FINLAND

Gulf of Bothnia

SWEDEN

Uppsala

Stockholm

BALTIC SEA

Peterhof □   □ St. Petersburg
                □ Tsarskoye Selo

Tallinn

ESTONIA

□ Novgorod

Volkhov

□ Rostov

Nizhny Novgorod □

Volga

Sergiev □   □ Vladimir

Moscow □   Klyazma

Oka

□ Pskov

□ Cesis

LATVIA

LITHUANIA

Dvina

RUSSIA

Nemunas

□ Kaunas

RUSSIA

BELARUS

Don

Gdańsk □

Malbork □   □ Biskupin

Warta

Pripyat

Desna

Gniezno □

Bug

POLAND

Warsaw □

Vistula

Kiev □

Dnieper

Odra

UKRAINE

Auschwitz
□ □ Kraków

Bug

SLOVAKIA

Dniester

MOLD.

Vienna

Danube   Tisza

□ Eger

Sirei

Esztergom □   □ Budapest

HUNGARY

Drava

Murescul   □ Alba Iulia

□ Sibiu

CROATIA

Sava

ROMANIA

BOSNIA -
HERCEGOVINA   YUGOSLAVIA

Velika

UKRAINE

Dnieper   Don

Volga

RUSSIA

CASP.
SEA

CRIMEA

GEORGIA

BLACK SEA

Tbilisi   □ Mtskheta

AZR.

ARM.

IRAN

TURKEY

IRAQ

INTERNATIONAL DICTIONARY OF
HISTORIC PLACES

INTERNATIONAL DICTIONARY OF HISTORIC PLACES

INTERNATIONAL DICTIONARY OF
HISTORIC PLACES

VOLUME 2

NORTHERN EUROPE

Editor
TRUDY RING

Associate Editor
ROBERT M. SALKIN

Photo Editor
SHARON LA BODA

FITZROY DEARBORN PUBLISHERS
CHICAGO AND LONDON

Copyright © 1995 by

reproduction in whole or in part in any form.

or
11 Rathbone Place
London W1P 1DE
England

Library of Congress Cataloging-in-Publication Data:

International dictionary of historic places / editor, Trudy Ring;
  associate editor, Robert M. Salkin.
       p.   cm.
     Essays on the history of 1,000 historic places.
     Includes bibliographical references and index.
     Contents: v. 1. Americas — v. 2. Northern Europe — v.
3. Southern Europe — v. 4. Middle East and Africa — v. 5. Asia and
Oceania.
     ISBN 1-884964-05-2 (set) : $125 (per vol.). — ISBN 1-884964-00-1
(v. 1). — ISBN 1-884964-01-X (v. 2). — ISBN 1-884964-02-8 (v. 3).
— ISBN 1-884964-03-6 (v. 4). — ISBN -1-884964-04-4 (v. 5)
     1. Historic sites. I. Ring, Trudy, 1955–   II. Salkin, Robert M., 1965–.
CC135.I585   1995
973—dc20   94-32327
                                                     CIP

British Library Cataloguing-in-Publication Data

International Dictionary of Historic Places, Volume 2—Northern Europe
I. Ring, Trudy   II. Salkin, Robert M.
970

ISBN 1-884964-01-X

First published in the U.S.A. and U.K. 1995
Typeset by Braun-Brumfield, Inc.
Printed by Braun-Brumfield, Inc.

Cover Photograph: Salisbury Cathedral
Courtesy of Salisbury Cathedral

Cover designed by Peter Aristedes, Chicago Advertising and Design

Frontispiece and endpaper maps by Tom Willcockson, Mapcraft

CONTENTS

HISTORIC PLACES BY COUNTRY

AUSTRIA
Enns
Graz
Melk
Salzburg
Vienna

BELGIUM
Bruges
Brussels
Ghent
Liège
Louvain
Tournai
Waterloo
Ypres

CZECH REPUBLIC
Karlovy Vary
Prague

DENMARK
Copenhagen
Odense
Slagelse: Trelleborg Viking Fortress

ENGLAND
Amesbury: Stonehenge
Bath
Bosworth Field
Cambridge
Canterbury
Chester
Chichester/Fishbourne
Cirencester
Colchester
Dover
Durham
Ely
Glastonbury
Hadrian's Wall
Hastings and Battle
Holy Island/Lindisfarne
King's Lynn
Letchworth
Lincoln
London: City of London
London: Greenwich
London: Hampstead and Highgate
London: Hampton Court
London: Southwark
London: Westminster Abbey/Parliament Square
Ludlow
Nantwich
Oxford
Runnymede
St. Albans
Salisbury

Stratford-upon-Avon
Winchester
Windsor
York

ESTONIA
Tallinn

FRANCE
Agincourt
Aigues-Mortes
Aix-en-Provence
Albi
Amboise
Angers
Arles
Avignon
Les Baux-de-Provence
Bayeux
Bordeaux
Carcassonne
Carnac
Chambord
Chartres
Chinon
Cluny
Compiègne
Dijon
Fontainebleau
Laon
Laval
Le Mans
Loches
Lourdes
Lyons
Mont-St.-Michel
Nancy
Nantes
Narbonne
Nîmes
Omaha Beach
Orange
Paris: Île de la Cité/Quartier Latin
Paris: The Louvre
Paris: Place de la Bastille
Paris: Place de la Concorde
Paris: Place de l'Étoile/Arc de Triomphe
Périgueux
Poitiers
Reims
Rouen
Saint-Denis
Saumur
Soissons
Strasbourg
Toulouse
Troyes

EDITOR'S NOTE

Fitzroy Dearborn Publishers' *International Dictionary of Historic Places* is designed to provide detailed and accurate information on places that have been the site of important events in human history and that have been preserved for the benefit of future generations.

The dictionary includes five volumes, which combined will cover nearly 1,000 sites worldwide. Volume 2 is devoted to Northern Europe, from the British Isles to Russia. The places included in this volume exhibit great diversity; some are prehistoric sites, while others were marked by events in Roman times, the Middle Ages, the Renaissance, or the modern era. It follows Volume 1, which covers the Americas. Future volumes in the series will be Volume 3, Southern Europe; Volume 4, Middle East and Africa; and Volume 5, Asia and Oceania. The dictionary is intended for the use of students, teachers, librarians, historians, and anyone interested in historic places.

The entry on each site includes a detailed essay explaining the events that occurred there and their historical significance, as well as providing information on what the site offers to contemporary visitors. Headnotes for each essay provide the site's geographic location, a concise description, and the address of an information office at the site, or, when no such office is available, the address of a central contact for the area. We expect this information to assist persons who are traveling to any of the places, as well as those who write or call any of the sites to request material. Each entry also includes a section on further reading. This section is a selective listing of relevant published works, as recommended by the author of the entry.

The entries were compiled from publicly available sources, including books, magazine and newspaper articles, and, in some cases, material supplied by the site offices. We thank the various site offices for providing this material, as well as illustrations. The site offices' assistance in no way constitutes their endorsement of the facts presented. Our contributors and editors, however, have made every effort to ensure the accuracy of each entry.

Obtaining up-to-date information on countries that have undergone major changes in the past few years required a great deal of effort. Helga Brenner-Khan of the German National Tourist Office was particularly helpful in providing Fitzroy Dearborn with information on Germany. Tracey Ryan of the magazine *Russian Government Today* and Benjamin J. Quick of General Tours were especially helpful with regard to Russia.

We dedicate this volume to the memory of Honor Levi, who died suddenly on November 6, 1994. She contributed four essays to this volume; her husband, Anthony Levi, and daughter, Claudia Levi, also contributed to Volume 2, while essays by another daughter, Clarissa Levi, will appear in Volume 3. We extend our condolences to the Levi family.

Finally, for their valuable editorial assistance, we thank Mary F. McNulty, Marijke Rijsberman, Paul E. Schellinger, and Randall J. Van Vynckt.

—Trudy Ring

CONTRIBUTORS

Sharon Bakos

Philippe Barbour

Bernard A. Block

Shawn Brennan

Elizabeth Brice

Elizabeth E. Broadrup

Monica Cable

Ron Chepesiuk

Olive Classe

Daniel D. Collison

Elizabeth Devine

Sina Dubovoy

Laura Duncan

Stephen Ellingson

Mary Feely

Jeffrey Felshman

John A. Flink

Lawrence F. Goodman

Richard Greb

Judith Gurney

Patrick Heenan

Jeff W. Huebner

Tony Jaros

Patrick Keeley

Linda J. King

Rion Klawinski

Andrew M. Kloak

Manon Lamontagne

Monique Lamontagne

Bob Lange

Cynthia L. Langston

Sherry Crane LaRue

Gregory J. Ledger

Anthony Levi

Claudia Levi

Honor Levi

Mary F. McNulty

Thomas Cermak McPheron

Kim M. Magon

Brent Marchant

Christine Walker Martin

Julie A. Miller

Phyllis R. Miller

Laurence Minsky

Bruce P. Montgomery

Michael D. Phillips

Jenny L. Presnell

Marijke Rijsberman

Trudy Ring

Robert M. St. John

Robert M. Salkin

June Skinner Sawyers

Paul E. Schellinger

Patricia Ann Shepard

James Sullivan

Hilary Collier Sy-Quia

Noel Sy-Quia

Elizabeth Taggart

Randall J. Van Vynckt

Aruna Vasudevan

Noelle Watson

Joshua White

Richard G. Wilkins

Thomas Wiloch

Beth F. Wood

Peter C. Xantheas

INTERNATIONAL DICTIONARY OF
HISTORIC PLACES

Aachen (North Rhine-Westphalia, Germany)

Location: In the state of North Rhine-Westphalia in northern Germany near the Belgian and Dutch borders, forty-five miles west of Cologne, and fifty miles southwest of Dusseldorf, at the edge of the Ardennes Mountains.

Description: Founded by the Romans, probably in the first century A.D., Aachen later became the unofficial capital of the Carolingian Empire. Aachen came under French control from 1794 to 1815, and was known as Aix-la-Chapelle. When the city came back into German hands its name reverted to Aachen—derived from the old German word *ahha* meaning water. Belgium briefly occupied the city after World War I. During World War II, Aachen became the first German city to fall to the Allies, on October 20, 1944. Much of the city was destroyed in World War II, but some historic sites remain; these include the cathedral and ancient mineral baths.

Site Office: Verkehrsverein
Information büro Eslisenbrunnen
Friedrich-Wilhelm Platz
52062 Aachen, North Rhine-Westphalia
Germany
(41) 180 29 60

The exact date of Aachen's founding and settlement remains a mystery. Most likely, the region that is today known as Aachen was settled in the first century A.D., by the Roman Empire during the rule of either Augustus or Tiberius. The Romans referred to the region as Aquae Granni. This remote post of the Roman Empire was used as a health complex for the empire's troops who were stationed nearby. The health complex consisted of a temple and two hot sulfur baths—the Buchelthermae and the Münster-thermae—utilizing hot spring water that flowed up to the earth's surface from a depth of 8,000 feet. Tired or wounded legionnaires were able to soak away their pains in some of the hottest spring waters found in all of Europe. Records indicate that the Romans continued to use the hot spring's facilities into the second half of the fourth century, when the Franks finally drove the Romans out of Aquae Granni.

Wave upon wave of barbarian invasions helped chip away at the authority of the Roman Empire and at its power to rule over much of western Europe. The weakened empire was divided among numerous kingdoms and principalities in western and northern Europe. Aachen eventually came under the control of the struggling Merovingian Empire.

Plagued by familial discord, the Merovingian Empire would never develop into a hegemonic power. As generations of Merovingian rulers fought among themselves attempting to increase their power, the day-to-day business of running the family's various castles was left to the mayor of each castle. As a result, nonfamily members were let into the decision-making process. One of the most important mayors in the Merovingian Empire during the late seventh and early eighth centuries was Pépin of Herstal, Charlemagne's great-grandfather. Pépin of Herstal and his son Charles Martel shrewdly worked at expanding their power within the empire. Their machinations helped pave the way for Pépin the Short, Charles Martel's son and Charlemagne's father, to seize control of the kingdom from the feuding Merovingians. Pépin received full cooperation from the church in his grab for power. He was thereafter crowned king of the Franks in 751 and was named Patrician of the Romans by Pope Stephen II in 754. The Carolingian Empire was born.

Charlemagne not only inherited an empire, he also greatly expanded it. After he came to power in 768, Charlemagne almost doubled the size of the territory he had inherited. He annexed Saxony and most of Italy, marched south of the Pyrenees, and destroyed the Avar Empire in the Danube Valley. Charlemagne was also able to establish close ties between the Roman papacy and the territory that later would become Germany. This relationship would last for centuries and would make Germanic kings the official protectors of the Roman Catholic Church.

Charlemagne ran his empire without a fixed capital during the early years of his reign. He ruled from a series of castles in Compiègne, Düren, Frankfurt, Attigny, Liège, Thionville, Ingleheim, and Aachen. The decision to make Aachen the unofficial capital of the empire most likely was made by Charlemagne in the winter of 788–89, while he stayed at a nearby royal villa originally constructed by the Merovingians.

Charlemagne was a fanatic about personal hygiene. This obsession with cleanliness was probably one of his reasons for choosing Aachen, with its hot springs, as the seat of the empire. Einhard, Charlemagne's biographer, gives credence to the argument that Charlemagne chose the city for its possessing hot springs; he writes that Charlemagne "loved the vapors of the naturally warm waters . . . and therefore constructed a royal residence" in the city. Aachen also presented other benefits to Charlemagne. The city was in the area of the empire from which Charlemagne's family originated, and this factor provided a sense of security. The city was also free from attack by hostile neighbors; the nearest neighbors were hundreds of miles away.

During Charlemagne's rule, the Mediterranean world was split into three distinct spheres of influence. Much of the Middle East and all of North Africa were under the Islamic sphere headquartered in Baghdad. Southeastern Europe, parts of Italy, and sections of Asia Minor constituted the

Aachen's imperial cathedral
Photo courtesy of German Information Center

second sphere of influence, called the Byzantine Empire, which was centered in Constantinople. Aachen, the seat of the Frankish Carolingian Empire, was the center of the third important sphere. This third sphere was to bring together the Germanic, Gallic, Teutonic, Latin, and Frankish worlds. The marriage of these different peoples and customs under one empire was an attempt to crush barbarianism in Europe.

Aachen was showered with praise by many of Charlemagne's contemporaries. The city was called "the new Jerusalem" and the "second Rome." Great sums of money were spent to build Aachen into an imperial gem. Charlemagne's biographers hint that much of the money came from Charlemagne himself. It probably would have been easy for Charlemagne to spend lavishly to construct Aachen without inconveniencing himself financially. Between 791 and 796, he amassed a huge fortune when he seized treasures from the Avars, enemies in the Danube River valley, as they retreated in defeat from the battlefield. It is believed that these treasures financed much of the construction in the city.

Distance did not appear to be a major concern when construction began on Europe's newest grand city. Glass, marble, metal, mosaics, paints, stone, and timber were transported to Aachen from across Europe, as were the various classes of laborers who were to work on the construction—the freemen, priests, peasants, monks, and professional artists. A note sent by Charlemagne to Pope Hadrian I testifies to the great efforts made in building the city; the king requested the pope's permission to remove marble and mosaics from church property in Italy, and bring it to Aachen to be used in various structures. Charlemagne also had stone and marble columns brought across the Alps from Rome and Ravenna, Italy.

Imperial Aachen had a center called the palatium. It consisted of Charlemagne's residence in the center surrounded by structures housing his entourage—many of whom were responsible for the upkeep of the main residence. Also located in the palatium was a grand sixteen-sided cathedral. It is unknown whether the palatium was enclosed by any type of gate or if it was guarded by a watchtower. Modern Aachen's city hall is located in the part of town once occupied by the palatium.

Two different architects built the palatium's two most important structures: the palace and the cathedral. Einhard, Charlemagne's biographer, was entrusted with laying out the design for the palace. His inspiration seems to have been models from the Roman architect Vitruvius. Einhard linked the castle to the city's hot springs, thereby providing heat to the structure. The layout of the massive complex dramatically altered the topography of Aachen. The complex was oriented from east to west, while the street system, set up earlier by the Romans, was oriented from north to south. The result was that many of the roads laid out in a grid pattern were turned into triangles.

Odo of Metz designed the grand cathedral in Aachen's palatium. The design was greatly influenced by the design of the church of St. Vitale in Ravenna. Ultimately, Odo's design was a distinctive mixture of Germanic, Celtic,

and Mediterranean elements; the greatest prominence was given to Mediterranean influences, a highly unusual development at that time in the Germanic world.

The structure's interior was truly complex, with an octagonal rotunda surrounded by vaulted aisles. Over the years the interior has been altered, but accounts written during the reign of Charlemagne provide insight on the church's original interior. They tell of precious metals, jewels, and geometric designs throughout; of lavish gifts received from Rome, Constantinople, Baghdad, Jerusalem, and Spain, all displayed prominently; and of paintings used to educate the illiterate masses on the teachings of the church.

The palatium complex was not militarily fortified. During the time of Charlemagne, there was no real need to construct a compound that would double as a fortress. Frankish society was stable and its enemies were hundreds of miles from Aachen. Such a militarily deficient complex differed greatly from the fortress-like castles constructed as soon as 100 years later.

Some historians have argued that the builders of Aachen simply imitated the past splendors created in other great cities like Rome. Other believe that Aachen was the result of blending a number of styles together to create a truly regal city. Aachen reached its greatest heights during the reign of Charlemagne from 768 to 814. Charlemagne attempted late in his reign to prepare his son Louis the Pious to rule. For a short while in 813, the two jointly ruled the empire. Despite the grooming by his father, Louis led the Carolingian Empire on a downward slide. Louis would later be forced to abdicate the throne by his three sons. The sons, in turn, would tear the empire apart. Charlemagne, however, had laid the foundations for the Holy Roman Empire, founded in 962 by the German King Otto I.

Aachen would remain an important city following Charlemagne's death. Both the French and the Germans attempted to use the glory of Charlemagne as an inspiration to later generations. Both saw Aachen's history as their own. The Germans crowned most of their kings and emperors in the city from 936 to 1562.

Holy Roman Emperor Frederick I (Barbarossa) went so far as to have Pope Pascal III canonize Charlemagne, in 1165, and claim him as a German, so as to refute any claims made by the French that Charlemagne was a Frenchman.

Aachen became a free imperial city around 1250, giving it political and commercial advantages over other cities in the Holy Roman Empire. It also was the site of several important church conferences in the Middle Ages. By the 1500s, however, it had declined because of its insecure position on the Franco-German frontier and its distance from the center of the Holy Roman Empire. The coronation site for Germanic kings was finally moved in 1562 to Frankfurt am Main.

Despite the decline of its political stature, Aachen maintained some distinction in subsequent centuries. Its hot springs made it a popular destination for prominent vacationers; it became known as the "Spa of Kings." Aachen was the site of the peace conferences that ended the War of Devolution in 1668 and the War of the Austrian Succession in 1748. The French occupied Aachen, which they called Aix-la-Chappelle, in 1794, during the French Revolutionary Wars. The city was formally annexed by France in 1801, but was given to Prussia in 1815 following the Congress of Vienna. It was a Prussian military base during the Franco-Prussian War in 1870–71 and a German air force base in World War I; after Germany's defeat in the latter conflict, the Allies occupied Aachen from 1918 to 1930.

World War II brought about the destruction of many of Aachen's historic structures. Aachen would become the first German city to fall to the Allied troops, but not without a vicious fight. A visitor to the city soon after it fell to American troops on October 20, 1944, would comment:

> The city is as dead as a Roman ruin, but unlike a ruin it has none of the grace of gradual decay. . . . Burst sewers, broken gas mains, and dead animals have raised an almost overpowering smell in many parts of the city. The streets are paved with shattered glass; telephone, electric light and trolley cables are dangling and netted together everywhere; and in many places wrecked cars, trucks, armored vehicles and guns litter the streets.

Six weeks of battle saw Aachen being pounded by 75 large-scale Allied bombing raids. In one day alone, 62 tons of bombs were dropped in one neighborhood, while another section of town bore the brunt of 169 tons of explosives. When the smoke cleared after Aachen had fallen to the Allies, more than 80 percent of the city's residential stock was either destroyed or badly damaged; 43 percent of the city's buildings were flattened; 40 percent were left standing but were structurally unsound. It was estimated that 3.5 million square yards of debris were heaped across the once glorious city. The city was without transportation, gas, electricity, water, post, or telephone service.

Ground troops left their mark on Aachen as well. Advances and retreats, by both the Germans and the Allies, were measured in mere city blocks or even yards. German soldiers and Aachen's few remaining residents often hid from the Allies in underground bomb shelters and in the city's sewer system. Capturing any sector of the city required that troops sweep through not only the streets, but also its buildings, bomb shelters, and sewers.

Capturing the city was essential to the Allies for two reasons. First, Aachen had to fall to ensure the safety of Allied troops advancing toward the Rhine River. The city was being used during the war as a German air base, which would have to be completely destroyed to prevent it from becoming operational again should the town fall once more into Nazi hands. Also, any German soldiers left behind in the city would be able to attack the Allies behind their own lines.

The second reason for striking at the city was symbolic. It was the first time an attack was launched on German soil in more than a century. Capturing the city was tantamount to ripping out the heart and soul of National Socialist mythology. Hitler preached that his rule was Germany's Third Reich and that it would last a thousand years. Germany's First Reich, he said, was the Holy Roman Empire, centered in Aachen for many years, and lasting until 1806. The second Reich, he said, began when Bismarck unified Germany in 1871 and lasted until 1918. By taking Aachen, the Allies undermined a symbol of National Socialism and captured an important city in Germany's history. Some said that Aachen was to Germany what Stalingrad was to the Soviet Union.

Its strategic importance and its history were not the only reasons why the German fought so hard to save Aachen. Many German soldiers feared the prospect of death at the hands of their own government, or of having themselves or their family imprisoned in German concentration camps if they refused to fight to the last man. German soldiers cornered in the ruins of a luxury hotel chose to pelt the advancing Americans with empty champagne bottles, rather than surrender when their ammunition ran out. When the German troops realized they were defeated, many succumbed to despair. Some took their own lives while others numbed themselves by consuming whatever alcoholic beverages were available.

After the Germans at Aachen finally surrendered on October 20, 1944, the rebuilding of the city began under Allied supervision. The effort did not last long, however. When Aachen fell Hitler attempted to turn the tide with one last great offensive—the Battle of the Bulge, near Bastogne, Belgium. The Battle of the Bulge delayed the end of the war in Europe until May 1945 and resulted in 80,000 additional casualties. Hitler's offensive proved disastrous to the already war-ravaged city of Aachen. Modest attempts at rebuilding the city were thwarted as Allied soldiers, concerned that the city might fall back into German hands, set fire to empty houses and caused havoc in Aachen as they retreated. When the war ended, Aachen became part of the British Occupational Zone. The rebuilding process quickly began.

Because of the wartime destruction, little is left of the historic beauty of Aachen. The cathedral, however, suffered only slight damage. It has been repaired and is today the city's most historic site. Also repaired are the ancient hot spring baths. The city hall, built on the site of the palatium, dates from the fourteenth century; a few other historic buildings still stand.

Further Reading: Richard E. Sullivan's *Aix-la-Chapelle in the Age of Charlemagne* (Norman: University of Oklahoma Press, 1963) provides information on Aachen before and during Charlemagne's reign. It is particularly informative about the city's architecture and the operations of Charlemagne's imperial court. Russell Chamberlin's *Charlemagne: Emperor of the Western World* (London: Grafton, 1986) is a comprehensive work, dealing not only with the reign of Charlemagne but with the rulers who preceded and followed him. *Bloody Aachen* by Charles Whiting (New York: Stein and Day, and London: Cooper, 1976) details the Allied capture of Aachen in World War II. The author makes use of interviews with American and German soldiers, as well as Aachen civilians, to provide a personalized view of the fall of Aachen.

—Peter C. Xantheas

Agincourt (Pas-de-Calais, France)

Location: In the *département* of Pas-de-Calais, the town Azincourt (as it is known in France) is located thirty-three miles west/northwest of Arras, off the main road from Hesden to Arras.

Description: Site of the famous battle on October 25, 1415, in which an English army led by Henry V defeated a French army reportedly four to five times its size.

Site Office: Musée d'art et d'histoire
Azincourt, Pas-de-Calais
France
21 04 41 12

Agincourt, or in France Azincourt, was the name given to the battle of 1415, in which an English army of 5,800 led by Henry V of England defeated a French force at least four times its size. One of numerous battles fought between the English and French during what would later be called the Hundred Years War (1337–1453), it is one of the better documented and studied medieval battles and is also one of the most legendary military victories in English history. Often characterized as a victory of the English archer over the French mounted knight, and thus a harbinger of modes of warfare to come, it has also come to symbolize the occupation of northern France by the English—the Lancastrian occupation (from the House of Lancaster, the family of Henry V's ancestors and successors).

There are divergent opinions among historians about the historical significance of the battle of Agincourt. For some, Agincourt represents the greatest military achievement of medieval England, with Henry V as one of its greatest heroes, and doubtless the stirring passages of Shakespeare's *Henry V* have inspired many people to reach just such a conclusion. French scholars, on the other hand, have generally tended to look upon Agincourt as a less than decisive event. For many, including Edouard Perroy, historian of the Hundred Years War, the battle is significant only because it symbolizes the misery and terror that Henry V's campaigns inflicted on the French countryside. However one may choose to look upon Agincourt, it has come to exemplify what historian John Keegan has called the "slaughter-yard behaviour" and "outright atrocity" of medieval warfare.

Agincourt happened during what historians now think of as the second phase of the Hundred Years War, a period in which the dynastic claims of the English kings over the French crown were once again voiced and, more importantly, acted upon. In the previous century, Edward III had put forward such a claim, challenging the right of the French king Phillip VI of Valois to seize the landholding of Guienne which belonged to the English king, who was also a prince of

France. After two victories at Crécy (1343) and Poitiers (1356), Edward rescinded his claim to the French crown in return for one-third of French lands that he held in full sovereignty under the terms of the Treaty of Brétigny (1360). Following the deaths of Edward and his son, the Black Prince, victor at Poiters, much of the territory ceded by the treaty was regained by the French.

Upon his succession to the English throne in 1413, Henry V renewed his great-grandfather's claim to the French throne. His motives were various. Chief among them was the hope that an invasion of France would still the discontent and instability which had marked the reigns of Richard II, successor to Edward III, and that of his father, Henry IV, who had usurped the throne from Richard in 1398. Aware that there remained many in England who questioned his legitimacy to wear the English crown, Henry also sought a continental victory that would show him to be a true successor of Edward III and demonstrate the righteousness of his ancestor's claim.

Political disorder in France provided a further incentive to the English to resume the Hundred Years' War. The King of France, Charles VI, was subject to bouts of madness. Civil war seemed endemic, and political power had fallen into the hands of great princes of the realm. Since 1411, France had been effectively divided between two great families, the Burgundians and Armagnacs. On becoming king, Henry promptly entered into secret negotiations with the Duke of Burgundy to secure the Duke's neutrality and opened negotiations with the Armagnacs to cloak England's preparations for war. In August 1414, an English delegation to Paris proposed that Henry would marry Charles VI's daughter, Catherine, and receive in full sovereignty Normandy, Touraine, Maine, Anjou, Britanny, and all of Aquitaine. The French refused this demand. Henry made a similar demand in June 1415, with the added threat that if the French refused he would recover these territories with the sword. The French again refused, whereupon Henry joined his troops at Southampton.

War with France proved to be popular in England. Spurred on by the promise of glory (and plunder), 2,000 men-at-arms and 8,000 archers were quickly assembled by August of 1415 for the invasion of France. The men-at-arms, for the most part, were well-equipped with head-to-toe steel-plate armor that weighed as much as sixty pounds. Besides their swords and lances, they carried such weaponry as the battle-axe, battlehammer, mace or flail (a spiked ball and chain attached, swinging, to a short handle), and perhaps the most lethal weapon of all, the pole-axe, a half-axe-, half-hammer-headed steel pole with a spiked shaft. By far the largest group within the invading army consisted of archers. Archers, with their English longbow, had formed an impor-

Monument at the Agincourt battlefield
Photo courtesy of Mairie d'Azincourt

tant part of the English army ever since the longbow had proven effective against the French at Crécy (1343). The best English bowmen could loose up to twelve arrows a minute, a rate far superior to that achieved by the Continental cross-bowmen, and could shoot from a range of 300 yards. The arrows were often steel-tipped and capable of piercing lesser-grade armor at a range of sixty yards, though it seems arrows seldom killed large numbers of well-armored knights. Archers were most effective in unseating a mounted foe, and the only real means of removing archers from a defensive position was a direct cavalry charge. Archers also performed a dual role in the English army. They often assisted men-at-arms in hand-to-hand combat when a melee ensued. Mounted archers, who accounted for two-thirds of Henry V's archers, could also, if needed, reinforce a mounted charge.

Having led the campaign against the Welsh revolt (1400–09), the twenty-seven year-old king was already a seasoned veteran of medieval warfare, and he now turned his experience and his mastery of detail and organization to the French campaign. His first objective was the city of Harfleur, an important port on the mouth of the Seine located near present-day Le Havre. From there, he planned to strike at

Paris and later to continue southward to Bordeaux. Unlike his predecessors, who had concentrated much of their efforts in the southern kingdoms of Aquitaine, Henry's strategy was to gain a foothold in Normandy within striking distance of Paris. The strength of carefully-sited fortresses was a lesson he had learned in his Welsh campaigns; he brought with his invading army a team of miners to tunnel beneath enemy walls. He brought, too, sixty-five gunners to man the English guns to be used in the siege.

The siege of Harfleur began on August 18, 1415, three days after the English landing. Although it had only a few hundred defenders, Harfleur had natural defenses that were quite formidable. They included twenty-six towers and a moat wide enough to prevent the use of battling rams; the moat could be crossed by only three drawbridges. Initial attempts to tunnel underneath the moat were repelled by the French. The English changed their strategy and turned to bombarding the walls with their twelve siege guns, which included the king's own cannon, capable of firing stone-shot the size of millstones against the walls. Despite the continual bombardment supervised by the king himself, the port town still held out. Dysentery soon spread among the English forces, depleting the army by almost one-third. At last the walls at the southwest barbican (gatehouse) was breached on September 16. After a series of sorties, Harfleur was taken, and the small town garrison agreed to surrender on September 22; no French force had arrived to come to its rescue. Thus Harfleur became the first French town ever captured by the English by means of cannon.

The depletion and exhaustion of his troops forced Henry to abandon his original plans of marching on to Paris, but he was not daunted; he decided instead to march north to Calais. This northern trek through the Normandy countryside was also intended to symbolize his claim to the territory.

Calais, a safe haven, lay some 160 miles away, and Henry had hoped to reach it in an eight-day march, achieving this great speed by avoiding confrontation and travelling light. He left Harfleur on the October 8, leaving behind his siege wagons, with reduced force of 900 men-at-arms and 5,000 archers. After a five-day march to the River Somme, he found his crossing at Blanche-taque thwarted by French troops on the other side. Determined not to retreat, he turned southeastward and followed the river until he found an unguarded crossing. The English troops finally crossed the Somme at Béthencourt but were well aware as they continued to march that they were being shadowed by a French force to the north. On the 24th of October they learned that their path to Calais had been intersected by a large French army estimated to be 25,000 men, half of whom were men-at-arms. After a sixteen-day trek, much of it through rain, the English troops reached the small hamlet of Maisoncelles located south of the field at Agincourt. The French were camped farther to the north at Ruisseauville, where, it is said, they played at dice for English captives.

At dawn, on October 25, both armies took to the field. It lay nestled in between the villages of Agincourt to the west and Tramecourt to the east. Lined on both sides by small woods, the field was very narrow, so much so (less than 1,000 yards at the centre) that the French were effectively prevented from making full use of their numerical superiority. Most of the French were dismounted men-at-arms in full plate armor, though two detachments of mounted knights were on the wings of the field, and the third line of attack remained mounted. Though the French had archers and crossbowmen on hand, these forces were deployed at the rear of the lines, there being only sufficient room on the field for men-at-arms.

The English 800 men-at-arms were all dismounted and, like the French, had sawn off their lances. Organized in three ''battles'' each flanked by archers, they covered the entire center of the field. On each side of the English line more archers were placed; each stood behind an eleven-foot-long stake, sharpened at both ends, set into the ground in front of him as a means of protection against enemy calvary. After a four-hour wait, during which both sides held their ground, Henry marched the English forces forward to within firing range of the French lines (about 300 yards); the English

archers replanted their stakes and loosed their arrows. Volley after volley harassed the French lines, and as the first line of French men-at-arms advanced, 500 mounted Frenchmen charged the English archers. The attack was repulsed: the French horses, stung by arrow fire, became unmanageable, and the line was thrown into confusion; horses that reached the archers were impaled on the stakes, and their riders killed.

The previous day's rain had turned the freshly-harvested field to mud, and weighed down by more than sixty pounds of armor, the dismounted French warriors trudged in knee-deep mud, while the threat of English arrows, which could pierce armor at sixty yards, pushed them towards the center. As both enemy lines clashed, the English, who had not advanced to meet the oncoming force, were thrown back a few feet but held their ground. In the fierce hand-to-hand combat that ensued, many French knights were unable to draw their weapons, their lines having become so compacted. The fighting became even more confused when the second line of French advanced into the fight, for the numeric superiority of French forces proved to be more of a hazard than an advantage. Thousands of French men-at-arms were knocked down, trampled, and even drowned in the mud of the battle. The melee was joined by lightly-clad English archers who, having spent their arrows, picked up sword, battle-axe, and other weapons. One group of French soldiers did manage to isolate the Duke of Gloucester, the king's younger brother. While rescuing his brother, Henry received a blow that dented his helmet; otherwise, he emerged unscathed.

By midday, the French had been routed. The English were in possession of the field. Mounds of French bodies lay scattered about. Thousands had been taken prisoner—perhaps as many as 3,000—and brought to the rear of the English line. Under the codes of medieval chivalry, knights who surrendered could expect to be ransomed. But, Henry ordered 200 archers to slay the prisoners, sparing only those princes of the blood who were likely to prove the most valuable. He feared a counterattack from the third French mounted line, and feared that the captives would break free to join their countrymen. The slaughter was stopped when the attack never materialized, but it is estimated that as many as 2,000 French prisoners had already been killed.

Four hours after the battle had started, the French herald, Montjoie, who had been watching the events of the day unfold with his English counterparts, rode in to the English encampment and announced that they had won the day. Henry named the battle Agincourt after the nearby castle to the west.

The list of the French dead filled ten pages. Against 500 English dead, the French army had lost nearly 10,000 men, including the slaughtered prisoners. After a brief rest at Maisoncelles, where the English dead were taken and burned, Henry continued his march to Calais, taking with him some 1,500 prisoners, most of whom were of noble birth. Some of the French dead were buried in the churchyards of Agincourt and Ruisseauville, but most, about 5,800, were buried in three large pits marked with a large wooden cross.

As stunning a victory as it was, the battle of Agincourt was not decisive. Even as Henry returned to a hero's welcome in England, the French forces tried in vain to regain Harfleur. Two years of diplomatic negotiations failed to provoke French concessions. In 1417, Henry returned to France and began a systematic conquest of Normandy, facilitated by the Duke of Burgundy's march on Paris, which fell to the Burgundians in 1418.

The Treaty of Troyes, negotiated in 1420, made Henry V of England heir to the throne of France. It was solemnized by the marriage of Henry to Catherine, daughter of King Charles VI. But rather than ending the fighting, the treaty succeeded only in further fragmenting France; the overlooked Dauphin (the French king's son) gathered his forces in the south of France to regain his right of succession. Henry spent the rest of his life fighting to consolidate his gains. In 1422, during the siege of Meaux, he contracted dysentery, to which he succumbed.

His premature death perhaps spared him many of the troubles that beset his son, Henry VI. Though Henry VI had succeeded Henry as not only King of England but also as King of France, the dual monarchy that Henry V had created by conquest was short-lived. English occupation had awakened a sense of French nationalism that burst forth in 1429 when a young peasant girl, Joan of Arc, rallied to the aid of the Dauphin. Her martyrdom at the hands of the Henry's successors spurred the French forces to unite under their own French King Charles VII and expel the foreign garrisons. By 1453, all English gains achieved under Henry V had been reversed.

Today, the field at Agincourt has changed little. Small clumps of woods, replanted by successive generations, still line the field as they did centuries before. A large crucifix now marks the battlefield. It was erected in 1870 by the Vicomte de Tramecourt in memory of the French men-at-arms who fell in defending their country in 1415. A chapel built on the site by the Marchioness de Tramecourt in 1734, to honor the French dead, was destroyed during the French Revolution. A small museum in the village of Azincourt now commemorates the battle and the arts of medieval warfare.

Further Reading: Among numerous books on Agincourt, three are notable for their exhaustive coverage of the battle and for their comprehensive explanation of its background: *The Hundred Years War* by Edouard Perroy, translated by W.B. Wells (London: Eyre and Spottiswoode, 1951; Bloomington: Indiana University Press, 1959); *The Face of Battle* by John Keegan (London: Penguin, and New York: Viking, 1976); and *Henry V as Warlord* by Desmond Seward (London: Sidgwick and Jackson, 1992).

—Manon Lamontagne

Aigues-Mortes (Gard, France)

Location: Near the Mediterranean coast in the *département* of Gard in southern France, approximately five miles south of Nimes, and nineteen miles east of Montpellier; to the east lies the 300-square-mile area of unspoiled salt marshes called the Camargue; to the west, the modern sea resort of La Grande Motte.

Description: A well-preserved fortification surrounding a medieval town, built in the thirteenth century under Louis IX; narrow streets form a grid plan surrounded by thick ramparts complete with towers; once a port and point of departure for two of St. Louis' Crusades; later deprived of access to the sea by the silting of the canals; known for the Tour de Constance, a thirteenth-century keep used as a prison for Huguenots during the Counter-Reformation in the eighteenth century; today a center for the salt industry with a population of approximately 4,000.

Site Office: Office de Tourisme Syndicat d'Initiative
Porte de la Gardette
B.P. 32
30220 Aigues-Mortes, Gard
France
66 53 73 00

The name Aigues-Mortes comes from the Latin name Aquae Mortuae ("dead waters"), first recorded in an eighth-century text. It seems apt for a location surrounded by miles of flat coastal plain punctuated by still lagoons and salt marshes. To some visitors the austere appearance of the medieval walled city with its quiet, narrow streets that permit little light or air to pass through may seem lacking in vitality, but the town has had an important and lively history.

Early in the Middle Ages, a Benedictine abbey was built here on the foundation of a Roman villa that had been turned into a Merovingian estate, the Villa Varanègues. The abbey was called *Insula Psalmodia,* perhaps because of the chants often sung by the monks. They built chapels, fished, raised cattle, and extracted salt. Their products were traded by boat with nearby Beaucaire and Saint-Gilles.

After the Saracens destroyed the abbey in 720, the monks collected the funds necessary to restore it from neighboring religious communities, the faithful, and the clergy. The abbey flourished under the protection of the popes for hundreds of years before Louis IX, the king of France, chose the site in hopes of changing the course of history.

The taking of Jerusalem by the Turks in 1244 prompted Pope Innocent IV to hold a council to plan its recapture for the Christian world. There would be a seventh crusade, led by the king of France, a holy man who had taken a vow, following a grave illness, to fight the infidels.

King Louis needed a Mediterranean port from which to launch his ships, but Marseilles belonged to Provence, and Montpellier had ties to Aragon. France had to have its own port. He had already chosen Aigues-Mortes as a location because of its natural harbor inside the lagoons. As early as 1240 he had begun negotiations with the abbot of Psalmodi to purchase the abbey. Plans for the crusade hastened the completion of the sale, and construction was begun.

The location had its drawbacks: strong winds jostled the ships, the inner harbor repeatedly silted, and there was little fresh water for the troops. Because Aigues-Mortes was not on a main trade route, concessions had to be worked out with Montpellier and Nîmes so that provisions could be secured. Workmen built a canal (the former Canal Vieil), which would allow small crafts to be loaded at the port and sent out to the sea, where the larger ships could anchor. Fresh water for the troops was shipped in.

Louis IX, later to be called St. Louis, put a great deal of effort into preparing his army of 15,000 for the Crusade of 1248. Ships were ordered from Marseilles and Genoa, the troops were assembled at Aigues-Mortes. In all, 36,000 men and women took part in the effort. Besides the soldiers and the knights, there were bishops and other clergymen, middle-class merchants, and peasants. Some of the noblemen brought their wives. Most came from France, but other European countries were represented as well.

At first the town was surrounded by wooden ramparts. Of the stone monuments remaining today, only the Tour de Constance was built before Louis's first crusade. There was not enough time in the three and one-half years between the beginning of the construction of the port in 1244 and the departure for Cyprus in 1248 to complete either the harbor or the ramparts; they were completed later under Louis's son and grandson.

While in Aigues-Mortes, the king and queen stayed at the Tour de Constance, built of Rhône valley stone from nearby Beaucaire. From the platform at the top there was a magnificent view of the Mediterranean. Massive with thick walls, the tower had to be built on a foundation resting on wooden pilings to keep it from sinking into the soft soil. The tower was later used as a royal prison.

With the king's departure in 1248, his mother, Blanche de Castille, was named regent. She was left with the responsibility for completing the negotiations for the provisioning of Aigues-Mortes and for continuing to collect revenues from churchmen to fund the crusade.

Upon his return in 1254, Louis acknowledged that in his haste to prepare for the crusade, he had exploited some of the other cities of the region. He made a tour of eastern

The ramparts of Aigues-Mortes
Photo courtesy of Office Municipal du Tourisme, Aigues-Mortes

Languedoc to mend relations and to give compensation to those cities that were unfairly treated. He restored municipal charters to Nîmes and Beaucaire.

In spite of the fact that the first crusade against Egypt was largely unsuccessful, Louis undertook preparations for a second crusade from Aigues-Mortes in 1270 to Tunis via Sardinia. His death from the plague only a few weeks later, however, brought an end to the spirit of the crusades.

The spirit of Aigues-Mortes lived on, however, and many pilgrims used the port to travel to the Holy Land or to Rome. It was also a convenient stop for Italians or Provençaux on the road to Santiago de Compostela. The spiritual environment of the Camargue region, with its medieval churches at Saint-Gilles and Saintes-Maries-de-la-Mer, made it a popular place for pilgrims from all over Europe.

Philip III, the Bold, the son of St. Louis, resumed the project of building the fortification at Aigues-Mortes. In 1272 he commissioned Guillaume Boccanegra, originally of Genoa, to build the ramparts and to repair the port. The purpose of the ramparts was to protect the inhabitants from intruders, strong winds, and blowing sand. Boccanegra died two years later. The work continued under local supervision, but was still far from complete in 1289 when an official report was filed with Philip the Fair, the son of Philip III. The second stage of construction was begun after bids from two Genoese entrepreneurs living in Aigues-Mortes, Cominelli and Boccuccio, were received. While it is not clear which of them was chosen to undertake the project, the work was completed by the beginning of the fourteenth century. Unified in style despite its two phases of construction, the well-preserved fortification is an excellent example of military architecture. Its imposing authority stands as a monument to St. Louis built by his successors.

The design of the fortification, with its twenty towers, surrounded in part by a moat, made it impregnable to attack, at least until the advent of firearms. Many have compared its style to other fortresses of the period at Antioch, Jerusalem, or Damietta. Its nearly rectangular shape extends for about one mile around its perimeter. Originally drawbridges crossed the moat, providing access to the gates. Each of the towers has a notched platform at the top intended for defense. Two chambers, one above the other, topped by vaulted ceilings and accessible by a spiral staircase, fill the interior space of the towers. A stone-paved walkway leads around the top of the ramparts, protected by a parapet with loopholes once used for shooting arrows.

Seventeen wide stairways provided access to the top of the ramparts from the inside, and ample space was left between the walls and the first row of houses to allow for soldiers to maneuver freely.

The walls are thirty-six feet high and more than eight feet thick, with small openings. There are five main gates: on the north side, the main entrance, called the Porte de la Gardette, and the Porte de Saint-Antoine; on the east side, the Porte de la Reine, named for Anne of Austria, wife of Louis XIII; and to the south, the Porte de la Marine (navy) and the

Porte des Moulins, so named because it once had a windmill atop each tower beside it. In addition, there are five smaller entrances, each cutting through the base of one of the five smaller square towers. On the south wall, the names suggest the sea: the Porte de l'Arsenal (dockyard), the Porte des Galions (galleons), and the Porte de l'Organeau (mooring). The eastern Porte de Cordeliers was named for the order of Franciscan friars who lived in Aigues-Mortes from the thirteenth to the eighteenth century. The western Porte des Remblais (filling in) may have been named for the construction method of the ramparts, which called for two parallel stone walls to be filled with rubble.

The towers, most of which are located at the gates, also have names closely tied to the site: the Tour du Sel (salt), the Tour de la Mèche (the axle of a rudder), and the Tour de la Poudrière (powderhorn), to name a few. The Tour des Bourgignons, located at the southwest corner, got its name in this way: after the Armagnacs massacred the Burgundians in 1421, the cadavers were thrown into the tower and covered with salt to protect the town from pollution. The Tour Carbonnière, an advance post on the road leading to the Porte de la Gardette, served as a tollbooth from 1409 until the French Revolution. Once a gate fitted with portcullises, it now sits between two roads built around it in the nineteenth century.

The port of Aigues-Mortes remained an important import-export center in the thirteenth and fourteenth centuries. The Canal Vieil was early abandoned in favor of an access to the Bay of Repausset. The gradual silting of the canals continued to change the navigability of the port, and eventually it fell into decline. While several attempts were made to restore the port, it never regained its age of glory. Its population, more than 15,000 at the time of the crusades, has decreased to about 4,000 today.

Most of the history of Aigues-Mortes recorded since the fourteenth century centers on the Tour de Constance, which served as a prison for enemies of the king. In 1307 forty-five Templar knights were imprisoned and questioned by royal commissioners who hoped to force a confession that might bring down the Order of the Templars, thought to have too much power and wealth to suit the king's purposes. From 1375 to 1377 Charles of Artois served a sentence for treason in the tower, and in 1455 the duke of Alençon did likewise for having sided with the enemy.

During the religious wars of the 1560s the Tour de Constance was used as a prison for Protestants. The first was Hélie Boisset, a Calvinist preacher, imprisoned along with Pierre d'Aisse, the governor of Aigues-Mortes who welcomed him. Although the two were later hanged, by 1575 Aigues-Mortes had become a Calvinist stronghold.

A second wave of persecution followed the revocation of the Edict of Nantes, which had granted religious toleration. Many who refused to give up their beliefs were imprisoned in the late seventeenth and early eighteenth centuries in the Tour de Constance. It was ninety-eight feet high, twenty-two feet wide, and nearly twenty feet thick. The three-arched bridge connecting the tower to the ramparts was added in the sixteenth century. There were two large circular chambers inside, one above the other. A round opening in the vaulted ceiling of each chamber let in a small amount of light (and rain in the upper chamber).

There was little hope of escape from the tower, but Abraham Mazel miraculously freed himself and sixteen other Camisard prisoners in 1705 after seven or eight months of planning. Seventeen others, afraid to try for fear of punishments, stayed behind. Those who escaped let themselves down by a rope, scaled the ramparts, managed to get past the guards, and crossed the wet marshes. Had they been caught, they would have been forced to become galley slaves.

After the escape of the Camisards, the narrow openings in the walls of the tower were sealed with iron bars. Starting in 1715, only women were imprisoned in the Tour de Constance. Two other towers also served as prisons during this period: the Tour de la Reine and the Tour Saint-Antoine.

When Cardinal Fleury took the place of the duke of Bourbon in 1726, persecution became even more severe. Attendance at Protestant assemblies was punishable by life imprisonment or slavery. The Tour de Constance began to fill with women whose fathers, husbands, and brothers were sent to the galleys. Some women were imprisoned for more than forty years. Many became ill because of cold winds, mosquitoes, and poor hygiene. Provisions were sparse, and help from family members was often thwarted by fines or other obstacles. The number of prisoners in the tower at one time varied from twenty-eight to thirty-eight. Some died while imprisoned, and a few abjured their faith to gain release, but most stayed, refusing to compromise their principles.

One woman, Marie Durand, was imprisoned at the Tour de Constance at the age of fifteen. Her brother, Pierre Durand, was a well-known preacher. Unable to capture him, the authorities, wishing to make an example of his family, imprisoned his seventy-two-year-old father in a fort at Brescou, along with Marie's fiancé, Matthieu Serres. Marie, who was young and pious, became the spiritual leader of the women in the upper chamber. She attended the sick, wrote letters dictated by others, read from the Bible, and led the women in singing Huguenot hymns. Even though her brother was hanged in 1732, two years after she entered the tower, Marie was not released until 1768, at the age of fifty-three. Visitors to the tower today can still see Marie's misspelled word scratched near the rim of the opening to the chamber below: "*Reçistez*" (Resist!).

Other horrifying stories have survived this period. Isabeau Menet entered in 1734 with a three-month-old infant. He was allowed to remain with her for six years before being sent off to Geneva. Isabeau was released after fifteen years, but by then imprisonment had driven her insane. A letter from 1730 reveals that Suzanne Daumezon gave birth to a child while living in the unhealthy tower.

Little was done to improve the conditions for the women. They remained miserable and forgotten until the prince of Beauveau, governor of Languedoc, took pity on them in 1767. Repulsed by their situation, he went against

Louis XV's wishes and ordered the women released. After that, the Tour de Constance was converted to a prison for criminals.

Today the Tour de Constance stands as a reminder of the religious intolerance of the past, from the crusades to the Counter-Reformation. The ramparts remain largely as they were in the thirteenth century, except for some modifications that have been made over the centuries. Openings have been made to allow for cannon, the bridges and portcullises have been removed, and the rear entrances to some of the towers used as prisons have been sealed off.

The moat was filled in during the eighteenth century at the time of the digging of the canal between Aigues-Mortes and Beaucaire. This canal, along with the channel leading to Grau-du-Rol, improved travel within the Rhône delta area, but trade with other Mediterranean ports declined steadily with the silting of the port; it finally came to a standstill after World War I.

Despite the privileges accorded the residents by the kings of France, the inhabitants of Aigues-Mortes lived a difficult and impoverished existence for hundreds of years. The lack of fresh water led to sickness, malnutrition, and early death. Malaria was a problem until conditions were improved and quinine began to be used. In the twentieth century residents of Aigues-Mortes have been able to supplement their livelihood in the salt and fish industries with the cultivation of grapes and rice. The extraction of sea salt, a tradition dating back to the monks of Psalmodi and perhaps even to the Roman times, remains a major industry in the area. Weather conditions are generally favorable to the process. The fall winds push the saltwater over the dunes into the low areas of the coastal plain. The summer heat and winds cause the water to evaporate from the ponds, leaving salt deposits.

Today the approach to Aigues-Mortes is a tree-lined street leading to the Porte de la Gardette. Inside the site, a statue of St. Louis dominates the square that bears his name.

The château, located in the north corner near the Tour de Constance, was built in the sixteenth century as a residence for the governors of Aigues-Mortes. It is protected by a wall at the front and by the ramparts at the side and back.

The Church of Notre-Dame-des-Sablons is a Gothic church begun before the Crusade of 1248 and completed by the end of the thirteenth century. Its bell tower collapsed in 1634 during the religious struggles, and restoration was begun in 1741. The changes made used elements from several periods subsequent to the thirteenth century, so the result is a hybrid of architectural styles. Following the Revolution, the church was successively used as a Protestant church, a barracks, and a salt warehouse before being returned to the Catholic Church in 1804. The altar tables are from the former abbey of Psalmodi.

A refreshing contrast to the walled city with its endless souvenir shops and cafés is afforded by the nearby Camargue, a 300-square-mile delta plain located just to the east, formed by the division of the Rhône into two branches just above Aries. Here among the salt marshes and the alluvial deposits, white Camargue horses, thought to be related to a Mongolian breed, have run wild and fed on grass since Roman times. Black cattle are tended by cowboys called *gardians*. Red flamingos and hundreds of other species of waterfowl come here to live in the national nature preserve. No visit to Aigues-Mortes would be complete without a ride through the Camargue, as open and untamed as Aigues-Mortes is rigidly planned and enclosed. And yet, despite the endless movement of the sea, the land, and the river that surround them, both remain remarkably unchanged.

Further Reading: *Aigues-Mortes* by Alain Albaric (Château-de-Valence, Gard: Editions du Vent Larg, 1967) is a concise history of the city in French. It includes photographs and maps of the region, as well as a bibliography of earlier histories of Aigues-Mortes. *Aigues-Mortes: Un port pour un roi* by Georges Jehel (Roanne/Le Coteau: Editions Horvath, 1985) is a thoroughly researched study of the development of the Mediterranean port of Aigues-Mortes as socially and economically important for France and the Capetian kings in the Middle Ages. *Aries, Saint-Gilles, Aigues-Mortes, la Camargue* by Jean-Max Tixier (Avignon: Aubanel, 1972) is a guide for travelers, also available in an English version, of the history and monuments of the Rhône delta. Although most of the book centers on Aries, the chapter on Aigues-Mortes is informative. *Louis IX and the Challenge of the Crusade: A Study in Rulership* by William Chester Jordan (Princeton, New Jersey: Princeton University Press, 1979) is a scholarly analysis of Louis IX's pursuit of an ideal, of his commitment to his country and his church. A complete account of the Counter-Reformation in France is provided in the two volumes of *The Huguenots and the Revocation of the Edict of Nantes* by Henry M. Baird (New York: Scribner's, 1895; reprint, New York: AMS, 1972). A brief history of the women prisoners at Aigues-Mortes is told in *Les Prisonnières de la Tour de Constance* by Samuel Bastide (Vennes/Lausanne, Switzerland: Augur, 1957). Although written in French, the book has engraved drawings of the Tour de Constance and its prisoners on each of its sixty-four pages. It also includes Huguenot poems written about the persecution. *Discovering the Camargue* by Monica Krippner (London: Hutchinson, 1960), while not flattering to the walled city of Aigues-Mortes, is nevertheless a delightful description of the way of life in the Camargue as observed by a British woman who lived there for a short time.

—Sherry Crane LaRue

Aix-en-Provence (Bouches-du-Rhône, France)

Location: A little more than fifteen miles northeast of Marseilles, at the southwest edge of the Provençal alps, where the foothills merge with the Rhone basin; on the main road from the Rhone valley to the Italian peninsula.

Description: Former capital of Provence, built on the site of thermal springs; the birthplace and home of Cézanne and still a gathering point for artists; the seat of a renowned annual music festival; university town, host to the arts section of the University of Aix-Marseilles.

Site Office: Office de Tourisme, Syndicat d'Initiative
2 Place du Général de Gaulle
13100 Aix-en-Provence, Bouches-du-Rhône
France
42 16 11 61

The town of Aix was founded by the Roman consul Sextus Calvinus, who gave his name to its thermal springs, the Aquae Sextiae. One and one-half miles to the north, a federation of Celto-Ligurian tribes had had their capital on the plateau Entremont before Calvinus destroyed it in 123 B.C. Aix was the first Roman settlement in Gaul, and under Augustus it became a Roman *colonia,* a status that carried rights to Roman citizenship. The general Marius won a huge victory over migrating tribes in the area in 102 B.C., perhaps killing as many as 100,000, and Aix acquired part of the territories of Marseilles when that town lost its independence in 49 B.C. When the empire became Christian, a church with a baptistry (the font of which still remains) was built at Aix. The town was capital of its province as early as A.D. 375, and became seat of a bishopric in the fifth century. Eventually, it ceased to be a garrison town and became a resort.

Aix was taken by the Visigoths in the fifth century and thereafter repeatedly plundered by Franks and Lombards until it is said to have been captured by the Saracens in 731, although evidence for the extent and duration of Saracen activity in Provence has increasingly been challenged. Very little is known of the history of the Aix region from the seventh century until the tenth. In the eleventh there is clear evidence of the emergence of powerful clans, whose heads were to become the regional counts. In 1145 the right to mint coinage at Aix was conceded by the German king Conrad III to Raymond des Baux, sometimes regarded on almost no real evidence as a hero of Provençal independence and who was formally ousted in 1150 by the Catalan Bérenger family. In the first half of the thirteenth century, Raymond Bérenger V was the first count of Provence to reside for much of his time at Aix.

The only surviving thirteenth-century Provençal Gothic buildings are in or near Aix: the nave of Saint-Sauveur, the church of Saint-Jean-de-Malte, and the apse and chancel of Saint-Maximin-la-Sainte-Baume about twenty-two miles to the east. The teaching of law, began at Aix in 1409, inaugurating what was to become the university.

One of the most popular counts of Provence, René I of Anjou, reigned from 1431 until his death at Aix in 1480. As "*le bon roi René*" (good king René) he has acquired a high but probably unwarranted reputation for learning and artistic patronage. He did write verse and greatly improved the region's economy through innovations in the management of estates and farms. Nevertheless, he was forced into a political arrangement destined to ensure that Louis XI would inherit Provence and incorporate it into France.

In the first days of 1482 Palamède de Forbin, the agent of Louis XI, took possession of Provence at Aix in the name of the king. Palamède was removed in 1483 and replaced by Jean de Baudricourt, governor of Burgundy, who more savagely completed the incorporation of Provence into France in spite of the protests of the estates, then meeting at Aix. Charles VIII (1470–98) attempted conciliation in order to ensure the possibility of a safe retreat through Provence should his campaign in the Italian territories go poorly. Outbreaks of anti-Semitic violence in Provence from 1484 to 1501 (when the Jews were expelled), made clear to Charles's successor, Louis XII, that the administration of the region needed reorganization.

A *parlement* intended for Aix was inaugurated in Provence, but because of the plague, it met at Brignolles in 1502. The successor of Louis XII, Francis I, having failed to secure election as Holy Roman Emperor, came almost inevitably into conflict with his successful rival, Charles V. Charles V supported the insurgency of Charles, duc de Bourbon, who carried Provence in 1524 and occupied Aix on August 7. Bourbon was defeated at Marseilles and Francis reentered Aix triumphantly on October 1. When Francis resumed hostilities with Charles V in 1535–36, the emperor laid waste to much of Provence, entering a deserted Aix on August 9, 1536. A truce was signed at Nice in 1538, when Charles V was threatened from the east by the Turks, now in alliance with France.

Aix remained the capital, the administrative center, and the seat of the *parlement* of Provence until 1790. Education flourished there as well. Courses in medicine and theology were added to those in law; the university was then upgraded in an attempt to rival Toulouse and Montpellier; and a secondary school was opened in 1543. Still, the town had a reputation for being dirty. One contemporary report complains of the lack of privies and ditches, so that citizens were compelled to use their roofs, and sanitation was by unreliable and sometimes infrequent rain. Sixty lanterns were

A cafe in Aix
Photo courtesy of French Government Tourist Office

installed to light the streets in 1568. By the end of the six-teenth century, Aix was at the center of a small but important literary group, led by Guillaume du Vair and including Fran-çois de Malherbe.

In 1562, the first consul at Aix led the anti-Protestant reaction to the attempt by Catherine de Médicis to introduce toleration for Huguenots. After 1628, when the Huguenots were defeated by royal forces at La Rochelle, the Protestants began to regroup in and around the area of Aix-en-Provence. Their presence brought threats of retaliation from Cardinal Richelieu. In response, unrest developed among Aix's local magistracy, who were threatened with new taxes, an exile of the *parlement*, and the loss of their hereditary offices. The unrest soon spread to the masses as well, especially as con-ditions deteriorated with a new epidemic of plague in 1629. Ultimately, tax reform was averted, the hereditary offices were not abolished, and the *parlement* was allowed to return; however, a new inner court, known as the Chambre des Requêtes, was imposed from Paris in January 1641. The Collège royal, founded by Henry IV in 1603, was confided to the Jesuits in 1621 and absorbed the nascent university.

The city then by and large reflected the wider history of provincial France during the seventeenth and eighteenth centuries, when the grander houses of Aix were built. At Paris the third estate of Aix was represented at the 1789 estates by the revolution's greatest orator, the Honoré-Gabriel de Mirabeau, who was disinherited and discredited among his fellow aristocrats. After the revolution, the polit-ical and economic standing of the city declined as that of Marseilles swiftly increased. Today, the only hint of the cen-tral role Aix once played in the affairs of Provence and France is a summer music festival held in the archbishop's palace.

Aix's wealth of historic architecture is a constant re-minder of the city's former glory, however. The Cours Mi-rabeau, running along the site of the old ramparts, ends to the west in the Place du Général de Gaulle, formerly Place de la Libération, with its splendid nineteenth-century Fontaine de la Rotonde, and to the east with another fountain, by Pierre-Jean David d'Angers and portraying King René holding a bunch of the muscatel grapes that he is said to have intro-duced to the region. In the center of the cours at the inter-section with the Rue Clemenceau is the less distinguished-looking Fontaine d'Eau Chaude, from which flows water from the thermal springs that had attracted Sextus Calvinus more than two millenia ago. Banks, cafes, shops, and hotels now occupy the ground floors, especially on the north side of the street, but some mid–seventeenth century houses and gateways remain. Among the more distinguished examples of domestic architecture in Aix are the carved seventeenth-century doorways of the 1675 Hôtel Boyer d'Eguilles, now the Natural History Museum, the 1707 Hôtel d'Albertas in the Rue Espariat, the Place d'Albertas itself, the Musée du

Vieil Aix, and buildings in the Rue Aude and the Rue du Maréchal Foch.

The seventeenth-century Hôtel-de-Ville is next to a sixteenth-century bell tower with its Provençal wrought-iron bell cage and has a fine ironwork balcony, while the old grain market, now the post office, has eighteenth-century carvings. From the courtyard of the Hôtel de Ville, there is access through a grand staircase to the Méjanes library. Aix contains other rich collections, the Musée des Tapisseries in the arch-bishop's palace, the paintings in the Musée Granet, the pic-tures and furniture of the Pavillon de Vendôme, and the regional collection and ceramics in the Musée Paul-Arbaud. Also, Paul Cézanne's former studio, Les Lauves, contains numerous mementos of the artist, who was born in Aix in 1839. The town's most important and impressive building, however, is doubtless the cathedral of Saint-Sauveur, whose medley of styles makes it more historically interesting than aesthetically attractive.

It is the simple Romanesque cloisters with their pairs of small columns that visitors find more attractive, although the cathedral itself has, together with its fifth-century font, a fine fifteenth-century triptych by Nicolas Froment, painter to King René, and a fifteenth-century stone retable. The cele-brated Brussels tapestries depicting the life of the Virgin and made for Canterbury Cathedral are now no longer on display, since three were stolen in 1977. The west front has sixteenth-century walnut carvings of twelve sybils and some of the Jewish prophets, and an unfinished, Gothic octagonal belfry from the fourteenth and fifteenth centuries. The architecture of the center is flamboyant sixteenth-century Gothic.

Aix is still a relaxed town of broad avenues, squares, and fountains. The center feels remote from the nearby con-gress hall, casino, university, and new housing and industrial developments. It is the heart of France's almond-growing industry, and produces a characteristic local delicacy made of diamond-shaped iced almond paste, called *calissons*.

Further Reading: In addition to city guides and articles in gaze-teers and encyclopedias, there is a rich literature, some of it in English, on the region of Provence and on Aix, for so long its capital. Sheer professional competence demands the mention of Jean-Jacques Gloton's *Renaissance et baroque à Aix-en-Provence. Recherches sur la culture architecturale dans le midi de la France de la fin du XVe siècle au début du XVIIIe siècle,* in 2 volumes. (Rome: French School at Rome, 1979) and Donna Bohanan's *Old and New Nobility in Aix-en-Provence 1600–1695. Portrait of an Urban Elite,* (Baton Rouge and London: Louisiana State University Press, 1992). Both works are understandably technical, but Boha-nan's is of particular interest because of the English-language con-tent of the bibliography. The best general introduction to Provence, including Aix, is the *Histoire de la Provence,* edited by Edouard Baratier (Toulouse: Privat, 1969). This work also contains a useful bibliography.

—Anthony Levi

Alba Iulia (Alba, Romania)

Location: On the terraces of the River Mureşcul, in the central plain of Romania known historically as Transylvania, 230 miles from Bucharest.

Description: Historic city with evidence of human settlements dating to the Neolithic era; closely linked with Romanian nationalist movements; today divided into a lower town and an upper town, known as the Citadel, which encloses numerous historic buildings behind a ring of fortifications.

Site Office: Tourist Information
Cetatae SA
Piata Iuliu Maniu 22
Alba Iulia, Alba
Romania
58 811195

Alba Iulia, long an occupied site, was known as Apulum by the Romans, as Bălgrad by the Slavs, as Karlsburg by the Austro-Hungarian aristocracy, and as Gyulafehérvár by the Hungarians. Archaeological excavations at the town have revealed settlements dating back to the Neolithic era. By the Bronze Age, the population of the region—which later became known as Transylvania—was divided into a number of tribes who formed a northern branch of the large ethnic group of Thracians. The ancient Greek historian Herodotus knew them as the Getae, "the most courageous and loyal of the Thracians." Roman historians identified the main tribe as Dacians. A historian writing prior to the second century recorded that "central Transylvania was inhabited by Appuli," doubtless a tribe associated with what is today Alba Iulia. More than forty Dacian urban settlements in central Transylvania, including one called "Apulon," were mapped by second-century historian, Ptolemy.

In 217 B.C. the Roman armies began to move north toward the lands of the Dacians, and Roman rulers increasingly dominated the region. The first and early second centuries A.D. saw frequent conflicts between Dacians and Romans. The native forces prevailed temporarily, and the Romans were forced to sign a treaty favorable to the Dacians. Still, the Romans gained a foothold in the region. Roman engineers and officers came to Dacia to build fortifications and advise the Dacian army. After renewed disturbances, Emperor Trajan subdued the Dacians in 102. A treaty was negotiated, but was later violated by the Dacians. This time a holy war was declared by the Roman Senate; it ended in the total annexation of Dacia and its transformation into a Roman province with a governor, Roman administration, and occupation by the Roman army.

The Dacians rapidly assimilated, adopting the Latin language, and Rome inducted many Dacian cohorts into the Roman army. Trajan resettled numerous people in Dacian towns and villages from all over the empire. Apulum was one of the chief sites of this colonization. It was also important militarily; from the start of the occupation, Apulum was the home of Legion XIII Gemina. The stone fortress built to house the legion would be the site of four successive citadels over the centuries. When Dacia was divided into three provinces in 159, Apulum became the capital of the province named Dacia Apulensis. Two Roman roads originating in the Danube region met in Apulum, then continued on to the north.

Various pieces of evidence suggest that, despite the extent of assimilation, the native population was able to retain many aspects of its identity. We know of the continuous presence of native Dacian culture from the names of settlements, including Apulum; from archaeological evidence, such as gravestone inscriptions; and from surviving chronicles. Perhaps the best evidence for the strength of the native culture is the continued presence of insurrections against the Romans.

As conflicts in the region escalated, the Romans began to withdraw. The province of Dacia Superior was abandoned by Hadrian after 117, and the legion, with one cohort still stationed at Apulum, became the only official Roman presence. The withdrawal led to further instability in the region. The area was repeatedly invaded by Germanic tribes, most notably the Goths. Ultimately, in 271 and 272, all of Dacia was officially abandoned as a Roman province.

Nevertheless, the Roman presence was not completely obliterated, even after the official evacuation. As with the continued presence of the native culture, the presence of a substantial Roman population during this later period is verified by a wealth of archaeological evidence. In the early 1980s, for example, excavations at the ruins of the former city of Apulum revealed a Roman cemetery with graves dating well into the Middle Ages; of the 450 graves, 75 contained remains from the second and third centuries, 195 dated from the seventh through tenth centuries, and 180 were from the eleventh and twelfth centuries. The discovery of dated bronze coins and other goods at excavation sites suggests that the settlement retained an urban character well into the fifth century. Even as Germanic, Slavic, and Magyar (Hungarian) invaders settled in the area, the Daco-Roman population remained.

The constant succession of invaders left little recorded history of the region over the ensuing years. In the ninth and tenth centuries, a palisaded earthwork citadel was erected on the site of the old Roman fortress at Alba Iulia. The earthwork was most likely built to protect the prince and court of the local *voivodate,* a political organization of tribes; according to different historical sources, these tribes were either

The fortress gate at Alba Iulia
Photo courtesy of Romanian National Tourist Office

Slavic, Slavo-Romanian, or Romanian in ethnic character. It was around this time that the city acquired the name Bălgrad (White Town). This was also the period that the Magyars solidified their authority in Transylvania. Between 1103 and 1113, the Magyars established a bishopric at Alba Iulia and introduced the religious orders, most notably the Cistercians, into the region.

The year 1241 brought renewed invasion to Alba Iulia, this time by the Mongols, who destroyed most of the town's major buildings. One of the casualties was the twelfth-century Romanesque cathedral. A French architect supervised the rebuilding of the structure, completed in 1300. The new Catholic cathedral followed the design of a traditional three-aisled Romanesque basilica, with early Gothic cross-vaulting. The fortress also needed to be rebuilt as a result of the Mongol invasion; the new structure featured a stone wall surrounding the cathedral and the bishop's palace.

As Romania struggled for independence from this long succession of invaders and conquerors, Alba Iulia became a symbol of nationalist aspirations. The city was first linked to the nationalist cause by Michael the Brave (Mihai Viteazul) at the end of the sixteenth century. After the Ottomans invaded the region in the fifteenth century and Hungary

fell to the Turks at the battle of Mohács in 1526, Transylvania was detached politically from Hungary and made a distinct vassal principality of the Ottoman Empire. Michael the Brave fought to throw off this vassalage, and in the process became the first native ruler of a united, autonomous Romania. Originally the ruler of Walachia (the principality south of Transylvania), Michael set up a series of shifting alliances with Transylvania, Moldavia (the principality east of Transylvania), and the Habsburgs of Austria; by the end of the sixteenth century he had not only driven out the Turks, but had conquered Transylvania as well, and would soon gain dominion over Moldavia, thereby uniting all three of the Romanian principalities.

On November 13, 1599, Michael entered Alba Iulia triumphant. Although he would rule there for only eleven months, he made a lasting impact both on the town and on the psyche of nation. As an ethnic Romanian, he enacted measures to alleviate the oppression of the Romanian population and clergy. He erected a cloister in Alba Iulia and took up residence at the fortress, which had been rebuilt again in 1561. (The ruins of this structure still stand and can be seen today behind the bishop's palace.) Michael's reign was short-lived, however. In September 1600 he was forced out of

Transylvania by the Habsburgs and the local noblemen, who feared his centralizing policies. On August 19, 1601, he was killed by imperial general Giorgio Basta. Transylvania was once more caught in the conflict between the Ottomans and the Habsburgs.

By this point, Alba Iulia had long been an engine of cultural advancement in Eastern Europe. In addition to being a Catholic bishopric, it was a Eastern Orthodox metropolitan's seat. During the reigns of Bishop Lepes (1427–42) and Prince N. Zápolya (1461–68), state and Catholic church schools were founded there. Many of the new schools established in the following centuries were founded to spread the ideas of the Reformation, in particular the theories of John Calvin, which attracted the support of the Transylvanian princes. Chancellor Gábor Bethlen (1642–1716), György Rákóczi I (1593–1648), and Michael Apaffi (1632–90) all advanced the cause of education by founding schools that provided instruction following the Calvinist catechism. Bethlen's foundation received scholars from humble backgrounds who were destined to become Calvinist preachers; his rule also saw the founding of an academy whose instruction included Romanian, Greek, and Latin courses. The historian and statesman Count Brancovic (1645–1711) was educated at Alba Iulia in the metropolitan's school; he went on to serve Prince Apaffi. From the town school in Alba Iulia founded by Bishop Lepes in the mid–fifteenth century, there grew the Theological Academy of Bishop Stoica. In 1560 there also existed in the town a school of ecclesiastical music; its director was a Frenchman. The Academium Colegium was founded by Bethlen in 1629; it boasted a number of illustrious scholars, including Martin Opitz and Piscator.

One of the most important results of this educational renaissance was a wealth of new publications, many in Romanian. Gábor Bethlen himself produced two notable works: a *History of Transylvania* (1650–1700) and an *Autobiography* (1708–10). During the reign of György Rákóczi I, a publishing venture was established in which Romanian priest Popa Dobre was employed by Matei Basarab to print, among other things, a Calvinist *Catechism* (1640). Under Michael Apaffi, further editions of the catechism were published, as well as the *Bucoavna* dictionary. In addition, an *Atlas* appeared in 1639, the *Gospels* in 1641, the *New Testament* in 1646, and a Latin *Catechism* in 1656.

The second half of the seventeenth century saw a decline in this provision of knowledge. In 1658 Alba Iulia was devastated by the Turco-Tartar invaders; one of the casualties was the print works, which was rebuilt in the town of Sebes, ten miles south of Alba. After 1700 Alba Iulia temporarily lost its place as a center of cultural influence, possibly because of the removal of the metropolitan's seat to Fagaras and the fact that the reputation of the Romanian school and print works was superseded by others. But by the end of the eighteenth century, Alba Iulia was once more in the vanguard of cultural studies in Eastern Europe. Ignatius Batthyány was bishop of Transylvania from 1780 to 1798 under Empress Maria Theresa. It was he who furthered the study of many subjects, most notably astronomy, by the foundation of the Batthyanaeum in Alba Iulia. This library, which may still be visited, housed a collection of books and manuscripts unique in eighteenth-century Transylvania.

A series of religious synods held in Alba Iulia in the late seventeenth and early eighteenth century gave further importance to the town. It was these synods, running from 1697 to 1701, that gave birth to the Transylvanian Uniate Church, which united the Catholic and Eastern Orthodox branches in Romania. Adherents to both churches were to be given equal rights, but eventually the Catholics came to dominate, a development that gave rise to Romanian resentment and fed the spirit of nationalism in the coming decades.

In the late seventeenth century the Habsburgs finally gained the upper hand over the Ottomans, after decades of warfare that devastated the Transylvanian countryside. Alba Iulia, for a time renamed Karlsburg in honor of Emperor Charles VI, saw major architectural improvements under the Habsburgs. The bishop's palace was renovated, and the church that eventually housed the Batthyanaeum Library was constructed. The Catholic cathedral, to which a Renaissance-style chapel had been added in 1512, was altered again: the west doorway was rebuilt to feature a triangular pediment with four baroque statues. Most notable, however, was the fourth and final restoration of the fortress, now officially the Karlsburg Citadel. The project lasted from 1715 to 1738 and employed 20,000 serfs. Giovanni Morando Visconti, an Italian architect, designed the structure, and Prince Eugene of Savoy supervised construction. A cross-shaped fort was constructed and surrounded by a massive brick wall with numerous bastions and a two defensive ditches. Perhaps the most striking features of the citadel are its elaborately carved, baroque entrance gates, which lead to a maze of tunnels and passageways.

Habsburg domination became increasingly resented by the Romanian peasantry. Preferential treatment and political representation was given only to the Magyars and Saxons, minority populations in the country. Conditions for the Romanian majority deteriorated, and revolutionary movements took hold. Once again, Alba Iulia became a focal point and symbol for these aspirations.

In 1784, three Romanians—Horia, Closca, and Crisan—came to lead a group of peasants who wrongly interpreted a recent conscription act as the first step toward the abolition of serfdom. They believed that Horia's earlier meeting with Emperor Joseph II gave them the right to arm the peasants, assemble them at Alba Iulia, and attack any who stood in the way of emancipation. When the authorities tried to arrest Crisan in the town of Zărand on November 1, 1784, a riot erupted and soon spread across the countryside. Before the imperial army finally put down the rebellion in mid-December, 230 manor houses had been sacked and 100 nobles murdered.

Following their capture, the three leaders were imprisoned in the citadel of Alba Iulia. After Crisan committed suicide in his cell, Horia and Closca received the full brunt of imperial retribution. On February 28, 1785, before a crowd of 2,500 peasants on a plateau still known as Horia's Field, they were broken on the wheel, disemboweled, and dismembered. The leaders are memorialized in the Alba Iulia Obelisk, at the entrance to the Citadel.

The peasants' desire for emancipation soon merged with nationalist desires for Romanian representation in the imperial government and for the preservation of Transylvania's status a principality, distinct from Hungary. Alba Iulia was central to this movement as well. In what was seen by the Romanians as a major victory, the emperor announced a meeting of Magyar, Saxon, and Romanian representatives in Alba Iulia to begin February 11, 1861. Once again, the Romanians' optimism was unwarranted. The very composition of the representative body was designed to thwart their goals: there were 24 members to represent the Magyar population of 500,000, 8 members to represent the Saxon population of 200,000, and 8 members to represent the majority Romanian population of 1.4 million. Furthermore, the conference was to be chaired by a Magyar. It soon became apparent that the Magyars wished only to discuss how to incorporate Transylvania into Hungary. After two days, the representatives saw that no agreement could be reached among the three groups, and they dissolved the conference, submitting three separate recommendations to the Transylvanian chancellor.

Although Moldavia and Walachia came together as a united Romanian kingdom in the late nineteenth century, Transylvania would not join them until the collapse of the Habsburg Empire at the end of World War I. Fittingly, the new union of Greater Romania would be declared in Alba Iulia. On December 1, 1918, an assembly of 1,228 elected delegates from all walks of life—clergymen, soldiers, students, even representatives of sports associations—gathered in the historic town. A crowd of 10,000 peasants, workers, and townspeople cheered their final resolution: "The National Assembly declares the Union of the Romanians living in Transylvania, Banat, Hungary and of their territories with Romania." The assembly also declared a number of democratic measures meant to protect the rights of ethnic minorities; sadly, these principles were quickly forgotten. The same ethnic and nationalist tensions that had existed before the war continued to plague the new country.

Soon after the creation of Greater Romania, Prime Minister Alexandru Averescu proposed a symbolic coronation for King Ferdinand and Queen Marie at Alba Iulia. The king and queen initially refused to participate, not only because of their political differences with Averescu, but also because the bad economic climate made a lavish celebration inappropriate. They finally gave their consent in 1922, when Ion I. C. Brătianu reclaimed the prime ministry. Ironically, Brătianu was extremely unpopular among Transylvanian politicians; ultimately, they boycotted the event meant to celebrate their admission into the Romanian union. Nor could Romania's religious conflicts be put aside for the sake of the coronation. Because the ceremony was to be performed by a Romanian Orthodox priest, Catholic authorities could not abide it taking place in a Romanian church as well, even though the church had been built expressly for that purpose, and at no small expense. The ceremony ultimately took place in front of the church, on the pretext that the event should be witnessed by all the Romanian peoples.

The October 15, 1922, coronation was a lavish affair, attended by scores of foreign dignitaries (if no local ones), and modeled after the coronation of Nicholas and Alexandra at Moscow. While planning the event, Queen Marie declared, "I want nothing modern that another Queen might have. Let mine be all medieval." She later described the coronation as "a modest ceremony in comparison to others I have known," but in private correspondences she is said to have expressed embarrassment at its pomp and opulence. The event cost approximately $1 million.

In 1948 Alba Iulia hosted another lavish celebration of Romanian unity; once again, it would only expose the nation's underlying conflicts. This ceremony was religious: October 7, 1948, marked the 250th anniversary of the Alba Iulia Union and the birth of the Romanian Uniate Church. Metropolitan Iustinian Marina, patriarch of the Orthodox Church, saw the anniversary as an opportunity to undo the very unification they were celebrating. The extravagant ceremony proved the culmination of his campaign for Romania to "return to the Faith and Church of [its] ancestors." In the Alba Iulia Cathedral he received the act of submission of 423 priests; absent, of course, were the four Uniate bishops, who were subsequently excommunicated. The conflict was soon pressed still further. The Orthodox Church had, in the words of historian Georges Castellan, fallen into "a workable *modus vivendi*" with the Communist regime, and their influence became apparent in the ensuing months; not only were the four Uniate bishops imprisoned in December, but the following October their bishoprics were abolished and their property transferred to the Orthodox clergy.

Aside from its use as a backdrop for historical ceremonies, Alba Iulia played only a minor role in the twentieth-century history of Romania. In the Communist era it was an industrial center for the production of leather, boots and shoes, building materials, and foodstuffs; the Romanian army still occupied part of the citadel during this period. In 1975, the city celebrated its 2000th anniversary. More recently, the town's association with Romanian nationalism was renewed, albeit rather ominously. In 1990, a Vatra Party rally in Alba Iulia sparked a violent reaction from the crowd when the speakers stirred up lingering anti-Hungarian sentiment.

Further Reading: *Apulum,* the journal of the Regional Museum of Alba Iulia, covers most aspects of the town's history. The most recent works on Romanian history are Keith Hitchins's *Rumania, 1866–1947* (Oxford: Oxford University Press, 1994) and Martyn C. Rady's *Romania in Turmoil: A Contemporary History* (London: Tauris, and New York: St. Martin's, 1992). Keith Hitchins's *Orthodoxy and Nationality: Andreiu Şaguna and the Rumanians of Transylvania, 1846–1873* (Cambridge, Massachusetts, and London: Harvard University Press, 1977) contains an excellent discussion of the February 1861 conference at Alba Iulia. A detailed, if at times sloppy, history of Romania from the Dacian period to the Communist era can be found in Georges Castellan's *A History of the Romanians,* translated by Nicholas Bradley (New York: East European Quarterly, 1989; as *Histoire de la Roumanie,* Paris: PUF, 1985). R. W. Seton-Watson's *A History of the Romanians: From Roman Times to the Completion of Unity* (Cambridge: Cambridge University Press, 1934; Hamden, Connecticut: Archon, 1963) is dated but helpful in its discussion of the medieval and early modern periods.

—Patrick Keeley and Robert M. Salkin

Albi (Tarn, France)

Location: Midi Pyrenees region, in the region of Languedoc, in southwestern France; capital of the department of Tarn; forty miles northeast of Toulouse.

Description: Albi is situated on the Tarn River, where the high and dominating Massif Central begins to give way to the Garonne Plain. A city of over 40,000 inhabitants, it is sometimes referred to as the ''red city,'' due to the color of the brick used in much of the city's construction. The oldest surviving parts of the city date from medieval times; an eleventh-century bridge spans the Tarn river. The thirteenth-century Cathedral of St. Cécile and the Palais de la Berbie (bishop's palace), also from the thirteenth century, are the city's most important historical buildings.

Site Office: Office de Tourisme et Syndicat d'initiative
Palais de la Berbie
Place Sainte Cécile
81000 Albi, Tarn
France
63 54 22 30

Albi, on the banks of the Tarn River, lies at the northern border of the province of Languedoc. The site of present-day Albi has been inhabited by humans since prehistoric times, and the Languedoc itself has seen the march of many armies over its lands.

In 219 B.C. Hannibal passed through the Languedoc region on his march from Africa to the Alps to do battle with the Roman Empire. At the beginning of the fifth century, Albi suffered a wave of invasions by Germanic tribes from the east, first by the Vandals, then the Visigoths. At the same time that these invasions combined conquest with an assimilation of the existing inhabitants and the invaders, they marked the beginning of the end of the Western Roman Empire. In 650, Albi became part of the independent region of Aquitaine, and in 760 Albi was again conquered, this time by Pepin the Short, father of Charlemagne.

At the beginning of the Christian era, Albi was part of the Gallo-Roman region called Albigenses. In the fourth century, Christianity came to Albi; its first known bishop, Diogenien, was appointed at the beginning of the fifth century. Through the succeeding centuries, Christianity maintained a strong presence in Albi, and there is mention of the existence of a cathedral there, dedicated to St. Cecile, as early as 920.

In the tenth century, a heretical belief that became known as Catharism began its spread westward over Europe from the east, eventually reaching the Languedoc. Those in the Languedoc who held these beliefs were given the name Albigenses, literally ''the people of Albi.'' The belief of the Albigensian heretics, essentially dualist in nature, maintained that the principles of good and evil, God and Satan, existed together in the world and were equally powerful. All material things belonged to the domain of Satan (God, so the heretics held, was not omnipotent in this realm), and salvation was to be attained only through an absolute rejection of the material world. They rejected the sacraments of the church, which drew on the earthly materials of bread, wine, and water, and they denied the power of priests as the essential mediators between God and man. The Albigenses refused to eat any food that was the product of sexual breeding. They rejected marriage and suppressed, as much as possible, every bodily need. Because very few documents composed by the Catharist heretics exist today, the exact nature of their heresy is unknown. Most of our present day knowledge of them comes from their enemies who destroyed them.

Although never great in number, the Albigensian ascetics preached their beliefs to the people of the Languedoc, and the power and influence of the heretics grew, under the protection of William IX, duke of Aquitaine, and other nobility of the Languedoc. In 1119, at the Council of Toulouse, Pope Alexander III was unsuccessful in gaining the aid of the secular powers to quell the heresy. The activities of the Albigensian heretics continued unobstructed until the ascension of Pope Innocent III to the papacy.

Although Innocent III was at first committed to a peaceful conversion of the Albigensian heretics, the murder of Peter of Castelnau, a papal legate, in January 1208 set him on a bloody course. The murderer was identified as a servant of Raymond VI, Count of Toulouse. Previously, the murderer had been witness to a bitter dispute between Raymond VI and the slain papal legate. The pope, when he received this information, proclaimed Peter of Castelnau a martyr and Raymond VI directly accountable for his murder. Innocent III ordered the Cistercians, the order to which Peter of Castelnau belonged, to preach a crusade to destroy the Albigensian heresy. The pope offered the same indulgences to the knights who participated in this crusade as to those who had gone to battle the infidel in the Holy Land. By 1209 the Albigensian crusade had been raised.

The crusade was led by Simon de Monfort, a veteran of the Fourth crusade to the Holy Land. By then in his late forties (old by medieval standards), de Montfort was tall and had remained physically strong. He was universally admired by his contemporaries and was genuine in his hatred of the heretics. In July 1209, he attacked the two strongest cities in the south, Beziers and Carcassonne, both under the control of Raymond-Roger Trencavel, viscount of Beziers. The well fortified city of Beziers fell to the crusaders in a matter of a few hours, due to the confusion and disorganization of its citizen defenders and the fierceness of the attack: crusaders

A view of Albi from the cathedral tower
Photo courtesy of French Government Tourist Office

looted and set the town ablaze, massacring 20,000 people, many at prayer in the cathedral. The sack of Beziers—more spontaneous than planned—had a chilling effect on the citizens of the Languedoc. As the crusaders marched toward the great walled city of Carcassonne, they encountered towns and villages whose citizens had fled before them into the countryside. As a result, the crusaders reached Carcassonne relatively quickly, having faced little or no resistance.

The siege of Carcassonne, strongest of Raymond-Roger Trencavel's cities, lasted several weeks. Simon de Montfort personally led the assault on the *castellare* of Carcassonne, a sort of fortified suburb, and, without assistance, carried a wounded knight to safety, away from the fortifications, all the while under attack by the defenders' arrows. Carcassonne fell in August 1209. Minerve and Lavour, two strongholds of the Albigensian heretics, were then taken, and in September 1209, Albi, too, fell to the forces of Simon de Montfort. Raymond-Roger Trencavel, imprisoned near Carcassonne after its fall, died of dysentery in November 1209, whereupon Simon de Montfort was given Trencavel's viscounty. Simon met his end on June 25, 1218, during the siege of Toulouse. He was struck and killed by a block of masonry hurled by female defenders from the city's walls. The news of his death resulted in a joyous celebration in the besieged city.

The crusade against the Albigensian heretics continued for another twenty years, until the signing of the Treaty of Paris in 1229 between Raymond VII of Toulouse, the son of Raymond VI, and King Louis IX of France, later canonized as Saint Louis. Although the crusade was officially ended with a pledge of fealty by Raymond VII to Louis IX, the Albigensian heresy lingered in pockets of the Languedoc. It was left to the Inquisition, following on the heels of the crusade, to break the power of the Albigensian heretics. The final pitched battle, a siege against the heretics at Montségur, lasted for eight months in 1244. Two hundred defenders, along with their families, were burnt alive. In 1300, the court of the Inquisition at Albi, led by the bishop of Albi, Bernard de Castanet, tried thirty-five nobles suspected of heretical beliefs. The nineteen that were found guilty were shackled for the remainder of their lives.

The Albigensian crusade is recognized as one of the most profound conflicts of the Middle Ages, not only for the monumental bloodshed and destruction which took place, but also for the role the crusade played in the shaping of modern-day France. Prior to the crusade, when one spoke of France one meant northern France. Albi and its province, the Languedoc, were separated from Paris and removed from the control of the Capetian king by the Massif Central, a large and sparsely populated area with only a few rivers and no roads. The Languedoc functioned as a semiautonomous province with its own language (*langue d'oc*). Albi had been under the direct control of the ill-fated Trencavel family, viscounts of Beziers and vassals of the count of Toulouse. It was the age of the troubadors, musicians and storytellers who ranged from court to court of the petty nobles in the Languedoc and the other southern provinces, singing songs and telling tales in their native *langue d'oc,* a language that Dante's contemporaries regarded as finer than either Italian or French. It is doubtful whether the troubadors contributed to the spread of the Albigensian heresy. In contrast to the Albigensian heretics, the troubadors, and their more motley companions the *jongleurs,* sang of war and of the pleasure of suffering for unrequited courtly love.

After the crusade came to an end, the manners and the civilization of northern France began to be felt in the south. The crusade not only broke the power and influence of the Albigensian heretics, it also established the influence of Saint Louis and future French kings to the Mediterranean and the Pyrenees. The petty nobility that had supported the troubadors was destroyed. Government was now centralized in Paris, and France was beginning to assume what Cardinal Richelieu was later to consider its natural boundaries—the Pyrenees, the Alps, and the Rhône River.

In 1264, the secular power of Albi was placed in the hands of its bishops. In 1277, Bernard de Castanet began construction of a new cathedral there. An earlier cathedral had existed in Albi since at least the tenth century. This new cathedral, dedicated like its predecessor to St. Cécile, would reflect the city's recent turmoil. Its red brick exterior resembles not so much a cathedral as a fortress. Although high gothic architecture had reached the south by the mid-thirteenth century the cathedral of St. Cécile is basically Romanesque, with sheer walls rising vertically, unbroken by ornament or window until its upper storeys. The design has no exterior flying buttresses that an enemy might scale or find shelter under during an attack. The cathedral was built as a stronghold by Bernard de Castanet, perhaps because one of his predecessors had been held captive in the old cathedral by the heretics. The austerity of the exterior is interrupted only by a fifteenth-century entrance on the cathedral's south side, covered by a fanciful porch carved from white stone, in the sixteenth century.

The cathedral's design is unique in that it has no interior transept and no side aisles. In contrast to the stern exterior, its interior is highly decorated. A fifteenth-century fresco of the Last Judgment that stands at the west end of the cathedral features such illustrious French leaders as Charlemagne, St. Louis, and Blanche de Castile situated beneath the angels and the apostles, reflecting the inseparability of the secular power of the French state and the church in the thirteenth century. Richly carved stone screens enclose the choir, whose niches were filled with statues of saints until Revolutionaries destroyed them.

In 1346, a famine ravaged Albi, followed in 1348 by the black plague, which arrived in France through the port of Marseilles in January 1348 and spread west to the Languedoc. The black plague reached Albi on July 8, 1348. In the smaller towns, the plague took from four to six months to take its victims before it died out. Estimates are that one-third of the population of Europe was killed by the black plague, with some areas losing 90 percent of their population.

Subsequent to the black plague, from 1361 to 1382, Albi, like many towns in southern France and Italy, suffered under the menace of the *routiers,* remnants of the free companies of mercenaries that fought on the continent of Europe in the various wars of the fourteenth century. Bands of trained, armed mercenaries, the *routiers* came from every occupation and class (including the church itself), ranging from street ruffian to knight. They formed under a captain and pillaged cities and towns. Those with walls could reasonably defend themselves, but the unfortified cities and the countryside fell prey to the *routiers,* paying a tribute to buy freedom from attack. The problem grew so serious that Pope Urban V, installed in the papal palace at Avignon, issued bulls of excommunication for anyone giving aid or supplies to the *routiers* and granting special indulgences to those who fought against them.

The secular rule of the bishops at Albi continued until 1678, when they were replaced by archbishops who governed the city until the French Revolution in 1789. The cathedral of St. Cécile, like many other churches in France, suffered during and after the Revolution. In 1794, the anti-christian Cult of Reason, an extremist revolutionary group organized by Jacques Hébert, took control of the cathedral and proclaimed it a Temple of Reason. Hébert, along with his Cult of Reason, were eventually eliminated by Robespierre; the Roman Catholic rite was reestablished at the cathedral in 1804.

In 1864, the same year the railroad came to Albi, Henri de Toulouse-Lautrec, the French painter who portrayed the nightlife of Montmartre, was born there at the Hôtel du Bosc. Albi was the family's ancestral home, and one of Henri's ancestors, named Baudoin, fought in the Albigensian crusade.

After Toulouse-Lautrec's death in 1901, a search was begun to find a suitable location for the works he had left to his mother. The Palais de la Berbie, which had been given to the town of Albi in 1907 by the Department of the Tarn on condition that it be turned into a museum, proved to be the perfect location. On July 30, 1922, the museum was opened by Léon Bérard, Minister of Public Education and Fine Arts. Over 600 works of art by Toulouse-Lautrec are housed in the museum today, including early sketches of his parents and the posters that brought him fame: Jane Avril at the Moulin Rouge and Louise Weber dancing at the Moulin de la Galette, among others. The works of art donated to the museum by his mother, the countess Adele, are owned by the city of Albi in perpetuity.

Today, Albi is the center of tourism for the gorges of the Tarn river. An industrial center, it manufactures cement, dyes, flour, synthetic textiles, and glass.

Further Reading: *Languedoc* by James Bentley (London: George Philip, and Topsfield, Massachusetts: Salem House, 1987) is a casual, informative guide, with photographs, to the history and small towns of the Languedoc. *The Albigensian Crusade* by Jonathan Sumption (London: Faber, 1978) is a scholarly but very readable history of the Albigensian crusade and its effects on the social, political, and ecclesiastical life of southwestern France. Detailed attention is also paid to individual battles and military strategy. Barbara W. Tuchman's excellent *A Distant Mirror* (New York: Knopf, 1978) vividly chronicles the events of the fourteenth century in France and England. *Toulouse-Lautrec* by Henri Perruchot, translated by Humphrey Hare (Cleveland: World Publishing Company, and London: Joseph, 1960), is an account of the artist's life and includes sketches by Toulouse-Lautrec at the beginning of each chapter, along with photographs of the artist and his family and associates. Finally, in French, the *Histoire d' Albi* (Toulouse: Editions Privat, 1983), compiled under the direction of Jean-Louis Biget, is a detailed history of Albi, including a chronological table of events. Unfortunately, it is unavailable in English.

—Rion Klawinski

Amboise (Indre-et-Loire, France)

Location: In west central France, on the left bank of the Loire River, about 15 miles east of Tours and 55 miles southwest of Orleans; 130 miles southwest of Paris.

Description: A small town dominated by the remains of a château that was, in the fifteenth and sixteenth centuries, the principal residence of the royal family, thus making Amboise the virtual capital of France. After the château ceased to be used by the royal family, much of it was demolished by its subsequent owners in the eighteenth and nineteenth centuries, and by bombardment during World War II. The north facade and several rooms of the château remain intact, however, as do its two imposing towers, the Tour Hurtault and the Tour des Minimes, and the St.-Hubert Chapel. Near the château is the Clos-Lucé, a manor house once occupied by Leonardo da Vinci.

Site Office: Syndicat d'Initiative
Bureau Municipal de Tourisme
37402 Amboise, Indre-et-Loire
France
47 57 01 37

Quite shallow archaeological investigation has demonstrated the existence of prehistorical settlers at Amboise and a later Celtic settlement at the highest point of the cliff on which the Amboise château now stands. In 56 B.C. the apparently flourishing Celtic village was turned by the legions of Crassus into the fortified town Ambactia. In the first decade of the sixth century the victory of Clovis and his Franks over Alaric II and the Visigoths at Vouillé ensured that Amboise and the Loire territory became subject to Frankish laws and customs.

According to Sulpicius Severus, St. Martin's regrettably unreliable hagiographer and disciple, the region had been christianized in the late fourth century by Martin, the bishop of Tours, who had a pagan temple replaced by a church dedicated to St. Denis. According to legend, a warrior named Ingelgar removed St. Martin's relics so they would not fall into the hands of Norman invaders in 853, and Ingelgar received the reward of suzerainty over the fief of Amboise, granted him by Charles the Bald. Ingelgar's descendants were the counts of Anjou, also famous as the Plantagenet family, who eventually occupied the English throne. Amboise remained a fief of the Anjou family until it became a French crown possession in 1431, when Charles VII confiscated it from Louis I d'Amboise, a member of the senior branch of the Amboise family, who had conspired against Georges La Trémoille, a favored member of Charles VII's court.

At Amboise, the Loire is at its narrowest and splits to flow by each side of the small Île d'or, so that it is also at its most easily bridged. It had long lain on an established trade route between the north of France and the Rhône, and the earliest bridges appear to have been well upstream from the castle. It was the best known of the medieval counts of Anjou, Fulk III Nerra, count from 987, who built a stone bridge at the most westerly point of the Île d'or, giving access from northern France to the road to Spain, and to protect it constructed a massive stone tower where a Carolingian wooden structure had stood. He also built a collegiate church nearby, to which a chapter was attached. Fulk, notorious for alternating periods of piracy and penance, left a man named Lisois as his lieutenant to govern the château, while the Buzançais seigneurs governed the town. Lisois married Hersende, daughter of Archambaud de Buzançais, and from that union derived the illustrious family known after the town as the "d'Amboise," who retained ownership of the château, and its younger branch, the "Chaumont d'Amboise," who were to produce key figures in the military and civil administrations of Louis XII.

When Charles VII acceded to the French throne in 1422, his effective control scarcely extended beyond the middle section of the Loire valley. It was there that he was always to feel most secure. He began to repair and refortify the medieval château at Amboise, which began its hundred years of glory in the mid-fifteenth century under his son Louis XI, who worked to unify France and strengthen the monarchy. It was at Amboise that Louis XI's son, Charles VIII, was born, brought up, and met the woman whom dynastic considerations dictated that he should marry, Margaret of Austria, daughter of the Holy Roman Emperor Maximilian. It was from Amboise in 1469 that Louis XI created the Order of St. Michael at the château's collegiate church of St.-Florentin to rally forces loyal to him in his attempt to enlarge and secure most of the territory constituting Renaissance France. In 1483 Louise of Savoy, mother of Francis I, was brought to Amboise to be brought up by her aunt, Anne de Beaujeu, daughter of Louis XI and later co-regent during the minority of her brother Charles VIII. Although Louis XI spent relatively little time at Amboise, he made it his family's home.

Louis XI installed cannon at Amboise and, in order to accommodate the royal household, built a wing running south from the western edge of the present royal lodging. To the west of this new wing another was constructed, running from the western end of the preserved ramparts and almost touching the St.-Hubert Chapel to create a western triangular courtyard. To the east of the wing running south from the lodging further wings, north-south and east-west, enclosed further large courtyards. Behind the wall to the east of the Tour des

The château at Amboise
Photo courtesy of Office de Tourisme, Syndicat d'Initiative, Amboise

Minimes an Italian garden was laid out, enclosed to the south by galleries forming yet another wing, behind which was a vast empty space. The collegiate church was absorbed in the new structures, and Louis XI had a new church built outside the palace-fort, near the edge of the river, today's St.-Florentin.

Charles VIII succeeded to the throne in 1484 at the age of thirteen. The leading nobles rebelled against the regency of Anne de Beaujeu, but were defeated at St.-Aubin-du-Cormier in 1488, and Charles VIII married not Margaret of Austria but, in December 1491, Anne of Brittany. Charles now devoted himself to the transformation of Amboise into a dream palace. In 1492 a work force of 300 began construction, carrying on by torchlight when it got dark, and with fires to make the stone workable when it froze. The château is essentially flamboyant French Gothic in style, with Italian decorative motifs, some of which apparently predate the Italian campaign from which Charles is known to have brought back a team of Italian craftsmen. The Italian Wars of 1494–95 in which France and other major European powers sought control of Italy, did not interrupt work at Amboise. They also started well, and Charles triumphed at Rome and Naples before an alliance formed behind him to cut off his retreat.

His son had been killed, and the baggage train of treasures he had assembled had been captured, but there is an account of the arrival at Amboise of "eighty-seven thousand pounds of tapestries, books, paintings, marbles, porphyry, and other furnishings" from Italy. In spite of his defeat, Charles had been impressed with Italian expertise in sculpture, painting, garden design, and the decorative arts. Italians had been at work in the Loire valley during the last quarter of the fifteenth century, and, by the time of Charles VIII's death in 1498, Italian decorative innovations had become well established at the French court. The Italian touches had been integrated with the French Gothic of the existing royal lodging to create a new French Renaissance style, grafting an Italian feeling for the use of space as well as Italian decoration onto the formidable French architectural expertise gained in building cathedrals. At least one of the Italian architects brought to France in the wake of the retreat of Charles VIII, Boccador, also worked on the new Paris town hall, as well as on the châteaux at Blois and Chambord.

Louis XII, Charles's cousin and successor, preferred Blois, but continued the construction work at Amboise. He had the ramparts strengthened to resist artillery, and the château's strong defenses were still evident hundreds of years later; the north facade was scarcely damaged when World War II bombing ruined much of the rest of the château in 1940.

In the interests of producing a male heir, and of preserving the unity of France, Louis XII had his first marriage, to the disabled Jeanne de France, annulled, and married Anne of Brittany, Charles's widow. The annulment, not achieved without difficulty at Rome, was pronounced in the church of St. Denis at Amboise on December 17, 1498. Until a direct

male heir could be produced, the heir presumptive was Francis, son of Louise of Savoy. In spite of eight pregnancies Anne produced only one son, stillborn, and Francis did in fact become king.

Francis acceded to the throne, as Francis I, on January 1, 1515, and like his predecessors made Amboise the capital of his kingdom. At his palace many royal events, births, betrothals, and marriages, became the excuses for elaborate festivities. Francis received the Holy Roman Emperor Charles V there in 1539, and for this occasion the Tour des Minimes was illuminated by torches. Francis brought Leonardo da Vinci from Florence to work in Amboise for the last few years of his life. Leonardo's house, the fifteenth-century Clos Lucé, some 650 yards from the present entry to the château, is preserved as a museum. He asked to be buried in the church of St. Florentin. The church was destroyed in the nineteenth century, and his remains have been reburied in the St. Hubert chapel at the château. Additions to the château continued into the middle of the sixteenth century. These included a tennis court, outbuildings, a nursery for the children of Henry II (Francis I's son and successor), and the pavilion of Catherine de Médicis, Henry's queen.

Amboise was the last of the French forts to double as a palace. A century later, when the focus of French political life had moved to Paris, Versailles was adapted from a hunting lodge into a sumptuous royal residence. The glorious period in the history of Amboise, in fact, did not by long survive the completion of its construction. The château, however, did see a few more incidents of historic significance. In March 1560, when construction was finished and palace-forts of the Amboise type were practically obsolescent, a Huguenot plot known as the "conjuration d'Amboise" was discovered. The plot was aimed at seizing Francis II, a weak and sickly king, and undermining the politically important Catholic Guise family, who actually held the power in Francis's administration. The court retired from Blois, which was undefended, to Amboise, which was fortified, and the conspirators, as they arrived, were arrested and summarily executed. Some were hanged from the north facade of the royal lodging overlooking the Loire, their corpses left as a warning to passers-by, while others were tied in sacks and drowned in the river, and yet more were beheaded and dismembered.

The French royal family did not again settle at Amboise, although a treaty with the Huguenots was signed there by Charles IX on March 18, 1563. It was less favorable to the Huguenots than the edict of the preceding January, but acknowledged their status as a separate religious group within France.

The châteaux of Amboise and Blois passed into the hands of Gaston d'Orléans, the rebellious, cowardly, and dissolute brother of Louis XIII, and from 1611 to 1638 heir presumptive to the throne. In 1631 troops loyal to Louis XIII and Richelieu took the château at Amboise and razed the external fortifications. In 1660 the château reverted to the crown, and became a prison. In the eighteenth century Louis XV gave the château to the duc de Choiseul, for whom Amboise was elevated to a duchy, carrying a peerage. Choiseul needed to create access to the enclosure of the château for much bigger carriages than those of the sixteenth century, and in creating it destroyed much of the newer construction. A great deal more demolition occurred in the first decade of the nineteenth century.

What was left of the château had been returned to the Orléans family in 1802, but reverted to the crown in 1830 and was sporadically used in the nineteenth century. Louis-Philippe, king from 1830 to 1848, used it as a summer residence, but the château was confiscated from the Orléans family during the 1848 revolution, and used to imprison Algerian leader Abdelkader and his Bedouin entourage, who had surrendered in 1847 after warring against the French for years. The château was sacked by Prussian troops in 1870, and returned to the duc d'Orléans in 1873. From 1900 to 1933 it was used as a retirement home for the households of the Orléans family. In 1940 the château, by then home of the French air ministry, was badly damaged by shelling.

The town of Amboise, beautifully situated and containing buildings from the same era as the château's great period, is little more than an extension, even a spill-over, from the château, but includes the restored early-sixteenth-century town hall, the clock tower built by Charles VIII, the church of St.-Denis, and a fountain by Max Ernst.

Further Reading: There are a large number of books in French about the Loire châteaux, many excellently illustrated, such as Claude Frégnac's *Merveilles des châteaux du val de Loire* (Paris: Hachette, 1964), but most of the publications in English are tourist handbooks. A worthwhile resource on the Renaissance art historical background is Anthony Blunt's *Art and Architecture in France, 1500–1700* (Harmondsworth, Middlesex: Penguin, 1953; second edition, 1970). Also, on Amboise generally, *Amboise* by Pierre Charles and Sylvain Knecht (Tours: Knecht, 1958), has an excellent summary in English of the informative text.

—Anthony Levi

Amesbury (Wiltshire, England): Stonehenge

Location: About eight miles north of the city of Salisbury and two miles west of Amesbury in the county of Wiltshire, on a broad flatland known as Salisbury Plain.

Description: A monumental structure of ancient origin consisting of a primary circle of large standing stones topped with horizontal lintels, most of which have long since fallen, leaving Stonehenge in ruins. The site includes several other features of note, such as the Trilithons, Bluestone Circle, Avenue, the Heelstone, and the so-called Aubrey Holes, all of which seem linked to the primary megalithic structure. Various theories have labeled Stonehenge a ritual site, an observatory, a calendrical device, and even the burial place of Celtic kings. While evidence exists to support all of these theories, none has ever been conclusively proven.

Site Office: Tourist Information Centre
Redwood House
Flower Lane
Amesbury, Wiltshire SP4 7HG
England
(980) 623255 or 622833

Like the pyramids of Egypt or the Colosseum of Rome, England's Stonehenge has aroused and inspired countless generations with its quiet majesty. The timeless megalithic monument stands proudly on Salisbury Plain in southern England, as much a part of the landscape as the grass that surrounds it and yet obviously, and eerily, man-made. Presented with a photograph or artist's rendering, few people would fail to recognize the monument, although no one in the world is able to explain its significance.

Stonehenge thus differs from its fellow ancient architectural wonders: the Colosseum was constructed with an obvious purpose in mind, and the names and dates associated with its construction are known with a great deal of certainty. The pyramids served primarily as tombs for great leaders, and although the chronology of their construction is, for the most part, vague, both their purpose and their occupants are generally known. But nobody knows who built Stonehenge, or why. Theories abound, and many are undoubtedly partially true. Some, like the theory that Stonehenge was used as a calendrical device, have been generally accepted by so many for so long that they have come to be accepted as fact. But when the subject is Stonehenge, the only "fact" is that it *is*. Just as it always has been.

Stonehenge actually has very little history associated with it, assuming "history" is defined in the conventional sense as the contextual analysis of prior people, places, events, or ideas. The only verifiable names, dates and events associated with Stonehenge are relatively modern and, at any rate, have nothing whatsoever to do with its construction or original purpose.

Because of its relative lack of conventional history, historical analysis of Stonehenge tends to concentrate on several specific areas: physical analysis of the monument itself, the chronology of its construction as borne out by archaeological evidence, the theories regarding its purpose, and, perhaps most important, the mythology with which it has been endowed.

The most obvious component of Stonehenge is known as the Sarsen Circle (all components of Stonehenge are generally capitalized). It is not known where the name "sarsen" originated, but the stones themselves are tertiary micaceous sandstone. The Sarsen Circle measures 97 feet in diameter and originally consisted of 30 upright stones each approximately 7 feet in width and 13 1/2 feet in height. These uprights were presumably capped with 30 lintels, producing a complete ring around the tops of the Sarsen Stones. The gently curved lintels were held in place not by mortar, but by knobs carved into the tops of the Sarsen Stones that fit into holes cut into the bottom of the lintels. The stones have been numbered 1 through 30 for research purposes. The space between stones 1 and 30 is a foot wider than the space between the other standing stones and is capped with a larger lintel, implying that the space was probably used as an entrance to the monument.

Of the thirty original standing stones, only seventeen remain standing today. Seven have fallen but are still located at the site and six are missing completely. Only six lintels remain in their original positions, and portions of two have fallen to the ground and are still at the site. Because only eight lintels can be accounted for, and because they would have been the last stones to be placed, it cannot be said with certainty that the other twenty-two were ever erected. Most archaeologists assume they were, but that they were scavenged by other builders at some point in time.

How or when most of the missing or fallen stones met their fate is unknown, but some mishaps have been very recent. Stone 22 fell in 1900. Its lintel came down with it and broke in two, but both Sarsen Stone and lintel were returned to their original positions in 1958. Stone 23 met a similar fate in 1963, lacking a lintel, and was repositioned in 1964.

Within the Sarsen Circle lies the much smaller Bluestone Circle. Named after the stone from which it was constructed, the circle measures 76 feet in diameter and consists today of 20 stones, four of which are actually not Bluestone at all, but rhyolite. It is estimated that approximately 60

Stonehenge at sunset
Photo courtesy of Salisbury District Leisure and Tourism Services

stones originally stood in the circle, based upon their average spacing of two feet and archaeological evidence of the stumps of less hardy stones found along the line of remaining stones. Interestingly, two of the stones in this circle, numbered 150 and 36, feature mortise holes similar to those found on the massive lintels of the Sarsen Circle, leading to speculation that they were brought to Salisbury Plain from a smaller monument elsewhere—probably Wales. Below the Bluestone Circle has been found evidence of another, apparently never completed, circle known as the Double Bluestone Circle.

The largest single stone to be found at the site, known as the Heelstone, sits 256 feet to the northeast of the center of the Sarsen Circle. Unlike the stones within the circle, the Heelstone is undressed—it is in its natural state—and has no direct relationship to any other stone. It stands alone, a few feet from what is now the A344 road.

Within the Sarsen Circle reside two stone horseshoes, one consisting of standing Sarsen Stones and a smaller, inner horseshoe composed of smaller Bluestones. The open ends of both horsehoes face northeast, aligned with the likely entrance of the Sarsen Circle and Bluestone Circle, and facing toward the Heelstone. The larger of the two is composed of trilithons: three-stone constructs consisting of two standing stones topped by a lintel. Unlike the Sarsen Circle, these trilithons stand alone, and are not connected to one another. The five separate structures that make up the horseshoe are graduated in height, with the trilithons at the open end of the pattern measuring 20 feet in height, the middle trilithons 21¼ feet, and the central structure reaching 25½ feet.

The trilithons have suffered a great deal of wear and tear over the eons. The right-hand stone of the central trilithon, number 55, toppled at an unknown time and lies on the ground in two pieces. Its lintel lies nearby on the so-called Altar Stone. Stone number 59, from the left-hand side, is broken into three pieces, as is its lintel. The other trilithon on the left, stones 57, 58, and 158, fell in 1797 and was replaced in 1958 when work was done on the Sarsen Circle. Stone 57 features carved grooves that form a quadrilateral symbol similar to symbols found on megalithic structures in Brittany, the northwestern portion of France once known as Less Britain.

The inner Bluestone Horseshoe, approximately thirty-nine feet in diameter, originally consisted of nineteen standing bluestones ranging in height from six feet at the open ends to eight feet in the center. Eight remain standing, four have fallen and seven are missing. All of the stones are free-standing.

The stones of the Bluestone Horseshoe provide ample evidence of previous use in another monument. Stone 68 has a long groove down the length of one edge which was found to fit perfectly into a ridge carved into stone 66, nearby. Because the stones were not erected adjacent to one another, but were obviously meant to fit together, there is great reason to believe that they were used together in an earlier structure. Stone 69 of the Bluestone Horseshoe still sports the remains of a knob on top found to match the two former lintels now serving as uprights in the Bluestone Circle. Stones 71 and 72 were once the same stone, split in half for the horseshoe.

The Altar Stone, which lies in front of the central trilithon, received its name because of its central position. Its table-like dimensions of sixteen feet long by three feet four inches wide and one foot nine inches thick certainly back up this line of speculation. The Altar Stone lies in two pieces and is partially obscured by fallen stone 55, of the central trilithon. There is also evidence of digging around the Altar Stone, most probably by fortune-seekers looking for treasure buried beneath the central stone.

Most archaeologists agree that the Altar Stone has probably occupied its central position since Stonehenge was constructed. There is, however, no definitive proof for this belief. An empty hole behind the central trilithon was once filled with a large stone. The Altar Stone may have stood alone and fallen, or it may have been one of a pair. Nobody knows, but the theory that it has always been in its current position is a logical assumption.

Another horizontal stone lies approximately midway between the Sarsen Circle and the Heelstone. This stone, known as the Slaughter Stone, received its misleading name as a result of the pervasive Druid myth which, depending upon the source, usually involved ritual sacrifice. There is no evidence to support this assumption, but it has endured for generations.

There are no natural sources of building stone within thirteen miles of Stonehenge, meaning its builders had to haul the imposing stones from many miles away. The Bluestone constructs at Stonehenge are of five types: spotted dolerite, rhyolite, volcanic ash (none of which survives above ground today), micaceous sandstone, and a greenish-gray sandstone that exists at the site only in small fragments.

The dolerite and rhyolite, by far the most common types of stone used in the Bluestone structures, came to Salisbury Plain from the Presely Hills in Dyfed, Wales. The Altar Stone is the only example of micaceous sandstone to be found at Stonehenge and was brought to the site from the Cosheston Beds of Mill Bay on the shore of Milford Haven, Dyfed, Wales. This has been borne out by the discovery of an identical stone at another Wiltshire archaeological site, the origins of which had already been traced to the same Welsh quarry.

The Bluestones were transported to Salisbury Plain in one of four possible ways. Dyfed is a distance of approximately 190 miles, which would not have made overland transport impossible, but would have made it less than effi-cient. Because one or more of the stones originated in Mill Bay, it is likely that at least part of the journey was made by water. One route would have been to the mouth of the Bristol Avon and then up the river as far as the draft of the boats would allow. It is conceivable that the stones could have been sailed around Land's End to the Salisbury Avon and then up the river, but the treacherous nature of the seas around Land's End makes this theory unlikely. The final possibility is by sea from Milford Haven to Hayle, in Cornwall, overland six to eight miles to Marazion in Mounts Bay, and then along the coast by sea to the mouth of the Salisbury Avon at Christ-church and up the river. Although complicated, this theory has been championed by many because of the discovery of Presely stone on the coasts of Devon and Dorset and near the mouth of the Salisbury Avon. The second possibility, up the Bristol Avon, is the most favored. The stones were likely slung between two canoes rather than shipped in a single boat, due to safety considerations.

The enormous Sarsen Stones that make up the largest constructs at the Stonehenge site were brought overland from a site somewhere between Marlborough and Newbury, the nearest location from which such stones would have been available. Exactly how the stones were moved those thirteen miles remains uncertain, but it is possible that the builders of Stonehenge possessed knowledge of the wheel, making such a journey much easier.

It has been proven through excavation that upon arrival at the site, the Sarsen Stones were slid into their holes via inclines which were quickly filled as soon as the stones were pulled upright. After being allowed to settle into place, the tops of the stones were dressed to a uniform height, knobs carved and lintels hoisted aloft, probably by means of wooden ramps.

In addition to the more obvious stones, Stonehenge also includes important earthworks which, though often overlooked, are integral to the site. Surrounding the monument is the Bank and Ditch earthwork. Actually, the Ditch shows no signs of being a planned earthwork, but was created solely in the course of providing material necessary to build the Bank, which has a diameter of 320 feet. Even though it is just a by-product of the Bank, the Ditch has yielded the lion's share of archaeological relics. Cremation burials have been discovered within its walls and many pieces of what can best be described as industrial refuse have also been found there, such as primitive tools, stone fragments, animal bones, and pieces of Neolithic pottery.

The Causeway and the Avenue lead into the monument from the northeast. The Causeway is simply the point at which the Avenue crosses the Bank and Ditch, and is narrower than the Avenue itself. The Avenue begins at Stonehenge, crosses Salisbury Plain, winds its way through the nearby valley and woods and terminates very near the bank of the River Avon, at West Amesbury. The Avenue was built in two sections, with the first radiocarbon dated to approximately 2100 B.C. and the second, to the river, at 1100 B.C.

Additional points of interest at the Stonehenge site

include the Aubrey Holes, named for the historian John Aubrey, who discovered them in 1666. Fifty-six in all, the Aubrey Holes ring the monument in a circle measuring 284½ feet across, and their consistent spacing of 16 feet apart has led archaeologists to believe that they were excavated for some particular purpose. That purpose has never been discovered, but the holes themselves have been found to contain all manner of refuse, from human bones to wood ash, and seem to have been put to a variety of informal uses after their original use was discontinued. On paper, a right triangle can be drawn connecting holes 7, 28, and 56, and each set of three thereafter. The sides of the triangle correspond to the dimensions of the Megalithic Yard (2.7 feet), the standard unit of measurement in the British Neolithic and Bronze ages.

Along the line of the Aubrey Holes are the Four Stations, consisting of two stones, numbers 91 and 93, and two mounds, numbered 92 and 94. Lines drawn from 91 to 93 and from 92 to 94 intersect at the center of Stonehenge. The Four Stations could very well have been used for purposes of timekeeping, as the sun rises directly over stone 91 as seen from mound 92 at the summer solstice. The sun also rises directly over stone 93 as seen from mound 94 at the winter solstice. Other alignments based on the Four Stations can break the year into quarters and eighths, as well. In 1963 it was discovered that a line drawn between stone 91 and mound 94 is in alignment with the most southerly moonrise. A line from stone 93 to mound 92 aligns with the most northerly moonset.

The monument is also encircled by a series of Y Holes and an inner ring of Z Holes. Presumably, there would have been 30 holes in each series, but only 17 Y Holes and 16 Z Holes were ever actually dug. Their contents have provided no clues to their purpose.

Despite its notoriety, Stonehenge is not, strictly speaking, unique. Henge monuments exist elsewhere in the British Isles, and Stonehenge is simply the largest and best preserved. The word "henge" means "hanging," and is taken from the lintels that top the trilithons in later henge monuments. Henges were not always constructed of stone, either. Not far from Stonehenge, also in Wiltshire, is Woodhenge, the site of a Neolithic henge monument constructed of wood. Woodhenge was discovered through aerial photography in 1925. The original wooden posts of the monument have long since disappeared, and modern archaeologists have marked their locations with stone markers. Adjacent to Woodhenge is Durrington Walls, another henge monument, excavated in 1967.

Henge monuments are closely related to the stone circles that can be found all over the British Isles. In fact, there are approximately 900 stone circles in Britain. Such circles consist of a circle of standing stones and, usually, an associated bank and ditch. The Avebury circle, also in Wiltshire, shows evidence of smaller stone circles within the outer ring, not unlike the inner constructions found at Stonehenge.

Most stone circles descend from earlier henge monuments, and appear to reflect new ideas in construction technology (i.e., humans discovered how to work wood long before they learned to shape stone). Stone circles that do not show evidence of an earlier henge monument tend to be both small and from late in the henge-building period. This period can be identified as between 3300 B.C., when the oldest stone circle on record was constructed, at Newgrange, Ireland, and 1500 B.C., the approximate construction date of the youngest known circle, at Perthshire, Scotland.

The construction of Stonehenge is divided into three periods: Stonehenge I, II, and III, with the third period consisting of three distinct phases, IIIa, IIIb and IIIc. It has been estimated that Stonehenge I required at least 30,000 worker hours to construct, meaning that the monument was not built overnight and required careful planning.

The construction of Stonehenge I has been dated, depending upon the source, to between 3100 B.C. and 2700 B.C., and consists of the Bank and Ditch, Heelstone, and Aubrey Holes. There is evidence that a wooden enclosure was constructed during this phase, as has been the case at most stone henges and stone circles. A total of fifty-three stake holes dug in eleven incomplete rows of six have been discovered near the likely entrance to the structure.

It is generally believed that Stonehenge I was built by the Beaker people, who arrived in Britain in approximately 3000 B.C. This culture or group of cultures has been named by modern historians after the distinctive pottery it left behind, as very little else is known about it. Archaeologists have identified two groups of contemporary Beaker people, one originating in northern Europe and the other on the Iberian peninsula. It is theorized that the Beakers, an advanced people who had mastered coppersmithing, and thereby ushered in the first metal age, the Bronze Age, were nomadic people who landed in Britain in search of new sources of ore for their metals. Their arrival in Britain fomented significant changes in the prevailing Neolithic lifestyle, with the locals quickly adopting not only Beaker technology, but ritual, as well.

Stonehenge II consists of the Double Bluestone Circle, which exists today only as archaeological evidence of excavation beneath the still-extant Bluestone Circle. An antler found at the site and probably used as a tool has been radiocarbon-dated to approximately 2100 B.C.

Phase IIIa, circa 2000 B.C., saw the removal of the Double Bluestone Circle (assuming it was ever completed) and the erection of the great Sarsen Circle and internal trilithons. The Four Stations and Slaughter Stone were also erected during this phase.

Stonehenge IIIb has been dated to approximately 1550 B.C., or perhaps slightly earlier, and includes the Y and Z holes and an oval of Bluestones that predated the existing Bluestone Circle. This oval follows the line of the Bluestone Horseshoe and may have included two trilithons, the lintels of which now reside in the Bluestone Circle. Phase IIIc was begun immediately following the completion of IIIb and consisted of the construction of the Bluestone Circle and Bluestone Horseshoe.

Stonehenge III was completed during an era in which the Wessex region was ruled by a powerful elite with the resources and ambition to make the most of the era's limited technology. The discovery in 1808 of a series of burials near Stonehenge revealed an advanced culture capable of advanced metalworking. The burials also display a definite Beaker heritage. Some archaeologists have added a fourth stage to the construction, that of the extension of the Avenue in 1100 B.C.

The theories surrounding the purpose for which Stonehenge was built range from probable to outlandish. Theories of the latter variety, involving pagan rituals of questionable morality, may be at least partly accurate when used to describe some of the likely happenings at Stonehenge since its construction, but have little relevance to its original purpose. Generally, Stonehenge and other henges and stone circles are considered to be astronomical devices and sites for the rituals associated with the seasonal festivals timed by the structures.

Many of the popular beliefs about Stonehenge are in error. For example, Stonehenge was *not* built by the Druids. While the ways of the Druids remain more mysterious than not, surviving evidence proves that the Druids were an Iron Age people who landed in Britain in approximately 300 B.C. Although it is possible that the Druids may have found a use for the impressive, ready-made monument they found on Salisbury Plain, they were in no way responsible for its construction. However, what little is known about the religion of the Iron Age Celts corresponds to a surprising degree with the religious practices of Neolithic and Bronze Age British tribes. The pattern of seasonal festivals was similar, and the same ritual sites were constantly reused from one era to the next. There were no mass immigrations to Britain between the time of the monument's construction and the arrival of the Druids, so if the Druids incorporated parts of what they found in Britain into their own culture, it stands to reason that many Druidic rituals would be similar to the native British rituals that preceded them.

Proving theories regarding the probable astronomical uses of Stonehenge is, strictly speaking, impossible. Many stones have moved over the centuries, still others are missing entirely, and the sidereal tilt of the earth, which creates a wobble in the planet as it spins on its axis, slowly alters humankind's view of the heavens, albeit very, very slowly. However, extant evidence does strongly suggest that astronomical observation was among the roles played by Stonehenge and similar monuments.

It was first observed in 1771, for instance, that sunrise on the summer solstice is directly over the Heelstone. On the other hand, the first flash of sunrise, which should be aligned with the center of the Avenue, appears today nearer to the Heelstone's peak, due to the inevitable shift of the earth in relation to the sun.

The theory that Stonehenge served a calendrical function is supported not only by its positioning but by history. Seasonal rituals have always played important roles in European culture (summer and winter solstice celebrations survived until the relatively recent past). Because the festivals could not be held on just any day, keeping track of the passage of time was important. Such observations as Stonehenge made possible were also important for agricultural purposes, marking the time for planting, for example.

Stone axes also seem to have ritual significance, and have been found at several stone circles. Axes were used as pagan symbols throughout northwest Europe and even in relatively recent times were placed in cattle troughs in western England to prevent sickness in the animals. During excavations at Stonehenge in 1953, small representations of bronze axes and daggers were found on stones 4 and 53, implying that the monument and the weapons were somehow linked together.

The oft-related subjects of fertility and dancing are also discussed as possible uses for henge monuments. The phallic shape of the standing stones themselves is self-evident, and they were viewed by some as ensurers of fertility until relatively recent times. As to dancing, which even today usually comes before concerns of fertility, there is ample European tradition linking ritual dancing to stone circles. In fact, some circles are reputed in legend to be the petrified forms of young women punished for dancing on Sunday. The stone circle in Cornwall called the Merry Maidens is attached to such a legend, as is another circle in Somerset.

Throughout recorded history the construction of Stonehenge has always been ascribed to the earliest culture known at any given time. Medieval writers bestowed the honor on Merlin. Seventeenth- and eighteenth-century antiquaries said the Druids built it. Modern archaeologists, who, unlike previous historians, have the distinct advantage of radiocarbon dating, attribute its presence to the Beaker people.

The story of Merlin, as chronicled by Geoffrey of Monmouth, was regarded as gospel truth until recent centuries and is still the most famous of the legends surrounding Stonehenge. Geoffrey's account of the origin of Stonehenge is found in *The History of the Kings of Britain,* an important work taken quite seriously for centuries. Completed in 1136, the history tells the story of English royalty from Brutus to the coming of the Saxons. However, to say that Geoffrey embellished the facts with more than a little fantasy would be an understatement. Still, on some matters, mention in Geoffrey's history is all that has survived.

According to Geoffrey, Aurelius, then king of England, defeated his arch enemy Hengist in battle sometime around A.D. 500 (Geoffrey included only three dates in his entire history). Upon returning home, Aurelius "collected carpenters and stone-masons together from every region and ordered them to use their skill to contrive some novel building which would stand for ever in memory" of the men who so bravely died in battle.

Mere mortals could not do justice to the men who had

died in service of their king, so Aurelius sent for Merlin the Magician, who was never at a loss for ideas. "If you want to grace the burial place of these men with some lasting monument," Merlin said to Aurelius, "send for the Giants' Ring which is on Mount Killaraus in Ireland. In that place there is a stone construction which no man of this period could ever erect, unless he combined great skill and artistry. The stones are enormous and there is no one alive strong enough to move them. If they are placed in position round this site, in the way in which they are erected over there, they will stand forever."

Aurelius was skeptical, noting that England was well-endowed with its own stones. Merlin swayed the king by telling him that the "stones are connected with certain secret religious rites and they have various properties which are medicinally important. Many years ago the Giants transported them from the remotest confines of Africa and set them up in Ireland at a time when they ruled that country . . . There is not a single stone among them which hasn't some medicinal value."

Merlin convinced Aurelius, who ordered his brother Uther, father of Arthur, to take 15,000 men to Ireland and return with the Giants' Ring. Uther was forced to do battle with the Irish leader, Gillomanius, but prevailed quickly, and returned to England with the stones. With great pomp and circumstance, Merlin re-erected the stones at their present location, surrounding the burial site of Aurelius's brave legions. According to Geoffrey, Aurelius and Uther were also buried beneath the Giants' Ring.

Stonehenge survives today primarily as a symbolic monument to something else. Exactly what that something else is, is up to the beholder. To the fictional musical group Spinal Tap, Stonehenge is the basis of a song about Druids and demons. To the re-founded Druidic order of Britain, Stonehenge is a ritual site, albeit an off-limits one. To many thousands of counterculturists, Stonehenge is the destination for the annual summer solstice celebration, often involving sex and drugs, and ultimately requiring intervention by police.

Today, visitors to Stonehenge are forced to stay behind a rope that protects both stones and earthworks from the millions of evidence-destroying feet that pass its way each year. Despite its location miles from the nearest city of substance, Stonehenge is the United Kingdom's second-most-visited tourist attraction, after only the Tower of London. Facilities around the monument have recently been upgraded and now consist of more than the obligatory gift shop. Stonehenge is under the management of English Heritage.

There is undoubtedly much more to be learned from Stonehenge, and however distressing the ropes around the monument may be to the interested observer, all that can be learned from simple visual inspection has been learned, and the vast majority of visitors to the site can learn all they care to know from a guidebook. Many an archaeologist has called for a complete excavation of the site, but others have spoken against such an action, noting that to remove all of the stones, no matter how carefully, would be to destroy the monument. It seems that humanity will never have all the answers.

Further Reading: *Stonehenge* (London: Hamilton, and New York: Macmillan, 1956; reprint, Harmondsworth, Middlesex: Penguin, 1990; New York: Viking Penguin, 1992) by the R. J. C. Atkinson is the standard one-volume source for researchers interested in Stonehenge. *The Origins of Britain* (London: Routledge and Kegan Paul, 1980; Chicago: Academy Chicago Publishers, 1983) by Lloyd and Jennifer Laing, the first installment in the five-volume series *Britain Before the Conquest,* covers British prehistory from the earliest known traces of civilization through the opening era of the Bronze Age and provides a wealth of detailed information that is still easily accessible to lay readers. *Stonehenge* (London: Her Majesty's Stationery Office, 1959; updated, 1981) by R. S. Newall is the official, government-sanctioned explanation of the monument. Geoffrey of Monmouth's *The History of the Kings of Britain* (1136), translated by Lewis Thorpe (Harmondsworth, Middlesex, and New York: Penguin, 1965) is less than reliable as history, but makes for great entertainment and is worthy reading as a study of British mythology, particularly that surrounding its most famous topics, such as King Arthur and, of course, Stonehenge.

—John A. Flink

Amsterdam (North Holland, Netherlands)

Location: On the south side of the IJ, a chain of lagoons, now mostly docks, that form the Markerwaard, a long inlet at the southwest corner of the IJsselmeer; connected with the North Sea by the Nordzee Kanal.

Description: Largest city and capital of the Netherlands (although The Hague is the seat of government); home to many of the country's cultural treasures; possessor of the largest freshwater harbor in Europe. A crescent-shaped paradigm of urban planning crisscrossed with canals and 1,000 bridges, Amsterdam forms part of the *Randstad* ring of cities that encircle the Netherlands' rural heart.

Site Office: VVV Amsterdam
Stationsplein 10
1012 AB Amsterdam, North Holland
Netherlands
(20) 5512512

Although Amsterdam's strategic economy necessarily has tied it to the turbulent political history of of Holland and, later, the Netherlands, for much of its history the city has enjoyed a great degree of independence. The Hague to the south eventually became the locus for the princely courts and intricacies of political administration, leaving Amsterdam free to develop as the Netherlands' economic powerhouse and cultural capital.

Until the thirteenth century the site of Amsterdam was occupied by a small fishing settlement of wood and mud houses leased by the lords of Amstel from the bishops of Utrecht. Though the nobility was based at Ouderkerk, five miles south of present-day Amsterdam's center, Gysbert II of Amstel built a castle on the site in 1204. Amsterdam's very name, however, was derived from a different structure: the dam that was built here in about 1270 to separate the mouth of the Amstel River from the branch of the Zuider Zee (now the IJsselmeer) known as the IJ.

Amsterdam was typical of the medieval dike-town: the river was dammed, its water subsequently apportioned into inner and outer harbors, and the excess diverted to the sea through alternative channels. The dikes served multiple functions, not only retaining the otherwise threatening water (most of the reclaimed land is below sea level), but also serving as the base for land routes and even for buildings that were related to trade. In Amsterdam, for example, the dam from early on was the chief public place. The town halls have always been located here, with the present one (now the Royal Palace) built in the seventeenth century, right next to the Nieuwe Kerk of 1308.

Though water abatement and reclamation of land began in the area in the eighth century, the need to build and manage dikes to keep the waters from encroaching led to the establishment of water catchment boards in the thirteenth century. These independent authorities still exist, as the management of the water-land balance in a city such as Amsterdam remains critical, especially as efforts to reclaim land continue.

Amsterdam was not the only town to grow through such reclamations, but from the earliest days it was able to parlay its waterways and access to the sea into a successful and long-lived trade-based economy. The dam had barely been built when, in 1275, Count Floris V of Holland acknowledged the importance of the fishing village of Amstelledamme by granting the villagers freedom from tolls on travel and on trade in their own goods within the county of Holland. This exemption marked the first document in which the town was mentioned. In 1300 Bishop Gwyde of Utrecht conferred on Amsterdam its city charter, and in 1317 the clergy withdrew completely when the bishopric of Utrecht transferred the city to Count Willem III of the House of Hainault, which ruled Holland.

Already in 1323 Amsterdam began to gain commercial stature, when the city became the point at which duties were levied on beer imported from Hamburg; this led to increased trade, particularly in grain, with the Hanseatic League, an active network of towns scattered mainly throughout central and eastern Europe. The Dutch succeeded in developing a competitive fleet that could offer transportation at a lower cost than that of the league members. The economic advantage allowed the Dutch not only to ship raw materials from the Baltic back to Holland, but also to sell Dutch goods and serve as distributors of other countries' commodities.

Toward the end of the fourteenth century, Holland began to challenge the monopolistic and nearly impregnable Hanseatic League, with the Dutch aggressively pursuing trade with England, the Baltic countries, and the west coast of Europe. Major commodities at stake included wheat, cloth, herring, beer, and fur. In the mid–fifteenth century Amsterdam competed especially with Lübeck, long the kingpin of the league thanks to its own strategic location on the Baltic and aggressive mercantile policies.

In later centuries Amsterdam was party to the contentious politics that occurred when the various foreign nations with interests in the Low Countries—the Spanish, English, Austrians, and French—vied for economic or religious control of the country or opposed Dutch intentions. But from about 1400, a governing system of four burgomasters elected annually by the Council of Elders afforded Amsterdam a degree of independence from whomever was ruling Holland.

Along one of Amsterdam's canals
Photo courtesy of Netherlands Board of Tourism

So even when the country was incorporated into the Duchy of Burgundy in 1428, Amsterdam continued to control its own economic expansion as well as its physical boundaries. In the fifteenth century, the physical city was extended to the east, west, and south, all this despite a fire in 1421 that nearly destroyed Amsterdam, and another fire in 1453.

Prominent among the infrastructural enhancements was the stone city wall begun in 1481, which supplanted the system of defensive canals and earthen ramparts between gates and towers. This wall protected the burgeoning city and its groundswell of immigrants. Amsterdam repeatedly benefited from its industrious immigrants, regardless of whether they were fleeing economic hardships in the surrounding countryside, religious persecution elsewhere, or resettling from the eventual Dutch colonies. Especially numerous were refugees from Spanish oppression elsewhere in Europe, including Protestant merchants from the southern Netherlands (now Belgium), Portuguese Jews, and French Huguenots.

Only in the sixteenth century did Amsterdam's economy begin to falter, with the onset of the Reformation. In

May 1535, Anabaptists—a group that espoused adult baptism and that was attracted by certain communistic and prophetic elements of Protestantism—sought to occupy the Town Hall. Many of the rebels were executed for this abortive insurrection. Further disorder plagued Amsterdam during a famine in 1566, when Reformers stormed Catholic churches and monasteries.

The Protestant theology that did eventually take hold in Amsterdam was that of the French theologian and reformer John Calvin, who proposed a radical but logical doctrine based on the belief in predestination. Amsterdam's leaders were inclined toward religious tolerance, but complications arose when the Holy Roman Emperor Charles V abdicated in 1555. He was succeeded as king of Spain and lord of the Netherlands by his very Catholic son, Philip II, who, unlike Charles, was raised in Spain and had no experience with the low countries far to the north. In May 1567 the Spanish government strictly forbade the practice of Calvinism, and to enforce the edict, in August, Philip sent to Holland the feared duke of Alba, who commenced a bloody rout of rebellious anti-Catholic elements.

Amsterdam remained pro-Spanish, however, even during an uprising in 1568 by the northern provinces of the low countries. The anti-Spanish movement gained serious momentum in 1572, however, with William the Silent, prince of Orange, firmly in charge of the rebellion against Spain. Rebels supporting William continued to attempt incursions—the French Huguenots from the south and the "Sea Beggars" from the coast—but Amsterdam resisted the attempts of the latter. The city finally surrendered to William's troops in 1578, becoming party to the northern Low Countries' war of independence from Spain.

In the spirit of tolerance that had reigned in Holland, the idealistic William, now as stadholder of the northern provinces, sought peaceful coexistence among Protestants and Catholics. Amsterdam, however, opted for the *Alteratie,* a political (but bloodless) revolution that led to the departure of all pro-Spanish civic leaders, clerics, and clergy from the city, in favor of a new civic administration comprising mostly immigrant Protestant merchants. As a gesture of renewed tolerance, however, the *Satisfactie van Amsterdam* legislated an end to persecution based on religious beliefs.

The Union of Utrecht in 1579 allied the seven northern provinces of the Low Countries into an effective United Provinces of the Netherlands (although Philip had not actually relinquished the area). This self-styled Dutch Republic comprised a bastion of Protestant practice to counter the staunch Catholicism of the ten southern provinces. The politics of this arrangement made coexistence of the faiths untenable, however, and in 1580 Catholicism was outlawed in Amsterdam; nonetheless, the occasional hidden church did allow services in the newly proscribed religion to proceed.

The fall of rival Antwerp to the Spaniards in 1585 was a windfall for Amsterdam, which quickly moved to take the former's place in maritime trade. Amsterdam's shippers sponsored new routes to the Mediterranean, Africa, the West Indies, and the Americas. In 1602 the United East India Trading Company was founded, and Amsterdam merchants—eager to control the exotic commodities of Asia—were among the major shareholders. The Dutch trade empire was at its peak at about this time. The country's ships were a ubiquitous presence on the seas: there were well over 4,000 Dutch vessels active, with most of the business expedited through Amsterdam.

By the seventeenth century Amsterdam matched its might in commodities transport with parallel control of the international banking business. Amsterdam's religious tolerance paid off handsomely in this case: the city's financial prowess resided largely with Jews who had found asylum from persecution elsewhere. In 1586 Portuguese Jews introduced what was to become a signature commercial venture for Amsterdam: diamond trading and processing. Amsterdam's commercial prominence developed further with the establishment of the Stocks and Commodities Exchange in 1611.

The fortunes associated with being the capital of the Dutch Republic, the greatest seafaring nation in Europe, provided a solid base for the city's Golden Age. This was when the most important port in the world also enjoyed primacy in the practical art of city planning and the fine art of painting.

The city's first task in about 1600 was to accommodate its ever-increasing population (100,000 by 1620 and 200,000 by 1660), which strained the confines of the tight medieval city. The result was a marvel of early town planning in Europe, with notable features being a governing width of a grand eighty feet for canals, individual lots, stipulation of minimum distances between the backs of buildings, and zoning to control the locations of commercial and industrial properties.

The "Plan of the Three Canals" by Hendrikje Staets, which was approved in 1607, expanded upon plans for a grand system of canals begun late in the prior century. In the new plan, three main ring canals arranged concentrically about the old city center—the Herengracht, Keizersgracht, and Prinsengracht—formed a monumental setting for the patrician houses and buildings related to the city's mercantile prominence. Buildings of lesser stature were relegated to radial canals that would connect the main canals. Logically, the harbor was reserved for major shipping activities, and another area to the west of the canals, the Jordaan, was set aside for artisans, industry, and charities. The critic Lewis Mumford calls this synthesis of planning and building "a miracle of spaciousness, compactness, intelligible order." The resulting urban fabric was indeed remarkably uniform, enhanced by trees and small-scale facades averaging 26 feet wide, but with rich variations in color, gables, and ornamentation. The provision for ample space between the backs of houses—the minimum distance was 160 feet—provided for amenities such as generous, well-lit gardens.

Construction of this ambitious "Venice of the North" began in 1613, with work continuing through the next several decades. Daniel Stolpaert, a surveyor-architect, was largely responsible for the ultimate realization of the canal system. Such a capitalistic plan was bound to have its drawbacks for the less fortunate, and in contrast to the showpiece ring canals, the Jordaan was cramped; the average passageway was a paltry eighteen feet wide instead of eighty, and the district became a ghetto for poor workers and immigrants.

Architects such as Hendrik de Keyser, city architect from 1612, complemented the city's plan with a series of new edifices. Working mostly in the Dutch Renaissance style, de Keyser designed the Zuiderkerk, the first Dutch Reformed church in Amsterdam, and then the influential Noorderkerk. In 1620 he also began the monumental Westerkerk, where Rembrandt was buried on October 8, 1669. Jacob van Campen, the Netherlands' leading classical revival architect, designed the new Town Hall (now the Royal Palace) according to the principles of the Italian Renaissance architect Andrea Palladio. In an acknowledgment of the structural realities of building on reclaimed land, the building rests on at least 13,659 piles.

Even more distinguished were the succession of painters who demonstrated magnificent skill in everything from genre paintings to formal portraits, still-lifes to landscapes, historical scenes to depictions of religion and mythology. This was the age of the master painter of light and shadow, Rembrandt Harmensz van Rijn, many of whose paintings—portraits of doctors, guild members, burghers—convey the genteel prosperity that characterized the Dutch Republic in its Golden Age. The prolific Rembrandt, who settled in Amsterdam in 1631, led a whole school of artists in the city, who enjoyed an enthusiastic reception among the city's burgher class. Notable counterparts in Amsterdam to Rembrandt—the portrait painter *nonpareil*—were Jacob van Ruisdael, often considered the greatest Dutch landscape painter, and his student Meindert Hobbema, a leading landscape painter of the following generation.

A country as prosperous as the Netherlands was bound always to have its challengers, and the Dutch Golden Age lasted only until England arose to challenge the country and gain commercial supremacy. The relationship of the Dutch and English had been tenuous for some time, with rivalry between London and Amsterdam reaching a crisis in the mid-seventeenth century. The Anglo-Dutch War of 1652–53 was only the first of several wars that the two countries conducted by the end of the century. At first, the Dutch held their own in these naval battles, with Admiral Michiel de Ruyter even sailing right up the Thames in 1667.

Continued problems with the English, new attacks from Louis XIV's France to the south, and intensified competition from other trading centers to the east such as Hamburg, combined to drain the vitality of the Dutch Republic and its capital. In addition to Amsterdam's economic travails, revolutionary upheavals rocked the city in the eighteenth century, with riots in 1747 against taxes and the entrenched ruling class.

The 1780 to 1784 war between the Netherlands and England disrupted Amsterdam's trade seriously enough to make definitive the city's eclipse by its rivals. Exacerbating Amsterdam's commercial eclipse throughout the eighteenth century was a simple, natural fact: the Zuider Zee, between the city and the North Sea, was becoming silted.

In 1795 Amsterdam was occupied by French revolutionary troops, becoming part of the Batavian Republic until 1806, and then capital of the Kingdom of the Netherlands under Louis Bonaparte, Napoléon's brother. The former's sympathies toward the Dutch led to his removal in 1810, however, and Napoléon annexed the entire country to France. Amsterdam's port declined precipitously as the rest of Europe sought to defeat Napoléon, especially through the Continental Blockade, which cut the city off from its traditional markets.

After the defeat of Napoléon and expulsion of the French in 1813, Amsterdam again became the capital of the Kingdom of the Netherlands, a constitutional monarchy from 1815 under William I, although the seat of government remained in The Hague. As in earlier times, the southern and northern provinces proved incompatible for cultural reasons, and the southern Netherlands gained independence as the Kingdom of Belgium in 1831.

Having survived the trials of revolution and occupation, Amsterdam began efforts to recoup its economic losses. One disappointment was the ambitious dredging of the fifty-mile Noordhollands Kanaal from 1819 to 1825, which failed to meet the city's shipping needs; the canal was simply too narrow and too shallow. However, one bright spot in the economy remained: the diamond trade. Amsterdam's diamond industry had received a boost early in the eighteenth century when Holland was granted a monopoly for handling the newly discovered Brazilian diamonds, and in 1867 another new market opened up with access to the South African mines. Shortly thereafter, Amsterdam finally succeeded in reviving its shipping trade by building a modern, deepwater route to the sea, the fourteen-mile Noordzee Kanaal dug in from 1865 to 1876.

The return of economic vitality brought further increases in population, accompanied by a new social awareness of popular needs such as adequate housing and improved public health. To address this new set of problems, the city began to fill in canals and widen streets in the inner city, and to plan again for peripheral growth.

As well, on the eve of the twentieth century, the brilliantly logical Stock Exchange (1898–1903) by H. P. Berlage provided Amsterdam with one of the seminal works of modern architecture. And an influential, if idiosyncratic, "Amsterdam school" of architects made its debut with the Scheepvaarthuis in 1916.

In 1913 the Social Democrats won a majority on the city council, and Amsterdam became a reliable stronghold of democratic socialism. The city's social progressiveness translated into design efforts such as Amsterdam's social-housing prototypes between the world wars.

While neighboring Belgium witnessed major battles

during World War I, the Netherlands stayed neutral; during the war, however, Amsterdam was plagued by severe unemployment, food shortages, and an influx of refugees, not so easily accommodated during this time of crisis as they had been during the city's more prosperous years.

The poignant story of Amsterdam's long-assimilated Jewish population exemplifies Amsterdam's much less fortunate situation during World War II. The Nazis nearly eliminated the city's Jewish community after German troops occupied the city on May 16, 1940, and began deportations to concentration camps. Despite such events as the "February Strike" of February 25, 1941, whereby workers of Amsterdam protested the deportations, and an intensified underground resistance movement, the Nazis succeeded in deporting about 100,000 Jews.

The diary of Anne Frank, the German Jewish girl whose family fled from Germany to Amsterdam in 1933, and who hid from the Nazis for two full years before being discovered and deported in 1944, vividly recounts the everyday sense of uncertainty and terror that prevailed throughout the years of occupation. The Anne Frank Huis on the Prinsengracht has become a point of pilgrimage as a reminder of this tragic episode in the city's history.

In addition to this human tragedy, Amsterdam's harbor and industrial district were destroyed. But after Canadian troops liberated the city on May 5, 1945, rebuilding began; additionally, in 1952 the Amsterdam-Rhine Canal opened, enhancing the city's access to the inland waterways serving the rest of Europe.

Continuing the tradition of quirky liberalism that had come to characterize the Dutch, and Amsterdam in particular, in the mid-1960s an anti-establishment group called the "Provos" (from *provocatie*) staged a series of publicity-generating actions. In 1970, heading toward temporary legitimacy but disavowing any notion of leadership, the successor "Kabouter" (Gnome Party) delegation won five seats on the city council.

The problem of lack of adequate housing in the city came to the fore even as Amsterdam celebrated its 700th anniversary in 1975. The police confronted angry residents of the Nieuwmarkt district, and protesters obstructed attempts to demolish housing to make way for the city's new subway system. By 1979, with more than 60,000 people on Amsterdam's waiting list for housing, many empty houses were occupied by squatters known as "Krakers." In April 1980, housing riots in Amsterdam marred the coronation of Queen Beatrix. Subsequent laws required that empty dwellings be registered, essentially outlawing squatters.

Continuing the marriage of commerce and culture, Amsterdam remains the Netherlands' largest industrial city and one of the world's leading financial centers, reigning nonetheless as "European City of Culture" for 1987. Architecturally, the city deftly blended its new City Hall, which opened in 1988, with the Muziektheater opera house in a building known as the "Stopera" (*stadhuis/opera*). Complementing the newer architectural forays are the approximately 7,000 houses—mostly from the seventeenth-century Golden Age—that have been classified as historical monuments.

An ongoing problem for Amsterdam is automobile traffic, which increasingly encroaches on the order inspired by the canals. But the "Venice of the North" remains defined by its waterways and the more than 1,000 bridges that span them, contributing to the allure that makes Amsterdam one of the most beloved urban ensembles in the world.

Further Readings: John J. Murray's *Amsterdam in the Age of Rembrandt* (Norman: University of Oklahoma Press, 1967) offers a concise, general introduction to the city's Golden Age. *Amsterdam: The Life of a City* (Boston and Toronto: Little, Brown, 1972) by Geoffrey Cotterell is an unabashedly subjective and anecdotal narrative that reveals a true enthusiasm for the city, and is nonetheless wide-ranging and informed. *Traders, Artists, Burghers: A Cultural History of Amsterdam in the 17th Century* by Deric Regin (Assen/ Amsterdam: Van Gorcum, 1976) is a well-informed, straightforward account of Amsterdam's Golden Age that discusses the relevant social, cultural, political, and economic topics. Two remarkable works by Simon Schama are in a class of their own, for both size and originality. *The Embarrassment of Riches: An Interpretation of Dutch Culture in the Golden Age* (New York: Knopf, 1987) registers nearly 700 pages, and establishes a national context for the particulars of Amsterdam. This massive and richly illustrated analysis of the Dutch character during the seventeenth century includes a delightfully interpretive "Bibliographic Guide to the History of Dutch '*Mentalité*,'" which delves into such arcania as food history, the comforts of a house, and marriage manuals. *Patriots and Liberators: Revolution in the Netherlands, 1780–1813* (New York: Knopf, and London: Collins, 1977) offers 745 pages of interpretive history of the revolutionary upheavals that occurred after the decline of the Dutch Republic, including the political intrigues affecting Amsterdam. In *Dutch Primacy in World Trade, 1585–1740* (Oxford: Clarendon Press, and New York, Oxford University Press, 1989), author Jonathan I. Israel does justice to the complex subject of trade economics with his exhaustive but eminently readable analysis of the ascendancy of the Dutch maritime provinces—including Holland and Amsterdam—after the fall of rival Antwerp in 1585. Maps, economic tables, and graphs complement the text. Renée Kistemaker and Roelof van Gelder's *Amsterdam: The Golden Age, 1275–1795*, translated from the Italian by Paul Foulkes, (New York: Abbeville Press, 1983; originally published as *Amsterdam, 1275–1795, Buono Governo e Cultura in una Metropoli di Mercanti*, Milan: 1982) is a beautiful, oversized text that broadens the concept of "Golden Age" to include the first several centuries of Amsterdam's history. The authoritative text is accompanied by hundreds of exquisite illustrations, maps, tables, and photographs, many in full color. The book includes a very useful prose chronology of the Netherlands and Amsterdam, from 1275 to 1795, and special essays about the Eighty Years War (1568–1648), the beginnings of social security in Amsterdam, and Amsterdam's civic orphanage. *The Making of Dutch Towns: A Study in Urban Development from the Tenth to the Seventeenth Centuries* (New York: Simmons-Boardman, 1960) by Gerald L. Burke explains in fascinating detail how Dutch towns developed, including the process of reclaiming land from the sea. A chapter on Amsterdam includes wonderful maps from various periods of the city's growth. Audrey M. Lambert's excellent *The Making of the Dutch Landscape: An Historical*

Geography of the Netherlands, second edition (London and Orlando, Florida: Academic Press, 1985) places the physical development of Amsterdam in a somewhat broader context of urban and economic issues, and includes a wealth of citations for further information.

—Randall J. Van Vynckt

Angers (Maine-et-Loire, France)

Location: In western France, along the Maine River about 5 miles above its confluence with the Loire, about 55 miles east of Nantes and 180 miles southwest of Paris.

Description: During the Middle Ages, Angers was the chief residence of the counts of Anjou, with whom the Plantagenet line of English kings originated. The area was a battleground in numerous wars, but Angers has retained much of its medieval architecture, including the Angers Castle and several churches. Angers was capital of the former province of Anjou and continues as capital of the modern department of Maine-et-Loire.

Site Office: Office de Tourisme, Syndicat d'Initiative
Place Kennedy B.P. 2316
49023 Angers, Maine-et-Loire
France
(41) 23 51 11

Angers, in the fertile Loire valley, is one of the most historic and beautiful cities in France. The region was originally settled by Celtic farmers in the Neolithic period, and later by a Gallic hunting tribe known as the Andecavi; the city was made capital of the state of Andes. The old town was originally a hill fortress situated on the left bank of the Maine River, on the borders of Brittany and the region that came to be known as Anjou. The Andecavi chief Domnacus fought against the Romans in 52 B.C., but the tribe was soon conquered by Fabius, a lieutenant of Caesar, and came under Roman rule. In Gallo-Roman times, the city was known as Juliomagus, and the surrounding country was called the Civitas Andegavensis or Civitas Andecavorum. Juliomagus was a well-developed city, with public baths and an arena.

During the reign of Valentinian III in the fifth century A.D., the territory broke away from the Roman Empire and became part of the Armorican Confederation. Its chief city was then renamed Andegavia. In 471, the city and region passed into the hands of the Merovingian kings, the first Frankish dynasty of ancient Gaul. The region became known as Pagus Andegavus under the Merovingians and Carolingians (the second Frankish dynasty); under the latter dynasty, founded in 751 by Pépin III, the region was organized as a province, or county, which was nominally administered by a count representing the king of France, or West Frankish Kingdom. The province soon became known as Anjou (which roughly approximates the modern department of Maine-et-Loire); its chief city became Angers. Robert the Strong was made count by West Frankish (or French) king Charles II, the Bald, in 860. The city suffered Norman invasions and occupations around this time.

Angers became the chief residence of the successive hereditary counts of Anjou (or Angevin counts) beginning in the late ninth century. Each of the counts defended and added territory to the feudal state. The first dynasty of counts began when one of the sons of Robert the Strong entrusted Anjou to a man named Ingelger, who became founder of the first Angevin dynasty. Ingelger's son, Fulk (or Foulques) I, the Red, drove the Normans out of the country and usurped the title of count. He enlarged his domain by seizing a part of neighboring Touraine. After his death in 942, he was succeeded by Fulk II, the Good, who repaired the destruction the province had suffered in the Norman invasions. Geoffrey I Grisegonelle, who succeeded Fulk II about 960, began the family policy of territorial expansion through conquest. Geoffrey I also helped Hugh Capet seize the French Crown, but died soon after the new king assumed the throne in 987.

Geoffrey's successor, Fulk III Nerra, was one of the most extraordinary and powerful members of the first dynasty of counts. "Black Fulk," who ruled from 987 to 1040, was one of the greatest castle builders of all time, and also one of the greatest war strategists. Fulk III spent practically his entire life fighting his more powerful archrivals, the counts of Blois, who were situated two territories to the east; they were separated by the county of Touraine. Although his army was usually much smaller than his enemy's, Fulk III kept the counts of Blois, especially Thibault I, from encroaching upon his lands by building an incredible number of strongly fortified castles along Anjou's borders; at least eighteen still survive. His fortresses were initially built of wood, with most later converted to stone, at strategic sites along rivers and roads, both north and south of the Loire. His small army could move quickly from one end of his domain to the other, as each castle was never more than a day's march from another. Meanwhile, Fulk III took time off from fighting to make several pilgrimages to Jerusalem; he died while returning from one.

In the eleventh and twelfth centuries, the city of Angers was the center of one of the most powerful fiefdoms in France. Fulk III's son, Geoffrey II Martel, ruled from 1040 to 1060 and realized his father's unfinished plans. He gained secure possession of Touraine from Blois, and also established his suzerainty over the Maine district: this latter annexation would bring the Angevin counts in conflict with the dukes of Normandy (after being dislodged from the area, the Normans settled in northwestern France). Geoffrey II died in 1060, leaving no sons, so two nephews—Geoffrey III, the Bearded, and Fulk IV, le Rechin—shared the succession. They had to contend with the dukes of Normandy—especially William the Conqueror—who captured Maine before invading England in 1066, and Fulk and Geoffrey were in conflict with one another as well. Fulk defeated Geoffrey in

Aerial view of Angers
Photo courtesy of French Government Tourist Office

battle in 1068, but he still had to relinquish most of the lands acquired by Fulk Nerra and defend his fiefdom against the claims of the Norman dukes. After Fulk IV died in 1109, his son Fulk V, the Young, tried to win back this territory. The long struggle with the dukes of Normandy ended in 1128, when Fulk V married his fifteen-year-old son Geoffrey V to twenty-seven-year-old Matilda, the daughter and heiress of Henry I of England, and granddaughter of William the Conqueror. The marriage gave Geoffrey V a claim to Normandy and England.

Geoffrey V, count of Anjou from 1131 to 1151, became known as "le bel Plantagenet" because of the sprig of broomplant (*plante de genet*) he wore on his helmet. After Henry I died in 1135, Geoffrey Plantagenet succeeded in occupying and strengthening Maine and Normandy, and pacifying his rivals in Anjou. He left Anjou to his son Henry in 1151; Henry became count of Anjou and Maine, and duke of Normandy. A year later, the eighteen-year-old Henry married the older and much sought-after Eleanor of Aquitaine, whose marriage to King Louis VII of France had recently been dissolved due to infidelity. He ascended to the English throne as King Henry II in 1154, starting the Angevin, or Planta-

genet, line of English kings. At its twelfth-century height, the so-called "Angevin Empire"—really a patchwork collection of feudal fiefdoms—stretched from the Scottish border to the Pyrenees. The house of Plantagenet retained the English crown until the death of King Richard II in 1399.

The Plantagenet kings of England kept Anjou in their possession until the early thirteenth century. Beginning in 1203, King Philip II (or Philip Augustus) of France conquered Normandy, Maine, and Touraine, and then seized Anjou from King John of England in 1205. The English tried to retake Anjou in 1214, but were defeated in battle. Anjou was formally ceded to France's royal domain by the Treaty of Paris in 1259. The Plantagenets no longer ruled Anjou.

Though the writer Henry James once remarked that Angers had been "stupidly and vulgarly modernized," the city is still rich in architecture dating from the twelfth and thirteenth centuries—unlike Nantes, 55 miles to the east, where many historic structures have been eliminated through a series of urban renewal programs. The Plantagenet rule of Angers was characterized by the construction of magnificent monuments; Geoffrey V's marriage to Matilda had not only led to a lasting Angers-England connection, but had also resulted in a distinctly Angevin style of architecture. Two examples are the former Hôpital de St. Jean, which was founded in 1175 by Henry II of England, as part of his penance for the murder of Thomas Becket, and completed in 1210; and the gothic-vaulted Cathédrale de St. Maurice, which was built in the late twelfth and early thirteenth centuries. The former was used as a hospital until 1865 and now houses a museum. The cathedral still retains its original stained glass, among the finest in France, and has a famous collection of Gothic tapestries. Also noteworthy is the church of St. Serge, which was built in the twelfth century and is renowned for its Angevin Gothic vaulting. Several other churches and abbeys in the city date back to the Middle Ages or earlier.

The most fantastic medieval edifice in the city, however, is the massive, moated Angers Castle (or Château d'Angers). It was built to guard a point at which the Maine River narrows, and was the last line of defense along an ancient route between the English Channel and the Loire, five miles upstream from the castle. Although the original wooden castle dates back to Count Fulk III Nerra's time, it was reconstructed in solid stone during the 1230s by King Louis IX of France, the future St. Louis, as a precaution against attack by the English. The young Louis had not yet been to Palestine, but he knew the style of the Crusaders' castles there. The formidable feudal fortress, arranged in a roughly pentagonal pattern with 1,025-yard-long curtain walls and 17 round bastion towers varying in height between 130 and 190 feet, could withstand any invasion.

Today, the castle is known throughout Europe as the repository of Angers's richest treasure and one of the greatest masterpieces of medieval art: the Apocalypse Tapestries (or Tenture de l'Apocalypse). The 70 pieces of flamboyant tapestry, made between 1373 and 1380, cover a distance of more than 110 yards and illustrate the biblical Book of Revelation. Based on the designs of court painter Hennequin de Bruges, the tapestries were woven from wool and gold thread in the Paris workshops of Nicolas Bataille, the greatest tapestry-weaver of his time, for Louis I, Duke of Anjou, who had ordered them for the walls of his castle. During the French Revolution, however, the tapestries were thrown into the streets, where citizens cut them up and used the pieces for carpets, bedspreads, horse blankets, cart covers, and fruit tree canopies. In 1843, religious authorities began to trace the pieces, and eventually recovered almost two-thirds of the original 180 yards.

The second Angevin dynasty of counts commenced in 1246 when King Louis IX gave Anjou as an appanage, or monarch-granted land endowment, to his younger brother Charles, count of Provence; in 1266, Charles assumed the crown of Naples and Sicily as Charles I, founding an Angevin dynasty of kings there. His son, Charles II, succeeded him as king of Naples in 1285; an earlier revolt had ended Angevin rule in Sicily. Five lines of European rulers descended from Charles II's five children; his daughter, Marguerite, inherited Maine and Anjou when she married Charles

III of Valois. Charles III greatly improved the economic and social conditions of the people of Anjou, who were largely family farmers. The son of Charles III, Philip of Valois, count of Anjou, became king of France as Philip VI in 1328, thereby reuniting the two provinces with the French crown until his death in 1350.

Philip VI's son, King John II of France, gave the countship of Anjou to his son Louis in 1354—beginning the third Angevin dynasty. When Anjou was raised to the rank of a duchy in 1360, Louis I, now duke of Anjou, also inherited the throne of Naples; Anjou was an appanage to this kingdom. At this time, during the Hundred Years War (1337–1453), English soldiers under the command of Sir Robert Knollys ravaged and devastated the Loire valley, including Anjou. Louis I and his son Louis II, however, were less interested in defending the duchy than securing the kingdom of Naples, and they spent most of their time fighting away from Anjou. After Louis II died in 1417, his widow, Yolande of Aragon, tried to protect Anjou from the English. Their forces had besieged the Loire valley by the late 1420s, causing great destruction to Angers, and would be driven back to England beginning in the 1430s as a result of France's renewed nationalism and the martyrdom of Joan of Arc.

Anjou's last and greatest duke was René I (René the Good, or René le Bon), born as the second son to Louis II in 1409. Through lineage and marriage, he eventually commanded an empire that included Provence, Piedmont, Lorraine, Naples, Sicily, and Jerusalem. René became king of Sicily in 1417 and became entitled to Naples in 1434. He set sail for Naples in 1438, but never took possession of it; it was captured by Alfonso V of Aragon after a 1441 siege. René returned to France the same year, and soon established a flourishing, almost fairy tale–like court life at Angers Castle; he constructed a chapel and most of the other buildings inside the ramparts and kept a menagerie of monkeys, leopards, and lions inside the dry moats. His daughter married King Henry VI of England. After his second marriage in 1454, René retired to Provence (where he was count) and painted pictures, composed poems, and wrote books of romance. René died in 1480. The duchy of Anjou was officially and permanently passed to the crown of France under King Louis XI, René's cousin, in 1480.

After René's death, the castle fell into disrepair. Toward the end of the sixteenth century, King Henry III ordered Governor Donadieu de Puycharic to raze the fortress; dismantling, however, proceeded slowly, and the king died before plans could be carried out, preserving the castle for posterity. In the seventeenth century, King Louis XIV turned it into a prison; he incarcerated his former finance minister Nicolas Fouquet there during his sensational trial for embezzlement in the early 1660s. In the nineteenth century, the castle became a prison again, this time for English sailors captured during the Napoleonic Wars, and it served as a primary munitions depot for the Germans in World War II. The castle was bombed by the Allies in 1944;

fortunately, the Apocalypse Tapestries had been temporarily removed.

Angers also retains many buildings from the Renaissance. One of the city's most beautiful mansions, the Barrault House, was built in 1487. Built for Olivier Barrault, the treasurer of Brittany, it today houses the city's public library, as well as galleries and museums including the Musée David, with works by noted native sculptor Pierre-Jean David, or David d'Angers. Most of David's studio works—statues, busts, bas-reliefs, and medallions—are permanently displayed in the museum, capturing the personalities of famous historical figures. David, who also designed the facade of the Pantheon in Paris, fashioned the big statue of Duke René that stands in the busy crossroads just south of the castle.

Angers was the scene of fighting during the Wars of Religion (1562–98), and Huguenots (French Protestants, or Calvinists) captured the city in 1585. It developed as a river trading center for fruits and vegetables shipped to England and the Netherlands; the manufacture of boat sails and hemp rope became important industries beginning in the seventeenth century. The citizens of Angers vigorously supported the French Revolution and, in 1793, put down an uprising of Vendean (counter-revolutionary) royalists against the revolutionary government. Angers's Universitas Andegavensis—the most famous Catholic theological school in France, dating to 1364—was suppressed during the revolution, but was reconstituted nearly a century later, in 1875, as the Université Catholique de l'Ouest.

In the twentieth century, the most historically significant event in Angers was its occupation by the Germans from June 1940 to August 1944. Although much of Angers was severely damaged during the war, its most historic buildings escaped serious injury, and the city has been considerably restored.

Angers is today the capital of the Maine-et-Loire department; much of the region of Anjou was incorporated into this department in 1790. The historic and cultural region of Anjou also includes parts of the Indre-et-Loire, Mayenne, and Sarthe departments. The region remains primarily rural, but the city of Angers has become heavily industrialized in recent decades; this change, along with the development of railway, has led to a large increase in the city's population since World War II. Nevertheless, Angers's traditional industries have adapted to changing times. As it has for centuries, the city serves as a large agricultural market, with the rich lowlands of the Loire and its tributaries producing wines, fruits, and flowers for the surrounding area, as well as for the markets of Paris. The city's market gardens have spawned many horticulture industries, such as nurseries, bulb plantations, and vegetable biotechnology. Another major activity, which had its roots in the thirteenth century and still provides employment for many Angers residents, is the extraction and export of slate from nearby quarries; this has earned Angers the appellation of the "Black City," referring to the shale and slate of which the city was once almost exclusively built. The nickname remains, even though stone replaced slate as the main construction material even before the Revolution. Textiles have been made in and around Angers since the fourteenth century; in more modern times, traditional weaving has developed into the manufacture of ropes and cables from hemp and jute. Since the 1970s, newer factories in the city's *zone industrielle* have produced computers, radio and television components, and camera equipment. An activity for which Angers has become world famous is the distillation of Cointreau liqueur. First produced in 1849 by the Cointreau brothers, the liqueur is still made at the family-owned distillery with alcohol from Angers and sun-ripened tropical oranges.

Further Reading: Much historical and geographical information about Angers and the surrounding region is contained in *The Loire* by James Bentley (Topsfield, Massachusetts: Salem House, 1986); *The Loire Valley: Plantagenet and Valois* by Henry Myhill (London: Faber, 1978); *The Loire* by Vivian Rowe (Washington, D.C., and New York: Luce, 1970); and *The Companion Guide to the Loire* by Richard Wade (Englewood Cliffs, New Jersey: Prentice-Hall, 1983; London: Collins, 1984). The area's role in national history is chronicled in Jean Dunbabin's *France in the Making: 843–1180* (Oxford and New York: Oxford University Press, 1985), which covers the period from the collapse of the Carolingian Empire to the rise of the French monarchy.

—Jeff W. Huebner

Anglesey/Holyhead (Gwynedd, Wales)

Location: An island off the northwest coast of Wales, Anglesey is approximately 260 miles northwest of London, to which it is connected by rail and road. Holyhead, its main port and town is located on Holy Island on the west coast, separated from Anglesey by a narrow channel.

Description: Anglesey has many remains of Stone Age settlements and was an important Celtic center in the Iron Age. More recently it was home to a significant copper mining and smelting industry. Holyhead prospered as a port during Victorian times. Today, the area is a popular tourist destination, not only because of its historical significance but because of its mild climate, beaches, and nature preserves.

Site Office: Economic Development and Tourist Department
Ynys Mon Borough Council
Llangefni, Anglesey, Gwynedd LL77 7JA
Wales
(0248) 810833

For centuries, men and women have crossed the Menai Strait to the Welsh island of Anglesey, which lies in the Irish Sea within sight of the dramatic mountains of Snowdonia on the mainland of Wales. They have come to seek out its mild climate, its fertile soil, its lakes and forests, its harbors and beaches, and, in recent times, its nature reserves and bird sanctuaries. Some have restricted themselves to the 275 square miles of Anglesey proper. Others have crossed to nearby Holy Island and its port of Holyhead. Still others have made pilgrimages to tiny Church Island or have attempted to visit the small group of bare, rocky isles of West Mouse, Middle Mouse, and East Mouse.

There is clear evidence that Anglesey was inhabited during the Stone Age, probably by people who came from the western Mediterranean as early as 4000 B.C. The island contains burial chambers typical of Stone Age communal graves found elsewhere in Europe. When constructed, these graves, called cromlechs in Wales and barrows in England, were covered with stones and mounds of earth with enormous boulders, known as megaliths, at the entrance. In 1910, a survey identified some fifty-four megalith graves in Anglesey; only about twenty of these can be located now.

Several of the Stone Age graves that have survived in Anglesey are elaborate structures composed of long passages with several compartments leading off to the side. One such chambered tomb at Trefignath, a few miles south of the port of Holyhead, consists of a forty- to forty-five-foot passageway with three, or perhaps four, compartments. Another chambered tomb near Lligwy takes advantage of a natural cleft in the rock of a hill, with a twenty-five-ton capstone, supported by a series of upright stones, marking the entrance. A third chambered tomb, Bryn Celli Ddu (the Mound of the Black Grove), near Llanfair, appears largely intact, although it has probably been restored in recent years. Within its passageway there is a central upright stone, six feet high, which was discovered lying on its side in the early 1920s. The whole grave was originally covered by stone slabs, of which several remain, and an early description mentions a stone bench around the sides; the bench is no longer present. Some believe that the tomb was at the center of an earlier stone circle, like that of Stonehenge or Avebury in England.

Another striking chambered tomb is Barclodiad y Gawres (the Apronful of the Giantess) located near the sandy beach of Cable Bay on the west coast of Anglesey. Considerably restored, this tomb was constructed some 5,000 years ago at the cliff edge, fifty feet above the sea. A mound covers a narrow, slab-lined passage about twenty feet long, leading to a series of burial chambers. Within, there are stones with inscribed patterns similar to those found in chambered tombs in Ireland, Spain, and Portugal. This suggests that there may have been communication and trade between Stone Age inhabitants in Anglesey and Ireland, where chambered tombs are relatively common, and also with prehistoric settlers in the European continent. Chambered tombs with similar cup-marked capstones and holed slabs have been discovered in Scandinavia, the Caucasus, and India. In Anglesey, as elsewhere, these are only found in districts bordering on, or easily reached from, the sea and are absent from interior sites, except those that are easily accessible from the coast.

Immigrants came to Anglesey from the Rhineland about 1800 B.C., during the Bronze Age. These new inhabitants are known as Beaker People, due to the distinctive waisted red and brown pottery vessels they used throughout Europe. There are few remains of the Beaker People in Anglesey, although it appears that, like their Stone Age predecessors, they preferred life by the sea.

Of far greater importance to the history and subsequent culture of Anglesey was the arrival, in about 150 B.C., of the first Iron Age Celtic tribes. There are abundant traces of these early Celtic inhabitants. The most dramatic find occurred in 1942–43, when the Royal Air Force base at Valley, some five miles south of Holyhead, was enlarged in order to serve as the transatlantic terminal of U.S. Air Force flights to Britain during World War II. A horde of Celtic Iron Age remains was recovered from a nearby bog when this was excavated in order to provide peat to serve as the base of a new runway. Excavators found 138 objects made of bronze, iron, and wood, including not only weapons such as swords, spearheads, and shield bases, but also vehicle parts, a bridle bit, sickles, currency bars, pots, rings, and even a trumpet.

Bryn Celli Ddu, top, and Holyhead Mountain, bottom
Photos courtesy of Ynys Mon Borough Council

One object, apparently used for captives or slaves, consisted of an iron chain more than ten feet long with five fixed iron neck rings. Most of these objects are in the National Museum of Wales at Cardiff, but some are in the Oriel Ynys Mon Heritage Gallery near Llangefni.

One interpretation of the valley find is that it represents votive offerings used during Druid rites. The Druids were Celtic priests about whom little is known except that they carried out rites in sacred oak groves, and attempted to forecast the future and placate their gods with animal or human sacrifices. Anglesey was an important Druid center. When a Roman army was dispatched to conquer the island in A.D. 61, the Roman historian Tacitus described its encounter with the Druids:

> The enemy army was ranged along the shore like a forest of weapons and soldiers among which women ran ceaselessly about like Furies, shrieking imprecations, with black robes and dishevelled hair and torches in their hands. All around stood Druids with their hands raised to the sky howling wild curses.

During the Roman occupation, the islanders lived on their own in the hills in small circular or rectangular huts. Remains of some of these hut settlements have survived to the present day. On the northeast coast, near the beach at Lligwy, outer walls and stone hut foundations attest to what was probably an important Celtic settlement that continued to be occupied for at least 250 years. There is evidence of a large settlement of more than fifty huts on Holy Island, and rectangular double banks and ditches at Caer Leb, near Bodowyn, suggest a farming community during Roman times. The Romans constructed several hill forts in Anglesey that can still be identified. The hillfort Caer Gybi at Holyhead may have served to protect the port against pirates sailing the Irish Sea.

Christianity had been established in Anglesey by the time the Romans left in A.D. 406, and hermits and monks established holy wells and churches on the island during the early Christian age. Several of these can still be identified, including St. Seriol's hut and well at Penmon on the eastern corner of the island, the site of a church dedicated to St. Patrick in Llanbadrig on the north coast, and the site of the chapel of St. Dwynwen on Llanddwyn Island off the south coast. St. Dwynwen is said to have been one of the twenty-four beautiful daughters of Brychan, a prince of fifth-century Wales. She died in 465, and her birthday is still observed on January 5. Traditionally, the legend goes, lovers made pilgrimages to the site of her chapel to discover their fortunes by watching the movement of an eel in the nearby freshwater spring.

After the Romans left, Anglesey was at the mercy of raiders from the sea, particularly Irish Picts who not only pillaged communities along the coast but also settled there. The Irish Picts dominated life in Anglesey until the middle of the sixth century, when they were defeated in battle by the Celtic tribe of Brythons, led by Caswallon Law Hir. It was during this period, as the Irish were leaving Anglesey, that the Celts in the north of mainland Wales united to form the kingdoms of Gwynedd and Powys.

The royal family of Gwynedd governed Anglesey from the sixth to the thirteenth century. They had an important palace at Aberffraw, on the southwest coast of the island, apparently constructed mainly of wood, but neither the building nor its foundations have survived. Some believe that stonework in a local Aberffraw church came from this royal palace. During the rule of the Gwynedds, Anglesey was known as Mon Mam Cymru—the Mother of Wales. Essentially flat, with a fertile soil and mild climate, the island served as the granary for the mountainous mainland. Giraldus Cambrensis, a twelfth-century chronicler, noted that, "when the crops have been defective in all parts of the country, this island, from the richness of its soil and abundant produce, has been able to supply all Wales."

The last Gwynedd monarch, Prince Llywelyn, was defeated in battle by King Edward I of England in 1282. Wales then came under English rule. When Edward I built a number of castles—his Ring of Stone—to hold the Welsh princes in check, he built two castles overseeing the Menai Strait, one at Caernarvon on the mainland, and the other at Beaumaris, which was then the principal town and chief port of Anglesey. Beaumaris came to serve not only as a garrison town for the English, but also as a naval station. Despite these precautions, or perhaps because of them, resistance to English rule continued in Anglesey, as in the rest of Wales. Following the rebellion of Owen Glendower in the fifteenth century, a stronger Act of Union in 1536 proclaimed English as the official language of all legal and government business in Wales, and disqualified any Welshman unable to speak English from holding public office in his own country. If anything, the Celtic and Welsh character of Anglesey was strengthened by this decree.

Throughout the centuries, the people of Anglesey have been primarily involved in agriculture; in modern times at least, barley and oats have been their traditional crops. Agricultural production on the island has been particularly intense during wartime, when imports of food from abroad into Britain have been severely curtailed. The economy, however, has not always been dependent entirely on agriculture or even, as in modern times, tourism. In the eighteenth century, copper mining on Parys Mountain in northeast Anglesey was a very important enterprise.

Parys Mountain apparently had been mined by the Romans, but there is little evidence of subsequent activity at the site until the great copper boom that began in 1768. The copper sheathing required by British warships resulted in a strong demand for copper at this time, and the discovery of ample deposits of this metal in Parys Mountain made it one of the most important opencast copper mines in the world. Production survived the criticism of the efficacy of copper sheathing on warships following the British naval losses in the American Revolution. Until the end of the eighteenth century, there were some 1,200 men, women, and children employed in the mines. A decline began shortly thereafter, and the work force had shrunk to 207 in 1806, and to 122 in 1808. By 1815, the copper industry in Anglesey had collapsed, mainly as a result of cheaper copper coming to Britain from Africa and the Americas.

The copper mine also spawned a smelting industry at Amlwch, up to this time only a small fishing port to the north of Parys Mountain. The smelting industry carried on for several decades after the decline of the Parys mine by importing smelting ore from other British mines and from abroad. A shipbuilding industry was established and the port grew rapidly; it was soon able to accommodate some forty vessels at anchor, albeit only small boats. With copper the outward cargo, ships often arrived with tobacco leaf that was processed in Amlwch factories. The smelting industry was severely constrained by the lack of nearby sources of the coal needed in the smelting process, and the prosperity of Amlwch declined in the second half of the nineteenth century. Today, it is one of Anglesey's larger towns with some small industry.

In modern times, Anglesey can be reached from the northernmost tip of Wales over two unusual bridges spanning the Menai Strait. Before these were built, it was only possible to reach the island on ferries or small boats that often foundered in the swift currents; it is recorded that some 200 persons drowned between 1664 and 1842 crossing the Menai Strait to Anglesey. The oldest of the bridges is the 600-foot Menai Suspension Bridge, the world's first large iron suspension bridge, designed by Thomas Telford in 1818 and completed in 1826. The Telford bridge is very tall as it had to conform to British Admiralty specifications that required at least 100 feet of headroom at high tide so that the great sailing ships of Her Majesty's Navy could pass beneath. The Menai Strait was a major shipping thoroughfare in Victorian times, and more than 700 vessels were registered at Menai ports in the 1870s.

The other, and newer bridge, is the double-decker Britannia, with cars above and trains below, the latter shuttling between London and the port of Holyhead, where ships await to carry passengers across to Ireland. In order to lure tourists to the railway line when it was first opened, the Victorians named the first town and railway station on the island side of the Britannia bridge Lanfairpwllgwyngyllgogerychwymdrobwlllantysiliogogogoch (St. Mary's Church in the hollow of white hazel near a rapid whirlpool and the Church of St. Tysilo near the red cave). The town is now commonly known as Llanfair.

Holy Island, one of several islands with the same name to be found off the coasts of Britain, is separated from Anglesey by a narrow channel, now spanned by road and rail bridges. From the top of Holyhead Mountain, although only 720 feet above sea level, it is possible, on a clear day, to see not only Snowdonia Mountain in Wales but also the Mountains of Mourne and the Wicklow Hills in Ireland. Holyhead, Anglesey's largest town, lies at the tip of Holy Island and, like Parys Mountain, it experienced its greatest prosperity in Victorian times.

Holyhead was probably first settled by the Romans, who constructed a fort above a creek on the site. When Edward I of England conquered Wales, Holyhead was described as an unimportant creek, although there was a church and probably a small monastery on the site. When Elizabeth I came to the throne of England in 1558, conditions in Ireland required a reliable method for sending packets of state papers between the Queen and her Lord Lieutenant in Dublin. A packet transmission system was devised for this purpose, by which state papers traveled from London by horse or coach across England and Wales to Holyhead and from there, by packet boat, across the Irish Sea, to Ireland. Individuals could also send papers by this system if they had royal permission to do so. In 1666, an act was passed establishing a national Post Office, and a contractor was appointed to oversee the packet boats in Holyhead. The crossing to Dublin was hazardous and boats were subject to raids by privateers as well as to severe weather conditions. Three packet boats capsized in storms in the late seventeenth century, one with the loss of 120 passengers, and this led to a decision to use a Holland-built packet boat, which was deemed better able to stand up to the stormy conditions. A lighthouse with a perpetual light was erected near the harbor in 1804; previously ships had to rely on a coal fire burning in a grate at the top of the cliff. There were restrictions on what could be brought from Ireland to Holyhead, and some goods required custom duties. In the early eighteenth century, a port "tide waiter" was appointed, whose job it was to wait for ships to come in on the tide so that he could inspect their cargoes and assess any duties payable.

As methods of travel across England and Wales improved, so did the fortunes of Holyhead. In 1810, a 1,000-foot pier was constructed and a lifeboat permanently positioned on the anchorage ground. Shortly thereafter, sailing cutters ceased to be used for the port and steam vessels were introduced. The mid-nineteenth century was a time of prosperity for Holyhead, as it was for Anglesey as a whole. The population of Holyhead increased from 3,869 in 1841 to 8,863 in 1851, mostly as a result of work begun on the railway station as well as on the massive breakwater, which took twenty-eight years to build. Stone for the breakwater was quarried from Holyhead Mountain with large quantities of explosives; on one occasion Queen Victoria and other members of the Royal Family traveled to the site to witness an explosion.

In 1848, the first train arrived in Holyhead and was feted at the new station, which, according to a contemporary account, was complete with "Refreshment Rooms . . . Waiting and Dressing Rooms . . . Telegraph Office and a well furnished Book Room." The railway company also built a large cattle shed and luggage station on the pier. Another sign of the prosperity was the illumination of the town in 1856 by gas lights while the town band played in celebration. Today, power is supplied to much of Anglesey and Holyhead by the Wylfa nuclear power station near Cemaes on the north coast of the main island.

Even in its heyday, Holyhead was essentially a railway town with a lot of primitive housing stock. Many residents relied on a man with a donkey cart to provide them with sand to cover the earthen floors of their houses. By 1920, the fortunes of the port had fallen into decline; the process was exacerbated by the Great Depression. Much of the traffic to Ireland moved to other ports, and the declaration of Irish independence in 1922 brought restrictions in trade between Britain and Ireland. Today, there is some light industry in the town, which has retained only a shadow of its Victorian energy. A considerable proportion of the population are descendants of Irish and other immigrants who came to work on the breakwater and the railway terminal. As a result, Holyhead does not have the same distinctive culture as the rest of Anglesey.

Further Reading: Few books on Anglesey and Holyhead are easily found outside Wales. *Mona, Enchanted Island* by Geoffrey Eley (Royston; Priory Press, 1968) is a chatty, romantic account somewhat lacking in detail. *Holyhead: The Story of a Port* by Lloyd Hughes and Dorothy M. Williams (published by the authors, 1981) contains vignettes drawn from local history sources. *Best Walks in Wales* by Richard Sale (London: Constable, 1988) contains excellent site descriptions, and *The Isle of Anglesey, North Wales* (Cardiff: Wales Tourist Board, 1994) is full of interesting information. The Anglesey Antiquarian Society in Llangefni has recently done a worthwhile series of specialized studies including *Copper Mountain* by John Rowlands (1966), *Prehistoric Anglesey* by Frances Lynch (1991), *A New Natural History of Anglesey* by Helen Ramage (1990), *Two Centuries of Anglesey Schools 1700–1902* by David Pretty (1977), *Medieval Anglesey,* by A. D. Carr (1975), and *Portraits of an Island: Eighteenth Century Anglesey* by Helen Ramage (1987). *Shorelands Summer Diary* by C. F. Tunnicliffe (London: Clive Holloway, 1984; Bridgeport, Connecticut: Merrimack, 1985) contains many fine drawings of birds found on Anglesey.

—Judith Gurney

Arles (Bouches-du-Rhône, France)

Location: In southeastern France, fifty-five miles northwest of Marseilles, along both banks (chiefly the right) of the Rhône; bordering the northernmost point of the marshy Camargue, where the Rhône delta spills into the Mediterranean.

Description: Arles was a strategic city for early imperialist peoples, capital of Provence under the Roman Empire, and seat of early Christian archbishops. Many ruins survive from these times. In the nineteenth century, Vincent Van Gogh spent two years in Arles and painted many local subjects. Although none of his work is displayed in Arles, the city does have four museums dedicated to the fine arts.

Site Office: Office de Tourisme, Syndicat d'initiative
35 place de la République
13200 Arles, Bouches-du-Rhône
France
90 18 41 20

Long ago an industrious commercial hub for a multitude of Europe's imperial states, the city of Arles today evinces a becalmed lack of urgency after centuries of relative dormancy. An ancient seat of government for the Greeks, Romans, and Gauls, and the site of an important early Christian bishopric, Arles fell into decline during the Dark Ages. In the tenth century, Arles regained much of its former splendor when it became the capital of the Kingdom of Arles. Today the city, no longer situationally imperative to European planners, is primarily a center of art, a role it began to develop during Vincent Van Gogh's prolific stay there in 1888–90. The artistic treasures of Arles are now maintained in four public museums.

The Greeks and the Phoenicians were most likely the first civilized inhabitants of Arles; evidence of their stay dates to the sixth century B.C. The Greeks provided the Provençal language with hundreds of words, and they brought to the region cherry trees, chestnuts, and olives. They also introduced worship of the sun to the inhabitants of Provence. Early immigrants came from Liguria; later came Roman colonizers. In 100 B.C., the Roman consul Gaius Marius built a canal linking Arles to the Mediterranean, and the city became a major port. Located at the southernmost bridgeable point of the Rhône, Arles also became a focal point for land traffic between Spain, Italy, and France. At the invitation of the Celts, the Roman emperor Julius Caesar annexed the region, ensuring the effectiveness of its government and commerce. By approximately 49 B.C., Caesar had a taken personal interest in the development of Arles, making the city the capital of Roman Provence.

Sometime between that year and the first century A.D., the Romans began construction of the impressive Arènes amphitheatre. Les Arènes, built to accommodate 26,000 spectators, was the site of many savage gladiatorial battles pitting men against beasts. The amphitheater remains standing today, supported by two tiers of sixty arches each (a third level has fallen to ruin). During the Middle Ages, Les Arènes was fortified and partitioned into dwelling and trading spaces. Restored in the nineteenth century, the venue is once again a public stadium, hosting Sunday afternoon bullfights beneath its weathered towers.

One hundred yards from Les Arènes lies the rubble of the Roman Théâtre Antique, a performance space of considerable grandeur that has suffered more extensive damage than its neighbor. Two surviving columns, nicknamed *deux veuves* (two widows), suggest the architectural beauty that preceded the theater's function during the Dark Ages as a rock quarry. Originally built to seat at least 7,000, the remains of the Théâtre's arcade, now able to accommodate an audience of only a few hundred, enjoys a revival each July during Arles's Music and Drama Festival.

Another Arlesian remnant dating to Roman times is Les Alyscamps (Provençal for "mythical burial ground," originally named Champs Élysées—Elysian Fields), a thoroughfare to the southeast of the town flanked on either side by tombs that once were breathtakingly detailed. Early Christians at Arles interred their dead in highly ornamental sarcophagi along Champs Élysées. The practice was abandoned toward the latter part of the Middle Ages, and the route was further marred when it was dissected by a railroad line a century ago. Most of the important finds from the site are now housed in the Musée d'Art Chrétien, but Les Alyscamps does retain some degree of ominous presence.

In the early 300s, the Roman emperor Constantine built a magnificent palace at Arles—standing in ruin today—that housed Provence's largest baths, the Thèrmes de la Trouille. Emperor Flavius Honorius added to the city's prestige when he made it the capital of the "three Gauls" in approximately 400, but by then the Roman Empire was faltering. Arles, however, had been permanently shaped by its Roman occupation: physically, by the development of public buildings and extensive roadworks, and culturally, by Christianity, the vulgar forms of the Latin language, and Roman literature and philosophies.

As Roman government fell into decline at Arles, the city began to rely more heavily on the influence of its Christian administrators. By the first century A.D., St. Trophimus had founded the bishopric of Arles; in 417, the city was made seat of the primatial see. In August 314, Arles hosted the first fully representative synod of the Western Roman Empire's bishopric. Constantine called forty-three members of the

The Roman Théâtre Antique at Arles
Photo courtesy of French Government Tourist Office

clergy to a meeting to attempt to repair the Donatist schism, a rift within the church, in which the followers of Donatus had rejected earlier decisions of the Christian synods in Rome and Africa. Although the Donatists continued to protest the findings of the church at Arles, Christian leaders considered them irrefutable.

The Visigoths, who had wrested control of the region from Rome, controlled Arles from the fifth to the eighth century A.D. During that reign, the city retained its significance to the Christian church, providing the site for St. Augustine's consecration as the first Archbishop of Canterbury in 597. Invading Saracens from Spain pillaged the city in the year 730, instigating a long period of instability. In the tenth century Hugh of Arles, count of Provence, and Rudolf II, duke of Burgundy, who had been rivals for some time, joined forces against the area's various invaders. Their alliance led to an agreement that united Provence and Burgundy as the Kingdom of Burgundy in 934. (The northwest portion of the kingdom, controlled by Richard the Justiciar, soon split off as the Duchy of Burgundy. In the late twelfth century the kingdom began to be called the Kingdom of Arles, with the name Burgundy reserved for the duchy.) The city of Arles regained some of its former splendor as the capital of the new king-

dom, ruled first by Rudolf II. His son Conrad succeeded him in 937.

Conrad's son Rudolf III became king in 993. He was childless, a factor that led to a rivalry for succession to the throne; the Germans in particular sought to rule the kingdom. The German king and Holy Roman Emperor Conrad II was a distant relative of the Burgundian rulers, but he made the relationship closer by marrying one of Rudolf's nieces. Therefore in 1032, Rudolf, on his deathbed, chose Conrad II as his successor. The Germans, with far-flung possessions in their Holy Roman Empire, devoted little time to the Kingdom of Burgundy and ruled it through viceroys, the result being that the inhabitants remained independent-minded and resistant to German authority. At the same time, the kingdom had internal problems created by ethnic and cultural differences among its residents. French influence in the area grew through marriages and intervention in political affairs, and French nobles gained countships in several parts of the kingdom. Therefore in 1378 Holy Roman Emperor Charles IV ceded the Kingdom of Arles to France.

The late Middle Ages saw the development of another of Arles's notable sites, the church and cloister of St.-Trophime. They were begun in the eighth century but built primarily during the eleventh and twelfth centuries. The well-preserved cathedral features noteworthy examples of sculptors' friezes, depicting scenes of the Old Testament for a largely illiterate congregation; surrounding the main portal are representations of the Last Judgment, the twelve apostles and the Nativity. The Romanesque cloister adjacent to St.-Trophime, with its statuesque saints guarding the four corners, is considered one of Arles's finest attractions.

Very few Arlesian records are available for the latter years of the Middle Ages, a time of occasional suffering and strife in the deteriorating city. Figures from 1437–38 reveal that two-thirds of the city's residents were laborers, farmworkers, shepherds, fishers, and hunters. The rest were vintners, completing the composition of an agricultural community. Not until the end of the fifteenth century did Arles partially recover, when immigrants arrived from the north of France to take advantage of the region's fertility, and when, too, it enjoyed some modest prosperity, the result of its situation on an important European trading route. Arles's place in the Christian hierarchy of Europe also slipped: by the end of the 18th century, its bishopric had been suppressed, and the remarkable St.-Trophime had ceased to function as a cathedral.

By the 1880s, Arles was a small town of 23,000 that had found a role in the modern era as the railroading center of the south of France. Arles was known primarily for its strikingly beautiful women, whose long, black, lace-trimmed costumes were made famous by the author Alphonse Daudet and by the composer Georges Bizet. Arles was also known as a home to a disproportionate number of mentally ill persons, mostly patients of the city's Hôtel-Dieu hospital.

Arriving at this setting in 1888, the painter Vincent Van Gogh rented a room above the Restaurant Carrel. Van Gogh's short and stormy stay in Arles would provide the town with an enduring mythic profile in the eyes of the world. During his fifteen months in Arles, he painted more than 200 canvases, including *Le Pont de Langlois, Cafe du Soir,* and a rendering of Les Alyscamps, even as that ancient installation was being overrun by the railroad. None of the other homes and cafes that served as subjects for Van Gogh's paintings in Arles remains today. Many were destroyed by bombing during World War II. Nonetheless, as Van Gogh biographer David Sweetman writes, "our image of the place has been . . . heavily conditioned by Vincent's re-creation of it," and suggests that Van Gogh "rigorously excluded much that displeased him."

Van Gogh's idyllic portraits of Arles belie another actuality: during his residence, he often was treated badly by the townspeople. In the hope of creating a fraternity of artists, a "studio of the South," Van Gogh invited Paul Gauguin to join him in Arles. Their association quickly became strained. Distressed, Van Gogh sliced off his own ear, wrapped it in newspaper and delivered it to a prostitute named Rachel at the Maison de Tolerance.

The city of Arles, while acknowledging its indiffer-

ence toward Van Gogh while he studied there, readily admits its debt to the artist for reviving its name. Four museums cater to patrons of the arts who are attracted to Arles for its role in Van Gogh's life, although none of them contains his works, which are now too prohibitively priced for small-town foundations.

Located on rue Balze, the Musée d'Art Chrétien displays an exemplary collection of religious art. Housed in a desanctified Jesuit church of baroque design, the museum features many Biblically decorated marble sarcophagi rescued from the tombs at Les Alyscamps. Below ground level lies the Cryptoporticus, a 300-by-500-foot cellar used as a granary 2,000 years ago, as well as a portion of the Roman sewerage system devised two centuries later.

The Musée d'Art Paien on place de la République resides in another desanctified church, the 17th-century Church of Ste.-Anne. Its collection includes Roman mosaics, busts, and sarcophagi that are engraved with mythological (rather than Biblical) scenes. Also included are the Greek sarcophagus of Hippolytus and Phaedra and two casts of the Venus of Arles, the original of which is in the Louvre.

The Musée Réattu, on rue du Grande Prieuré, is housed in a fifteenth-century palace of the Knights of Malta. Although Jacques Réattu was never considered one of France's finest painters, the repository that takes his name offers canvases by such French masters as Gauguin, Léger, Vlaminck, and Dufy, as well as a playful cluster of fifty-seven drawings by Picasso. The Musée Réattu also features a collection of photography which encompasses some of the field's greatest practitioners.

Lastly, on rue de la République, the Musée Arlaten displays period rooms portraying life in Arles through the ages. This folk museum, founded in 1896 by the poet Frédéric Mistral, combines furnishings and decorative items with properly attired mannequins to provide still-life images of Arlesian social development and Provençal customs.

Like the rooms of the Musée Arlaten, the legacies of Arles are firmly rooted in the past. Once a key location for the imperial and religious authorities of Europe, this city on the Rhône is virtually no larger today than it has been for centuries. As a result, the Old World Arles exudes a languid eccentricity befitting an artwork in progress.

Further Reading: *The Splendor of France* by Robert Payne (New York and London: Harper, 1963) provides a well-crafted portrait of Provence, focusing on the peoples of the region and their customs. *Van Gogh: His Life and His Art* by David Sweetman (New York: Crown, and London: Hodder and Stoughton, 1990) retells in meticulous detail the artist's life in Arles, and in the process illuminates an Arles arriving at the modern era. *The Penguin Guide to France,* edited by John Ardagh (New York: Viking Penguin, and London: Collins, 1985) and *France At Its Best* by Robert S. Kane (Lincolnwood, Illinois: Passport Books, 1991) are the most informative of the travel guides.

—James Sullivan

Armagh (Armagh, Northern Ireland)

Location: South of Lake Neagh; bordered by the Dungannon district on the northwest; Craigavon district on the northeast; Banbridge district on the east; Newry and Mourne districts on the southeast; and the Republic of Ireland on the southwest; thirty-three miles southwest of Belfast, Northern Ireland; eighty-nine miles northwest of Dublin, Republic of Ireland; in administrative County Armagh, Northern Ireland.

Description: Diocesan headquarters for Church of Ireland and Roman Catholic Church in Ireland; city, seat, and urban district (established 1973), and administrative county (established 1974, comprising northern portion of former traditional county of Armagh); 261 square miles. Founded, according to tradition, in A.D. 450 by St. Patrick.

Site Office: The Old Bank Building
40 English Street
Armagh, Armagh BT61 9DB
Northern Ireland
(861) 527808

Armagh, a city, seat, and urban district in County Armagh, Northern Ireland, has been the ecclesiastical capital of Ireland for over 1,500 years and was the site of a reknowned medieval school of theology and an important intellectual center of the western world from the fifth to ninth centuries. According to tradition, Armagh was founded in the mid–fifth century by St. Patrick, the patron saint of Ireland, although this is a matter of dispute among historians. Patrick, however, is inseparably linked with Armagh and it has long claimed him as its founded. Armagh is also the primatial see of Ireland.

Armagh has adopted for itself the nickname, "the Irish Rome," for its religious significance and because, like Rome, it is positioned among seven small hills. In his writings, Patrick called Armagh "my sweet hill," and his original hill fort and stone church are believed to have been built at the site where St. Patrick's Church of Ireland Cathedral now stands. Rebuilt at least seventeen times, little remains of the original medieval church. Although its gargoyle-guarded exterior and Norman tower are essentially twelfth century, its present appearance owes much to eighteenth century rebuilding and nineteenth century restoration. On the opposite hill is the twin-spired St. Patrick's Roman Catholic Cathedral designed by neo-Gothic architect James McCarthy. It was begun in 1840 and dedicated in 1873. Considered McCarthy's masterpiece, the curvilinear decorated cathedral features statues of Sts. Patrick and Malachy and a Byzantine-like interior.

The work of eighteenth-century native architect Francis Johnston is evident in Armagh. Among his most famous creations are the Mall (a rectangular green surrounded by town buildings), the Royal School, the Bank of Ireland building, Charlemont Place, the courthouse, the market house, and the 200-year-old observatory. These structures were built with a local yellow limestone. The Mall is paved with Armagh marble, a polished, red limestone. Many of Armagh's streets have preserved their historic flavor, and their pattern follows the lines of ancient fortifications. The city is currently involved in a regeneration plan to preserve and present Armagh's historic past.

Armagh is Ireland's oldest recorded settlement. Traces of occupation dating as far back as the Late Stone Age (about 2000 B.C.) and the Late Bronze Age (about 700 B.C.) were excavated at Emain Macha, an eighteen-acre circular site located on a drumlin across the River Callan, two miles west of what is now Armagh. Both Emain Macha (Macha's Palace) and Armagh (derived from *Ard-Macha,* meaning Macha's Height) are believed to have been named for Macha, wife of Nevry, a semimythical Celtic war queen who arrived in Ireland 608 years after the biblical flood and founded the palace of Emania about 300 B.C. She is believed to be buried in the side of the hill where the Church of Ireland Cathedral now stands. However, some sources hold that these settlements were named after two different goddesses, both named Macha.

Although archaeological research suggests that the site may have been destroyed and abandoned by the second century B.C., Emain Macha, now more popularly known as Navan Fort, was Ireland's principal settlement at the time of Christ. Two centuries later, it appeared on a map of the known world by Alexandrian geographer Ptolemy, who depicted an area in Ireland identified as *Isamnion,* considered by scholars to be a form of Emain Macha. Armagh may have been a secular site, possibly even Ulster's capital, before it became a religious center, while Emain Macha flourished nearby as a military and ceremonial center for the kings of Ulster. On the other hand, many historians believe that Emain Macha was the prehistoric capital of Ulster from the fourth century B.C. until the fourth century A.D.

Emain Macha was also the setting of the Ulster Cycle, the ancient record of the epic and heroic tales of Cuchulain and the Red Branch Knights, Ireland's order of chivalry. The cycle of songs and poems has been compared to the epics of Homer. The Ulster Cycle is believed to be over 2,000 years old, but was only recorded in the seventh century A.D. It includes Ireland's version of the *Iliad,* the epic known as *Táin Bó Cúailgne* or "*The Cattle Raid of Cooley,*" a mythical tale of a battle between two divine bulls with a pseudohistorical setting depicting warfare between the Connachta and the

St. Patrick's Church of Ireland Cathedral
Photo courtesy of Bord Failte, The Irish Tourist Board

Ulaid, two powerful Irish dynasties. The heroic legend of the Ulster Cycle undoubtedly has some historical base. In the first century A.D., King Conchobáir (Conor Mac Nessa) ruled Ulster from his palace in Emain Macha with the Red Branch Knights, whose headquarters and training school were also based in Emain. His death in A.D. 33 marked the passing of the old pagan order, as according to Ulster legend, King Conchobáir died in reaction to the news of the death of

Christ, and was therefore regarded as the first Irish martyr and the first Irish pagan to have reached heaven.

The ancient kingdom of Emain Macha fell in either the fourth or fifth century A.D. The date is disputed among historians. However, most historians agree that the fall of Emain Macha coincided with the demise of the ancient kingdom of Ulster. According to folk tales, in 332 A.D., three Collas brothers from Connaght killed Fergus Fogha, the last of the traditional kings of Ulster, in the battle of the Black Pig's Dyke. They destroyed his palace, burned Emain Macha to the ground, and moved the capital to where Armagh now stands. The Red Branch Knights were driven eastwards into Down and Antrim. It is also believed that in the fifth century, Niall Noígiallach, a prince who claimed descent from Cormac mac Airt, founder of the Connachta, waged war with Ulster, drove the Ulaid (Ulster inhabitants) northeast out of the province, and conquered and destroyed Emain Macha, leaving it deserted by the middle of the fifth century. Niall of the Nine Hostages, as he was known, died in approximately 450 A.D., and his descendants took the dynastic name of Uí Neíll. Still others hold that Emain Macha was destroyed by the Connachta as late as A.D. 450 and that it was still the most important political center when Patrick founded his church at Armagh in 444. Regardless of whether Emain Macha was a vital political center or the deserted site of ancient ruins when Patrick founded his church two miles away in Armagh, the missionaries undoubtedly came to the area impressed by Emain Macha's enduring fame, past glory, and early Christian connection. Its political importance was also a factor. In time, Armagh replaced Emain Macha as the political center.

Interestingly, Patrick did not intend Armagh to be a permanent administrative center. At first denied the hilltop site he had selected in Armagh, Patrick later acquired the land after converting King Daire, a local chieftain descended from the Collas brothers. After establishing his Armagh headquarters, the itinerant missionary moved on to found over seven hundred Christian communities across Ireland, making him the most important and venerated figure in Ireland's history.

Most of the known details of Patrick's life are drawn from his *Confessio,* a testimony of his life in which he vindicates himself against critics' charges and bequeathes his legacy to those whom he had Christianized. Patrick is also credited with bringing literacy to Ireland. A few of these writings, some of which are preserved in Armagh, are the earliest surviving documents of Irish history.

Born in Roman Britain, the son of a Roman official who was also a minor deacon, Patrick was captured by Irish raiders at the age of sixteen and brought to Ireland as a slave where he herded sheep or swine in County Antrim, less than fifty miles northeast of Armagh. During his six years of captivity, Patrick, an unconfirmed Christian in pagan Ireland, underwent a spiritual transformation, according to his *Confessio,* and resolved to devote his life to God. Guided by a voice heard in a dream, Patrick made a 200-mile journey leading him to a seaport, where he escaped from Ireland and returned to his home in Britain. Called back by "the voice of the Irish," Patrick returned as a consecrated bishop to Ireland where he claimed to have baptized thousands of native pagans and set up Christian churches "beyond where no man lived and where nobody had ever come to baptize or to ordain clergy or to confirm the people." According to legend, Patrick drove the snakes out of Ireland, perhaps a metaphor for his saving of the pagan Irish souls.

Indeed, the legendary nature of the historical figure known as Patrick has long confounded scholars. Because the *Confessio* is not an autobiography and is chronologically imprecise, and other accounts of Patrick's life are equally obscure, historians have encountered difficulties and discrepancies regarding details and time frames of the missionary's life, particularly the dates of his arrival in Ireland and death. These chronological problems have given rise to the theory of "the two Patricks": one a Roman missionary believed to be Palladius, the first bishop to the Irish, who came to Ireland in the 430s and was known there as Patrick; the other a British missionary who arrived in Ireland some decades later. Whether the patron saint of Ireland was a Briton named Patrick or a Roman named Palladius, or an historical confusion of both, St. Patrick has long been celebrated by the Irish as the man who Christianized Ireland and founded the primatial see of Armagh.

The Patrick of the *Confessio* is believed to have died in Ireland in 492. A few early sources claim that Patrick was buried somewhere on the main hill at Armagh. Others cite Downpatrick as his burial place, and a stone in a cemetery behind a church further asserts this. However, Armagh continues to challenge Downpatrick's claim as St. Patrick's burial place.

After Patrick's death, a trend toward monasticism within the Irish church gathered impetus. Within a century, new monasteries had replaced old Patrician foundations as the important religious and learning centers. In the western Christian world, Ireland was unique in that most of its chief churches were headed by monastic hierarchies, rather than bishops. Even the church of Armagh, Patrick's own, eventually accomodated itself to the system. At Armagh, Patrick's immediate successors had been bishops. Before the end of the fifth century, however, Cormac became Armagh's first abbot and Patrick's religious center at Armagh became a monastery that flourished and became a school of theology famous throughout Europe. For the next two centuries, the head of the church of Armagh served as both bishop and abbot.

During this time, Ireland's Roman church organization, in an effort to counter pressure from new ecclesiastical groups, established Patrick's church of Armagh as the "primacy of all-Ireland," using Patrick's historic writings as supporting evidence. In the seventh century, the *Liber angeli* (Book of the Angel), also known in Armagh as *Canoín Phátraic* (Patrick's Testament), the *Vita Patricii* (Lives of Patrick), by Muirchú and Bishop Tírechán, and *Collectea,* also by Tírechán, all claimed that Patrick's church at Armagh was the primatial church and held appellate jurisdiction over Ireland. The *Liber angeli* further asserted that the supremacy of the Irish church had been invested in Patrick and that his successors must retain this position.

Trends toward secularization in Ireland gained momentum in the eighth century, countered by a reform movement aimed at fortifying the Christian community. By this time, Armagh's abbot no longer served as bishop. Instead, this role was now reserved for a subordinate community member for the administration of episcopally ordained sacraments.

Armagh was now the most important church in Ireland. It had jurisdiction over a large number of churches in Northern Ireland and other areas, and, as Ireland's primatial see, its head took precedence over all the clergy of Ireland.

From 750 to 850, Armagh grew and gained significance under the Uí Néill (O'Neill) dynasty. It was ruled during this period by families of the Airgialla, a satellite state of nine kingdoms dependent upon the Uí Néill and known as Niall's "hostage-givers." The Uí Néill ruler, while not a king, was the most powerful secular leader in Ireland. The bishop of Armagh held an equally powerful position within the Irish church. The two forces united in 804 when Aed Oirnide, the overking of the Uí Néill, was ordained king by Armagh's abbot and acted as protector of Armagh's interests.

Three years later, Armagh again reinforced its claim of primacy in a small gospel book, *The Book of Armagh*. Commissioned by Patrick's successor, Torbach, abbot of Armagh, and written by Armagh scriptorium master Ferdomnach, the *Book* contains almost the complete text of the New Testament; the *Life of St. Martin of Tours;* the *Vita Patricii,* an abridged version of the *Confessio;* and the *Liber angeli.*

Although the primacy of Armagh was generally recognized by the end of the seventh century, it was by no means a controlling force. Several provinces, most notably Dublin, Cashel, Tuam, Emly, and Kildare, continued to operate independently.

Armagh's primatial claim took on a political importance as Ireland's kings proclaimed "Patrick's Law," a sort of Armagh church tax. The Uí Briúin (O'Brien) kings proclaimed the law in Connacht between 783 and 836, as did Munster's king, Feidlimid mac Crimthainn. Mac Crimthainn also interfered in Armagh's political affairs and was the first king to wage war against Ireland's monasteries.

Armagh served as a leading intellectual center of the western world, with some 1,200 scholars based there by the ninth century. This was followed by several centuries of war with Viking armies from Denmark. The city was attacked five times by Danes between 839 and 1092. During the late tenth and early eleventh centuries, Brian Boru, the self-proclaimed "emperor of the Irish," became a national hero when he attempted to drive the Vikings out of Ireland. In 1001, he attacked Uí Néill territory; in 1002 and 1005 he made expeditions to the northern Irish states to take hostages in order to demonstrate his authority as high-king of Ireland. During 1005, when he was about sixty years old, Brian Boru spent a week in Armagh, where he made an offering of twenty ounces of gold to the Irish Church, confirmed Armagh's apostolic see of St. Patrick as the supreme ecclesiastical city of all Ireland, and recorded the decision in Latin in *The Book of Armagh*. In the latter he described himself as *imperator scotorum* ("emperor of the Irish"). Brian Boru was killed at the Battle of Clontarf in Dublin on Good Friday, 1014, and was buried in Armagh behind St. Patrick's Cathedral along with his son, who was also killed in the battle.

Brian Boru ranks with St. Patrick as one of Ireland's legendary figures. In breaking up the Uí Néill monopoly, Brian Boru succeeded in creating a high-kingship for Ireland, a dream envisioned for over two hundred years. More importantly, he brought about a unified kingdom of Ireland, the struggle for which continues today.

After the Viking Age, from the eleventh century onward, the Irish church underwent a reformation that reflected a new leaning towards Rome. In Armagh, a series of lay abbots from a local family, the Clann Uí Sínaigh, dominated the church from 965 to 1129. The Uí Sínaigh abbots had not taken holy orders and some of them were married. Armagh joined the reform movement when the seventh in a series of members of the Clann, Cellach Uí Sínaigh (St. Celsus), inherited the position of abbot in 1105 and had himself consecrated bishop as *comarba Pátraic* ("heir of Patrick') the following year. For the first time in many years, Armagh, the primatial see of Ireland, was under the rule of a bishop who was also an heir of Patrick.

The next year, Gilla Espaic, or Gilbert, who had spent time at the monastery in Rouen, France, was made bishop of Limerick and appointed papal legate. Gilbert put forth a plan for a diocesan and parochial organization for Ireland and a uniform liturgy within the church. In 1111, a national synod at Raíth Bresail was assembled, attended by 3,000 clerics, 300 priests, and 50 bishops, and presided over by Gilbert, Cellach, and Muírchertach Ua Briain of Munster, the Irish high-king descended from Brian Boru. At the synod, the Irish church was organized into two provinces with archbishoprics in Cashel and Armagh. Ireland was divided into twenty-four sees, replacing the old monastic system. Armagh was assigned what is approximately its present jurisdiction in counties Armagh, Tyrone, and Derry in Northern Ireland and Louth and Meath in the Irish Republic. The reorganization was successfully implemented within forty years due in large part to the efforts of Cellach's hand-picked successor, Máel Maedóc (St. Malachy). Maedóc succeeded, against great opposition from the Uí Sínaigh, in resigning the see and becoming head of the church of Armagh. He traveled to Rome to seek the pope's approval for a reorganization of the Irish church along the lines of the Raíth Bresail synod and to obtain grants of *pallia,* insignia for the two proposed archbishops. On the way to Rome, Maedóc stayed in Clairvaux, Burgundy, at a leading Cistercian monastery as the guest of St. Bernard. He was so impressed with the order that he introduced the Cistercians to Ireland, where, in 1142, they began their first settlement at Mellifont on land granted by the king of Airgialla. Maedóc died six years later in Clairvaux, on his second journey to Rome, but his design for the Irish church was realized four years later at the synod of Kells-Mellifont. A new church organization was formed, dividing Ireland into thirty-six sees with four metropolitans—Armagh, Cashel, Dublin, and Tuam—under the now-officially confirmed primacy of Armagh. The Raíth Bresail and Kells-Mellifont synods essentially implemented the diocesan organization the Irish church maintains today.

Armagh reached its pinnacle between the Battle of Clontarf in 1014 and the Anglo-Norman invasion of 1170. In 1169, Ruari O'Connor, Ireland's last high-king, founded a professorship for Irish and Scots in Armagh. But Armagh's triumphs of the twelfth century were offset by even greater challenges. The Anglo-Norman invasion led to struggles for the see and precipitated conflicts with Dublin over the issue of primacy. Ireland came under English rule in the twelfth century, and Armagh suffered in the religious wars waged between the English and the Irish. Armagh was also devasted by a series of earthquakes in 1112, 1121, 1150, and 1166.

In 1556, Shane O'Neill destroyed many of the town's buildings in an attempt to prevent its occupation by the English. The Protestant English forces were victorious, nonetheless, and brought many changes to Ireland.

In the first decade of the seventeenth century, a plantation scheme was prepared in London and put into effect to assure that planters would outnumber the native Irish. The Irish inhabitants were segregated in designated parts of the colonized areas so that a new Protestant community could be created. Land from Counties Armagh, Cavan, Coleraine, Donegal, Fermanagh, and Tyrone was confiscated and granted out again in one- to two-thousand-acre lots at easy rents, largely to English and Scottish settlers, on the condition that the new landowners or "undertakers" bring in Protestant tenants. The segregation plan was abandoned within twenty years, as the native Irish were found to be a valuable labor force and were not moved off the land in advance.

In 1642, Armagh was again destroyed, this time by Phelim O'Neill. Armagh regained some of its prosperity and enjoyed a period of peace from the mid-eighteenth to nineteenth centuries, when it saw a final architectural flourish. The wealth of the Protestant clergy is evident in Armagh's Georgian buildings, many of which survive today. In 1840 Armagh's St. Patrick's Catholic Cathedral was built. St. Patrick's Church of Ireland Cathedral was also rebuilt and restored during this period.

In 1801, Ireland became a part of Great Britain. In 1920, the Government of Ireland Act partitioned Ireland into two parts: Northern Ireland (Ulster), consisting of the counties of Armagh, Antrim, Down, Londonderry, Tyrone, and Fermanagh; and Southern (Republic of) Ireland, consisting of the remaining twenty-six counties. Although most of Ireland won independence as a self-governing dominion through this act, the northern one-sixth remained under British rule.

During this decade an Irish independence movement devoted to a united Ireland was born, giving rise to the Irish Republican Army (IRA). Throughout the 1920s, the group was involved in bombings, raids, and street battles on both sides of the border. The Republic of Ireland severed ties with Britain in 1949 and the IRA was eventually outlawed by the Republic.

It remained active, however, and in 1969, the IRA split into two groups. The majority advocated a united Ireland through peaceful means. The minority, the provisionals, insisted on a terrorist campaign. Four years later, Armagh county was abolished as an administrative unit when Northern Ireland's local government was reorganized and a district was established. The following year, the northern part of the traditional county of Armagh was designated as an administrative county.

The predominantly Catholic Armagh has had a leading role in the "Troubles" in Northern Ireland and is known for its cross-border incursions. Southern Armagh district and adjacent areas near the border were hotbeds of continued sectarian violence well into the 1990s. In late August of 1994, after twenty-five years of terrorism, the IRA called a ceasefire, proclaiming an end to violence in favor of political solutions toward a united Ireland.

Further Reading: Four excellent sources on the history of Armagh include *The Oxford Illustrated History of Ireland* (Oxford and New York: Oxford University Press, 1989), edited by R. F. Foster; *The Peoples of Ireland: From Prehistory to Modern Times* (Notre Dame, Indiana: University of Notre Dame Press, 1986) by Liam de Paor; *The Course of Irish History* (Cork: Mercier Press, 1984), edited by T. W. Moody and F. X. Martin; and *Medieval Ireland: The Enduring Tradition* (London: Macmillan, and New York: St. Martin's, 1988) by Michael Richter. The Bord Fáilte's *Ireland Guide* (New York: Irish Tourist Board and St. Martin's, and London: Macmillan, 1993) is an authoritative, up-to-date travel guide that features notes on Armagh's history.

—Shawn Brennan

Augsburg (Bavaria, Germany)

Location: At the river confluence where the Wertach joins the Lech, 42 miles northwest of Munich, 93 miles southeast of Stuttgart, and 50 miles east of Ulm.

Description: Oldest town in Bavaria, originally a Roman settlement; a prosperous business center in the late fifteenth and early sixteenth centuries, with much architecture dating from that period; city restored after extensive damage by bombing in World War II.

Site Office: Touristik- und Kongress-Service
Bahnhofstrasse 7
86150 Augsburg, Bavaria
Germany
(21) 50 20 70

Augsburg, with more than a quarter of a million inhabitants, is the third-largest as well as the oldest town in Bavaria. Unlike the old centers of most older German towns, Augsburg's Altstadt dates overwhelmingly from the huge surge in the town's prosperity during the brief period from about 1470 to 1525, and in outward appearance retains in consequence an unusual stylistic unity. The dominant architectural style relies on the forms earlier adopted south of the Alps, less angular and more graceful than the late medieval Gothic characteristic of Nürnberg at the same period.

The old town is oblong in shape, about 1,300 yards from north to south and 875 yards from west to east. The principal street, the Maximilianstrasse, forms the north-south axis. Very broad and lined with flamboyant fifteenth- and sixteenth-century dwellings, sometimes half-timbered, and sometimes with Renaissance fronts, it is the main shopping street and leads from Augsburg's principal church at the south end, past two fountains by Adriaen de Vries, toward the Rathaus and square just beyond its northern end, and farther still, to the cathedral. Toward the east of the old town lies the former Jakob suburb, the site of the 1519 *Fuggerei*, eight streets with fifty-three gabled houses built by banker Jakob Fugger to house Augsburg's poor, and let to them at a token rent and the undertaking to pray for the souls of the founders. The broad avenues around the old town replaced the fortifications removed in 1703.

The history of Augsburg is intimately connected with the commercial activities of two great banking families, the Welsers and the more important Fuggers, between 1495 and 1525. After the decline of the Medicis, they became the principal bankers to Europe's rulers outside France, including the pope and the emperor, and financed much of the trade across the Alps, within the empire and as far east as Novgorod and Budapest. Jakob Fugger once had to remind the Holy

Roman Emperor Charles V that he had financed the bribes that had secured his election as emperor in 1519. Including a trivial amount for fees, they amounted to the equivalent of more than 4,600 pounds of gold. The Welsers lent Charles V a quarter as much again, with three Italian banks lending together a little more than the Welsers. The Habsburgs never settled a debt estimated at 4 million ducats with the Fuggers. The fortunes of the town declined, as they had risen, with those of the Fuggers, whose business was finally destroyed by the consequences of the discovery of the New World with its precious metals and of the sea routes to the east.

A Roman camp of one or two legions with auxiliaries was founded by Augustus after the region, in which the Romans established the province of Raetia, had been conquered in 15 B.C. by Tiberius and his brother Drusus, Augustus's stepson, and considered by him to be his best general. Tiberius established the civilian settlement known as *Augusta Vindelicorum,* principal town of Raetia, which was granted the status of *municipium* by Hadrian. Augsburg became the center of Rome's road network north of the Alps.

After the fall of Rome the settlement was quickly overrun by Germanic tribes and then came under the power of the Frankish kings. In the late eighth century Augsburg was almost entirely destroyed in the war of Charlemagne against Tassilo III, duke of Bavaria from 748, and finally fell into the hands of the kings of Swabia. Augsburg recovered and grew around a church that had been consecrated in 807, and Bishop Ulrich surrounded the settlement and church with walls in 955. This enabled the town to withstand an attack by Hungarian tribes, who were defeated at nearby Lechfeld by Otto I on August 9, 955.

The town was early awarded the right to mint its coinage, confirmed again in 1521, to hold a daily market, and to exercise independent jurisdiction. In the first half of the eleventh century a settlement developed around the Perlach watch tower on the present Rathausplatz. In 1084 the Holy Roman Emperor Henry IV occupied the city, shortly after his submission to Pope Gregory VII at Canossa, implicitly acknowledging the primacy as well as the independence of spiritual jurisdiction. In 1156 Frederick I Barbarossa, crowned Holy Roman Emperor the previous year, nonetheless issued a constitution for Augsburg that limited episcopal jurisdiction. Augsburg's history is important for what it shows about the struggle for primacy in Europe between military and sacerdotal sovereignties.

About 1200 the cluster of domestic dwellings was enclosed together with the cathedral and two monastic foundations by a communal wall. In 1276 Augsburg rewrote its own constitution, which was confirmed by Rudolf I of Habsburg, who by this date was king of Germany, had the support of Pope Gregory X against Ottokar of Bohemia, and

The Schaezler Palace in Augsburg
Photo courtesy of German Information Center

was on the point of establishing Habsburg authority in Austria-Hungary. In 1316 Emperor Louis IV the Bavarian formally confirmed Augsburg's full status as a free city of the empire, a status already contained in the 1276 constitution. Dues paid to the emperor were trivial. Although from the time of Henry IV the emperors favored the rise of the political power of the merchants, the town was governed by a patrician council until 1368, when the craftsmen succeeded in acceding to power under a constitution that lasted until 1548. In practice it was the interest of the larger merchants that the fiscal regime most favored.

The emperors' preferred route for their armies across the Alps was the Brenner pass, eighteen miles south of Innsbruck, opening the way through Brixen, Bolzano, and the Adige River to Venice, and Augsburg's commercial rise was due to its controlling position on the routes between Milan, Venice, and everywhere north of the Alps and east of the Rhine, as well as to the Rhône valley. It was the established route to Italy that enabled Augsburg merchants, exploiting the south German mines and the new technology for separating out the silver found in copper ore, to transport iron, copper, silver, tin, lead, and arms to the Italian peninsula. The town's resulting prosperity attracted the German princes to borrow money from Augsburg merchants, individually or corporately, as early as the thirteenth century. Gradually such transactions came to involve credit.

Augsburg's dominant banker, Jakob Fugger, was born in 1459, the last of his parents' ten children, and was originally destined for the church. All the males except one entered the family bank, managing trading desks in Augsburg, Venice, and Nürnberg. One brother, Marx, acquired connections useful to the bank in an administrative post at Rome from 1471. The Fugger bank, for instance, collected in Hungary, Bohemia, and Poland the indulgence proclaimed by Julius II for the building of St. Peter's, and became papal agent for the collection of these indulgences in Germany. The character of Augsburg, as it developed, was determined by the riches of many of its merchants, and by the immense fortune of its richest banker.

From 1488 to 1534 Augsburg was a powerful member of the defensive Swabian league of free cities. Although the town accepted the Lutheran creed in 1534, from 1548 it demanded parity of treatment for Roman and Lutheran institutions. When Martin Luther was first summoned to Rome, the elector intervened to demand a trial on German soil. Cardinal Tommaso de Vio, also known as Cajetan, was acting as papal legate to the last diet of Emperor Maximilian I at Augsburg, and Luther appeared there before him three times in October 1518. It was in the bishop's palace at Augsburg that the famous *Augsburg Confession,* drawn up by Philipp Melanchthon and signed by seven princes and two cities, was presented to Charles V on June 25, 1530. The document became one of the definitive statements of the Lutheran faith. When King Henry VIII of England and King Francis I of France both died in 1547, a year after Luther, Charles V, having won his military victory at Mühlberg, obtained the

diet's approval for another effort at religious compromise, the *Augsburg Interim* issued on May 15, 1548. Maurice, Duke of Saxony, drove the emperor out of Augsburg in April 1552 and the peace negotiated that year became in September 1555 the Peace of Augsburg, conceding legal recognition although not toleration within the empire for Lutheranism.

Early in the period of rising prosperity Augsburg's cathedral was rebuilt. The Gothic chancel to the east was consecrated in 1431, but the western portion, with red and white inner walls, was originally an eleventh-century basilica, rebuilt in the fourteenth century. The eleventh-century bronze door panels, with thirty-two bas-reliefs, have been retained in the portal of the south aisle. The cathedral has twin slender Gothic towers with sharply pointed steeples and contains four magnificent altarpieces by the elder Hans Holbein and some twelfth-century stained glass windows depicting the prophets. Although half of Augsburg was destroyed toward the end of World War II, including the now reconstructed Rathaus, the cathedral escaped damage.

There are two churches, St. Ulrich and St. Afra, adjacent to one another on the Ulrichsplatz at the south end of the Maximilianstrasse. One is the former late neo-Gothic church of the Benedictine abbey, begun in 1467 on the site where St. Afra was buried in A.D. 304. The other church, Protestant and completed in 1710, now shares a crypt into which St. Afra's tomb has been moved and where the tomb of the tenth-century St. Ulrich is to be found. The larger church, with tower and onion-shaped dome, was consecrated in 1500 and contains three important renaissance altars and the tomb of the Fugger family. The church is historically important for evidence of Augsburg's advanced role in importing Italian decorative and architectural style into present-day Bavaria. The wrought-iron work and woodcarving in both churches is normally described as baroque, as sometimes are the altars. At the crossing of the transept there is a bronze *Crucifixion* dating to 1607.

In the league of Swabian towns Augsburg rapidly overtook Ulm between 1490 and 1510 as the most significant city. Its wealth increased fourfold in the fifty years following 1475; its population doubled to about 40,000 between 1450 and 1520; and its great artists, the Holbeins and the Erhart father and son, Michael and Gregor, were poached from Ulm. It was from Michael Erhart that Ulrich Fugger, Jakob's elder brother, commissioned a carved wooden altarpiece, and there is a carved statue by Gregor Erhart, dating from about 1500, in the church of St. Ulrich and St. Afra.

In the fourteenth-century Annakirche, where Luther's meeting with Cajetan took place, less than 200 yards west of the Rathausplatz and now Protestant, Ulrich and Jakob Fugger agreed with the Carmelite prior in 1509 to build a large funeral chapel for themselves and their late brother Georg. An extension of the church to the west, the chapel was consecrated in 1518. It has a ribbed vault, an organ, a patterned marble pavement, stained-glass windows, choir stalls, pearwood busts, marble mourning poutti, and figures and reliefs in marble and limestone. Michael Baxendall, the leading authority on south German sculpture of the period, has called it "the first High German essay at a monumental complex of art in a more or less Italianate, or Renaissance, style."

The church itself, which contains the *Jesus, the Friend of Little Children* by Lucas Cranach the Elder, was rebuilt in the closing years of the fifteenth century, and given its baroque decoration between 1747 and 1749. The early sixteenth-century Dominikanerkirche and the Moritzkirche were also redecorated in baroque style during the eighteenth century. The original Rathaus, built by the town architect Elias Holl from 1616 to 1620, has been rebuilt since World War II. Its famous Golden Hall, so called from the profusion of its gilding, was renovated in 1984. Elias Holl also remodeled in 1622 the group of sixteenth-century fortified buildings with two vaults and a central courtyard known as the Red Gate at the extreme southeast corner of the old town. It is largely Holl's buildings that give the town its present decorative and Italianate early baroque architectural character.

Augsburg suffered greatly in the Thirty Years War of 1618 to 1648, when it was besieged and taken by Gustavus Adolphus in 1632 before surrendering to the imperial forces in 1635. The population declined from 48,000 to 16,000. In 1703 it was bombarded by the electoral prince of Bavaria, and forced to pay a large ransom, and if suffered again in Napoléon's campaigns. The town was finally incorporated into Bavaria in 1806. It had already begun to recover in the eighteenth century, and in the mid-nineteenth century the town's economy again became buoyant, relying on metal products and textile production.

During World War II some 12,500 dwellings in Augsburg were destroyed and 32,000 seriously damaged. The modern town, mostly rebuilt by 1963, preserves the Catholic and imperial Fugger inheritance, and its center is still redolent of a rich merchant town. Although most of the building is post-Renaissance, the important architecture ranges from late Gothic to baroque. As in so many German cities with historic pasts to preserve but modern economies to build, Augsburg has a compact, preserved, restored, or rebuilt center that exploits the tourist potential of the city's past, with a large number of museums. Bertolt Brecht was born here in 1898 in a house that is now a museum. The Holbeins' house has been reconstructed and turned into a small art gallery. The former family house of the Welsers has been turned into the Maximilian museum; there is a Roman museum for the archaeological relics of the Roman period; and there is a Mozarthaus, devoted to Wolfgang Amadeus Mozart's father, Leopold, in the house where he was born. The city's main art gallery is in the Schaezler Palace, where there is also a state collection, including Albrecht Dürer's portrait of Jakob Fugger. Since 1970 the city has also had a university.

Further Reading: For Augsburg's general history there is little in English specifically devoted to the town apart from the usual brochures, guides, gazetteers, and encyclopedia articles. In German the most reliable modern work on Augsburg's history is to be found in the second edition of the *Handbuch der Bayerischen Geschichte,* edited by Max Spindler, fourth volume in six (Munich: C.H. Beck'sche Verlagsbuchhandlung, 1968). On the art history of southern Germany, Michael Baxendall's *The Limewood Sculptures of Renaissance Germany* (New Haven and London: Yale University Press, 1980), offers the best treatment, and Augsburg figures largely in appropriate political, art, and economic histories. On Jakob Fugger and his commercial dealings, there is an excellent two-volume biography in German by Götz von Pölnitz, *Jakob Fugger: Kaiser, Kirche und Kapital in der oberdeutschen Renaissance* (Tübingen: J.C.B. Mohr, 1949), and an informative single-volume work in French by Léon Schick, *Un grand homme d'affaires au début du XVIe siècle: Jacob Fugger* (Paris: S.E.V.P.E.N., 1957).

—Anthony Levi

Auschwitz (Bielsko, Poland)

Location: Town in southern Poland, approximately thirty-three miles west of Kraków, on the River Sola.

Description: The most imfamous of World War II concentration camps; also the site of a museum and memorial to its more than 2 million victims.

Site Office: Katowice/Auschwitz Tourist Office
Ul. Mlynska 11
Katowice, Katowice
Poland
59-64-18

There is no site in the world that holds a distinction so grisly as that of the Polish town of Oświęcim, better known by its German name of Auschwitz. Following the Nazi invasion of Poland in September 1939, German leader Adolf Hitler stepped up his efforts to eliminate groups he deemed "undesirable," namely Jews, Romanies (gypsies), Slavs, homosexuals, and the mentally retarded, a plan that later came to be known as the "Final Solution." Throughout Europe, small, obscure towns on rail lines were converted into concentration camps for the most horrifying atrocities the modern world has known. Auschwitz was the largest of these camps; in its existence, it witnessed the systematic elimination of more than 2 million Europeans, mostly Poles and Jews. Although the camp at Auschwitz has long since been abandoned, tourists walking through its ruins today can still feel the terror that will forever linger in the Polish countryside.

Auschwitz has a history that goes back to the fifteenth century, the height of Polish power in Europe. The town has served in many capacities, most notably as the capital of a small feudal duchy that was incorporated into Poland in 1457. Its actual naming stretches back to the age of the Piasts, when it served as a stronghold for the dukes of Oświęcim. It is the site of many historical buildings, including a parish church built in the fourteenth century and refurbished in Renaissance style during the sixteenth century.

Until 1941, Auschwitz played only a minor role in the destructive Nazi empire. But following relatively easy victories in Poland, Czechoslovakia, Austria, and France, Hitler felt sufficiently secure to turn his attention to other matters, namely the destruction of the those whom he did not consider part of his "master race." In March 1941, SS (Schutzstaffel) commander Heinrich Himmler ordered camp leader Rudolf Höss to begin the implementation of new plans for Auschwitz. Himmler explained that the easy access by rail, the isolation of the town, and the general topography of the area, which made the camp easy to hide from outsiders, made Auschwitz a prime candidate for a elimination site. In a sum-

mer meeting with Höss, Himmler reported that plans had been laid for the Final Solution, and that Auschwitz would play a vital role.

Hitler's orders were also sent directly to Hermann Göring, head of Hitler's state apparatus. Göring legitimized the construction of the death camps, making them one of the state's top priorities in his four-year plan. Between Himmler, Göring and Höss, Hitler had a triumvirate of men hard at work building the house of horrors at Auschwitz. In speeches, Himmler proudly told his men that all Jews must be removed from the occupied territories in Eastern Europe.

A few weeks after his meeting with Himmler, Höss received a visit from SS officer Adolf Eichmann. Eichmann came to plan actual details of elimination with Höss, namely how and when the killings would begin. Eichmann was impressed with the results of using carbon monoxide gas to eliminate mental patients; he suggested they use the same tactics at Auschwitz, only on a grander scale. By late summer of 1941, employees were being trained in the use of cyanide gas, better known to most of the German population as Zyklon B. In September, the Nazis conducted the first gassing tests at Auschwitz, as construction on main elimination camp at nearby Birkenau began. Some 250 patients from a nearby hospital and 600 Russian prisoners of war were killed; the Nazis deemed the test a success.

The sprawling Auschwitz-Birkenau complex began receiving trainloads of prisoners by March 1942. During that year, Jews from Slovakia, France, Norway, Belgium, and Croatia all arrived, most destined to face the gas chambers. For the previous two years, these Jews had been forced out of their homes and crowded into makeshift ghettos in the major population centers of Europe. These ghettos were filthy places that managed to strip most of the dignity from their inhabitants even before the ghettos were "liquidated" and their citizens sent to the concentration camps; thousands died before they could even leave the ghettos.

When their ghettos were destroyed, the Jews were told they were being sent east for resettlement and work. Such cover stories were told to quell the rampant rumors of mass death camps in eastern Germany and Poland; the lies were believed because they were what the captives wanted to hear. Prisoners boarded boxcars headed for various installations, crowded together with barely room to breathe. If they were fed at all, it was sparingly. One trainload of Greek Jews traveled twenty-seven days from the island of Corfu, without once being fed. When their train finally arrived at Auschwitz, half were dead and the other half were comatose, unable to even stand for processing; they had to be carried off to the gas chambers.

Upon arrival, prisoners were separated into two lines. The elderly, young, sick, or retarded went into the left line,

Monument to those who perished at Auschwitz

which proceeded immediately to the gas chambers. A mere 10 percent qualified to stand in the right line; these were men and women judged by the Nazi guards to be the most fit for hard labor. Men immediately separated from their wives and children often watched them marched away to the chambers, never to be seen again.

Even the strongest men or women generally lasted only a month under the tortuous conditions in the camp. Any food given them was either rotted or devoid of any nutrition, leading to extreme malnourishment, dysentery, and other disease. Prisoners who looked more ghostly than human struggled to stand under their own power. When the Nazis found evidence of contagious disease in one of the compounds, their remedy to quell further infection was immediate elimination of all compound members.

Prisoners were constantly humiliated, tortured and forced to live like animals. Anywhere from 800 to 1,000 people were crammed into a barracks at night, fighting for an inch of space so they could sleep. Guards arbitrarily shot, beat, or hanged prisoners in front of family members as they watched. They would make prisoners strip down and squat for hours in the freezing cold and rain while they were counted. In the end, everyone who came to Auschwitz was destined to die in the gas chambers—it was simply a matter of when; those in the left line were at least spared the suffering of those in the right line left to struggle on.

Many of those sent to the chambers had little idea of their fate, even up to the time of their death; they needed to believe that what was happening to them could not possibly be real. Following their arrival, those sent to the left were lined up and marched into a large courtyard, where faucets used to water the grass jutted out of the ground. Since most of the prisoners had not had water for many days, they rushed toward the faucets, falling out of order. The Nazi guards waited patiently, letting the thirsty people drink and fill the pots and pans they carried with them. They knew that the prisoners would be easier to handle if they had been given water first. After a while, the lines began to reform.

The prisoners then marched another 100 yards, where they were met by a set of concrete steps leading underground with a sign above that read "Baths and Disinfecting Room." Some of the filthy prisoners were even happy at the sight, thinking they were finally going to have a shower. A long, brightly-lit room awaited them at the bottom of the stairs; benches and coathooks jutted out from the walls. Signs posted all over instructed the prisoners to tie their clothes and shoes together, and remember which coathook they were hung upon in order to avoid confusion after the bath. In actuality, the Germans wanted this done so it would easier for them to sort and pack up these items for shipping to Nazi soldiers and citizens.

The prisoners waited for their next order. Nazi soldiers then entered the room and, with megaphones, told the people they had ten minutes to strip naked and organize their belongings. Shy men, women, and children were apprehen-sive, but threatening gestures from the soldiers made them comply. Another SS officer opened the final door, leading to another large room that had pipes from floor to ceiling with holes that looked like shower heads. When the room had filled, all personnel were ordered out. The door closed, and the lights turned off, leaving the frightened prisoners in total darkness.

Outside, the shower room pipes extended above ground, where a "health officer" prepared cyanide granules to be dropped down. Upon release, the gas began seeping into the chamber, killing everyone inside within five minutes; the guards waited patiently for another five minutes to ensure everyone inside was dead. Twenty minutes later, giant ventilation fans began expelling the gas so that the process of cremation could begin.

The site inside the chamber was horrifying. Some 3,000 distorted, bloated, ghastly bodies were piled on top of one another, a grim sign that some of the prisoners had tried to clamber to the top of the pile in order to escape the rising gas. Employees then began removing the clothes and shoes from the hooks for disinfection as the bodies were being hosed down and dragged away for processing. From the death chamber, employees loaded between twenty and twenty-five bodies onto an elevator headed for the incineration room. The bodies were then placed on chutes that deposited them next to the fires, where they would face their final humiliation before being reduced to ashes.

For the Nazis, both hair and gold were valuable commodities; the piles of dead bodies gave them an opportunity to collect both. Gold's value was obvious, but hair was just as important: it expanded uniformly in varying humidity, making it good for use as a trigger in delayed action bombs. Before the bodies were thrown into the fire, crematoria employees shaved the heads and pried the dead mouths open, extracting gold fillings, teeth, and bridges. After they were pulled, the teeth were dropped into an acid solution to strip away the bone and flesh; the gold was then taken from the solution. Each crematorium collected on average about twenty pounds of gold per day.

When this process was complete, three bodies at a time were put onto metal sheets and placed in the ovens for cremation, which took about twenty minutes. All that would remain of the bodies was ash, which was picked up by trucks and dumped a mile away in the Vistula River. There were four crematoriums, each with fifteen ovens; several thousand bodies could be cremated each day, demonstrating the extreme size and scope of the complex. After the war, Rudolf Höss testified that at its height, the camp could hold more than 140,000 prisoners at a time. He also testified that while he was the camp commandant between June 1941 and August 1943, he alone ordered the deaths of more than two million Jews. To this day, there is no exact count of how many were killed at Auschwitz, but best estimates put the figure between 2.5 million and 4 million.

Auschwitz was also well-known for its grisly human experimentation overseen by Dr. Josef Mengele, better known as the "Angel of Death." Occasionally, as the trains were being emptied and the prisoners sorted, Mengele made a general search for Jews skilled in either the medical or dental professions. The people he chose came to be known as the Sonderkommando, or SK. Sonderkommando members were well fed and well treated in comparison to the rest of the camp, but they were forced to work in the crematoriums in various capacities. Men who had been dentists were made to extract teeth for gold; former doctors were members of Mengele's death team. Mengele was fascinated by dwarfs and twins; whenever they arrived in the camp, he sought them out for either horrible live experiments or had them shot for dissection and analysis.

Sonderkommando members suffered from extreme depression; many killed themselves because of the brutality of the jobs they were made to undertake, or because they knew that the rest of their family members had been gassed. Many SK members were forced to participate in the killing of groups that included their own loved ones. The average life of Sonderkommando members was four months; after that, Mengele would order their deaths and choose a new unit in order to keep his team fresh and productive. The new unit's first job was the cremation of the old guard.

As the Nazis gave more ground to the advancing Russian Army in late 1944, steps were taken to "clean out" the camps, first by stepping up the killing and then later by attempting to destroy them to hide the evidence. In October 1944 alone, more than 33,000 Jews were killed at Auschwitz. By early November, Nazi officers were marching prisoners out of the camps eastward to slave labor camps such as Bergen-Belsen in Germany. An earlier prisoner revolt had destroyed one crematorium; the others were either set afire or dynamited in haste, but some of the camp was not destroyed. On January 27, the Red Army liberated what was left of the camp, and the first concrete evidence of the terror that occurred there was released to the world.

Visitors to the site today will note that the camp has been left in much the same state that it was fifty years ago. The site has been meticulously cleaned in order to facilitate easier preservation, and a museum helps tell the story of Auschwitz-Birkenau; a motion picture filmed by Soviet soldiers on the day they liberated the camp is shown several times a day. In the barracks hang horrific photos of life in the camp, along with lists of victims categorized by nationality.

From the Auschwitz concentration camp, visitors can walk the 1.5 miles down a road to the ruins of the gas chambers and crematoriums at Birkenau; at the back of the camp is the Monument to the Glory of the Victims, designed by local artists and erected in 1967. To the right of the monument behind a group of trees sits a pond, still murky and grey with the ashes of the victims, a dark reminder that the stain of Auschwitz on humankind will never disappear.

Further Reading: Lucy S. Dawldowicz's *The War Against the Jews* (New York: Bantam Books, 1986; London: Penguin, 1990) details the entire Nazi campaign to eliminate the Jewish population of the world. *Auschwitz: A Doctor's Eyewitness Account* (New York: Arcade Publishing, 1960; London: Panther Books, 1962) by Dr. Miklos Nyiszli is a harrowing tale of a Sonderkommando doctor forced to follow Josef Mengele's orders in order to survive at Auschwitz. *Auschwitz and the Allies* (New York: Holt, Reinhart and Winston, and London: Joseph, 1981) by Martin Gilbert is an analysis of how the Allied forces learned of the Nazi extermination of the Jews, and what they did about it. Finally, a good travel guide with details on sights, food, and lodging in Poland is *The Berkeley Guides: On the Loose in Eastern Europe* (New York: Random House, 1992).

—Tony Jaros

Avignon (Vaucluse, France)

Location: In Southern France along the Rhone River on the shoulder of Provence and just south of Orange.

Description: Seat of the Holy Roman Church from 1309 through most of the Great Schism; site of papal offices; currently site of world-famous theater festival.

Site Office: Office de tourisme et Syndicat d'initiative
41 Cours Jean Jaures
84000 Avignon, Vaucluse
France
90 82 65 11

Until recently history has been unkind to Avignon, the central city of the County Venaissan (Comtat Venaissan), sitting along the Rhone River between Dauphine and Provence. Petrarch described the city as an unholy Babylon, hell on earth, sink of iniquity, and a cesspool of the world. In modern times a fairer assessment has been made.

The reason for antiquity's disgust is that Avignon was for seventy years the seat of the Holy Roman Church, from 1309 through the Great Schism. To the minds of medieval historians and their successors, especially those employed by the Italian papacy, the only seat of the Catholic Church could be in Rome, where both Peter and Paul were martyred. In this view, the Avignon papacy could only be seen as a scandal, especially since the majority of Avignon popes were French.

But the popes had reasons for settling at Avignon during the fourteenth century. First, Rome and its surroundings were unsafe due to civil strife and war. Second, popes had, for centuries, spent much of their time away from Rome, as it becomes unbearably hot in the summer. Reasons for choosing Avignon were many: it is protected by the Rocher des Doms (Cathedral Rock) on one side, and by the Rhone River on the other, making it quite safe. The people were friendly, unlike the famous Roman mob, and the temperature was clement, save the occasional cold, dry gusts of the French *mistral*. Finally, the County Venaissin was a papal state with a university, a cathedral (the Notre Dame des Doms), monasteries, a Templar commandery, and most important, a bishopric complete with a bishop's palace.

Little is known about early Avignon that is not caught up with religion, superstition, and myth. One historian dates the town back to the Deluge. St. Martha, sister of Lazarus, having been cast out of the Holy Land by the Jews, began her cult in a cave near Avignon, establishing a church and a nunnery. Southern France fell to Rome in 125 B.C.; Avignon became a Roman colony with Latin as its official language. Then history falls silent until the advent of the Barbaric Wars, in which Avignon figured prominently. In the eighth century

the Saracens conquered the Rhone Valley; they, in turn, were conquered by the Franks in 739.

In the treaty of Verdun (843) ending the war between the grandsons of Charlemagne, the Rhone Valley was given to Lothair, son of Louis the Pious. During the next century, the valley was attacked repeatedly by pirates. Next followed a period of rule determined by marriages, inheritance, and finally internecine war. In 1185 Alphonso, count of Toulouse, and Raymond Bérenger, count of Barcelona, made a truce by which Avignon was dually owned. In the latter half of the twelfth century, Provence became known for its troubadours, one of whose favorite subjects was Adelais, countess of Avignon.

In 1206, William, count of Avignon, confirmed a charter whereby the bishop of Avignon was given control over the city. Meanwhile, Little Benet, the chief of the Order of Hospitallers, built the Pont St. Bénézet, the famous bridge across the Rhone, with a bridge chapel and a nearby hospital for the poor. According to legend, Benet, when still a child, was directed to build the bridge by Jesus Himself.

Avignon became embroiled in the Albigensian Wars of the late twelfth century. In 1165, certain sectaries of the church were tried for heresy, convicted, and ordered to return to the orthodoxy of the church. This occurred at Lombers, near Albi, and the sectaries came to be known as the Albigenses. Pope Lucius III decreed the Inquisition in 1183 to end the heresies; all of Christendom was to come to the aid of the church against the sectaries, including all counts. The count of Toulouse, however, favored the heretics, and the citizens of Avignon favored Toulouse. The Albigensian Wars between Toulouse and the kings of France ended in devastation and misery for the heretics, in disgrace for Toulouse, and in capture for the citizens of Avignon, which was given over to Raymond, count of Provence. In 1271, the County of Venaissin became the possession of the Holy See, and Avignon came under the jurisdiction of Charles of Anjou, king of Naples and Sicily.

The rule of the Avignon Popes began as a result of the political and religious schemes of King Philip IV, the Fair, of France. In 1303, Guillaume de Nogaret was sent by Philip into the Italian town of Agnani, in order to charge Pope Boniface VIII with heresy. Assisted by the Colonna family, Nogaret captured, insulted, and imprisoned the pope, who died of shock and exhaustion within a month. The period following that of Boniface's short-lived successor, Benedict XI (1303–04), marks the beginning of a new organization of the Catholic Church, which included a constant, stable residence for the pope and the centralization and control of all activities by the Holy See.

At the conclave of Perugia in 1304, the college of cardinals was severely divided. Philip the Fair was still de-

Pont St. Bénézet
Photo courtesy of French Government Tourist Office

manding the posthumous trial of Boniface VIII. The French faction of cardinals supported Philip, while the Italians demanded the outrage be atoned. It was decided to look outside the college for an impartial figure, which they found in Bertrand de Got, who became pope in 1305, taking the name Clement V. Philip the Fair persuaded Clement to be crowned at Lyons instead of Rome, and, with news of Italy and Rome deteriorating in civil war and class revolt, Clement decided to stay in France, thus beginning the seventy years of the Avignon Popes.

Clement V was a compromiser and in bad health. He was easily swayed by Philip, who not only wanted Boniface VIII tried, but was accusing the Knights Templar of incredible evil. In 1312, Clement organized the General Council to decide these questions, and things largely went Philip's way. The outrage at Agnani was pardoned and the Templars were arrested. Despite being under Philip's thumb, Clement proved an able politician in other areas; for instance, the choice of bishops and list of benefices came directly under the influence of the pope. He completed the last of the great code of ecclesiastical law with the publication of the seventh book of *Decretals*. His good nature led him, however, into nepotism and his excessive generosity led to high taxes on the

religious and a devaluation of the treasury. He died attempting to return to Gascony on April 6, 1314.

The election of Clement's successor was no less difficult than Clement's own. This time the college was divided between Italians, Gascons, and Provençals. Sequestered in Carpentras, the Gascons helped incite a riot that resulted in the murders of Italians, attacks on the Italian cardinals' homes, theft and plunder of townsfolk and curia dwellings, and even a siege of the conclave with death threats against the Italian cardinals. The Italians broke through a wall and fled, some to Avignon. Eventually, under pressure from the count of Poitiers, the college assembled in Lyons and, after some difficulty, elected Jacques Duèse, who at the age of seventy-two took the name John XXII.

Duèse had been a bishop at Avignon, so it was natural for him to want to wait out the Roman difficulties at the bishop's palace there. At Avignon, the centralization of the Catholic Church was expedited. John improved the treasury through taxing ecclesiastical benefices and abolishing excessive spending. He took upon himself the bestowment of bishoprics, creating a number of clients for his affection, among them the nobility.

Following John's death in 1334, the college elected

Jacques Fournier, an orthodox Cistercian monk, who took the name Benedict XII. Benedict attacked nepotism and favoritism, allegedly declaring that the pope is to be "like Melchisidek, having no father, no mother, nor kin." He ordered bishops and ecclesiastes to return to their benefices or incur sanctions. He laid out many more reforms in great detail—in fact too much detail to be really effective. He was also a terrible politician, unable to prevent the outbreak of the Hundred Years War.

Benedict's most important accomplishment was in the rehabilitation of the bishop's palace, turning it into the Palais des Papes, the normal residence of the Holy See. High walled, bereft of decor, it served as a cloister, fortress, and administrative offices for the papal curia. The palace helped to effect the administrative centralization of the church, as more and more power was overseen from a singular purview, Avignon.

Benedict, under strain from the Hundred Years War, soon died of exhaustion in 1342. The election of Pierre Roger was made unanimously by the cardinals conclave. Roger, who chose the name Clement VI, held doctorates in theology and law, and was a friend of both Edward III of England and Philip VI of France. He had tried to prevent war's outbreak from the beginning.

Clement VI was vastly different from his predecessors. He was more than generous, giving away so many benefices that even bishops complained. He lived like a nobleman, and his taste led him to deplete the treasury in an effort to enlarge the palace, which he did with flair. He also purchased Avignon from Joan of Naples, reconstructed the abbey of Chais-Dicu, and loaned much money to kings and lords of France. He held numerous festivals and entertained the brightest minds and talents of his day.

He took care to mind the spiritual aspects of his people, preaching and presiding at religious rites. His charity was boundless during the Black Plague, which killed more than half of Avignon's population. He offered sanctuary to Jews, who, accused by superstitious mobs of creating the plague, were sent by thousands to their deaths.

Clement attempted a crusade, developing an alliance between the papacy, the Venetians, Cyprus, and the Knights Hospitallers. He also took pains to end the Hundred Years War and was successful in obtaining a few, albeit short-lived, cease fires.

Clement's extravagance made Avignon the light of the world, yet it nearly bankrupted the church. Thus, when Clement died in 1352, the sacred college looked toward a more frugal successor. Étienne Aubert, who took the name Innocent VI, was an elderly man in poor health, with a tendency to vacillate and become depressed. The college may have intended to control him, but they were soon disappointed. Innocent resurrected Benedict XII's austerity and reforms. He set the Inquisition upon the Franciscan heretics without mercy, so much that St. Bridget declaimed against him.

Innocent appointed an accomplished warrior named Gil Álvarez Carillo de Albornoz, who was also a cardinal of the college, to be legate for all of Italy, charging him with reestablishing the authority of the Holy See. By force, Albornoz conquered Viterbo and Orvieto. Next he called a parliament of the pope's vassals, which brought them under papal rule. He then reconquered Spoleto, Ancona, and Romagna. Albornoz founded the Papal State and his constitution regulated its laws until 1816. He defeated the Visconti with the help of Charles IV, the Holy Roman Emperor.

Innocent had his hands full with the Treaties of Calais and Bretigny, which ended the Ango-French conflict. Mercenaries who had been hired to fight in the war were left with no other employment than plunder and murder along the Rhone River. Avignon was looted on many occasions and besieged until the payment of a ransom, which the papal treasuries could not afford. Innocent set to fortifying the city with new walls and gates. Then another plague beset the city, killing 17,000. The stress was too much for Innocent, and his health quickly failed. He died on September 12, 1362.

Guillaume de Grimoard was elected pope and installed on October 27, 1362, taking the name Urban V. He was the only one of the Avignon popes to be canonized by the church. The reasons are numerous. First, Urban had never been a part of the hierarchy and was never involved politically with kings or lords; he was a monk and kept the monastic habit the rest of his days. He was also a university teacher and scholar of canon law. His love of education led him to found a studium at Trets, universities at Orange, Kraków, and Vienna, the school of music at Toulouse, and the Colleges of St. Benedict and the Twelve Doctors at Montpellier.

Art and architecture flourished under Urban's guidance. He added a peace garden to the papal palace, fortified Avignon, and restored the abbey of St. Victor at Marseilles, endowing its church jewels, tapestries, ornaments, and other gifts. He built a cathedral at Mende, renovated the priory of Chirac, founded collegiate churches, and raised a bridge over the Lot at Salmon.

In the late 1360s, Urban undertook the first of several unsuccessful attempts to move the papacy back to Rome. England and France were at peace for the time, and Avignon, under frequent attack by mercenaries, was now more dangerous than Rome, where the citizen mobs had been pacified by a new constitution. Most important, Urban wished to return to the site of Peter's and Paul's martyrdoms and the traditional home of the Catholic Church. The location would give him greater authority to organize a new Crusade and negotiate possible reunification with the Byzantine Church under Emperor John V Palaeologus. The move would also add weight to Urban's call for absentee bishops to return to their dioceses.

Urban set sail for Rome in April 1367 and arrived, after several brief stops, in October. In order to prevent a disruption in the day-to-day affairs of the church, Urban retained much of the administrative apparatus at Avignon. For the duration of his stay at Rome, therefore, church affairs were administered concurrently from two sites. The pope took with him to Rome the college of cardinals, the papal chamberlain, the chancery, and the penitentiary. At Avignon there remained the papal treasury, the papal library, and a host of clerks, notaries, and employees of the Camera Apostolica (Papal Chamber). The majority of these employees were under the authority of the papal treasurer, since the most important administrative responsibilities left to Avignon were financial. Avignon had long been an important commercial center, and was particularly well positioned to act as intermediary in papal interactions with northern Christendom.

The continued existence of these administrative facilities would prove fortuitous; in 1369 King Charles V of France renewed the conflict with England, and Urban returned to Avignon to better negotiate an end to the hostilities. Soon after his return to the city, he fell ill and died.

Pierre-Roger de Beaufort, Clement VI's nephew, was unanimously elected pope in 1370, taking the name Gregory XI. He was a learned man, both in civil and canon law, and has been highly praised for his piety. He inherited a huge deficit from Urban's spending; nevertheless, he gave generously to Avignon's convent of St. Catherine and house of the Penitents, and funded the construction of an orphanage, a clock for the city tower, and the repair of the Pont St. Bénézet. He added to the papal library and helped restore Avignon's Roman palaces. A staunch supporter of the Inquisition, he overcrowded the prisons with heretics.

Gregory, like his predecessor, attempted to return the papacy to Rome, though the move preceded the rift of the Great Schism. Despite advice to stay in Avignon, Gregory felt he could rule the Papal State better from its center—Rome. A visit from Catherine of Siena helped solidify his decision to leave Avignon. Once again, the papal administrative offices were divided between France and Italy, with the bulk of the financial administration remaining in Avignon. The pope and his curia arrived in Rome on January 13, 1377. Gregory's health, unfortunately, could not withstand both the stormy trip and the subsequent experience of Rome's climate, and he died of exhaustion on March 27, 1378.

With the Holy See back at Rome, the mobs of citizens demanded a Roman or, at the very least, an Italian, to be named as pope. The nervous cloister of cardinals, having received news of the threats being made against them, gave in easily, naming Bartolomeo Prignano, the archbishop of Bari and a former vice-chancellor of the curia. Prignano took the name Urban VI and was crowned in April 1378. He immediately proved himself brazen and rude, abusing the emperor's favorite cardinal and sending the bishops back to their benefices. Thirteen French cardinals decided to declare the election null and void, explaining that it was made under pressure from the Romans. After fleeing to Agnani, they chose Robert of Geneva at the conclave at Fondi. This second pope took the name Clement VII.

This was the beginning of the Great Schism. Half of Christendom supported the Italian Pope; half supported the French one. The lines of division were purely political: France and its allies against France's enemies. Clement at first attempted to depose Urban by force, but failed. He returned to Avignon in 1379.

Clement maintained a huge, lavish court at Avignon. He proved a humanist, supporting the arts and literature, and he held frequent festivals. Much of this courtly style was supported through loans and heavy taxation—the papal treasury was still burdened with Urban V's deficit. Clement felt this lavish spending necessary, however, to impress upon the northern nobility that his was the true papacy. Clement also looked for religious signs that his rule had been ordained by God. He found such proof in the miracles at the tomb of Peter of Luxembourg, one of the first cardinals Clement had appointed. Armed with this proof, he resumed his struggle to unseat his rival, Urban VI. He employed all his wit and charm to elicit aid in conquering Rome, but once again, he failed. When Urban died in 1389, the Roman cardinals elected another pope, Boniface IX, to take Urban's place.

When Clement VII died in 1394, even the French nobles were anxious for the schism to end. But the French cardinals ignored their petitions and elected as Boniface's rival Pedro de Luna of Aragon, who took the name Benedict XIII. In response, the University of Paris proposed that both popes resign; the proposal was seconded by an assembly of prelates and lay clergy. Finally, the king of France supported the proposition. But Benedict refused to step down. As a result, the French clergy denounced him and the French king withdrew his support. Most of his cardinals deserted him, the curia abandoned the city, and the citizens of Avignon rose against him. The king even hired mercenaries to besiege the Palais des Papes. But Benedict refused to give up, working tirelessly to keep the morale of his supporters. A compromise was reached in which Benedict stayed in the palace under the guard of the duke of Orléans—in essence, under house arrest. But the plucky Benedict eventually escaped in 1403, and his tenacity elicited the obedience of a few kings. When he was deposed formally by the Council of Constance in 1415, he retired to a hermitage in Catalonia and died, still "pope," in 1422.

Since that time, no Frenchman has ever been elected to the Throne of St. Peter. Until 1693 Avignon was ruled by a series of cardinal legates, among them Cardinal Guiuliano della Rovere, who restored the episcopal palace and convinced his uncle the pope to make Avignon an archbishropic. In the sixteenth century Avignon became involved in the Wars of Religion, as nearby Orange was taken over by Protestants. Avignon remained tolerant toward heresies, but its Catholicism was reinforced by the arrival of the Jesuits in 1564.

The embers of French nationalism were already glowing in Avignon, whose citizens despised their Italian papal governors. At times the crown had occasion to seize the papal state in order to extort agreements from Rome; the people welcomed these occupations, often embellishing them with public humiliations of the vice legates. Such acts were viewed as seditious by the aristocracy, including the king, even though the shouts of the citizens were "Vive le Roi!"

The eighteenth century brought Avignon famine in 1721, a plague in 1759, and the annexation to France in 1791 under Napoléon Bonaparte. Mistral became its most famous voice since the time of the Troubadors in the fin de siècle, and Avignon is today the site of a world-renowned theater festival.

Avignon, until this century, has been reviled in the annals of history. Yet, it was in Avignon that the church realized the centralization of its administration and its power, a legacy that still remains. The architectural and artistic contributions of the popes continue to attract tourists. The Palais des Papes consists of two palaces: one, rebuilt by Benedict XII, is austere; the other, built by Clement VI, is lavish, with high ceilings and frescoes by Giovanetti. There is the Cathedral of Notre-Dame-des-Doms, where John XXII and Benedict XII are buried. The Musée du Petit Palais, which served as a guest house to papal visitors, was refurbished in the 1970s to house over 300 pieces of medieval Italian art. Also standing are vestiges of the Pont St. Bénézet, and the Musée Lapidaire shows artifacts of Avignon's Roman period.

Further Reading: *The Avignon Papacy* by Yves Renouard, translated by Denis Bethell (Hamden, Connecticut: Archon, 1970; London: Faber, 1971) is a short book covering the seventy-year period of the Avignon popes and the antipopes of the Great Schism. Canon Mollat's *The Popes at Avignon,* translated by Janet Love (London: Nelson, 1963; San Francisco: Harper and Row, 1965) is a Catholic history of the same period, which ignores the antipopes, yet is quite comprehensive and detailed. Finally, *The Story of Avignon* by T. Stokey (London, 1911), while rather old, is marvelously filled with renderings of ancient texts, and includes mentions of other characters such as Petrarch and Rienzi.

—Gregory J. Ledger

Baden-Baden (Baden-Württemberg, Germany)

Location: On the northern edge of the Black Forest in the state of Baden-Württemberg, southwestern Germany; 69 miles west of Stuttgart, 108 miles south of Frankfurt, and 40 miles from the French city of Strasbourg.

Description: City famous for its warm mineral springs, casino, luxurious hotels, ancient castles, and ruins.

Site Office: Kurverwaltung
Augustaplatz 8
76530 Baden-Baden, Baden-Württemberg
Germany
(7221) 27-52-00

Baden-Baden is in the Schwarzwald, the legendary Black Forest, in the German state of Baden-Württemberg. About ninety miles long and twenty-five miles wide, the forest is filled with streams, lakes, fir tree groves, and mountains covered with pine and birch trees. There are small villages, farms, vineyards, ancient castles, parks, spas, and ski resorts.

The scenic Schwarzwald Hochstrasse (Black Forest High Road), running through the Rhine Valley, begins at Baden-Baden. The Badische Weinstrasse (Baden Wine Road) passes just southwest of Baden-Baden, running through many of the area's vineyards, which are known for their fine white wines.

Baden-Baden is close to major railway lines connecting it with Stuttgart, Munich, and Frankfurt. Each day about ninety trains stop at Baden-Oos, the railroad station just north of the city.

The Oosbach River, often called the Oos, flows through the center of Baden-Baden. The city's downtown area, on the east side of the Oos, is the commercial and residential section. The city's uptown area, on the west side of the Oos, contains the mineral spring baths, casino, and hotels. The Lichtentaler Allee, a park along the west bank, is the focal point of the city's activities. The walkway is lined with shrubs, trees, and a profusion of roses, azaleas, rhododendrons, and zinnias.

Baden-Baden is a spa, a health resort featuring thermal springs whose mineral waters contain chlorine, salt, and sulphur. Since the time of the Roman Empire, visitors have come to the city to bathe in and drink the mineral waters, believing they could prevent illness and cure ailments.

The Romans built baths in all the lands they conquered. They especially prized the mineral waters of natural hot springs, which required no heating and were recommended by Roman doctors for their therapeutic qualities. The Roman Emperor Vespasian, who brought troops to the Black Forest about A.D. 74, built the town's first baths. The em-

peror Caracalla (Marcus Aurelius Antoninus), who had built Rome's famed Thermae Antoninianae, also called the Baths of Caracalla, conducted an expedition into Germany in A.D. 212. In 213 he visited the baths to seek a cure for his arthritis. Caracalla named the town Civitas Aurelia Aquensis and made lavish improvements. He leveled a cliff and resurfaced the baths with white marble and green granite. Inscriptions on the buildings honored Diana Abnoma, the bath goddess, and Apollo Grannus, the water god.

Despite the natural warmth of the waters, some Roman baths had radiant heating. Wood or charcoal fires produced heat that passed under the bath's cement floor. As the temperature changed, the cement expanded or contracted without cracking.

In 233 native Germans, the Alamanni, took over the area. In the early Middle Ages, the Christian Franks took the town and built its first castle. In the eleventh century the castle became the residence of the Dukes of Zähringer, who already ruled the town of Freiburg on the southern edge of the Black Forest. When one branch of the family died out in 1218, another branch of the same family succeeded them. Adopting the title of margrave, they called themselves margraves of Baden. (They were made grand dukes of Baden in 1806, a title they retained until 1918.)

In 1245 the Margravine Irmengard founded the Kloster Lichtental, a Cistercian convent in the town's Lichtental quarter. It still functions as a convent, and the main convent building has a retail shop selling liqueurs and handiwork made by the nuns.

The Stiftskirche, a Catholic parish church, also was built about 1245. A collegiate order took over the Stiftskirche in 1452 and continued to operate the church until 1806. The order added a Gothic crucifix in 1467 and a Gothic tabernacle about 1500. The church also contained tombs of the margraves. In 1689, the Stiftskirche was damaged by fire and was partly rebuilt in the baroque style. Between 1953 and 1956 modern stained-glass windows were installed.

The Yburg, a castle four miles south of Baden-Baden, was mentioned in documents dated 1245, although the exact date it was erected is unknown. The castle was ruined in the Peasants' War of 1525, but was later restored by one of the margraves of Baden.

In 1437 the margraves of Baden built a Renaissance-style residence that was destroyed by fire in 1689. With the coming of the Reformation in the 1500s, the margravial family split into the Catholic Baden-Baden line and the Protestant Baden-Durlac line; the town took the name of the Catholic line. Later this name helped distinguish Baden-Baden from the old state of Baden in which it was located, and from cities called Baden in Switzerland and Austria.

The Hohenbaden Castle, which the margraves took

The baths quarter in Baden-Baden
Photo courtesy of German Information Center

over in the eleventh century, was damaged by fire in the fifteenth century. Tourists can still visit the ruined castle, now called the Altes Schloss (Old Castle). Its tower offers a view of all of Baden-Baden, the Rhine plain, and the Vosges Mountains.

The Schloss Favorite, a baroque-style summer palace northeast of town, was built in the early eighteenth century for Margravine Sibylla Augusts, the widow of the Margrave Ludwig Wilhelm, called the ''Türkenlouis'' because of his victories in Turkey. The palace, standing in a park with a Magdalene chapel, had a hall of mirrors. Its state rooms were decorated with stucco, wood, and marble, and artisans employed such techniques as mother-of-pearl inlaying, agate and ivory painting, lacquering, wax modeling, and pearl embroidery. Today the palace displays a fine porcelain collection and the original kitchen.

By the late sixteenth century Baden-Baden was a recognized spa resort. But the town, like most of Germany, suffered extensive damage during the Thirty Years War in the first half of the seventeenth century. In the eighteenth century, when the English town of Bath achieved fame as a watering hole, Baden-Baden was still being rebuilt. By the early 1800s, Baden-Baden was once again an elegant spa resort, and its new casino was a major attraction. With the new railways giving those who could afford it greater mobility, Baden-Baden attracted visitors from Britain, France, the Low Countries, and the major German cities. The town was in vogue with Europe's nobility, and French was its official language. Its distinguished visitors included Napoléon III and Empress Eugénie, Queen Victoria and the future Edward VII, the writer Fyodor Dostoyevsky, and the composer Johannes Brahms. It also attracted the prosperous middle class, eager to mix with celebrities.

The white-colonnaded Kurhaus, the famous spa at the north end of the Lichtentaler Allee, was built in 1824. The neoclassical building replaced an eighteenth-century structure housing an earlier spa and casino. The new Kurhaus and the Spielbank (Casino) also shared the same building.

The Spielbank, Germany's oldest casino, was designed by Charles Sechan, a French theatrical set designer, and was commissioned by the Benazet family, which came from France to manage the casino in 1838. The gaming rooms were based on decorative styles from the periods of Louis XIII through Napoléon III. Its famous patrons included Emperor William I, Chancellor Bismarck, the Aga Khan, and actress Marlene Dietrich. The establishment still operates today—serving an equally elite clientele. Patrons are required to show their passports, and Baden-Baden residents need a letter of permission from the Burgomaster to enter the gaming rooms. Guests can play blackjack, baccarat, roulette, or poker, but there are no slot machines.

The nearby Trinkhalle (Pump Room), where patrons drank the mineral waters, was built in the nineteenth century. Its design was based on the old Roman style, and it was decorated with frescoes of Black Forest legends. Other buildings in the area housed an internationally respected art gallery and a theater.

The Friedrichsbad, also called the Old Baths, was under construction from 1869 to 1877. It was one of the city's major spas, commissioned by Grand Duke Friedrich von Baden. Mark Twain was a guest and claimed that the baths cured his rheumatism. Today the establishment offers several therapeutic bath programs, including a two-hour program consisting of a shower, two saunas with temperatures ranging from 130 to 160 degrees Fahrenheit, a brush massage and soaping, a thermal steam bath, three freshwater baths at up to 60 degrees Fahrenheit, and a thirty-minute rest. The spa also offers electrotherapy, hydrotherapy, and massage. Excavation for the Friedrichsbad's bath building revealed the Römanische Badruinen, ruins of a Roman bath installation, still on view today.

One of the margravial palaces, destroyed by fire in the seventeenth century, was replaced with a baroque structure completed in 1847 and called the Neues Schloss to distinguish it from the Altes Schloss ruins of the Hohenbaden Castle. Today the building houses the Zähringer Museum, which contains historical nineteenth-century rooms that recreate the atmosphere of the palace when it was the summer residence of the grand dukes of Baden. It also contains historical documents, works of art collected by the margraves over the centuries, and margravian family portraits. The outbuildings house a museum devoted to the town's history.

The Brahmshaus is located in the Lichtental quarter of Baden-Baden, once a separate town. Johannes Brahms spent each summer at this residence from 1865 until 1874. Here he worked on various compositions, including his *Lichtentaler Symphony* (No. 2). The building has become a museum featuring Brahms memorabilia, and a music festival, called the Baden-Baden Brahms Days, held in May of alternate years. Each festival consists of about eight concerts.

Baden-Baden was never bombed in World War II. After the war, the city became headquarters for the French Command in Germany and the site of a Royal Canadian Air Force base. Russian, British, and Soviet army personnel were also stationed there.

In 1990 the population of Baden-Baden was 50,000, including the surrounding incorporated villages. The city has more visitors than any other city in the Black Forest, attracting about 275,000 people a year. Although Germany has more than 250 registered spas and health resorts, Baden-Baden is considered the country's most elegant, sophisticated spa resort.

Recently, many of the city's trees, like those elsewhere in the Black Forest, have begun to show signs of blight. This environmental problem, caused by acid rain resulting from automobile emissions and nearby industries, is the subject of great concern in Germany.

Further Reading: *Romans on the Rhine* by Paul MacKendrick (New York: Funk and Wagnalls, 1970) discusses the early cultural history suggested by archeological evidence. In *Germany: A to Z Guide* (Chicago: Rand McNally, 1980) Robert S. Kane treats the history, architecture, and tourist attractions of forty German cities. The book is illustrated with maps of major cities and sections of the country.

—Phyllis R. Miller

Basel (Switzerland)

Location: On the Rhine River near the meeting point of France, Germany, and Switzerland.

Description: Intellectual capital of Switzerland, with its oldest university; home of the oldest publicly owned art collection in Europe.

Site Office: Verkehrsbüro Basel
Schifflande 5
4001 Basel
Switzerland
(61) 25 50 50

Basel—often spelled "Basle" in English, following the French style—is the Swiss city that, with its suburbs, makes up the half-canton of Basel-Stadt, now a separate political unit from the neighboring half-canton of Basel-Land. The military and commercial advantages of Basel's geographical site have attracted settlers for at least two thousand years. The city stands between the last section of the Rhine River, before it breaks into tributaries to the north, and the Birsig River, on a narrow piece of land that is both easily defensible and yet accessible from the territories that were to become France, Germany, and Switzerland.

The first known inhabitants of the area were members of the Celtic tribe known to the Romans as the Helvitii and which Julius Caesar's armies defeated in battle in 58 B.C. One of the Romano-Celtic towns established in this period was Augusta Raurica, now Augst, seven miles from modern Basel. Much of this town, including its theatre, can still be seen today, along with a hoard of Roman treasure excavated in 1962 and displayed in the Römermuseum Augst. Some time before A.D. 200, another town was established by the Romans on the riverbank where Basel's cathedral now stands, although the first written reference to this town as "Basilia" dates from 374. Only twenty-seven years later the Germans, who had been making inroads into the Roman Empire for generations, seized the town and expelled the Roman soldiers. As the Germans, mingling with the Franks, gradually organized the federation that became the Holy Roman Empire, Basel was drawn into closer trading contact with its neighbors. It was still not secure, however. In 916, for example, the church that had been built on the site of the Roman town was demolished during an attack by Magyar warriors (forebears of the modern Hungarians).

In 1006 the Holy Roman Emperor granted control over the town and the surrounding countryside to the bishops of Basel, who remained princes of the empire and rulers of the city for more than 500 years, though they faced increasing challenges from the merchants and their city council. Their cathedral, the Münster, was rebuilt and reconsecrated in 1019; rebuilt once again, this time in red sandstone, in the following century; and partially rebuilt and enlarged on several occasions since; most extensively after 1356, when the city was struck by an earthquake. Four other Basel churches also survived the earthquake. The Leonhardskirche is a Gothic church built on top of a Romanesque crypt and attached to part of the remaining city wall; the Martinskirche, begun in 1225 and completed in 1451, has been deconsecrated and made into a concert hall; the Barfüsserkirche, founded by the barefoot Franciscan friars in 1250 but later rebuilt, became a museum of the city's history in 1894; and the Peterskirche, also founded in the thirteenth century, contains frescoes painted around 200 years later.

From 1431 to 1448 the city was the meeting place for the last in a series of Church Councils convened to reunite and reform European Christianity, following the schism between Catholics and Orthodox in the eleventh century and a further schism between popes in Rome and "antipopes" in Avignon, which had started in 1378. The conciliar movement, which had developed through the councils at Pisa in 1409, at Constance from 1414 to 1418, and at Pavia in 1423, now seemed on the verge of making significant changes that might have averted some of the bitterness and violence of the Reformation in the following century. But the council's suggestions were opposed and disregarded, not only by the papacy, reunited since 1418, but also by the monarchs of western Europe and the leaders of the eastern churches, who preferred to make their own direct arrangements with the pope. In 1460 Pope Pius II forbade any further councils outside papal control and marked this reassertion of supremacy by founding the University of Basel. Its first building was on the Rheinsprung, a street that slopes up toward the Münster, with which its teachers and students were closely associated.

During the fifteenth century Basel had to defend itself in three wars, first between 1424 and 1428; then against the Habsburg dynasty, in the War of St. Jacob, between 1443 and 1450; and, lastly, against the growing power of Burgundy, between 1473 and 1476. In all three conflicts the city's survival depended at least to some extent on the relative security and sophistication of its finances, indicated by the willingness of its creditors to continue lending sums that they knew would be paid back promptly. In 1499 Basel was the neutral location chosen for negotiations between the Holy Roman Emperor Maximilian I and the group of ten Swiss cantons that had just defeated his Schwabian League of South German States and thereby prevented him from realizing his plans for a supreme court and a uniform tax for the whole of his empire. The resulting Treaty of Basel is now regarded as the foundation for the separation of the Swiss Confederation from the empire in fact though not yet in theory. In the same

The Holbein fountain in Basel
Photo courtesy of Swiss National Tourist Office

year the ten cantons invited Basel to join them. The city took up the invitation in 1501, becoming the wealthiest city in Switzerland and the only one to have a university.

Basel marked its newly secure status by replacing the guard on its gate with a woman who spun cloth while collecting the tolls. Between 1504 and 1514 the city's council built a new Rathaus (city hall) on the marketplace. Its printing and publishing industry led the country, issuing many books recognized as landmarks of the Northern Renaissance, such as those illustrated by Albrecht Dürer and Hans Holbein. (A fountain based on Holbein's work now stands in Basel.) The single most influential of these books was the Greek version of the New Testament issued in 1516, not because it became wildly popular (it did not), but because of the idea it represented, that it was possible to return to and reexamine the origins of Christianity. Its editor, the Dutch scholar Desiderius Erasmus, was already renowned both for his satirical work *The Praise of Folly* and for advocating pacifism, denouncing colonization as theft, and demanding state intervention in the economy.

The tolerant atmosphere of Basel was soon put to a severe test, for in Switzerland as elsewhere in northern Europe the Protestant Reformation led to political disruption and violence. But in Switzerland the Reformation took a unique course; unlike in Germany or Scandinavia, the movement here was initiated not by Martin Luther or his disciples but by a native of the country, the priest Ulrich Zwingli, who had been born in 1484 and had graduated from Basel's university. His campaign against corruption in the Catholic Church, stimulated by Erasmus's work, began in Zurich in 1519 and quickly went beyond Luther's, rejecting almost all decorations in churches as encouragements to idolatry, interpreting the communion service as a simple commemoration of the death of Jesus without any real presence of his body or blood, and demanding justice for the oppressed peasants whom Luther feared and denounced.

From 1523 Johan Hussgen, a German theologian also known as Ioannes Oecolampadius (the Greek form of his name), who had been an assistant to Erasmus, used his new post as a professor of theology at Basel University to help promote Zwingli's ideas. Also in 1523 the city council removed the vestiges of the prince-bishop's rights as overlord, inducing him to move to Porrentruy. Erasmus, now living in Basel, avoided theological disputes until Luther's attacks on his efforts to find a middle way between Catholics and Protestants forced him to respond by declaring his loyalty to the pope; at the same time, however, he advised the city council of Basel to tolerate both kinds of Christianity. In 1528 a Zwinglian uprising in Basel led to the banning of Catholic services, the destruction of holy images, and the rise of Hussgen to a position of decisive influence over the city council. All that remained of the medieval political and religious regime were the church buildings and the image of the prince-bishops' staff, which still appears in the coats of arms of both Basel-Stadt and Basel-Land. (Having sold their rights in Basel to the city council in 1579, the bishops retained their

titles and their control of the valleys of the Jura until the time of the French Revolution.)

The unity of the Swiss was now threatened by the religious wars among the cantons which broke out in 1529 and 1531. In the latter year Zwingli was killed in battle and Hussgen died either of a fever or of grief at Zwingli's death. Eventually it was agreed that each canton would choose which version of Christianity to adopt. Basel joined with Zurich, Bern, and Schaffhausen in upholding the Reformation but they were in the minority against seven Catholic and two mixed cantons. Religious toleration did not disappear from Basel. Erasmus, who in 1529 had departed for the German city of Freiburg forty miles away, was allowed to return in 1535. He died in the following year, still a Catholic, and was buried in what had become a Protestant cathedral. Also in 1536 John Calvin, having fled from France, wrote and published his *Christianae Religionis Institutio* (*Institutes of the Christian Religion*) in Basel before moving on to power and fame in Geneva. Many reformers in Basel, most notably Sebastian Castellion, professor of Greek at the university, came to oppose Calvin's intolerance just as they had opposed the pope's, even after the unification of the Calvinist and Zwinglian movements in 1549. The city council was even tolerant enough to permit the printing of the first European edition of the Koran, in 1543.

The achievement and maintenance of even these limited measures of tolerance in the Zwinglian areas of Switzerland contrasted sharply with the experience of most European countries. During the seventeenth century the Swiss succeeded in elevating their traditional preference for negotiation over fighting—sometimes disregarded, as in two further internal wars over religion in 1656 and 1712, but always present—into a basis for their full independence from the Holy Roman Empire. This was at last accepted by the empire and the other European powers at the peace talks of 1648, which concluded the Thirty Years War. The Swiss delegation at these talks was led by Johann Rudolf Wettstein, Burgomeister (mayor) of Basel. Switzerland's neutral position in international wars was first officially asserted in 1674 and has been upheld ever since. The canton of Basel at this time still comprised both the city itself and the countryside around it. In 1652 many of the canton's farmers joined with those of Bern and Lucerne in the Peasants' Union led by Nicholas Leuenberger, which fought and lost three battles against soldiers employed by the cities. The urban elite, organized in restrictive guilds based on crafts, rested their political power on their economic control of the villages. This control was intensified as the production of silk ribbons, the main industry of the canton, expanded with the invention of the bar loom and the participation of Calvinist refugees fleeing from Catholic cantons and from France, where King Louis XIV had rescinded the toleration of the Protestant minority. Partly through their presence, partly through the influence of the general Francophilia sweeping Europe at that time, buildings in the French style, and even the use of the French language, became marks of elite status in Basel and other Swiss cities, while the silk ribbon industry, known as the Posamenterie, was bifurcated between the workers who made the ribbons in their own homes in the countryside and the so-called Bän-delherren ("ribbon lords"), the financiers who lived inside the city. Among the latter were Lukas and Jakob Sarasin, whose luxurious houses on Martinsgasse, built in the 1760s and now called the Blaues Haus ("Blue House") and the Weisses Haus ("White House"), give some indication of the wealth that the industry generated.

In 1792 the bishopric of Basel was conquered and annexed by France. Switzerland as a whole remained aloof from entanglements with France for a few more years, but the French army's need to maintain contact with its conquered territories in Italy meant that sooner or later Napoléon Bo-naparte would move against the Swiss, who stood in the way. In 1798 the radical Peter Ochs, a member of the Basel government who had been born in France, tried to persuade his colleagues to ally themselves with the French Revolution. Their refusal to do so prompted France to take direct action, attacking Bern, occupying Switzerland, and establishing a "Helvetic Republic" led by Ochs and other "patriots." They became increasingly unpopular as long-serving officials were dismissed, supplies and conscripts were seized, and it became clear that the regime was no more than a puppet of French interests. In 1803 they were replaced by a "reformed" Swiss Confederation, firmly under Napoléon's control as "Mediator", in which Basel was one of the six "director-cantons" with slightly greater notional powers than the others. This system was no more popular or effective and collapsed in December 1813, when the army surrendered and the 200,000 troops assembled by the Allies at the border entered Basel.

The return of the old order meant the revival of its political divisions. The residents of the urban section of Basel canton still denied any political participation to residents of the rural section, who outnumbered them by around two to one. The Swiss Diet sent soldiers into the canton but could not persuade either side to compromise, and in 1833, after a battle in which the farmers were victorious, the canton was divided between the 14 square miles of Basel-Stadt (city) and the 165 square miles of Basel-Land (country). Even today, each of these half-cantons sends only one delegate to the Ständerat ("Council of States"), the upper house of the Swiss Parliament, instead of the two delegates sent by full cantons.

Just as the various cantons had been split into two groups by the rise of Protestantism in the sixteenth century, so now they were bitterly divided by the rise of liberal dem-ocratic movements. By 1847 liberal governments controlled the majority of the cantons, including Basel-Stadt. Basel now had around 27,000 inhabitants, who were to witness and take part in its transformation into the center of the Swiss chemical industry. In 1845 Christian Friedrich Schönbein, a professor at the university, had applied nitric acid to cotton to make guncotton, the first synthetic material ever created; in the same year the first train to enter Switzerland arrived in the city from Strasbourg. Twelve years later J. R. Geigy and W. Heusler began to manufacture natural dyes to supply to the 4,000 or so ribbon makers of Basel-Land. This combination of aca-demic research, efficient transportation, and commercial en-terprise has helped the industry to grow into a major employer and source of wealth, producing plastics and drugs as well as dyes; the industry is now dominated by three main corpora-tions: Sandoz, Hoffmann-La Roche, and CIBA-Geigy.

In 1914 Jean Jaurès and August Bebel, the leaders of the French and German socialist parties, met at Erasmus's tomb in Basel to symbolize their joint commitment to peace. This gesture failed, of course, to prevent World War I, during which Switzerland remained neutral. After hyperinflation had caused the collapse of the German currency in the chaotic aftermath of the war, the leading states of Europe established at Basel what is now the Bank for International Settlements (BIS), the worldwide club of the heads of central banks. Since their monthly meetings are held in secret it is difficult to assess just how much influence they can exert over the national governments that appoint them.

Switzerland remained neutral throughout World War II as well, though its intelligence agents shared information with those of France until the Nazi invasion of that country, its air force attacked German airplanes that entered Swiss airspace until ordered not to, and, in response to rumors of a German invasion in 1940 and 1941, the general staff planned to abandon Basel, Bern, and other lowland cantons and fight on from redoubts in the Alps.

Basel retains its crucial role in Swiss commerce to-day. The trains that connect it to all three countries in its vicinity, along with its quays on the Rhine, transport around half of all goods imported into Switzerland. It is still also a center of Swiss cultural activity. Its Kunstmuseum (Art Museum), which houses one of the best collections of paint-ings in the country, was begun in 1661 when Basilius Amer-bach gave the city his own collection. It is the oldest publicly owned art gallery in Europe. The city's international atmo-sphere is perhaps best indicated by the unique quirks of its transportation services: Basel airport is in France, not Swit-zerland, while in the city itself there are three railroad sta-tions, belonging to the French, German, and Swiss national networks. Such arrangements seem somehow appropriate to the city which once gave refuge to the Dutch humanist Er-asmus, the German theologian Hussgen, and the French re-former Calvin.

Further Reading: Among the general histories of Switzerland that provide a great deal of information about Basel, Charles Gilliard's *A History of Switzerland* (Westport, Connecticut: Green-wood, 1955) and Georg Thürer's *Free and Swiss: The Story of Switzerland* (London; Oswald Wolff, 1970) stand out as concise yet comprehensive, while Wilhelm Oechsli's *History of Switzerland, 1499–1914* (Cambridge: Cambridge University Press, 1922) re-mains readable and thought provoking after more than seventy years. Jonathan Steinberg's *Why Switzerland?* (Cambridge, Lon-don, and New York: Cambridge University Press, 1976) covers many aspects of Swiss history, economy, and society in a shrewd and entertaining manner. George Faludy's *Erasmus of Rotterdam* (London: Eyre and Spottiswoode, 1970; New York: Stein and Day, 1972) is an interesting biography of the great humanist, placing his life and work in the broad context of the north European Renaissance and Reformation.

—Patrick Heenan

Bath (Avon, England)

Location: One hundred and fifteen miles west of London, thirteen miles east of Bristol.

Description: Spa with preserved Roman baths and many important examples of Georgian architecture.

Site Office: Bath Tourist Information Center
The Colonnades
11-13 Bath Street
Bath, Avon BA1 1SW
England
(225) 462831

The story of Bath is largely the story of four men. King Bladud, according to legend, founded the city in the ninth century B.C. Almost three millennia later, in the eighteenth century, three men turned Bath, then a sleepy provincial town, into the most popular resort for the British nobility and the country's wealthiest and most fashionable citizens. Richard (Beau) Nash served as arbiter of manners in Georgian Bath; the architect John Wood designed much of the city's Palladian architecture; and Ralph Allen, financier and philanthropist, funded several of Georgian Bath's major enterprises.

According to tradition, Bladud, though heir to the British throne and father of King Lear, was stricken with a skin disease so severe that his father, the king, sent him away from the court. Eventually, Bladud became a swineherd, herding his pigs—which also developed the skin disease—in the area around Bristol. One day, he drove the pigs across a stream in the belief that there were more acorns on the other side; from a hillside, he observed the pigs charging down into a swamp. As he tried to get them out, he waded into the swamp himself. There he observed that the pigs' sores were healing, and soon both the pigs and Bladud were cured. He returned to his kingdom, reclaimed his rightful place, and chose the healing swamp with its hot spring as his royal seat. The spring waters have come at a constant rate (half a million gallons a day) and a constant temperature (118 degrees Fahrenheit) for thousands of years.

In the first century A.D., the Romans reached the area and built a settlement with baths called "Aquae Sulis" ("Waters of Sul," after the Celtic deity presiding over the waters). How the Romans discovered the mineral springs is unknown, but the settlement was a spa first and foremost, even though it was established as part of Rome's military conquest of Britain. According to their habit of integrating local deities with Roman deities, the Romans also constructed an elaborate temple dedicated to Sul-Minerva (Minerva being the Roman goddess of wisdom and healing). Into this temple, the Romans threw silver, bronze, and pewter coins and other artifacts as offerings to Minerva. When the baths themselves were uncovered in the eighteenth and nineteenth centuries, archaeologists found a sophisticated plumbing system of lead pipes and drains, and a central heating system utilizing hypocausts (flues and hollow bricks that conduct hot water through the walls and floors).

A new period of activity began in the seventh century, when the Abbess Bertana was given a grant of land for a convent "near the city called Hat Bathu." Offa, King of Mercia, also constructed a minster at Bath, and so began the city's 500 years as an ecclesiastical center largely ruled by Benedictine monks.

By the tenth century, Bath was enjoying both fame and prosperity. There were four parish churches, three springs, a mint, and a renowned monastery. It was sufficiently prominent to be chosen as the site for the coronation of Edgar, first king of all England, in 973.

The Norman Conquest affected the city very little, but the local population rebelled against Norman rule in 1088, leaving the city in ruins. William Rufus, son of William the Conqueror, made a grant of the abbey of Bath together with its property to John of Tours, royal physician and chaplain and thenceforth Bishop of Bath. Reportedly, John paid the king £60 for the city.

Bishop John expanded the city, undertaking several construction projects and a restoration of the baths. According to the description of the author of the *Gesta Stephani* (Works of King Stephen, compiled about 1138), "Through hidden channels are thrown up streamlets of water, warmed without human agency, and from the very bowels of the earth, into a receptacle beautifully constructed with chambered arches. These form baths in the middle of the city, warm and wholesome and charming to the eye." He added, "sick persons from all England resorted thither to bathe in these healing waters and the strong also, to see those wonderful burstings out of warm water and bathe in them." The bishop also began construction on a vast Norman cathedral on the site of the current Bath Abbey. Though only small pieces remain, its size can be inferred from the fact that all of the present Bath Abbey occupies only the area of its nave. Many well-educated monks enriched the community with their learning.

The prosperity of medieval Bath depended on the wool trade, and the city became a community of weavers. Geoffrey Chaucer immortalized this aspect of the city in his portrait of the Wife of Bath: "For weaving she possessed so great a bent/She outdid the people of Ypres and of Ghent."

The fortunes of the city turned in the late Middle Ages. The Black Death of 1349 devastated the priory, bringing the number of monks down to twenty-one. The religious community never recovered its former strength. When

The baths of Bath, with the abbey in the background

Bishop Oliver King visited in the late fifteenth century, he found an undisciplined community—the monks idle and women present where they should not have been. King reported having a dream which he took as a divine command to rebuild the crumbling abbey and reinvigorate the monastic community. The bishop obeyed and began building the current abbey, a notable example of English Gothic architecture. It contains the first English fan vault and the huge expanses of glass typical of the Gothic style. Bath Abbey was the last great church to be built in England before Henry VIII broke with Rome. The monks, however, did not seem to have recovered their inspiration, to judge by reports of dissolute living in the monastery.

Seventeenth-century Bath faced other problems—both economic and social. Many parishes throughout England sent those who were poor and sick to spas such as Bath. Often these people stayed on, established residence in Bath, and thus became eligible for poor relief from the city. They lived by begging from visitors to the spa, in competition with the professional beggars who already crowded Bath.

Another question arose over the city's morals in connection with mixed nude bathing, and the city's reputation also suffered from doubts about the hygiene of masses of sick people bathing in the same water.

Nevertheless, the city was honored with four visits by Queen Anne (two when she was a princess, two when she was queen) to alleviate her gout. Where the queen went, there went the court and those who moved in court circles. What they found was a dirty city without adequate lodgings and with unpaved streets, a place infested with beggars, pickpockets, and various types of con men. Typical of the lack of civility was the rudeness of the sedan chairmen reported by Celia Fiennes after a 1690 visit to Bath:

There are chairs as in London to carry the better sort of people, but no control is exercised over them; they imposed what fare they chose and when these were disputed would not let their customers go. If it was raining they would open the top and let him or her, often an invalid, be exposed to the wet until in despair the charges were met.

The city fathers made just two concessions to the expanding number of visitors. One was to ban nude bathing. The other was to appoint a master of ceremonies. The first to hold this office, a Captain Webster, was addicted to both drink and gambling. When he was killed in a duel, the mantle fell to a young man—Richard (Beau) Nash—who had arrived in Bath in 1705, made himself agreeable to the right people, and won consistently at gambling (apparently without cheating).

As master of ceremonies, Nash set about cleaning up the mess—both physical and social—that he found in Bath. He arranged for the streets to be paved and cleaned and for the sedan chairmen to be licensed. Nash responded to the requests of local doctors for the construction of an indoor space where their patients could come to take the waters by building the Pump Room. The first Pump Room was built in 1706. Nash engaged musicians to play there, assessing visitors a fee of a guinea for this attraction.

Nash's next targets were the visitors themselves. With his passion for propriety, he cast a cold eye on the manners and dress of Bath's guests. He issued a code of behavior, including the rules

1. That a visit of ceremony at first coming, and another at going away, are all that are expected or desired, by ladies of quality and fashion,—except impertinents.

4. That no person take it ill that any one goes to another's play, or breakfast, and not theirs,—except captious by nature.

5. That no gentleman give his ticket for the balls to any but gentlewomen.—N.B. Unless he has none of his acquaintance (see Oliver Goldsmith, *Life of Nash*, 1762).

Additional Nash directives covered clothing, banning, for instance, riding boots on men and aprons on women. He also outlawed the wearing of swords in the city and subsequently prohibited dueling.

Nash's strictness controlled amusement as well. Balls were to begin at exactly 6:00 P.M. and to end exactly at 11:00 P.M. Nash enforced his rules and once turned a deaf ear to the plea of the Princess Amelia for one more dance after the 11:00 P.M. deadline. The balls proceeded in a rigidly prescribed order: approximately two hours of minuets, during which each gentleman danced with two ladies, then country dances for an hour, then an interval "for the gentlemen to help their partners to tea," then "amusements" until Nash called a halt at 11:00 P.M.

Visits to Bath followed a set routine. Fashionable newcomers were greeted by the sound of the abbey bells and music from the "city waits." After the newcomers had paid the appropriate subscriptions (assembly rooms, booksellers, etc.), they prepared to use one of the city's five baths. People commonly bathed between 6:00 A.M. and 9:00 A.M., with ladies coming in dresses so copious as to hide their shapes. After the bath, visitors convened in the Pump Room, where those who had come to take the waters began their regimen of three glasses a day. The day passed with visits to coffee houses, to church, and to the parade for a walk. In the evening, the company reconvened in the Pump Room, and they concluded the evening at the assembly houses with balls or plays, or a visit to the gambling houses. Gambling was one of the major evening amusements for visitors to Bath at the time.

Seven years after Nash arrived, an enterprising eighteen-year-old from Cornwall, Ralph Allen, came to the city. He built his first fortune by ridding the postal service of its corruption and, in the process, turning a handsome profit for himself and for the government. With capital realized from his involvement with post office, Allen bought the quarries at Combe Down, near Bath, in anticipation of a construction boom in the city. The Bath stone was too soft to withstand the rigors of London's climate, but local demand for the stone was such that Allen amassed a second fortune.

Enter John Wood, a twenty-year-old surveyor and builder. Wood's dream was no less than to make Bath a second Rome. He envisioned an assembly hall to be called the Royal Forum of Bath; a sort of sports arena to be called the Grand Circus; and the Imperial Gymnasium of the City. All three were to be modeled on classical Roman examples. Though Wood's scheme was never realized (and one wonders how he meant to turn a small, provincial city into a replica of a world capital), enough of his work remains for the visitor to Bath to appreciate his genius. In 1729, work began on Wood's first great work, Queen's Square, with its Palladian facade designed to resemble a palace. Though work went on for seven years, Wood's vision for the west side of the square was never realized.

Meanwhile, leading citizens of Bath decided that a new hospital was needed for the many poverty-stricken patients who came to the city. Dr. William Oliver (inventor of the Bath Oliver biscuit, designed for those on a restricted diet) and Beau Nash were among the fund raisers. Ralph Allen donated stone from his quarry, and John Wood served as the architect, without pay, for the Mineral Water Hospital, which was built between 1738 and 1742.

In the year of his death, 1754, Wood embarked on another plan—the Circus, a grand arrangement of residential buildings around one of the major inter-sections of Bath. Architectural critics agree that Wood showed his genius in the layout and design of the Circus. The Circus is approached from three streets, so that the viewer does not look across into another street but to the facade of one of the buildings, all with three tiers, one upon the other, and each crowned with a stone acorn, in tribute to the legend of King Bladud. Wood did not live to see the Circus completed, but his work was carried on by his son, John Wood, Jr.

The careers of Allen and Wood converged again when the financier commissioned the architect to build a country home for him on a slope outside Bath. For ten years, from 1735 onwards, Wood worked on the project, but then the cantankerous architect fell out with his patron. For the twenty years that he lived in Priory Park, Allen extended his hospitality to writers and artists. Henry Fielding was a frequent guest and immortalized Allen as the benevolent Squire Allworthy in *Tom Jones,* while Alexander Pope paid tribute to Allen's charity and modesty.

The three creators of Georgian Bath died within a decade: John Wood in 1754, Nash in 1761, and Allen in 1764. In 1771, John Wood, Jr., who had carried out his father's plan for the Circus, showed his own architectural gifts in the Royal Crescent, only a short distance from the Circus. In *The City of Bath*, the historian Barry Cunliffe describes the Royal Crescent as "being perhaps the most perfect essay in urban architecture in the western world," going on to compliment the younger Wood for integrating his and his father's designs. Another innovative design of the end of the eighteenth century was Robert Adam's Pulteney Bridge, which spans the river Avon and is the only bridge in England with shops.

From about 1800, Bath declined as a gathering center for the affluent and titled. The increasingly stern strictures on gambling—the lifeblood of amusement in Bath—together with the increasing popularity of other spas, such as Cheltenham, contributed to this decline. Bath appealed to those in search of urban pleasures, while the nineteenth century saw an increasing interest in visits to "nature". Bath lost out not only to other spas but to such nature areas as the Lake District.

In the infamous Baedeker Raids of 1942, in which the German Luftwaffe targeted places in England of cultural and historical interest, Bath experienced severe damage (though most of the Georgian buildings were spared). After damage to major structures was repaired, property developers succeeded in destroying many eighteenth- and early nineteenth-century buildings. However, in 1966, Bath was chosen by the British Government as one of four historic towns to serve as case studies for the reconciliation of old towns with twentieth-century urban requirements.

The latter part of the twentieth century finds tourists in large numbers visiting Bath, stopping at the Pump Room to sample the waters, and testing the opinion of Charles Dickens's Sam Weller (from *Pickwick Papers*), who decided that the water had "a wery strong flavour o' warm flat irons."

Further Reading: *The City of Bath* by Barry Cunliffe (New Haven, Connecticut: Yale University Press, and Gloucester: Sutton, 1986) is a thorough examination of the architectural history of Bath. David Gadd's *Georgian Summer: Bath in the Eighteenth Century* (Bath: Adams and Dart, 1971; Park Ridge, New Jersey: Noyes, 1972) is an entertaining and well-researched social and architectural history of Bath from its founding by the Romans.

—Elizabeth Devine

Les Baux-de-Provence (Bouches-du-Rhône, France)

Location: Les Baux-de-Provence lies in the southwest of Provence, not far east of the Rhône River and two historic cities built on its banks, Avignon (less than twenty miles to the north) and Arles (about twelve miles to the southwest).

Description: On a rocky outcrop in the Alpilles chain of small mountains, Les Baux-de-Provence was built on a strategic plateau. The massive medieval castle was the scene of many intrigues and dominated the village below it for eight centuries. The castle was torn down in the seventeenth century, but some ruins remain visible. The village contains some fine restored Renaissance houses as well.

Site Office: Office Municipal de Tourisme
Hôtel de Manville
13520 Les Baux-de-Provence, Bouches-du-Rhône
France
90 54 34 39

The medieval lords of Les Baux-de-Provence (Les Baux), their campaigns, and their courts made the fame and infamy of this dramatic hilltop site. The first castle was built on the rocky promontory at the end of the tenth century, and the noble family ruled its changing Provençal domains until the 1420s, battling for power in the region, gaining and losing territories through marriages, allegiances, and war.

There is little certainty about the history of the site before the tenth century, although prehistoric weapons found in the area have been dated back to around 2000 B.C., and artifacts showing influences of Italian, Greek, and Sicilian civilizations have been unearthed. A Ligurian Celtic tribe, the Desuviates, may well have made use of the many natural caves in the vicinity, and had a settlement on the Bringasses plateau adjoining Les Baux.

The Romans built a fortification at the foot of Les Baux. A stela carved with three figures was found at the site in the nineteenth century, and for a time was thought to represent three St. Marys and therefore named the *Trémaïe*. It more likely depicts a noble couple overseen by a pagan goddess. A further Roman relief sculpture, called *des Gaïe*, was subsequently found nearby.

As Roman domination faded, the Visigoths swept down through Europe, and Euric, seventh king of the Visigoths, took the town of Arles in 480. For a time he persecuted the Christians there, and some of them are said to have fled to the rock of Les Baux and begun to build there. Various stories about the derivation of the place's name have circulated. In the past it has been claimed to descend from a lord of the house of Balthe, a Gothic word meaning courageous, while for a long time the lords of Les Baux claimed the name

derived from that of Balthazar, one of the magi who, according to one legend, traveled as far as Les Baux, establishing a castle there—the Star of the East significantly featured on the family's coat of arms. The most likely explanation, however, is that the name comes from a Ligurian word meaning an elevated site.

The line of the lords of Les Baux has been traced back to one Leibulfe at the end of the eighth century, but the family's origins are unclear, and Leibulfe may have been a mythical figure. At the beginning of the tenth century, the archbishops of Arles gained considerable sway over the region and surrounded themselves with powerful men. One of these was Isnard, who took charge of numerous ecclesiastical estates in the area; in the 970s the archbishop Massanes gave him land in the vallis Felauria, now the valley of Les Baux. The first castle was built on the site at this time. One of Isnard's relatives was Pons the Elder, whose son Pons the Younger married Profecta, daughter of the powerful head of the Marignagne family, and came to hold a position of importance among the lords of Provence. The first man to carry the name of the place as his family name was Hugues des Baux, son of Pons the Younger, born in 981. Guillaume Hugues, who took over from Hugues, is said to have played a large role in the first crusade of 1095.

It is the time of Raymond I des Baux (1095–1150) that marks the family's great struggle for power in Provence. Raymond married Etiennette, daughter of Gilbert, count of Provence. Gilbert's elder daughter, Douce, had married Raymond Bérenger IV, count of Barcelona. Gilbert died without a male heir in 1109, and Douce inherited eastern Provence, assigning control in 1113 to her husband, who already ruled the western part of Provence. Raymond des Baux would come to resent the possessions and power of the house of Barcelona. For a time, however, he helped Raymond Bérenger to defend the region against the advance of the Spanish Moors. But when Raymond Bérenger's son Bérenger Raymond came to power in Provence, Raymond des Baux rebelled, starting a Provençal civil war that raged for fourteen years. Already a powerful man in the region—tradition has it that from the time of his marriage he owned seventy-nine towns, villages, castles, and estates (called the *terres baussenques*) in the region and owned the formidable castle of Trinquetaille, next to Arles—Raymond des Baux wished to redress the daughters' unequal inheritance and claimed Etiennette's right to Provence. Sixty-three barons and knights took Douce's side, sixty-four Etiennette's. The Genoese supported Bérenger Raymond, only to betray and kill him in 1144. He left a seven-year-old son, another Raymond Bérenger. This boy's uncle and guardian carried the same name as he, and was count of Barcelona and King of Aragon. He came to Provence to oversee the boy's possessions and fight

Les Baux-de-Provence
Photo courtesy of French Government Tourist Office

Raymond des Baux, and during an assembly of the Estates in Tarascon in 1146, he forced the majority of the lords opposed to the count of Provence to swear loyalty to his nephew. This was the first meeting of the Estates to be mentioned in Provençal history.

Raymond des Baux sought the support of the Holy Roman Emperor, Conrad III, also king of Arles (the castle of Les Baux was attached feudally to the kingdom of Arles) Conrad confirmed Raymond's rights to the fiefs owned by his father or brought into his family by his wife. These nominal confirmations of rights were of little practical use though, and by 1150 Raymond des Baux was defeated and many of the *terres baussenques* ravaged. Raymond went to Barcelona to submit to his adversary and even hand over control of the castle of Trinquetaille; he died soon after.

Etiennette sent her two eldest sons to Barcelona to swear allegiance to Raymond Bérenger the Younger of Barcelona. Five years later, though, Etiennette's son Hugues des Baux rose against the house of Barcelona. With the reconfirmation of his rights by the new Holy Roman Emperor, Frederick Barbarossa, and the backing of the Arlesians, his cousin the count of Toulouse, and other lords, he rekindled the war and refused to give up Trinquetaille. A second treaty

was made in 1156 confirming the first and limiting Hugues's family to the pastures and castle of Les Baux. But Trinquetaille was not given up. The count of Barcelona besieged it but failed to take it. In revenge, he ravaged the territories of Les Baux and Arles, and captured Les Baux. Returning in 1161 with much more powerful troops, he destroyed Arles and rapidly won Trinquetaille, razing it to the ground. A new treaty put Etiennette in control of the family possessions on the condition that the sovereignty of the count of Provence be recognized. Etiennette died soon thereafter, and Hugues once more resumed hostilities, but in 1162 he was defeated.

The count of Barcelona arranged for the marriage of his nephew, the count of Provence, to Frederick Barbarossa's niece, bringing the Holy Roman Emperor's support to his side. The family of Les Baux was given back the lordship of Trinquetaille, and peace was restored to Provence.

In 1172 the lordship of Les Baux fell to Hugues's brother, Bertrand. He had married Tiburge, daughter of Raimbaud III, prince of Orange, who died without a male heir in 1175, leaving the principality of Orange to Bertrand. The Holy Roman Emperors confirmed and strengthened Bertrand's rights, including the right to display his flag from the Alps to the Rhône, and from the Isère River to the Mediterranean. Bertrand completed the nearby abbey of Silvacane, founded by his father, and gave endowments to various churches in the region, but he was assassinated in 1181.

Of Bertrand's three sons, all became heads of powerful families: the third, Guillaume, became prince of Orange after Bertrand; the second, Bertrand, lord of Berre; and the eldest, Hugues IV des Baux. Hugues revolted against his overlord, Alphonse, count of Provence, who ordered his capture dead or alive. He was imprisoned, but freed on paying a ransom. A treaty of 1206 settled the disputes, confirming various rights for Hugues to lands around Les Baux. Debts, though, forced him to sell other territories, including in 1225 the huge lake of Vaccarès. Through marriage to Barrale, he had come into possession of part of the viscounty of Marseilles, but gave back the Marseilles territories to its inhabitants to settle his debts with them; in 1226 he tried to reclaim the territories. The dispute was sorted out by a papal legate who arranged for payments to Hugues.

The so-called Albigensian heresy was splitting southern France at this time, the counts of Toulouse supporting the Albigensian side, the counts of Provence, traditional Catholicism. Hugues des Baux took the side of Toulouse against his natural overlord. The latter, Bérenger, reacted by dispossessing him of his towns and castles, and imprisoning him. However, he then employed Hugues to negotiate a successful peace with the count of Toulouse and restored his territories to him.

The family and court of Les Baux would not be associated only with fighting through its history. Poetry counted among the most highly considered courtly pursuits in Provence, and the traditions of the troubadours flourished at Les Baux. One Fouquet celebrated Adelasie, wife of Berald des Baux, and among his compositions was the *Complaint of Berald* on the death of his wife. The troubadours also recounted the nobles' achievements in war, and Fouquet followed Guillaume des Baux on Simon de Montfort's crusade against the Albigensian heretics. Perdigo was another troubadour who sang the praises of Guillaume and the deeds of Hugues des Baux. When Gui de Cavaillon attacked Guillaume in his compositions, the prince of Orange replied in verse—some of the lords of Les Baux themselves gained a reputation for their poetry. The century saw the flowering of the courts of love: select noblewomen judged questions of gallantry and apparently awarded a crown of peacock feathers and a kiss to the finest poet. The troubadours and the so-called *val d'Enfer* (vale of Hell) next to Les Baux are said to have inspired Dante's *Divine Comedy*.

Before Barral, Hugues's son, came into his inheritance in 1240, he was already serving the count of Toulouse as senechal, or steward, and married the latter's niece, Sibylle d'Anduze. He was excommunicated for a time by the archbishop of Arles for conquering the Venaissin territories to the north. In 1240 the count of Toulouse decided to take Arles and put it under the control of the Holy Roman Empire; Barral appears to have ceded Trinquetaille to assist the plan. Louis IX of France intervened, placing the castle back under

Barral's control; the Arlesians, displeased with Barral's role, seized Trinquetaille and other of Barral's territories to put an end to his rule. Peace and the status quo were restored by a treaty of 1245.

The powerful Charles of Anjou, brother of Louis IX, became count of Provence, and had plans to take control of the three self-governing towns of the region: Avignon, Marseilles, and Arles. This time the Arlesians found for a short time a defender in Barral. However, when Charles laid siege to Arles, Barral withdrew, and Arles lost its independence. Barral then came to an agreement with Queen Blanche, mother of Charles and Louis and regent during their absence on a crusade, to submit to their designs on Avignon and Arles in exchange for forgiveness of his past deeds. In 1252 his fiefdoms in the Venaissin were returned to him for a promise that he would serve in the Holy Land. From then on relations were good with the new count of Provence. Barral was even given charge of suppressing a Marseilles rebellion. He was an ostentatious lord, who surrounded himself with poets, including the loyal Bertrand d'Alamano. Fighting and its glorification in poetry went hand in hand in those times, and when Barral was fighting in Marseilles, he fought Boniface de Castellane, reputed as a warrior and troubadour. Barral's daughter Cecile entered Provençal legend for her beauty, bearing the name of Passe-Rose.

In 1264, when Charles of Anjou was granted the crown of Naples by Pope Urban IV, Barral joined the expedition to secure these new territories. Barral's son Raymond was in the front guard at the Battle of Benevento against Manfred, who had been king of Naples and Sicily, and was granted the county of Avellino as a reward, while his other son Bertrand was given the task of dividing up Manfred's treasures. Barral was appointed grand justiciary of Sicily in 1261, and podestat of Milan.

Of Barral's two sons, Raymond became grand senechal of Provence, and later commander of the royal fleet and captain general of the cavalry, but he was killed by his own men in a skirmish in Calabria. Bertrand, who also acquired many other positions of importance, married Philippine de Poitiers, daughter of the influential count of Valentinois. He was renowned for his courage both on the battlefield and in tournaments, as is recounted in the chronicle of the first tournament held in Provence, which claim that he brought down twenty horses without receiving any wounds himself. His style of living led him into heavy debt, however, and he even had to sell the prized Trinquetaille to the archbishop of Arles in 1300. He died in 1305 while visiting Palestine.

His grandson Hugues Raymond became a close friend of King Robert, the new count of Provence, and became embroiled in the machinations and murders of Robert's granddaughter, Queen Joan I of Naples. She received one of the finest European territorial inheritances of the times, including southern Italy and Provence, but brought renewed turbulence to the region. On behalf of Queen Joan, Hugues Raymond received the homage of the lords of Provence, and went to the pope in Avignon as her ambassador. He became grand admiral of her kingdom. Joan sought his support in Provence when king Louis I of Hungary came to avenge the death of Andrew of Hungary, her first husband, whom she was strongly suspected of murdering. Joan tried to flee Provence, but was kept prisoner by Hugues Raymond, who feared she might try to rid herself of her Provençal territories. She was set free only when the pope had secured assurances that she would not relinquish Provence to any other power.

Hugues Raymond would play an extraordinary role in the siege of Aversa in Italy in 1350 by the king of Hungary. He brought help to the besieged, and negotiated a truce with the Hungarian, while arranging for Joan and her second husband, Louis of Taronto, to be taken to Gaeta. In the meantime he took the castle where Joan's sister Marie, widow of the duke of Duras and, via her mother, titular empress of Constantinople, was held. He forced Marie to take his son Robert in marriage, and once sure the marriage had been consummated, placed Marie and her treasures on his ship bound for Provence. He stopped outside Gaeta, and was ordered to see Louis of Taranto. He feigned gout, but Louis came to him and, accusing him of betrayal, assassinated him, taking his two sons Robert and Raymond prisoner. Pope Innocent VI tried to have Robert des Baux freed, but in 1354 Marie went to his cell and had him assassinated.

Back in Provence, Robert of Duras was wreaking havoc. In 1355 he took the castle of Les Baux and captured Raymond des Baux's brother Antoine. He amassed considerable support from those wishing to take revenge on Queen Joan. Civil war was brewing. Queen Joan and the pope jointly organized troops to defeat Robert. The major lords of Provence rallied brigands. A siege was mounted in June, with a fort built facing the castle of Les Baux, including a tower for catapulting rocks onto the stronghold. By August, Robert admitted defeat. Some argue that he surrendered because of large payments promised him by the pope. The brigands went on the rampage after the siege, and had to be paid to leave the area by the pope. But then Raymond des Baux appears to have attracted them back to Provence in his own violent schemes. Queen Joan's troops managed to take many of Raymond's strongholds, and Count Jean d'Armagnac protected the threatened Marseilles and went to besiege Les Baux. Antoine joined his brother, and with the brigands pillaged and burned Aix-en-Provence. By the end of the 1360s, however, Raymond and Antoine had lost most of their gains, and in 1371 Raymond died.

Raymond left an only daughter, Alix des Baux. Her grandfather, the vicomte de Turenne, was appointed her guardian. He was part of an extremely influential family, two members of which became Avignon popes. Turenne is legendary in Provence for the anger he unleashed concerning disputes over territory with the house of Anjou, and over inheritance with the papacy. Furthermore, he usurped Alix's power. Known as the Scourge of Provence, he brought terror to the region from 1388 to 1399. Based at Les Baux, he defied Marie of Anjou, and flew the flag of the Roman pope

from the castle. He is best remembered for stories of the gruesome way he treated prisoners for whom no ransom was paid, simply having them thrown off the precipitous rocks of Les Baux. A General Assembly held in Aix-en-Provence in 1391 ordered the besieging of the places held by Turenne. Pope Clement VII also put together an army, with Adon de Villars, made husband of Alix by Turenne, as one of its leaders. Turenne was first offered a sum of 20,000 florins if he gave up arms; this offer seems only to have incensed him, and he arranged for pirates to attack some of the coastal cities. The militia proved insufficient to combat Turenne. Foreign fighters were recruited, but they, too, set about pillaging the region; they had to be paid large sums to disband. Eventually the intervention of the French king Charles VI brought about a change in the situation. In 1397 a force of 3,000 men amassing to lend further support to Turenne was stopped and disbanded by royal order. By 1399 Turenne's strength was sapped, and he agreed to a treaty. Soon breaking it, he was hounded to Tarascon, where he was surrounded by royal troops. Taking to the river, he is believed to have drowned. Alix was able to assume power in Les Baux.

On Alix's death in 1426, however, the Baux family line ended. Until 1481 the town came under the direct rule of three counts of Provence: Louis III, René, and Charles III. These counts preserved and reconfirmed the special privileges from which Lex Baux's inhabitants had benefited under their own lords, exempting them from various taxes, and allowing them a good degree of independence. Charles III died without heir, leaving his cousin Louis XI, king of France, to inherit his possessions and assume the title of count of Provence. So Provence was reunited with the kingdom of France.

The French kings' policies were aimed at bringing Provence under tighter central control. Louis XI ordered the destruction of the fortifications of Les Baux, with others in the area, to reduce the refuges for warring factions. The next French king, Charles VIII, placed Les Baux under an administrator with the title of baron. The second baron was Anne de Montmorency, King Francis I's favorite, who repelled the Holy Roman Emperor Charles V from Provence in 1536. Perhaps to recognize Montmorency's success, Francis I visited Les Baux in 1538. By letters patent of 1539 the town was given the right to hold two fairs a year and a market once a week, while in 1545 money was donated for the rebuilding of the ramparts. The command of the place was put in the hands of a noble in the absence of its baron.

In these times, Renaissance styles of architecture from Italy were influencing Provence, and in Les Baux many fine noble houses were built by wealthy families. One still-standing and particularly admired piece of architecture, a pavilion ordered by Jeanne de Quiqueran, baroness of Baux, has become famous as the erroneously named pavilion of Queen Joan.

The Reformation brought strife to Les Baux as elsewhere in France. Protestantism flourished in Provence. Reformers seized power in Les Baux in 1561, destroying religious images and burning the castle furniture. Many Catholics fled to Arles. Captain Gauchier de Quiqueran besieged Les Baux successfully in the same year, and executed several reformers. After a period of persecution, the treaty of Amboise of 1563 restored the right to free worship to French Protestants, and they reemerged in Les Baux. A fine window in the town still bears an inscription of 1571, *Post tenebras lux* (After the darkness, light), the motto of the Calvinists. The baron des Baux Jacques de Bauche and one of his captains, Jacques de Vérassy, brought back a harsher attitude toward Protestantism early in the seventeenth century. In 1619 the latter refused the Protestant minister entry into the town. The Protestants of the town brought the case to the court of Grenoble, but it turned down their complaint. The power of Protestantism in Les Baux declined rapidly, due to the absence of a minister and mass recantations.

Antoine de Villeneuve, sieur de Mons, succeeded de Bauche as the fifth and final baron des Baux. Close ally of the Duke of Orléans, enemy of the French king Louis XIII, he left the country with the duke. Cardinal Richelieu ordered the capture of Les Baux by Charles de Grille, whose attempt, which involved disguising some of his number as women, was inept. The town council insisted that the king himself should write confirming his orders. The angered governor of Provence sent his captain Saucourt to lay siege to the town. Eventually the king's letter arrived and after a twenty-seven-day siege, the town allowed Saucourt to enter.

An act drawn up by the royal court in 1631 was then forced on the inhabitants of Les Baux, by which the ramparts would be destroyed at their expense. They were also forced to pay 100,000 pounds to buy back their territories, effectively being made to pay for Mons's actions and the ensuing siege. The castle, which had stood 800 years, was brought down in a month under the direction of the master mason from nearby Tarascon. Some ruins of the castle, however, remain visible today. The town was crippled by its enforced debts, but by 1642 it was arranged that the king should buy back the territories of Les Baux. In that year, he gave Les Baux to the prince of Monaco of the Grimaldi family, as a reward for loyal allegiance to the French crown after the Spanish had tried to establish a territorial stronghold in Monaco. So the Grimaldis became marquises of Les Baux, and their coat of arms was soon placed on the entrance gate to the town. Six marquises ruled Les Baux, from 1643 to 1795, but the population declined, many departing to live in the valleys. The first marquis had a nobleman's house built on the plateau; the last was paid around 200,000 pounds in cash by the French nation for Les Baux, and the Grimaldi house was destroyed in the French Revolution.

In the early nineteenth century, the place gave its name to bauxite, the chief ore from which aluminum is made, after a chemist discovered its presence in the area. From the middle of the century, it was the writer Frédéric Mistral who revived the fame of Les Baux in his poems in Provençal, which are steeped in the history of the location and recount tales of the house of Les Baux, whom he famously described as a "*race d'aiglons*" (race of eagles). He won the Nobel Prize in 1904. His poems were part of his unceasing and romantic attempts to revive Provençal culture, and when blonde hair of extraordinary length was found in excavations in Les Baux's oldest church of St. Vincent, Mistral took it for the Provençal museum he founded in Arles. Another celebrated local poet of Les Baux, Charles Rieu, who scarcely left the parish in his life, is commemorated with a statue at the top of the village. Les Baux became something of an artists' colony at the turn of the century. Now the romance of the location, the ruined castle and restored Renaissance architecture, and the medieval tales and legends associated with the place draw mass tourism, and make Les Baux among the handful of most visited sites in France.

Further Reading: By far the most detailed history of Les Baux-de-Provence, entitled *Les Baux et Castillon* (first published 1902; reprint, Raphèle-les-Arles: C. P. M. Marcel Petit, 1987) was written by abbé L. Paulet. Although the author's biases of religion and local birth are evident, he consulted archives and researched the history in depth. He not only relates the dramatic history of the lords of Les Baux and their adventures; he also presents in some detail the history of the community more generally. Not all points and dates are clear, however. Paul Pontus, in his book *Les Baux* (Paris: Nouvelles Editions Latines, 1971), gives a lucid, shorter account. Lucien Bély has written several small but reasonably helpful tourist guides, even if some of the details are left blurred. They contain roughly the same text and are published by Ouest France, located in Rennes. Their titles are *La Gloire des Baux* (1981), *Les Baux en Provence* (1982; English translation available) and *Aimer le Baux de Provence* (1987).

—Philippe Barbour

Bayeux (Calvados, France)

Location: In the Calvados department of the Normandy region, northern France; 20 miles west of Caen; 150 miles west-northwest of Paris.

Description: Originally a Gaulish settlement, then a Roman town, Bayeux became an agricultural center with a lace-making industry. It was the first French town to be liberated by the Allies in 1944 and is famous for the Bayeux Tapestry, presents the story of the Norman Conquest of England of 1066.

Site Office: Office du Tourisme et Syndicat d'initiative
1 rue des Cuisiniers, B.P. 200
14403 Bayeux, Calvados
France
31 92 16 26

Situated on the River Aure five miles from the Channel coast, Bayeux was the capital of the Bajocasses (or Baiocasses) tribe of Gauls, and was written about by Pliny the Elder around A.D. 50. The Romans named it Augustodorum. Subsequently captured by Bretons and Saxons, it was finally taken over by the Vikings who invaded Normandy and devastated the town in 858. It had already become a walled city with baths and temples by the third century, and a bishopric was established in the following hundred years. Rollo, the Viking chief, went on to marry the governor of Bayeux's daughter, Popa. After the treaty of St-Clair-sur-Epte with the king of France in 911, Rollo was baptized, and made the first duke of Normandy. The citizens continued to speak the Norse language, however, rather than French.

Rollo's great-grandson, William, who became duke of Normandy in 1035, received the Saxon, Harold Godwinson, in Bayeux in 1064, two years before King Edward the Confessor of England's death. Against his will, Harold solemnly swore that Duke William would accede to the English throne. It was an act that would lead directly to the conquest of England by the Normans. King Edward the Confessor died, childless, early in January of 1066 and Harold was offered the throne by the Saxon nobles. When he accepted, William set about organizing a fleet and invasionary force, which was ready within seven months. It set sail on September 27, arriving the following day in Sussex. The Norman army occupied Hastings, and on October 14 William's and Harold's armies engaged in battle a few miles outside the town. Harold was killed when he was pierced in the eye with an arrow, and William's army emerged victorious. William, who became known as William the Conqueror, was crowned king of England in Westminster Abbey on Christmas Day.

The precise origins of the Bayeux Tapestry, which shows the saga of the conquest, are not known. At one time it was believed to have been the work of William's wife, Queen Matilda, and court ladies. Now it is generally believed to have been commissioned by William's half-brother, Odo de Conteville, count of Kent and bishop of Bayeux, and to have been embroidered (it is not technically a tapestry at all) by Saxon workers or nuns in England over a ten-year period. This attribution is strengthened by the inclusion in it of three of Odo's followers who are mentioned in the Domesday Book, an inventory of English lands and landholdings ordered by William and completed in 1086. These three warriors are among the very few named figures on the tapestry. Worked in eight colors of wool on a linen background, the tapestry is some 20 inches wide by about 230 feet long. It was used on certain feast days, particularly on July 1, the Feast of Relics, to decorate the Cathedral of Notre Dame in Bayeux. The end part of the tapestry has been destroyed or has decayed, but nevertheless it is a unique documentary account of the conquest, and of the ships, armaments, clothes, and buildings of the period. Another similar embroidery done for William's daughter, Adela, described in Latin verse by Abbot Baudri, has not survived.

The central part of the tapestry can best be described as a sort of strip-cartoon, with scenes showing the events that led up to the conquest, including Harold's visit to France, the oath he made to William in Bayeux, the funeral of Edward the Confessor at Westminster, the coronation of Harold, the embarkation of the French fleet and the crossing to Sussex, the cavalry and foot-soldier attacks at the battle of Hastings, and the death of Harold. The tapestry also depicts the appearance of Halley's comet in 1066, which was taken to be an omen by both sides in the dispute, and which is, incidentally, one of the first visual records of this comet. Latin inscriptions above the various events illustrated are stitched in the Saxon manner of writing. The earlier scenes are divided by stylized trees, but the second section runs continuously. The narrow friezes that decorate the top and bottom along the length of the tapestry occasionally show additional details of the battle, such as Norman soldiers stripping the English dead of their clothes, armor and weapons, or the rape of English women. Elsewhere they contain stylized birds and beasts, and scenes from Aesop's fables. It is a work of an impressively documentary and realistic nature.

The tapestry is mentioned in the inventory of the cathedral treasury in 1476. From 1803 to 1804 it was exhibited in Paris on Napoléon's instructions, its historical importance having been realized with the publication of articles by two archaeologists in 1724. It was returned to Bayeux in 1815, where constant rolling and unrolling caused great damage to it. The tapestry is now housed in the Centre Guillaume le Conquérant, an eighteenth-century building opposite the cathedral that was a seminary until 1970.

Details of the Bayeux Tapestry
Photo courtesy of Office de Tourisme, Bayeux

Of the original Cathedral of Notre Dame, completed in 1077 by Bishop Odo, only the crypt and the Romanesque bases to the towers on the west front remain. The cathedral was severely damaged by fire in 1106 at the hands of the army of Henry I of England. Henry had it restored; the forces of his grandson Henry II were to ruin it again in 1159, and the cathedral was rebuilt again; much of the present structure dates from the thirteenth century. It was Henry II who was instrumental in the murder of Thomas Becket, the archbishop of Canterbury. The tympanum of the south porch has this story carved on it, ending at the top with the scene of Becket's assassination at the high altar of his own church. The exterior of the cathedral is supported by flying buttresses, part of the the thirteenth-century construction. The central tower over the crossing dates from the fifteenth century. The original cupola was removed in 1857 because of structural weakness, and replaced with a cast-iron "bonnet." It is one of the most beautiful of Romanesque-Gothic cathedrals in France. Inside there is both light and lightheartedness.

The nave has sturdy, wide-arched pillars from the twelfth century, with a band of carved foliage and four-leaf clovers beneath the clerestory level, where the windows are of lightly colored stained glass. The carvings between the apex of each arch and the pillar are sometimes quite irreverent; apes, dragons and jugglers mingle with bishops and saints. The clerestory, the vaulting of the nave and the chancel are from the thirteenth century, with the chancel dating from about 1230, on the outlines of the previous apse. It has tall, slender pillars and is lit by rose windows. The ornately decorative choir stalls were carved in the late sixteenth century by an artist from nearby Caen. The Chapter House, a lovely late-twelfth century building, has a tiled or glazed brick floor incorporating a maze design, and it was at the desk here that the beatification of St. Theresa of Lisieux was signed. She was canonized in 1925.

Although the town of Bayeux is small, it is both well preserved and well restored. Half-timbered houses in the Normandy style, sometimes turreted, sometimes with overhanging upper stories, and dating from the fifteenth century onward, remain in the streets surrounding the cathedral. An eighteenth-century mansion, the Hôtel du Doyen, is now home to the Bayeux lace workshop, and has examples of the characteristically floral patterned lace on show. More lace is exhibited in the Musée Baron Gérard, as well as porcelain from Bayeux and Rouen; Italian, Flemish, and French paintings; and seventeenth-century tapestries bearing the arms of Louis XIII and Anne of Austria. In the Place des Tribunaux, stands a plane tree planted in 1797 and called the Tree of Liberty. A statue of the poet and political writer Alain Chartier, born toward the end of the fourteenth century in Bayeux, is also in the town. He is best remembered for his poem *La Belle Dame sans Merci*, and the local secondary school has been named after him.

After the Normandy beach landings in 1944, which initiated the liberation of France from the German occupation of World War II, Bayeux was the first town to be declared free. Charles de Gaulle, leader of the Free French Army and later the first president of the Fifth Republic, made a speech there on June 14, 1944. A column in the square is named for him and marks the occasion. The fifteenth- to seventeenth-century former governor's mansion in the Rue Bourbesneur leading eastward from the square, honors de Gaulle's memory and his association with the town by housing the Musée mémorial de la Bataille de Normandie, which tells the story of the two-and-a-half months of the operation. Tanks of the period stand outside the museum. Inside there are military uniforms and equipment.

The D-Day invasion was code-named Operation Overlord, and despite a small delay because of bad weather, was launched on June 6, 1944. British, American, and Canadian troops earmarked the Utah and Omaha beaches with British and Commonwealth forces acting as a spearhead on the Sword, Juno, and Gold beaches. It is the Omaha and Gold beaches that are closest to Bayeux, and one of the many war cemeteries in the area lies on the southwestern outskirts of the town. Bayeux itself was never of strategic importance during the war, and escaped serious damage. For this reason, it has remained one of the most intact and enchanting small towns in northern France.

Further Reading: Many encyclopedias and guidebooks provide valuable information on Bayeux, but a more personal view is available in *Portrait of Normandy* by D. Pitt and M. Shaw (London: Hale, 1974). The tapestry has had several major monographs devoted to it: *The Bayeux Tapestry: the complete tapestry in colour, with introduction, description and commentary* (London: Thames and Hudson, 1985) by D.M. Wilson, and *The Mystery of the Bayeux Tapestry* (London: Weidenfeld and Nicholson, 1986; Chicago: University of Chicago Press, 1987) by D.J. Bernstein are two recent ones. In French, *La Tapisserie de Bayeux* (Rennes: Ovest-France, 1977; English translation, 1985) by the then-curator of the museum, S. Bertrand, contains an interesting section on life in general in eleventh-century France.

—Honor Levi

Berlin (Germany)

Location: In the eastern part of Germany, about 365 miles southeast of Hamburg and 35 miles from the Polish border. The city covers an area of 341 square miles and lies in the valley of the Spree River, which runs through its center.

Description: Germany's capital and largest city (1992 estimated population: 3,376,800). For much of its long history (it celebrated its 750th Anniversary in 1984), Berlin has been an important industrial and cultural center; despite the ravages of World War II, and the difficulties of the Cold War, of which Berlin was the symbolic center, many historic sites remain intact or have been restored.

Site Office: Berlin-Touristen-Information
Europa-Center
Budapester Strasse
10820 Berlin
Germany
(30) 2 62 60 31

Berlin's origins can be traced to the twelfth century, when two towns called Berlin and Cölln occupied the banks of the Spree River. The name ''Cölln'' first appeared in documents in 1237, and ''Berlin'' seven years later. Located on the northeastern outskirts of the Holy Roman Empire, the two towns were a convenient stopping point for traveling tradesmen, who would often visit for several days to sell to local merchants a wide assortment of goods, including wool, oak, rye, hides, and furs.

The two towns thrived and agreed to cooperate on defense and trade matters, although they had little enthusiasm for union. Nevertheless, in 1307 the towns merged under the name of Berlin, and built a town hall. Soon after, the city joined the Hanseatic League, a federation that provided commercial and protective advantages to free towns located around the Baltic Sea.

During the next few centuries, Berlin grew and continued to prosper, a fact that made it vulnerable to attacks by the robber barons that roamed the areas. In 1414, however, Frederick from the house of Hohenzollern defeated the most prominent of the outlaws; the following year, he declared himself prince elector of Brandenburg. The Hohenzollerns would rule the empire from the city for the next 500 years.

The emerging Hohenzollern dynasty worked to exert its control over the independent-minded citizens of Berlin, who had enjoyed special privileges under the houses that preceded its takeover. The inhabitants revolted several times, but the Hohenzollern family crushed these uprisings, and Berliners lost many of their privileges. In 1470 Berlin became the official residence of the prince elector of Brandenburg.

Berlin became the Hohenzollern dynasty's administrative center, and people flocked to Berlin in search of work and better living conditions. Between 1450 and 1600 the city's population swelled from 6,000 to 12,000 residents. In an effort to extend its control over the Berlin citizenry, the Hohenzollern adminstration issued numerous decrees to curb such social problems as drunkenness and public rowdiness, but these laws were hard to enforce. After all, the elector's court and bureaucracy set the standard, which was one of excess.

In the 1500s living conditions were wretched for many residents. Garbage rotted in the streets, and the drinking water was often contaminated. It was not surprising that in 1576 the plague devastated Berlin, killing almost 5,000 people.

No sooner did Berlin begin to recover from this catastrophe than it was again hit by hardship. Between 1618 and 1648 the city was caught up in the events of the Thirty Years War, one of European history's most devastating conflicts. Elector George William, who ruled Berlin during this period, was not one of the strongest members of the house of Hohenzollern. He tried to play both sides of the conflict, and ended up having to pay indemnities to each side. Berlin's economy and people suffered; the population was reduced once again to a mere 6,000 people.

Fortunately for Berlin, the great Hohenzollern ruler Frederick William succeeded his weak and incompetent father in 1640. Under Frederick William's rule, which lasted until 1688, the city once again became prosperous. The ''Great Elector'' encouraged the arts and industry, sponsored many building projects, and oversaw development of the Oder-Spree Canal, which made Berlin an important port.

An enlightened ruler for the times, the elector Frederick William welcomed religious refugees, most notably the French Protestant Huguenots, whose skills helped make for important contributions to Berlin's economic development. The population of Berlin increased during his reign to more than 20,000, and by 1700 an estimated one in five Berliners was of French extraction.

In 1701 the Great Elector's successor, Elector Frederick III (since 1688), became Frederick I, the first king of Prussia. He made Berlin the capital of his growing fiefdom, seeking to rival Paris through patronage of the arts and sciences. Among his accomplishments, King Frederick designated Berlin the center of the newly founded Prussian Academy of Sciences, with the brilliant philosopher and mathematician Gottfried Wilhelm Leibniz as first president. Frederick's son, in turn, was primarily a soldier. As King Frederick William of Prussia from 1713 to 1740, he strengthened the army and made the administration more efficient.

Frederick William I's son, Frederick II, who ruled

Scenes at the Brandenburg Gate before (top) and after the Berlin Wall's fall
Photo courtesy of German Information Center

from 1740 to 1786, built on the achievements of his predecessors. Known as Frederick the Great, Frederick II was much like his great-grandfather, the Great Elector. He supported the arts and culture, and made Berlin an intellectual center that attracted such luminaries as the author Gotthold Ephraim Lessing and the philosopher Moses Mendelssohn. The king himself was an able composer and prolific writer.

While Berlin prospered under Frederick the Great, the king was never popular with the people, largely because he led Prussia into several wars. The conflicts did not devastate Berlin as the Thirty Years' War had done, but Austrian and Russian forces managed to occupy the city during the Seven Years' War. Still, by the end of his reign, Frederick had made Berlin the center of a great European power.

Frederick the Great's nephew, Frederick William II, ruled for only a decade, and it was under his son Frederick William III, who reigned from 1797 to 1840, that Berlin's fortunes changed. A few years after the overthrow of the king of France in the French Revolution of 1789, Napoléon Bonaparte, an obscure Corsican soldier, came to power and soon embarked upon a campaign to conquer Europe. On October 27, 1806, after the Battle of Jena, French troops occupied Berlin, forcing Frederick William III to transfer his government to Königsberg.

Entering the city, Napoléon's troops marched triumphantly to the Brandenburger Tor (Brandenburg Gate), the city's symbol of independence. To break the Berliners' proud spirit, Napoléon had the Quadriga (a sculpture of a chariot drawn by four horses) removed from the top of the gate and shipped to Paris, the center of his empire. When Napoléon finally was defeated a few years later, happy Berliners recovered the Quadriga and returned it to its proper place.

The French occupation of Berlin ended in 1809, but it had weakened the absolute power of Frederick William. Inspired by the ideals of the French Revolution, many Berliners became involved in the wars of liberation of 1813 and 1814. After Napoléon's defeat in 1815, however, the king regained his authority. Frederick William III had promised that the people would have a "state constitution," but that promise was not fulfilled duing his lifetime.

Frederick William's long reign was a period of economic and cultural expansion. The city became an important center of trade and industry, producing silk, porcelain, machinery, and woolens. In 1809 the philologist and statesman Wilhelm von Humboldt founded Friedrich Wilhelm University (later renamed Humboldt University), and it quickly grew into a great center of learning, attracting many scholars of international note. In keeping with Berlin's status as a major European power, the king commissioned the construction of many fine public buildings.

The early nineteenth century marked the beginning of the Industrial Revolution, and thousands of peasants flocked to Berlin looking for work. Many ended up laboring in factories under brutal conditions and at low pay. Meanwhile, the city's growing middle class began to demand more political

power to match its newly gained social status. Hopes for positive change were high in 1840, the year Frederick William IV succeeded his father, but he proved to be a reactionary ruler who had no intention of changing the status quo.

The political situation came to a head in 1848 when Berlin became the center of revolution in Prussia. The February uprising in Paris sparked protests and rallies in Berlin, and bloody clashes broke out between protesters and the police. The king quickly surrendered to the demands of the rebels. He granted a new constitution that included the right to vote, freedom of assembly, and freedom of the press. But the victory was short-lived. By the end of 1849, the king had crushed the revolution in Berlin and regained control of the city.

Although the failure of the revolution dashed Berlin's hopes for political freedom, the city did embark on a period of steady economic growth. In fact, between 1850 and 1900, Berlin became Europe's second-largest industrial center. By 1870 more than 1,000 new factories had been founded, and the urban area almost doubled. Meanwhile, by 1900 the population had increased from 900,000 in 1871 to more than 2.5 million.

In 1866 Berlin became the capital of the North German Confederation, a political association of several German states dominated by Prussia. Four years later, under the brilliant leadership of Otto von Bismarck, Prussia went to war with France and won. The following year, Prussian King William I, Frederick William IV's brother and successor, was crowned German emperor at Versailles.

During William I's rule, the city's appearance changed remarkably. Many new buildings were constructed, including the noted Reichstag or lower chamber of the German parliament. But Berlin's growth had another side as well. As one writer described it, "The whole area between the centre of the city and the suburbs was literally crammed full with large tenement houses put up at an alarming rate. In the north and east of the city downright slums with dreary streets and gloomy inner courtyards sprang up."

Prussia would have two more emperors after William I. The first, Frederick III, was dubbed the "three month king" because he ruled for only a short period before his death in 1888. The second, William II, was the last of the Hohenzollern emperors. During his reign, Berlin's growth was prodigious. The city's population, for example, doubled from 2 million to 4 million between 1900 and 1920.

When World War I broke out in 1914, Berliners, like most Germans, were enthusiastic. By 1918, however, the war had gone badly for Germany, and Berliners turned against it. William II, who, during World War I, was called "Kaiser Bill" by his enemies, fled to the Netherlands after the Allies defeated Germany in 1918, ending Hohenzollern rule.

After World War I, the Prussian monarchy and the German Empire dissolved, and Berlin became the capital of a new German republic. The city experienced the ruinous inflation of the 1920s, which dislocated its economic life and weakened the status of its prosperous middle class.

But Berlin continued to grow, increasing a staggering tenfold in geographical area when most of the surrounding suburbs were incorporated into the city in 1920. In all, Berlin incorporated seven formerly independent towns, twenty-seven landed estates, and fifty-nine rural communities to form Greater Berlin, which was then divided into twenty precincts. The move made Berlin Europe's biggest industrial city, as well as a major banking and stock-exchange center. By the mid-1920s the population of Berlin stood at 4.5 million.

In the 1920s Berlin became one of Europe's leading centers of culture. Music, theater, art, and film all flourished. Actresses Greta Garbo and Marlene Dietrich and the director Fritz Lang began their careers in Berlin; Wilhelm Furtwängler conducted the Berlin Philharmonic Orchestra; architect Ludwig Mies van der Rohe and dramatic producer Max Reinhardt were prominent figures in Berlin. Architectural giants such as Martin Gropius and Hans Scharoun left their imprint on the city. Before World War II, Berlin boasted 7 opera houses, 35 theaters, and more than 20 concert halls, as well as 150 daily and weekly newspapers.

By the 1930s, however, economic conditions in Germany had worsened, and the country came under the sway of Adolf Hitler and his National Socialist (Nazi) Party. Interestingly, the people of Berlin never strongly supported the Nazis, whom, in the words of one writer, they "regarded as gangsters from southern Germany." Hitler did gain power, though, becoming Germany's chancellor on January 30, 1933. Soon after, he embarked upon a campaign of terror against his political opponents. In November 1938, in what became known as *Kristallnacht* (Crystal Night, the Night of Broken Glass), Jewish-owned businesses and synagogues in Berlin and throughout Germany were destroyed.

In September 1939 World War II began. It went well at first for Germany, but the Allies eventually turned the tide of battle, and Berlin ended up paying a terrible price for the Führer's aggressive ambitions. War reduced Berlin's population by a third from its 1939 level of 4.5 million, while destroying nearly half of the city's buildings, including 85 percent of its industrial plants and 75 percent of its cultural heart—the museums, theaters, and concert halls.

After the war Berlin was split into two parts, reflecting the divisions of the Cold War. The Soviets controlled the heart of the pre-1945 city in eastern Berlin, and the western powers, led by the United States, exercised supreme authority in their sector, western Berlin.

The city became a constant source of tension between the two sides for forty years. In 1948 the Soviet Union blocked land routes into the western sector and in turn into Western Germany, controlled by the Allies, from which West Berlin drew its sustenance. Numerous crises and confrontations followed the blockade, which inspired the massive Allied airlift of supplies to West Berlin. In 1961 the construction of the Berlin Wall cut off virtually all contact between East and West Berlin.

The Allies generously financed western Berlin's reconstruction, and this part of the city rose from the ashes to become prosperous and thriving. The East, on the other hand, never recovered from the devastation of World War II, and millions of people fled to West Berlin and West Germany.

In the 1980s a movement for freedom took hold in the Soviet Union and its satellites that led to the end of Communist rule in East Germany. In 1990 Germany was reunited into a single, non-communist nation, with a united Berlin to be reinstated as its capital. The Berlin Wall was destroyed, and once again Berlin stood poised to take its place as one of the world's most dynamic and prominent cities.

Today Berlin's history is evident in the Brandenburg Gate, the Reichstag, medieval churches, and other preserved and restored buildings, including the Charlottenburg Palace, once the summer residence of Prussian kings and the last remaining Hohenzollern Palace. Reunited Berlin also has a thriving cultural scene, encompassing opera, symphony, and theatrical groups and a wide variety of art museums.

Further Reading: The reader may consult the following books for further information about Berlin's history: *Berlin* by Richard Stein (New York: Blackbirch Press, 1991); *Berlin: In Brief* (West Berlin: Press and Information Office of Land Berlin, 1972); *The Berlin Crisis* by Robert M. Slusser (Baltimore: Johns Hopkins University Press, 1973); *The Weimar Chronicle: Prelude for Hitler* by Alex de Jong (New York and London: Paddington Press, 1978); *The First Book of Berlin: Tale of a Divided City* by David Knight (New York: Watts, 1967); *Before the Deluge: A Portrait of Berlin in the 1920s* by Otto Friedrich (New York: Harper, 1972); and *Imperial Berlin* by Gerhard Masur (New York: Basic Books, and London: Routledge, 1971).

—Ron Chepesiuk

Biskupin (Bydgoszcz, Poland)

Location: Fifty-five miles northeast of Poznań and thirty-one miles southwest of the city of Bydgoszcz in the Bydgoszcz province of north-central Poland near the town of Gaşawa.

Description: Originally an ancient fortified settlement on an island in Lake Biskupin; over time that island became a peninsula, and today Biskupin is a small Polish village on the Biskupin peninsula.

Site Office: Biskupin Museum
88410 Gaşawa, Bydgoszcz
Poland
253-53

In north central Poland, on the banks of a medium-sized lake called Lake Biskupinskie, lie the remains of the ancient town of Biskupin. Although its modern counterpart stands a few miles to the east, the excavated remnants now an archaeological park, of ancient Biskupin, attract all the attention of visitors to this region.

Biskupin, which means "bishop's village" in Polish, is a village more famous for its prehistoric past than its present. In 1933 while leading students on a field trip, a local schoolmaster, Walenty Szwajcer, noticed a number of oak poles sticking up in the middle of Lake Biskupin. Thinking that these poles might be of some historical significance, Szwajcer called the Museum of Poznań. Shortly thereafter, the museum sent Professor Józef Kostrzewski to investigate. Kostrzewski authenticated the poles as relics from the prehistoric era and, in conjunction with Zdzisław Rajewski, began directing Poznań University's excavation of the site in 1934. The excavation was disrupted by the Nazis during World War II, but it began again in 1946.

The excavation findings at Biskupin cover 10,000 years, from the late Paleolithic Age to the Middle Ages. More than five million fragments and articles including wood, horn, bone, pottery, and metalware, have been excavated. The excavated relics from the Middle Ages suggest that a feudal order existed earlier than had previously been thought. The most significant findings at the Biskupin site, however, relate to prehistoric times.

The area was populated several times during prehistory, first between 9000 B.C. and 8000 B.C., again at the end of the Neolithic period (1800 B.C.), and once again during the early Bronze Age (c. 1500 B.C.). The first settled population, however, dates from the early Iron Age, approximately 550 B.C. At this time a fortress was built on an island in Lake Biskupin. This fortified settlement lasted roughly 140 years. Fortunately, after the settlement was abandoned, the rising water level preserved the submerged fortress, and provided

archaeologists with vast amounts of information pertaining to the prehistoric Biskupin settlers.

These settlers comprised one of the larger groups of the Lusatian culture. Lusatian culture refers collectively to the Lausitz Urnfield groups who covered large areas of middle and east-central Europe. The eastern zone of the Lusatian culture, where Biskupin was located, was heavily influenced by the Carpathian trade route originating in the town of Hallstatt. The excavation at the Hallstatt mines has been so instrumental in the study of prehistory that the years from 1200 B.C. to 475 B.C. are referred to as the "Hallstatt period." Traditionally, the period is broken down into four smaller periods: Hallstatt A (1200 B.C.–1000 B.C.); Hallstatt B (1000 B.C.–800 B.C.); Hallstatt C (800 B.C.–600 B.C.); Hallstatt D (600 B.C.–475 B.C.). Thus, the settlement at Biskupin occurs in the context of Hallstatt D.

Prior to the excavation at Biskupin, little was known about the Lusatian culture. Most of the societies that comprised the culture were transient, which made traces of them difficult to find. Archaeologists differ in the precise dates that they attribute to Lusatian culture; however, using the widest parameters they place the Lusatian culture somewhere between the seventh and fifth centuries B.C. Pollen records from forest reclamation studies of middle and east-central Europe confirm that Lusatian settlements were ultimately abandoned, leading to the collapse of the culture.

The settlement at Biskupin is noted for the sophistication of its architecture, which, according to Rajewski, has "not been equalled by anything from that period in Europe." The fortress is surrounded by a belt of oak and fir poles driven into the lake bottom at a forty-five-degree angle. The belt is three to nine rows deep, with 35,000 poles used in the entire belt. This belt served three purposes. The poles provided a first line of defense against enemy attacks, acted as a breakwater that helped prevent the island's shoreline from eroding, and served as an ice-breaker during the winter months. Inside this protective belt, along the shoreline, were defensive ramparts. The base of the ramparts consisted of stacked logs; on top of this base were timber crates filled with earth and stone that had been plastered with clay. This created a fortress wall 19 feet high and 1,500 feet in circumference.

Inside the wall there was a village, complete with wooden houses, storerooms, a street system, and public spaces. There were slightly more than 100 houses and 3,599 feet of roads. The total area of the settlement was roughly 4.9 acres. In all, more than 8,000 cubic yards of timber were used in the construction of the settlement. Experts estimate that this made the Biskupin settlement three times larger than its neighbors. Proportionately, 35 percent of the usable area of the settlement was used for defensive and safety purposes,

Restoration of the ancient settlement at Biskupin
Photos courtesy of Polish National Tourist Office

23 percent was covered by streets and public spaces, and 42 percent was devoted to housing. Archaeologists believe that the total population in the settlement was between 1,000 and 1,250 people. Peter Wells, however, in his book *Farms, Villages, and Cities: Commerce and Urban Origins in Late Prehistoric Europe* claims that this number is much too high. He believes that many of the houses were storage facilities, not residences, and therefore places the range of inhabitants closer to 50 to 200.

The work within the settlement appears to have been divided along gender lines, with the women doing the farming work and the men tending to the livestock. The women cultivated the soil with small wooden plows pulled by animals. Plows had not yet been introduced in much of Europe, but in the larger Hallstatt communities like Biskupin, plowing was replacing the digging method of cultivating the soil. The women also made furrowing hoes out of antlers soaked in vegetable acids. The crops they planted included four species of wheat and two species of barley, millet, peas, lentils, and beans. The livestock that the men kept included pigs, short-horned cattle, sheep, and small horses. Because the ratio of excavated domestic animal bones to excavated wild animal bones was roughly three to one, experts assume that hunting did not play a large part in either the economy or the culture. The excavation also revealed local production of bronze implements. However, there is no convincing evidence that iron-working techniques, which were still relatively new, had been mastered.

The houses in the Biskupin settlement were all built in accordance with a uniform plan. Each house was thirty-two feet long and twenty-six feet wide, and was divided into two rooms: a large main room and an anteroom. The main room covered 645 square feet with the hearth on the right and the sleeping quarters on the left. It was separated from the anteroom by a wicker door. The anteroom covered the remaining 215 square feet and seems to have been used to quarter livestock during extreme weather. Kitchen pots of two to four gallons were common in most houses. The fact that all of the houses were built from one plan and outfitted in uniform fashion demonstrates a high level of social organization and distinguishes Biskupin as one of the stellar examples of Lusatian culture.

Further indication of the high level of social organization at Biskupin can be found in the sophisticated manner in which goods and property were distributed. While soil, woods, water, and underground raw materials belonged to the tribe or sometimes to a particular clan, the livestock was considered private property. Likewise, looms, canoes, and nets were owned jointly, while bronze and iron goods belonged primarily to the chieftains. Furthermore, burial patterns indicated that shamans and metalworkers were accorded a higher social standard than the average settler.

This first settlement was destroyed by intertribal warfare. A second fortress was immediately built squarely upon the old fortress's foundation. The new fortress was virtually the same as the older one, with the exception that the wall on the ramparts was higher. The second settlement was destroyed by fire, and burial patterns confirm that there was a significant decrease in population after the settlement burned. The joint life span of the two Biskupin settlements was approximately 140 years.

One wonders, though, why a third settlement was not erected after the fire. While any answer must remain hypothetical, it is likely that the shifting climatic conditions of Europe at the time were at least partly responsible. In order to feed its population, the Biskupin development depended on high-yield leguminous crops such as peas and lentils. Legumes grow best in lowland areas near bodies of water, hence Biskupin's main food source would have been located close to the water's edge. During the Hallstatt D period, the climate in central Europe changed significantly. New weather patterns increased the annual precipitation dramatically. Consequently, the level of Lake Biskupin would certainly have been rising dramatically as well. This would, in turn, ruin the leguminous crops and seriously curtail Biskupin's ability to feed its inhabitants, prompting, perhaps, the abandonment.

After the second Biskupin settlement was burned, the area was largely deserted. It was settled again, first by small groups in the first century A.D., and then later in the fourth and fifth centuries during the late Roman period. At the start of the seventh century, there was once again a settlement fortified by an oaken palisade and abatis at Biskupin. The island where the old settlement had been no longer existed, and a peninsula had taken its place. A castle occupied the peninsula, while most of the villagers lived on individual plots spread along the lakefront.

The castle was built on two-thirds of an acre, while the remaining two and one-half acres were covered by the suburban development that was, essentially, agricultural. This appears to be an early feudal structure, and suggests that feudal societies existed well before historians and archaeologists had previously suspected. The castle was destroyed between the ninth and tenth centuries, but the feudal distribution appears to have remained intact, with the feudal lord living elsewhere.

Near the end of the eleventh century, the prince of Gniezno, a mere twenty miles from Biskupin, "gave" the village to the archbishop of Gniezno. Biskupin was one of many villages that were given to the archbishop at the time, and it was then that the name "Biskupin" was used for the first time. In 1325, the archbishop moved the village slightly more inland on the peninsula to its present location.

Although very little has occurred of any note in Biskupin in modern times, many scientists and tourists visit the archaeological park on the site each year. In the park the original gateway is on display, as well as fragments of the defensive wall and two rows of houses that were reconstructed in the precise manner in which they were originally built. A five-minute walk from the entrance to the park is a museum that houses a scale model of the entire fortification.

Further Reading: *Biskupin Polish Excavations* by Zdzisław Rajewski (Warsaw: Polonia Publishing House, 1959) is the most comprehensive source on the excavation at Biskupin. The material is presented clearly and his illustrations are helpful. In addition, there are two texts that pertain to late prehistoric Europe and discuss the Biskupin settlement in helpful ways. *Tribe and Polity in Late Prehistoric Europe,* edited by D. Blair Gibson and Michael N. Geselowitz (New York and London: Plenum Press, 1988) contains the essay ''Demographic and Economic Changes in the Hallstatt Period of the Lusatian Culture,'' by Janusz Osotoja-Zagórski, which deals with Biskupin extensively. However, the article is fairly theoretical and often difficult to read. *The European Iron Age* by John Collis (New York: Schocken Press, and London: Batsford, 1984) also discusses the Biskupin settlement, although to a significantly lesser degree than either of the other two sources.

—Lawrence F. Goodman

Blenheim (Bavaria, Germany)

Location: Thirty miles northwest of Munich, on the upper Danube river, in the German region of Bavaria; 120 miles north of German-Austrian border at Füssen.

Description: Small Bavarian village; to the north are wooded hills, to the south is marshland; in between is a plain traversed by slow-moving streams on their way to the Danube; site of a major battle in the War of the Spanish Succession in 1704.

Contact: Tourist Information
Bahnhofstrasse 10
D-89420 Höchstädt, Bavaria
Germany
(9074) 4412

Established in 1600, Blenheim is a typical Bavarian village, located along that part of the Danube that flows through southern Germany. Like most of the small farming villages along the banks of the Danube, Blenheim is a quiet town in which the inhabitants routinely go about their work from one season to the next, working outside in the fields in spring and summer, and then moving the work inside in the colder fall and winter months. Blenheim is situated in between the region of Franconia to the north and the town of Heidelberg to the south. From all outward appearances, there is little to distinguish Blenheim from the many other villages along the Danube; however, in the annals of history, the events that transpired in and around Blenheim in August of 1704 forever save it from the anonymity of its neighbors.

Blenheim (spelled *Blindheim,* in German) is historically significant as the site of the battle of Blenheim during the War of the Spanish Succession. The battle of Blenheim was a great victory for the allied forces of England, Holland, and Austria over the Franco-Bavarian armies, and clearly marked the beginning of the end of France's military dominance of Europe. Just as clearly, the battle marked the beginning of England's rise to military prominence. The context of the battle is as much political as military.

By the end of the seventeenth century there was a precarious balance of power in Europe. The Austrian Empire, ruled by Leopold of the Habsburg family, controlled most of eastern Europe, but the empire was in a state of decline. To the west was France which, under the rule of Louis XIV, had become more and more aggressive and whose military power had been increasing steadily for forty years. To the north of France was Holland, a major maritime and economic power. Holland was wary of France becoming more powerful than it already was. Fortunately, Spain's holdings in the so-called Spanish Netherlands (roughly equivalent to what today is Luxembourg and Belgium), created a buffer zone between Holland and France.

This delicate balance of power was threatened by the fact that Charles II of Spain was close to death and had no immediate heirs. The three claimants to the Spanish throne were Philip of Anjou (grandson of Louis XIV), Archduke Charles (Leopold's younger son by his second marriage), and Joseph Ferdinand (Leopold's grandson by his first marriage). Because of the size of Spain's holdings, the question of who would succeed to the throne had enormous implications for Europe. Either France or Austria would double in size. Further, if the Spanish Netherlands were to fall to the French, the balance of power between France and Holland would be disrupted. After a flurry of diplomatic activity and the sudden death of Joseph Ferdinand, Charles II bequeathed his entire kingdom to Philip of Anjou.

As soon as Philip assumed the throne, Louis seized the Spanish fortresses of Mons, Namur, Ostend, Oudenarde, and Antwerp in the Spanish Netherlands. Alarmed by France's aggressiveness, England, Holland, and Austria formed the Grand Alliance in September 1701 in order to protect themselves. From that point forward, the allies began arming themselves for war. On May 15, 1702, war was declared on France in London, Vienna, and the Hague. To further complicate matters, in September of that same year Bavaria joined forces with France. The ruling family of Bavaria, the Wittelsbach family, agreed to help France defeat the allies in return for a promise of huge territorial gains at the expense of the Habsburgs in the event of a Franco-Bavarian victory. Thus, the war of the Spanish Succession began.

Sir John Churchill, Duke of Marlborough and a direct ancestor of Winston Churchill, was placed in command of the English army. Prince Louis, Margrave of Baden, was his counterpart commanding the imperial forces of Austria. The allies' initial plan was for Baden to conduct the campaigns in the south, and for Marlborough to beat back the French in the northern Dutch territories. By the end of the campaigns of 1703, however, it became apparent that the northern territories were too strongly held by the French. The allies would have to change their strategy to be victorious.

By the spring of 1704, the allies decided to coordinate their efforts and concentrate on engaging the Franco-Bavarian forces somewhere along the Danube. Historian Hilaire Belloc claims that Prince Eugene, one of the Austrian Imperial Army's finest generals, conceived of the plan and wrote to Marlborough about it. Major-General J.F.C. Fuller, on the other hand, claims that it is unlikely that Eugene thought of the plan. Instead, he suggests that Count Wratislaw, the Imperial Envoy Extraordinaire, was initially responsible for the plan. While there is considerable ambiguity as to who

Blenheim Palace, Sir John Churchill's monument to his victory
Photo courtesy of Blenheim Palace

Diorama of the Battle of Blenheim
Photo courtesy of Höchstädt Tourist Information

first conceived of this plan, the conditions certainly warranted it. Austria was crumbling economically and was increasingly susceptible to a Hungarian attack from the east. This made Vienna ripe for the picking, should the French capture and control the Danube. The reports were that Marshal Marsin of the French army and the Elector of Bavaria were going to join Marshal Tallard, supreme commander of the French forces, and his 36,000 men and proceed up the Danube, with Vienna the obvious target. The threat to Vienna (and hence the empire), in conjunction with the strength of the French position in the northern territories, convinced the allies to coordinate their efforts along the Danube. Thus began Marlborough's famous "march to the Danube" in the spring of 1704.

By all accounts, Marlborough's famous march was one of the most successful and well-orchestrated marches in military history. He began in Roermond in mid-May, proceeded southeast to Bonn, followed the Rhine south to Wiesloch, and then turned away from the Rhine valley and marched to the Danube, arriving at Amerdingen on July 1. In all, Marlborough's forces marched more than 340 miles. It is significant to note, however, that for most of this time the French generals could not be sure of Marlborough's destination and thus had to wait for him to commit himself before reacting. Initially, the French supposed that Marlborough's goal lay along the Moselle River. When he turned away from the Moselle at Coblenz, the French believed that Strasbourg and Alsace were his destinations. It was not until Marlborough turned away from the Rhine valley in Wiesloch on June 7 that the French could be certain that he was heading for the Danube.

On June 10, Marlborough reached Mondelsheim and met with Baden and Prince Eugene. Their strategy was for Eugene to wait along the Rhine and guard against Tallard and Villeroy (another of the French marshals), while Marlborough continued marching toward the Danube, combined with Baden's forces already in place just north of Ulm. Together, Marlborough and Baden planned to engage the Elector and Marshal Marsin somewhere along the Danube. This is just what they did, and while Eugene returned to the Rhine, Marlborough kept moving his troops southward, reuniting with Baden at the town of Launsheim on June 22.

Nine days later, the combined forces arrived at Amerdingen, and Marlborough was in position for the first offensive of the campaign along the Danube: the taking of the fortress known as the Schellenberg in the town of Donauwörth.

Marlborough wanted to take Donauwörth because of its strategic importance. Whoever held Donauwörth had, in the Schellenberg, a strong position from which to control the Danube and all points east. Thus the town was important for the allies to control if they wanted to protect Vienna. Moreover, from Donauwörth Marlborough could have access to the supply-rich region of Franconia to the north. Donauwörth was fifteen miles west of Marlborough's camp at Amerdingen. On July 2, he marched his troops, under the command

of Lieutenant-General Goor, to Donauwörth and ordered them to attack the Schellenberg that same afternoon.

Marshal Count D'Arco of the French army, who had recently been sent to protect the Schellenberg, never expected Marlborough's men to attack on the same day that they had marched fifteen miles. His preparations against the attack were not scheduled to be completed until the next day, and his reserves were not scheduled to arrive until then. When Marlborough's forces attacked at 5:00 P.M. on July 2, then, they were able to carry the day. His losses, however, were heavy: 1,400 killed and 3,800 wounded. The French retreated to Augsburg, where the Elector of Bavaria had entrenched himself and his army upon hearing of the defeat at the Schellenberg.

With the Elector isolated from Tallard and Villeroy, who were still in the Rhine valley, Marlborough attempted to entice the Elector to switch sides. When diplomacy failed, Marlborough resorted to intimidation. He authorized the use of nearly 5,000 troops to set about ravaging Bavaria, burning towns and brutalizing the inhabitants. Marlborough's hope was that the Elector would switch his loyalties in order to save his people. The attempt, however, was unsuccessful. The Elector held out until Tallard arrived in Augsburg with his forces on August 3. The mere presence of Tallard was enough to force Marlborough to recall his troops and concentrate on the threat posed by Tallard and the Elector.

When Tallard marched south to reinforce the Elector, Eugene followed him so stealthily that neither Tallard nor Villeroy realized that Eugene was heading toward a rendezvous with Marlborough. Eugene's arrival was doubly helpful to Marlborough. With Eugene's army, the allies' strength was roughly equivalent to that of the Franco-Bavarian forces. Moreover, Eugene's arrival enabled Marlborough to separate himself from Baden. Marlborough and Baden had been disagreeing constantly on strategy. Marlborough wanted to proceed westward on the Danube, while Baden wanted to move eastward and take the town of Ingolstadt. With Eugene's forces at his service, Marlborough could let Baden go to Ingolstadt with a small task force to besiege the town, thereby relieving himself of Baden's intervention without significantly reducing his own fighting force.

Tallard moved his forces from Augsburg to Lauingen on August 8, and on to the Blenheim plateau on August 11. His plan was to move down the Danube and break down the Allied forces. On August 12, Marlborough's forces joined Eugene's in the town of Munster, just ten miles northeast of Blenheim. Marlborough knew that Villeroy was not far behind Tallard and that his only chance of beating the French was to attack before Villeroy could provide reinforcement. He therefore decided to attack on August 13.

The town of Blenheim, situated on the northern bank of the Danube, consisted of a central church, roughly 200 houses, and one big manor house. To the left of the town was the plateau where Tallard, Marsin, and the Elector were encamped, and three miles farther to the left was the town of

Oberglau. Marlborough's plan called for Eugene to attack at Oberglau, Lord Cutts (nicknamed "the Salamander") to attack at Blenheim, and for Marlborough himself to attack the plateau at the center. Movement in the allied camp began at 2:00 on the damp Wednesday morning of August 13. Marlborough and Cutts were in place by 7:00 A.M., but Eugene's forces were not in place until 12:30 P.M.

The French, meanwhile, were surprised when they awoke to discover the allies preparing for an attack. Because the French enjoyed a slight edge in forces (most estimates place the French forces at 60,000 and the allied forces at 56,000) and because their geographical position was strong, the French were confident of victory. Marsin and the Elector fought on the left near Oberglau, Tallard in the center, and a garrison of troops under the command of Lieutenant-General the Marquis de Clérambault de Palluau was sent into the village of Blenheim. There was a fatal flaw in the French plan, however. They fought as two separate armies, Marsin and the Elector forming one army, and Tallard's forces forming another. There was almost no coordination between them, which, in the heat of the battle, cost them dearly.

When Eugene's troops were finally in place, Marlborough gave the order to attack. On the allied right (Oberglau), Eugene's troops were repulsed as they tried to cross the Nebel, the stream that separated the allied forces from the Franco-Bavarian forces. They fought hard all day, but were never able to overcome the enemy. In Blenheim, Lord Cutts, with 11,000 men and a battery of guns, ordered the first assault on the village. Tallard, recognizing the strategic importance of the town, had garrisoned Blenheim with nine battalions and placed seven battalions behind the village in reserve, with eleven more battalions as well as twelve squadrons of dragoons within reach. Cutts was repulsed, and it is estimated that he lost one-third of his men in this first assault.

As he was retreating, Cutts was pursued by the Gendarmerie, the elite of the French infantry. Cutts called for reinforcements and was sent five squadrons. One of these was Colonel Palmes's squadron, whose spirited fighting routed the Gendarmerie, and allowed Cutts's men to prepare for a second assault. This assault also failed, but other events all but assured the allies of victory. During the course of the assault, the French had set some houses on fire to halt the enemy gunfire. Unfortunately for them, the wind changed and the fire came into the village. Clérambault panicked, calling up the seven battalions in reserve *and* the eleven behind them, which Tallard had wanted to keep on hand to cover a retreat. There were now so many French troops in Blenheim that it was difficult for them to move. More importantly, these troops were isolated from Tallard. As long as they were contained in the village, they were useless.

Marlborough observed the situation and ordered Cutts to desist before he could mount a third assault. With the French left engaged by Eugene's troops and the right contained in the village, Marlborough realized that Tallard was now susceptible to a concentrated attack on the center. At roughly 3:00 P.M., Marlborough reformed his troops into two

lines and prepared to attack Tallard head-on. At this point Marlborough marshalled ninety squadrons and twenty-three battalions, as compared to Tallard's fifty to sixty squadrons and nine battalions. Unlike Eugene's troops, who were attacked while crossing the Nebel, Marlborough's troops were not attacked until they had already crossed the river. Disorganized as they were, Marlborough's troops withstood the attack and were able to regroup. This was due in large part to Eugene's prompt response to Marlborough's request for reinforcements. Although he could scarcely afford to give up any troops, Eugene never hesitated in aiding his commander.

Tallard, on the other hand, enjoyed no such luxury. Because the Elector and Marsin were not coordinated with him, Tallard was unable to rely on them for reinforcements. To make matters worse, the troops he had set aside at Blenheim to cover a retreat were trapped in the village, due to Clérambault's panic. Consequently, when Tallard requested reinforcements from Oberglau and Blenheim he received none. In the end, Marlborough crushed Tallard. Marsin and the Elector, hearing of Tallard's fate, retreated westward. While Eugene pressed the retreating Franco-Bavarian forces, Marlborough joined Cutts's men and surrounded Blenheim. Inside, the French were leaderless, for Clérambault had disappeared. He was later found to have drowned in the Danube, but whether his death was a suicide or further evidence of his panic has never been determined.

At 7:00 P.M. the allies advanced on the village. Blansac, the maréchal de camp, was now in command inside the village. He wrote several times to Tallard asking for permission to withdraw, but received no reply. This was because Tallard himself had been captured by Marlborough's men. Finally, Blansac surrendered and the worst defeat in French military history came to an end.

Had Marlborough not defeated Tallard at Blenheim, the French would certainly have taken Vienna and destroyed the Habsburg Empire. Most of Europe would have fallen under French rule. Instead, the French were forced out of range of Vienna and began losing battle after battle. Hilaire Belloc writes, "Blenheim must always be remembered in history as the great defeat from which dates the retreat of the military power of the French in that epoch, and the gradual beating back of Louis XIV's forces to the frontiers which may be regarded as the natural boundaries of France." France's military losses, moreover, turned into British military gains.

After the events of August 13, 1704, serenity and anonymity once again descended upon Blenheim. The roar of heavy artillery was replaced by the gurgling sounds of the Danube. Although the fields were ravaged by the battle, the Bavarian farmers living at Blenheim were able to repair the damage over the years. The nearly 300 intervening years between then and now, moreover, have passed rather uneventfully. The battle is commemorated by a memorial stone placed on the battlefield on August 13, 1954, the 250th anniversary of the action. In the nearby town of Höchstädt, a museum features a huge diorama portraying the battle. Marlborough, upon returning to England, built what may be the most famous "monument" to the battle—Blenheim Palace, in the town of Woodstock. The palace is open to visitors, who can view tapestries that tell the story of the battle of Blenheim. The palace also is the birthplace of Marlborough's famous descendant, Winston Churchill.

Further Reading: *The Battle of Blenheim* by Peter Verney (London: Batsford, and New York: Hippocrene, 1976) is the most objective and historically informative account of the battle of Blenheim. In addition, the book explains the political background for the War of the Spanish Succession, the allied effort leading up to the battle, and the aftermath of the battle. *Six British Battles* by Hilaire Belloc (Bristol: Arrowsmith, 1931) contains a chapter on the battle of Blenheim that is less academic and more readable than Verney's book. Belloc's pro-British bias does not hinder his account of the battle, and his commentary helps explain the course of the battle. *A Military History of the Western World, Volume Two* by Major General J.F.C. Fuller, three volumes (New York: Funk and Wagnalls, and London: Eyre and Spottiswode, 1954–56) contains a chapter on Blenheim that combines the historical detail of Verney's account of the battle (although to a lesser degree) with Belloc's narrative skill.

—Lawrence F. Goodman

Bordeaux (Gironde, France)

Location: In the Aquitaine region in southwest France, situated on the banks of the Garonne River. The Bordeaux wine region includes all the vineyards within the administrative *département* of the Gironde.

Description: Site of important events in almost every period of French history; famous for its wines and its grand architecture, with many notable buildings dating from the eighteenth century and concentrated on the left bank of the Garonne.

Site Office: Office de Tourisme, Syndicat d'Initiative
12 cours de 30 Juillet
33080 Bordeaux, Gironde
France
56 44 28 41

A Gaulish Celtic tribe, the Bituriges Vivisques, were the first known settlers on the left bank of the Garonne River, next to the Devèze tributary, around the third century B.C. The settlement grew at an important point on the Celtic tin route from the north, connecting the Atlantic with the Mediterranean. In return, wine exports from Italy flowed northward from the end of the second century B.C. Bordeaux would have to wait until the Roman occupation to begin cultivating the vine; before the beginning of the Christian era, the area was famed for its wheat.

The Bituriges Vivisques acquiesed peaceably to the Roman conquest by Crassus in 56 B.C., being concerned primarily with free trade. Known to the Romans as Burdigala, the town became part of the province of Aquitania. Roman Bordeaux was established along two axes, the *cardo* and the *decumanus*, now the rue Ste. Catherine and the rue Porte-Dijeaux. Provincial Romans were soon allowed by imperial decree to start planting their own vines, around A.D. 50. After some experimentation, the basilica variety was cultivated successfully, and the Bordeaux wine trade was born. The timing was fortunate, as the northern sources of tin had been depleted.

Burdigala was to become capital of Roman Aquitaine, certainly by the time of the Aquitainian governor and short-lived Roman Gaulish emperor, Tetricus, in the 270s. He succeeded the Roman Gaulish emperor Victorinus, backed by the Roman troops on the Rhine, and by Victorinus's mother, to whom he was related. His investiture took place in Bordeaux. The Roman emperor Aurelian brought this division of the empire to an end in 274.

Gallo-Roman Burdigala was one of the liveliest centers of commerce in Gaul, attracting many foreigners, including Greeks and Celts. Roman traditions fused with Celtic ones, while Roman monumental buildings were constructed, including the Palais Galien, thought to have been an amphitheatre, and the Pillars of Tutelus, possibly part of a forum.

Threatened by Germanic invasions, the town was enclosed toward the end of the third century behind walls forming an almost perfect rectangle. Within, ships docked in the protected port. Burdigala became the metropolis of the second Aquitaine after the province was split in two by Diocletian's reorganizations.

The Roman university came into prominence at roughly the same time. From 335 it was attracting scholars from across the Roman world. Ausone, famed for the insights his writings and poetry gave into fourth-century Aquitaine, was a university rhetor. He would become tutor to Gratien at the imperial court at Trier. From 333 pilgrims traveled from Bordeaux to Jerusalem, and from some time in the fourth century Bordeaux had a cathedral church.

The first northern invaders had attacked Bordeaux in 276. Waves followed in subsequent centuries. In 406 much of Bordeaux was burned. Vandals passed through in 408, shortly before the Visigoths in the 410s. When the Visigoth kingdom became an independent state under Euric, the court often stayed in Bordeaux.

In 498 the Franks arrived, occupying the town briefly. Clovis established a permanent Frankish force there after his victory at Poitiers in 507. At the close of the century Bordeaux came under increasing threat of attack from the older Gascon tribes. By the end of the century power was, in reality, in the hands of the Gascon dukes.

The Moslem invasion from north Africa reached Aquitaine in 732. Charles Martel defeated the Berber cavalry between Tours and Poitiers, and Bordeaux opened its gates to him. In 768 Pépin the Short won the town from the Visigoths, and in 769 Charlemagne had the Château of Fronsac, to the east of Bordeaux, built to protect the limits of Carolingian territory.

A series of counts with the title Duke of Gascony ruled Bordeaux in the ensuing period. Then, in 844, Vikings attacked. In 848, Bordeaux was ransacked. Gregory of Tours had described Bordeaux as a place dotted with churches and monasteries in the seventh century, including the basilica of St. Seurin and the abbey of St. Croix; but all was swept away by the Viking invasion.

Little is known of Bordeaux's history from the Viking attacks until the end of the tenth century, when it briefly became the principal town of the grand duchy of Gascony under the Sanche family. The last of the family died without heir in 1032, leaving the Gascon succession open. By 1058, the count of Poitiers had gained control of Aquitaine, and Bordeaux was given the role of capital of Aquitaine and

St. André Cathedral
Photo courtesy of French Government Tourist Office

Gascony, despite Poitiers's older standing. The new ruling family frequently stayed in Bordeaux, but stationed officers there to represent it permanently.

The house of Poitiers maintained firm control over the Bordeaux church. In 1089–90 Duke William IX granted his son special privileges for the building of St. André, to be-

come Bordeaux's cathedral. In 1096 William IX received Pope Urban II in Bordeaux. Conflict between Aquitaine and Rome caused division, though, when William X appointed a new archbishop who supported Anaclet, rival to Pope Innocent II, at the double papal election of 1130. As penance, William X undertook a pilgrimage to Santiago de Compostela. He died on the journey in 1137. His young heiress was Eleanor of Aquitaine.

The French king Louis VI organized an extravagant retinue to accompany him to Bordeaux for the marriage that same year of Eleanor to his son, the future Louis VII. The town saw the king imposing his officers in place of local ones; the city's revolt of 1147–48 was apparently in protest of this intervention. But the connection of Bordeaux and Aquitaine to the French throne was curtailed. In 1151–52 Eleanor and Louis VII returned to Bordeaux, and their marriage was dissolved.

In 1152, Eleanor was to find a new husband, Henry Plantagenet. Their marriage would profoundly affect the course of Bordeaux's history, of Anglo-French relations, and of European history. Henry Plantagenet was count of Anjou and duke of Normandy, and in 1154, he acceded to the English throne as Henry II. The union between England and Aquitaine, with Bordeaux as its capital, was to last three centuries. Richard the Lion-Hearted, son of Eleanor and Henry, twice held court at Bordeaux, in 1174 and 1176. When Richard's brother John became king of England, the French king Philip Augustus declared the confiscation of John's continental possessions in 1202. In 1204 the French army laid siege to Bordeaux. Two years later, Alfonse VIII of Castile, Eleanor's grandson, besieged the city as a pretender to Aquitaine. Both sieges were unsuccessful.

The end of the twelfth century saw a surge in Bordeaux's population, and in 1227 a second town wall was completed to include the St. Eloi district. The St. Eloi gate became the seat of the *Jurade,* the body of town councillors. The clergy, which had dominated the city's life until the twelfth century, saw its political influence diluted.

From the start of the thirteenth century, wine exports to England and northern Europe began to dominate Bordeaux's trade to an extraordinary degree. With the French king Louis VIII's conquest of La Rochelle, Bordeaux remained as the Plantagenets' sole great continental Atlantic port. In 1214, the Bordeaux bourgeois were exempted from paying duty on wine from their own vineyards. Then they won the right to control the time of release of wines from farther inland.

The mid-thirteenth century saw popular uprisings within English Aquitaine. Henry III had to cross from England in 1242, and then in 1253–54, to deal with the unrest. On the second occasion he brought his son, Prince Edward, to help appease the discontent caused by the brutal government of the king's deputy, Simon de Montfort. Edward, who became King Edward I in 1272, tried to improve relations, but conflict between rulers and the merchant class continued. The thirteenth century also witnessed the organization of the town's parishes and the establishment of large religious orders. Five mendicant orders built important centers from around 1228 to 1287.

In 1305, the archbishop of Bordeaux, Bertrand de Got, was elected Pope Clement V. This position brought the Gascon church great privileges. As pope, Clement V so-

journed three times in Bordeaux before removing the papal seat from Rome to Avignon in 1309. He was unabashed about showering his family and Gascons with ecclesiastical posts, and St. André benefited from special subsidies.

The size of the wine trade by this time was enormous by medieval standards, bringing great prosperity. From 1305 to 1309 annual exports of wine from Bordeaux (principally to England) have been estimated between 22 million and 24 million gallons, comparable to the amount being exported in the best years of the nineteenth century. The city expanded extraordinarily, a third wall being completed in 1327. Wine storehouses began to be built on the river's edge, and the town was by now surrounded by ecclesiastical vineyards. Most of the land around the southern suburbs was cleared at the start of the fourteenth century, and the new vineyards were the favored form of investment for the bourgeois. Virtually all the production was of red wine, or claret.

Further local conflict over rulership of the city led to the appointment of a series of English mayors after 1325. Conflict with France was to intensify with the start of the Hundred Years War in 1337, and, together with outbreaks of plague, would reduce the wine trade quite substantially. The town itself was rarely under threat, but it was the important center of English Aquitaine. After the battle of Maupertuis in 1356, Edward, the victorious Black Prince (son of Edward III), captured King John II of France, and brought him as prisoner to Bordeaux for a time, ending the first stage of the great war.

From 1362 Bordeaux became the capital of the short-lived Principality of Aquitaine, granted by Edward III to the Black Prince, with Aquitaine's administration no longer answerable to England. The constable became the treasurer of Aquitaine, and the prince had his own special coinage struck. His military campaigns were expensive, though, and led to increased taxes. The Bordeaux townspeople did not protest, but two important vassals, the count of Armagnac and the lord d'Albret, appealed to the French king for help, and the war was rekindled. The prince renounced his principality in 1372.

The beginnings of English king Richard II's and French king Charles VI's reigns brought a period of peace toward the end of the century. But by 1400, the troops of Louis I, duc d'Orleans, were closing in on Bordeaux, and Henry IV, who had succeeded to the English throne the previous year, was too involved with events in England to assist. However, the *Jurade*'s forces managed to destroy the French fleet at St. Julien du Médoc. After 1420, Bordeaux's urban militia made successful forays into the country, reconquering much land. By 1438, the French war front had moved into Gascon territory, though. Charles d'Albret attacked Bordeaux, and while the town resisted, the suburbs were sacked.

The final French offensive began in autumn 1450. The Bordeaux mayor Shartoise attempted a counterattack, but the Bordelais were massacred in a disastrous day known in Bordeaux as the *male jornade,* which translates roughly as "painful day." In 1451 French armies beleaguered and won the town. The townspeople were granted favorable conditions by the victors, but a group still resisted, led by Captal de Buch, who called on the English to rescue the town. English general John Talbot entered Bordeaux in 1452, but Charles VII's troops encircled the area. Talbot advanced a little east, and clashed with the French at the battle of Castillon, where he was killed. This date in 1453 marked the end of the Hundred Years War and of English rule, with Aquitaine and Bordeaux falling under French control.

During this period of anguish, Bordeaux's Archbishop Pey Berland was an outstanding figure. Surrounded by warfare, he fought for the founding of the medieval university, which Pope Eugene IV granted in 1441. Pey Berland's spiritual guidance also gave him a saintly reputation on his death in 1458.

The French kings ordered massive new defenses for Bordeaux—the Château Trompette and the Fort du Hâ. In 1462 Bordeaux was granted its own *parlement*. Charles VII imposed new export taxes, and while English ships were not barred from the port, trade diminished. In the 1460s Louis XI reduced the export tax, declaring that "without its communication and trade with the English kingdom, Bordeaux cannot be Bordeaux." The French conquest reopened trading links with the interior, while another boom for Bordeaux came from the export of Toulouse and Languedoc pastel for dyeing.

Bordeaux's activity expanded greatly along the river. The town's nickname of the *port de la lune* (port of the moon) probably originated from the name of the Château de la Lune given to the royal fortress, but would come to refer to the grand crescent sweep on the left bank of the Garonne.

In some years of the sixteenth century, more boats exported pastel than wine from Bordeaux. Competition from New World indigo and disruption caused by the French Wars of Religion spoiled the trade. The wine trade continued relatively steadily through the century, with England remaining a reasonably strong partner until the wines of Portugal became the object of English attentions at the century's close.

Sixteenth-century Bordeaux, strongly interested in spiritual matters, was greatly encouraged by the founding of the Collège de Guyenne in 1533. (Aquitaine was also known as Guyenne or Guienne.) The college taught in the best Renaissance humanist tradition of Desiderius Erasmus. Michel de Montaigne was the school's most famous pupil. He resigned in 1570 from his administrative duties at Bordeaux's *parlement* to dedicate himself to his writings. The extraordinary *Essais,* analyzing his times and his character, and from which the genre of the essay derived, were first published in Bordeaux in 1580.

Famous scholars from throughout Europe taught at the college, early disseminating evangelical, even Lutheran ideas. But the pyres were burning early, from the 1540s, for some Protestants. The *parlement* pressed home Bordeaux's role as a Catholic bastion. In 1572, the Bordeaux Protestants were subjected to the horror of a St. Bartholomew's Day

massacre in their town. After that, they lay low, unlike the feuding parties in the countryside.

Conflict with the centralizing state increased, and Bordeaux townsmen participated in the popular revolts. The town was quite slow to join in the 1548 revolt against the *gabelle,* or salt tax, until it lost its privileges. The king's army came in force to discipline the town. Through the Wars of Religion, the Bordelais supported the king of the moment. Louis XIII was married with great pomp in Bordeaux in 1616. Whenever their rights were threatened, however, the Bordelais reacted. This occurred in 1635, when an attempt to put a tax on innkeepers' barrels of wine sparked a riot. The tax was suspended.

The *Fronde,* a nationwide uprising in 1648 against centralizing power, saw the town turn against the governor of Aquitaine, the duc d'Epernon. The Château Trompette yielded to the townspeople. Royal troops surrounded Bordeaux, which was granted an amnesty by Cardinal Mazarin in December 1649. In each of the two subsequent years violence flared up again, first in support of the rebellious royal princes, second, from autumn 1651, with the Ormistes' unrest. This royalist group played on the mystical bond between the monarch and the people, which it said was being broken by the intervention of the royal powerbroker Mazarin. The Ormistes supported monarchy, but not the monarch's appointed bureaucrats. The Ormistes also encouraged the townspeople to turn against the Bordeaux *parlement;* peace and the *parlement* were only restored in 1653–54. The last riot for more than a century came in 1675, in protest of the *papier timbré,* tobacco and pewter taxes. Once again royal forces came in great number. Another royal fortress, Fort Louis, was erected.

The church in seventeenth-century Bordeaux became aggressively reforming, with the powerful cardinal François de Sourdis—an extravagant patron of the arts as evidenced in Bordeaux's St. Bruno church and sculptures by the Berninis—encouraging the way for his successors.

Through the seventeenth century England and Brittany remained Bordeaux's traditional trading partners. A major trading change at the end of the sixteenth century, however, had been the arrival of ships from the United Provinces (now the Netherlands). By 1700–01, they would take more than two-thirds of Bordeaux's wine exports. In the seventeenth century, Aquitaine's plums, too, were bought in quantity by the Dutch to help their sailors combat scurvy on their trips to the Americas.

The New World would become an extremely important trading partner for Bordeaux. The new mercantile explosion came with the establishment of the sugar industry after part of Santo Domingo came under French rule in 1697. Bordeaux would become the largest port in France in the eighteenth century, and the gateway to Europe, re-exporting the goods arriving from the West Indies, while products from its own region, such as wines, flour, plums, and heavy equipment were shipped to the expanding Indies. From 1770 to 1774, an average of 225 ships per year left Bordeaux for the

Indies. From 1786 to 1789 the average was 265 ships, 23 of these being for slaves. Vast numbers of African slaves were shipped to the Indies in this triangular trade. Much of Bordeaux's wealth and glorious golden architecture was built on income from the slave trade.

The architectural splendor of Bordeaux dates primarily from the eighteenth century. The new royally appointed governors of the region, the *intendants,* wished to enhance the town's name. The *jurats* for almost a century opposed the expense of embellishment, and the dangers of opening up the city walls, feeling the town could rest on its wines' glory. The breakthrough occurred in 1729 when the king's architect, Jacques Gabriel, realizing the magnificent potential of the site, wrote of it to the king. Progress was slow, as Louis XV's typical method was to make the inhabitants pay subscriptions on a series of forced loans, but by then Bordeaux could easily afford the transformation. Under the *intendant* Boucher, the Place Royale came into existence. In the center stood an equestrian statue of Louis XV. The style followed that of Versailles.

The *intendant* Louis-Urbain Aubert, marquis de Tourny, succeeded Boucher and continued construction in the city. New tree-lined allées circled the old town; squares were planted on the medieval edge to open up trade and fine vistas; ornamental gateways ennobled the main entrances into the old town; and a grand public garden was created, although reserved for the elite. Travelers to Bordeaux considered it one of the most magnificent cities in Europe, François de la Rochefoucauld referring to a luxury of buildings as widespread as in Paris.

In the century of the Enlightenment, the Académie royale de Bordeaux was founded in 1713 and became the center of the town's intellectual life. The most influential Bordelais thinker was Montesquieu, a political philosopher whose major work *L' Esprit des lois,* analyzing different forms of government, published in 1748, greatly influenced political thought in Europe and America.

One of the grandest pieces of Bordeaux's neoclassical architecture, the Grand Théâtre, was designed by Victor Louis in the 1770s for the Marshal de Richelieu. Another monumental building erected in great part for personal glory was the Rohan Palace, ordered by the archbishop Mériadeck de Rohan. On the outskirts, from the 1760s on, local craftsmen built rectangular stone bungalows, called *echoppes,* which typify outlying Bordeaux. As one historian puts it: "The Bordelais of the eighteenth century lived in the midst of a perpetual building site."

The French Revolution would damage Bordeaux's trade greatly. Among the Bordeaux professional class, there was wide support for moderate, federalist republicanism, a movement that gained the name of the Girondins because of the number of supporters from the Bordeaux area. In the Terror of 1793, the Girondins were hounded by the radical Jacobins. That summer many of these so-called slaves of federalism were rounded up. Of 200 arrested, most were treated relatively leniently, while 18 were condemned to death. In fact the Terror, administered in Bordeaux by a former teacher from the Toulouse area, by the name of Lacombe, appeared fairly indiscriminate in the people it condemned. Many merchants had retreated to the countryside for safety.

Under the new national government known as the Directory, which held power from 1795 to 1799, there was some improvement in trading conditions, a number of American merchants establishing themselves in Bordeaux. But Napoléon's belligerence led to a damaging Continental blockade of France.

Bordeaux stagnated in the nineteenth century. At least a bridge, the Pont de Pierre, was finally built in 1822 to span the Garonne. The forts of the Château Trompette, the Fort Louis, and the Fort du Hâ, symbols of past royal military control, were dismantled by 1835. Much of medieval Bordeaux was condemned. The writer Victor Hugo loved the town. "Take Versailles and mix it with Antwerp and you have Bordeaux," was his description of it before the destruction of the medieval, which he lamented: "you are tearing up, one after the other, all the pages of your old book, only to keep the last page. You are chasing out of your town and wiping out of your history Charles VII, the kings of England, the dukes of Guyenne, Clovis, Gallien and Augustus, and in their place you are erecting a statue of M. de Tourny. You are knocking down something quite magnificent to put up something quite minor." In place of the Château Trompette, the vast esplanade of the Place des Quinconces was developed through the nineteenth century, with its mass of statues to civic pride, culminating at the end of the century with the Monument aux Girondins.

At least the lack of entrepreneurial spirit saved Bordeaux from the worst results of the industrial revolution. But its shipping and maritime trades dwindled. The wine trade maintained its position through the first half of the nineteenth century, which saw the continued growth in prestige of individual wine châteaux, or vineyards, as opposed to generic wines blended by merchants. The famed classification of the vineyards of the Médoc (on the spit of land north of the town) was judged in 1855, taking into account the great importance of each vineyard's site, as well as an element of social standing. The Second Empire, from 1852 to 1870, was a period of great prestige and prosperity for the region's wines, with the Americas representing important markets. This boom occurred despite an attack of oidium fungi on the vineyards from 1852 to 1860. Far more serious was the spread of phylloxera after 1875; the insects devastated the vineyards, and from 1880 Bordeaux was importing Iberian and Italian wines. Only the grafting of sturdy unaffected American stock would bring about the recovery of the Bordeaux vineyards.

In 1870 the Franco-Prussian War began. The government moved temporarily to Bordeaux in that year. In the two subsequent great twentieth-century wars with Germany, in 1914 and 1940, it would do the same, and caused Bordeaux to be given the name of the "tragic capital" of France. In

1870 and 1914, the Grand Théâtre became the brief home of the French National Assembly.

Two extremely powerful mayors have dominated the evolution of Bordeaux in the twentieth century. The first was Adrien Marquet, who was elected mayor in 1925, and remained until 1945, encouraging large loan-financed building projects for the town.

By June 1940 France was defeated by Germany. Marshal Pétain announced the armistice on the Bordeaux-Lafayette radio station. General Charles de Gaulle flew off immediately from Bordeaux's Mérignac airport to London to organize the French resistance. Marquet briefly became minister of the interior. Bordeaux was designated as being within the German-occupied France, and the French government moved to Vichy. The Germans built an important submarine base north of Bordeaux. The city was bombarded several times by the Allies, although spared the devastation of many other towns.

About 1,000 Girondins were deported; over half never returned. Some 300 were shot by the Germans, 50 in one hostage camp after the assassination of the German administrator Reimers. The area's resistance suffered from a lack of unity after it was infiltrated by an influential collaborator, known as Grandclément, in 1942; he was shot by the Bordeaux resistance in July 1944. Two generals, de Gaulle and Jacques Chaban-Delmas, came to restore the hierarchy. The last Germans left Bordeaux at the end of August 1944, but only surrendered the port of Verdon to the north in April 1945. Secrecy and silence still surround the degree of collaboration in Bordeaux as elsewhere in France. After the war, Marquet was incarcerated on a charge of crimes against the French nation. In Bordeaux's war trials, 66 people were condemned to death (15 of these sentences were carried out), and 1,303 people were sentenced to terms of "national indignity."

The major personality to emerge in postwar Bordeaux was Chaban-Delmas, the general of the resistance, a national rugby player and French tennis champion, soon mayor of Bordeaux, then holding various ministerial posts, becoming French prime minister from 1969 to 1972. His many positions of power at the municipal, regional, and national levels allowed him to encourage a number of economic developments in Bordeaux and Aquitaine, generally maintaining a cooperative balance between the city, to the right politically, and the region, to the left. The aeronautical industry replaced shipping in importance after the latter's difficult decline. Two new bridges were opened over the massive Garonne in the 1960s. The new Mériadeck quarter brought urban regeneration to the center, while to the north Bordeaux-Lac was developed into a vast exhibition area, now hosting the Bordeaux International Fair, the internationally important book fair (the Salon du Livre), and the massive biennial wine fair (Vinexpo).

Another name to bring honor to the town was the writer François Mauriac, whose works are steeped in the atmosphere of Bordeaux and the nearby Landes pine forest. He was awarded the Nobel Prize in 1952. Cultural events increased after the war, the first Mai de Bordeaux arts festival taking place in 1950, the more controversial avant-garde festival Semaine Sigma launched in 1965. Museums opened in the 1980s include one on regional ethnology and one on World War II resistance. The Cité Mondiale du Vin, a permanent international wine center, was completed in 1991.

Plans have been afoot to build a subway and develop Bordeaux's rather deprived and tatty right bank. The architect Ricardo Bofill has drawn up extravagant plans. It remains to be seen whether the economic climate allows these bold projects to progress and become a worthy counterpart to Bordeaux's glorious left bank.

Further Reading: The great work on Bordeaux's history, *Histoire de Bordeaux* (Bordeaux: Fédération Historique du Sud-Ouest, 1962–74), was compiled in seven volumes under the direction of Charles Higounet. Much more approachable is the one-volume work he also edited, *Histoire de Bordeaux* (Toulouse: Privat, 1980; second edition, 1990). In English, the tourist office sells an illustrated guide with surprisingly learned historical detail, even if the translation is rather quirky and at times inaccurate. It was compiled by curators at the Musée d'Aquitaine, Délie Muller and Jean-Yves Boscher, and is called simply *Bordeaux* (La Brède: Sud-Image, 1993).

—Philippe Barbour

Bosworth Field (Leicestershire, England)

Location: Leicestershire, the Midlands, England; located 97 miles from London; 29 miles from Birmingham; 16 miles from Leicester; and 2 miles from Market Bosworth.

Description: Site of Bosworth Battle, 1485, including the battle trail—1¾ miles long around the battlefield—and King Richard's Well.

Site Office: Bosworth Battlefield Visitor Centre and
    County Park
Sutton Cheney
Market Bosworth, Leicestershire CV13 0AD
England
(455) 290429

On August 22, 1485, a battle took place at Bosworth Field in Leicestershire, England. Traditionally this event has been held to mark the end of the Middle Ages in England, the end of the thirty-year Wars of the Roses, the end of the Plantagenet dynasty of kings and the beginning of the Tudor dynasty.

Certainly it is true that the last Plantagenet died on the battlefield that day, that he was the last English king to die in such a way, and that the first Tudor king took his place on the throne. While the Middle Ages can hardly have ended at Bosworth on one summer morning, any more than they could have begun on October 14, 1066, at Hastings, as is often claimed, the 419 years between 1066 and 1485 are the period generally regarded as the Middle Ages in England. The three medieval Gothic styles of art and architecture—Early English, Decorated and Perpendicular—are broadly contained within the period when England was ruled by the fourteen Plantagenet kings.

The name Plantagenet comes from the Latin *Planta genista*—a sprig or flower of the broom. This emblem of the Angevin kings—the kings of Anjou in France—was used by Henry II, son of Geoffrey Anjou, who became the first Plantagenet king of England. Throughout the period, from 1154 to 1485, these kings of England were intermittently at war with Scotland, Wales, Ireland, and France. They also had to fight to remain on the throne themselves.

The first seven Plantagenets include many famous names: Henry II (1154–89) is well known for his turbulent relationship with Thomas Becket. Richard I (1189–99) was the Lionheart, who spent more time on the Crusades than in England. It was in John's reign (1199–1216) that the Magna Carta was signed. The first "parliament" was held in the reign of Henry III (1216–72) and strengthened in Edward I's time (1272–1307). Edward II (1307–27) was a weak king, succeeded by Edward III (1327–77), the seventh Plantagenet,

in whose reign the Black Death struck, in 1348, wiping out some one-third of England's population. Geoffrey Chaucer, author of the *Canterbury Tales,* was known at Edward III's court, and at his son's. It was Edward III who, having failed to unite Scotland with England, turned his attentions to France and so began the Hundred Years' War between the two countries.

Edward III's son Edward (the Black Prince) preceded him in death, so he was succeeded by his grandson, Richard II (1377–99), the eighth and last of the direct line of Plantagenets. Richard's successor, Henry IV (1399–1413), was a cousin from the Lancastrian branch of the line; Henry, surnamed Bolingbroke, was a son of John of Gaunt, the fourth son of Edward III and Duke of Lancaster. Not only did Henry IV do battle with Scotland, Wales, and France, but he also had to deal with the opposing offshoot of the Plantagenet line, the Yorkists. His son, Henry V (1413–22), was particularly successful in France, regaining much of the French land previously lost, and on his death he left England as the dominant power in France. His son Henry VI (1422–61 and 1470–71) came to the throne as king of both England and France. Unfortunately, he was only nine months old.

The prospect of a nine-month-old king of England and France posed many problems. In England there were always ambitious royal relatives and nobles eager to seize any possible opportunity to gain the throne for themselves—the problem of the "over-mighty subjects." Medieval kings had no regular army or police force to support their authority. In times of trouble a king had to summon the armies belonging to the great lords of the land. The other side of the coin was that these same armies could be turned against the king himself, or against other lords, in the pursuit of power and land.

To prevent any such power struggles developing, the lords of England devised an original solution for the government of England during the king's minority: a protectorate, consisting of a council of leading men of the realm. The device was successful, and the child was crowned King Henry VI of England in November 1429 at the age of eight and King Henry II of France in December 1431. By the age of sixteen he had taken up his duties as a gentle and peace-loving king, who symbolically ended the Hundred Years' War with France by his marriage to Margaret of Anjou in 1445.

In 1453 Margaret gave birth to Henry VI's heir, Edward, but Henry had sunk into his first bout of insanity by then and was unaware of the event. For the second time in his life a power vacuum occurred, reactivating the old blood feud between the Yorkists and the Lancastrians. The result was the series of battles that came to be known as the Wars of the Roses, the first of which was the Battle of St. Albans on May 22, 1455.

King Richard's Well
Photo courtesy of Leicestershire County Council

The name "Wars of the Roses"—not generally used until the eighteenth century—was inspired by the symbol of the white rose used by the Yorkists and the red rose later adopted by the Lancastrians. In essence, the term covers thirty-two years during which the two branches of the Plantagenets struggled against each other to gain the throne of England. The fighting was not continuous, but consisted of a series of short, sharp battles, and generally affected the ordinary people of England very little. It was a political struggle that released old antagonisms between noble families from Cornwall to Wales to Northumbria. Across the country, lords and earls took sides, and sometimes changed sides, under the Yorkist or Lancastrian banners, as a means of avenging old grievances and in a bid to gain power for themselves. In the process an unprecedented number of nobles were killed in battle, or as a result of battles.

The first phase of the Wars of the Roses ended in 1461 with the succession to the throne of Edward IV (1461–83), the first Yorkist king. Edward was the son of the Duke of York, who was the richest landowner in the country and who had been "protector" to Henry VI in his first period of mad-

ness. Henry VI was still alive, albeit in the Tower of London, alternating between sanity and madness, and he was briefly restored to the throne, from October 1470 to March 1471. However, Edward IV won two decisive victories, at Barnet and Tewkesbury, and thus returned to the throne. He quickly had both Henry VI and his own treacherous brother, George, Duke of Clarence, put to death in the Tower; the latter reputedly was drowned in a vat of wine.

Edward IV had married "beneath him," a widow by the name of Elizabeth Woodville. She bore him ten children, including his heir, Edward, born in 1470. King Edward enjoyed the good things in life: food, wine, and plenty of mistresses. Perhaps it was an over-indulgent lifestyle which led to his early death at the age of forty. His son, Edward V (1483), was only twelve years old when his father died.

Unlike brother George, Edward IV's other brother, Richard, Duke of Gloucester, had been totally loyal throughout the reign. Richard's power base was in Yorkshire, where he had gained a great deal of respect both as a soldier who had brought peace and stability to the area, and as a generous man. There is much debate about Richard's motives upon the death of his brother, Edward IV. In his play *Richard III,* William Shakespeare paints a picture of a ruthless, ambitious man, determined to be king at all costs and all too ready to seize the opportunity presented by a defenseless, twelve-year-old heir. Others take the view that Richard genuinely wanted only the role of protector to the boy-king, in order to reduce the influence of the child's mother's scheming and ambitious family, the Woodvilles. Some say that Edward IV had wanted Richard to become his son's protector, but that the royal council opposed this choice; they wanted to see the boy Edward safely crowned immediately.

Whatever the truth, Richard took young Edward into his charge and, as his protector, led him to London, arriving there on May 4, 1483. Already alliances had formed, with Richard and the dukes of Hastings and Buckingham—the three great landowners and Yorkists—opposing the Wood-villes. On the arrival of the Yorkists in London, Edward V was taken to the Tower—for his protection—to await his coronation, and Richard was made "protector" by the council; this title was to last only until Edward V's coronation.

The sequence of events from May to July 1483 varies from account to account, largely depending on the writer's attitude toward Richard. Plans for Edward V's coronation at first proceeded, but his mother was in sanctuary, having taken with her her younger son, Richard, and daughter, Elizabeth, and the late king's treasure. Tension built up between the council and Richard, and on Friday, June 13, Richard suddenly accused Hastings and other members of the council of treason. Hastings was dragged out of the Tower, where the meeting was being held, and beheaded. Many others were imprisoned, and Edward V was joined in the Tower by his younger brother, Richard, duke of York.

What treason Hastings had committed is not clear, but his summary execution had an unsettling effect on those who saw or heard about it. With a powerful Yorkist army to back him up, Richard had more "conspirators" put to death, and by the end of June charges of illegitimacy were made against Edward V and his brother. A petition asking Richard to become king was delivered on June 26; unsurprisingly, he acquiesced, and was crowned Richard III on July 6, 1483.

Perhaps Richard could have become a strong and successful king, despite the blood that had been let in the course of his usurpation; there are many who feel he would have made an excellent monarch. But there were two young princes in the Tower, and as long as they remained there, alive, Richard knew he would never be secure on the throne.

No one knows exactly how or when the princes died, or who killed them or issued the order for them to be killed. Although Sir James Tyrell is said to have confessed to having murdered them, others have pointed the finger at the Duke of Buckingham. But as doubt turned to certainty concerning their deaths, it was Richard who was held responsible, an accusation he never denied. No bodies were ever displayed, and no funerals took place, but in 1674 two small skeletons were discovered in the Tower.

After his coronation, Richard III spent the summer of 1483 buying support, making generous gifts of lands and privileges, giving pardons and remissions of taxes. By September, however, the south and southeast of England, including London, were rife with rumors about the young princes' presumed fate; when Richard's right-hand man, the Duke of Buckingham, publicly turned against Richard and gave his support to the Lancastrian pretender, Henry Tudor, instead of to the princes, it was finally assumed that they were, indeed, dead. Richard's disaffected subjects then looked to Henry Tudor to rid them of the "child murderer."

It was Henry Tudor's mother, Lady Margaret Beaufort, who was the direct Lancastrian heir to the throne of England, through John of Gaunt; she accepted that the country was not ready for a female monarch, however, and did all she could to promote her son. The Tudors had been a lowly Welsh family until Henry V's wife, Catherine de Valois, married Owen Tudor after the death of the king. This marriage had been something of a scandal at the time, but Henry VI had acknowledged the three sons and one daughter produced by his mother and Owen Tudor as his half brothers and half-sister, and had given them titles. To Edmund Tudor, the firstborn, he gave the title of Earl of Richmond. After the death of Edmund, his wife Lady Margaret Beaufort married for a second time, this time to Lord Stanley. So it was that Lord Stanley became stepfather to Henry Tudor, Earl of Richmond, a fact which did much to determine the outcome of the Battle of Bosworth.

Henry Tudor himself was relatively unknown until this time, having lived in exile (as a Lancastrian) all his life, first in Wales, then in Brittany. But in 1483, as the new hope for Richard's enemies, he twice set sail for England as Buckingham led an uprising. The rebellion failed, and Buckingham and others were beheaded. Those conspirators who could do so escaped to Britanny to join Henry Tudor. At Rennes on December 25, 1483, he took a vow to marry Elizabeth of York, sister of the two dead princes, so uniting the houses of York and Lancaster and bringing peace to the land.

Throughout 1484 Richard worked hard to establish himself as a just and worthy king, with some success; but in April of that year his son and heir died, and in March 1485 his wife, Queen Anne, died too. Richard had Elizabeth of York at court, and a rumor soon spread that he had poisoned his wife in order to marry his niece, and so thwart Henry Tudor's plans. Richard denied this change, but his support in England was diminishing, while in France Henry Tudor had gained the blessing of, and financial support from, the king of France. Conflict appeared inevitable.

In early June, Richard left London to take up residence at Nottingham Castle. The Midlands were the least stable part of his realm, and it was a good central base from which to await the threatened invasion by Henry Tudor. On August 7 the pretender landed at Milford Haven in Wales, the land of his father's family. By August 11 Richard had heard the news and sent urgent messages to loyal (he hoped) lords and earls, demanding their immediate presence, with their armies. Lord Strange, the son of Lord Stanley, was taken hostage by Richard in the hope that this would prevent Lord Stanley and his brother, Sir William, from supporting Henry Tudor.

Henry had arrived with some 2,500 men, composed of supporters plus Scottish and French mercenaries. He moved slowly from the Welsh coast, collecting fewer extra troops than he had hoped. Richard moved from Nottingham Castle to Leicester on August 20, and by August 21 he had amassed an army of 10,000 to 15,000 men. They set out from Leicester in style, camping that night near Sutton Cheney, north of the battlefield. Henry's army, which had grown to about 5,000 men, camped three miles south of Richard's, with Ambion Hill between them.

Richard had planned to place his men on top of Ambion Hill, from where they would look impressively menacing. At the foot of the hill was marshy land, which would present Henry's army with difficult terrain from which to mount an uphill challenge. In the event, it seems that Henry Tudor began to advance sooner than the king had expected, and Richard's army was rushed up the hill. The story goes that Richard missed both Mass and his breakfast on that morning of August 22, making do with a drink of water from the well now named after him. The king, the loyal Norfolk, and the Duke of Northumberland commanded the royal troops. Between the two armies the Stanley troops of 5,000 to 6,000 men remained in two groups. Both sides were expecting their support, but neither felt sure of it.

The Earl of Oxford led the Tudor troops against Norfolk's royal army. The first phase of the battle lasted about one hour, during which Henry's men held their own. At this point Richard, surveying the scene, saw the Stanley troops still immobile, and behind him the Duke of Northumberland's men hung back. Richard must have decided it was time for him to take a decisive lead, so, in true warrior-king fashion, he led a cavalry charge down the hill, straight toward his enemy, Henry Tudor.

Unfortunately for Richard, it seems likely that his horse got stuck in the mud at the foot of the hill, on the plain. He fought valiantly, but all around him those who had charged with him were being cut down, and it was at this moment that the Stanleys decided at last to choose their side. It was not the king's side they came in on, and Richard was hacked down—probably by Stanley's Welsh soldiers—crying out against the traitors who had deserted him. His body was stripped and mutilated, tied to a horse and dragged to Leicester, where it was put on display for two days. The battle was over within two hours, and the victorious army moved south to Crown Hill; there Henry Tudor was proclaimed King Henry VII, and the coronet Richard had been wearing in the battle was placed on the new king's head. He was crowned formally on October 30, 1485, and married Elizabeth of York on January 18, 1486. The Tudor dynasty had begun.

In fact Bosworth was not the last battle in the Wars of the Roses. That took place at Stoke two years later, and was the Yorkists' last attempt to oust the Lancastrian. The battle was bitter, lasting longer than the battle at Bosworth, but it ended the Wars of the Roses forever, leaving the Tudors more securely on the throne. The thirty-two years of wars exhausted the battling aristocracy, leaving many lords and their heirs dead and their funds severely depleted by the expense of war and loss of lands. The Black Death of 1348 had already triggered the breakup of the feudal system, with the labor shortage increasing the survivors' wages and the demand for workers freeing the serfs from their ties to the land of one lord. These two factors were pivotal in leading to the growth of the new merchant class, which by Henry Tudor's time was already developing into a new gentry. Henry VII was to restrict the power of the remaining aristocracy in favor of the emerging gentry, so ridding the monarchy of the scourge of the Plantagenet kings, the over-mighty subject. In this way Henry VII prepared the ground for England's future expansion and prosperity during Tudor times.

The battle site is now a public park administered by Leicestershire County. A self-guided trail around the battlefield is open year-round. There are periodic re-enactments of the Battle of Bosworth Field and other historical events. At the Visitor Centre, open April 1 through October 31, there is an exhibition that re-creates medieval times, and a theater that shows excerpts from a film of Shakespeare's *Richard III*.

Further Reading: *The Battle of Bosworth* by Michael Bennett (Stroud, England: Alan Sutton, and New York: St. Martin's Press, 1985; corrected paperback edition, 1993) was originally published for the battle's quincentenary in 1985. It is now the standard work on the battle itself and the events leading up to it. *The Field of Redemore: the Battle of Bosworth 1485* by Peter J. Foss (Leeds: Rosalba, 1990) offers a "definitive" reconstruction of the battle site and action. *Bosworth Field and the Wars of the Roses* by A. L. Rowse (New York: Doubleday, and London: Macmillan, 1966) still provides a very readable and reasonably accurate account of the time from Richard II to the Tudors. *The Chronicles of the Wars of the Roses*, edited by Elizabeth Hallam (London: Weidenfeld and Nicolson, 1988) is the third in a series of three books that cover the Plantagenet monarchs. (*The Plantagenet Chronicles* and *Chronicles of the Age of Chivalry* are the first two, both also edited by Hallam). Heavily illustrated, these books provide a very accessible way of reading extracts from the primary sources of contemporary chronicles, supplemented by short but informative articles written by historians, and including a good bibliography. *The Making of the Tudor Dynasty* by R. A. Griffiths and R. S. Thomas (Gloucester: Sutton, 1985) is a companion volume to Bennett's book above. *Richard II; Henry IV, Part 1* and *Part 2; Henry V; Henry VI, Parts 1, 2,* and *3; Richard III,* all by William Shakespeare, are available in a variety of editions. Although Shakespeare's famous history plays are based on historical fact, as far as it was known, they were written in the 1590s, a century after the Battle of Bosworth, and being dramas for the stage, Shakespeare allowed himself a certain amount of artistic license. There are a great many books on Richard III, and the Richard III Society publishes a journal, *The Ricardian*.

—Beth F. Wood

Bruges (West Flanders, Belgium)

Location: In the northwest corner of Belgium, approximately six miles south of the North Sea, sixty miles northwest of Brussels, and fifty miles west of Antwerp; capital of the province of West Flanders.

Description: Named for the bridges spanning its old canals; one of most prosperous cities in the Middle Ages, owing to its location on the Zwin estuary and its thriving cloth industry; much of its medieval appearance has endured to this day.

Site Office: Dienst voor toerisme
Burg 11
B-8000 Bruges, West Flanders
Belgium
(50) 44 86 00

Although mentioned in chronicles as having existed as early as the seventh century, Bruges's development really began in the year 865, when Baldwin of the Iron Arm, the first count of Flanders, built a fortress at the head of the gulf of the Zwin on the site of an old Norwegian port. Northern Europe was under constant threat from Norse robber bands, and many people sought the protection and rule of feudal lords rather than risk murder and rape at the hands of marauders. A small settlement grew in the shadow of the fortress, made up at first of a church, a prison, and an inn.

In the middle of the twelfth century a road was built from the settlement, now called Bruges, east to Cologne. A canal to Ghent was built in the late part of the century, and trade to the south as well as the east was opened. Linked to the North Sea and thus to England by the Zwin, the settlement grew so rapidly that by the thirteenth century Bruges had become one of the most important trading centers in Europe.

Shiploads of wool from England arrived daily in Bruges, where the wool was woven into cloth and lace that were among the finest in the world. Boatloads and cartloads of Bruges cloth were sent to the rest of Europe and beyond. The medieval world bought all the cloth from Flanders that it could get, and large numbers of weavers from the Flemish countryside came to Bruges to work. After them came the other skilled workers who were necessary to the growth of a new urban market. These workers formed associations called guilds, which were the forerunners of modern trade unions.

The guilds resolved to take political action to ensure their private liberties as well as the success of their members. Though Bruges was owned by nobles and merchants, guild members expanded the walls around the city first built around the small settlement by Baldwin. The fortified town was a "borough," and its inhabitants were called "burghers."

The nobility granted autonomy to the borough of Bruges by signing into law the *Charte de commune,* which contained freedoms and rights that were unmatched anywhere else in Europe, except in the other Flemish cloth-making cities of Ghent and Ypres. Anyone able to get inside the walls and live within the borough for a year and a day could become a member of the commune. The only obligations owed to the lord by burghers of the commune were loyalty and service in his army.

The combination of free trade and powerful labor made Bruges rich, and nobles took notice. Sturdy and independent, the burghers of Bruges in 1298 roused the interest of Guy of Dampierre, count of Flanders, whose attempts to control Bruges (as well as Ghent and Ypres) were unsuccessful. But the count did succeed in giving the king of France, Philip IV, the Fair, an excuse for meddling in Flanders. Philip pushed Dampierre aside and annexed the province in 1301.

The upper classes in Flanders generally identified themselves with the French and were called Leliarts, in honor of the French fleur-de-lis. The common people spoke Flemish and called themselves Clauwerts, or Men of the Lion's Claw, in reference to the lion on the flag of Flanders. Power struggles under these banners marked most of the fourteenth century in Flanders.

In the early 1300s, the Bruges guilds were united; they represented thousands of workers. The largest and most powerful guild was the Weavers Guild, headed by Pieter de Coninck. Known for being anti-France by both the citizens of Bruges, and the French themselves, de Coninck was asked to leave the city, which he did.

When King Philip sent his emissary, Jacques de Chatillon (soon to be the ruling governor of Flanders), to Bruges, the townspeople admitted him inside the walls, taking him at his word that only 300 soldiers had accompanied him there. But de Chatillon lied. With him were 2,000 French knights, whose general demeanor convinced the people that trouble was ahead for them. It was rumored that the knights were going to massacre Clauwerts—men, women, and children—and the whole city of Bruges was terrified.

In the meantime, de Coninck was not idle while outside the walls, but spent his time gathering citizens from nearby Damme for an assault on the Frenchmen who held Bruges. Most nights, de Chatillon and his men feasted and celebrated into the late hours. At dawn on Friday, May 18, 1302, de Coninck and Jan Breydel, leader of the Guild of Butchers, brought a small army through the gates. The townspeople within had been waiting for their arrival. They armed themselves, joined with de Coninck's army, and took to the streets. Anyone who could not answer the Clauwert challenge "Schild en Vriend" (shield and friend) in appro-

Market Square in Bruges

Another view of the city
Photos courtesy of Belgian National Tourist Office

priate Flemish, or pronounce the slogan in the proper accent, was slaughtered. The Flemings massacred the drowsy French, in what came to be called the "Bruges Matins."

King Philip was livid. Several weeks later an army of 40,000 knights arrived in Flanders to take revenge. They were met by men from every part of Flanders, and on July 11, Flemish workers met French knights on a field near Kortrijk. Though the French had superior numbers and skill in warfare, the Flemings knew the terrain. They made traps for the French by covering marshland with brushwood. Many French knights and their horses drowned, were injured, or became fixed targets for the arrows of the Clauwerts. Flanders won a complete victory. All that was recovered from the French were the little gold spurs they wore on their boots, which were hung the next day from the ceiling of the basilica in Bruges. The Battle of the Golden Spurs marked the greatest victory of Flanders over France.

In 1340 there was yet another battle between Flanders and France. When King Edward III of England declared himself the king of France in 1337 (which marked the first year of the Hundred Years War), Philip of Valois (who also declared himself king of France) persuaded the count of Flanders, Louis I de Nevers, to arrest Englishmen in Flanders. Edward retalitated by arresting Flemings in England. The Flemings knew that good relations had to be maintained with England or their source of wool would be lost. With their livelihoods at stake, the burghers of Bruges, Ghent, and Ypres succeeded in securing, with the help of Philip, the neutrality of Flanders for a short time. However, Flanders was eventually forced to choose sides, and chose England.

In June of 1340, Philip sent 300 warships to anchor in the Zwin. Edward sent a fleet to meet them. It was not concern for the communes of Flanders that prompted Edward's action as much as his fear of losing a pivotal trading partner. In the resulting Battle of the Sluis, 30,000 French died, and 4,000 of the victorious English, but Flanders had once again resisted French domination.

Later in the century, the communes rebelled against Louis II de Male, count of Flanders, over his extravagance and harsh taxation; he sought alliance with the French. In 1382, Philip van Artevelde (the son of guild leader Jacob van Artevelde of Ghent) led a Flemish army against Louis's French-supported army, and took Bruges. Late in the same year, Louis and the French took up arms at Roosebelee, but this time, on November 27, the French inflicted more than 60,000 casualties on Flanders. Artevelde was one of those killed. The communal government at Bruges was never the same.

Bruges suffered less at the hands of the French than the rest of Flanders, thanks to the intervention of the duke of Burgundy and the support of some nobles whose goodwill the burghers had bought with heavy bribes. Nonetheless, the French were on a rampage and spared no one outside the city.

Nearly constant fighting among the powerful Flemish cities of Bruges, Ghent, and Ypres had fractured Flanders's ability to fight, and recognizing this weakness, these cities resolved to put aside their differences. But they did not stay united for long. In 1383, Louis de Male died and his son-in-law Philip II, the Bold, duke of Burgundy, inherited Flanders. He incorporated the region into his already sizable holdings. The rule of the House of Burgundy marked the end of Flemish independence.

Despite the loss of privileges, Bruges became wealthier than it had ever been. The people of Flanders may have been weakened, but their enemies were not in much better shape. France and England were still fighting the Hundred Years War, and French leadership was weak. With power up for grabs, the House of Burgundy had grand plans, and Philip the Bold missed no opportunity to expand and consolidate his powers. He restricted the rights of the guilds in Bruges, but did not crush them. He needed the guilds' manpower and their taxes.

Philip's son, John the Fearless, established bad relations with Bruges early on. In 1411 John was fighting the Armagnacs with the contracted assistance of Bruges. He found himself in the embarrassing position of waiting for the rest of his men, who had been delayed. The Bruges soldiers honored their contract up to the minute that it expired; after it had done so, all of John's promises and prayers could not induce them to continue fighting on his behalf. John the Fearless rode alone to Paris, after virtually begging the soldiers for a hand.

Bruges again refused to assist him in 1414, this time against the English, and, disgusted, John retired to Ghent, where he spent lavishly on a residence, bought off local burghers, won the support of the general population, and eventually signed a commercial treaty with the English. Then, with nothing to fear, he turned his attention back to Bruges.

He marched into town one morning with his men and took control. Sixteen officers were deprived of their appointments, and their homes were given to some obscure citizens on whose loyalty the duke could depend. The town charter was revoked, and in its place a document called a *Kalfel* was signed. Liberties were curtailed and new taxes were imposed, many restrictive laws were put into force, and the guilds were humiliated.

After John's death in 1419, however, Philip II, the Good, became count of Flanders, and with him came a golden age for Burgundian rule in Flanders. Bruges became the chief market of the Hanseatic League, an association made up primarily of German cities, and its trade expanded to include practically all of Europe. The population of Bruges grew to 80,000. The arts flourished as never before. The master painter Jan van Eyck lived in Bruges, as did Hans Memling. William Caxton, an Englishman living in Bruges, set up a printing press with a Bruges printer, Collaert Mansion. They put out the first book printed in English, *Recuyell of the Historyes of Troye*.

The Flemish painters devoted their skills to the faithful rendition of God's world. An apprentice would have to master the creation of some 800 colors made by methods as obscure as follows (according to Anselm Boodt, author of *The Perfect Jeweler*, 1644): "Take a new-laid egg and pierce a small hole in it. Suck out the egg-white, replace it with quicksilver and ammonia; seal the hole and place the egg in a heap of warm horse dung for forty days. All this must be done at the height of summer. Finally break the shell of the egg and mix the contents with gum arabic." Such recipes for paint were jealously guarded and meticulously executed, for the artisans who painted the banners and the shields for the noble families also made the portraits of the various merchants. Colors and objects were assigned symbolic meanings beyond themselves, and entire stories could be told at a glance.

Bruges was then a drinking town, surpassed by none in its consumption of wine and beer, and it hosted the largest festivals in Europe. But all was not well. Jan van Eyck was paid some 100 livres annually for his work, and that was considered generous. His patron had an annual budget of 400,000 livres, with an additional 200,000 for travel and wearing apparel. And though Philip established the "Order of the Golden Fleece" in Bruges in 1430 to honor the weavers, eight years later he was revoking privileges as had his father and grandfather before him. Also in 1438, he prohibited English wool from being imported to Bruges. While this helped fill his coffers at first, it later helped to bring about disaster for the city.

During the fourteenth and fifteenth centuries, Bruges was the premiere city of northern Europe in commerce, politics, and art. But in the late fifteenth century England developed a cloth-spinning technique of its own. The English industry rose while the Flemish one declined. The powerful guilds of Flanders, however, refused to give up control of trade and instituted harsh protectionist measures against their former trading partners. Geographical factors also contributed to Bruges's problems; the Zwin, which had gradually been silting up, was now impassable by all but the smallest boats. Finally, political developments put an end to Bruges's period of prominence.

Philip the Good's son, Charles the Bold, became duke of Burgundy in 1467, but was killed in battle ten years later. All of the Burgundian lands were in turmoil, and the rule of Flanders passed to the Habsburgs in 1482—exactly 100 years after the House of Burgundy's accession. Bruges fell under the power of the Habsburg monarch, Maximilian of Austria.

Bruges was against him from the start, and even went as far as putting him in the town prison for three months in 1488. Maximilian was named Holy Roman Emperor in 1494. He ruled most of Europe, but this did not stop Bruges from continuing to fight him. As a result of the loss of the Zwin and constant power struggles, most of Bruges's trade moved to Antwerp. Bruges was not Rome; it could not and did not fare well against the power of a continent. Eventually, Bruges came to be called "Bruges la Morte."

Bruges was not quite dead, but as a world market, it may as well have been. The city began a long decline. Much

of Flanders was converted to Protestantism in the early part of the sixteenth century. Rule of Flanders passed to Spain (also controlled by a branch of the Habsburgs) in the latter half of the century, and while trade picked up slightly with the importation of Spanish wool, Bruges also had to contend with another Spanish import: the Inquisition. Nearly constant religious fighting and persecution eventually led to the split still seen in Belgium today—southern Belgium is mostly Catholic and northern Belgium is mostly Protestant. Spanish rule ended in 1713, after the Treaty of Utrecht ended the War of the Spanish Succession and transferred Flanders to the Austrian branch of the Habsburgs.

Bruges was controlled by Austria until 1794, when rule over Flanders fell once again to the French as a result of the French Revolutionary Wars. French rule continued under Napoléon I; with his eventual defeat, the Treaty of Paris of 1815 amalgamated Belgium and Holland into the United Kingdom of the Netherlands. The Belgians rebelled against Dutch rule, however, and the Kingdom of Belgium became a nation in 1830 and was formally recognized the following year. Under the reign of King Leopold II from 1865 to 1909, Belgium steered clear of tumult in Europe and became a small power in its own right.

The Flemish language was officially recognized by the Belgian state in 1886, and a few years later, a new canal from Bruges to the sea was built. It opened in 1895. Though the new canal and establishment of a port at Zeebrugge (Bruges-on-Sea) in 1907 were breathing new economic life into Bruges, occupation by the Germans during World Wars I and II put the town's revival on hold until the middle of the twentieth century.

Tourism plays a large part in its prosperity in the 1990s, but not as large a part as new industry, drawn by access to the sea. The newer industries are kept outside central Bruges, where the modern descendants of the original burghers have voted to keep the city much as it was when the rest of the world said it was dead.

Among the well-preserved sites in the center of Bruges are the Beguinage, dating from 1245, once home to the Princely Beguinage of the Vineyard, an order of women dedicated to asceticism and philanthropy. It is now occupied by Benedictine nuns. Tiny houses are laid out gracefully on a large lawn. One of them, the Beguine's House, has exhibits illustrating the life of the medieval inhabitants. Bruges also boasts many historic homes and churches, the latter including the Basilica of the Holy Blood, which contains the Relic of the Holy Blood, a vial containing fluid reputed to be some of the water in which Joseph of Arimathea washed the blood-stained body of Christ. The older parts of the basilica are in the Romanesque style, dating from the twelfth century, while the newer parts are Gothic, from the fifteenth century. Bruges's numerous museums include the Memling Museum, housing major works by Memling, including *Reliquary of Ursula* and *The Marriage of St. Catherine*.

The geographical center of Bruges is the Market, a square lined with seventeenth-century houses. In the center of the Market is a monument to Jan Breydel and Pieter de Coninck. Also in the Market is a belfry, the lower portion of which dates to 1296 and the upper portion to 1487. In the belfry's 266-foot tower is a 47-bell carillon. Near the Market is another square called the Burg, on the site of the ninth-century fortress. Adjacent to the Burg is the City Hall, dating to 1376, one of the oldest in the low countries. It features exhibits on the history of Bruges.

Further Reading: *The Story of Bruges* by Earnest Gilliat-Smith (London: Dent, 1905) is packed with names, facts and figures. *A Short History of Belgium* by Leon Van Der Essen (Chicago: University of Chicago Press, 1915) is also very thorough. *Bruges: The Cradle of Flemish Painting* by François Cali, translated by Dennis Chambelin (Chicago: Rand McNally, 1963; London: Allen and Unwin, 1964; as *Bruges, au berceau de la peinture flamande*, Grenoble and Paris: Arthaud, 1963) is capricious, otherworldly, and does much to disorient the reader. The overall effect is somewhat maddening, but can also be charming at times. *Bruges and West Flanders* by George W. T. Ormond, illustrated by Amcdée Forestier (London: A. and C. Black, 1906) is a series of historical vignettes describing the main historical events of Bruges and West Flanders.

—Jeffrey Felshman

Brussels (Brabant, Belgium)

Location: On the Senne River in the Brabant province of central Belgium; 155 miles from Bonn, 185 miles from Paris, and 190 miles from London.

Description: Capital city of Belgium and of Brabant province; center of commerce and textile production in the Middle Ages and early modern period; present-day international economic, tourist, and cultural center and site of the headquarters of the European Economic Community (EEC) and the North Atlantic Treaty Organization (NATO).

Site Office: Tourist Information Brussels
Brussels City Hall
Grand-Place/Grote Markt
1000 Brussels, Brabant
Belgium
(2) 513 89 40

Brussels began in the late third century as a small trading settlement called Bruocsella (a place in the marshes) on the shores of the Senne. In the seventh century its inhabitants built a small chapel on Mount Michael. The Cathedral of St. Michael would later stand on the same site.

In 977 Charles, duke of Lower Lotharingia, built a stone castle on the Island of Saint Géry in the Senne River. Charles acquired his duchy after he broke with his French royal family and gave his allegiance to the Holy Roman Emperor. Lower Lotharingia included the area that would become the Duchy of Brabant, and Charles's descendants would rule the region for centuries as the dukes of Brabant.

Charles completed his castle and took up residence there in 979, now officially recognized as the year of the town's establishment. Almost immediately, the castle became a meeting place for craftsmen and traders. Access to the castle could be gained across several bridges, one of which, the Jodenbrugge, became home to a Jewish ghetto.

Construction of a new, Romanesque church on the site of the old chapel began in 1047. The timber structure, known as the Church of St. Gudule, burned down in 1072, however. Work on its replacement, the Church of St. Michael, began in 1266 and continued into the seventeenth century. The building has two square Gothic towers, a royal crypt, and stained-glass windows by the sixteenth–century court painter Bernaert van Orley and other artists. St. Michael's was designated a cathedral in 1961.

By the thirteenth century, the dukes of Brabant had built a new castle on Coudenberg Hill, the site of the present-day Place Royale. The town at this time was composed of a series of winding narrow streets—the widest was but fifteen feet across—and it is not surprising that a fire in 1276 swept rapidly from structure to structure, ultimately destroying one-third of all houses.

Brabant, for reasons of geography and economy, was spared much of the warfare and social unrest that gripped Europe in the fourteenth century; as a result, it prospered as a center of commerce and textile manufacture. The town soon covered more than 1,000 acres, and the streets were paved with cobblestones in 1334. In 1379, a second line of fortified ramparts enclosing the inner city was built to reinforce the earlier, twelfth-century fortifications. The marketplace, called Grand' Place, became the center of activity in town.

Brussels could not escape all of the strife in Europe, however. While it was beyond the reach of the Hundred Years War, local conflicts soon overtook it. In the fourteenth century the count of Flanders occupied the town, giving rise to the city's greatest hero, Everard t'Serclaes. T'Serclaes drove out the Flemish forces with a mere 100 men; he was later ambushed by the count's men and mortally wounded.

Brussels's growth continued in the fifteenth century. Construction of the Gothic-style Town Hall (Hôtel de Ville) on Grand' Place began in 1402 and was completed in 1480. Its central tower is flanked by two wings. The right wing was added after the rest of the structure was completed, when the original space proved inadequate. Because the land available for the second wing was limited, the addition left the tower off center. The Town Hall houses the offices of the city's aldermen and the city council's meeting chambers.

In 1430 the dukes of Burgundy took over most of the area that would become Belgium and eventually moved their capital from France to Brussels. The city grew prosperous and Grand' Place became an elegant square where the dukes held tournaments and jousts.

In 1455 the town's population was more than 43,000. Most inhabitants worked for the ruling family or for wealthy merchants who exported the locally produced goods. Brussels was known for the luxurious cloth, tapestries, and lace produced by cottage industry and exported to other countries. Other exports included weapons, armor, and fine paintings. Spurred on by this prosperity, the craftsmen, organized into guilds, had struggled since the thirteenth century to break the power of the nobility and the wealthy merchants who controlled the city and its industry. Whereas such movements proved successful much earlier in Bruges and Ghent, the guilds of Brussels made little progress until the fifteenth century.

Notre-Dame du Sablon, a Gothic-style church known for its stained-glass windows, was built in the fifteenth and sixteenth centuries by the city's crossbowmen and replaced a small chapel they had built earlier. It contains a statue of St. Hubert taken from the Antwerp Cathedral.

In the first half of the sixteenth century, Charles V,

Guild House on the Grand' Place
Photo courtesy of Belgian National Tourist Office

duke of Burgundy and later Holy Roman Emperor, frequently visited Brussels, and the town became his unofficial capital. The city continued to grow, and the annual festival, the Ommegang, began. The second half of the century was a time of conflict between Protestants and Catholics. Brussels was occupied by Spanish troops led by the duke of Alva. In 1568 the duke had the counts of Egmont and Hornes beheaded in Grand' Place because of their attempts to soften the policies of Spain's Catholic king, Philip II, which repressed religious and political dissent.

The Spanish branch of the Habsburg family controlled Brussels until 1695. Charles II was reaching old age (he would die five years later) without an heir. The French king, Louis XIV, sent troops to take Belgium. Between August 13 and 15, 1695, French troops bombarded Brussels with cannonballs, largely to divert attention and reinforcements from the French seige of Namur. Almost 4,000 buildings were destroyed, and another 500 were damaged. In Grand'Place only the Town Hall remained standing. Charles was compelled to choose Louis's grandson, Philip of Anjou, as his successor. Philip's ascent to the throne triggered the War of the Spanish Succession.

By 1698 the craftsmen had rebuilt their guild houses in Grand' Place. All the buildings are designed in an Italian-Flemish baroque style, but each is different. Six connected guild houses at the square's upper end are called the House of the Dukes of Brabant because of the nineteen busts of the dukes displayed at ground level.

One of the results of the War of the Spanish succession was that the Austrian branch of the Habsburgs took over the rule of the low countries, including the city of Brussels, in the early eighteenth century. The Brussels guilds continued to press for municipal rights, and in 1719 one dissenter, the craftsman François Anneessens, was beheaded. The accession of Maria Theresa to the Habsburg throne in 1740 touched off the War of the Austrian succession, with various European powers supporting or opposing Maria Theresa's right to rule. Brussels became a battleground, and between 1746 and 1748 the city was occupied by the French. The 1748 Treaty of Aix-la-Chapelle ended the war and left Maria Theresa in control. The Habsburg-appointed governor of Brussels, Charles of Lorraine, resumed his office in the city.

The Park of Brussels (Parc de Bruxelles) on Coudenberg Hill, originally the pleasure garden and hunting grounds of the dukes of Brabant, became city property in 1776 and was laid out as a French garden in 1787. Today, the Palace of the Nation stands at one end and the Royal Palace at the other.

In January 1790, influenced by the French Revolution, Brussels participated in a proclamation creating the United Belgian States. Austrian troops briefly occupied the city, and then all of Belgium was annexed by the Revolutionary French government. When Napoléon was defeated at nearby Waterloo, the 1815 Congress of Vienna awarded Belgium to Holland.

But the citizens of Brussels longed for Belgian self-

rule, and, in August 1830, an audience attending an opera at the Théâtre de la Monnaie near Grand'Place responded dramatically to the aria "Amour Sacré de la Patrie" (Sacred Love of Country). Their response was taken up by others outside and spread throughout the country. Soon a volunteer militia was formed; a new Belgian flag was raised; and crowds of people began singing a new national anthem, "La Brabançonne."

On September 23, 1830, 14,000 Dutch troops invaded Brussels to put down the uprising. Villagers from Liège joined the rebellion, and the Dutch were defeated in a four-day battle in the Park of Brussels. The Kingdom of Belgium was established on September 27, 1830, and Brussels, which then had a population of 100,000, became its capital. The Congress Column, a monument erected in 1859, commemorates the National Congress of 1830 that had proclaimed Belgian independence. (An eternal flame at its foot honors unknown soldiers of World Wars I and II.)

The following years saw great progress. The Free University of Brussels (Université Libre de Bruxelles) was founded in 1834, and in 1835 Brussels was the first city on the European continent to build a railway, which connected Brussels and Mechlin. In 1847 a covered pedestrian thoroughfare, the St. Hubert Arcades (Galeries Saint-Hubert), was constructed of glass with iron girders. It was Europe's first covered shopping gallery and is still known for its fashionable shops and gathering places.

In the mid–nineteenth century the city built the neo-classic building housing both the Museum of Ancient Art (Musée de l'Art Ancien) and the Museum of Modern Art (Musée de l'Art Moderne). The Museum of Ancient Art occupies the ground floor and upper floors and displays paintings by major Belgian painters of the fourteenth through eighteenth centuries. It is especially known for the works of the sixteenth-century painter Pieter Bruegel the Elder, including the famed *Landscape and the Fall of Icarus* and the work of his sons, Pieter Bruegel the Younger and Jan Bruegel. The Museum of Modern Art, which shares its entrance with the Museum of Ancient Art, has seven subterranean levels and displays works of nineteenth- and twentieth-century Belgian painters, including surrealists René Magritte and Paul Delvaux.

One of Brussels's famed attractions is the Manneken-Pis, a small bronze statuette depicting a little boy urinating into the fountain below. The statue, affectionately called "Little Julian," stands at the top of a fountain near Grand' Place. Originally made of sugar, the figure was copied in stone, probably in the fourteenth century. It was replaced by a bronze figure created by the sculptor Jerome Duquesnoy in 1619, which was stolen by the English in 1745 and the French in 1747. Louis XV, after returning the statuette, made amends for the French kidnapping by giving costumes to it. So began the tradition of dressing the statue; today, Little Julian has over 300 costumes, and wears a costume on each holiday. In 1817 Little Julian was stolen again, and this time the figure was shattered. But the pieces were fitted together to make a mold, and the figure still poised on the fountain was cast from that mold.

The Neo-Gothic House of the King (Maison du Roi) was originally built in the 1500s by Holy Roman Emperor Charles V to house various community functions. Damaged by fire in 1695 and imperfectly restored, it was rebuilt in its original form in 1873. It houses the City Museum (Musée Communal), which exhibits relics of the city's archaeological and historical development. The wardrobe of the Manneken-Pis is stored on the top floor.

Brussels was the site of World Expositions in 1880, 1888, and 1897. The Park of the Cinquantenaire (Fiftieth Anniversary) and its museums were constructed for the 1880 exposition, celebrating Belgium's first fifty years of independence. The park's museums include the Museums of Art and History and the Museum of the Army and Military History, which are connected by a triumphal arch. The collection of the Royal Museum of Art and History includes artifacts from ancient Egyptian, Greek, Roman, near Eastern, and South American civilizations. The Military Museum is devoted to Belgium's martial history, and its displays include warplanes from both World Wars.

The Greco-Roman Palace of Justice (Palais de Justice) houses the city's law courts. Designed by Joseph Poelaert, it took twenty years to build and was completed in 1883.

Place du Petit-Sablon, a square designed by the architect Beyaert and laid out in 1890, is surrounded by a wrought-iron fence bearing forty-eight small bronze statues representing the guilds of medieval Brussels. Statues of the counts of Egmont and Hornes are in the center of the square.

The Royal Palace (Palais Royal) at the south end of the Park of Brussels was built early in the twentieth century. In the late twentieth century King Baudouin, the constitutional monarch, lived at Laeken Palace on the city's outskirts, but came to the Royal Palace for state occasions.

In 1914, at the start of World War I, Germany invaded Belgium. Brussels fell to the Germans and was occupied on August 20. The Burgomaster of Brussels, Adolphe Max, led a protest against the German invaders, but the city would not be freed until the armistice was signed. On November 22, 1918, King Albert and Queen Elisabeth returned triumphantly to the city.

In May 1940 the Germans again invaded and occupied Belgium. Brussels was one of the centers of the World War II resistance movement, and it has been said that more captured British servicemen escaped to Britain through Brussels than any other European city. In 1940, inspired by the action of Adolphe Max in World War I, citizens of Brussels held a protest against the German invaders. Brussels was liberated by the advancing Allied forces on September 4, 1944. Four days later, the government-in-exile returned.

The postwar years saw Brussels once again become a leading city in Europe. As before, a world's fair was held in Brussels, this time in 1958. The Atomium, built for the exposition, is a 400-foot-high building designed to represent the nine atoms of an iron molecule enlarged 165 billion times. It can be seen against the skyline from any point in the city. Bruparck, a 12-acre leisure park next to the Atomium, features about 400 models of Europe's buildings, landscapes, railways, and roads and a recreation of Brussels as a medieval village.

Brussels was chosen as the headquarters of the European Economic Community (EEC) in 1959. The EEC is headquartered in a twelve-story structure with glass-sided slabs that is built in the form of a four-pronged star. In 1967 the city also was designated as headquarters for the North Atlantic Treaty Organization (NATO). Soon many international organizations and business firms established their headquarters in Brussels, and modern buildings were constructed to house them throughout the 1960s.

In 1979 Brussels celebrated its thousandth birthday, based on the date that Duke Charles of Lower Lotharingia occupied the castle he built on the site. By this point, Brussels was a major international capital, easily reached by roads, railways, and airlines, and had more international organizations than Paris, Geneva, or London.

Further Reading: *Belgium and Luxembourg in Pictures* by E. W. Egan (London: Oak Tree, 1967; New York: Sterling, 1977) includes photographs of many of Brussels's famous buildings. In *A Masterpiece Called Belgium* (Englewood Cliffs, New Jersey: Prentice-Hall, 1984), Arthur Frommer devotes more than 100 pages to the history and attractions of Brussels.

—Phyllis R. Miller

Budapest (Pest, Hungary)

Location: In north-central Hungary, on the banks of the Danube River, about 135 miles southeast of Vienna.

Description: Capital of Hungary and center of its economic, political, and cultural life; often described as one of the world's most beautiful cities, rebuilt several times after wartime destruction. The city acquired the name Budapest in 1872 when the communities of Buda, on the west bank of the Danube, and Pest, on the east side, were united.

Site Office: Tourinform
Suto u. 2
1052 Budapest, Pest
Hungary
(1) 117-9800

The city of Budapest did not exist until 1872, the year the towns of Buda and Pest amalgamated; however, as early as the first century A.D., the site of the present-day city served as a Roman outpost called Aquincum, a name that refers to the area's thermal waters. Aquincum was the capital of the province of Lower Pannonia in the second and third centuries A.D. and served as a commercial center and an important military post, protecting the border of the Roman Empire. Early in the fifth century, the Huns, under the leadership of Attila, defeated the Romans and took control of the settlement. It is believed that the word "Buda" is the name of Attila's brother.

Various Germanic tribes occupied the site until the ninth century, when the Magyars arrived from the area between the Volga River and the Ural Mountains. The Magyars had the reputation of being fierce warriors, and they terrorized their enemies until 955, the year Holy Roman Emperor Otto I defeated a Magyar army at Augsburg. According to legend, Otto's troops blinded seven Magyar soldiers and sent them back to Hungary as a warning, should they ever think about fighting him again.

The Magyars constructed a settlement on the west side of the river, and another community developed on the east; these became the towns of Buda and Pest. The Magyar rulers converted to Christianity, but they still paid homage to their old pagan gods. Christianization intensified after the Magyar Duke Vajk was crowned King Stephen I on Christmas Day 1000. He became a Christian zealot, seeking out and killing nonbelievers and breaking the power of the local pagan chieftains. For his "services" to the Christian faith, he was canonized as St. Stephen in 1083, forty-five years after his death. Stephen was also instrumental in politically consolidating Hungary and strengthening the central power of the crown. He was the first king of the Arpád dynasty and is considered the founder of Hungary.

The strong kings who followed Stephen, László I and Könyves Kálmán, imposed and strengthened a feudal system that increased the crown's power and forced many Hungarian peasants to become serfs. After Kálmán's death in 1116, however, anarchy ensued. The successive rulers curried favor with feudal lords by making large land grants to them. Meanwhile, the kings lived in luxury while the peasants endured extreme poverty. In the thirteenth century a nobleman named Regent Bánk led a revolt against King Andrew II; this led to the implementation of a document known as the Golden Bull in 1222. This seminal document recognized the rights of the nobility and gentry and laid the basis for Hungary's constitutional development, but it also marked the decline of the power of the crown and Hungary's role as a major force in central European affairs.

The weakened state was subject to invasion, and in 1241 a Mongol army destroyed Pest, then crossed the frozen Danube to lay waste to Buda. Another major setback for the Hungarians was their defeat by the Mongols in the major battle of Muhi, near the present city of Miskolc. The invaders terrorized Hungary for another year, killing more than half the population, but they departed as quickly as they had come. After they left, the Hungarian king Béla IV embarked on social, economic, and military reforms, and ordered the construction of Buda Castle as a fortress against an anticipated return of the Mongols. They did not return, but the town of Buda developed anew around the fortress. What had been Aquincum became known as Óbuda (Old Buda), and Pest was rebuilt as well. Buda, however, was the focus of development.

One of Béla IV's major achievements was to begin construction of a magnificent Gothic church (later known as Matthias Church) in Buda; it is still one of Budapest's most impressive structures. Until the fifteenth century, every king added to the church's architecture. In 1310 Charles I, or Charles Robert of Anjou, became the first king crowned in the church.

Charles faced the difficult task of uniting the country and ending infighting among the Hungarian nobility. He worked hard to win the support of nobles from across the country and was aided by his enemies' quarrelling among themselves. Militarily, he fought to reclaim territory from recalcitrant feudal lords within Hungary, but sought peaceful relations with the country's neighbors. By 1323 he had completed Hungary's reunification. Charles died in 1342 and was succeeded by Louis I, the Great, who ruled until 1382.

In the late fourteenth and early fifteenth centuries, King Sigismund, who came to the throne in 1387, continued

Belvarosi Templom, an inner-city parish church in Budapest
Photo courtesy of IBUSZ Travel Inc.

developing the capital of Buda by ordering the construction of a new palace. The town attained special status, becoming one of a handful of Hungarian cities to be designated a royal free city. One could not judge Buda's importance by the size of its population, however; by the end of the fourteenth century, the town had a mere 8,000 residents.

In the second half of the fifteenth century, King Matthias Corvinus brought Renaissance humanism to Hungary. Education and the arts flourished at his royal court in Buda. He also lent his name to the Buda church in which he was twice married. In addition to building a library and putting a great number of book copiers to work, Matthias brought economic prosperity to Hungary and strengthened its defenses by organizing Hungary's first permanent professional army, known as the Black Army. He expanded Hungary's territory and made his kingdom the most powerful in central Europe. He imposed high taxes that burdened the country's peasants, but he earned the title of "Matthias the Just" when he promulgated the highly progressive 1486 Code of Laws. Matthias died in 1490.

In the sixteenth century, the Ottoman Empire, seeking to expand its boundaries, attacked Hungary. The Ottoman Turks briefly occupied and looted Buda in 1526, and they returned in 1541 for a much longer occupation of Buda and other Hungarian territory. For the next century and a half, Buda was a provincial capital within the Ottoman Empire. Some aspects of Ottoman rule were harsh and repressive; many Hungarians were sold into slavery, and others were sent to Turkey. The conquerors did practice religious toleration, however; although some of Buda's and Pest's churches were converted to mosques, Christians and Jews were allowed to practice their religions. The primary marks of Ottoman rule remaining on Budapest today are the city's Turkish baths.

The Habsburgs, rulers of Austria, captured Buda from the Ottoman Turks in 1686, and by the turn of the century they controlled nearly all of Hungary. Buda had suffered significant damage in the fighting of 1686, and the city's historic structures, including the castle and Matthias Church, had to be almost completely rebuilt. The Habsburgs turned out to be absolutist and repressive rulers; they also taxed the citizens heavily and plunged the peasants ever more deeply into poverty. This led to a succession of Hungarian independence struggles, the most successful being that of the *kuruc* (cross) army, led by Ferenc Rákóczi II, which succeeded in occupying upper Hungary and some of Austria before being defeated by the Austrians in 1711.

The Hungarians gradually acquiesced to Habsburg domination, but an independence movement emerged again in the 1820s, with Pest and Buda at its center. This culminated in an 1848 rebellion against the Habsburgs. The rebellion began peaceably on March 15, when independence leader Lajos Kossuth met with the emperor in Vienna and won recognition for an independent Hungarian government. He made Pest, which had surpassed Buda in population, his capital. Later in the year, however, the Habsburgs changed

their minds, and Austrian troops, aided by Russia, marched on the capital and reconquered Hungary.

During the same era, István Széchenyi, a Hungarian count, had been working to gain greater autonomy for Hungary by compromise with the Habsburgs, but many of his supporters eventually gave up on his moderate approach and allied themselves with the more radical Kossuth. As Hungary's Austrian rulers attempted to strengthen their hold on the country after the 1848 rebellion, they became suspicious of Széchenyi, had him followed, and finally in 1860 gave him the choice of being committed to an insane asylum or committing suicide. He chose the latter. He did leave his mark on the BudaPest area; he had overseen the building of roads and the first suspension bridge in the twin cities, initiated projects to improve navigation on the Danube, encouraged industry, and imported English thoroughbred horses, the ancestors of today's *puszta* horses. He also had founded the Hungarian Academy of Sciences in Pest in 1825.

By the 1860s, the Habsburg empire had begun to weaken, due in part to Austria's defeat in the Austro-Prussian War of 1866. As a means of keeping the peace domestically, Austria in 1867 agreed to recognize Hungary as an equal partner in the empire. While the Habsburg emperor, Francis Joseph I, retained the crown, Hungary received the authority to manage its internal affairs. In 1872 Buda and Pest were united as Budapest, and in 1893 the city was granted the status of a royal capital.

Budapest prospered under the new arrangement, and expanded rapidly in the late nineteenth and early twentieth centuries. Many of the city's major buildings—the Opera House, the Parliament Building, the Vigadó concert hall, and luxurious hotels—date from this period. Budapest's road system was expanded and improved, and the city built the first underground railway in continental Europe. The city's second bridge, the Elisabeth Bridge, was completed in 1903.

As part of the Austro-Hungarian empire, Hungary entered World War I on the side of the Central Powers. Although the war took a heavy toll on Hungary—it lost two-thirds of its territory in the postwar Treaty of Trianon—the defeat of the Central Powers and the collapse of the Austro-Hungarian monarchy provided the opportunity for Hungarian independence. In 1918 Hungary proclaimed itself an autonomous republic, and a revolutionary group called the National Council assumed control of the government, with the liberal Count Mihály Károlyi as president and prime minister. The count was forced out in 1919, however, as Communists headed by Béla Kun took power. Just four months later, though, the Communists were driven from power when Kun's forces were defeated by a Hungarian army formed by the landowner class, led by Admiral Miklós Horthy de Nagybánya and aided by Romanian allies. Horthy became regent of Hungary in 1920 and held that position until 1944. Under his regime, Hungary remained something of a feudal society, dominated by a landed gentry, with a large and impoverished peasant class.

Budapest had suffered little physical damage in the

war, but had to deal with a great influx of refugees afterward. Major public works projects, however, alleviated homelessness, and industrial production recovered quickly, soon reaching prewar levels. Budapest between the wars was an attractive and cosmopolitan city.

In World War II, Hungary chose the losing side again, becoming Germany's ally after Hitler invaded the Soviet Union. Horthy attempted, too late, to pull out of this alliance, and he was removed from power by the Nazis in 1944. Ferenc Szálasi, a Hungarian Fascist, was installed as the country's premier. Szálasi intensified the process of deporting Hungary's Jews to death camps; the deportation of some rural Jews had taken place under Horthy, but the Budapest Jewish community had remained intact. (Jews in Budapest, and in Hungary as a whole, had enjoyed a great degree of acceptance from the Middle Ages until the 1930s, when Fascism began to attract sympathizers.) Only about 150,000 of Hungary's prewar Jewish population of 600,000 survived the war. The Jews in Budapest, however, fared better than those elsewhere thanks to the intervention of the Swedish diplomat Raoul Wallenberg. Wallenberg and the Swedish embassy in Budapest arranged to have 20,000 of the city's Jews treated as Swedish citizens and housed in buildings that were declared branches of the embassy. The move prevented their deportation and extermination. Wallenberg disappeared while on a trip to the Soviet Union in 1945; his fate remains a mystery, although the Soviet government announced that he had died there in 1947.

Allied bombing raids and the ground combat that ensued in the fight for Budapest in 1945 destroyed much of the city. While retreating, the Germans blew up all of the bridges over the Danube. Immediately after the war, however, rebuilding began.

Politically, Budapest and Hungary were in turmoil. In 1945, in the first free elections to be held for many years, the social reform–oriented Smallholders Party, which represented the interests of the middle class and small farmers, won control of the national government. The Communist Party, however, with control of the police force and assistance from the Soviet Union, worked behind the scenes to build up its power, using such tactics as having leaders of other parties arrested on suspicion of being spies for the United States. By 1949, Hungary's Communist Party had seized control of Budapest and the country.

The Communist government carried on the rebuilding of the country, fostered the growth of industry, and improved transportation systems. The government also imprisoned its critics and outlawed all organizations it did not control. By October 1956, dissatisfaction with the government culminated in a general uprising in Budapest. Demonstrators assembled beneath the city's statue of Sándor Petófi, a poet who had been involved in the 1848 revolution, and issued demands for freedom of speech and other reforms. The rebellion spread throughout the country, with workers' councils taking over many factories. Even the army took the side of the reformers. Imre Nagy, who had been Hungary's premier

from 1953 to 1955, was recalled to that position around the time of the revolt. Although a member of the Communist Party, he was in favor of reform and critical of Soviet influence. During the uprising, he was supportive of the rebels' goals but also attempted to maintain peaceful relations with the Soviet Union. In early November, however, the Soviet Army invaded the country with the intention of crushing the rebellion. On November 4 the army surrounded the Parliament Building in Budapest. Nagy was removed from office and a new Soviet-sponsored government installed. Nagy was executed in 1958 after a sham trial; many other supporters of the rebellion were put to death as well. In all, an estimated 20,000 Hungarians died and another 200,000 fled the country.

The Soviets put János Kádár at the head of the new government. Kádár was loyal to the Soviets in foreign policy matters, sending troops to help crush the 1968 uprising in Prague. Domestically, however, he supported economic reforms and civil liberties. By the 1980s, Hungary had one of highest standards of living in the eastern bloc. Hungarians could own property, buy a large selection of consumer goods, and travel to the west. In 1990, under a new coalition government, Hungary began further moves toward capitalism, enthusiastically privatizing its industry. The transition proved economically painful, however, and in 1994 the former Communist Party, now the Socialist Party, won a majority of seats in Parliament. The Socialists emphasized their commitment to forging a democratic government and a market economy, while trying to ease the greatest pains of the transition—unemployment and lowered living standards.

Budapest has remained the jewel in Hungary's crown, with its fine hotels and shopping areas attracting tourists and giving it an advantage in the new capitalist era. Budapest's numerous historic sites, rebuilt again and again after wartime ravages, also draw numerous visitors. The castle, Matthias Church and other historic churches stand in the Buda section of the city, as do the Old Buda Town Hall (used from 1710 until 1872) and the Király Turkish Baths. On the Pest side, the Danube Promenade was home to exclusive hotels and gourmet restaurants in the late eighteenth and early nineteenth centuries; since the area's destruction in World War II, it has been returned to only a hint of its former glory, but it does include the Vigadó concert hall and an attractive park area. The Pest side of the city also holds the Opera House, dating from 1884, where Gustav Mahler made his debut as chief conductor and Giacomo Puccini supervised the first performance of *Madame Butterfly*. Budapest's Jewish community is concentrated in Pest; its landmarks include the Great Synagogue, the Jewish Museum, and the Weeping Willow monument, honoring the victims of Fascism.

Further Reading: Much information about Budapest is contained in the following books: Ivan Vilgyes's *Hungary: A Nation of Contradiction* (Boulder, Colorado: Westview, 1982); *A History of Hungary,* edited by Peter F. Sugar (Bloomington: Indiana University Press, and London: Tauris, 1990); Miklos Molnar's *Budapest: A*

History of the Hungarian Revolution (London: Allen and Unwin, and New York: Crane-Russak, 1971); Jorgt Hoensch's *A History of Modern Hungary* (London and White Plains, New York: Longman, 1988), and Stephen Brook's *Vanished Empire: Vienna, Budapest, and Prague* (New York: William Morrow, 1988).

—Ron Chepesiuk and Trudy Ring

Caernarvon (Gwynedd, Wales)

Location: Northwestern Wales, at the southwestern end of the Menai Strait; on Route A487 eight miles southwest of Bangor and nineteen miles northwest of Porthmadog, or on Route A4085, thirteen miles northwest of Beddgelert.

Description: Thirteenth-century castle and town on the site of Roman and Norman forts; additional attractions include the Segontium Roman fort, the Museum of the Legions, the Royal Welsh Fusiliers Regimental Museum, and the thirteenth-century Llanbeblig Church.

Site Office: Caernarvon Tourist Information
Oriel Pendeitsh
Castle Street
Caernarvon, Gwynedd LL55 2ND
Wales
(286) 672 232

The centuries-long effort of three great states to conquer the Welsh is memorialized nowhere better than in the medieval walled town of Caernarvon (in Welsh, Caernarfon), on the Menai Strait separating Anglesey and Snowdonia in northwestern Wales. Here can be seen the remains of the long reach of imperial Rome, and close by is the magnificent castle and town built by Edward I of England on the site of its Norman predecessor. For each conqueror the strategic usefulness of Caernarvon was seen and utilized with increasing technical and architectural skill, and today Caernarvon remains a place of royal as well as military significance.

The first inhabitants of the area were Neolithic people who dwelled there about 2000 B.C. Five hundred years later the area was settled by the "Beaker People," so called for the pottery they made. During the Bronze Age extensive trading was apparently carried on with people as far away as the Mediterranean. Between 500 and 300 B.C., the Celts arrived, and a Celtic tribe called the Ordovices were living in northwestern Wales when the Romans arrived around A.D. 61.

Although never as fully subjugated as southern England, Wales was a place of interest to the Roman legions who were conquering the known world for the empire. They utilized their famous expertise to construct a system of military roads and strategically placed forts. The fort built at Caernarvon, called Segontium, was founded about A.D. 78 and manned by auxiliary troops. It was constructed to guard the west end of the Menai Strait and the land adjacent to it. Situated 150 feet above sea level, the fort was built in a rectangle with four gateways. A wall and the southwest gate can still be seen from the outside. This gate had two passages each nine feet wide, flanked by twin towers sixteen feet square. The internal buildings are the only ones remaining

exposed in any Roman fort in Wales and date to no earlier than A.D. 150. The headquarters included an enclosed courtyard, an assembly hall, and the regimental chapel where the regimental standard was kept. A second, smaller fort was later built near the first; both had been abandoned by the time the Romans left Britain in the early fifth century. Sir Mortimer Wheeler excavated the site, on the southeast edge of town, in the 1920s. The Museum of the Legions, a branch of the National Museum of Wales, contains Roman artifacts recovered from the site. After being abandoned by the Romans the fort was claimed by Welsh chieftains, who gave Caernarvon its name: the Welsh phrase *Caer yn Arfon* described "the fort in the land over against Mona," Mona being an ancient name for Anglesey. Eventually the area became a part of the Welsh principality of Gwynedd.

Segontium had an influence on later Welsh history far greater than its size or remote location might suggest. Legend reports that Magnus Clemens Maximus was declared emperor of Rome by the Segontium garrison troops; this association with the imperial throne was incorporated by the Welsh into the Mabinogian, the collection of Welsh folktales, in which Magnus was called Macsen Wledig. Further legends claim that the future emperor Constantine the Great was born at Segontium. These imperial associations established Caernarvon as a place of divine appointment and political significance, a significance not lost upon later rulers.

In 1066 William the Conqueror crossed the English Channel from Normandy and took all of England by defeating the Saxon King Harold II in the Battle of Hastings. But the Normans spent the next two centuries trying to subdue Wales. Evidence of their efforts is visible in the remains of the many castles they built throughout the country, especially in the south. Late in the eleventh century the Normans pushed into Gwynedd in northwestern Wales, and there Hugh of Avranches, the first earl of Chester, built a crude motte-and-bailey castle near the site of Segontium. Motte-and-bailey castles were timber structures built upon a motte, or mound, twenty to thirty feet high, either shaped from a natural hill or artificially constructed. A palisade surmounted the motte, and inside was a two- or three-story wooden tower. Connected to the motte was the bailey, a structure where the castle living quarters were located. When under attack the residents retreated into the palisade on the motte. Although some of these timber structures were later rebuilt in stone, most were allowed to decay, and such was the fate of the Caernarvon castle. The motte, however, remained.

The extensive and expensive castle-building programs of the Normans and later of the English were a response to the continual threat of Welsh uprisings. Though lacking in central organization, the Welsh successfully exploited both the strengths of their own leaders and the weaknesses of their

The investiture of Prince Charles at Caernarvon Castle in 1969
Photo courtesy of Arfon District Council

opponents. Thus the strong Welsh prince Llywelyn ap Gruffudd (Llywelyn the Last) of Gwynedd obtained an advantageous treaty from Henry III of England in 1267 that granted him the title Prince of Wales and made him overlord of most of Wales. This stilled the waters for only ten years; Llywelyn's relentless and impolitic efforts to increase his power enraged the new English king, Edward I, one of the strongest monarchs to ever sit on the English throne. Edward attacked Llywelyn, and the treaty resulting from this war stripped Llywelyn of all but his title. Edward began at once a monumental program of castle-building to maintain his conquests, employing an initial workforce of 1,845 diggers, 790 carpenters, and 320 masons.

This all too concrete evidence of English dominion and Edward's insistence on the primacy of English law and customs spurred the enraged Welsh to rebellion. In the ensuing war both Llywelyn and his brother Dafydd, who had succeeded him as prince, were killed, and the rebels were scattered. Edward then determined that Gwynedd, the rebel stronghold, should be subdued and English government firmly established. Edward's capture of Caernarvon in 1283 proved to be one of the turning points of the war. He understood the political value of Caernarvon's legendary imperial

associations and set out to exploit them. He determined to build there not merely a fortress but a palace, a castle that would evoke the past and establish the legitimacy of his rule even as its strength and advanced design defended it.

The architectural genius behind Edward's castle-building campaign was Master James of St. George, whose innovative military architectural design in continental Europe had brought him to Edward's attention. Edward himself had been on the Crusades and had seen first-hand the new approaches to castle design in Europe and the Middle East. One spectacular result of that knowledge was Caernarvon Castle, begun in 1283. "If it be well manned, victualled and ammunitioned," wrote seventeenth-century observer John Taylor, "it is invincible." Constructed on the rocky peninsula between the mouths of the Seiont and Cadnant Rivers, the castle commands the Menai Strait. It forms the southern section of the town defenses, whose walls form a large D, with the curve facing the strait. Shaped roughly like an hourglass, the castle is 600 feet long, its curtain wall thickest and highest along the exposed southern face. The original plan, never completed, called for a central wall between its two wards. The eastern of these wards was built around the old Norman motte. Though the motte has since been leveled, a difference in the elevation of the two wards remains noticeable.

One of Europe's great medieval castles, Caernarvon Castle dominates its town and its surroundings. The curtain walls are guarded by an assortment of thirteen polygonal towers, each different from the rest, and two gatehouses. The Eagle Tower, which housed the generous private quarters of the castle's governor, Sir Otto de Grandson, first justiciar of North Wales, boasts an unusual cluster of three turrets. At one time each bore the statue of an eagle; now only one remains. Largest of all the towers, the Eagle Tower was designed to serve as the last post of defense in the castle; it has an additional exit so that defenders could supply it separately and escape as a last resort. The Queen's Tower is connected to the Eagle Tower and also contained lavish accomodations. Today it houses the regimental museum of the Royal Welch (sic) Fusiliers.

Four of the towers guard the two gatehouses, the King's and Queen's Gates. The King's Gate, main entrance of the castle, is filled with an awesome variety of defensive measures. The gate faces north into the town, protected by its own set of walls. Once inside the town, attackers had to bypass the drawbridge that pivoted between the moat (now the modern roadway) and a pit, in full range of the arrow-loops in the gatehouse walls. The gatehouse's sixty-foot long passage was defended by six portcullises and five thick doors, all under view from arrow-loops along each side and "murder holes" in the passage roof. At least some of these measures may have been added after Welsh rebels sacked the uncompleted castle in 1294 and held it for six months. It would not prove so vulnerable again. The Welsh rebel leader Owen Glendower lost 300 men trying to penetrate the gatehouse in 1401; he repeated the attempt in 1402 and again in 1403, when his forces were turned back by a garrison of only 28 defenders.

Caernarvon Castle was established to serve as the seat of Edward's new government in North Wales, and as such was an important administrative center. In the fourteenth century the lower ward was lined with timber-frame buildings where the daily business of the castle was conducted. Today only foundations remain. The Great Hall, a 100-foot-long structure on the south wall between the Queen's Tower and the Chamberlain Tower, was the government office and the central dining hall. Opposite it were the kitchens, located separately to reduce the risk of fire. Edward did not overlook any opportunities to take advantage of Caernarvon's symbolic usefulness. Edward arranged for his wife to give birth at Caernarvon in 1284, so that he might offer a native-born prince of Wales to the Welsh people. The grand style of the castle, the Roman eagles on the turrets, and the bands of colored stonework on the facade, designed to recall Constantinople and Emperor Constantine, were all intended to connect English rule with Welsh history. But Edward's son, Edward II, never returned to Caernarvon. Edward II's finances could not keep pace with his ambitious building plans, and Caernarvon Castle, like so many of his others, was never finished. The royal residence had little chance to fulfill its palatial function.

For centuries the castle and its town led a quiet existence. During the English Civil War, Caernarvon Castle served as a refuge for the Royalists and was besieged and surrendered to Parliamentary forces. From then on it was largely neglected, although it was still imposing enough to inspire Samuel Johnson, who viewed it in 1774 and described it as "an edifice of stupendous honesty and strength."

Restoration of the castle began in the 1840s, when Anthony Salvin was commissioned to repair the stonework. David Lloyd George, the British prime minister, recommended that Caernarvon Castle be the site for Prince Edward's investiture as prince of Wales in 1911, encouraging further efforts to preserve it. Lloyd George was of Welsh parentage and had represented Caernarvon in Parliament. The attention both before and after the investiture provided a long-overdue refurbishing, and since then members of the royal family have made Caernarvon a regular stop on their tours. Both George VI and Elizabeth II stayed there during tours following their coronations, and in 1969 millions of people around the world witnessed the ceremony televised from the castle ward investing Charles as prince of Wales. Today the 700-year-old walls and towers stand in excellent condition, an irresistible draw to tourists.

The current town of Caernarvon was born with the castle; its walls were begun in 1284. Within those walls, which run about a half mile in circumference, the old town still exists, the maze of streets containing houses, shops, and inns, though the town has expanded well beyond its medieval defenses. The old county hall, built in 1863, lies within the walls, and the guildhall, dating to 1874, is on the site of the town's east gate. Two medieval churches still exist, the garrison church of St. Mary's, dating to the fourteenth century and housed in a corner of the city walls, and Llanbeblig Church, built in the thirteenth century on the old Roman cemetery. Llanbeblig is dedicated to St. Peblig, the uncle of Constantine the Great, according to Welsh tradition. A statue of David Lloyd George overlooks the weekly market held in Castle Square. Caernarvon serves as the seat of Gwynedd County.

During the nineteenth century the local slate industry generated great activity in the harbor, and a small museum in the ex-dredger *Seiont II*, docked at Victoria Dock, explains the town's maritime past. Tourism and recreation are important industries here. The Royal Welsh Yacht Club and the Caernarvon Sailing Club are headquartered here. Sea fishing in the Menai Strait and salmon and trout fishing in the Seiont River are popular activities. The beauties of the surrounding landscape make a fitting backdrop for the splendor of castle and town. The rugged terrain is dominated by the Snowdon Mountains to the east, including Mt. Snowdon, the highest mountain in Wales at 3,560 feet; nearly half of the county is part of Snowdonia National Park. To the west, across the Menai Strait, lies the equally picturesque land of Anglesey. An excellent vantage point for viewing the scenery is a rocky outcrop called Twthill, half a mile northeast of the old town and once home to Bronze Age residents.

Further Reading: *Castles in Wales: History, Spectacle, Romance* by Roger Thomas (Basingstoke, Hampshire: Automobile Association, and Cardiff: Wales Tourist Board, 1982) offers well-illustrated and detailed accounts of the many remaining castles in Wales, including Caernarvon, along with a helpful capsule history. Providing a somewhat broader context are Paul Johnson's *The National Trust Book of British Castles* (London: Weidenfeld and Nicolson, 1978; New York: Morrow, 1980) Johnston's and James Forde-Johnston's *Great Medieval Castles of Britain* (London: Bodley Head, 1979), both also well illustrated. *The Mabinogion,* the ancient collection of Welsh folktales, provides an inside look at a largely forgotten culture. It is available in many editions.

—Elizabeth Brice

Cambridge (Cambridgeshire, England)

Location: Cambridge lies on the right bank of the Cam River in the southeast central part of England, fifty-six miles north of London and 40 miles south of King's Lynn.

Description: One of the most beautiful cities in Europe, Cambridge is economically as well as architecturally dominated by its internationally known university. The town, which has now spilled over on to the Cam River's left bank to the north and west, is situated on a low plain at what used to be a ford over the river before the land rose slightly to the northwest up Castle Hill. With the gentle hills to the south and southeast, the site provided an easily defendable fort for the Romans.

Site Office: Tourist Information Centre
Wheeler Street
Cambridge CB2 3QB
Cambridgeshire
England
(223) 322640

Cambridge's river was originally called the Granta, from which derived the seventh-century Grantacaestir. At the time, Bede called it "a desolate little city." The city became Grantebrycge, then Cantebrig in the twelfth century. In the fifteenth century the "Cant" became "Cam," and from about 1600 that is what the river was called. The ford by Castle Hill offered the best route from east Anglia to the north and west, and the river gave access to the sea. The Romans built a twenty-five-acre fort where the land rises on the west bank of the river and two roads intersected, Akeman Street running from southwest to north-east and the Via Devana from southeast to northwest. The fort lasted from the late first century until the early fifth, after which the site was resettled by Angles, then in the late ninth century by Danes, and finally by Normans.

The Roman bridge was rebuilt in the eight century, and in spite of raids, the town prospered. The Danish settlement filled out the site of the present town center, up the river from the bridge at Castle Street to Mill Lane, and along the King's Ditch. The latter was probably dug in the late ninth century to rejoin the river, east of Magdalene. The Saxon conquest in approximately 917 resulted in a development south of the Danish settlement. Cambridge had a mint, and St. Bene't's church must have been built about the year 1,000. The pre-conquest tower survives.

William the Conqueror's victory in 1066 led to the establishment of the castle where the mound now stands. Its erection required the destruction of twenty-seven houses. The Domesday account of 1086 shows a prosperous town governed by William's sheriff, Roger Picot, who pulled down more dwellings and built a third mill. However, the townspeople objected to the size of his levies.

Under William's fourth son, Henry I, the city acquired monopoly rights for the landing of goods in the shire and the levying of tolls on them. The annual September Stourbridge fair, outgrowing rivals in the nearby meadows to become one of the most important in England, was established to the northeast of the city at Barnwell. A large number of churches, ecclesiastical foundations, charitable institutions, and religious houses were established within the city, together with guilds and confraternities. Two of these helped to found one of the earliest colleges in 1352, Corpus Christi, unique in having been established by townspeople. The town was moved from the diocese of Lincoln to Ely in 1109, and its first charters date from 1201 and 1207, in which the existence of a Jewish quarter was recorded.

Cambridge University's origins are traceable to the arrival from Oxford of a group of regent masters who were fleeing the hostility of the Oxford townspeople between 1208 and 1209. They rented lodgings near Great Saint Mary's Church for themselves and their quasi-apprenticed students. John Grim, the highest ranking academic of the group, seems originally to have come from the town on the Granta. Oxford teachers had also dispersed elsewhere, but outside Oxford only Cambridge was allowed to flourish. By 1225 the bishop of Ely was treating the scholars as a separate canonical society, and by the following year the new Franciscan order had established a foundation in the town. Other orders soon followed, including Dominicans and Carmelites by midcentury. In 1231, Henry III granted substantial privileges to the scholars of Oxford and Cambridge, implicitly defining the degree of civil autonomy they were to enjoy. By 1233, Gregory IX issued a bull addressed to the chancellor and university of Cambridge confining jurisdiction over the scholars to their own chancellor or the bishop. The first statutes date from about 1250 and, as at Oxford, a series of disturbances, some resulting in numerous deaths, led to the strengthening of the university's privileges by the king.

The Cambridge student hostels were less numerous but much bigger than those at Oxford. They numbered about twenty, each with some twenty-five or thirty students, although the six fifteenth-century hostels for lawyers averaged eighty students each. University teachers and students from the monastic and other religious foundations, whose relationship with the university was sometimes bitter, were also in residence.

The first Cambridge college, Peterhouse, modeled on Walter de Merton's Oxford foundation, was established in 1284 by Hugh de Balsham, bishop of Ely, under royal li-

Trumpington Street and Pembroke College, Cambridge, in 1815
Print courtesy of Cambridgeshire Libraries

cence. The scholars were at first housed with the Austin canons, but strained relationships led to their removal outside the Trumpington Gate. They were given two hostels and the church of St. Peter's, now Little St. Mary's, next to the present Fitzwilliam museum.

Most early Cambridge colleges were built around a quadrangular court with a hall, rooms, and sometimes a chapel. This model was shared by many monasteries, inns, and country houses. Chapels posed problems, since they inevitably diminished the attendance and income of the parish churches. At Peterhouse, the college was attached to the parish church by a gallery, and Corpus Christi was physically attached to St. Bene't's. Later colleges often have imposing gate towers, as if they were castles. Peterhouse has a partly original thirteenth-century hall in a mostly fifteenth-century court. The old court at Corpus Christi still retains its 1352 aspect, although it has been restored and expanded. Christopher Marlowe was an alumnus, and the college has the best collection of silver plate in Cambridge.

Most of the older buildings of the older colleges such as Clare, founded in 1326, and Pembroke, founded in 1347 by the widow of the earl of Pembroke, have been destroyed. However, the seventeenth-century court of Clare is extraordinarily elegant. Pembroke's chapel by Christopher Wren, a gift to the college from his uncle, the bishop of Ely, is of historical importance as the first ecclesiastical building in Cambridge based directly on an antique model. The other two fourteenth-century foundations, Gonville and Caius, founded in 1348, and Trinity Hall, founded in 1350, have few ancient buildings, although Trinity Hall's seventeenth-century library retains the authentic atmosphere of its period. Like their Oxford counterparts, Cambridge colleges do not display their identities on the their street fronts.

No new colleges were founded in Cambridge between Corpus Christi in 1352 and Godshouse in 1439, which was later refounded as Christ's in 1505. During that period, Oxford's lead in size and reputation continued to increase. Then, for reasons that almost certainly have to do with suspicions of Oxford's lingering inclination towards the doctrinal attitudes of John Wycliff, a leader of the Protestant Reformation, Cambridge began to attract new foundations. No new colleges were founded at Oxford between Magdalen, established in 1458, and Corpus Christi, in 1517. This marked the period of Cambridge's most startling growth with King's (1441), Queens' (1448), St. Catherine's (1473), Jesus (1496), Christ's (1505), and St. John's (1511).

The three large-scale architectural gems among the Cambridge colleges are King's, St. John's, and Trinity. Visually most impressive of all is King's, primarily because its magnificent frontage on King's Parade and its chapel, which was not intended to serve merely as a college chapel for the corporate worship of some seventy college members. In fact, the great church served no ecclesiastical purpose. Although conceived by Henry VI, it actually became a willful, not to say arrogant, display of Tudor political power erected on the wrecked ambitions of the defeated, deposed, and deranged

last Lancastrian king, Henry VI. Henry had at first envisioned for his foundation only the small court that now houses the university registry, and only later developed triumphalist architectural plans. The grandiose project for a university college, linked to his founding of Eton, was executed in his then-obsolescent perpendicular Gothic style by the Tudor successors who had defeated him. The unfortunate Henry, put to death in 1471, had dedicated much of his young adulthood to establishing the twin institutions. He had laid both foundation stones, but got little further in Cambridge than acquiring the ground for King's, clearing it, and planning the chapel.

Nonetheless, the chapel has been considered the finest Gothic building in Europe. Huge in size, nearly 300 feet long, 80 feet high, and 40 feet wide, its extraordinary system of spatial relationships instantly overwhelms the viewer, whose eyes are immediately forced upwards towards the intricate and delicate fan vaulting. Its curves are echoed in the 1606 organ case, and set off by the magnificent sixteenth-century woodwork in the screen and choir stalls. The fine stained glass windows, dating from 1515, constitute what is probably the largest complete set of medieval windows left in Europe. The other buildings at King's are mostly from the eighteenth and nineteenth centuries. The magnificent lawn sweeping magisterially down to the river was once intended to be the site of a further quadrangle.

The building of St. John's, established in 1511 and the second college founded by Lady Margaret Beaufort, countess of Richmond and Derby and mother of Henry VII, was largely accomplished by her confessor and executor, St. John Fisher, bishop of Rochester. Lady Margaret, who had also established divinity professorships at both Oxford and Cambridge, left the bulk of her wealth to the new foundation. There are four courts east of the river. The easternmost, fronting St. John's Street, has an imposing three-story Tudor brick gatehouse with rich fan vaulting on the ground floor and an elaborate heraldic decoration over the gate, which includes two yales supporting the Beaufort arms. (Yales are mythic beasts with antelope's bodies, goat's heads, and swivelling horns.) The figure of St. John, to whom an ancient hospital on the site had been dedicated, is a 1662 replacement of the original, removed during the civil war. The second court is one of the finest in Cambridge, and the third leads to the famous Bridge of Sighs over the Cam, modeled on that in Venice. The nineteenth-century chapel is by the Gothic revival architect, George Gilbert Scott.

Trinity College was founded by Henry VIII in 1546 and reflects in its scale and design the flamboyance character of that monarch. It is larger than any other collegiate establishment in Cambridge and larger than Henry VIII's other grandiose academic foundation, Christ Church in Oxford. Like Christ Church, it incorporated existing institutions, notably Michaelhouse and a number of hostels. The original King's Gateway, dating from the sixteenth century, is preserved from the old King's Hall, although altered. It opens onto Great Court, the largest in Cambridge, with buildings of different dates. The sixteenth-century gate, chapel, and hall have been embellished, but not enough to disguise their graceful proportions. The rather heavily neoclassical fountain dates from 1602. In the second court, the superb library, designed by Wren in 1676, contains magnificent carvings by Grinling Gibbons. Wren wanted to build a domed free-standing library, but in the end settled for the present two-armed neoclassical building, with the library on the court side raised to first-floor level over a low round-arched arcade and adapted to the height of the other already existing ground floors.

Much of the building was carried out under the mastership of Thomas Nevile, whose name is commemorated in the Jacobean second court, which was open toward the river to its west until Nevile closed it with the library and Cam blocks. New Court is a nineteenth-century extension to the college, but opens onto the fine grounds leading down to the river. The best that can be said for the twentieth-century buildings is that they exploit constricted space with ingenuity.

When it comes to Trinity, the famous 1911 edition of the *Encyclopedia Britannica* majestically declares, ''The eminent alumni of this great college are too numerous to admit of selection.'' In the university *Prospectus,* the college itself mentions half a dozen prime ministers; Herbert, Marvell, Dryden, Byron, Tennyson, and Housman among the poets; Macaulay, Acton, and Trevelyan among the historians; Newton and Rutherford among the scientists; and Russell and Wittgenstein among the philosophers.

The concentration of great names is not surprising since England, unlike the countries of continental Europe, and at whatever social cost, gave a monopoly of university education to only two institutions until the nineteenth century. What cannot be doubted is the consequent intensity of historical interest attached to Oxford and Cambridge, where Christ's has the mulberry tree under which Milton is reputed to have written *Lycidas,* and Pepys's library, in the original cases, is at Magdalene. Pembroke and Emmanuel have chapels designed by Wren. Both cities have for centuries attracted the most advanced and munificent forms of architectural and artistic patronage, and have in the twentieth century not only become conscious of the richness of their own histories, but have also come to take seriously their social roles as custodians of that history, the display of its visual relics, and the moral obligation to make the cultural inheritance of which they are custodians accessible to any who want to examine it, whether in books or in buildings.

As a result, at Cambridge as at Oxford, immense ingenuity has been exercised to combine a large modern university, functioning in often ancient buildings in crowded and twisted streets, with a viable modern market town and county center that itself has to cater to the preservation of ancient buildings and modern gardens, and to provide the usual amenities of modern urban living. Cambridge is not merely a museum, and not merely an academic society. Cambridge is the hugely successful product of the attempt to combine its various custodial functions with all the others essential to it. Its most impressive achievement today is the success with which its different and apparently incompatible functions are daily reconciled. Other ancient European cities certainly face similar problems, but few have solved them more successfully.

Further Reading: There are innumerable guides and handbooks to Cambrige, catering to every sort of interest, as well as the normal treatments of town and university in encyclopedias, gazetteers, and historical surveys of all kinds. There are memoirs and formal histories of all the older colleges, and many formal historical and archival records have been published. Seriously recommended is *A History of the University of Cambridge* in four volumes, edited by Christopher Brooke (Cambridge and New York: Cambridge University Press, 1988), with copious bibliographies. For a general guide, F. A. Reeve's *Cambridge* (London: Batsford, and New York: Hastings, 1964) is reliable and readable, if more interested in the university than the town. For a readable account of the changing historical relationship between town and university, see Rowland Parker's *Town and Gown: The 700 Years' War in Cambridge* (Cambridge: Cambridge University Press, 1983).

—Claudia Levi

Canterbury (Kent, England)

Location: Canterbury is situated in the county of Kent, approximately ninety miles south east of London. Easily reached via the M2 motorway, it is only ten to fifteen miles north of the port cities of Dover and Folkestone.

Description: Founded during Roman times, Canterbury is dominated by its Norman/Gothic cathedral, seat of the Archbishop of Canterbury, the first one of whom was St. Augustine in 597. Centuries of tumult followed that first appointment, but the cathedral managed to remain preeminent in English life, particularly after Thomas Becket was murdered at its altar and Canterbury became the most important pilgrimage destination in northern Europe. During Dissolution in the reign of Henry VIII, the shrine of St. Thomas was destroyed, and the interest in pilgrimages diminished, never again to reach the same levels. The cathedral continued to be the archbishop's seat, and during Elizabeth I's reign, from 1538 to 1603, Canterbury became an important city in the growth of Protestantism. Today, Canterbury remains one of the great Cathedral cities of Europe.

Site Office: Canterbury Tourist Information Centre
Department SF3, P.O. Box 198
Canterbury, Kent CT1 1UX
England
(227) 766567

The city of Canterbury, nestled in the verdant Kent countryside, has endured centuries of war and peace, scandal and honor. Its historical importance, as the seat of the Archbishop of Canterbury, is signified by its magnificent cathedral, a blend of Norman, Gothic, and Perpendicular styles. The Norman Conquest saw its grand rebuilding, after a fire in 1067. The city, by then, was home to a large Benedictine monastic community. It was in the Cathedral, a few years later, that Thomas Becket was murdered by knights loyal to Henry II, and became enshrined there as the martyred St. Thomas. His martyrdom led to centuries of pilgrimages to the church, which inspired Chaucer's *Canterbury Tales,* and to its reputation as the greatest pilgrimage church in northern Europe. After another fire in 1174, the restored cathedral was the first major building in England to incorporate Gothic architectural elements. It is as a medieval city, then, that Canterbury is most famous. But one must go back to the Romans, who founded Canterbury as Durovernum Cantiacorum, to find the town's roots before the Saxons, then Normans, took over.

Evidence has been found of human habitation dating back to the Iron Age, and a small settlement already existed on the site, a throwback to Caesar's attempts to conquer the island, when in A.D. 43 Romans came to settle Durovernum Cantiacorum. The Roman emperor, Claudius, looked to Britannia, then a fragmented land of feuding kings, to bolster his weakening reputation and power. He ordered 43,000 Roman soldiers to Britain; they landed at Richborough, Kent, and quickly secured the southeast of England, having accepted the surrender of twelve English kings. London (Londinium) and York (Eboracum) became the two premier cities, though Canterbury grew into a small center of trade and agriculture as well as iron production. It was also an important junction of major Roman roads, at a point between the Roman port towns and London. Evidence of Roman occupation was buried under the settlements of subsequent societies, and it was not until after the bombings of World War II that archaeologists began in 1944 to excavate in earnest. They discovered a Roman street plan that was completely different from the modern one; the buildings of Durovernum were widely spaced. Signs of opulence were apparent in much of the recovered materials, which included marbles, mosaics, and fine silver.

Constantine the Great, best remembered as the first Roman emperor to convert to Christianity, presided over the "Golden Age" of Roman power in Britain, between 306 and 337. Pagan practices, previously tolerated, were made illegal, and trade flourished. But with the death of Constantine in 337, Britannia was plunged into tumultuous times. Military rebels proclaimed themselves emperors and feuded, invasions from Scotland were a constant threat, but the area still seemed to enjoy relative prosperity in the last half of the fourth century. The greatest threat to all of England came from the increasingly powerful Saxon pirates. When Roman soldiers were recalled to the continent, as the Saxons invaded the Roman empire there, the depleted militia left in Britain struggled to cope with fierce Saxon raids. By 410, their forces were in disarray, and the Saxons had succeeded in separating much of Britain from Roman rule. The last known appeal from Britons for military help from the Romans came in 446, but the plea went unheeded.

Of the period that followed, from 450 to 595, little record remains. A vacuum of power created by the retreating Romans led to struggles between three European tribes then in England: the Angles, from Schleswig-Holstein in Germany; the Saxons, from the Saxony region of Germany; and the Jutes, from Jutland. The Jutes settled in Kent, while the Anglo-Saxons settled further to the north and west, moving inland via the river systems. This period, sometimes called the Dark Ages, is not wholly dim. It is from this period that the Arthurian myth arises, and during it that such intellectual giants as Bede and Gildas lived. Only a few of the 200-odd kings from this era stand out as memorable figures, King

Canterbury Cathedral
Photo courtesy of Canterbury City Council

Ethelbert of Kent among them. Ethelbert attempted to establish order and laws "after the manner of the Romans" when he came to power in 561. His capital was Canterbury. His French queen, Bertha, was a Christian, and the king gave her the church of St. Martin for worship. In 597, the Benedictine monk Augustine led a missionary expedition from Rome to England, though the group was at first reluctant to cross the Channel because of the Britons' reputation as savages. Eventually, they made the crossing and were graciously welcomed into the court of King Ethelbert, who himself was then won over to Christianity and granted Augustine authority to preach and convert. The king also gave him a church, which would in time become Canterbury Cathedral. Augustine also founded the abbey of St. Peter and St. Paul, later known as St. Augustine's; its ruins still exist. Augustine, proclaimed archbishop, set up his official seat, or "cathedra," in what is now the Cathedral Precincts. Until 758, however, archbishops were buried in the abbey; it was considered more important than the cathedral. With the growth of the cathedral and its priory, rivalry inevitably developed between the two religious institutions, the Cathedral and the abbey. Celtic missionaries from Ireland, jealous of the Benedictine's preeminence, competed for souls to convert. But regardless of such enmities and differing theologies, Augustine's mission had been a success: Christianity had gained a solid foothold in Britain, and England's awe-inspiring cathedrals, of which Canterbury is the most renowned, are its triumphs.

In 978, Ethelred, an Anglo-Saxon descendant of Alfred the Great, began his reign as king. During this time, the Vikings, or Danes as they came to be called because of their involvement with the Danish king, resumed their attacks on the coasts of Kent and East Anglia. Disaster struck Canterbury when, in 1011, it was sacked by Dane invaders; the archbishop was taken prisoner and watched his cathedral burn to the ground. Fortunately, in 1016, King Canute, a Dane, began his rule of England and succeeded in creating a short period of peace and prosperity. Known not only for his strength and military prowess but also for his piety, King Canute raised and gave money to restore the Christian churches that had been destroyed during the years of pillaging, including Canterbury Cathedral, which now grew and prospered until it was destroyed once again during the Norman Conquest. When Canute died in 1035, a confusing dance of inheritance began, ending with the victory of Edward, son of ex-king Ethelred, who had fled to Normandy with his family in 1013. Edward the Confessor (as he came to be called), more French than English by then, came to power in 1042, and after a relatively peaceful reign, he died in 1066. In that year the course of English history was changed forever. Three claimants fought for the throne: William, Duke of Normandy; Harold, Earl of Wessex; and Harold Hardrada, King of Norway, who was killed in battle soon thereafter. William collected an army in Normandy and sailed to England, marching inland unopposed in September 1066, while Harold of Wessex marched his army south to meet him. Their armies met outside the town of Hastings, where a fierce and now famous battle was fought. Near the end of the day, the Normans mounted a final attack, killing Harold, and allowing William the Conqueror to claim the crown and found a new royal dynasty.

The conquest of the Normans presented some dramatic changes, most obviously in the introduction of fortified castles, one of which was built at Canterbury by William, who recognized the town as of strategic importance. No longer were towns laid out within walls; now, only the Norman lord and his family lived within the safety of walls— those of his castle. The Normans also established the social and military system of feudalism, whereby a person's rank was determined by the amount of land that he held. Language began to change as well, as Norman French words intermingled with Anglo-Saxon, the amalgam slowly evolving to modern English. From 1066, the king, the archbishop, and the abbot were the authorities in Canterbury. The clergy, who were the only people who were literate, filled the important posts in church or state. They were also large property owners, involved in trade and commerce; they built numerous buildings out of the local stone, flint, many of which still survive throughout the city and countryside. Canterbury's church was destroyed by fire in 1067. In 1070, William the Conqueror appointed a Norman, Lanfranc, to be Archbishop of Canterbury. Lanfranc built a huge new cathedral on the site of the former church. Lanfranc also built up the monastery to become one of the largest Benedictine establishments in the country. For centuries to come, the Archbishop of Canterbury would be not only the most significant religious figure in the country, but also one of its most prominent statesmen.

The immense wealth and properties of the Church soon became a point of intense bickering and envy between church and state. The quarrels came to a head in the reign of Henry II, which began in 1154. Henry II, a powerful, shrewd ruler with a legendary temper, married Eleanor of Aquitaine, who brought with her as dowry the French territory running from the Pyrenees to Bordeaux: Henry now controlled land running from the Scottish border, through England and France, to the Pyrenees. Despite such power, Henry bridled at the Church's authority and independence and strove to put in place a more cooperative, subservient Archbishop of Canterbury, an ally—and to do so without offending the papacy. Thomas Becket, his loyal adviser and chancellor, seemed the perfect candidate. When the position of archbishop was vacated in 1162, Henry saw his chance and appointed Becket to

the post. Much debate rages among historians over the rest of the story, but virtually everyone agrees that Becket's attitudes changed when he came to power and that the king resented him for what he regarded as his disloyalty. Becket suddenly became pious, a prophetic spiritual leader committed to the Church, not the crown, and the more Henry pushed for cooperation, the more Becket reacted against him, and the more heated their arguments became. By 1164, the exasperated Henry was determined to break the resistance of his stubborn archbishop, and he produced the Constitutions of Clarendon, outlining the customs and privileges of the king over the Church. At first, Becket gave in to the pressure, but soon after he denounced the Constitutions and fled to Normandy, in exile. Henry promptly attempted to crown his son as successor to Becket in 1170, but Becket swiftly reacted by issuing an order to close all churches in the kingdom. His orders were followed, enraging Henry. Becket then returned to Canterbury, punished Henry's supporters, and further alienated himself from the king. Some accounts hold that, in a fit of temper, Henry steamed, ''Will no one rid me of this turbulent priest?'' The four loyal knights in attendance took the statement literally and rode off to Canterbury Cathedral. They found Becket in front of the altar of St. Benedict and killed him with blows to the head. Outrage, shock, and horror from Christians far and wide followed. In an attempt to quell the uproar, and as penance, Henry walked barefoot through Canterbury and submitted to a flogging by monks. He also paid a visit to Pope Alexander III, to mend fences with an angry Church. In the meantime, miracles were said to be taking place at Becket's tomb in the cathedral, and it fast became a place of pilgrimage in England. In 1172, the pope canonized Becket. With a martyred saint in residence, Canterbury's supremacy over the rival town and minster of York was soon sealed. The likes of Chaucer's pilgrims in his mid-fourteenth century work, *Canterbury Tales,* flocked to Canterbury Cathedral to prostrate themselves in front of the shrine of St. Thomas. Gifts presented in honor of St. Thomas soon increased the Cathedral's wealth and influence, despite Henry's determination to pursue his arguments with Becket's successors. When the east side of the cathedral was destroyed by fire in 1174, many of its Gothic features were incorporated into its new choir, including a magnificent place for Becket's shrine over a lofty crypt.

Canterbury became second only to Jerusalem and Rome in importance to pilgrims, who brought wealth and trade with them. The town, full of artisans, churchmen, tradesmen, and great landowners, flourished, as did the innskeepers. Many pilgrims in the Middle Ages combined their secular and religious holidays and considered a pilgrimage a good way to spend time away from home with a group that would create enough variety to suit everyone's needs. Others were sick or vagrants, and Canterbury boasted several hospitals founded specially to minister to their needs. Almost all of the successive kings of England, as well as dignitaries, visited the shrine, though only one, Henry IV, is buried there. In 1376, the popular English warrior of the Hundred

Years' War, the Black Prince, was buried there as well. But eventually the interest in pilgrimages, relics, and indulgences began to decline. The seventh and last celebration of the Translation of St. Thomas took place in 1520, with King Henry VIII, Holy Roman Emperor Charles V, and Cardinal Wolsey in attendance. Only eighteen years later, in November 1538, the once-devout Henry VIII declared St. Thomas a traitor, had his shrine pillaged (it was never completely found again), and all references to him destroyed. The Dissolution (and, with Martin Luther's prodding in Wittenburg, the Reformation) was underway, and with one order Henry had destroyed Canterbury as a pilgrimage site. Henry VIII had already broken with the Roman Catholic Church in his great ambition to become head of the English Church and Defender of the Faith—and in order to divorce wife Catherine to marry pregnant Anne Boleyn. He ordered the destruction of all monasteries in the land, most of which order was planned by Wolsey's successor, Thomas Cromwell, and carried out by his supporters. The destruction of the great monasteries of Canterbury further diminished the city's ecclesiatical fame. Ironically, it was also Henry who unwittingly gave Canterbury another saint and made the city once again a destination for pilgrimage—though that result would come only centuries later. Henry beheaded his once-favored chancellor, the Catholic Sir Thomas More, who resisted Henry's break from Rome. More's daughter, who lived in Canterbury, secured his head and brought it to Canterbury to be buried in the family vault there. When the Pope beatified Sir Thomas More in the 19th century, Canterbury once again became a pilgrimage site for Roman Catholic pilgrims.

After Henry's death in 1547, a series of struggles for power ensued, ending in 1558 with the accession to the throne of Henry VIII's daughter, Elizabeth I, who ruled until 1603 and brought strong, shrewd leadership to her country. England had veered toward Protestantism under her father and his successor, Elizabeth's half-brother Edward VI, then lurched towards Catholicism under her half-sister, Mary, and Elizabeth recognized the need for greater political stability and less confusion in official religious doctrine and certainly less religious persecution. Canterbury once again became a center for Protestantism, this time for those suffering persecution during religious wars in France and the low countries, the Walloons and the Huguenots. After the massacre of St. Bartholomew in Paris in 1572, the numbers of Huguenot refugees in Canterbury increased, and they were given use of the crypt in Canterbury Cathedral; a small French Protestant group still worships in Black Prince's Chantry there. The main occupation of these refugees was weaving, and wool-weaving became a chief industry of the town. Though Canterbury was not a port city, it lay on a direct route from the port of Dover to London, and it evolved into a center for the marketing and distribution of woollen merchandise. Many of the buildings still standing in the city's center were built by Huguenot weavers, in a distinctive style of high, gabled fronts with loft doors leading to storage areas for wool and other merchandise. At the same time, a group of radical

Protestants, called Puritans, was becoming more extreme and vociferous in their rebellion against the heavy hand of the Church of England. Two Puritans from Canterbury, James Chilten and Robert Cushman, fled on the *Mayflower* to Massachusetts. It was also during Elizabeth I's reign that what is now called English Renaissance, in art, architecture, and music, flowered. The best poets of the day dedicated their works to her, Edmund Spenser began publishing his epic poem, *The Faerie Queene,* in 1590, and Shakespeare wrote his greatest works during the last decade of her reign. Canterbury's most famous son, Christopher Marlowe, was part of this illustrious company. His life was brief (he died in a tavern brawl in Deptford in 1593, at age 29), but during that short life he wrote such poems as "The Passionate Shepheard to His Love" and such now well-known plays as *Tamburlaine the Great, Dr. Faustus,* and *Edward the Second.* Most literary scholars now agree that Marlowe's mastery of dramatic blank verse served as a foundation for Shakespeare's greatest triumphs.

Following the Civil War, in the mid-1600s, when the city proved a hotbed of unrest that sparked several larger conflagrations, life in Canterbury quieted down. The Cathedral continued in its role as seat of the archbishop, the most important religious figure in the country; there were the usual kinds of scandal; trade grew, as did the population, at a moderate rate. The Industrial Revolution left Canterbury behind, when mill and coal towns further north gained the ascendancy, and Canterbury's industrial manufacturing concerns foundered. In 1830, the Canterbury-Whitstable Railway opened. Whitstable was a seaside fishing village not far to the east, and the Canterbury Railroad Company pioneered, as one of the first ten passenger railways in the world, a rail link between the two towns using the locomotive, the *Invicta.* And thus the first railway-passenger season ticket in the history of the world was issued at Canterbury, though it was not good for very long: the *Invicta* never worked well, the rail link never reached its full potential, and the railroad was soon a memory.

The 1940s brought World War II to Canterbury. The Battle of Britain began in the spring of 1940, and German bombers flew over Canterbury en route to London during the Blitz. The Baedeker Raids—so named for the series of tourism books that Adolf Hitler used to devise the targets of his deadly raids—levelled the Luftwaffes' greatest destruction on the city. Much of Canterbury was destroyed, though the Cathedral, miraculously, made it through the raids relatively unscathed, as did many of Canterbury's historic monuments. Now, modern Canterbury is a populous, lively market town, still dominated by its glorious Cathedral. The city boasts numerous events at the Cathedral, as well as museums, centers for the arts, schools (including the University of Kent and the prestigious King's School founded by Henry VIII), shops and restaurants. Some of the Norman castle's keep still stands, as does the West Gate, the remaining gate of several that locked out intruders during medieval times. Its surviving architecture mixes medieval with post-modern, lending the town an eclectic, modern but impressively historic air for the many tourists who arrive each day—an aura similar perhaps to the one Chaucer's pilgrims might have noted centuries earlier.

Further Reading: A good, concise guide to the long and confusing history of England's rulers is presented by Christopher Daniell in A *Traveller's History of England* (New York: Interlink Books, and Moreton-in-Marsh, Gloucester: Windrush, 1991). Though Canterbury is rarely a focus for the events he relates, Daniell captures the overall historic goings-on and trends from Roman to modern times. For an interesting portrait of medieval times and Chaucer's life see Derek Brewer's *Chaucer and His World* (New York: Dodd, Mead, and London: Eyre Methuen, 1978). A close scrutiny of Canterbury's historic buildings and monuments is provided by John Boyle in *Portrait of Canterbury* (London: Robert Hale, 1974). Having served for thirty years as Town Clerk of Canterbury, Boyle intimately knows the plan and buildings of the city, and its sometimes scandalous, sometimes humorous history.

—Christine Walker Martin

Carcassonne (Aude, France)

Location: In the Languedoc-Roussillon region of France, 350 miles south of Paris and 35 miles east of Toulouse, on the Aude River.

Description: Carcassonne, capital of the govermental department of Aude, has two distinct sections, the Cité and the Ville Basse. The Cité rests atop a hill above the Aude River; reconstructed in the nineteenth century, it is the most complete example of a fortified medieval town that exists in France. It was a key site in the Albigensian crusade in the thirteenth century. The Ville Basse, or lower town, lies on the plain that stretches to and beyond the Aude River, and throughout its history it has experienced the growth that was unattainable for the heavily fortified, and consequently confined, Cité.

Site Office: Office de Tourisme de Carcassonne
15 boulevard Camille-Pelletan
11012 Carcassonne, Aude
France
68 25 07 04

Carcassonne stands at the crossroads of both historic trade routes and routes of invasion, running both east-west and north-south. Even its name is ancient. Two hundred years before the time of Christ, the Celts called this area "Carcaso." Since the time of the Roman occupation of Gaul, there have been walls at Carcassonne.

A fort, built by the Romans on a hilltop that would eventually become the fortified Cité, guarded the route of wine merchants between Narbonne and Toulouse. Between 413 and 435, the Cité was alternately occupied by the forces of Rome and by the Visigoths, a Germanic tribe from the east, who eventually wrested control of the entire Languedoc region from the Romans. Portions of the fortifications of the Chateau Comtal, still extant, date to the fifth century and are credited to the Visigoths. Carcassonne was invaded and occupied by Saracens from Spain in 725. They called the city Karkashuna. Charlemagne later captured the city, and it become strongly Christian, resisting Moslem influence.

In the twelfth century, Carcassonne came under the control of the Trencavel family, viscounts of Béziers and vassals of the Count of Toulouse. The family developed and enriched the walled city. At mid-century, a cathar bishop, Guiraud Mercier, was installed at Carcassonne. The Cathars were a religious group considered heretics by the Catholic Church. They believed that the principles of good and evil were equally powerful, and that all earthly things existed in the domain of Satan. From the tenth century, the beliefs of the Cathars had been spreading westward from eastern Europe and had taken firm root in the Languedoc. Among these beliefs was that salvation was to be found only in a total rejection of all things material. This belief led the Cathars to reject, among other things, the sacraments of the church, as they involved the use of the earthly products of bread, wine, and water. In the Languedoc, the Cathars were called Albigensians (after the town of Albi, a stronghold for the Cathars), and they found acceptance among the ruling family. Roger II Trencavel was excommunicated for a brief period for sheltering the heretics. The Cathars were so protected by the Trencavel family that, in 1204, it was possible to hold an open religious debate in Carcassonne between Catholics and Cathars.

In the early thirteenth century, the walls of Carcassonne were sorely tested. In 1208, Peter of Castelnau, a papal emissary, was murdered. Word reached Pope Innocent III that the murderer was a servant of Count Raymond VI of Toulouse, who was sympathetic to the Cathars. Innocent III raised a crusade to combat the heresy.

In 1209, the Albigensian Crusade, under Simon de Montfort, began its campaign in the south. On July 22, 1209, the city of Béziers was sacked and burned by the crusaders. Some 20,000 of its inhabitants were killed. After Béziers, the crusaders began their march on Carcassonne. Meeting little or no resistance en route, the advance guard of the crusaders arrived at the city's walls on July 28, 1209. The citizens of Carcassonne braced for the worst. Carcassonne was much better fortified than Béziers, and the defenders were led by Raymond-Roger Trencavel, their viscount. The suburb to the north, called the bourg, and the one to south, known as the castellare, were also fortified, surrounded by ditches and walls. Between the Cité and the Aude River lay the unfortified suburb of St. Vincent. The Cité, if cut off from the river, had to rely on deep wells for its water supply.

The main force of the crusaders arrived at the walls of Carcassonne on August 1, 1209. The suburb of St. Vincent was occupied first, effectively cutting off the Cité from the river. On August 3, the fortified bourg was stormed and taken. On August 7, the far stronger castellare was won by the crusaders, but retaken by Carcassonne's defenders the next day.

Despite several attempts at a negotiated surrender by the forces of Simon de Montfort, Trencavel rejected all terms. By August 14, conditions within the Cité had deteriorated badly. Extreme summer heat had dried up the wells. Mosquitoes fed off the carcasses of dead animals in the street. A stench filled the city, and disease was spreading rapidly.

Trencavel agreed to be escorted by the crusaders to a meeting where he would negotiate the surrender of Carcassonne. Once outside the Cité walls, Trencavel was immediately seized and put into chains, in violation of the terms of

A view of Carcassonne
Photo courtesy of French Government Tourist Office

the escort. The remaining defenders did negotiate a surrender and were allowed to leave Carcassonne peacefully on August 15.

The crusaders ransacked the city for valuables, and elected Simon de Montfort Lord of Carcassonne. Raymond-Roger Trencavel was imprisoned near Carcassonne and died in November 1209.

The Albigensian Crusade continued for another twenty years, until 1229, and eventually broke the power of the cathars in the south of France. The crusade also played a role in the consolidation of France under the Capetian kings in Paris and in the establishment of the modern-day borders of France.

Between 1270 and 1290, two French kings, Philip III the Bold and Philip IV the Fair, added to the defenses of the already well-fortified city. Carcassonne remained a major outpost of defense against Spain. Its reputation for being impregnable earned Carcassonne the name "maid of the Languedoc."

In 1659 the Peace of the Pyrenees resulted in the annexation of the territory of Roussillon to France. As a result, the Pyrenees, farther to the south, became the firmly established border between France and Spain, and the need to maintain the strong walls at Carcassonne diminished. Over the next two hundred years, the walls of the Cité fell into disrepair and crumbled. The deterioration continued until the mid-nineteenth century when, under the supervision of architect Eugéne-Emanuel Viollet-le-Duc, the restoration of the walls of Carcassonne was begun.

Viollet-le-Duc, born into a wealthy family, refused to be trained at the École des Beaux-Arts but chose instead to study architecture in Italy. His association with Prosper Mérimée, the French author and an inspector for the French Commission des Monuments Historiques, eventually turned Viollet-le-Duc's interest and energy to the architecture of the French Middle Ages.

History has found little merit in Viollet-le-Duc as an architect. His original buildings are considered commonplace. As a scholar of medieval architecture, however, he was both consistent and daring. Viollet-le-Duc saw the Gothic style of architecture as the beginning of the ascendancy of the lay world over the religious domination of the Middle Ages. Although Gothic cathedrals reflected the glory of God and the power of the church, Viollet-le-Duc saw them as structures based on the rational principles of a construction system that consisted of ribbed vaults and buttresses. He compared the skeletons of Gothic cathedrals of the Middle Ages to the skeletons of nineteenth-century iron buildings being built in his own age. He also defended the use of modern techniques and materials in restoration.

His first restoration work was in the French city of Vézelay in 1840. Then came restorations of the churches of Sainte Chapelle and Notre Dame in Paris. In 1844, Viollet-le-Duc had begun restoration work on the basilica of Saint-Nazaire in Carcassonne when, with local archaeologist Jean-Pierre Cros-Mayreveille, he decided to attempt a restoration of the entire medieval Cité. With the backing of Mérimée, the restoration began in 1855. The work was eventually completed by pupils of Viollet-le-Duc in 1898.

Critics of his restoration question whether his work is an authentic reflection of medieval architecture in southern France. They claim that perhaps, as an architect from the north of France, he never understood the medieval Languedoc. However, he did make use of local masons, stonecutters, blacksmiths, carpenters, and sculptors in his restoration of Carcassonne. Balancing the criticism of the work is the comment of American author Henry James, who even though he preferred ruins over restoration, described Carcassone as "a splendid achievement."

Nearly two miles of ramparts, in a double wall, surround the hilltop Cité. The outer ramparts are built downhill from the inner, allowing for defense from both levels at the same time. Nineteen of the thirty-five defensive towers protecting the ramparts are located on the outer wall. Between the inner and outer ramparts are the *lices*. Used for tournaments in times of peace, these open spaces afforded no protection for the invader who breached the outer defenses. Two gates allow entry to the city. The *porte Narbonnaise* is on to the east and the *porte d'Aude* on the west. The very heavily fortified *porte Narbonnaise* was constructed by order of King Philip the Bold as protection for the least defensible approach to the Cité. Within the fortified walls is a fortress castle, the Château Comtal, defended by its own moat and gates. Viollet-le-Duc also reconstructed the château's "hoardings," spaces with trap doors for dropping missiles on any invaders who had made their way across the ring of walls.

Also within the walls on the Cité is the basilica of Saint-Nazaire, the first stones of which were blessed by Pope Urban II and laid in 1096. The tomb of Simon de Montfort is located in the basilica, although his body is not. It has been removed from the tomb and placed at Montfort l'Amaury, west of Paris.

The Ville Basse was built in 1260 by King Louis IX, known as St. Louis, and supplanted the already existing suburbs of the Cité that had fallen with such relative ease to Simon de Montfort. The Ville Basse had its own walls and was much better fortified than its predecessors. The increased defenses did not, however, stop the Black Prince, Edward of England, from razing the Ville Basse in 1355, during the Hundred Years War. It was quickly rebuilt.

Viollet-le-Duc also conducted restorations in the Ville Basse, most notably on the thirteenth-century Gothic cathedral of Saint-Michael. The cathedral was incorporated into the defensive ramparts of the Ville Basse in the rebuilding that occurred subsequent to the ravages caused by the Black Prince. As a result, the exterior of Saint-Michael itself resembles a fortress. Former entrances, now blocked up, can be easily traced.

An essential factor in the growth of Carcassonne and the Ville Basse is its connection to the Canal du Midi. The canal, which traverses the southwest corner of France between Toulouse and Agde, connects the Atlantic and the Mediterranean.

In Roman times, Augustus Caesar wished to build a canal across the isthmus of Gaul. Charlemagne shared this desire. Through the years, plans for such a canal were proposed by, among others, the French kings Francois I, Henri IV and Louis XIII. None of these plans was brought to fruition, due to the engineering problems involved in providing a year-round water supply for such a canal and the complications in financing the project.

In 1660, however, a solution to the problem of the water supply found. Pierre-Paul Riquet, the Baron of Bonrepos, was in charge of collecting the salt taxes in Languedoc-Roussillon. He also was a contractor for the royal army, which at that time was engaged in nearly constant warfare against Spain. Riquet was interested in a canal as a means of easing the difficulty of both travel and supply in the region. After consulting with, among others, a fountain maker and a mathematician regarding the amount of water necessary to supply a canal 52 feet wide and 6.5 feet deep, Riquet proposed to divert streams from the Black Mountains to the pass at Naurouze, a point 620 feet above sea level and higher than any point on his proposed canal. From this location, water could be directed either to the Garonne or the Aude River, thereby ensuring a year-round water supply.

Riquet sought assistance for his plan from the Archbishop of Toulouse who, in turn, put Riquet's plans in the hands of one of Louis XIV's more powerful ministers, Colbert. Colbert called Riquet to Paris. Riquet's plan was compatible with Colbert's desire to expand French trade.

At his own expense, Riquet dug a trial canal between Naurouze and Revel. In October 1666, Louis XIV signed an order for the construction of a "navigable canal between the Mediterranean Sea and the Ocean."

Construction began in 1667, with 12,000 workers using picks and spades to dig the canal. Riquet, exhausted from his long work in realizing the dream of his canal, died in October 1680. Paul Mathias, another engineer, completed the work begun by Riquet. The first cargo, twenty-three barges of grain, was transported to Beaucaire from the Garonne River on May 15, 1681. Expansion of the canal has continued into the nineteenth and twentieth centuries. Its length has reached nearly 150 miles. Through the years, the principal cargo on the canal has been grain.

Although it remains the preeminent example of a fortified medieval town, Carcassonne has marched with the times. In 1847, gas illumination was installed in the city. In

1857, railroad service began. In 1928, the city celebrated the 2,000th anniversary of the founding of the Cité, and in 1942, 2,000 of Carcassonne's inhabitants demonstrated in support of the French Resistance against the Nazis. On August 20, 1944, the German army left Carcassonne, executing members of the Resistance and burning the *quai Riquet* in its wake, calling to mind the actions of the Black Prince in the Ville Basse 600 years earlier.

Today, although some of its economy is based on leather tanning, wine production, and the manufacture of hosiery, Carcassonne is primarily a tourist center. In France, only Paris and Mont-St.-Michel attract more visitors each year. The Canal du Midi remains important to the Languedoc, used chiefly by vacationers for recreational cruising.

Further Reading: *Languedoc* by James Bentley (London: George Philip, and Topsfield, Massachusetts: Salem House, 1987) is a casual, informative guide, with photographs, to the history and the small towns of the Languedoc-Roussillon region of France. *The Albigensian Crusade* by Jonathan Sumption (London: Faber and Faber, 1978) is a scholarly but very readable history of the crusade and its effects on the social, political, and ecclesiastical life of southwestern France. Detailed attention is also paid to individual battles and military strategy. In French, the *Histoire de Carcassonne* (Toulouse: Editions Privat, 1984), compiled under the direction of Jean Guilaine and Daniel Fabre, details the history of the city and includes a chronological table of important events. It is unavailable in English.

—Rion Klawinski

Carnac (Morbihan, France)

Location: On the Atlantic coast at the north end of the Bay of Quiberon in the department of Morbihan, Brittany, in western France, southwest of the village of Auray.

Description: Village of approximately 4,200 inhabitants; coastal resort; prehistoric site of more than 3,000 granite monuments, mostly menhirs (single stone monuments), arranged in four systems of uneven rows, or alignments, erected during the Neolithic period, probably in the fourth millenium B.C.; contains the passage grave of Kercado and the tumulus of Mont-Saint-Michel within a ten-square-mile area; the largest area of stone rows in Brittany.

Site Office: Office de Tourisme, Syndicat d'Initiative
74 avenue des Druides
56340 Carnac, Morbihan
France
97 52 13 52

In the village of Carnac sits a seventeenth-century church dedicated to St.-Cornély, the patron saint of horned cattle in the province of Brittany. A popular legend holds that St.-Cornély, finding himself pursued by Roman soldiers, turned the columns of soldiers into stone, forming the alignments that remain at Carnac today.

The mystery surrounding the origin and purpose of these stones has given rise over the last two hundred years to many theories, some of which have been discounted and others which continue to be debated. The stones were believed at various times to be altars of the Druids (pagan priests), statues of deities, or megalithic calendars. Some early visitors to the area thought them to be phallic symbols and sacrificial stones left by the Gauls.

Before 1800 the sites were viewed by many as territorial markers, or landmarks. In 1483 Pierre Garcie dit Ferrande recorded that he had used the mounds at Locmariaquer, a few miles east of Carnac, as guideposts for sailing his ship. In 1805 Count Maudet de Penhouët claimed that the stones represented records of battles fought in the area. Jacques de Cambry called them astronomical observatories for Druids.

By the middle of the nineteenth century, it became clear that the megaliths preceded the Druids by several millenia. Egyptologists, seeking to link the cultures of western Europe with those of the Mediterranean, assumed that the megaliths were copies of Egyptian tombs. In 1870 the Reverend W. C. Lukis was the first to suggest that the alignments may have been connected with funeral rites.

More recently Alexander Thom, a Scottish engineer together with his son, Archie, spent some forty years visiting stone rows in France and England, making accurate maps of each site. Thom believed that there was a standard unit of measure used in Brittany and Britain, equal to 2.72 feet in length, which he called the megalithic yard. Between 1970 and 1974, the Thoms plotted 3,000 stones in the Carnac region. From his research, Alexander Thom deduced that the rows had been erected to mark the extreme risings and settings of the moon. While his theory remains highly controversial and unaccepted by most archaeologists today, his accurate maps of Carnac sparked a renewed interest in the rows, which had been neglected by archaeologists for some time. While Neolithic settlers of Carnac would have been likely to observe the movements of the sun between the two solstices, lunar movements were quicker and more complex, waxing and waning each month, and completing a cycle of positions along the horizon once every 18.61 years. It would have taken generations of observations to chart the extremes.

Recent advances in radiocarbon dating and tree-ring calibration have shown that the megaliths in Brittany existed long before any in the Mediterranean, and much earlier than was previously believed. Dating the monuments has always been problematic, due to the influence of the relationship between the monuments and the communities of people who exist side by side with them. Some monuments have been given supernatural names or assigned supernatural powers. Even in Christian times, childless women would rub up against the Giant Menhir of Kerderff, near the Ménec alignments, hoping for fertility. While in some cases this sense of awe has helped to preserve the monuments, it has also helped to perpetuate the myths surrounding the origins of the stones. Seen as holy places, some of the tombs and menhirs were Christianized with crosses and other decorations added by later settlers of the region.

Surveys and excavations began in 1820. The early explorers recorded some information about the megaliths that was helpful to later archaeologists who found some of the sites disturbed by development of the communities. For example, it is now known that between 1826 and 1835, builders tore up hundreds of stones from the Carnac area to build a lighthouse on Belle-Île, an island to the south of Quiberon.

Another cause of the confusion in dating was the practice of reuse of some of the passage graves at Carnac by later inhabitants of the region. The chambers were closed, similar to Bronze Age vaults, and a few metal objects were found in them, associating them with the Bronze Age. Radiocarbon dating has placed the Tumulus Saint-Michel at Carnac in the fourth millenium, i.e., before 3000 B.C. French archaeologist P. R. Giot has dated other sites in Brittany in the fourth or fifth millenium B.C., placing the region firmly in the Neolithic Age.

Prehistoric stone monuments at Carnac
Photo courtesy of French Government Tourist Office

James Miln, a Scotsman eager to do research, began excavation of Carnac in 1874. After three years, he uncovered a Roman villa at Le Bossenno, which has since been destroyed. Upon his death he left all of his archaeological findings to the village of Carnac. This was the beginning of the Miln-Le Rouzic Museum that bears his name today.

In 1904 the Société Préhistorique Française was founded to control the investigations of prehistoric monuments. In Brittany, Zacharie Le Rouzic, a student of Miln, dominated the exploration of the region from 1909 to 1939. He re-erected many of the fallen stones, placing a red concrete plug at the base to show that they had been moved. His classification work has been continued by Glyn Daniel, P. R. Giot, J. L'Helgouach, and others. The interpretation of data has moved away from a focus on the ritual monuments to a broader view of archaeological evidence.

Who were the prehistoric people who built these rows, and why were the stones important enough to invest so much labor to erect them? The people were mostly farmers who grew crops, built homes, raised cattle, made tools, and built their own tombs. The coast of Armorica was a fertile region, but as the population grew, the descendants of the first settlers of Carnac had to move on to less fertile lands.

While excavation has revealed very little about the purpose of the rows, they seem to be associated with cromlechs, or stone "circles," some of which are square or horseshoe-shaped. Le Ménec and Kerlescan each have two cromlechs, one on the east and another on the west. The stones increase in height as they approach the cromlechs, suggesting that they may have provided avenues leading to the cromlechs. The cromlechs may have been used as places of assembly, but it is unclear how many people gathered there at one time. They are wide enough to squeeze in 7,000 people, but they may have served an elite of only 250 or so.

Judging by the charcoal discovered during excavations, it is possible that the cromlechs were used for bonfires to celebrate the solstices. (The axes of the eastern and western cromlechs at Ménec are lined up with the midwinter sunrise and the midsummer sunset, respectively.) Or perhaps the cromlechs were related to funerary rituals, given that the Carnac region has such a large concentration of passage graves and *tertres tumulaires,* or funerary mounds. Recent study of some of the depositions found in the tombs of the Carnac region (fibrolite and jadeite axes, variscite necklaces) have led some archaeologists to speculate that this area must have been inhabited by an elite.

With the shift from a hunter-gatherer society to an agricultural economy in the fourth millennium, a new social order would have been necessary. There would have to have existed an interdependency between generations because of the time lapse between preparation of the land and its productive return. As the elders gained control of religious practices, stone axes became an important element in burial rituals. Carvings found on some of the menhirs include motifs of axes, along with horned animals and shepherds' crooks, echoing the importance of axes in an agricultural society. The deposition of fibrolite axes at the base of the decorated menhir at LeManio 1 implies a relationship between the long mounds and axe exchange.

Competition between social groups gave rise to larger passage graves, requiring greater numbers of people to build them. The jadeite axes excavated at Carnac are rare, suggesting a greater social differentiation in this area. Moreover, the variscite necklaces found in the *Grand Tumulus* monuments can be considered to be symbols of power belonging to a social elite.

The rich resources of this area—fertile soil, marshlands, and proximity to the sea—meant that this society would have had abundant production for some time. Its power seems to have been short-lived, however. By the end of the fourth milllennium B.C., the large *Grand Tumulus* monuments gave way to new monuments which showed little or no social differentiation. The reason for the decline of the elite is unclear. It may have been internal conflict or overuse of resources caused by increasing competition, or perhaps a combination of factors.

During the conquest of Gaul by Julius Caesar in 56 B.C., the Veneti, the tribe who inhabited the region at that time, rebelled against the Romans from the fort-like dwellings in which they lived. Finding it hard to attack the forts, Caesar engaged the Veneti in a battle at sea, where he was able to immobilize their ships by breaking their halyards.

Many years later, after the French Revolution, the same Bay of Quiberon was the scene of another engagement. In 1795 6,000 French *émigrés,* mostly noblemen, joined by groups of peasant royalists called Chouans, tried to invade the Quiberon peninsula under protection of the British navy. They were driven back by the French and forced to surrender.

The stone rows of Carnac attract thousands of visitors each year. In fact, some of the rows have been closed off to protect the ground from further erosion. But the nearby Miln-Le Rouzic Museum contains excellent artifacts, including many examples, mostly casts, of megalithic art from all over France. Of particular interest are the decorated stones in the inner room moved here to be protected.

There are also many examples of the reddish-brown beaker bell pottery collected from the tombs that come from a later period. A bead of faience (a blue-green, glazed glass material) which was discovered by Le Rouzic in 1926 at Paro Guren II is now part of the exhibit. It is believed to have been made by the Egyptians in the thirteenth or fourteenth century

B.C., suggesting that burial chambers were used in several phases spanning hundreds or thousands of years.

A tour of the alignments covers several miles. The Ménec alignment group is first, about one-half mile northwest of the village. There are eleven or twelve rows extending a length of more than 3,000 feet from east to west, each side easing uphill slightly toward a cromlech.

The space between the rows varies from six to twelve meters, and the rows are far from straight. It is likely that the rows were added to in later phases. There are more than 1,100 standing stones in the group, ranging in size from two to thirteen feet in height. The eastern cromlech is now part of the little hamlet of Ménec.

The Kermario group is located 370 yards northeast of the eastern end of the Ménec alignments. Its name means "the place of death." A view facing east is possible from the viewing platform located at the west end of the rows. There are seven nearly parallel rows and three rows unrelated to the others. The rows contain 982 menhirs, ranging in size from fifty centimeters to six and one-half meters. At the east end, the rows pass over LeManio, a *tertre tumulaire,* or long mound. It was found to contain a large menhir decorated with a serpentine motif at its base and five stone axes. This passage grave may be the earliest tomb on the site.

The next group of alignments is located about 440 yards northeast of Kermario. It is called Kerlescan, "the place of burning." The rows come closer together and even cross one another as they move east toward the hamlet of Kerlescan. To the west is the best preserved of the cromlechs at Carnac, a rectangle open on the north side. The Kerlescan alignments have 540 menhirs in thirteen lines, plus thirty-nine more forming the cromlech. To the north is the Kerlescan tomb, a long, chambered mound excavated by Lukis in 1868. It is 150 feet long, and it has two compartments separated by two upright stones with a "porthole" carved out between them.

The Kercado tomb lies to the southeast of the Kermario alignments. It is a circular chamber about 100 feet in diameter. It was excavated in 1863 by René Galles, and a second time in 1924 by Le Rouzic. Its contents from the second excavation—beaker pottery, arrowheads, gold plaques, and variscite beads—can be seen at the Carnac museum. The excavated material discovered by Galles, including jadeite axes and human bone fragments, is on exhibit at the archaeological museum in Vannes. Kercado is an example of an early passage grave, dating from the fourth millenium B.C.

The last group of alignments, Petit Ménec, is made up of two or three damaged lines 250 yards east of Kerlescan. Here many of the stones were removed to build a lighthouse on Belle-Île.

To the east of the village of Carnac, toward the sea, is a large, long mound called the Tumulus Saint-Michel. On top of the monument sits the chapel of Saint-Michel, added during the Christian era. This tumulus is one of the longest in Europe, approximately 370 feet long, 200 feet wide, and 35 feet high. It was excavated by Galles in 1862, and later by Le Rouzic. It appears to have been constructed in several phases. The first phase was probably the construction of the main chambers, covered by a circular cairn. The second phase involved the construction of small cists and the enclosure by a long mound. In the final phase a small passage grave was constructed at the eastern end. The *Grand Tumulus* monuments had no access to the outside, so that excavation had to be done by a series of tunnels. There are six other monuments in the *Grand Tumulus* series, all within twenty kilometers of Carnac. The dating of these monuments is not exact, but some archaeologists believe that they are mid-Neolithic, overlapping with the passage graves.

As the passage graves evolved, there came to be less of a distinction between the chamber and the passage leading to it. The Mané Kerioned C monument is trapezoidal in form, with the passage and the chamber merged. Older graves, such as Kercado, continued to be used in the late Neolithic at the same time that new tombs were being built. Although most burials appear to have been by inhumation, there is evidence of cremation at Kercado and at Mané Lud.

It would be impossible to describe the rituals practiced by these early people from the evidence gathered from excavations. It seems clear, however, that there was not just one ritual, but a sequence of rituals associated with each site.

While the form of the burial monument underwent many transformations during the Neolithic period, it is evident that these monuments played a crucial role in the lives of the people of this region, important enough that they continued to build them and reuse them over a period of 2,000 years.

Further Reading: *From Carnac to Callanish: The Prehistoric Stone Rows and Avenues of Britain, Ireland and Brittany* by Aubrey Burl (New Haven, Connecticut, and London: Yale University Press, 1993) is a comprehensive study of stone rows and stone circles found in prehistoric sites in western France and the British Isles. It includes excellent photographs of many of the sites, as well as maps of some of the alignments at Carnac as recorded by Alexander Thom. *Lascaux and Carnac* by Glyn Daniel (London: Lutterworth, 1955) is a delightful guide to both prehistoric regions. Mr. Daniel's descriptions of the monuments, while lacking the advantage of radiocarbon dating, are detailed and personal accounts of his travels there. *Statements in Stone: Monuments and Society in Neolithic Brittany* by Mark Patton (London and New York: Routledge, 1993) relates the stones to interactions of a living society, offering explanations for what was left behind by the prehistoric people of Brittany. Patton includes many interesting drawings of artifacts and maps showing distribution of various kinds of monuments. *Ancient France: Neolithic Societies and Their Landscape*, edited by Christopher Scarre (Edinburgh: University Press, 1983) is a collection of articles on the Neolithic period in various parts of France. The article ''The Neolithic of Brittany and Normandy'' by James Hibbs provides information on Carnac. *The Bronze Age in Barbarian Europe: From the Megalithic to the Celts* by Jacques Briard, translated by Mary Tarton (London and Boston: Routledge and Kegan Paul, 1979) is primarily a study of a later period, but it does offer some reasons for the confusion on dating the Carnac monuments.

—Sherry Crane LaRue

Cashel (Tipperary, Ireland)

Location: Forty-eight miles northeast of Cork in County Tipperary.

Description: Cashel rises nearly 360 feet above the surrounding plain. The site is dominated by the ruins of a Gothic cathedral, a Romanesque chapel, and the churches of two monastic orders.

Contact: Tourist Information
41 The Quay
Waterford, Waterford
Ireland
(51) 75788

Rising high above the Tipperary plain, the Rock of Cashel dominates the landscape of southern Ireland just as its past masters once dominated the political and ecclesiastical landscape of medieval Ireland. In the fourth and fifth centuries, Cashel (also the generic Irish name for a stone or earthen ring-fort) was established as the home of the Eóganacht Dynasty—the kings of Munster—and it remained the seat of political power in southern Ireland until the site was given to the Catholic Church in 1101 by King Murtagh O'Brian. Cashel became the second archbishopric of Ireland ten years later, and in the following century Dominican, Cistercian, and Franciscan orders founded houses near the archbishop's cathedral. It continued to be a center of episcopal and monastic activity until the mid-sixteenth century when the monasteries were dissolved under the reign of Henry VIII.

The center of Cashel has always been the 360-feet high outcropping of limestone that today is topped by the ruins of a twelfth-century Romanesque chapel and round tower, a thirteenth-century Gothic cathedral (rebuilt in the sixteenth and seventeenth centuries), the main hall of a secular college built in the fifteenth century, a high cross dating to about 1150, and several intricately carved tombs. Below the rock to the southeast are the remains of the town of Cashel. It was a walled city of roughly twenty-eight-acres that probably provided goods and services first to the Munster kings, and then to the archbishopric. The remains of the town wall, the church of the Dominican friary, and the Cistercian abbey's chapel and cloisters are the only visible reminders of the medieval town.

According to one legend, Cashel was founded by Conall Corc, leader of the Eóganacht tribe upon his return to Ireland in the fourth century after an extended exile in Britain. Another legend claims Cashel was established after two swineherders were visited by an angel who revealed the future history of the kings of Munster to them. They reported the vision to the area's tribal king, who gave the land to the swineherder, who subsequently sold it to Conall Corc. His-

torians have tried to reconstruct the founding and early history of the Munster kings, but the extant annals are largely silent and there is scant archeological evidence. The story that has emerged from legend and tradition is one of political ambition, inter- and intratribal warfare, and the gradual transformation of a political center into a religious one.

By the mid–fifth century the Eóganacht dynasty had returned to southwestern Ireland, established their rule in the northern areas of the province of Munster, and built a fortress at Cashel. Through military conquest and a series of alliances with tribes to the south, they became the dominant power in the southern half of the island by the middle of the sixth century. Yet compared to the northern dynasty of the O'Neill (Uí Néill) and the kings of Leinster (the province lying to the northwest) and Connacht (the province lying to the northeast), the Eóganacht were a weak house both militarily and politically. Unlike the O'Neill, who had ambitions to unite all of Ireland under their centralized rule, the Munster kings generally ruled over a loose federation of southern tribes and seemed content to avoid the constant intertribal warfare of the seventh and early eighth centuries. Instead, Cashel and Munster were becoming key centers of Christianity. Near the end of the fifth century, St. Patrick allegedly visited Cashel and baptized the king, Aenghus (Cengus). Thereafter there was a close connection between the church and the kings at Cashel—several were bishops or abbots and others were instrumental in bringing ecclesiastical reform to the Irish church throughout the Middle Ages. The first effort at reform, the Céli Dé movement in the eighth century, was based in Munster and perhaps owes its origins to the political and religious environment, created by the Munster kings, that encouraged the development and extension of Christianity in the south.

The relative peace of Munster would come to an end in the eighth and ninth centuries as internal and external enemies brought war to the kings of Cashel, and as the Eóganacht made a bid for the high kingship of Ireland. By the early decades of the eighth century, the O'Neill dynasty had consolidated its power in the north and the eastern midlands, and was pushing to control the south. Munster, which had become the dominant kingdom in the south, responded by attacking the O'Neills and their allies in the 720s and 730s under Cathal mac Finguine, and then again under the powerful reign of Feidlimid mac Crimthainn between 820 and 840. Although both Munster rulers rebuffed O'Neill attacks and successfully raided the monasteries in their territory, the O'Neill proved stronger and by mid-century had defeated the Munster forces in several battles, carried off hostages from Cashel, and won the fealty of the king of Munster. During this same period, the Eóganacht rulers were confronted by increasingly frequent raids by Vikings, who established a

The Rock of Cashel
Photo courtesy of Bord Failte, The Irish Tourist Board

permanent presence at Waterford and Limerick. Although these defeats weakened the Eóganacht dynasty, it may have rallied if not for the rise of the Dál Cais, a tribe from eastern Munster with aspirations to replace their overlords. The Dál Cais effort to depose the Eóganacht and take Cashel began in the 940s, and they finally took control of Munster under Brian Boru (or Bóroime) in the early 980s. Unlike the earlier kings of Cashel, Brian Boru tried to obtain the high kingship of Ireland. Over the next thirty years Brian battled the Kingdoms of Leinster and Connacht (usually winning the battles, but unable to ultimately control the territory), routed the Norse of Dublin, and forayed repeatedly into O'Neill territory in the north. By 1006, Brian was the recognized king of Ireland, having defeated the island's major kingdoms and extracted their oaths of loyalty. During his reign, Brian began to rebuild Ireland's infrastructure (e.g., roads and bridges) and its religious and educational institutions that had been damaged or destroyed by the years of war. This work was cut short, however, when he died putting down a revolt by Leinster and the Dublin Norse at the Battle of Clonfort in 1014.

The death of Brian opened the competition for the high kingship to all of Ireland, and it also signaled the beginning of the end for Munster as a political power in Ireland. Internecine struggles for the throne and challenges to it from Leinster and several minor kingdoms to the north and east of Cashel divided and weakened Munster during the eleventh and twelfth centuries. Yet, the kings of Cashel continued to wield ecclesiastical power during the medieval period. For example, Fedlimid mac Crimthainn (king from 820 to 846) was a zealous advocate of the Céli Dé reform movement, and he tried to use his military power to enforce a more rigorous set of rules to guide the spiritual practices in his kingdom's monasteries (e.g., simplifying dietary regulations but not promoting fasting). He also tried to influence the appointment of abbots by the monks at Clonmacnois and attacked monasteries in O'Neill territory not only to enrich his treasury but to put them under O'Brian rule. Brian Boru was also active in church affairs. In 1005 he visited Armagh where, according to author Brian Ó Cuív, "he confirmed to the apostolic see of St. Patrick the ecclesiastical supremacy over the whole of Ireland." One historian notes that Brian was probably responsible for bringing English mass-books and service-books to the Irish church. Several of Brian's descendants and successors to the Munster throne continued to be strong patrons of the church.

Brian's great-grandson, Murtagh O'Brian was probably the most capable and influential of the Munster patrons during the remainder of the medieval period. Murtagh frequently wrote to Anselm, archbishop of Canterbury, who was one of the Catholic Church's leaders in instituting the Gregorian reforms. Undoubtedly the two discussed the many problems facing the Irish church, including simony (the buying and selling of ecclesiastical benefices), the ownership of sacred property and offices by laymen who passed them down to other family members, and clerical marriages. Together they formulated a plan to reorganize church practices and policies to make them more congruent with those used on the Continent. In 1101 Murtagh called the First Synod of Cashel, at which clerical and lay leaders from southern Ireland rewrote the rules of the church. A papal legate presided over the synod and it passed eight decrees that were intended to purify the church and make it independent from secular powers. The first outlawed simony and the eighth narrowed the range of morally acceptable marriage partners (e.g., a man could no longer marry his stepmother), and condemned divorce. The remainder addressed the question of ecclesiastical autonomy. The second decree excluded churches and monasteries from having to pay rent or tribute to secular authorities, the seventh exempted clergy from secular jurisdiction, and the third forbade lay ownership of ecclesiastical benefices (i.e., a church office that includes an income).

At the end of the synod, Murtagh gave Cashel, his seat of government, to the church in perpetuity. This gift may have been motivated by political ends—namely to deprive the Eóganacht (one of the claimants to the Munster throne) its ancestral home and thus minimize its power, and to encourage the church to establish a second episcopate in the south to rival Armagh in the North, and thus bring some power back to Munster. Ten years later (1111), Murtagh was instrumental in calling a second synod at Rath Breasail (near Cashel). This time clergy from the whole of Ireland attended, and the synod successfully reorganized the structure of the Irish church. The extant annals provide little information about the decrees that were passed or the nature of the discussions. However, they do report that the church in Ireland was to be organized along the lines of the English church—two archbishoprics with twelve bishops each. Armagh was named the northern and Cashel the southern archbishopric.

A final church synod was held at Cashel in 1272, although this one was convened by the conquering Henry II of England rather than by one of the O'Brian kings. In 1169 the Norman monarch of England commissioned a small army to help a Leinster king win back his wife and lands, as well as plant the Norman flag on the island. The Normans met resistance and by 1172 had established themselves only along the eastern coast at Dublin and south to Waterford. Henry was determined to visit his new territory and sailed to the island with an army of 4,000 in October of 1172. The Irish kings of the south recognized the power of the English king

and offered their loyalty to him throughout the fall. During this time Henry visited Cashel, and as part of his colonization project, called a national council of Irish bishops to reorganize church practices according to the rules and norms of the church in England. In the existing accounts of the discussions and decrees issued by the council, it is recorded that the Irish bishops agreed to be guided by the rules and practices of the English church, and that they pledged, as the Lord of Ireland, their loyalty (and thus that of the Irish church) to Henry II. The synod also legislated on matters of personal morality—marriage and divorce, the necessity of baptism, holding masses for the dead, and tithing to the church. Just as the Norman invasion and Henry's 1172 visit ended Ireland's political independence, so too did the Second Synod of Cashel end the independence of the Irish church. It had finalized the reforms that had already occurred on the Continent and aligned its rules and practices with those sanctioned by Rome and England.

This final church-wide meeting marked the climax of Cashel's importance as a political and religious center in Ireland. From the eleventh through thirteenth centuries the archbishops of Cashel and the kings of upper Munster (Thomond) and lower Munster (Desmond) embarked on an ambitious building program. Between 1101 and 1111 the first cathedral was built atop the Rock of Cashel, probably under the guidance of Murtagh O'Brian. In 1134, the king of Desmond, Cormac Mac Carthy (Mac Carthaig) added a small Romanesque chapel to the cathedral. Known as Cormac's Chapel, it is probably one of the finest examples of this type of architecture in Ireland, and it is noteworthy for its steep stone roof, which contains a vaulted chamber, as well as the intricate carvings around the doorways and windows. The cathedral was enlarged in 1169 by the Thomond lord, Donal Mor O'Brian (Domnall Mór O'Brien) and then again by three successive archbishops in the thirteenth century. As an episcopal see, Cashel also attracted a number of religious orders that established houses near the cathedral. Benedictine monks built a monastery just west of Cashel near the beginning of the twelfth century and may have been responsible for bringing the German workers or at least the expertise necessary to build Cormac's Chapel. In 1243 the Dominican order founded a priory near the cathedral, and English Franciscans established a house in 1265. The last monastic order to build at Cashel were the Cistercians, who were brought by the Cistercian archbishop of Cashel, David Mac Carwill, in 1272. Hore Abbey took over the lands of the Benedictine monastery as well as the supervision and income from St. Nicholas hospital in the town. Finally, a college for vicars was built on the Rock in 1406.

As political power slowly shifted to the north and east (i.e., Dublin and England) and religious power to Canterbury and three other Irish archbishoprics, Cashel settled into the quiet life of a provincial religious town. The English Reformation under Henry VIII hurried Cashel's slide into desuetude, first by stripping Cashel of its status as an important Roman Catholic center and locating a Church of England

Center at the site, and second by dissolving the monasteries surrounding the Rock and giving or selling their rich lands to secular landowners. The cathedral and buildings atop the Rock of Cashel fell into disuse and eventually into the ruins that remain today, while the town under the Rock became, in the words of historian Kevin Whelan, "a wretched shanty town which straggled into existence along its approach roads." Today Cashel remains a site of magnificent religious ruins, set amid the green farmlands of county Tipperary. The quiet among the ruins belies the noise and activity of kings and clerics that once filled the churches and cloisters at Cashel.

Further Reading: In 1972 the Gill and MacMillan publishing house of Dublin produced a series of books on Irish history. They are well-documented, concise accounts of the political, military, and religious events of ancient and medieval Ireland. They include *Ireland before the Vikings* by Gearóid Mac Niocaill; *Ireland before the Normans* by Donncha Ó Corráin; and *Anglo-Norman Ireland, c. 1100–1318* by Michael Dolley. Micheal Richter's *Medieval Ireland: The Enduring Tradition* (London: Macmillan, and New York: St. Martin's Press, 1988) is a more comprehensive history of ancient and medieval Ireland that places the development of Irish society, arts, and education in the context of its religious and political history. *The Irish Church in the Eleventh and Twelfth Centuries* by Aubrey Gwynn (Dublin: Four Courts Press, 1992) and *The Church and the Two Nations in Medieval Ireland* by J. A. Watt (Cambridge and New York: Cambridge University Press, 1970) provide detailed narratives of the relationship of church and state prior to the Reformation and outline the development of Christianity in Ireland.

—Stephen Ellingson

Cesis (Latvia)

Location: At the base of the Vidzeme (Livonia) highlands, approximately fifty-five miles northeast of the Latvian capital of Riga, straddling both banks of the River Gauja.

Description: Ruled for centuries by a variety of foreign powers; center of the crusading Livonian Order, created by Pope Innocent III to convert the residents of the Baltic area during the Middle Ages; today a historical city protected by the Latvian government.

Site Office: Cesis Tourism Bureau
3 Uzvaras bulvaris
Cesis LV-4101
Latvia
371-241-2224

Populated since early medieval times, Cesis is Latvia's second oldest city after Riga. Throughout its history it has also been known by the German name Wenden and the Russian name Tsesis. The first mention of the town of Cesis is found in the Livonian Chronicle, written in 1225. Composed in Latin by Latvian chronicler Henricus de Lettis, also known as Henry Litovsky, the Livonian Chronicle is considered the single most important account of early Latvian history. The area was inhabited even earlier than the chronicle, however, for a castle known as Latgale Fortress stood on Riekstu Hill as far back as the eleventh century. This fortress would later become the stronghold for the crusading Fratres Militiae Christi.

In the Middle Ages, the Baltic region was eyed as an important supplier of food and industrial raw materials for surrounding countries. German merchants were especially aggressive in their attempts to establish favorable trade with Latvia. Following the German merchants came the missionaries of the Catholic Church, charged with converting the natives of Latvia by nearly any means necessary. By 1201, Livonia (the Latin term for Latvia and a portion of Estonia) had a permanent bishopric. In 1202, the bishop of Livonia, Albert, had convinced Pope Innocent III to issue a bull authorizing the creation of the crusading *Fratres Militiae Christi*, or Livonian Order.

In 1204, the pope consecrated the Livonian Order. The dress worn by these crusaders included a white cloak marked by a red cross on the shoulder and a sword. From the costume came the popular name of the order: the Knights of the Sword. The newly formed order quickly adopted the early German crusaders' standard of the day: convert or die. The Livonian Order was similar in structure and intent to other contemporary crusading orders. Members of the group were required to take vows of poverty, celibacy, and obedience. It was the duty of all knights to convert unbelievers using force if necessary, and above all to defend the Catholic Church.

Members entering the group were divided into three classes: those chosen as knights; those entering as clerics, or *fratres capellani;* and the *fratres servientes,* or common soldiers and artisans. Only members of noble descent could be chosen as knights. Knights were stationed by district in castles constructed next to the castle-mounds of defeated Latvian nobles. Each castle was self-governed by a council, headed by a *commendator* appointed by the master of the order, an elected official who usually ruled for life.

The Livonian Order had three masters during its existence from 1202 to 1237. The third master, Volquin, was given the site of the vanquished Latgale Fortress in Cesis to build a monastery and residence for the order. Volquin is reported to have been a ruthless crusader, eager for battle and anxious to expand the territory under his control. As he and his order attacked tribe after tribe, Lithuanians began to sympathize with their troubled neighbors. Volquin died in battle in 1236 when two tribes—the Zemgallians and the Lithuanian Samogithians—put aside their differences in order to defeat the raiding and looting crusaders.

The order's numbers dwindled after this defeat. Angered by the brutality of their efforts, Pope Gregory IX sharply criticized the *Fratres Militiae Christi*. However, determined to save the church's influence in the region, the pope in 1237 combined the Livonian Order with the Order of Teutonic Knights, an order based largely in Prussia. Hermann Balk was declared provincial master of the Livonian branch and took up residence in the fortress at Cesis.

The former Knights of the Sword shed their original costume and adopted that of their new order, wearing a white mantle with a black cross. Many of the Knights continued to wear black tunics under their new mantles and the order was subsequently nicknamed the Black Knights.

The Teutonic Order continued battling the local tribes, particularly the Zemgallians and the Samogithians of Lithuania. The tribes attacked and defeated a number of the order's castles, and by 1345 occupied a number of former order strongholds, marching as far as Cesis. Cesis, declared a town in 1323, had a stubborn patriotic spirit despite the presence of the Order's fortress, and is credited with flying the Latvian flag as early as the thirteenth century. As legend has it, the maroon-white-maroon flag of Latvia was first created in Cesis.

With the crusading orders clearing the way for German economic expansion in the Baltics, the Hanseatic League (a mercantile association of northern European towns) was able to gain a firm foothold in Latvia. The league was a vital economic and political force in many parts of

Cesis Manor, top; Cesis Medieval Castle, bottom
Photos courtesy of Latvian Honorary Consulate, Chicago

northern Europe from the thirteenth through fifteenth centuries. The league's name is thought to have come from the medieval German word *Hanse,* meaning "guild" or "association." The profitable group of merchants grew as they protected their new trading routes and towns by ridding the areas of pirates and robbers, building lighthouses to aid in safe navigation, and establishing a highly protective trade guild eager to establish monopolies. By 1282, Riga, the capital city of Latvia, became a Hanseatic member. Cesis followed, becoming part of the league by the end of the fourteenth century. The Hanseatic League soon had over 100 member towns, most of them within Germany.

During the fifteenth century, the town of Cesis grew. Master of the Teutonic Order Walter von Plettenburg occupied the fortress at Cesis and enlarged it substantially during his reign from 1491 to 1535. One notable addition to the castle was the Western Tower, one of few parts of the fortress still standing today.

In the mid-sixteenth century, Latvia became the battleground of the Livonian War, fought from 1558 to 1583 between Russia and Poland. Cesis—then called Wenden—fell to the Poles under Sigismund II in 1561, along with three other districts including Riga, Treiden, and Dünaburg.

Russia under Czar Ivan IV, known as Ivan the Terrible, sought to wrest Livonia from Poland's control. In 1570, Ivan presented Duke Magnus of Denmark with the Livonian crown, a rather odd act since Livonia was not Ivan's to give. Magnus had long had his eye on the Livonian throne. A decade earlier during a visit to the country, Magnus had been graciously received by a people who hoped his stewardship might free Livonia from both Polish domination and Russian aggression. But now Magnus came with a large force of Muscovite soldiers; Russian occupation had merely been substituted for Polish rule. The new king was widely regarded by both native and foreign nobles as a mere puppet of the czar, and the Latvian people were dismayed to find their homeland newly allied with their ancient foes, the Russians. But the Livonian-Teutonic Order found itself ineffective against the powerful occupying force, and in 1561 it dissolved itself.

The alliance between King Magnus and Czar Ivan was short-lived, however, and by 1577 Magnus was eager to escape the czar's control. That year, seeking new support in the area, Magnus turned to the recently nominated king of Poland and engaged him in secret talks. At the same time, Magnus encouraged the Livonian nobility, who had previously resisted his authority, to accept his aid in fighting the Russian troops. Furious at what he considered treason by his vassal, Ivan began a terrible assault on Livonia, leveling the country's castles and strongholds until he had chased Magnus to the fortress at Cesis.

Ivan followed Magnus to Cesis and began a ferocious assault on the town. Hoping to spare Cesis and its population, Magnus surrendered and was imprisoned by Ivan. The garrison at the fortress was determined not to be captured by the czar, and rather than surrender, destroyed themselves. Furi-

ous, Ivan commanded his forces to raze Cesis and torture the town's inhabitants.

Battered but undestroyed, Cesis was recaptured by the Poles in a bloody battle in 1578. King Stephen Báthory of Poland founded a Roman Catholic bishopric in Cesis, presenting the town and its fortress as a residence to Bishop Patricius Nidecki in 1583. Stephen paid close attention to the internal affairs of the newly taken Livonia. The civil restructuring that took place during his rule gave Cesis the same trading and coinage rights as Riga and allowed local businesses to trade along the Gauja River.

Peace under Polish rule was brief, however, and the area was again torn apart during the Swedish-Polish War of 1600 to 1629. In 1600, angered at the treatment of his envoys at the hands of the Poles, Charles IX of Sweden attacked Livonia, capturing several cities including the stronghold of Cesis. Under Swedish rule, the town of Cesis and its fortress were presented in 1622 to the king's close friend and trusted counsellor, Axel Oxenstierna. In 1671, the battered town was nearly destroyed by a disastrous fire.

The Swedish rule of Latvia proved beneficial; Livonia was able to sustain economic and cultural growth under this peaceful master. Once again, however, the respite proved short-lived, and Livonia once again became a battleground, this time for the Great Northern War, fought from 1700 to 1721.

The weary Livonian-German nobles soon sought to end Swedish domination of their lands, turning to other neighbors who would defeat the Swedes but allow reasonable independence and autonomy. The nobles first turned to the Polish throne, at the time occupied by the German Augustus II. When he proved to be weak and inefficient, the Latvians turned to Russian Czar Peter I, the Great.

Unfortunately for Livonia, the nobles had underestimated Peter's desire for total conquest. Determined to rule Latvia once again, Peter ordered the entire land destroyed in an attempt to drive out Sweden. Under the Russian commander Sheremetiev, a number of towns including Cesis were completely destroyed, displacing thousands of citizens. The extreme tactics proved effective, however, and Cesis once again reverted to Russian hands.

With the death of Czar Peter II in 1730, Peter the Great's niece, Grand Duchess Anna, maneuvered herself into the position of empress of Russia. The same year, the empress presented the town of Cesis as a gift to her favorite counselor, Minister of State Ernst Biron, making the former town the private property of Biron. The town changed hands again in 1747 when Empress Elizabeth presented Cesis to Count Bestuzhev-Ryumin. Fire again ravaged the fortress in 1748, then held by the local count Sievers. In 1759, the status of town was once again bestowed upon Cesis.

The nineteenth century saw great industrial development in Latvia, primarily in the capital city of Riga. A railway was started and a branch line connected with Cesis in 1889. Progress, however, was again halted by war; this time the fall of the Russian czar and the ensuing Russian Civil War tore apart the land of Latvia. The Latvian Rifles, a military group formed during World War I to protect the homeland against the advancing Germans, supported the Bolsheviks. The Latvian Rifles had no great love for the Bolsheviks, but were encouraged and ultimately misled by Lenin's promised goal of self-determination for all nations. The Latvian Rifles stormed the city of Cesis and secured it for Lenin and his Bolsheviks on November 9, 1917.

Latvian Rifle support for the Bolsheviks, however, was not as sound as the Russians would have liked, and it was quickly discovered that the Rifles' first allegiance went to Latvia. In response to this unflinching nationalism, the Bolsheviks nearly engineered the destruction of the Rifles by throwing them into battles against enormous odds. Another fiercely patriotic group, the Brigade of Northern Latvia under the command of Jorgis Zemitans, captured Cesis in the name of Latvia on May 30, 1919, reuniting it with the rest of the country.

On June 17, 1940, Latvia once again lost its freedom, invaded by the neighbor its citizens had regarded warily for so many centuries, this time not called Russia, but the Soviet Union. Latvia's Baltic neighbors Lithuania and Estonia were also invaded and incorporated into the Soviet Union at the same time.

On September 6, 1991, Latvia—along with its Baltic neighbors—gained its long-sought freedom from foreign rule. It was officially recognized by the new Soviet provisional body, the State Council, as an independent country. The ancient city of Cesis, one of Latvia's oldest, is today proudly recognized by the Latvian Tourist Board as an important site in Latvia's long and rich history.

Further Reading: Readers interested in tribal and crusading history in Latvia will enjoy Alfred Bilmanis's *A History of Latvia* (Princeton: Princeton University Press, 1951; Oxford: Oxford University Press, 1952). *Latvia in the Wars of the Twentieth Century* (Princeton Junction, New Jersey: Cognition Books, 1983) by Visvaldis Mangulis concentrates mainly on the role of Latvia and Latvians in the unrest that has shaken their homeland in the past century.

—Monica Cable

Chambord (Loir-et-Cher, France)

Location: Nine miles northeast of Blois on the Cosson River, in the department of Loir-et-Cher, in central France; approximately ninety miles south/southeast of Paris.

Description: Château, with large forested domain; one of the most famous of the Loire region châteaux; commissioned by Francis I of France, 1519; construction largely completed by 1547, the year of Francis's death; designed and built under the direction of a succession of architects, on a model by Domenico da Cortona.

Site Office: Château de Chambord
Chambord, Loir-et-Cher
France
54 20 40 18

The brainchild of Francis I of France, Chambord stands as perhaps the culmination of the Renaissance ideals associated with that king. The sprawling château and its enormous, mostly forested, grounds suggest an atmosphere of high leisure—a rich playground for an art- and sport-loving king. French royalty after Francis, notably Louis XIV, maintained the estate as a country retreat, altering the building according to their tastes; but it is as an embodiment of Renaissance architectural style, and as a reflection of the king for whom it was built, that Chambord continues to attract interest. Indeed, architectural historian Ralph Dutton considers Chambord "the most splendid example of French Renaissance architecture in the country."

Twentieth-century historians have not been the only observers to discuss Chambord in superlatives. Jerome Lippomano, Venetian ambassador to Henry III, describing the château and its setting in 1577, wrote: "I have seen during my life many magnificent structures, never one more beautiful or sumptuous [than Chambord]. The interior of the park, within which the château is situated, abounds with forests, lakes, and brooks, with pastures and hunting grounds; and in the center rises this beautiful edifice with its gilded crenels, its leadwork-covered wings, its pavilions, terraces, galleries. . . . We left the place astonished. . . .'' That the Italian ambassador should have been so taken with the forested park surrounding the château is no accident: Francis wanted the grounds to constitute a major feature of the estate, defining the edifice itself in terms of both form and function. Unlike the châteaux at Blois or Amboise, Chambord was to be for Francis strictly a *maison de plaisance,* a retreat from affairs of state where he could pursue his favorite activity: the hunt. Because he wanted to situate his new château far from all commerce (the châteaux at Blois and Amboise were sur-

rounded by towns), he chose a site by the banks of the Cosson River, bought up much of the surrounding marsh-land, and began construction on a wall ten feet high and some twenty-two miles in circumference to surround the estate.

Built on the site where previously had stood the fortified structure from which Joan of Arc set out for the siege of Orleans, the new edifice was commissioned by Francis in 1519, four years into his reign. Little information survives about the genesis of the building, the designer, or the extent of Francis's involvement in the planning. Sources indicate that Francis de Pontbriant was appointed superintendent of buildings on September 6, 1519. What is known after that is only a succession of names of master masons, to no one of which can historians with certainty attach any particular work: Jacques Sourdeau, then his son Denis, then Pierre Trinqueau (until 1538), then Jacques Coqueau (who appears to have been there from 1538 through 1569). A model thought to have been made by Domenico da Cortona (preserved through drawings made by André Félibien) gives some indication of the present structure, although clearly the plans were altered by French masons (perhaps at times even by Francis himself) as the château took shape. Whereas da Cortona's model is a fairly typical Italian plan, Chambord emerged as anything but typical—and in a style that came to be regarded as essentially French.

Doubtless the most striking and distinctive feature to the approaching observer is the château's roof with its fantastic array of chimney stacks, cupolas, turrets, and dormers. The appearance of the roof recalls a cityscape or, perhaps more in keeping with the intention of the place, an overgrown forest to rival the surrounding woods. The central cupola rises above all other structures and acts as a focal point in the composition of the roof; beneath it, and in a sense presupposed by it, lies the great central staircase, which along with the roof is the other outstanding feature of Chambord. This freestanding staircase, consisting of twin sets of stone steps that intertwine with each other around a single shaft but never meet, was not part of da Cortona's original plan. Many have considered the design for this double spiral staircase to be by Leonardo da Vinci, whom Francis had invited to France in 1517, but this attribution cannot be verified.

The main section of Chambord is a square-shaped *donjon,* or central keep, at each corner of which stands a large tower. From this keep extend two long gallery wings that form the front section of the château's outer quadrangle, and at each of the four corners marked by this quadrangle stands a tower similar to those of the *donjon.* Thus the layout of Chambord resembles in some respects that of a medieval fortress. But the design of the keep incorporates a striking innovation: the double spiral staircase ascends from the center of a Greek cross made by the four *Salles des Gardes.* This

The Château de Chambord
Photo courtesy of French Government Tourist Office

Greek cross divides each of the keep's three floors into *appartements*—a feature that, appearing for the first time here at Chambord, would become widely adopted in Renaissance château construction for the next two centuries.

Construction proceeded slowly, all work having stopped between 1524 and 1526 when Francis was in captivity in Italy. By 1526, however, 1,800 workers were on the job, and by 1537 the *donjon* was being roofed. Thus when Francis received the Holy Roman Emperor Charles V at Chambord in 1539, he was able to dazzle the emperor in high style among the most sumptuous of surroundings. Work continued on the wings of the quadrangle, and by 1540 two floors of the king's *appartement* had been built. In 1547, the year Francis died, the *appartement royal* was finished. That so much of the work should have been complete by the time of Francis's death is astonishing given the scope of the project. The west wing was completed in 1550, and some works were still in progress in 1571. No significant alterations have been made since then, except some additions made by Louis XIV to the quadrangle surrounding the court and to the northern front of the *donjon*.

Chambord's history has been checkered since the death of Francis. The château was never inhabited in the manner he had hoped it would be, and despite varying degrees of attention from successive kings, its course soon became one of disrepair. Under Henry II, Francis's son, much of the remaining work was completed. However, with the death of Henry II in 1559, and with the Wars of Religion (1562–98) straining the royal purse, further progress and repairs were interrupted. Catherine de Médicis, the wife of Henry II, visited Chambord often with her sons Charles IX and Henry III; but after the death of Charles IX in 1574, the royal retreat lay mostly dormant for the better part of half a century.

It wasn't until the middle of the seventeenth century that Chambord was brought back to anything like the life for which Francis had intended it. Louis XIV repaired there often to hunt, especially in the early years of his reign, and he apparently maintained his court with accustomed flair and *richesse;* Molière's "Monsieur de Pourceaugnac" and "Le Bourgeois Gentilhomme" were given their first performances at Chambord in 1669 and 1670 respectively. For Louis XIV the first floor of the north front of the *donjon* was altered to create a splendid suite of rooms; the northern arm of the cross was closed off, thus joining two of the corner *appartements* into a grand *appartement royal*. Other changes were made, as well as some much-needed reconstruction of the vaults in the central court, by Jules Hardouin-Mansart, who would later become the king's architect at Versailles.

After Louis XIV's attention was taken up completely by the new royal residence at Versailles, Chambord again fell into disuse and disrepair. His successor, Louis XV, took little interest in the massive Renaissance structure, preferring the more delicate rococo style popular during the early eighteenth century. In 1725 he offered the château to his father-in-law, Stanislaw I Leszczyński, the ousted king of Poland, who remained there until 1733. The ex-king's exile at Chambord was mostly unhappy owing to his ill health and the generally dilapidated conditions of the building. After another period of abandonment lasting fifteen years, the château was again occupied, this time by the illustrious military figure the Maréchal de Saxe, who received in 1748 the offer of Chambord from Louis XV in payment for his services to the throne. The general, whose residence there marked the last time Chambord was ever to witness the life of the Court, died in 1750, having done nothing to improve the estate. Since that time Chambord has never again been fully occupied.

The history of Chambord from the mid-eighteenth century was sadly one of continued neglect and gradual decay. After the death of the Maréchal de Saxe, the estate returned to the Crown and fell under the slack supervision of a series of governors. During the French Revolution, most of the movable possessions were either sold or burned. Napoléon gave the estate to Maréchal Berthier in 1809, along with the title Prince de Wagram. This prince confined his efforts at refurbishment to the task of having sculpted for him some ornamental medallions with the letter "W," which he hoped to see replace those bearing the "F" left behind by Francis I. After the prince's death in 1815, his widow, unable to raise the money necessary for maintaining the château, obtained permission from Louis XVIII to sell the estate. Chambord was purchased by public subscription in 1821 and bequeathed to the heir of Charles X (the latter's grandson, who it was hoped would ascend the throne of France as Henry V), the young Duc de Bordeaux, later to be called the Comte de Chambord.

The Duc de Bordeaux received the estate after the Duchesse de Berry, his mother, together with Comte Adrien de Calonne, worked to save Chambord from being sold stone by stone at the hands of the "Bande Noire," which since the French Revolution had seized other châteaux in the country. During the revolution of 1830, the young prince fled France with the rest of the royal family, living exiled in Austria for forty-one years. When the law of exile against the Orleans family was finally repealed, the Comte de Chambord returned to France and spent his one and only night in the château whose name he now bore, on July 2, 1871. From Chambord is dated his *Manifeste* in which he states his allegiance to the white flag of the Bourbons. His refusal to accept the French Tricolor put an end to all hopes of a restoration, and three days later the prince left France, never to return.

Several items that had been prepared for the Comte de Chambord's triumphal entry into Paris as Henry V, including the satin-lined coach in which he was to ride, remain housed at Chambord.

Meanwhile, the estate itself had continued to decay. About the time Chambord was placed on the list of *Monuments Historiques* in 1840, the novelist Gustave Flaubert described the deplorable conditions of "ce pauvre Chambord." After the Comte de Chambord's death in 1883, his nephews the Duc de Parme and the Comte de Bardi oversaw a few minor repairs, but any further efforts to restore the estate were interrupted by World War I. Not until after the state had acquired Chambord from the Bourbon-Parme family in 1930 for 11 million francs did any serious restoration get underway. The French Republic committed the combined efforts of the Administration des Domaines, the Ministère de l'Agriculture, and the Secrétariat aux Beaux-Arts to improve the château and its environs. In January of 1947, Michel Ranjard was named chief architect in charge of the restoration. He began his enormous task by taking stock of the innumerable flaws wrought by time, disuse, and misconception in the original construction, noting "the flagrant lack of coordination among the various teams of workers" alongside the other countless instances of disrepair.

Chambord has no doubt benefited enormously from the attention it has received in recent decades. Along with the numerous structural repairs, much of the original detailing and sculpting has been copied and replaced. The château now perhaps approaches in appearance the level Francis had envisioned for it. The *appartements* of Francis and Louis XIV have been restored and decorated according to their original plans, and great pains have been taken throughout to retain the period flavor. The park has been restored to evoke its original hunting-ground aura: numerous deer and wild boar roam the woods contained within the reconstructed wall surrounding the estate. The château contains a hunting museum, and in summer the inner courtyard surrounding the *donjon* is host to musical and theatrical performances.

Further Reading: *Chambord* by Claudine Lagoutte, from the series Ouest-France (La Guerche-de-Bretagne; Raynard, 1984), offers good photographs and detailed analysis of the château's layout, construction, and successive alterations, as well as some informed accounts (or creative guesswork) of the various inhabitants. The analysis, however, too often gives way to nationalistic fervor; and, lacking a translation, Lagoutte's book will be of limited value to the English-speaking world. Less detailed, but still very useful, is the chapter on Chambord by Ralph Dutton in *Great Houses of Europe*, edited by Sacheverell Sitwell, with several fine photographs by Edwin Smith (London: Weidenfeld and Nicolson, 1961). Dutton provides a sound overview of the architecture as well as the history of the château. A specialized treatment by Anthony Blunt in *Art and Architecture in France 1500–1700* (London: Penguin, 1953; New York: Viking, 1954, fourth edition, London: Penguin, 1980) offers a specialized treatment focusing on architectural problems and influences.

—Paul E. Schellinger

Chartres (Eure-et-Loir, France)

Location: Approximately fifty miles southwest of Paris.

Description: A small city with a population of about 40,000 and home to the queen of Gothic churches. The Cathedral of Chartres is one of the great examples of the Gothic style with its beautiful stained glass, ornate sculpture, flying buttresses, and soaring towers.

Site Office: Chartres Tourist Office
Place de la Cathédrale
28005 Chartres, Eure-et-Loir
France
37 21 50 00

Chartres was once a small rural district far from the bustle of Paris. The growth of the great metropolis has closed the distance between the two worlds and now Chartres seems to be just a bit beyond the outskirts of Paris. Nonetheless, it still keeps some of its rural flavor and in spite of the thousands of tourists who visit the city each year, Chartres maintains a more leisurely pace than its larger metropolitan neighbor.

What attracts so many visitors to Chartres each year is the famous Cathedral of Notre Dame at Chartres, commonly referred to simply as Chartres. A stunning example of Gothic architecture, Chartres was an important pilgrimage destination during the medieval period because of its shrine to the Virgin Mary. Thousands made the annual journey to Chartres to celebrate the feast of the Virgin. Things are not all that different today as thousands make the journey to experience this impressive structure, perhaps not so much for religious reasons as for aesthetic ones.

The loftiest expressions of the medieval period are seen in the spires and stained glass of the Gothic cathedral. The cathedral itself represented a composite of the anonymous work of stone cutters, masons, carpenters, and metalworkers. It was the greatest product that a town and its craftsmen could produce. At the time, the importance of a town was in large part measured by the size and the height of its cathedral. The growth of the two, town and cathedral, was indistinguishable just as was the interweaving of affairs both spiritual and worldly among the people. During this period the cathedral was more than simply a spiritual center, it was also the geographical heart of not only the town but of people's lives as well.

As with so many other church sites in Europe and indeed throughout the world, the structure that now stands in Chartres is but a relatively modern version of the many structures that have stood at the site dating back to at least 743, when there is record of a cathedral being destroyed by fire. A later construction was also destroyed by fire in 858, when the

town was sacked by the invading Danes. In 962, 1020, 1134, and 1194 other fires either damaged or destroyed the cathedral standing on the site now occupied by Chartres Cathedral. The building that we know today dates from the early thirteenth century. Unlike so many other medieval churches, which were constructed over decades and sometimes centuries, the most of the construction at Chartres took place during a relatively brief period. Construction began immediately after the fire of 1194 and was essentially complete by 1222. The Gothic structure was not dedicated until 1260 after most of the decorative elements, including the stained glass, were in place. Because of the time in which it was built, Chartres is unique in that it is the culmination of the early Gothic period and the pacesetter for what was yet to come. It was the first cathedral built in the high Gothic style. Later cathedrals are taller and bigger, but none has the same unified feeling as Chartres and none has such a complete and beautiful collection of stained glass. There are no examples of a truly complete Gothic cathedral. The ideal plan has seven towers; Chartres was once planned with nine. The complete Gothic cathedral exists only in drawings.

Medieval Chartres was an important pilgrimage church in large part due to one particular relic, a piece of cloth known as the "tunic of the virgin," which was thought to have been worn by Mary. The story of how the tunic came to Chartres is an interesting and oft-repeated tale. Legend has it that when Mary felt that she was soon to die, she gave her clothes to an honest widow who had served her since the time of her son's death. The clothing was responsible for many miracles in Palestine, and those who were charged with its care had beautiful reliquaries made for the clothes. The tunic in question is said to have eventually come into the possession of a Jewish woman. Two Christian pilgrims who happened to stay with the woman were able to steal it from her through trickery. They then took the relic back to Constantinople where they attempted to hide it. However, the tunic made itself known by several miracles and it was taken by the emperor who built a domed temple for it. The reliquary became known as the protector of the empire and remained in Constantinople until it was given to Charlemagne when he passed through the city returning from Jerusalem. He had the relic taken to his chapel at Aachen and eventually it was given to Chartres by Charlemagne's grandson, Charles the Bald. The tunic became the most important relic at Chartres and was credited with many miracles. The cult of Chartres and the cult of the Virgin depended in large part on this relic. The tunic was kept in the grotto or crypt of the cathedral along with a statue of the Virgin, which was also an important item in the development of the cult of Chartres. In addition to these two items there was also a well in the grotto, and the water from the well was also an important element in

Chartres Cathedral
Photo courtesy of French Government Tourist Office

Chartres's appeal as a pilgrimage site, for the water was thought to have curative powers. The constant stream of pilgrims into Chartres was essential for the small town's economic survival.

The cathedral at Chartres that originally housed the tunic of the Virgin was essentially destroyed by fire in 1194. Miraculously, the tunic survived, a feat that spurred the people in their resolve to rebuild. Chartres Cathedral as we know it today was reconstructed in the record time of only twenty-seven years, incorporating some of the surviving elements from the earlier structure. The pilgrims provided much of the funding for the construction of the new cathedral and even carried wood and stone to the construction site and contributed other forms of unskilled labor.

Chartres today is unique among medieval cathedrals in that it remains essentially unchanged from the period in which it was built. Constructed of a very durable limestone, the structure has required very little repair, and few alterations and changes have been made to the building in the past 700 years. Furthermore, it has been spared the ravages of war.

As the finest existing example of a Gothic cathedral, Chartres has been the subject of endless studies. One of the things that makes the structure so interesting is the almost complete anonymity of the builders. Volunteers, paid craftsmen, and guild members certainly all worked on the cathedral, but nothing is known of specific individuals or even master craftsmen. It is not even clear exactly how the construction was supervised. Indeed, a close examination of the structure reveals inconsistencies throughout. However, particularly from a distance, the overall effect of the cathedral is of a unified, planned whole.

A fine example of this is the west facade, which is made up of two very different towers as well as a triple portal and rose window that had survived from the previous Romanesque church. The towers were completed some 400 years apart, thus accounting for the difference in style. The west entrance is the most diverse at Chartres, while the north and south porches were created during the same period and hence are more unified in style.

Upon entering the cathedral, one is struck by the absence of walls. The nave and the aisles are delineated only by columns and the outer walls are filled with stained glass. The nave itself is 53 feet wide, 130 feet long, and 122 feet high. Beyond the transept one enters the choir and apse with its circulating apsidal chapels, another distinctive feature of the Gothic style. The most striking element in the interior of Chartres Cathedral is the stained glass windows, which bring in the natural light while still maintaining a mystical, otherworldly atmosphere within the cathedral. In contrast to the preceding Romanesque style, in which thick masonry walls did not allow for many windows, Gothic was light and airy. The development of flying buttresses, which removed the support functions of the wall to detached piers, allowed for a much lighter and thinner wall, which in turn could house stained glass windows. The stained glass of the Gothic ca-

thedrals replaced the murals and mosaics of earlier styles and gives these structures their most distinctive quality. The two basic styles of windows are the rose, or circular, windows and the lancets, or tall, arched windows. The stained glass in Chartres was donated, and in most cases there is an indication within the window itself of who the donor was. The guilds, for example, are represented in at least nineteen of Chartres's windows, and in each case the signature of the guild is represented by a craftsman engaged in the typical work of the guild. Royal families also donated windows and usually had them marked with their coat of arms.

Unlike the monasteries of an earlier period, the Gothic cathedral was meant for the general populace, and thus the greatest amount of sculpture was on the outside, generally clustered around the portals. While there is also sculpture in the interior, particularly on the tops of the columns, it is at the entrances to Chartres Cathedral that one finds the greatest abundance of sculpture. The figures presented are closely related to a new iconography that developed during the medieval period and depended on a logical, chronological sequence of presentation.

Each of the porches at Chartres contains some 700 carved figures combining biblical scenes and lives of the saints with mythical figures, fantastic animals, and contemporary history. The doorways of the west facade are called the royal portals, and the sculpture there concentrates on the life of Christ. The central tympanum presents Christ in his majesty surrounded by symbols for the four evangelists and the twenty-four elders of the Apocalypse. On either side of the central tympanum Christ is presented at the start and the close of his earthly life. This portal dates from the earlier structure and the sculptures have a certain static quality characteristic of the Romanesque style. As one would suspect, the sculptures in the north and south portals are somewhat more lifelike and freer in their execution. These portals were completed later than the west portal, and their sculptures more clearly represent the qualities of the Gothic. The portals are equally filled with figures, but there is a greater concentration on Mary and her earthly mission.

The Gothic period was characterized by an attempt to find a synthesis between the cares of this world and those of the world beyond. The nearly complete melding of the secular and spiritual life of the people is symbolized by the Gothic cathedral, which served both as a house of worship and a civic center. Brilliant sunlight streaming through stained glass windows brought together the tangible and the ethereal in the creation of a new type of interior space nowhere better executed than at the finest example of high Gothic architecture, the Cathedral of Notre Dame at Chartres.

Further Reading: There is a great deal available on Chartres Cathedral, including texts that deal separately with the sculpture or the stained glass. Among those that attempt a comprehensive look, Jean Favier's *The World of Chartres* (New York: Harry Abrams, and London: Thames and Hudson, 1990) is the most lavishly illustrated with drawings and photographs. This oversized volume is a wonderful introduction to this impressive structure. A more in-depth account of the actual construction of the cathedral can be found in John James's *Chartres: The Masons Who Built a Legend* (London and Boston: Routledge and Kegan Paul, 1982). James's text is filled with drawings and black-and-white photos that clearly explain Gothic building techniques and gives clear explanations and examples through the use of cross sections and diagrams. A more academic approach is *Chartres Cathedral*, edited by Robert Banner and a part of the Norton Critical Studies in Art History (New York: Norton, 1969). Perhaps the most famous description of Chartres is in Henry Adams's *Mont-Saint-Michel and Chartres* (Boston and New York: Houghton Mifflin, 1905; London, Constable, 1914).

—Michael D. Phillips

Chester (Cheshire, England)

Location: In the county of Cheshire, in northwest England, on the right bank of the River Dee, seven miles above its estuary, eighteen miles south-southeast of Liverpool.

Description: England's only walled city, Chester is a major tourist center on the Welsh-English border. Its two and one-half mile promenade along the walls, distinctive "Rows," timber houses, and medieval cathedral make it one of the most medieval-looking cities in England.

Site Location: Tourist Information Centre
Town Hall
Northgate Street
Chester, Cheshire CH1 2HJ
England
(244) 317962

One of the principal gateways to north Wales, Chester is best known for its city walls. The north and east sides of the walled circuit incorporate much of the original Roman masonry while the west and south sides are medieval extensions to the Roman structure. Part of a chain of fortresses built by the Romans in their conquest of Britain, Chester was known in Roman times as Deva, after the River Dee. It derives its modern name from the Anglo-Saxon word *Ceaster,* meaning Roman fort.

Discovery of cremation burials in urns suggests that present-day Chester may have been the site of a wooden fort or temporary base as early as A.D. 60. The stone fortress was built from A.D. 76 to 79, in time for Julius Agricola's conquest of the native Ordovices of northern Wales and the Brigantes of northern England. During the Roman occupation of Britain, Deva became the chief western legionary fortress. It was ideally situated on one of the crossing points of the river, and its deep-water harbor provided a base for the Roman fleet. It originally housed the second Adiutrix Legion from A.D. 76 to 86, and from A.D. 88, the famed Twentieth Valeria Victrix Legion, responsible for patrolling north Wales, Cheshire, and the frontier with Scotland.

The Roman fortress at Deva was similar to most of those found throughout Britain. It was rectangular in shape with the praetorium (commandant's quarters) in the center. It included such facilities as an aqueduct and a lead-piping water system that serviced a large bathhouse and exercise hall. Other structures included barracks, granaries, ovens, and a large amphitheater. Many of the buildings were initially timber constructions that were gradually replaced with stone ones in the succeeding century. Civilian settlements developed outside the eastern and western defenses, but it would appear that these areas did not grow very large.

The Roman garrisons abandoned Chester in the fifth century. It was frequently raided by neighboring tribes from Wales, Scotland, and Ireland. In 615, Aethelfrith, king of Northumbria, defeated a Welsh army, and it is believed that the fortress was then ruined but not totally deserted. In the ninth century, Chester fell victim to the Viking raids, as did much of Anglo-Saxon England. In 894, Danes captured it but were eventually defeated by troops of Alfred the Great, king of Wessex, who had begun to fortify his northern defenses by building a series of fortresses or *burhs,* garrisoned enclosures.

This system was extended to Chester in the tenth century when Anglo-Saxon influence reached Mercia. Aethelflaed, daughter of Alfred the Great, was responsible for much of the refortification of Mercia. She married Aethelred, ealdorman of Mercia, and when he died in 911, she continued to govern as "Lady of the Mercians." She continued the Anglo-Saxon policy of establishing fortified settlements, building a chain of fortresses from Edissbury and Runcorn to the Pennines.

After her death in 918, Mercia passed to her brother, Edward, king of Wessex, thus uniting the two most powerful kingdoms in Anglo-Saxon England. Throughout the tenth century, Chester became a thriving port and a prosperous community. In 973, the West Saxon king Edgar came to Chester, where Celtic kings had gathered to swear their allegiance to him as their overlord.

The death of the last Anglo-Saxon king, Edward the Confessor, and the Norman invasion that soon followed in 1066 brought an end to the peaceful prosperity that Chester had enjoyed under Anglo-Saxon rule. It was the last English town to yield to William the Conqueror. It fell to him in 1070 after he laid waste to much of the Cheshire countryside in the winter of 1069–70. He granted the town and many of the surrounding lands to his nephew Hugh of Avranches, nicknamed Lupus (the Wolf), as a palatine earldom.

Hugh Lupus, with the help of St. Anselm, founded a Saxon monastery and built the Benedictine abbey of St. Werburgh. A church had been founded on the site in the tenth century, as a resting place for St. Werburgh, a Mercian princess who died in the early eighth century. Her burial place became a shrine and a focus for pilgrims during the Middle Ages. According to tradition, the bones of St. Werburgh were brought to Chester from Hanbury to protect them from Viking raids. The earl of Chester richly endowed the monastery with lands and privileges. A great rebuilding of the monastery was undertaken in the thirteenth and fourteenth centuries and included the rectory, chapter house, Lady Chapel, the choir and south transept. The abbey was dissolved in 1540 by Henry VIII. Chester was made a bishopric, and the abbey was changed to the Cathedral of Christ and the

The Chester city walls and water tower
Photo courtesy of Chester City Council

Virgin Mary. The present cathedral buildings are those of the Benedictine abbey.

Chester Castle was also begun in this period. Little of the original medieval construction remains; only Agricola's Tower stills stands to this day. Chester Castle became the seat of government for seven successive powerful earls from 1077 to 1237. The earls of Chester were among the most powerful noblemen throughout medieval England and meddled in many affairs, among them the succession to the English throne. Subsequently, Henry III annexed the earldom to the crown in 1237 and in 1254, and it was conferred on the future King Edward I. Since 1301, the title has been held by the heir to the English throne and has been one of the titles held by successive princes of Wales.

Annexation of Chester by the crown assisted Edward I's conquest of Wales. Chester became a vital supply center throughout the wars with the Welsh in the late thirteenth century. It was also the starting point for Edward's expeditions into Wales. Edward was at Chester when Wales conceded defeat in 1282. Welsh chieftains came to Chester in 1300 to do homage to the prince of Wales.

Spurred by its increasing role during the Welsh campaigns, Chester became the most important seaport in the northwest of England during the Middle Ages. Trade flourished particularly in the thirteenth and fourteenth centuries, as candles, salt, and cheese were exported to Ireland in exchange for grain, hides, and fish. Merchant ships from other European countries, notably Spain and Germany, would also call into port.

As trade expanded, the social and political structures of Chester became more complex. Much of the political gain had been granted through charters either by the king or the earl of Chester. The office of sheriff is one of the earliest in England and dates from the 1120s. The first royal charter, granted by Henry II, reestablished trading rights in Dublin. Chester received its first mayor, William the Clerk, in 1228.

Trade guilds also became increasingly important. As in other medieval cities, guilds were created to regulate and protect local trades and craftsmen. One of the most unusual efforts of the Chester craft guilds was the annual performance of the Corpus Christi plays, better known today as the Chester Plays. These recreations of the story of Christ's life were in fact large processions. Large wagons served as movable stages and were pulled through the streets from one appointed station to another. Each guild was responsible for all or part of a play, including the construction and design of the sets and props. During the Reformation in England attempts to end the annual performances were often ignored. In 1575, Sir John Savage, mayor of Chester, was summoned before the Privy Council to justify his approval of the annual pageant despite the disapproval of the archbishop of York. After this, the plays were never again performed.

Prosperity in medieval Chester also encouraged the development of another distinctive feature of the community. ''The Rows'' represent a particular urban landscape that evolved in the medieval period. While visiting England in 1870, the American novelist Henry James described the Rows as ''an architectural idiosyncrasy which must be seen to be appreciated.'' They are buildings that consist of covered galleries at the first-floor level. Reached by steps from the streets, the galleries are public rights of way on top of shops at street level. Since medieval times, separate owners possessed different stories, which may explain the complex structure of the buildings. Today, the buildings are of varying architectural styles and dates, but common features are the ''seldae'' or stone-built stalls at street level. Above this level, the buildings tend to be timber-framed. The Rows appear to be a method developed to maximize the number of shops on the street. Despite the loss of some of the Rows over the centuries, they were spared total destruction by Chester's economic decline as a major port city in later centuries.

In the sixteenth century, Chester benefited from a revival of trade with Spain and Portugal, importing wine, oil, fruit, and spices. Increased imports of Irish rawhides spurred tanning and other leather crafts. In 1506, Henry VII granted the city its ''Great Charter'' permitting the citizens to elect their own governing body, a local assembly consisting of the mayor, two sheriffs, twenty-four aldermen, and forty common councilmen. This privilege was seldom seen elsewhere in England. The city was to be governed in this way until 1835.

During the English Civil Wars (1642–48), Chester remained loyal to Charles I. The king visited Chester in 1642, and later in 1645, while the city was intermittently besieged by the Parliamentarians. The king is reported to have watched the defeat of the Royalist troops at the Battle of Rowton Moor, fought about three miles southeast of Chester, from the tower on the city walls. Charles I fled into Wales following the battle, eventually giving himself up in March 1646 at Newark to the Scots, who handed him over to Parliament in 1647. Chester capitulated in February 1646.

In the eighteenth century, Chester's economy declined. There was an attempt to improve the navigation of the River Dee. A new channel in the river, the New Cut, and a canal were dug, but the improvements were ineffective and Chester eventually lost its position as a major port center to Liverpool. The new channel, however, attracted some new industry to the city.

The city was also undergoing considerable aesthetic changes throughout the eighteenth and nineteenth centuries. Chester Castle was extensively rebuilt between 1789 and 1813. Grosvenor Bridge was built in 1832. Until its construction, the only local bridge had been the thirteenth-century Dee bridge. The Grosvenor Bridge brought about the first change in the city's street pattern, which had remained consistent since the Roman period. Grosvenor Street, the new thoroughfare, cut diagonally from Bridge Street and is the main route to north Wales across the Dee.

British railway expansion brought further changes to Chester's landscape. The first railway lines came to Chester in the 1840s. The General Railway Station, built on the northeastern limits of the city in 1847–48, led to the construction of smaller terraced houses in the vicinity. City Road, a direct route between the railway and the city center, was built in the 1860s, further disrupting the ancient streetscape.

As was the case with so many British cities and towns during the Industrial Revolution, Chester witnessed many social changes. Slums and courts along Foregate, behind the Town Hall, housed the city's working poor. Cramped and unsanitary housing often led to outbreaks of typhus and other such diseases. At the same time, new industrial wealth encouraged a rebuilding of the city center in the second half of the nineteenth century. Victorian architects such as John Douglas and Thomas Meakin Lockwood designed their own distinctive buildings, which added further to Chester's architectural riches.

In the twentieth century, a diverse manufacturing base grew in Chester and the surrounding area. Companies such as British Aerospace, Sharp, Brother, General Motors, Marconi, and Shell have been attracted to Chester, but it remains principally an international tourist center.

Visitors have been attracted to Chester since the medieval pilgrimages to St. Werburgh, and since 1781 guide-books have encouraged the development of a tourism industry. An estimated 1.5 million tourists visit Chester annually. Henry James called Chester ''probably the most romantic city in the world,'' and despite successive changes through centuries, it has retained its medieval character where, as James said, ''every shade and degree of historical colour and expression'' has been preserved.

Further Reading: Chester's importance as a Roman frontier outpost has been thoroughly studied by V. E. Nash Williams in *The Roman Frontier in Wales* (Cardiff: University of Wales Press, n.d.; revised by Michael G. Jarret, Cardiff: University of Wales Press, 1969). Geoffrey Barraclough's ''The Earldom and County Palatine of Chester'' in *Transactions of the Historic Society of Lancashire and Cheshire* (Chester), volume 103, 1951, pages 23–57, examines the political power wielded by the earls of Chester in the medieval period. Patrick Ottaway's *Archeology in British Towns: From the Emperor Claudius to the Black Death* (London and New York: Routledge, 1992) discusses the development of Chester's unique architecture following the departure of the Romans from Britain. Henry James's description of Chester can be found in *English Hours* (1905), now available in a variety of editions.

—Manon Lamontagne

Chichester/Fishbourne (West Sussex, England)

Location: On the southern coast of England, sixty-two miles southwest of London and eighteen miles east of Portsmouth.

Description: Chichester is the county town of West Sussex, built on the site of a Roman fort that later evolved into a ruling center for the sub-Roman native princes. It has remained an important commercial center since the first century A.D. Among its attractions are Roman ruins, a Norman cathedral, and fine examples of Georgian architecture. The adjacent village of Fishbourne was the site of the Roman port and the Fishbourne Palace, a Roman-era dwelling.

Site Office: Chichester Tourist Information Centre
29A South Street
Chichester, West Sussex PO19 1AH
England
(243) 775888

Like many other places in England, the city of Chichester and the neighboring village of Fishbourne rest upon the rubble of the Roman Empire. Unlike many other places, some excellent Roman remains have been uncovered here. Roman tiles, coins, and inscriptions have been found periodically for centuries, but only recently has the extent of Roman involvement in the area begun to be understood.

In the first century B.C. the southeastern section of England was occupied by warring tribes who had considerable contact with peoples across the English Channel. After Gaul was conquered by Caesar, some of his opponents fled to Britain. Among them was Commius, a Belgic leader who had at first supported Caesar. He established a kingship in what is now Sussex, Surrey, and eastern Hampshire. His son Verica ruled the local tribe, known as the Atrebates, for many years while maintaining a friendly relationship with Rome. When hard pressed by a rival tribe to the north, Verica fled to Rome seeking help.

The political unrest in southern Britain was well timed to suit the plans of the new emperor Claudius, who was looking for a low-risk military venture to prove himself. It was not the first time Rome had thought to conquer Britannia since Caesar's campaigns there. In A.D. 40 an invasion had been planned, but the legions had mutinied and the plan was abandoned. Now Claudius was ready to try again. In A.D. 43 four legions in three divisions set sail and landed somewhere in Kent. With the support of Verica's tribe, the Romans defeated their British adversaries on the banks of the Medway, driving them back across the Thames. The future emperor Vespasian led the Second Legion into the west to secure the Isle of Wight and further territory.

Vespasian's campaign required a stable camp and a supply base, preferably in a well-harbored, friendly territory. The territory along the southern coast occupied by the Atrebates answered very well; the Romans built their port facilities and storehouses where the village of Fishbourne now stands and placed the camps for the legions and their auxiliaries to the east, at the present site of Chichester. Verica may or may not have returned with the Romans; in any case he would have been too old to have served their needs for a client king. Another man was elevated in his place, a mysterious figure named Tiberius Claudius Cogidubnus, and assigned the management of the local tribes under Roman supervision. Although little is known of him, his life was intertwined with the development of Chichester and especially of Fishbourne.

The first timber structures built convenient to the harbor at Fishbourne were granaries, clearly designed to contain the stores of food needed to support the 8,000 to 10,000 men under Vespasian's command. The camp itself was a mile to the east, between two streams. Because the site lies under the modern town, its original extent is not known, but timber structures in a typical barracks style have been found. It would have been standard procedure to have defensive works, such as banks and ditches, surrounding the camp; although unconfirmed, some evidence of such works has also been found. The presence of soldiers also encouraged the trade that southern Britain and Rome had practiced for decades; craftsmen and tradesmen began to congregate on the outskirts of the camp. A town, called Noviomagus Regnensium, was born.

By A.D. 47 the Romans had accomplished their objective: southeastern Britain had been annexed to the empire. A new network of forts and communication roads had been constructed further inland, and the supply post at Fishbourne and the base camp were no longer needed. The soldiers departed, leaving behind Roman culture and a new center of trade. Rather than reverting to its pre-Roman roots, the town flourished as a Roman settlement. Where the barracks had stood, metalworking and pottery were carried out. Streets were laid out in best Roman style, in a plan that guides the city layout to this day. Buildings were constructed, mostly in timber, but evidence for at least one masonry structure dating to A.D. 58 was found in 1740, when ruins and an inscription dedicated to Emperor Nero were uncovered.

At Fishbourne there were also dramatic changes that may well have been due to the comfortable position of the Roman favorite Cogidubnus. On the site of the former granaries, two new timber buildings were constructed, sited next to one another, end to end. The northernmost was apparently intended to serve as a workshop with living space attached, while the southern building was designed as a dwelling for

The Chichester Market Cross
Photo courtesy of Chichester District Council

the upper class. Other buildings in the area of the harbor may also have existed beneath the position of the current village. While Chichester was developing into an active trading community, a large and, for the time, luxurious dwelling was constructed near its harbor. Its owner must have been a person of wealth, and Cogidubnus is the obvious possibility.

Then in the early 60s, the timber structures at Fishbourne were replaced by something entirely new: a substantial Roman-style masonry building that has become known as the protopalace. It was 190 feet long by more than 150 feet wide, and included a large colonnaded garden, baths, dwelling rooms, and an area for servants to work and live. The remains of elaborately carved Corinthian capitals, possibly from the garden colonnade, imply the employment of continental craftsmen in the building's construction. In its design, its additional luxuries, and the quality of its decoration, the protopalace was a significant advance over its predecessor, implying a corresponding advance in the life of its owner.

One more leap made by the owner was reflected in the site, and this was the greatest leap of all. Between 75 and 80 a true palace was built on a scale sufficient to impress even a Roman. The project took five years to complete and must have been a huge enterprise, involving the coordination of vast quantities of building supplies and large groups of skilled craftsmen. When completed, the palace was composed of four wings arranged in a square around a central garden measuring 250 feet by 320 feet, laid out in a formal Roman style and watered by an elaborate underground irrigation system. It included impressive guest suites with small private gardens and a magnificent audience chamber, in addition to private living quarters. Its walls were plastered and painted to imitate a variety of colored marbles, and its floors were laid in mosaics. The protopalace that preceded it was incorporated into one corner. To the visitor approaching from Chichester, the palace presented a monumental pedimented facade. Having passed through the large entrance hall in the east wing measuring 80 feet by 105 feet, and then across the garden, the visitor mounted steps to the west wing, seated 5 feet higher than the rest of the palace. Within its pedimented entrance supported by four large columns was the audience chamber, where the master of this palace awaited.

Presumably, that master was the ever-more-prosperous Cogidubnus or his heir. One of the few things known of him is that he was given the title *legatus augusti,* a rare honor for a non-Roman. It is possible that the honor was a reward from Vespasian, who in the year 69 became emperor, and whose favor might well have resulted in sufficient wealth for Cogidubnus to build the Fishbourne palace.

Benefits of the emperor's favor would not have been limited to the king alone. During the years between the late 50s and the late 80s, Chichester also saw steady and prosperous growth. The town probably continued to serve as supplier of some goods to the army, as well as developing into an agricultural market center and exploiting its trading position with the continent. The same craftsmen who worked on the Fishbourne palace left evidence of their work in Chichester as well, and it seems reasonable that continental craftsmen brought to work on the palace might have stayed to participate in Chichester's boom. They helped to build the public baths and probably other public buildings. In a Roman town the public baths were the center of social activity. The Chichester baths were located on the site now under Chapel and West Streets. They were probably used and maintained throughout the years of Roman occupation, possibly until the early fifth century.

Client kingships such as Cogidubnus's were not hereditary, and upon his death sometime late in the first century the territory reverted to the empire. The kingdom was divided into three *civitates,* with the town of Noviomagus as the capital of a new administrative district that included Sussex and part of eastern Hampshire. The transfer of power would have required some legal delicacy, and it is known that two prominent Roman lawyers were in Britain at the end of the century, probably serving as advisers to the Roman governor, Agricola. Noviomagus's new status as a *civitate* resulted in the construction of both a forum and a basilica. It was ruled by a council called an *ordo,* made up of 100 councillors who were property holders. The *ordo* met in the basilica, next to the forum. The location and design of the Chichester forum is not known, but it was probably similar to that at Silchester, with a courtyard that served as a market, the basilica along one side, and shops and offices along the other three. One possible location for the Chichester forum is on the corner of West and North Streets, where a large masonry wall with a colonnade has been found beneath the streets.

By the end of the second century, construction of the central part of the town was complete, with the major public buildings and large town houses occupying the prime sites, and the city walls finished. While the major public buildings were masonry, most of the houses would have still been built of timber with wattle-and-daub walls. The exact boundaries of the town are not known, but some of it lay outside the walls, particularly on the south side, where there is evidence of occupation back to the first century. The four gates, one at each compass point, provided egress and guaranteed the survival of the two main thoroughfares that intersect at right angles in the center of town. Chichester's ampitheatre was located to the southeast of the town and was built about A.D. 70.

During the second and third centuries, the palace at Fishbourne underwent some drastic changes. The modifications made involved both a reduction of size and an effort to modernize and extend the private living quarters. The most famous of the mosaics in the palace, called the Dolphin Mosaic, dates from mid-second-century renovations. From a palace intended to serve some official capacity, it was transformed into a luxurious villa. Then, in the late 290s, a fire swept through the villa, destroying enough to make the building unlivable and ensure its demolition. The rubble that remained was gradually reclaimed by nature, until the medieval village that developed around the ruins turned the old palace grounds into farm fields and ran ploughs through the old mosaic floors.

The fortunes of Chichester also changed dramatically with the fortunes of the empire. The influx and attacks of various tribes upon Britain in the late fourth century caused a temporary state of chaos throughout much of the country. In 368 Count Theodosius arrived with Roman reinforcements and a plan to shore up the country's defenses. The bastions added to Chichester's walls may date to this era. Order was restored for a time. In Chichester the public baths were still being repaired and maintained around 370. Sometime after that Noviomagus began a long, slow decline. By 410 the Roman army had left Britain entirely to concentrate its attention on the continent. The suburb outside the Eastgate was deserted by the late fourth or early fifth century. The land outside the walls may have become more important for agricultural purposes, as the area reverted from a money economy to one of subsistence and barter.

In 457 the Saxon raider Aelle arrived in Britain and with his sons attacked a number of British settlements. One of Aelle's sons was called Cissa, and it is from his name that the modern name of Chichester derives: Cissa's Castra, from the Latin word for camp. But Cissa's relationship to the town is obscure, and the reason for the attachment of his name unknown. By the late seventh century, when the conversion of the southern Saxons by St. Wilfrid is recorded, it is likely that the town and surrounding country had been fully occupied by them.

The crowning glory of Chichester today is the great Norman cathedral that was begun in 1091 by Bishop Ralph de Luffa and continued by Seffrid II. Although damaged by several fires through the centuries, the church remains substantially as it was built in the twelfth and thirteenth centuries. The interior length of the cathedral is 393 feet. It is 131 feet across at the transepts and 90 feet across at the nave. Its spire is 277 feet high, easily visible for miles at sea. The nave and choir are Norman, with the early English style apparent in other areas. The cathedral's original tower and spire collapsed during a storm in 1861, but were successfully rebuilt. It is the only English cathedral to have a detached bell tower, built in the fifteenth century.

During the Middle Ages Chichester prospered as an important trade center first for wool, then for wheat. In 1501 Bishop Edward Story gave the city its unusual market cross, which stands fifty feet high at the intersection of the two old Roman streets. A number of other buildings dating from the medieval period remain, including the twelfth century St. Mary's Hospital, the Old Guildhall, and part of Greyfriars Monastery. Much of Chichester was torn down and rebuilt during the eighteenth century, and a number of Roman remains were discovered at this time. The town reappeared in Georgian dress, and much of its fine Georgian architecture remains, particularly in West Street and the Pallants. Three miles to the northeast is Goodwood House, seat of the dukes of Richmond, an unusual triangular house designed by James Wyatt in 1780. The third duke laid out the famous Goodwood Racetrack on the estate in 1801.

Two events have contributed to the town's fame in this century. One was the establishment of the Chichester Festival Theatre in 1962 in a new facility located just outside the city walls. Sir Laurence Olivier served as its first artistic director, and it has gained an international reputation. The other event was the accidental discovery of the remains of the Fishbourne Palace in 1960. Workmen laying water pipe stumbled onto a mosaic floor, and the resulting excavations revealed the largest single Roman building known in Britain. A museum at the site houses the many artifacts recovered, and the formal garden has been replanted.

Excavations in both Chichester and Fishbourne continue as development, demolition and excavation permit, and archaeologists continue in their efforts to fill in the picture of Chichester's Roman past.

Further Reading: The story of archaeological work in Chichester and Fishbourne can be found in the following works: *Fishbourne: A Roman Palace and its Garden* by Barry Cunliffe (Baltimore: Johns Hopkins, and London: Thames and Hudson, 1971) and *Roman Chichester* by Alec Down (Chichester: Phillimore, 1988). Both are well illustrated, detailed accounts written by the archaeologists in charge.

—Elizabeth Brice

Chinon (Indre-et-Loire, France)

Location: Chinon is 28 miles southwest of Tours, 18 miles southeast of Saumur, and 173 miles from Paris via Orléans. It lies wholly on the right bank of the Vienne as the river flows northwest to join the Loire upstream from Saumur, and separates the river from an outcrop of rock at the northwest edge of the town on which stand the ruins of two châteaux and a fortress.

Description: An important town in the Middle Ages, used as a residence by kings of France and England; the site where Joan of Arc called upon Charles VII in 1429 and persuaded him to fight to reclaim the French crown; birthplace of author François Rabelais; also noted for its wines. Portions of medieval château-fortress complex and numerous other historic buildings have been preserved.

Site Office: Office de Tourisme, Syndicat d'Initiative
12 rue Voltaire B.P. 141
37501 Chinon, Indre-et-Loire
France
47 93 17 85

A settlement existed at Chinon before the Roman occupation of Gaul, and the site was occupied by the Visigoths before the land became the subject of great feudal disputes. It passed first to the counts of Blois, and then in 1044 to the counts of Anjou. The town as it exists today is dominated by the great spur of rock containing the ruins of the great châteaux. The whole ensemble covers an area some 440 yards by 75 yards.

The first part of the château complex to be constructed was the westerly, the relatively small Château du Coudray, on the left when seen from the river, and containing the ruins of the tenth-century Tour du Moulin. It is connected by a stone bridge where once there was a wooden drawbridge to the largest central ruin, the Château du Milieu, where the old *logis royaux* (royal lodgings) form part of the southern wall. The Fort St.-Georges to the east was the last to be constructed and was named after the patron saint of its chapel and of England; the English king Henry II had it built in the twelfth century and died there in 1189. Henry was one of the Plantagenet kings of England; this dynasty originated with the counts of Anjou and controlled substantial portions of France. The fort was connected to the central château by a wooden bridge ending with a drawbridge until the whole structure was replaced by the present stone bridge. The fort, of which only the foundation of the outer wall remains, was built simply for defensive purposes. The terrain on which the two châteaux had been built was protected to the south by the river, and by ravines to the west and north; the fort guarded the château complex against attack from the plain to the east.

The earliest buildings in Chinon of which vestiges remain are the Tour du Moulin of the Château du Coudray and the Church of St. Mexme, a tenth-century basilica on the edge of the town due east of the châteaux. Only the nave and the facade remain. Most of what is left of the Château du Coudray dates from the thirteenth century. During the early years of that century, French king Philip Augustus (Philip II) built the tower in which Joan of Arc was later to lodge. After the deaths of Henry II of England and his successor Richard the Lionheart, as well as the victory of the Capetians over the Plantagenets, Philip Augustus succeeded in reclaiming Maine, Touraine, Anjou, Normandy, and Poitou from the English crown in 1204 and, after a long siege, took possession of the fortified châteaux at Chinon.

It was in the Château du Coudray that 140 members of the order of Knights Templar were imprisoned by Philip III (the Bold) in 1307. The order, founded in 1119 as a religious insititution intended for the protection of pilgrims to the Holy Land, was exempt from taxation and had become rich from banking and other financial transactions during the Crusades. It had reached a membership of 15,000, but had lost its high religious standing and acquired a generally unsavory reputation. Philip sought to appropriate its funds and, to the displeasure of Pope Clement V, who regarded the matter has coming under his own spiritual jurisdiction, moved legally against the order. Arrests were made in October 1307 and confessions exacted under torture. The order was suppressed in May 1312; 54 of the 140 imprisoned at Chinon were burned to death in Paris in 1308, as were others elsewhere.

It is for its fifteenth-century history, however, that Chinon is best remembered. Much of the Loire valley as well as the part of France to its north had come under English dominion in the second half of the fourteenth century and, with their Burgundian allies, the English were gaining ground in France. Events early in the following century strengthened English control. By the Treaty of Troyes of May 1420, Henry V and his heirs were to succeed to the throne of France. On the French king Charles VI's death in 1422, his son, the dauphin Charles VII, had himself proclaimed king of France, but did little to redress the territorial losses. Henry V's infant son Henry VI was titular king of France.

The Château du Milieu, first built in the eleventh century, was rebuilt by Charles VI and Charles VII, who held his court here. In 1399 the "Marie Javelle" bell was installed in the strangely shaped Tour de l'Horloge at the east entrance, 115 feet high but only 16 feet wide. Of the royal lodgings only a few steps and the west gable remain, but it was here, on the first floor, that Charles VII was summoned by Joan of Arc to send relief to Orléans, which was under siege by the English, and reconquer France. Joan, not without difficulty, had persuaded the king's local representative,

The château complex at Chinon
Photo courtesy of Office de Tourisme, Chinon

Baudricourt, to have her escorted from Lorraine to Chinon, dressed in male attire and carrying a sword. Charles VII kept her waiting two days before seeing her, and is said to have made himself indistinguishable among the courtiers to test her ability to pick him out. Finally won over to belief in her mission, he set in motion the measures that led to the lifting of the siege of Orléans and to his own coronation at Reims on July 17, 1429. Chinon commemorates Joan of Arc with a museum dedicated to her in the Pavillon de l'Horloge. Its chief contents are a wide variety of paintings that portray or feature her.

Later in the century the chronicler Philippe de Commynes became governor of Chinon. He rebuilt the church of St.-Etienne in the space of ten months during 1480 and his coat of arms can still be seen on the facade. By the late fifteenth century, however, the town had lost its importance, and the kings of France were building more architecturally sophisticated châteaux and hunting lodges elsewhere on the Loire. Chinon retains its historical interest on account of François Rabelais's birth there about 1483 and his description in his fiction of the detailed topography of the region.

Rabelais's father, Antoine, was a substantial landowner, dean of lawyers, occasional judge, and owner of one of the large townhouses. François Rabelais's birthplace—his family's country home, just outside of Chinon—is now a museum dedicated to the writer and his career.

Old Chinon, below the châteaux on the right bank of the river, retains its medieval aspect, with a "Grand Carroi" (from *carrefour,* or crossroads), where once Joan of Arc rode into town, and its ancient houses, in one of which (now a bakery) Richard the Lionheart is said to have died in 1199 after his defeat at Chalus. In 1321 the city's Jews, accused of poisoning the water supply, were rounded up and burned alive on the island in the Vienne.

In 1633 the châteaux passed into the ownership of Cardinal Richlieu, whose family retained the complex until the 1789 revolution. Stones from the ruins are said to have been moved to build cardinal's new château at the nearby town of Richelieu. It was only in 1855, when the writer Prosper Mérimée was inspector general of historic monuments, that plundering of the site ceased and a move toward conservation began. Today the complex, partially restored, is a showcase of medieval architecture and history.

Further Reading: There is no single work on Chinon in English apart from small guidebooks, and further reading is to be found chiefly in works on French history or on the Loire châteaux or on the history of French architecture. For a general, well illustrated work in English, Jean Martin-Demézil's *The Loire Valley and its Treasures* (London: Allen and Unwin, 1969) can be recommended. It is a translation of *Trésor du Val de Loire* (Paris and Grenoble: Arthaud, 1987). The *Histoire du pays de la Loire* by François Lebrun in the series issued by Privat of Toulouse (no date) offers a professional introduction to its subject.

—Anthony Levi

Cirencester (Gloucestershire, England)

Location: On the River Churn in Gloucestershire, England, on the eastern edge of the Cotswold Hills, approximately twelve miles southeast of Gloucester where the A417 road intersects with the A429.

Description: Cirencester was the second most important town in Roman Britain after London. The town was also prominent in the Middle Ages as a wool trading center. The town's architectural beauty has earned it the honorary title of Capital of the Cotswolds.

Site Office: Tourist Information
Corn Hall Building
Market Place
Cirencester, Gloucestershire GL7 2NW
England
(285) 654180

Cirencester's name holds a number of indications about its past. Its last two syllables derive from the Old English *ceaster,* an Anglicized version of the Latin *castrum,* meaning camp. "Ciren" is a remnant of the town's Roman name Corinium. In the Domesday Book, a statistical survey conducted during the reign of William I, the town appears as Cirecestre, a spelling that may account for the pronunciation "sissiter", which was widely used by the town's locals until the nineteenth century and is still occasionally heard today.

Archaeological findings have established that the town site was occupied by a farming settlement in pre-Roman times. The Roman conquest of A.D. 43, during the reign of Emperor Claudius I, prompted the erection of a Roman military encampment here. In approximately A.D. 49, this camp became an auxiliary fort, later to develop into a civil settlement that grew to become the capital of the Dobunni tribe after A.D. 60. The Dobunni had originally been centered at the confluence of the Rivers Avon and Severn (approximately twenty miles north of Cirencester, near present-day Tewkesbury) and surrendered to Claudius in the conquest of A.D. 43. As was the pattern with other Roman conquests, Roman forms of administration were quickly imposed on newly gained territories.

Centers of population could be classified into three different categories, each of a different rank. Of highest rank were the *coloniae,* city-states that were created as legionary settlements by allotting plots of land to retired legionnaires (those who had performed twenty-five years of service with a legion). The purpose of setting up such colonies was the strategic establishment of local, self-sustaining sources of military recruits for the legions, which could consist only of Roman citizens. There were four of these *coloniae* in Britain,

at present-day Colchester—the original capital of Roman Britain—Lincoln, York, and Gloucester. The next-ranked settlement category was the *municipium.* Unlike the inhabitants of *coloniae,* those of a *municipium* would not automatically have Roman citizenship. This could, however, be acquired through public service: magistrates and their families obtained Roman citizenship at the conclusion of their term of office. The third-ranked category was that of capital of a *civitas.* Britain was divided into *civitates,* administrative regions that typically replicated the tribal territories prior to Roman occupation. The administration of each region was handled from a capital, the organization of which followed the model of a *municipium,* though Roman citizenship was not available to its magistrates. All three categories shared the same form of local government, which took the form of an *ordo,* a council made up of 100 councillors and several magistrates, often drawn from the local elite. The distinctions in rank among the different types of settlement were based on the extent of inhabitants' access to Roman citizenship, which conferred a number of privileges. Most importantly, it gave hereditary social status. More pragmatically, it also offered protection from summary flogging or execution, as well as the right to trade under Roman law.

Corinium Dobunnorum, "Corinium of the Dobunni," was founded as a *civitas* capital. During Claudius's reign, and prior to the extension of Roman rule into Wales and northern England, Corinium stood at the western limit of the new imperial province of Britannia, and thus of what the Romans regarded as the civilized world. The Roman imperial frontier extended roughly along a line running southwest to northeast, from the mouth of the River Axe to Humber. This line coincides with the Fosse Way, one of Britain's major Roman roads; the route is followed today by the A429 road.

The Romans' primary purpose in laying roads was to make possible the rapid mobilization of troops—particularly important during the years that immediately followed the Claudian conquest, for the Roman army had to remain mobile to maintain possession of newly secured territories. No doubt the Fosse Way was built to facilitate this process. It is significant to the development of Corinium that the Fosse Way ran through the town. Even more significant was the fact that another major Roman road, the Ermine Way—the modern-day A417 road—intersected with the Fosse Way at Corinium. Yet another Roman road, known as Akeman Street, joined this intersection. These major roads were supplemented by a network of side roads, or *deverticula.* Corinium's remarkable accessibility has endured to this day: the modern town can be reached from ten different directions by road. The road system of the Roman Empire, with its 50,000 miles of hard-surfaced highway and numerous feeder roads, was so exten-

The Roman Wall at Cirencester
Photo courtesy of Corinium Museum, Cirencester

The Parish Church of St. John the Baptist
Photo courtesy of Woodmansterne Limited

sive that it led to the saying "All roads lead to Rome." In the Cotswolds, it seems that all roads led to Corinium.

The presence of Roman troops must have been a major spur to Britain's economic growth. They were the only significant group of wage earners in Britain at the time, and their demand for goods and services would have been considerable. Corinium's position within the road system was possibly a strong incentive for many traders and craftsmen who catered to the needs of Roman soldiers (they had to follow their customers wherever they went) to set up in the town on a permanent basis. Corinium's growth was given further impetus during the period from 75 to 80, when Governor Agricola, in an attempt to encourage large-scale Romanization in architecture, dress, and education, developed a town center in Cirencester. Such urban development would include the building of a forum, a public building that would also serve as a symbol of civic organization. It would include a market area and house offices for the magistrates and the *ordo,* as well as a basilica for the conduct of law tribunals and public business. In the second and third centuries a channel was dug, into which the River Churn was diverted, as part of the town's defenses. Corinium's significance as a commercial center is confirmed during Hadrian's reign, when it was one of the towns to build a *macellum,* or regular market-hall. The town's public buildings also included an amphitheater, built during the third century, which was discovered southwest of the present-day town. Most archaeological findings in Circencester were made accidentally; archaeological discoveries seem to be the unavoidable byproducts of civil engineering work in this town. Two famous mosaics, known as the Hunting Dogs and the Four Seasons, were discovered in 1849 in Dyer Street while workers were digging for drains. More recent large-scale development has yielded numerous findings, including the discovery of some 450 Romano-British burials that have provided a great deal of historical detail.

In Roman times, economic activity centered on agriculture, and the *civitas's* great villa estates are proof of the area's extraordinary wealth. Villas were Roman-style countryside houses and generally served as working farms that stood in clusters around towns. In the case of Corinium such villas were spread over an approximate thirty-mile radius. Particularly fine examples are North Leigh, Ditchley, and Stonesfield, east of Cirencester. They were constructed by urban craftsmen, and their level of decoration was an indication of their inhabitants's wealth. Despite two centuries of excavations, little knowledge has emerged regarding daily life in a villa estate. The villas varied in size, but followed a basic structure: a rectangular main building with a corridor (which could also be in the form of a veranda) that linked two short side wings. In larger villas, the side wings tended to be longer, and were often linked by a wall that formed an enclosed front yard. Outlying buildings could also be built for various agricultural uses. The villa would have at least one set of bath chambers, served by plumbing that was also linked to kitchens and latrines. The central structure normally contained several rooms that were heated by wall or underfloor ducts. A large room could serve as the main reception room and possibly also as the main dining room, or *triclinitan.* This room would also contain the house's finest mosaics, essentially large-scale decorative items for which Corinium-based craftsmen were particularly renowned. They profited from the flowering of what was Britain's richest villa area at the time and, in turn, must have contributed greatly to Corinium's economic growth. Fine examples of these mosaics can be seen today in Cirencester's Corinium museum, along with full-scale reconstructions of Roman interiors.

Roman culture also brought greater gastronomic diversity to Britain. The new ingredients that the Romans introduced, taken for granted today as staples of British fare, included carrots, peas, beans, apples, and pears. Another glimpse of Roman urban life is offered by the museum's description of Roman cuisine. The recipes were recorded by one Apicius and have been compiled into a museum booklet by Marian Woodman. Delicacies of the time included boiled ostrich (served with its feathers intact), milk-fattened snails, and stuffed dormice. A popular condiment was *liquamen,* extracted from barrels of salted anchovies that had been left to rot in the sun.

Corinium's emergence as Roman Britain's second most important administrative center was in large part due to a number of changes made to the island's gubernatorial arrangements. From A.D. 43 until the third century, Britain was ruled as a single province by a governor, who also acted as commander in chief of a sizable military contingent, which constituted 10 percent of the entire Roman army spread throughout the empire. During the period 195 to 197, Governor Albinus abused his command over the Roman legions based in Britain to seek the Roman throne. In an attempt to avoid future repetition of such rebellion, the emperor diminished the British power base by splitting Britannia into two provinces in 213. The northern province was named Britannia Inferior, "Lower Britain," with its capital at York. The southern province was given the name Britannia Superior, "Upper Britain," with London as capital. This new arrangement did not, however, prevent further threats to imperial authority. In 287 Carausius, a general appointed by Emperor Maximian to deal with Saxon attacks on northwestern Gaul, seized Britain instead. For a brief period (287–96), Britain became a separate empire. Maximian attacked Britain in 288–89 but had to make peace. Carausius was murdered in 293 by his subordinate Allectus, who took over the throne. He kept it for three years before Britain was brought back into the Roman empire in 296 with a two-pronged invasion by Constantine and Praetorian Prefect Asclepiodotus. The repeated power struggles over Britain had highlighted two issues. The first of these was that anyone allowed to abuse the command of Britain was difficult to displace militarily. The second issue was the growing importance of Britain within the empire, especially at a time at which other parts of the empire were being badly affected by barbarian incursions and economic decline. It was against this backdrop that Britain was subdivided yet again, this time into four provinces, to which a fifth, Valentia, was added later on. The first four new provinces were Maxima Caesariensis, Flavia Caesariensis, Britannia Secunda, and Britannia Prima; their respective capitals were present-day London, Lincoln, York, and Cirencester. It is curious that Corinium should become the provincial capital of Britannia Prima, especially in view of nearby Glevum's (present-day Gloucester) higher rank as a *colonia*. Corinium is not known to have received any honorary rank or status, and it seems that its logistical advantage and prominence as a trading center were sufficient justification for choosing the town as a gubernatorial seat. It soon became the second most important Roman center, after London, in Britain.

Corinium's status as Britain's second town continued into the early fifth century, but soon waned with the decline of Roman rule. Increasing barbarian invasions into Roman territory on the European Continent had forced Rome to withdraw military forces from Britain, which in turn left the island increasingly exposed to Saxon raids. Frustrated with such imperial neglect and the continued burdens of Roman taxation, the Britons expelled Roman officials in 409. Although central imperial control was never reestablished, Romanized life continued into the middle or even late fifth century. Beyond this period, Corinium begins to figure less prominently in British history, and the town entered a phase of long-term decline, until the capture and subsequent development of the town in 577 by Saxons.

Unlike other significant Roman towns, Cirencester's outward appearance displays little of its centuries under Roman rule. Most of it is displayed in the Corinium museum, or remains buried under the town's current surface. The present-day layout is mainly the product of medieval development, when Cirencester was effectively rebuilt. Thus, the present-day marketplace does not mirror the square dimensions of its predecessor, the Roman forum. Overlooking the market is the Parish Church of St. John the Baptist, hailed as one of England's great wool churches. Its construction dates back to Norman times, when Henry I ordered the construction of an abbey in 1117, a factor that stimulated new urban growth. Prosperity returned to the town with wool production, for which the rich pastures in the surrounding area were particularly suitable. The abbot played a prominent role in this trade, having obtained royal charters in 1215 and 1253 to conduct what became famous wool fairs. The wool produced in the area was of the finest quality, and was commonly referred to as the "Cotswold Lion." The town's medieval reliance on wool is still acknowledged in the names of two of the town's inns, the Golden Fleece and the Ram. Most of the construction that shaped the church's appearance today took place in the fifteenth century, funded by the wealth of the town's wool merchants. Their munificence resulted in a church that is larger than some English cathedrals, and which would have been even more impressive if original designs had been followed to build a spire over the tower.

A remarkable feature of St. John the Baptist is its three-storied south porch that faces the marketplace. It was built in 1490 to provide a meeting place at which to conduct public, non-ecclesiastical business. The abbey was destroyed following Henry VIII's Dissolution of the Monasteries, a result of the 1534 Act of Supremacy under which Parliament denied papal authority by establishing the Church of England with the monarch as its head. The church of St. John the Baptist and the Spital Gate are remainders of the great compound that constituted the abbey, and the church's south porch became the Town Hall. Although the porch was returned to the church in the seventeenth century, it is still referred to today as the town hall by locals.

The town is bordered on the west by the 3,000-acre Cirencester Park, owned by the Earl of Bathhurst's family since the late seventeenth century. Cirencester House, inside the Park, built between 1714 and 1718, is the Earl's seat. The house is closed to the public, and its extraordinarily high horseshoe-shaped yew hedge keeps it invisible at street level. The public is admitted in the Broad Walk, at the end of Cecily Hill, a five-mile-long avenue that stretches into Cirencester Park.

Cirencester House, the Church of St. John the Baptist, and most of the town's buildings are built in oolite limestone, in plentiful supply in the area. Just below the surface lies the first layer of limestone, which is used mostly for dry-stone walls that line field boundaries. Below this layer is the oolite, which is carved and sawed with ease, hardening with exposure to the elements to become extremely durable. It is the same material that the inhabitants of Corinium used in their constructions 2,000 years ago. Remains of quarrying activity have been discovered southeast and northwest of the amphitheater, where side roads linked the quarry to the Fosse Way.

Further Reading: *The Oxford Illustrated History of Roman Britain* by Peter Salway (Oxford and New York: Oxford University Press, 1981; second edition, 1993) is an authoritative work on Romano-British history. *The Cambridge Historical Encyclopedia of Great Britain and Ireland,* edited by Christopher Haigh (Cambridge and New York: Cambridge University Press, 1985) offers conveniently arranged chapters on Roman Britain.

—Noel Sy-Quia and Hilary Collier Sy-Quia

Clonmacnois (Offaly, Ireland)

Location: In the center of Ireland, thirteen miles south of Athlone, on the River Shannon between Lough Ree and Lough Derg.

Description: Ancient monastic site with numerous medieval ruins; begun A.D. 545 by St. Ciaran the younger; destroyed and rebuilt many times until 1552 when it was destroyed by the English; now a National Monument site in the care of the Commissioners of Public Works.

Contact: Midlands-East Region Tourism Organisation
Dublin Road
Mullingar, County Westmeath
Ireland
(044) 48650

In early times, Clonmacnois was one of the most influential religious communities in Ireland. Founded by St. Ciaran in A.D. 545, it was not only a large monastery, but also a thriving city situated in the middle of Ireland on the banks of the River Shannon and beside the Eiscir Riada, the most important roadway of early Christian Ireland. In the sixth century the Shannon was the chief waterway and the safest route in Ireland. Ultimately its location may have contributed to its downfall since marauders found it so accessible.

It's hard to imagine that this was once a thriving city. There are scattered homes in the gently rolling countryside, and many sheep. A few pleasure boats rock on the River Shannon. The only remains are of a stone church and related buildings in a graveyard located several miles on country roads from the main Dublin-Galway highway. The Shannon is no longer a major thoroughfare. Clonmacnois is now a National Monument site protected by the Commissioners of Public Works.

St. Ciaran was born in Roscommon, just north of Clonmacnois, in A.D. 516. His father was Beoit, of noble birth, although a carpenter by trade. His mother was Darerca, also of a high-ranking family. He was one of five sons, four of whom became priests. Two of his three sisters became nuns. He was entrusted to the care of Deacon Justus, who trained him from an early age. Ciaran was credited with miracles while still a student. Later he studied with St. Finian of Clonard, where he became famous for his sanctity. He was so highly regarded that St. Finian made Ciaran his deputy when he was away.

Ciaran then went to the monastery of St. Nennidius on an island of Lough Erne. After that Ciaran proceeded to Aran to study with St. Enda. While he was there he had a vision of a large fruitful tree transplanted to the middle of Erin. It sheltered the whole island, and its fruit was carried beyond the sea. According to historian John Canon Monahan, St. Enda explained the vision to him as follows: "The great tree which thou hast seen is thyself, Erin shall be full of thy honour, and the shadow of they grace shall protect the whole island, and multitudes shall be satiated with the fruits of thy fasting and prayers; go, then, in the name of God, to the banks of the Shannon, and found thy church there." The two saints then erected a cross on the coast of Aran as a pledge of their undying spiritual friendship.

Ciaran erected his first church at Isell-Ciaran. From there he proceeded to Inis-Ainghin (Hare Island) in Lough Ree, and many disciples joined him. After living there for three years and three months, he proceeded to Ard-Mantain, near the Shannon. It was very beautiful, but after resting there for a while, he said "If it is here we remain we shall have much indeed of the riches of this world, but the souls sent to heaven from it shall be few." He then went on to Clonmacnois, which was called Ard-Tiprait at that time. "Here will we stay," he said, "for numerous will be the souls that will ascend to heaven from this spot."

According to legend, when Ciaran was placing the post in the ground for the first building, he was joined by Diarmaid Mac Cerbhaill, who at the time had been banished by King Tuathal Maelgarbh. Ciaran blessed Diarmaid, and told him he would be king the next day. That night the king was killed, and Diarmaid was summoned to assume his place. In gratitude, King Diarmaid gave Ciaran and his brethren gifts of land and many churches. Several different names for the monastery were used in its early years, but eventually Clonmacnois, meaning "meadow of the descendants of Nós," became universally accepted. Ciaran died at age thirty-three, only a few months after the monastery was begun. It is said that before he died, he prophesied that his monastery would be attacked by "wicked people."

In spite of his death, Clonmacnois grew and prospered, and the monastery became famous for piety and learning. A large community was formed of monks, clerics, anchorites, bishops, scholars, and students, plus lay attendants and workers. Through the centuries the great Uí Neill dynasties of the south were strong supporters of the monastery. In addition, King Diarmaid and his descendants (sixteen of whom became kings) were settled within easy reach of Clonmacnois, and their connection with the monastery was intimate. By the middle of the seventh century it was in a top position among the monasteries of Ireland. It always had a national, rather than a regional character because its monks and abbots came from all parts of the country. Because of its central location it was the burial place of many of the kings of Connacht and Tara.

Unfortunately, Ciaran's prophesy came true. Clonmacnois was repeatedly destroyed and rebuilt for centuries. It

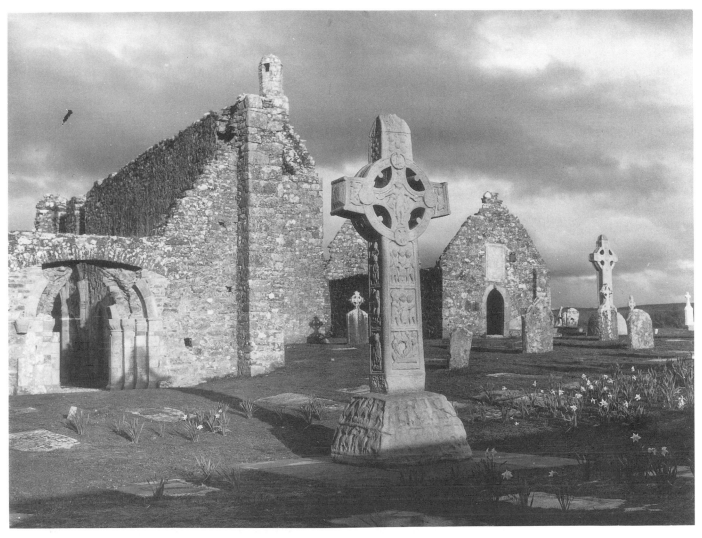

The Cross of the Scriptures at Clonmacnois
Photo courtesy of Bord Failte, The Irish Tourist Board

was subject to both burnings and plunderings. Since the early buildings were of wood and thatch, they burned and were rebuilt easily. Until the year 833 the recorded burnings were probably accidental. Then everything changed. Féidlimid of Munster burned the town to the door of the church and massacred the townspeople. During the next two centuries, Clonmacnois was plundered or burned eighteen times, including nine times by the Norse, and four times by the Munstermen. It was plundered by the O'Briens and the O'Farrells (separately) in 1044, and again later that century by the O'Rourkes, Foxes, McAuleys, MacCocklains, O'Kellys, O'Maddens, and by Niall MacLochlainn.

In 1178, Hugh de Lacy, constable of the English king, sacked the whole settlement; in 1179 there was an attack when 105 houses were burned to the ground; in 1202 and 1204 it was plundered by William de Burke, who led an English-Irish group. In addition, there are records of battles between Clonmacnois and Birr and Clonmacnois and Durrow

in the eighth century. At that time the monasteries possessed extensive lands occupied by a lay population that could provide fighting men. The destruction that took place at Clonmacnois was not connected with religion; the motives were political. The leaders were therefore always willing to make generous reparations, which is one reason Clonmacnois repeatedly recovered from destruction, and its possessions grew. In 1216, for example, the English king ordered compensation for the lands occupied in fortifying the castle recently built there. The bishop was to be paid for the orchards cut down, for the cows, horses, oxen and utensils confiscated by the king's men.

Clonmacnois ceased to function as a monastery after the death of Abbot Maelciaráin Ó Maeleóin in 1263. It survived as the seat of an impoverished bishop until the final destruction of the churches and other buildings by the English in 1552. After that complete destruction, periodic attempts were made to repair some of the buildings, and two structures were used for worship in the seventeenth and eighteenth centuries. In 1880, Clonmacnois was turned over to the Commissioners of Public Works. However, until 1955 only the buildings were owned by the state. The Representative Body of the Church of Ireland finally presented the grounds to the commissioners in 1955, after which point much work was done to improve the grounds and buildings. The remains at Clonmacnois are now an impressive open-air mueum of ancient Irish Christian art. Nothing remains of the original buildings because they were wood and thatch. The existing ruins are of stone buildings, crosses, and monuments built from the ninth to the thirteenth centuries.

The Cross of the Scriptures (King Flann's Cross), a ring-headed, intricately carved, sandstone cross, dates from the tenth century. The head, shaft, and base consist of carved panels; some are difficult to understand. The Last Judgment on the east face of the head and the Crucifixion on its west face are clear. The lowest panel on the east face of the shaft shows a monk on the left and a warrior on the right, pressing a long pole into the ground. This could be St. Ciaran and King Diarmaid planting the first corner-post of their church, but it could also be Abbot Colman and King Flann, either building the cathedral or setting up this cross. The inscriptions on the east and west faces of the foot of the shaft are unclear, but appear to commemorate King Flann, who died in 916, and Colman, who died in 921. The cross is now housed in the Visitors' Center.

The South Cross, from the early ninth century, is a Celtic cross. A carving of the Crucifixion is on the west face of the shaft. The North Cross, from about the year 800, is a shaft. It is decorated with tail-biting lions and a curious figure with crossed legs, which may be the Celtic god Cernunnos.

Temple Kelly may have been founded in 1167 by Conor Kelly on the site of St. Ciaran's hospital. Only the foundations remain. Temple Doolin (or Dowling) most likely dates from the ninth century, and was repaired in 1689 by Edmund Dowling as a family mausoleum. Apparently Temple Hurpan (also called MacLaffey's church) was added in the seventeenth century and is joined to Temple Doolin. Temple Meaghlin (or Temple Ri) dates from the early thirteenth century; it is a plain rectangular building, with attractive east windows. Temple Ciaran (Eaglais Bheag) was apparently begun in the ninth century, with additional work at various later dates. It is only twelve and one-half by eight feet internally, and is named after St. Ciaran, who is believed to be buried in the northeast corner. Temple Finghin (or Finian) dates from the twelfth century and has a small attached round tower fifty-six feet high. It is also known as MacCarthy's Church. There appear to have been modifications made to the Romanesque chancel-arch in the seventeenth century to hold it together. Temple Connor was founded in 1010 by Cathal O'Connor, king of Connaught. Today it is perfectly pre-

served, and since the mid–eighteenth century it has served as a place of worship for the local Protestant community.

The Nun's Church dates from the tenth century. It is some distance from the other buildings and there are traces of a causeway leading to it. It closely resembles Temple Finghin in its Irish Romanesque style, which is especially apparent in the chancel arch and the west doorway. Some restoration work was done in 1867 by the Kilkenny Archaeological Society.

O'Rourke's Tower was built in the tenth century, probably by Fergal O'Rourke, and repaired in 1120 by O'Malone, a bishop and abbot. It is a good example of the detached belfry common in early Irish monastic sites, and is complete, except for the conical cap.

The Cathedral was probably begun in the tenth century by Flann Sinna, king of Tara, and the abbot Colman. Modifications were made in the twelfth, thirteenth, fifteenth and seventeenth centuries. Figures of St. Dominic, St. Patrick, and St. Francis adorn the north doorway; these carvings were erected by Dean Odo, who died in 1461. A stone-vaulted chancel was inserted into the existing cathedral during the fifteenth century. In 1647, Charles Coghlan, the vicar-general, made an attempt to repair the cathedral, which had been roofless since the English raid in 1552. It is believed that many of the ancient High Kings of Ireland were buried here, including Turlough Mor O'Connor and his son Ruairi.

The Castle located beside the Shannon dates from the early thirteenth century. It was probably built by John de Grey, bishop of Norwich, to guard the line of the Shannon, along with Athlone Castle. It was partially destroyed in the seventeenth century to prevent its being used by Cromwellian forces, and today portions of the walls lean out at various angles.

St. Finian's Well is about two hundred yards northwest of Temple Finghin (Finian) and St. Ciaran's Well is about a quarter of a mile southwest of the castle to the right of the Shannonbridge road. According to Irish folklore, both are believed to have healing powers.

Originally, the cemetery contained numerous monumental grave slabs, laid flat on the graves of those whom they commemorated; the 240 surviving slabs are now displayed in the visitors' center. They date from the seventh to the twelfth centuries. Irish inscriptions cover many of them and their diversity and design eclipse those from other sites.

Further Reading: *Clonmacnoise, Co. Offaly.* (Mullingar, County Westmeath: Midlands-East Tourism, n.d.) is a very brief history and description of the ruins written for tourists. *Clonmacnois: A Historical Summary* by John Ryan (Dublin: Office of Public Works, 1973) and *Records Relating to the Diocenes of Ardagh and Clonmacnoise* by the John Canon Monahan (Dublin: Gill and Son, 1886) are both detailed accounts of the history and lore of the site.

—Patricia Ann Shepard

Cluny (Saône-et-Loire, France)

Location: Sixty miles north of Lyons, in the Saône-et-Loire Department.

Description: Cluny was a Benedictine monastery founded in 909. It became a center of monastic reform during the 10th and 11th centuries, and its abbots created a centralized network of monasteries across Europe (known as the Cluniac Order) that were owned by Cluny and whose monks owed their allegiance to its abbot. Cluny once was home to over 400 monks and the largest basilica in Christendom.

Site Office: Office de Tourisme
Tour des Fromages
71250 Cluny
France
85 59 05 34

Located in southern Burgundy and about sixty miles north of Lyons, Cluny was one of the most important monastic centers in western Europe during the tenth and eleventh centuries. In two centuries Cluny grew from a small monastery in an isolated French valley to become the mother-house for 1,450 priories of the Benedictine order that were scattered across France, Germany, Spain, Italy, England, and Switzerland. The abbots of Cluny reestablished Benedictine discipline, reorganized the liturgical and administrative practices of the order, developed Romanesque architecture, and laid the foundations for the Gothic style of the late medieval period. The Cluniacs were also instrumental in effecting religious reform and strengthening the power of papacy during the central Middle Ages.

The story of Cluny's rise must be grounded in the context of feudal Europe because its development was a response to the political chaos and religious abuses that were endemic in the ninth and tenth centuries. In the Carolingian Empire, monasteries were incorporated into the feudal system of government, and they had become the property of and vassals to secular rulers. Kings and princes appointed abbots or sold monastic offices for profit (i.e., simony), and the income from these religious houses were placed in the royal treasury. Lay control over many of the monasteries in Europe led to a loosening of spiritual discipline that was reflected in the failure to faithfully observe the liturgical and ascetic practices of the Rule of St. Benedict, which governed monastic life in the middle ages. (The Rule of St. Benedict organized monastic life around a set of eight religious services of prayer, scripture reading, and music interspersed with periods of manual work. The Rule mandated a simple diet for the monks, imposed silence in their day-to-day lives, and placed them under the prescriptions of poverty, charity, and celibacy.) As historian Raffaello Morghen notes, religious life had been reduced to ''an outward show of ritual and cere-

mony,'' and matters of the spirit had been become matters of secular wealth and power. H.E.J. Cowdrey reports that reform and spiritual revival had begun late in the ninth century, but few expected it to come from a small and relatively new abbey in central France.

In 909 the aged and heirless Duke William III of Aquitaine established a monastery on his lands at Cluny as a means to ensure his eternal salvation. In the charter, William decreed that the monks of the new house would strictly observe the Benedictine Rule. More importantly he freed Cluny from control by all secular powers (i.e., the nobility), placed the new monastery under the ownership of St. Peter and the protection of the papacy, and allowed the monks to freely elect their abbot. Although these provisions were not unique to the period, Cluny's abbots and several supportive popes would use this charter to free the church from lay control and unify monastic life under Cluniac administration and practice.

Duke William appointed Berno, abbot of Baume, to build the new abbey upon the strict observance of the Benedictine Rule. Little is known about his reign (909–926), except that the new house was poor and remained relatively undeveloped. The monks inhabited the Roman villa that had served as William's hunting lodge, erected a small chapel, and brought the land under cultivation. Cluny's second abbot, Odo (926–944), established the Cluniac liturgical tradition and began the work of reforming monasteries in France and Italy that would eventually create a centralized Cluniac order. Odo was a scholar and reformer who was passionately concerned with the devotional life of Benedictine monasteries. He greatly expanded the amount of time the monks spent in prayer and psalm-singing and hence altered the traditional Benedictine custom of balancing spiritual and manual labor. At Cluny, lay brethren and serfs did the majority of the agricultural and artisanal work, which freed the monks to devote their time to worship, prayer, copying manuscripts, and studying. Odo justified this shift in monastic duties in his writings, especially *Occupatio*. They focus on the necessity of observing the Rule, of renouncing the world and serving God through meditation and worship, and uniting the earthly church and the heavenly one in the mass.

Odo reestablished Benedictine orthodoxy at Cluny, and both secular and ecclesiastical leaders took notice. In 927 King Rudolf of France affirmed Cluny's independence and granted it the right of coining its own money. Twelve years later King Louis IV issued a similar royal charter. In 931 Pope John XI affirmed Cluny's autonomy and endowed it with the unprecedented right to receive monks from other houses who wished to live in an environment that strictly kept the Rule, and commissioned its abbot to reform any monastery given to him for that purpose. Pope Leo VII renewed these rights in 938. These affirmations and support of Clu-

Spires rising above Cluny
Photo courtesy of Office de Tourisme, Cluny

ny's rights, the immunity from interference, as well as Odo's program of renewal provided Cluny with a basis for its future expansion and marked it as the model for monastic life. During his abbacy, Odo reinstated the Rule at several monasteries in central and southeastern France, introduced the Cluniac spirit to monasteries in Spain, Switzerland, and Italy (the latter by the invitation of Alberic, the sovereign of Rome). At the same time the nobility of Burgundy and the papacy began to entrust parcels of land, churches, and even monasteries to Cluny, which enriched its treasury and reputation.

Under the next two abbots, Aymard (944–965) and Mayeul (965–994), Cluny's lands increased and were brought under more efficient cultivation, and the number of dependent houses in France multiplied. Again, the papacy stepped in to affirm Cluny's rights, protect its possessions, and encourage Cluniac reform. Mayeul became the confidante of Otto I, the Holy Roman Emperor who entrusted several Italian houses to Cluny's care, and thus continued Odo's practice of cultivating secular and ecclesiastical relationships.

Many historians believe that Cluny reached the height of its power under the rule of Odilo, who served as Abbot from 994 to 1049. Near the end of his life Odilo allegedly boasted, "I found an abbey of wood, and I leave an abbey of marble." This is an apt summary of Odilo's accomplishments, both literally and figuratively, for he was a master politician who used the support of kings and popes to expand Cluny's sphere of influence, an adept administrator who centralized Cluniac government, and the first great builder who enlarged the monastery and built a church that would serve as a model for many other Benedictine houses.

The aging Mayeul appointed Odilo as his coadjutor, and the young monk was thrust into the world of ecclesiastical politics when he was sent to defend Cluniac claims before the pope. In 990, with the support of Emperor Otto III, he obtained a papal bull from Gregory V that confirmed the abbey's possession of thirty-seven monasteries in twelve French dioceses and threatened excommunication to any lord, secular or religious, who invaded and plundered Cluniac churches, houses, or lands. The same bull gave Cluny the right to choose any bishop to consecrate its churches and ordain its priests, thus effectively freeing the monastery from the control or interference by any ecclesiastical officer except the pope. Several French bishops contested their loss of power and potential revenue, and many local nobles in central and southern France tested papal oversight by trying to extract tolls or taxes from Cluniac houses or by building on Cluniac land. As Cluny's abbot, Odilo secured papal confirmation of exemption from all interference and of its new rights from Benedict VIII in 1016 and from John XIX in 1024 and 1027. He also secured the protection and patronage of the Holy Roman Emperors (Otto III, Conrad II, and Henry III), the king of France, and the rulers of Burgundy. Those who tested the papal threat of excommunication failed in their bid to profit at Cluny's expense and in many ways served to strengthen the power of the Holy See. Cluny and its dependent houses were now free from interference or control by local bishops and were free to serve as the Church's (and the papacy's) emissaries-at-large.

Papal oversight and secular protection encouraged Odilo to engage in reform, to centralize the administration of Cluniac houses, make them subject to the authority of one abbot, and to institute the observance of a common set of practices. The idea of a "Cluniac Order" was born with Odilo; it refers to a situation in which other monasteries, priories, or deaneries (i.e., granges) became the property of Cluny and the abbot of Cluny served as the sole abbot of all other houses. Odilo frequently visited the fifty or sixty houses placed under Cluny's direct control during his abbacy to exercise jurisdiction and ensure that the liturgical and devotional customs of the mother-house (codified between 996 and 1030) were followed. This was a revolutionary mode of monastic government; previously, European monasteries existed as separate entities—free to create their own customs and subject only to the authority of their own abbot. (This movement towards centralization reflected the consolidation and centralization of secular rule occurring across Europe.) Odilo was also an energetic reformer, and he was entrusted with monasteries in Burgundy, Provence, Auvergne, and Poitou. Several more Italian houses were reformed during his nine trips to Italy, and he trained Spanish monks, at the request of Spain's rulers, to bring the Cluniac spirit to the Iberian Peninsula.

Cluny's prestige and power were enhanced under Odilo by his engagement in secular politics and his ambitious building projects. The war between the dukes of Burgundy, Nevers, and Bourbon in the late tenth and early eleventh centuries threatened the security of monastic estates and the lives of their servants. Odilo sponsored the Pax Dei (Peace of God) at the council of Anse in 994, in which he threatened to excommunicate any soldier who brought war to monastic holdings. The Truga Dei (Truce of God) grew out of this, and by 1042 it had been proclaimed in all the dioceses of France and introduced in Italy.

An increase in the number and size of gifts (in land, buildings, mills, fisheries, and money) provided a substantial income that allowed Odilo to renovate Cluny and several of its priories (Charlieu, Sauxillange, Payerne, Nantua, and Romainmotier), as well as to build several new monasteries (Saint-Flour, Paray, Souvigny and Ambierle). Cluny bene-fitted from a theology and popular belief in the necessity of public penance. The overriding religious concern of eleventh-century Europeans was ensuring their salvation. The aristocratic class looked to the monasteries and secured the intercessory prayers and absolutions of monks through gifts of lands and estates. (In addition, many aging nobles joined monasteries either as monks or as lay brethren to earn their salvation.) At Cluny, Odilo used the income from these gifts to build a new and larger church, enlarge most of the monasteries' living and work quarters, and erect a magnificent Roman-style gate for the abbey. The church, Cluny II, was built between 955 and 1000, and it established the Cluniac school of architecture. It was characterized by tunnel vaulting (which greatly improved the acoustics for chanting), horse-shoe-shaped chapels, and an aisled nave that ended with multiple apses.

Odilo's successor (and the first abbot freely elected by the monks rather than appointed by the dying abbot), Hugh was also a builder. He held the abbacy from 1049 until 1109, and during this period he doubled the number of houses directly dependent on Cluny, tripled the number of monks living at the monastery (from 100 to 300), significantly enlarged the living quarters, infirmaries, workshops and church at the site, and established important patronage relationships with the kings of Spain and England. Hugh's building program is best known for the new basilica he commissioned in 1088 and finished twenty years later. Cluny III was the largest church in Christendom until St. Peter's Basilica in Rome was rebuilt during the sixteenth century, and it reflected the Cluniac preoccupation with worship. Its nave was 260 feet long (120 feet longer than Clunny II) and over 400 feet from end to end. The nave was crossed by two transepts (an unusual architectural feature of the period) that were surmounted by bell towers. It soared to a height of 92 feet, and its vaults were supported by intricately carved columns. The nave ended in a rounded apse that had five horseshoe-shaped chapels radiating from it. It was funded by annual subsidies of Alphonso VI of Spain and Henry I of England. Leading architectural historians identify Cluny III as one of the precursors of the Gothic tradition.

Yet Hugh built more than churches. Like Odilo he was active in secular and ecclesiastical politics and developed strong relationships with Popes Leo IX, Alexander II, Stephen IX, Nicholas II, Gregory VII, Urban II, and Paschal II. He served as a papal legate in France and throughout the Holy Roman Empire—settling disputes between nobles and the church, adjudicating at church councils, and eventually helping Gregory free the church from control by the emperor during the "Lay Investiture Crisis" of 1076 to 1083.

With the death of Hugh in 1109, Cluny entered a period of disorder, financial crises, and dissent that arose internally and from changes in European monasticism. As a result, during the twelfth century Cluny slipped into a decline from which it never recovered. Cluny's internal problems began in the previous century with the expansion of the abbey, and the frequent absences of the abbot. Gradually the discipline of the Rule was relaxed to accommodate building schedules and the increasingly large numbers of monks (especially those who entered late in life) who were incapable of fully participating in the daily liturgical round. It took many years to master the difficult liturgy of the Cluniac services, and as the number of older men entering the novitiate increased Cluny began to relax its standards or did not allow them to participate fully in the chanting. Many of its brightest and most holy monks were sent to establish or run dependent houses.

The real trouble began in 1109 when Pons was elected abbot. He lacked his predecessors' religious virtue, zeal for reform, and administrative ability. One historian calls him a "splendor-loving" feudal baron who unwisely exhausted the abbey's dwindling income on renovating the cloister, buying relics, and adorning himself with the trappings of a secular lord. In 1119 some monks at Cluny lodged a formal protest with the pope that their abbot had betrayed the monastic vows of humility and poverty. Pons voluntarily resigned in 1122 amid great dissension within the order. In 1125 he forcibly tried to regain the abbacy by attacking the new abbot, Peter the Venerable, at the Swiss monastery he was visiting. For a year Pons plundered the abbey's granges and priories—forcing monks to pledge their loyalty to him, burning and killing at will. The pope intervened, imprisoned Pons, and restored the abbacy to Peter. But by this time the authority of the office was severely weakened, distrust and rebellion were rife throughout the order, and local bishops began attempting to regain control of Cluny and its dependencies. Its treasury was depleted and its most generous benefactor, Alphonso VI of Spain, was dead.

Eventually Peter restored stability and discipline to the Cluniac Order. The basilica was finally finished and consecrated in 1131, and he increased the number of monks at Cluny to 460 (although this number made it difficult for the abbot to personally influence his flock). Peter also wrote new statutes for the order during the 1130s as a way of reinstating uniform observance of Cluniac practices.

The primary external force that weakened Cluny's influence and prestige came from changes within European monasticism. Cluny established the idea of a monastic order, and new orders multiplied during the late eleventh and twelfth centuries. Its greatest and most vocal rival was the Cistercian order, founded in 1098 in nearby Burgundy at Cîteaux. The Cistercians represented a revival of the eremitic tradition in which solitude, austerity, humility, and a literal observance of the Rule of St. Benedict were championed. The Cistercians, especially its most famous leader, Bernard of Clairvaux, castigated Cluny for its failure to live according to Benedictine orthodoxy and charged the Cluniac monks with greed and worldliness. They claimed that Cluniacs offended God and were distracted from their true monastic duties by the requirements of managing their estates and by their ostentatious and unnecessary building programs and liturgical practices. Although Peter the Venerable answered these charges (many leveled during the abbacy of Pons) and tried to bring Cluniac

practices into closer agreement with the Rule, he was unable to stem the church-wide movement towards more simple forms of religious devotion and organization.

The Cistercian order eclipsed the Cluniac in France, England, and Spain during the twelfth century, and Cluny occupied a respectable but less vital role in the affairs of church and state thereafter. The fourteenth century witnessed a marked decline in devotion, and the number of monks at Cluny dropped to eighty. In the mid–eighteenth century, the community razed the medieval lodgings, and the last mass was celebrated in 1791. A revolutionary mob defaced the basilica and burned many of its furnishings in the same year. The abbey was sold to a private party in 1798 and was slowly demolished during the nineteenth century. Today a single transept of Hugh's magnificent church is all that remains of this medieval center of European monasticism.

The history of Cluny is a story of its remarkable set of abbots and their commitment to restoring discipline to monastic life. Cluny provided a new model for monastic practice and administration that would be extended or adopted throughout the Middle Ages. It was a product and example of the union of ecclesiastical and secular politics that characterized feudal society. In the words of H. E. J. Cowdrey,

> The Cluniac system . . . was a conflation of two ideas: the community of the Rule under an abbot, and the dominion and control of a king over his tenants-in-chief and sub-vassals. . . . Cluny thus made use of the two most powerful ideas in early medieval society, that of the religious obedience of a monk to his abbot, and that of the fidelity and mutual obligations of vassal and lord. The pivot of the whole system was the abbot of Cluny, who was at once in his spiritual capacity the father and sovereign of all Cluniac monks and in his forensic capacity, as *persona* or lord of the church of Cluny, the overlord to whom all the churches linked to her owed fealty.

Further Reading: A concise but comprehensive book on the history and customs of Cluny is Joan Evans's *Monastic Life at Cluny, 910–1157* (Oxford: Oxford University Press, 1931; reprint, Homden, Connecticut: Archon, 1968). Readers interested in the architecture of the middle ages should examine Evans's detailed history of Cluny's architecture, *The Romanesque Architecture of the Order of Cluny* (Cambridge: Cambridge University Press, 1938; reprint, Godstone, Surrey: Gregg, 1972), and Kenneth John Conant's *Carolingian and Romanesque Architecture, 800–1200* (London and Baltimore: Penguin, 1959). *The Cluniacs and the Gregorian Reform* by H. E. J. Cowdrey (Oxford: Clarendon, and New York: Oxford University Press, 1970) and *Cluny under Saint Hugh, 1049–1109* by Noreen Hunt (London: Arnold, and Notre Dame, Indiana: Notre Dame University Press, 1967) and provide excellent overviews of the history of Cluny and well-documented accounts of Cluny's role in the ecclesiastical reforms and political affairs of the tenth and eleventh centuries. Hunt also edited a collected volume of papers by leading Cluniac scholars that explore the devotional and artistic life at Cluny, *Cluniac Monasticism in the Central Middles Ages* (London: Macmillan, 1971; Hamden, Connecticut: Archon, 1972), which includes articles by Hunt, Raffaello Morghen, and others.

—Stephen Ellingson

Colchester (Essex, England)

Location: In Essex County, England, approximately fifty-five miles northeast of London.

Description: First Roman colony in England, making it the country's oldest recorded town.

Site Office: Tourist Information Office
1 Queen Street
Colchester, Essex CO1 ZPJ
England
(206) 712920

Colchester was known during ancient times as Camulodunum, meaning "fortress of Camulos," the British god of war. The modern Colchester is from the Old English for "Roman station on the river Colne," and points to the town's prominence in Britain's pre-Roman and Roman history. Camulodunum became part of the Roman empire in A.D. 43 and was the first English town to appear in historical records, but the area had been settled long before. Recent archaeological findings point to the existence of a settlement at the head of the estuary of the River Colne around 1100 B.C. Details of this earlier period are scant and can be drawn only from archaeological research; it is Roman contact with Britain that forms the beginnings of recorded British history. It is known that Belgic tribes, originating from an area bounded by the rivers Seine, Maine, and Rhine in western Europe settled in southeast England. They included the two tribes that were to play significant roles in Colchester's early history: the Catuvellauni, who settled northeast of the Thames, and the Trinovantes, who occupied the area of present-day Essex. During the period between 80 and 50 B.C., a second wave of Belgic invasion took place. Contact with Rome in connection with trade is known to have occurred before Julius Caesar's expeditions, which took place a century before the Roman conquest of A.D. 43.

Julius Caesar made two expeditionary invasions of Britain in 55 and 54 B.C. They took place in the middle of Caesar's major undertaking at the time, the invasion of Gaul, present-day France, during the period between 58 and 51 B.C. The Roman empire preferred to take military action in response to a *casus belli,* an act or situation that justified war. The reason Caesar gave for the first expedition was that Gallic tribes based in Britain had supported the enemies he had fought in Gaul. The real intention was reconnaissance, in preparation for a potential conquest. Caesar sailed with two legions—each consisting of approximately 5,000 soldiers—to Britain, probably from Boulogne, France. He landed August 26 in 55 B.C. at a spot between Deal and Walmer on the Kentish coast in southeast England. This main force of infantry waited at anchor for the cavalry transports.

They had been delayed at their departure, and were carried back to Gaul by the tide. Despite this setback, Caesar engaged the British in battle, and they were quickly beaten. During the days that followed, Caesar received in his camp many of the British chieftains, who were seeking peace. On August 30 and 31, however, a storm broke out that again prevented the arrival of cavalry support and damaged the infantry ships. The Britons quickly realized the weakness of the Romans' position and acted quickly to exploit it. Using war chariots, they attacked the Romans, but were eventually pushed back from the beach by the Romans. Again the British chiefs came seeking peace. But, lacking the necessary forces and supplies for a prolonged presence on British soil, Caesar returned to Gaul, planning to return the following year.

The 54 B.C. expedition was carried out with five legions and 2,000 cavalry, with the aim of conquering Britain. On this occasion, the British tribes hostile to Rome had united under the leadership of Cassivellaunus. The Roman campaign began well with the defeat of Cassivellaunus's military coalition. The Romans, with help from the Trinovantes, who feared Catuvellaunian hegemony, stormed and took Cassivellaunus's base, thought to be present-day Wheathampstead in Hertfordshire. Despite these successes, Caesar did not follow through with the planned conquest of Britain. Fearing an uprising in Gaul, he was forced to leave Britain. The previously hostile Britons were left with a token reminder of Rome: an obligation to provide hostages and an annual tribute.

Britain had developed into a potential new addition to the Roman empire on several counts. First and foremost, Britain's value to Rome lay in the efficiency of its inhabitants' farming methods, which yielded a cereal surplus that could be put to good use within the empire. Britons had also achieved a measure of technological skill and social organization. These attributes would make them more receptive to Roman forms of government.

In the century that followed Caesar's abortive conquest, Catuvellaunian influence established itself over most of southern Britain. The Catuvellauni created an empire during this time that included the Trinovantes. The latter's territory, centered around Camulodunum, was seized by Cunobelinus of the Catuvellauni in about A.D. 7. Coins minted with that date show that he had moved the new Catuvellaunian capital to Camulodunum. At his death in A.D. 40, Cunobelinus was succeeded by his sons Togodumnus and Caratacus.

A rival kingdom to the Catuvellauni had been established by Commius in present-day Hampshire and Sussex. Commius had moved from Gaul, where as a chieftain he had dealt with Caesar. His subjects belonged to the Atrebates

Colchester Castle, top; St. Botolph's Priory, bottom
Photo courtesy of Colchester Borough Council

tribe who, as rivals to the Catuvellauni, seem to have enjoyed the backing of Rome. This rivalry between the Commian kingdom and the Catuvellaunian empire resulted in the expulsion of Verica, a descendant of Commius, by one of Cunobelinus' sons. It was the event that gave the Roman emperor Claudius his *casus belli* for his invasion of Britain in A.D. 43. The actual reason for the invasion was Claudius's need for glory and popularity at home. Claudius had been made emperor by the Praetorian Guard, following Caligula's assassination in A.D. 41. Although supported by the military, he was considered incompetent by the Senate, and his hold on the throne was not secure. In A.D. 42, the governor of Dalmatia attempted a coup, for which he received significant backing from his colleagues in the senate. The enterprise failed when the two legions under the governor's command withdrew their support for the uprising. Claudius's need for credibility had become urgent. The conquest of Britain, unsuccessfully attempted by Julius Caesar, would greatly bolster his domestic standing.

For the invasion, Claudius put together a main force made up of four legions, led by Aulus Platius. The sea crossing culminated in three separate landings, to make British resistance more difficult. It is believed that one landing was made near Richborough in northeastern Kent, where a camp has been discovered in recent times. Another landing was made near Chichester, possibly to allow the Romans to meet up with forces loyal to Verica. Nothing is known about the third landing. Aulus Platius first defeated in battle Caratacus, then a force led by Togodumnus. These successes prompted a section of the Dobunni tribe (based in western England, around present-day Cirencester) to surrender to the Roman forces. Despite this setback, the Britons fought on with determination, but they were eventually overwhelmed by their Roman opponents, who enjoyed the advantages of superior equipment and more disciplined military training. The propaganda effect of the conquest was certainly exploited to full effect. When Claudius concluded the military campaign, he did so in the company of a large number of distinguished senators. Many historians believe that the seizure of Camulodunum was actually delayed until Claudius could appear in person. One historical source mentions that the military force accompanying Claudius included a number of elephants, on which he and the senators are supposed to have entered Camulodunum. There the emperor received the submission of eleven British kings, thus beginning three and one-half centuries of Roman rule in Britain.

Having been the capital of Britain's most dominant tribe, Camulodunum was now transformed into the capital of the new Roman province, Britannia. To underline his military triumph, Claudius named the town Colonia Victricensis, "Colony of the Victory." A legionary fortress was built on the site now occupied by the present-day town center. The organization of Roman rule in Britain followed the pattern established in other parts of the empire. In A.D. 49, the legionary fortress became a civil settlement named Colonia Claudia. The term *colonia* was applied to cities of the highest rank within the Roman Empire. As in the case of Camulodunum, the colonia were often created at legionary fortresses. Plots of land were given to retired legionaries—those who had served twenty-five years in a legion—to create a provincial settlement of Roman citizens. These settlements would, in due course, provide new recruits for the legions, which could enlist only Roman citizens. Colonial government was in the form of an *ordo,* an elected council made up of about 100 *decuriones,* councillors, drawn from the local pre-Roman aristocracy. The organization of these councils was in fact based on that of the Roman Senate. Camulodunum was the first *colonia* in Britain, and was followed by Lindum and Glevum (present-day Lincoln and Gloucester, respectively) toward the end of the first century. A temple was erected to Claudius, who was deified after his death in A.D. 54. Although apparently a religious building to serve the imperial cult—for which Camulodunum was the provincial center—it actually had a political purpose. Sacrifices were made here on behalf of living emperors, and as such the temple served the purpose of promoting public displays of loyalty to the imperial power of Rome. When performed by the local aristocracy, they were designed to set an example to local inhabitants of lower rank.

The government of other parts of conquered Britain was often in the form of a client-kingdom. Client-kings were bound to Rome by treaties that ensured a favorable disposition toward the empire. In return, they received the benefit of Roman protection. It was a common arrangement in the first three decades following the conquest. Upon the death of the first client-king, the Romans normally imposed an *ordo*-type administration, as was the case when Prasutagus, king of the Iceni (based in present-day Norfolk and northwest Suffolk) died.

The Iceni figure prominently in early Romano-British history, for they revolted twice against Rome. The first occasion was in A.D. 47, when Britannia's governor attempted to disarm them under the *Lex Iulia de vi publica,* the Julian Law on public violence. Designed to facilitate imperial control of Roman subjects, it prohibited the bearing of arms, and was one of the laws enacted throughout the Roman Empire from the end of the first century B.C. to the second century A.D. The second Icenian revolt resulted in the destruction of Camulodunum in A.D. 60.

With the death of the Icenian client-king Prasutagus, the Romans moved to install their own administration. Prasutagus had bequeathed half of his kingdom to his daughters and the other half to Emperor Nero, in the hope appeasing the Romans and thus keeping his kingdom safe. But the governor's centurions and the slaves of the provincial procurator (a senior official in charge of the estates of defeated tribes and taxation) actually took the whole kingdom with force. The Icenian nobles were treated badly. Prasutagus's daughters were raped and his widow Boudicca (she is also referred to as Boadicea) was flogged. Further grievances were incurred when the procurator Catus Decianus tried to recover the grants of money that Claudius had made to the native leaders, and Icenians were evicted from their estates. The enraged Icenians were joined by other tribes, unspecified in historical records, and by the erstwhile friends of the Romans, the Trinovantes. They had also been aggrieved by the appropriation of their lands for the settling of legionaries.

The Roman forces at Camulodunum were ill prepared to face the wrath the British tribes. The defenses of the legionary fortress had been leveled to make way for the civil development of the town, and the city was quickly overwhelmed and destroyed. A small number of soldiers managed to hold out for two days in the Temple of the Imperial Cult. The temple was viewed as a particularly loathsome symbol of Roman imperialism, and Boudicca was determined to destroy it. It was surrounded and burned; those seeking refuge inside were massacred by the British, who quickly moved on to Londinium (present day London), which was also sacked. A similar fate befell Verulamium, present-day St. Albans in Hertfordshire. Eventually quelled by the imperial army, the revolt ended with the suicide of Boudicca and heavy casualties on the British side.

Londinium became Britannia's new capital in the wake of Boudicca's uprising. It took Camulodunum fifteen years to recover from the event. Though diminished in its urban status, the town was still important enough to serve as the center of the Imperial Cult. The temple was rebuilt and wall fortifications were erected, which stretched for some 9,000 feet around the town. Much of them can still be seen today, especially along Balkerne Hill to the west of the town center, and along Vineyard and Priory Streets. Balkerne Gate is the largest surviving Roman gateway in Britain. One pedestrian gate and two guardhouses can still be seen today. A large part of the gateway now lies under a pub, the wittily named Hole-in-the-Wall.

Other evidence of the town's Roman history, the remains of a Roman amphitheater, lie near the town's castle, in Maidenburgh Street. The curve of the foundations is indicated by dark paving stones on the road. Parts of the walls and floor have been preserved inside a modern building, where they can be viewed through a display window under 74 Maidenburgh Street.

Toward the end of the fourth century, the Romans pulled troops out of their distant provinces in response to barbarian invasions closer to home. Barbarian groups near Britain began taking advantage of the Romans' absence. Finally in 409, the British elite, vexed at the continued need to pay taxes to an empire that could no longer defend them from attack, expelled all remaining Roman officials.

When the Normans reached the city in the eleventh century, Colchester was a flourishing town; the Roman walls and many Roman ruins were still visible. Approximately ten years after the invasion, the Normans began construction of Colchester Castle upon the foundations of the Roman Temple of Claudius. It was completed in 1120 and contains the tallest Norman castle tower, or keep, in Europe. Colchester was a suitable site for a Norman fortress for a number of reasons. It had a harbor, which could be used as a warship base from which to defend the eastern coastline and the Thames estuary to the south. The town remains strategically important even today: there is still a military garrison here, and a tattoo (a military spectacle of marching and music) takes place here in even-numbered years. The town was also commercially active in the twelfth century, and its existing infrastructure represented another advantage to the Normans. The Roman walls, erected a thousand years before, were still in place. The stone base of the Temple of Claudius was still surrounded by the stone walls of its court, and there was a plentiful supply of building materials nearby in the form of Roman ruins. It was an ideal spot for a fortress. The ditch that can be seen today was dug at this time, and the earth was thrown onto the Roman walls to form a defensive mound. The castle originally stood two stories higher than it does today.

Another Norman structure, St. Helen's Chapel, stands above the remains of the Roman Amphitheater. The first written record of the chapel was made in 1097, but the present structure dates from the thirteenth century. St. Helen's stands at the edge of the Dutch Quarter, so called because some 500 Flemish refugees settled here during the sixteenth century to engage in the town's cloth trade.

An obelisk on the north side of the Castle commemorates Colchester's brief involvement in the English Civil Wars of 1642 to 1651. King Charles I, crowned in 1625, had increasingly alienated Parliament with his beliefs in absolute government by the crown and his sympathies toward Catholicism. Matters came to a head when Parliament attempted to curb the king's powers in 1640; this eventually led to the outbreak of civil war two years later, in which royalist Cavaliers fought the parliamentarian Roundheads. The second Civil War began in April 1648, when royalists captured the towns of Berwick and Carlisle. On June 4, a royalist uprising took place in Colchester. Roundheads attempted to storm the town on June 13 but were repelled by Sir Charles Lucas, a cavalry commander. The following day the roundheads began a siege of the town, which was to last seventy-five days until Colchester's surrender on August 28. Sir Charles Lucas and two brothers-in-arms, Sir George Lisle and Sir Bernard Gascoigne, were imprisoned in the castle for a short period. After being court-martialed, Lucas and Lisle were executed. The obelisk that can be seen today marks the spot where they died.

The ruins of St. Botolph's Priory Church also bear witness to the fighting that took place in 1648. They stand just outside the Roman walls at the southeastern corner of the town, and show a great deal of recycling of Roman building materials. Built on Saxon foundations—which places it in the period between the Roman withdrawal from Britain in the fifth century and the Norman conquest in the eleventh century—St. Botolph's became England's first Augustinian priory in about 1100. It stood for five and one-half centuries before its destruction in the second Civil War. Near St. Botolph's and along Priory Street, which follows the Eastern stretch of town wall, one can see how the wall was reinforced with bastions, which were put in place in the fourteenth century. Brick repairs are also visible, made after damage of the 1648 siege. Continuing along Priory Street and up East Hill, one arrives at St. James's Church, the town's largest medieval church. This is also the spot where the town's East Gate stood until 1674. Diagonally opposite St. James's stands Colchester's finest Georgian house. Built in 1718, it was owned by Charles Gray, Member of Parliament for the town. The house is known as Hollytrees and today houses a museum of toys, costumes, and decorative art items.

Colchester is often mentioned in connection with the paintings of John Constable (1776–1837), who, along with contemporary John Turner (1775–1851), is hailed as one of Britain's finest landscape artists. Colchester has traditionally served as the base from which to explore the landscapes that Constable liked to paint. They are to be found in the area that runs northeast and west of the town. Off the A12 road five miles to the northeast of Colchester is East Bergholt, the artist's birthplace. Other destinations in the area include Dedham: the town provided the inspiration for many of his works;

and Flatford: the mill and the thatched-bridge cottage feature in Constable's painting *The Hay Wain.*

Colchester also figures in the literary arts. 11 and 12 West Stockwell Street was the home of Ann and Jane Taylor from 1796 to 1811. The sisters are remembered today for the following lines, written by Jane:

"Twinkle, twinkle little star,
How I wonder what you are!
Up above the world so high,
Like a diamond in the sky!"

Further Reading: *The Cambridge Historical Encyclopedia of Great Britain and Ireland,* edited by Christopher Haigh (Cambridge and New York: Cambridge University Press, 1985) provides a readable introduction to Britain's early history, with conveniently presented explanations of different aspects of Roman culture. *The Oxford Illustrated History of Roman Britain* by Peter Salway (Oxford and New York: Oxford University Press, 1981; second edition, 1993) is an excellent, detailed work covering Roman-British history, from Caesar's expeditions to the end of Roman rule.

—Noel Sy-Quia and Hilary Collier Sy-Quia

Compiègne (Oise, France)

Location: Situated at the confluence of the Oise and Aisne Rivers, some fifty miles northeast of Paris.

Description: Ancient city in northeastern France, site of the signing of the Armistice of 1918 and also of the capitulation of France to Nazi Germany in 1940.

Site Office: Office de Tourisme, Syndicat d'Initiative
Hotel de Ville B.P. 9
60321 Compiègne, Oise
France
44 40 01 00

The history of the city of Compiègne begins in Roman times, when it was a settlement named Compendium. Little is known about the place, hidden in dense forests, until Merovingian rulers in the early Middle Ages held councils and assemblies there. At an uncertain time, a palace was built that served primarily as a hunting lodge. Charlemagne is reputed to have favored the palace at Compiègne, gathering his most favored knights there to hunt in the 35,000-acre forest. Legend has it that he decided at Compiègne on the ill-fated military intervention in Spain, which ended with the tragic battle at Roncevaux sung in the *Chanson de Roland.*

Throughout much of the Middle Ages, Merovingian and Carolingian kings had battled internal divisions in the realm. The reign of Charlemagne's son, Louis the Pious, was marked by greater Frankish unity and considerable political and military expansion. Charlemagne's grandson, Charles the Bald, attempted to consolidate these gains. He saw Compiègne as one of the major centers of his kingdom and tried, unsuccessfully, to rename the city after himself—as Carlopolis. In about 850, he donated the original palace at Compiègne to a religious order. The building then survived for almost a millennium as the abbey of St. Corneille, but it was completely destroyed during the French Revolution. Charles built another, grander royal residence on the bank of the Aisne River. The building also served as a fortification against the invading Normans and is distinguished by a massive defensive tower. For about 500 years, Frankish kings added to and embellished this palace on the river.

Charles the Bald had been unable to root out internal rivalry altogether, however, and further squabbles ensued upon his death. The Catholic Church capitalized on the confusion and extended its worldly influence in the late ninth and tenth century. During this time the abbots of St. Corneille controlled the local government of Compiègne. When the last Carolingian, Louis V, died in 987, a group of nobles and bishops met at Compiègne and Senlis and decided to offer the crown to Hugh Capet, who was then count of Paris, Senlis, Dreux, and Orléans. This was the beginning of Capetian rule.

During several centuries of chaos in the realm, Compiègne remained with the French crown. The palace built by Charles the Bald was visited frequently. Charles V, however, envisioning greater splendor, gave the structure to the Jacobins in the mid–fourteenth century and started over on the edge of the forest with a palace that remained a hunting lodge and a favored spot of French monarchs until the mid–nineteenth century. The residence was altered frequently to suit the individual requirements of a long succession of kings and queens. The most thorough alteration, by Louis XV, was begun in 1728 and left very little of the original structure.

Compiègne entered another turbulent phase in its history with the reign of Charles VI (1380–1422), who is variously known as Charles the Mad or Charles the Beloved. Charles VI has chiefly gone down in history for his rapacious greed and the endless taxes he levied on his subjects. He needed the money, in part, for his campaigns against the Burgundians, who were making encroachments against the French. Compiègne, tired of the heavy taxes, rebelled and sided with the Burgundians—an ill-advised decision. Charles VI immediately marshaled an army and marched on the city. The Burgundians failed to defend their new allies, and the rebellion was quickly suppressed.

Having learned its lesson, Compiègne swore fealty to Charles VII when he acceded to the throne in 1422. When the Burgundians and the English laid siege to the city in 1430, Compiègne defended itself under the leadership of William de Flavy. Joan of Arc, then a national hero after having won a series of military victories for the French king, was dispatched to the city to help raise the siege. On May 25, 1430, she led a raiding party of about 500 men outside the city gates to attack one of the encampments of the Burgundians. Her party ran into more resistance than anticipated and beat a hasty retreat. Joan herself made up the rear and was taken by the pursuing Burgundians just outside the city gates. John, Duke of Luxembourg, imprisoned her in his castle at Beaulieu and subsequently sold her to the English. Taken to Rouen, she was tried, convicted of heresy and sorcery, and burned at the stake in 1431.

Although France lost Joan of Arc at Compiègne, the city successfully withstood the siege. After six months, the Burgundians and the English retreated. However, still under the capricious and cruel government of the infamous William de Flavy, the citizens of Compiègne found little relief after their victory. De Flavy was removed from office for egregious misrule when the people of Compiègne petitioned the crown to intercede on their behalf. Unfortunately, he managed to get himself reinstated shortly afterward when he promised to curb his debauchery and willfulness. In spite of the promise, even greater misdeeds followed. When he imprisoned several members of notable French families and

LA MARNE
L'YSER
BELGIQUE
ARTOIS
LA SOMME
L'OISE
EMIN des DAMES
CHAMPAGNE

VERDUN
ARGONNE
LES EPARGES
LORRAINE
ALSACE
VOSGES
ITALIE
ORIENT

The memorial flame at the Clairière de l'Armistice
Photo courtesy of French Government Tourist Office

ordered their execution, he overstepped his limits, however. One of his notable victims was the father of his wife, who then plotted to have him killed. In 1449, de Flavy's wife found his barber willing to cut de Flavy's throat. When the barber somehow failed to execute the plan efficiently, de Flavy's wife herself accomplished the murder. Upon de Flavy's death, the relatives of another victim, Pierre de Rieux, Maréchal de France, filed a lawsuit that sought to hold de Flavy's family responsible for his crimes. The suit resulted in 1513 in a parliamentary decree that ordered the erection of a monument in the Rue du Pont at Compiègne. It bore a public indictment of de Flavy for his crimes against de Rieux. The stone cross does not survive, but the text of the inscription is perpetuated in the historical record.

During the Wars of Religion in the sixteenth and seventeenth centuries, Compiègne remained staunchly Catholic in all struggles against the Protestant League. The city remained relatively calm and stable amid the upheavals in the country. In 1624, the first of a series of signings of historic international treaties took place at Compiègne, when Cardinal Richelieu signed a treaty that ended the war with the Dutch.

In 1728, Louis XV completely reconstructed the residence at Compiègne originally built in the fourteenth century by Charles V. Unhappy with the inconveniences of medieval architecture, he commissioned a more ''livable'' hunting lodge. Louis XVI first met Marie-Antoinette there, and it was their efforts that transformed the structure into a palace second only to Versailles and Fontainebleau. Unlike its owners, the palace survived the French Revolution virtually unscathed.

The site of lavish balls and parties under the Napoléons, Compiègne has in recent history more often found itself a theater of war. During the Franco-Prussian War, Compiègne served as headquarters of the invading German army. Like most of northern France, Compiègne was laid waste by the ravaging trench warfare of World War I. But early in the morning of November 11, 1918, in Marshal Ferdinard Foch's rail car, placed on a siding in the forest near Compiègne, Germany signed its unconditional surrender to the allied forces. Foch's rail car was left on its siding as a memorial of allied victory, and a commemmorative marble slab was erected nearby. The slab bore the inscription, ''Here on the eleventh of November, 1918, succumbed the criminal pride of the German empire . . . vanquished by the free peoples which it tried to enslave.''

During World War II, Compiègne again suffered much damage, though its famed buildings survived. On June 17, 1940, Marshal Henri-Philippe Pétain asked the Germans for an armistice, one day after replacing Paul Reynaud as the head of the French government. One June 22, 1940, the armistice was signed by the French delegation and by Adolf Hitler and other members of the German party. The signing took place in the same rail car as the signing of the armistice that ended World War I. This new armistice called for the disarming of the French forces and the surrender of 60 percent of France to German control. The remainder of the country, while it remained under French government, was required to collaborate with the Nazi regime. Before the signing, a triumphant Hitler, bent on revenge for Germany's disgrace in 1918, stood before the commemmorative marble slab. Three days later, he had it blown up. Foch's rail car was also removed by the Germans and taken to Berlin, where it was reputedly destroyed in the Allied bombings. The site of the historic signings, known as Clairière de l'Armistice, is still marked in a clearing in the forest by a replica of Foch's rail car.

From 1941 to 1944, Compiègne served as one of the deportation centers for thousands upon thousands of the victims of the Holocaust. It was one of a ring of such centers around Paris, together with Pithiviers, Beaune-la-Rolande, and Drancy. To promote their ultimate goal of Aryanization, the Germans planned what they called "the great Parisian action" of 1941, which called for the deportation to the concentration camps in Germany of 4,000 Jews per week, 1,000 from each French camp "serving" Paris.

Decades later, Compiègne became a fashionable resort for Parisians. Its architectural riches include, besides the palace of Compiègne, a medieval Benedictine monastery and the former abbey church of Morienval. The church, built in the eleventh century in the northern Romanesque style, contains the oldest ogival vault in France. The Hôtel de Ville, completed in the sixteenth century, now houses a museum of early Roman art.

Further Reading: Clara E. Laughlin's *The Martyred Towns of France* (New York and London: Putnam, 1919) gives a somewhat breathless but otherwise serviceable account of Compiègne's history up to the end of World War I.

—Kim M. Magon and Marijke Rijsberman

Copenhagen (Denmark)

Location: One of Denmark's easternmost points, on the islands of Sjælland and Amager, directly across the Øresund from Sweden.

Description: Capital city of the Kingdom of Denmark and the country's chief port (Greater Copenhagen's 1988 est. pop., 1,343,916). In the nineteenth century, home of Hans Christian Andersen and Søren Kierkegaard.

Site Office: Copenhagen Tourist Information
Bernstorffsgade 1
DK-1577 Copenhagen V
Denmark
11 33 25

For centuries Copenhagen has been an important trading center for the Baltic region because of its port, situated on the narrow waterway between the islands of Sjælland and Amager, which is the largest in Denmark even today. Earlier still there had been human activity and settlement in and around the two islands for thousands of years. Remains have been found indicating the presence of hunting tribes in what is now central Jutland from about 14,000 B.C., and they may well have reached the islands, which lie to the east of Jutland. In about 4,000 B.C. the first permanent agricultural settlements were established. As the area became more densely populated, territorial conflicts developed, out of which the tribe to be known as the Danes emerged in about A.D. 500. They controlled a territory, already known as Danmark, which included what is now southern Sweden. Their small settlement of Havn (harbor) was to develop into Copenhagen.

Although most Danes were farmers—and this corresponds with archaeologists' discovery of a farmers' village on the site from about the tenth century—fishing was probably the leading activity there throughout the six centuries that saw the voyages of the famous Viking ships as far as North America; the conversion of the Danes to Christianity, starting with their King Harold Bluetooth in 960; the expansion of Danish power over England and the North Sea under his successors; and the collapse of this short-lived empire in civil wars between 1131 and 1157. It was at this point that Havn changed from a small fishing village to a focus of political and economic power. Having reunited the Danes, King Valdemar I, the Great, who ruled from 1157 to 1182, granted control of the settlement to Absalon, the bishop of Roskilde and archbishop of Lund (later archbishop of Denmark), a faithful adviser who built a large castle on the tiny island of Slotsholmen in 1167. Up until his death in 1201 Absalon, regarded as the founder of the city, went on to serve as chief adviser to Valdemar's successor Canute, king from

1182 to 1202, overseeing the expansion of Danish power over the southern and eastern coasts of the Baltic Sea. This policy was continued until 1223, when King Valdemar II, captured by Count Henrik of Schwerin, secured his release only at the cost of giving up most of these conquests. For Copenhagen the collapse of this second Danish empire had the beneficial side effect of encouraging Valdemar II and his successors to turn their attention to internal affairs. In 1254 a charter of municipal rights was granted to the town of Havn; from 1290 its character was altered when moats and ramparts were constructed around it. By 1300 a growing and prosperous *borger* (merchant) class had developed, and Havn had became known as Købmandens Havn (merchants' harbor), from which the city's modern Danish name, København, originates. Absalon's castle survived at the heart of the town until 1369, when opponents of King Valdemar IV, who was then engaged in a costly and disastrous war with the Hanseatic League, attacked and destroyed it.

The Union of Kalmar in 1397 formally brought Denmark, Norway, and Sweden together under the authority of Erik VII, known to European history as Erik of Pomerania. The castle in Copenhagen was rebuilt for Erik and became the official residence of the Danish monarchs from 1416.

Under Erik's rule Danish merchants took over the trade in foreign goods entering Denmark from the former monopoly exercised by the merchants of the Hanseatic League. Copenhagen, one of the towns that increased its wealth through expanding its participation in this trade, was granted royal trading privileges in 1417 in recognition of its key role in the Baltic region. Erik, like Valdemar IV, found his plans obstructed by the Hanseatic League, which attacked Copenhagen in 1428; the Hanseatic League was badly affected by Erik's imposition of the Sound Dues, a toll on all ships entering the waters off the coast of Helsinger, in 1429. In alliance with other commercial interests and with Danes who objected to Erik's heavy taxation (increased in order to pay his troops), the Hansa saw to Erik's replacement as monarch by Christopher III of Bavaria, in 1439. He exempted his Hanseatic allies from the Sound Dues, and established his court at Copenhagen, which now had about 10,000 inhabitants, confirming its status as capital in succession to Roskilde.

After Christopher's death Christian I became the first Danish monarch to be crowned in Copenhagen, in 1449. Christian founded Copenhagen University in 1479: the city was to be a cultural capital as well as a political and economic one. As it turned out, however, it was to be the capital, not of the Union of Kalmar, but of Denmark alone (though its kings also ruled Norway). Sweden broke away from the rule of Christian II, who was overthrown by Danish nobles in 1523 and replaced by Frederick I. (Christian fought to regain

The Copenhagen skyline
Photo courtesy of Danish Tourist Board

his throne for nine years, until he was captured and imprisoned.) The succession of Frederick's Lutheran son Christian III, in 1533, confirmed the adherence of Denmark to the Protestant Reformation.

Under Christian IV, who was king from 1596 to 1648, there was a further shift in power and wealth away from the majority of Danes, who were farmers, toward the monarch and other great landowners, along with their allies, the townspeople of Copenhagen and elsewhere. Estates increased in size, and the burden of rents and taxes grew heavier, but Denmark's chief export, the thousands of cattle taken to Germany every year, also increased in numbers and value; and thus the nobles and the king acquired a larger surplus to spend in the capital. Christian IV's lasting contribution to the development of Copenhagen, forming the foundations of the modern city, is still very much reflected in its architecture today. His enlargement and reconstruction of Copenhagen involved creating the seamen's district of Nyboder (which has recently been restored), reclaiming and laying out the district of Christianshavn (Christian's harbor) on the island of Amager, and extending the city's fortifications to the north. His legacy of important public buildings, most of which still carry his "C IV" monogram, includes the Rosenborg Slot, the suitably rose-colored palace where he died, which has been a museum since 1830; the Royal Arsenal, now a museum of military history, which contains the longest vaulted hall in Europe; the Børsen (Exchange),

topped by an extraordinary spire of dragons' tails intertwined, which today houses the city's Chamber of Commerce and related bodies; and the seamen's church known as Holmens Kirke (the church on the island). But perhaps the most distinctive building put up by Christian IV is the Rundetårn (round tower) of the Trinity Church, completed in 1643. This was originally an observatory but is now popular with tourists who can walk up its 230-yard spiral ramp (there are no stairs) to survey the city from the top. Unlike the Russian emperor Peter the Great, however, they are not allowed to drive a coach drawn by six horses up the ramp.

At Christian's death, after a series of defeats in the Thirty Years War, which raged across Germany from 1618 to 1648, Denmark was weakened just as Sweden was entering its "Age of Greatness." In 1657 the occupation of Jutland by Swedish troops and the threat that they might advance on the capital forced Denmark to surrender all its remaining territory in what is now southern Sweden. Three years later, after the Swedish siege of Copenhagen in 1658 had been ended through the intervention of warships sent by the Dutch, the Treaty of Copenhagen confirmed the loss, and King Frederick III gave the city equal status with the nobility in recognition of its citizens' loyalty. Since Denmark became a hereditary monarchy in the same year, 1660, this status in itself might seem little more than a symbolic gesture, but it did reflect the influence of the city's leaders, who had advocated the change out of resentment of the nobles' former rights to elect kings and to avoid taxes. Some of these citizens were then appointed as royal officials, serving a king who by 1665 had made himself absolute ruler of Denmark. Under his patronage the city's population rose by about 100 percent, to reach 60,000 by 1700, and Copenhagen's merchants accounted for about one-third of Denmark's trade with other countries.

During the eighteenth century Copenhagen lost about one-third of its population to a plague, which struck in 1711–12, and many of its people and buildings to two great fires, in 1728 and 1795. But the city recovered each time, partly through the economic stimulus of royal building programs, through which the absolute monarchs both maintained Christian IV's legacy of public buildings and extended it. In 1733, for example, Christian VI began the Christiansborg Slot (palace) on the site of the castles of Bishop Absalon and King Erik of Pomerania. After twice burning down and undergoing several alterations and restorations, the palace has housed the Folketing (parliament) since 1918. In 1749 Frederick V began the building of the Amalienborg Slot in honor of his wife, Queen Sophie Amalie. This enormous rococo palace was originally four separate buildings owned by nobles but became a royal residence in 1795 after Christiansborg first burned down. They were designed by Niels Eigtved as the centerpiece of the new district of Frederikstad, which also contains other residences he designed for leading nobles, the gigantic domed Frederiksidrke or Marmorkirke (marble church), and the Prince's Palace, completed in 1744 and now part of the National Museum.

Denmark remained neutral throughout the American Revolution and the European wars of the late eighteenth century, preferring to prosper from free trade with all the various combatants. By 1801 Copenhagen, still dominating the Danish economy, had more than 100,000 inhabitants. But neutrality was not permitted in the continental struggle between Britain and France. In 1801 and again in 1807 British warships attacked the Danish fleet docked in the city's harbor, as punishment for its role in facilitating trade between the neutral states (Denmark, Russia, and Sweden) and the French Empire and for its failure to join the alliance against Napoléon. On the second occasion large areas were destroyed, hundreds of citizens were killed, and the entire Danish fleet was taken to Britain. The Danes then allied themselves to the French just as Napoléon's luck was about to turn for the worse; the outcome was the bankruptcy of the state in 1813 and the loss of Norway to Sweden in 1814.

By 1831, when many of the people of Copenhagen welcomed the news of another French revolution, the absolute monarchy was clearly under threat. King Christian VIII who came to the throne in 1839, presided over agricultural reforms, the introduction of railways, and, in 1848, the end of absolute rule and the introduction of a liberal constitution. By 1850, Copenhagen, with a population of 130,000, had a new cathedral, the Vor Frue Kirke (Church of Our Lady), a new city hall (now law courts), and a new prison, all designed in the classical style by C. F. Hansen, as well as new university buildings in the neo-Gothic style by Peder Malling. The city's residents could spend their free time in the Tivoli Gardens, the amusement park, opened in 1843, which continues to provide a variety of restaurants and many outdoor entertainments. Nineteenth-century Copenhagen was home to Nicolei F. S. Grundtvig (1783–1872), who became bishop of Copenhagen in old age after having founded the "folk high schools" that were to inspire adult education movements all over Europe; to Hans Christian Andersen (1805–75), whose novels are as well known within Denmark as his fairy stories are elsewhere; and to Søren Kierkegaard (1813–55), the melancholy philosopher of Copenhagen University whose work was to be taken up by the existentialist movement in the twentieth century.

Meanwhile the city was growing rapidly. Its ancient fortifications were torn down in 1867, and the effective absorption of the suburbs of Frederiksberg and Gentoffe extended the capital north and west; the population rose to 261,000 by 1880 and to 454,000 by the beginning of this century. Industrialization lay behind this expansion, bringing with it a number of slums on the outskirts of the city, a militant Marxist movement, which first organized strikes in 1872, and as a side effect, a small group of wealthy philanthropists. This last group was typified by Carl and Ottilia Jacobsen of the Carlsberg Brewery, who donated their collection of Greek, Roman, and Etruscan sculptures to Denmark in 1888; soon after that they created the Carlsberg Foundation, which continues to contribute to the museum Ny Carlsberg Glyptotek, which opened to the public in 1897. In 1899 the increasing polarization between labor unions and employers, which had led to damaging strikes and lockouts, subsided with a kind of truce between the two sides. Its terms—recognition of unions, binding agreements, strike ballots—still govern industrial relations today, though they did not avert the month-long general strike of 1920 called to protest the economic depression of the time.

In 1905 this modern industrial capital city, no longer a small town on the northern edge of Europe, celebrated the opening of Martin Nyrop's new Rådhus (City Hall), with its 370-foot tower, the tallest in the country. The building's main embellishments are inside a large and complex astronomical clock designed by Jens Olsen and, outside, a fountain erected in 1923 which is surmounted by Joachim Skovgaard's vivid sculpture of a bull fighting a dragon. Eight years later the head of the Carlsberg Brewery commissioned the famous bronze statue of Hans Christian Andersen's Little Mermaid, sculpted by Edvard Erichsen—whose model was Copenhagen ballerina who had danced the role—and placed on the seawall as if gazing on her former home. Some residents, and some visitors too, prefer the Gefion Fountain nearby, which depicts the goddess Gefion with her four sons who have been transformed into oxen.

During the 1930s Copenhagen began to expand further to the west as the city council built apartment blocks in what had once been farmers' villages. In 1940 the Grundtvigs Kirke, perhaps Copenhagen's most distinctive modern building, was completed on the city's highest point. This yellow brick church, designed by P. V. Jensen Klint to resemble a pipe organ, is a memorial to Bishop Grundtvig. It was in the same year, on April 9, that, with German troops approaching the city, the Danish government was told by the German ambassador that it must accept a German army base or face the consequences. Denmark, a neutral state since 1814 (apart from the brief war of 1864 in which it had lost Schleswig-Holstein to Prussia and Austria), surrendered and was occupied for five years, up to the liberation of Copenhagen on May 4, 1945, which had been preceded by the evacuation of Danish Jews to Sweden and a general strike in Copenhagen. The Museum of the Danish Resistance Movement, in Churchill Park, now commemorates the tragedies and heroism of the war years.

Denmark became a founder-member of the United Nations and then of the North Atlantic Treaty Organization (NATO), abandoning neutrality but forbidding the importation of nuclear weapons. In 1973 Denmark became the first Scandinavian country, and so far the only one, to join the European Community, now the European Union, which it briefly disrupted by rejecting the Maastricht Treaty, aimed at greater unity, in a referendum in 1992. The capital of this cosmopolitan nation retains some of the atmosphere of the fishing village it once was long ago, and much of the grandeur of its past kings, especially Christian IV and Frederik V, though little of their violence and autocracy. These kings surely would not have welcomed the fact that their royal capital now has the Little Mermaid as its best-known symbol.

Further Reading: Just as Copenhagen dominates the economy and society of Denmark, so it plays a large role in general histories of the country, such as W. Glyn Jones' *Denmark: A Modern History* (London, Sydney, and Dover, New Hampshire: Croom Helm, 1986). More than half a century after it was first published, Monica Redlich's *Danish Delight* (London: Duckworth, 1939) remains a vivid account of one year in the life of a foreigner growing accustomed to Copenhagen.

—Monique Lamontagne

Culloden (Highland, Scotland)

Location: Six miles east of Inverness between the Moray Firth and the River Nairn.

Description: Site of the April 16, 1746, Battle of Culloden; now under the jurisdiction of the National Trust for Scotland.

Site Office: Culloden Visitors' Centre
Culloden Moor, Highland
Scotland
(436) 790607

On April 16, 1746, the dream of returning a Stuart to the British throne died on blood-soaked Culloden Moor. The last battle fought on British soil, Culloden ended in the defeat of the Highland army led by Charles Edward Stuart, the Bonnie Prince Charlie of legend, at the hands of government forces led by the Duke of Cumberland.

In 1688, King James VII of Scotland and II of England, an unpopular Roman Catholic monarch, attempted to secure some measure of freedom for his fellow Catholics. Government leaders, critical of James's pro-Catholic policies, encouraged William of Orange, James's Dutch Protestant son-in-law, and his daughter Mary to come to Britain to rule jointly. James fled to France. Yet there were many in Scotland who remained loyal to the exiled king. These supporters were called Jacobites (from *Jacobus*, Latin for James).

An early attempt to restore James to the throne occurred the following year when John Graham of Claverhouse, Viscount Dundee (the Bonnie Dundee of popular song), defeated government forces at the Battle of Killiecrankie on July 27, 1689. But Graham was fatally wounded, and, with their leader gone, the Jacobites soon foundered. Although James died in 1701, the hope of a Stuart revival lived on through his son, James Edward, known as the "Old Pretender" (claimant to the throne), and his son, Charles Edward Stuart, the "Young Pretender." The death of the last Stuart monarch, Queen Anne, James VII's half-sister, in 1714 and the swift succession of George of Hanover, a German Protestant, set in motion a series of Jacobite uprisings. Two major attempts to restore the Stuart monarchy to the British throne took place during the first half of the eighteenth century: the 1715 Rising and the 1745 Rising.

In 1715 the Earl of Mar raised the standard at Braemar on behalf of the Old Pretender. Mar then proclaimed the exiled king as James VIII of Scotland and III of England. Mar was not much of a military leader. On November 13 he fought an indecisive battle at Sherriffmuir against the army of George I. Rather than taking advantage of the uncertainty of the situation, he withdrew to Perth and his support evapo-

rated. The Jacobite cause was once again in disarray. The government, however, wasted little time by launching a systematic program to pacify the Highlands. Attempts were made to destroy the Gaelic language with the establishment of English-language schools. Further, General George Wade, the commander in chief for Scotland, built military roads intended to help the government keep a watchful eye over the Highlanders.

Thirty years later, in 1745, another and more dramatic Jacobite Rising took place when Prince Charles Edward Stuart accepted the challenge to regain his ancestral throne. The prince and seven supporters landed on the Island of Eriskay in the Outer Hebrides on August 2, 1745. His initial reception was less than encouraging, however. When confronting the young prince for the first time, Alexander Macdonald of Boisdale promptly told him to go home. With that the prince reportedly responded, "I am come home, sir" and sailed for Moidart on the Scottish mainland, landing at Loch nan Uamh, near Arisaig. Seeking support, he distributed letters to the clan chiefs, appealing to their sense of loyalty and duty to the crown. Subsequently, the standard was raised at Glenfinnan on August 19, 1745, where some 900 members of Clan Cameron and Clan Macdonald had gathered. The Forty-Five had officially begun.

Initially the prince's heather army consisted of a small force of about 1,200 men. Yet the more he traveled across the countryside, the more support he received, so that by the time the prince made his triumphant march to Edinburgh his ragtag army was nearly 3,000 strong.

Charles met with some early triumphs. His army routed government forces under Sir John Cope on September 21, 1745, at the Battle of Prestonpans. Then the slow march toward London began. On November 16, the northern English town of Carlisle surrendered. Flushed with success, the army moved on to Manchester. By early December, they had reached as far south as Derby, less than 150 miles from London. The presence of an enemy force so menacingly close placed the capital city on alert.

But internal conflict within the Jacobite army threatened to undo the hard-fought victories of the campaign. While Charles and his field commander Lord George Murray quarrelled over strategy, a thousand or so homesick Highlanders returned to their glens. Worse yet was the news that the government army, led by the Duke of Cumberland, George II's young son, was heading in the prince's direction. Charles wanted to march all the way to London. But his advisers won the day, and on December 6 the decision was made to retreat to Scotland and launch a new campaign in the coming year. The prince used Inverness as his winter base from which he continued the fight with attacks on Fort Augustus and Fort William.

The Culloden Memorial Cairn
Photo courtesy of National Trust for Scotland

Meanwhile, Cumberland had already reached Aberdeen. Gaining strength and momentum with each day, Cumberland made plans to advance on Inverness. By April 14 he was in Nairn. His speed caught the Jacobites off guard and ill prepared to wage war. Yet the battle drums were beating, and the inevitable moment of conflict was about to arrive. All that was needed was a site to make their stand.

Lord George Murray preferred ground where the likelihood of better defensive positions would be available, but John William O'Sullivan, the quartermaster general and the prince's personal favorite, disagreed. Instead, he recommended the flat unprotected terrain of Drummossie Moor, near Culloden House, an area some six miles east of Inverness. It was a fatal mistake. The wide space would prove no obstacle to the skills of Cumberland's expertly trained and disciplined army: it would be a showcase for their superior cavalry and, more significantly, offer great opportunity for their artillery. But it was not suited to the guerrilla tactics of a Highland army. The Highland army depended on the success of its Highland charge, a chaotic and mad charge of men against the enemy that was followed by hand-to-hand combat using broadsword and dirk.

Murray knew about Cumberland's upcoming birthday celebration and so proposed a surprise night attack on Cumberland's camp. That night at eight o'clock the Highland army set out in Cumberland's direction. They marched in the dark, in two columns, but made little progress. Dawn was approaching and still they had not reached Nairn. Delayed repeatedly by the muddy terrain and unable to see in the pitch darkness, they realized that by the time they would arrive at their destination it would be close to daylight. Frustrated, angry, and on the verge of exhaustion, they returned to Culloden, arriving around 7:00 A.M. They were also hungry since their provisions had been left behind in Inverness. In the meantime, Cumberland was already on the attack. By the time he reached Culloden, many of the Highlanders had gone off to the hills in search of food; others were asleep.

At 11:00 A.M. the two armies confronted each other. It was to be the first time that the prince would act as commander in a battle during the Rebellion. Charles carried a light broadsword and wore a tartan jacket and buff-colored waistcoat. Many people from the surrounding countryside, anxious to see the outcome for themselves, came to watch.

The battle began with particularly deadly results for the Highlanders. The initial artillery fire devastated the Highland ranks. Despite the onslaught, the prince, unable to see what was happening at the front, did not issue an order to charge. Finally, an exasperated Lord George Murray requested that the order be given. The prince did so and sent a young Highlander named Lachlan MacLachlan to deliver the message to the front lines. But MacLachlan was struck down by cannonball, causing further delay and countless deaths. By the time the Highlanders did charge, the battle was virtually lost. The methodical routine of the government forces—while one line fired with deadly precision, the other reloaded—plus the introduction of powerful grapeshot had already taken its toll. Further aggravating the situation was the weather. Sleet, wind, and rain blew in the Jacobites' direction, while smoke from their own guns virtually blinded them.

In less than a hour the battle was over. Charles fled the scene on horseback, led by one of his officers. Some say he was in tears. The casualties tell the story: the government dead numbered less than 400, but Jacobite losses were more than 1,000. Meanwhile, Cumberland entered Inverness and took up residence at the house where Charles had stayed only days before.

After the carnage on the battlefield ended, more bloodshed occurred as British troops went on a killing frenzy, indiscriminately slaughtering fleeing clansmen and innocent bystanders, many of them women and children. They tortured, killed, and mutilated their victims. Cumberland ordered his men to visit all the cottages in the vicinity of Culloden and in the nearby hills and farmhouses to search for the rebels. The road to Inverness was littered with the victims of what is now considered one of the most disgraceful episodes in British military history.

The killing continued unabated for days. Some government soldiers set fire to a barn containing thirty Highland officers and men. Rebels and rebel sympathizers were, in the duke's view, traitors and thus deserved no mercy. He even suggested a permanent solution: transporting entire clans to the colonies.

In London, preparations were being made to welcome the duke. But any Scot living in the British capital was on guard. Indeed no Scot felt safe in the capital—whether Jacobite or Tory, humble working man or man of letters. Conditions were not much better in Edinburgh, where mobs searched for suspected Jacobites in the wynds and lanes of the old town.

Cumberland, christened the savior of the Union, returned to England a national hero. Handel composed "The Conquering Hero" in his honor, and a flower was named "Sweet William" after him. In Scotland, however, they renamed a weed "Stinking Willie." A new dance was invented in honor of the battle, the "Culloden Reel."

Meanwhile, Charles remained a fugitive and a reward of £30,000 was put on his head. Thousands of troops searched for him. For five months Charles wandered the lonely western isles, taking refuge in South Uist, Skye, and back on the mainland. The government soldiers became more desperate as the search went on. Houses were burned, and the level of brutality intensified.

In September 1746 Charles was picked up by a French frigate at Loch nan Uamh, the same site where he had arrived with so much hope only fourteen months earlier. He would never again return to Scotland. He died in 1788 in Rome, a drunken shadow of the idealistic youth who captured, for a brief moment, the spirit of his time.

Meanwhile, the courts were making every effort to crush the Highlanders and their way of life. The Privy Council, in violation of the 1707 Treaty of Union that united England and Scotland, announced that all Jacobite rebels would be tried in England, not Scotland, as provided by law. Trials were held in London, York, and Carlisle. Ultimately, 120 prisoners who were captured at Culloden were executed; more than 1,100 were banished or transported. Many were sent to the North America and Caribbean colonies as indentured servants.

The Disarming Act of 1746 forbade the carrying of weapons, prohibited the wearing of the kilt, and outlawed the playing of the pipes. An equally devastating piece of legislation was the Heritable Jurisdictions (Scotland) Act of 1747, which stripped the clan chiefs of their hereditary power, effectively destroying the ancient patriarchal system of the Highlands. By the end of the eighteenth century, the Jacobite era had become nothing more than a cue for a sad old song.

The Jacobite cause may have died on the fields of Culloden but the spirit of Jacobitism survived in the form of romantic literature (Sir Walter Scott, Robert Burns, James Hogg) and in the songs of Lady Nairne and Sir Harold Boulton. Indeed, by the mid–nineteenth century, Jacobite songs were second only to love songs in popularity in Scotland, and the myth of Bonnie Prince Charlie had found its way into the national consciousness. Bonnie Prince Charlie had become the last true Scottish hero and Culloden was where he had met his tragic—but noble—end. Jacobitism, a political movement that had threatened to destroy the very existence of the fragile pact between England and Scotland, was miraculously transformed into pure nostalgia, with its sentimental pull of exile and loss. Widespread distrust of the Highlander turned into fascination. The post-Culloden generation embraced the Jacobite era as one belonging to Scotland's heroic past. Charlie became a hero for all seasons, and the Highlander, who only decades before had been an object of ridicule and derision, became the national symbol of Scotland. Finally, in 1822 King George IV, dressed in full Highland regalia, made his famous visit to Edinburgh, giving the official stamp of approval on all things Highland.

Today, Culloden Moor is one of the most easily accessible of British battle sites, containing a number of monuments and memorials, buildings and stones. The Old Leanach Cottage, a thatched farmhouse, is the only building to survive from the Culloden era. Inhabited until 1912, it was taken over by the Gaelic Society of Inverness until 1944, when it was given to the National Trust for Scotland. In 1959, it was converted into a visitors' center. A new and larger center opened in 1970.

A twenty-foot-high memorial cairn was erected in 1881. Casualties of Cumberland's army were allegedly buried under the English Stone, although no evidence to substantiate such a claim exists. The body of Alexander

MacGillivray of Clan Chattan was found under the Well of the Dead. The Graves of the Clans lie on either side of a nearby road, while the Keppoch Stone marks the spot where Alasdair MacDonell, a clan chief, reportedly fell. The Irish Memorial is dedicated to the Irish soldiers in the French army who fought for Bonnie Prince Charlie. According to legend, the Duke of Cumberland surveyed the battle scene while standing on the Cumberland Stone. The Culloden Visitors' Centre offers exhibitions on the battle and the clan system. It also contains a bookshop and tearoom.

Further Reading: In *Scottish and Border Battles and Ballads* (New York: Potter, 1975), Michael Brander describes the battle-fields that inspired some of Scotland's finest traditional song, while William Donaldson in *The Jacobite Song: Political Myth and National Identity* (Aberdeen: Aberdeen University Press, and New York: Pergamon, 1988) explores the special place Jacobite song has in the Scottish political tradition. These general histories are recommended: *A History of Scotland* by J.D. Mackie (London: Pelican, and New York: Penguin, 1964), and *A Concise History of Scotland* by Fitzroy Maclean (London: Thames and Hudson, and New York: Viking, 1970). John Prebble has written several books on Scottish history, always with great sympathy and fierce passion. In *Culloden* (London: Penguin, 1967), he tells the story of the ordinary men and women involved at Culloden who fought for a cause that never really belonged to them.

—June Skinner Sawyers

Dachau (Bavaria, Germany)

Location: Town in Bavaria, twelve miles north-northwest of Munich, on the Ammer River.

Description: Small industrial area and artists' colony, most notable as the site of a Nazi concentration camp during World War II; the camp was liberated (and more than 30,000 prisoners freed) by the Allied army in late April 1945.

Site Office: Verkehrsverein Dachau
Konrad-Adenauer-Strasse 3
85221 Dachau, Bavaria
Germany
(8131) 84566

To the casual visitor, the German town of Dachau seems serene and friendly enough, nestled on a hilltop, dotted with quaint old homes and populated by friendly residents. Tourists wander through the streets, marveling at the architecture and the views of Munich and the Alps sparkling in the distance. For many, however, the mere utterance of the word Dachau is enough to conjure up horrible memories. Splendid scenery and cultural distinction aside, the town will always be remembered primarily as the birthplace of Nazi concentration camps during World War II and the training ground for the terror that ran rampant across the European continent for more than a decade.

In the twelve years from 1933 to 1945, the atrocities committed at Dachau were horrendous enough to effectually erase any other prominent historical memories of a town that is older than nearby Munich, dating back to the days of Charlemagne in the ninth century. Even today, Dachau is filled with landmarks and old homes that display its long legacy. The most notable of these structures is the Renaissance Palace, originally constructed in the sixteenth century and renovated for the ruler Max Emanuel in 1715. The castle has served many purposes during its 450 years of existence, including use as a field hospital in the Battle of Austerlitz during the Napoleonic Wars. Bombings and invasions have taken their toll on the castle; an ornate ballroom that serves as a concert hall during the summer months makes up most of the one remaining wing. In the center of the town stand the Parish Church of St. Jacob, a large, yet carefully crafted structure dating back to the sixteenth century, and the Schlossbraueri, a 250-year-old brewery. Dachau has always been a prominent vacation site, attracting residents from Munich and artists wanting to draw and paint the surrounding natural wonders.

The town's dark period began shortly after Adolf Hitler took power in 1933. In March, Munich police commissioner Heinrich Himmler announced the construction of what he called internment camps, designed to hold political prisoners and others whom the state considered threatening or undesirable; Dachau was the first of these camps. Ironically, the camp was constructed on the site of an old explosives factory that had been dismantled in accordance with the terms of World War I's Treaty of Versailles.

The camp, of course, was only the beginning of what would come to be known as Hitler's "Final Solution," the attempt to eliminate all persons whom he considered threatening to his perfect society. To carry out this goal, the German leader needed a strong, organized system of killing, which was spearheaded by members of the SS, or secret police. At roughly the same time Dachau was being converted to its new purpose (June 1934), control of the camps was transferred from private owners concerned with financial gain to SS members concerned only with inflicting the greatest amount of suffering on the greatest number of people. Over the next twelve years, prisoners were exterminated in any and every way imaginable.

Soon after construction ended at Dachau, hushed reports of torture and death at the camp began circulating among the German citizens, as an increasing number of government detractors disappeared. During the summer of 1933, the Nazis ran unopposed and sailed to easy victory in nearby Munich's political elections; many who offered resistance were sent to a camp for "reeducation." And although Dachau was not equipped as an extermination camp, reports of Jews being slain there surfaced as early as 1934. Of the more than 10,000 Jews in Munich, only a few hundred would survive the holocaust that was to come.

The first mass shipment of Jews to Dachau came in 1938; most of these 13,000 Jews were released after bribing the guards and promising to leave the country. Those who had no money ended up as victims when they were dispersed to newer and larger camps designed for mass killing. And many died in Dachau itself. During their three-month internment there, roughly 700 of these Jews died from beatings, hangings, and the harsh camp conditions. Unfortunately, this number would soon seem merciful; the comparatively gracious practice of selling prisoners their freedom was soon replaced by the systematic orchestration of their deaths.

Random acts of violence, the basis on which all of the camps essentially ran, are incomprehensible by basic human standards. Men and women, young and old, healthy and strong, were regularly kicked, beaten, and stripped of all dignity. Regardless of their strength or ability, they were put on a gruelling work detail that would last eighteen or more hours a day. Meals were scarce, devoid of nutrition, and often interrupted by shootings or beatings. The SS officers were obsessed with order, forcing the prisoners to stand or squat in line for hours each day, often unclothed in freezing

Dachau concentration camp, with memorial to "The Unknown Prisoner"

temperatures. If an escapee was found, horrible beatings were administered to everyone in the camp. When prisoners were finally sent off to the barracks, anywhere from six to ten individuals were crammed into each bed for a few hours of tortured sleep.

In the large and complex network of killing and torture centers that developed over the course of the war, Dachau was older, smaller, and less severe in comparison to the other camps. A three-tiered classification system was developed to distinguish the sites from one another, and Dachau was listed as a I, or mildest on the scale. Camps in the II classification had worse conditions, while camps classified as III sites were death camps; few who were sent there lived to tell their tale. As the years passed and the system became disorganized, the camps became more alike and the classification system became meaningless.

Exact statistics of the number processed and killed in the camp are not available: many of the records were destroyed by fleeing Nazis at the end of the war. Roughly 250,000 are said to have been imprisoned at Dachau during its twelve-year existence, resulting in 50,000 deaths in the camp itself. Two million people were killed at Auschwitz, the most grisly camp. At its peak, the gas chambers at Auschwitz could process in only two days the number of prisoners killed in all the history of Dachau. The Nazis never regarded Dachau as an extermination center; they constructed only a small gas chamber and crematorium there, in 1942, both of which were never used to full capacity. Sheer numbers do not tell the whole story of the camp, however; Dachau served many other vital roles that enabled the other camps to form and grow as they did.

Size and death totals aside, Dachau was different from other camps. It was part of a "base camp" system, which also included the camps at Buchenwald and Sachsenhausen. Though these base camps seldom housed more than 100,000 prisoners, they sat at strategic rail hubs, convenient for allowing officers there to organize countless boxcars of prisoners, systematically sending thousands to their deaths. Dachau also served as a temporary home and processing center for prisoners being sent to slave labor camps in the east, and an evacuation center when these same eastern camps were abandoned late in the war.

The camp was also widely known for the beginnings of brutal human experimentation, which spurred larger testing centers at other sites. The "research" took place under the guidance of Dr. Sigmund Rascher and usually resulted in horrendous pain and death for those who unknowingly volunteered or were chosen to participate. The various studies conducted there included an examination of the symptoms and spread of malaria: 1,100 prisoners (many of them Polish clerics) were injected, studied, and sacrificed, their bodies racked by fever and chills until their deaths. Another test sought to discover the effects of ingesting sea water. The most grotesque of the experiments was a study of extreme and varying levels of temperature and pressure on the human body, in the hopes of helping German pilots survive bailouts

at high altitudes. Parts for a decompression chamber were shipped to Dachau and assembled; one experiment involved the use of human guinea pigs to discover the boiling point of blood in relation to simulated altitudes as high as 70,000 feet.

By virtue of being the first of the camps, Dachau was the prototype and training center for all other sites, and that is its most infamous distinction. Under the leadership of the initial camp commandant Theodore Eicke, a streamlined plan for the annihilation of an entire race was developed, polished and later implemented elsewhere. Like Dachau, camps that followed were well-concealed, near a large city (to provide supplies and entertainment for the SS officers), and serviced by a rail line. Even the way the buildings at Dachau were laid out and structured served as the blueprint for other camps: every site had a headquarters, an SS residential area, and a prisoners' compound, which was constructed only after the guards and operators were satisfied with their accommodations. Prisoners were usually shipped to the camps before they were finished and forced to live without any substantial shelter (regardless of season) until building was completed.

Theodore Eicke's brutal legacy also included the formation of a personnel training center; before an SS officer was given any position of responsibility at another camp, he had to take a battery of courses at Dachau. When a guard or operator was well-schooled in the management, torture, and extermination of prisoners, he was transferred into the main system. This centralization of officer training ensured that a cohesive effort was taking place throughout the empire. Oddly enough, the only time the Dachau complex was not used as a prison camp was during late 1939 and early 1940, when all of the prisoners were temporarily transferred to other sites so a group of SS men could complete their schooling.

Despite its "mild" characterization by the Nazis, Dachau was a brutal and horrible place. Starvation, disease, and death raged uncontrollably, although there were at times some half-hearted efforts to stave them off. After 1939, a program was developed to immunize camp residents against typhoid fever and dysentery, but it failed miserably: the population was transient, and malnutrition, overwork, and generally deplorable living conditions continued to be simultaneous parts of camp policy. Prisoners were deprived of sleep and forced into grueling slave labor. One job was known as "gardening," the carrying of incredibly heavy amounts of stone at a breakneck pace. Prisoners who crumbled under the strain were usually shot, beaten, or mauled by dogs. And although relatively few gassings took place, there are countless stories of hangings, shootings, and even immolations occurring at Dachau.

As the war progressed and Nazi fortunes soured, the once highly organized prison system, with Dachau at its center, began to crumble. Hitler's order to escalate the extermination caused numerous overloaded trains to flood into the camp. Because Dachau lacked the means to process and organize the influx of bodies, the prison population soared to unmanageable levels. Guards grew more impatient and, consequently, the camp became more brutal. As the Soviet Army liberated the eastern camps, more prisoners were evacuated to camps like Dachau in the west. There were numerous stories of trains passing through with the majority of passengers dead from exposure or starvation; near the war's end, a train arrived from Buchenwald with only 1,200 of 5,000 passengers still living.

On April 29, 1945, Dachau and more than 30,000 prisoners were liberated by the advancing American units; in spite of the death that surrounded it, many soldiers commented on the how peaceful and beautiful the town still seemed. One week earlier, Himmler had ordered the entire prison population of Dachau eliminated, but for some reason, his orders had not been heeded. Even so, what was left of the camp was enough to bring even the most hardened of war veterans to tears. The soldiers found forty open boxcars crammed with 2,000 bodies, and even more bodies were stacked neatly next to the crematorium. Prisoners who had been forced on death marches from other camps in the weeks prior to liberation poured out of the camp buildings to receive medical attention or roam free.

Throughout the months that followed, Allied journalists and statesmen toured the camps throughout Germany, returning to confirm the destruction and death that had been rumored for years. To make sure the German citizens knew what they had either actively or tacitly supported, Allied commanders forced large segments of the population to visit either Buchenwald or Dachau.

Dachau remains one of the most famous of German camps not only for its pioneering spirit of death, but for its close mirroring of the rise and fall of Nazi society. The camp's birth nearly coincided with the rise of Hitler, and, like that of the German leader, Dachau's reign of terror lasted for twelve years. The day after the camp was liberated, Hitler took his own life in a Berlin bunker. In its existence, it never housed a majority of Jews like other camps; instead it more closely represented the entire range of people whom Hitler hated, including communists, gypsies, and homosexuals. Dachau was a society based on fear, hatred, and authoritarian rule, a symbolic representation of the conditions under which the whole of Europe was forced to live.

Visitors to the site today are faced with stark reminders of the grisly events that took place there. What remains of the cell blocks, guardhouses, and crematorium are all open to the public for viewing, and some 500 exhibits help to document life and death at Dachau.

Further Reading: *The Holocaust: The Fate of European Jewry* by Leni Yahil (New York: Oxford University Press, 1990) is a complete look at the attempted destruction of the Jewish people. *Deliverance Day: The Last Hours at Dachau* by Michael Selzer (Philadelphia: Lippincott, 1978; London: Sphere, 1980) traces the last days of the camp before it was liberated, as does *Inside the Vicious Heart* by Robert H. Abzug (New York: Oxford University Press, 1985). Other good comprehensive works of holocaust history and camp life that make frequent mention of Dachau are *Holocaust* by Martin Gilbert (New York: Holt, Rinehart and Winston, 1985; London: Collins, 1986); *Atlas of the Holocaust* by Martin Gilbert (London: Joseph, 1982; as *Macmillan Atlas of the Holocaust,* New York: Macmillan, 1988) and *Theory and Practice of Hell* by Eugen Kogon (New York: Octagon Books, 1973).

—Tony Jaros

Deventer (Overijssel, Netherlands)

Location: Northwest Netherlands, in the province of Overijssel, on the Ijssel River, approximately fifty miles northwest of Utrecht.

Description: A center of trade since the early Middle Ages and headquarters of a religious movement, the Brethren of the Common Life, in the late Middle Ages; has three surviving medieval churches and many buildings from the Dutch Golden Age in the seventeenth century.

Site Office: VVV-Deventer
Brink 55
7411 BV Deventer, Overijssel
Netherlands
(31) 5700-16200

The present site of the city of Deventer has had some form of settlement since the Bronze Age (between 1750 and 500 B.C.), although archaeological evidence suggests that continuous habitation probably began only after 500 B.C. and developed over the next few centuries near what are now Bursesteeg (Bourse Lane) and Kranensteeg (Cranes Lane). Little is known of these early centuries.

By the late eighth century A.D., Deventer had become part of the empire conquered by the Franks, which included most of modern France, Germany, the Low Countries and northern Italy. The Franks introduced the feudal system of landholding into the region and it was under their protection that Sts. Willibrord and Boniface led campaigns to convert their subjects to Christianity. In 765, an Anglo-Saxon monk, Lebuinus (Liafwin), arrived in Deventer as a missionary and settled on a hill called the Berg, overlooking the Ijssel River, where he erected a wooden chapel. It was apparently destroyed in a raid by Saxons within thirty years after his arrival. The earliest written reference to "Daventre Portu" (Deventer) dates from 806 and indicates that the town already had some economic significance as a port for trade along the Ijssel, having grown up around a fortified manor established by the Franks.

Between 843 and 870, Deventer was ruled by the Kings of middle Francia, one of three kingdoms eventually established after the division of the Frankish empire in 814. Middle Francia was then fought over by West and East Francia (roughly corresponding to modern France and Germany), until 925, when the East Franks won control. By then, Deventer had become a center of both religious and political power as the main residence of the Bishops of Utrecht, who had been driven from their own city by Viking invasions around 900. (The Vikings had also attacked Deventer in 881). The bishops were to maintain residences in both cities for about 100 years, from 918 onward.

Between 953 and 962, the bishops and their territory, including Deventer and the rest of the province of Overijssel, were under the control of Bruno, Archbishop of Cologne and brother of Otto I, who was crowned Holy Roman Emperor in the latter year. He and his successors then granted more lands and wealth to the bishops of Utrecht, so that by 1040 they governed Overijssel, Gelderland and Drente, and the town of Groningen. The bishops were also the leading landowners in the northern Low Countries, now the Netherlands. Such Deventer street names as Korte Bisschopstraat (Short Bishop Street) and Hofstraat (Court Street) commemorate the connection.

Deventer also began to flourish as a commercial center during these years. The Viking raids had put Dorestad and many other rival ports out of business. The merchants of Deventer, who had access to relatively secure trading routes between Flanders and northern Germany by land and between the North Sea and the Rhine along the Ijssel, were able to attract new settlers even after the Viking attacks ended at the beginning of the eleventh century. The merchants of Deventer administered estates on behalf of the kings of Germany and their successors, the Holy Roman emperors, as well as for the bishops of Utrecht and a number of wealthy monasteries; they extended their trade as far as Riga in Latvia and Bergen in Norway, a connection marked by the figure of a crowned fish on the wall of the Atheneum, the city library; and they maintained close relations with Cologne, Münster, Leipzig, and other German cities, as is shown by the large amounts of German pottery excavated in recent years. Their resources included a mint and a toll station on the Ijssel. Like the merchants of Tiel, Utrecht, and other trading cities, they frequently clashed with the bishops over their respective rights and duties in governing the city, and raising and spending funds, but did not try to achieve complete independence.

The prosperity of early medieval Deventer permitted the building of four churches for the steadily expanding population. The Lebuinuskerk or Grote Kerk (Lebuinus Church or Great Church) was started in 1040, on the initiative of Bernold (Bernolphus), bishop of Utrecht, near the site of Lebuinus's wooden chapel. Its Romanesque crypt has remained unaltered to this day. The construction of the Bergkerk (Hill Church) between 1198 and 1209, on a site to the northeast of the original center of the city, was a result of the growth of the population, as was the extension of the city's fortifications to take in more and more land. The Mariakerk (St. Mary's) was in use some time before 1334, but was in a decayed state by 1591 when religious services ceased, and in 1647 the remains of the building were converted into an armory. The fourth church, the Broederenkerk (Brothers' Church) was a Franciscan foundation from the early fourteenth century and became a place of worship for

A farmers' market in Deventer
Photo courtesy of Netherlands Board of Tourism

French-speaking residents in the seventeenth century, but has belonged to the Roman Catholic Church since 1803.

Two great fires broke out in 1235 and 1334 and destroyed much of Deventer, although it was quickly rebuilt each time, and increasingly in stone and brick rather than wood, another sign of prosperity. By this time, Deventer, Zutphen, Kampen, and other *Ijsselsteden* (cities on the Ijssel), had begun collaborating against the pirates who frequently raided ships on the river. Deventer also became a member of the Hanseatic League, an association of trading cities along the coasts of the North and Baltic Seas which, from its origins some time in the late thirteenth century, rapidly came to dominate northern European commerce.

The city then became famous throughout northern Europe as the headquarters of an austere religious movement known either as the *Devotio Moderna,* in its Latin form, or the *Moderne Devotie,* in Dutch. The movement's followers traced their emphases on scholarship and on an ascetic way of life to the writings and example of religious reformer Geert Groote, 1340–1384. Many of them lived communally in groups known as the *Broeders des Gemenen Levens* (Brethren of the Common Life), which organized orphanages and homes for the aged. The fifteenth-century mystic, Thomas à Kempis, was a member of the movement and his book, *The Imitation of Christ* is generally thought to exemplify the philosophy of *Devotio Moderna.* Another follower was the theologian, satirist, and pacifist thinker Desiderius Erasmus. Both men studied at the Latin School in Deventer.

However, while Deventer's cultural importance was increasing amid the religious upheavals that would culminate in the Protestant Reformation of the sixteenth century, the town's commercial importance was declining. The growth of Amsterdam and other coastal ports in the provinces of Holland and Zeeland undermined the power and cohesion of the Hanseatic League, and trade along the Ijssel was disrupted both by the gradual silting-up of the river and by a series of wars between the Emperor Charles V and the dukes of Gelders over control of the province of Gelderland to the southwest. Even so, Deventer's city walls were extended one more time, at the turn of the fifteenth and sixteenth centuries (parts are still visible today), and the two city churches were rebuilt and expanded. The lower half of the Gothic tower of the Grote Kerk was completed in 1499 and the conversion of the Bergkerk from a decaying Romanesque building into a Late Gothic magnificence was finished in 1500. De Waag, a Late Gothic building completed in 1528, with bricks recycled from two abandoned fortresses, was both the weigh-house for the city's merchants and the headquarters of its Civic Guard militia. It now houses the Municipal Museum. Two private houses from about the same time, Vogelsang (singing of the birds) and Kroonenberg (Mountain of the Crowns), now contain the Speelgoed- en Blikmuseum, or Museum of Toys and Tin.

By the middle of the sixteenth century, Deventer was home to approximately 11,000 people, among them many of the first professional printers in the Netherlands. Their trade is commemorated by a stone book placed near the city hall. It was partly through their influence that the city opened one of the first public libraries in the Netherlands, the Atheneum, for Athena, the Greek goddess of wisdom, in 1560.

Deventer suffered more than it benefited from the Dutch Revolt, which erupted in 1568, following several years of growing protests against the persecution of Protestants and the centralization of government by King Philip II of Spain, the ruler of the Low Countries at the time. Because it was the headquarters of the largest Spanish garrison in the province of Overijssel, Deventer was a principal target for the rebels. In 1578, the rebels seized control and expelled the Spanish troops. Five years later, the city council, which the rebels had established, placed iron chains across the main roads. (The stones in which their end-links were embedded can still be seen in the walls of many old houses today.) This precaution proved useless when, in 1587, Deventer was betrayed to the Spanish by Sir William Stanley. Stanley was an Englishman who had been sent with 1,200 English and Irish soldiers to govern the city during the brief term of Robert Dudley, Earl of Leicester, as regent of the Netherlands. The rebels, under Prince Maurice of Nassau, the Captain-General of the Netherlands, finally took back the city after a siege lasting ten days in 1591. Deventer had been repeatedly damaged and looted during the revolt and its population fell by as much as one-third. Most of the remaining Catholics left after 1591, when Calvinism became the official religion of Deventer.

While many cities took over educational institutions from the Catholic Church or developed their own universities after they were given religious freedom under the Union of Utrecht, the constitution of the Dutch Republic proclaimed in 1579, Deventer did not erect the Atheneum (or Illustre) School until 1630. The Atheneum did not have the status of a university and therefore it was unable to compete against the universities of Utrecht, Amsterdam and other rival cities as a place of education for the sons of the city's élite.

After the revolt, political and economic leadership in the Dutch Republic of the United Provinces, which was finally declared independent in 1648, was firmly in the hands of the merchants of Amsterdam and other western cities in Holland. Ruined by war and economic competition, Deventer took nearly two centuries, from 1645 to 1830, to recover its earlier peak population of 13,000. Yet trade revived, as Deventer became an entrepôt for goods from the Ruhr district sent through Cologne, and the rebuilding of the city after the years of revolt went ahead. By 1613, the Grote Kerk's tower was made higher and given an octagonal dome, in which a carillon was installed, between 1647 and 1664. A Latin inscription runs around the dome; *Fide Deo, Vigila, Consule, Fortis Age* (Trust in God, be watchful and cautious, act courageously). The city mint, which existed until the end of the eighteenth century, was given a large brick building called the Muntentoren (mint tower). It was damaged in World War II but has since been restored. The many houses and shops dating from this period, most in the "Dutch gable" style called Gelders-Overijssel, include a pharmacy which still carries its own Latin motto, *Deo Fidendum, Mediis Utendum* (Trust in God but take your medicine). The palace of the bishops of Utrecht, abandoned after 1527 when their secular powers and properties were sold to Emperor Charles V, was finally demolished and replaced by an open square, the Nieuwe Markt (new market).

Deventer endured yet another siege during a civil war that broke out in the province of Overijssel in 1660. A costly war between the Dutch and the English had just ended in an English victory and the country was bitterly divided between supporters of William III, Prince of Orange, and supporters of Jan De Witt, the Grand Pensionary (chief executive) of Holland. In Overijssel, De Witt's party, based at Zwolle, launched a series of attacks on Deventer, the Orangist center. The war continued over the next three years, until De Witt negotiated a settlement between them, only to be overthrown and murdered by an Orangist mob in The Hague in 1672. In the same year, Overijssel, Gelderland, and Utrecht were overrun by the armies of King Louis XIV of France and his ally, the bishop of Münster. After their ejection in 1675, the three provinces were allowed back into the Dutch federation only after they agreed to allow the Prince of Orange a veto over elections and appointments both at the provincial and local levels. The Stadhuis (town hall), which the newly tamed city councillors were to occupy, was put together from the remains of an older town hall; of the Wanthuis, the headquarters of the clothmaker's guild; and of the Landshuis, the hall of the medieval states assembly of Overijssel. It was completed in 1694.

During the eighteenth century, Deventer was noted for its large numbers of skilled workers in various processing industries, such as the making of linen clothing and wooden furniture, leatherwork, and metalwork. These industries depended on the surrounding villages for materials and on the Dutch and North German markets for customers. In 1703–1704, the leaders of the guilds managed to take control of the city government but were quickly expelled and suppressed by the wealthy Orangist merchants. Their descendants faced a new challenge eighty years later, in the form of the radical Patriot movement, which circulated petitions demanding democracy and mounted protest demonstrations in several Dutch cities. The Patriots' militia was suppressed in 1787 by soldiers sent by King Frederick William II of Prussia (the brother-in-law of William V, Prince of Orange), and their many supporters in Deventer were dismissed from their jobs and expelled from the city. The surviving patriot leaders went on to cooperate with the French revolutionary army, which invaded the Netherlands in 1795 and created the Batavian Republic. Its Grand Pensionary (president) was Rutger Jan Schimmelpenninck, a native of Deventer. The Republic was replaced by the Kingdom of Holland, ruled by one of the brothers of the French Emperor Napoléon, in 1808. A new Kingdom of the Netherlands, ruled by the Orange family, took over in 1815.

Nineteenth-century Deventer was one of the strong-

holds of the Separatists (*Afgeschiedenen*), a group led by liberal theologians which broke away from the state-sponsored Calvinist Reformed Church in 1834. They were forbidden to meet outside private homes until 1838 and could not provide teaching for children until 1842, but their church has since become one of the leading religious groups in the Netherlands. The city also became a center of light industry, linked to Utrecht and beyond by the Overijssel Canal, opened in 1855, and the state railroad system completed in 1889. One factory turned out velocipedes and other versions of the bicycle. An iron foundry, J. L. Nering Bögel, produced the musical clock placed in the tower of the Grote Kerk in 1866, the fountain erected in 1898 to mark the coronation of Queen Wilhelmina, and the cast-iron roof and pillars of a market building, the Botermarkt, opposite the Grote Kerk. The Botermarkt is embellished with yet another sardonic Latin inscription about trust: *Fide sed cui vide* (Be trusting, but see who it is).

By 1879, there were 19,000 people living in Deventer. Over the course of the next 100 years or so, this figure grew to more than 65,000 as Deventer, still a commercial and industrial center, weathered the storms of World War I, the Great Depression, and World War II. During the Nazi occupation the local police earned the honor of being called "slack" by the occupiers, and many of them collaborated with the Resistance. The city was liberated by Canadian troops in April 1945. (One of the bridges at Deventer was later to "stand in" for the bridge at Arnhem, the site of one of the major battles of World War II, when the British director Richard Attenborough made his film about the battle, *A Bridge Too Far*, in 1977.

Deventer has led the way among postwar Dutch cities in trying to reverse the flow of population out of the city center into the suburbs by redeveloping the Bergkwartier (the Berg district) and the Nordenbergkwartier (the northern Berg district), which together form the historic heart of the city. The program of the NV Bergkwartier, a special corporation set up in 1960, has included the restoration of the Bergkerk, finished in 1970, and the placing of five stone tablets on the walls of buildings associated with the Brethren of the Common Life, in 1981.

Further Reading: There is a great deal of useful and interesting information on Deventer in such general histories of the Netherlands as Bernard H. M. Vlekke's *Evolution of the Dutch Nation* (New York: Roy Publishers, 1945; London: Dobson, 1951). Audrey M. Lambert's *The Making of the Dutch Landscape: An Historical Geography of the Netherlands* (London and San Diego, California: Academic Press, 1985) includes an examination of the city's development. Wayne Ph. Te Brake's *Regents and Rebels: The Revolutionary World of an Eighteenth-Century Dutch City* (Oxford and Cambridge, Massachusetts: Basil Blackwell, 1989) provides a detailed survey of life in Deventer during a turbulent period in its history.

—Monique Lamontagne

Dijon (Côte d'Or, France)

Location: In northeastern France, region of Burgundy, valley of the Ouche river; accessible by road, rail, and canal.

Description: Settled in 300 by Gallic tribes; historical capital of the Duchy of Burgundy, occupied by the Valois line of nobility (1364–1477); current residence of the prefect of the Côte d'Or; current home of several important museums, including the Musée des Beaux-Arts.

Site Office: Office de Tourisme, Syndicat d'Initiative
34 rue des Forges, B.P. 1298
21022 Dijon, Côte d'Or
France
80 30 35 39

Dijon, the capital city of Burgundy, is today most famous for its mustard. Though situated in the wine country, Dijon is not the center of its industry, which is in Beaune; it is known for its *cassis,* a blackcurrant liqueur, of which Dijon produces 85 percent of France's consumption. As far as historians are concerned, Dijon's success (if longevity is the equal of success) is not explained by its geography or its climate. It was the site of the medieval capital of the Duchy of Burgundy, and the four dukes of the Valois line, Philip II (the Bold), John the Fearless, Philip III (the Good), and Charles the Rash, made it the axis mundi of a glorious dynasty in the fourteenth and fifteenth centuries.

Human remains have been discovered in the Dijon region dating back as far as 15,000 B.C., but it was the Greeks and Etruscans involved in the tin trade who established the area in 600 B.C. It was fortified as a Roman settlement around A.D. 273 by Aurelian. After the decline of the Roman empire in the west, a group of Germanic invaders known as the Burgundiones appeared, taking as their possessions an area including eastern France, Switzerland, and the Rhône Valley. The reign of Burgundiones, noted for their long hair, which they scented with rancid butter, lasted until the death of Gundebald, sometime after 500.

According to the Treaty of Verdun of 843, the region of Burgundy was given to Charles the Bald, heir of Charlemagne. The region evolved into a French duchy. During the ninth century, Christian monks turned Burgundy into a safe haven for themselves and for the relics of certain saints. Protected from the raids of the Norse, the monks were free to concentrate on intellectual and religious pursuits.

In the eleventh century, the bishop of Langres surrendered Dijon to King Robert II, the Pious, of France. Dijon came to be ruled by a succession of Capetian dukes. By 1328, the Capetian line had died out and Philip of Valois (Philip VI) became king of France. By this time England and France

were at odds with each other over territory; the tension broke out into the Hundred Years War in 1337. At the military defeat of Poitiers, King John the Good of France was taken prisoner by Edward the Black Prince. John's son Philip, who had stood by his side in the midst of battle shouting encouragement, accompanied his father to England as a prisoner, but was treated as a member of the royal household. While at a banquet, Philip slapped a butler who had offered a dish to the king of England before he had offered it to the king of France, and the action earned him the sobriquet, "Philip the Bold." As a reward for his steadfast loyalty and spirit, John gave Philip the duchy of Burgundy in 1363. So dawned the era of the Valois dukes.

Philip the Bold, an unattractive man, married the equally unattractive Margaret, countess of Flanders, in 1369, thereby securing the counties of Flanders, Artois, Rethel, and Nevers, after the death of Margaret's father in 1383. Philip had long since determined that Dijon was to be the capital of his dynasty. He immediately began rebuilding and refurnishing the ducal palace of the Capetians. He had a tower built at the palace, which is known as the Tour de Bar, after René de Bar, who had been incarcerated in the tower for several years during the reign of Philip the Good, Philip the Bold's grandson. In 1377, Philip the Bold began plans for a monastery outside of Dijon, and in 1385 the charter of the Carthusian monastery of Champmol was made. The architect was Drouet de Dammartin, and the whole project called upon the talents of Flemish artisans, among them Claus Sluter, whose sculpture is today recognized as the work of genius.

Philip the Bold was extravagant and ambitious, but no more so than his contemporaries. For a visit from King Charles IV of France on his way to see the pope at Avignon, Philip arranged a reception that surpasses the imagination. The garden at the St. Etienne abbey was destroyed in order to build the lists for tourneys and jousts. He dressed the knights alike in red or white velvet, and the ladies in gold cloth at his own expense. When the visitors arrived, Philip gave the king and his brother two horses each; the king's sister-in-law received a golden goblet and a gold water pitcher studded with precious jewels. Charles IV was given to paranoia and delusions (he once believed that he was made of glass). Philip was, as a result, co-regent of France, and this enabled him to pilfer from the royal treasury, which helped him pay for such elaborate spectacles. The king, in this way, paid for his own gifts.

Philip the Bold died in 1404. He was succeeded by his son John the Fearless. In 1396, John had been part of another, failed crusade against the Turks. Unlike the many who died at the hands of the Turks, John had been captured and held for ransom. This trial earned him his sobriquet, "the Fearless." He is remembered for ordering the ambush assas-

The Cathedral of St. Bénigne
Photo courtesy of Mairie de Dijon

sination in 1407 of his cousin the duke of Orléans, who had taken his father's place as co-regent. The murder plunged France into turmoil. Burgundy went to war with the Armagnacs, those who had sided with Orléans and were under the leadership of the count of Armagnac. The Armagnacs even issued a piece of propaganda, a letter written to John the Fearless by Satan, thanking him for his faithful service.

In 1418, John won Paris in a coup that involved the wholesale slaughter of Armagnacs. He then hoped to force the Dauphin to accept his "protection." But at a meeting held on the bridge of Montereau, he was cleft through the skull with a hatchet by one of the Dauphin's knights. History has not been kind to John, but he was no more ruthless than his contemporaries. He was a patron of the arts, a prudent financial strategist, and a great military leader.

Philip the Good, heir of John, is credited with creating Burgundy's Golden Age. During his reign of forty-seven years, Philip enlarged the Burgundian state to include Holland, Hainaut, Zealand, Brabant, Limburg, and Luxembourg. He exacted vengeance for his father's murder by forming an alliance with Henry V. After Henry's victory at Agincourt in 1415, the French were forced to agree to the Treaty of Troyes, which declared the Dauphin to be a bastard and Henry the true heir to the French throne. It was Philip who handed Joan of Arc to the English. The Treaty of Arras gave Philip independent rule over Burgundy, and thereafter Philip did not seek power in France. Rather he concentrated his imperialism on the Low Countries. Although he continued to claim Dijon as his capital, having a tower built for him, he spent more and more time in his palaces away from Dijon.

Philip's lasting legacy is the court lifestyle he created during his tenure. On the occasion of his marriage to Queen Isabel of Portugal, he created the Order of the Golden Fleece, which lasts to this day, based on the faded ideals of chivalry. One unfulfilled aim of the order was the launching of a crusade to free Constantinople from the Turks. Another aim was to revive the spirit of chivalry among knights, which had been in decline since the thirteenth century. Philip informed the pope of the order's intentions; the pope was quick to issue three bulls endorsing the movement. But the order was more than just a spiritual affair; politically, it coalesced the power of the knights, who numbered more than thirty, under the leadership of Valois Burgundy. Since Philip, unlike the kings of France or England, had no natural or nationalistic rights to fealty from the subjects of his far-flung array of territories, the unity invented by the oaths of the order placed him finally at the center of a powerful oligarchy.

Philip also hosted extravagant fetes and banquets at his court, collected the works of great artists such as Jan van Eyck (who was trusted with political duties as well), patronized writers, and had perhaps the first amusement park created at his palace in Hesdin, complete with distorting mirrors, bridges that collapsed when stepped upon, and amazing water-powered automatons. Philip was known both for his piety and for his immorality. He is known to have taken thirty mistresses and produced seventeen children out of wedlock, yet at other times he would fast on bread and water and spend hours in his vespers. He died in 1467.

The portrait history draws of Philip the Good's heir, the last duke of Burgundy, is not flattering. During his life, Charles's sobriquet was "le Travailleur"—the Industrious. Later, after his death, he was called "le Temeraire," which can be translated as "the Foolhardy," "the Rash," or "the Bold." He was known to be impetuous and had a violent temper. His goodness died with him: he was faithful to his wife, and was driven by a strong sense of justice. His foolish political and military policies are his lasting legacies, however; they cost Burgundy their duke and Charles his life.

Charles's court was as accomplished as his predecessors'. He loved Italy: he wore Italian-style clothing, spoke Italian, became close friends with Italian princes, and filled his army with Italian soldiers. He attended his meals with great ceremony and was a generous patron of the arts. He was also an avid reader and musician.

One of Charles's aims was the unification of his divided Burgundy through the annexation of new territories. In 1471, he laid claim to the throne of England. Lorraine was annexed in 1475. He entertained the idea of being crowned Holy Roman Emperor. His archenemy, Louis XI of France, was a cunning diplomat who realized that power resided in money and not in military strength. Louis subsidized Charles's enemies, whom Charles made with ease. He bribed England's Edward IV to abandon the notion of invading France, and paid the salaries of Swiss mercenaries, who defeated Charles at Grandson and Morat. In a desperate attempt to save Lorraine, Charles laid siege to Nancy. With the aid of the Swiss, the defenders of Nancy were able to overcome the Burgundians. The duke was killed, his body discovered later in a frozen pond, half-devoured by wolves. Since Charles had left no male heir, the duchy reverted to France.

In the five centuries since, Dijon became a subject of the French state, and the middle class rose in power, building many fine homes. The Parlement of Dijon assumed the station of the higher classes, patronizing the arts and sciences and producing a saint, Jeanne de Chantal. In 1650, the leadership of the Fronde, a revolt against the rule of Cardinal Mazarin, was taken by the governor of Burgundy, but quickly fizzled. Burgundy and Dijon were spared the worst of the Terror during the French Revolution, but suffered from symbolic protests that saw the destruction and dismantling of great works, especially those that had to do with religion. For instance, only the base of Claus Sluter's *Well of Moses* survives. After the revolution, the country was divided into ninety departments, and Burgundy lost its identity. Not until Charles de Gaulle reformed the departments in 1960 for economic reasons did the name Burgundy return to the region; Dijon became the residence of the prefect.

But the majesty of Valois Burgundy remains, in the towers of the ducal palace, now converted into the town hall and Musée des Beaux-Arts, one of the greatest museums in the world, housing the tombs of Philip the Bold and John the Fearless, as well as a fine-arts collection spanning hundreds of years. Elsewhere in Dijon, the triumphal arch Porte Guillaume graces the Place Darcy. The Cathedral of St. Bénigne, exemplary of the Gothic style contemporary with Valois Burgundy, also houses a 1,000-year-old Romanesque octagonal sepulchre. The Church of Notre Dame, with its flying buttresses and gargoyles, is an architectural masterwork; the Church of St. Michel, built in the early sixteenth century, offers one of the best Renaissance facades in France.

Further Reading: *Dijon and the Valois Dukes of Burgundy* by William R. Tyler (Norman: University of Oklahoma Press, 1971) is a short, comprehensive study of Dijon in the time of the Valois Dukes, including habits of the ordinary Dijonnais along with the fêtes of the extraordinary Valois; *Burgundy* by Anthony Turner and Christopher Brown (London: Batsford, and New York: Hippocrene, 1977) includes a highly readable history of Dijon and Burgundy, but its purpose also includes the evaluation of wine and cuisine. See also the works of Richard Vaughn, which include Valois *Burgundy* (London: Allen Lane, 1975).

—Gregory J. Ledger

Dortmund (North Rhine-Westphalia, Germany)

Location: In the eastern Ruhr District of the state of North Rhine-Westphalia, at the southern end of the Dortmund-Ems Canal, thirty miles south of Münster.

Description: A major industrial and commercial center since the nineteenth century, with some buildings and institutions dating back to the thirteenth century.

Site Office: Verkehrsverein
Königswall 20
44135 Dortmund, North Rhine-Westphalia
Germany
(02 31) 5 02 21 74

Dortmund today is the seventh-largest city in Germany, with a population of approximately 600,000. It is a center of industry and trade, part of what the Germans call the *Ruhrstadt* (Ruhr city) or *Ruhrgebiet* (Ruhr district), the continuous, built-up area of high population density and industrial production which occupies 2,200 square miles along the valley of the Ruhr River. Accordingly, it can be difficult to look back, beyond Dortmund's steelworks and shopping centers, beyond the Florianturm, the 695-foot-high television tower dominating the Dortmund skyline, to the beginnings of settlement in the area 2,000 years ago. That is when the Celtic people known to the Romans as the Germanii arrived in the Ruhr valley.

Such an effort of imagination is made even more difficult by the sheer lack of tangible information about more than 800 of those 2,000 years, for the migrating peoples who passed through the valley in successive waves—the Germanii, the Vandals, the Huns, the Alamanni, the Franks, and others—left only fragmentary archaeological evidence of their existence. The earliest written references to what is now Dortmund indicate that its development as a trading post, during the ninth century, depended on its location at an intersection between a north-south road and the Hellweg, the principal east-west road of the Carolingian empire. The empire was a loose political structure established by the Franks under their leader, Charlemagne, about 800 and divided up among his successors. The settlement, recorded as Throtmanni in 880, served the economic needs of a nearby castle, Schloss Hohensyburg, the ruins of which can still be seen. A regular public market existed from 900 onward, connected by trade to other towns all over Europe.

During the following centuries, the town's merchant leaders exerted their control over the surrounding villages whose produce they handled and took a leading role in the *Kreis* or district centered on the town. By 1152, when the town's name was recorded as Tremonia (its Latin form), this district was under the overlordship of the various rulers of Westphalia, but in 1220, Dortmund's ruling group were powerful enough to escape the overlords' control by applying for and receiving recognition as a *Reichsstadt,* an imperial city, subordinate only to the Holy Roman Emperor. It was about this time that the city fathers built the fortified walls that are marked today by a ring of streets—Königswall, Ostwall, Hoher Wall, and others—around the city center.

Dortmund's independent status was confirmed in 1332 by a decree known as the Privilegium Ludovicum (the Privilege of Emperor Ludwig). Even so, the city was happy to oblige the local rulers in some matters, for example in 1293, when King Adolf of Nassau granted its citizens the right to brew beer and thus inaugurated what is still one of Dortmund's leading industries. As the only imperial city in Westphalia, Dortmund (a name first recorded in 1222) was able not only to withstand the pressure of local rulers but also to carve out an independent area from the surrounding countryside known as the Palatinate of Dortmund. This area, established in 1343, contributed to the city's privileged status during the Middle Ages when it was chosen as one of the locations for meetings of the Reichstag, the parliament of the Holy Roman Empire, and, from around 1400, for an imperial court of justice known as the *Westfälische Feme* or the *Dortmunder Freistuhi.*

In the middle of the fourteenth century, Dortmund joined with the nearby towns of Bochum, Duisburg, and Essen, under the name of the "associated merchants of Westphalia" to seek and achieve membership in the Hanseatic League of trading cities, with members primarily in Germany, but also the Low Countries and on the Baltic Sea. The League wielded decisive influence over trade in northern Europe from the late thirteenth century until the late fifteenth century, when its rivals, the coastal cities of the Netherlands, began to expand their commercial activities.

Four churches remain as monuments to this imperial and Hanseatic city of Dortmund. The city's principal church, dedicated to its patron saint, Reinold, was begun in the thirteenth century and restored in 1956. It is noted for its west tower, completed in 1701, its Late Gothic altarpiece, and its eagle-shaped pulpit, built in approximately 1450. The Marienkirche (St. Mary's), founded in 1220 and restored in 1957, has a fourteenth-century choir stall and an altarpiece painted by Konrad von Soest in the early fifteenth century. The thirteenth-century Petrikirche (St. Peter's), restored in 1963, is famous for its enormous altarpiece made by Adrian Averbecke in Antwerp in 1520 and containing 633 gilded figures. Finally, the fourteenth-century Propsteikirche (Priory church), founded by Dominican friars and restored in 1965, has an altar triptych painted by Derick Baegert in about 1480.

Dortmund, featuring the Reinoldikirche, left center, and Marienkirche, right center
Photo courtesy of German Information Center

All four churches became Protestant establishments when the city council declared Dortmund a "reformed" city in 1523, only six years after Martin Luther had initiated the Protestant Reformation. An Archigymnasium (high school) was founded in 1543 to teach the new faith to the sons of its leading families and in 1570 the city formally subscribed to the Confession of Augsburg, the declaration of belief drawn up on behalf of the Lutherans by the theologian Philipp Melanchthon in 1530.

From the middle of the sixteenth century onward, the economic and political status of Dortmund declined as civil and international conflict increased, disrupting trade, and as local rulers, pursuing the mercantilist doctrine of seeking maximum self-sufficiency, invested in developing their own trading towns as rivals to "free" cities such as Dortmund. The city retained sufficient prestige to be the location for the diplomatic negotiations which resulted in the Dortmund Recession of 1609, an agreement, signed at the Rathaus (city hall), between John Sigismund, the Elector of Branden-

burg, and Philipp Ludwig, the Count Palatine, which settled their dispute over the province of Jülich-Cleves. However, the Thirty Years War, which began soon afterward, in 1618, was especially disastrous for Dortmund. While the population of Germany as a whole is believed to have decreased by between one-third and one-half, owing to the violence, disease, and famine of those years, Dortmund lost 5,000 of its approximately 7,000 inhabitants. An unknown but probably large proportion of the 5,000 were lost through emigration to more peaceful areas.

Dortmund, now little more than a village with a marketplace, retained the formal status of an imperial city, but its medieval glory was long gone. In 1803, under pressure from the French ruler Napoléon, the imperial authorities reformed the system, so that all but six of the fifty-one remaining imperial cities were "mediatised," that is, absorbed by neighboring larger states and deprived of their access to the Reichstag and the emperor. In any case both the parliament and the position of emperor were abolished only three years later. Dortmund was taken over by Nassau in 1803, by the Grand Duchy of Berg in 1808 and then by the French in 1809, as the capital of their *département* (district) of the Ruhr. This arrangement, in turn, lasted only six years, until the final overthrow of Napoléon in 1815, after which Dortmund went with the rest of the Ruhr valley to the Kingdom of Prussia, the second largest of the states, after Austria, which formed the new German Confederation in the same year.

In 1818, Dortmund's population was 4,300, less than it had had 200 years before. Within a generation it was to be transformed virtually out of all recognition. Beginning in the middle of the nineteenth century, the opening of deep coal mines—a total of twenty at their maximum extent—and the building of iron and steel works changed Dortmund from a minor commercial town to a major industrial city. It was linked to the rest of the rapidly expanding northern German economy, first by the railroad between Cologne and Minden, financed and constructed by the Prussian state and opened in 1847, and later by the Dortmund-Ems Canal, opened in 1899. This waterway, 167 miles long, with nineteen locks, still connects Dortmund to the port of Ems on the North Sea, serving the same function for the eastern Ruhr district as the Rhine River does for the western Ruhr. It is connected to the Rhine itself through the Rhine-Herne and Wesel-Datteln canals and with the Weser river through the Central and Coastal canals. Dortmund, the leading port on this canal network, has the largest harbor on any canal in Europe.

While the deep coal mines and the iron and steel firms, led by Hoesch AG, accounted for most of the industrialization in Dortmund, other industries also developed, notably the mechanized production of lager beer for mass markets. The leading traditional brewers of the city founded the Dortmunder Union Brauerei AG in 1873 as a consortium. It still flourishes in Dortmund today, having merged with a Berlin rival in 1972 to form Dortmunder Union-Schultheiss Brauerei AG, the largest beer-making enterprise in Germany. There are five other breweries in Dortmund, where 1.14 billion pints of beer are produced every year, making Dortmund the second largest center of beer-making in the world (after Milwaukee, Wisconsin) and the largest in Europe.

The clearest indicator of the rapid changes which Dortmunder underwent was the exponential growth of its population. From 44,400 in 1871, the year in which Prussia led the formation of the German Empire, it grew to 110,000 in 1895 and then to 260,000 in 1910. These totals are a little misleading because part of this astonishing expansion was due to repeated extensions of the city boundaries to take in more and more of the suburbs that had so recently been quiet villages. However, the significant point is that there must still have been people in the city, at least up until around 1910, who could remember when Dortmund itself had been a quiet village too.

During World War I, Dortmund benefited economically from the enormous increase in the demand for coal and steel, but by July of 1917, with the economy faltering and the war effort clearly turning out to be in vain, it became necessary to curb the growing dissatisfaction by placing the city under martial law. In November 1918, just before the end of the war, the abdication of Emperor William II and an uprising by sailors, marines and harborworkers in the northern city of Kiel sparked a wave of unrest across the country. Workers' and Soldiers' Councils seized control of Dortmund and most other German cities. On November 10, a crowd of up to 60,000 people gathered in Dortmund to celebrate what seemed to be the start of a new society.

They celebrated too soon. Even as the mayor and the police chief departed, the moderate Social Democratic Party took a majority of the seats on the revolutionary council and ensured that the existing social order did not break down. Just two policemen were dismissed from their posts, while the 2,000 or so members of the supposedly "revolutionary militia," the Red Guards, were used mainly to prevent looting. On January 7, 1919 it was announced that Communists had attacked the state bank, the Social Democratic newspaper office, and the security guards' barracks, but had been easily defeated. However, the Social Democrats had prevented a visiting Communist from even speaking in Dortmund and the alleged ringleaders of the attacks were discharged from prison, without trial, soon afterwards, which suggests that the affair may have been arranged by the Social Democrats themselves. Certainly they benefited from it: the mere threat of extremism was sufficient to calm the atmosphere and Dortmund's brief revolution was over before it began. Three months later, when nearly 80 percent of the Ruhr coal miners went on strike, the government declared a "state of siege" and a Social Democratic journalist, Carl Severing, was appointed as Reichskommissar (state commissioner) for the Ruhr district, with his headquarters in Dortmund. The strikers were forced back to work, many at gunpoint, by the end of April 1919.

With the collapse of both the revolution and the general strike, Dortmund ceased to be a focus of political activity, weathering the hyperinflation of the early 1920s, the Great Depression, and the Nazi era by concentrating on producing coal, iron, steel, beer, and machinery in peace and war alike. The people of the city paid a high price for their relative passivity: Allied bombing of Dortmund during World War II is estimated to have destroyed 65 percent of the city's buildings, and 93 percent in the historic city center. The citizens were left with 600 million cubic feet of rubble to clear away before reconstruction could begin. Most of the area formerly enclosed by the medieval city walls could not be restored, and is now occupied by a large pedestrian shopping precinct. However, as was previously mentioned, the four medieval churches were all eventually rebuilt, as was the complex known as Westfalenhalle (Westphalia Halls). Westfalenhalle is a group of sports facilities, exhibition buildings and conference halls originally built in the Volkspark (People's Park) on the edge of the city in 1925. New buildings include the Paul Gerhardt Church, built in 1950; the church of St. Boniface, completed in 1954; the City Theater, completed in 1965; and the University, built in 1966 on a campus two miles southwest of the city center. It opened in 1968, and now has more than 23,000 students.

As much as 49 percent of the area within the city boundaries consists of parks and other open spaces. Dortmund's numerous museums include two major buildings in the central European form of Art Nouveau known as Jugendstil. One is a colliery machine hall, Zollern II/IV, built in 1903 in the suburb of Bövinghausen, and now part of the Westphalia Industrial Museum. The other, a house near the city center, was built in 1923 and is now the Museum of Art and the History of Culture. Dortmund also is home to museums dedicated to the history of beer-making, to the local schools, and to the victims of the Nazi regime, as well as the German Museum of Cookery Books (Deutsches Kochbuchmuseum) and the Fritz Hüser Institute, a library and museum of literature written by and for the industrial working class.

Following the postwar "economic miracle" (Wirtschaftswunder), during which the economy of West Germany rapidly expanded and diversified, Dortmund was transformed yet again, although not so drastically as in the nineteenth century. Just as Throtmanni, in the year 880, was at an intersection of road traffic, so contemporary Dortmund is a major interchange on the German railroad network. More than 1,000 passenger trains pass through the city every day, including 100 international expresses. The steel industry is still led by Hoesch, now Krupp-Hoesch, producing both basic steel and cold rolled strip steel. Machine-building and

metal processing also continue, but all the coal mines have been worked to exhaustion and closed down. The manufacture of computer software has become a leading activity.

Nowadays, two decades after the economic miracle ended with the petroleum crisis of the early 1970s, more than two-thirds of the working population are in the service industries, notably in insurance. However, with an 11 percent unemployment rate, Dortmund is closer to recent British or French levels than to the historically low rates of western Germany. Yet in a sense even these conditions are not new to Dortmund, a city which has risen and fallen and risen again in response to changes in the patterns of production and consumption in Germany, in Europe and now in a wider world, its people always seeking to balance local autonomy with due regard for the pressures and demands of political and economic powers far beyond the valley of the Ruhr.

Further Reading: Regrettably, there is little in English on the history of Dortmund itself, as distinct from general histories of Germany, or works by economic and social historians on the Ruhr district as a whole. The events of 1918 and 1919 figure largely in Jürgen Tampke's book *The Ruhr and Revolution: The Revolutionary Movement in the Rhenish-Westphalian Industrial Region 1912–1919* (Canberra: Australia National University Press, 1978; London: Croom Helm, 1979). Unfortunately, it is not easy to obtain. The Information and Press Bureau of the City of Dortmund issues a number of pamphlets, in several languages, mainly on the present situation of Dortmund.

—Patrick Heenan

Dover (Kent, England)

Location: Dover is situated on the southeastern coast of England, in the county of Kent, approximately 100 miles from London. Set in a narrow cleft at the foot of chalk cliffs, the town sprawls below its castle which towers 300 feet above. The coast of France lies 20 miles away, across the English Channel.

Description: The coastal community has a 2,000-year history as a trading center and a point for invasion and defense of the island of Great Britain. The earliest recorded military action there was the Roman invasion of 55 B.C.; the most recent was Operation Dynamo of 1940, which involved the rescue of more than 340,000 British and French troops pinned down by the advancing German army on the coast of France at Dunkirk. Modern Dover is a bustling port town, terminus for ferry service to and from the European mainland, as well as rail links to and from London.

Site Office: Dover Tourist Information Centre
Townwall Street
Dover, Kent CT16 1JR
England
(304) 205108

Before Roman times, there were numerous settlements along the white chalk cliffs where Dover now stands. Its strategic position on the English Channel, as the nearest point between Britain and the rest of Europe via northern France, meant that Dover fast became the front door of Britain, as a port and center of commerce and industry. As early as 54 B.C., when Caesar invaded Britain along its south coast and established Roman influence throughout Kent, Dover's reputation as a port began to evolve. Despite political and social unrest over the ensuing centuries, Dover's role—and prosperity—steadily increased. After the Norman Conquest, it became the seat of government for the Cinque Ports, a confederation of coastal towns charged with providing ships for defense when needed, in exchange for various economic advantages. Dover Castle, still impressively seated on the cliff above town, benefited from Norman fortifications that proved to be an effective defense against assault. Hitler's bombardment of Britain was the last challenge that the castle endured, though it served as a garrison until 1958. A steady stream of sovereigns have been welcomed to Britain at Dover. Now, as the busiest passenger ferry port in Europe, offering services between the European mainland and Britain, Dover is a landing or embarkation point for thousands of tourists each year.

Dover was known by the Romans as Dubris, a word probably derived from the Britons' name for the town, Dwf-fyrrha (Steep Place): the high chalk cliffs presented a formidable barrier to invasion. Julius Caesar knew the place as a permanent settlement set in a slight gap in the cliffs, with a small, natural harbor. Caesar apparently felt that Dubris was significant enough to start his invasion of Britain there. Thus, in late August, 55 B.C., he sailed across the channel from present-day Boulogne to land on Dover's shore, with approximately 6,000 Roman soldiers. The resistance from the Britons proved to be so fierce there that Caesar was not able to land his troops and had to sail seven miles farther up the coast, to try to land on flatter shores. Thus, Dover might take credit for repulsing the first historical invasion of British shores. The Romans finally did overcome the Britons north of Dover, but heavy fighting continued between the Britons and the dismayed Romans, who had not expected to meet such energetic defense. Caesar's troops were better armed and organized, but even after they overcame the Britons they were continually pestered by regrouped armies of the enemy. Caesar decided to return to Gaul with his troops before winter and later regain a foothold with more soldiers and ships. The following year, 54 B.C., Caesar returned to successfully defeat the British in numerous skirmishes in the southeast and west of England; this time, he made Britain a tributary of the Roman Empire. Though Caesar and his troops returned to Gaul shortly after their victories, the southern areas of Kent, in particular, would remain a part of the empire for the ensuing century. During this time, trade grew with the Continent, there was a steady flow of visitors from Italy, and Roman influences became increasingly prevalent in many British tribes. A main road, now called Watling Street, was established by the Romans between Dover and London, and the remnants of this road as well as numerous coins unearthed by modern archaeologists indicate that the port of Dover became an important center of commerce for the Romans.

Increased British hostility toward the Romans led to another invasion in A.D. 43 by the Emperor Claudius, whose troops landed along the Kentish coast. This time, the Romans stayed to occupy their northern frontier and built fortifications as well as buildings and towns of more permanence. Dover maintained its importance as a port city and center of trade and distribution, though the main economic and military focus of the Romans had shifted farther north, to London and York. Troops were housed atop the Dover cliffs, near the present site of a lighthouse, built by the Romans to guide merchant vessels to port. Although no trace of other structures has been found there, in all likelihood the Romans also built a castle, barracks, or some fortification on the site.

The gradual decay of the Roman Empire meant that Britain again was vulnerable to fresh invasion, both by the powerful Saxon pirates from the east and the Picts and Scots

The White Cliffs of Dover
Photo courtesy of Dover District Council

from the north. Amid power struggles between rebel Britons, southern Britain began to break up into small tribal states, and by the mid-400s, Roman rule had effectively ended. Much of what was left by the Romans was destroyed, including towns, churches, public baths, and even roads; the ruins would remain for later societies to piece together. In Dover, the lighthouse on the cliff beside the castle, mosaics, and a painted house have survived.

It was not until the Saxon King Ethelbert of Kent came to power in 561, with his capital at nearby Canterbury, that civilized order returned to the area. As trade with the Continent was renewed, so Dover's position as a key port and seat of commerce was enhanced. King Ethelbert converted to Christianity, a faith first brought to the island by the Romans then somewhat forgotten, and he allowed St. Augustine, a missionary from Rome, to start a mission in Kent. Augustine's efforts were centered in Canterbury, but his influence spread to other towns; Priory of St. Martin-le-Grand was established in Dover in 691. The Vikings, or Danes as they came to be called, resumed their intensive attacks on the south coast during the late tenth century, and much of the area around Dover was again sacked and burned.

In 1066 came the Norman Conquest, when William the Conqueror claimed the crown after winning a fierce battle near Hastings, thus founding a new royal dynasty and social order. The Normans brought new systems to Britain, most notably feudalism; they introduced fortified castles, one of which looms on the cliffs over Dover's city center. Anyone familiar with military strategy knows that the clifftop site, with its view across the English Channel, would be an ideal place from which to repulse invasions. (Only twenty miles from the Continent, Dover is also an ideal place from which to launch invasions.) The castle is still Dover's most imposing feature, situated 300 feet above the town and occupying thirty-five acres of land. Though the Romans probably established a fortification on the site, it is the Norman castle that endures. The keep at the center of the upper court and several of the castle towers are constructed with massive stones, set in even, Norman masonry.

Acknowledgment of Dover during Norman times first appears in the Domesday Survey, carried out by William I, which described all of Britain's towns, economic activities, and populations in amazing detail. When Dover was overrun by troops loyal to William, soon after the battle at Hastings, they almost completely destroyed it. William, understanding the importance of earning the support of the population, especially in key coastal areas, ordered punishment for his soldiers and compensation for the sufferers, and the town duly supported its Norman conquerors. Dover's prosperity increased over the ensuing years, most likely due to nearby herring fisheries, easy access to continental trade, and free government, whereby freemen were able to carry on their mercantile pursuits without interference.

William's reign also saw the creation of a confederation of five coastal towns that would become known as the Cinque Ports. These towns received special privileges from the Crown in exchange for "ship service"—the provision of warships and men to protect against the inevitable skirmishing that occurred among English merchants, marauding pirates, and French forces. Dover was one of those initial five towns and became the principal station and seat of government of the confederation. Eventually, most of the towns between Sandwich and Hastings joined the confederation, providing a powerful defensive or offensive tool, as needed. Dover normally supplied more than twenty ships for the various opportunistic raids and counterattacks against France throughout the thirteenth century.

In retaliation for one such raid, the French in 1293 counterattacked at Dover, taking advantage of the absence of King Edward I, who was then fighting in Scotland. The French burned and sacked the town, but could not succeed in taking the castle. Not long after, they were repulsed, suffering great losses, but the town was essentially destroyed and would spend many years rebuilding. The next English monarch, Edward II, walled in the city to protect it from a similar fate.

In the sixteenth century, King Henry VIII ordered additional fortifications to the castle at Dover. Because Dover's famed harbor had for years been silting up, Henry ordered a major renovation project costing thousands of pounds. Between 1533 and 1535, several breakwaters were installed to allay the silting problem. Unfortunately, Henry's ideas did not work; by 1541, the breakwaters had crumbled in the face of heavy storms and further blocked the mouth of the harbor; at certain points, the harbor was only three feet deep, rendering it almost useless. The harbor fell into disuse, and Dover's trade decayed as Henry lost interest in his project and later monarchs failed to find the funds to continue it. It was not until 1580 that any steps were taken to repair the harbor, funded by licensing fees, shipping taxes, and donations.

The early seventeenth century saw Dover ascend to significant economic importance to England, as an international entrepôt, or central trading and market center. During England's ensuing series of depressions and crises, including its Civil War, Dover remained the main source for much-needed economic activity in the form of import and export of foreign commodities. In addition, Dover had always been a place where sovereigns were received, due in part to its proximity to the Continent. Henry VIII greeted the Holy Roman Emperor Charles V in Dover in 1520; in 1625, the British King Charles I received his queen; in 1660, Charles II landed to begin his reign.

By the eighteenth century, France and England were again at war, and the city again became a key defensive stronghold. The castle was reinforced and barracks were excavated in the solid rock cliffs underneath the castle, large enough to house 2,000 men. Batteries of heavy guns were placed on top of the cliffs, especially around the area known as the Western Heights. Subterranean tunnels and stairways connected the castle with the town below. The French never came close enough to directly threaten Dover, but the major earthworks produced in anticipation of their attack would be put to the test more than a century later.

During the nineteenth century, a large pier was completed, along with upgraded port facilities, which ever since have served as a major landing place for ferry and merchant traffic with ports in France, Belgium, and other continental countries. By 1900, Dover had settled into its role as a lively port town. Local industry consisted of shipbuilding, sail and rope making, and other light industrial activities. With its large, fine hotels and homes along the waterfront, Dover also became a resort town for summer beach-goers.

But Hitler changed all that. After war had been declared on Germany by France and Britain in September 1939, the Allied armies waited for action, and waited for almost a year, during what became known as the "Phony War." About 390,000 British troops were stationed in northern France, together with allied French troops, in anticipation of an attack by German troops. When at last Hitler began his attack, his *Blitzkrieg* (Lightning War) stunned the allies. Using massed tanks from the powerful Panzer divisions and rapid movement—the total opposite of the strategy employed during World War I—Hitler easily succeeded in splitting the Allied armies and reaching the English Channel. The Blitzkrieg began in early May 1940; by the end of the month, the German army had pushed bewildered British and French troops to the beaches at Dunkirk, a large French port. Pinned to the English Channel at the edge of the Continent, with nowhere to move, the English and French seemed to face certain annihilation.

It was Vice-Admiral Bertram Ramsay who instigated Operation Dynamo, based in Dover, for the rescue of the troops. Within a matter of days, Ramsay had organized a flotilla of naval ships as well as hundreds of small private boats to run between Dover and Dunkirk, between May 26 and June 4, to transport the men from the French beaches back to England. In the end, more than 340,000 allied troops had been plucked from the beaches under heavy German fire: the operation was hailed as both a victory and a miracle. Troops streamed into Dover, the injured filling every hospital. The British forces established a headquarters at Dover Castle, with an intricate communications operation beneath the ground, in the honeycombed chambers that were the legacy of a previous generation's efforts to oppose Napoleon. But the "victory" was short-lived: from these cliffs, British Prime Minister Winston Churchill contemplated the glare of fighting on the distant French shore; soon, France had surrendured to Germany. Britain now stood alone, the only country in Europe still capable of opposing Hitler.

As the war intensified, the inhabitants of Dover suffered greatly: German troops assaulted their shores with heavy artillery; German aircraft flew overhead. Dover was heavily damaged, particularly by the long-range artillery that the Germans had placed at Cape Griz Nez, directly across the channel. Those watching from Dover could see the flash of cannon in France, then minutes later feel the thunder of falling shells. Ninety-six percent of the houses in Dover were damaged, and many residents fled the town. A hospital to accommodate 500 patients was dug into the chalk cliffs; it proved to be the only safe haven in town. Singer Vera Lynn comforted the thousands of displaced families with her hopeful song of peace, "Bluebirds over the White Cliffs of Dover."

Dover was rebuilt after the war, as it had been following the devastations it suffered in prior centuries. After the war's end, a regional headquarters was prepared below Dover Castle, in case of a nuclear war; its existence was a secret until the mid-1970s.

Dover's dominance as a port and economic center has now waned. It continues to welcome thousands of visitors each year who ride the ferries to and from France and Belgium on holidays, though a large proportion of these visitors do not stay and instead use Dover as a link to other parts of the country. Even now, its significance as a ferry port is under threat from the development of the "Chunnel," the tunnel under the Channel linking Britain with France, and hence the rest of Europe. The Chunnel passes through nearby Shakespeare Cliffs, named after the playwright's depiction of the cliffs in King Lear, but Dover is not a terminus for the service. However, Dover's skill at survival, its docks and quays, shops and hotels, and its indestructible castle and cliffs, all ensure that Dover will not fade away.

Further Reading: For a concise guide to England's turbulent political history, see Christopher Daniell's *A Traveller's History of England* (New York: Interlink Books, and Moreton-in-Marsh, Glos.: Windrush, 1991), which provides an overview from Roman to modern times. For a detailed historical account of the town of Dover, see Rev. S. P. H. Statham's *The History of the Castle, Town and Port of Dover* (London: Longman, 1899). While it is a period piece and can be a complicated read in places, it discusses Dover's early economic and political fortunes in great depth. Robert Carse provides an interesting, pertinent account of Operation Dynamo and the beginnings of Hitler's Blitzkrieg in his book *Dunkirk: 1940* (Englewood Cliffs, New Jersey: Prentice-Hall, 1970).

—Christine Walker Martin

Dresden (Saxony, Germany)

Location: In the southeast corner of the former East Germany, in the Saxony region. Situated on both banks of the Elbe River.

Description: Founded about 1200 and named capital city of Saxony about 1500; became center of artistic and architectural treasures and producer of Dresden china; site of one of Napoleon's last great victories in 1813; ravaged by Allied bombing during World War II, the city has since been extensively restored.

Contact: Tourist Information
Pragerstrasse 8
01069 Dresden, Saxony
Germany
(351) 4955025

The city of Dresden, Germany, stands as a two-pillared monument to the most creative and the most devastating impulses of humankind. As the capital city for the rulers of Saxony during the Renaissance, Dresden became one of Europe's most prominent centers for the fine arts. The city's baroque and rococo architecture was lavishly ornamental, and its museums housed a priceless collection of paintings by the old masters and sculptures by the ancient Greeks. Porcelain was developed for the first time in Europe in Dresden in 1708, and the city lent its name to the decorative style known worldwide as Dresden china. In the nineteenth century, Dresden's new opera house, designed by Gottfried Semper, provided the city with a hub of musical activity, with performances of works by prestigious composers such as Bach, Handel, Strauss, Telemann, and Wagner.

Meticulously cultivated over centuries, Dresden's aesthetic essence and architectural beauty were reduced to rubble in two days in February 1945, as Allied forces bombed the city relentlessly. Because Dresden's Altstadt and Neustadt (Old Town and New Town) areas were not involved in industrial production—military or otherwise—many believed the Allied raid on the city was unjustifiably ruthless. British and American documents on the attack were classified for decades, leaving Allied motives and German casualties to speculation. Various sources dispute the number of civilian deaths in the bombing of Dresden, with estimates ranging from 8,000 to 250,000. The true figure is impossible to calculate, given that the streets of Dresden had become a haven for European war refugees whose identity was never discovered.

Today many of Dresden's finest buildings have been carefully restored. Since the reconstitution of a unified Germany, Dresden's hotels have multiplied to accommodate the burgeoning tourism industry, and the city, with a population of more than 500,000, is once again one of the largest in the country. However, reminders of the devastation of World War II dot the cityscape: the Frauenkirche, once one of Europe's most magnificent cathedrals, has been left in charred ruins, and parts of the downtown area remain vacant, awaiting construction.

Geographically the Dresden Basin lies at a sharp bend in the Elbe River. Settlers building a fortress here in about 1200 found that the region—unusually temperate for northern Europe—was well suited for agriculture. Fruit trees and vines thrived on the rich topsoil.

By the year 1500 the leaders of Saxony had chosen Dresden as their seat of government. To ensure its safety, the city was surrounded by strong lines of fortification. With expansion confined by these city limits, Dresden's masons and carpenters built upward; by the eighteenth century, much of the city stood five stories tall.

The arrival of the Saxon rulers created a class system in Dresden that revolved around the wealthy. The courts of the nobility employed servants and secretaries, and craftspeople lavishly appointed the majestic houses of their commissioners. From the latter half of the 1600s into the following century, the elector Frederick Augustus the Strong led Dresden's transformation into a capital city erected in the baroque style. The architects Matthäus Daniel Pöppelmann and Georg Bähr, among others, were instrumental in creating the visual splendor of detail and symmetry which earned Dresden the nickname "the Florence on the Elbe."

Given the cultural passions of its upper class, Dresden quickly established itself as a notable center for the fine arts. By the mid-seventeenth century Dresden (along with Munich and Vienna) was a permanent home for the Italian opera. The composer Heinrich Schütz, the first to merge Italian lyricism with the vigor of the German Protestants, was the director of Dresden's music from 1615 until his death in 1672. Schütz campaigned for the salvation of the city's musical institutions throughout the religiously charged Thirty Years War (1618–48), and his three works titled "Passions," published in 1666, were inspirations for the later oratorios of Bach and Handel. Schütz's successor, Johann Adolf Hasse, Dresden's court conductor well into the 1700s, was considered the finest master of German-Italian composition.

In 1707 ground was broken for the impressive Gemäldegalerie of the Saxon Electoral Court. The vestibule of the palace, known as the Zwinger, housed an initial collection of 535 "royal paintings" drawn from the private collections of a succession of Saxon rulers. The display included offerings by masters of the Italian and Dutch schools of painting, as well as a fledgling representation of Germanic work. Elector Frederick Augustus's active pursuit of great works of art

Dresden's Frauenkirche, undergoing restoration

The opera house, designed by Gottfried Semper
Photo courtesy of German Information Center

provided a precedent for future government officials in Dresden; shortly the collection covered the walls from floor to ceiling not only of the Zwinger but also of several other facilities.

In Dresden in 1708 the chemist Johann Friedrich Böttger, while staying at his summer house on the Brühlsche Terrasse (called the "Balcony of Europe") overlooking the Elbe, became the first European to discover the porcelain-making process. By 1710 the ruler Augustus had established a porcelain factory in Meissen, to the north of Dresden, placing it under Böttger's supervision. Originally producing red stoneware, the factory in 1713 began making the delicate white porcelain known today as Dresden china. The factory directorships of the painter Johann Höroldt (from 1720) and the sculptor Johann Kändler (from 1733) combined to create a world-renowned style of ornamental, Oriental-influenced china that featured elaborate decorations and painstaking designs. Kändler designed lifelike porcelain floral arrangements as well as the famous raised figures of tableware called "Swan Service," each of which contributed to the distinction of Dresden china. The Meissen factory continues to manufacture chinaware today.

The establishment of the porcelain trade marked the beginning of an industrial era in Dresden; at the same time, the city was embarking on a period of frequent unrest. During the 1740s the Catholic Saxon dynasty lost the sympathies of its Protestant population to the post-Reformation leaders of neighboring Brandenburg and Prussia. Before a military union between Saxony and Austria could take place, the Prussian army attacked and soundly defeated the Saxons, and Frederick the Great occupied Dresden in 1745. The Second Silesian War ended with the signing of the Peace of Dresden on December 25 of that year, and the next day Frederick celebrated Christmas at the magnificent Frauenkirche cathedral.

Despite the turmoil, the Saxon successors to Augustus the Strong remained committed to the enthusiastic support of the fine arts in the city. By 1754 the Dresden collection had amassed 1,446 paintings, including signature works by Canaletto, Rubens, Rembrandt, and van Dyck. Its most recent masterpiece at the time, Raphael's Sistine Madonna, was acquired by Augustus III in 1753.

During the Seven Years' War, begun in 1756, Dresden came under siege in the midst of a fierce battle for the disputed territories of northern continental Europe. Foreshadowing the events of World War II two centuries later, the city's Frauenkirche withstood three days of bombardment from enemy fire; and remained standing.

Dresden also was the site of an important military engagement in the nineteenth century. By 1813 the French emperor Napoléon had regrouped his troops in Germany after losing half a million men in battle in Russia. Europe was in a period of armistice, but no one expected the peace to last: both Napoléon and the forces of a united Europe were preparing for battle. In August of that year, Napoléon's commanding officer in Dresden, Saint-Cyr, reported that the Russian-led allied forces were preparing to attack Dresden.

Napoléon, who considered Dresden to be of utmost importance to his lines of communication in Germany (and who had come to know the city while quartered at the Marcolini Palace during the armistice), accepted this challenge in what would prove to be the last great battle of his reign.

On August 25, 1813, before Napoléon had arrived back in Dresden, a Russian light division foray to the suburbs of the city revealed to Saint-Cyr the gravity of the impending battle. But, rather than storm the city, the Russian command chose to retreat and wait for Austrian reinforcements. This hesitation proved costly: Napoléon's troops received word of the peril and hurriedly marched on Dresden.

By the time Napoléon arrived in Dresden on August 26, the French had amassed ten battalions and thirty cannon in the urban Altstadt (Old Town). Dresden's 30,000 residents were alarmed at the proximity of the enemy forces, fearing retaliation for their ill treatment of prisoners of war. Advancing on the Grosser Garten from the south, columns of Russian and Prussian recruits captured several public buildings and set fire to many more homes, and they threatened to overtake the densest part of Dresden's center. But a resurgent French army signaled an abrupt turnaround when it repelled an attempt to take Prince Anton's garden. By nightfall the French had recaptured virtually all of the buildings they lost earlier in the day, and Napoléon inspected 700 Austrian prisoners at the Royal Palace. The people of Dresden were encouraged by the performance of Napoléon's 70,000 troops against enemy forces nearly double in size. Throughout the night, additional French soldiers streamed into the city over the bridges crossing the Elbe.

In the heavy rains of the following morning, Napoléon's men drove the Russian forces from the Blasewitz wood, and the Austrians were driven into the villages surrounding Dresden. Wet muskets would not fire, and the French killed or captured as many as 15,000 to 25,000 enemy soldiers in gruesome hand-to-hand combat on civilian streets. Such was the exuberance of their victory that French soldiers in Dölzschen broke open the town's wine cellars, engaging in a "Bacchus feast" before the battle was decisively settled.

Napoléon's bold decision to defend Dresden in spite of a shortage of men made the Battle of Dresden one of his most compelling triumphs, but his conquests in northern Europe were nearing their close. Two months later, Saint-Cyr's men in Dresden were blockaded by the Russian-Austrian alliance. This time, starving and exhausted, they vacated Dresden on the condition that if they lay down their arms, they would be allowed to return to France; having agreed, the French were promptly taken as prisoners.

During the nineteenth century Dresden moved toward a position of increased commercial significance. The area's economy received a major boost in 1839, when Frederick List built the Saxon railroad linking Dresden and Leipzig; the following year, the run was extended to Prussian Magdeburg. Further gains came when Saxony joined the German Empire in 1871. With growing industry, however, came a growing number of factories, and the city spread to its outskirts, sacrificing some of its architectural finery. By the beginning of the twentieth century, that part of Dresden that had spread across the Elbe, Neustadt (New Town) was itself an old district.

In May 1849 the city experienced an uprising of revolutionaries, among whose ranks were counted the architect Gottfried Semper and the composer Richard Wagner. Wagner, raised in Dresden, saw the insurrection put down within a week, and he was consequently barred from Saxony. He had done some of his most important work in Dresden, serving as the opera's second conductor from 1843 to 1849, and first producing *Rienzi, Der Fliegende Holländer,* and *Tannhäuser* there.

Spared any warfare during World War I, Dresden saw another revolution take place after that war, and this one was successful. The Workers' and Soldiers' Council forced the abdication of the king of Saxony, after which the red flag of Germany flew overhead.

During World War II Dresden operated as an "open city," providing a workplace for Allied prisoners of war and a haven for refugees from war-torn areas of Europe. Dresden was equipped with just two shelters (located under the Goehle Werk and Zeiss Ikon factories) that could withstand the force of 1,000-pound bombs, and the city had forfeited all of its antiaircraft artillery to the front line, replacing the guns in some cases with wooden replicas. Despite these vulnerabilities, the people of Dresden took comfort in Marshal Hermann Göring's claim to have erected a "roof over the Reich," and many were convinced that Dresden was too beautiful to be attacked.

Still, the 1943 Allied bombing of Hamburg led officials in Dresden to order the construction of an interlocking network of underground cellars, and to allocate three "safe open spaces" in the event of attack—the Grosser Garten, the meadows of the Elbe, and the Grosses Ostragehege, site of the enormous Friedrichstadt hospital complex. Dresden had also begun to prepare for potential disaster to its priceless stores of art. As early as 1942 the local commander, the Gauleiter Martin Mutschmann, had ordered the dispersal of the city's treasures to private castles and country homes, a haphazard process that left the collection damaged and in disarray.

In January 1945 the British Royal Air Force began to reconsider the proposal nicknamed "Thunderclap," first put forth a year earlier, which would ravage Berlin, Munich, and Breslau with 25,000 tons of bombs over the course of four days. On January 25 Sir Arthur Harris of the RAF Bomber Command suggested supplementally raiding the cities of Chemnitz, Leipzig, and Dresden, presumably to disrupt German efforts to accommodate evacuees from the east; two days later, he was given authorization to do so.

On February 3 the U.S. Allied Air Force launched more than 1,000 bombers on Berlin, killing 25,000 people within minutes; the international press, stationed in that city, was outraged. Many saw the raid, which took place one day before the opening of the Yalta conference, as a display of might to the newly allied Russians.

The night of February 13–14, 1945, was Fasching night in Germany, the carnival celebration during which children dress in costume to conclude the season before Lent. Due to inclement weather, the first scheduled wave of U.S. bombers to fly over Dresden was postponed. As a result, the much more forceful British RAF attack, scheduled to be the second, became the first. At about 10 P.M. 244 British four-motor Lancaster bombers descended on the city, followed three hours later by another 529 planes. In the daylight of February 14, the USAAF attacked with 316 B-17 Flying Fortresses; the 527 originally scheduled were reduced in number because of the almost total demolition already accomplished. Bombing was confined to the civilian Altstadt area; left untargeted were possible strategic locations such as the Friedrichstadt complex, the railroad station linking the Hauptbahnhof and Neustadt station, and the autobahn bridge over the Elbe.

The havoc wrought by the Allies continued: two months later, on April 17, the USAAF sent 572 bombers carrying 1,690 tons of bombs over the city's marshaling yards, and another eight planes targeted industrial facilities. Two days later, on April 19, the air force bombed points on the Dresden railway and refugee camps in suburbs such as Pirna. Dogfights with Luftwaffe Messerschmidt pilots unwilling to admit defeat saw the downing of several U.S. planes.

Allied forces apparently chose to launch the relentless air raid on Dresden because its old, wooden, densely populated center would provide a horrific spectacle of flames certain to devastate an already weakened German spirit. Additionally, some theoreticians have suggested that the British and American allies, already committed to redrafting Europe and assigning 120 million Europeans to Stalin's jurisdiction, may have wanted to leave the tenuously-allied Russian premier with crippled spoils. Stalin had indeed requested Leipzig, which was equipped with military industries; Chemnitz, too, had factories involved in tank production. But Dresden was home to no such industry.

At the time of the bombing, the American novelist Kurt Vonnegut was serving as a prisoner of war producing malt syrup for pregnant women in an underground factory near the Friedrichstadt known as Slaughterhouse Five. Ironically, he survived the attack under shelter of the factory; he preserved his recollections of Dresden's destruction in his 1969 novel *Slaughterhouse Five,* considered a masterwork of antiwar sentiment.

A specially commissioned U.S. report on the bombings of Dresden, released from classification in 1978, claimed that the city suffered 8,200 to 16,400 casualties. Alexander McKee, author of *Dresden 1945: The Devil's Tinderbox,* puts the figure at between 35,000 and 70,000, and cites the widespread German belief that as many as 250,000 civilians lost their lives.

On May 8, 1945, the last day of World War II in

Europe, Russian forces occupied Dresden and began their "treasure hunt" for the hidden art of Dresden's remarkable museums. Before taking refuge, the Gauleiter Mutschmann had abided Hitler's so-called Nero Command, which instructed German leaders to sabotage their own food supplies and roadworks in defeat, actions which further complicated Russian efforts to resurrect the city and its collections.

More than 18 million cubic meters of rubble had to be cleared from the ruins of Dresden. From the Pragerstrasse promenade, one mile of the Altstadt in any direction was leveled to the ground, and the area of devastation has been calculated at fifteen square kilometers—more acreage than the entire city claimed in 1890. Of 220,000 dwellings in the city limits, 80,000 were reported completely destroyed, with only 45,000 unharmed.

About 200 precious paintings that had not been removed from the Schloss Pilnitz in Dresden perished in the bombings. Scouring crates and searching basements in the German countryside, the Russian government eventually recovered most of the works that had been removed from Dresden. For a time, 762 of the paintings were displayed in the Pushkin Museum in Moscow, and hundreds more in Kiev. From 1955 through 1958, the Russian government returned 1.5 million items to the German Democratic Republic; after a triumphant showing in the Berlin National Gallery, many of the works were hung in the rebuilt Zwinger. In 1979 a loan show called *The Splendors of Dresden* visited the museums of New York, Washington, and San Francisco, heightening American awareness of the city's artistic richness.

In addition to the Zwinger, the city of Dresden has restored the Schloss Pilnitz gallery and the Albertinum gallery at Neumarkt. The latter features the fabled Grunes Gewölbe (Green Vault), which, dating to the reign of Augustus the Strong, houses a world-class collection of objets d'art. Further restoration efforts include the historic Altmarkt and Neumarkt shopping areas that comprise Dresden's traditional center; the Kreuzkirche cathedral, with its fire-stained rendering of the Crucifixion; the Rathaus meetingplace; the Semperoper, the opera house designed by the celebrated architect Gottfried Semper; and the sixteenth-century Johan-neum, once the stables of nobility, today the home of a fine collection of vintage engines and automobiles.

Other attractions in Dresden include the Katholische Hofkirche (also known as the Cathedral of St. Trinitas), which shelters a crypt containing the tombs of forty-nine Saxon rulers, and a 250-year-old organ said to be one of the finest made by the renowned Silbermann family; the Porcelain Museum; the Armee Museum of ancient military history; and the Buch Museum, which recounts the history of books and publishing from the Middle Ages. Dresden's business focuses on industrial research and development, and the city's chief products are medicines, electronics, and optical and precision instruments. The city is home to the Central Institute of Nuclear Physics and the Technical University at Dresden, as well as the prestigious Palucca School of Ballet, and one of Germany's most highly-regarded art academies.

Tourism in Dresden has been rising steadily since the dismantling of the Berlin Wall, and the subsequent economic boost has encouraged a refinement of the modern, boxy architecture to which builders resorted in the aftermath of World War II. The hotel industry is building facilities at a rapid rate, anticipating a full return of the exquisite elegance the city knew before the bombs of 1945.

Further Reading: *Dresden 1945: The Devil's Tinderbox* by Alexander McKee New York: E. P. Dutton, and London: Souvenir, 1982 is a graphic account of the city's demolition told in part by firsthand accounts of survivors of the World War II bombing raids. *Napoleon's Last Campaign in Germany–1813* by F. Loraine Petre (London: Arms and Armour Press, and New York: Hippocrene Books, 1974) is a military chronology originally written in 1912. Portions of *A History of Modern Germany 1648–1840* by Hajo Holborn (New York: Knopf, 1964; London: Eyre and Spottiswoode, 1965) are helpful, as are entries in *Germany at its Best* by Robert S. Kane (Lincolnwood, Illinois: Passport Books, 1985) and in *Fodor's 1994 Germany* (New York). Finally, *Dresden Gemäldegalerie,* part of the Great Galleries Series, written by Gertrud Rudloff-Hille (London: Oldbourne Press, 1961), provides a comprehensive history of the city's art.

—James Sullivan

Drogheda (Louth, Ireland)

Location: Republic of Ireland, eastern region, about twenty miles north of Dublin.

Description: Small town near one of the most impressive pre-Christian burial places in Europe; mentioned in many ancient Irish tales, particularly those pertaining to the mythic hero Cuchulainn; closely associated with the story of St. Patrick, who brought Christianity to Ireland; the site of a notorious massacre in the seventeenth century; near the battlefield where dreams of a Catholic, Gaelic Ireland were defeated by Protestant, English forces.

Contact: Midlands and East Regional Tourism Organization
Dublin Road
Mullingar, Westmeath
Ireland
(44) 48650

The Irish name for Drogheda is *Droichead Átha,* Bridge of the Ford. The name presumably comes from the fact that Drogheda lies on the River Boyne, about four miles from the mouth. Historians believe that the first people to live in Ireland came from the European mainland about 6000 B.C. and settled on the northeast coast, somewhat to the north of present-day Drogheda.

About two miles east of Drogheda, also on the Boyne, is Brugh na Boinne (Palace of the Boyne). One of the most interesting pre-Christian burial places in Europe, the site is mentioned in many ancient Irish stories. Brugh na Boinne actually consists of three communal graves some distance apart, at Dowth, Newgrange, and Knowth. Dowth is circled by stone curbing and has a twenty-seven-foot-long stone passage running to a central chamber. In this chamber rest the Bronze Age kings of Ireland. Their graves date back to about 2000 B.C.

Newgrange was built as a communal tomb around 4000 B.C. This neolithic passage grave is one of the most spectacular in Europe. Originally thirty-five standing stones surrounded the grave, but today only twelve remain. Newgrange consists of a cairn (mound of stones) 280 feet in diameter and 44 feet high. Some of the stones are decorated with spiral and zigzag patterns. A sixty-two-foot stone passage leads to a central chamber. The domed, twenty-foot-high chamber is roofed with beehive vaulting and has three recesses carved with elaborate patterns. The chamber contains several stone basins, but today their use is unknown.

Newgrange was constructed so that on the winter solstice (the shortest day of the year) the rising sun shines through an opening above the passage entrance. Slowly, the sun stretches down the passage, eventually lighting the central chamber for fifteen to twenty minutes. Because the stones are carved with intricate patterns, they glow eerily in the light. Then the sun retreats, not to touch the central chamber for another year.

Historians believe that Vikings plundered much of the royal remains and treasure buried in Newgrange. One pagan king, Cormac, escaped this fate of grave desecration. He did so by insisting that he should be buried across the River Boyne in a wood now called Rosnacree. According to legend, Cormac's descendants made many attempts to bring his body to Newgrange, but the waters of the Boyne always rose in fury and prevented them from crossing.

The third burial mound, Knowth, has yet to be entirely excavated. A summit chamber and surrounding curb were discovered in 1942; the central burial chamber has yet to be found. Excavators hope that the Vikings missed the burial chamber, thus leaving valuable material yet to be uncovered.

Between 600 B.C. and 400 B.C., Celtic tribes from Britain and mainland Europe invaded Ireland. Celts were the first people in Europe known to make iron. Their society was divided into three classes: rulers; common people; and educated people, who included poets, or *filí,* and priests, or *druids.* Celtic culture spread across Europe, but retreated before an expanding Roman Empire between about 300 B.C. and A.D. 100. Celtic culture was preserved in Ireland, however, and in other remote regions such as Scotland, Wales, the Cornwall region of England, and Brittany in northwest France.

About twenty miles north of Drogheda is Dundalk, believed to be the scene of the birth and death of the legendary Cuchulainn. A hero of Irish folklore, Cuchulainn was a powerful warrior for the province of Ulster, and the son of the Celtic god Lugh. He obtained his name from the Irish word *cu,* meaning hound, after he killed a ferocious watchdog belonging to a chief called Culain. Dundalk at that time was known as Dun Dealgan, or Dealga's Fort, and stood on the inland edge of today's town. Dun Dealgan was the site of the Cattle Raid of Cooley, or *Táin Bó Cúailnge,* the story of which is central to the Ulster cycle of ancient Irish tales. The tale is recounted in western Europe's oldest epic in a native language, which has been referred to as the Irish *Iliad.* If the raid took place in fact, it probably did so around the time of Christ. According to the tale, Queen Maeve of Connaught coveted a great bull that belonged to the ruler of Ulster. She sent warriors to steal the bull, leading to a four-day battle. Cuchulainn defended the area against the raiders and survived, although many warriors on both sides were killed. Cuchulainn is believed to have met his death in Dun Dealgan while defending it on the road to the north. The mythical

St. Lawrence's Gate at Drogheda
Photo courtesy of Bord Failte, The Irish Tourist Board

giant Finn McCool's mastiff Bran also is rumored to be buried in the Dundalk area.

About fifteen miles southwest of Drogheda is the Hill of Tara, where in ancient times a royal court was held. According to archaeologists, Tara was grand even in the Bronze Age (about 2000 B.C.). Originally Tara was home to the kings of Meath (once a province, now a county of Ireland). Later, Tara became home to the Irish high king. Every three years provincial rulers, warriors, historians, and poets came to a *feis,* a hosting, at Tara to hear laws made and judgments passed. A glorious time at Tara was the reign of Cormac mac Airt (227–266), who built a banquet hall 760 feet long by 90 feet wide to hold 1,000 guests for one *feis.* Today, all that remains of the hall are parallel ridges.

At Tara today is a circular fortress 850 feet around, known as Rath na Riogh (the Kings' Rath, or fortress). Within that rath are two other mounds, the Dáil, or assembly place, and Cormac's house. On the summit of Cormac's house is a statue of St. Patrick and the Lia Fáil, or Stone of Destiny. According to legend, Ireland's ancient kings were crowned on this stone. However, some claim the original Lia Fáil was stolen and carried first to Scotland and then to England's Westminster Abbey. Numerous other raths also remain, including one known as the Rath of Leary, named for the king who ruled when St. Patrick brought Christianity to Ireland in the fifth century.

According to legend, St. Patrick in 433 walked to Drogheda, where he took a boat along the River Boyne to the village of Slane. At Slane is a hill that commands a wide view stretching southwest to Tara. The night that Patrick arrived was Easter Eve, when it was the custom throughout Christian lands to light the Paschal Fire to commemorate Christ's resurrection. According to Ireland's ancient pagan religion, however, that was the one night of the year when no light must shine. Patrick lit a fire at the top of the hill. From Tara, the king saw the flames. The druids warned, "If that fire is not quenched now, it will burn forever and consume Tara." The king rode the ten miles to Slane with his warriors. Patrick and his followers went down the hill to greet them, but were seen as a herd of deer and passed by the warriors unharmed. From that day, the king put no obstacles in Patrick's way as he converted the Irish to the Christian faith.

One of the most beloved saints of the new religion was St. Bridget, or Bríd. Now one of the patron saints of Ireland, Bridget probably was born at Faughart, to the north of Drogheda, in the fifth century. Five miles south of Drogheda, in Duleek, stands what is believed to be Ireland's first stone church. It was built by St. Patrick, who placed it under the direction of his disciple, St. Cianan (Keenan). The church was followed by a priory in 1182, of which the ruined church remains.

After 795, the Vikings began raiding the east and south coasts of Ireland. They settled near harbors, where they established towns. From these towns the Vikings raided the countryside, destroying monasteries and robbing them of their treasures. The Danish raider Turgesius, or Thorgestr,

took Drogheda in 911 and established a settlement there. In 1014, Brian Boru, the high king of Ireland, defeated the Vikings at the Battle of Clontarf (now part of the north side of Dublin). The Vikings were allowed to remain in Ireland and through intermarriage merged with the Celts.

Near Drogheda today stand the ruins of Monasterboice, a monastery founded by St. Boetius that was a great seat of learning until the twelfth century. The ruins include a tower, which at a height of about 100 feet is reputed to be the highest in Ireland, and two churches. The burning of the tower in 1097 is the last recorded event in the history of the abbey. The Muiredach Cross, which dates from the tenth century, is considered one of the finest in Ireland. It is fifteen feet high, with twenty-two panels depicting the fall of man, the crucifixion, and Christ at Judgment Day. It supposedly was erected by the tenth–century abbot Muiredach. The West Cross also is considered very fine; it stands more than twenty-one feet high and consists of about fifty sculpted panels, although many of them are worn.

About four miles south of Monasterboice is Mellifont Abbey, which was the first Cistercian house and the first abbey founded by a foreign religious order in Ireland. It was founded in 1142 by St. Malachi O'Morgair on the site of a seventh-century convent.

In the 1160s, the high king of Ireland, Turlough O'Connor, deposed the king of Leinster, Dermot MacMurrough. Dermot went to England to ask King Henry II, a Norman, for help. Dermot promised Norman soldiers a share of any land they helped him reconquer. With Norman help, Dermot recovered his kingdom in 1170. After Dermot's death, a Norman named Strongbow declared himself king of Leinster. Other Normans seized Irish land, and Henry II forced them to recognize him as lord of Ireland. The expanding Norman grip on Irish land soon led to tension. North of Drogheda, at the present village of Lungangreen, Norman John de Courcey battled O'Hanlon, chief of Armagh, in 1180. There was no clear winner, and both sides retired with heavy losses.

By the fourteenth century, the Normans held nearly all land in Ireland, but the Normans slowly became Irish as they married Irish people and adopted Irish customs. English control dwindled to only a small part of Ireland around Dublin, known as the Pale. Drogheda was an important center of the Pale. The town frequently was the rendezvous of English armies sent to put down rebels in Ulster. In 1395, the four chief Irish princes submitted to England's Richard II in the Magdalen monastery at Drogheda.

Numerous parliaments were held at Drogheda. Of particular historic note was the parliament of 1494, which passed Poynings Law. Under Poynings Law, no law passed by the Irish parliament could go into effect until it was ratified by the English privy council.

To the north of Drogheda, Edward Bruce, who was crowned king of Ireland in 1315, conquered Dundalk the following year. Bruce was slain in nearby Faughart in 1318, during a battle against Sir Richard Bermingham. Dundalk was surrounded by walls during the time of Henry IV and underwent six sieges between 1560 and 1650. Most of the walls were removed in 1747.

In 1541, Henry VIII of England forced Ireland's parliament to declare him king of Ireland. He extended English law to Ireland and tried to introduce the Protestant religion. Although Henry's heir and daughter Mary I was a Catholic, she continued her father's policies in Ireland. Mary seized Irish land and awarded it to English settlers. This practice was known as the plantations. Her heir and sister, Elizabeth I, made Catholic religious services illegal in Ireland. A series of rebellions were put down by Elizabeth in 1603.

As the plantations continued, Irish Catholics feared that they would lose their land. They also worried that the Puritans, who were gaining power in England, would persecute them. In 1641, they rebelled. Drogheda was besieged, but the English Sir Henry Tichborne and Lord Moore held off the Irish Phelim O'Neill. The town held for the English and was relieved by James Butler, the earl of Ormonde. About two miles north of Drogheda, on the site of a medieval castle used by O'Neill as his headquarters during the siege, Tichborne and his son built the mansion Beaulieu between 1660 and 1667.

Unrest continued, however. In 1649, Oliver Cromwell was sent to Ireland as commander in chief and lord lieutenant. Cromwell's mission was to safeguard continuing Protestant plantations. In September, after a siege, Oliver Cromwell conquered Drogheda, and his troops massacred the garrison and townspeople. According to Cromwell's account, 2,000 people were killed. Others say 3,000 died. Included among the victims were women and children. Other Drogheda residents were transported to Barbados in the West Indies. The massacre at Drogheda, followed by one in Wexford in October, persuaded many Irish towns to surrender to Cromwell. Cromwell awarded more Irish land to English Protestant settlers and deprived Catholics of political rights. He departed from Ireland after only nine months, leaving a notorious reputation behind. Today, a fragment of Drogheda's defenses remains at St. Lawrence's Gate, which consists of two towers joined by a blocked arch.

In 1669, St. Oliver Plunkett, the Catholic bishop of Armagh, became primate of Ireland. Plunkett was arrested in 1678 during the Popish Plot, an imaginary Catholic conspiracy invented by Titus Oates and Israel Tonge. Oates and Tonge spread a rumor that the Jesuits were plotting to assassinate England's Protestant Charles II and replace him with his Catholic brother James. About thirty-five innocent people were tried and executed for treason during the Popish Plot, including Plunkett. In 1678, Plunkett was hanged, drawn, and quartered at Tyburn (London). Today Plunkett's embalmed head is housed in a shrine in Drogheda's St. Peter's Church, which was erected in his memory. The door of Plunkett's cell is preserved beside the church. A shrine to Oliver Plunkett also stands on the spot where he served Mass and ordained priests, north of Drogheda and about three miles

south of the village of Blackrock. August 14, the feast of the shrine, is a time of religious gathering there. Oliver Plunkett was beatified by the Catholic Church in 1920 and canonized in 1975.

After Charles II's death in 1685, his Catholic brother became King James II of England. James abolished many anti-Catholic laws in England. In 1688 the English people, most of whom were Protestant, forced him from the throne. William of Orange—a Dutch, Protestant ruler married to James's daughter Mary—became William III of England. James went to Ireland, where he raised an army to fight William. James also borrowed troops from Catholic France. Protestants in Ulster supported William and helped him defeat James in the Battle of the Boyne on July 11, 1690. The battle took place in King William's Glen, just north of Drogheda. According to legend, King James fled the field when it was clear his forces had lost. He arrived at an Irish castle, where he told the woman of the house about his defeat. He said sarcastically, "Madame, your countrymen are fine runners." She replied, "But I notice your Majesty has outstripped them all." Drogheda held for James II under Lord Iveagh but surrendered the day after the Battle of the Boyne. The war ended with the Treaty of Limerick in 1691. The Battle of the Boyne was decisive in consolidating English, Protestant control over Ireland. Today, the anniversary of the battle is commemorated by Protestants in Northern Ireland. After William's victory, the Irish lost more land to the English. By 1704, Irish Catholics held about one-seventh of Irish land. They were forbidden to purchase or inherit land, barred from the Irish parliament and army, and restricted in the practice of their religion.

In 1791, Theobald Wolfe Tone, Thomas Russell, and Napper Tandy founded the Society of United Irishmen. The United Irishmen aimed to unite Irish Protestants and Catholics in order to free Ireland of British control. In 1798, they rebelled in various parts of the country. Napper Tandy landed at Donegal, on the northwest, with French soldiers. He became drunk, was carried back to his shop, and later was captured in Hamburg. In other parts of the country, the rebellion was put down by Lord Cornwallis. In 1800, the Act of Union joined the parliaments of Ireland and England. Tandy was released through the efforts of Napoléon in 1802. Today Napper Tandy is buried at Castlebellingham, near Drogheda.

In Slane, the soldier poet Francis Ledwidge was born in 1887. Ledwidge was killed fighting in World War I in 1917, and his home is now a museum. Today Slane village is overlooked by a hill on which lie the ruins of an ancient Abbey Church and College, which were rebuilt in 1512. Nearby stands Slane Castle, built in the late eighteenth century on the site of a former castle. The castle's grounds form a national amphitheatre. In the 1980s, such performers as Bob Dylan, the Rolling Stones, and Bruce Springsteen performed there. The castle was damaged by fire in the late 1980s. In 1983, the site of the Battle of the Boyne was purchased by an American, James Delaney. The land is now held in trust and a park of reconciliation may be developed there.

Further Reading: *The Celts* by Gerhard Herm (London: Weidenfeld and Nicholson, 1975), describes Celtic history, culture, and myths. *Modern Ireland, 1600–1972* by R. F. Foster (Harmondsworth, Middlesex: Penguin, 1988; New York: Lane, 1989) is a thorough account of modern Irish history, including events at Drogheda.

—Mary Feely

Dublin (Dublin, Ireland)

Location: On Dublin Bay on the east coast of Ireland.

Description: Capital city of the Republic of Ireland; home to Swift, Yeats, Joyce, Beckett, and other writers.

Site Office: Bord Failte Eireann (Irish Tourist Board)
Baggot Street Bridge
Dublin 2, Dublin County
Ireland
(1) 676-5871

The city of Dublin takes its English name from the Dubh Linn (dark pool), which used to lie where the Liffey and Poddle rivers met, and its Irish name, Baile Atha Cliath ("town of the hurdle ford"), from a ford over the Liffey close by. The bay on which Dublin stands was known to the Greek geographer Ptolemy in the first century A.D., and the city's harbor was fought over by local rulers. It is possible that St. Patrick, the semi-legendary missionary who brought Christianity to Ireland, conducted baptisms at a well where the cathedral named for him was to be built. The Irish culture of these early centuries is well represented in modern Dublin. The Old Library of Trinity College displays the Book of Kells and other religious texts illuminated by Irish monks; the National Museum of Ireland includes among its wide-ranging collections several pre-Christian Ogham Stones (with inscriptions in the old Ogham script), the Tara Brooch, and the Bell of St. Patrick. However, these remnants come mostly from elsewhere on the island, and Dublin itself appears not to have become an important settlement until after the arrival of Viking settlers, who began to call the place "Dyflin" in about A.D. 836.

A first wave of mainly Norwegian invaders was driven out, but was soon followed by Danes, who defeated the Irish in 919, built a town at what is now Wood Quay, and began holding their meetings on a hill known as the "Hoggen." The two sides at the Battle of Clontarf in 1014 indicate how far assimilation between the two peoples proceeded, for it was a conflict between Sitric, the Danish king of Dublin, allied to the Irish king of Leinster, and Sitric's father-in-law Brian Boru, the high king of Ireland. In 1038 Sitric established Christ Church Cathedral as the seat of the bishops of Dublin, who became archbishops from 1152.

The continuing division of Ireland into several kingdoms made it impossible for the Irish to resist the conquest of the island by the Norman rulers of England. Dermot Mac-Murrough, king of Leinster, decided to seek help from Henry II, king of England, in a dispute with other Irish kings that had led to his exile to Bristol in southwest England. In 1169 the Norman commander Richard de Clare, Earl of Pembroken, nicknamed "Strongbow," married Dermot's daughter; in 1170 he conquered Dublin, and in 1171 he succeeded Dermot as king of Leinster. King Henry then assigned Dublin to the control of the merchants of Bristol and received the surrender of the Irish kings. Strongbow died in 1176, having become balliff (royal governor) of Ireland and begun the rebuilding of Christ Church.

Medieval Dublin was controlled by a series of lords deputy of Ireland, who took their oaths of office in Christ Church Cathedral, which was also the meeting place of the parliament of Ireland. In practice they governed only the city itself and the area around it known as the Pale, and not even all of that. After St. Patrick's church, beyond the city walls, was designated a cathedral—in rivalry to Christ Church—early in the thirteenth century, the archbishops of Dublin used both cathedrals, while also governing the areas around them, known as the "Liberties" because they were free from control by the city government. Although life was not easy for most Dubliners—for instance, 4,000 of them died as the Black Death struck Ireland—the city's merchants mostly prospered, developing separately from the rest of the largely agricultural country.

Under the Tudor dynasty Ireland became much more closely bound to England. The Tudor era opened in Dublin in 1487 with the coronation in Christ Church of Lambert Simnel, a baker's son who claimed to be the heir to the Plantagenet dynasty, but he was swiftly defeated and made a servant in the household of King Henry VII. In 1534 "Silken" Thomas Fitzgerald began his rebellion against English rule at St. Mary's Abbey (now in ruins), but he found little support, his siege of Dublin Castle was soon defeated, and he was executed, with five of his uncles, in London in 1537. Neither event had anywhere near the impact of the Reformation, initiated in Ireland by George Browne, who was archbishop of Dublin from 1536 to 1554. It was resisted by the priests of St. Patrick's Cathedral, which was stripped of its revenues and treasure and allowed to fall into disrepair, but not by most Dubliners, who once again proved to be very different from most Irish people of the time, who never accepted Protestantism.

In 1565 Henry Sidney became the first lord deputy of Ireland to take up residence in the castle, which had been begun in about 1204, and from which he and his successors were to govern the whole island on behalf of the English (later British) crown through to 1922. The late sixteenth century was another period of contrasts, between the military repression of the Irish countryside by English troops and the commercial success of Dublin which provided many of the supplies for these troops. Dublin's loyalty was rewarded with Queen Elizabeth's foundation of the University of Dublin, usually referred to as Trinity College (or TCD), in 1592.

Dublin played no part in the Irish rebellion of 1641,

St. Patrick's Cathedral
Photo courtesy of Bord Failte, The Irish Tourist Board

except as a safe place for English Protestants to escape to as the war erupted in the countryside, and during the English Civil Wars, between 1642 and 1649, it was a stronghold of supporters of the parliamentary cause while most of Ireland supported King Charles I. Following the king's execution in 1649, James Butler, Duke of Ormonde, failed to capture the city for King Charles II and fled to France. Later the same year Oliver Cromwell, chosen by Parliament as lord lieutenant of Ireland, arrived in Dublin with 12,000 troops. His cavalry used St. Patrick's Cathedral as stables for their horses, but otherwise Dublin avoided the harsh measures imposed on the island.

By this time there were about 9,000 people in Dublin, almost all living south of the Liffey. Ormonde, appointed in 1662 to govern Ireland after the restoration of Charles II, encouraged the development of the north bank, starting with the stone quay which is still named for him. Throughout the city, churches were restored and new roads laid out. What

is now Phoenix Park, the largest public park in Europe, was enclosed and opened to the public; the common land in south Dublin, near St. Patrick's Cathedral, was also enclosed and given the name St. Stephen's Green; and the Hoggen was removed in 1682 when College Green was laid out on its site. Ormonde also commissioned the Royal Hospital at Kilmainham as a home for retired soldiers. (It has recently been restored and opened to the public.) By the time Ormonde left office in 1689 Dublin had about 60,000 inhabitants. Once again, however, Ireland was caught up in English conflicts. James II, the last Roman Catholic monarch of England, Scotland, and Ireland, held his last "parliament" in Dublin in 1689, after he had fled from London and the Protestants William III and Mary II (his son-in-law and daughter) had been invited to take his place. After his defeat at the Battle of the Boyne in Northern Ireland in 1690, James retreated to Dublin, where he boarded a ship for France, leaving behind him the decisively defeated Irish Catholic population under the thumb of a self-confident Protestant ''Ascendancy'' of Dublin merchants and Anglo-Irish landowners. Their power was symbolized by a statue of William put up outside Trinity College in 1701 (and blown up in 1925).

However, the Irish Parliament, from which Catholics were excluded, was seriously restricted in its authority by the claims of the British Parliament to overrule its decisions. Among those who protested against this restriction was the anonymous author of the *Drapier Letters,* pamphlets issued in Dublin in the 1720s, who was in fact Jonathan Swift, now famous as the author of *Gulliver's Travel.* Swift had been born in Dublin in 1667 and had been a student at Trinity College; he was dean of St. Patrick's Cathedral from 1713 to his death in 1745.

The Ascendancy had a lasting impact on the appearance of Dublin through the architecture and town planning it sponsored. The Mansion House, the official residence of the lord mayor of Dublin, was constructed in 1710, though its front was added in the nineteenth century; the first of the superb Georgian townhouses of Merrion Square was built in 1726; and the construction of a new Parliament building, opposite Trinity College, began in 1729. Leinster House, originally the residence of the dukes of Leinster and since 1922 the seat of the modern Irish Parliament (Oireachtas), was built in 1745. The classical front of Trinity College was completed in 1759—the same year that Arthur Guinness began brewing the beer that would make his name famous throughout the world. But the architectural masterpieces of Georgian Dublin were both built later in the century, both designed by the Englishman James Gandon: the Custom House, on the north bank of the Liffey, then the largest building in the city, decorated, by Edward Smyth, with imaginary faces personifying the four principal rivers of Ireland; and the Four Courts, completed in 1802, another domed building on the river bank.

By 1800 the population of Dublin was about 200,000. The Act of Union in that year was Britain's response to the failed Irish rebellion of 1798 in which, characteristically, Dublin had not been involved, apart from a hopeless attack on the castle by ill-organized peasants. The Irish Parliament was bribed to vote for its own abolition, its building was sold to the Bank of Ireland, and most of the landowners and nobles who had been prominent in it soon departed.

In 1798 the Irish had their last rising led by Protestants, for the Irish nationalism of the nineteenth century would be a more popular but also more definitely Catholic movement, in which Dublin would be more prominent. Its first great leader, Daniel O'Connell (1775–1847), known as the Liberator, lived in Dublin while campaigning for the lifting of legal restrictions on Roman Catholics. That was achieved for most purposes, in 1829, throughout the United Kingdom, though, like Oxford and Cambridge universities, Trinity College was to remain restricted to members of the established church for another four decades. O'Connell went on to become lord mayor of Dublin from 1841 until he was jailed three years later after moving a resolution calling for the restoration of the Irish Parliament.

In 1882 a monument to O'Connell was put up on Sackville Street, the main thoroughfare of the city (which was renamed O'Connell Street forty years later). By 1882 Dublin had lost much of its wealthier population to the suburbs, linked to the city by train since 1834, and gained thousands of poor residents, many of them originally peasants seeking to escape the Great Famine of the 1840s; many of the townhouses of Dublin had become slum dwellings.

The rise of nationalist militancy culminated, also in 1882, in the assassination of the chief secretary, Lord Frederick Cavendish, and his undersecretary, Thomas Burke, who headed the British administration in Ireland under the viceroy.

The National Gallery of Ireland, opened in 1864, the National Library and Museum, added in 1890, and other institutions helped to foster the modern Irish culture typified by such Dubliners as the playwrights Oscar Wilde (1854–1900) and George Bernard Shaw (1856–1950), the poet W. B. Yeats (1865–1939), and the novelist James Joyce (1882–1941). It is striking that all four men, like Joyce's disciple, the playwright Samuel Beckett (1906–89), spent large parts of their adult lives, achieved their fame, and died outside Ireland, though Yeats was one of the founders of Dublin's Abbey Theatre (in 1904) and became a senator of the Irish Free State.

After centuries of unsuccessful struggles against British rule, most of Ireland was to achieve independence through the medium of Sinn Féin, the nationalist movement founded in Dublin in 1905. Its military wing, the Irish Volunteers, was formed in 1913 in response to the creation of the Protestant and Unionist Ulster Volunteers in Nothern Ireland. Sinn Féin did not officially support those who mounted the rebellion in 1916 that has since come to symbolize the cause of Irish freedom. The Easter Rising, which lasted six days, was carried out by an uneasy alliance between Irish Volunteers led by the Romantic nationalist Patrick Pearse and the Irish Citizen Army of the Marxist labor activist James Connolly. The rebels attacked the castle, where about fifty of them died, and occupied the General Post Office where they were defeated. The Post Office now contains a statue *The Death of Cuchulainn* (a hero in ancient Irish legends), commemorating the Rising whose leaders were executed.

Where the Rising had failed, protest and negotiation succeeded, only to be followed by partition and civil war. The Anglo-Irish Treaty of 1921 effectively implemented the British Government of Ireland Act of 1920, dividing the island between the twenty-six countries of the Irish Free State, governed from Dublin, and the six counties of Northern Ireland, then controlled by Unionist politicians and armed Ulster Volunteers determined to resist incorporation into the majority-Catholic state. Antitreaty forces led by Éamon de Valera seized the Four Courts and were removed, in June 1922, only after the building had been severely damaged by shells fired by the protreaty government forces, which also destroyed most of the public records stored there.

De Valera led most of his followers back into peaceful politics in 1927 and soon came to dominate the Free State, writing the Irish Constitution, in force since 1937, and serving as prime minister for many years and as president for the fourteen years up to his death in 1975. The Free State, redesignated the Republic of Ireland in 1949, was characterized by rising levels of unemployment and emigration, and Dublin stagnated along with the rest of the country. However, by 1966, when the fiftieth anniversary of the Easter Rising was marked officially by the opening of a Garden of Remembrance and unofficially by the blowing up of Nelson's Pillar on O'Connell Street, De Valera's protectionist policies were being abandoned and the economy was beginning to expand.

The current "Troubles" of Northern Ireland began in 1968 with the Unionists' deployment of armed police against peaceful campaigners for civil rights for the Catholic minority. Though more than 3,000 people have died in the violence that has followed the re-creation of the "Irish Republican Army" and the intervention of the British Army, few have been citizens of the Irish Republic. The Troubles first affected Dublin in February 1972, when the British Embassy in Merrion Square was burned down. But this was hardly an expression of popular feeling, for the political wing of the IRA, which calls itself "Sinn Féin," receives less than 2 percent of the votes cast in the Republic.

In 1973 Britain and Ireland both joined what was then the European Community (or Common Market), renamed the European Union in 1993. Not only does the Republic receive about six times more in European Union funds than it pays in, but its trade has been reoriented away from dependence on Britain toward free exchange with all the member states. However, in spite of the resulting overall rise in prosperity, Ireland remains relatively poor in comparison with most of northwest Europe, some areas of Dublin and other cities having 70 percent or more of their adult residents unemployed. In a city of more than 920,000 people, where current eco-nomic and social problems necessarily take priority, it is not always easy to find the political will, and the funding, needed to protect Dublin's heritage. Those buildings that are in constant use have of course been well maintained—notably the Four Courts and the Custom House (now housing the Ministry of the Environment), once they had been restored after the civil war—but much else has been lost through neglect in the lean years and development in the fat ones. Thus, for example, between 1978 and 1980 the construction of a controversial new city hall went ahead at Wood Quay despite the discovery on the site of thousands of Viking artifacts; many of the objects were rescued, but many others were lost.

Nowadays Dublin is being renovated and embellished, using funds from public and private sources including the European Union, and the once-notorious levels of air pollution are also being sharply reduced. Viking Dyflin, medieval Dublin, the Dublin of the Protestant Ascendancy, and the Dublin of the Easter Rising and the civil war all now have their memorials in a city that has nevertheless left them far behind.

Further Reading: Desmond Clarke's *Dublin* (London: Batsford, 1977; New York: Hippocrene, 1978) is perhaps the best of the many books surveying the whole of the city's history. Thomas and Valerie Pakenham's *Dublin: A Travellers' Companion* (London: Constable, New York: Macmillan, 1988) is an anthology of eyewitness accounts of Dublin events and places from Viking times to the twentieth century. *Dublin Through the Ages*, edited by Art Cosgrove (Dublin: College Press, 1988), is a collection of academic essays on the development of the city. Thomas Farel Heffernan's *Wood Quay: The Clash Over Dublin's Viking Past* (Austin: University of Texas Press, 1988) is a vivid and thought-provoking account of the most celebrated dispute over preserving Dublin's past. Finally, after eighty years the short stories in James Joyce's *Dubliners* are still powerfully evocative, while his *Ulysses*, depicting Dublin on June 16, 1904, in enormous detail, will have its status as one of the greatest novels of the century confirmed when John Kydd's definitive edition is published.

—Patrick Heenan

Durham (Durham, England)

Location: On the River Wear, 16 miles south of Newcastle upon Tyne on the A177 road.

Description: Medieval city with walls, castle and cathedral, and a nineteenth-century university.

Site Office: Tourist Information Centre
Market Place
Durham, County Durham DH1 3NJ
England
(91) 384-3720

Durham was originally known as Dunholm, from an Anglo-Scandinavian word meaning "hill island." The rocky hill that is the focus of the old town is in fact on a peninsula, surrounded on only three sides by the River Wear. Little is known of the early history of the area, although the remains of prehistoric settlement have been uncovered, and it appears that a bishop was consecrated in the area now called Elvet (originally Aelfet ee, "swan island") in 762.

The origins of the modern city can be traced back only to 995, when the remains of St. Cuthbert were brought to Dunholm. Cuthbert, the prior (head) of the monastery on the island of Lindisfarne (also called Holy Island) and briefly a bishop, had died more than 300 years earlier, in 687. It is said that when his coffin was opened in 698, his body was found not to have decayed, a phenomenon taken to indicate his holy status. The monastic order that he had led left Lindisfarne in 875 after raids by the Danes. It was not until eight years later that they were able to establish a new monastery, when the Danish King Guthred granted all the land between the Wear and Tyne rivers to the Church of St. Cuthbert in the town of Chester-le-Street (now in County Durham), where a bishopric was established. As raids by both Danes and Scots continued, the monks removed the saint's remains to safety at Ripon in Yorkshire. In 995 they decided to return to Chester-le-Street but, according to legend, when they had stopped for a rest on a hill en route, they found that it was impossible to move St. Cuthbert's coffin. The coffin's intransigence was taken as a message from St. Cuthbert himself that he desired to be buried nearby, at Dunholm. A church was built near the present site of Durham Cathedral to house the new tomb. The site was almost certainly chosen by the monks as being well placed for a new settlement, being defensible, since there was water on three sides and a small piece of land on the north side of the church. Three years later Aldhun, Bishop of Chester-le-Street, ordered the construction of a stone cathedral, and fortifications were also built, including a fortress on the northern, landward end of the peninsula, in time to fend off a Scottish invasion in 1006. The cathedral was completed around 1020 and soon became a repository for treasures, including the remains of the Venerable Bede, the first historian of England.

It was probably after the Norman Conquest that the name of the town changed from Dunholm to Durham. In 1068 Saxons rebelling against the Norman King William I, who had conquered England two years before, seized the town and killed the Earl of Northumberland and 700 soldiers who had been sent to suppress their uprising. While the monks took their treasures (including St. Cuthbert's coffin) off to Lindisfarne, King William re-established order in the region, appointing Walcher of Lorraine as Bishop of Durham while the Saxon Bishop Ethelwin became an outlaw. In 1070 the remains of St. Cuthbert were returned to Durham permanently. In 1072 a new Earl of Northumberland ordered the construction of a castle and of the fortifications around the peninsula, though the only part of the castle still surviving from this date is the Norman Chapel. The castle was one of the main residences of the Bishops of Durham until 1837, which is why the grassed area in front of the castle is known as Palace Green.

Bishop Walcher himself became the next Earl of Northumberland, having paid £400 for the title and the earl's lands, and thus founded what by the thirteenth century was being called the Palatinate. Like the earls of Chester in Northwest England and certain magnates in continental Europe, the prince-bishops, as they were known, exercised a combination of spiritual and political powers that made them sovereigns within their own territory. This special status did not save Walcher from being murdered in 1080 by rioters at Gateshead who were protesting at his involvement in the death of the Saxon nobleman Liulph. His successor, William of St. Calais, another Norman, replaced the Saxon monks of the Priory with Normans in 1084 and began to rebuild the Cathedral in 1093.

The city walls which still surround almost all of the old town were constructed by Ralph Flambard to replace the old fortifications. Flambard was bishop from 1099 to 1128. In 1104 he had the tomb of St. Cuthbert placed behind the high altar, plainly marked as "Cuthbertus." He also saw to the digging of a moat around the castle, the creation of Palace Green, and the founding of the parish church of St. Giles in the northeastern part of the town. He may well have founded the suburb of Framwellgate, to which the Framwellgate Bridge was built in the last year of his episcopate. It was rebuilt in the fourteenth century and is still standing today. The oldest bridge in the city is Elvet Bridge, which also dates from the twelfth century.

Because of its location near the Scottish border, Durham played a crucial role in the conflicts between England and Scotland. In 1136 King Stephen of England whose claim to the throne was being challenged by the Empress

Durham Cathedral
Photo courtesy of Durham City Council

Maud, came to Durham to persuade King David I of Scotland to stop supporting her. In 1139 Durham was the site chosen for a peace conference between the two kings, which resulted in the Treaty of Durham, under which the Scots were given the countryside around the city. Between 1140 and 1144 the bishopric was disputed between a Scots claimant, William Cumin, and the English claimant, William de St. Barbara, who received the support of the monks and the citizens and eventually drove Cumin out.

Hugh Pudsey, who became bishop in 1153, granted a charter to the city of Durham around 1180, creating the corporation that, in various forms, was to govern the citizens on the bishops' behalf for more than 650 years. Having restored the castle and other buildings that Cumin's supporters had destroyed or damaged, Pudsey also completed the Galilee Chapel in the Cathedral, which was to be used by women, who were forbidden to approach the tomb of St. Cuthbert. After his death in 1195, Pudsey's successors frequently quarrelled with the monks of Durham Priory over their respective rights, until Bishop Richard Poore reached an agreement with them, known as Le Convenit, in 1229. The increasing power of the bishops over the monks and the citizens alike was symbolized in 1284 by Bishop Anthony Bek's placing of not one but two thrones in the hall which he had built inside the castle—one to represent his spiritual leadership, the other his political control. Bek took more than 1,600 soldiers with him when he joined King Edward I's invasion of Scotland in 1296, and he was eventually rewarded for his loyalty to the king by being made an honorary king himself, his realm being the Isle of Man in the Irish Sea.

In 1346, while Edward III of England was attacking the city of Calais in Northern France, the French King Philip II persuaded his ally David II of Scotland to invade England. But the Scots were defeated at Neville Cross (now part of the Durham suburbs) by an army raised by Edward's wife Queen Philippa, in which one of the regiments was led by Thomas Hatfield, Bishop of Durham, and Lord Ralph Neville of Raby Castle, south of Durham, who was a member of the most prominent family in the area. Neville was the first layman to be buried in the cathedral, in 1367, and the Neville Screen behind the main altar is named for him.

The history of medieval Durham is not entirely about warrior-bishops and their magnificent buildings. Life for the ordinary people of the city was much less grand and much more dangerous. Many were killed in their homes by floods in 1316; many more died during outbreaks of the Black Death in 1349 and of another plague in 1416, and the thunderstorm of 1429 which caused the collapse of the Cathedral tower must also have caused extensive damage and death in the crowded streets of the poor. Death was not always the result of natural causes. The level of violence in what was still a small town by modern standards is suggested by the entries in the Sanctuary Book of the Cathedral, which show that, for example, between 1464 and 1525, of the 331 people who sought refuge there, 283 were said to have committed murder. Sanctuary was claimed by touching the frightening image of a lion's face holding the knocker on the north door of the Cathedral.

By 1536, when the English Reformation was under way, the prince-bishops' special powers seemed anomalous, since the only other palatinates in the kingdom, Chester and Lancaster, had reverted to the Crown many years earlier and had anyway been held by laypeople. Accordingly, King Henry VIII took away the bishops' judicial powers, though he left their other privileges unaltered. Later that year the "Pilgrimage of Grace," a rising of Northern Catholics against the changes being decided in faraway London was joined by people from Durham who took St. Cuthbert's banner from the Cathedral and marched south behind it. Their rebellion failed, however, and in 1538 the Cathedral's treasures were confiscated, the shrine of St. Cuthbert was dismantled, and the pilgrimages to it, which had been a major source of income for many people in the city, were brought to an end. In 1541 the Priory was converted into a Chapter, a group of priests, led by a Dean, who were placed in charge of the Cathedral administration.

Under the boy-King Edward VI, Bishop Cuthbert Tunstall, who had been appointed in 1530 and remained in office throughout the upheavals of the Reformation, was jailed on charges of treason, and the Duke of Northumberland plotted to take over the Palatinate. His larger plot, to make his daughter-in-law Lady Jane Grey Queen after Edward's death in 1553, was swiftly defeated; the new Catholic Queen Mary I restored Bishop Tunstall, and St. Cuthbert's shrine was reconstructed. But Tunstall remained in London, and five years later, when Elizabeth I became queen and Protestantism was triumphant again, he was removed from office and replaced by James Pilkington.

In 1565 Bishop Pilkington issued a new charter for the city corporation, appointing an alderman and twenty-four burgesses whose decisions were to be subject to his veto. Four years later the bishop and his appointees briefly lost control of the city when another Catholic uprising brought the Latin Mass back to the Cathedral for about ten days. Pilkington was thus forced to attend to the enforcement of religious and political order and could not devote much time or money to the Cathedral buildings, which were in a state of disrepair by the time he died in 1575. Under Bishop Tobias Matthew, a mayor was appointed for the town in a charter of 1602. He and the 24 common councillors of the corporation, chosen by the twelve trade guilds, had a certain degree of autonomy from the bishops, which they and their successors managed to maintain through to the abolition of their posts in 1835.

During the period of the English Civil Wars (between 1642 and 1648) and the Commonwealth (the republic that replaced the monarchy from 1649 to 1660), Durham was under Parliamentary control, following yet another Scottish invasion, in 1640, which had driven Bishop Thomas Morton out of the city. With the bishopric effectively vacant the palatinate was abolished in 1646. In 1650 the cathedral was used as a prison for more than 4,000 Scots captured by Oliver

Cromwell's troops during the Battle of Dunbar. Much of the woodwork inside the cathedral was destroyed when it was used as firewood to keep the men warm during the winter, and the Scots, who had never forgotten their hatred of that family, damaged the Neville tombs in the nave. John Cosin, bishop for twelve years after the Restoration of King Charles II in 1660, received back the palatine rights over the city, as well as the castle, both of which had been bought by the city corporation during the years of the Commonwealth. He did much to restore the Cathedral and castle to their former glory; he also built almshouses (in 1666) and a library (in 1669), both of which now belong to the University of Durham.

Durham experienced several decades of slow but steady growth in the peaceful conditions of the eighteenth century and early nineteenth century, during which many of its older buildings were given architectural facelifts, their Georgian fronts belying their real ages. Then the façade of power was also changed, for the old corporation of Durham was replaced by an elected body under the 1835 Municipal Reform Act, and in 1836, after the death of Bishop William Van Mildert, the last vestigial powers of the Palatinate reverted to the Crown. The word "Palatine" survives now in the County Palatine of Durham, which includes the city in a local government area much smaller than the original territory of the prince-bishops.

Van Mildert's memorial is the University of Durham, which he helped to establish in 1832 and to which he gave the ruins of Durham Castle. Like the medieval universities of Oxford and Cambridge, the University of Durham has no single campus, being organised as a group of colleges. The oldest of these is University College, which has been accommodated inside the Castle since it was restored in 1840. For more than 100 years the University also had departments located in Newcastle upon Tyne, sixteen miles to the north, until they were reorganized as the University of Newcastle in 1963.

The Industrial Revolution had a great effect on Northern England, especially where coal and iron were predominant natural resources, as in County Durham. The coal miners from the villages around the city, who formed the Durham Miners' Association in 1870, have paraded through the city every July, with their brass bands and banners, during the festival known as the Durham Miners' Gala, which has also been a celebration of the nationwide labor movement. A large

black oak plaque inside the Cathedral pays tribute to those many mine workers who died while digging for the mineral that fueled the British economy until very recently. While the mining areas were expanding, Durham itself fell behind other towns in the county, moving from being the fourth largest in 1800 to being the seventh largest in 1900, its population oscillating around 18,500.

Eighty-one years later the population of the city was only around 26,500, and most of its citizens were older and in more prestigious occupations than those of other towns in the county, for Durham had retained its local government and cultural functions even as the industrial economy around it had grown, flourished and then decayed. The city's authorities have managed to keep the construction of modern buildings away from the old town and preserved most of the medieval and Georgian architecture, with the result that Durham is still strikingly unlike any other city in the Northeast of England. One result of the city's unique development is that it is now unable, as a small market town, to compete effectively with the large modern cities of Teesside, Newcastle, and Sunderland, all of which are within twenty miles of Durham. Another result is that the legacy of its prince-bishops, especially the Castle, the Cathedral and the University, continue to attract the visitors (including the university's thousands of students) whom Durham requires to sustain its relative prosperity.

Further Reading: Gaillard Thomas Lapsley's *The County Palatine of Durham* (New York and London: Longman, 1900) is a comprehensive academic study of just what a palatinate was, written in a style that now seems somewhat quaint but is at least unpretentious. Arthur Mee's *The King's England: Durham,* revised by B. Bennyman, (London: Hodder and Stoughton, 1969), covers the whole county and includes a detailed guide to the history and the buildings of the city. Nikolaus Pevsner's *County Durham* (London: Penguin, 1953; revised edition, by Elizabeth Williamson, 1983) is marked, like all the volumes in his "Buildings of England" series, by a combination of fascinating architectural information with vague and unreliable historical passages, all delivered in a tone of assumed infallibility. Margaret Bonney's *Lordship and the Urban Community: Durham and Its Overlords 1250–1540* (London and New York: Cambridge University Press, 1990) offers a wealth of information on the medieval city, its bishops and its priory, presented, like so much contemporary academic historiography, in lifeless prose.

—Patrick Heenan

Echternach (Luxembourg)

Location: In northeastern Luxembourg, on the Saure River, which forms the border between Luxembourg and Germany.

Description: Medieval monastic town and religious center famous for the Benedictine Abbey of Echternach and its basilica, containing the remains of St. Willibrord; the Golden Gospels of Echternach, an illuminated manuscript; and the Church of St. Peter and St. Paul, the oldest Christian sanctuary in Luxembourg. Each spring a dancing procession occurs on Whit Tuesday, the Tuesday after Pentecost.

Site Office: Syndicat d'Initiative
Porte St. Willibrord (Basilica)
Echternach L-6401
Luxembourg
7 22 30

Like its more influential neighbor Trier, Echternach was once home to Roman residents, as evinced by the ruins of a large Roman palace and fortress under the crypt in the Church of St. Peter and St. Paul. A Roman road connected the town to the main Roman highway from Treves to Bitburg and Cologne.

By the late seventh century, Echternach had an almshouse or hospice for itinerant Scottish monks, who had been sent to establish religious communities throughout Western Europe in an attempt to strengthen Christian practices and combat rising pagan influences there. St. Willibrord, a Scottish Benedictine monk who had previously established abbeys at Utrecht and throughout Frisia (Germany), settled in Echternach and established an abbey that became a regional center for his missionary work. The abbey was located on an estate awarded to Willibrord in 698 by St. Irmina and her daughter Plectrude, consort of Pepin II. He died in 739 at the age of 81, and was buried in the crypts below the church.

St. Willibrord provided the impetus for the next two centuries of the abbey's growth and prosperity. Not only did the abbey serve as a center to transfer Anglo-Saxon culture to Frisia, but it also became a center of religious and secular learning. A seminary prepared monks to be missionaries as Willibrord had been. An agricultural school taught peasants crop rotation and other agricultural techniques, some of which remained in use well into the twentieth century. The abbey also operated a school for artists, who produced many illuminated ivory sculptures.

The Springprozession, a dance festival for which Echternach has become famous, was first mentioned in eighth-century records. The procession probably has ties to spring rites in honor of the goddess Diana. One legend of the dance's origin cites a strange illness that afflicted the town's cattle, causing them to wander out into the fields and dance themselves to death. The townspeople formed a procession beginning at the river and danced to the tomb of St. Willibrord; their procession cured the cattle and saved the livelihood of the town. Another legend dates the procession to a time when Willibrord was still living in Echternach. According to the story, a wife of a townsman named Veit died while in the Holy Land on a pilgrimage. Her family accused Veit of her murder and had him arrested and convicted. As his last request on the scaffold, Veit asked for a fiddle. The music he played caused his false accusers to confess their crime and his innocence and caused the townspeople to dance to exhaustion and near death. At dusk, the fiddler Veit disappeared into the hills. Only the arrival of St. Willibrord stopped the dancing and saved the lives of the townspeople. This legend may be tied to lore surrounding St. Vitus's dance, a disease characterized by loss of control of the limbs (Veit, the fiddler's name, is close to Vitus). The dance also resembles other dances meant to improve the health of participants.

The dance begins some time around eight or nine o'clock on the morning of Whit Tuesday and lasts much of the morning. The procession is led by the priests, followed by the children, and then other pilgrims making their way to St. Willibrord's tomb. The dance has a strange pulsating rhythm. The walkers take five steps forward and two or three backward while chanting, "Holy Willibrord, founder of churches, light of the blind, destroyer of idols, pray for us." The procession has ended at the basilica since 1906, when Willibrord's remains were moved there from the Church of St. Peter and St. Paul. The character of the dance has also changed since its origins. The fifteenth-century proclamation of the archbishop of Treves forbade dancing; hence, the dance became more of a slow walk.

With the rise of Holy Roman Empire in the ninth century, the abbey at Echternach underwent organizational and administrative changes. Under the reign of Charlemagne, the abbey prospered and expanded its role as a regional learning center, administered by Benedictine monks. In 847, the Benedictines were replaced by secular abbots and canons, who plundered the cloisters and neglected their educational and ecclesiastical duties. In 950, the abbey was acquired as part of an imperial fief by Sigefrid, a regional prince who had ties to what would become the ruling house of Luxembourg. Sigefrid became the abbey's "advocate," and in 973 he requested reform monks from Trier to replace the secular abbots, thereby restoring the abbey's original Benedictine ties.

It was around this time that work began on the Golden Gospels of Echternach (Codex Aureus Epternacensis), a lavish, illuminated production of the Gospels with a golden, jewel-encrusted cover. This codex was commissioned in

A view of Echternach
Photo courtesy of Luxembourg National Tourist Office

parts by Holy Roman Emperors Otto the Great and Henry III. The illuminated manuscript was produced in Echternach at the abbey, still a center for illumination. The ornaments on the cover were a gift from Emperor Otto III to the abbey. After the production of the Gospels, however, Echternach began to decline in importance as a center of religious art, as styles produced elsewhere came into favor.

The eleventh century not only brought the abbey fame as a center for manuscript illumination, but a new basilica as well, built in front of the tenth-century Church of St. Peter and St. Paul. The construction was prompted by a fire that destroyed the original basilica and cloister in 1017; most rebuilding was completed by 1031, when the shrine was reconsecrated by Archbishop Poppo of Stavelot. Around this time frescoes were painted on the walls of the crypt below the basilica. Gothic additions to the original Romanesque structure were made in the thirteenth century and included a new roof and larger windows. The eighteenth century brought new wealth to the town, and much of the old abbey complex was torn down and rebuilt in the French baroque style.

When Napoléon invaded Luxembourg, centers of religious art and learning such as Echternach were early targets of the revolutionary fervor. In 1794 the French army plundered the basilica and ransacked the crypts, scattering the remains believed to be those of Willibrord. The basilica was initially used as a barracks and stables by the troops; following its sale at auction in 1795, it was transformed into a faience (pottery) factory. Meanwhile, the Benedictine monks fled to St. Peter's in Erfurt, taking with them a wealth of manuscripts, including the Golden Gospels. At Erfurt, the monks acquired the services of Maugérard, a monk involved in the buying and selling of manuscripts. (He later had ties to the Bibliothèque Nationale.) Maugérard arranged for the Golden Gospels to be purchased by Duke Ernst II of Saxe-Gotha-Althenbourg, one of the few German princes who valued medieval art and culture and sought to preserve it. The Gospels were passed down through the family until the dissolution of ducal property before World War I. The current duke established a foundation to preserve the Gospels and other family treasures. Financial losses in World War II forced the foundation to sell the Golden Gospels to the Germanisches National Museum in Nürnberg, Germany.

After the revolution passed, the townspeople placed what was left of Willibrord's scattered dust into a Roman sarcophagus in the Church of St. Peter and St. Paul. They also raised money to restore the basilica. In 1906, Willibrord's remains were moved to the crypt of the basilica itself and placed in a Carrara marble sarcophagus. The buildings of the abbey complex, which, even in the early twentieth century, still comprised a quarter of the town's structures, were put to use housing a wide variety of public and private institutions, such as governmental offices, barracks, a dairy, and a school for girls. The Hospice of St. Willibrord stood well into the twentieth century and continued to function as a rest home for the elderly; it was the second-oldest surviving hospice in Europe, after the Hôtel Dieu in Paris.

Much of Echternach was seriously damaged in World War II, particularly during the German counterattack led by General Carl Rudolf Gerd von Rundstedt in late 1944. The fifteenth-century Guild Hall was practically the only historic building to escape unscathed. Much of the town has been restored, however, including the basilica. In 1950 the baroque buildings of the monastery complex were restored and reopened to house a school for Echternach children.

Further Reading: Little has been written in English solely on Echternach. Readers should therefore consult histories of the Grand Duchy of Luxembourg or church histories. One of the newer English-language histories of Luxembourg is *The Grand Duchy of Luxembourg: The Evolution of Nationhood, 963–1983* by James Newcomer (Lanham, Maryland, and London: University Press of America, 1984). Newcomer draws on such earlier publications as *History of the Grand Duchy of Luxemburg* by A. H. Cooper-Prichard (Luxembourg: Linden, 1950) and *Luxemburg in the Middle Ages* by John Allyne Gade (Leiden: Brill, 1951). Peter Metz's *Golden Gospels of Echternach: Codex Aureus Epternacensis*, translated by I.S. and P. Gorge (New York: Praeger, and London: Thames and Hudson, 1957) offers a fascinating examination of the history of the codex and its representation of medieval Christianity as well as wonderful black-and-white and color plates of the work. Anecdotal information about Echternach and the dancing procession abounds. Beryl Miles's *Attic in Luxembourg* (London: John Murray, and Hollywood-by-the-Sea, Florida: Transatlantic, 1956) provides an eyewitness account of the dance. In *Land of Haunted Castles* by Robert Casey (New York: Century, 1921; London: L. Parsons, 1924) and *Luxembourg, Land of Legends* by W. J. Taylor-Whitehead (New York: Macmillan, 1950; London: Constable, 1951) the myths surrounding the dance are artfully told.

—Jenny L. Presnell

Edinburgh (Lothian, Scotland)

Location: Situated immediately to the south of the Firth of Forth, and occupying some five miles of north-facing slope between the Pentland hills and the Forth estuary, Edinburgh lies on the east coast of Scotland, slightly more than 40 miles west of Berwick-on-Tweed, near the easternmost point of the English border, 40 miles east of Glasgow, and 377 miles northwest of London.

Description: Edinburgh is the capital of Scotland and a center for medical education. At its core is the Old Town, built on top of a cliff of black basalt jutting 250 feet above sea level and, across the valley of the North Loch, the New Town with its well laid-out avenues, crescents, terraces, and streets. Edinburgh Castle overlooks the city from a height of 443 feet above sea level. Edinburgh is the birthplace of writers Sir Walter Scott and Robert Louis Stevenson.

Site Office: Edinburgh and Scotland Information Centre
3 Princes Street
Edinburgh, Lothian
Scotland
(31) 557-1700

Edinburgh's early history has to be reconstructed from a few archaeological finds and much informed academic speculation. The earliest postglacial settlers may have arrived as early as 7000 B.C., and there is some evidence of Stone Age exploration along the Forth. By 1500 B.C., swords were being used, and the area around the crag known as Arthur's Seat was being defended. By 800 B.C., there were hill forts near Edinburgh and south as far as the present border with England. Late in the first millenium B.C., Craiglockhart Hill, Blackford Hill, and the present site of Holyrood were occupied, ringing the present city from southwest to east. In the second half of the second century A.D., the Romans built the Antonine wall between the Clyde and the Forth, their northernmost defense. One of the Romans' major roads terminated at Cramond, within the modern city limits.

Members of the dominant Celtic tribe, the Wotadini, having maintained a stable relationship with the Romans, appear to have moved their settlement to the castle site after the withdrawal of the Romans from Britain around A.D. 500. There are legends of subsequent wars, perhaps carried as far south as Yorkshire. By 854 Din Eidyn (Eidyn's hill fort) was called Edwinesburh (Edwine's stronghold) by Simeon of Durham, who puts it in the diocese of Lindisfarne, and continuous occupation of the castle site is to be assumed, although there is no evidence until the eleventh century. The oldest surviving building in the city, now incorporated into

the castle, is the twelfth-century Chapel of St. Margaret, grandniece of Edward the Confessor and queen of Malcolm III Canmore, son of Duncan I and victor over Macbeth in 1054. After the Scottish invasion by William the Conqueror in 1072, Malcolm made peace with the English king, but continued to raid England repeatedly even after being forced to submit to William Rufus in 1091. David I, in whose reign the chapel was probably built, first declared the castle a royal domain in the twelfth century.

Three of Malcolm's sons reigned from a residence on the castle site, and the upper town rapidly expanded. The Church of St. Giles is believed to have been built about 1110 during the reign of Margaret's son, Alexander I, and Holyrood Abbey was founded in 1128 by his younger brother David I. Parts of the original St. Giles nave still remain, although the surviving arcade is from the early thirteenth century. The church was retained as a parish church after the dissolution of the abbey, but its vault collapsed in 1768. Meanwhile, in 1367 a huge L-shaped keep was built at the southeast corner of the castle. It was later reduced in size and only a fragment still survives, incorporated into the 1573 Half Moon Battery. The castle as it now stands dates primarily from sixteenth and seventeenth centuries, and the Palace of Holyroodhouse, unfinished when James IV was killed in 1513 at Flodden, is, in its present form, largely the work of the seventeenth-century architect William Bruce.

Under David I, Edinburgh had become a burgh in 1130, a status that allowed it to act as a market and as a manufacturing center, particularly for cloth. The city became the site of the first Scottish mint. Holyrood became a burgh of its own, with its own jurisdiction, and was called the Canongate after its resident canons. The crossing known as Netherbow just east of the present intersection of North Bridge and High Street marked the limit between the jurisdictions. The town charter granted by Robert the Bruce in 1329 ensured the growth and prosperity that would eventually make Edinburgh the undisputed capital of Scotland, a position for which it had been rivalled first by Perth and then by Stirling. The transfer of the royal seat from Perth to Edinburgh dated from the assassination of James I in Perth in 1437.

James II was crowned in Holyrood, and lived with the widowed queen in the castle. Eight of his fourteen parliaments and all those of James III were held on the site of the new Tolbooth, next to the Cathedral of St. Giles. Under James II, Edinburgh was described in a royal charter of 1452 as "the principal of our kingdom."

Until as late as 1681, all water had to be pumped from wells south of the Canongate, and surprising number of texts describe how far social life and street design were dictated by the legal right to empty slops into the street after 10

The Scott Monument
Photo courtesy of Edinburgh Tourist Board

o'clock in the evening from what were still mostly unglazed windows. In 1482, the Golden Charter of James III granted the magistrates the hereditary office of sheriff, with power to levy duties on all goods landed at Leith.

James IV did much to enhance the city's standing before his death in 1513 fighting the English at Flodden. He established the supreme court and began the construction of Holyrood palace. During his reign, the city was partly enclosed by walls. Fragments of the wall still stand, forcing expansion into the tall buildings of the narrow, cramped streets at right angles with the present Royal Mile.

It was in the sixteenth century that the principal sites of the Old Town were first developed. The College of Surgeons was established in 1505, and the College of Justice endowed in 1535. In 1582, young James VI granted a charter to encourage the provision of buildings to house the teaching of "humanity, philosophy, theology, medicine, and laws," the source of the future university. The Old Town had spread southward down the hill from the castle into what is now the Grassmarket. The architecture was predominantly small-scale and domestic, although some of the landed families were building their townhouses in the Canongate.

After Flodden a second line of wall was built, protecting the area on which Heriot's School and Greyfriars Church now stand. The Canongate, however, was still exposed. The new fortifications did not prevent the Earl of Hertford from ravaging the city on behalf of Henry VIII in 1544.

The preaching of Protestant reformer John Knox and the relationship of Mary, Queen of Scots with Elizabeth of England kept Edinburgh at the center of Scotland's political and religious affairs during the sixteenth century. The town's fortunes diminished, however, with the departure of James VI to occupy the English throne in 1603. Grass grew even in the High Street. Nonetheless, the cathedral was built, and, although the stonework is much restored, there is a fine Gothic nave with a central tower crowned by a lantern supported by eight flying buttresses, and a steeple. In 1559, John Knox became minister of St. Giles, which was designated a cathedral twice, under Charles I and Charles II. Robert Lorimer's Thistle Chapel was added in 1911.

Parliament House was erected behind the cathedral between 1632 and 1639, and the Scots parliament met there from 1639 until 1707. Next to it is the National Library. Opposite St. Giles are the City Chambers facing the Mercat Cross, the hub of the old city. South of the castle, Heriot's Hospital was begun in 1628. The mercantile sector of the town was clustered around the cathedral. Oliver Cromwell occupied the city in 1650, and in 1745 Prince Charles Edward briefly held court in the Palace of Holyroodhouse.

It was not until the late eighteenth century that Edinburgh regained its elegance, cultural life, and political importance. It became technically possible to drain the north Loch, and, soon thereafter, to bridge the valley. The North Bridge, sixty-eight feet high and 1,128 feet in length, was completed in 1772, allowing the Canongate area to expand

northwards, and in the next fifty years, four more bridges (South Bridge in 1788, Waterloo Bridge in 1820, King's Bridge in 1833 and King George IV Bridge in 1834) were created, giving access to the city from all directions and allowing the creation of the New Town.

In 1767, plans for the New Town as a residential district were approved. The plan set out a grid five streets deep and seven streets wide, with a broad central axis ending in the present grand squares, St. Andrews Square to the east and Charlotte Square to the west. It was completed by 1833. The west portion of Princes Street Gardens, created in the drained loch bed, was once the private amenity of Princes Street owners. It was separated from the public eastern portion by the causeway of rubble now known as The Mound, which resulted from the construction of the New Town. Two monumental neo-Greek buildings stand on the Mound and now house the Royal Scottish Academy, built in 1832, and the National Gallery of Scotland, built in 1859. Both are the work of William Playfair. Nearby, an 1844 neo-Gothic spire supports a statue of Sir Walter Scott.

The railway tracks leading to Waverley Station pass underneath the Mound. Their construction in 1847, brought the inevitable destruction of Princes Street's domestic dwellings and the development of the street into a shopping center. A few eighteenth-century buildings still remain in Princes Street. The Register House, built between 1772 and 1792, at the east end facing the North Bridge, is by James and Robert Adam. In nearby St. Andrews Square, the Royal Bank of Scotland is installed in an eighteenth-century mansion built for Sir Lawrence Dundas. Edinburgh has enshrined its eighteenth-century elegance in the Georgian House, a fully furnished town residence in Charlotte Square preserved by the National Trust for Scotland.

It was largely the economic need for access to England's colonial markets that drove Scotland into the Act of Union with England, signed in Edinburgh in 1707. The Act brought little immediate prosperity to Edinburgh, since it had virtually no goods to offer. However, several years of intense building and rebuilding occurred. Hospitals and schools were established, and the inchoate university was reorganized into faculties. George Heriot's hospital school, begun in 1628 and completed in 1693, was built across the Grassmarket from the Edinburgh Castle. In the 1730's Heriot's standing was challenged by George Watson's Hospital.

The intellectual life of the city flourished during the eighteenth century in what has been called the Edinburgh Enlightenment. From it emerged such thinkers as David Hume and Adam Smith. James Boswell came from Edinburgh, which also nourished the novelist Tobias George Smollett and the poet Robert Burns. *Blackwood's Magazine,* the *Edinburgh Review,* and William Smellie's *Encyclopedia Britannica,* founded in 1771, were all launched in the city and were a product of the intellectual and literary vigor of the city's life.

In 1831, with the New Town all but completed, Telford's Dean Bridge was erected over the Water of Leith at its west end to provide an imposing approach to the New Town houses. The creation of the Old Town caused the relatively swift evacuation of the professional classes from the Old Town to the New, and Edinburgh began quite suddenly to experience a degree of social segregation, all the more damaging because the city establishment had become authoritarian in the wake of the French Revolution.

The development of the New Town dwindled together with Edinburgh's cultural vigor early in the nineteenth century. In 1833, the town went bankrupt. It creditors, forced to accept an issue of 3 percent annuity bonds, lost some 25 percent of their money.

As it stands today, the city's most attractive feature is its Georgian architecture, crowned by the Charlotte Square of Robert Adam. Adam also designed the university's old buildings, which were completed in 1827 by William Playfair, the architect of the 1818 City Observatory.

No doubt more imposing is Edinburgh Castle, spectacularly impressive when seen from below its cliff. It was severely damaged in 1573 when James Douglas, fourth earl of Morton, took it from Mary, Queen of Scots. It is now an historically interesting building housing a museum. The Palace of Holyroodhouse still has the rooms associated with Mary and her second husband, Lord Darnley. It was the site of Mary's interview with Calvinist John Knox and where David Rizzio, her favorite secretary and confidante, was stabbed. It was also here that she married the Earl of Bothwell in 1567. The state apartments are still in occasional use.

The most historically interesting section of the town is the Royal Mile, which descends Castle Hill from the castle Esplanade. Along its route are the Cathedral of St. Giles, the Parliament House and the John Knox house.

It is scarcely possible to depict Edinburgh without mentioning the famous Forth rail bridge some eight miles away, a stunning feat of engineering, built between 1883 and 1890, to carry the railway from London to Aberdeen. The bridge, a steel cantilever structure, is one-and-a-half miles long, with two main spans of 1,700 feet. Just upstream there is now also a suspension road bridge opened in 1964 with a center span of 3,300 feet and two side spans of 1,338 feet.

Edinburgh is fighting hard to retain its position as Scotland's capital against the challenge from Glasgow, the cultural life, economic prosperity, and industrial base of which are increasingly regarded as superior. However, in medicine, Edinburgh remains unchallenged. The city is still the seat of the British government's Scottish administration, of Scotland's principal legal institutions, of the headquarters of the Scottish clearing banks, and of most of the principal Scottish providers of financial services.

Further Reading: There are numerous guides and encyclopedia articles devoted wholly or partly to Edinburgh, but there are no modern and reliable general histories of the city. The Scottish Development Department's *Buildings of Special Architectural or Historic Interest: City of Edinburgh* (Edinburgh, 1971), is excellent, but for specialists, and Alexander John Youngson's *The Making of Classical Edinburgh 1750–1840* (Edinburgh: Edinburgh University Press, 1966) is also formally academic. There is a profusion of memoirs, guided tours, and histories of particular institutions, buildings, and aspects of Edinburgh life, but the most generally recommendable works must include Ian Nimmo's *Portrait of Edinburgh* (London: Hale, 1975), concentrating on the city's way of life; George Scott-Moncrieff's illustrated *Edinburgh* (Edinburgh: Oliver and Boyd, 1965); the four-volume *Historic South Edinburgh* by Charles J. Smith (London: Skilton, 1978–88); and the very informative *Edinburgh Past and Present* by Maurice Lindsay and David Bruce (London: Hale, 1990), containing many ancient and modern photographs taken from the same spot. The best all-round introduction to the city's life is Allan Massie's *Edinburgh* (Edinburgh: Mainstream, 1994).

—Claudia Levi

Eger (Heves, Hungary)

Location: Northeastern Hungary; situated in the Eger River valley between the Mátra and the Bükk Mountains.

Description: One of the first Magyar settlements in eastern Europe and later a flourishing medieval town; location of several sieges by the Turks; site of archiepiscopal see and many ecclesiastic institutions; seat of ecclesiastic learning in the eighteenth and nineteenth centuries; known today for its vineyards.

Site Office: Tourinform
Dobó tér 2
3300 Eger, Heves
Hungary
36-321-807

The Eger River valley is thought to have been occupied as early as the Neolithic Era. The Magyars, a group of Finno-Urgic tribes from central Russia, settled in the region around the ninth century. The town itself first rose to prominence as one of the five bishoprics created by Stephen the Great, first king of Hungary, following his coronation by Pope Sylvester II on Christmas Day, 1000. In exchange for the pope's recognition of his preeminence among the Hungarian Magyar chieftains, Stephen agreed to Christianize the Magyar and Slavic population, donate land to the church, establish bishoprics and cathedrals, and collect tithes for their upkeep.

Stephen allowed Benedictine monks to enter Hungary, and it was they who built the monastery five miles north of Eger, at the foot of Bélkö Hill in the village of Bélapátfalva. It is the only monument of Cistercian Romanesque architecture still standing in Hungary. The monastery was built in 1232, when the bishop of Eger, Kilit II, moved the French Order of Cistercians to this site from their abbey in Belharomkut. The structure is a basilica, with a nave, two aisles, and transepts forming a Latin cross. Below a detailed rosette, semicircular arches span the main entrance. The exterior of the church and its doors and windows are Romanesque, but the vaulted ceiling and arches of the aisles, renovated in the thirteenth century, are Gothic.

By the end of the twelfth century, Hungary was among southeastern Europe's leading powers and traded gold, silver, copper, and iron with western Europe. Eger, along with the rest of Hungary, continued to develop a feudal economy with the crown wielding absolute power. The crown's power began to diminish in the thirteenth century, however; in 1222 the Golden Bull was signed, a Hungarian document comparable to the Magna Carta, enumerating the rights of the nobility. Their new authority led to frequent conflicts among the nobles, and by 1241 their loyalties were so badly divided that Hungary's king Béla IV barely commanded enough forces to meet the Mongols as they swept through Hungary. With little or nothing to stop them, the Mongols razed Hungary, burning to the ground towns and villages and slaughtering their inhabitants. Half of Hungary's population died during the invasion. Eger was not spared; few of its buildings survived.

During the years of reconstruction that followed, immigrants from Germany and Italy helped repopulate the country. Béla IV distributed crown monies and lands among his nobles to redevelop Hungary's towns, stipulating that the nobles build and fortify castles. Because of the inability of Hungary's towns to defend themselves against the Mongols, careful thought was given to new strategies of defense. With the exception of royal residences such as Esztergom and Székesfehérvár, the country's castles had been earthen structures fortified by stockades. The new garrisons, among them Eger Castle, were stone-and-mortar constructions, and it was hoped these would better thwart enemy attacks. In Eger's case, this degree of planning was prophetic; Eger owes its historical significance to the role Eger Castle played in a number of battles.

To ensure the castles would be adequately manned, Béla offered shelter to those peasants living outside the castle towns in exchange for their help with the castles' defense. With feudal lords governing the garrisons and, consequently, wielding the power that those garrisons symbolized, the king's role became less and less significant.

The House of Árpád died out in 1301 with the death of Andrew III, and a succession of foreign kings came to the Hungarian throne. The country's economy steadily developed around foreign trade and gold mining, and Hungary enjoyed a golden period during the reign of Matthias Corvinus. Renaissance art and culture flourished in the second half of the fifteenth century, and Eger was one of Hungary's main centers of culture. Although Matthias also established a university and a printing press, his lasting contribution to the arts was the Bibliotheca Corviniana. At his death, Matthias's library had amassed over 2,000 manuscripts and printed volumes. Only a fraction of the library's vast holdings survive today, but these are treasured by libraries the world over.

During the sixteenth century, in the war for decisive control of Europe, Hungary served as the battleground between the Austrian Habsburgs and the Turks. Hungary's people and lands suffered, exploited by both sides. Fearing capture by Asian slave traders, most of Hungary's population fled remote villages for the relative safety of larger, fortified settlements, such as Eger.

One of the bright spots in this period came in 1552, when the 2,000 inhabitants of Eger defended the castle

Turkish Minaret in Eger
Photo courtesy of IBUSZ Travel Inc.

against a siege by a Turkish army of 150,000. Outnumbered but tenacious, the citizens of Eger fought the Turkish forces for five weeks until, remarkably, the Turks retreated. The castle's captain, István Dobó, distracted the enemy's rear guard and side flanks with raiding parties that slipped outside of the fortress through a maze of underground tunnels. Meanwhile, men and women poured boiling water and tar on enemy soldiers trying to scale the bastion walls. News of the

victory at Eger spread across Hungary and Europe, making heroes of Dobó and the residents of Eger and proving the Turks were not invincible. Géza Gárdonyi's famous novel, *Egri csillagok* (*The Stars of Eger*), details the courageous battle that prevented the northeastern section of Hungary from coming under the rule of the Ottoman Empire.

Fifty years later, another battle in Eger yielded different results. Not trusting Hungarian forces, the Habsburgs had stationed non-native mercenary soldiers in the castle. Turkish forces under the leadership of Sultan Mehmed III surrounded Eger Castle and promised the mercenaries trapped inside safe retreat if they surrendered. The garrison capitulated, and what had been one of the key fortresses safeguarding northern Hungary fell to the Ottoman Empire. Eger was occupied by the Turks from 1596 until 1687.

During its capture and occupation, the town of Eger was destroyed by the Turks and established as a Turkish province. Several structures from the Turkish occupation remain. Today, the Hospital Church, Kórháztemplom, occupies the site of a Turkish mosque that was demolished in 1841. Nearby rises the slender spire of the former mosque's minaret; forty meters high, the fourteen-sided landmark contains a spiral staircase leading to the roof balcony from where the muezzin called Moslems to prayer. Built at the turn of the seventeenth century, the minaret is the northernmost Islamic monument still standing. The Turks also made famous the natural warm springs in Eger. Although bathing facilities there date to the fifteenth century, it was only during the Turkish occupation that a large number of bath houses were built and bathing activity began to flourish at Eger. Today, the baths district of the city includes thermal and medicinal spas; a restoration of the Turkish bath houses began between World Wars I and II.

In 1683 Emperor Leopold I established the Holy League of Austria, Poland, and Venice and undertook a great campaign to oust the Turks, a campaign bolstered by a huge number of Hungarian soldiers. These allied forces reclaimed Buda in 1686 and Transylvania in the following year. Gradually, the Turks were expelled from Hungary. The allied forces blockaded Eger Castle in the second half of 1687 and forced its Turkish occupants, soldiers and civilians, to withdraw; 600 Turkish citizens remained, converted to Christianity, and assimilated themselves into the culture. In 1702, Eger Castle shared the fate of a number of other Hungarian castles when Emperor Leopold I ordered that it be blown up. The Habsburgs feared that these fortresses might be used as strongholds for Hungarian factions battling for independence from Vienna. In the case of Eger Castle, the emperor's primary objective was the demolition of the casemate network that István Dobó had used to such great advantage during the victory in 1552. Several stories deep, lying one on top of another, the casemates consisted of barrack and cannon chambers and shaft observation corridors. Recent excavation has discovered an underground well, 100 feet deep, that provided a reliable water source for the castle's defenders during enemy attacks. The casemates' strategic significance was great; the miles of underground tunnels allowed the castle's troops to move undetected, open sniper fire on their attackers, or surprise the enemy by suddenly appearing from unlikely positions. The casemate system was successfully buried by the Habsburgs' explosions, but in recent years nearly all of the passageways have been cleared and much has been excavated, including gun barrels and other ammunition from the bottom of the well, thrown there by those last to surren-

der. The St. John's Cathedral was also destroyed by the Habsburgs; only the foundations of this Romanesque and Gothic cathedral remain. Today The Castle Museum, located in the former Bishop's Palace, contains a collection of documents, old weapons, and other relics, as well as István Dobó's tombstone.

Despite the Habsburgs' destruction of its castle, Eger served as one of the central battlegrounds for the Rákóczi War of Independence, a huge peasant uprising from 1703 until 1711, when their commander, Sándor Károlyi, surrendered.

For Eger the eighteenth century was a period of reconstruction. The town developed rapidly and in direct correspondence to its growing trade along the Danube with Austria and the increasing number of craftsmen residing there. In 1777 there were more than 1,000 craftsmen in Eger.

The wealthy bishopric in Eger was made into an archbishopric in 1804, and a number of ecclesiastic institutions were established there. Stylistic trends of the eighteenth century, namely baroque, dominated most of the town's buildings. Local craftsmen gave such a unique stamp to their handiwork that a kind of "Eger baroque" developed, which differs strikingly from the baroque creations of other Hungarian cities. Perhaps the finest example of "Eger Baroque" is the Minorite Church, built between 1758 and 1783 by local craftsmen to the plans of a Prague architect. The altarpiece depicting the Virgin Mary and St. Anthony is the work of local painter Johann Lucas Kracker. Kracker also painted a fresco in the library of another outstanding work in the baroque style: the Lyceum. The library's collection contains over 700 manuscripts and 80,000 volumes, including 87 incunabula (books printed prior to 1501). The Lyceum was commissioned as a university in the second half of the eighteenth century by the bishop of Eger, Count Károly Esterházy, but the Habsburgs opposed the establishment of a university in Eger, and it became an archiepiscopal secondary school for girls instead. The complex is among the many buildings in Eger done in the Louis XVI style by Jakab Fellner, and its large inner courtyard is thought to be one of Hungary's finest enclosed architectural spaces. Of special interest is the eighteenth-century observatory built by Henrik Fazola, a master ironsmith, to the specifications of the renowned astronomer, Miksa Hell. Fazola also created many of the spectacular wrought-iron gates and other ornamentation for castles and palaces throughout Eger and Hungary. One of his masterpieces is the spectacular baroque wrought-iron gate on Eger's county hall.

The U-shaped, baroque structures that comprise the archiepiscopal palace were built in stages during the eighteenth century by several architects, including Jakab Fellner. After the Cathedral of Esztergom, Eger's neoclassical cathedral is Hungary's largest ecclesiastical structure, measuring 300 feet long and 173 feet wide. Designed by József Hild, the cathedral was built between 1831 and 1839 on a site once occupied by a medieval church. Statues of Hungary's eleventh-century kings, St. Stephen and St. László, flank the bottom of the stairs at the entrance and are the work of the Italian sculptor Marco Casagrande. From the sacristy a staircase leads up to the cathedral's treasury, where many examples of religious art are displayed, among them a fifteenth-century chalice with enamel and filigree decoration, an eighteenth-century, cross-shaped reliquary in silver-gilt, an eighteenth-century ciborium with a coat of arms, and a pair of eighteenth-century altar cruets with their golden tray.

The wealth generated by the archbishopric continued to feed the economy into the nineteenth century. Wine production had been one of the city's leading industries since the thirteenth century. The locally produced Egri Bikaver, or "Bull's Blood," named for its deep red-black color and full body, continues to be internationally famous. Heavy industry would not arrive in Eger until after World War II. Under Communist authority, there developed a gear factory, sheet-iron works, technical industry factory, and furniture factory. The mid-twentieth century also saw a large number of modern buildings erected in the city center and the outskirts of town.

Further Reading: *A History of Hungary* by Peter F. Sugar (Bloomington: Indiana University Press, and London: Tauris, 1990) is an excellent and articulate study of the nation's history, as is Ernst C. Helmreich's *Hungary* (New York: Praeger, 1957). Specific attention is focused on the different regions of Hungary in Zoltán Halász's *Hungary,* translated by Zsuzsa Béres and J.E. Sollosy (Budapest: Corvina Kiadó, 1980), including the most detailed description of Eger's famous siege in 1552. *Hungary: A Comprehensive Guide,* edited by Gyula Németh, translated by Charles Carlson and George Maddocks (Budapest: Corvina Kiadó, 1981) and *Guide to Ten Hungarian Towns* by István Wellner, translated by Katalin Pogány and Brian Mclean (Györ: Panorama, 1986) provide detailed maps and descriptions of the ruins and sites in Eger.

—Elizabeth Taggart

Eichstätt (Bavaria, Germany)

Location: On the Altmühl River in West Bavaria, about sixty-seven miles north-northwest of Munich.

Description: Eichstätt was founded as a bishopric in 741 by St. Boniface. It is the site of numerous historic structures and is surrounded by the 1,875-square-mile Altmühl Nature Park. In or near the city are several museums with fossils from the Jurassic Era; more than 300 caves, some of which have yielded prehistoric artifacts; ruins of Roman fortifications, dating back to the first century A.D.; and more than 60 castles or ruins.

Site Office: Verkehrsbüro
Kardinal-Preysing-Platz 14
85072 Eichstätt, Bavaria
Germany
(84 21) 79 77

Eichstätt derives its reputation as a "city of clerics" from the fact that it was founded as a bishopric in the mid-eighth century and later ruled by prince-bishops for nearly 900 years. Although the tradition of prince-bishops existed throughout Europe in the Middle Ages, it is most strongly associated with Germany, and is exemplified by Eichstätt. As the name implies, prince-bishops acted not only as the spiritual leaders of their sees, but also were noblemen answerable, in many cases, directly to the king. Thus, they often became rivals in spiritual, economic, and political arenas. One offshoot of this competition was the building of beautiful cathedrals, decorated with the finest sculptures and paintings. Eichstätt, though never a major German city, is known for its well-preserved Romanesque, Gothic, and baroque architecture built by its prince-bishops during the Middle Ages.

Although Eichstätt began as a Roman station in the first century B.C., evidence shows people had lived in the area since prehistoric times. The Altmühl Valley National Park, which surrounds Eichstätt, contains more than 300 caves, many of which have provided archaeologists with evidence of Neanderthal man, and of old-stone-age, bronze- and iron-age settlements.

The Romans, however, provided the first recorded history of the area that now is Eichstätt. In 15 B.C., Caesar Augustus sent an army to the Alpine area of Germany, where it founded the province of Raetia. When the Roman troops marched through Gaul, they found seminomadic tribes of hunters and herdsmen. Though primitive, these early Germans could be fierce and cunning adversaries, as the Romans learned when they tried in vain to vanquish them.

The most famous example of this fierceness occurred in A.D. 9. Hermann (or Armin), chief of the Cheruschi tribe, lured a Roman field army into a trap. In the battle of the Teutoburg Forest, near what was to become the border of Prussia, the Germans annihilated 20,000 Roman troops. In spite of the fact that the Romans later defeated Hermann, they made few further attempts to conquer territory in eastern Germany.

Having thus been discouraged from eastward expansion, Rome secured its western territories using the Rhine and Danube as natural borders against the German invaders. The upper reaches of these two rivers, however, had no such natural protection. In this area, now northern Switzerland and the Alpine area of Germany, including Bavaria where Eichstätt is located, the Romans built a 300-mile-long ditch-and-bank fortification called the Raetian Limes, or simply the Limes. The Limes connected the Rhine, south of what is now Bonn, to the Danube, southwest of Regensburg.

Today, remnants of the Roman conquest can still be found near Eichstätt, including ruins of the Limes and a Roman road, running just north of the town, from Preith toward Fort Biriciana, and across much of the region covered by the national park. While most of this ancient fortification has collapsed, some portions still stand, and others have been restored, along with gates, towers, and forts. Roman estates behind the Limes served as supply stations for Roman troops for the first two centuries A.D., and numerous ruins of these estates still exist.

When the Roman Empire began to crumble in the early fifth century, tribes such as the Visigoths, Vandals, and Huns overran and settled the frontiers of Germany, including the areas around Eichstätt. By the late fifth century, many of these tribes had settled into communities or formed small kingdoms under Frankish rule. Most of the population consisted of serfs and free peasants, dependent on landed nobles. By the seventh century they had adopted many Roman customs, such as law, military organization, and agriculture, including the cultivation of grapes to make Rhenish and Mosel wines.

Although many of these tribes were adopting Christianity as well, others clung to their native faiths. In the mid-eighth century, St. Boniface, an English missionary, traveled to Germany to convert the pagans. He subsequently founded numerous bishoprics, including the episcopal seat in the Nordgau at Eichstätt. He named St. Willibald the first bishop of Eichstätt in 741. In 761 St. Willibald's sister, St. Walburga, became abbess at the double monastery at Heidenheim, where she promoted the education of German women and became known as a patroness against hunger and plague. St. Walburga's remains are enshrined at the church of the former Benedictine abbey in Eichstätt. The Gothic-style shrine was added to the baroque abbey by architect Martin Barbieri and, in keeping with Benedictine tradition, it is still visited every year by pilgrims.

The statue of St. Willibald at Eichstätt
Photo courtesy of Informationszentrum Naturpark Altmuhtal

Charlemagne became the first Frankish king of the Carolingian dynasty in 768. He quickly conquered most of western Europe, from the Atlantic to the Adriatic, and from the North Sea to the Pyrenees, including what is now Bavaria and Austria. Part of Charlemagne's effort to increase trade in his new empire included plans for a canal between the Altmühl and Danube Rivers, which would have improved river traffic for Eichstätt and other towns on the waterways. The project, begun in the autumn of 793, was called Fossa Carolina, or Canal of Charles. Weather and geological obstacles defeated Charlemagne's efforts, however, and construction was halted. Remnants of the project still exist.

Charlemagne made his most lasting mark on Eichstätt through his attempts to reestablish the Roman Empire and link papal authority and political rule. Charlemagne continued the work of St. Boniface in strengthening the authority of the church, but, in turn, he also tightened his own grip on the church's activities. He maintained control of episcopal elections and threw enormous financial support to the church by way of monetary endowments and large fiefs. Some of these fiefs included manors, making the bishops of those sees among the wealthiest landholders in Germany. In the tenth century all the bishops were elevated to the rank of noblemen. By 908, when Eichstätt was chartered, the concept of prince-bishops had been established, and the town and diocese would remain under this form of rule until 1802, when Eichstätt was secularized.

Wielding both secular and religious authority, prince-bishops could mobilize considerable resources as patrons of education and the arts. Their accomplishments are particularly evident in the magnificent cathedrals, churches, and monasteries in prince-bishoprics such as Eichstätt. The town's cathedral, for instance, was founded at the time St. Willibald became bishop in the mid-eighth century, but alterations and additions were made to the structure for the better part of a millennium. The history of this work can be traced in the cathedral's varied architectural styles: its two spires are Romanesque; its nave, gallery, and mortuarium are Gothic; the west facade, built in gratitude for the church having been saved from destruction during the Thirty Years War, was created by Eichstätt architect Gabriel de Gabrieli in the baroque style. The cathedral contains such artistic masterpieces as the Pappenheim Altar, mortuarium windows by Hans Holbein the Elder, and sculptures by Renaissance artist Loy Hering.

Other famous structures include the Residenzplatz (Residence Square), built for the clerics by architects Gabrieli, Jakob Engel and Maurizio Pedetti; another Gabrieli project, the Hofgarten (Court Gardens), the former summer residence of the clerics, which now contains the administration offices of the Catholic University; Willibaldsburg, a castle overlooking the city, which was the seat of prince-bishops; the church of the former Benedictine abbey, with the tomb of St. Walburga; and the Baroque Capuchin Church, founded in 1189, which contains a reproduction of the Holy Sepulchre of Jerusalem.

Ironically, the opulent splendor of cathedrals and ecclesiastic holdings such as these contributed to the groundswell that shook the church to its very foundations: the Reformation. Even predominantly Catholic enclaves like Bavaria were threatened. In Eichstätt, a see still ruled by a prince-bishop, the town and aristocracy sided with the Protestants. Although Germany would remain more than 50 percent Protestant, the Catholic Counter-Reformation, beginning around 1560, effectively restored much that was lost. Through a decree from Rome reaffirming Catholic absolutism, reforms in the church, and heavy-handed treatment of Protestants by Albert V, ruler of Bavaria, Catholicism was reestablished in Eichstätt.

By the end of the seventeenth century, rule by prince-bishops in Eichstätt had lasted nearly 900 years, surviving not only the Reformation, but plague, wars, and political intrigues. The prince-bishop of Eichstätt had even outlasted secularization of church property in the late eighteenth century by Emperor Joseph II.

Eichstätt's centuries-old tradition of prince-bishop rule, however, did not outlast Napoléon. During his rise to power in the early eighteenth century, he reorganized Germany to make the larger secular states dependent on France and to isolate Austria. To gain cooperation from German princes for his reorganization, he promised them ecclesiastical holdings. In 1802, Germany was completely secularized, with only parts of the Eichstätt see remaining in Bavaria. Four years later, Napoléon effectively dissolved the already moribund Holy Roman Empire and formed most of the German states into the Confederation of the Rhine. By 1809, his troops occupied Austria.

After Napoléon's defeat and the subsequent Congress of Vienna in 1815, the German states formed the Germanic Confederation, a loose collection of states based on Napoléon's reorganization. This organization, coupled with reactionary forces opposing stronger unification, resulted in tougher monarchical control throughout Germany. In 1817, the king of Bavaria gave Eichstätt, with the landgraviate of Leuchtenberg, to his son-in-law, Eugène de Beauharnais.

Monarchical rule in the early and mid-nineteenth century came under pressure from rebellious Germans, inspired by the revolution in France. Many pushed for a new constitution and a unified Germany. In 1821, the see in Eichstätt was reestablished. By mid-century, after Eichstätt had returned to the political jurisdiction of Bavaria, Germany had begun to evolve from an agrarian country to a center of industry. The nation's first rail line was completed from Nürnberg to Fürth in 1835, and by 1870, a line ran along the Altmühl between Eichstätt and Gunzenhausen. Besides promoting rapid industrialization, the rail lines also helped unify the country.

Today, Eichstätt's businesses include textiles, paper manufacturing, metalworking, and printing. It has a Catholic university and hosts thousands of religious pilgrims and tourists who come to visit the city of prince-bishops.

Further Reading: *A Concise History of Germany* by Constantine Fitzgibbon (London: Thames and Hudson, 1972; New York: Viking, 1973) offers an interesting, but personal, interpretation of the German psyche and the historical events that shaped it from antiquity through the early 1970s. Mary Fulbrook's book, also entitled *A Concise History of Germany* (Cambridge and New York: Cambridge University Press, 1990) ties together a broad range of complex issues and material, while following the main lines of German historical development. Fulbrook's book is more subjective than Fitzgibbon's, and reaches from antiquity through the unification of Germany in 1990.

—Richard G. Wilkins

Eisenach (Thuringia, Germany)

Location: In the heavily forested state of Thuringia in eastern Germany; 30 miles west of Erfurt, 150 miles southwest of Berlin; at the confluence of the Nesse and Hörsel Rivers.

Description: City dating to the eleventh century; site of one of the most famous castles in Germany, the Wartburg, situated on a hill 600 feet above the town; as the seat of the medieval landgravate of Thuringia, famous for its court poets, immortalized in Wagner's opera *Tannhäuser,* and for the charitable works of Landgravin Elizabeth; refuge for Martin Luther and place where he translated the New Testament into German; traditionally considered the birthplace of Johann Sebastian Bach; transformed into a symbol of German nationalism, especially after the university student rally of 1817.

Site Office: Eisenach-Information
Bahnhofstr 3
99817 Eisenach, Thuringia
Germany
(36 91) 7 61 62

Nestled at the foot of the Thuringian Mountains and surrounded by dense pine forests, tiny Eisenach has had an exceptional impact on Germany's spiritual and cultural history. The exact date of Eisenach's founding is unknown, but the town likely developed during the construction of Wartburg Castle on a hill to the south. As indicated by carbon dating of the building's oldest remains, construction most likely began in 1086. Hundreds of workers were needed for the project, which undoubtedly took decades to complete. Eisenach came into existence just below the castle, along an important trade route at the confluence of the Hörsel and Nesse Rivers.

Upon its completion, the Wartburg towered above the surrounding Thuringian forest. The fortress and its outbuildings were enclosed by formidable walls, from which knights surveyed the neighborhood beyond. The castle had several huge, drafty rooms: the Rittersaal, or knights' chamber; the dining hall; the landgrave's (lord's) chamber; and a chamber for his consort. Most famous of all, however, was the singers' hall, where the era's wandering minstrels performed.

Late in the eleventh century, the Wartburg became the property of a local lord, Ludwig der Springer (the Jumper). In 1131 the Holy Roman Emperor elevated the tiny state of Thuringia to an independent landgravate, and Ludwig's heir, also named Ludwig, became its first landgrave. It was during the reign of Landgrave Hermann I (1190–1271) that the Wartburg would achieve true fame for its rich cultural life.

Hermann's father, Ludwig III, wanted his sons to be educated and cultured. Accordingly, he sent them to Paris, where they studied at the relatively new University of Paris, the Sorbonne. Late medieval European culture was at its height, and Hermann fell quickly under its influence. Upon his succession to the landgraviate in 1190, he threw open the Wartburg's gates to the poets and musicians of the day—the wandering minstrels, or Minnesänger. Thanks to his patronage, the most famous poets in German medieval history, including Wolfram von Eschenbach, Heinrich von Ofterdingen, and the most famous of all, Walther von der Vogelweide, found a welcoming refuge at the Wartburg. Their songs and love poems, with their gallant knights and lovely ladies, established the Wartburg as the easternmost outpost of high medieval culture.

Six centuries later, composer Richard Wagner encountered the beauty of the Wartburg's dramatic setting and heard the ancient legends surrounding the fortress. The most famous legend, put in writing in the thirteenth century by poet and minstrel Wolfram von Eschenbach, had as its theme a *Sängerkrieg,* or minstrel competition, that allegedly took place at the Wartburg at the time of Hermann's reign. How much of Eschenbach's ballad is fact can never be ascertained, but Wagner seized on this tale and made it the centerpiece of his epic opera, *Tannhäuser.* Today, a portion of the museum dedicated to Eisenach author Fritz Reuter, himself a Wagner enthusiast, has been transformed into a small Wagner museum, displaying Reuter's huge collection of Wagneriana.

The fame of the Wartburg's Minnesänger was soon rivaled by the renown of another of its residents: the saintly landgravin Elizabeth of Hungary. Having married Landgrave Ludwig IV in 1221, when she was at most fifteen years old, Elizabeth and her ladies in waiting went daily into Eisenach on errands of mercy to the town's poorest. Moved by the plight of indigent men and women too sick to care for themselves, she established a hospital for them, the remains of which can still be seen at the foot of the Wartburg. She often tended the sick personally and performed other acts of mercy. The wealthy landgrave tolerated her generosity, even her relinquishing of personal jewelry and expensive clothing. But the local nobility was incensed and decried her extravagance. Concern for his reputation motivated Ludwig to forbid his wife's activities, especially her daily trips with baskets of provisions for Eisenach's poor. Catching her by surprise one day as she was making her way down the Wartburg, he demanded to know what she was carrying in her large covered basket. She stammered "roses," and when he uncovered the basket, there were indeed beautiful roses in it. Upon her husband's death, Elizabeth was compelled to leave the Wartburg. A mere four years after her untimely death in 1231, Pope Gregory IX canonized her. Elizabeth left an in-

The Wartburg
Photo courtesy of German Information Center

delible mark upon the Wartburg, despite her short life and the castle's long history since then. In the nineteenth century, when the Wartburg underwent a major restoration that transformed it into a romanticized version of a medieval castle, Eisenach native Moritz von Schwind painted a series of sentimental frescoes, still on display today, depicting stages in her life.

After Elizabeth's departure, the fortress declined as a cultural mecca. But by then high medieval culture, in the form of beautiful church architecture, had disseminated widely in the region. Eisenach already was full of impressive churches and monasteries, including the late twelfth-century St. George's Church, with its dozens of altars, and the thirteenth-century St. Nicholas Church, today home of the town's museum of medieval art. There were also the imposing monasteries of the Dominican, Franciscan, and Carthusian religious orders. By 1283, Eisenach had received a city charter and supported a number of schools.

The Wartburg continued in the possession of the local rulers, but they lived there less and less often. It was a manned fortress when Martin Luther came to Eisenach as a teenager to continue his education. He arrived in the small medieval city of 4,000 inhabitants in either 1498 or 1499, and lived there for three years while he attended the Latin school. Corporal punishment was the rule in schools in those days, and was applied often enough to give Luther painful memories late into adult life. But these experiences did not affect his positive memories of the town, which became one of his favorites. The house in which he lived as a student, called simply the Lutherhaus, has been preserved and contains a small museum. Later, when critics accused Luther of being a foreigner and introducing foreign ideas into Germany, he retorted that he had lived and been educated in Eisenach, understood to be one of the most German of cities.

Luther's education prepared him to enter the university at Erfurt, thirty miles away. His return to Eisenach twenty years later was far more dramatic than his departure had been. By late 1521, he was a notorious and wanted criminal, forced to flee to safety after having refused to recant his beliefs before Holy Roman Emperor Charles V in the city of Worms. The emperor declared him an outlaw, which meant anyone could seize him and deliver him up for execution.

Nonetheless, Luther survived, in large part due to his official protector, Frederick III, the elector of Saxony. Frederick secretly sided with Luther's doctrines and intended to defy the emperor. Consequently, he ordered that Luther be "abducted" after he fled Worms, and be brought in secret to the Wartburg, where he could remain in hiding until he was safe (he was expected to grow a beard and shed his monk's clothing for the attire of a local knight). Luther agreed to the plan and soon found himself in the castle. The Wartburg was barely inhabited but quite livable, and Luther relished the opportunity for intellectual work that seclusion offered him. He threw all of his efforts into his forbidden work: a German translation of the New Testament. His room in the Wartburg

is today open to the public, although nothing original remains in it.

Luther stayed a year in the Wartburg. Meanwhile, word had gotten to his friends in Wittenberg of his whereabouts, and he took the risk several times of visiting them clandestinely. According to legend, he also made several secret visits into the city of Eisenach where, in disguise, he delighted in questioning the citizens about the "heretic" Luther and eliciting their views. He was always moved by their positive support of his doctrines, or what they understood of them. Eisenach would become one of the most thoroughly Lutheran of German cities.

In part because of Luther's doctrines, some of which carried a radical social message, the peasantry throughout much of eastern Germany rose up in rebellion against their lords, igniting the Peasants' War in 1524. Their defeat the following year was inevitable. Seventeen local rebels were brought to the marketplace in Eisenach, where they were executed. A simple cross in the pavement still marks where their executions took place.

Perhaps because of the seeds planted in the heyday of the Minnesänger, or because of the Protestant emphasis on hymn singing, the town of Eisenach again became renowned for its music. As long as it remained the capital of Thuringia, the landgrave maintained a court orchestra. Visitors observed that music was played everywhere in the town. One family in particular, the Bachs, served as court musicians in Eisenach. For this reason, it is presumed that famed composer Johann Sebastian Bach was born in the family home there in 1685, when his father was a court musician. There is, however, no record to confirm whether the house that is now the Bachhaus museum was in fact the family home and, consequently, the birthplace of the composer. Nevertheless, in 1906 the Bach Society, which purchased the house, refused to entertain the possibility that the composer might have been born elsewhere in Eisenach. The Bach festival in Eisenach did not get under way until 1950; it takes place annually on the composer's birthday, March 21.

In 1741 Thuringia was absorbed into its larger neighbor, the state of Saxony-Weimar, and Eisenach ceased to be a state capital. The Wartburg, meanwhile, deteriorated, soon becoming a ruin of its former self. But as the castle physically declined, its symbolic value as a cultural treasure and monument to Germany's past increased. Germany's most famous writer, Johann von Goethe, visited it in the late eighteenth century and was so taken by its romantic locale and its ancient history that he made numerous sketches of it. His carefully preserved drawings are among the few records of the fortress's appearance in the late eighteenth century, before it was renovated and altered in the nineteenth.

The Wartburg's symbolic value increased still further with the rise of modern nationalism. Napoléon Bonaparte's conquest of Germany and his consolidation of the German states into a confederation gave birth to widespread national feeling. Eisenach had been occupied, vandalized, and set afire by French troops. While the town itself had no university, in 1817 members of the nationalist student organizations known as Burschenschaften gathered from eleven of the German states at the one monument that symbolized German unity—the Wartburg—ostensibly to commemorate the 300th anniversary of the Reformation. Wittenberg might have been a more appropriate place for this festival, but it lacked a powerful symbol. The festival of hundreds of students, visitors, and Eisenach residents lasted two days at the Wartburg, and concluded one evening with a massive book burning of works considered antinational in character or content.

Few events catapulted the ancient fortress into national prominence as much as this festival, which focused attention on the woeful state of the Wartburg and its dire need for repair. Funds were finally obtained and the work commenced in 1838; it would last until 1890. Unfortunately, the project included a great deal of renovation that was not historically accurate. Rather than returning the fortress to its original appearance, the project's planners sought to make it conform to romantic, nineteenth-century notions of what a medieval castle should look like. A moat with a drawbridge was added because castles were supposed to have moats and drawbridges, although as far back as anyone could ascertain, the Wartburg had had neither.

The Wartburg continued to be romanticized throughout the nineteenth century, by Wagner, Franz Liszt, and numerous German writers. The nationalism evident in the Burschenschaften rally of 1817 culminated in the political unification of the German states in 1871. The religious and spiritual legacy of the Wartburg receded behind its importance as a symbol of German nationalism, symbolism it

retained into the twentieth century and that the Nazis exploited.

World War II took a heavy toll on Eisenach and its fortress. Small as the city was, its railroad was of strategic importance; all of its important buildings—the medieval churches, Luther's student residence, the Reuter house with its famous Wagner collection, Bach's birthplace, and the town hall—suffered irreparable damage. The Wartburg survived, but was battered once again into disrepair.

From 1952 until 1983, under the Communist regime, the Wartburg underwent its second major restoration. This time, however, every attempt was made to reestablish its original twelfth-century appearance. Meanwhile, Eisenach's historic buildings were carefully restored as well. A new monument that had escaped previous notice became of central importance: the site of the founding in 1869 of the German Worker's Party, ancestor of the German Communist Party. It remains to be seen what will become of this historic shrine in the aftermath of the unification of East and West Germany, and the collapse of Communist rule.

Further Reading: There is nothing currently in English on Eisenach or Wartburg castle. *Frommer's Comprehensive Travel Guide* of Germany for 1991 does contain a short summary of the most important historic monuments of the town, including Wartburg castle. Fascinating accounts of the Wartburg and Eisenach in the time of Luther can be found in Martin Brecht's *Martin Luther*, volume one, *His Road to Reformation, 1483–1521*, translated by James L. Schaaf (Philadelphia: Fortress, 1985; as *Martin Luther: Sein Weg zur Reformation, 1483–1521*, Stuttgart: Calwer, 1985).

—Sina Dubovoy

Ely (Cambridgeshire, England)

Location: In Cambridgeshire, fifteen miles north of Cambridge and approximately seventy miles northeast of London. Cambridgeshire is one of four counties that make up the region known as East Anglia.

Description: Originally a monastic community, Ely has played a part in English history from Anglo-Saxon times. The cathedral at Ely is a towering example of Romanesque and early English architecture that rises up from land that was once an island until the surrounding marshes were drained. This feature of its location earned the cathedral the name Ship of the Fens. At 517 feet, the nave of the cathedral at Ely is the fourth longest in England.

Site Office: Tourist Information Centre
Oliver Cromwell's House
29 St. Marys Street
Ely, Cambridgeshire CB7 4HF
England
(353) 662062

Much of the history of Ely has to do with its religious associations. Its name (eel-lee) comes from a legend according to which St. Dunstan punished the local monks for impiety by turning them into eels. St. Augustine (first Archbishop of Canterbury, 601–604) is supposed to have consecrated a church (later destroyed) in the area, and in 673 the Queen of Northumbria, St. Etheldreda, founded a religious community at Ely—then more commonly known as the Isle of Ely. According to the Venerable Bede, this community consisted of a house for women and another for men, both overseen by the Abbess.

St. Etheldreda, the daughter of the king of the East English, had been married early to a prince named Tondbert. When he died, she married Egfrid, King of Northumbria. Although married to Egfrid twelve years, Etheldreda, according to legend, remained a virgin. After convincing King Egfrid to allow her to enter a convent, Etheldreda joined a community at Coldingham. She fled to Ely (at the time an island surrounded by marshes) when the king changed his mind and attempted to force her back home. The Archbishop of York, St. Wilfred, in 673 officially installed St. Etheldreda as Abbess of Ely—a post she held for six years before dying in 679. Her story is told in a series of eight carvings held within the cathedral at Ely, and her remains were kept in a shrine near the high altar until they were destroyed during the Reformation. Throughout the Middle Ages, St. Etheldreda was the most popular female saint in England.

St. Etheldreda had two sisters who also founded religious communities. One of them, Sexburga, from the Isle of Sheppey, came to Ely to become the community's second abbess until her death in 699. The third abbess, Sexburga's daughter and Etheldreda's niece, was, like her aunt and mother, an Anglo-Saxon queen who had entered the religious life after having been widowed.

The only artifact at Ely that dates from Saxon days is called the Ovin stone. It is really the base of an early eighth-century cross bearing a Latin inscription that translates as "Thy light to Ovin grant, O God, and rest, Amen." Ovin, Etheldreda's steward while she was queen, managed her estates before he joined the monastery at Ely.

Because of its remote location, the community at Ely flourished quietly until 869 when Danes invaded the area. The monastery was left in ruins and its inhabitants killed. For the next century, Ely was cared for by secular clergy until in 970 the monastery was refounded, this time as a male only community under the Rule of St. Benedict. The abbey received a royal charter from King Edgar—a document that is kept today at Cambridge.

Under the Abbot Britnoth in the tenth century, Ely flourished. The local gentry helped protect the monastery from continuing Danish invasions and also contributed to its support. The great Saxon king Canute, according to a poem still extant, celebrated Candlemas (February 2) at Ely one year during his reign (1016–1035).

When William the Conqueror invaded England, Ely was the site of the final resistance by the Saxons. The Saxon commander Hereward the Wake put up a brave fight in the area and had much support; however, the monks at Ely became tired of housing his troops. They gave the Normans the information they needed to mount a successful invasion and the area fell in 1071.

The first Norman abbots at Ely did much to recover the lands lost to Danish invasions and restore the estate to its original boundaries. Abbot Simeon, appointed by William in 1081, began construction of the present cathedral, which was completed in 1189. This long construction period can be appreciated today by comparing the mixture of artistic traditions found in different parts of the building.

In 1109, the pope gave his blessing to the creation of the diocese of Ely and appointed the abbot Hervé le Breton as bishop. The cathedral until then had been under the jurisdiction of the diocese of Lincoln. This event is significant in that the church at Ely now became a cathedral—the seat of a bishop. Much of the money that had previously gone to the monastery was now used in constructing and adorning the church building.

Improvements continued to be made to the cathedral during the Middle Ages. In the early thirteenth century the bishop added the Galilee Porch—one of the earliest examples

Ely Cathedral
Photo courtesy of East Cambridgeshire District Council

of early English architecture extant. In 1252 a new presbytery that replaced the original Norman east end of the building was consecrated. King Henry III and his son were among those present at the event. This rebuilding program became necessary because over the years the cathedral at Ely had become a popular place for pilgrimages. To take advantage of this flourishing pilgrim traffic, fairs were held at the cathedral twice each year. These fairs were called St. Audrey's fairs, and the word ''tawdry''—a corruption of ''St. Audrey''—came to be used for the souvenirs sold at them and eventually for anything flimsy or cheap.

In 1321 a Lady Chapel was founded between the presbytery and the north transept. It was during this chapel's construction in 1322 that the central Norman tower of the cathedral collapsed. The Norman choir was destroyed and the nave and transepts severely damaged. Experts, including King's Master Carpenter William Hurley of London, were called in by the bishop to determine what was to be done with the damaged area, measuring seventy-two feet in diameter. The result was what is now considered one of the masterpieces of medieval architecture—the octagonal lantern tower. Rising sixty feet, the tower has a total weight of 400 tons supported by a marvelous scheme of pillars and posts. The tower is considered by many to be the most outstanding architectural feature of the cathedral.

As the lantern tower was being constructed, the Norman choir was restored and the building of the Lady Chapel completed. One item of historical note from this era involves the first record of the purchase of bricks anywhere in England; the order came from Ely during the years 1335 and 1337. These bricks can still be seen in the walls of the Lady Chapel, which itself makes a unique historical claim in that at forty-six feet, its roof is the widest medieval stone vault in England.

Between 1321 and 1349, the period during which Black Death decimated the population of England and Europe, some of the cathedral's most significant construction took place. The last phase of construction on Ely Cathedral probably took place around 1392, when an octagonal belfry was built over the west Norman tower. Considerable effort has been required ever since to keep the weight of the belfry from causing the tower's collapse. The tower was strengthened by construction of a stone casing in 1476. With the construction of two small chapels, Ely Cathedral by the beginning of the sixteenth century was complete. Ely, however, did not escape the religious turmoil during the Reformation unscathed.

In 1539 King Henry VIII dissolved the monastery at Ely, although the episcopal see remained. Two years later, the king established a college for choristers at Ely; thus, although the monks were gone, the pattern of religious life and worship in the community continued. Bishop Goodrich was responsible for carrying out the Reformation edict of removing all images and other such artifacts from the cathedral. St. Etheldreda's shrine was destroyed and most of the decoration in the Lady Chapel dismantled. Most of the monastic buildings were spared, however, and are still in use today.

The great English reformer Oliver Cromwell lived from 1636–46 in Ely on an estate he had inherited from his father. His house still survives as a tourist attraction. Tradition holds that Cromwell ordered the cathedral clergy to stop chanting services and to concentrate instead on preaching. His command was supposedly ignored until Cromwell had the doors forcibly locked.

The years that followed were impoverished ones for English religion in general and also for Ely Cathedral, which remained closed for seventeen years. Little was done to maintain the building, although in 1699 the northwest corner of the transept was restored. This work was supposedly done by the great architect Sir Christopher Wren, whose uncle was bishop of Ely from 1638–67. Overall, however, the condition of the cathedral deteriorated well into the next century. In 1750, an architect from Cambridge, James Essex, undertook restoration and some redesigning that likely saved the cathedral.

It wasn't until 1839 when a Dr. Peacock became dean of Ely that a significant period of restoration began. Among the changes were a new floor for the nave and new carvings for the choir stalls. A boarded ceiling was installed in the nave and painted in 1858—a significant example of Victorian church art. New windows were also installed during this time, most notably the great east window portraying events from the life of Jesus. Prime Minister William Gladstone in the 1870s held up Ely as an example of vital cathedral life.

During the 1950s further restoration was undertaken on the cathedral, especially to the timbers of the lantern tower. The altar was moved to its present position underneath the octagon, creating a flexible liturgical space. The west tower, always a problem, received attention during the early 1970s with a near total renovation using the latest engineering techniques. In 1974 the original 1907 pipe organ was rebuilt. Ely still supports a choir of men and boys, keeping alive a tradition begun by Henry VIII. His King's School still operates, with some 800 boys and girls studying on the grounds of the community originally established more than 1,300 years ago. The cathedral caused a stir in 1986 when it became the first cathedral in England to charge admission. These funds go toward the continuing restoration of the building.

Further Reading: A loving and detailed description of Ely Cathedral and its history is found in *Ely Cathedral* by Michael S. Carey, dean emeritus of the cathedral (London: Pitkin Pictorials, 1973). Although meant as a guidebook/souvenir, this book contains a wealth of factual as well as anecdotal information. *British Cathedrals* by Paul Johnson (New York: Morrow, and London: Weidenfeld and Nicolson, 1980) focuses on the cathedral's architecture and gives very little historical context. It is excellent for its description of the cathedral and its ornamentation. *Let's Go: The Budget Guide to Britain and Ireland 1994* (New York: St. Martin's Press, and London: Pan, 1994) has the most interesting description of Ely of any travel guide and gives historical facts not found elsewhere. Another travel guide with an excellent section on Ely is *Fodor's 94 Great Britain* (New York: Random House, 1993).

—Linda J. King

Enns (Upper Austria, Austria)

Location: On the west bank of the River Enns, just south of the Danube. Approximately 60 miles east of the German border; 25 miles south of the border with the Czech Republic; 100 miles west of Vienna.

Description: A small city in a valley by the confluence of two rivers, which was first established as a military outpost of the Roman Empire, Enns has been continually inhabited for at least 2,000 years. Enns is peaceful in the 1990s, belying a violent past, which can be glimpsed in bits and pieces: Parts of a medieval wall, a lookout tower built in 1568, and a notable museum containing numerous Roman artifacts remain from the times when armies passed through during wars of conquest, and religious outlaws of a variety of denominations were drowned in the river or burned at the stake.

Site Office: Enns Touristeninformation
im Sparkassenreiseburo
Linzer Strasse 1
Enns, Upper Austria
Austria
72 23/32 61

Enns was granted a municipal charter by Duke Leopold VI, of the ruling house of Babenberg, on April 22, 1212, making it the first city in Austria to be legitimized by noble fiat. Enns was granted its charter ahead of Linz, Salzburg, and Vienna. Though Enns was an important port in the Middle Ages, located at the end of the ancient "iron road" from Steyr and the port at which salt and iron from the southern regions were transhipped by way of the Danube to the east and west, Enns, even then, was not the wealthiest or largest city in Austria. But it was then, as it is today, located on a natural border to two frontiers. The River Enns has marked the spot at which marches of invaders from the east have stopped, a natural border between the Slavs in the east and the Germans in the west. In 791, again in 901, and as recently as 1945, the river has been a dividing line. Though the oldest known settlement here was Roman, placed on the south bank of the Danube as a bulwark against the barbarians in the north, the place was probably occupied long before the Romans arrived.

The earliest known inhabitants of Upper Austria were probably Thracian-Illyrians, who lived south of the confluence of the Enns and the Danube, in Hallstatt, around 1000 B.C. Evidence of their habitation comes from their word for salt—"Hal"—similar to "Hall," the word used in the Danube Valley to the north. Salt was then (and is now) a major product of this region. These people lived in peace for more than 500 years. No weapons were found in their tombs, but amber and art objects were found in abundance. Then, roaming into the region from the north, came the Celts, and peace became a memory.

The Celts were a warrior tribe who performed human sacrifices. They ruled the area, which they called Noricum, for several hundred years—their funeral mounds, called *Tumuli*, were found throughout the Danube Valley. Rome and the Celts had many contacts over those years, not always peaceful. As Roman power grew, Celtic influence waned. The Roman legislature passed a law in 97 B.C. forbidding human sacrifice; however, Celtic cults continued this custom over one hundred years later. Even by the year A.D. 200, the municipal temple in Lauriacum (just outside of modern Enns) was designed to accommodate worshipers of both Celtic and Roman gods. Roman rule over Noricum, politically, became official in 16 B.C.

The Romans built a road from the southeast that ended just past the present-day site of Vienna, and stationed their largest garrisons to the east of it. Noricum's first imperial capital, Carnuntum, built entirely from stone, was approximately twenty-six miles east of present-day Vienna. Smaller Roman garrisons were stationed at strategic points to the west; one of these was at Lauriacum-Lorch, the site of present-day Enns. The Romans built a temple, a forum, and baths there; they probably had several stone houses, and chances are good that the Romans intermarried with the local populace.

Living north of the Danube were Germanic tribes, while the Romans established the empire south of the river, as wall as on the river itself. The Romans built a large fleet to patrol the river between the garrisons, both for military purposes and for trade. Though there were occasional embassies between the Germans and the Romans, hostilities between the two were never far from the surface. Vienna replaced Carnantum as Noricum's leading city in 249, and vineyards along the Danube produced a fine wine which drew nobles from Rome to this northern border.

Diocletian became Roman emperor in 284, and in 301 issued an edict against Christianity. It did not take long for Roman authorities to create the first Austrian Christian martyr. Florian, a Roman army veteran, was an administrator at Lauriacum who had converted to Christianity. He was condemned for excessive zeal by a Roman court; a millstone was hung around his neck, and he was thrown into the River Enns, where he drowned. He was canonized some 380 years later, and is invoked against fire. One local poem goes like this:

> O thou Holy Florian,
> spare our cottage,
> burn the other one.

Christians were persecuted in the Roman Empire until 313, when Constantine I issued the edict of Milan.

Statues of Pilate and Christ at Enns's basilica
Photo courtesy of Enns Touristeninformation

While Roman administrative power did not last past the fourth century, Roman Christians established outposts which continued some of Rome's traditions. In Enns, a Christian basilica was built around 360, and in the middle of the fifth century, St. Severinus, the "Apostle of Noricum," took up residence there. Severinus roamed the length of the Danube, warning the people of an approaching barbarian invasion. Few listened, but Severinus was right. After Severinus died, his body was removed to Italy by his followers. They were fleeing from the invasion of the Huns, and believed that Severinus's corpse would be desecrated if it were left to rest in Noricum. This action signalled others of the end of the imperial Roman presence in Noricum, and when Huns plundered much of the land east of the Enns, Noricum itself ceased to exist. The Huns probably did not reach the Enns, and only occupied Noricum for about fifty years—their destination was Rome itself. But when the Huns were gone from the valley, another barbarian horde, the Avars, rose from the east to take their place.

The Avars were probably of Turkish origin, as opposed to the Huns, who had wandered to central Europe from the Siberian plain. They arrived in the early sixth century, settling mostly in the area east of the Enns, which was known west of the Enns as the "Avar desert," because the Avars destroyed everything there that they had not built themselves. The Avars ruled the area directly across the river from Enns until Charlemagne, the Christian, Frankish king took his armies east to restore Christianity to the Avar territory. Bavarians, meanwhile, had settled on the site of Enns.

In 791, Charlemagne's armies camped near Enns, using the town as a base before riding to the war against the Avars, and by 803, his armies had expelled the Avars. Charlemagne then established in the area a frontier district, first called Ostmark, renamed Ostarricchi, and later Osterreich. The frontier district marked most of the boundaries of present-day Austria.

One of the signs of the importance of Enns to Christendom was that soon after Charlemagne's campaign against the Avars, a synod was held there. A Carolingian church was built in Enns at roughly that time. Charlemagne began a tradition of establishing monasteries throughout Upper Austria. After his death in 814, his three sons carved up the Holy Roman Empire and one of them, Louis I (the Pious), became the ruler over Ostmark. He made the church responsible for tax collections. Tax revolts and other disturbances marked his reign, but these were nothing compared to yet another threat from the east—the Magyars. The Magyars reached the Enns in 900 and practically leveled the St. Florian Monastery. An army made up of Bavarian and Ostmark soldiers was formed. They met the Magyars in a battle near Enns. Some 1,200 Magyar bodies were recovered after the battle; the rest of the horde withdrew to the east.

In 901, the victorious Bavarians built a fortress at Enns and restored the St. Florian Monastery. They believed there would be no more trouble from the Magyars. None came for another six years, but then the Magyars returned to the River Enns, only much father south, decimated an army of Bavarian knights, and crossed into Bavaria itself.

Though the war continued for another half century, trade along the Enns and the Danube never stopped. After the Magyars were effectively destroyed in a battle at Lechfeld by the German emperor Otto the Great in 955, Otto appointed Leopold of Babenberg as margrave of Ostmark in 976. This inaugurated the Babenberg period, which lasted 270 years. Though the Babenbergs ruled as representatives of the German kings, the family ruled with such skill they became virtually independent. Leopold I began a policy of land settlement by Germans, which was continued by his successors. Boundaries established during the Babenberg period lasted until 1918. Much of Austria saw wealth it had never seen before, and Enns saw its fortunes rise to perhaps the highest level in its history.

The Babenbergs actively sided with the Catholic Church in most matters, and resolved territorial disputes peacefully by turning them into domestic quarrels. They built monasteries all over Austria and married into the powerful houses of Europe. In the tenth and eleventh centuries, Enns was one of the largest and most famous of the Ostarricchi cities. The Babenbergs lived first in Melk, not far from Enns, but they gradually moved east, and eventually settled in Vienna. Enns benefited when thousands of Crusaders streamed down the Danube on their way to holy war in the east. They left silver and gold in the cities along the river. In 1212, when Leopold VI granted the first town charter in Austria to Enns, a near-mania for charters among Austrian cities ensued. Vienna received one in 1237, but many other Austrian cities had to wait until the Babenberg dynasty ended. With a new threat in the east (from Mongolians this time), and a bank collapse precipitated by a church ruling against charging interest (of which no such rule existed among Jews, who filled the role of financiers), the last male of the Babenberg line, Frederick II (also known as "The Quarrelsome") died in 1246 without leaving an heir. No one was left in charge.

Because the imperial line in Germany also came to an end at approximately the same time, there was no immediate successor appointed to rule Austria after the extinction of the Babenbergs, and dynastic havoc ensued. The king of Bohemia, Otakar II, stepped into the power vacuum and attempted to establish authority over Austria. He did so in great part by supporting the towns over the nobility. Where he could buy a town by granting it a charter, he did, and where a charter was not enough, he paid cash. As to the nobility, his army conquered castles wherever they found them. But the Germans settled their dynastic troubles in 1272. They elected Rudolf of Habsburg as their king, and four years later, Rudolf declared war on Otakar. Slowly, a German army marched east to Vienna.

Sitting behind the fortified walls in Vienna, Otakar counted on an Enns squire, Konrad von Summerau, to stop Rudolf's army before it could cross the River Enns, but the Germans found the gates of the city of Enns wide open. Rudolf of Habsburg defeated Ottokar in 1278 in the Battle of the Marchfeld, and began a Habsburg reign over Austria which lasted into the twentieth century.

It began as a bloody reign. Though the battle was won, the following century and a half in Austria were filled with violence, both man-made and natural. Constant infighting between nobles was exceeded only by a succession of natural disasters. A plague of locusts hit Austria in 1338 and 1339, great floods in 1340, and there was even an earthquake in 1348. And in 1349, bubonic plague, the Black Death, swept through the Danube Valley. Dazed, the people asked what could have caused these troubles. Someone, somewhere, gave an answer that resounded for more than half a century, from the lowest peasant all the way up to Albert V, the Habsburg rule in 1420: the Jews.

Pogroms were common occurrences through the later half of the fourteenth century, but in 1420, the institutionalized pogrom became the law of the land. It started with a rumor in Enns: The Church there had been broken into (probably the Basilica of St. Lawrence, built over the remains of the Carolingian church, the early basilica, and the Celtic-Roman municipal temple), the wafer smeared with blood and the host desecrated. The Jews were blamed. Austrian ruler Albert V ordered all Jews in Austria arrested and had 270 burned at the stake. He expelled the rest. Jews named Austria "Erez Ha Damim"—"The Bloodstained Land."

The Habsburgs followed Babenberg policy in one respect—they married into other imperial houses all over Europe, and eventually became the most powerful dynasty in Europe. But all of the centuries that followed were marked by turmoil, both religious and political. Sometimes the struggles were resolved by persuasion, other times by arms and flame. Enns fortified itself through the fifteenth and sixteenth centuries, and continued to profit off the salt and iron trade. A lookout tower, which still stands, was constructed in the fifteenth century, and many solid houses were built in the sixteenth century. Though religious strife brought about by the Protestant reformation affected all of Europe, Enns, like most of Austria, remained mostly Catholic.

In the eighteenth century, Austria became ruler of a large empire, but the Austro-Hungarian empire was finished by the end of World War I. When the Nazis overran Austria in 1938, Austria itself ceased to exist. Renamed Ostmark, the country became a province of the Third Reich. Austrians, accustomed to invasion, adjusted easily. Only the flag at Enns changed, and the only smoke seen in the flames of war curled from the concentration camp in Mauthausen, a couple of miles to the north.

Ostmark was divided into four parts by the allies in 1945—the British took one part, the French another, and the Americans split the rest with the Russians. The Enns River was the dividing line between the American and Soviet sections. Americans occupied Enns. In 1955, Austria was granted its independence in return for permanent neutrality. Since then, Enns, like Austria, has become a milder version of what it always was—a host to visitors from all over the globe, peacefully picking up the gold and silver left behind by tourists passing through the area where the River Enns meets the Danube. Notable sites include the Basilica of St. Lawrence, which houses numerous art treasures, including statues of Pilate and Christ, with Pilate portrayed as a Turkish grand vizier. The Municipal Museum displays artifacts from Roman times and Enns's city charter. The city tower and several historic town houses still stand.

Further Reading: Very little is available about Enns specifically. A few good general histories of Austria are available in English, but for an overview of the history of the entire region, *The Danube* by Erwin Lessner, with the collaboration of Ann M. Lingg Lessner (Garden City, New York: Doubleday, 1961), is lively, authoritative, and engaging. Because its subject is the river, a more static boundary than those marked by law, the book includes much that is left out of other histories of Austria.

—Jeffrey Felshman

Erfurt (Thuringia, Germany)

Location: On the Gera River, sixty-two miles west-southwest of Leipzig, in what used to be Prussian Saxony; between Gotha and Weimar, about twelve miles from each.

Description: Medieval commercial center at the point of intersection between an important north-south trade route and the *via regia* (king's route), the principal trade route in the high Middle Ages between the Rhineland towns, Poland, and the Ukraine; home of a university founded in the fourteenth century and later attended by Martin Luther.

Site Office: Erfurtinformation
Bahnhofstrasse 37
0-5020 Erfurt, Thuringia
Germany
(61) 2 62 67

Modern Erfurt has since 1990 been the capital of the Federal German Land (state) of Thuringia. Since 1950 a town of 41 square miles and a population of some 220,000, it is also the local government center of the "Land" of Erfurt, with the publicly stated intention of regaining some of the importance it enjoyed during the millenium preceding 1800. The colorful and carefully restored old town (Altstadt)—little more than one mile across and immensely rich in historical associations is partly ringed by a road; the Altstadt forms the core of a city that is economically dependent on horticulture and light industry, including the traditional manufacture of clothing and footwear, but now also on microelectronics, optical engineering, food processing, and the production of office equipment. The reopening in 1993 of the university that was suppressed by Prussian decree in 1816 may well modify the character of the town, as undoubtedly will the completion of restoration and rebuilding in the back streets of the Altstadt. This latter effort should bring these buildings into harmony with the Altstadt's late-medieval style of pointed gables, pitched or swept dormer windows, steeply pitched roofs, often half-timbered, plastered, and painted in pastel colors, but mostly without pillars, rounded arches, or other Italianate architectural and decorative motifs.

The association of Martin Luther with Erfurt has tended to determine the town's historical image and to overshadow other important aspects of its past, often still discernible through the layers of its surviving medieval architecture. A Stone Age settlement, the site was in the hands of the Franks by the sixth century, and was evangelized by Celtic and Anglo-Saxon monks under St. Boniface, later archbishop of Mainz, who had Erfurt erected into a bishopric in about 742, although the diocese was soon afterward reabsorbed into Mainz. Under Charlemagne Erfurt was in 805 designated an imperial center for trade with the Slavs. However, the settlement later became subject to the overlordship of the archbishops of Mainz, who had recourse to charters issued by the emperor Otto I and succeeded in enforcing their claim on Erfurt; they did this even during a period in which the overlords were the landgraves of Thuringia in the early twelfth century, and despite a charter granted by the Holy Roman Emperor Frederick II.

Frederick I Barbarossa held at least five imperial diets at Erfurt. By the end of the twelfth century, a score of diets and synods are known to have taken place there; the settlements had meshed into a larger unity; and in 1162 a surrounding wall was built to protect the enclosed area, divided in 1182 into twenty-four parishes. The wall, partly destroyed in 1165, was rebuilt with six gates in 1168. By 1170 Erfurt was referred to as a *municipium,* or municipality, and in 1196 as a *civitas*. It had slowly bought rights of independence from needy archbishops of Mainz and was in practice becoming a free city. Thirty-six churches existed there by the end of the twelfth century. By that date trade, particularly in woad (a blue dye) and dyed cloth, had begun to make the town wealthy, and staple rights conferred to the town a monopoly over trade in woad. Many of Erfurt's richer merchants were now having their houses built.

Ruins of the Romanesque royal monastery of St. Peter, built between 1103 and 1147 and destroyed during the Prussian siege of 1813, can still be seen on the rise northwest of the hill, which dominates the town center and on which now stand next to one another its two magnificent principal churches, the cathedral (Marienkirche) and the Church of St. Severus (Severikirche). Broad steps, seventy in all, now lead up to the churches from the large market square. By 1154 work had begun to replace a collapsed former minster with a Romanesque basilica, predecessor of the present cathedral. The original Severikirche was probably consecrated in 1148.

Provision for school teaching and care of the sick existed from the twelfth century, but Erfurt's great expansion started in the thirteenth century, when a number of new monastic foundations were established. The first town hall was begun in 1258, by which date thirty-seven routes emanating from Erfurt are known; in 1268 the town made the first of the territorial acquisitions that would cover 350 square miles, as much as a small principality, by the end of the Middle Ages. By the opening of the fourteenth century the cathedral, in use since 1170, lacked only its central tower. A third of the town had been destroyed by fire in 1291, and the Severikirche must have been substantially completed by 1332.

From 1303 to 1311 the most important of the northern mystics, Meister Eckhart, whose taste for expressing his spiritual teaching in paradoxical formulations led to a partial and posthumous condemnation in 1329, lived in Erfurt as supe-

View of Erfurt, from its cathedral
Photo courtesy of German Information Center

rior of the new Dominican Saxon province. The six stone arches of the famous Krämerbrücke (Merchant's Bridge), built over the ford in the Gera in 1325, replaced the previous wooden structures that had often burned down and secured the trade route through the town. The bridge was later widened with the support of the wooden beams that can still be seen, making it possible to build workshops, trading houses, and dwellings on each side. It is now the only bridge north of the Alps to support houses still occupied by residents. There were originally sixty-two units, now combined to make thirty-four, half-timbered, plastered, and mostly built between the sixteenth and eighteenth centuries. A recent U.S. guidebook has called it "possibly the most beautiful shopping avenue in the world." The bridge's short length of 410 feet encouraged tall buildings of up to 85 feet; the bridge is 62 feet wide.

Later in the fourteenth century the town was caught up in Saxony's internecine war and was unsuccessfully besieged by the Wettin faction, abetted by the Holy Roman emperor and king of Bohemia Charles IV. Erfurt had developed four *hohe Schulen* (advanced schools), and now became, with Cologne, the only German town to contain a

university that was a purely civic rather than a princely foundation. The papal bull authorizing the university dates from 1379, and the institution opened in 1392, in size second only to the University of Prague, and unique in containing all four graduate faculties—theology, medicine, civil law, and canon law—in addition to the arts. Luther studied there from 1501 to 1505, and joined the Erfurt Augustinian choir monastery in 1505. Ordained in the cathedral in 1507, he remained attached to the Erfurt monastery until 1511, when he definitively left for Wittenberg. His cell was destroyed by fire in 1872, but the convent near the right bank of the river a few minutes' walk downstream from the Krämerbrücke has since been rebuilt.

The apogee of the town's prosperity was reached in the fifteenth century when Erfurt became a member of the Hanseatic League. Although in practice Erfurt was a free city, its overlordship was officially transferred by the Treaty of Amorbach in 1483 from the archbishops of Mainz, who remained feudal superiors, to Saxon "protectors" after both the archbishop of Mainz and the Saxon Wettins had again attacked the town. Erfurt had to pay compensation to both. It had grown from a trading post to a thriving market and mercantile town, and had become a cultural center, but was not independent of either Mainz or Saxony, which had been hostile to one another even before the sixteenth-century schism. Gutenberg studied at the university, and the first printing press east of the Rhine was established in Erfurt. Merchants from all over Europe used Erfurt's weighing house and assaying facilities.

The university in the first decade of the sixteenth century was dedicated to the fashionable *"via moderna,"* insisting on the transcendence of God to the point of making his law arbitrary rather than rational, but it also became the focus of an advanced antischolastic reaction; this antischolasticism was "humanistic" more in its resolute insistence that religious perfection was not separate from but integral to a moral achievement based on rational norms rather than any attitude toward the literature of Greek and Latin antiquity. In Luther's time Erfurt had only one humanistic professor, Nikolaus Marschalk, who published the first Greek book in Erfurt in 1501, but moved to Wittenberg a year later. The cultivation of an antique style came comparatively late, leaving Erfurt a score of years behind Heidelberg and Basel. The circle of young humanists which coalesced around Crotus Mutianus Rufus at Erfurt did not come together until about 1512.

The prevailing attitude allied exterior conformity to scholastic educational practice with an undercurrent of anticlerical feeling, supporting longer than most institutions the antipapal council at Basel convoked in 1431 and favoring the conciliarist view that the plenitude of ecclesiastical teaching authority and jurisdiction lay not with the pope but with a general council. Characteristically, the Erfurt theological faculty joined Cologne and Mainz in demanding the suppression of Hebrew books in the interest of converting the Jews, and then joined Cologne, Louvain, Paris, and Mainz by mildly condemning the *Augenspiegel* (*Eye Mirror*) by Johannes Reuchlin, who found the use of Hebrew spiritually elevating. Nonetheless, the two authors of the celebrated antischolastic satire ridiculing the official position, the *Letters of Obscure Men* (1515–17)—Crotus Rubeanus, Erfurt's rector from 1520, and the anticlerical but superficial polemicist Ulrich von Hutten—were both connected with Erfurt. After a decade of religious conflict and a peasant revolt of 1525 that made all churches evangelical, the churches were equally divided in 1530 between Lutheran and Catholic communities; both the cathedral and the Severikirche remained Catholic, although membership of the governing council entailed annual attendance at the Lutheran church from 1559, and the links with Saxony strengthened.

By the late fifteenth century the tide was also turning against Erfurt's commercial prosperity: Saxony was building up Leipzig as a trading center, Erfurt's defense against the archbishop-electors of Mainz and against the Saxons had been expensive, and the value of money was sharply declining. In addition the two-thirds of the town's dwellings that had burned down in 1472 needed to be rebuilt, and the town's defenses required the construction of a fort on the site of the former royal monastery. Erfurt's earlier refusal to win for itself formal status as a free imperial city had left it with debts that became unaffordable with the decline in trade. Interest on a debt of 550,000 gulders was nearly 6 percent per annum, and the city's insolvency led to a revolution of the merchants and traders against the governing council, which had become a clique that favored Saxony and exploited the situation to its own advantage. In what is known as Erfurt's *"tolles Jahr"* (mad year) of 1509, the council was effectively deprived of power by the broader community of merchants and artisans. The situation was resolved in favor of Mainz, and the student riots of 1510 were largely pro-Saxon demonstrations. Only a diminished prosperity was reestablished as Erfurt became dependent on what was already virtually the market gardening for which the town later became famous. Despite the town's economic decline, some of the grand houses still standing today were built in the sixteenth century.

During the Thirty Years War Erfurt was under attack from Swedish Protestant as well as Catholic imperial military forces. The Swedes occupied Erfurt from 1631 to 1649 when Gustavus II Adolphus, victorious over the Catholic forces of Tilly at Breitenfeld, decided to pursue his defeated enemy westward rather than take Vienna, not yet the essential seat of Habsburg power. After the end of the war Erfurt was finally forced in 1664 to accept the renewed dominion of Mainz, which kept twelve commissioners in the city until Erfurt was finally ceded to Prussia in 1802. Cheap indigo had replaced Erfurt's expensive woad by the end of the seventeenth century.

After Napoléon's victory at Jena in 1806, Erfurt surrendered to the French, and was later besieged by Prussian forces who caused substantial damage to the town in 1813 before capturing it in the first week of 1814. The Prussians secularized most of its religious institutions and closed the university. The fortifications were dismantled in 1873, allowing the town to develop the industrial potential of its position and resources. Industrialization came quickly with connection to the railway system in 1847, and, although Erfurt is often said to have suffered relatively little damage during World War II, the air raids of 1944 did destroy buildings in the Altstadt. In 1945 nearly 10,000 buildings were destroyed or severely damaged.

The largely pedestrianized Altstadt retains the late-medieval layout of a busy merchant town. Meandering backstreets enhance the atmosphere generated by the fine renaissance and baroque facades of the important buildings on the main streets. The two principal streets are short and lead between the three major squares, the Anger, (also the name of a street), the Fischmarkt, and the Domplatz, which lie in an almost straight line. The Markstrasse connects the Domplatz to the Fischmarkt and was once part of the royal route leading from the Fischmarkt over the Krämerbrücke. The Schlösserstrasse leads south from the Fischmarkt to the Anger, once the woad market, where there are still outstanding buildings, such as the 1577 three-story "Zum schwarzen Löwen," and near which is the fourteenth-century Kaufmannskirche (Merchant's Church), which replaced an eleventh-century Friesian merchant's church. Lutheran from 1521, it has a rich, early-seventeenth-century baroque interior. Bach's parents were married there, and three of their sons in succession were directors of the governing council's orchestra.

Nearby is the governor's palace (Statthalterei) of 1711, with an imposing white baroque facade accented by dull red borders whose white plaster decoration contains elements, by that date universal, of Italian derivation. In 1800 a landscaped garden was laid out in front of the building. It was here that Goethe met Napoléon and that the governor from 1772 until 1802, Karl Theodor von Dalberg, later archbishop-elector of Mainz, received Goethe and Schiller—the latter of whom he inspired to write *Wallenstein;* as well, Dalberg became friendly with Wieland, who—with Goethe, Schiller, and Herder—was one of the principal poets of the Weimar group.

The architectural and decorative styles of Erfurt therefore reflect different periods, although late-medieval styles predominate and unify the Altstadt, where Saxon baroque is lightened by the pastel colorings and occasional Italianate ornamentation. Much medieval stonework remains, and, although the most imposing buildings are the later ones, the style of the Altstadt is still medieval, with clusters of little streets surrounding the squares, which now let in light and air, and adjacent terraced houses of different heights. The Fischmarkt, badly damaged by the wartime air raids, has been meticulously restored. Its sixteenth-century house "Zum roten Ochsen" is a rich merchant's dwelling with a facade made increasingly splendid in different styles as the centuries passed, but unified by the way the architectural features from different centuries have been arranged to echo one another. Built in 1562 for a chairman of the council and woad trader, it was rebuilt as a trading house with a dwelling on top in 1737, and as a cinema in 1913. Rebuilt again in 1979, it now contains a gallery of modern art. A frieze over the ground floor shows figures depicting the seven days of the week and the nine Muses. The heavier baroque facade of "Zum breiten Herd" opposite it has a history similarly characteristic of Erfurt. This building was commissioned by a high civic dignitary in 1584 with a frieze above the ground floor depicting the five senses, and the neighboring inn was acquired and rebuilt to give a single, architecturally unified facade. The joined exteriors were restored in 1969 and the interiors in 1984.

The dominating complex of the cathedral and the Severikirche with the few houses on the hill overlooking the town is imposing in detail. The Domplatz beneath the two churches was the site specified by Charlemagne as the original trading place between Germans and Slavs. It was Boniface himself who took the first steps toward building on the site of the later royal monastery the church "*beatae Mariae virginis*" (of the blessed virgin Mary), whose convent of Benedictine nuns acquired the relics of St. Severus from Ravenna in 836. The nuns were moved to the site of the Severikirche in 1123. The church burned down and was rebuilt in the twelfth century, again in the thirteenth to fourteenth centuries, and was extensively restored after a fire in 1472. The three towers were renewed in 1494–95, when it is possible that the middle tower was added. The whole church was renovated from 1845, having been used as a hospital by the Prussian occupying administration from 1813. The interior was restored from 1979 and 1982.

The five naves are created by slender pillars supporting simple vaulting that starts on each pillar at exactly the same height. The central nave is double the width of the other four, and its vaulting also reaches higher. The church is architecturally imaginative and its ground plan unique. Topping the eastern facade are three towers, the central one taller than the flanking pair, each surmounted by a tall, pointed, and copper-clad spire, with a small ball of copper on top. The spires are elegantly echoed in miniature at the outer edge of the copper-clad, gabled dormers protruding from the roof and on the chapel of Mary to the north. The slenderness of the columns supporting the vaulting of the church's interior gives an impression of great height to the building. It seems likely that the outer naves were added late to the ground plan, which then became block-shaped, without a transept.

The baroque high altar of 1670–80, with its three-tier structure and renaissance inspiration, and the extravagantly decorative organ of 1714 magnificently enhance the visual effectiveness of the building. The church also contains splendid examples of late-medieval sandstone sculpture, an alabaster relief of St. Michael from 1467, and an intricate housing for the baptismal font of the same date—slender, elegant, and very tall. An impressively carved sarcophagus of St. Severus depicts scenes from his life, including one of the saint between his wife Vincentia and his daughter Innocentia.

The cathedral, built to the south of the Severikirche, was originally a late-twelfth-century Romanesque basilica, with two of its towers finished in 1201 and 1237, and a central tower completed before 1307. Parts of the twelfth-century construction are still visible. By 1329 the huge vaulted supports for the extension of the chancel to the east known as the "cavata" were built out from the hill, allowing the ninety-eight-foot chancel to be added in the fourteenth century, with an axis some eight degrees askew from that of the nave, presumably due to an error when the cavata were built. While the cavata were being built, a large, new crypt was installed, the massive arches of which lift the chancel almost forty feet above the Domplatz. The aisles of the fourteenth-century structure were made wider than the nave, to leave the pillars supporting the nave's vaulting in the same positions as before.

Almost 895 panes of medieval stained glass remain from the original 1,100, and thirteen of the fifteen windows are almost as intact today as they were when Luther was ordained beneath them. The fourteenth-century portal, a progressively recessed series of pointed arches with statues of the apostles surmounted by a tall triangle filled with stone tracery, is not elaborate, but again creates a stylish repetition of the same sharply acute angle, as if each element in the design were exerting itself to extend upward. Aesthetically the effect created by the cathedral is much reinforced by other echoes in the stonework.

During the Napoleonic wars both churches were turned into fortresses, and the French stabled their horses in the cathedral, which is now fully restored. The 1697 high altar includes a tabernacle that repeats in miniature the eight twisted columns entwined with vine leaves supporting the baldachin. Other important furnishings in the interior include the twelfth-century stucco retable of the enthroned madonna and child, and the freestanding "Wolfram" of the same date, an almost-life-size bronze figure holding a candlestick in each hand, with a third candlestick at the back of his neck. There is a fine, octagonal sandstone font, and some medieval and sixteenth-century paintings of variable quality. No description of Erfurt could omit reference to the "gloriosa," the medieval bell nearly seven feet high and eight feet across, finally recast in 1497 by a bell and cannon maker from the Netherlands. Two earlier bells survive, as well as one from the seventeenth and two from the eighteenth century. Others have been replaced.

Erfurt remains essentially a city of merchants' houses, converted warehouses, and trading floors, not of palaces and monuments. The construction of grand dwellings extended into the governorship of Philipp Wilhelm von Boyneburg from 1702 to 1717, after which, despite the efforts of Dalberg, Erfurt entered a period of decline. The postwar restorations have been extraordinarily successful in balancing the needs of a working city against the obvious desirability of restoring as much as possible of what remains from each of eight or nine centuries. The use of color has unified the renovated facades. It may have obscured something of the architectural diversity extending over so many centuries, and in an understandable desire to recreate a workable but aesthetically pleasing ensemble to have favored the tourist at the expense of the cultural historian. At no previous date did the currently restored buildings all look new together. No solution could have been perfect, but that adopted shows a degree of sensitivity and taste at least equal to the best shown in other European cities with similar restoration problems, such as St. Petersburg and Prague.

Further Reading: There is very little literature in English devoted specifically to Erfurt, but in the excellent Art Guide series issued in translation by Schnell of Munich there are illustrated and informed pamphlets on the cathedral (1991), the Severikirche (1993), and, for its Lutheran interest, the Augustinian monastery (1991). All are highly recommended and have detailed lists of fuller treatments. The fundamental work is the six-volume (in nine) *Geschichte Thüringens,* edited by Hans Patze and Walter Schlesinger (Cologne and Graz: Böhlav, 1967–79), with a single-volume bibliography. Friedrich Benary's *Zur Geschichte der Stadt und der Universität Erfurt am Ausgang des Mittelalters* (Gotha: Friedrich Andreas Perthes, 1919) analyzes the town's financial administration from 1473 to 1531, and indicates where the source materials are published or are to be found. A lengthy, annotated reading list is to be found in the publication of the Hessische Landeszentrale für politische Bildung, *Hessen und Thüringen. Die Geschichte zweier Landschafter von der Frühzeit bis zur Reformation,* edited by Achim Güssgen and Reimer Stobbe (Büdingen: Hessische Landeszentrale fur politische Bildung, 1992). Gerd Schöneburg's *Erfurt. Führer durch die historische Altstadt* (Erfurt: published by the author, 1992) is a handy, informative, and mostly accurate guide to the history of the town's buildings, with a formal chronology.

—Anthony Levi

Esztergom (Komárom, Hungary)

Location: Across the Danube River from Slovakia, forty-three miles northwest of Budapest.

Description: Capital of Hungary in the Middle Ages, with the ruins of a fortified royal palace; headquarters of the Catholic Church in Hungary and the seat of an Archbishop, with the country's largest cathedral.

Site Office: Komturist (Tourist Information Office)
13 Széchenyi tér
Esztergom, Komárom
Hungary

Esztergom, now a quiet border town of 33,000 people, played a prominent role in the creation and consolidation of the former Kingdom of Hungary, a multiethnic state considerably larger than the modern-day republic, which existed from around 1000 until its disintegration in 1918. The area had been inhabited since at least Neolithic times: the Castle Museum (Vármúzeum) in Esztergom displays archaeological finds covering thousands of years, up to and including remnants of the Celts, who settled along the Danube River around 2,400 years ago, and of the Romans, who conquered the Celts and held Esztergom as an outpost of their empire between the first and fifth centuries A.D. It was the Romans who discovered the thermal waters, which are still used in the city's public baths, and who first planted the vines from which the local wines are still made. Esztergom, then called Solva Mansio, was one of the places where the philosopher-emperor Marcus Aurelius wrote parts of his *Meditations,* widely regarded as a classic of Stoicism.

During and after the collapse of the Roman Empire, the Carpathian Basin, the geographical feature which corresponds to most of the territory of contemporary Hungary, was frequently traversed by migrating peoples such as the Avars and the Huns, who wished to escape population pressure in Western Asia and to seek new lands and wealth in Europe. However, the history of the Hungarian city of Esztergom cannot be said to begin until the late tenth century, with the conversion to Christianity of many of the Hungarians, the people who called themselves the Magyar and who had arrived in the Carpathian Basin, under the leadership of their semilegendary ruler Arpád, around the year 896. His descendants were to rule Hungary until 1301. One of them, Géza, founded a stone castle at Esztergom as the main residence of his royal household and began the building of a cathedral there for one of the two archbishops appointed to organize the spread of Christianity into the region. Since then every archbishop of Esztergom has also been primate (head) of the Catholic Church in Hungary. After Géza's death his son Vajk, who is said to have been born in Esztergom in 970,

was baptized as Stephen (István) and received what is now called St. Stephen's Crown from the pope, in return for his promise to Christianize all his subjects. The crown was placed on his head by the archbishop in Esztergom on Christmas Day, 1000. Soon afterward, Stephen founded a new cathedral next to his father's castle on the hill, now known as Várhegy (Castle Hill), that dominates Esztergom. In the Middle Ages this structure came to be known as the Cathedral of St. Adalbert, in honor of the priest who is said to have come to Esztergom from Prague to baptize Géza and/or Stephen.

In 1018 Stephen built a second royal capital at Székesfehérvár, which, unlike Esztergom, was on the main road taken by pilgrims making their way from western Europe to Jerusalem. From then until the thirteenth century the two capitals performed different functions: kings usually were crowned and married in Esztergom, and the royal mint was located in the city, but Hungary was mainly governed from Székesfehérvár, the site of the royal tombs. The two cities stood out among the forty or so designated as capitals of counties; their connection to royalty attracted merchants from what are now Belgium and Italy, and they tended to be under the control of the church rather than of the secular nobility. However, while they could satisfy the demand of the royal household for luxuries, during the eleventh and twelfth centuries the produce of the countryside around each of the royal castles was insufficient to meet its demand for food, fuel and materials; ultimately, the household had to move from estate to estate around the kingdom.

Esztergom, then also known by the Latin form of its name, Strigonium, reached its peak of wealth and importance under King Béla III, who reigned from 1172 to 1196. Béla was unusually open to influence from both the Western and the Eastern traditions of the continent. On the one hand, he had been impressed as a child by the splendor of Byzantium, the capital of the Eastern Empire; on the other, he had married two French wives in succession, introduced the Cistercian and Premonstratensian orders into Esztergom and other Hungarian cities, and made contact with Western European nobles passing through Hungary on their way to Crusades in Palestine. He invited French architects to improve his own capital at Esztergom by building a new cathedral and a new castle (often referred to as a palace), both in the early Gothic style.

Most of Hungary was occupied by the Mongols during 1241 and 1242. The city of Esztergom was attacked and the bathhouses founded by the Romans were damaged, but the Mongols could not capture the castle. After their retreat, King Béla IV tried to reorganize the political and economic framework of the country along Western European lines, making Buda (now part of Budapest) the only permanent

The cathedral at Esztergom
Photo courtesy of IBUSZ Travel Inc.

royal capital and converting Esztergom and other castle towns into commercial centers by giving them single administrations uniting both Hungarian residents and foreign merchants. Esztergom, however, was a special case; the archbishops retained some control over what continued to be the religious center of the kingdom. They took up residence inside Béla III's castle from 1249.

After the Arpád dynasty ended in 1301, the monarchy became elective, which in practice meant that whoever paid the highest bribes or made the biggest concessions to the nobles would become king, whether he himself was Hungarian or not. The archbishops, who served as regents in the months, sometimes years, between the death of each king and the election of the next, continued to exercise considerable political power, especially whenever a king spoke no Hungarian or even stayed away from the kingdom for years on end. But they were unable to reverse the trend toward frag-

mentation; by the time of King Károlyi I, who reigned from 1308 to 1342, many of the castle towns had become the nuclei of virtually independent regions governed by the major barons, who used their monopoly over local agriculture and commerce to resist royal control. In 1311 the most powerful of these rebellious lords, Máté Csák, launched an attack on the lands of the archbishop, failing to dislodge him but causing considerable damage in the city and the surrounding countryside before being defeated.

Matthias Corvinus, a leading member of the noble Hunyadi family, was the only Hungarian elected to the throne during this period. During his reign, from 1458 to 1490, he managed to rein in the power of the barons, if only temporarily, and cooperated with Archbishop János Vitéz in trying to make Buda and Esztergom centers of Renaissance scholarship and culture. The partnership ended in 1472, when he discovered that Vitéz was conspiring with rebel barons against him; Matthias Corvinus replaced him with Ippolito d'Este, an eight-year-old nephew of his Italian wife Beatrice. During the reign of King Ulászló II, the power of the barons increased to such an extent that some of them effectively took over the administration of the kingdom while others formed a resentful opposition to them. One of their leaders, Tamás Bakócz, who had risen from a peasant background to become chancellor of Hungary, had himself declared archbishop of Esztergom in 1497, partly in order to gain access to royal revenues, which he happily supplemented with the proceeds of bribes and of the embezzled property of King Matthias's widow. Some of the fortune he amassed was spent on rebuilding and expanding the castle at Esztergom, while the city itself became so impoverished that it could not maintain its former tradition of self-government and allowed him to seize control.

The weakness of King Ulászló II gave the Habsburgs of Austria, the most powerful dynasty in the neighboring Holy Roman Empire, the opportunity to increase their influence over Hungarian affairs. Archbishop Bakócz, campaigning to become pope in 1513, sought to win favor with the Habsburgs, who were prominent defenders of Catholic tradition against the first stirrings of Protestantism. When he lost the election, he announced a crusade against the Turks in 1514. It was ostensibly for this purpose that he raised a peasant army that spread rapidly across the country and turned against him. The revolt was ultimately put down by the other barons, who executed the peasant leadership and changed the laws to remove even the minimal rights that the peasants had enjoyed until that point. Archbishop Bakócz, politically defeated and restricted to Esztergom, died in 1521. He was buried in a chapel inside the cathedral that he had had built in red marble by Italian craftsmen, between 1506 and 1511, and completed in 1519 with an altar of white marble designed by the Italian sculptor Andrea Ferrucci da Fiesole. When the old cathedral was replaced in the nineteenth century, the chapel was disassembled into 1,600 numbered pieces and reassembled inside the new cathedral.

In 1526 King Louis II and many of the Hungarian nobles died fighting the invading Ottoman Turks under Sultan Süleyman I, the Magnificent, in the Battle of Mohács. After a period of confusion and continued fighting, involving rival claims to the throne and Habsburg intervention, the country was divided in 1552. The Turks ruled the center from Buda; a group of nobles formed the Principality of Transylvania, under Turkish protection, in the east; and the Habsburg archduke Ferdinand I, who was already king of Bohemia and who would eventually become Holy Roman Emperor, ruled a much-reduced Kingdom of Hungary in the west. As for Esztergom in particular, the Turks conquered it and other cities on the military highway to the capital in 1543. It was they who reopened and expanded the thermal baths in the city.

In the course of subsequent wars between Habsburg emperor-kings and Ottoman Sultans, Esztergom came under an unsuccessful siege by Habsburg armies for two months in 1594. One of the Hungarian soldiers who died during the siege was Bálint Balassi, whose love poems were later to influence many other Hungarian writers. The local history museum in Esztergom is named for him, and he is also commemorated by a sardonic line from one of his poems, which has become proverbial in Hungary, to the effect that if people did not commit sins God would have nothing to forgive.

Esztergom was taken by Habsburg forces in 1595, was taken back by the Ottomans in 1605, and was then surrendered to the Habsburgs under the Treaty of Zsitvatorok in 1606, only to be passed back and forth, though remaining mainly under Turkish control, until 1683. During the long years of war, its archbishops, clergy, and other leading residents were often forced to escape from the city and its cathedral was damaged, its economy ruined, and its population drastically reduced by famine, disease, and emigration. The Catholic faith, tolerated by the Muslim Ottomans to an extent that no Christian country then would have tolerated Islam, was maintained in Esztergom under the leadership of its archbishops, especially the Jesuit scholar Péter Pázmány, who was archbishop from 1616 to 1673 and founded the university in the city.

In 1683 the Ottoman armies were turned back from the gates of Vienna and the Habsburgs launched their largest counteroffensive so far, taking Esztergom and other northern cities and pressing on to Buda, which was seized in 1686. Once the Turks had been expelled, a process completed by the 1699 Treaty of Karlowitz, the Habsburgs revived the ancient tradition of governing Hungary through collaboration with the leading landowners. In the years up to 1848, they granted limited rights of self-government to only eight "free royal cities," including Esztergom. Esztergom, known as Gran to the German-speaking imperial officials and its German residents, and as Ostrihom to its Slovak residents, was to be under Habsburg rule until 1918.

Recovery from the wars was stifled by an outbreak of plague in 1739, as well as by the refusal of the archbishops to return to the city. Nevertheless, new buildings were constructed in the eighteenth century, many in the elaborate and expensive baroque style then in favor. These include the Town Hall, originally built in 1698 for János Bottyán, a leader of the abortive independence movement of 1699, expanded in 1729 and reconstructed in the 1770s; the parish church of the Víziváros district at the foot of the castle hill, completed in 1738; the Old County Hall, built in 1747; the Franciscan church attached to a new theological college and opened in 1750; the so-called Öreg-templom (Old Church), built in 1762 to replace a church destroyed during the Ottoman occupation; and yet another church, since demolished, built on the site of the former cathedral on the orders of Empress-Queen Maria Theresa (who reigned from 1740 to 1780). The statues of the royal saints Stephen and László I that stood in front of it are still to be seen today, although the church itself has been replaced by a new cathedral, the largest church ever built in Hungary, on the hilltop site of the former cathedral and royal castle.

Work on the new cathedral (sometimes referred to as a basilica) began in 1822, two years after Archbishop Sándor Rudnay had arrived to reestablish Esztergom's position as the center of the Catholic Church in Hungary. Its first architects were Pál Kühneland and János Packh, although later stages were the responsibility of a third architect, Jószef Hild. Its consecration ceremonies, in 1856, included the world premier of the Esztergom Mass by the Hungarian composer Ferenc (Franz) Liszt and the installation of the main altar, a marble structure that includes a reproduction of Leonardo da Vinci's painting *The Last Supper* and, above it, a depiction of the *Assumption of the Virgin Mary into Heaven* by Michelangelo Grigoletti, one of the largest oil-on-canvas paintings in the world. The building was finally completed, under Archbishop János Simor, in 1869. From this monumental neo-classical church, with its portico of eight columns, its dome 236 feet high and 85 feet wide, and its twin belfries with their own smaller domes, there are views over the Danube River into Slovakia.

Archbishop Simor, who died in 1891, founded two of Esztergom's museums. The Christian Museum (Keresztény Múzeum), inside a new Archbishops' Palace in the city below the cathedral, was opened in 1875 to display the medieval and later works of art collected by archbishops of Esztergom over the centuries, including pictures by the Flemish Renaissance painter Lucas Cranach the Elder. The Cathedral Treasury (Kincstár), opened in 1886, contains religious objects such as a ninth-century cross made of crystal and an ornate, bejeweled baroque chalice (communion cup) donated by Maria Theresa.

In the parliamentary Kingdom of Hungary, created by the liberal revolution of 1848 and the Compromise of 1867 (which led to the Austro-Hungarian Empire), Esztergom retained its role as the religious capital for the majority of the kingdom's inhabitants, whether they were Hungarians, Slovaks, Germans, or any of the several other minority nationalities. After World War I, however, these minorities formed their own countries or, as in Transylvania, took Hungarians with them into Romania and other existing states. Esztergom, now just across the Danube from the new state of Czechoslovakia, became a focus of nationalist feeling and of archaeological work partly inspired by it. The remains of King Béla

III's castle were first excavated in the 1930s, but work was interrupted by World War II, during which Hungary became an ally of Nazi Germany, in the hope of retrieving its lost territories. Esztergom was damaged by war in 1944 and 1945, when the Hungarian government drove out the Germans, only to be overwhelmed by the arrival of the Soviet Red Army: the ruins of the bridge over the Danube, built just after World War I but destroyed at this time, still stand, unrepaired, as a memorial. The Communist regime installed in the aftermath of the war at first persecuted the Catholic Church, imprisoning the archbishop, Cardinal Mindszenty, but during the 1960s some liberalization took place, allowing, among other changes, the resumption of work on the castle remains and other remnants of pre-Communist times. The artifacts and sections of buildings discovered over these years—notably King Béla's chapel, which was filled with soil during the Habsburg-Ottoman wars as an extension of the castle defenses—form the basis of the modern Castle Museum.

The historic buildings of Esztergom have been visibly polluted in recent years by emissions from the factories of Slovakia, the focus of heavy industrial development during the four decades of Communist rule. It is an open question whether Hungary, which abolished its own Communist regime in 1989, will be able to afford the financial and political costs of trying to counteract or eliminate this pollution. In the meantime the cathedral, the castle museum, the baroque churches and the ancient baths, in their varying states of repair, go on bearing witness to the long history of Esztergom, city of kings (some of them saints) and archbishops (some of them politicians).

Further Reading: The history of Esztergom is covered in some detail in *A History of Hungary*, edited by Ervin Pamlényi (London and Wellingborough: Collet's, 1975) and another book with the same title, edited by Peter F. Sugar (London and New York: Tauris, 1990). It is striking that while the former was compiled under the Communist regime and the latter after its collapse, both have contributions from the same Hungarian scholars. There is a vivid description of Esztergom in the 1930s in Patrick Leigh Fermor's travel book *Between the Woods and the Water* (London: Murray, 1986; New York: Penguin, 1987).

—Patrick Heenan

Fontainebleau (Seine-et-Marne, France)

Location: Northern France, thirty-five miles southeast of Paris.

Description: Town famous for a beautiful château that stands in a nearby forest.

Site Office: Office de Tourisme, Syndicat d'Initiative
31 place Napoléon Bonaparte, B.P. 24
77302 Fontainebleau, Seine-et-Marne
France
64 22 25 68

The charming château of Fontainebleau—built and rebuilt, beautified and restored by French monarchs over the centuries—shares its name with a nearby town and surrounding forest. As a royal residence, Fontainebleau eventually was overshadowed by the fame and splendor of Versailles. Yet it was a favorite home of French leaders ranging from the Renaissance's Francis I to the Revolution's Napoléon Bonaparte. Bonaparte fondly referred to Fontainebleau as Le Maison des Siècles, or the House of the Centuries.

Early in the twelfth century, a man called Fontaine de Bliaud reportedly built a manor where the château now stands. Because the manor stood beside a spring, it eventually became known as "the Blue Fountain," or Fontainebleau. Encircling the manor was a forest that today comprises 42,500 acres. This forest, and the excellent hunting it offered, is probably the reason that Fontainebleau became the home of royalty.

Later in the twelfth century, Louis VI built a tower near the edge of a pond in the forest. In 1137 Louis VII added a chapel to his father's tower. The chapel was consecrated by Thomas Becket in 1169, while he was in exile from England.

A monastery was established at Fontainebleau in 1259 by Louis IX, also known as St. Louis. According to legend, Louis opened the monastery to soothe his conscience for having expanded the château. The Trinitarian monks who lived there ran a hospital for the local people. The monastery, located west of the present château, remained in existence until the time of the French Revolution.

Philip IV, known as Philip the Fair because he was regarded as the most handsome man of his day, was born at Fontainebleau in 1268. He met his end there in 1314 after his horse threw him while he was riding in the forest.

Fontainebleau remained essentially a hunting lodge until the reign of Francis I. Francis, a cultured man of the Renaissance who loved the arts, found the château almost a ruin. His immediate predecessors had preferred the châteaux of the Loire region. Francis decided to transform Fontainebleau into a place of beauty and elegance. In the late 1520s he tore down the original building, leaving only the tower.

In its place was built the first part of the château that exists today: the pavilions facing the Cour du Cheval Blanc (White Horse Court), also known as the Cour des Adieux (Court of the Farewells). A long gallery known as the Galerie François I was added to connect the château with the two-story Trinity Chapel on the Cour Ovale (Oval Court).

Work on the château was supervised by master mason Gilles le Breton, who probably designed the alterations as well. Very much a building of the Renaissance, Fontainebleau marked the departure of French architecture from Gothic design. Le Breton's Porte Dorée (Golden Gate), a mixture of new and old styles, introduced a new kind of architecture to France. This style of architecture was part of a wider art movement known as mannerism. Mannerist artists valued artistic invention and imagination, and their creations were elegant and courtly. Le Breton also added an elaborate external staircase, which has since been destroyed.

As a cosmopolitan connoisseur of the arts, Francis turned to Italian artists to decorate his new palace. The masters who did so made the palace a center of artistic and architectural innovation. They were known as the School of Fontainebleau.

Two important members of the school were Giovanni Battista Rosso and Francesco Primaticcio. These painters and stucco workers further established mannerism in France. Their style—influenced by such Italian artists as Michelangelo and Raphael—dominated not only interior design but French painting and sculpture for decades.

Rosso used paint and stucco in the Galerie François I to create figures that seemed to reach from the wall. His technique gave the gallery's 210 feet a sense of drama. Primaticcio's chief works at Fontainebleau were the Chambre de la Duchesse d'Etampes and the Galerie d'Ulysse. The figure drawing and sculpting he accomplished became very influential.

The painter Nicolò dell'Abbate, who arrived at the château in the 1540s, is sometimes described as the third Fontainebleau master. Dell'Abbate was noted for the illusions he created with paint. The architect Sebastian Serlio came to Fontainebleau, where his ideas carried great weight. He was joined by the important classical architect Giacomo Barozzi.

The brilliant sculptor Benvenuto Cellini worked on the château between 1540 and 1545. Among other achievements, he carved a new entrance that featured a reclining nymph surrounded by deer, stags, and greyhounds.

Francis delighted in his "delicious deserts of Fontainebleau." He kept much of his extensive art collection, as well as his library, at the château. The keeper of his library was the poet Mellin de Saint-Gelais, and the humanist Pierre Duchâtel also found a home at Fontainebleau.

Fontainebleau Palace
Photo courtesy of French Government Tourist Office

The Appartements de la Reine were built between 1545 and 1565. The bedroom was to be occupied by many French queens over the centuries: Marie de Médicis, Marie-Thérèse, Marie Leszczyńska, Marie-Antoinette, Josephine, Marie-Louise, Marie-Amélie de Bourbon, and the Empress Eugénie.

Fontainebleau ceased to be a center of artistic innovation in the mid-1560s, yet magnificent additions continued. Built by Francis but decorated under his heir Henry II, the Galerie Henri II is regarded as the most splendid room in the château. It overlooks the Cour Ovale and is topped by an elaborate ceiling carved from walnut. Paintings designed by Primaticcio and painted by dell'Abate depict mythological scenes. The room is festooned with the monograms of Henry and his mistress, Diane de Poitiers, and emblems of the mythological goddess Diana.

Two of Henry's sons who in time also became king, Francis II and Henry III, were born at Fontainebleau. Neither did much to beautify the palace; their brother Charles IX added the Salle des Gardes during his reign.

It was the popular Henry IV who loved Fontainebleau as much as had Francis I. Even though he moved the royal library from Fontainebleau to Paris, Henry spent a fortune on almost doubling the château in size. His additions included the Galerie de Diane, which is more than 263 feet long. Henry also added the Galerie des Cerfs (of the Stags), decorated with dozens of stags' heads. According to legend, Henry was galloping through the forest of Fontainebleau one day in 1610. He heard hunters approach. A huge and hideous Black Huntsman appeared, uttered a cry of warning, and vanished. Shortly afterward, Henry was assassinated by a religious fanatic.

During the seventeenth century, a group of artists known as the Second Fontainebleau School came into being. Less influenced by Italian art than those of the original Fontainebleau School, the Second Fontainebleau School never gained much artistic influence.

After Henry's assassination, Louis XIII replaced the external staircase built under Francis I with the present horseshoe staircase. Louis XIV, known as the Sun King, concentrated on building his own palace at Versailles. Nevertheless, he added to Fontainebleau the group of buildings known as the Petits Appartements.

In 1657, Queen Christina of Sweden came to stay at Fontainebleau as the guest of Louis XIV. While there, she accused her secretary, the Marquis de Monaldeschi, of treason. She ordered her guards to execute him in the Galerie des Cerfs. Because he was wearing a coat of mail, de Monaldeschi had to be stabbed in the throat. The execution shocked the French royal family, and Christina was asked to leave France. Today, de Monaldeschi's coat of mail is displayed in the gallery where he died. Other visitors to Fontainebleau included the exiled King James II of England in 1688 and Russia's Peter the Great in 1717.

In 1685, Louis revoked the Edict of Nantes at Fontainebleau. The edict, introduced by Henry IV in 1598, guaranteed French Huguenots (Protestants) freedom of conscience, social and political parity with Roman Catholics, and some freedom of worship. Louis's action launched a period of government persecution of Huguenots. About 200,000 Protestants fled the country as a result.

In 1725, the marriage of Louis XV to Marie Leszczyńska, daughter of the exiled Polish king Stanisław I, took place in the gallery of the Chapelle de las Sainte-Trinité. It had been built for Henry II and decorated under Henry IV. The bride later requested that nude statues in the palace, allegedly carved by Primaticcio during the reign of Francis I, be veiled. Her request was granted. Another change in decoration made under Louis XV was the removal of three of the thirteen paintings by Ambroise Dubois in the salon Louis XIII—the doors had to be widened to make way for the wide skirts then worn by aristocratic women.

Louis later demolished the south wing of a group of buildings around the Cour du Cheval Blanc to make way for more apartments. Admirers of the earlier buildings severely criticized the demolition. Louis XVI added another long gallery to the château in 1785, but occupied the building only once before the French Revolution took his throne and life.

Louis XVI's gallery marked the end of the construction of Fontainebleau launched by Francis I more than 200 years before. Subsequent rulers of France continued to renovate and redecorate the palace, however. During the French Revolution, the furnishings and artwork of Fontainebleau were stolen. Satyrs flanking the fireplace of the Galerie Henri II, for example, were melted down. Nothing but the walls and ceilings remained. The building itself, however, was left intact.

Napoléon Bonaparte became leader of France in 1799 and crowned himself emperor in 1804. Napoléon summoned Pope Pius VII to France to oversee his coronation, and received the pope at Fontainebleau. Pius was housed in the Appartements des Reines Mères, which once had been occupied by Catherine de Médicis, the wife of Francis I.

As Napoléon I, the emperor spent 12 million francs restoring Fontainebleau, where he held his royal court. Napoléon and his first wife, Josephine, occupied the Petits Appartements erected by Louis XIV. In October 1809, Napoléon returned to Fontainebleau after defeating Austria in battle at Wagram, near Vienna. Josephine hurried to the door that connected their bedrooms to welcome him home, but found the door walled up. Napoléon had decided to divorce Josephine because she had failed to give him an heir. When

Pope Pius refused to recognize Napoléon's divorce from Josephine, the furious emperor annexed the Papal States in retaliation. Pius promptly excommunicated those who took part in the annexation. Napoléon had Pius arrested and brought to Fontainebleau. The pope remained a prisoner in the Appartements des Reines Mères, where he had previously stayed as a guest, until 1814.

In 1810, Napoleon married Marie-Louise of Austria, who then took Josephine's place at Fontainebleau. The couple's son, also named Napoléon, was born in 1811 and baptized in the gallery of the Chapelle de la Sainte-Trinité.

In 1813, Napoléon's forces were defeated in the Battle of Nations by an alliance of Great Britain, Austria, Prussia, Russia and Sweden. Napoléon retreated into France with his foes on his heels. In March 1814 the allied nations occupied Paris. In April Napoléon relinquished his throne. In a room at Fontainebleau today called the La Salle d'Abdication (Abdication Room), he signed a document that read: "The Allied Powers, having declared the Emperor Napoléon to be the only obstacle to the reestablishment of peace in Europe, the Emperor, faithful to his oath, declares that he renounces for himself and his successors the thrones of France and Italy, and that there is no personal sacrifice, even that of his life, which he is not ready to make in the interests of France." The allies sent Napoléon into exile on the island of Elba. Napoléon bade farewell to his guards from the horseshoe staircase overlooking the Cour du Cheval Blanc. He told them,

> Soldiers, my old companions in arms, whom I have always found on the road to honor, the time has come for us to part. I might have stayed longer among you, but it is useless to lengthen the struggle. It might even lead to civil war, and I will not again tear the bosom of France. Be happy and enjoy your well-earned repose. As for me, I desire no pity. I still have a mission to perform, to make known to posterity the great things we have done together. I should like to embrace you all, but as that is impossible, this flag shall take your place.
> (translated by Frances M. Gostling)

He embraced the flag of the regiment, got into his carriage, and drove into exile. Since then, the Cour du Cheval Blanc also has been known as the Cour des Adieux (Court of the Farewells).

Napoléon stayed on Elba for less than a year. In February 1815 he set sail from the island with supporters who had followed him into exile. He landed in Cannes and began marching toward Paris. Troops who were sent from Paris to arrest him instead hailed him as their emperor and joined his ranks. Napoléon stopped at Fontainebleau, where he reviewed his soldiers before leading them into Paris on March 20.

Napoléon's second reign ended with his defeat at the Battle of Waterloo on June 18. He abdicated for a second time on June 22. He was banished to the island of St. Helena in the South Atlantic Ocean, where he died in 1821.

Today, the rooms Napoléon used at Fontainebleau—

his study, bedroom, map room; Josephine's bedroom, boudoir, sitting room; the music room and throne room—are decorated with such relics as a lock of his hair, the cradle of his only son, the hat he wore when he returned from Elba, and a painting by Girard of Napoléon in his coronation robes. In La Salle d'Abdication is a copy of his first abdication, lying on the table on which he signed it.

After Napoléon's final abdication, further restoration was carried out in turn by Louis XVIII, Charles X, and Napoléon III. Napoléon III, in particular, spent large sums of money on the palace. He enlarged the library at Fontainebleau to 50,000 books, which now are stored in the Galerie de Diane, sometimes called the Library of Napoléon III.

During the 1830s, a group of artists known as the Barbizon school developed in the village of Barbizon near Fontainebleau. Members of the school—such as Jean-Baptiste-Camille Corot, Jean François Millet, and Théodore Rousseau—painted scenes of rural beauty in soft, muted colors. During his twenty-seven years in Barbizon, Millet painted such well-known works as *Angelus* and *The Reapers*. He and Rousseau are buried in the graveyard of a church seen in *Angelus*.

A frequent visitor to Fontainebleau during the 1870s was Scottish writer Robert Louis Stevenson, who lived nearby. During one visit to the palace, he met Fanny Osbourne, who later became his wife.

During World War II, Fontainebleau was used as the headquarters of German General von Braunitsch. It was liberated by American general George Patton in August 1944 and was the military headquarters for the Allied forces in Europe for several years afterward. During the 1960s and 1970s, the Galerie François I was restored, and today the château remains one of the great showplaces of France.

Further Reading: A good general history of the château up to 1950, in French, is Albert Bray's *Le Château de Fontainebleau* (Paris: Vincent, 1956). In English, there are sections on Fontainebleau in Frances M. Gostling's *The Lure of French Chateaux* (New York: McBride, 1926) and F. J. Forster's *Excursions Out of Paris* (New York: Pageant, 1965). Some information on Fontainebleau also is available in a biography of Francis I, *Francis in All His Glory,* by Burke Wilkinson (New York: Farrar Straus, 1972). On specific features of the château, "The Iconography of the Galerie François I at Fontainebleau" by Erwin and Dora Panofsky in *Gazette des beaux-arts* (Paris), series 6, volume 22, September 1958, pages 113–190, is worth reading, and Kenneth Woodbridge's *Princely Gardens: The Origins and Development of the French Formal Style* (London: Thames and Hudson, and New York: Rizzoli, 1986), provides information on the Fontainebleau Gardens that is not available elsewhere.

—Mary Feely

Fribourg (Fribourg, Switzerland)

Location: The canton, or state, of Fribourg covers about 645 square miles in western Switzerland, bordered on the north and east by the canton of Berne, on the south and west by the canton of Vaud, and on the west and north by Lake Neuchâtel. The city of Fribourg, capital of the canton, is located on a peninsula on the Saane River, about twenty-two miles southwest of the city of Berne, and thirty-three miles northeast of the city of Vevey. It is on the main railway between Zurich and Berne, and Lausanne and Geneva.

Description: Fortress town built by Duke Berthold IV of Zähringen in the mid-twelfth century; medieval center of trade; bastion of Swiss Catholicism; center of renewed trade after the introduction of rail transport in the nineteenth century.

Site Office: Office du Toursime Région Fribourg
Square des Places 1
CH-1700 Fribourg-Suisse, Fribourg
Switzerland
(37) 81 31 75

The city of Fribourg was founded in the twelfth century as a military outpost protecting a ford across the Saane River. The region, which now includes both the city and the canton, was settled much earlier, however. Artifacts found near Lake Neuchâtel, the northwestern border of the canton, indicate that prehistoric people lived there during the Neolithic Era. In the latter part of the Bronze Age (1200–900 B.C.) Celtic tribes crossed the Alps, settling parts of what is now Switzerland. These were the forebears of the Celtic-Helvetian tribes Julius Caesar's troops conquered in 58–57 B.C. to control the passes leading through the Alps.

In the middle of the third century, as the Roman Empire began to decline, Germanic tribes invaded and, over the next two centuries, settled the area east of the Aar River, northwest of Fribourg. Burgundians also were settling along the middle Rhine area and west of the Aar, and later the Franks followed. By 507, the Franks had conquered the region west of the Rhine, western Switzerland, and Burgundy. In the mid–seventh century, Frankish rulers had established themselves as kings; local political power was exercised by counts acting under their authority. After the death of the Frankish king Charlemagne, the empire was divided under the Treaty of Verdun in 843, making it vulnerable to Viking marauders from the north and the Magyars from the east. What is now Switzerland eventually passed in 1033 to the Holy Roman Empire, a loose confederation of European states ruled by Emperor Conrad II. The pope assumed religious leadership of the empire.

Because of the inability of the Frankish monarchy to protect its empire from invaders, feudalism became the expedient form of rule. Under feudalism, landless freemen offered their services to wealthy, powerful landowners in return for shelter, money, and protection. Powerful European families such as the houses of Habsburg, Savoy, Zähringen, and Kyburg came to acquire immense holdings in and around modern-day Switzerland.

The Zähringens, for instance, founded a number of towns strategically located for military advantage, one of which was Fribourg. In 1157, Berthold IV, duke of Zähringen, had substantial power as a landowner, and governed other areas as rector of Burgundy. He sanctioned the building of Fribourg or, in German, Friyburg (Free Town) on a steep hill overlooking the Saane River as a stronghold to protect the ford at that site, and as a secure place for freemen in the area against powerful neighbors. Some of those neighbors, higher nobles, saw this as a threat and tried to intervene. So, while building the town walls, the workers had to hire mercenary soldiers to protect them and the outpost. The Church of Notre Dame also was constructed about this time.

Although the stronghold was for the protection of all persons under Berthold's authority, numerous lesser nobles soon established an aristocratic hierarchy. For instance, they obtained permission from the bishop of Lausanne for the right to be buried in specific monasteries, apart from the other townspeople. This artistocractic reputation followed Fribourg into the nineteenth century. The city's diversity of language—part of the town still speaks German, part French—is another vestige of this period, emphasizing the fact that it was founded by Burgundians and Germans.

Berthold IV died in 1185 and was succeeded by his son Berthold V, who left no heirs. As a result, the Zähringen holdings, including Fribourg, passed to the house of Kyburg, related by marriage, when Berthold V died in 1218. The Kyburg dynasty, itself, came to an end in 1264, and much of central Switzerland, including Fribourg, passed to Habsburg ruler Rudolf IV in the 1270s. Rudolf IV had been elected king of Germany and Holy Roman Emperor in 1273.

While Swiss communities such as Uri, Schwyz, and Unterwalden moved toward independence in the thirteenth and fourteenth centuries as the Everlasting League (forerunner of the Swiss Confederation), Fribourg continued as an Austrian possession under Habsburg control. At the behest of Austria, Fribourg attacked Savoy in 1447, drawing retaliation from Bern, its ally. Bern invaded and defeated Fribourg, forcing the city to pay damages to Savoy. The citizens of Fribourg subsequently revolted, seceded from Austria in 1452, and two years later received a limited status in the Swiss Confederation, under the protection of Berne.

In 1476, Fribourg joined the Swiss Confederation in

Tympanum over the entrance of St. Nicholas Cathedral, Fribourg
Photo courtesy of Swiss National Tourist Office

the Burgundian War against Duke Charles the Bold, who wished to expand and restore Burgundy to the status of a sovereign kingdom. In early 1476, Charles captured a Swiss garrison in a castle at Grandson, near Lake Neuchâtel, and executed all but two men, who were spared to help with the executions. Two days later, on March 2, soldiers from Fribourg joined a force totaling 18,000 men gathered from all parts of the confederation and, shouting the war cry, "Grandson!" they routed the Burgundians. Although Charles soon returned with 20,000 men for an extended campaign, the Swiss ultimately prevailed. In January 1477, they drove the Burgundians from Nancy, killing Charles in a marsh as he tried to escape. These victories over one of the world's greatest armies secured Switzerland's reputation as a military force, and its independence from countries that otherwise might have considered a military invasion.

Due in part to Fribourg's assistance against Charles the Bold, the city was admitted into the Eternal Pact, or Swiss Confederation, in 1481, along with Solothurn. These were the first additions to the pact in 128 years, since Berne in 1353. Historically, the rural cantons had feared that the confederation, composed mostly of urban cantons, would become too powerful, posing a threat to their own independence. Therefore, they opposed expansion of the confederation, and the inclusion of Fribourg and Solothurn nearly triggered a civil war. Nicholas of Flue, a hermit canonized in the twentieth century, arbitrated an agreement in which the two towns would be admitted on the condition that they relinquish other outside alliances.

Although Fribourg, like most of the Swiss cities, enjoyed a democratic form of government at this time, a few wealthy families gathered an increasing amount of power. By the sixteenth century, democracy had given way to oligarchy throughout Switzerland. The dominant families increased their power through alliances, marriages, purchases, and military takeovers. Their tactics, combined with growing unrest in the underclass, fueled rivalries between cities and cantons.

The tenuous unity was put further at risk by the Reformation. The Reformation movement was led in Switzerland by Huldrych Zwingli at Zurich, Desiderius Erasmus at Louvain, and Frenchman John Calvin at Geneva. While the specifics of their beliefs differed, they all inveighed against corruption in the church. Zwingli and Calvin attacked church doctrine where they saw it conflicting with the Bible and challenged authorities on such issues as eating meat during Lent and the right of clerics to marry. Erasmus protested abuses in the church and wanted to return to what he saw as the high standards of the days of the church's beginnings. This, he would do through the study of the Bible. The Swiss Reformation began in 1523 with Zwingli's preaching in Zurich and quickly spread, mostly through the urban cantons. Erasmus, after being forced out of Louvain, settled first in Basel, then in Fribourg from 1529 to 1535.

Although the Reformation established a small beachhead in Fribourg, and even while its neighbors were turning Protestant, the city ultimately remained Catholic. This outcome was largely the result of efforts by Peter Canisius, a Jesuit sent by the church to combat Protestantism with public debate and education. St. Canisius, the first German Jesuit (canonized in 1925), traveled throughout Europe preparing Catholics to meet Protestant arguments. He founded twenty Jesuit colleges from 1549 to 1580, the last of which was in Fribourg. He also served as cathedral preacher in Fribourg. The tomb of St. Canisius is located in the former Jesuit Church of St. Michael, built from 1604 to 1613.

The Reformation continued for centuries to have a divisive effect on Switzerland. As the oligarchies amassed more wealth and power, rebellions by the underclass, inspired by revolutions in France and America, became more frequent. Some Swiss firebrands were forced to flee to France, where they tried, and ultimately succeeded, to convince the French Revolutionary government to intervene. In 1798, a French army under General Guillaume-Marie-Anne Brune invaded Switzerland and captured Bern, Solothurn, and Fribourg, as well as other cantons. A year later, Napoléon Bonaparte, then a brigadier general, came to power. He emptied the coffers of Bern, one of the richest Swiss cities, unseated the aristocrats of Fribourg and the other cantons, and established the Helvetic Republic.

The failure of this attempt to centralize the Swiss government resulted not only from internal strife, but also from the fact that, in the late eighteenth century, Russia, Austria, and France were fighting the war of the Second Coalition on Swiss soil. In 1803, Fribourg was re-established under Napléon's Act of Mediation, which created a new Swiss Confederation. At this time, Morat, now Murten, was added to Fribourg, enlarging the canton to its current size. The new confederation established Switzerland as a neutral state made up of cantons governed by limited self-rule. Napoléon declared himself Mediator of the confederation, making Switzerland subservient to France. With the defeat of Napoléon at the "Battle of Nations" in Leipzig in 1813, and his subsequent abdication as emperor a year later, an Austrian army entered Switzerland. Austria, England, Russia, and Prussia brokered the Pact of Confederation, a loose federation of self-ruled cantons and a federal assembly. In 1815, the Treaty of Paris guaranteed a permanent neutrality to Switzerland.

Following the signing of the treaty, the Swiss oligarchies resurfaced. Their grip on authority, however, began to slip over the next fifteen years, as rural cantons rebelled against being dominated by the richer city cantons. The July Revolution in Paris in 1830 encouraged citizens in Switzerland to demand freer trade, a freer press, more religious tolerance, and a more democratic form of government. In 1833 Basel divided, with half the canton seceding from the confederation.

In 1841, the secularization of the monasteries in Aargau further aggravated religious tenisons between Catholics and Protestants. Four years later, a civil rebellion erupted over the question of whether to allow Jesuits in Lucerne into the confederation. Protestant volunteers marched on Lucerne and, although they were easily repelled, the action prompted Fribourg and the six other Catholic cantons to form a union for mutual protection—the Sonderbund. The federal diet outlawed the Sonderbund in 1847 and sent an army to put down the rebellion, which it did within a month, first by taking Fribourg. Subsequently, the Jesuits were expelled and the seven cantons of the bund fined.

The most important result of the war was the creation of the 1848 Federal Constitution, which led to the present Swiss democracy. It established the current political structure, relegating foreign relations to the federal government, guaranteeing freedom of press, equality before the law, freedom of movement, association, and religion. The first rail system was built in 1847 and, after numerous disputes over control, contracts, and use, it was nationalized. Berne was declared the capital in 1855 and the Swiss Federal Institute of Technology was created in Zurich.

Today, the canton of Fribourg remains primarily agricultural. The capital city, a center of commerce in the Middle Ages, regained that reputation with the advent of the rail system. Fribourg is on the main railway from Lausanne to Berne and connects with Morat and Payerne. Beginning with the nineteenth century, it also enjoyed a resurgence in weaving and tanning enterprises, which had faltered after the fifteenth century. The city now also produces chocolate, cardboard boxes, machinery, electrical equipment, clothing, and chemical products.

Fribourg continues to reflect Swiss parochialism, or *Kantonligeist*—"little-canton-mindedness." Even the city, itself, is subdivided, with the oldest part of the city, the Bourg, located well above the river bank and the rest of the city located on the lower section of the bank. A third area, a modern development, is located on a still higher area, west of the Bourg. Although there are four national languages in Switzerland—French, German, Italian, and Romansh, a Latin dialect—French and German predominate, with about two-thirds of the people speaking German. In Fribourg, however, two-thirds of the populace speak French, the rest, mostly German. About 57 percent of Swiss are Protestant,

with 41 percent Catholic. Fribourg, however, which resisted the Reformation, is staunchly Catholic.

The city's medieval architecture, particularly in the Bourg, has earned it the reputation of being "the flower of the Gothic age." This section includes many buildings and homes with Gothic facades, as well as towers, walls, and gateways dating from the thirteenth to seventeenth centuries. Medieval structures include the thirteenth- to fifteenth-century St. Nicholas Cathedral; the twelfth-century Church of Notre Dame; thirteenth-century Église des Cordeiliers, a Franciscan church; the convent of Maigrauge with a restored thirteenth-century church; another thirteenth-century church, the former Augustinian Church of St. Maurice; the former Jesuit Church of St. Michael dating from the seventeenth century and containing the tomb of St. Peter Canisius; and the town hall, dating from early sixteenth century, with a tower built in 1642. Other notable structures include the nearby Hauterive Abbey, about four and one-half miles southwest of Fribourg, a Cistercian abbey dating from the twelfth century, and several bridges on the Saane River, including the seven-arched Pont de Zähringen bridge, and the Fribourg Bridge, built from 1830 to 1834.

Further Reading: *A History of Switzerland* by James Murray Luck (Palo Alto, California: Society for the Promotion of Science and Scholarship, 1985) offers a comprehensive look at Switzerland, with a fair amount of detailed information about cities and cantons, such as Fribourg. *A Short History of Switzerland* by E. Bonjour, H.S. Offler, and G.R. Potter (Oxford and New York: Oxford University Press, 1952; Westport, Connecticut, and London: Greenwood, 1985) presents another straightforward look at Switzerland, including maps dividing the nation linguistically and others potraying the country at various periods in its history. *The Swiss Without Halos* by J. Christopher Herold (New York: Columbia University Press, and Oxford: Oxford University Press, 1948; reprint, Westport, Connecticut, and London: Greenwood, 1979) provides a subjective and often humorous look at Swiss history. *The History of the Helvetic Confederacy* by Joseph Planta (London: John Stockdale, 1807) is a detailed, though dated and somewhat subjective, look at Switzerland, written by the principal librarian of the British Museum at the turn of the nineteenth century.

—Richard G. Wilkins

Fulda (Hesse, Germany)

Location: In the state of Hesse, sixty-two miles northeast of Frankfurt.

Description: Lying on opposite banks of the River Fulda, the city of Fulda was founded in 1114. An important center of Christian history since the foundation of a Benedictine monastery in the eighth century, Fulda's association with the Catholic Church has left a heritage of magnificent medieval and baroque buildings.

Site Office: Verkehrsbüro
Schlossstrasse 1
36037 Fulda, Hesse
Germany
(61) 10 23 45

Fulda grew up around its abbey, a Benedictine monastery with a continuous history of almost 1100 years from 744 to 1803. The founder of the Abbey of Fulda and its first abbot was Sturm. He was a pupil of St. Boniface, the Anglo-Saxon missionary and martyr who preached Christianity and organized its dissemination among the pagan Saxons of the mid-eighth century. Sent as an oblate at the age of five to a Benedictine monastery in Exeter (southwest England), Boniface rejected offers of preferment in the Anglo-Saxon church in order to evangelize Germany. In 723 Pope Gregory II made Boniface peripatetic bishop of all Germanic Christians, and in 747 Boniface became archbishop of Mainz. Fulda was Boniface's favorite retreat towards the end of his life. Using the papal connection that supported the efforts of Boniface, Abbot Sturm placed Fulda at the center of a great missionary effort.

Early monasticism cultivated protective connections with princes and their courts, a virtual necessity given the circumstances of those violent times. The cooperation between church and princes shaped the political organization of northwest Europe, once the barbarians who had invaded the former Roman Empire had been either beaten or accommodated. The successful expansion of central power produced kingdoms that lasted, governed by rulers who could marshal sufficient support to usurp the traditional rights of local potentates. Royal power in turn shaped the ambience of monastic communities. Fulda became the mother house to a number of monasteries and nunneries that adhered to the *Regula* (rules) and teaching of St. Benedict, and that were settled partly by monks and nuns from England. Successful missionary efforts had to be cemented by the piety of local nobles who could endow churches and monasteries. The ties between the missionaries and the local aristocracy often aroused the resentment of the local population, however.

Relations between Fulda Abbey and the secular authorities in the town and surrounding areas were sometimes stormy, due to the numerous exemptions from local aristocratic and ecclesiastical control mutually conferred by church and nobles throughout the vast possessions of the religious community. So there was a need for tact during periods of reform, since members of the cloister were often connected with powerful families; these formed a vested interest through their ownership of property and local or national offices of government. The last thing a reformer needed was an area full of aggrieved and powerful men, such as had appeared during England's ninth-century ecclesiastical reforms. However, the cloister continued to provide missionaries to the Saxons, which was especially encouraged by Charlemagne. Fulda and other church institutions in return helped bring the former Saxon lands "firmly within the Romano-Frankish cultural orbit," as historian Roger Collins puts it.

Fulda grew wealthy from gifts of land and money by greater and lesser aristocratic families from all over the eastern half of the Frankish realm. The first buildings to comprise Fulda Abbey were consecrated on March 12, 744, the feast of St. Gregory the Great, pope, and inspiration behind England's conversion from Celtic to Roman Christianity more than a century earlier. A basilica dedicated to the Holy Savior, built over the tomb of Boniface, was completed in 818. The Michaelskirche, an imitation of the Holy Sepulchre in Jerusalem and finished in 822, is the principal physical monument to the activities of Fulda's early abbots, and it is still in use. Placing Fulda under the special protection of the papacy ensured exemption from future episcopal interference, and from early in its history Fulda was granted special privileges, or "liberties," by the Carolingian kings. A branch of Charlemagne's family settled east of the Rhine. Pépin III, the Short, father of Charlemagne, made Fulda a royal abbey in 764; Carloman, Pépin's brother and co-ruler, added further liberties, and Charlemagne continued the patronage begun by his father. The favored position of Fulda depended on the continued goodwill of the ruling dynasty. The abbots cultivated the royal connection and were rewarded by a stream of charters of gift, confirmation, and exemption. When the desired charters were not forthcoming, Fulda, not alone in this practice, forged them. Fulda's abbots sought to fend off intrusions from other laymen, since they were opposed to lay domination. To manage the influence of different secular authorities they encouraged the notion that anointed kings were closer to divine power than other worldly potentates.

The huge landholdings of such monasteries as Fulda were the source of heavy civic responsibilities as well as great economic rewards. Monastic ideals included poverty and seclusion from the world. But there is nothing quite so expen-

Statue of St. Boniface at Fulda's cathedral
Photo courtesy of German National Tourist Office

sive as supporting royal favor. Fulda, being a royal monastery, took up a rather different position from cloisters patronized by less elevated mortals. Perpetual prayers for the king and kingdom were the least of the monks' duties. Visits from the royal family to Fulda or its dependent houses in other parts of the realm meant feeding many dozens of extra mouths. Payments of agricultural produce had to be made to the royal palaces. Administrative and clerical services were due to the king. The provision of military equipment for the imperial treasury was not unknown; indeed, one former scholar of Fulda, Servatus Lupus, when Abbot of Ferrières, was at one point required by Charlemagne to lead a contingent of the imperial army into battle—refusal to do so would have cost more than the principle was worth.

So important a place as Fulda was bound to suffer from intrusions of the world beyond the walls of the cloister, which could have unsettling effects. Fulda's history is not free from violent eruptions of both clerical and lay opinion as well as deviations from Benedictine rule. The abbots' responsibilities were particularly vulnerable to pressures from outside the community. The needs of the church and the monks had to be satisfied from the landed possessions that the monks' families had given to the monastery to support them for life. But there remained great swaths of property that the abbot, as head of the community, could put to what use he liked. As early as 812, the monks at Fulda complained directly to Charlemagne that their abbot was using monastic property for other purposes. A few years later, Louis the Pious heard a similar complaint from Fulda. He established some rules concerning the management of Fulda's holdings and, in 817, sent St. Benedict of Aniane, "the second Benedict" as he became known, to oversee the reforms. In the early tenth century the cloister had again to be reformed by the introduction of monks from Scotland, sent to revive the true spirit of the Rule of St. Benedict. By 1013 Fulda's eminence among the imperial German abbeys was confirmed when it became the center of a reform of the province of Lorraine instigated by Henry II.

The achievement of the great early reformers must not be underestimated. Reform meant that communities of laymen became communities of committed spiritual brethren. Freedom from local overlordship had first to be sought; then a self-contained religious community had to be built under royal or episcopal protection. Younger children sometimes posed problems to aristocratic families who did not wish to divide their property. These families often sought to place their children in monasteries, which needed them to maintain a supply of literate monks; but monasteries had to resist accepting children sent to them because they were in one way or another unfit for aristocratic responsibilities or for life in general. It is possible that Charlemagne's biographer Einhard was sent to Fulda, of which his parents were benefactors, because his very small stature meant that he would be useless as a member of a warrior class. Fortunately for all concerned, Einhard also turned out to be an exceptional scholar.

The impact of Christianity on pagan Europe is seen in the immense importance of organized religion at this time. In the Middle Ages, Christians built up institutions of a kind that today we would associate exclusively with the state, and religious loyalties were often more strongly felt than political allegiance. Religious communities played an important part in secular as well as religious affairs. The peculiar characteristics of the western church, notably the position of the papacy and the leading part played by monks, can be seen as a response to the conditions created by the decline in secular government and education after the fall of the Roman Empire. By about 850 the foundation of the medieval church had been laid, not so much the result of deliberate planning by anyone, but often the product of the response of a number of individuals. Popes, bishops, monks, emperors, kings, and laypeople contributed to a result that no one could have foreseen. Fulda became a major political center.

Fulda's intellectual achievements are as impressive as its political status. During the ninth century the abbey had more than 400 monks in its cloister. A school for the training of secular priests was attached to the abbey, thanks to Charlemagne's legislation. Fulda's scriptorium was responsible for the survival and transmission of texts by Greek and Roman authors, including Tacitus, Suetonius, Ammaianus, Vitruvius, and Servius. The practice of conserving old texts was inherited from Bede of Northumbria and the English School of York; it was carried on by Alcuin, who was at York before Charlemagne asked for his assistance at Aachen. There were problems to be overcome, however, before such work could succeed. The Anglo-Saxon monk Alcuin recognized the importance of overcoming the insular practices of particular scriptoria and established a common scribal standard. It was under Hrabanus Maurus (822–842), Fulda's greatest medieval abbot and builder in 838 of the Peterskirche, that the necessary training in grammar and reading was given, since texts were dictated to the monk-copyists.

Fulda's Anglo-Saxon influence also inspired the production of theological works. These works of the Carolingian schools have long been recognized as altered transmissions of earlier texts, but they were not simply "home-made quilts." Alcuin's tract on the Holy Spirit, *De fide Sanctae*, completed in 802, was thought important enough to be copied regularly up to the sixteenth century. By an editorial technique that resulted in "a kind of saturated paraphrase," Alcuin sought "to select, combine, juxtapose, add to, adjust, comment upon, and otherwise utilize a large body of literary fragments," as historian John Cavadini describes the technique. One of the legacies of the work of preservation was the development of new avenues of study, including the disciplines of the jurists of canon and civil law. The nascent German national literature was also developed at Fulda.

The pressure of renewed barbarian attacks and invasions from the north and the east rent western Europe from the end of the ninth century. Charlemagne's empire gave way to a number of successor kingdoms. His educational achievements were not lost. Fulda's reputation was enhanced in 968 when the pope declared its abbot primate over all Germanic and Gaulish abbots. For some time the abbey remained a model monastery responsible for transplanting its habits of observance and discipline to others throughout northern Europe. During the tenth and eleventh centuries such monasteries as Fulda remained the dominant centers of learning. They became the focus for a renewal of learning in the eleventh century. This period was perhaps the most brilliant in terms of the monks' production of illuminated manuscripts, murals, gold-work and sculpture. Teaching in monastic schools ensured this continuity of learning until the universities of Europe developed, principally from cathedral schools.

The Abbot of Fulda assumed the office of arch-chancellor to the empress and was granted pontifical rights in 1133. The enormous privileges granted the abbey had naturally led to its becoming a territorial power. In the early thirteenth century the abbot was elected a prince of the Holy Roman Empire. In 1294 the abbot's household was separated from the monastery proper, though this resulted in tensions among the cloister's noble occupants. On the abbey's estates, disaffection among tenants was not unknown, with particularly recalcitrant rebel vassals active in 1353 and 1395. It was the weakening of these bonds that eventually allowed the incursion of secular princes, to the detriment of the spirituality and the economy of Fulda.

The unity of Carolingian Europe had been maintained through the collaboration of princes and monks. Part of the bargain struck by the princes in return for their overlordship of monasteries was that the world would be kept out of the cloister and that an abbey's landed possessions would be sacrosanct. This arrangement was inimical to many a greedy local magnate at the best of times; it began to be questioned more closely in the sixteenth century, as princes kept reverting to type and compromising the governance of monastic empires, rivals to their own hegemony. A challenge to the huge liberties of religious landlords was posed when princes sought to cancel claims to land that generations of abbots would have meant to uphold. Strong abbots could resist such challenges; under them existing arrangements only gradually unraveled. One of the results of this struggle was the Peasants' War (1524–25), which should be seen against the background of strong princes at odds with Rome. The war was a temporary and exaggerated expression of what had been going on for centuries—a seesaw struggle between peasant communities and lords about their respective rights.

By the Reformation, Fulda Abbey had begun to lose some of its lands, and its position of authority in the Germanic church was compromised by the penetration throughout northern Europe of Protestant reformers. Abbot von Henneburg, who served from 1529 to 1541, showed some sympathy with the Reformation theology. However, the influence of the Counter-Reformation was ultimately stronger than that of the Protestant Reformation. In the sixteenth and seventeenth centuries Abbots von Dernsbach and von Schwenisburg led the recovery—with the assistance of reform monks from the great monastery of St. Gall—and resisted the dissolution of Fulda. The confirmation of Catholic supremacy in the region was also marked by the foundation in the early seventeenth century of a Franciscan monastery, the Frauenberg. In 1604 the town of Fulda was granted quasi-episcopal jurisdiction in the teeth of opposition from the neighboring cities of Wurzburg and Mainz. The see of Wurzburg included that part of Fulda that lay on the right bank of the river; the see of Mainz embraced the part of the town that lay on the left bank. Valuable revenues would be lost to the two original bishoprics. The grant of quasi-episcopal jurisdiction was confirmation that the exemption from control by local bishops granted to Fulda by the papacy in the ninth century was still a force to be reckoned with.

Secular ambitions again raised their head when Gustavus Adolphus (1594–1632) gave Fulda Abbey as a principality to the local magnate William, Landgrave of Hesse; but his sovereignty lasted only ten years. The violent upheavals of the Thirty Years War (1618–48) severely damaged both the abbey and the state of Hesse. Perhaps the greatest damage done was the scattering of the manuscript collection and the library of Fulda Abbey. (What few volumes remain of the formerly extensive collection are on display in the State Library in Fulda.) By the late seventeenth century, however, Abbots Kircher, Brouwer, and von Spee had accomplished a revival of Fulda's fortunes, and set the stage for further development. In 1727 the abbot of Fulda was elevated to the rank of auxiliary bishop. The University of Fulda, founded by the abbot, combined Benedictine and Jesuit schools from 1732 to 1802. In 1752 the town of Fulda became a full bishopric (the abbot a prince-bishop) under the Holy See; the monastic cloister kept their rule but became the cathedral chapter. In 1755 the diocese of Fulda became suffragan to the metropolitan bishopric based in the city of Mainz, though the cloister and Fulda's daughter houses still remained outside the bishop of Mainz's jurisdiction. Under Prince-Bishop Heinrich von Bibra in the eighteenth century, the monastic state saw improved educational and pastoral care, and the town of Fulda underwent an economic revival and considerable urban development. The manufacture of porcelain was a particularly important activity, the factory flourishing between 1761 and 1789. It was during the eighteenth century that the magnificent baroque quarter of Fulda was realized. The cathedral (1704–12), the castle (1706–21) and the Orangery (1722–25) are the main ornaments of this development.

After the Treaty of Paris (1802) the diocese of Fulda and the abbey were secularized and suppressed. Its possessions fell successively into the hands of the royal families of Orange and France and then those of the grand duchy of Frankfurt. Fulda's landed wealth was still coveted. This we see from the fact that the bishopric of Fulda remained vacant and in the hands of secular princes until 1829. In 1816 most of its possessions were ceded by Prussia to Hesse-Cassel and Bavaria. But in 1866 the lands of Fulda were annexed entirely by Prussia. The see was also kept deliberately vacant between 1873 and 1886, during the period of the "Kulturkampf'' (cultural struggle), when Otto von Bismarck waged a cam paign against political Catholicism in Germany. The intellectual life of Catholic Germany was, in part, still sustained by activities of the bishops of the diocese of Fulda. Important episcopal conferences took place near the tomb of St. Boniface from 1867.

The emergence of industrialization in Germany in the late nineteenth century led to the establishment of several industries in and around Fulda, including weaving, dyeing, and linen and other textile manufacture. Brewing and railway engineering were also important economic activities that started then. Agriculture was still important; Fulda was the local market for cattle and grain. The town has long benefited from the phenomenon of tourism. Many of the medieval and baroque buildings that were the background to the historic events of Fulda have become museums, such as the Schloss Fasanerie, which houses an important collection of Roman statuary.

Further Reading: A multitude of books on the Middle Ages cites the importance of Fulda and its abbey. These include *Early Medieval Europe: 300–1000* by Roger Collins (London: Macmillan, and New York: St. Martin's, 1991) and *The Barbarian West: 400–1000* by John M. Wallace-Hadrill (London and New York: Hutchinson, 1951; third edition, 1967). Also helpful are Fulda's own records, translated and annotated by Timothy Reuter as *The Annals of Fulda* (Manchester and New York: Manchester University Press, 1992). Various aspects of the abbey's history are highlighted in other works, including John Cavadini's ''The Sources and Theology of Alcuin's 'De Fide Sanctae' '' in *Traditio* (New York), volume 46, 1991; and David Ganz's ''The Preconditions for Caroline Minuscule'' in *Viator* (Berkeley and London), volume 18, 1987.

—Patrick Keeley

Gdańsk (Gdańsk, Poland)

Location: Gdańsk (or in German, Danzig) is a major north-central Polish port city on the Bay of Gdańsk, about three miles from the Baltic Sea. It is situated on the Motława River, a branch of the Vistula (two miles inland), Poland's chief waterway. Gdańsk, located about 170 miles northwest of Warsaw, is also linked with two smaller neighboring port towns, Gdynia and Sopot, in a conurbation called the Trojmiasto (Tri-City).

Description: Leading cultural, historic, and industrial center of northern Poland, and capital of the province of Gdańsk; site of buildings dating from the Middle Ages and Renaissance; important port city for nearly 1,000 years and for a time the busiest port in Europe; its shipyards, among the largest in the world, were the birthplace of the Solidarity trade union movement.

Site Office: Centrainy Osrodek Informacji Turystycznaj
Ulica Heweliusza 27
Gdańsk, Gdańsk
Poland
(58) 31-43-55

The city of Gdańsk, formerly known as the Free City of Danzig (its German name) between World Wars I and II, has long been one of the most important commercial ports on the Baltic Sea: as such it has served as a cosmopolitan, fiercely independent flashpoint for a series of historically significant struggles to define the Polish nation-state. Gdańsk has changed hands many times in its 1,000-year history, when Poland's boundaries shifted all over the eastern European map and sometimes vanished completely. More than that of any other city of eastern Europe, Gdańsk's history has reflected the millennia-long Polish resistance to the German colonizing movement to the east; the Poles traded claims to Gdańsk with the Teutonic Knights, the Prussians, and finally the Nazis. Poland's refusal to surrender Gdańsk to Germany led to Hitler's invasion of Poland in 1939, precipitating World War II. After the war ended in 1945, the heavily damaged city became part of Poland again. Gdańsk's shipbuilding industry is one of the largest in the world; workers strikes at the city's Lenin Shipyards in 1980 led to the creation of Solidarity, a trade union movement that helped bring about free elections in Poland—and the creation of a non-Communist government.

Gdańsk was founded in the tenth century, a time when west Slavonic tribes began to establish relatively stable kingdoms throughout eastern Europe. The Polish nation originated from five main tribal groups, two of which were the Polanie and the Pomorzanie. The local Kaszub-speaking Po-

morzanie, or Pomeranians, were living along the seacoast. The Polanie had settled farther south, in an area to be called Great Poland. "Poland" is held to have begun with a tribal union under the Polanie Piast dynasty in the mid-ninth century. In 966 Mieszko I solidified the state by adopting western Christianity, placing his lands under protection of the papacy. The Polanie extended its influence as far north as the Vistula River estuary on the Baltic Sea, where Gdańsk is now situated; in time, the Vistula became Poland's main arterial trade route, eventually making Gdańsk an international port.

Originally the site of a fishing village, Gdańsk existed as a fortified marine trading center on the left bank of the Motlawa, a branch of the Vistula (two miles inland), as early as 970. The city was first mentioned as *urbs Gyddanyzc* in 997, when it was an important port on the *Via mercantorum* (the merchant's route). The Piast dynasty ruled a strong and united Poland until 1138—though Piasts would remain in power in various parts of the nation until 1370. By the beginning of the thirteenth century, a complicated system of successions led to the establishment of separate provincial dynasties and independent principalities, which considerably weakened the state. Gdańsk was controlled by the Polish princes of eastern Pomerania, and became capital of the Pomerania principality. (In 1148, under Pope Eugenius III, *castrum Kdanzc*, along with eastern Pomerania, was part of the Polish diocese of Wloclawek; the first parish church was founded in 1185.) One of the Pomeranian princes, Duke Swientopelk, granted Gdańsk its municipal autonomy in about 1260. Before Mestwin II, son of Swientopelk, died in 1294, he had bequeathed the principality to cousin Przemysław II, duke of Great Poland, who held it until 1308.

Gdańsk developed into a rich trading center in the thirteenth century. In the late 1200s the city, already heavily "Germanized" through shipping and trade with Lübeck, became an important member of the Hanseatic League. The league was a merchants' guild, a federation of free towns in northern Germany and adjoining countries formed for the purpose of conducting maritime trade—and promoting economic advancement—between eastern and western Europe. Throughout the medieval period, the league brought many German burghers (middle-class merchants), traders, craftsmen, scholars, and artisans to Gdańsk and Poland. It was also at this time that the Gdańsk area, because of its coveted location, began to suffer a series of foreign invasions that have plagued Poland ever since.

In the thirteenth century, the military and religious Order of the Teutonic Knights, idle since their return from the Crusades, extended the German *Drang nach Osten* ("thrust toward the east"). In 1226 a powerful Polish duke felt threatened by the Prussians, a pagan tribe of Baltic origin who inhabited a region on Poland's northeast frontier, and he

Memorial to Gdańsk shipyard workers who died in 1970 protests

invited the knights to help him. By the end of the century, the knights had succeeded in conquering and Christianizing the Prussians, killing many in the process. The Teutonic Knights, however, proved more dangerous than the Prussians; they were soon warring with Poles along the Baltic coast. In 1308 the knights invaded Pomerania, which was still outside Poland and which was also being claimed by the dukes of the neighboring region of Brandenburg. The knights were called upon to expel the Brandenburgers, but they went farther than expected, capturing Gdańsk for their own and massacring many of its inhabitants. In 1309 the Teutonic Knights moved their headquarters from Venice to the magnificent Malbork (German, Marienburg) Castle, just south of Gdańsk; they had begun building the Gothic fortress in 1275. The order grew more powerful, and would remain in possession of what is now northern Poland for the next century and a half, a period in which German colonists greatly increased the region's prosperity. During this time, Gdańsk's grid-patterned Main Town (Glowne Miasto, now the center of modern Gdańsk) grew up perpendicular to the medieval port town. Products from the Vistula River basin, like rye and timber, were shipped to the west from quays along the Motlawa.

Despite the Teutonic Order's aggressive presence, Gdańsk flourished in the fourteenth century. It was a period of great building activity. The buildings included many big brick churches, most notably that of St. Mary, a Flemish-Gothic edifice that was begun in 1343 and completed in 1502. St. Mary's (known in Polish as Kosciol Najswietszej Marii Panny, or Church of Our Lady) is not only the largest church in present-day Poland and the entire Baltic basin; it is also the largest brick Christian church in the world, capable of accommodating 25,000 people. The dominating Town Hall (Ratusz Glowny), originally built in 1357–82, was destroyed during World War II but has been rebuilt into the Gdańsk Historical Museum.

The Polish kings, however, never renounced their rights to Gdańsk and Pomerania. Casimir III the Great (Kazimierz Wielki), who ruled Poland from 1333 to 1370, welcomed thousands of persecuted Rhineland Jews into the tolerant state, and granted them the right to be governed by their own laws and elected officials. In the late fourteenth century, Poland formed a historic 400-year union with Lithuania, the only pagan nation left in Europe, the frontiers of which extended north and east toward Moscow. The union of crowns—which became the Polish Commonwealth of the Kingdom of Poland and the Grand Duchy of Lithuania in 1569, making it a remarkably multinational state of many different ethnic origins and languages—began in 1386 when the Lithuanian prince Jagiełło married Jadwiga, Casimir III's young princess daughter. After ordering his people to adopt Roman Catholicism, Jagiełło was crowned king a year later as Władysław II, and ruled until 1434. The Jagiełłonian dynasty ruled into the late sixteenth century.

King Władysław II, who called himself *Pomoraniae dominus et haeres,* and the combined Polish, Lithuanian, and Tatar armies stopped the Teutonic Knights' expanionist drive by defeating them at the Battle of Grünwald in 1410. But the king failed to secure Gdańsk and Pomerania. In 1454 the Pomeranian nobility, along with Gdańsk and Prussian cities, renounced their allegiance to the knights, and appealed to Poland for aid in overthrowing the Teutonic Order (which had been excommunicated by the pope); in return, they offered their submission to the Polish crown. King Casimir IV (Kazimierz Jagiełło), son of Władysław II, who ruled from 1447 to 1492, proclaimed the incorporation of Gdańsk and Pomerania (as well as most of Prussia) into Poland in March 1454. This was followed by the Thirteen Years War (1454–66), when Polish/Lithuanian forces defeated the Teutonic Knights. The knights formally recognized the 1454 proclamation with the October 1466 Peace Treaty of Toruń; they retained a part of Prussia. The Germanized Gdańsk became a part of Poland—which it would remain for more than three centuries. The city of 30,000 inhabitants was granted local

autonomy by the king in gratitude for its loyalty, and Malbork Castle became a Polish royal residence.

Gdańsk reached its productive peak—its "Golden Age"—during the Renaissance. It became the major international port on the Baltic, as well as the greatest grain exporter in Europe. Gdańsk continued to prosper into the eighteenth century, when it was the most populous city in eastern Europe, with nearly 80,000 people—and when Poland was first "partitioned" off the map and Gdańsk became part of the Prussian Kingdom.

With the Teutonic Knights' stronghold over Gdańsk broken, Poland gained control over the all-important Vistula river trade—and the Baltic seaboard. As a democratic, independent port city without which the landlocked kingdom could not prosper, Gdańsk received many royal freedoms and privileges not granted to other cities; it was not compelled to share tax revenues with the crown, for instance. Gdańsk was, for all intents and purposes, a city-state within the kingdom. In the fifteenth century, the number of ships entering the port greatly increased, leading to the construction of towering grain silos and harbor cranes, some of which remain. The Gdańsk shipyard launched its first warship in 1568, and the Polish Maritime Commission was begun in 1572. At just the time when northern Europe's soaring population needed more grain supplies, wheat and rye were floated up the Vistula from the fertile plains of central and southern Poland to be sold at Gdańsk to German, Dutch, and Scottish merchants; between 1541 and 1754 Gdańsk's grain exports amounted to between 150,000 and 200,000 tons annually. Poland fed the growth of western Europe in the Renaissance just as North America's wheat fed the Industrial Revolution in Europe three centuries later. Gdańsk also gained fame during this period as a great European center of furniture making, weaving, and silver artisanship.

Gdańsk's "Royal Route," which dates from the sixteenth century, runs from the west end of Main Town down Long Street (ulica Długa) to Long Market (Długi Targ), the town square at the very heart of old Gdańsk. When kings entered the city on their annual visits, they would pass through a series of Renaissance gates which led to the medieval port. The first, at the entrance of Main Town, is High Gate (Brama Wyzynna), built in 1576. Next is Golden Gate (Brama Zlota), built in 1614. At the east end of Long Market, by the old harbor, is Green Gate (Brama Zielona), built in 1568, which doubled as a royal residence. The route is lined by stately old houses that once belonged to rich burghers who headquartered their powerful merchants' guilds in the nearby Artus Mansion, built in 1616.

The Jagiełło dominion greatly expanded to almost all of central and eastern Europe in the fifteenth and sixteenth centuries through dynastic marriages. Besides Poland and Lithuania, the kingdoms of Hungary and Bohemia soon owed allegiance to the Jagiełło crown. Bohemia brought the sixteenth century Reformation to Poland; Gdańsk, due to its strong German cultural influence, became largely Protestant. Also during this time the Polish nobility and gentry "magnates'' (the latter benefiting from the agricultural boom, and often wealthier than the kings) became the most powerful social class—a ''gentry democracy.'' While this class prevented the development of an absolute monarchy (unlike the autocratic systems rising on either side of Poland in Russia and Prussia), it also made the commonwealth difficult, almost impossible, to govern. When the last of the Jagiełłos died in 1573—a time when Poland was the largest kingdom on the European continent—the nobles and gentry won the unprecedented right to elect monarchs. The result: the magnates restricted royal power, and Gdańsk, along with the rest of Poland, became increasingly vulnerable to outside encroachments.

The seventeenth and eighteenth centuries saw numerous conflicts. The 1655–1660 Swedish wars left much of Poland destroyed, halting Gdańsk's economic growth. Subsequent wars with the expansionist Russian Kingdom and the Ottoman Empire further weakened the commonwealth, and Poland fell into slow, near-chaotic decline. In 1764 Stanisław II Augustus Poniatowski was elected to the throne, the last king of an independent Poland until the state was restored in 1918. Meanwhile, to the west, the Hohenzollern dynasty had renounced its allegiance to the Polish crown and became kings of Prussia (and later of the German Empire). In 1772 Frederick the Great (who ruled 1740–86) forcibly partitioned one-third of Poland's territory among Russia, Prussia, and Austria, leaving Gdańsk cut off from Poland and surrounded by Prussian territory; the Prussians established a customs house at the mouth of the Vistula. With the Second Partition of 1793, Gdańsk was invaded and came under Prussian rule. Port trade rapidly dropped, and the city lost much of its wealth. The Germans now called the city Danzig. With the Third Partition of 1795, the Polish state was completely abolished, and the commonwealth disappeared from the map of Europe. At this time, the Catholic Church became the sole preserver and defender of Polish culture and identity against foreign oppression.

French Marshal François Joseph Lefebvre besieged and occupied Danzig during the Napoleonic Wars in 1807, when it was first granted the privileges of a ''free city'' by the Treaty of Tilsit. The city was protected by two Polish regiments from the Grand Duchy of Warsaw, a short-lived puppet state that Napoléon carved out of the Polish territories annexed by Prussia. But Danzig's economic stature greatly diminished because of its territorial separation from Poland. After Napoléon's retreat from Moscow in 1812, Russians briefly captured Danzig. The city's senate sent a delegation to Paris and London in 1813, and to Vienna in 1814, to appeal for reunification with Poland. Instead, with Napoléon's defeat, the 1815 Congress of Vienna repartitioned Poland among Russia, Prussia, and Austria again, and Danzig became the chief town of the West Prussia province (what was nominally the Kingdom of Poland was under Russian patronage, and lost any semblance of independence from Russia after an unsuccessful 1830 revolt by the Poles). Though Danzig became industrialized, especially in ship-

building, it failed to regain its status as a great Baltic trading port.

Following 1830, the partitioned nation of Poland was subjected to systematic Russification and (Prussian) Germanization, both of which effectively suppressed Polish culture, language, and Catholicism. Four major insurrections by occupied Poles in the nineteenth century all resulted in heroic defeat. This led to both an intense nationalism and a mass emigration of political and cultural leadership, and, later, of the peasantry. After 1871—when Germany and Prussia united into the German Empire under the Prussian leadership of Otto von Bismarck, first chancellor of the Hohenzollerns—Danzig was a typical eastern German city. At the outbreak of World War I (1914–18), for example, only about 4 percent of Danzig's inhabitants were Poles or of non-German stock.

Czarist Russia fell in March 1917, and Poland declared its independence November 11, 1918, Armistice Day, which ended World War I. But new frontiers were in dispute. German troops were disarmed and driven out. Russia's new government recognized the Polish people's right to a free and reunited state, and U.S. President Woodrow Wilson urged the creation of ''an independent Polish state'' with ''a free and secure access to the sea.''

Under the June 1919 Peace Treaty of Versailles, when the map of Europe was redrawn by the Allied Powers, Pomerania and much of West Prussia were restored to Poland. The recognized state was also granted independent access to the Baltic Sea, the so-called ''Polish Corridor,'' which separated East Prussia from the rest of Germany. Danzig—natural port for the Vistula, as well as for most of Poland—was inside the corridor. Since the Allies did not want the overwhelmingly German city under direct German control, they declared Danzig a ''Free City.'' Under the supervision of a resident high commissioner appointed by the newly created League of Nations, however, Poland was to have free, unrestricted use of the port, and could collect customs duties. Danzig also had its own legislative assembly, or Volkstag, but Poland had the authority to conduct the city's foreign relations. The Free City of Danzig had a territory of 731 square miles and a population (1929 census) of 407,500, including 15,890 Poles.

Danzig's German-controlled assembly and the city's Polish administrators, however, were often in conflict over certain Versailles Treaty provisions. Since Poland had failed to secure outright possession of the port, it was prohibited from using Danzig as a naval base or munitions depot. In 1920 Danzig longshoreman refused to unload munitions destined for Poland during the three-year Polish-Soviet War, in which Russia took large Polish territories to the east in 1921.

The Polish government finally decided to build a port in its own territory, on the narrow corridor to the Baltic. Beginning in 1924, Poland poured millions of dollars into the construction of Gdynia, a small fishing village about fifteen miles north of the Free City on the Gulf of Danzig. The harbor was deepened, and docking facilities were built. In addition, Poland built fortifications to protect Danzig, including Westerplatte. Danzig Germans, who had been helping to boycott Polish exports, protested at the expected loss of trade. Though much of Danzig's former trade was channeled through the new port, Danzig also continued to prosper. Between the wars, it was mostly a port for bulk cargoes (coal, iron ore, timber); from 1913 to 1938, for example, the tonnage of ships entering port increased fivefold, and the annual amount of imports and exports rose more than threefold. Yet Gdynia rapidly became Poland's main seaport and naval base for commercial and military purposes, and soon surpassed Danzig in the volume of goods handled. Gdynia served as the major Baltic port until the outbreak of World War II.

Gdynia's remarkable growth became a serious threat to the Free City of Danzig, and the League of Nations Council tried for more than a decade to settle the controversy. Political and economic disputes between Poland and Danzig intensified with the rise of the German National Socialist (Nazi) Party. In March 1933, during the "Danzig Crisis," Germany's two-month-old Nazi government attempted to reduce the rights of the Polish harbor police; the attempt was repelled when Poland sent a regiment to reinforce the harbor's Westerplatte garrison—in defiance of both Germany and the League of Nations. When the Nazis gained a majority of seats in Danzig's Volkstag two months later, they formed a new city government. New senate President Hermann Rauschning took a conciliatory approach to Poland, but he was forced to resign in favor of hardliners who wanted a "nazification" of Danzig's political regime. The Nazis gained more seats in the 1935 elections. In October 1938 Hitler first demanded the return of Danzig to Germany, and territorial rights in the Polish Corridor. Poland refused. In March 1939 Hitler issued an ultimatum to Poland concerning Danzig and ordered the construction of a German-controlled road through Polish territory to East Prussia. The Germans provoked incidents in Danzig.

The first shots of World War II were fired on September 1, 1939, when the German battleship *Schleswig-Holstein* bombarbed Westerplatte's military depot for twenty-one days. On that same day, Hitler proclaimed Danzig a part of his greater Third Reich; German troops invaded the Free City and, despite heroic fighting by the Poles, captured it. Britain and France declared war on Germany. After the Soviet army invaded Poland on September 17, the state was partitioned between Nazi Germany (1939–45) and Soviet Russia (1939–41). Danzig was formally annexed to the Third Reich in October. Since Danzig was heavily German, it largely escaped the nazification policy being carried out elsewhere in Poland, which involved the mass removal, labor deportation, and extermination of Poles, as well as the repression of the Polish language and the Catholic Church. These Poles were replaced by German colonizers. During the war, some six million Poles were killed by Nazis, about half of the victims Jews.

At the February 1945 Yalta Conference, with the Soviet army occupying Poland and Danzig, the Allied Powers (Roosevelt, Churchill, and Stalin) created a Soviet-sponsored Polish Provisional Government; the Soviet Union took some eastern Polish territories. In March Danzig was the scene of heavy fighting between the German and Soviet armies. At the August 1945 Potsdam Conference, Polish lands previously held by Germany—Pomerania and most of Prussia—were returned once again to Poland; these territories included Danzig. Poland was now one nation again; its borders were shifted into the shape it has today—and roughly what it was in the tenth century. Germans were expelled, and Communists were elected in 1947.

Danzig, however, had been renamed Gdańsk. Severely bombed, the city was in ruins: 55 percent of its buildings were destroyed, and the medieval center of the city suffered great damage. The Soviets had dismantled industrial plants. Population had declined from 250,000 in 1939 to 118,000 in 1946. But reconstruction was undertaken, and completed by the 1960s. Historic buildings—St. Mary's Church, the Town Hall, old merchant houses on the Długi Targ, and other buildings of Gothic, Renaissance, and baroque architectural styles—were restored to their original conditions at high cost.

In 1970 Gdańsk was the center of protest against the Communist government. Steep, sudden food price rises and new incentive wage rules just before Christmas led to massive occupation strikes at shipyards along the Baltic coast. Thousands of workers from Gdańsk's Lenin Shipyards (now the Gdańsk Shipyards) marched in protest to the local Communist Party headquarters, eventually setting it on fire. Fighting broke out, and police opened fire on unarmed demonstrators, killing twenty-seven. The militia spilled more blood at other Baltic shipyards. A general strike movement began in factories all over Poland, with workers demanding wage increases and a price freeze. The 1970 workers' revolt resulted in a revocation of price rises and an increase in real wages.

A decade later, with Poland in economic crisis and workers better organized, Gdańsk once again was historically destined to serve as a flashpoint for a popular uprising, which transformed the nation—and the world. The industrialized port city, long a fiercely independent symbol of Poland's struggle to be free, gained international recognition as the birthplace of Solidarity and the root of Poland's—and Communist east-central Europe's—new movement toward democracy. The election of Karol Wojtyła, archbishop of Kraków, as Pope John Paul II in October 1978, and his visit to Poland a year later, also set the stage for the extraordinary events of the 1980s.

In August 1980 striking Lenin Shipyard workers, led by labor organizer Lech Wałęsa (who had taken part in the 1970 protests), demanded wage rises, lower prices, and other labor and political reforms. Other Baltic shipyard and Polish factory workers joined in the occupation strikes. With the August 30–31 Gdańsk Agreement, the government granted workers twenty-one concessions, including the right to form independent trade unions and the right to strike—unprecedented political developments in a Soviet bloc country.

Wałęsa was elected chairman of the national coordinating committee of independent labor unions, or Solidarity. By 1981 nearly ten million workers nationwide had joined Solidarity, which rapidly swelled into a popular revolutionary movement for democracy. Spurred by the fear of Soviet intervention, however, prime minister General Wojciech Jaruzelski imposed martial law in December 1981; Solidarity was outlawed, and demonstrations and strikes were banned. Wałęsa and other Solidarity leaders were arrested, but trade union members continued to meet secretly. Wałęsa was freed in November 1982, and martial law was suspended a month later. Solidarity was legalized in 1988. Poland's first free elections since the World War II Communist takeover brought Solidarity victory in June 1989. In August Tadeusz Mazowiecki became prime minister, the first non-Communist to head an Eastern bloc nation. After the fall of Communism in 1989, Wałęsa became president and head of state, in December 1990.

Today, Gdańsk remains one of the world's largest shipbuilding centers and the home of many architectural treasures that have survived the city's turbulent history. Many historic mansions still line the "Royal Route." The fourteenth-century church of St. Mary and numerous museums also are among Gdańsk's attractions.

Further Reading: *Hippocrene Companion Guide to Poland* by Jill Stephenson and Alfred Bloch (New York: Hippocrene Books, 1991) is an introduction to contemporary Poland, its people, culture, and history. It contains a chapter on "Gdańsk and the Beauty of the Baltic Sea." *Poland* by Marc E. Heine (New York: Hippocrene Books, 1987) is similar to the *Hippocrene Companion Guide,* but less travel-oriented; it contains more historical and cultural information region by region. Gdańsk is included in "The North." *The Struggles for Poland* by Neil Ascherson (New York: Random House, 1987; London: Pan, 1988) shares the experiences of a British journalist who had covered Poland for thirty years. It is an insightful examination into the complex, chaotic historical forces that have shaped a state, a nation, and its people, from 966 to 1986.

—Jeff W. Huebner

Geneva (Geneva, Switzerland)

Location: At the southwestern corner of Lake Geneva in the Rhone valley, in the Geneva canton of Switzerland, 100 miles from the French city of Lyons.

Description: Capital of Geneva canton and headquarters for many international organizations; frequent site of international political conferences; and center of theological, intellectual, and scientific activities.

Site Office: Office du Tourisme de Genève
1, rue de la Tour de l'Ile
Case Postale 5230
CH-1211 Geneva 11, Geneva
Switzerland
(22) 28 72 33

The city of Geneva, located in the French-speaking portion of Switzerland, is a center of European cultural life and international diplomacy. With a tradition of independence, it has provided a refuge over the centuries for persons whose ideas made them unpopular elsewhere—persons as diverse as the Protestant theologian John Calvin and the Enlightenment satirist Voltaire. Its independent spirit, in addition to its neutrality in times of war, also has made it a natural home for international organizations, about 200 of which maintain headquarters in Geneva, including the International Red Cross, the World Council of Churches, the World Meteorological Organization, the European Center for Nuclear Research, and the World Health Organization.

The area's first settlement occurred at the end of the Ice Age, at the foot of Mont Saleve. About 2500 B.C. a large lake village, consisting of dwellings built on piles standing in the water, was at Lake Geneva's western end, in what is now the modern port. Later the inhabitants moved to a hill on the left bank, now occupied by the city's old town district. The first fortified settlement was built on the hill by a Celtic tribe, the Allobroges, who were conquered by the Romans in 120 B.C.

The town was first called Geneva in the writings of Julius Caesar, who described a walled town on the Rhone occupied by Gallic people, through which he passed in 58 B.C. Caesar ordered the bridge over the Rhone destroyed to delay the advance of the Helvetii into Gaul. By A.D. 379 Geneva was a Christianized Roman city and the seat of a bishop's office. As the Roman Empire declined and other peoples began to invade outlying areas, Geneva was taken by the Burgundians, who ruled it from 443 to 534. In 534 the Franks, in turn, defeated the Burgundians, and ruled Geneva for a time. It eventually became part of the Frankish kingdom of Lotharingia, when the Carolingian lands were divided up in 843.

The area was contested for years, however, and in 888, as feudalism was dawning, Geneva passed to the second Burgundian kingdom. In 1033 both Geneva and Burgundy became part of the Holy Roman Empire, as a result of a bequest by Burgundian king Rudolf III, who died without an heir, to Holy Roman Emperor Conrad II, who had married one of Rudolf's nieces. In 1162 the empire granted Geneva's bishop the right to rule the city, and thereafter the bishops were prince-bishops of the Holy Roman Empire.

One of Geneva's most notable structures, the Cathédrale-St-Pierre (Cathedral of St. Peter), dates from this period, having been built from about 1150 through 1232. The building has been remodeled over the centuries and today displays a mixture of Romanesque, Gothic, and neoclassical styles. During the Reformation the Calvinists would change it from a Catholic cathedral to a Protestant church, the Temple de Saint-Pierre, and it was the site of preaching by John Calvin. Archaeolgical excavations have revealed foundations of other churches from as early as the fourth century under the cathedral.

At the close of the thirteenth century the people of Geneva, tiring of rule by the prince-bishops, tried to establish their own municipal government. They sought the aid of the count of Savoy, Amadeus V, against their bishop. But the count not only seized the bishop's castle; he declared himself the official through whom the bishop exercised his judicial powers. The people had traded rule by the bishop for rule by the counts of Savoy. In 1401 Amadeus VIII of Savoy, elected pope as Felix V, named himself to the see of Geneva, and through 1522 this seat was frequently occupied by a member of the house of Savoy.

Geneva began to grow in the fifteenth century. Commercial establishments were built along the banks of Lake Geneva and the Rhone, and the town expanded north into the suburb of Saint-Gervais. Trade fairs flourished in Geneva at this time, with cloth from neighboring Fribourg being among the commodities sold there.

In 1530 the Genevese appealed to the cantons that formed Switzerland for help in eliminating the house of Savoy's domination. Swiss troops intervened and forced the duke of Savoy to sign a treaty in which he agreed to give up control of Geneva. A system of fortifications was built around the city at this time and the citizens, cut off from territories with which they had traded, began to concentrate on banking and such trades as watchmaking, which they learned from French Huguenot refugees.

The Genevese welcomed the Protestant Reformation, which rejected or modified some of the doctrines and practices of the Roman Catholic Church. Geneva became a haven for refugees fleeing persecution in France, Italy, and England, including many writers, intellectuals, and politicians.

295

Cathédrale St. Pierre, Geneva
Photo courtesy of Swiss National Tourist Office

In 1535 Geneva officially became a Protestant city and declared itself a republic. A city council took over the municipal powers once held by the bishop.

John Calvin, a French Protestant refugee, came to Geneva in 1536 and joined forces with Guillaume Farel, a reformer who began preaching the new faith in Geneva in 1532. Calvin's doctrines included the rejection of papal authority and the concept of salvation by faith alone. Calvin was not immediately successful in winning over the Genevese to his way of thinking, and was driven out of the city in 1538. He returned triumphantly in 1541, however, and strongly influenced the city's ecclesiastical and state affairs, in effect establishing a theocracy based on strict church discipline. Many people fled Geneva to escape the austerity of Calvinism; many others, however, were drawn to the reformer, and the cathedral became Calvin's church and was expanded to accommodate the throngs who came to hear him preach.

In 1559 Calvin founded the Academy of Geneva, which would later become Geneva University. Founded chiefly to train theologians, the institution also gave the citizens of Geneva an interest in intellectual matters. Calvin used a former Catholic chapel, next to the cathedral, as his Auditoire, a lecture hall where he taught his doctrines to missionaries. John Knox, founder of the Church of Scotland, preached in the lecture hall from 1556 to 1559, and Calvin encouraged Protestant refugees to hold services there in their native languages.

Calvin, who had become the virtual ruler of Geneva, died in 1564, and Geneva was surrounded by hostile neighbors who threatened its independence. Charles Emmanuel, the duke of Savoy, hoped to restore Geneva to Catholicism. On December 11 and 12, 1602, he sent troops on a surprise attack, prepared with collapsible ladders to scale the city walls. But they were defeated when Mère Royaume, wife of the director of Geneva's mint, seeing the Savoyards climbing the walls, emptied a pot of hot soup over their heads and then gave the alarm. The duke's army was driven out with considerable losses. The Genevese celebrated their victory by gathering in the former cathedral to sing the 124th Psalm. The victory is commemmorated by the annual Festival of the Escalade, when the battle is reenacted. Many citizens wear costumes, and children are treated to chocolate versions of the soup pot filled with marzipan vegetables. On July 11, 1603, the duke of Savoy signed the Treaty of St. Julien, which ended the house of Savoy's attempts to take over Geneva and recognized the city's independence.

The city grew prosperous and gained more land through agreements with France and Sardinia. Geneva's banks established branches throughout the world, the city constructed beautiful public buildings, and a powerful patrician class developed. A further influx of religious refugees came after France revoked the Edict of Nantes in 1685; the edict had granted French Protestants freedom of religion since 1598. Geneva nurtured the Enlightenment thinkers of the eighteenth century as well; the philosopher and writer Jean-Jacques Rousseau was born there, and the satirical author Voltaire, having made enemies in France, England, and Prussia, found a haven there.

In 1798, however, during the French Revolutionary Wars, Geneva was annexed by the French and was reduced to being a French departmental capital. Geneva's economy suffered, and the city's population ceased growing. The Genevese allied themselves with France's enemies and thus were in a good bargaining position after the fall of Napoléon. The republic of Geneva was restored in 1814 and on September 12 of that year became the twenty-second canton of the Swiss Confederation.

Middle- and working-class Genevese began to demand more power in the city's government and obtained some concessions in 1841. In 1846 the conservative government was overthrown and a radical group wrote a new constitution in 1847 establishing a democracy with universal male suffrage. (Women would not win the vote until 1960.) Under the new leadership railway lines were built and the Bank of Geneva was established. In addition, the fortifications surrounding the city were torn down, allowing urban expansion. Development of hydroelectric power in the late nineteenth century helped industry grow.

Geneva's international role continued to grow, and on August 22, 1864, the first Geneva Convention was signed by sixteen countries in the Alabama Hall of the City Hall. This agreement laid the foundation for the International Red Cross, and was the first in a series of treaties calling for the humane treatment of soldiers and civilians in wartime. In 1871 and 1872 an international tribunal met in Alabama Hall to settle a dispute between the United States and England, resulting from England's shipyards supplying the Confederate navy with ships during the U. S. Civil War.

Geneva, with its tradition of independence and neutrality, became the site of the headquarters of the League of Nations in 1919. The League met in the Palais Wilson, once a major hotel, from 1925 to 1936. The Palais des Nations was built between 1929 and 1937 to house the organization. It is the European continent's second-largest edifice, only exceeded by the Palace of Versailles. In 1940, when the League was dissolved, Geneva's economy suffered severely.

After World War II Geneva's population increased and employment grew. In 1945 the Palais des Nations became the European headquarters of the United Nations, and many international organizations established their headquarters in the city. The World Health Organization was established in Geneva in 1948. In 1958 the city was the site of negotiations by the United States, the Soviet Union, and the United Kingdom on ending nuclear weapons testing. In 1985 the cornerstone of the International Museum of the Red Cross and Red Crescent was laid in the presence of Nancy Reagan, wife of U.S. president Ronald Reagan, and Raisa Gorbachev, wife of Soviet leader Mikhail Gorbachev.

Modern Geneva has an economy based on services such as banking, insurance, securities trading, and tourism, in addition to the city's role as a headquarters for international groups. Manufacturing emphasizes precision machinery and instruments, and the chemical industry produces medicines and fragrances.

Like many old European cities, Geneva has a concentric arrangement of neighborhoods forming belts around the original nucleus. The city's old town section is on a steep hill on the Rhone's left bank. It is known for its old streets, stairways, fountains, and for its historic buildings, including Geneva's oldest private house, the Maison Tavel. The area is surrounded on three sides by buildings and broad streets following the line of the city's old fortifications. The cathedral stands on the highest point of the hill.

Business is concentrated in the section below the old town to the north and in Saint-Gervais. The area where most international organizations are headquartered is far north of the town's center.

Geneva's unusual features include an artificial island originally built to defend Geneva's harbor from possible attacks by the dukes of Savoy. It was named Rousseau Island (Île Rousseau) in the 1800s, and became a public park featuring a bronze statue of the writer and philosopher.

The City Hall (Hôtel de Ville) was built from the fifteenth through seventeenth centuries. The Palace of Justice (Palais de Justice), standing on the site of a Roman forum, was built from 1707 to 1712. Although it was originally a hospital, it has housed the law courts since 1860.

Duke Karl II of Brunswick, who lost his German dukedom, left his great fortune to Geneva on the condition that the city erect an elaborate monument to him. After he died in 1873, the city built the Monument Brunswick.

Geneva's most famous monument, the memorial to the Reformation (Monument de la Reformation), was started in 1909 on the 400th anniversary of the birth of Calvin and finished in 1917. The 300-foot-long granite wall features statues of Reformation leaders and bas-reliefs showing scenes from the history of the Protestant Reformation.

Other famous sites are the harbor with its Jet d'Eau, the world's highest fountain; the three connected parks, Mon Repos, Perle du Lac, and Villa Barton; the English Garden, featuring a floral clock; Place du Bourg-de-Four, used for markets and fairs in the Middle Ages, which still has some of its original buildings; and the Russian Orthodox Church (Eglise Orthodoxe Russe), with its golden domes.

The city's museums include The Museum of Swiss Citizens Abroad (Musée des Suisses à l'Etranger), containing exhibits dealing with Switzerland's relationships with other countries; the International Museum of the Red Cross (Musee Internationale de la Croix-Rouge), dramatizing humanitarian acts performed in times of natural and manmade disasters; the Museum of Art and History (Musée d'Art et d'Histoire), with its collections of fine and applied art and archaeology; and the Horology and Enamels Museum (Musée de l'Horlogerie et de l'Emaillerie), which treats the science of horology from its origins to the present and also illustrates Geneva's art of ornamental enameling.

Further Reading: In *Switzerland* (New York: Henry Holt, 1994) Paul Hofmann treats the country's three major cities: Zurich, Basel, and Geneva. It provides much detail on the history of Geneva and how that history is illustrated by the city's major sites. For discussions of John Calvin's place in the history of Geneva, consult *Calvin's Geneva* by E. William Monter (New York and Chichester: Wiley, 1967); and Richard C. Gamble's *Calvin's Work in Geneva* (New York: Garland, 1992).

—Phyllis R. Miller

Ghent (East Flanders, Belgium)

Location: Northwestern Belgium, where the Lys River meets the Scheldt; approximately fifteen miles south of the border with the Netherlands and twenty miles south of the North Sea; the provincial capital of East Flanders, forty miles northwest of Brussels, capital of Belgium.

Description: City founded sometime in the tenth century, one of the most important in the world during the Middle Ages and the Renaissance. More of Ghent's medieval buildings have survived to the present day than those of any other city in modern Belgium; they include the Castle of the Counts of Flanders (1138); The Cathedral of St. Bavo, which houses *The Adoration of the Lamb* by Hubert and Jan van Eyck; the Cistercian Abbey of Bijloke (1228), now the Museum of Archaeology; the Belfry (one of three medieval towers in the city), built by the freedmen of the Guilds in the fourteenth century; the Town Hall; and the "Friday Market" (Vrijdagmarkt), center of the medieval city.

Site Office: Stedelijke Dienst voor Toerisme
Belfortstraat 9
9000 Ghent, East Flanders
Belgium
(91) 25 36 41

*T*he Adoration of the Lamb, the great fifteenth-century polyptych by Hubert and Jan van Eyck, bears an inscription in Latin that reads, "Behold the Lamb of God, which taketh away the sins of the world." After being stolen, chopped up, and spirited all across Europe by armies and thieves for nearly 500 years, it seems something of a miracle that *The Adoration of the Lamb,* completed in 1432 by Jan van Eyck, is on display in the chapel of Joos Vyd, in the Cathedral of St. Bavo, in the center of old Ghent, the city in which the work was executed.

One of the greatest and most mysterious of European artworks, *The Adoration of the Lamb* is comprised of twelve interlocked oak panels, eight of which bear paintings on either side. The panels depict saints, angels, prophets, sybils, Adam, Eve, God, and the Madonna, alongside contemporaneous persons: the donor and his wife, pilgrims, and merchants. Their faces are painted with intricate detail, and the work as a whole provides a vivid portrait of fifteenth-century Ghent. All but one of the panels are original. Since their return to Ghent by the American army after World War II, they are together in the same place for the first time in several hundred years.

A quatrain on the front claims that Jan van Eyck's older brother Hubert, "the greatest painter who ever lived,

began the work, which his brother Jan, the second in art, finished at the instigation of Jodocus Vijdt. With this verse on 6 May he invites you to look at his work." While other works by Jan van Eyck are known throughout the world, nothing else by his older brother Hubert has been seen anywhere since the fifteenth century. The details of the paintings are rendered so precisely that, even after centuries of study, no one can tell for sure which were painted by Hubert and which by Jan. Some scholars speculate that Hubert never existed, that he was an invention of Joos Vyd, Jan's patron, a bourgeois of Ghent jealous of Jan's identification with the city of Bruges, Ghent's great rival. Hubert's existence remains as mysterious as the meaning of the story told in the twenty paintings of *The Adoration of the Lamb.*

At the time of the polyptych's first unveiling, Ghent was still young, rich, armed, and dangerous. No one then living in the city would have thought that Jan van Eyck's noble protector, Philip the Good, Duke of Burgundy, would be an instrument of Ghent's downfall. Ghent's citizens, descended from the first entrepreneurs of Europe, had built a thriving city-state, 80,000 people strong, which would be dictated to by no one. Within the city itself, more than fifty guilds were in near constant conflict—with the merchants, with dukes and counts and other feudal lords they had expelled more than two centuries earlier, and with each other. Ghent was always boiling, often boiling over.

In the tenth century, however, Ghent was relatively peaceful. Elsewhere in northern Europe, armies were continually on the march, villages were routinely plundered, but the Abbey of St. Amand, established in the seventh century at the point where the rivers Scheldt and Lys came to a head, was a safe haven, as were other abbeys in the Lowlands. These abbeys were the only safe places in northern Europe. Near the abbey, the counts of Flanders had built a fortified castle, and by the wharf on the banks of the River Lys a small settlement was taking shape. Two hundred years later the settlement was anything but peaceful.

During the tenth and eleventh centuries, Ghent was one of a number of villages in Flanders that were nominally vassals of the King of France, but a succession of weak kings neglected the region. On the other side of the Scheldt, the German king Otto I ruled his vassals by force and edict, but the counts and dukes of Flanders paid scant tribute or attention to France. They presided over extensive lands, serfs, and a few trading posts populated by a small group of free merchants.

The territory lay on the route of the Norman conquests; Norman soldiers had left behind ruined monasteries and a few free people who, while obliged to the lord of the land, were the property of no one. Some of the Flemings who had gone to England with the Normans ended up staying

The Castle of the Counts of Flanders
Photo courtesy of Belgian National Tourist Office

there, forming their own associations, while in Ghent the free folk grouped in the shadow of the castle, trading up and down the river in the summer, staying put in the winter. As the counts in the castle instituted the feudal system on the surrounding lands, the traders on the river below its windows invented capitalism.

Though sheepherding had been a way of life in Flanders for over a thousand years (the quality of Flemish wool had been noted by Caesar), the world didn't take note of the Flemish weaver's skill until traders began taking samples north and south along the Scheldt and east along the Lys. After that, money literally flowed into the settlement. The popularity of the Flemish cloth led to construction of an east-west "high road" in the mid–twelth century from Cologne to Bruges, with Ghent a major stop on the route. Now wealth came to Ghent from all directions, and after came people, though not all of them to perform labor or engage in trade.

Around 1200, the countess of Flanders, Jeanne of Constantinople, and her sister Marguerite started the first of three *beguinages* in Ghent. Named after the hood (or beguin) worn by nuns, these were self-contained retreats populated entirely by women who formed their own communities but

did not become nuns or take any religious oath. The women of the *beguinages* performed works of charity, and the Grand Beguinage of Ghent grew into a small, separate city with institutional buildings, eight streets, and eighty houses.

With the increased demand for Flemish cloth came a corresponding demand for rights in Flemish cities. In Ghent, as in other cities in Flanders, the merchants asked for, and were granted, latitude from the counts of Flanders in determining their own destiny. The merchants joined with wealthy landowners to set up a Council of Aldermen to administer the rules of the town. They raised the funds to build a wall, a belfry, and a town hall. The walled city was called a *bourg* or borough, the residents, *bourgeois,* or burgesses.

The wealthy landowners and merchants of Ghent did not act alone in securing the town from feudal obligations. The laborers in the fledgling industries formed organizations among themselves. Their guilds were the forerunners of modern labor unions, comprised of craftsmen who, in the beginning of the twelfth century at least, were mostly subservient to the merchants. But when princes, dukes, and counts opposed expansion of the rights of the town, the guilds led the fight against them. When merchants and landowners opposed expansion of the rights of their laborers, the guilds rose to fight them as well.

The law, amended by the counts on behalf of the merchants, also worked in favor of the laborers by making provisions for increasing their numbers. Called "*Keure*" in Flemish and "*Charte de Commune*" in French, individual rights were granted to any person who managed to enter the borough and remain within its boundaries for a year and a day. That individual could then call him or herself a resident of the *commune* who owed the lord nothing more than an oath of allegiance and fealty in time of war. As more Europeans sought to enter the *communes* and become bourgeois, the Flemish towns that had won these rights grew into cities. The power of the guilds grew also, as did the resistance of the merchants.

While most of the European population languished in feudal obligations, ruled by a few individuals graced by god with the accident of birthright, in Ghent the struggles that began with the birth of free trade made occasional allies of nobles and guilds or nobles and merchants; more often, however, these conflicts set guild against merchant, guild against guild, city against city, and Ghent against the world. Inside Ghent, the merchants were more powerful than nobles. The guilds sought a share of their wealth, and looked to get it through their numbers and growing political power.

The count of Flanders, Guy of Dampierre, seeking to conquer the merchants once and for all, allied himself with the guilds of Ghent, Bruges, and Ypres. By the beginning of the fourteenth century, approximately 60,000 people lived in Ghent, and the Weavers Guild there was one of the strongest organizations in Europe. Comprised of freedmen whose productivity had out-paced the production of local wool, the guild had ties to the monarchy in England, from where most of the wool was now coming. This business alliance with England did not escape the notice of England's enemy, King Philip of France. The merchants of the three cities called for help from King Philip, who was more than willing to send the royal French army into battle against his nominal vassals, the count of Flanders and his subjects.

On July 11, 1302, on a Flemish field, the ragtag fighters from the *communes,* called "men of the lion's paw" because they carried the gold and scarlet flag of The Lion of Flanders, battled France, the most powerful army in Europe (who fought under the *fleurs-de-lis* and were called "men of the lily"). In what would later be named the Battle of the Golden Spurs, the knights of France were butchered by the workers of Flanders. Golden spurs were all that remained of the flower of France after it met the lion of Flanders.

The guilds of Bruges and Ypres took command of their cities, but Ghent's were divided against themselves, giving the merchants an opening to power. The guilds built the belfry, where the city charter guaranteeing the rights of its citizens was kept under lock and key. The guilds administered justice in the Friday Market, sentencing their own members without interference from the nobility. The guilds built and ran city hall. But the most powerful guild of all, the weavers, jealously guarded their position against the others, even to the point of engaging in a bloody battle against the Fullers Guild (who shrank and thickened woolen cloth) in the Friday Market.

An influential member of Ghent's merchant class temporarily resolved those differences. Jacob van Artevelde has been called one of the first labor leaders in the world, if not the first. Named captain of the municipal army in 1338, he became a more important figure than the count in the war brewing between England and France. As leader of the army, he resolved to unite the guilds against the merchants and the nobility, tempering the power of the weavers without usurping it. While van Artevelde connived to keep Ghent out of the conflict between England and France, the guilds ran the city. Ghent had already gone to war with Bruges, for the good of no one. Kings and counts still sought to impose their will upon the *communes.* While van Artevelde was alive, the united guilds of Ghent were the supreme power in Flanders, but their sway didn't last long. Falsely accused of favoring English interests, van Artevelde was murdered by a mob in 1345.

His son, Philip, took over leadership, but in the swirl of patricians, *communes,* guilds, dukes, and clergy, all fighting for a piece of the pie, the son was unequal to the task. Attacked from within and without, Philip van Artevelde died in 1382 in battle against the new count of Flanders, Louis de Male, who was allied with France. While the rest of Flanders fell in line behind the count, Ghent alone continued free for the next two years before it, too, fell. Louis de Male died in the same year, and Flanders came under the rule of the dukes of Burgundy.

By 1430, Philip the Good ruled over practically every city and principality in the area that would later become Belgium. Philip set up a new arm of authority called the States General, which exalted the power of the clergy and noblemen at the expense of the *communes.* The duke of Burgundy's reach was regal, his armies battled to the edge of the Mediterranean, and his territories were the equal of France and England in wealth and power. In 1450 he imposed a new tax on Ghent, but Ghent refused to pay.

In 1453, led by the weavers, Ghent went to war just outside its city walls against the armies of Burgundy. Twenty thousand free citizens of the *commune* were slaughtered there, and the city ceased to be free. Succeeding Philip in 1467, his son, Charles the Bold, used the Burgundian armies to crush any and all Flemish resistance to his policies. But in the fifteenth century, Flanders was experiencing problems beyond the brutality of dukes.

To create its world-renowned cloth, Belgium had been dependent on shipments of English wool. Since they had learned how to make cloth themselves, however, the English were no longer shipping much wool to Flanders. While Bruges's outlet to the sea, the Zwin, had silted over, stifling trade there for good, and the Flemish cloth industry as a whole was using an inferior grade of Spanish wool, Ghent was just barely hanging on by virtue of its grain trade when one of its own, a native son, dealt the city a crushing blow.

Charles the Bold died, and his daughter, Mary of Burgundy, married Maximilian of Habsburg, son of German Emperor Frederick III. Mary died in 1482, but not before producing a son, another Philip, who was a young child at the time of her death. General revolts broke out once again in Ghent, but by 1492 the Habsburgs had established themselves there. Philip the Fair, no longer a child, moved into a castle in the city, where his son, Charles, was born in 1500. He was one of approximately 100,000 people who lived in Ghent at the time.

Charles became King Charles I of Spain in 1516 and Holy Roman Emperor Charles V in 1519. While Charles kept his court in Spain, Flanders was his special interest, and Ghent one of the objects of his royal focus. As an agent of the Inquisition, Charles V brought a special ruthlessness to his role as defender of the faith against the Protestant Reformation. But money, not religion, was at issue between Charles and Ghent; Charles wanted more of it, and Ghent didn't want to give it to him. In the end, Ghent paid.

After his representatives were thrown out of the city, Charles returned to his hometown, bringing his army with him. Ghent, unable to resist the superior might of the empire, saw its charter revoked, its guilds crushed, and its leading citizens hanged as an example to their fellows. The Abbey of St. Bavo was torn down and a fort erected in its place, with an army garrison left behind. Ghent was forced to pay cash as well: in addition to a tribute of 500,000 florins, the city was made to provide an annual fee of 6,000 florins. One of the first free city-states had been cast back into a vassalage it had barely known. The *commune* was dead. The city would not recover for another three centuries.

Although the rule of its native son had proved disastrous for Ghent, the rule of Charles's son Philip was, in some respects, even worse. When Charles left his throne in 1555 to join a monastery in Spain, Philip II took over where his father had left off. Now the Inquisition's torture of heretics was extended, and Philip II tolerated absolutely no opposition to his rule of Spain and Rome. Local bishops would adhere to Philip's rule, while the Inquisition controlled Ghent both spiritually and politically. To enforce his rule, Philip appointed the Duke of Alba, who convened the "Council of Blood." Thousands were put to death, and Protestants and Catholics had a common enemy.

In 1575, a few rebels under the leadership of William the Silent assisted by the States-General called a meeting in Ghent. The meeting produced a treaty, called the Pacification of Ghent, which declared independence for Belgium and the Netherlands. The rebels called for the restoration of their privileges and an end to Spanish rule. As such, the Pacification of Ghent was a declaration of war. More significantly, however, it was also a peace treaty between the Catholics and Protestants of the lowlands, who agreed to set aside their differences and put an end to religious persecution. Yet the declaration of religious tolerance failed to hold, and the land split in two—Catholics to the south in what would become Belgium, and Calvinists up north in the Netherlands. While Ghent remained in Catholic Belgium, it was also a Calvinist stronghold. Helped along by the English defeat of Spain in 1588, the treaty did succeed in breaking Spain's grip on the region, but at the same time it opened the door to invasion by other powers.

Ghent remained a fractured city. Its cloth industry was moribund. The greatness it had once achieved was treated with reverence by its citizens, cavalierly by outsiders. At the order of Emperor Joseph II, who was offended by their nakedness, the panels showing Adam and Eve on the polyptych of *The Adoration of the Lamb* were removed in 1781. In 1794, the lamb itself, along with three other panels, was removed to Paris. Then in 1800, a new industry rejuvenated Ghent. Related to the old, it came, as the wool for Flemish cloth had once come, from England.

A citizen of Ghent, Lieven Bauwens, stole a model of the first spinning jenny (for cotton) from Manchester, and just a few years later Ghent was making cloth in mass quantities again. Factories attracted thousands of workers; the Industrial Revolution had arrived. The grime was as real as nineteenth-century Manchester's, and so was the effect the new industry had on the city's coffers, which bulged once again. In 1816, the van Eycks' Lamb returned to Ghent, and in 1830 statehood was conferred on Belgium. The panel of Adam and Eve, however, was still missing.

In 1814, Britain and the United States met in the city to sign the Treaty of Ghent, which ended the War of 1812. But this act had little effect on the city. Ghent's renewed cloth industry was the city's defining feature throughout the nineteenth century, and as it grew, so too did other industries, as well as the town itself. Once again, labor movements formed to fight the cloth merchants, electing socialists to the Belgian parliament in 1894, and eventually winning rights for working people again.

During the twentieth century Belgium was constantly invaded, rarely independent, until after World War II. By the 1970s, the country was a center of European capitalism, though it was not itself rich. All but one of the original panels of the *Adoration of the Lamb* had been returned to Ghent, that of Adam and Eve having been returned in 1861. The missing panel, with *St. John the Baptist* on one side, and *The Just Judges* on the other, was held for ransom in 1934. The thief separated the painting of John the Baptist and returned it as proof that he held the remaining panel, but the Belgian government refused to pay for the return of *The Just Judges*. It has never been recovered; a copy now stands in its place. The other panels, after traveling the world, were restored in 1950. The figures in the paintings, fifteenth-century citizens of the *commune,* are home again under the flag of Flanders in a quieter Ghent.

Further Reading: *A Short History of Belgium* by Leon Van Der Essen (Chicago: University of Chicago Press, 1920) is good, informative, and not so compressed as one would expect from its title. More current, livelier, but less informative is *Belgium: The Land and the People* by Donald Cowie (New York: A.S. Barnes, 1977;

London: Yoseloff, 1978). Cowie's history is highly subjective, and his section on Ghent is mostly a gloss. Despite its title, *A Masterpiece Called Belgium* by Arthur Frommer (Englewood Cliffs, New Jersey: Prentice Hall, commissioned by Sabena, Belgian World Airlines, 1989) is a very nicely done guide-book written by a man who obviously knows and loves his subject. Frommer's ardor is almost embarrassing at times, but his enthusiasm for the land, and for what he calls "Turbulent, Seething Ghent," is communicable. *The Complete Paintings of the Van Eycks* by Robert Hughes and Giorgio T. Faggin (London: Penguin, and New York: Viking, 1986) provides good sections on the panels of *The Adoration of the Lamb*.

—Jeffrey Felshman

Glastonbury (Somerset, England)

Location: Approximately 130 miles west-southwest of London and 25 miles south of the west-coast city of Bristol in the Mendip district, county of Somerset.

Description: One of the oldest continually inhabited sites in England; legend claims that St. Joseph of Arimathea visited the site shortly after the death of Christ, bringing with him the chalice from the Last Supper, popularly known as the Holy Grail; reputed to be the final resting place of King Arthur and Queen Guinevere, and was, for a time, said to be the final resting place of St. Patrick of Ireland.

Site Office: Tourist Information Centre
The Tribunal
9 High Street
Glastonbury, Somerset BA6 9DP
England
(458) 832954

The name Glastonbury means many things to many people. Continually inhabited since at least 300 B.C., Glastonbury figures in such diverse historical episodes as the Roman occupation of Britain, the Norman Conquest of 1066, the dissolution of the monasteries of Britain in 1539 and the lives of St. Patrick, King Arthur, and Jesus Christ. The site has been used for religious purposes probably from its very beginning and has been the site of Christian gatherings almost since the time of Christ. Glastonbury also was once home to the wealthiest monastic order in Britain.

Centuries of being at the center of often-tumultuous affairs in Britain have taken their toll on Glastonbury. The once-proud abbey now sits in romantic ruin and the modern town of Glastonbury, while economically vital, retains little of its fomer prestige. Residents, however, are well-versed in the lore of their home town, and bookshops specializing in matters related to Glastonbury abound.

The distinctive topography of Glastonbury and the surrounding countryside handily lends itself to the area's association with the legendary and ethereal. Glastonbury in particular is topographically extreme, with Glastonbury Tor rising 522 feet from a bed of low-lying marshland that, although drained centuries ago by its monastic inhabitants, is still prone to flooding, recreating, however briefly, the ancient tableau that has led modern scholars to identify the area as the fabled Isle of Avalon, of Arthurian legend.

The marshes, properly known as the Somerset Levels, have been drained and otherwise altered many times over the centuries, but never completely. In 1618, for example, authorities attempted to drain the Levels, but were prevented from accomplishing very much by the protests of the locals,

who stood to lose their way of life if the marshes were destroyed; the periodic flooding brought deposits of silt that enriched the local soil, and the marshes also yielded fish and plant life for area residents. Even by the end of the seventeenth century Somerset Levels remained only partially drained despite nine special commissions for the work issued by Charles I and an Act of Parliament in 1669.

Archaeological excavation has also been especially fruitful at Glastonbury, with successive projects since 1892 adding significantly to modern knowledge of British history in general and Glastonbury in particular. Fascinated by discoveries of relics of lake-dwelling societies in Switzerland in the nineteenth century, Somerset-based antiquary Arthur Bulleid began to search for evidence of similar habitation in Britain. In 1892 Bulleid discovered pottery fragments at Glastonbury that warranted full-scale excavation of the area. Five years later he also discovered the Iron Age village of Meare, located nearby.

The village at Glastonbury, which has been almost totally excavated, shows evidence of occupation since as early as 300 B.C., although the site's population was probably at its greatest approximately 200 to 250 years later. Discovered at the site were the remains of 90 primitive huts extending over a triangular area of about 135 feet by 100 feet; about 20 to 30 of these huts were probably occupied at any given time. The huts were round, constructed of vertical timbers, and graced with clay floors and hearths, many of which bore decorative features.

The nearby village of Meare was composed of approximately 12 small houses dating from 300 B.C. and found beneath a layer of refuse and clay deposited there after Glastonbury had become the dominant settlement of the area.

A significant hoard of relics was unearthed at Glastonbury and Meare, including rare wooden items such as intricately worked beams and planks, kitchen utensils and tools, and what may have been part of a loom. The skill of the Iron Age carpenters was noteworthy, as evidenced by the presence of such techniques as mortise-and-tenon jointing, and the use of dowels to hold together larger pieces. Also found were dugout canoes and items of iron, lead, bronze, copper, and glass. Wheat, barley, peas, and beans were the staple crops of the area, and the residents also raised sheep, horses, dogs, pigs, goats and cattle, as borne out by the variety of bones found at the sites.

Five dice were also found at Glastonbury, as well as a box for their storage and several other playing pieces fashioned from glass. Pottery, the stalwart of archaeological relics, decorated with incised curvilinear ornamentation, was also found at Glastonbury and Meare.

Many legends are associated with Glastonbury and, as is usually the case, most of the area's tales came into being

The Glastonbury Holy Thorn, top; St. Mary's and St. Joseph's Chapels, bottom
Photos by Mike Zapp, Glastonbury, Somerset, England

long after the often-sketchy facts upon which they were based. One legend in particular, that of Joseph of Arimathea's visit to Glastonbury shortly after the death of Christ, proved to be fertile ground for the area's myth-makers, providing the basis for such famous tales as the legend of the Holy Grail and the identification of Glastonbury as the burial place of King Arthur and Queen Guinevere.

Although mentioned only briefly in the bible, Joseph of Arimathea was, in Christian legend, the wealthy follower of Christ who obtained his body from the Romans and had it buried in the tomb from which he arose three days later. Briefly imprisoned by the Jews, Joseph was set free by the risen Christ, who entrusted him with the safekeeping of the Holy Grail. The grail, often known as the chalice, was the drinking vessel used by Christ and his disciples at the Last Supper.

Along with his brother-in-law Bron, Jospeh formed a society of the grail to ensure its safekeeping. Together they traveled to Glastonbury and, according to the poet Robert de Boron, founded the first Christian church on the site:

> That Joseph came of old to Glastonbury,
> And there the heathen prince Arviragus
> Gave him an isle of marsh whereon to build,
> And there he built with wattles from the marsh
> A little lonely church in days of yore. . . .

The legendary wattle structure built by St. Joseph is reputed by some to be the world's first Christian church. Certainly, if factual, the structure would have been the first church in Britain, if not the world. The legend alone is enough to make Glastonbury the honorary cradle of Christianity in Britain.

That the grail was an important part of Joseph's contribution is explained later in the legend:

> The cup, the cup itself from which our Lord
> Drank at the last sad supper with his own.
> This, from the blessed land of Aromat—
> After the day of darkness, when the dead
> Went wandering o'er Moriah—the good saint
> Arimathean Joseph, journeying brought
> To Glastonbury, where the winter thorn
> Blossoms at Christmas, mindful of our Lord.

It is said that Joseph, not willing to allow the grail to be put to regular use for fear of theft or destruction, hid the grail at the foot of what is now known as Chalice Hill. Some say that it was not the grail, but two vials of Christ's blood that were buried near the hill, and that the Grail was stored in a great, no longer extant, tower upon a hill.

The legend also asserts that due to their proximity to the grail, or the vials of Christ's blood, the two streams that flow from Chalice Hill are especially holy, and run with the blood of Christ. It is not known exactly when this addition was made to the legend, but it is unknown prior to the dissolution of the monasteries in 1539. Such suppression has often ended in results completely unlike those hoped for by the suppressors and, in this case, the legend eventually was put to commercial use. In 1750 the residents of Glastonbury,

in an effort to stimulate the local economy, declared that the waters of the streams of Chalice Hill possessed miraculous powers of healing. Asthma sufferers, who made up a very large potential market, were especially encouraged to partake of the healing waters. A pilgrimage route was created, and records exist declaring that on May 5, 1751, approximately 10,000 people came to Glastonbury seeking of the holy waters.

The legend of St. Joseph of Arimathea, however, does not end with the world's first church, healing waters, and the Holy Grail. Joseph is also reputed to have planted the first Glastonbury Thorn, sometimes called the Holy Thorn. Native to the area around Glastonbury, the Glastonbury Thorn is a hardy plant that, fitting for a legend, blooms only at Christmastime and thrives nowhere but Glastonbury. If planted elsewhere, it is said, Glastonbury Thorn will die within a few years. According to one version of the legend, the first plant arose at the foot of Wirrall Hill when Jospeh struck his staff to the ground while resting. Other versions say that the plant sprang from one thorn dropped from Christ's Crown of Thorns, while still others, such as Richard Pynson's story of the life of St. Joseph, published in 1530, say that it was a walnut tree that grew from the place where St. Joseph's staff struck ground, while still mentioning the miraculous thorns:

> Great mervaylles men may se at Glastenbury,
> One of a walnut tree that there dooth stande
> In the holy grounde called the semetery
> Harde by the place where Kynge Arthur was founde
> South fro Joseph's chapell it is walled in rounde . . .
> Three hawthornes also that groweth in Werall
> Do burge and bere grene leaves at Christmas
> As fresh as other in May, when the nightyngals
> Wrestes out here notes musycall as pure as glas.

In the year 166, according to legend, the first abbey church was founded at Glastonbury by Roman missionaries who came at the behest of King Lucius, perhaps the first Christian king of Britain. The fact that Christianity was not even a legal religion in Roman Britain until 313 makes this legend impossible to verify with any certainty. It is known from a variety of sources, however, that a Celtic monastery did exist on the site prior to the Saxon conquest. After conquering Somerset in 658, the Saxon chieftain Cenwealh took a special interest in Glastonbury, turning it into a model of cooperation between the Saxons and the Celts. Saxon monks gradually replaced Celts, but, unlike most places of new Saxon domination, the Celtic tradition was never forcibly stamped out by the Saxons; it was instead incorporated into what would become the new British tradition. Toward the end of the seventh century King Ine of Wessex enlarged the abbey at Glastonbury, making the new cooperative religious bastion an enlightened model. Excavations from 1924 to 1926 revealed the remains of Ine's church under six feet of clay dumped on the site in the thirteenth century for the rebuilding of the nave of the later church.

Among the great variety of archaeological remains

uncovered at Glastonbury are two mausoleums within the boundaries of the ancient Celtic cemetery that forms one of the site's earliest settlements. Both are low and wide, probably marked with crosses at one time, and very rare in Britain. It is assumed that they were reserved for persons of great importance, and it has been suggested that they were the first resting places of St. Patrick and St. Indracht, a lesser-known Irish saint. It is known that in the original Church of St. Mary, as it existed before the fire of 1184, the altar was flanked by a pair of pyramidal shrines that were reputed to contain the relics of the departed saints. The constant presence of Irish monks at Glastonbury and its popularity as a destination for Irish pilgrims lend credence to the notion that the saints were interred in what would now be considered an unthinkable spot for an Irish notable.

In 943 St. Dunstan was appointed abbot of Glastonbury by his friend King Edmund I and refounded Glastonbury as a grand Benedictine abbey based on scholarship and worldly productivity. This precedent was to stand the abbey in good stead and become an important part of the development of tenth-century Britain. The monks engaged in copying books, learning and playing music, and gardening. They also reclaimed marshland for agricultural purposes, adding their part to the continual efforts to drain the marshy ground, and even established a glassmaking factory, of which remains have been found. The revival of Glastonbury under Dunstan was carried on by his successors, resulting in four Glastonbury monks rising to the level of archbishop of Canterbury before the Norman Conquest.

The successful monastery at Glastonbury continued to grow through the eleventh century, continuing to thrive and becoming one of the wealthiest abbeys in Britain. According to the Domesday Book, the taxation survey compiled by the Normans in 1086, the abbey held lands in five counties. The county of Somerset, too, had grown in wealth and importance; the Domesday Book lists approximately 600 villages and small hamlets throughout the county that are still in existence today, representing almost all of Somerset's current settlements. Unlike most counties, which have grown considerably since 1086, Somerset was almost fully developed by that time.

In 1184 fire destroyed the mostly timber-and-wattle compound at Glastonbury. Although the monastery's wealth and power allowed it to initiate an immediate program of rebuilding, the plan was fated never to be completed. Bickering and dissention within the church, leading up to the dissolution of 1536 to 1539, worked against the reconstruction, which would have required at least a century or two under the best of circumstances. Following the dissolution the abbey became a quarry for other construction projects, a fact that accounts for its current ruined state more than ravages of time.

Pivotal though Glastonbury's place may be in the grand scheme of British history, especially the years prior to and immediately following the Norman Conquest, it is the town's association with the legend of King Arthur that has brought it most of its modern fame and, not surprisingly, modern tourists, as well.

According to the quasi history recorded by Geoffrey of Monmouth, Arthur's fabled sword Excalibur, originally known as Caliburn, was forged at Avalon. More importantly, Avalon is where the king was taken following his battle with his illegitimate son Mordred, in which he sustained mortal injuries: "Arthur himself, our renowned king, was mortally wounded and was carried off to the Isle of Avalon, so that his wounds might be attended to."

Although Geoffrey, who wrote in the mid-twelfth century when Glastonbury was well established, makes no attempt to describe the location of the mystical Avalon, persistent legend and archaeological evidence found at Glastonbury Tor, which some say may have been the site of Camelot, and in the vicinity of the abbey have led scholars to equate Avalon with Glastonbury. Furthermore, a lead cross found at the burial site alleged to be that of King Arthur and Queen Guinevere read, according to contemporary accounts, "Here lies Arthur, the famous king, in the island of Avalon" providing a convincing argument for Glastonbury as Avalon.

The legend of Arthur as the "Once and Future King" proved to be a mixed blessing for the rulers of England after the conquest. Legends that Arthur was not actually dead, but sleeping until such time as he was needed again, wrought havoc with kings' ability to govern, particularly in western Celtic regions that were loathe to form a united kingdom. Such legends have persisted even into modern times. For example, it was suggested during the Battle of Britain in 1940 that Winston Churchill and the Royal Air Force pilots who miraculously fought back superior German forces were really King Arthur and his Knights of the Round Table, awakened in time of need.

Henry II, in attempting to subdue the troublesome, and fervently Celtic, Welsh decided that the Once and Future King needed to be laid to rest forever. While traveling in Wales he learned from a loose-lipped bard the secret that was widely known, but assiduously guarded in Celtic circles that Arthur was, in fact, dead and buried between two pillars in the cemetery of Glastonbury Abbey. Henry suggested that the monks look for the burial, but nothing was done. When the structure burned in 1184 the opportunity presented itself and, in 1190, one year after Henry's death, the area between the pillars was excavated.

At approximately seven feet below the surface the diggers found a slab of stone and the lead cross previously discussed. Nine feet lower they discovered a large coffin fashioned from a hollowed-out oak log. Inside, the coffin contained the skeleton of a large man and some smaller bones with a tuft of yellow hair remaining, presumably that of Guinevere. A sword was also found during excavation, which was presented by Richard I to Tancred of Sicily in 1191 as nothing less than the fabled Excalibur. The bones were gathered into a single casket and stored with the abbey's other treasures. For the Plantagenets, the discovery of

Arthur's very dead bones made certain that Arthur would not become the king of the future.

There is no record of the location of Arthur and Guinevere's bones for at least forty years following their exhumation, but it is assumed that they lay in the abbey's treasury. At some point they were interred in another tomb on the abbey's grounds. In 1278, following a successful offensive in Wales, Edward I ordered this tomb opened. The event was chronicled by the contemporary historian Adam of Domerham:

The lord Edward . . . with his consort, the lady Eleanor, came to Glastonbury . . . to celebrate Easter. . . . The following Tuesday . . . at dusk, the lord king had the tomb of the famous King Arthur opened. Wherein, in two caskets painted with their pictures and arms, were found separately the bones of the said king, which were of great size, and those of Queen Guinevere, which were of marvellous beauty. . . . On the following day . . . the lord king replaced the bones of the king and those of the queen, each in their own casket, having wrapped them in costly silks. When they had been sealed they ordered the tomb to be placed forthwith in front of the high altar, after the removal of the skulls for the veneration of the people.

The tomb to which the bones were moved was discovered in 1931 in the western portion of the choir near the high altar. Only the base of the tomb's cavity remains, as the tomb itself was destroyed at the Reformation and the bones dispersed.

Henry VIII's reign spelled the end for Glastonbury's power as a monastery. When Rome refused to accept the validity of Henry's marriage to Anne Boleyn and, therefore, Elizabeth's status as heir to the throne, Henry responded with the Act of Supremacy, which confirmed his status as the one and only leader of the Church of England. The subsequent Act of Succession confirmed Elizabeth as heir, but was not accepted by Pope Clement VII. In 1534 Henry insisted that all leaders of religious houses take an oath accepting the Act of Succession, thereby acknowledging the validity of Henry's marriage to Anne Boleyn and the right of Elizabeth to succeed her father to the throne, and to deny the validity of Henry's previous marriage to Catherine. Henry and his advisers suggested that such an oath would not present a problem to any loyal Englishman, but on March 23, 1534, Clement issued a verdict that Henry's marriage to Catherine had been lawful, thereby implicitly denouncing both Elizabeth and Anne Boleyn.

Most clergymen assented to the act, reasoning that Henry was the most immediate threat. Those who could not take the oath met with exile, torture, or execution. It was the Act of Supremacy that fomented the dissolution, however. Henry, given to a lavish lifestyle, needed money. Parliament refused his requests for a larger budget for the crown, so Henry, now sovereign head of the English church, began to look elsewhere for money. Parliament granted the crown a tax of 10 percent of the annual incomes of all of Henry's new holdings, but nobody knew how much this was worth. In 1535, Henry's chancellor Sir Thomas Audley sent out a team to gauge the value of all ecclesiastical property in England. Together, the 650 monasteries and nunneries and 200 friaries and hospitals reported annual incomes in excess of 200,000 pounds, which would yield 20,000 pounds a year to the crown.

Attempts to collect the new tax did not go very well, and many religious houses rioted rather than give up a portion of their incomes. Those that resisted were dissolved in fact simply because their leaders were executed and their possessions confiscated. The actual Act for the Dissolution of the Greater Monasteries, which passed both houses of Parliament in May 1539, not only replaced the 10 percent tax with complete submission, but also made legal the actions involved in the supression of the recalcitrant houses. The abbots of Glastonbury resisted to the very end, but were executed for treason. Abbot Whiting, leader of the abbey, was hanged, drawn and quartered on the Tor, possibly in the presence of Henry himself, who journeyed to the George and Pilgrims Inn, which still stands, to witness the dissolution of England's most powerful monastery. In one sweeping act, the religious landscape of England was changed forever.

The site of Glastonbury's abbey lapsed into such mundane use that it was actually sold into private hands, and was not returned to ecclesiastical duty until 1908, when the Church of England repurchased it. In 1965 the bishop of Bath and Wells, who has jurisdiction over the abbey, allowed Catholics to celebrate Mass on its grounds for the first time since the Reformation. The crypt of the old church has also been restored as a nondenominational chapel.

Further Reading: *The Quest for Arthur's Britain* (London and New York: Praeger, 1968) by Geoffrey Ashe, a noted Arthurian scholar, details not only the legend of Arthur, but also the reality as borne out by historical record and archaeological excavation. Glastonbury figures prominently in both aspects of such study. *Highways and Byways in Somerset* (London: Macmillan, 1924) by Edward Hutton puts the history of Glastonbury, and other notable Somerset locales, into a modern context by relating the stories behind what has survived to the twentieth century. Hutton concentrates on religious matters. *A Documentary History of England* (London, and Baltimore, Maryland: Penguin, 1966), edited by J. J. Bagley and P. B. Rowley, contains complete renditions and brief analyses of the documents most pivotal to English history between the years 1066 and 1540, including the Act of Supremacy and the Act for the Dissolution of the Greater Monasteries.

—John A. Flink

Gniezno (Poznań, Poland)

Location: In a rich, lake land agricultural region of west-central Poland, 145 miles northwest of Warsaw, 30 miles east of Poznan, along the Piast Route, Poland's historic "Memory Lane."

Description: Poland's first capital and the place of coronation for Polish kings until 1320; site of the first independent Polish archbishopric in the year 1000 and burial place of St. Adalbert, Poland's first patron saint; variously under the authority of Greater Poland, the Polish province of Poznań, Prussia, the Grand Duchy of Warsaw, and Germany; one of the two capitals of the Warsaw-Gniezno Roman Catholic Archdiocese.

Contact: Provincial Information Center
Stary Rynek 59
Poznań, Poznań
Poland
52-61-56

The origins of Gniezno, one of Poland's oldest cities, predate by centuries the country's recorded history. According to tribal legend, the three Slavic brothers Lech, Czech, and Rus created, respectively, the Polish, Bohemian, and Russian peoples. Lech, the mythological founder of Poland, is said to have established his city at Gniezno because he found white eagles nesting there. The settlement was named Gniezno, meaning in Polish "nesting site," and the white eagle was adopted as the new nation's symbol, which it remains. According to historical conjecture, it was actually in the seventh and eighth centuries, that groups of western Slavs first settled the area between the Oder and Vistula Rivers, in what is now west-central Poland. The most prominent of these tribes called themselves the Polanie (or Polanians), literally "people of the open fields." Their country, called Polska, was centered on the lake region around Gniezno; it later became known as Wielkopolska (Greater Poland) to distinguish it from the southern part of their realm, called Malopolska (Lesser Poland). Archaeological evidence indicates that the Polanie had established a fortified stronghold at what is now Gniezno in the eighth century.

Piast, reputedly a peasant, links Poland's legendary past to its recorded history. He became Polanian ruler about the mid-ninth century and launched a dynasty that was to last five centuries; the Piast, or Polanian dynasty, would eventually unite neighboring tribes into one Polish kingdom by 1370. Piast's first notable descendant—the first historically recorded Polish ruler—was Mieszko I, prince of the Polanians, chief and warlord of the tribe, who lived from approximately 922 to 992. He was in possession of Greater Poland and surrounding areas by about 960, making Gniezno the first capital of the Piast state. The town formed one of several garrisons in the region, heavily fortified with earthworks, palisades, and moats, and guarded by several thousand men.

The years 965 to 966 marked a momentous turning point in Polish history. Although Mieszko's activities were first recorded by medieval Latin chroniclers in the 960s, his marriage to Dubravka, a Christian Bohemian princess, in 965, is generally regarded as the point at which Poland entered European history. A year later, as part of the marriage agreement, Mieszko renounced the pagan religion of his ancestors and was baptized into the Christian faith, converting to Roman Catholicism. These acts brought Mieszko and his people into the world of western culture and religion (though many Slavic tribes continued practicing paganism until 1226, when the Order of the Teutonic Knights began subjugating them to Christianity). Greater Poland was now formally allied with the Holy Roman (German) Empire, which had for many years been trying to convert the Poles by force, rather than with the Byzantine Empire, which embraced the Eastern Orthodox form of Christianity. Mieszko's baptism didn't immediately lead to the creation of a formal church hierarchy, but it prepared the way for the foundation of an ecclesiastical province of Poland, with its see at Gniezno.

Though the imaginary line between East and West—Roman and Byzantine—was drawn at Gniezno, Mieszko eventually made the neighboring city of Poznań his capital and official residence in 968; the so-called Christian prince also established Poland's first local bishopric there. In the following years, Mieszko fought to establish his realm's independence from the Holy Roman Empire. After Dubravka died, Mieszko married a nun named Oda. By the time of his death in 992, he had succeeded in uniting and controlling most of the territories that make up present-day Poland. In 991, the year in which Latin chroniclers first described a "realm of Gniezno," Mieszko asked that his lands be placed under the direct protection of the papacy, in order to prevent the Holy Roman Empire from extending its imperial sovereignty over Greater Poland. It is interesting to note that much of this information about the birth of Catholicism in Poland was first documented by a Jewish trader, Ibrahim-ibn-Jakub, who probably spent some time in Gniezno in the late tenth century.

Mieszko was succeeded by his son, Bolesław I Chrobry (the Brave), who would become the first crowned king of Polania. Though he, too, realized the importance of establishing an independent religious state and expanded his father's territories—capturing Prague, for example, in 1003, and Kiev in 1018—most of his domain had been either divided or conquered by the Germans or Bohemians by the time of his death in 1025. (During the pagan rebellion of 1035 to 1037, Czech king Bretislav captured Gniezno.) It was

Twelfth-century door at Gniezno's cathedral
Photo courtesy of Polish National Tourist Office

Bolesław, however, who sealed Gniezno's destiny as an important Catholic center. In 996, he received a mission from Rome headed by Vojtěch (in Polish, Wojciech), the exiled bishop of Prague. Vojtěch, a Czech, had studied at Magdeburg (Germany), where the local archbishop christened him Adalbert, the name by which he is more generally known in the West. The devoted missionary stayed among the Polanians for a while before sailing from Gdańsk to the land of the pagan Prussians, in what is now northeast Poland. Though initially successful, Adalbert was murdered by the Prussians in 997. Bolesław is said to have ransomed the missionary's mutilated corpse for its weight in gold; Adalbert's remains were brought back to Gniezno and buried before the altar of the town's cathedral. Rome quickly proceeded with his canonization; St. Wojciech, or the church dedicated to St. Adalbert, the nation's first patron, soon became the object of a popular Polish cult.

In the year 1000, the pope urged Holy Roman Emperor Otto III to make a pilgrimage to Gniezno, to visit St. Adalbert's tomb and to create a metropolitan see. Bolesław and papal representatives entertained the emperor, who ceremonially established a Roman Catholic archdiocese, the first in Poland independent of Germany, headed by an archbishop and primate at Gniezno. Otto nominated St. Adalbert's half-brother Radim, or Gaudentius, to be the first archbishop. Also during these festivities, the emperor recognized Poland's political and ecclesiastical independence by placing his crown on Bolesław's head, dignifying the Polanian ruler as an "elder of the Roman nation" and a "brother and aide in the Empire." Bolesław's coronation as Poland's first king took place, with Rome's blessing, in 1025, the year of his death. To this day, Gniezno–which obtained town rights in 1243–remains the official administrative center of Poland's Catholic Church and the residence of the Cardinal Primate of Poland, although the primatial see is now also linked with Warsaw. It has been said that the long road to a Polish pope— Pope John Paul II, the former Cardinal Karol Wojtyla of Kraków, was elected to the Vatican throne in 1978—began a millennium ago in Gniezno.

The ecclesiastical province of Poland, See of Gniezno, would continue throughout the Middle Ages, as the diocesan structure expanded and local bishoprics were established in cities and provinces coming under Polish domain. The church hierarchy created a strong ecclesiastical unity that managed to endure even when the Piast state—threatened by descendant rivalries and foreign invasions by Germany, Bohemia, Kievan Russia, the Teutonic Order, and Tatars—disintegrated into a number of independent principalities and provincial dynasties in the twelfth and thirteenth centuries. The period of fragmentation, or regionalization, began with the death of Bolesław III, Krzywousty (the Wry-mouthed), in 1138; it essentially ended when Piast prince Władysław I, Łokietek (the Short), reconquered Polish lands by 1314, and Greater Poland once again shared the fate of the united federal Polish state. Władysław was crowned king in Kraków (capital of Lesser Poland) by the archbishop of Gniezno in 1320; Kraków became the capital of the newly unified realm that same year. Władysław and his son, Kazimierz III, Wielki (Casimir the Great), who assumed the throne in 1333, are considered the real founders of the Polish monarchy. Casimir, one of Poland's most enlightened statesmen, more than doubled his inherited territory and restored the kingdom to political, commercial, and cultural prominence by the 1360s.

Very little remains of Gniezno's medieval past. But the Cathedral of St. Wojciech still stands, and is considered the most imposing Gothic cathedral in Poland. The original cathedral was built by Mieszko before 977; the current structure is the third to occupy the spot, its twin-towered, tall-naved Gothic exterior dating to the mid-fourteenth century. Its interior was remodeled in the baroque style during the eighteenth century, but the cathedral was rebuilt and re-Gothicized after World War II; the occupying Germans had turned it into a concert hall in 1942, and it was fire-bombed in 1945. Many of the cathedral's treasures were plundered by the Nazis—but luckily not the masterpiece Gniezno Door, a pair of magnificent cast-bronze Romanesque doors that depict eighteen scenes from the life and death of St. Adalbert. The bas-relief panels begin with the martyr's birth, and conclude with the laying of his remains at Gniezno. The doors, dated to approximately 1175 and believed to be of Flemish origin, are among the finest examples of early medieval Polish art and are also among the oldest works of their kind in Europe. St. Adalbert's remains are contained in a silver sarcophagus, supported by four silver pallbearers, above the cathedral's main altar; the reliquary, created by baroque master Peter van der Rennen of Gdańsk, is topped by a reclining figure of Poland's first saint.

The dynastic union of the Kingdom of Poland and the Grand Duchy of Lithuania lasted from 1386 to 1572, establishing a bulwark of Roman Catholicism on the eastern frontier; Lithuania had been the last pagan nation in Europe. Their combined armies conclusively defeated the Teutonic Order (which held most of northern Poland) in the Thirteen Years War of 1454 to 1466; by the end of the fifteenth century, Poland-Lithuania controlled a vast empire that stretched from the Baltic to the Black Sea. In 1569, the two states became a commonwealth, initiating a period of elective monarchy and multinational democracy that lasted until 1795. But in a destructive period known as the Deluge, which lasted from 1654 to 1660, Russian and Swedish armies overran the commonwealth; the Poznań, or Posnania province (formed out of Greater Poland), including Gniezno, capitulated to the Swedes in July 1655.

The Polish-Lithuanian Republic was partitioned three times among its more powerful neighbors between 1772 and 1795, the last time vanishing completely. In the First Partition, Russia, Prussia, and Austria all annexed parts of the commonwealth in 1772, leaving only a small Polish state that included Greater Poland. In the Second Partition of 1793, Poland was further divided between Russia and Prussia; Gniezno passed to Prussia, which intensified a Germanization process that had begun as early as the thirteenth century with the arrival of the Teutonic Knights. In the Third Partition of 1795, Poland's last remaining territory was claimed by three powers again; the state was abolished from the map of Europe, until it regained its independence after World War I. During this intensely nationalist period, the state was, in effect, the church: the deeply entrenched Roman Catholic archdiocesal system became the sole preserver and defender of Polish culture and identity against foreign aggression.

In 1807, after the Treaty of Tilsit, French emperor Napoléon created a small, weak Polish state called the Grand Duchy of Warsaw as a reward to Poles who had emigrated to France after the last partition and formed Napoléon's Polish Legions. The duchy incorporated Poznań, as well as Gniezno. The puppet state, ruled by the king of Saxony, was set up as an instrument of Napoléon's foreign policy—a check to Austria, Prussia, and Russia. After Napoléon was ultimately defeated by the Russians at Waterloo, the 1815 Congress of Vienna (which redrew Europe's map in the post-Napoleonic period) reverted Poznań to Prussia as the Grand Duchy of Poznań; it also established the so-called Congress Kingdom, a czarist-ruled Polish state of limited territory and sovereignty that lasted until 1874. The Poles, however, did not submit peacefully to the partitioning powers. After a great national insurrection in 1830 was crushed by the Russians a year later, the Congress Kingdom virtually became a cultural and political extension of the Russian Empire. Another uprising of 1863–64 eventually led to the extinction of the Congress Kingdom of Poland as a political unit.

Meanwhile, in Prussian-controlled Poznań, the experiment of settling Germans on Polish soil (begun by Prussian king Frederick the Great after the First Partition) was resumed as policy at the end of 1830; the Polish language was no longer used equally with German in government offices, law courts, and schools. After Poznań revolted unsuccessfully against Prussian rule in 1848, a stricter policy of Germanization was applied. In the Kulturkampf period of 1872 to 1874, instituted by Otto von Bismarck, former Prussian chancellor and founder of the newly created German Empire, German became the language for all teaching, except for religious instruction. Polish was completely eliminated in 1900, which led to the famous schoolchildren's strike of 1903–04. Poznań had been renamed, in German, Provinz Posen.

During World War I, Russia fought Austria and Germany on Polish territory, with Polish leaders supported by the Allies, especially France. The chaotic Soviet Revolution of 1917, as well as the defeat of the Central Powers, enabled the Poles to form a state of their own in 1918. In December 1918, shortly after Armistice Day, the majority Poles of Poznań successfully rose against German domination, and Gniezno became part of the newly independent Poland. A peace treaty signed by the Soviets and Poland in 1921 (after a 1920 war) gave Poland substantial territories in the east.

In September 1939, at the beginning of World War II, Nazi Germany and Soviet forces invaded Poland, each an-

nexing parts of Polish territories. Once again, from 1939 to 1945, there was no independent Poland, and Gniezno was in German hands; Poznań was called Gau Wartheland, because it was centered around the Warta River area. The Nazis established one of their first forced-labor camps in Poland at Gniezno. Heavily destructive fighting and systematic Nazi terror, especially directed at the nation's Jews, left Poland in ruins and rubble. At the Allied conference in Potsdam, Germany, in 1945, the Soviets retained territories in the east, while Poland regained large former German territories in the west; the province, or voivodship, of Poznań was more or less restored to its prewar boundaries.

During the German occupation, the Nazis suppressed the structure of the Polish Church and attacked its clergy; this was especially true in territory incorporated into the German Reich, like the Warta area, which included the archdiocese of Gniezno-Poznań. Though the primate of Poland, Cardinal August Hlond, left the archdiocese under pressure from the Polish government, the Catholic Church continued to exist, because many Germans practiced the religion. (The Reich contained 30 million Catholics who recognized the Pope and the traditions of the Holy Roman Empire. Moreover, the Vatican had also made a concordat with Nazi Germany in 1933, and the new Pope, Pius XII, elected in March 1939, had been Papal Nuncio in Germany for many years.) Up to 1,300 Polish churches and chapels were closed in the Warta region; an additional 500 were turned into storehouses or given to the German Evangelical Church. The Gniezno Cathedral was just one of many churches stripped of its artworks and later devastated. There was a mass arrest of Polish priests from 1939 to 1941; out of 2,500 priests working in the Warta region before the war, some 1,550 were sent to concentration camps, of whom 752 lost their lives.

Poland came under Communist control in 1949, when large-scale industrialization and reconstruction programs were begun. Gniezno, long a local trade and railway center, developed industries related to the manufacture of engineering equipment and chemicals, as well as the processing of local agricultural produce. A series of workers' strikes and riots due to price increases in Poland in the 1970s and 1980s led to the creation of the Solidarity Party. Ultimately, the Communist government fell in 1989.

Further Reading: Norman Davies's *God's Playground: A History of Poland,* volume 1, *The Origins to 1796* (New York: Columbia University Press, 1982; Oxford: Oxford University Press, 1983) is the definitive account of the nation's early history. *Poland* by Marc E. Heine (New York: Hippocrene, 1987) is a compact account of the nation's history and culture, on a region-by-region basis. Jozef Garlinski's *Poland in the Second World War* (London: Macmillan, and New York: Hippocrene, 1985) discusses the Nazi and Soviet occupations of Poland. The reader might also wish to consult Neil Ascherson's *The Struggles for Poland* (New York: Random House, 1987; London: Pan Books, 1988), an insightful examination of the complex, often chaotic historical forces that shaped a state, a nation, and its people.

—Jeff W. Huebner

Goslar (Lower Saxony, Germany)

Location: Situated 45 miles southeast of Hannover, 120 miles southeast of Bremen; in the heart of the Harz Mountain range.

Description: Town founded in 922; made famous in 968 for the discovery of silver in nearby Rammelsberg; in 1056, Emperor Henry III built his famous fortress palace, the Kaiserpfalz, which is the largest intact secular structure in Europe in the Romanesque style; in the sixteenth century the town became renowned for its "Gose" beer; untouched during World War II, Goslar is a popular tourist attraction and beautifully preserved historic medieval town.

Site Office: Tourist-Information
Markt 7
38640 Goslar, Lower Saxony
Germany
(53 21) 28 46

Goslar has been continuously inhabited for more than 1,000 years. Although it never achieved the size or distinction of London or Paris, cities of similar age, one can come to Goslar and find, still standing and inhabited, 168 buildings constructed prior to 1550, and another 144 built between 1550 and 1650. These structures, moreover, are not hard to find amid towering skyscrapers; they compose the town's core. Add to this profusion of intact medieval splendor the Kaiserpfalz, finished in 1056 and still standing sound and unaltered, and one has a living museum.

Goslar appears to be relatively remote, the nearest city being more than forty miles away. Moreover, Goslar is enveloped by thick forests in the heart of the Harz Mountains of northern Germany. This mountain chain is rich in lore and mythology. Its highest peak, the Brocken, was immortalized in Goethe's *Faust;* the range was also the home of the pre-christian Teutonic deities, as well as Little Red Riding Hood and Sleeping Beauty. But if Goslar seems rather out of the way today, it was never cut off from larger happenings outside its borders. In fact, it was often the site of dramatic events: the visit of the pope in the eleventh century, the Reformation, the Thirty Years War, and the rise of the Third Reich.

According to oral tradition, a tiny settlement existed at the site in the early tenth century, possibly earlier. A small river, the Gose, provided the food, water, and transportation needed for life in those days, while the thick spruce and fir forests provided game and fur. By 922, the German emperor Henry I officially designated the town as "vicus Goslariae," or Goslar, after the river. Nothing was remarkable about the place until 968, when veins of silver were discovered by accident in nearby Rammelsberg, a mountain in the Harz.

This is commemorated in the thrice-daily playing of the glockenspiel in the town square, where life-size figures re-enact the scene of Ramm, vassal of Emperor Otto I, accidentally noticing veins of silver uncovered by the scratching of his horse's hooves. The discovery marked the start of a silver rush, and Goslar turned overnight into a mining town, as rough and ready as any in the American West.

Some individuals, including the German rulers, began to grow fat on the profits of mining, and Goslar became a favorite spot of the emperors. In the mid-eleventh century, Henry III developed a personal fondness for Goslar that resulted in his ordering the building of an imposing palace for his royal visits. This became known as the Kaiserpfalz, an immense structure built in the Romanesque style of the day, with a chapel the size of a church in a separate, attached building, known as St. Ulrich's. Goslar by then was a walled and fortified town. Like the Tower of London built decades later, the Kaiserpfalz was meant to overawe the inhabitants and local princes; in that day and age, it must have succeeded.

Even by today's standards it is a large building, comparable in size to a cathedral, and it is considered the oldest surviving secular medieval structure in Europe. It consists of two stories, divided into immense "halls" or rooms, with walls that were more than twenty feet thick. When royalty was in residence there, important matters of state were discussed, important visitors received, and much medieval pageantry was witnessed by Goslar's residents, including the visit of Pope Victor II in the 1050s.

Inside the Spartan-looking St. Ulrich's Chapel (originally only one of several chapels within the Kaiserpfalz) is buried the heart of Emperor Henry III himself, although his body lies buried in the city of Speyer. In his lifetime he had expressed the wish to have his heart laid to rest in his beloved Kaiserpfalz, where his son and successor, Henry IV, was born. Nearly 100 gatherings of the early Reichstag (or parliament) took place at the Kaiserpfalz. Miraculously, the Kaiserstuhl, or bronze throne of the emperors in which they presided over these meetings, is still preserved and on display.

In 1200, Goslar was one of the largest towns in north Germany, with 5,000 inhabitants. It was also quite wealthy, enabling the burghers to maintain the Kaiserpfalz at their own expense. In the thirteenth century, when the German emperors lost interest in Goslar, the wealthy citizens of the town took over the Kaiserpfalz and repaired it substantially, altering the interior to conform to the popular Gothic style of the day. The townsfolk also removed the beautiful fountain from the imperial compound in order for it to grace the town square, where it still stands today. Clearly the decline of imperial interest in Goslar did not harm the city; if anything,

A street scene in Goslar
Photo courtesy of German Information Center

its residents, many of them grown wealthy from the nearby silver, gold, copper, and lead mining, could afford to assert their independence. In 1268, Goslar joined the prestigious Hanseatic League, a union of northern European cities. In 1340, wealthy burghers bought the rights to an imperial charter, recognizing Goslar as a city, and guaranteeing its virtual independence. Along with a city charter, they procured the all-important rights to the mines in nearby Rammelsberg, although not the mines themselves. This allowed them to mint their own coins, many of which are on display in the civic museum.

Wealth led to more building. In 1200, there were only a handful of churches in the town. By 1500, the number of churches had climbed to forty-seven, endowing the city with the nickname "northern Rome." Of these, only five are still standing: the oldest, St. Jakobi, finished in 1079; the church of St. Cosmas and St. Damian; St. Stephan's; the Neuwerk-kirche; and the Frankenberg Kirche. In 1450, the beautiful Rathaus, or town hall, was built, complete with expensive

stained glass windows and an inviting covered shopping arcade on the ground level. In the early 1970s, the Rathaus underwent a major restoration, enabling the visitor to appreciate what one of Germany's oldest town halls looked like in the mid-fifteenth century. The town hall still contains the lavish paintings, upwards of fifty-five, that adorned the assembly room of the burghers in the early sixteenth century. These paintings depicted stages in the life of Christ and other biblical figures, but they bore likenesses to the inhabitants of Goslar, who sat as models for them.

Mining, while the most lucrative enterprise, did not become the sole major business in the town. Brewing beer for thirsty miners became big business also, and "Gose" beer became famous throughout the area. Goslar also became an important producer and exporter of roofing tiles, which adorned the houses of the well-to-do. Many of the elite lived within the city's twenty-three-feet-thick town walls, with their 182 fortified towers. Only four of these interesting towers are left standing today, and no two of them look alike.

These wealthy burghers belonged to or headed the many guilds in the city. The guild houses are almost all gone, but for a few—richly decorated and imposing—that convey some impression of Goslar's former commercial glory. The city's decline began slowly in the sixteenth century. There was much tension between townsmen and the local ruler of Brunswick (the name of the principality in which Goslar was situated). A fire, possibly a result of arson, burned down most of the town's churches in 1527. At this time, a new doctrine was beginning to gain popularity in the city, emanating from the university town of Wittenberg. Goslar's churches were soon adopting the doctrines of Martin Luther, and the old monasteries and convents began to shut down. Goslar's citizens defied their unpopular local duke and sided with the anti-Catholic League of Schmalkald. When the League was defeated in 1552 by the Catholic forces of Holy Roman Emperor Charles V, the city was punished by the duke, who rescinded Goslar's rights to the nearby mines, including the use of most of the surrounding forests.

Deprived of the source of its commercial wealth, the town's fortunes sank. While beer became the most important export, it did not make up for the loss of the mines. To make matters worse, Goslar did not escape the brutalities of the Thirty Years War that followed in 1618. Then bubonic plague raged in the town for two years, from 1622 to 1624, claiming 3,000 lives out of a population that was no more than 10,000. The few Jews who were permitted to live in the town were driven out, blamed for the pestilence. With peace in 1648 they returned and reclaimed their small synagogue, only to have it closed in 1670 because of rumors that they were blaspheming Christ. It was later reopened, but few Jews remained in Goslar thereafter, as the town stagnated and its population declined.

The town's architecture suffered as well during the economic downturn. A portion of the Kaiserpfalz burned to the ground, and the rest fell eventually into decay and disrepair. In time, the palace's surviving furnishings were sold to raise money, St. Ulrich's Chapel was turned into a prison, and the Kaisersaal, or Imperial Hall, was used as a granary. The town's impoverishment ultimately saved much of Goslar's architecture, however; while they put their once-grand buildings to rather undignified uses, the townspeople were too poor to raze the structures and erect new ones. Eventually, with the rise of nationalism during the drive for German unification, the people became concerned with restoring and preserving their architectural heritage. In 1868, restoration began on the Kaiserpfalz, turning it into a national shrine. A statue has been erected there in honor of William I, the last emperor to visit the Kaiserpfalz and the most generous donor to the restoration project.

With the birth of a new, centralized German Reich in 1871 the economy of the town began to move forward. The railroad had just recently come to Goslar, and some buildings were torn down, as were most of the lovely medieval towers. But Goslar remained an agricultural town, with the Industrial Revolution merely passing it by. Tourism, in the form of German travelers visiting the city from spas in the outlying area and from nearby cities, also began to make its mark.

Goslar might have become then what it is today—an intimate, preserved medieval town that beckons to the native as well as to the foreign visitor—had it not been for the ravages of the first half of the twentieth century. Its economy was destroyed in the hyperinflation of the early 1920s; its gradual recovery afterward was interrupted by the Depression and the coming of dictatorship. A local son—Erwin Rommel—became a prominent Nazi officer (although he eventually became opposed to Hitler) during World War II; yet Goslar's residents were not much attracted to Nazism, even though their town represented to the regime everything that was Teutonic and German. Adolf Hitler paid a visit to Goslar in 1934 and designated it the official site of the agricultural ministry of the Third Reich. Nazi rearmament brought an airfield and training camps to Goslar, and an unpopular stridency and militancy in a city whose residents could not bring themselves to participate in the destructiveness of Kristallnacht in 1938 (the synagogue was not closed down until the 1950s, when the town's newspaper was installed there). Fortunately, Goslar was spared from damage during World War II despite the existence of the airfield and the fact that Goslar had become a major hospital center.

After the war, occupation by British and Danish troops followed, as did the restoration of the town. During the renovation of the early 1970s, dozens of beautifully written letters, dated in the year 1400, were discovered accidentally. Goslar is a tourist attraction, a living museum, and a starting point for excursions to the nearby Harz Mountains.

Further Reading: Currently, the only sources of information on Goslar in English are guidebooks, which highlight the major historic sites within the old town district. In German, several books were published in the late 1960s and 1970s that furnish an overview of Goslar's medieval past and its major attractions: Peter Lufft's *Schönes Goslar* (Hannover: Fackelträger Verlag, 1969); Klaus Krause's *Goslar: Über den Dächern einer 1000jährigen Stadt* (Goslar: Verlag Goslarsche Zeitung, 1976); and *Goslar, So Wie Es War* (Düsseldorf: Droste Verlag, 1978).

—Sina Dubovoy

Göttingen (Lower Saxony, Germany)

Location: Sixty-five miles south of Hannover and one hundred forty-five miles south of Bremen, in the north German state of Niedersachsen (Lower Saxony); situated on the Leine River, and southwest of the Harz mountains of Grimm fairy-tale fame.

Description: Göttingen is famous for Georgia Augustus University of Göttingen, established in 1737, the first university in the German states to allow complete freedom of thought and research. In the 1920s and 1930s, the university was the world center for the study of mathematics and physics. Its faculty members have won thirty Nobel Prizes. Göttingen was also a commercial center in the Middle Ages and a member of the Hanseatic League; in Göttingen's Old Town district, several medieval buildings, including churches and the town hall, have been preserved.

Site Office: Fremdenverkehrsverein
Markt 9
37073 Göttingen, Lower Saxony
Germany
(5 51) 5 40 00

The town of Göttingen is so indistinguishable from its world-renowned university, it is difficult to imagine that the city existed at least 500 years before the university ever was established. This milestone did not occur until 1737, long after Harvard University had been founded, to say nothing of such venerable centers of learning as Oxford or the Sorbonne. The opening of the University of Göttingen (officially, the Georgia Augusta University of Göttingen) that year, amid much festivity, injected new life into a dying city. The sleepy, moribund town of Göttingen was selected as the site of a new, major university precisely because of its small population, abundance of land, and relative remoteness from major cities. The low cost of living in this area, compared to Berlin or Bremen, was a selling point in attracting students.

Up to that point, there seemed to be nothing at all remarkable about Göttingen. Architects arriving in town to design the university disdained what they beheld: the medieval spires of the Johanniskirche and the Albanikirche and the fourteenth-century, unfinished Rathaus (town hall). The Gothic-style buildings surrounding the town square, which did not even boast a fountain or statue, were not deemed worthy of notice. Not until the rediscovery of the Middle Ages in the nineteenth century did Göttingen's Old Town achieve some distinction as a gem of medieval architecture, with its ancient fortified walls and crooked cobblestone

streets, dominated by the ornate Rathaus. In 1901, the town square finally was graced by a charming fountain with its whimsical statue, of the goose girl of Grimm fairy-tale fame.

In fact, the not so unremarkable city had led a long and embattled existence by the time the university came to endow it with lasting fame. Like so many medieval German cities, its beginnings are shrouded in the murky, remote past. A settlement above the town square had borne the name "Gütingi" in the tenth century; all that was remembered of Göttingen's predecessor was that the hamlet had boasted a church, and that both disappeared in time. The new settlement that replaced it evolved further south and took on its name, which evolved into Göttingen by the time this successor town obtained a city charter in 1251.

Medieval Göttingen was well situated for that time on a major north-south artery, the Leine River, nestled in the picturesque, fertile "Leinetal," or valley. Goods could be hauled on barges all the way north to Bremen and the North Sea. Göttingen flourished, becoming a center of trade in the region known as Hannover, which was an independent political entity under the Welf family dynasty. Agriculture blossomed as well in the hilly, fertile valley outside the city walls, with wool and woolen textiles becoming the mainstay of Göttingen's economy. As a major river port, Göttingen gained entry in 1351 into the prestigious, wealthy Hanseatic League, which was dominated by German commercial giants such as Lübeck, Danzig, and Bremen, remaining a member until 1572. During this time, the beautiful town hall or Rathaus was built and nearly finished, as well as the city's oldest church, the Johanniskirche (Church of St. John's). Göttingen was virtually independent—the norm for many towns in the Middle Ages because of the dearth of centralized rule, a situation lasting far longer in the German states than anywhere else in Europe, including Italy.

A city with a prominent bourgeoisie needed a university for its youth. In the 1540s the city council petitioned the Holy Roman Emperor Charles V for permission to charter a university. However, waves of religious and social discontent, emanating from the university town of Wittenberg, crested in civil and religious war, in which Göttingen sided with the Protestant-led armies of the League of Schmalkald. The League's war against the Catholic Holy Roman Emperor's forces ended in defeat. Göttingen was fortunate in its wealth, for its citizens put up the unheard-of sum of 10,000 gold florins in 1548 as a penalty for having suborned the crown.

Thereafter followed an outbreak of plague and the horrendously destructive Thirty Years War, the brutalities of which made all previous warfare seem tame by comparison. When peace finally was signed by Protestants (supported by unlikely Catholic French allies) and Catholics in the German

A view of Göttingen
Photo courtesy of German Information Center

city of Worms in 1648, two-thirds of the German states lay in ruins; it would take 100 years for Göttingen to recover.

Göttingen became a neglected backwoods river town, whose commercial life centered around local agriculture, at a time when momentous developments were occurring that would bring the city back to life. As it happened, in 1714, the Electorate of Hannover, as the political state was called, in which Göttingen was situated, became ruled by Great Britain, an anomaly that lasted until 1837. This happened partly because of much intermarriage among the European royalty, and partly because the only successor to the English throne upon Queen Anne's death in 1714 turned out to be her relative, the Elector of Hannover, who became George I that year.

The town had fewer than 20,000 inhabitants when the English king George II, also known as George Augustus, agreed to establish a university there in his own honor. The year was 1734, and it would take three more years for the university to admit its first students. One means of promoting the university to prospective students was to advertise the fact that it boasted a riding hall and stables, so students could comfortably bring their horses along, a thing impossible in a major city. Georgia Augusta University was the first German university to boast such a novelty, and it worked.

Göttingen in a short time would be transformed from a sleepy rundown backwater to a brilliant intellectual center. To offset what the court architects thought was the town's dismal, uninspiring architecture, no expense was spared erecting university buildings of dazzling neoclassical elegance, which subsequent buildings continued to imitate. It was to be the most modern university in the German states, its charter specifying complete freedom of inquiry, even allowing students the unheard-of privilege of using the university library (hitherto always restricted to faculty). By the turn of the twentieth century, the small university could boast the third-largest library in the German Empire.

Freedom of inquiry and hospitality toward foreigners did not extend to Hannover's oppressed Jewish minority, however; as to Göttingen, Jews were not even permitted to reside within the town's walls. Excluding Jewish citizens from study was so commonplace throughout the German states, however, that it attracted no attention.

The combination of unrestricted intellectual freedom and a generous endowment soon attracted the leading thinkers and researchers to the university. By the time Benjamin Franklin went out of his way to visit Göttingen in 1766, he had heard so much about it that he hoped to find in it a model of just what a modern university should be for his native city of Philadelphia; the new University of Pennsylvania was founded on the same principles as its German counterpart. The University of Göttingen turned out to be the first German university to confer the doctoral degree on Americans, and remained until World War I the most popular choice for Americans studying in Germany.

The list of famous scholars and students who came to the university reads like a "who's who" of German intellectual history. Nor were Americans excluded from the roster of notables: Henry Wadsworth Longfellow, historian George Bancroft, Linus Pauling, and J. Robert Oppenheimer studied there. Among Germans, the famous Grimm brothers were students there, and in the 1830s, Jakob Grimm was a philologist on the faculty at Göttingen. The Prussian Otto von Bismarck studied there, while the creator of nuclear fission, Otto Hahn, was only one of the university's thirty Nobel Prize winners. Others included the famous Max Born and James Franck. The principle of the telegraph was first elaborated at Göttingen; the element "aluminum" was discovered in a laboratory there, as was the TB bacillus, Vitamin D. The foundations of modern aerodynamics and hydrodynamics were established at Göttingen in the early 1920s. In short, the University of Göttingen would become world-renowned, attracting students and faculty from all over the world, and would retain its prominence until the Nazi takeover of Germany in 1933.

The Industrial Revolution with all its grimness bypassed the town, despite the coming of the railroad in 1854. Higher learning and intensive scholarship did have its lighter side: the university from the outset had a strict code of behavior as well as a punishment bloc for students who failed to live up to the rules, especially sobriety. The "cells" were always packed, and not surprisingly, in time their walls were covered with graffiti. However, the university's reputation for academic achievement imprinted itself on town life. A spartan, almost monastic lifestyle was expected of faculty and students alike. Ostentation and conspicuous consumption in turn were frowned upon, as the wife of one professor in the late nineteenth century discovered, to her chagrin, when she was informed that professors' wives were expected to wear the same hat to market as well as on formal visits.

With the city's new lease on life, the town resumed participating in larger events bestirring the German nation, which was still divided into dozens of political entities. Seven professors, known as the "Göttingen Seven," were expelled by the kingdom's ruler in 1837 because of their open protest against the revocation of Hannover's liberal constitution of 1833. It was the heyday of German nationalism, quite different from the strident chauvinism of the twentieth century. Besides an ardent desire for political unification, German nationalists in the first half of the nineteenth century were opposed to autocratic rule and called for the emancipation of the oppressed German Jewish minority.

These goals conflicted with the aims of Hannover's autocratic ruler, King Ernest Augustus. In 1837, the year Queen Victoria ascended the British throne, Hannover became an independent kingdom, in the shadow of its overpowering, aggressive Prussian neighbor. In 1866, war erupted between the two states, initiated and planned by Prussian head of state, Otto von Bismarck, a former law student at Göttingen. Hannover was forcibly incorporated into Prussia until Bismarck completed political unification of the German states in 1871, establishing Berlin as the capital of the new German Empire.

The city and university were swept up in the tremendous elation at Germany's political unification and the resulting German "Fatherland." The German Reich was the strongest military power on earth, next to the Russian Empire, and the biggest political state in western Europe.

A new mayor was elected to the town council in 1870, the remarkable and indefatigable Georg Merkel. Until his resignation as Lord Mayor in 1895, he overcame fiscal conservatives and personal opponents as he embarked on a sweeping urban renewal program that transformed Göttingen into a modern community. The streets were entirely paved for the first time in the town's history, and new and modern drainage and sewage systems, as well as a water purification plant, were erected. Aware of the fact that the city's young people never bathed, but only washed their hands and faces, Merkel overcame puritanical resistance to the installation of bathing facilities in the city's four new schools. To his lasting credit would go the complete reforestation of the denuded Hainberg on the outskirts of town, which took eleven years of arduous planning and digging. Aware of the need for cultural improvements, he backed the establishment of the city's first theatre, and saw to it that new buildings in Old Town conformed to the city's historic architecture. His successor as Lord Mayor, Georg Friedrich Caslow, continued his modernization program, especially electrification. Not surprisingly, these urban reforms were supported consistently by the university.

By 1900, Göttingen had a population of slightly more than 30,000 inhabitants and had acquired the reputation of a pleasant, progressive, and enlightened community. Perhaps for these reasons, the town attracted a sizable Jewish minority. In the new German Empire, German Jews were granted full citizenship and all careers were open to them. Centuries of discrimination, however, made it difficult for Germany's largest minority to accommodate itself to the new changes. There were only a trickle of Jews studying at the university, even by 1900. Jews were allowed to live in Göttingen, but prejudice even in this enlightened town was marked: very few of the city's clubs and organizations admitted Jews. The new synagogue strikingly resembled a church building, and the town's Jews strove to assimilate themselves, even if this meant rejecting their own faith. Except for a handful of very conservative Jews, the majority of them, approximately 700, adopted German names and lived unobtrusively, reluctant even to establish a Jewish school so as not to raise eyebrows. It had not been more than a half century since Jews were forbidden to live in Göttingen.

The start of World War I in 1914 was a turning point for the city, as for the rest of Germany. The university kept its doors open, despite the dearth of students, all called to the army. The town waxed patriotic, and blindly supported the Kaiser and his aggressive aims. After the war, returning veterans made up the majority of the student body, mostly a disgruntled lot for whom Germany's defeat was viewed as some kind of betrayal. The new Nazi Party gained strong support in Göttingen long before it did elsewhere in Germany, and from the 1920s onwards, more than half of the town council were Nazi Party members. Overt persecution of the Jewish minority was impossible under Germany's new and much chastened postwar government. However, the town's Jewish population began a conspicuous decline, perhaps because there were greater opportunities elsewhere in the new democratic German republic, or perhaps because anti-Semitic articles, appearing as early as 1920 in the town's daily newspaper, the *Göttinger Tageblatt,* made them feel unwelcome.

The university in the 1920s and early 1930s blossomed as never before: Nobel prizes in medicine, physics, mathematics and chemistry were regularly—an average of every other year—bestowed on faculty members. More than 200 plaques commemorating notable professors and students—including Bismarck—dotted residences throughout the city where they had once lived. Paul von Hindenburg of World War I fame, president of the Weimar Republic from 1925 to 1932, paid a ceremonial visit to the city, as did Adolf Hitler, for whom the townsfolk had voted overwhelmingly in elections in 1932 and 1933.

The university's decline as an intellectual center was swift under the new Nazi regime. Virtually all the members of the mathematics faculty had been Jewish, and they were forced in 1933 to take extended leaves of absence. To the greater glory of American mathematics, these professors were fortunate to find refuge in the United States. Never again would Göttingen be the world center of mathematics as in its pre-Nazi heyday. By the time of the university's 200th anniversary in 1937, it had lost its academic freedoms, and was often the site for Nazi ceremonies and speeches. The town zealously carried out official persecution of the Jewish minority, which culminated in the November 10, 1938, *Kristallnacht,* with the destruction of all remaining Jewish-owned enterprises and the torching of the synagogue. The Jewish population disappeared in the final deportation of 1942; only a handful live in Göttingen today.

The university town remained unscathed during World War II, and emerged with all of its buildings intact. One can to this day see the original, still unfinished, fourteenth-century Rathaus, the whimsical statue of the goose girl in the center of the town square, and quaff good German beer in the many long-established standing *Kneipen,* or pubs, throughout the Old Town. The university, with its student body of 30,000, is located largely outside of the town, for the sake of expansion. In the new reunited Germany of the 1990s, Göttingen may yet rise to the illustrious heights it once attained before world wars and political upheaval cast their long, dark shadows.

Further Reading: There is nothing in English on Göttingen apart from travel guidebooks, which highlight the major historic sites within the Old Town district. There are, however, two excellent sources in German. *100 Jahre: Göttingen und sein Museum,* written by multiple authors (Göttingen: Stadtisches Museum Göttingen, 1989) was published to commemorate the centenary of the town's museum but actually provides much information on Göttingen as well as the museum. A great deal of local history also is contained in *Göttingen: So Wie es War* by W. Nissen and W. R. Rohrbein (Düsseldorf: Droste Verlag, 1975).

—Sina Dubovoy

Graz (Styria, Austria)

Location: On the banks of the River Mur, in the province of Styria, southeast Austria; eighty-seven miles south-southwest of Vienna.

Description: Graz is the second-largest city in Austria. In existence since at least 1115, it became in the fourteenth century the seat of the Leopoldine branch of the Habsburg imperial family, thus initiating a long association with the Habsburgs, who contributed significantly to the city's growth and importance within Austria. In the twentieth century Graz has become an important commercial center for Austria.

Site Office: Grazer Tourismus
Hans-Sachs-Gasse 10/4
A-8010 Graz, Styria
Austria
(316) 835 241

The city of Graz is dominated by the Schlossberg (Castle Hill)—the name Graz comes from *gradec*, Slavonic for fortress. The town emerged on a strategically significant site from which to guard the southern end of the Mur Valley, which forms an important approach to Vienna. The area has probably been the site of some sort of fortification since the ninth century, though the area's real importance as a defensive site emerged during the centuries of invasions from the east by Hungarians and Turks. By the thirteenth century the city of Graz was ringed with walls against enemies. The settlement is first mentioned by the name of Graz in 1115 and achieved the status of city in 1240. The city's rise to prominence within Austria began in 1564. In that year, the three sons of Emperor Ferdinand I divided Austria among themselves. The youngest son, Charles, received the territories of Inner Austria, including the provinces of Styria, Krain, Görz, Carinthia, Istria, and Western Croatia. He took up residence in Graz, thereby turning the city into a capital. The Habsburgs built several important baroque churches in the city and also established a university. The Schlossberg last served as a defense in the early nineteenth century against French forces.

The area was held by the Romans from about 15 B.C. until the fifth century A.D. The Romans began to conquer the eastern Alps during the reign of Augustus and established three Roman provinces in what is now Austria. Graz was in the province of Noricum, which included Styria, Carinthia, eastern Tyrol, Salzburg, and Upper and Lower Austria south of the Danube. Communications with Rome were maintained via the existing trade routes through the Alps, which had been used before military occupation to supply Rome with timber, wool, iron, and lead. The Romans' hold on the area was weakened by the end of the second century, and despite a triumphant campaign led by Marcus Aurelius, they were harassed by continuous attacks by tribes such as the Marcomanni. When the Roman Emperor Theodosius died in 395, troops were recalled to the heart of the Roman Empire to defend Rome itself against the marauders, who eventually sacked the city in 410. Rome's decline left the frontier camps unmanned and the Alpine provinces open to the invasions of local tribes.

The area around Graz was subsequently settled by southern Slavonic tribes but was conquered by Charlemagne, marking the beginning of the area's long association with the Holy Roman Empire and the later Habsburg Dynasty. Charlemagne's successors lost large tracts of land to the Magyars, who conquered modern-day Hungary, leading Emperor Otto I to establish the Bavarian Ostmark or "Eastern March" in 955 to defend against further Magyar attacks. In these tumultuous circumstances the Austrian nobles appealed to the king of Bohemia, Przemysl Otakar II, for protection. In 1276 Rudolf I, a Habsburg who had been elected king of Germany, occupied Graz. The king of Bohemia and his supporters held the castle in Graz and would not recognize Rudolf's claims on the region, but they were defeated and Przemysl Otakar II was killed at the Battle of Dürnkrut in 1278.

Rudolf I made Styria a hereditary Habsburg possession in 1282, but after the Treaty of Neuburg in 1379 the Habsburg possessions were divided between the Albertine and Leopoldine branches of the family, and the city became the residence of the Leopoldine branch of the Habsburgs. Its significance as a strategic defense site and its long association with the Habsburgs accounted for Graz's importance as a city until the nineteenth century.

The Habsburg dynasty, founded by Rudolf I, is of crucial importance to the history of Europe. It provided Austria with its rulers for nearly 700 years, from 1278 until 1918. Through the Holy Roman Empire Habsburgs ruled most of Europe for more than 300 years, from the late fifteenth to the early nineteenth centuries, and also provided the rulers of Spain from 1516 to 1700, thereby extending their influence to the Spanish expansion into the Americas. The Holy Roman Empire was founded by Charlemagne but became properly established in 962, when Otto I, king of Germany, was crowned emperor at Rome by the pope. The Holy Roman Empire, modeled on the Roman Empire, was designed to be a universal monarchy. Until the time of Maximilian I, the last emperor to be crowned by the pope, there was a powerful allegiance between the Vatican and the Holy Roman Empire, which formed a crucial part of the balance of power for much of Europe's history. At its height in the tenth and eleventh centuries, the Holy Roman Empire encompassed Germany, the Netherlands, Austria, parts of the Czech and Slovak Re-

The Mausoleum of Emperor Ferdinand II
Photo courtesy of Grazer Tourismus

publics, Switzerland, eastern France, and northern and central Italy. From the eleventh century, however, the emperor and the pope were in continuous conflict over who held the supreme position within the allegiance. Emperors were henceforth no longer crowned by the pope, though the institution of the Holy Roman Empire survived, albeit in a diminished form, until the nineteenth century, when Napoléon, who crowned himself emperor of the French, ceased to recognize the Holy Roman Emperor, even as a symbolic figurehead. Francis II of Austria abdicated the imperial title soon afterward, in 1806.

Maximilian I, Holy Roman Emperor from 1493 to 1529, succeeded in reuniting the possessions of the two branches of the Habsburg family and, by marrying his son Philip to Joanna of Castile in 1496, obtained the newly united Spain for the Habsburg empire. The empire's influence grew until the rule of Maximilian's grandson, Charles V, saw a Habsburg empire that included all of Europe except England, France, Scandinavia and Portugal. Charles V was Holy Roman Emperor from 1519 to 1556, and king of Spain, as Charles I, from 1516 to 1556. He granted the German and Austrian territories of his vast empire to his brother Ferdinand I, and when Charles died the imperial title was inherited by Ferdinand, thus splitting the Habsburg dynasty once again, this time into the imperial and the Spanish lines. Ferdinand married into the Hungarian royal family and thereby inherited parts of Hungary and Bohemia, thus laying the foundations of Austro-Hungarian power, which was strengthened considerably in 1699, as a result of the Treaty of Karlowitz. The imperial line outlived the Spanish line, and in the War of the Spanish Succession (1701–14), fought against Louis XIV of France, it gained many possessions previously in the hands of the Spanish Habsburgs. In the eighteenth century the Habsburg throne passed to Maria Theresa, who ruled from 1740 to 1780, but her entire reign was spent in constant conflict with the emerging kingdom of Prussia. This conflict continued into the nineteenth century, when several wars were fought with Prussia, a factor that led to the decline of Austria's influence over Germany. The last Habsburg ruler, Charles I, emperor of Austria and king of Hungary, abdicated in 1918, leaving Hungary to declare its independence. Both Austria and Hungary declared themselves republics.

The Habsburg dynasty thus played an essential part in the evolution of Europe: Austria in particular made a vital contribution to European history through its role as bulwark against the Turks. Graz figured prominently in this protracted battle to preserve Catholic supremacy in Europe. The Ottoman Empire, founded in 1288 with its center in modern Turkey, took over in the minds of contemporary Europeans as the source of the expansionist Moslem threat. Ever since the eighth century the Saracens (the Christian term for the Arab warriors of the time) had established footholds in and around the Mediterranean, including Sicily and several cities along the southern coast of Spain. In 720 Moslem armies has crossed the Pyrenees into France, but in 732 the tide had turned with the defeat of the Moslems at Poitiers, France.

Nevertheless, part of the passion with which the Crusades of the eleventh to thirteenth centuries were fought was a result of the deep fear held by many Catholic Europeans of the expansionist Moslems. This fear was rejuvenated in the sixteenth and seventeenth centuries, when the Turks posed a constant threat to the rest of Europe. Austria bore the brunt of this aggression, with persistent attacks and invasions. In 1529 and again in 1683, the Turks stood before the gates of Vienna, but were both times repulsed. Graz, as a strategic point between the Turks and Vienna, thus assumed historic importance for Europe. A reminder of this significance is the town's most visited landmark, the Landzeughaus, or provincial arsenal. It provides a particularly vivid impression of the town's former exposure to the threat of Turkish invasion. Today, the Landzeughaus serves as a museum; in the past, it was a comprehensively equipped armory. Before its construction in 1642, weapons were stored in several places, including cellars in the city wall and in spare space in the Landhaus (the Styrian provincial parliament). The building was designed by the Italian architect Santino Solari, whose work can be seen in other parts of Austria, most notably the Dom, or cathedral, of Salzburg. Serving as a central armory, at the time the biggest in the world, the Landzeughaus enabled the city's defenders to arm themselves on short notice. Its contents could arm 28,000 men. Its four floors have remained virtually unchanged and still contain the original weaponry, arranged as if prepared to defend the town. Approximately 30,000 items are displayed, among them more than 1,300 suits of armor and thousands of firearms, cannon, swords, and halberds. The risk of Turkish invasion receded at the beginning of the eighteenth century, and in 1757 the national government tried to take possession of the Landzeughaus, against the will of Graz's inhabitants. They appealed directly to Empress Maria Theresa for permission to keep the armory as a gift for the courage they showed in the defense of the country's border against Turkish incursion. The empress granted the request, and ever since the people of Graz have dedicated resources to the preservation of the Landzeughaus.

Preservation of other defensive features of the town was not so easy. The ramparts that once stood on the Schlossberg were taken down on Napoléon's instructions in 1809, after he had defeated Austria. A large ransom was paid by the town to preserve two of the castle's towers. These are the Uhrturm (Clock Tower) and Glockenturm (Belfry). A short distance from the funicular station on the Schlossberg, the famous octagonal Glockenturm, built in 1588, houses *Liesl,* Styria's largest bell. A short downhill walk leads to the Clock Tower, built in 1561, which now houses the Schlossmuseum. Its clock dates from 1761. The land formerly occupied by the ramparts was made into a public park in the nineteenth century.

In addition to the prolonged threat of Turkish invasion, there were other difficulties that the people of Graz had to endure. These are represented in the Landplagenbild, or Picture of the Land's Calamities, on an outside wall of the southern end of the Cathedral, facing a small square. A 1481 fresco describes the misfortunes of the plague (or Black Death), Turkish invasions, and locusts, all of which had befallen Graz in 1480. Graz was to suffer another outbreak of plague in 1680; one-fourth of the city's residents died in that epidemic. The Cathedral was built in 1438 on the site of an earlier church. With Graz's elevation to status of capital and seat of royal residence in 1564, the Cathedral became the Court Church. In 1786 it became the seat of the Bishop of Seckau. Adjoining the Cathedral to the south is the Mausoleum of Emperor Ferdinand II, who died in 1637. It contains the tombs of the emperor and his first wife Anna Maria. Its baroque interior and the high altar of its Chapel of St. Catharine were designed by Johann Bernhard Fischer von Erlach, an architect who added to the town's renown.

Graz is the birthplace of Fischer von Erlach, who along with his contemporary Johann Lukas von Hildebrandt is hailed as one of the most outstanding and influential Austrian architects of the baroque period. The house in which the architect was born in 1656 stands in a street that is today named after him. Fischer von Erlach studied in Rome, where exposure to the works of Giovanni Bernini and Francesco Borromini was to influence his own creations in Austria. Characteristic features of the baroque style in general and Fischer von Erlach's work in particular are elaborate, fluid forms in undulating facades and rich ornamentation. Their combination was designed to represent and evoke in the beholder's contemporary consciousness a glorious vision of Heaven. Fischer von Erlach's most celebrated buildings include the Dreifaltigkeitskirche (Church of the Holy Trinity) in Salzburg, which is reminiscent of Borromini's Saint Agnese Church in Rome, and the Karlskirche in Vienna. His secular buildings include two Vienna landmarks: Schönbrunn Palace and the Hofbibliothek (Imperial Library). The latter was constructed after Fischer von Erlach's appointment as surveyor general of imperial buildings in 1705.

Opposite the Graz Cathedral in the Hofgasse stands the Burg, an imperial palace constructed by Duke Frederick V (later Holy Roman Emperor Frederick III) in 1438 that now houses government offices. Few features of the original building remain, with most of the structure dating from the nineteenth and twentieth centuries. A grand courtyard contains a double spiral staircase built in 1499. A smaller courtyard displays busts of distinguished Styrions. Another older feature is the Burgtor, a gate that leads into the Stadtpark, the town's municipal park.

Graz is also the birthplace of Archduke Francis Ferdinand, the heir to the Austro-Hungarian empire. He was born in 1863 in the town's Palais Khuenburg, which today functions as the Stadtmuseum, or city museum. Francis Ferdinand took the unusual step of marrying a woman of lesser rank, Sophie Chotek. This constituted a morganatic marriage, whereby Francis Ferdinand's descendants were denied any rights of succession upon his death. His death itself has figured largely in the history of Austria, and that of the world. He was assassinated in Sarajevo on June 28, 1914, by a Serb nationalist, an event that triggered World War I. Austria had been eager to suppress Serb nationalism, and the archduke's assassination was more than adequate justification for a declaration of war on Serbia. The conflict escalated to include two power blocs, led respectively by the Central Powers of Austria-Hungary and Germany and the "Allied Powers" of Britain, Russia, and France. In 1917 the United States entered the conflict on the side of the Allied Powers, and the war ended in 1918.

During the twentieth century Graz has continued its development as an industrial and cultural center. This development had begun in the seventeenth and eighteenth centuries and intensified under the leadership of Archduke John from 1811 to 1859. Among the accomplishments of his administration was the founding of Graz Technical University. Industries in Graz manufacture a wide variety of products, and the city also is a center for commerce in the agricultural products of the surrounding countryside.

Graz's industrial significance, however, made it a target for Allied bombings during World War II (Austria had been annexed by Nazi Germany in 1938). Some 15 percent of the homes in Graz were destroyed by air raids in 1945. After the war's end, Graz was occupied by British forces until 1955. Since then, Graz has refurbished and added to its cultural institutions. Its Academy of Music and Performing Arts was founded in 1963. Today the historic and cultural sites in Graz draw many visitors, adding tourism to the mix of the city's diversified economy.

Further Reading: *Austria and the Austrians* by Stella Musulin (London: Faber and Faber, and New York: Praeger, 1971) contains a chapter on Styria and provides a descriptive account, written in an engaging style, of Graz's surroundings.

—Noel Sy-Quia and Hilary Collier Sy-Quia

Haarlem (North Holland, Netherlands)

Location: Western Netherlands, on the Spaarne River.

Description: Capital of North Holland, one of the country's twelve provinces. Industrial and market city; part of the northern wing of Randstad Holland, the urban fringe around the agricultural heart of the Netherlands.

Site Office: VVV Haarlem
Stationsplein 1
2011 LR Haarlem, North Holland
Netherlands
(23) 319059

A river town that began on high ground but grew to engulf surrounding marshes, Haarlem is part of the Randstad, the urban core of Holland that includes Amsterdam, Rotterdam, and The Hague. Haarlem's history draws from industry on the one hand and from the fine arts on the other, with one unhappy interlude thrusting the town into the fray of Counter-Reformation violence.

Activity in the fine arts, especially in painting, was so prominent during the seventeenth century that several towns—Amsterdam, Haarlem, Delft—could claim their own "schools" of artists. Amsterdam had Rembrandt, Delft had Jan Vermeer, and Haarlem boasted the portraitist Frans Hals.

The prosperity that allowed the luxury of arts patronage was founded on a solid industrial base. Haarlem was a center of linen production, becoming especially prominent in the late sixteenth and early seventeenth centuries after wars in the southern Low Countries against the Spanish rulers interfered with the cloth trade in rival cities such as Ghent and Antwerp. In addition to the many rural Dutch who became urbanized by moving into towns such as Haarlem beginning in the Middle Ages, there was plenty of industrious immigrant labor from the southern provinces and other areas of Europe where conflicts between non-Catholics and the Spaniards made life difficult.

Haarlem was not always so open to outsiders, however. In fact, the first development on the site was the Gravensteen, a stronghold built on the Spaarne River in about 977 in an effort to repel Danish invaders. The defensive site took on a commercial aspect when the counts of Holland determined that they could levy tolls from the spit of land in the river known as the Bakenes. Strategically located on the main route into northern Holland, Haarlem served as a market center for the Kennemerland, a large agricultural area to the north. The town grew rapidly, profiting by being party to the Dutch economic juggernaut that was expanding to all parts of Europe and well beyond.

By the twelfth century, Haarlem had become one of the chief residences for the counts of Holland, a role it filled until the thirteenth century. Count Willem II granted town rights to Haarlem in 1245, about which time the knights' houses clustered around an open field were converted into shops and the count's own house became the town hall.

Manipulation of the landscape began early on, when as a defense strategy the counts dug a canal—the Bakenessegracht—alongside their castle during the thirteenth century. Haarlem's earliest expansions proceeded in the manner of a traditional town: there was ample high, dry ground just west of the Bakenessegracht, and the town expanded into that area in the mid-thirteenth century. The Grote Kerk (Bavokerk, or Church of St. Bavo), Town Hall, and Market Square all were built in the new section, in effect shifting the nucleus from the castle area. In the fourteenth century expansion occurred across the river, east and southeast of the castle area. Major additions of land in about 1426 continued Haarlem's growth to the southwest. In the fifteenth century, however, Haarlem began to expand into low-lying areas, taking on the character of a *grachtenstad*—a water town. It grew in this manner again in the seventeenth century, with transportation becoming increasingly water-based until many canals were later filled in.

The physically expanded town's chief economic activities in medieval times were beer brewing, textile-related industries (such as weaving, bleaching, and dyeing), and shipbuilding. Industrial prosperity allowed the town to provide services for the less fortunate, such as almshouses, begun in 1395, with numerous examples extant. The fifteenth century witnessed considerable industrial expansion accompanied by a rapid increase of population. Haarlem had 5,000 inhabitants in 1400, 20,000 in 1500, and 40,000 in 1600.

The existence of many Dutch towns has been contingent upon the technology of reclaiming arable land from swamps, marshes, and, most dramatically, from lakes or the sea and its inlets. Haarlem poses an interesting hybrid of the town-making process in Holland: the city began on naturally dry, high ground, but then expanded to include low-lying, marshy territory, making Haarlem a *grachtenstad* as well. These divergent geographies resulted in a formal duality in the town plan. Notable in the oldest, highest part of Haarlem is the irregular development characteristic of many medieval town centers, while the more tenuous, low-lying areas feature careful planning that allows the control necessary to maintain a delicate balance between the town and the water.

Haarlem, as Amsterdam, has for centuries been a center for printing. The degree of the historic merits of the printing industry in Haarlem are disputed, however. There have been attempts to claim efforts on the part of Laurens Janszoon Coster (c. 1370–1440), a Haarlem town official, that would make him a contemporaneous inventor with Johannes Guten-

St. Bavo church in Haarlem
Photo courtesy of Netherlands Tourist Board

berg of movable type. But scholars discount this notion due to lack of documentation. In any event, the establishment of a newspaper at Haarlem in 1656—the *Oprechte Haerlemse Courant*—reinforced the importance of printing in Haarlem. The publication survives as the *Haarlems Dagblad*, ranking among the oldest European newspapers.

Haarlem's primacy in the horticultural trade is perhaps less debatable. The town's important market-gardening and horticultural (bulb-growing) interests did not develop until the seventeenth century, when the sandy soil that was so abundant in the region proved of primary benefit to tulip cultivation. The tulip, a flower native to the northern and eastern Mediterranean, had made its way to the Netherlands from Constantinople and the Levant during the late sixteenth century. As demand for the flowers and bulbs grew during the 1620s and 1630s, the Haarlem area introduced increasingly sophisticated cultivation and varieties. In 1634 a financial frenzy took over the tulip market, with rampant speculation controlling the value of certain varieties of bulbs until the market crashed in 1637. But the local growers south of Haarlem continued to be the primary suppliers of the exotic bulbs for the next two centuries.

As in the rest of Europe, the Reformation introduced to Haarlem a religious conflict that became inextricably linked to the city's politics and occasionally interfered with the town's prosperity. The Protestant theology that took hold here was that of the French theologian and reformer John Calvin. The Calvinists came to power in Haarlem, and the Catholics quickly were put on the defensive. When the Holy Roman Emperor Charles V abdicated in 1555, his Catholic son, Philip II, succeeded as king of Spain and lord of the Netherlands. In May 1567 the Spanish government outlawed the practice of Calvinism, and to enforce the edict in the Low Countries, in August, Philip sent the feared duke of Alba, Fernando Álvarez de Toledo, to Holland, where the latter began a "Council of Blood" to expunge the heretics.

Much to the detriment of Haarlem and its allies throughout Holland, Amsterdam collaborated with the Spaniards, serving as a supply depot. From December 1572 Haarlem was under siege by Alba's son Frederick of Toledo, and famine set in. While Haarlem struggled mightily to withstand the siege—in the winter at least, skaters could smuggle in supplies over frozen waterways—the Spaniards had uninterrupted access to relief through Amsterdam. Not only that, but they could move their fleet freely from the North Sea through the Zuider Zee, past Amsterdam and into the Haarlemmer Meer, blocking Haarlem's main supply route. Despite attempts by Prince William of Orange and various groups of townspeople to defend Haarlem, the besieged Calvinist town capitulated to the Spaniards on July 12, 1573.

As a condition of surrender, Haarlem had agreed to deliver to the victors fifty-seven of the town's leading burghers in exchange for a general amnesty. Frederick initially agreed to this request, but three days later the Spaniards commenced a massacre in which 1,800 members of the defending garrison of 4,000 were executed, along with all Calvinist ministers. The governor himself, Wigbold van Ripperda, was not spared.

The Spaniards, however, were ultimately frustrated in their efforts to re-Catholicize Holland, and Alba departed the country in December 1573. In the wake of the devastating siege, a great fire swept Haarlem in 1576, destroying much of the inner city. Just two years after that, on May 29, 1578, a number of Catholic churches and religious establishments in the area were plundered during a famine.

The northern provinces, including Holland, eventually formed a union at Utrecht under William of Orange, in 1579. The idealistic stadholder sought to establish harmony among the Protestants and Catholics, but this was problematic in the cities, where political and religious sentiments were especially strong. The urban areas championed a Protestant Dutch Republic to counter the entrenched Catholicism of the ten southern provinces, where the Spaniards continued to hold sway.

The political and religious situation within the Dutch Republic stabilized as the economy boomed, and Haarlem recovered from the dreadful events of the 1570s along with the rest of Holland. By the early 1600s the town was executing a master plan for rebuilding, with a number of attractive new buildings and orderly streets replacing the more haphazard development of the prior century.

Well before Rembrandt began his career capturing burgher life on canvas in Amsterdam, Haarlem was thriving as a mecca for painters. In fact, Haarlem was home to Frans Hals (c. 1581/85–1666), a painter who ranks with Rembrandt and Vermeer as one of the leading Dutch masters. He migrated to Holland from Antwerp with his parents after that city fell to the Spaniards in 1585, and probably moved to Haarlem in 1591. Primarily a portraitist, Hals ushered in the greatest age of Dutch painting with his life-size group portrait called *The Banquet of the Officers of the Haarlem Militia Company of St. George*, painted in 1616; in this landmark work he brought to the canvas with unprecedented vivacity the remarkable vigor that characterized the Dutch Republic at the height of its Golden Age. Here Hals conveyed with unconventional expressiveness the confidence of twelve men who epitomized prosperity and civic pride.

Hals later developed the sense of spontaneity and joviality for which he is best known, in such paintings as *The Laughing Cavalier* (1624), *The Merry Drinker* (c. 1628–30),

and *Malle Babbe* (c. 1630–33). Among Hals's students were five of his sons, his brother Dirck, Judith Leyster and her husband Jan Miense Molenaer, Adriaen van Ostade, Philips Wouwerman, and Adriaen Brouwer.

Other important artists who worked in Haarlem during the Dutch Republic's Golden Age included Hendrik Goltzius, an outstanding etcher and woodcut artist who founded the etching school in Haarlem in 1582; Hercules Pietersz Seghers, a painter and etcher of fantastic landscapes usually of ominous mountain scenes with little sign of civilization; Willem Claesz Heda, a master of texture and surface in his still-life canvases; Pieter Jansz Saenredam, a pioneer of realistic "portrait" paintings of churches. One particularly notable family included Salomon van Ruysdael, who painted Baroque landscapes primarily of village, canal, and winter subjects, and his nephew Jacob van Ruisdael (who used an alternate spelling of the family name), the premier landscapist of the next generation who eventually moved to Amsterdam.

Haarlem's architecture—marked by its picturesque almshouses and austerely graceful brick churches, and crowned by the delectable commercial buildings Vleeshal and Waag—offered likely subjects for painters who sought beauty in the details of the everyday environment. Largely responsible for this picturesque distinction was Haarlem native Lieven de Key (1560–1627). As Haarlem's first town architect, he redesigned the square around the medieval Grote Kerk—the enlargement of which in the late fifteenth century had created circulation problems—and Town Hall to accommodate his Vleeshal (1603). The latter building, originally Butchers' Hall, features stepped gables, lavish ornamentation, and a copper roof.

Offsetting the triumphs of artistry have been problems associated with being a town more or less dependent on industry, but these have been occasional. Tax riots that plagued a number of towns in 1624 were especially serious at Haarlem; the town militia fired upon the mob, killing five people and wounding many more.

Haarlem's course of prosperity continued through the seventeenth century, but during the eighteenth century the city suffered along with other older industrial towns. Its population declined from an estimated 47,000 inhabitants in 1732 to just 21,000 in 1795, when the Dutch Republic became the Batavian Republic under the auspices of the French. (After the collapse of Napoléon's empire, the Netherlands and Belgium became a united kingdom in 1814.) Haarlem nonetheless remained enough of a cultural center to attract visits from the likes of Handel and Mozart, both of whom played the 5,000-pipe organ at the Grote Kerk, the latter musician doing so at the precocious age of ten in 1766.

While much serious land drainage was under way in the eighteenth century in regions near Haarlem, the town's greatest challenge in terms of controlling nature involved the struggle to conquer the Haarlemmer Meer, just to the southeast of the city. This body of water had grown from only 7,500 acres in 1550 to 46,250 acres in 1839, due largely to the cutting of peat in the lowlands, which allowed smaller areas of water to merge. Storms on the lake created conditions that turned it into a so-called "waterwolf," threatening the various proximate towns, including Haarlem, Leiden, and Amsterdam. The scope of the project was perplexingly huge, thwarting at least fifteen schemes during the seventeenth through early nineteenth centuries. Dutch engineers were able to drain the lake only in the mid-nineteenth century, after steam power had become available to drive the massive equipment that such a large *droogmakerij* (making dry) required. It took twenty-seven steam pumps well over a decade to drain the nearly 47,500 acres. As was typical in such reclamations, a dike 37½ miles long was constructed around the perimeter of the low land, with an adjacent canal serving as buffer between the new polder (reclaimed land) and the higher ground. Although the massive feat of draining the Haarlemmer Meer was completed in 1852, the resulting land was considered undesirable, with uneven drainage and cholera plaguing the pioneers. The private investment in infrastructure that usually invigorated such areas was not forthcoming. Activity did not begin in earnest there until about 1900, when large-scale agricultural operations began to occupy the zone. Since then—despite being thirteen feet below the level of the North Sea—the former lake has become the site of Schiphol, the busy airport for nearby Amsterdam.

As the threat of the unruly lake abated, Haarlem rebounded as a locus of economic vitality. In 1824, for example, the Nederlandsche Handel Maatschappij, an organization set up to dominate the export trade from Holland to the Dutch East Indies, took advantage of the city's expertise in textiles by locating cotton weaving, bleaching, pressing, and dyeing operations there. Even after the near-war conditions that led to the split in 1830 of the greater Netherlands into the Netherlands in the north and Belgium in the south, several major Belgian industrialists were persuaded to locate their production facilities for Java-bound exports in Haarlem. So pronounced was this spate of new industrialization that Haarlem earned the moniker "Manchester of the Netherlands."

The town's economy was further boosted by the opening of the Amsterdam-Haarlem railway line in 1839, which facilitated subsequent industrialization. In the ensuing century and a half, the textile industry has given way to metal manufactures, food processing (especially the luxury cacao and chocolate trade), service functions, and tourism; the city continues to thrive as a printing center, too. Vast bulb fields stretch to the north and south of Haarlem.

Haarlem today shows little of its water-based heritage. Many of the formerly distinctive canals—including the prominent Oude Gracht, Raamgracht, and Voldersgracht, have been filled in to serve as roadways. But the city's artistic heritage is abundant: Lieven de Key's decorative Vleeshal now houses art exhibitions. And a former almshouse houses the Frans Hals Museum, with the world's premier collection of work by Hals and other Haarlem artists.

Further Reading: *Blue Guide: Holland* (London: Black, and New York: Norton, 1989) by John Tomes provides a brief history of Haarlem, with map, but is most notable for its information about the city's art and architecture. It also includes well-informed introductory essays about Dutch history and art in general. *Dutch Art and Architecture: 1600–1800* (Harmondsworth, Middlesex, and Baltimore, Maryland: Penguin, 1966; third edition, 1977) by Jakob Rosenberg, Seymour Slive, and E. H. ter Kuile, part of the thorough Pelican History of Art series, features an extensive discussion of the Haarlem painter Frans Hals and his protegés. *The Embarrassment of Riches: An Interpretation of Dutch Culture in the Golden Age* (New York: Knopf, and London: Collins, 1987) by Simon Schama offers a national context for a detailed look at the seventeenth-century Dutch character; but his rich text is relevant to any Dutch city, including Haarlem. *The Making of Dutch Towns: A Study in Urban Development from the Tenth to the Seventeenth Centuries* (New York: Simmons-Boardman, 1960) by Gerald L. Burke offers a detailed account of how Dutch towns developed, including the process of polderization, whereby wetlands are drained for other purposes. Concise discussions of Haarlem include information about the town's various physical expansions. A broader account of Haarlem may be gleaned from *The Making of the Dutch Landscape: An Historical Geography of the Netherlands* (New York: Academic Press, and London: Seminar Press, 1971; second edition, London and Orlando, Florida: Academic Press, 1985) by Audrey M. Lambert, which provides numerous references to the city in the context of the Netherlands' physical and economic development.

—Randall J. Van Vynckt

Hadrian's Wall (Cumbria and Northumberland, England)

Location: The wall originally extended from the Firth of Solway in the west of England across the Tyne-Solway isthmus (in the counties of Cumbria and Northumberland) to the mouth of the Tyne River on the eastern coast. Most of the remains visible today are the central and eastern sections.

Description: Commissioned by the Roman emperor Hadrian in A.D. 122, the wall was probably completed by A.D. 128; it marked the frontier of the Roman Empire in Britain and served military and administrative purposes until approximately A.D. 410. Along the remains of the wall, and its forts and turrets, are several visitors' centers and museums. The wall has been designated a UNESCO World Heritage Site.

Site Office: Carlisle Visitor Centre
Old Town Hall, Greenmarket
Carlisle CA3 8JH, Cumbria
England
(228) 512444

According to the sole surviving Roman commentary on Hadrian's Wall, from the *Scriptores Historiae Augustae, Vita Hadriani, 11 2,* the structure was built to separate the Romans from the barbarians. Underlying this simple assessment was a set of complex political changes regarding the goals of the empire, the role of the army, and the limits of Rome's power. During the early years of his reign, the Roman emperor Hadrian who, ruled from A.D. 117 to 138, visited the frontiers in Germany and Britain, and in 122 ordered his three legions in Britain to build a wall of stone and turf across the seventy-mile (seventy-six Roman miles) isthmus between the mouth of the River Tyne on the eastern coast of Northumberland and the Firth of Solway on the western coast of Cumbria. For nearly 300 years Roman soldiers and auxiliary troops drawn from Britain, Gaul (France), and Germany that manned the wall effectively controlled access through the northern frontier and checked the incursions of the hostile native tribes in what is now Scotland. In order to understand why the wall was built and its importance in Roman and British history, it is necessary to place it in the context of the Roman invasions of Britain and the political changes in the empire during the end of the first and beginning of the second centuries.

In 55 and 54 B.C. Julius Caesar invaded Britain, and several of the southeastern tribes sued for peace once they had witnessed the power of Caesar's army. Troubles in Rome and in the recently conquered province of Gaul forced Caesar to return to the Continent with his legions, and he was prevented from returning to continue the conquest of the island and establish a provincial government. Britain remained free of Roman occupation for nearly 100 years as the succeeding emperors looked eastward for military glory. However, in A.D. 43 the new and insecure emperor Claudius, who ruled from 41 to 54, decided to invade Britain as the means to secure the loyalty of the army and the Senate. The initial campaign was successful, and for the next forty years the Roman military extended its rule from the southeast, first to Wales and then north to the Scottish highlands. The decisive northern campaigns began with the governorship of Agricola. In A.D. 78 he defeated the most dangerous tribe in the northern and western counties (i.e., Cumbria, Yorkshire, and Northumberland), the Brigantes, and established a strong presence at the Tyne-Solway line. Between 79 and 83 Agricola pushed north past the Clyde-Forth isthmus (subduing the Votadini, Selgovae, and Novantae) and won a decisive battle against the highland tribes in 83 at Mons Graupius, located at the northwestern edge of Scotland.

Although defeated, the northern tribes were far from subdued, and Roman policy called on the army to push forward until all of Britain (and possibly Ireland) was under Rome's control. However, during the closing decades of the first century and opening decades of the second, the conquest was stopped, troops were withdrawn from Scotland to the Tyne-Solway line, others reassigned to the Continent, and a series of forts, watchtowers, and military roads, called the "Stanegate" system, were established along the line of what would become Hadrian's Wall. The construction of some form of frontier defense in Germany, begun by Domitian (who ruled from 81 to 96) in the 90s and continued by Trajan (who ruled from 98 to 117) during the first two decades of the second century, reflected military setbacks along the Rhine and Danube, and the acknowledgment by Rome that the empire's frontier (which heretofore had been fluid and in a state of continual expansion) had reached its limits.

Hadrian came to power in 117 determined to stabilize the empire's borders and bring peace to the provinces. Apparently the Brigantes and tribes of lower Scotland were creating serious problems along the frontier, revealing the weakness of the Stanegate system of defense. Without a natural barrier, such as a river, to demarcate the frontier, Hadrian decided to erect a wall running the entire length of the Tyne-Solway gap that would enable the army to monitor any sort of movement, including that of small-time raiders, or anything suggestive of a larger assault. The goal was to foster exploitation of the island in a peaceful manner, all the way to the frontier.

Hadrian's plan called for a stone wall, ten Roman feet thick and fifteen feet high, to be erected from Newcastle on the east coast to the river Irthing (forty-five Roman miles), and a turf wall of similar dimensions to Bownes-on-Solway at

A section of Hadrian's Wall
Photo courtesy of Carlisle City Council

the Cumbrian coast. Small forts (housing twelve to thirty-two soldiers) with gates to the north and south were placed at each Roman mile. They were intended to serve as bases from which to conduct regular patrols in the Scottish lowlands and to control access through the frontier. Two observation turrets filled the gap between the milecastles. A ditch (roughly ten feet deep and from twenty-six to forty feet wide) fronted the wall, and a second, twenty-foot wide ditch called the "Vallum" was dug behind the wall in order to secure the wall from attacks from the rear (especially by the Brigantes), channel pedestrian traffic, and demarcate the end of the military zone.

The bulk of the Roman army, comprising auxiliary companies (who were recruited from the native Britons and the other defeated "barbarians"), was originally to be garrisoned at large forts (three to nine acres) a few miles south of the wall, but this plan was altered soon after construction had begun. The wall and forts were largely built by soldiers of the three Roman legions in Britain—at hand for major campaigns and for construction work—who quickly realized that in case of an attack, the small patrol forces on duty at the milecastles would probably be overwhelmed before troops could reach them. The existing forts from the Stanegate system were moved or extended to the wall, and new forts were built astride it. The eighteen forts allowed the Roman authorities to place a formidable army on the wall, and they improved the

army's ability to react to disturbances along the border and more effectively patrol the roads and countryside of the Scottish lowlands. The threat of attacks by the Novantae and Selgovae in the northwest led the wall builders to extend milecastles and observation turrets, enclosed between two ditches, along the Cumbrian coast, and to construct three outpost forts (Birrens, Bewcastle, and Netherby) north of the western sector of the wall. By 132 the wall was substantially complete, although a few forts were completed and the turf wall rebuilt in stone during the end of Hadrian's reign in the late 130s.

Upon Hadrian's death in 138, the new emperor, Antoninus Pius, who ruled until 161, reversed his predecessor's frontier policy. Between 139 and 143 Antoninus's army reoccupied Scotland. By the end of the 140s a new frontier wall (made of turf) and system of forts and milecastles were erected nearly one hundred miles north of Hadrian's Wall along the thirty-seven-mile (or forty Roman miles) gap between the Firth of Forth and the Firth of Clyde (the "Antonine Wall"). It is unclear why Hadrian's Wall was abandoned by the new emperor, but historians suggest several possible reasons: disturbances by the northern tribes (the Selgovae and Damnonii in the lowlands, and the Caledonians and Maeatae north of the Antonine Wall) forced a punitive campaign, or Hadrian's wall was simply too far from the major centers of resistance to be effective; the British army, displeased by its generally administrative role under Hadrian, pushed for a greater military role in exchange for its loyalty to the new emperor; Antoninus lacked military prestige, and a short, but successful, war in Britain (considered the easiest province in which to win battles) would secure his position. Whatever the reason, the Antonine Wall had a short life, as military disturbances within Britain and along the continental frontiers made a permanent occupation of Scotland impossible.

The last four decades of the second century were turbulent ones in the north of Britain. Literary and archeological evidence suggests that the Brigantes (a tribe within the Roman borders) revolted in the early 150s, possibly destroying the forts at Birrens, Brough-on-Noe (just south of Yorkshire), and Newstead (which was situated halfway between the two walls, along the major north-south military road—Dere Street). During this period, German reinforcements for Britain's three legions landed at Newcastle-on-Tyne, suggesting that troop strength was low and that all three legions were probably operating in the north. Further evidence of the Brigantian revolt comes from the orderly abandonment of the Antonine Wall and destruction of its forts during 154 and 155. These troops would have been sent south to the Hadrian Wall system to restore order, and parts of the wall were repaired by the end of 158. Before he died in 161 Antoninus ordered Roman troops back to the Antonine Wall, but this move was premature. In 162 and 163 renewed uprisings in the Pennines and Scottish lowlands forced the Roman army to retire to Hadrian's Wall under the command of the province's new governor, Calpurnius Agricola. This second return to the original frontier may have been caused by both the difficulty of securing the hilly terrain of the Clyde-Forth isthmus and bogs of the lowlands from an increasingly hostile group of native tribes, and the unavailability of reinforcements because Rome was also busy putting down insurrections along its eastern frontier. Calpurnius Agricola was charged with the task of refitting the Hadrianic defenses, and inscriptions from the forts at Carvoran, Chesterholm, and Corbridge attest to this work. His successor continued this work in the late 160s: the forts at Stanwix and Great Chester underwent extensive renovations. During this period of repair, the Vallum at Hadrian's Wall was cleaned out, and many of the gates at the milecastles were narrowed or closed, thus restricting access across the frontier to the more heavily supervised forts. In addition several forts along Dere Street in the Scottish Lowlands continued to be occupied as the army's advance surveillance posts.

Northern Britain remained unsettled throughout the 170s, and in a new wave of attacks in the early 180s, the Scottish tribes overran the wall (suggested by the levels of destruction at the Halton Chesters, Rudchester, and Corbridge forts, and several smaller forts in the Scottish hinterland), killed a Roman general and many of his troops, and wreaked havoc in the northern frontier area. According to the Roman historian Dio, a new general, Ulpius Marcellus, ruthlessly put down the insurrection by 184, but abandoned the

Scottish lowland forts and retired to Hadrian's Wall. This course of action suggests the continued inability of the Roman army to pacify the northern tribes, and the necessity of halting imperial expansion. The politics of Rome had serious consequences for Britain during the 190s. A power struggle followed the murder of the emperor Commodus (180–192) in 192, and the governor of Britain, Albinus, hoped to become the next emperor. He led a significant number of the province's troops to Gaul in 196 to engage another rival, Septimius Severus, but was defeated in 197. The withdrawal of Albinus's troops left the border understaffed, and some literary and archeological evidence suggests that the Caledonians and Maeatae overran Hadrian's Wall and that the new governor, appointed by the emperor Severus (192–211), sued for peace with the Maeatae.

Peace was ephemeral. Punitive campaigns into the lowlands and wall rebuilding occupied the Roman army during the opening decade of the third century, but the Scottish tribes remained ungovernable. In 207 the province's governor asked for reinforcements or an imperial expedition to help him subdue the fractious tribes. Severus and his army arrived in 208, rejected the Caledonian and Maeatae requests for peace, and fought a difficult but successful series of campaigns until the emperor's death in 211. Although he had hoped finally to crush the northern tribes and incorporate the entire island within the province, Severus died before he had accomplished this task, and his oldest son abandoned the campaign, signed a new treaty with the northern tribes, and returned to Rome. However, the Severan campaign brought a long peace to the frontier that lasted until the turn of the century.

Little is known about the years of peace in Britain. During the early decades many fort structures (such as administrative buildings, granaries, barracks, gates, and aqueducts) were built or repaired, and with stability in the north, the military's attention turned to fortifying the southeastern coast against piratical raids. Throughout the period, northern troops were transferred to the Continent or the southeastern coast, and a number of hinterland and wall garrisons were abandoned.

The long peace also encouraged the expansion of civilian settlements (the *vici*) outside of the wall forts and the development of a few independent cities near the frontier (Corbridge and Carlisle). The large Roman army at Hadrian's Wall, ranging from roughly 35,000 men during the Hadrianic era to about 15,000 to 16,000 during the late second and third centuries, required a steady supply of provisions and manufactured goods. The *vici* attracted a variety of merchants and artisans, who provided the army with bronze, iron, leather goods, pottery, glassware, food, pack animals, and building materials. Veterans often settled in the *vici*, as did the wives and children of the soldiers on duty. The *vici* also provided what little entertainment (e.g., taverns, gaming establishments, prostitution) was allowed in the spartan life of the frontier soldier.

Many of the *vici* occupants were immigrants (from

southern Britain, Wales, or other provinces in the empire)—the army had displaced many of the native inhabitants—although the pacified tribes on the frontier supplied the army with cereals (corn, oats, hay), beef, seafood, poultry, milk products, and beer. The native peoples also provided the army with a rich religious life and a new pantheon of deities, which many soldiers adopted or incorporated into their romanized belief systems.

A second phase of rebuilding the wall and its fortifications, as well as erecting new coastal defenses, occurred following the successful campaign of the emperor Constantius Chlorus (who ruled from 292 to 306) in 296 to unseat a British pretender to the throne, Allectus, and the emergence of new enemies who often skirted the wall by attacking from the sea: the Picts from Scotland, the Scots from Ireland, and the Saxons. Constantius and his son Constantine returned to northern Britain in 306 to put down damaging invasions of the Picts, and Constantine (emperor from 306 to 337) ordered the reoccupation of several hinterland forts north and south of Hadrian's Wall during the opening decades of the fourth century. The frontier remained relatively quiet, perhaps the result of peace treaties with the Picts and Scots. However, both groups, along with the Saxons, Franks, and Atacotti, raided Northumberland and Yorkshire repeatedly during the 360s. Their targets tended to be the wealthier civilian communities south of the frontier, although several wall forts evince damage from military attacks at this time. In 367 the army's advance scouts (the *areani*) failed to warn of an impending invasion by the "barbarian" alliance (because they had been bribed to stay quiet), and the two senior military/administrative officers of northern Britain were ambushed and killed during the incursion. The loss of the army's leaders and the ferocity of the attack caused a severe breakdown of military discipline, with many soldiers and officers deserting or taking extended leave. Rome sent a new general, Theodosius, to restore military and civil order and punish the invaders. Theodosius successfully repelled the barbarians, but did not counterattack. Instead, he disbanded the *areani*, and strengthened the Scottish forts and the defenses along Hadrian's Wall.

These actions did little to deter the Picts and Scots, and the second phase of "barbarian" raids began about 380 and lasted until the end of Roman control of Britain in 410. Although additional armies were sent in 382 and 398, they were withdrawn to the Continent after the campaigns in which they repelled but failed effectively to defeat and pacify the Picts and their allies. Imperial power struggles, hatched by several British leaders in the 390s and early 400s, drained the province of troops as battles for control of the armies and the emperorship took place on the Continent. In 409 the southern government of Britain collapsed in the face of an internal revolt that eliminated the administrative apparatus to supply and pay the northern army. Already weakened by troop withdrawals, constantly harassed by the Picts, and faced with debilitating supply shortages, the Roman army units stationed at Hadrian's Wall most likely disbanded themselves and turned to civilian pursuits between 410 and 420. In the end, Roman control over Britain ceased because power struggles at the center of the empire, and the military demands of protecting a vast frontier, left Britain with too few troops adequately to defend the province.

Although Hadrian's Wall survived 300 years of barbarian attacks, it fared less well during the ensuing centuries as monks, nobles, and farmers borrowed stones from the wall to construct their churches (e.g., Lancaster Priory, Carlisle Cathedral, and the Venerable Bede's monastery at Monkwearmouth and Jarrow), castles, and field walls. The mid-eighteenth century rebellion by the Scottish Prince Charles revealed the necessity of cross-country roads in the North (the English army hoped to stop Prince Charles in Newcastle, but the lack of east-west roads prevented this when he took an alternative north-south road through Carlisle in 1745). The British Army destroyed most of the western wall from Newcastle to Carrawburgh in order to build the Military Road that follows the route of the wall. Fortunately, the army chose to take the road south of the rocky, central ridge of the wall, and the central section of Hadrian's Wall remains relatively well preserved. In addition, large sections of the Vallum and northern ditch are still visible.

A great many archaeological investigations have been conducted at Hadrian's Wall during the twentieth century, primarily at several wall forts (Housesteads, Chesterholm [Vindolanda], Chesters, Corbridge, and Carrawburgh) and nearby milecastles and turrets. The Roman Army Museum at Carvoran, the Museum of Antiquities at the University of Newcastle, and the fort at Vindolanda house many of the Roman artifacts found during the digs, and are the best local sites to learn about Hadrian's Wall. Many tours of Hadrian's Wall begin at Carlisle, where there is also on award-winning museum of Roman and prehistoric artifacts.

Further Reading: *Hadrian's Wall* by David J. Breeze and Brian Dobson (London: Allen Lane, 1976) is an excellent and comprehensive historical account of how and why Hadrian's Wall was built, and ably discusses the military and political importance of the wall; Breeze and Dobson are leading experts on the wall. *Roman Britain, 55 BC–AD 400: The Province Beyond Ocean* by Malcolm Todd (Brighton, Sussex: Harvester Press, and Atlantic Highlands, New Jersey: Humanities Press, 1981) and *The Northern Counties to AD 1000* by Nick Higham (London and White Plains, New York: Longman, 1986) are sound histories that place Hadrian's Wall in the broader context of ancient Britain. While all three books discuss the archaeological evidence underpinning the histories of the wall, three other texts are detailed archaeological analyses of the structures built at the wall that also offer reassessments of the wall's history: *Hadrian's Wall Bridges* by P. T. Bidwell and N. Holbrook (London: Historic Buildings and Monuments Commission for England,

1989); *The Roman Fort at Vindolanda, at Chesterholm, Northumberland* by Paul T. Bidwell (London: Historic Buildings and Monuments Commission for England, 1985); and *The Roman Military Defence of the British Provinces in Its Later Phases* by Derek A. Welsby (Oxford: B.A.R., 1982). *Hadrian's Wall: A Personal Guide* by Robin Birley (Newcastle upon Tyne: A.R.P., 1990) and *The Building of Hadrian's Wall* by Robin Birley (Newcastle upon Tyne: A.R.P., 1991) are very readable overviews of the history of the wall, although they are directed at a tourist audience.

—Stephen Ellingson

Hastings and Battle (East Sussex, England)

Location: Hastings lies on the south coast of England, sixty-two miles southeast of London. Battle is situated seven miles northwest of Hastings.

Description: Hastings, once one of the Cinque Ports, is a seaside resort with an old quarter of fishermen's houses and much elegant nineteenth-century architecture, although some of it was destroyed by bombing in World War II and replaced by modern buildings. A promenade runs along the seafront with its pier. On one of the hills onto which the town has spread lie the ruins of the Norman castle of Hastings. Battle was the site of the Battle of Hastings in 1066, in which William the Conqueror's victory led to the Norman conquest of England. Battle Abbey, which dominates one end of the town's main street, was built on the battleground.

Site Offices: Hastings Tourist Information Centre
4 Robertson Terrace
Hastings, East Sussex TN34 1EZ
England
(424) 718888

Battle Tourist Office
88 High Street
Battle, East Sussex TN33 0AQ
England
(424) 773721

The name of Hastings is inextricably linked with undoubtedly the best known date and certainly one of the most important battles in English history, fought on October 14, 1066. In that year, William, Duke of Normandy, landed with his troops on the southern English coast to claim the crown of England, commanded the defeat of Harold Godwinson's army, and took the throne as William I—he is often popularly referred to as William the Conqueror. Harold was to be the last Saxon king of England. The Norman Conquest of England initiated by the Battle of Hastings marks a crucial turning point in English history, government, and society. William and his troops did not land at Hastings, however, but at or near Pevensey, a few miles west of Hastings; and despite the name given to the battle, it did not take place on the site of the town of Hastings, but some seven miles northwest, at a place subsequently, and aptly, named Battle. The Battle of Hastings gives the two towns of Hastings and Battle their primary importance in English history.

The name of the town of Hastings comes from the times when the Jutes and Saxons invaded and settled in England in the centuries after the Romans' departure from Britain early in the fifth century; it is said to derive from the clan of the Hastingas, or sons of Haesten, who made their home there. Mention is made in Simeon of Durham's chronicle recounting King Offa of Mercia's conquest of the kingdoms of Sussex and Kent in 771 of "the men of Hastings" being among those subdued. A mint was in existence in Hastings by 928, a sign of a settlement of some importance.

Hastings is also well known as having been one of the original Cinque (pronounced "sink") Ports, along with Romney, Hythe, Dover, and Sandwich. This unique medieval confederation's duty was to offer ship service to the king to protect the southeast coast in exchange for privileges, and it has sometimes been popularly and rather exaggeratedly described as the precursor of the English Royal Navy. It is not clear when the organization was officially formed, but the ports might have served Edward the Confessor. The organization does not appear to have been in operation when William invaded.

Edward the Confessor died in January 1066, without a direct heir to his throne, and it was the dispute over his succession that led to William's expedition and the Battle of Hastings. Edward's background is important in understanding the reasons for William's invasion. Edward was the son of Aethelred II and Emma (herself daughter of the third Duke of Normandy, Duke Richard II, and so a relative of William). He had spent his early life in exile from England at the Norman court during the reign in England of the Danish king Canute. His Norman connections were therefore strong, and when he became king of England he gave some Norman nobles positions of importance both in the church and the state.

On Edward's death, there were three contenders to the throne: Harold Godwinson, Earl of Wessex; Harold III, King of Norway; and William, Duke of Normandy. Edgar Aetheling had a better hereditary claim than all three, being the grandson of Edmund II Ironside, the natural son of King Aethelred, but he was still a minor at the time of Edward the Confessor's death, and therefore not chosen. The English throne was not regarded as strictly hereditary in those times; in fact, a new king was chosen by the Witan—a council of advisers—of the deceased king. Illegitimacy was not a bar to kingship either, and William was a bastard.

The reasons for William's invasion and the events of the expedition and battle were recounted in several medieval chronicles, some written just a few years after the battle, others at a later date. Together with the Bayeux tapestry, dating from the late eleventh or early twelfth century, they are used as the main sources of information on these historic events, but any serious historical analysis should bear in mind that all but one of these accounts gave the view from a Norman perspective and all were composed within the highly stylized, glorifying, and romanticizing forms of the era.

Hastings Castle, top, and Senlac Hill, bottom
Photos courtesy of Hastings Borough Council

None of them was written by participants or eyewitnesses, and speculation plays a part in any interpretation of the events.

Edward had certainly become attached to William during his time in Normandy, and it is likely that he nominated him his heir, a promise given formal ratification, it seems, in 1051 by Robert the Norman, archbishop of Canterbury. Harold Godwinson journeyed to Normandy in 1064. His motive for the journey is disputed. Norman records claim that it was to confirm William's succession. An alternative claim has been made that he went to try to release his brother and nephew, held hostage in Normandy since the family's temporary banishment from England by Edward. At all events, Harold was made to swear on holy relics to William's rightful succession. Harold III of Norway claimed the throne of England on the basis of a treaty of 1038 concluded by his father, Magnus, with Harthacnut, the Danish King of England from 1035 to 1042.

Harold Godwinson was not of royal descent, but he already commanded the king's army and was Edward's brother-in-law. Despite the supposed promise to William, Edward and his advisers, as Edward lay dying, chose Harold Godwinson as successor. He was crowned on January 6, 1066. In May, Earl Edwin repelled the first attack on Harold's rule, by his brother Tostig in eastern England. Then in September, Harold III of Norway landed on the northeast coast. Harold Godwinson marched up to rout the Norwegians at the Battle of Stamford Bridge on September 25; Harold III was killed in this battle. But two days later William and his fleet were setting sail to attack the southern coast, backed by papal approval and a fair wind.

William's invasion campaign had been carefully planned. After a diplomatic mission to Harold Godwinson had failed, William amassed his army, ably gaining the support of his Norman barons and then making concerted efforts to secure support from the powers of northern Europe. Although he was not directly aided by France and Flanders, a strong contingent of volunteers from those states fought for William. The largest additional forces, however, came from Brittany. Furthermore, William sought the pope's religious approval through the archdeacon of Lisieux. Pope Alexander II declared Harold a usurper and William the rightful successor to the English throne; he sent William a banner he had blessed and a ring supposedly enclosing some of St. Peter's hair. This papal support gave the invasion the aspect of a religious war.

The building of William's fleet lasted into the summer. It has been estimated at 700 ships, transporting perhaps 11,000 men and more than 3,500 horses. The commonly told story is that William was prevented from invading earlier by adverse winds. No evidence has emerged that Harold III and William planned their invasion attempts together; the difficulties of communications of the times and the two men's competing claims make it unlikely. It is possible, however, that William knew about Harold III's intentions.

William's armada met no opposition on landing.

Harold Godwinson knew of William's preparations much earlier in the year, and through the summer of 1066 the coast was guarded. But the ships on guard apparently either ran out of supplies, or had their crews paid off and scattered before the invasion. The fyrd, militia that had provided a coast guard during the summer, also seems to have fulfilled its annual obligations by the beginning of September, and broken up. William's landing was certainly timely, so soon after the Saxon defenses had disbanded and hot on the heels of Harold III's invasion attempt in the north.

William directed his troops to make camp at Hastings very soon after their arrival. He first received news of Harold Godwinson's movements from the messenger of a Norman settled in the area. His troops laid waste, as is testified in the Domesday Book, to numerous neighboring villages. These actions have been interpreted as part of a strategy to force Harold down south to fight. Harold, in York, received news of William's invasion at the beginning of October, and sped back south with his elite troops, the housecarls. By the evening of October 13, after a halt of a few days in London, he had covered the roughly 260 miles to the area where the battle would be fought, "at the hoary apple-tree," as the Anglo-Saxon Chronicle laconically recorded it.

Harold's plan, the chroniclers and most historians agree, was to surprise the Norman invaders and attack them before they had become too firmly settled. So he traversed the forest of the Andredsweald separating London from Hastings as quickly as possible, there to collect his troops on Caldbec or Senlac Hill, where the open down started, before launching his attack on William's encampment. William appears to have received intelligence of Harold's movement south, and decided to advance and engage in battle with Harold on the morning of October 14, before all the latter's forces had assembled and rested. William's men, in contrast, would be well prepared.

Chroniclers tell of William's men in prayer the night before, while the English feasted; of Bishop Odo of Bayeux presiding over Mass for the whole Norman force before leaving Hastings; of William giving a stirring speech to all his troops in preparation for the battle; and of personal challenges to one-on-one combat between Harold and William. These stories are doubtless embellishments of events, but ones that capture the imagination of those susceptible to historical romanticism.

The chronicles refer to the battle being fought *in planis Hastinges,* meaning "in the unwooded country [not plain] of Hastings" in medieval Latin, and it is from these references that the name of the battle no doubt emerged. It is also known as the Battle of Senlac. Most modern estimates of the sizes of the each army are of 6,000 to 7,000 men; some, calculating by the numbers of French who came over in the ships, reckon up to 9,000 men may have made up William's forces, and that on the Saxon side, with the arrival of more militia to join Harold's ranks in the course of the day, the English forces may have risen to as many as

10,000. The Norman chronicles state that Harold's army was much greater in number than the French. Harold's position on top of Senlac Hill was surely a strong defensive one. The enemy would have to climb the ridge; below, the ground was possibly marshy around the Senlac brook.

On the battlefield, William's forces approaching from the southeast divided into three columns. On the left, the column was mostly of Bretons; in the center, William led the main, Norman force; on the right were French and Flemish forces. The columns were further divided into three, the archers at the front, followed by infantry, and behind them mounted knights. To the north, Harold's forces consisted of two elements, housecarls and fyrd. Less is known about their deployment. It has been suggested that Harold stretched out his flanks to make the most of the natural protection of the Senlac ridge at both ends. A shield wall of closely linked soldiers was formed. The Saxon army had no or very few archers. The armies were therefore different, and although the numbers were not dissimilar, William's forces included a large component of archers and of cavalry, which gave him greater flexibility in the fight.

The battle began in mid-morning, and seems to have involved several waves of attack by William's forces. First, archers sent low volleys of arrows into the English, with little result. The attack that followed was not successful in breaching the English line, and the Breton left flank is reported to have retreated in disarray, prompting the English to follow it down the hill. Panic ensued among the Norman troops, with rumours circulating that William had been killed. He reacted swiftly, apparently removing his helmet to show his troops he was alive, and commanding his cavalry in the center to circle left to mow down the descending English infantry.

A lull ensued, but with more English militia arriving at the site all the time, William was no doubt keen to attack again. The next attempt was more successful, breaching the English line in places. Once again, part of William's force was followed down the hill by seemingly overeager Englishmen who met the same fate as the previous band. Historians have argued over whether or not the second flight was a tactical feint, as the chroniclers present it, to tempt the English to descend from their strong position. Whatever the case, it was effective.

The Norman attack that followed was decisive. During the assault, the archers sent their arrows high into the air. These volleys were followed by several mounted attacks. Harold's two brothers were killed in the fighting and the English forces began to break up. Then, as so famously depicted in the Bayeux tapestry, Harold was struck in the eye by an arrow. Housecarls carried on defending him, but were eventually overpowered by French knights who delivered the final death blow to the Saxon king. Towards dusk the battle was lost for the English. The death toll was high on both sides, the invaders possibly losing about a quarter of their number and the English even more. Two additional incidents feature in certain accounts, first of William commanding the pursuit of the fleeing English and second of a band of French knights attacked by survivors from Harold's army and sent to their deaths down a ravine named the Malfosse.

Historians have speculated variously on the reasons for the outcome of the battle. On the English side, they have drawn particular attention to Harold's overhastiness, the possible weariness of his housecarls after Stamford Bridge, and the fact that the English troops had not all arrived when the battle commenced; and, on William's side, they have cited the freshness of the Norman and French troops and their use of more advanced techniques of warfare. Whatever the reasons for the outcome, it was a battle that raged most of the daylight hours, and was bitterly fought.

The historical impact of the Battle of Hastings would be far reaching. Through that victory and the Normans' subsequent conquest of the whole of England, the country was turned from a land largely under Nordic influence to one transformed by Norman rule and culture. French became the language of governance, and a rigid feudal system was imposed, diminishing the federal nature of the earldoms of England and making for more centralized government. In the longer term, as a consequence of the conquest, the houses of Normandy, Blois, and Anjou ruled England for more than 400 years, shifting territorial interests to the continent, and creating the great rivalry and conflict between the English and French crowns that lasted for centuries.

William stayed for a while around Hastings to see how the other English leaders would react. The building of the first Norman castle in England, at Hastings, was under way, on a site probably previously occupied by a Saxon fortification. The Witan in London nominated Edgar Aetheling king, but resistance to William quickly crumbled, and on Christmas Day, 1066, he was crowned in Westminster Abbey.

The construction of Battle Abbey was soon ordered by William as penance for his warring in England. Popular tradition has it that he vowed before the battle to erect a monastery on the site were he victorious. The earliest record of this story, however, derives from a charter from 1154 produced, or forged, by the monks of the abbey. In fact, it is likely William made his vow in 1070, the year in which papal legates formalized their recognition of his right to the throne of England. He wished the high altar to be on the marshy spot where Harold fell. Four monks from the Benedictine monastery of Marmoutier on the Loire, who had been sent to form the community, began to build the abbey at a more suitable location. When he discovered the change, William ordered them to follow his wishes. In 1076 the eastern wing of the abbey church was consecrated. William's treasury funded the entire building, and the abbey stood outside normal episcopal jurisdiction. At the time William died, it was richly endowed, the fifteenth wealthiest religious foundation in England. The privilege of a *leuga,* or land within a league of the high altar, was granted to the foundation, while William bequeathed it his royal cloak, a collection of relics, and a portable altar he had used during his military campaigns. The town of Battle can claim the oldest street directory in the country, compiled forty years after the battle, showing just how fast it grew up to serve the abbey.

William II of England, or William Rufus, third son of the Conqueror and his successor as king of England, came with his army to Hastings Castle in 1094 in preparation for an expedition against his oldest brother, Robert of Normandy. Detained by bad weather, he was able to preside over the consecration of the whole abbey church.

Through strong abbots, the foundation grew in wealth and protected its privilege of exemption from episcopal control, although it was challenged repeatedly by bishops of Chichester. In May 1157, the king ruled in the abbey's favor, but difficulties with royal administrators led the monks to pay 1,500 marks to King John in 1211 to avoid further interference, leaving the abbey free to elect its own abbot, but ending its royal connection and protection.

Hastings Castle continued to be used as a practical stopping point for royalty. Its College of St. Mary, founded in 1090 within the walls by the Norman Count of Eu, had as its dean for a time Thomas Becket, but there is no evidence he spent any time there. King John did issue his ordinance of the sovereignty of the seas from there in 1201, forcing all ships to lower their topsails to the Lieutenant of the King or his Admiral, or be treated as foe.

Hastings's historical prominence in the century following the Norman Conquest centers on its role as one of the Cinque Ports. It played a fair part in national defense and attack on the seas, but its sailors also frequently pursued the more maverick policy of piracy. The earliest charters treating the ports as a confederation date from 1260 and 1278, but reference was made in these documents to earlier charters from Henry II of around 1155 and 1156, in which conditions had been stipulated; the ports were to put at the king's disposal fifty-seven ships, each manned by twenty-one men and a boy, for fourteen days each year. In turn, Henry II's charters probably merely restated services and rights agreed some time in the eleventh century. Because of their strategic and maritime importance, the Cinque Ports could wield great influence. With the loss of Normandy to the French king in 1204, King John issued a string of charters to strengthen their loyalty. Once, however, in 1216, they actually joined the French side, but the king paid for their renewed support, and in 1217 the confederation won an important sea battle off Sandwich, the Ports' twenty fighting vessels beating a French force over twice their number. In 1242 the fleet was deliberately engaged by the king to raid the French coast, while in 1277 it was central in capturing Anglesey. In 1282 it fought against the Welsh, and later in the century against the Scots. With the establishment of Simon de Montfort's Parliament in 1265, the importance of the Cinque Ports was evident in that they were each asked to send four men, while other towns only had the right to two representatives. An ugly incident occurred in 1290, the year of the Jews' expulsion from England, when some of their number were attacked, robbed, and massacred by Cinque Port sailors. In fact, Hastings would be infamous for its piracy for centuries.

From the fourteenth century the story of Hastings is one of decline. Its harbor became increasingly vulnerable to the encroaching sea and began to silt up; it suffered from French raids, which in 1339 and 1377 led to the sacking and burning of the town; and through the century it was badly affected by sea floods. At this time, the abbots of Battle commanded the defense of a stretch of the coast; in 1377, Abbot Hamo led a particularly successful counterattack, repelling the French. With the growth of the Royal Navy in the fifteenth century, the influence exerted by the Cinque Ports declined further. The confederation continued to defend its rights to privileges, even though by 1450 its maritime power had effectively disappeared.

Almost a century later, in 1538, Battle Abbey lost its religious power with the dissolution of the monasteries under Henry VIII. The king donated the abbey to his Master of the Horse, Sir Anthony Browne. Browne destroyed the church, the chapter house and a section of the cloister, turned the west range into his home, and built a magnificent tomb for himself and his wife in Battle church. The abbey would thereafter serve as a country estate to three succeeding families.

A Crown Commission discovered in 1581 that some Hastings land that should have been handed to the crown on the dissolution of the monasteries had been taken by local families. The problem was solved by having the town pay a yearly lease for the disputed lands and having Hastings issued a new charter of self-governance under a mayor in 1588. A new harbor was built in 1597 to try to restore the town's fortunes, but it was severely damaged by a high tide in the same year. Impoverished, the town sent the king a petition in 1636 calling for help as the remains of the harbor structure were in danger of being destroyed by the sea. That destruction happened in 1656. The town survived as a minor fishing port.

The fashion for sea air and bathing led to a revival of Hastings's fortunes from the end of the eighteenth century. In 1771, the new landlord of the Swan Inn, Thomas Hovenden, was the first to grasp the potential, drawing tourists to the town. By 1815 it was hard to find places to stay. In the 1820s James Burton came to the area and decided to design and have built a whole new resort town in the valley next to Hastings. So fashionable St. Leonard's was born, attracting many more people. By 1881 Hastings was the second largest English resort after Brighton, adorned with a typical oriental-inspired pier, which burned down in 1917 but was replaced in 1922. St. Leonard's was later to merge with Hastings. Despite suffering from bombing raids in World War II and from the growth of package holidays abroad, Hastings has managed to keep tourism as the mainstay of its economy. Visitors are drawn to the sea, the lovely nineteenth-century architecture, and the ruins of Hastings Castle. Battle Abbey, which also attracts many visitors, is now partly a private school and partly owned by English Heritage—in 1976 a donation from citizens of the United States allowed Britain to buy the ruins of the abbey and the site of the battle, a battle that marked the last successful invasion of England and changed the course of English history so profoundly.

Further Reading: There are numerous publications on the Battle of Hastings. For a modern explanation of the Norman invasion and the Battle of Hastings, R. Allen Brown's chapter on the Norman Conquest of England in his *The Normans and the Norman Conquest* (London: Constable, 1969; New York: Crowell, 1970; revised, Woodbridge, Suffolk: Boydell and Brewer, 1985) is a well-sourced, scholarly exposition. The information in this chapter is condensed in a booklet by the same author, *The Battle of Hastings and the Norman Conquest* (London: Pitkin Pictorials, 1982), a good introduction for the general reader; it was republished as *William the Conqueror and the Battle of Hastings* (1988). Another recommended book on the invasion and battle is Brigadier C. W. Barclay's *Battle 1066* (London: Dent, 1966), offering a clear account of events leading to the battle and reasonable speculation on the possible course of events of the conflict. Of equal interest, and enjoyable to compare with Barclay, are two pieces by Lieutenant-Colonel Charles Lemmon, one a chapter in *The Norman Conquest: Its Setting and Impact* (London: Eyre and Spottiswoode, 1966), and the other a booklet, *The Field of Hastings* (St. Leonards-on-Sea: Budd and Gillatt, 1957), which is a useful practical guide to the battle and its battle-field. As to sources from the period of the Norman conquest, of the Norman chronicles, that by William of Poitiers, William the Conqueror's chaplain, is regarded as the most authoritative. The other main source of the period is the Bayeux tapestry. For books specifically concentrating on the history of Hastings, J. Manwaring Baines's *Historic Hastings* (Hastings: F.J. Parsons, 1955) is the most often cited work and offers useful, well sourced details. A more recent publication, *Hastings: A Living History* (Hastings: Hastings, 1987) is an easy-to-read, illustrated account of Hastings and St. Leonards by a former mayor, David Thornton, helpful in filling in information on the town in the last century. On Battle and the site of the Battle of Hastings, the slim volume *Battle Abbey* by J.G. Coad (London: English Heritage, 1984) offers a clear, reliable introduction to the history of Battle Abbey, and provides a good summary of the causes of the Norman invasion and the battle. Eleanor Searle's *Lordship and Community: Battle Abbey and its Banlieu 1066–1538* (Toronto: Pontifical Institute of Medieval Studies, 1974) will be of interest to those wanting greater detail not just on life at the abbey, but on the structure of English medieval society. Eleanor Searle also edited and translated an edition of the Chronicle of Battle Abbey, *Chronicon Monasterii de Bello* (Oxford: Clarendon Press, 1980).

—Philippe Barbour

Heidelberg (Baden-Württemberg, Germany)

Location: Fifty-six miles south of Frankfurt, in the state of Baden-Württemberg along the banks of the Neckar River.

Description: For centuries the home of the powerful Palatinate, the town was badly damaged during the Thirty Years War, and again in the Palatine War of Succession, but efforts have been made throughout the centuries to preserve what remains of historic Heidelberg Castle and rebuild other town buildings. Germany's first university, the 600-year-old Heidelberg University continues to draw high-caliber students and faculty.

Site Office: Verkehrsverein Heidelberg
Friedrich-Ebert-Anlage 2
69115 Heidelberg
Germany
(62 21) 1 08 21

Heidelberg has been inhabited for hundreds of thousands of years, as documented by the finding there of the oldest human bone in Europe: the jawbone of pre-human *homo heidelbergiensis,* or Heidelberg Man, judged to be over 500,000 years old. Discovered in 1907 in the nearby town of Mauer, the jawbone is today on display in Heidelberg at the Kurpfälziches Museum.

Around 400 B.C., the Heidelberg area was settled by Celtic tribes, later by Teutons, and finally by Romans. In A.D. 40, across the Neckar from today's Heidelberg, the municipal district of modern-day Neuenheim was the site of a Roman fort, manned by the twenty-fourth Roman cohort and the second Cyrenaican cohort. The fort remained active for several centuries until it was conquered in 260 by the Alemans.

During the first millennium, a number of religious holdings were established in the Neckar Valley near modern Heidelberg. In 764, Lorsch Monastery was founded, followed in 870 by Michaelskloster, a monastery founded by the abbey of Lorsch on the Heiligenberg, or "Holy Mount," inside an ancient double Celtic rampart estimated to date from 5 B.C.

These two monasteries were joined by the Neuberg Monastery in 1130, and later by the Schönau Monastery, established by the bishopric of Worms in 1142. Schönau was the first instance of Worms's influence in the Neckar Valley, and it is here that the first mention of Heidelberg is found. A document written in 1196 refers to Heidelberch, a settlement located close to a Worms castle and under the castle's protection. This settlement became modern-day Heidelberg.

The settlement of Heidelberch grew rapidly and by 1214 had become the political center of the powerful Palatinate. The Golden Bull of Emperor Charles IV in 1356 pro-

claimed that in the absence of the Holy Roman Emperor, secular princes were to be given the title of Count Palatine and thus authorized to govern a territory in the name of the church. The Palatinate was divided into two areas: the Upper Palatinate ruling northern Bavaria along the Naab River, and the Lower, or Rhenish, Palatinate governing the areas along the middle Rhine River. It was this Rhenish Palatinate that made its capital at Heidelberg. The Golden Bull was also the first effort to create a written constitution for the growing empire and gave rise to the Electoral College, a group composed of archbishops and lay princes empowered to elect the German king (usually also crowned emperor). The power of the Counts Palatinate was so great that for nearly 500 years they controlled the Electoral College.

In the middle of the fourteenth century, Germany began to establish universities, ostensibly to diminish the influence of long-established foreign universities on future German clergy and court members. Rupert I, count Palatine at Heidelberg, is credited with founding the first such university, at Heidelberg in 1386.

At its founding, Heidelberg University, like other new German universities, concentrated on the study of canon law. In keeping with its model the Sorbonne, however, the university also offered studies in law, medicine, and liberal arts, although these faculties were sparsely represented. The newly created university gained popularity some years later when the emperor supported an Italian candidate for pope rather than a French one, causing German intellectuals to leave the Sorbonne in great numbers. The Palatinate, under Rupert I, offered minor financial support for the fledgling university, with most of the funding coming from Pope Urban VI.

As the number of students continued to increase, Rupert II eased the faculty's housing shortages by driving Heidelberg's Jewish population from their houses and synagogues and converting their property into university living quarters. The majority of the school's buildings, however, continued to expand outside the city's original boundaries, as marked by the historic square Universitätsplatz. The Alte Universität, or Old University, the original site of the University of Heidelberg, still lies to the north of the square and is marked by the Lion's Fountain, a representation of the heraldic animal of the Palatinate.

Germany's oldest university has played a major part in the religious, cultural, and political life of the country. In its early days, the University of Heidelberg and its faculties played a large role in the tide of humanism sweeping the nation. In April 1518, Martin Luther entered the university to defend his recently written ninety-five theses. The growing popularity of Luther's beliefs, combined with the acceptance of the Reformation within the Palatinate by Prince Elector

A view of Heidelberg
Photo courtesy of German Information Center

Ottheinrich, caused the evolution of the university from staunchly Catholic philosophies to equally firm Protestantism. A mere forty-five years after Luther began his protest, the Heidelberg faculty gathered together to draft a definition of the ideals of the Reformation; and, in 1563, the Heidelberg Catechism, a guiding formulation of Reformation beliefs, was composed.

In 1803, Grand Duke Charles Frederick of Baden proclaimed Heidelberg University to be the state's first national university. Charles Frederick proceeded to invite a number of scholars to Heidelberg to study, work, and teach, thereby increasing the school's renown. The University of Heidelberg has maintained its prestige to the present day. During World War II, the Nazis recognized its importance and influence in Germany and swiftly brought the school and the faculty under their sphere of influence. Withstanding several wars, the school has produced a number of Nobel Prize-winning scientists and continues to be one of the foremost universities in the world.

Equally important to the city of Heidelberg is the Heidelberg Schloss, or castle. Now in ruins, the once-majestic castle had humble beginnings as a stronghold overseeing the Neckar River ford and ferry. As Heidelberg grew, so did the castle. The electors of the Wittelsbach dynasty's Palatinate quickly claimed the fortress as their primary residence, adding to it as needed, enlarging the humble fortress into a sprawling palace.

The Heidelberger Schloss's numerous tenants each added their section according to the style of the time. The near-continuous additions for 500 years produced a delightful architectural eclecticism. The largest part of the complex, built in the Renaissance and baroque styles, dates to the sixteenth and seventeenth centuries, a period in which the Palatinate electors held court at the Schloss.

The most famous part of the castle was built from 1556 to 1566 during the reign of Count Ottheinrich. Appropriately named Ottheinrich's Building, its facade is one of Germany's finest surviving examples of Renaissance architecture. The facade is covered not only with images of Christian saints, but members of the Roman pantheon as well. Jupiter and Mars stand among the five virtues of strength, faith, love, hope, and justice. Count Ottheinrich himself is depicted above an elaborate doorway styled after a classical triumphal arch.

The castle continues to house the world's largest wine vat, the Heidelberger Fass. This barrel was originally constructed from 130 oak trees and held 55,000 gallons of wine, paid as tax by the Palatinate's wine growers. One famous legend tells of the Tyrolean dwarf Perkeo, court jester in the eighteenth century during the reign of Elector Carl Philip and reportedly drinker of frightening quantities. It is rumored that when asked if he could consume the Fass's wine in a single gulp, the jester replied in his native Italian, "Perchè no?" (Why not), and was thereby given his nickname. The dwarf dutifully guarded the Fass until his death; according to legend, he died when he drank a glass of water by mistake.

Begun at nearly the same time as the town's Schloss, the Heiliggeistkirche (Church of the Holy Ghost) was constructed from 1399 to 1441 at the request of Elector Rupert III in celebration of his claiming the German throne. Despite his uneventful reign as King Rupert I, his church has stood commandingly on the Marktplatz for nearly six centuries. The base and walls of the church are Gothic and the roof and steeple are clearly baroque; the top of the church was rebuilt in the mid-sixteenth century, accounting for the major differences in styles. The vendors' stalls lining the Gothic walls remain from the church's medieval days.

The Church of the Holy Ghost's galleries were built in 1440 to hold the Palatinate's growing library. The first manuscripts were those taken from the early abbey of Lorsch, and by the late sixteenth century, the Bibliotheca Palatina had become one of Europe's most important collections. The library of 5,000 books and 3,500 manuscripts was commandeered in the Thirty Years War by General Johann Tserclaes von Tilly and presented as a gift to the Vatican.

Another famous portion of the church suffered at the hands of the French in the late seventeenth century. Heiliggeistkirche was the resting place for more than 50 prince-electors of the Palatinate until the destruction of the tombs and desecration of the bones in 1693. The only tombs left intact are those of the church's founder, Rupert, and his wife Elizabeth.

Despite Martin Luther's activity at Heidelberg University in the early 1500s, it was not until the reign of Calvinist elector Frederick III in the 1560s that the Palatinate became the driving force behind Germany's Protestant movement. As the Reformation swept through Germany, the gap between the Catholic Church and German nobility widened. The church rapidly formed alliances with the lesser nobility in hopes of regaining its diminishing authority, but by the end of the Reformation the nobility—including the Palatinate—had become the real ruling power in the country.

The division of the empire into Lutheran, Calvinist, and Catholic factions made unity increasingly difficult. Rising unrest throughout Europe heightened the growing tensions in Germany. In the early 1600s, the Spanish general Ambrogio Spinola cautiously planned his attack on the Dutch, counting on alliances from German Catholic strongholds for his success. The missing piece in Spinola's puzzle was backing from the Palatinate, at this time strongly Calvinist.

The rising diplomatic intrigues and foreign threats on Germany persuaded a number of princes to cast aside their differences and enter into an alliance called the "Union." Led by Elector Frederick IV, the Union professed to represent Protestants generally, despite most of its members being Calvinist. The Union was not only well received among those Germans opposing the House of Habsburg and its monopoly of the imperial throne, but obtained the support of the Venetians and the Dutch, and King James I of England gave his only daughter Elizabeth to Frederick IV's son, Frederick V, to wed.

But the political and military manipulations of the Union, which thrust Elector Frederick V onto the throne of Bohemia for a few short months, ultimately backfired. Frederick not only failed to loosen the Habsburgs' hold on the imperial throne, but in the process he lost the electorship to the Catholic Maximilian of Bavaria. This loss marked the beginning of the Thirty Years War.

The town of Heidelberg came under attack in the summer of 1622, the siege led by Count Johann Tserclaes von Tilly, general of the army of the Catholic League, and the Spanish general Gonzales de Córdoba. After defending their city for eleven weeks, the garrison of Heidelberg admitted defeat, surrendering on September 19, 1622. Tilly's troops plundered the fallen city, with the general taking the Bibliotheca Palatina and presenting it to the pope as a gift.

In 1649, Frederick's son Charles Louis was allowed to return to the Palatinate's former seat of power. Seeking peace after years of turmoil, he attempted to forge an alliance with France in 1671, offering his daughter, Elizabeth Charlotte, to the brother of Louis XIV, the Duke of Orleans. For the next fifty years, the princess Liselotte, as she was called, wrote over 4,000 letters, detailing her homesickness and the days she spent as a "royal slave" in Paris.

In 1685, after the heirless death of Liselotte's husband, Louis XIV claimed the princess's right to power in the Palatinate and invaded the Rhineland. From 1688 to 1696, for the second time that century, Heidelberg fought bitterly against invaders in the Palatine War of Succession. Again, the city fell, and the triumphant French troops plundered the city in 1689 and again in 1693, all but leveling the castle. In 1774, a fire raged through the historic Schloss, completing its destruction.

The rule of the Palatinate from Heidelberg ended in 1720 when Prince Elector Charles Philip left the city. At odds with the citizens of Heidelberg over religion, he moved the Palatinate to Mannheim, the new home of the ruling party until 1777, when Charles Theodore became elector of Bavaria and set up court in Munich.

Heidelberg's citizens had seen their city destroyed and rebuilt a number of times, and in the eighteenth century they tried once again to restore their historic town. Much of today's Heidelberg dates from that reconstruction, as shown by the numerous baroque buildings standing upon older Gothic foundations.

Peace was not to last in the Rhine and Neckar Valleys, and early in the nineteenth century the Palatinate's lands were once again invaded. The wars of the French Revolution and the Napoleonic invasion cost the Palatinate its sovereignty, and by 1795 the French had control of the entire Rhineland west of the Rhine. After Napoléon's defeat at Waterloo, the former lands of the Palatinate were given to Bavaria. It was not until 1838 that the Palatinate was officially reunited.

In 1848, a group of liberal German nationalists met at what would later become known as the Heidelberg Committee, or Vorparlament. At the Heidelberg assembly, an electoral law was produced that called for the governments of each state to organize elections for the National Assembly at Frankfurt. For a brief time, German nationalists were united in thought and goal. This unity was short-lived, however, and by June 1849 the Frankfurt Assembly had been dissolved.

By October 1848, the Palatinate-Baden rebellion had broken out in both Baden and Saxony. During this period, Heidelberg was the seat of the revolutionary army that would ultimately lose the war. Once again, Heidelberg was invaded, this time by Prussia. The town remained occupied by Prussian troops until 1850.

Throughout its six centuries of rich history, the town of Heidelberg had seen its share of strife and destruction. Its citizens had time and again rebuilt the city, attempting to recreate the town's romantic aura and rich cultural heritage. Fortunately, the city escaped this century's greatest challenges: the widespread destruction of World War I—after which the Rhineland was briefly occupied by France—and the bombing raids of World War II.

Today, Heidelberg is a popular tourist spot for millions of German and international visitors. The town's university is considered one of the best in the world and maintains excellence in both its students and faculty. Once the glittering residence of Palatinate electors, the Heidelberg Castle lies in ruins above the city, continuing to exert its Romantic influence over the Neckar River region and this age-old town.

Further Reading: Readers interested in the role of Heidelberg in the Reformation will enjoy Hajo Holborn's book *A History of Modern Germany* (New York: Knopf, 1959; London: Eyre and Spottiswode, 1965), a detailed account of Germany from the fifteenth century until the mid-seventeenth century and the Peace of Westphalia. *The Thirty Years War* by C. V. Wedgwood (London: Jonathan Cape, 1938; New York: Anchor, 1961) is a comprehensive study of the international intrigue giving rise to the devastating Thirty Years War in Europe. Two general books on Germany and Heidelberg are *Modern German History* by Ralph Flenley (London: Dent and Sons, 1959) and Malcolm Pasley's collection of essays in *Germany: A Companion to German Studies* (London: Methuen, 1972; second edition, London and New York: Methuen, 1982). Pasley's essays include one by Agatha Ramm entitled "The Making of Modern Germany, 1618–1870," discussing the Heidelberg Committee of the mid-nineteenth century. Walter Marsden's *The Rhineland* (London: Batsford, and New York: Hastings House, 1973) details the history and culture of Germany's Rhine River Valley, the home of Heidelberg.

—Monica Cable

Holy Island/Lindisfarne (Northumberland, England)

Location: Just below the Scottish border, off the Northumberland coast of England, approximately twelve miles south of Berwick-upon-Tweed and two miles from shore, connected to the mainland by a causeway that is passable at low tide.

Description: Small island that was an early center of Christianity, named Holy Island in the eleventh century; contains the village of Holy Island (also called Lindisfarne), ruins of the eleventh-century (second) priory, and nineteenth-century restorations of the twelfth-century St. Mary's Church and the sixteenth-century castle. The eighth-century illuminated manuscript called the Lindisfarne Gospels, now in the British Museum, was created here.

Site Office: Holy Island Village Information
Holy Island
Berwick-upon-Tweed, Northumberland, TD15 2RX
England
(289) 89253

Across the shire of Northumberland extends a huge swath of volcanic upthrusts called the Great Whin Sill. Among the islands it spawned off the northeastern coast lies the small isle originally known as Lindisfarne and now called Holy Island. Composed partly of sand dunes and marsh grass, the island is bounded by cliffs on the northern and eastern coasts. Two prominent outcroppings of the Sill's dolerite rock guard the southern coast: one called the Heugh, between the village and the sea, and the other Beblowe Crag, topped by Lindisfarne Castle. Twice each day the tide floods the modern causeway to the mainland as it does the old Pilgrim's Way. Only two square miles, with a village of 200 people, Holy Island has nonetheless played a role far greater in the religious history of England than its size would suggest. Nine saints and sixteen bishops were residents of Lindisfarne, and from the island the conversion of northern England to Christianity was begun in earnest.

In A.D. 634 Oswald, the nephew of Northumbria's first Christian king, defeated pagan usurpers to regain the throne. Oswald had been educated by the Irish Christian monks of Iona, and he quickly asked for their help in converting and leading his subjects to the faith. After an unsuccessful first effort, St. Aidan was sent to Northumbria; he chose the island of Lindisfarne as the site for his new monastery and base of operations for his missions. By the year 651, when Oswald and Aidan both died, the preaching of the monks and the support of the king had solidly established Christianity among the Northumbrians, and Lindisfarne was known as the mother of monasteries.

The most famous of the nine saints associated with Lindisfarne was St. Cuthbert, devoutly honored even in that age of saints. Cuthbert was called to the priesthood by a vision of St. Aidan's soul ascending to heaven at the moment, as Cuthbert later learned, of Aidan's death. He was trained at the monastery of Melrose in Scotland before beginning an illustrious career of preaching and proselytizing, becoming as well loved for his humility as he was well known for his oratory. Cuthbert helped found a monastery at Ripon and then succeeded his old prior at Melrose before being sent to Lindisfarne.

Christianity in England had just emerged from the confrontation between the Celtic Church and the Roman Church, which established, through the Synod of Whitby in 664, the authority of Rome over English Christians. Lindisfarne as a center of Celtic Christianity had played its own role in this struggle, and one of the consequences of the Synod was the resignation of the current abbot from his Lindisfarne monastery. But he managed to ensure that a Celt would take his place, and in due course Cuthbert arrived to lead the monastery as it sought to deal with the new order imposed by Roman rule.

At Lindisfarne, a windswept and lonely place, Cuthbert began to seek respite from his responsibilities in even greater solitude, hoping to devote himself to prayer and meditation. A tiny island a hundred yards off the southwest corner of Lindisfarne served him at first, as it had served St. Aidan as a hermitage. It is now known as St. Cuthbert's Island, or with familiar fondness as Cuddie's Isle. A chapel was later built and dedicated to him that was excavated in the 1880s. But this retreat was not far enough from the pilgrims who would come seeking him out. In 676, after twelve years of service as Lindisfarne prior, Cuthbert was granted permission to retire to the uninhabited island of Inner Farne, within view of Lindisfarne to the south, where he farmed enough ground to sustain himself. His hermitage was still not solitary, for the Lindisfarne monks and other visitors called upon him even here.

In 684, much against his wishes, Cuthbert was persuaded to accept a bishopric, but only after the combined supplications of King Ecgrith of Northumbria, Bishop Trumwine of Abercorn, and a host of illustrious churchmen, who had journeyed to Lindisfarne to seek him at his hermitage. Cuthbert was consecrated Bishop of Lindisfarne the following year. After two years he retired again to his hermitage due to declining health. The abbot of Lindisfarne, Herefrith, ministered to him there during his last days until his death on March 20, 687. Cuthbert was buried under the floor of the monastery church in a stone sarcophagus. Eleven years later the coffin was opened so that the bones could be "elevated" as a sign of sainthood. To the amazement of the monks, Cuthbert's body and shroud remained undecayed. As news of

Lindisfarne Priory
Photo courtesy of Berwick Borough Council

the miracle sped through Northumbria, a cult of followers sprang up, perpetuating his memory.

To the monks at Lindisfarne Monastery, Cuthbert remained a powerful patron. Even before the opening of his tomb, a magnificent illuminated copy of the Four Gospels had been executed in his memory. This book, known as the Lindisfarne Gospels, is one of the most superb examples of medieval manuscript illumination and of Celtic art in existence. Now in the British Library, its origins were detailed in a colophon added two and a half centuries after its creation by a priest named Aldred, who added an Anglo-Saxon interlinear translation of the Latin. In his colophon, Aldred credited Eadfrith, later a Bishop of Lindisfarne, with the writing. Another future Lindisfarne Bishop, Ethelwald, had bound it. A hermit named Billfrith had added the gold and silver and jeweled ornamentation to the original binding, long since lost. Aldred's Anglo-Saxon translation is the earliest surviving English version of the Gospels.

Close inspection of the manuscript indicates that a single scribe, presumably Eadfrith, wrote the entire book and did the elaborate and integral decorations. A work of this type would have taken a scribe working without interruption a minimum of two years to complete; it is therefore believed to have been accomplished before Eadfrith's election to the bishopric in 698. The script used is called insular majuscule and was developed in Ireland. Through the Irish monks at Iona it spread into Northumbria and the rest of England. It was a formal script, entirely appropriate for a book intended as an altar Bible for ritual use. The book consists of 258 leaves, made from at least 129 large pieces of vellum, similar to parchment, produced from calfskin. The elaborate decorations were colored with pigments made from red and white lead, verdigris, yellow ochre, indigo, yellow arsenic sulphide, and blue lapis lazuli, among other materials. Gold paint was used sparingly on the manuscript itself, but gold lavishly ornamented the binding.

The peaceful existence of the Lindisfarne monks was destined to change. In 793 the Anglo-Saxon Chronicle, an annual compilation of events, noted: "In this year dire portents appeared over Northumbria and sorely frightened the people. . . . A great famine immediately followed these signs and a little after . . . on 8 June, the ravages of heathen men miserably destroyed God's church on Lindisfarne, with plunder and slaughter." The Vikings had arrived and had made Lindisfarne the target of one of their earliest attacks. But the true extent of the damage is uncertain, since the abbey's most cherished relics were untouched. After the Norsemen had departed, the brothers returned to restore their island to order, living now under a constant threat. The Vikings had learned that monasteries made easy and profitable targets; one after another fell to their attacks.

The Lindisfarne Stone, discovered in the 1920s by workers renovating the Priory, was long believed to have been carved as a memorial of the Viking raid. On one side it depicts a fierce line of warriors, and on the other is shown two worshippers before the cross. Recent scholars have questioned this interpretation, suggesting that the stone was instead a gravemarker, illustrating the Day of Judgment. After several years on exhibit in the British Museum, it now resides in the more modest museum on Holy Island.

In 875 Bishop Eardulf decided the monks could no longer risk keeping such important relics as the bones of St. Cuthbert and the Lindisfarne Gospels; taking the house's treasures he set out in search of a safer location, accompanied by the entire island populace. After roaming for seven years, they finally settled at Chester-le-Street, a few miles north of Durham. In 995 they fled again in fear of a Viking threat, this time to Ripon; as they returned home they received a sign from St. Cuthbert to re-establish the monastery in Durham. The body of St. Cuthbert rests in the cathedral there today.

The Lindisfarne Gospels disappeared from the historical record during the late Middle Ages, when it may have been kept in the Durham Cathedral. But it was very likely sent to London by the commissioners of Henry VIII during the dissolution of the monasteries in the early 1500s; in 1567 it was in the possession of William Bowyer, the Keeper of the Records in the Tower of London. It had probably been dispatched there because of its jeweled binding. From Bowyer's son it was purchased by Sir Robert Cotton in the early seventeenth century and was among the treasures deeded to the nation by his heirs in 1703. It is now on display in the British Library. The church on Holy Island contains a facsimile of the book and a handmade carpet, embroidered by local women, which replicates the elaborate decorations of the manuscript.

Meanwhile Lindisfarne was apparently deserted. There is no mention of the island in historical records for two and a half centuries, except for a brief sojourn there by the last Anglo-Saxon bishop of Durham, hiding from the expected retribution of William the Conqueror. When the Normans invaded the English churches as well, they set about reviving Benedictine monasticism. William of St. Calais, the first Norman bishop of Durham, established a Benedictine community on the island again, as a dependency of Durham, and around 1093 work began on a new Priory Church for Lindisfarne—now called Holy Island in honor of its venerable past. The new building, entirely Norman in design, was stylistically similar to Durham Cathedral, built at the same time, and designed on a plan similar to Whitby Abbey. It was largely completed by 1300.

It is the ruins of this priory that the visitor to Holy Island sees today; all traces of the original monastery have long since vanished. The west front church entrance retains its elaborate doorway and part of one of its flanking twin towers. Part of the north wall and two of the eastern bays remain. The vaulting is gone except for one rib and its keystone, locally known as the Rainbow Arch. To the south are the remains of the priory building.

After the Priory of Lindisfarne was dissolved by Henry VIII, the buildings fell into disrepair and were periodically stripped of materials for other building. In 1888 the then owner of the island, Major-General Sir William Crossman, excavated the abbey site and made a plan of the ruins. In 1913 he transferred the property to the government. Further excavations were carried out during the 1920s, and the ruins were restored and stabilized. Today the ruins are under the protection of the Department of the Environment.

Next to the Priory sits the island church, St. Mary's, founded in the mid-twelfth century. In 1860 it was extensively restored, and little of the original exterior is visible. The interior dates to the twelfth or thirteenth century, and features alternating courses of red and white stone in the north arcade piers, and a trefoiled piscina in the south aisle.

With the Reformation, Holy Island seemed destined to sink back into oblivion, but it became important in a different sphere: military strategy. As Tudor Kings and Queens quarreled and warred with Scottish Stuarts and their French allies, the little island not far from the Scottish border took on a new significance. In 1544 the Earl of Hertford, having landed 2,200 troops and anchored 10 battleships in the harbor the year before, used Lindisfarne as a naval base for his devastating attack on Edinburgh. The military value of Holy Island was not forgotten, and soon the island was fortified. Upon the bulbous outcropping called Beblowe Crag that guards the sea lane was built a stone fort, now called Lindisfarne Castle. It never boasted a large garrison, and when James VI of Scotland became James I of England, Holy Island's strategic importance faded as well.

Religious tensions brought the site back into prominence when Scottish Presbyterians resisted Charles I's efforts to establish an Episcopal form of worship throughout both kingdoms. In 1639 Charles made Lindisfarne his naval base for an abortive advance toward the recalcitrant Scots waiting north of the Tweed, but both sides withdrew without fighting. Perhaps the most memorable circumstance about Lindisfarne during this period was the commander of its garrison, Captain Robert Rugg, as famous for his generous hospitality as for his spectacularly large nose. Amiable and patient, Rugg was moved to surrender the fort to Parliamentary forces during the Civil War, perhaps partly because of the large arrearage of back pay he had accumulated serving the king. In spite of Parliament's promises, he had no better luck collecting from his new masters, and died still unpaid. After the Restoration, the garrison dwindled. The fortifications were reinforced in the 1670s during the Dutch Wars, and the fort was briefly seized by a sympathizer of the Old Pretender, James Stuart, in 1715. The garrison was finally withdrawn in

1819, after which the fort was used briefly as a coast guard station before falling into disuse and disrepair.

In 1901 Edward Hudson, editor of the weekly magazine *Country Life*, visited Holy Island and at once fell in love with the tumbledown fortress perched on its rock. Hudson bought the property and retained the brilliant architect Edwin Landseer Lutyens to renovate it as a country house. Finding the fort still structurally sound, Lutyens modified its line while retaining its Tudor aspect and revamped the rambling garrison quarters into a comfortable Tudoresque Revival home. Lutyens also designed some of the furniture and fittings, and seventeenth-century antiques furnished the rest. His friend Gertrude Jekyll designed the gardens. Hudson entertained the royal, the rich, and the famous at Lindisfarne Castle before selling the property in 1921. It was donated to the National Trust in 1944, and house and garden are restored to their designers' plans. Each year the Castle hosts 50,000 visitors.

In 1964 a large area of the island was made a National Nature Reserve, although regulated hunting is allowed. Thousands of migrating wading and water birds come to Holy Island annually, including some endangered species. The island offers 8,000 acres of saltmarsh, mudflats, dunes, and rocky cliff faces, waystations for dozens of species of migrating birds.

Today the inhabitants of Holy Island live in the village, not a monastery, and make their living through fishing, farming, and the tourist trade.

Further Reading: A detailed and affectionate account of Holy Island's past and present is provided by Magnus Magnusson in his *Lindisfarne: The Cradle Island,* illustrated by Sheila Mackie (Stocksfield, Northumberland: Oriel Press, 1984). Excellent color plates illustrate *The Lindisfarne Gospels* by Janet Backhouse (Oxford: Phaidon, 1986), which describes the history and production of this and similar medieval illuminated manuscripts. For more information about the Viking raids on England the *English Heritage Book of Viking Age England* by Julian D. Richards (London: B.T. Batsford/English Heritage, 1991; New York: Hippocrene, 1992) offers a balanced and well-illustrated account, with many useful maps.

—Elizabeth Brice

Iona (Inner Hebrides, Scotland)

Location: A small island in the group called the Inner Hebrides, off the west coast of the Highlands of Scotland, in Strathclyde (previously Argyll). The island lies off the southwest tip of the Isle of Mull, in the north Atlantic Ocean.

Description: Site where in 563 St. Columba, a Christian Irish prince and monk, landed and established Celtic Christianity, first in Scotland (Northern Pictland), spreading via Lindisfarne in Northumbria into England. Previously inhabited by Picts and/or Celts, it is now the home of The Iona Community.

Site Office: The Iona Community
The Warden
The Abbey
Iona, Inner Hebrides PA76 6SN
Scotland
(681) 7404

Iona is formed from some of the oldest, and most stable, rock structures in the British Isles, dating from the pre-Cambrian age and composed of ancient Lewisian gneiss, or volcanic rock. Some bands of limestone in this rock were recrystallized as they went through processes of heat and pressure, producing a green-veined marble, particular to Iona. This was quarried until the early twentieth century. Today a few multicolored stones, called Iona pebbles, can still be found near St. Columba's Bay in the south east of the island.

The island's original name was Hy or Hí, meaning "yew-tree," and the yew was one of the trees sacred to the Celtic Druids. The name, Iona, came about because of a misreading of the name, "Ioua Insula," (Yew-Place Island). Although the island was probably occupied by one of the Celtic tribes at some time, it lay in the territory generally belonging to the northern Picts, the indigenous inhabitants of northern Scotland or Alba, who may have originated in the Bronze Age or even the late neolithic period. The word Pict is from the Roman, *Picti* meaning "Painted Ones," referring to the Picts' custom of tattooing and painting their bodies, using a blue dye obtained from the woad plant. No one is quite sure who these people were; some think they may have been the ruling class of some ancient British Celtic tribe or tribes. Their language, remains of which survive on sculptured stones and metalwork and in personal and geographical names, has yet to be deciphered, but may in some way be related to the Welsh language. Whoever the Picts were, they were pagan, and it is the northern Picts whom Columba set out to convert to Christianity in 563.

When the first Irish-Scots, the Dalriadans, first colonized Argyll is a matter of debate. While it is possible that the process started as early as the third century A.D., most schol-

ars agree that the main colonization took place at the end of the fifth century and into the sixth century. When Columba arrived on Iona, the island, and the surrounding area, including islands and the mainland of Argyll, were under the control of the Irish Dalriadans, not the Picts, though the northern Picts controlled a neighboring area. The Dalriadans were relatives of Columba, and they were Christian.

In 312 the Roman Empire had embraced Christianity as its religion and from that time England, as part of the empire, came under the influence of the new religion. However, in 410 the Romans abandoned the country and by 550 England had been taken over by the Anglo-Saxons, who destroyed Christianity there. The northern Picts, however, had never been subjected to the Romans, and continued in their pagan Celtic traditions. It seems likely that the Picts invited, or at least welcomed, the Dalriadan-Scots, who helped them to defend the area first against the Romans, and then the Anglo-Saxons. Ireland had never been part of the Roman Empire either, and it was the arrival of Saint Patrick about 432 that started the conversion of the Irish.

Columba was born in Ireland on December 7, 521, in Gartan in County Donegal into a royal—and a Christian—family. He was the grandson of Conall Gulban, who founded one of the branches of the Northern Uí Néill dynasty in Northern Ireland, and the great-grandson of Niall of the Nine Hostages, the High King of Ireland, who ruled from 379 to 405. Although born a prince, Columba seems to have been given to the church almost immediately, in the care of his foster father, Cruithnechan, the priest who baptized him Colum (in Latin, Columba, meaning The Dove). In Ireland he became known as Columcille, Colum of the Church. After studying with two St. Finians, St. Finian of Molville and St. Finian of Clonard, and a Christian bard named Gemman, he was ordained a priest in 551, at the age of thirty.

For the next ten years he worked in Ireland, a strong leader and innovator, founding several churches and two monasteries, Daire Calgaich (Derry) and Dairmagh (Durrow). During this time he built up a formidable reputation as a priest, a man of God, and a worker of miracles. Why he left such a successful career in Ireland is not really clear, and the stories concerning the reason for his departure are many and varied. Most of them relate a tale about a dispute that arose between Columba and his old teacher, Finian of Molville, who had a rare copy of the Vulgate, St. Jerome's Latin translation of the four gospels, which Columba copied without permission. Finian discovered this and somehow, directly or indirectly, events escalated, developing into the Battle of Culdreihmne (Cul Drebene) in 561, between Columba's Uí Néill tribe and the pagan Irish king, Diarmaid, in which 3,000 of the king's men died.

Some say that Columba was held responsible and ex-

Interior of Iona's abbey
Photo courtesy of Argyll and Bute District Council

communicated, subsequently pardoned but sent into exile as a penance. Others say he chose the penance himself, out of remorse at the death of so many. (In the Irish Celtic Church, ''white martyrdom'' or exile was approved but ''red martyrdom'' or death was not.) The popular version of the story tells only of Columba's fervent religious zeal and desire to convert the northern Picts to Christianity. Others attribute his going to political, as much as religious, motives. An Irish prince, Columba was the cousin of the Dalriadan King Conaill, King of the Scots in Dalriada (Argyll). The dynasty was newly established there and certainly a person of royal blood and spiritual standing could be helpful to the Scots in their dealings with the surrounding Picts and their leaders.

Wherever the truth lies, in 563 Columba crossed from Ireland to Alba in a small boat, a coracle, with twelve companions. He is reputed to have landed in the south at Port na Churaich (The Bay of the Coracle) or St. Columba's Bay with the twelve monks. After landing, he is said to have walked up the nearby hill, to the west, Carn Culri Eirinn (The Hill of the Turning Back to Ireland) and as he could still see Ireland from there, decided to establish his monastery on the east side, away from the sight of his beloved homeland. The island was empty, save for the ruins, still to be discerned today, of a dun, *Dun Bhuirg,* presumably built by pagan predecessors. These duns, or hill forts, were first built and occupied in the first millennium B.C.

Whether Iona was given to Columba by the Picts or by his cousin, King Conaill, is disputed, but the community he established there seemed to be quite secure. The monks set about building their simple cells and monastic buildings, of wood, wattle, and clay. A site on the hill, *Tor Abb,* has revealed the foundations of what may have been St. Columba's cell, and parts of the monastery's boundary walls have been excavated. The *Reilig Oran* or *Odhrain* (Oran's graveyard) was in use from 563. Oran is thought to have been one of the twelve monks who accompanied Columba to Iona, and the first to die. It is possible that he gave himself as the sacrifice that was necessary in those days when a new religious sanctuary was established. While modern Christians may shrink from such a notion, it must be remembered that the Celtic Christians of the sixth century were newly converted from their indigenous religion in which sacrifice, both animal and human, was an important element. Indeed, the worst punishment given by the Druids, the aristocratic priests of the Celtic religion, was to be banned from attending sacrifices.

The Ionan monks' daily life has left few visible traces, but an account was written by Adomnán, abbot of the community from 679 to 704, about a century later, and there is information in Bede, too, which indicates that the monks settled into a life of prayer, farming, wood-gathering, stone-carving, and other such day-to-day jobs and devotional activities. They grew wheat and barley and cared for sheep and cattle, the latter being kept on nearby islands. They ate beef, mutton, venison, pork, fish, and seal meat, and used seal oil for their lamps. They also grew holly, used to produce ink.

Little is known of Columba's mission to the Picts, but certainly Columban monasteries were established in Pictland, in the area to the north and east of the Dalriadan-Scots' lands in Argyll. It seems that Columba met King Bridei (Brude), king of the Picts, perhaps in his dual role of prince and priest, again combining the political with the spiritual. Some say that the saint converted the king, but there is no real evidence for this. Iona became the mother church of both the Irish and the Scottish Dalriada, with Columba recognized as their religious leader.

From Iona what is known as Celtic Christianity spread, reaching not only into Pictland but down into England and over to Luxueil in Gaul and to Bobbio in the Apennines. The period of greatest expansion came about as a result of King Oswald of Northumbria's request for an Ionan monk to convert his kingdom, Oswald himself having already become Christian. In 634, St. Aidan responded to this request and established his base at Lindisfarne, an island which at low tide is linked to the mainland.

The Celtic church had reached Britain in the second and third centuries, establishing Celtic Christian communities in Cornwall, Wales, the Pictish lowland areas, and Ireland. The Celtic Christianity that Columba and his successors established differed from Roman Christianity in that it emphasized communal living, individual prayer, asceticism, everyday work, and penance. The Celtic Christians were austere in lifestyle, practicing what they preached. There were differences in the baptism ritual and the way the date of Easter was established. Perhaps they drew more on their Celtic, and Druidical, heritage, although the Roman church too found it had to be adaptable in England, allowing the retention of certain elements of the old religion in their churches and practices.

In 597 Columba died, having predicted the event well in advance. A few months later, St. Augustine was dispatched to England by the pope, to bring both the English and the Irish-Scots back to the church of Rome. Augustine established his base at Canterbury in Kent, and slowly the influence of his Roman church moved northward in England. At the same time, the influence of the Celtic church, based at Lindisfarne, was moving southward in England and conflict became increasingly likely. Things came to a head because King Oswiu (Oswy), son of Oswald of Northumbria and a Celtic Christian, was married to a princess of the Roman faith. They found themselves celebrating Easter on different days, so in 664 Oswiu called a synod to the monastery of Streaneshalch, 'The Bay of the Lighthouse,' now known by its Danish name, Whitby. The outcome of the Synod of Whitby was in favor of the Roman church, spelling the beginning of the end for Columba's Celtic Christianity. Some monks remained on Iona, which became famous as a center of learning. It is said that the Book of Kells was written, or started, on Iona. Columba himself was a scribe of some renown and his followers carried on the tradition, being instrumental in the great flowering of Celtic art in the form of manuscript illumination.

Viking raids began at the end of the eighth century

and continued through to the end of the tenth century. The monastery on Iona was plundered many times, 806 being the year in which sixty-eight monks were killed on the island. In 804, after the first raids, land was secured for a new monastery in Kells, Ireland, and in 807 building there began, and most of the monks moved to Kells. A few remained on Iona until the twelfth century, despite the dangers.

Meanwhile, the Dalriadic Scots had been in conflict with their neighbours, the Picts. In 843, Kenneth MacAlpin became King of the Picts and the Scots, marking the end of the Picts and the birth of the nation that was to become Scotland. From this time until the end of the eleventh century the Reilig Odhrain cemetery on Iona became the burial place of all the Scottish Kings, including Macbeth and Duncan, whom he murdered. Donald Bane was the last MacAlpin to be buried there in 1097. It is said that there are forty-eight Scottish kings, eight Norwegian Kings and four Irish Kings buried there. The graveyard is still used by the islanders and occasionally permission is given for others to be buried there, as was the case in 1994 when John Smith, the leader of the Labour Party in Britain, died suddenly and was buried on Iona. Although Malcolm III had a royal vault built at Dunfermline, refusing burial on Iona, his wife, Margaret, is believed to be responsible for the building of the pink granite Chapel of Saint Onran in 1080, probably on the site of an older chapel. This is the oldest complete building to be seen on Iona today.

By 1203 the last few Celtic monks had left the island, and a Benedictine Abbey was built, at the instigation of Reginald, the famous laird (lord) of the Clan (family) Donald. Parts of the abbey remain today. An Augustinian nunnery followed, marking the first time women were allowed on the island. The nunnery is now a ruin. The burial mound at Martyr's Bay, once thought to contain the bodies of the sixty-eight monks killed in 806, recently has been found to hold female bodies only; it was probably the nuns' burial ground. In 1450 there was some rebuilding of the abbey, in the style of the original construction, and in 1507 the Benedictine Abbey became the Cathedral of the Western Isles. It was in the fourteenth and fifteenth centuries that the West Highland School of Carving reached its zenith at Iona. There are some 600 stone crosses and graveslabs from the school, the work of four different workshops; Iona was probably the earliest and seems to be the best. There were once 350 crosses on Iona, but now there are only 3, thanks to the ravages of the Reformation zealots in the sixteenth century. The three that remain are now known as St. John's, St. Matthew's and St. Martin's, the last one being the oldest.

The island belonged to the Macleans of Duart until the late seventeenth century, when it became part of the Duke of Argyll's estates. In 1773 Dr. Samuel Johnson went to Iona, a visit documented by James Boswell in his biography of Johnson and marked by plaques on the island. In 1897 a cairn, or mound of stones, was built to mark the 1,300th Anniversary of Columba's death, and in 1899 the Eighth Duke of Argyll gave the abbey to the Church of Scotland for the use of all Christian denominations. The church formed the Abbey Trustees and restoration work was begun. On June 9, 1905, St. Columba's Day, the restored abbey was rededicated, and it was opened for public worship in 1912. The chapel is said to contain the relics of St. Columba.

In 1938 Iona took on a new lease of life. The Reverend George Macleod, later Lord Macleod of Fuinary, was a minister in the Govan district of Glasgow. It was the time of the Great Depression, with enormous numbers of people unemployed, and he wanted to revive a sense of community and mutual support. To that end he created a fellowship of six young ministers in training, along with some plumbers, masons, and carpenters, and set out to rebuild the monastery buildings on Iona, with the trustees' blessing. Macleod felt he was following in the footsteps of Columba, a saint who had been able to combine both spiritual and political concerns. Despite the standstill imposed by the World War II The Iona Community accomplished the complete restoration of the abbey outbuildings.

Today The Iona Community is an ecumenical center, under the jurisdiction of the Church of Scotland. There are some 150 full members who stay at the abbey while training but who then go out to places all over the world, specializing in community development and church renewal. The rules of the community involve praying for each other, working for world peace, using prayer for healing, and giving part of their income to a charitable fund, most of which is used in the Third World. The community is funded by private contributors. All those involved get together whenever possible, meeting on Iona where there is a small permanent staff.

This small Scottish island, with a climate warmed by the North Atlantic drift, has white beaches and a clear, green sea, and has inspired painters, in particular a group called the "Scottish Colourists," over the centuries. The work and influence of the Iona Community is growing all the time and the island draws a large number of visitors each year.

Further Reading: A key work is *Adomnán's Life of Columba* by A.O. and M.O. Anderson (London: Nelson, 1961). It is because Adomnán wrote his Life of Columba that we know so much about the saint, although fact needs sifting from fiction. The three parts deal with his prophecies, his miracles and his visions of angels. *Columba* by Ian Finlay (London: Victor Gollancz, 1979) is an interesting account by a specialist in Celtic art, casting new light on the saint's life, drawing on the art as well as on original sources. *Iona* by the Eighth Duke of Argyll (1884) is a small volume that may prove hard to find. *The Picts and the Scots* by Lloyd and Jenny Laing (Phoenix Mill, England and Dover, New Hampshire: Alan Sutton, 1993) reports on the latest archaeological discoveries, including Iona, and their implications in a very readable way. *Celtic Britain* by Charles Thomas (London and New York: Thames and Hudson, 1986) deals with British history from when the Romans left to the late seventh century in particular, drawing on the latest historical and archaeological research. It includes chapters on Celtic Cornwall, Scotland, Wales, Britain in general and the church and monasteries, as well as Celtic art.

—Beth F. Wood

Jarlshof (Shetland, Scotland)

Location: Shetland consists of about 100 islands, 150 miles north of the Scottish mainland. Seventeen of these islands are inhabited; the population of Shetland is about 23,000. Jarlshof is located on the southern tip of the Shetland island of Mainland, just west of Sumburgh airport.

Description: Jarlshof is considered one of Shetland's most important archaeological sites, and has been occupied for some 3,000 years from Neolithic to Viking to medieval times. Jarlshof is under the care of Historic Scotland.

Contact: Jarlshof
Historic Scotland
20 Brandon Street
Edinburgh, Scotland EH3 5RA
(31) 556-8400

Jarlshof, an ancient historic site on the edge of Europe in the Shetland Islands, has seen continuous settlement for some 3,000 years. Three main stages of settlement are represented at Jarlshof: the late Bronze Age, the late Iron Age, and the Viking Age.

The first inhabitants of Shetland were groups of Neolithic farmers who settled around what is now Sumburgh and the coastal town of Lerwick 5,000 to 6,000 years ago. Some historians theorize that these farmers were probably descended from immigrants who had settled in the northern isles about 1,000 years earlier. Such peoples, a mixture of hunter-gatherers and farmers, had established themselves in Deeside in mainland Scotland by the fourth millennium B.C., before finally reaching Orkney and then Shetland. They brought livestock with them and grew barley.

Around 1,500 B.C., Shetland experienced a climate change, as did much of Scotland. Its weather turned less hospitable and became wetter and cooler. Peat began to accumulate on higher ground, forcing the farmers to move to the crowded inlets or *voes*. By the late Bronze Age (1000 B.C. and later), Shetland was on the verge of a population explosion, with many more people than it could comfortably accommodate. The combination of climatic change and population increase led to the development of a more competitive society that tended to form small groups rather than live in isolated farmsteads. These settlements, such as the one at Jarlshof, were often located near the shore. As the climatic changes continued, these settlements shifted from farming to husbandry. The earliest houses at Jarlshof belong to the late Bronze Age. The oldest building is a Bronze Age smithy.

The Iron Age in Shetland occurred from around 600 B.C. The inhabitants bred sheep and maintained short-horned cattle, pigs, ponies, and dogs. They made their living largely from the sea, catching seabirds and hunting grey seal and walrus for food and oil. Like their forebears, they grew sturdy crops such as barley and wheat.

As Shetland society grew more complex, the social fabric began to change. This societal upheaval eventually led to feuds and necessitated the building of fortresses called *brochs*. Brochs are the most visible evidence of the Iron Age in Shetland. A broch is a tall circular fort with a thick hollow wall and a single entrance. About seventy-five brochs are in Shetland. When not being used for defensive purposes, these sturdy structures doubled as farmhouses. One of the best-preserved brochs in Shetland is at Jarlshof.

In the centuries before the arrival of the Norsemen, very little changed in Shetland. A different type of housing structure did develop, however, called the *wheelhouse*. A wheelhouse was a dwelling with a turf roof supported by piers that radiated from the center of the structure like spokes in a wheel. It was believed that wheelhouses were made from the stones of brochs and even could have formed, at one point, a part of the broch. Eventually, the wheelhouses were either abandoned or modified. The inhabitants of the area at that time were called Picts by the Romans, and Péttar by the Norsemen. They were an early, enigmatic Celtic people who had already been converted to Christianity by the time of the Viking invasion.

Viking raids began as early as the eighth century. A Norse earldom developed in the eleventh century in Shetland as well as in neighboring Orkney, in Caithness on the Scottish mainland, and in the Outer Hebrides. The Norse seem to have been able to bring the Picts under their control with no great difficulty and built on the sites that they had once occupied. Estimates of the population of Shetland during the Viking Age range from 20,000 to 22,000 people.

The Viking Age lasted roughly from A.D. 800 to 1100. Most of the Viking settlers came from the western fjords of Norway, an area around present-day Trondheim. The Norsemen found the Shetland landscape similar to their Scandinavian homeland. They didn't ask for much: easy access to the sea, land fertile enough to grow grain or other sturdy crops, and pasture for the livestock.

The first important Norse ruler was Sigurd, who ruled toward the end of the ninth century. It was Harold Hårfager (Fairhair), however, who began the process of unification among the various Norwegian tribes. From his base in western Norway, in the Hardanger area, Harold made forays into the Scottish Isles. Historian Johannes Brøndsted believes that the early Viking raids in Shetland and Orkney were primarily motivated by piracy, not settlement, or that they were meant to establish bases for future attacks on the Scottish mainland. Only later, in the last twenty-five years of the ninth century,

A view of Jarlshof
Photo by Harvey Wood

did Harold establish an earldom in the Orkneys, with Shetland also under his domain.

The northern isles were stepping-stones for the Norse to other, more distant places—the Faroes, Iceland, Greenland, and, finally, Newfoundland or "Vinland" in the New World. The reasons that scholars provide for the Viking invasions vary. Some say an increase in population at home persuaded the young and aggressive to look elsewhere for opportunities. Others say the expansion of trade opportunities and the riches that they promised was an important factor behind the raids. Whatever the Vikings' motives, the ultimate success of the Viking excursions relied on the existence of the Viking longship, an agile creation that allowed the Norse to navigate the cold northern waters with speed and agility.

The first Scandinavian settlement at Jarlshof consisted of a small farming community, dating back to first half of the ninth century. Unlike other parts of Shetland, it appears that the people who settled there originally came from southern Norway. Approximately six to eight generations of Norse people lived at Jarlshof.

Jarlshof during the Viking Age contained a fairly extensive Norse village. Indeed, evidence of Norse settlement is clearly visible. The main dwelling house at Jarlshof, for example, is based on Norse design. It was a typical Norse farm that was modified over a 400- to 500-year period, from the ninth to the thirteenth centuries. Divided into two rooms on either side of main entrance, with a large living room and small kitchen, it conformed to the construction of a typical Norse drinking hall. Over the years this living arrangement was modified to include a byre, or barn, at one end of the structure and smaller rooms along the walls. Originally a single dwelling, it grew until more houses were built and a fairly significant farmstead developed. Extra units were added to the end or side of existing houses as required.

Norse dwellings in Shetland were fairly simple structures. The longhouse (usually seventy feet by twenty feet) was the main building. A typical longhouse had a byre, a living area that contained a fireplace, and sleeping quarters. Remains of traditional longhouses are found today throughout Scandinavia. The typical longhouse in Scandinavia had a porch, a large main room with a hearth, a bedroom, and a dairy. The longhouse style of architecture remained quite common throughout Shetland until fairly recent times. Unlike its Scandinavian counterpart, the Shetland longhouse was made of stone, not timber, because Shetland is treeless.

In addition to farming, textile making was also an important activity at Jarlshof. The community had an artistic side, too, as evidenced by the existence of bone combs and carved pins. The backbone of Norse society in Shetland at the time was the *bondir,* or peasant-farmer. As head of the Shetlandic household, the peasant farmer retained *odal rights,* a form of land ownership that traditionally provided ample rights to the eldest son.

The language spoken both in Shetland and Orkney at the end of the Viking era was Norn, a west Scandinavian form similar to Norwegian. Shetlandic Norn, in particular, bears a strong resemblance to the dialects of southwestern Norway, especially in the area around Stavanger. As late as the 1890s, many Scandinavian words and phrases lived on in Shetland, especially in the more remote islands. Today, the Shetland accent retains a distinctive Scandinavian infection. Scandinavian street names are common. Frequently, Shetlandic children bear Scandinavian names or Shetlandic variations.

In 1195, Shetland was placed under the direct rule of Norway as a *skattland,* that is, a Norwegian tax-paying tributary. Around this time, both Shetland and the Faroe Islands to the north were placed under joint administration. Within two centuries, however, Norway had lost much of its power because of economic problems and the country's meager natural resources. But especially devastating was the effect of the Black Death, which ravaged the population and thwarted any Norwegian attempt to extend control beyond its own backyard. In 1397, Norway was united with Denmark under Queen Margaret. In 1468, King James III of Scotland married Margaret, a Danish princess, daughter of King Christian I. The following year, Shetland was pledged to Scotland as a dowry for the king's daughter. Two years later, both Shetland and Orkney were annexed to the Scottish crown.

Conditions at Jarlshof reflected the change in power. The farmhouse and barn at Jarlshof that were built in the thirteenth century were modified over the years. A new farmstead was built in the fourteenth century and eventually abandoned in the sixteenth century.

It was at this point that the first laird's house, now one of the most prominent ruins still standing, was first erected. In 1564, Mary, Queen of Scots, gave both Orkney and Shetland to her half brother Lord Robert Stewart, the illegitimate son of James V. The local magistrate met at Jarlshof in a structure called the New Hall. By the early 1570s, Lord Robert had given the office of chief magistrate of Shetland to his half brother Laurence Bruce of Cultmalindie. In 1581, King James VI made Robert Earl of Orkney.

In 1592, William Bruce of Symbister leased the house from Earl Robert. Earl Robert was eventually succeeded by his son Earl Patrick Stewart. By the early years of the next century, Earl Patrick had built a grand house near Sumburgh Head on the southern tip of the Shetland's Mainland, which he called "the Old House of Sumburgh." The New Hall was converted to a kitchen. When Bruce regained possession of the holding, Earl Patrick refused to relinquish his dwelling rights and broke into the property. For this and other trans-

gressions, Earl Patrick was imprisoned in 1609 and executed in Edinburgh in 1615. Three years earlier, in 1612, Orkney and Shetland were again annexed to the crown.

Sir Walter Scott, the famous Scottish novelist, visited Shetland in 1814 and found inspiration in the wildness of the Shetland landscapes. He based the exploits described in his novel, *The Pirate* (1822), on the story of John Gow, an Orkney pirate, but, with artistic license, he set his tale 100 years earlier than it historically would have occurred. One of the scenes took place at Sumburgh Head, where the pirate was saved after a shipwreck by a character named Mordaunt Mertoun, the only inhabitant of Jarlshof. Scott named the old laird's house Jarlshof. The medieval name for Jarlshof is Svinaborg. The modern name Sumburgh is derived from this word, which means either "Svein's Fort" or "Fort of Pigs."

By the end of the seventeenth century, Jarlshof was in ruins, its proud and historic past left to the wind, the sand, and elements. It wasn't until the late nineteenth century that violent storms revealed the existence of earlier settlements. John Bruce, the owner of the Sumburgh estate, noticed that the storms had exposed some massive stone walls. Curious, he started to do some investigating of his own. Finally, in 1925, he placed the site under the jurisdiction of the government. Excavations at Jarlshof began in the 1930s. Most of the artifacts, which include cooking utensils, spinning, weaving and sewing materials, and knives and jewelry, are in the Royal Museum of Scotland in Edinburgh.

Jarlshof remains one of the most important and architecturally complex historical sites in Britain. Few other sites in Britain offer the breadth of history or the opportunity to experience first-hand so many phases of human settlement. A small museum is on the grounds and is open to the public.

Further Reading: A scholarly yet readable account of the Viking Age in Britain is H. R. Loyn's *The Vikings in Britain* (London: Batsford, and New York: St. Martin's Press, 1977). There are several good guides available that examine the architectural heritage of Shetland. They include Patrick Ashmore's *Jarlshof: A Walk Through the Past* (Edinburgh: Historic Scotland, 1993); Mike Finnie's *Shetland: An Illustrated Architectural Guide* (Edinburgh: Mainstream Publishing, 1990); Noel Fojut's *A Guide to Prehistoric and Viking Shetland* (Lerwick: Shetland Times, 1986); and Anna Ritchie's and David J. Breezes *Invaders of Scotland* (Edinburgh: Historic Scotland, 1991). *Scotland and Scandinavia,* edited by Grant G. Simpson (Edinburgh: John Donald, 1990), is a collection of essays that explores the historical links between Scotland and Scandinavia, including the changing nature of Shetland's role. For a history of the Vikings, see *The Vikings* by Johannes Brøndsted, translated by Kalle Skov (London and New York: Penguin, 1965). The best general portrait of Shetland is Liv Kjorsvik Schei's and Gunnie Moberg's *The Shetland Story* (London: Batsford, and New York: Hippocrene, 1988).

—June Skinner Sawyers

Kalmar (Kalmar, Sweden)

Location: On the southeastern shoreline of Sweden, in the region of Småland, 244 miles south of Stockholm; connected by the longest bridge in Europe to the vacation island of Öland, across Kalmar Sound.

Description: Port town the location of which was strategically vital to Sweden's lines of communication, commerce, and warfare in the Middle Ages and Renaissance; site of the signing of several historical pacts and laws, most notably the Kalmar Union of 1397, which joined Sweden, Norway, and Denmark into a single Scandinavian kingdom.

Site Office: Kalmar Tusistbyrå, Sverigereson Sydost
Larmgatan 6, Box 23
391 20 Kalmar, Kalmar
Sweden
(480) 153 50

As a port city with ready access on the Baltic Sea to Denmark, Germany, Poland, the Baltic states, and northern Sweden (including Stockholm), Kalmar came to be known as "the lock and key to Sweden." Its castle, an imposing fortress protecting the city's harbor, was frequently home to Sweden's kings and regents. Because of these distinctions, however, the city was repeatedly a battleground during conflicts among the Scandinavian powers and rival Swedish factions. Today, Kalmar rests in relative serenity, a small city that can focus its attentions on its local heritage and maritime industry.

Because of its strategic locale, the port of Kalmar was an important axis along the routes of the Vikings, whose period of domination reached its peak in the eleventh century. Archaeologists have uncovered Viking ships in the bogs of Kalmar's waterways; it is unclear whether the vessels were sunk by storm or warfare or, as some scholars believe, as a routine part of shipbuilding designed to harden the wood of the ships. During the decade of the 1120s, King Sigurd of Norway, crusading in the name of Christianity, laid waste to Kalmar, an early example of what were to be frequent raids.

Kalmar joined the group of Swedish cities—including Stockholm and Söderköping—that increasingly owed their wealth to the successes of the Hanseatic merchants, a league of German peoples hailing mainly from the regions of Westphalia and the Rhineland. The Hansas were such a factor in the development of Kalmar and Stockholm that, in the fourteenth century, the two cities passed legislation requiring that at least one-half of their town councils be composed of Swedes.

As Viking influence waned and was replaced by that of the free Hanseatic associations of the Baltic Sea, Sweden found itself seeking political guidance and military aid. From 1307 until 1612, the Hanseatic League besieged Kalmar no less than twenty-three times, indicating the magnitude of the port's importance to the region. The protection of the "lock and key to Sweden" was but one reason that the Swedes sought aid; they found it in the person of Queen Margaret.

Margaret was the daughter of the Danish king Valdemar and the wife of the Norwegian king Haakon. Valdemar died in 1375, Haakon, five years later, events that effectively made Margaret the ruler of both Denmark and Norway. When, in 1386, the life of Swedish Lord High Steward Bo Jonsson Grip came to an end, Margareta realized her opportunity to unify the three Scandinavian territories. Grip, an aggressive landholder who had systematically seized the property of German noblemen in Sweden and Finland, had instructed his surviving administrators to keep his German widow from his holdings. Intervening on behalf of the widow was Albert, king of Mecklenburg, who promptly appropriated Swedish properties for himself, an act that compelled the Swedish Council to seek the help of Margaret. The queen accepted the crown in Sweden in 1388, and for the first time the three Scandinavian powers (Finland was occupied by Sweden at the time) were united under one ruler. Under Margaret, the Mecklenburg army was driven from most of Sweden (although they did occupy Stockholm for nine more years).

Margaret's son and only heir, Olav, had died in infancy, but she assured the continuation of the Scandinavian union by selecting her great-nephew Erik of Pomerania as her successor. The fifteen-year-old Erik, bearing a name suitable for a king in all three nations, had his coronation at Kalmar on Trinity Sunday in 1397. It was here at Erik's inauguration that the Kalmar Union, the most sweeping pact in Scandinavian history, was drafted.

The original document of the Kalmar Union exists today in the Royal Archives in Copenhagen. Of the seventeen seals that were required to officially approve the agreement in all three participating countries, the Union is affixed with only ten; this apparent oversight has been the subject of debate since the signing of the pact. The creation of the document has been called the most intensively discussed event in Nordic history. The provisions of the Kalmar Union were broad in scope: a single king would rule the three Scandinavian provinces, which would share policies of defense and diplomacy but would retain their own judicial and administrative systems at home.

Historians are undecided whether Queen Margaret indeed sought to forge ties that would be mutually beneficial to each of the Scandinavian nations, or whether she wished to use her position to further the imperialist aims of her father, the Danish king Valdemar. Margaret sometimes implemented her plans for Scandinavia at the expense of the land-

Kalmar Castle
Photo courtesy of Swedish Travel and Tourism Council

holding Swedish noblemen and the church leaders who commanded the respect of Sweden's population; in many cases, she installed German and Danish nobles in Swedish castles. Yet she also spent a considerable amount of time in Sweden and tried not to appear overly generous to Danes living there. By all accounts, Margaret was a skillful diplomat with an impressive resolve, reigning during an era when Europe was overwhelmingly ruled by men.

Queen Margaret succumbed to the plague in 1412, at which time Erik realized the full role of the kingship to which he had been named fifteen years earlier. Erik was to grapple with the Hanseatic League in Sweden throughout his tenure, making favorable trade concessions with Denmark, which in turn fueled a Hanseatic trade embargo against Sweden. Swedish miners, whose exports were threatened by the Hansa blockade, took the country from Erik, an occasion that heralded the beginning of a new era of nationalist sentiment in Sweden—one that would not bode well for the Kalmar Union.

In 1470, Christian, Erik's successor in Denmark, brought troops to Stockholm to administer the Kalmar Union, but he was defeated by the newly elected Swedish regent Sten

Sture. Sture's victory sparked fifty years of patriotism among Swedish commoners supportive of the Sture family. In 1483 the *råd*, Sweden's aristocratic council of state, drew up the Recess of Kalmar, a piece of legislation designed to safeguard the many interests of the wealthy class, ensuring that no "base-born" men could be elevated to positions above the "good men" of the aristocracy. The terms of the Recess of Kalmar were not approved, but its tone was to affect Sweden's internal class relations for centuries.

Ascending to the regency of Sweden in 1512, Sten Sture the Younger was excommunicated by the Catholic Church and attacked by Denmark's Christian II in the names of both the church and the Kalmar Union. Sten the Younger died in battle in 1520; during the infamous Stockholm Massacre of Swedish leaders, his exhumed corpse was torched as part of the Danes' fierce display of might.

Gustav Eriksson, nephew of Sten's widow, had been taken hostage by Christian's forces, only to escape to Sweden to instigate an uprising that swiftly took the country from Christian's Danes. In June 1523, Gustav was elected Gustav I Vasa, king of Sweden, and the Kalmar Union had come to an end.

Although sections of Kalmar Castle date as far back as the eleventh century, it was Gustav Vasa who created many of the impressive Renaissance features of the building that remain today: its moat, its defense towers, its grand courtyards, and its dungeons, which lie below sea level to make escape by tunneling impossible. Kalmar Castle sits on a promontory overlooking Kalmar Sound, an ideal location for the defense of the town. With Swedish independence, Kalmar was even more strategically located than ever. It was the nearest city of significance to the Danish border, and the Kalmar Channel created a crucial passageway from the south to Stockholm. Of course, Vasa bolstered Kalmar Castle not only to create the cornerstone of Swedish defense, but also to create an impressive monument to his rule. Swedish legend has it that, during the thirteen-week celebration of his third wedding, he and his guests drank 59,700 gallons of beer behind the walls of Kalmar Castle.

Shortly into Sweden's new period of nationalism, Kalmar again became the scene of strife, this time of an internal nature. In 1525, council member Berend von Melen was at odds with Vasa, who happened to be his wife's cousin. Summoned from his castle at Kalmar to Vasa's court in Stockholm, von Melen was persuaded to turn his defiant city over to Vasa's army. But in Kalmar, von Melen's brother refused. When Berend von Melen was sent with Vasa's army to defuse the situation, he instead escaped to his castle, and later to Germany to the service of the Elector of Saxony. Gustav Vasa saw no recourse but to storm the von Melens' castle, an action that led to considerable bloodshed in Kalmar.

Yet another piece of legislation took the Kalmar name in May 1587: the Statute of Kalmar, a controversial ordinance that assured Swedish independence and security in anticipation of the coronation in Poland of Sigismund, grandson of Gustav Vasa. In September, the document was signed by Sweden's John III (son of Gustav Vasa) and by his son Sigismund; it once again reserved the authority of the *råd,* the aristocracy in Sweden, this time in the event of Sigismund's absence at his new post in Warsaw. Sigismund had been offered to Poland in the futile hope that the Poles would come to Sweden's aid against a threatening Russian nation.

In the meantime, Sigismund lost favor in his homeland, quarreling bitterly with John's brother Charles, who despised the aristocratic backers of his nephew. In 1598, Sigismund held fortresses in Kalmar, Stockholm, and Finland, but he was losing his hold on power. On May 1 through 2, 1599, seven years after Johan's death, Charles stormed Kalmar, which had been left unattended by an increasingly helpless Sigismund. In the melee, Charles ordered the execution of three senior officers of the castle at Kalmar, and their heads were hung on spikes in vengeance. The castle surrendered on May 12; Karl went on to vanquish Finland.

By 1600, Sigismund and Charles—now king of Sweden—were constantly pitted against one another. Charles's forces were no match for the mighty Polish army, but he was able to prolong the conflict due to an internal Polish rebel-

lion. Christian IV, the king of both Denmark and Norway, saw this conflict as the means by which to resurrect the Kalmar Union; he led his Danish troops into Kalmar in May 1611. The town, surprised and unprepared, fell almost immediately, and its castle surrendered two months later.

The Kalmar War lasted until the following year. As the Danish fleet pressed on and reached Stockholm's archipelago, King Charles, now aged and in poor health, proposed that he duel Christian to the death for the war's outcome. Christian vehemently declined, but he soon became disillusioned with his undertaking in Sweden. After Charles succumbed to a stroke, his seventeen-year-old son Gustavus II Adolphus successfully drove the Danes out of the city, back to the southern province of Skåne, and, by 1619, out of Sweden entirely.

Now free once more, the city of Kalmar thrived commercially. In addition to its waterways, which provided its navy, its commercial traders, and its fishermen with their livelihood, Kalmar was the home of a brisk business in firearms, as well as a manufacturing plant that produced uniforms for the Swedish army (Sweden was otherwise dependent on textile imports from western Europe.) But Kalmar was forced to rebuild once again in 1647, when a fire devastated much of the city.

It was during this period, in the mid–sixteenth century, that much of Kalmar's most impressive architecture was erected. Gustavus II Adolphus's daughter and successor, Christina, was well educated in Latin traditions. During her reign, she led Sweden into a period of architectural splendor dominated by baroque styles. Even after she abdicated in 1654, the building continued. In 1660, the well-traveled builder Nicodemus Tessin the Elder began his plans for the Cathedral at Kalmar, modeling it after the Jesuit Church at Antwerp. The structure is the foremost example of Baroque cathedrals in Sweden.

Given Sweden's longstanding peace and the displacement of shipping by air freight, Kalmar has lost much of its strategic importance. It still serves, however, as the administrative and commercial center of southeastern Småland. Many of the town's gabled homes and public buildings date to the seventeenth century, such as the governor's residence built in 1674 on the square Lilla Torget, where King Louis XVIII of France lived for a time in the early nineteenth century. The Kalmar Court House, built between 1684 and 1690, and the art museum called the Konstmuseet are other attractions, as is the Secondary School, designed by the creator of Stockholm City Hall.

After undergoing an intensive restoration program, Kalmar Castle today is one of Sweden's chief tourist attractions. Among its most magnificent rooms are the Banqueting Hall (Unionssalen), the Queen's Hall (Drottningsalen), and the Golden Hall. Erik XIV ordered the intricate decoration that graces the castle's State Chamber, with its ornate carvings and stuccowork. The castle's rococo chapel is still used for Sunday services, and its museum houses much of the finest art and artifacts Kalmar has to offer, including an exhibition of the salvaged effects of the Swedish man-of-war *Kronan,* which sank off the coast of Öland in 1676 with 800 men aboard. Raised in 1980, the ship yielded bottles of 300-year-old brandy and invaluable coins and weapons.

The longest bridge in Europe, spanning more than three and one-half miles over the Kalmar Sound, connects Kalmar to Öland, a popular resort island where the Swedish Royal Family spend their summer season. Kalmar is also within an hour's drive of some of the world's finest glass works, at Boda, Orrefors, Kosta, Strömbergshyttan, where artisans run communities in the woods of Småland centered on their workshops. These local settings reflect the tranquility of their chief hub of activity, Kalmar, a city that once weathered some of the greatest storms of northern European history.

Further Reading: Michael Robert's *The Early Vasas: A History of Sweden, 1523–1611* (Cambridge and New York: Cambridge University Press, 1968) and *Gustavus Adolphus: A History of Sweden, 1611–1632* (London and New York: Longmans, 1953) stand as definitive works. Good corroboration is provided by Irene Scobbie's *Sweden* (London: Benn, and New York: Praeger, 1972), part of the Nations of the Modern World series. For an evocative recreation of Sweden in its formative years, the books of Swedish author Vilhelm Moberg are exemplary.

—James Sullivan

Karlovy Vary (Czech Republic)

Location: Western Bohemia, Západočeský region, Czech Republic, at the confluence of the Teplá and Ohře Rivers; north/northwest of Pizen; seventy miles west of Prague.

Description: Resort town and health spa; founded by King Charles of Bohemia in 1349; site of Carlsbad Decrees, 1819.

Site Office: Kurinfo
Vřídelní Kolonáda
360 00 Karlovy Vary
Czech Republic
17 20 3569 or 17 24 097

Karlovy Vary, known in German as Karlsbad (also Carlsbad, anglicized), on the Teplá River in the Západočeský region of western Bohemia, is one of the best-known resort towns in Europe and the oldest and largest health spa in the Czech Republic. It is the site of the drawing up of the Carlsbad Decrees of 1819, designed to suppress civil liberty in Europe. Karlovy Vary hosts the International Film Festival in the summer of even years, an annual summer festival of tourist films (Tourfilm), and the Dvórák Autumn Music Festival.

Although most of Karlovy Vary's buildings are less than 150 years old, the spa town retains a distinctive Victorian character. Tall nineteenth-century houses line the streets of Old Town, and Belle Époque mansions decorate the banks of the wooded Teplá Valley. In the southern section of the town, modern buildings, a product of the "socialist functionalism" of the 1950s and 1960s, are juxtaposed with turn-of-the-century blocks. Highlights include the town's colonnades. The neo-classical, corinthian-columned Mlýnská (Mill) Collonade (later known as the Colonnade of Czechoslovak-Soviet Friendship), on the left bank of the Teplá, was built between 1871 and 1881 by Prague architect Josef Zítek. Situated on the Rusalka, Libussa, Prince Wenceslas, and Millpond springs, it is the heart of the spa town. Karlovy Vary's most famous spring, the Vřídlo, runs through the modern Vřídnelní (Spring) Colonnade, built in 1975. It was originally named Yuri Gagarin Colonnade after the world's first astronaut, who visited the spa in 1961 and 1966. It replaced the original structure built by the Viennese architects Helmer & Fellner. The white and grey wrought iron-trimmed Savodá (Park) Collonade, also by Helmer and Fellner, was built between 1881 and 1882. The following year, the Viennese duo completed the wooden, Swiss cuckoo clock-inspired Trźní Colonnade, in the center of Karlovy Vary on Charles IV and Market springs. The current structure replaced the original bronze one, which was destroyed. Rising

about the colonnade is Zámecká Věž (Castle Tower), built in 1608 on the oldest site in town. The colonnade closed for renovation in the early 1990s.

Another Helmer and Fellner design is the white, wedding cake-style Grandhotel Moskva, which reverted to its original name, Grand Hotel Pupp, after the 1989 revolution. Named for its founder, eighteenth-century confectioner, Johann Georg Pupp, the hotel is the longest-running establishment in Karlovy Vary. Founded in 1793 and rebuilt in the late nineteenth century, the 500-bed hotel is almost 300 years old with wings added some fifty years ago. Also notable is the twin-towered Church of St. Mary Magdalene, designed by Kilian Dientzenhofer in the 1730s. It is considered a Counter-Reformation masterpiece and one of the best of Karlovy Vary's few baroque buildings.

Today the town is essentially neo-classical. There was little reinvestment during the years of Communist control, and many buildings of Karlovy Vary are collapsing about its crumbling streets. However, the Czech Republic's Velvet Revolution of 1989 inspired reconstruction of some of the town's buildings, beginning in the early 1990s.

The area may have been known to the Romans, but King Charles of Bohemia (later Holy Roman Emperor Charles IV) is credited with the development of the spa town. According to legend, in 1347 Charles stumbled upon the Vřídlo (Czech for the "bubbly one"), Karlovy Vary's hottest (163 degrees Fahrenheit) spring, while on a royal hunting expedition near his castle, Loket. It is believed that the king's hunting dog fell into the hot spring while chasing a stag and was scalded. Charles investigated the situation, and, in consultation with his personal physician, declared that the spring, which gushes a forty-seven-foot geyser of hot water, possessed healing qualities and ordered baths to be established at the site. Two years later, Charles founded a settlement there, and in 1358 he built a hunting lodge at the site where the Trżní Colonnade now stands and named the town Karlsbad, German for "Charles's spa." (Karlovy Vary is Czech for "Charles's boil.") A dozen years later, the emperor granted the town its municipal charter. For the next two centuries, the town had only about 100 buildings.

Many medieval rulers, princes, clerics, and noblemen visited the spa, and by the sixteenth century Karlovy Vary had become a famous watering place. During the period of Swedish occupation in the Thirty Years War (1618 to 1648), Karlovy Vary was devastated by fires and war-related damage. By the latter part of the seventeenth century, under the patronage of the Habsburgs, word spread of the town's curative waters and Karlovy Vary's reputation grew internationally. A bathing establishment was built in 1704. Three years later, Karlovy Vary became a royal free town. After a fire in

A view of Karlovy Vary
Photo courtesy of Czech and Slovak Service Center

1759, many of the town's buildings were reconstructed in the baroque style.

Three years later, the first sizable bath house was set up in Karlovy Vary. The most popular treatment at this time was soaking in the hot water for two days and two nights until the skin was sore. But by 1766, a drinking cure of fifty to seventy cups per day of the hot sulphurous alkaline liquid was the preferred treatment. (It was also believed at the time that

drinking up to 500 cups per day would cure the "disease of poverty.") In 1789, a local physician, Dr. David Becher, carried out the first scientific analysis of Karlovy's spring waters and declared the drinking cure excessive. Instead, Becher prescribed his own drinking cure, an herbal liqueur called Becherovka, after himself, which is available today at bars and restaurants throughout Karlovy Vary. Although the brew contains no water from any of Karlovy Vary's twelve springs, it has been affectionately labeled Karlovy Vary's "thirteenth spring," and is jokingly proclaimed to be the only one that really works. Today, the spring water is recommended in smaller doses of five to seven cups per day.

Karlovy Vary's springs, which were formed 500 million years ago during a geological shift, are forced up from hot subterranean granite rocks more than a mile below the surface. There are actually about sixty springs in Karlovy Vary but only twelve are used for treatment. These twelve high-pressure (660 gallons per minute) alkaline-saline hot springs range in temperature from 109 to 163 degrees Fahrenheit and contain forty chemical elements that have been used in the medical treatment of metabolic disorders, diarrhea, constipation and other digestive complaints, liver and gall bladder ailments, gastric and duodenal ulcers, urinary tract disorders, diabetes, infectious hepatitis, and skin disorders. Besides drinking and bathing in the thermal waters, spa treatments have included exercise, walking, swimming, diets, acupuncture, pneumopuncture, electromassage therapy, carbon dioxide baths, and gas inhalation and injections.

The watering place maintained its popularity throughout the eighteenth century and reached its peak in the nineteenth century, when its chief baths, sanatoria, colonnades, and hotels were built. In its heyday, the spa was frequented by Europe's royal, rich, famous, and fashionable, and its guest book, dated back to 1569, boasts visits from Beethoven, Bismarck, Brahms, Chopin, Liszt, Paganini, Janáček, Dvórák, Empress Maria Theresa of Austria, Czar Peter I the Great, Frederick I of Prussia, Leibnitz, Prussian Field Marshal General Blücher, Augustus of Poland, Albrecht von Wallenstein, Kurt Waldheim, Haile Selassie, Francis pbxJoseph, Friedrich von Schiller, Aleksandr Pushkin, Bedřich Smetana, Tolstoy, Gogol, Goethe (thirteen times), Adolf Hitler, and Karl Marx, who is said to have written several chapters of *Das Kapital* during spa visits between 1874 and 1876.

In 1819, in the aftermath of the French Revolution and the Napoleonic Wars, the Congress of Vienna was organized. During this year, Austria's chancellor, Prince von Metternich, invited ministers of the major German states—Austria, Prussia, Bavaria, Saxony, Mecklenburg, Hanover, Württemberg, Nassau, Baden, Saxe-Weimar-Eisenach, and electoral Hesse—to Karlovy Vary to join in the creation of an agreement to repress civil liberty throughout Europe. The conference, which took place from August 6 to 31, 1819, was precipitated by the murder that year of German dramatist August Kotzebue by radical student organization member Karl Sand. Metternich capitalized on the concern over the climate of revolutionary outrage caused by this event to persuade the German governments to unite for the suppression of nationalism and liberalism within their states. Metternich's reactionary measures called for uniform censorship by the Diet of the German Confederation (Bund) of all periodical publications; the disbandment of recently formed nationalist student clubs (Burschenschaften); the placement of school

and university faculties under supervisory curators; and the establishment of a central investigating commission at Mainz to track down conspiratorial organizations. The repressive Carlsbad Decrees, as they were known, were agreed upon by the representatives of the German states on September 20, 1819. Over the next decade the decrees were enforced to varying degrees in the German states, and liberal political activities were temporarily curtailed. Although the decrees ultimately failed to suppress liberal development or stifle nationalism in the German states, they contributed in large part to the tension that led to the European upheaval of 1848.

A railway opened in Karlovy Vary during 1870 and 1871, which facilitated the flow of more visitors to the town and led to a flurry of art nouveau construction. Karlovy Vary's original colonnades were completed during this period. The town continued to attract 50,000 to 60,000 wealthy and aristocratic visitors a year until World War I. At war's end, Karlovy Vary became part of the new nation of Czechoslovakia.

On April 23 and 24, 1938, Sudeten-German politician Konrad Henlein made a speech at an SdP (Sudeten German Party) rally at Karlovy Vary in which he asserted that all territory owned by Germany in 1918 belonged to it by right and demanded that the fledgling state surrender its defenses. Henlein made eight specific demands to be fulfilled by Czechoslovakia before he would discuss peace. Some European leaders believed that the demands were mock negotiations designed to make a peace agreement impossible. Indeed, a month earlier, Hitler had ordered that "demands should be made which were unacceptable to the Czech Government." Hitler's Nürnberg speech on September 12, 1938, triggered rebellion in Czechoslovakia's German areas near the German border in western Bohemia. By the next day unrest had penetrated into the country as far as Karlovy Vary and martial law was ordered in some western Bohemian districts.

In 1938, after the Munich Pact, Karlovy Vary was taken by the Nazis with the Sudetenland, and its Czech population left the town. The following year, the Nazis dismantled the Vřídnelní Colonnade and melted down the copper roof and the floor for armaments. After World War II, Karlovy Vary was restored to Czechoslovakia and Czechs moved back into the city as its German inhabitants were transferred back to Germany in 1945. The city's sanatoria and hotels gradually reopened and that year medical facilities affiliated with Prague University were established in Karlovy Vary.

Over the next two decades, Czechoslovakia continued to struggle for political reform and liberalization. In April of 1967, East German State Council Chairman Walter Ulbricht visited the spa town to attend the conference of European Communist Parties. Nearly a year later, the Warsaw Pact conference was held in Dresden. In 1968, Warsaw Pact troops seized Czechoslovakia. In July, *Pravda,* the Soviet Party newspaper, charged that U.S.-made arms and a secret "subversive" American manual on preparation methods for the liberation of East Germany and Czechoslovakia were discovered near Karlovy Vary.

Ulbricht returned to Carlsbad in August 1968 for a roundtable with Czechoslovak leaders and declared "significant successes in the economic development of the country," even though Ulbricht himself opposed Czechoslovak liberalization. The following year, Karlovy Vary was among several areas in Czechoslovakia in which student protest demonstrations and hunger strikes for political reform took place.

For another two decades, Czechoslovakia struggled for political and economic reform. In November 1989, the "Velvet Revolution" brought a transformation of society and democratization of life to the country. The government-controlled economic system was abandoned, opening the door for the adoption of market-economy principles.

Since the Velvet Revolution, the spa towns of western Bohemia have become even more popular with visitors for their nostalgic atmosphere of the vanished era of the Austrian empire. Karlovy Vary, as the oldest of the Bohemian spa towns, is still the most popular, and its old German name, Karlsbad—as it is known outside the Czech Republic—particularly evokes images of an aristocratic Europe of the past. Each year Karlovy Vary's visitors double the local popula-

tion, and thousands more, especially neighboring Germans, come to the town in the summer.

Despite Karlovy Vary's popularity, the environmental health of the spa town is currently in jeopardy. Expansion of open-cast brown coal mining in north Bohemia has produced a "moonscape" of industrial air pollution, which severely threatens Karlovy Vary.

Further Reading: The Blue Guide's *Czechoslovakia* by Michael Jacobs (London A & C Black, and New York: Norton, 1992) provides notes on Karlovy Vary's history in addition to its usual travel information. Insight Guide's *Czech and Slovak Republics* (Boston: Houghton Mifflin, 1993), edited by Alfred Horn, recounts several lively tales from the town's early days. The Economic Commission for Europe's Paper, *Economics and Environment in the former Soviet Union and Czechoslovakia* (New York: United Nations, 1993) by Viktor Ivanovich Danilyan et al., edited by Aleksandar M. Vacić, examines the post-Velvet Revolution economic and environmental situation and describes the pollution problem affecting north Bohemia, including Karlovy Vary. *Czechoslovakia: Crisis in World Communism* (New York: Facts on File, 1972), edited by Vojtech Mastny, details the 1968 struggle for political reform in Czechoslovakia, while *Czechoslovakia Before Munich: The German Minority Problem and British Appeasement Policy* (London and New York: Cambridge University Press, 1973), by J. W. Bruegel, highlights events of 1938.

—Shawn Brennan

Kaunas (Kaunas Region, Lithuania)

Location: About sixty miles west of the Lithuanian capital of Vilnius, at the junction of Lithuania's two largest rivers, the Nemunas and the Neris.

Description: Lithuania's second-largest city; the city grew around Kaunas Castle, constructed in the eleventh century; an industrial and commercial center.

Site Office: Tourist Information Bureau
Laisves al. 88
3000 Kaunas
Lithuania
(7) 20 49 11

As the second-largest city in Lithuania, Kaunas has played a major role in the country's history since the Middle Ages. First mentioned in a chronicle of the Crusades in 1361, the site of Kaunas was inhabited as far back as several centuries B.C. It is Lithuania's industrial center, an important trading center, and served as the nation's provisional capital during the 1920s, when the capital city of Vilnius was taken by the Poles.

Archaeological remains of a prehistoric settlement have been discovered at Kaunas. These remains indicate that the area was first settled by a people known as the Proto-Balts, the ancestors of contemporary Lithuanians, Latvians, and other Baltic coast peoples. Fragments of Proto-Balt ceramic bowls and goblets—decorated with corded impressions, rows of indentations, or bands of slanted lines—have been found in Kaunas, along with stone battleaxes, hoes, flint broadaxes, arrowheads, and small tools.

The Proto-Balts built their settlements along waterways. In addition to the Kaunas site, located at the junction of the Nemunas and Neris Rivers, Proto-Balt settlements have been found on the Baltic seacoast at Gdańsk, Poland, and Sventaji, Lithuania, and along the Nemunas and Merkys Rivers. The Proto-Balts were farmers and fishermen who practiced a form of nature cult in which the stars, sun, and other natural objects were worshiped. They also practiced ancestor worship. These pagan religious beliefs were still practiced by Lithuanians into the Middle Ages, bringing them into open conflict with the crusading Teutonic Knights. The Lithuanians were the last European people to adopt Christianity.

The modern city of Kaunas grew around a castle built by a Lithuanian leader named Kaunas in the eleventh century on the peninsula formed by the Nemunas and Neris Rivers. Kaunas Castle's twenty-seven-foot high walls were ten and one-half feet thick with defensive towers in the four corners. Located on the eastern bank of the Neris River, the castle was a focal point for much of the military conflict between Lithuania and the Teutonic Knights of neighboring Prussia. The knights wished to convert Lithuanians and other pagan peoples in the Baltic region to Christianity by force.

In 1362 the Teutonic Knights and their allies from western Europe captured and plundered Kaunas. In retaliation for this attack, Lithuanian Duke Kęstutis launched raids into Prussia, the Teutonic Knights' land, but was ambushed and taken prisoner. Because of his noble status, Kęstutis was treated as a guest rather than a prisoner and was even allowed several private meetings with the Teutonic Grand Master Winrich. The kind treatment did not sway Kęstutis's loyalties, however, and in the fall of 1362 he made his escape and resumed attacks on Prussia. Attempting to find him, the Teutonic Knights and their allies from neighboring Livonia again besieged and captured Kaunas Castle, finding not Kęstutis but one of his sons. The castle changed hands several times over the years of struggle against the Teutonic Knights. The Lithuanians, believing they could not defend the castle against further attacks, demolished most of it in 1394.

In the fourteenth century Vytautas the Great made his residence near Kaunas during his decades-long fight against the knights and his rival for political power, the Polish king Jagiełło. About 1399, according to legend, he constructed the Church of Vytautas on the Nemunas River shoreline to commemorate his appointment by Jagiełło as vice regent of Lithuania. Another version of the legend claims that he built the church as a thanksgiving for having survived a ferocious battle with the Tartars. The Church of Vytautas is one of the city's oldest brick buildings. It houses the seventeenth-century painting *Christ Crucified* by Pranciskus Smuglevucius, a prominent Lithuanian painter, and a 1930 bronze medal of Vytautas by sculptor Petras Rimša. The church is still used today for marriage ceremonies.

The Basilica of Kaunas was begun in the early fifteenth century during the reign of Vytautas the Great. Reconstruction occurred several times over the next few centuries, with the latest reconstruction in the early part of this century. The basilica is the largest Gothic building in all of Lithuania.

In 1408 Kaunas acquired the Magdeburg charter, guaranteeing the city self-rule. During the fifteenth and sixteenth centuries the city became a primary trading center for Lithuania, Poland, and Russia. The Hanseatic League established an office at Kaunas as well. By the sixteenth century Kaunas was Lithuania's second most prosperous city, just behind the capital city of Vilnius.

Kaunas's rise as a commercial center paralleled its decline as a military outpost. With the defeat of the Teutonic Knights by a joint Lithuanian and Polish army in 1410 at the Battle of Grünwald (also called the Battle of Tannenberg), outside threats to the independence and safety of Lithuania

The Church of Vytautas, top, and Kaunas Castle, bottom
Photos courtesy of City of Kaunas

were less severe. Kaunas Castle was no longer vital to the security of the region, and the remains of the castle decayed over the next several centuries.

Kaunas's prosperity was undermined by wars and a series of other problems, beginning in the seventeenth century and continuing until the early twentieth century. The city was the scene of fighting in the Thirty Years War, the Seven Years War, and the Napoleonic Wars, and suffered disastrous fires in 1731 and 1800. Also, Lithuania, bound to Poland by a common ruler since the fourteenth century, was more formally united with that country by the Union of Lublin in 1569, and thus was involved in the partitioning of Poland by its more powerful neighbors in the eighteenth century. In the third such partition, in 1795, Kaunas was annexed by Russia. The city was involved in the unsuccessful rebellions by the Poles against Russian rule in 1831 and 1863. The Russians, for their part, attempted to make Kaunas into a western defensive outpost against possible German aggression. This strategy failed during World War I, when the country was occupied by German troops. With the chaos of the Russian Revolution of 1917 and the subsequent Civil War, the Lithuanian capital of Vilnius became part of an independent Poland, and Kaunas was made the provisional capital of a newly independent Lithuania in 1920.

The period of Lithuanian independence saw Kaunas undergo a remarkable transformation. The city became, Albertus Gerutis writes in *Lithuania: 700 Years,* "the educational and cultural center of Lithuania." Educational facilities were greatly expanded. Kaunas University was established in 1922 and renamed Vytautas University in 1930, on the 500th anniversary of Vytautas the Great's death. In Vilijampolė, a Kaunas suburb, an Academy of Veterinary Sciences was established in 1936. The Kaunas Art School, a conservatory, the Lithuanian Catholic Academy, and other educational institutions were founded during the 1920s in Kaunas.

The arts also flourished during this time. In 1924, Kaunas held the first Lithuanian song festival since the festivals were prohibited by czarist decree in 1904. Some eighty choirs participated in the first such festival; later festivals drew even more participants. In 1933, a women's song festival was held, drawing 3,000 participants from all over Lithuania. A state opera was founded in 1920 by Kipras Petrauskas, with performances not only in Kaunas but in a number of provincial centers as well. A symphonic orchestra broadcast performances from Kaunas on a radio station established in 1926.

Museums were also opened in Kaunas at this time. The Čiurlionis Gallery was established to house the works of noted Lithuanian painter Mykolas Čiurlionis. The Kaunas Military Museum opened in 1921. A Liberty Bell, donated by Lithuanian-Americans, was installed in the museum's tower in 1922.

A building boom was spurred by the construction of a number of new government buildings in the city. The Central Post Office and the Palace of Physical Culture were constructed in 1932, the Pienocentras Dairy Marketing Board and the Vytautas the Great Cultural Museum in 1934, the Chamber of Commerce and Industry in 1938, and the Officers Club in 1939.

Independence for Lithuania lasted only until 1940, when Soviet troops conquered the country during World War II. Soviet occupation was resisted by many Lithuanians, and Soviet measures to make the country into a Communist state were harsh. Religious teaching was suppressed. The theology department at Kaunas University was shut down soon after the Soviet occupation. The Kaunas seminary was turned into Red Army barracks, the Church of the Garrison became a museum, and two Lutheran churches were made into warehouses.

In 1941, when Nazi Germany invaded the Soviet Union and the Baltic States, the soldiers' arrival in Lithuania was at first greeted as a liberation. On June 22, 1941, when German troops entered Lithuania from the west, Lithuanian insurgents captured the radio station in Kaunas and broadcast an announcement of national independence. It was assumed by many Lithuanians, and by many people in other Baltic countries, that the Germans would restore their independence once the Soviet Army was forced out. Kaunas became a center for anti-Soviet activity. Of the estimated 2,000 casualties suffered by Lithuanian insurgents in their fight against the Soviet Army, at least 200 were from Kaunas. The provisional government that had ruled Lithuania before 1940 reestablished itself in Kaunas on June 25, 1941. Six weeks later, under intense pressure from the Nazis, who had no intention of allowing an independent Lithuania, the government disbanded.

During the German occupation, a Nazi task force of 1,000 men was assigned to locate and execute Jews in the Baltic Countries. Kaunas was the first center of their operations. Beginning June 28, 1941, this task force rounded up and executed Jews in Kaunas and surrounding areas. By July, a confidential German report claimed that 7,000 Jews had already been killed in Kaunas. The Ninth Fort, a fortress built in 1882 during czarist rule, became under Nazi occupation an execution site for Jews in Kaunas. Some 80,000 people in all were executed at the site.

Lithuania was reconquered by the Soviets in the summer of 1944. Kaunas was evacuated by the German Army by August 1. The removal of the German Army left the city in disorder. A wave of rapes and looting followed in the lawless environment until Soviet troops restored order once more. One of their first actions in Kaunas was the execution of at least 400 alleged criminals. Other Lithuanian cities saw similar mass executions.

Opposition to Soviet rule was widespread throughout the Baltic States. Kaunas saw a number of protests during the fifty years of Soviet rule. The most dramatic of these occurred on May 14, 1972, when a nineteen-year-old student named Romas Kalanta stood before the theater in Kaunas where the Lithuanian People's Assembly had approved the country's incorporation into the Soviet Union. Kalanta poured gasoline over his body, lit a match, and set himself on fire in protest of the loss of his nation's independence. The suicide provoked immediate reaction from students in the city. Several thousand of them took to the streets, fighting the police and shouting "Freedom for Lithuania!" Local police had to call in help from Soviet paratroopers and KGB units to quell the disturbances. Three other self-immolations, inspired by Kalanta, took place elsewhere in Lithuania within days.

In March 1990, Lithuania was the first of the Baltic States to declare its independence from the Soviet Union. Soviet president Mikhail Gorbachev fought this move by cutting off the nation's oil and other supplies. By that summer, the Soviets had agreed to negotiate the demand. But in January 1991, Soviet tanks rolled into Vilnius to crush the independence movement, seizing the radio and television stations and killing fourteen Lithuanian citizens. In August of that year, when a military coup against Gorbachev failed and forced him to look for help from reformist elements within his party, Lithuania again declared its independence. This time Gorbachev was too weak to reimpose Communist rule. By September 1991, Lithuania had been recognized as an independent nation for the first time in more than fifty years.

Kaunas today has a population of about 430,000 people and enjoys a status as Lithuania's leading commercial and industrial center. Trade between other Baltic and Eastern European countries flows through the city either by river or via an extensive system of railroads and highways. Little damaged by the fighting of World War II, Kaunas's Old Town area contains preserved examples of architecture dating back to the fourteenth century. Historic buildings include the Church of Vytautas, the cathedral, and Kaunas Castle, which was restored in the mid-twentieth century.

Further Reading: *Lithuania: 700 Years* (New York: Manyland, 1969), edited by Albertas Gerutis, is the standard history of Lithuania. *Lithuanian Towns: Past and Future* (Vilnius: Gintaras, 1970) by A. Spelskis is a more specialized source on the architectural history of the country but is written from a pro-Soviet viewpoint. *A Guide to the Baltic States* (Merrifield, Virginia: Inroads, 1990), edited by Ingrida Kalnins, provides a detailed guide to the historic sites to be found in Kaunas. *The Livonian Crusade* (Washington, D.C.: University Press of America, 1981) by William Urban gives an account of Kaunas in the Middle Ages during the battles with the Teutonic Knights, while *The Baltic States: Years of Dependence, 1940–1980* (Berkeley: University of California Press, and London: Hurst, 1983) by Romuald J. Misiunas and Rein Taagepera tells the history of Lithuania under Communist rule. *Lithuania* (Minneapolis: Lerner, 1992) contains recent information about the country since it regained independence.

—Thomas Wiloch

Kiev (Ukraine)

Location: On the banks of the Dnieper River in north-central Ukraine, 470 miles southwest of Moscow.

Description: Ancient city, now the capital of independent Ukraine. During the medieval age, Kiev was the capital of Kievan Rus, the Eastern Slavic state. After Kievan Rus's decline, the city was controlled successively by the Mongols, Lithuania, Poland, and the Russian Empire. In the early twentieth century, Kiev joined the Soviet Union as the capital of the Ukrainian Soviet Socialist Republic, which declared its independence in 1991.

Site Office: General Agency on Tourism
12 Hospitalna Street
Kiev, Ukraine
(44) 225 30 51

Kiev, the ancient city built on the sloping banks of the Dnieper River, has a turbulent history of independence and invasion, self-government and foreign control. Since its early prominence as the capital of Kievan Rus, the medieval Slavic state, Kiev has been subject to a Mongol invasion; Lithuanian, Polish, and Russian rule; a German occupation during World War II; and incorporation into the Union of Soviet Socialist Republics for most of the twentieth century, until regaining independence in 1991.

Kiev's urban landscape bears witness to centuries of violent political struggle. The domes of ancient churches and monasteries still crown some of Kiev's steep hills, but as much as the city is known for these, it is also known for the architectural treasures it has lost, those destroyed during invasions, occupations, civil unrest, and the transition to Communist rule. The city embodies the destruction, reorganization, and reconstruction such changes bring.

Kiev's mythic origins were recorded in the *Primary Chronicle,* an ancient document written by the monk Nestor at Kiev's Monastery of the Caves in the eleventh and twelfth centuries. According to Nestor, three brothers from a Slavic tribe settled on the steep hills of the Dnieper's west bank. The town they founded, Kiev, was named for the eldest brother, Kii. Although the roots of settlement at Kiev are undoubtedly ancient, scholars have not established Kiev's actual origin conclusively. Dendrochronological analysis of wooden remains in Kiev's Podil district date settlement there only to 887, although other evidence suggests settlement as early as the fifth century.

Whatever the date of first settlement, by the ninth century Kiev had grown into a center for international trade on the routes between Europe, the Christian Byzantine Empire, Baghdad, and the Khazar state of the lower Volga and

northern Caucasus regions. Kiev became the capital of Kievan Rus, an Eastern Slavic state credited as the birthplace of Russian civilization. The largest organized political state in medieval Europe, Kievan Rus extended throughout most of modern Ukraine and Belarus as well as parts of northwestern Russia, including present-day St. Petersburg and Moscow. Scholars attribute the founding of Kievan Rus to the Varangians, warrior-merchants known in the west as Vikings. Evidence suggests that a complex interaction between the native Slavic tribes and the Germanic-Scandinavian Varangians who gained control of the region led to the establishment of a strong Slavic state.

Kievan Rus flourished from the ninth through the twelfth centuries. After Varangian warrior Rurik founded a dynasty in 862, a succession of Varangian-Slavic princes acquired more territory and established a more unified state. Sviatoslav, who reigned from 962 to 972, united all East Slavic peoples under Kievan Rus by gaining control of the Volga River. Vladimir the Great, whose long reign lasted from 980 to 1015, initiated a change that further unified his peoples: he converted to Christianity. His subjects accepted the religion, and many of Kiev's residents were baptized in a mass ceremony in the Dnieper in 988. Christianity came to Kievan Rus by way of Byzantium, not Rome. The subsequent split between Roman Catholicism and Eastern Orthodoxy thus led to the historic conflicts between Kiev and its Roman Catholic neighbor Poland.

Vladimir's son, Yaroslav the Wise, presided over Kievan Rus during a period of peace and great prosperity in the early to mid–eleventh century. Two of Kiev's most important architectural treasures, the St. Sophia Cathedral and the Monastery of the Caves, were built during his reign. Vladimir's death, however, signaled the beginning of Kievan Rus's decline. Important trade with Byzantium and the Abassid Caliphate was lost after attacks on those regions by crusaders and Mongols, respectively. Kievan Rus's own base of power was reduced when more than twelve outlying Slavic principalities (including those at Novgorod and Vladimir-Suzdal) rebelled and gained independence in the twelfth century. Mongols invaded Kiev and destroyed the city in 1240, bringing Kievan Rus to a violent end.

Kiev continued to pay tribute to the Mongols until the fourteenth century, when the city came under the control of Lithuania. Nevertheless, Kiev remained a target of the Tatar raiders, who attacked the city periodically through the seventeenth century. Lithuanian rule, however, hastened Kiev's economic recovery by granting Kievan burghers unrestricted trade in the Grand Principality of Lithuania. As a result, merchants from across Europe, the Middle East, and Asia congregated in Kiev's Podil district to trade silks, rugs, spices, wool, iron, furs, salted fish, and more. Craftsmen in

Protective pavilion at Kiev's Golden Gate
Photo courtesy of Mystetsvo Publishers

Kiev profited by selling locally manufactured weaponry such as rifles, bows, and arrows.

In 1494 Podil's burghers even gained the right to govern themselves as a semiautonomous body under an elected burgher magistracy. These freedoms were part of the Magdeburg Rights, which also granted the burghers tax exemptions and judicial authority. These rights were upheld through Lithuanian and Polish rule, giving Kiev's merchants a freedom that helped sustain the city economically through the turbulent political periods to come. When Ukraine joined the Russian Empire in 1654, the czars affirmed the Magdeburg Rights, which survived until Czar Nicholas I rescinded them in 1835.

Ukraine's expanses of land, coupled with the rich trade routes along the Dnieper, prompted Polish kings to begin a campaign for possession of the Ukraine in the fourteenth century. In 1385, Lithuania and Poland formed a dynastic union, and Lithuania converted to Roman Catholicism. A Catholic church was constructed in Kiev in 1433, and the Polish king ordered a ban on new Orthodox church construction in 1455. Kiev was officially brought under joint Lithuanian-Polish control in 1471. When Lithuania and Poland merged into a single state in 1569, Poland took control of the Ukrainian territory, including Kiev.

Religious divisions soon split Kiev's population. Faced with pressure from Poland's Roman Catholic ruler, the vast majority of Ukrainian Orthodox bishops established the Uniate Roman Catholic Church at the synod of Brest-Litovsk in 1596. In this compromise decision supported by Poland, the bishops accepted the full primacy of the Holy See in doctrine but did not give up Orthodox ritual or Slavonic liturgical language in practice. Although some Kievans—generally the lower classes—retained the Orthodox faith, many of the Ukrainian elite converted to the Uniate church for fear of being labeled "backward" by the ruling Poles. Many Orthodox monasteries in Kiev were controlled by Uniate authorities, although one of the oldest and most significant, the Monastery of the Caves, remained firmly Orthodox. In 1615 Kiev's Bogoiavlensk Monastery founded the Kiev Orthodox Brotherhood, an organization opposed to Polish rule.

The Cossack army entered the conflict over Kiev in the seventeenth century. The Cossacks, an organized band of free men (most from Ukraine and including former serfs and slaves), governed themselves from a region south of Kiev called the Zaporozhian Sich. In the early seventeenth century, the Cossack Army joined the Kiev Orthodox Brotherhood to fight for the Orthodox population against the Catholics and Uniates. Cossack hetman (leader) Bohdan Khmelnytsky brought the struggle to a climax in 1648, when he led a major rebellion that forced most Poles out of Ukraine and left Kiev's Catholic monasteries in ruins. Polish landowners, Catholic and Uniate priests, as well as scores of Jews, were murdered. After victorious battles in Poland, Khmelnytsky returned to Kiev in 1649, where he was greeted as the city's heroic liberator.

The Poles, aided by the Lithuanian army, exacted revenge on Kiev in 1651, when they regained the city and caused extensive damage. The besieged Cossacks looked to Russia for protection. Khmelnytsky took an oath of loyalty to Russian Czar Alexis I Mikhaylovich in 1654, submitting the Ukraine to his control. Russia declared war on Poland, and the conflict raged until 1667, when Russia and Poland agreed to split the Ukraine. Poland gained the "right bank" (western Ukraine), and Russia controlled the "left bank" (eastern Ukraine) and Kiev. After formal annexation in 1686, the Ukraine became the Russian Empire's "Little Russia."

Despite the presence and power of the Cossacks, Kiev began to lose some of its self-governing rights to the Russian Empire. In 1654 Russian troops quartered in Kiev assumed the responsibility of defending the city, historically the duty of the burgher magistracy. Catherine the Great abolished the Cossack Hetmanate and appointed her own government, the Little Russian College, in 1764. A new town charter in 1785 granted governing power and management of the police to Kiev's governor, a Russian appointee. Catherine also took control of the Ukrainian Orthodox Church. By the end of the eighteenth century, Kievan autonomy was in jeopardy.

The Magdeburg Rights of self-government granted to Kiev's burghers by the Lithuanians were finally revoked by Czar Nicholas I in 1835. This action signified a broader effort by the Russian state to make Kiev—whose diverse population included Lithuanians, Poles, Jews, Armenians, Tatars, Germans, and Turks—a "Russian" city. Russian became the language of administration and education and, as such, the key to upward mobility. In 1834 Nicholas established St. Vladimir University, a Russian university in curriculum as well as language, in Kiev.

Such changes were seen as attacks on Ukrainian language, traditions, and identity. The Kievan impulse to fight for self-government resurfaced, and the embryonic Ukrainian national movement began. In 1846, about thirty Ukrainians founded the secret Brotherhood of Saints Cyril and Methodius at St. Vladimir University. Former serf Taras Shevchenko, an artist and poet revered as one of Ukraine's national heroes, was a founding member of the group. His poem "The Dream" denounces Peter the Great for "crucifying" the Ukraine and Catherine the Great for "finishing off his victim." When Russian authorities learned of the group, its members were arrested and deported to distant parts of Russia. This repression of a Ukrainian national organization led to further actions against Ukrainian intellectuals. Czar Alexander II's decrees of 1863 and 1876 severely limited the use of the Ukrainian language, making it illegal in schools and publications. As the twentieth century approached, the groups championing Ukrainian nationalism remained small and largely uninfluential.

The growing labor movement, however, sparked Kievans' interest more successfully. As early as 1879, protestors denounced workers' low wages, long hours, and poor working conditions at the Main Railway Shops (the largest commercial enterprise in the city) and in the Podil commercial district. The first strike in Kiev occurred that year, when workers at the Southwestern Railway's Main Shops walked out and were consequently fired. In the summer of 1903, workers' strikes that began in Odessa and Baku spread to Kiev. Workers struck at the railway shops, machine shops, print shops, and bakeries, among others. Violence erupted when Russian troops fired on an apparently unarmed crowd. The bloodshed drew more strikes in protest, and a confrontation in the Podil caused more casualties. Kievans' dissatisfaction with the Russian government grew.

The "Bloody Sunday" massacre of civilian protesters in St. Petersburg on January 9, 1905, prompted angry demonstrations by citizens across the Russian Empire. In Kiev, strikes lasted for several days and remained peaceful. The root cause of these actions against authority was economic rather than political. Wage hikes, shorter work days, and state-provided insurance were among workers' demands. Calls for political change emerged later that year. When the imperial government returned control of the Kievan universities to faculties in 1905, St. Vladimir University became the site of open "people's meetings." As workers' attendance grew, discussions became more political. Some Ukrainians used the forum to call for an independent Ukrainian state.

The massive Russian railway strike of October 1905 undermined the authority of the imperial government and prompted Bolshevik cries for a general strike in Kiev. Southwestern Railway workers struck in mid-October, demanding a parliament elected by universal and direct suffrage. The city was soon virtually shut down by a short general strike. This civic unity of purpose was rewarded by news of Nicholas II's October Manifesto, which declared the institution of civil liberties (freedom of speech, conscience, and assembly) and the creation of a lawful parliamentary system. The use of Ukrainian in books and journals was again permitted.

Shortly after the decree, Kievan socialists formed a local soviet with delegates from the city's major industrial plants. A month after the October Manifesto, however, civil unrest and military mutinies sparked by socialist agitation caused the imperial government to declare martial law in Kiev. The promise of the October Manifesto ended abruptly. The revolutionary movement crumbled, and the Kiev Soviet, weakened by arrests, lost its authority. The imperial government remained in control in the coming years, despite occasional strikes.

The Russian Revolution of 1917 toppled the czarist government. As Communists came to power across the former empire, the Ukrainian nationalist movement gained strength. In April 1917, a National Ukrainian Congress gathered in Kiev, where it elected a Rada (council) that declared the birth of an autonomous Ukrainian nation two months later. In November, the Rada convened a democratically elected Ukrainian Constituent Assembly. Asserting its power in the Ukraine, Russia's Communist government created a Ukrainian Soviet seated in Kharkov, an industrial Ukrainian city with strong Bolshevik support. The nationalist Rada fol-

lowed this action by reaffirming Ukraine's status as a "free and sovereign" republic.

From 1918 to 1920, Kiev was beset by violence and pogroms. Ukrainian nationalists fought the Communist Red Army, as well as Polish, German, and Austrian forces, for control of the region. By December 1919, the Red Army occupied Kiev and most of the Ukraine. The treaty of alliance devised by Vladimir Lenin formally incorporated the Ukraine into Communist Russia in 1920. Ukraine joined the Union of Soviet Socialist Republics (USSR) in 1922. Kiev, site of nationalist fervor, was not returned to its traditional status as Ukraine's capital until 1934, when it replaced Kharkov. Ukraine would not exist again as an independent democratic nation until the end of the twentieth century.

Although deprived of their political autonomy, Ukrainians were assured of the primacy of their own language by Soviet decree in 1923. Five years later, however, Joseph Stalin revoked this policy and mandated Russian as the Ukraine's second official language. This action presaged the Communist government's attack on Ukraine's intellectual, cultural, and even architectural heritage.

Countless Kievans lost their lives in this conflict. Stalin's artificial famine in the early 1930s caused millions of Ukrainian deaths by starvation. Between 1936 and 1941, his political police (the NKVD) routinely murdered Kievan citizens (intellectuals, political dissidents, and others), burying many in mass graves in a forest near Bykivnia outside Kiev. As many as a quarter of a million victims may be buried there.

Kiev's new role as capital of the Ukrainian Soviet Socialist Republic prompted a campaign of architectural destruction. The Communist government in Moscow used the transfer of the capital to Kiev as an opportunity to suppress both religion and Ukrainian nationalism. Communists had moved against religion in the 1920s by converting St. Sophia Cathedral and the Monastery of the Caves, both preeminent Ukrainian institutions, to Soviet state museums. Citing plans to rebuild Kiev in model Soviet style, Russian authorities in 1934 proceeded to demolish historical and architectural landmarks, which they perceived as symbols of Ukraine's dangerous nationalist tendencies.

Stalin's wrecking ball toppled St. Michael of the Golden Domes, an ancient church (built 1108–13) that had survived Mongol assaults and Russian civil war. Scores of secular and religious buildings across the city were razed to make way for Soviet administrative complexes—some of which were never built. Kiev lost irreplaceable parts of its streetscape and history during the 1930s, a "peaceful" period as destructive to the city as the German occupation during World War II.

Nazi forces began attacking Kiev in June 1941 and succeeded in occupying the city by autumn. In late September, Nazis ordered Kiev's Jewish population to assemble with their belongings at a particular location and time. Believing this was a step toward resettlement, Jewish citizens marched to a ravine near the outskirts of town. There, at Babi Yar,

33,700 Jews were executed by Nazi firing squads. The German forces used the site as a mass execution ground throughout Kiev's occupation, leaving over 100,000 Kievan Jews and other victims buried in mass graves. A memorial now marks the site.

Kiev was finally liberated by the First Ukrainian Front on November 6, 1943. Almost 800 days of occupation had ravaged the city. Some 6,000 buildings were demolished, and Kiev's prestigious main boulevard, the Khreshchatyk, and surrounding areas suffered extensive damage. Much of Kiev's historical architecture that had survived Soviet demolition efforts in the 1930s was lost during the war.

Soviet authorities blamed the Nazi invaders for inflicting this devastation, but modern scholars argue that the government itself was responsible. As a Nazi invasion became imminent, Soviet government officials made plans to destroy any Kiev site that might prove useful to the enemy. Before the Soviet army evacuated Kiev, numerous buildings and streets were mined. Kiev's civil home guard, which also retreated, set fire to the city's power station and destroyed bridges over the Dnieper. After Nazis gained control of the city, Soviet partisans began a campaign of demolition, setting fires and triggering explosions that left the central business district in ruins.

Kiev was rebuilt and expanded in the postwar years, largely transformed into a city of "Mussolini modern" edifices where ancient architectural treasures once stood. Tragedy struck again in 1986, when the Chernobyl nuclear reactor just sixty-five miles north of the city exploded. Radioactive particles leaked into the air, soil, and water, causing long-term contamination in the region.

Political revolution came to Kiev most recently in the early 1990s. With Communism in collapse throughout the USSR, Ukrainian nationalist fervor erupted again. On this occasion, nationalism triumphed. The Ukrainian people declared their sovereignty on July 16, 1990. In early 1991, the Ukrainian parliament instituted laws protecting freedom of conscience and religion. After a Communist coup against Soviet President Mikhail Gorbachev failed in August 1991, Ukrainians voted for national independence and outlawed the Communist party. An independent Ukraine, with Kiev as its capital, was reborn.

The city itself reflected this political transition. Monuments to Lenin, erected in the early twentieth century, were now removed from prominent city squares. Many streets previously renamed in honor of Communist leaders and martyrs now reverted to their pre-Communist nomenclature or received new names appropriate to the new political environment (October Revolution Square became Independence Square).

Despite the political changes that have continually remade Kiev, today's visitor can see much in the city that has remained constant. Kiev traditionally comprised three districts: Pechersk, Old Kiev (also called High City), and Podil. These three settlements, separated by hills and ravines, remained geographically isolated from each other until the

1830s, when architectural development joined the distinct "villages" into a single city.

Pechersk, a hilltop settlement in southeastern Kiev, developed as a site of both religious and military importance. The region derived its name from the Ukrainian word *pechery,* or "cave," for the natural caves found there. In 1051 Orthodox monk Anthony and his follower Theodosius founded the Monastery of the Caves (Kievo-Pecherska Lavra) on the site. The monks lived and worshipped in underground cells carved from the caves, which also served as burial chambers. For centuries pilgrims have journeyed here to experience the underground churches and view the monks' naturally mummified remains.

The Monastery of the Caves has, like Kiev, survived through periods of prosperity and destruction. In the Middle Ages, the monastery became the center of Orthodox Christianity in Kievan Rus. The monastery also played a significant cultural role in the Slavic state. Monks Nestor and Nikon wrote the historical *Primary Chronicle* here in the Middle Ages, and, in 1615, the Ukraine's first printing house was founded here. Although much of the monastery was destroyed in the Mongol invasion, it was rebuilt, expanded, and fortified in the following centuries, particularly during the reign of Peter the Great.

The monastery now includes not only the underground chambers but also the Trinity Gate Church, the Assumption Church (Uspensky Sobor), All Saints' Church, and Refectory Church, built largely in Byzantine and baroque styles. Another monastery landmark is its Great Belfry, a bell tower built from 1731 to 1744. Its four stories, crowned by a cupola, reach 317 feet into the air, earning its former distinction as Russia's tallest building.

After the Russian Revolution, the monastery was closed by the Communist government, which converted it into the Museum of Cults and Customs. During World War II, Nazi troops occupied the Monastery of the Caves, installing anti-aircraft weaponry within its well-fortified walls. In November 1941, a bomb exploded in the Assumption Church, leaving it in ruins. Although the Soviet government blamed German forces for the incident, scholars speculate that the bomb may have been planted by Soviet troops before the invasion.

Influenced by the religious revival that accompanied perestroika, the Soviet government returned the monastery to Ukraine's Orthodox Church in 1987. Today, the Monastery of the Caves is a place of worship and site of such cultural institutions as the Museum of Historical Treasures, Museum of Ukrainian Decorative and Applied Art, and the Museum of Ukrainian Books and Printing.

Pechersk commands an excellent view of Kiev and the curving Dnieper from its location high on the hills of the river's western bank. Both the Cossacks and the czarist Russian government utilized this strategic location by transforming it into a citadel. Cossack hetman Samoilovych initiated construction of a hilltop fortress in 1679. Less than twenty years later, the Cossacks had enclosed the monastery grounds with eighteen-foot-high defensive walls. In the eighteenth and nineteenth centuries, Pechersk became an administrative and residential center for czarist Russian officials and military commanders. A newer fortress to protect czarist interests was built in the 1830s. Pechersk's importance faded during the nineteenth century, as industry and business relocated elsewhere.

Old Kiev, which stands on a steep hill northwest of Pechersk and inland from the Dnieper, thrived as the center of medieval Kievan Rus. A medieval fortress built by Prince Vladimir protected the settlement, then known as Vladimir's Town. The original moat and earthen ramparts were supplemented by new ramparts with three massive gates during Prince Yaroslav's reign. The main entry, called the Golden Gate, stood forty feet high. Built in 1037, this famous gate was covered with beaten gold and other precious metals. Legend has it that Mongol Batu Khan entered the city through the Golden Gate in 1240, after Kiev's fall to his raiders. Victorious Cossack hetman Bohdan Khmelnytsky, hailed as Kiev's liberator, returned from Poland in 1648 through the Golden Gate. Two of the gate's parallel support structures have survived to the present day, and a partial reconstruction was completed in 1982.

St. Sophia Cathedral, built in Old Kiev in 1037, is the city's oldest church. Prince Yaroslav the Wise ordered its construction to give thanks for his victory over the Pecheneg tribe, who had attacked Kiev from the east. In addition to its religious role, St. Sophia's played an important part in the political and cultural life of Kievan Rus. The veche, or popular assembly, met here to debate; Western European diplomats were received here; and here Kievan Rus's princes were crowned. St. Sophia's also housed Ukraine's first library.

Like the Monastery of the Caves, St. Sophia's suffered severe damage during the Mongol invasion of 1240. Fighting during the periods of Lithuanian and Polish rule further ravaged the cathedral's facade. Restoration efforts began in 1654, when Ukraine came under czarist Russian rule. Originally, the church featured a main cupola encircled by twelve smaller cupolas, symbolizing Christ and his twelve apostles. During Czar Peter the Great's reign, six cupolas were added to St. Sophia's already distinctive silhouette. The cathedral's ancient Byzantine exterior was also transformed by a new baroque facade. The Communist government converted St. Sophia to an architecture and history museum in the 1930s. Today St. Sophia is particularly notable for the extensive eleventh-century mosaics and frescoes that grace its interior, including *Christ the Pantokrator* and the *Orant Virgin* (also called *Indestructible Virgin*), and scenes of ordinary medieval life (hunters, musicians, jesters, and more).

Just to the northeast of St. Sophia once stood Old Kiev's other medieval religious landmark, St. Michael of the Golden Domes. The church, whose main gold-covered dome gave it its name, was built by Prince Svyatoslav II between 1108 and 1113. St. Michael's survived the Mongol invasion and subsequent political violence largely intact. While most of Kiev's Orthodox clergy and monasteries converted to the Uniate Church in the seventeenth century, St. Michael's retained its Ukrainian Orthodox doctrine and practice. Like many of Kiev's Byzantine churches, it was remodeled with a baroque exterior in the seventeenth and eighteenth centuries. In the 1930s, St. Michael's was targeted for destruction by the Communist government. The church was torn down between 1934 and 1936 to clear the site for construction of a Soviet Capital Center. This complex was never built, and the site remains vacant.

The Mongol invasion left Old Kiev in ruins; it remained largely unsettled until a new wave of development in the nineteenth century, including the establishment of St. Vladimir University in 1834, revitalized the area. In the meantime, most of the city's residents and businesses moved north to the lowlands bordering the Dnieper. This thriving commercial district, home to Kiev's river port and marketplace, is called Podil, or "Lowertown." Settlement in Podil has been dated to the ninth century, although it was probably first inhabited many centuries before. Podil was home to craftsmen, farmers, fishermen, and merchants of many nationalities: Greek, Armenian, Italian, Jewish, and Slavic, among others. The burgher magistracy that enjoyed self-rule for many centuries was based in this part of the city.

While Pechersk and Old Kiev boasted grand churches and palaces, Podil remained architecturally unsophisticated. Narrow streets of one-story thatched wooden houses characterized Podil until a three-day fire destroyed the district in 1811. Podil was reconstructed and continued to be the economic heart of Kiev well into the nineteenth century. Yet Podil did not escape the Communist wrecking ball that claimed so many Kievan landmarks. Soviet authorities ordered the destruction of the medieval Church of the Pyrohoshcha Madonna, as well as two seventeenth-century churches: the Church of Saints Peter and Paul (Petropavlivska), and the Bratskyi Monastery of the Epiphany (Bohoyavlennya).

The building of the former Kiev Academy, one of the most prestigious educational institutions in czarist Russia, was spared, as was St. Andrew's Cathedral. This baroque structure, topped by five striking green, blue, and gold cupolas, towers above Podil on Starokievskaya Hill. It was built in 1752 by noted Italian architect Bartolomeo Rastrelli and is now open as a museum.

Kiev's transformation into a more modern city began in the nineteenth century. As the city grew into a commercial, educational, and transportation center, its population increased and spread outward from the Dnieper. Within the city, development erased the ancient divisions between Pechersk, Old Kiev, and Podil. A bustling new boulevard, the Khreshchatyk, emerged in what had once been a valley connecting Pechersk to Podil's river port. During the nineteenth century, the Khreshchatyk's small wooden shacks and distilleries were replaced by prestigious hotels, shops, banks, and businesses. After most of the area was demolished in the Nazi occupation, the Khreshchatyk was rebuilt in a uniform, modern style. Today it is a cultural center, home to the Museum of Ukrainian Art, the Taras Shevchenko State Museum, the Russian Art Museum, and the Museum of Western and Oriental Art.

Further Reading: Michael F. Hamm's *Kiev: A Portrait, 1800–1917* (Princeton, New Jersey, and Chichester, West Sussex: Princeton University Press, 1993) offers an extensive, well-researched account of Kiev as it developed under czarist rule. Hamm recounts Kiev's political history and also offers a portrait of daily life in the city, with special attention to minority populations. The book also provides an overview of Kiev's ancient and more recent history. For a detailed and extensively illustrated record of Kiev's lost architectural treasures, Titus D. Hewryk's *The Lost Architecture of Kiev* (New York: Ukrainian Museum, 1982) is excellent.

—Elizabeth E. Broadrup

Kilkenny (Kilkenny, Ireland)

Location: In southeastern Ireland, in the valley of the River Nore; sixty-five miles south-southwest of Dublin.

Description: Ancient capital of the kingdom of Ossory; site of religious school founded in the sixth century; center of Anglo-Irish parliamentary maneuvering that led to the Kilkenny Statutes of 1366, which forbade interactions between Irish natives and English colonists; capital of Roman Catholic Confederacy from 1642 until Oliver Cromwell's conquest of Ireland forced the city's surrender in 1650.

Site Office: Kilkenny Tourist Office
Shee Alms House
Rose Inn Street
Kilkenny, Kilkenny
Ireland
56/21755

Kilkenny is one of Ireland's oldest towns. Prior to the Anglo-Norman invasion, it was the capital of the kingdom of Ossory. The town's Gaelic name, Kil Cainneach (Church of Canice), was given in honor of the religious seminary founded in the sixth century by St. Canice. The monastery school soon gained fame, and students traveled there from across Christendom. Even prior to the founding of the school, Kilkenny was a market town dealing in dyes and woolen goods.

Kilkenny came under English rule quite early, and the repercussions of this rule would influence the bulk of Kilkenny's history. In 1172, Norman military leader Strongbow built a simple castle on the site which, most likely, had previously been occupied by the kings of Ossory. The castle burned down in 1174, and in approximately 1192 Strongbow's son-in-law, William Marshal, contracted to have Kilkenny Castle refashioned into a stone fortress. Around the same time or shortly thereafter, work began on St. Canice's Cathedral, on the site of the original sixth-century monastery. The diocesan see of Ossory, founded by St. Canice in the town of Aghabo in the sixth century, was moved to Kilkenny in 1202. Marshal's interest in Kilkenny County heightened when he was blocked from his feudal lands in Normandy by King Philip II (Augustus) of France. In 1207, Marshal came to Ireland for the first time to settle his Kilkenny territorial possessions. In addition to the stone castle, Marshal most likely built the timber and earthen stronghold Callan Motte as well. Marshal also established Englishtown, the colonial settlement in Kilkenny that would exist separately from the native Irishtown section until 1843.

Initially, the Anglo-Norman rulers of Ireland were unable to guide their scattered kingdoms in a unified direc-

tion. From the thirteenth until the end of the fifteenth century, climatic changes that damaged crops, foreign conflicts such as the Hundred Years War, and devastating epidemics hindered any real growth in the colonial Irish economy. In the fourteenth century, for instance, the bubonic plague swept across the Irish towns, wiping out nearly one-third of the colonial population. Many of the remaining colonists began to assimilate into the Gaelic culture, intermarrying with the natives and adopting their customs. These older colonists became jealous of the holdings of newly arrived English planters and sided with the natives against them. The native Irish took advantage of the weak and divided English presence by reclaiming abandoned or poorly defended lands. Their incursions, in turn, fueled the fears and resentments among the colonists. These resentment only grew when the English Crown imposed new taxes to compensate for the colony's shortfalls in revenue.

England finally turned its attention to the problems in Ireland after the 1360 Treaty of Bretigny provided a break in the Hundred Years War. King Edward III sent his son Lionel to confront the growing Irish unrest. Conceding that military force was not an option, Lionel turned to the authority of the Anglo-Norman parliament that had been meeting in Kilkenny since 1293. Under his impetus, the Statutes of Kilkenny were enacted in 1366. The acts were designed to safeguard the newly arrived English colonists not from native encroachments, but from assimilation into the Irish culture and clashes with the older, assimilated colonists. Intermarriage was forbidden between the Irish and English, as was trade; the use of the Gaelic language was outlawed among the colonists and even among Irish loyal to the Crown; only English names could be given to children; only Englishmen or Anglo-Irish could be granted ecclesiastical benefices; and colonists were forbidden to have contact with Irish musicians, poets, or singers. The level of tension between the older and newer colonists is evident in one particular provision forbidding the two groups from hurling insults at one another (the epithet "Irish dog" is singled out). Ironically, these statutes, meant to protect the English culture, were written in French, then the official language of civil documents.

The statutes proved to be largely ineffective, however, as the English turned their attention back to wars with the French and the Scots. English landowners found a variety of ways to circumvent the statutes, including bribery. As the English tried to legislate their authority from a distance, the revival of Gaelic culture and self-sufficiency continued.

At the end of the fourteenth century, Kilkenny became the seat of the powerful Butler family, who had been named earls of Ormonde by the English Crown. The third earl of Ormonde purchased Kilkenny Castle in 1391, and the structure would remain in the Butler family for almost 600

Kilkenny Castle
Photo courtesy of Bord Failte, The Irish Tourist Board

years. As the seat of this powerful family, Kilkenny became the most important inland town in Ireland by 1600. King James I made the town a free borough in 1609.

With the English Reformation started by Henry VIII in the 1530s, the Crown began a systematic seizure of church lands and closure of monastic orders. In response, the expelled priests and monks took their message of Catholicism—and resistance—to the Irish heartland. A renewed spirit of insurgency took hold across the countryside.

In October 1641 plans were discovered indicating that Dublin was to be the focus of an armed rebellion. While parties of rebellious native Irish were unable to take the symbolic city of English domination, they did drive Protestant settlers from their farms in the Ulster countryside. In England, the event was depicted as a calculated massacre and became a touchstone for English leadership, who would thereafter refer to the Protestant victims as "slaughtered saints." The king's emissaries in Dublin moved to declare all Catholics enemies of English authority, and the Parliament in Westminster enacted provisions to confiscate all lands belonging to rebellious Catholics. Ireland was soon plunged into civil war, with native Irish Catholics and Anglo-Irish Catholics uniting against the Protestants.

The united Catholic forces formed a Catholic Confederacy in 1642, at a national Catholic synod called in Kilkenny on May 10. There they created an executive branch (the twenty-four-member Supreme Council) and a parliamentary body (the General Assembly of the Confederate Catholics) that rejected the authority of both the English Parliament and the colonial Irish Parliament, which continued to meet in Dublin. The Confederacy threw its support behind King Charles I, who was already threatened by Parliamentary forces at home and thus more likely to meet the Catholics' demands. For his part, Charles was happy to allow the Catholic Confederacy to threaten the pro-Parliamentary Dublin government just enough for him to maintain control in Ireland. At the same time, he wanted to monitor the Confederacy so that it could not become too strong.

Toward these ends the king entered into protracted negotiations with the Kilkenny Confederacy. King Charles put James Butler, twelfth earl and first duke of Ormonde, in charge of his operations in Ireland, both on the field and at the negotiating table. Although a Protestant, Ormonde was fiercely loyal to the king and had strong connections to the Catholic forces in Kilkenny, the seat of his ancestral line. A truce came in September 1643. In England, the Parliament seized on these negotiations as proof of the king's Catholic sympathies and tried to parlay the treaty into greater support among Ireland's Protestants.

By 1645 the king was more in need of Irish support than ever, and consequently he ordered Ormonde to negotiate a permanent settlement with the Confederacy. The talks stalled, however, and in August Charles sent the pro-Catholic earl of Glamorgan to Kilkenny to break the impasse. Glamorgan concluded the negotiations on terms very favorable to the Confederacy. In return for full freedom to practice the Catholic faith and full control over their own churches, the Irish were to provide an army of 10,000 men for the besieged king. The treaty was to remain a secret until the army was assembled, but in October the archbishop of Tuam was killed in a battle near Sligo, and a copy of the document was found on his person. When Ormonde heard of the treaty, he demanded that the king disown it; Charles complied. Ormonde then began his own negotiations with the Anglo-Irish faction of the Confederacy. On March 28, 1646, he finalized a treaty with Confederate delegates in Dublin. The treaty did not guarantee free practice of the Catholic faith and was concluded without the assent of the native Irish faction or the knowledge of Archbishop Giovanni Battista Rinuccini, the strongly pro-native papal nuncio sent by Rome to aid the Catholic cause.

From the beginning, the alliance between the native Irish and the Anglo-Irish had been tenuous. The native Irish viewed loyalty to the king as a stepping-stone to greater independence; the Anglo-Irish owed their very positions to royal grants, and cared more about preserving their status than winning freedom for the people whose lands they now occupied. The Dublin treaty helped blow the Kilkenny Confederacy apart.

Since his arrival in Kilkenny in November 1645, Rinuccini had opposed dealing with Ormonde, whose Protestantism and strong ties to the Crown cast his motives into doubt. In June 1646, after victories by native military leader Owen Roe O'Neill, a new synod was called, in which the Confederate delegates who had met with Ormonde were excommunicated. In September, O'Neill, Rinuccini, and military leader Thomas Preston entered Kilkenny, dissolved the Supreme Council, and formed a new one with Rinuccini as president.

But the Confederacy was still divided over tactics and goals. Jealousy between Preston and O'Neill prevented coordinated military actions that might have taken Dublin, and by the time the General Assembly met again in Kilkenny in February 1647, a unified front was beyond hope. (It has been said that the phrase "fighting like Kilkenny cats" was coined in reference to the disputes in the General Assembly.) Eventually, in 1648, O'Neill separated his army from the Confederacy altogether. But by now it was too late. Ormonde returned to Kilkenny in September 1648 to head the sinking Confederacy; he dissolved the existing government and began reassembling members to rule as Commissioners of Trust. Late that year, O'Neill died in command of his forces in Ulster. In January 1649, King Charles was beheaded and Rinuccini departed Ireland, his mission hopeless.

Back in England, Oliver Cromwell had long viewed the situation in Ireland with building anger. He now planned an expedition to settle matters there, but made sure that the victorious Puritan Parliament provided adequate money and supplies for his troops before he set sail. He arrived in Ireland in midsummer 1649 with a force of 8,000 ground troops and 4,000 cavalry. After capturing Drogheda in the north, he divided his army, keeping one portion in the north and sending another south. He captured Wexford, then positioned himself for a spring offensive against Kilkenny.

Cromwell reached Kilkenny on March 22, 1650. He asked the leading town officials to surrender, but the city governor, Sir William Butler, sent word that he would not capitulate. Cromwell then commanded his artillery units to bombard the city. On March 26, 1,000 English infantrymen captured the Irish section of the city, but Englishtown, protected by defensive walls, held out for a few additional days before surrendering. Kilkenny Castle fell as well, and Ormonde went into exile on the Continent. The citizenry of Kilkenny paid £2,000 to protect their lives and property. During the campaign, Cromwell used St. Canice's Cathedral to quarter 500 horses for his cavalry detachments. In order to air out the makeshift stables from the intense smell, his Protestant troops smashed the cathedral's stained glass windows.

Cromwell was recalled to England two months later, but his troops continued their work until 1653, by which point they had defeated the last remnant of resistance. All Irish and Catholics, with the exception of forty laborers and skilled craftsmen, were ordered to leave Kilkenny by May 1654. Of course, every Irish Catholic in the former Confederate capital asserted that he was one of the forty laborers in question, and the town's English Protestants refused to support the statutes because the Irish workers were essential to maintaining their standard of living. New anti-Catholic laws were instituted, but they mostly affected Catholic landowners. On May 15, 1655, it was proclaimed that the abandoned houses in Kilkenny were only to be awarded to ''English and Protestants and none others.''

Kilkenny was never again so prominent in Irish history. It has had its share of famous residents, however, including Jonathan Swift, William Congreve, and Bishop George Berkeley, all of whom attended St. John's College, a Protestant school established in the sixteenth century. Kil-

kenny continued as a market town, especially after a rail connection to Dublin was established in 1852 and connections to Waterford and Maryborough were added in 1880. By the twentieth century, however, the town's importance had declined.

Today, Kilkenny is not only home to agricultural markets, breweries, and craft industries, but is also a popular tourist destination, known for its wealth of Norman and Tudor structures. These include the Tudor-period Rothe House, built in 1600 by merchant John Rothe and now home to the County Kilkenny Museum; and Kilkenny Castle, which the Butler family donated to the town, along with the surrounding fifty-acre gardens, in 1967. The restored St. Canice's Cathedral houses one of Ireland's finest collections of sixteenth-century tomb sculptures. Next to the cathedral is the 102-foot Round Tower, which, according to various sources, was either part of the original sixth-century monastery or added in 847 by King O'Carroll of Ossory.

Further Reading: *Medieval Ireland: The Enduring Tradition* by Michael Richter, translated by Brian Stone and Adrian Keogh (Dublin: Gill and Macmillan, and New York: St. Martin's, 1988; as *Irland im Mittelalter—Kultur und Geschichte,* Stuttgart: Verlag W. Kohlhammer GmbH, 1983) does a good job of revealing the early Viking and Norman influences on the developing Irish culture and power structure up to 1485. Liam De Paor's *The Peoples of Ireland: From Prehistory to Modern Times* (London: Hutchinson, and Notre Dame: University of Notre Dame Press, 1986) presents a comprehensive examination of Irish history. *The Curse of Cromwell: A History of the Ironside Conquest of Ireland, 1649–53* by D. M. R. Esson (Totowa, New Jersey: Rowman and Littlefield, and London: Cooper, 1971) gives a dramatic, if opinionated, outline of how the Catholic Confederacy bungled military and diplomatic opportunities through internal quarreling.

—Andrew M. Kloak

King's Lynn (Norfolk, England)

Location: In Norfolk County, on the east coast of England. King's Lynn stands at the mouth of the River Ouse, forty-two miles west of Norwich, on the A 47 road.

Description: Bustling seaport and agricultural town; a major trading center in medieval England.

Site Office: Old Gaol House
Saturday Market Place
King's Lynn, Norfolk
England
(553) 763 044

King's Lynn's present-day activities in marine and agricultural commerce are direct reminders of the town's prominence as a major trading center in medieval England. It was originally founded as Lynn, a Celtic word meaning "poor," referring to the mouth of the River Ouse on which the town stands. King's Lynn owes much of its emergence as a medieval trading town to the ancient process of saltmaking. The site of the present-day town and its surrounding area, like other locations along the Wash (the large bay in the eastern coast of England), met two important criteria that enabled the production of salt from the sea. The first of these was an inlet that allowed sea water to flow inland, beyond the coastal limits set by the high tide; the second requirement was the availability of burning fuel, such as peat or turf. Salt was extracted by first collecting sand deposited between high and low tide levels into clay- or wood-lined containers with a perforation at the base. Sea water was then poured through the container to extract the salt from the collected sand. The water was allowed to filter through the perforation at the base into pans which were placed on turf fires, where the salt turned to crystals. The sand thus used was then discarded onto heaps behind the salt-making spot. With the passing of time, these heaps grew larger in an unintentional process of land reclamation, eventually becoming mounds. It has been estimated that the building up of these waste mounds had taken place since prehistoric times and continued until the fifteenth century. Waste mounds more than twenty feet high and over sixty feet across are visible today in fields now a mile away from the sea. Lynn was constructed on land reclaimed in this way. Proof of the origin of Lynn's foundations could be seen in the nineteenth century, when a mound named Belasis or Rhondil stood near the city's east gate. The earliest documents dealing with land refer to salt-works, known as salterns, and their sand areas, known as *greves* to the north-east of the town. A twentieth-century architectural report mentions that the foundations of St. Margaret's Church were likely to have been built on a number of salt makers' mounds.

Salt, used extensively in fish and meat curing during the Middle Ages, was always in high demand. Through the history of its salt production one can trace the earliest beginnings of Lynn as a trading center: where the salt was made, it was also bought. Wool, according to documents dating from the eleventh century, was also traded here. Large flocks of sheep were apparently taken to King's Lynn by salt-makers moving to the coast for the summer, when weather conditions were most suitable for salt production. Trade in the area was further diversified with the growing of corn on land reclaimed from the area's marshlands, known here as the Fens. Commercial activity took place where traders landed along the coastline of the Wash. As time passed, the volume of trade increased, and with it arose the need for a harbor that provided a sheltered anchoring point as well as an organized market site. Lynn's location near the mouth of the River Great Ouse made it an ideal site for a harbor. Whoever organized such a market place would be able to generate significant revenue by collecting tolls from traders. One of the local lords first known to have profited from such an opportunity was Herbert de Losinga, Bishop of Norwich, who arrived in England from Normandy toward the end of the eleventh century, twenty-five years after the Norman Conquest of England by William I (the Conqueror). Based across the English Channel in Normandy for the greater part of his reign in England, William delegated a large portion of the government of England to bishops brought in from the European continent. These bishops were under tight royal control: they received their appointments and lands directly from the king, to whom they owed knight service. It was under this policy, continued by William I's successor, William II, that Bishop Herbert took up his appointment in 1091. A market, held on Saturdays, and a fair held on the feast of St. Margaret's, were already well established by the end of the eleventh century. This fact was recorded in 1096, when Bishop Herbert began the construction of his cathedral at Norwich and founded a community of Benedictine monks to serve it. Records of the time show that the endowments he gave to these monks to support their activities included the profits earned from his Saturday market. The monks were also given the Church of St. Margaret, at the time still under construction.

Episcopal jurisdiction over the town was so extensive that it was renamed Bishop's Lynn. Tolls, collected by the Tolbooth [sic] and then split between the bishop and the Earl of Arundel, were levied for the use of market and mooring facilities. Bishop Herbert was also a large landowner in the town, and earned substantial rental income. As the wealth of individual townspeople grew, repeated attempts were made to restrict the extent of the bishops' powers over the affairs of the town. In September 1204, the town received its first

The Customs House, on Purfleet Quay

The Guildhall, left, and St. Margaret's Church
Photos courtesy of Tourist Information Centre, King's Lynn

charter from King John I, granting it the status of free borough, which allowed for municipal, rather than episcopal, organization. The charter also provided the town a merchant guild and exempted it from having to pay customs to the bishop. King John marked the occasion by giving the town a sword, which is today displayed in the Guildhall, along with the charter itself. In spite of this charter, the bishop retained extensive rights over the town. His jurisdiction still covered its quays, waterways, and unbuilt plots of land, which yielded him substantial profits. Episcopal authority continued as well to hold sway over most legal disputes that arose in the town.

Tensions between civil ("town") and ecclesiastical ("gown") authority continued throughout the Middle Ages. In 1299, for example, a disagreement arose as a result of the bishop's insistence on his rights over the town's watercourses and streams. The town's civil authorities had built a basin, controlled by sluices, with which to maintain the water level in the ditches around the walls of the town. The bishop's servants broke this basin, and the town's grievances were augmented by the imposition of a tallage. Agreement on the issue was not reached until 1307. In 1309, feeling his authority threatened, the Bishop charged thirty-three of his tenants, all leading members of the community, with violating his jurisdiction by not observing the conditions stipulated in the contracts they held with him. This move was in great part aimed at curbing the growing commercial power of the merchant guild. In the late thirteenth century, complaints had been lodged that outsiders—those who spent less than a year and a day in the town—were obtaining, through membership in the guild, exemption from the town's tolls. The bishop not only recovered the revenue lost by curtailing this practice, he also extracted a charge of £40 as payment for the provision of his two courts and established his power to authorize the building of new quays.

Attacks on the bishops continued, and these were not always aimed at formal ecclesiastical authority: in 1377, a group of disgruntled artisans physically assaulted Bishop Despencer. In 1402, attempts were made to force Despencer's successor to repair the bishop's staith, a stretch of land by the water that served as a landing point for ships. Numerous threats of violence were also made to the bishop. In 1404, the Tolbooth clerks tried to collect a stallage fee from stall holders in the Tuesday Market, much to the citizens' anger. The townspeople rioted, and control of the Tolbooth eventually passed into their hands. These power struggles continued until King Henry VIII incorporated the town in 1525. Bishop's Lynn became King's Lynn, and the bishop's liberties and courts were transferred to the town in 1537.

Situated at the south end of present-day Queen Street is the Saturday Market Place, which, together with St. Margaret's church, formed the original site of the town. The church's proportions, at 230 feet in length and with the two towers of its west end, were unusually large for a parish church of the time. Large-scale construction appears to have been a fondness of Bishop Herbert: the spire of his other

construction, Norwich Cathedral, is the second highest in England after that of Salisbury Cathedral. In addition to functioning as a parish church, St. Margaret's also served as the church of a priory built by the Norwich monks. This double function was reflected in the division of responsibilities with regard to the upkeep of the church. The parishioners were responsible for the nave and the western towers, while the monks were in charge of the maintenance of the chancel (except for its north side, where the Holy Trinity Guild had its own chapel). Enlargements and improvements of the nave took place in the fourteenth and fifteenth centuries, reflecting the growing prosperity of the town's citizens. The south choir aisle housed the prior's own chapel. The church's property today includes two ten-foot-tall fourteenth-century brass tablets, considered to be among the most important in England. One of these, dedicated to ex-Mayor of Bishop's Lynn Robert Brauche, who died in 1349, depicts a peacock feast that was held in honor of King Edward III, who visited the town in 1349. The other brass, dedicated to Adam de Walsokne, who died in 1349, depicts contemporary country life.

The other building overlooking the Saturday Market is the Guildhall, built with a striking checker design on its façade in 1421 for the Guild of the Holy Trinity. Guilds performed an important civic function in the medieval town. Founded essentially as religious fraternities that promoted members' interests and performed charitable deeds, they were formed by groups such as seafaring merchants, shipmen, clerks, and coifmakers. Their activities were financed with money raised by membership dues, fines levied for absence from meetings, and members' funerals as well as bequests. The Guild of the Holy Trinity represented the town's traders, and was thus the most dominant of the town's guilds. It had a monopoly on the trade in millstones and marble. In 1318, it was agreed that the profit from this trade should not go to individuals, but to the guild as a whole, a move that further boosted its resources. The guild contributed to various projects in the town, such as the building of a charnel chapel at the north end of St. Margaret's in 1324, and the cleaning of the town dike in 1377. Its municipal and commercial activities were of great importance to the town: the Guild of The Holy Trinity was the only guild to have its own chapel at St. Margaret's, where it employed six chaplains. In 1391, when the ringing of the bell at St. Margaret's was banned due to damage to the fabric of the towers, exceptions were made for funerals of guild members and their wives. All religious guilds were dissolved in 1545, and the building became the property of the Corporation of King's Lynn. Today it serves as the center of the town's government. A specially constructed museum in the crypt displays regalia of public office.

The town's first market place occupied the land between Millfleet and Purfleet, two streams flowing into the Ouse; a second market developed at about the same time north of the Purfleet stream, with nearby mooring facilities for ships. The division in the town's commercial sites has continued to this day with the Saturday and Tuesday Market Places. These two markets are reached by a street, lined with magnificent merchants' houses and warehouses, known as King Street at one end and Queen Street at the other. St. George's Guildhall, once the site of the town's other significant guild, stands in King Street. Restored to its original condition, it is known today as the Fermoy Centre. Its perfectly preserved hammer-beam roof provides a setting for numerous social events, one of which is the King's Lynn music festival, held in July of each year. Also situated in King Street is the Customs House, which was built in 1683 by Henry Bell, mayor of King's Lynn and a renowned architect. The Duke's Head Hotel in the Tuesday Market was also built by Bell in 1689. To the north of the square lie the docks on the Great Ouse. The church on the Tuesday Market Place, St. Nicholas, was made into a parochial chapel, or subsidiary church to St. Margaret's, in the middle of the twelfth century by Bishop William Thurbe. St. Margaret's remains the principal church of the parish.

King's Lynn has functioned as a significant domestic and international commercial hub. Its inland waterways bore goods from and to the interior, while the harbor witnessed flows of goods to and from towns along England's eastern coastline, as well as abroad. The beginnings of international trade can be traced back to the area's salt-making industry, which at first attracted Norwegians and Flemings, who later were also attracted by the availability of wool. The diversification of trade can be seen in records of purchases made in the town for the royal household. During the period 1190–1200, these included hawks, falcons, spices, and wine. There was further diversification in the thirteenth century, when the traffic in goods included wines from France; furs and wax from northern Germany; hawks, fish, and timber from Norway. Domestic goods shipped included mineral ores, corn, ale, wool, and (later) cloth.

The fourteenth, fifteenth, and sixteenth centuries saw growing sophistication in the traded goods and deepening interaction with traders from abroad. Many of these traders settled in King's Lynn. In 1388, a number of Prussian merchants settled in the town and plied their trade to and from the Baltic, Spain, and Portugal. A group of Lynn merchants tried to be part of the vigorous Baltic trade by setting up their own base of operations at Danzig (present-day Gdansk in Poland), thus emulating their Prussian counterparts. Lynn's significance to the merchants of the Hanseatic League was made clear in 1475 with the construction of a series of warehouses (known today as the Hanseatic Warehouses), linked by a narrow cobbled alleyway—previously a canal—to the Satur-day Market Place. In addition to trade, merchants were also attracted to Lynn as a shipping center, where ships could be hired or chartered with ease. In 1393, Norwegian naval authorities hired three large warships from Lynn men. King's Lynn's national importance at the height of its economic power can be seen in the level of annual contribution (known as aid) paid to the Crown's Treasury. In 1182, King's Lynn paid an aid of 80 marks, while Norwich paid 100.

King's Lynn is also renowned as the birthplace of the religious mystic Margery Kempe, the daughter of John Brunham, once mayor of Bishop's Lynn. She married John Kempe in 1393 and bore fourteen children, then gave up married life to devote herself to religious activities, sailing from the town's port on her pilgrimages to Jerusalem, Rome, Germany, and Spain. She described her spiritual journey through life in *The Book of Margery Kempe,* which is remarkable for being one of the earliest English autobiographics, with its vivid representation of medieval life and engagingly colloquial style. She is known to have received communion in the prior's chapel of St. Margaret's.

Through the ages King's Lynn has also been attractive to treasure hunters, who have been brought here by legends about King John's Great Treasure. Four months after putting his seal to the Magna Carta at Runnymede, King John I was entertained at a feast given at Bishop's Lynn by the local nobility. On the following day he left for Newark Castle in Lincolnshire to the north. He sailed across the Fens, which in those days were not entirely drained; the marshland was navigable, but at great risk. His baggage ship was swept up by the tide. King John reached his destination, but his belongings did not. He died soon afterwards, in 1216.

Further Reading: A very helpful work is *The Making of King's Lynn: A Documentary Survey* by Dorothy M. Owen (London and New York: Oxford University Press for the British Academy, 1984). The introduction of this book provides reconstruction of the early history of King's Lynn, drawn directly from historical documents, dating from the eleventh century. Reproduced documents provide a mass of detail in their original wordings in Latin and medieval English. *The Berlitz Travellers Guide to England & Wales* (New York and Oxford: Berlitz Publishing Company, 1994) contains a summary of King's Lynn's landmarks and history, as well as useful information for the visitor. *Village England* edited by Peter Crookston (London: Hutchinson, 1971) and *English Market Towns* by Russell Chamberlin (New York: Harmony, 1985) both include illustrations and descriptions of King's Lynn.

—Noel Sy-Quia

Kinross (Tayside, Scotland)

Location: Twenty-five miles north of Edinburgh, on the banks of Loch Leven.

Description: Loch Leven Castle was the site where Mary Queen of Scots was kept prisoner from June 17, 1567, until she escaped on May 2, 1568.

Site Office: Kinross Service Area
Junction 6, M90
Kinross, Tayside
Scotland
(577) 863680

Kinross is a pleasant little town just east of the M90 motorway on the banks of Loch Leven, within easy reach of Edinburgh (an hour by bus, less by car). The Cleish Hills are to the south, the Ochils to the north and west, and the Lomond Hills to the north and east. Its main claim to fame is Loch Leven and its castle, where Mary Queen of Scots was imprisoned between 1567 and 1568. This 4,000-acre loch (about eight miles in circumference) is the principal loch of the Scottish Lowlands. It contains a bird sanctuary of international importance and is also famous for trout fishing. The loch contains several islands; the largest is eighty-acre St. Serfs near the eastern shore. Castle Island is a short distance from the town of Kinross, and can now be reached via a ferry boat.

Kinross County is mentioned in historical records as early as 1252, the time of Sheriff John of Kinross. Originally it was part of the Fife area, but about 1426 it was separated and the parishes of Kinross, Orwell, and Portmock became the Shire of Kinross. In 1975, the county was combined into the District of Perth and Kinross, within Tayside Region. The first Kinross village began near the loch on a promontory and gradually spread west. At one time it was the capital of the county. The name in Gaelic means "head and promontory."

The presence of a castle in Loch Leven dates back to the Picts. There is evidence that it was the home of Congal, son of Dongart, king of the Picts, in the sixth century. The present square tower was probably built in the eleventh or twelfth century; the ramparts and round tower are more modern. In 1328, Loch Leven was listed as one of the royal castles, along with Edinburgh, Stirling, and Dumbarton. Until 1368, there were various occupants and many battles for control of the castle. From then on, it was used as a prison for important political offenders. In 1390 King Robert gave the castle by charter to Sir Henry Douglas. His family controlled it for the next three centuries, although it was really state property; the family home was on the mainland, to the northeast of the present Kinross House. They began to rebuild it in 1400, but even though they lived there as well as on the mainland, it was still used as a prison for notable convicts such as Patrick Graham, the first Bishop of St. Andrews; Robert, the steward of Scotland, who later became King Robert II, and his son; and Archibald, Earl of Douglas.

There are records of sporadic violence on the site, such as when King Alexander III and his queen were taken from the castle on the island and carried off to Stirling to be kept safe from the English invaders in 1256. The castle was besieged by the English in 1301, but was retaken by Sir John Comy. In 1335, Sir Allan Vipont held the castle on behalf of the Scottish King David II, when Sir John Stryvelyn (Stirling), unsuccessfully besieged it with the help of English soldiers. That siege lasted nine months with a great loss of lives.

Originally the level of Loch Leven was four and one-half feet higher than it is now, and therefore Castle Island was much smaller. The tower was the original castle for about 200 years; other buildings and high ramparts were added in the early sixteenth century. Inside the parapet there was an elevated walkway so that sentinels could keep watch in all directions. Since the island was then so small, there was a little garden and only a few additional buildings of wood or stone.

The castle played an important part in the dramatic life of Mary Stuart, Queen of Scots. Early in her reign, she visited Loch Leven occasionally on her many travels around the country and enjoyed taking part in falconry there. Her first recorded visit was April 9 to 15, 1563, when she went hawking. She also visited the castle in July 1565 with her husband Lord Darnley.

Mary was a controversial queen of Scotland from the very beginning. She was born in 1542, a tumultuous time in Scotland, crowned queen after her father's death when she was only nine months old, and raised by her mother's family in France from 1548 because the political situation in her homeland was so volatile and unsafe. After many happy years in France, she was married to the young dauphin of France at the age of 15, became queen of France at 17, and was widowed just a year later—a dowager queen at 18.

In 1561 Mary returned to Scotland to take up her duties as queen. Still very young (only 19), she walked into the center of intrigue, plots, and counterplots led by the deceitful Scottish nobles. In addition, Queen Elizabeth I of England, her cousin, felt threatened by Mary's attempts to be declared heir apparent to the English throne; Elizabeth therefore attempted to exacerbate political tensions between the two countries. Also, Mary was a devout Catholic, and Scotland was by that time Reformed. The fact that she insisted on continuing to practice her religion was a sore point with her subjects.

A new husband for Mary was considered very important for political reasons, and after much plotting, Mary

Loch Leven Castle at Kinross
Photo courtesy of Perthshire Tourist Board

chose Henry Stuart, Lord Darnley. They married in 1565 and Mary gave birth to a son, James, in 1566. Unfortunately, the marriage did not go well. In 1567, Lord Darnley was murdered, and Mary was accused of being a party to the crime. She was never completely cleared of this charge. A few months after Darnley's death, Mary was abducted by Lord Bothwell and taken to his castle at Dunbar while on her way back from visiting her son at Stirling Castle. Bothwell and Mary were soon married. Although the nobles had assured Mary they would support this marriage, Bothwell had also been implicated in the death of Darnley. Most nobles soon turned against both Mary and Bothwell, and the common people were angry and suspicious. After a battle at Carberry Hill in June 1567 when the army of Bothwell and Mary was defeated, Bothwell fled, never to return. Mary was taken to Edinburgh, and then on to be imprisoned in the castle at Loch Leven.

Sir William and Lady Agnes Douglas, along with the Dowager Lady Margaret, were Mary's "hosts" (jailors) from June 17, 1567, until she escaped on May 2, 1568. The Douglas family felt they had as much right to the throne as the Stuarts, making them ideal for the job of keeping Mary a prisoner. She was first imprisoned in the Glassin Tower, a small round tower in the ramparts, with an entrance only from the courtyard. Although Mary was pregnant and very ill, the campaign began to force her to abdicate in favor of her son James. She soon gave birth to still born twins, and continued to be very ill. On July 24, 1567, after great pressure from Sir Robert Melville and several others, Mary finally signed the Deeds of Abdication of the Scottish Throne in favor of her son James, then one year old. She named her half brother, James Stewart (the Earl of Moray), as regent.

George Douglas, son of Sir William, was a member of an "Escape Committee" that had been formed to assist Mary. For some reason, he was banned from the island, which greatly decreased his ability to be helpful. Before leaving, he outlined the details of a proposed escape plan to Mary, and then told her that Willie Douglas would take his place on the island. Willie was only sixteen years old, and was known as "the Foundling", as it was said he had been abandoned as a baby beside the castle. (He may have been an illegitimate son of Sir William.)

After she signed the abdication papers, Mary was transferred to the solar story (above the main hall) of the main tower. She had more room, but much less privacy. Gradually she began to recover her health. The only activity she was allowed was embroidery; she was not even permitted to write letters. Many escape plans were considered and discarded, and some were attempted unsuccessfully. Sir William became suspicious of Willie's activities and banned him from the island for a time. He was allowed to return to the island, however, on April 30.

Finally, an elaborate scheme involving a party to distract everyone was successful, and Mary escaped by boat. Mary mounted her horse and road through town to the cheers of the common people. She went to Hamilton Palace where

she was joined by many supporters. After declaring her abdication invalid because it had been signed under duress, she demanded Regent Moray resign his position. He refused, and civil war began. Mary's forces were defeated quickly, and she fled to England. She was convinced, against all evidence and advice, that Queen Elizabeth would help her. Quite to the contrary, Mary spent the next seventeen years a prisoner in various English castles. In 1587, she was found guilty (although she proclaimed her innocence) of taking part in a plot to murder Queen Elizabeth, and she was beheaded.

Loch Leven Castle, where Mary was held by the Douglases, is a good example of the type of building called a "Tower Keep," which was a style of fortified home common in Scotland in the tumultuous sixteenth century. It is oblong in shape and originally had five stories, including a kind of penthouse under the steep roof, which no longer exists. A circular staircase is in the southeast corner. The upper windows formerly had external shutters. The original doorway was high up in the east wall on the second story, reached by a removable gangway, which made it easier to defend the castle.

The basement was used both as a prison and for storage. The only entrance was through a hatchway from the floor above; ventilation came from narrow openings in the walls. There was also a well in the center of the floor. So that the castle can now be visited, the south window was enlarged to become the present entrance, and a stone stairway was added up to the kitchen.

The dark kitchen with very small windows is above the cellar. Two of the windows have window seats; the one in the south wall has a sink with a runnel to carry dirty water outside. A salt box is at the back of the fireplace. A primitive toilet, called a "garderobe," is in a corner.

The huge main hall is above the kitchen. The original entrance was in the east wall of this room. Just inside this entrance is a trap door used to lower provisions to the kitchen and cellar. A large fireplace is in the west wall. The windows are deep-set with window seats, allowing views of the loch and countryside. The floor above the main hall is called the solar, and appears to have been divided into a sitting room and bedrooms.

There were various other buildings on the site. The most important was named the Presence Chamber, which was more than thirty feet long, eighteen feet wide and fifteen feet high. This was probably built for the use of Queen Mary. In her time it contained a crimson- and gold-covered throne and gold and silver silk curtains.

The town of Kinross lies between Highway M90 (the route to Edinburgh) to the west and Loch Leven to the east, along the north-south High Street. It is a pleasing mixture of ancient and new buildings. A small museum with exhibits from Kinross's eventful past is located in the High Street in the center of town.

This is fertile farm country, but from the late seventeenth century Kinross was also known for its cutlery. That trade was killed by competition from Sheffield during the early nineteenth century. The town developed as a settlement of wool and linen weavers; fishermen caught eels and trout in the loch.

The Kinross Weavers Society was founded in 1756 with some interesting criteria for membership. An annual fee of four shillings was charged, and only weavers "of good moral character, healthy constitution and visible means of support" were admitted. A Deacon presided over the meetings. He kept order with his baton and imposed fines for swearing or interrupting a speaker. Five shillings a week benefit would be paid if a member became ill and could not work; if he died, thirty shillings were paid for funeral expenses, and his widow received one pound a year.

By 1793 the most important cloth woven in Kinross was Silesia linen, which was used for window blinds. In addition, cloth was woven for private use by people in their homes from flax raised locally.

Today Kinross's Todd and Duncan Factory is the world's largest spinner of cashmere. It stands beside the loch on the site of an old woollen mill and a linen factory. Kinross Woollen Company ran the original woollen mill in the nineteenth century. It went bankrupt, and the empty mill was purchased by William Todd and James Duncan in 1895. They retained the original mill, and it is part of the complex today. The Kinross Linen Works operated from 1873 until 1963, when Todd and Duncan took over.

After having been lost, restored, and moved a couple of times, the Old Mercat Cross is now surrounded by flower beds in the small recreation park off the south High Street. An iron collar and chain, called the Jougs, are now attached to the cross. These were used in medieval times to collar lawbreakers and expose them to scorn. The Auld Manse, beside the park, is on the site of priest's cottages from Queen Mary's time. The present structure was built in 1769. It is believed by some to be haunted; it is said that ghostly footsteps and a baby's crying have been heard during the night. The Myre is a playing field across from the recreation park. It used to be an area where everyone could graze his cattle or dig peat. There was also a communal washing-house for the townspeople. The Old Tolbooth (or County House), near the entrance to Burns Begg Street, can be identified by its rounded gable. It was the town jail from about 1600 until 1826, when the new County Buildings were built. The dungeon-like cells were on the ground floor, the middle floor was the sheriff's court, and the keeper's house and debtors' room were on the top floor.

All that remains of the old parish church built in 1741 is the Steeple. The Town Hall now occupies part of this site. The Fountain beside the Steeple was built over the Cross Well, the public water supply, in 1886. An old section of Kinross dating from the early eighteenth century is behind the Steeple and Fountain. At that time the town was developing as a weaving center and a staging-post for horse-drawn

coaches. Some of the surviving little streets are Parliament Square (where local politicians gathered), Brewery Lane (from 1894 to 1896 there was a brewery here run by the Forbes Brothers, probably closed down by the Temperance Society), and School Wynd (the site of Blacklock's School, which was thriving in 1791).

Several important buildings were built in the nineteenth century, including the County Buildings built in Adam's style in 1826. The original interior included the old prison, which was closed in 1878. The cells have now been turned into stores and the Community Council meets here monthly. The Burgh Chambers were built across the road about 1835. The offices of the Town Clerk were here as well. The Bank of Scotland was built about 1835 and is still in use.

Kinross had become an important coaching station on the Edinburgh to Aberdeen route by the end of the eighteenth century. Several inns were opened to take care of travelers, and stables were also established. At that time, the leading inns were Kinross Green Inn and Kirkland's Inn, which were extremely competitive. In fact, at one time Kirkland's postboys linked arms in a barrier across the road to persuade travelers to stay at their inn rather than the Green. Both of these inns are still in operation.

Kinross House is a beautiful home of soft, mellow stone from the Cleish Hills built between 1679 and 1693 by Sir William Bruce, who was King Charles II's surveyor general and master of the works for Scotland. The house is currently occupied by Sir David Montgomery and is not open to the public; however, the gardens are open every afternoon from May to September. They contain colorful herbaceous borders, smooth lawns, and clipped yews. There is a wide-bordered avenue from Avenue Road to the house.

Kirkgate, in Gaelic, means the road to the church. The monks of St. Serf's Island built the first church here on the banks of the loch; there are no remains. Another church was built in 1675 and used until 1742. The only remains are the original central aisle, which has been reconstructed as a chapel for descendants of Sir William Bruce of Kinross House, now the Arnots of Arnot Tower near Scotlandwell. The ferry to Castle Island is located here, and it is the site of the first village buildings.

The Market Park is a lovely area with many fine old trees opposite the Green Hotel. Over the years many fairs and other events have been held there.

Further Reading: *Historical Guide to the County of Kinross* by the Kinross Antiquarian Society (Kinross, 1980) is a useful compilation of facts and suggested tours of the area. *Lochleven's Royal Prisoner: Mary Queen of Scots* by N. H. Walker (Kinross: author, 1983) contains the story of Mary Queen of Scots, including details of her imprisonment in Castle Loch Leven. A good biography of Mary is Antonia Fraser's *Mary Queen of Scots* (New York: Delacorte, 1969; London: Panther, 1970).

—Patricia Ann Shepard

Kraków (Kraków, Poland)

Location: On the Vistula River in southern Poland, 160 miles south of Warsaw.

Description: Former capital of Poland and present capital of the province of the same name; site of many historic buildings, including former imperial castle, several long-established churches, and second oldest university in central and eastern Europe; important in Polish nationalist movement but subject throughout its history to foreign invasions, including German occupation during World War II.

Site Office: Kraków Tourist Office
Ulica Pawia 8
Kraków, Kraków
Poland
22-04-71

The exact date of Kraków's founding is unknown. Archaeological excavations in and around Kraków have uncovered pottery fragments, weapons, stones, metal tools and coins dated from approximately A.D. 600. The people who left these relics were the first in a succession of inhabitants of the area.

A Slavic tribe called the Wislandie chose the area of Malopolska, or Little Poland, which encompasses modern-day Kraków, as its stronghold in the late eighth or early ninth century. The Wislandie did not dominate the area for long. The Great Moravian Empire, which originated in what are today the Czech and Slovak republics, overtook the Wislandie some time in the ninth century. The Moravian Empire was likely responsible for introducing Christianity to its pagan subjects.

In the late ninth or early tenth century, the Moravian Empire crumbled while under assault by the Magyars (Hungarians). The Magyars allowed Kraków to become an autonomous state under Wislandie control. The Magyars lost control of Kraków to an invading people called the Bohemians. The Bohemians' reign in Kraków would be a short one. The end of the tenth century found them pushed out of Poland by Mieszko I, leader of the Polanie tribe.

From 988 to 990, Mieszko I united the various tribes within the north and south of Poland and brought about the formation of a Polish state, into which Kraków was subsequently incorporated. Mieszko was the first leader of the Piast dynasty, which would rule Poland for more than three centuries. A convert to Catholicism, he linked Polish culture to that of western Europe.

Kraków experienced substantial growth and became a thriving town thanks to its important position along the European trade routes. The city was a crucial stopping point along the Kiev–Prague–Regensburg route. Items traded along this route, inevitably making their way through Kraków, included amber, salt, and slaves.

The first settlements constructed in Kraków were on Wawel Hill. One of the most important structures on the hill was a church dedicated to the Virgin Mary, the first of many important churches constructed in Kraków. Fragments of the original tenth-century church have survived subsequent building on the site. A castle also was built on the hill.

Settlements began to spread beyond Wawel Hill in the late tenth century. Structures built beyond the hill included additional churches and homes along with a marketplace, which was located half a mile down from Wawel Hill on a flat plain.

Bolesław I, the Brave, became the ruler of Poland following the death of his father, Mieszko I. Under his rule Kraków was made the seat of the Polish bishopric. A giant cathedral was constructed on Wawel Hill in the early eleventh century upon his request. After his death in 1025, however, Poland was in disarray and many hostile outsiders sacked both the nation and the city of Kraków, destroying the cathedral and other structures.

King Casimir I, the Restorer, who assumed the throne in 1034, was responsible for bringing stability to the Polish countryside and to Kraków. He ordered that towns and churches be rebuilt, and chose Kraków as the capital of Poland.

In 1072 Stanisław, a future patron saint of Poland, was named bishop of Kraków. With the capital being in Kraków, this position held political as well as religious significance. In 1079 Stanisław was accused of treason against King Bolesław II, the Bold. A royal court found Stanisław guilty; the sentence was the amputation of his limbs. The knights ordered to carry out the sentence, however, were reluctant to do so. Bolesław, impatient with their hesitation, killed the bishop in St. Michael's Chapel in Kraków. The facts of the case against Stanisław remain uncertain, but many historians believe he was indeed involved in a plot to depose Bolesław and to place the king's brother, Władysław Herman, on the throne. Bolesław was a capable ruler, but was considered by some to be cruel and immoral. In any case, Stanisław became a martyred hero to Polish Catholics, and in 1253 he became the first Pole to be canonized. St. Stanisław's Day is observed on May 8 in Poland, May 7 elsewhere.

The mid-twelfth century saw two important churches constructed within Kraków—St. Andrzej's (known in English as St. Andrew's) and St. Wojciech's (known in English as St. Adalbert's). St. Andrzej's, with its fortress-like exterior, is still standing and is considered to be one of the finest examples of Romanesque architecture in Poland. St. Wojciech's is

Wawel Castle at Kraków
Photo courtesy of Polish National Tourist Office

also still standing but has undergone a series of alterations over the years.

In 1241, Kraków was invaded by the Mongols. The city, at that time consisting mostly of unfortified wooden buildings, was easily sacked and almost completely burned to the ground. Half the population died, and only the few stone structures in the city remained unscathed.

The Mongols' retreat from Kraków signaled a new beginning for the city and for the Polish state, which had fragmented into numerous independent principalities during the twelfth and thirteenth centuries. A local duke granted Kraków a formal charter in 1257, and it would be confirmed as the capital of the reunified Poland in 1320. In the meantime, the residents of the city went about rebuilding. Unlike the first time the city was built, construction this time followed a clearly thought-out plan.

A new market square, Rynek Glowny, was established on the western outskirts of the city. By medieval standards it was massive, measuring more than 2,000 square feet. Streets were laid out in a grid pattern. The current design of the city has its basis in this plan. Houses along the newly constructed roads were built largely from one material—stone. Wawel Castle was left untouched, but the nearby cathedral was reconstructed; St. Wojciech's was rebuilt, drastically altering its Romanesque style into Gothic; St. Andrzej's was incorporated without change into the layout of the city; the parish church of St. Mary's was redone in Gothic style, giving it massive height and width; while Sukiennice, or Cloth Hall, previously made of wood and destroyed by the Mongols, was rebuilt as a glorious brick hall. Cloth Hall's new location was in the center of the marketplace.

The population of Kraków grew both during and after reconstruction. The population boom resulted, in part, from a large influx of Jews fleeing persecution in other parts of Europe. At a time when anti-Semitism was rampant throughout most of the continent, the Polish government's attitude was one of tolerance. Suburbs developed outside Kraków's city walls to accommodate the newcomers. The population expanded to an estimated 12,000 people.

A new ruler, Casimir III, the Great, controlled Poland from 1333 to 1370. Casimir the Great founded the University of Kraków (now called Jagiellonian University), the second oldest university in central and eastern Europe after the University of Prague. The university's charter was issued by the papacy on May 12, 1364. It was modeled after the University of Bologna in Italy and had eleven academic chairs—one in the arts, two in medicine, three in canon law, and five in civil law. Following its inception, the university experienced myriad problems, the most important being its difficulty in obtaining quality faculty from abroad. This led to its closure in the 1370s, but it reopened in 1400 under the new Jagiellonian dynasty.

The founder of the Jagiellionian dynasty was Jagiełło, a Lithuanian who had married Jadwiga, the last of the Piast rulers. Jadwiga died childless in 1399 and her husband, who outlived her by thirty-five years, assumed the throne. The Jagiellonian dynasty ruled Poland and Lithuania until 1572. Under the Jagiellonians, Kraków flourished—Christians and Jews lived in harmony, and there were good relations with the rest of Poland, the Black Sea peoples, and western Europe.

The Jagiellonian rulers worked hard to fortify the capital city and improve its appearance in the sixteenth century. The wall encircling the city was extended and a dozen watchtowers were constructed along the perimeter of the city to ensure its safety. Entrance into the city was permitted through only seven gates. A moat was constructed around the city's exterior walls by channeling the Rudawa River. The royal castle on Wawel Hill was enlarged, as was the city hall. The official population doubled to approximately 25,000 by mid-sixteenth century, with 18,000 residents living inside the walled city and about 7,000 more residents located in the largest suburbs. These numbers were probably misleading because there were people living in and around Kraków who were not granted citizenship in the empire and were thus not counted with the rest of the population. Those not counted included all Jews and some suburban residents. With the addition of these people, the actual population of Kraków and its suburbs was probably between 28,000 and 30,000.

Kraków also experienced industrial growth in the early to mid-sixteenth century. It became an important production center in eastern and central Europe, but growth did not occur without grime. Copper, gold, and silver smelting, as well as iron foundries and lead processing, caused much pollution.

Aside from industrial growth, the Jagiellonian dynasty also left an architectural mark on the city of Kraków. When one of the Jagiellonian kings, Sigismund I, the Old, married Bona Sforza, an Italian princess from Milan, she arrived in Kraków with an entourage that included many renowned Italian Renaissance architects. These architects successfully blended Polish Gothic with Italian Renaissance styles. One structure in Kraków that illustrates the marriage of these two styles is the imperial residence, Wawel Castle, rebuilt during Sigismund's reign in the first half of the sixteenth century.

The university had a period of great achievement from 1400 to 1550. At its height, the university was a strong influence on the character of the city, giving it an international atmosphere. More than 18,000 students attended one or more years at the university during the fifteenth century. At times, nearly half the student body was drawn from other countries. The curriculum was expanded to include theological, rhetorical, and astronomical studies, and a building was constructed to house the university, which had previously held classes in the castle, churches, and private homes.

The university's courses in astronomy became especially respected. Among the more famous graduates of the university was Nicolaus Copernicus. He came to study at the university in 1491 at the age of nineteen and graduated four years later in 1495. Copernicus's work forever altered the study of astronomy, by stating the earth was not the center of the universe, but rather one of several planets that revolved around the sun.

The second half of the sixteenth century witnessed the steady decline of Kraków's university. The Reformation and humanist movements sweeping across Europe hastened the university's decline, as more countries sought to establish their own universities, and the Kraków faculty was slow to adopt new ideas. Increasing ethnic hostilities made the city less attractive both to prospective international students and faculty. Wars in various parts of Europe also kept foreign students away. The university continued to operate, but did not begin to regain its former status for many years.

The Jagiellonian dynasty came to an end in 1572, with no male heir to ascend the Polish throne. The end of the Jagiellonian dynasty was one in a series of events that led to Kraków's overall decline. Changing trade routes, along with a pro-rural policy adopted by the Polish gentry, greatly reduced the importance of the city. In 1596, King Sigismund III Vasa moved the capital to Warsaw. Despite the change in capitals, Polish kings continued to be coronated and buried in the cathedral on Wawel Hill until 1764.

The change in status did not signal the end of beautiful architectural creations in Kraków. Between 1600 and 1800 the ornate baroque style left its mark on the city. Introduced to Kraków by Jesuit priests, most of the baroque architecture in the city took the form of ornamentation on existing buildings. These elements included facades, portals, courtyards, decorative sculptures, and ornamental stucco designs. Buildings fully constructed in the baroque style were usually churches financed by the Catholic hierarchy or wealthy aristocrats.

Poland was involved in war with Sweden in the seventeenth century, and the Swedes attacked and looted Kraków, leaving the city physically and economically devastated. Subsequent wars and internal weakness led to the disappearance of Poland as an independent nation during the eighteenth century, when it was partitioned, in three successive steps, among Prussia, Russia, and Austria. There was some Polish nationalist resistance to outside rule; on March 24, 1793, Thaddeus Kosciuszko stood in Kraków's market square and called for an uprising of the Polish people against the Russians, who at that time controlled the city. The rebellion failed, however, and Kraków was passed to the control of Austria in 1795.

The Napoleonic Wars brought more changes to Kraków. From 1807 to 1815 it was part of Napoléon's Grand Duchy of Warsaw. Then the Congress of Vienna, which reapportioned the territories of Europe after Napoléon's defeat, granted Kraków free city status. This status meant that the Polish nationalist movement took hold strongly in Kraków; activists even declared the establishment of the "Republic of Kraków" early in 1846. During its period as a free city, however, foreign powers, especially Austria, made persistent attempts to undermine Kraków's autonomy. The nationalist declaration of 1846 gave Austria even more reason to suppress Kraków, and the city became part of the Austro-Hungarian Empire in 1846. This time Austrian rule was more relaxed, though, than it had been previously, and the new rulers contributed to the beauty of Kraków by ordering that the city walls be torn down and replaced with an encircling belt of green space, which remains in place today.

Kraków finally returned to Polish rule in 1918, when a newly independent Poland was created at the end of World War I. The nations that had previously occupied Poland were in disarray, the Poles had risen up for independence, and peace terms called for the creation of an autonomous Poland. Polish rule was ruptured, however, when World War II began and Poland was quickly defeated by Hitler's German army. The German invasion began on September 1, 1939. On September 6, 1939, the Germans captured Kraków and chose it as their headquarters for the duration of the occupation. Wawel Castle became the home of the German governor-general, Hans Frank.

The Germans quickly worked to oppress the residents of Poland, especially the nation's Jewish population. In Kraków, the Germans ordered that all Jews move to the suburb of Podgórze or leave the Kraków area altogether. Podgórze became one of the largest Jewish ghettos in occupied Europe. Initially consisting of 16,000 people crammed into 320 buildings, the ghetto expanded as more Jews were shipped into it. The ghetto was sectioned off from the rest of the city by walls and had only three exits. Forced into the ghetto, Jews saw their synagogues desecrated, educational and religious practices forbidden, and their businesses and valuables stolen from them.

Outside of the city the Germans created the Kraków-Piaszow concentration camp, the second largest camp of its kind in Poland after Auschwitz. The camp was unique because it was one of only a few concentration camps to be located near a major city. Its original population consisted of 2,000 Jews. The population ballooned to more than 25,000 when the Jewish ghetto in Kraków was ordered liquidated in 1943. Overall, as many as 55,000 Kraków Jews either died in the ghetto or in the Kraków-Piaszow concentration camp, or were deported elsewhere and later perished in other concentration camps.

The Germans had every intention of destroying Kraków if forced to surrender it to the Allies. Their plans included dynamiting most of the city's important structures, among them Wawel Castle. The Soviet Red Army, in a surprise move, encircled the city so quickly that all the Germans could do was to flee. Kraków was liberated by the Soviets on January 17, 1945.

In 1946 yet another illustrious graduate came out of Jagiellonian University. This was Karol Wojtyła, who twenty-two years later would become the first Polish pope as John Paul II. Wojtyła, born in Wadowice, an industrial community fifty miles southwest of Kraków, had come to Kraków and entered the university in 1938. When the occupying Germans closed down the university, Wojtyła and his schoolmates continued their education through an underground study network. With the liberation of Poland and the reopening of the university, they were again able to attend classes openly, and Wojtyła graduated with distinction in theology in August of 1946. He subsequently was a parish priest in Kraków and elsewhere, and during the 1950s returned to Jagiellonian University to study philosophy and to serve as a lecturer in social ethics.

Wojtyła was named archbishop of Kraków in 1963, and, with his criticism of the Communist government that had ruled Poland since the end of World War II, he quickly became recognized as a political as well as a religious leader. He and the primate of Poland, Cardinal Stefan Wyszyński, pursued a largely successful campaign for toleration of the Catholic Church by the Polish government. When Wojtyła was named pope in 1978, he succeeded John Paul I (Albino Luciani), who had died only three weeks after his inauguration as pope. As pope, John Paul II has traveled widely, survived an assassination attempt, and emerged as a charismatic figure, although his conservative stands on many issues have aroused opposition from liberal Catholics. In 1979 his travels took him back to Poland for the observance of the 900th anniversary of the martyrdom of St. Stanisław. His tour of the country climaxed with an address to a huge gathering in Kraków. His appearance fed nationalist as well as religious feelings at a time when Poland was on the verge of great change.

Kraków had become increasingly industrialized under the Communist government, with a large steel foundry and a planned industrial suburb, Nowa Huta, developed just east of town. Kraków factory workers, like others throughout Poland, became involved in the Solidarity labor movement and were critical of the government. They participated in the various strikes and demonstrations that chipped away at the government's power. During the 1980s the activism on the part of the Poles, plus the influence of the progressive policies of Soviet Premier Mikhail Gorbachev, led to the collapse of Poland's Communist regime and the creation of a more democratic government and market-oriented economy.

Today Kraków is a center of Polish history, culture, and industry, although pollution from industry is posing some threat to its historic structures. These preserved structures include the Wawel Castle and cathedral, St. Mary's Church, Cloth Hall, and Jagiellonian University. Jagiellonian is one of twelve universities in Kraków. The city also boasts numerous museums, theaters, and opera houses.

Further Reading: *Saint Stanisław: Bishop of Krakow,* edited by Doyce B. Nunis Jr. (Santa Barbara, California: Saint Stanisław Publications Committee, 1979), provides extensive information on the history of Kraków and the life of its patron saint. *Cracow and its University* by Józef Dużyk and Stanisław Salmonowicz, translated by Marianna Abrahamowicz, (Kraków: Wydawnictwo Artystyczno-Graficzne, 1966) is a detailed, extensively illustrated treatment of the city and Jagiellonian University. A pro-Communist stance is evident in some portions of the book, written during the Cold War years. *The Kraków Ghetto and the Piaszow Camp Remembered* by Malvina Graf (Tallahassee: Florida State University Press, 1989) is an informative and interesting account of the author's struggle to survive the ghetto and the camp.

—Peter C. Xantheas

Landshut (Bavaria, Germany)

Location: On the banks of the Isar River, forty miles north of Munich.

Description: Often called the best preserved Gothic city in Germany; first capital of Bavaria under the Wittelsbach family; site of quadrennial festival reenacting the 1475 wedding of Wittelsbach heir Georg to Princess Jadwiga of Poland.

Site Office: Verkehrsverein
Altstadt 315
84028 Landshut, Bavaria
Germany
(871) 92 20 50

Forty miles north of Munich in the flatland of southern Germany lies Landshut, the first capital of Bavaria. Far from Bavaria's picturesque, gently rolling Alpine foothills, Landshut is not a common tourist destination. But whatever the town lacks in natural scenery it makes up in well-preserved historic beauty, with churches, castles, and gabled housed lining the cobblestone streets. The town and its two unusually wide parallel main streets—Altstadt and Neustadt (Old Town and New Town)—offer a striking glimpse of Germany's Gothic past.

The first mention of Landshut in a historic document dates from 1150. The town received its charter in 1204, when a profitable toll-collecting crossing of the Isar River was moved upstream, closer to the protection of Landes Hut, a stronghold built on a strategic bluff by Bavaria's second ruler from the house of Wittelsbach, Ludwig-der-Kehlheimer.

A branch of the Wittelsbach family ruled from Landshut until the twentieth century. At their height, the Wittelsbachs controlled not only Bavaria, but the Palatinate of the Rhine as well, and placed kings on the thrones of Germany (Rupert) and Sweden (Charles X, Charles XI, and Charles XII). Holy Roman Emperor Louis of Bavaria, himself a Wittelsbach, amassed great wealth during his reign in the early fourteenth century and worked steadily to heighten the glory of the Wittelsbach capital at Landshut. The town was perhaps at its prime under the "rich dukes" of Wittelsbach—Heinrich, Ludwig, and Georg—who ruled from the fourteenth to the early fifteenth centuries.

The castle of Landes Hut accommodated the ruling family almost continuously for 300 years, undergoing near-constant renovations, additions, and improvements as each new inhabitant saw fit. The Wittelsbach court left Landshut in 1503 but returned in 1516. They would remain in the castle for another twenty years, after which time they moved to the newly built Stadtresidenz, a palace located on Altstadt in the town of Landshut itself. The stronghold of Landes Hut then became known as Burg Trausnitz.

The sprawling fortress stands in testimony to the undertakings of the Wittelsbachs. The oldest parts of the castle are Romanesque: the belfry (commonly referred to as the Wittelsbach Tower), the inner gateway and its two round towers, the main building of the residence, and the two-storied court chapel. Expansions were undertaken in the fourteenth and fifteen centuries, with the addition of a number of defensive structures and residential buildings. The elaborate entrance to the castle, with its numerous gateways, watchtowers, and covered passages, originated during this period.

After the Wittelsbach royalty vacated the fortress in favor of the new Stadtresidenz, renovations began to transform the former Landes Hut into a *Lustschloss,* or pleasure palace. Prince Wilhelm, the future Duke Wilhelm V, lived in the palace in the mid–sixteenth century, overseeing much of the remodeling. Enamored with the splendor of the Renaissance style, Wilhelm hired Italian artisans recommended by Hans Fugger, a wealthy Augsburg merchant, to add new wings, ornate staircases, fanciful gardens, and mural-bedecked chambers.

The most famous surviving addition from the years of Wilhelm's patronage is the Narrentreppe, or Fools' Staircase, an elaborate painting along all levels of a stairwell in the Fürstenbau, or Princess' Wing. Alessandro Scalzi lined the staircase walls with life-size *comedia dell'arte* figures in 1578, making the Narrentreppe paintings Germany's earliest known portrayals of professional stage art. A number of characters from this art of improvisation can be found, including Pantalone and his servant Zanne, the courtesan Cortigiana, and the star himself, Harlequin, accompanied by his lover Colombine. Bringing up the rear of this elaborate procession are the duke and duchess of Bavaria, the parents of Wilhelm, who owed over 300,000 florins as a result of his patronage of the arts.

The final renovations of the castle took place in the mid–nineteenth century under King Louis II of Bavaria. Finding Burg Trausnitz as appealing as his predecessors, Louis had a number of chambers on the second floor outfitted splendidly.

Catastrophe struck the elaborate Burg Trausnitz on October 21, 1961. Fire raced through the historic landmark, and hundreds of years of Landshut's heritage went up in smoke. The Narrentreppe miraculously survived, and recent renovations have restored the venerable charm of this centuries-old site. Today, the former Landes Hut draws visitors both to the fortress itself and to the castle's hilltop perch for a wonderful panorama of the town, with its baroque façades and striking red roofs.

In the town itself stand two magnificent churches, the Holy Ghost Church at one end of Altstadt, and the Church of St. Martin at the other. The Holy Ghost Church dates from

Participants costumed for Landshut's Fürstenhochzeit festival
Photo courtesy of German National Tourist Office

1407; construction of the huge Church of St. Martin took place from 1389 to 1500. Both churches were designed by Master Hans von Burghausen, and St. Martin's has the distinction of being the only Gothic church believed to be designed by a single architect.

St. Martin's, with its brick tower soaring 436 feet, is one of the finest example of *hallenkirchen,* a Gothic-style church having both nave and aisles of the same height. The tower itself reigns as the highest brick church tower in the world. The three naves and brick tower of St. Martin's dominate the Landshut's skyline.

Von Burghausen's crusade to increase the height of his churches challenged the architectural possibilities of the time. The piers of St. Martin are a mere 39 3/8 inches across and rise a wondrous 72 feet into the air. The huge halls of the church are no less spectacular, with delicate pillars more than 90 feet high framing a stunning collection of fifteenth- and sixteenth-century artwork, seen in the exquisitely carved stone portals and altars. The pulpit and the high altar are especially grand. The renowned Landshut Madonna, carved in 1518 by native sculptor Hans Leinberger, is a brilliant masterpiece exhibiting the Bavarian style of sculpture during the transition from Gothic to Renaissance styles.

Outside St. Martin's runs Altstadt, the wide thoroughfare of the Old City. This picturesque boulevard is lined with stately gabled houses once used by the burghers of Landshut during the late Gothic and early Renaissance period. Past the burghers' dwellings is the Stadtresidenz, the city palace built by the Wittelsbachs after they left the Burg Trausnitz. Constructed from 1536 to 1543, the Stadtresidenz was the first Italian Renaissance–style structure built north of the Alps; artists and craftsmen were brought up from Padua to fashion the palace in the manner of the Palazzo del Té in Mantua.

Not only does the palace's exterior mimic the Italian style of the period, but the interior, with its rich staterooms covered with mythological murals, transports today's visitors over the mountains and back through time to Italy. Today, the impressive Stadtresidenz houses two museums: the Stadt und Kreismuseum with its collection of eighteenth-century applied art, and the Staatliche Gemäldegalerie's exhibit of sixteenth- to eighteenth-century European paintings.

While the majority of historic buildings found in Landshut are built in Gothic and Renaissance styles, baroque and rococo sites can be found as well. Baroque arrived in Landshut in the mid–seventeenth century along with the Jesuit order. The building of the Jesuit church from 1631 to 1641 coincided with the renovation of a number of late Gothic burghers' houses into a similar style. A century later, the Dominican Church and the Seligenthal Abbey, founded in 1232, were richly remodeled in the ornate Bavarian rococo fashion.

During the Thirty Years War (1618–1648), Landshut was the site of strategic battles between Germans loyal to the approaching Swedes, and those remaining loyal to the Bavarian and imperial crown. In the summer of 1634, Bernard, duke of Saxe-Weimar, and the Swedish marshal Gustavus Horn converged at Augsburg. Together, their combined force of 20,000 marched on Landshut, at that point held by the Bavarian and imperial cavalry. Imperial general Johann von Aldringer rushed to the town's rescue, but came too late to organize an effective resistance, and perished in Landshut's fall.

The capture of Landshut was not the final goal, but instead was an attempt to show the advancing Emperor Ferdinand II that aggression against Swedish interests in Germany would not be tolerated. The warning went unheeded, however, and further bloodshed ensued.

Although it played an important role in Bavarian government and culture, the town of Landshut was home to a major university for only a short time. The Bavarian University was transferred from Ingolstadt to Landshut in 1800, and then moved again to Munich in 1826 as part of the reorganization of German universities. A medical school specializing in surgery and a religious facility for training teachers remain today.

Modern-day Landshut usually lacks the throngs of visitors found in better-known German towns, but every four years the city's stately palaces, fabulous art, and imposing churches become the backdrop for the Fürstenhochzeit, or Prince's Wedding, sometimes called Europe's greatest folk festival. The festival commemorates the wedding of Georg, son of Ludwig the Rich, in 1475. To find a bride for his son, Ludwig sent Bishop Henry of Regensburg to Poland, where he negotiated with King Casimir IV for his eldest daughter's hand in marriage. The king consented and thus Jadwiga, princess of Poland, was to be wed to Georg. With the impending marriage announced, the townspeople began elaborate preparations. In autumn of 1474, the eighteen-year-old Jadwiga came to Landshut with her parents, accompanied by a splendid court of 642 persons. The royal wedding took place at St. Martin's on November 14, 1475, and was officiated by Archbishop Bernhard of Salzburg. The number of guests totaled more than 9,000, and included such nobility as Emperor Frederick III, 25 princes, a number of bishops and abbots, and countless nobles and emissaries from around the continent.

The town of Landshut was flooded with guests for the eight-day celebration. Duke Ludwig proclaimed that all merrymakers were his guests, and no shopowner or hotelkeeper in the town was allowed to accept money for services during this time. Preparations for the festivities were made well beforehand, with droves of livestock, vats of wine, and bags of exotic spices brought into Landshut from all over the world. In the course of the celebration, the duke's guests consumed 333 oxen, 490 calves, 1,133 sheep, 1,537 lambs, 684 pigs, 11,500 geese, 40,000 chickens, 14,000 pounds of raisins, and countless barrels and vats of wine and beer; the festivities were lit by candles made from 11,000 pounds of wax. The price of this lavish affair? An astonishing 55,700 Rhenish guilders—at a time when an elegant mansion could be bought for a mere 500 guilders.

The people of Landshut strive to keep the memory of the event alive. In 1882, when the Landshut Town Hall underwent renovations, artists were brought in from Munich to paint murals recreating the wedding. In 1902, the people to Landshut formed Die Förderer, a society committed to reenacting the lavish affair. The first procession was staged in 1903, and Die Förderer continues the tradition every four years in June, putting on Germany's largest historical festival.

The town's landmarks—St. Martin, Burg Trausnitz, and the old, gabled houses—provide an accurate backdrop for the performance. Eager to adhere as closely as possible to the original wedding, the people of Landshut take great pains in preparing their costumes. Some 1,300 residents take part in the Fürstenhochzeit, copying their attire from pictures and written records of the 1475 wedding. The festivities begin with the thunder of cannon and the pealing of St. Martin's bells. Princess Jadwiga and her cortège of over 500 are welcomed into the town and admired by Landshut natives and visitors alike. The reenactment continues along Altstadt, with the bride being welcomed by wedding guests and residents of the town to the sound of drums and fifes.

Turning now from Altstadt to the field at Wiesmahd, the reenactment portrays the camp life of the medieval wedding, with the aroma of roasting meats wafting over the galleries and tents. The nobility are seated in the galleries, ready to receive the knights on the jousting field. The knights proceed attired in full armor, lances at the ready, to joust to the end for the honor of first place.

Although the original wedding took place on November 14, the modern-day reenactment occurs in the summer in order to accommodate visitors on holiday. And even though the pageant successfully draws an increasing number of visitors, the wedding has escaped becoming an event for tourists; it remains a festival of the residents of Landshut and an expression of their pride in their lovely town and its rich history.

Further Reading: Max Ammer's article on Landshut published in the German Central Tourist Association's magazine *Deutsche Revue* (Hamburg), volume 1, 1965, provides an in-depth look at Landshut's history, architecture, and culture. The article is published in German, English and French. Two books including Landshut in their general discussion of Germany are Hajo Holborn's *A History of Modern Germany, 1648–1840* (New York: Knopf, 1964) and *Gothic Architecture* by Louis Grodecki, translated by I. Mark Paris (Milan: Electa Editrice, 1976; New York: Abrams, 1977). Readers interested in the tangled affairs of Europe during the Thirty Years War will enjoy C. V. Wedgwood's *The Thirty Years War* (London: Cape, 1938; New York: Anchor, 1961).

—Monica Cable

Laon (Aisne, France)

Location: Capital of Aisne Department in Northern France, seventy-seven miles northeast of Paris and twenty-nine miles northwest of Reims.

Description: Ancient city in northeastern France and former capital of the kingdom of France. Laon served as an episcopal see from the fifth century and was home to Carolingian kings and various bishops from the tenth century. The Notre-Dame Cathedral of Laon is a notable structure in early Gothic architecture.

Site Office: Office de Tourisme
Hôtel-Dieu
Parvis de la Cathédral
0200 Laon
France
23 20 28 62

Laon's history goes back to Roman times, when the city was known as Laudunum. It is one of few medieval cities almost totally preserved to the present day. Located on a rocky ridge about 330 feet above a surrounding plain that includes Picardy and Champagne, Laon was originally built behind nine fortified gates; three have survived through the centuries.

Because of its location atop steep slopes and a C-shaped rocky butte, Laon was a natural fortress. Attempted invasions rarely met with success. In its earliest days, Laon withstood invasions by Caesar, Attila the Hun, and the Northmen. In addition, the city's proximity to the border of Austrasia and Neustria further developed its reputation as a place of refuge.

Laon was an episcopal see from the fifth century until the French Revolution. From the end of the fifth century, Laon was designated a bishopric. Bishops here were dukes and attended the monarch during coronation ceremonies.

Also at the end of the fifth century, the first permanent structure was built at Laon: a church, initiated by the Archbishop of Reims. It was at this time that Remigius, the archbishop of Reims and later known as St. Rémi, baptized Clovis, king of the Franks. The conversion of the Franks, as well as donations from Clovis and Frankish nobility, made Remigius a wealthy archbishop. In fact, the territory of Reims grew too large for Remigius to manage, so he divided it into new dioceses. Laon was one of the bishoprics that resulted from the reorganization. Laon's cathedral chapter was a powerful one. By 871, during the reign of Charles the Bald, it had gained 2,000 manses by royal favor alone. The bishops were quite influential in Charles's kingdom.

During the violent period that shook Carolingian society in the latter half of the ninth century, Laon served as a refuge. In 881 and 883, it harbored the bodies of Sts. Quentin, Victoricius, and Cassian. St. Vincent, a monastery in Laon, also sheltered refugees and important relics, despite the fact that it was located outside the city's protective ramparts. Founded by Brunhilda, a Merovingian queen who lived from 534 to 613, St. Vincent came under several attacks in the ninth century. The bishops of Laon tried hard to restore the ancient abbey. It was during Bishop Roric's tenure (948–76) that St. Vincent finally found stability and became the second seat of the bishopric and the final resting place of Laon's bishops and canons.

Notre-Dame-la-Profonde was a monastery for women founded in 640 by Salaberga. Her first community of noble women, established near Langres and Luxeuil, was forced to flee to Laon during the Merovingian wars. Bishop Attilo welcomed Salaberga and her followers, who eventually numbered 300 women, including the wives and widows of Merovingian royalty. In addition to Notre-Dame-la-Profonde, Salaberga built six more churches in Laon.

The ninth and tenth centuries probably contributed the most to Laon's history, as it was during this era that Laon was the capital of France and home to the Carolingian kings and their castles. Laon was greatly influenced by the Carolingians, whose kings ruled the West Franks from about 843 to 977.

Laon's first Carolingian ruler was Charles II, the Bald, who ruled from 843 to 877 and served as emperor from 875 until his death in 877. Charles, a son of Louis the Pious, became king following a three-year war in which he allied himself with his brother Louis the German against their eldest brother, Lothair I. In 843 the ending of the war resulted in the Treaty of Verdun, which created boundaries that continue to influence the political map of twentieth-century Europe. According to the treaty, Charles the Bald's kingdom included Western Francia, or what is today modern France, plus additional territory in the north, south, and east. The treaty allotted the kingdoms of Italy and Central Francia, which included the Low Countries, Lorraine (the province later named after Lothair), and Provence to Lothair, who, as Louis the Pious's eldest son, ruled as emperor from 840 to 855. His brother Louis became king of Eastern Francia, or Germany.

Despite the Verdun treaty, invasions, revolts, and internecine warfare continued. In an effort to defend their kingdoms from each other and invading Normans, Arabs, and Hungarians, the kings built various castles. Charles the Bald divided his territory into small feudal units; by the time of his death, the kingdom had lost much of its strength and influence. Pope John VIII therefore passed the title of emperor to Charles the Fat, son of Louis the German. Back at Laon, Charles the Bald was succeeded as king of France by his son, Louis II, the Stammerer, who reigned from 877 to 879.

The Abbey of St. Vincent at Laon
Photo by Claude Jacquot, courtesy of Office de Tourisme, Laon

In 879, Louis II's son, Louis III, became the next king of France at the age of 16. Though he ruled with his brother Carloman because of his young age, Louis III successfully faced a revolt of the nobles and an invasion by Normans at Saucourt in 881. In 882, young Louis III died in a hunting accident.

From 893 through 923, Charles III, the Simple, another son of Louis II, ruled from Laon. The tenure of Charles III, who has been called "the last Carolingian with any real authority in France," included five years of civil war. After unsuccessfully trying to expel the Northmen from the mouth of the Seine, Charles granted a large part of what later became the province of Normandy to Rollo, the Northmen's king.

In 978, Lothair, grandson of Charles III, attempted to gain control of Lorraine, an effort that resulted in an invasion by Emperor Otto II. Hugh Capet, who had succeeded his father, Hugh the Great, as Duke of Franks, in 956, in effect brought Lothair's rule to an end. Capet, along with Emperor Otto III and Gerbert of Reims reduced Lothair's rule at Laon to a nullity. In 987, Hugh Capet was crowned at Noyon after winning an election organized by Adalbero, bishop of Reims, and Gerbert of Reims. Hugh Capet moved the capital to Paris in 987; he granted the royal palace at Laon to the bishop, who ruled the city as count. Under the bishop's rule, Laon became the center of an ecclesiastical seigniory.

Although the period from the tenth through twelfth centuries was marked by feuds and weakening kingships, it also resulted in the birth of French culture and civilization. For the most part, this culture was highly influenced by the church, including the abbey and cathedral schools. Laon flourished, as did French scholarship, especially in such subjects as philosophy and theology.

Louis VI, also known as Louis le Gros, or Louis the Fat, began his reign during the lifetime of Louis V, son of Lothair. He was the first French monarch whose ideas formulated a system of policy. He also influenced an important chapter in the civil history of France: the revolutionary idea of the enfranchisement of commons, or early municipalities, including Laon. These ideas would lead to violent conflict centuries before the French Revolution. On April 25, 1112, one of the bloodiest battles of the Middle Ages was fought in the city. The battle began when the commune of the bourgeoisie and merchants revolted against Bishop Gaudri for annulling their charter. The church and the house of the cathedral's treasurer were burned. After the King helped to recapture the town in a counterattack, the nobility declared revenge, leading to more bloodshed; then the peasants began their own attack.

Within three months, Laon was devastated, and its people began to rebuild the city and its church. They started on a pilgrimage throughout France to raise money for the project. According to legend, they performed miracles and carried such relics as a piece of the True Cross, a part of the Virgin Mary's dress, and a part of the sponge of the Passion.

When the new church was consecrated on August 29, 1114, it could not accommodate Laon's cathedral school or its growing population. Laon's cathedral school had been gaining in popularity throughout western Europe, and the city's population was expanding rapidly, thanks to the prosperous cloth industry. Clothing made in Laon was becoming well known—items were exported as far away as to Egypt. Laon's convenient location on the main route connecting Flanders and the Île-de-France was also good for business. The commune of Laon was reestablished in 1130. As a result of these developments, residents of the city decided a larger cathedral was necessary.

In the late 1150s, Bishop Gautier de Montagne undertook the work of establishing a new cathedral, the Notre Dame Cathedral of Laon. It was a project that would involve three different architects and take nearly seventy years to complete. Construction of the Laon Cathedral started with the choir and south transept. By 1170 these areas were nearly finished. Eight years later, a second architect helped to construct the two eastern bays of the nave. From 1190 through 1215, a third architect and building campaign helped bring the project to completion, with construction of the rest of the nave and the facade. In 1205 the original choir, a simple polygonal mass with a single ambulatory, was replaced with the long, flat-ended choir that still stands today.

The Gothic structure featured seven towers, two of which fell during the French Revolution. Two of the remaining towers feature statues of oxen. According to legend, during the cathedral's construction, a strong and helpful ox came to the aid of a team of workers who had met with trouble.

In 1225, the façade of Laon's cathedral was completed. It features three deep porches and impressive statuary. Because it is one of the few Gothic designs fully completed as originally intended, the cathedral is a noted architectural work. The cathedral school was a center of the Liberal Arts revival of the twelfth-century Renaissance.

In the thirteenth century, flying buttresses were added in an effort to modernize the cathedral with a High Gothic style. Small chapter houses were also added. These additions fit between the choir and the chapels off the transept's eastern sides. In the late thirteenth and early fourteenth centuries, private chapels were built between the pier buttresses. The ecclesiastical structures built around the cathedral include the Bishop's Palace, which is now the Palais de Justice. The Notre Dame Cathedral of Laon is regarded by many as the most impressive early Gothic cathedral.

Templars' Chapel, erected in Laon about the same time as Notre Dame, is a small chapel, unusual because of its octagon shape. A gallery near the Templars' Chapel houses paintings by the three LeNain brothers, who were born in Laon during the seventeenth century and established themselves in Paris in the late 1620s. Antoine, the eldest brother, was known primarily for his miniatures and small portraits. Louis excelled in small pictures of nature; Mathieu, the youngest, painted portraits and larger pictures, such as battle scenes and the mysteries and martyrdoms of saints.

The Church of St. Martin, built in the twelfth and thirteenth centuries, was formerly a Premonstratensian abbey. Although the church burned down in 1944, its abbatial buildings, added during the twentieth century, now house the city's library.

Laon has witnessed several wars. It changed hands several times during the Hundred Years War, which began in 1337. During the Napoleonic Wars following the Battle of Craonne on March 6, 1814, the Russians retreated to Laon and reunited with the Prussians. In a treaty signed with Austria on March 1, the forces agreed to join as allies in war against Napoléon if he refused to consent to their conditions. Although Napoléon did send word to consent, the message arrived too late and on March 9 and 10, 1814, the allied armies joined forces to defeat Napoléon, in a conflict known as the Battle of Laon.

Laon was occupied by the Germans several times over the next hundred years, first during the Franco-Prussian war of 1870–71. Though Germans occupied Laon during World War I, the city managed to survive without much damage. Unfortunately, World War II left many scars on the city.

Several historic buildings, including the Church of St. Martin, were badly damaged during bombing just prior to the liberation of the city on August 30, 1944.

Further Reading: *Monastery and Cathedral in France* by Whitney S. Stoddard (Middletown, Connecticut: Wesleyan University Press, 1966) provides background information and illustrations on medieval architecture, sculpture, stained glass, manuscripts, and other church treasures in France. *Art and Architecture in France 1500 to 1700* by Anthony Blunt (London: Penguin, 1954; New York: Viking, 1954; fourth edition, London: Penguin, 1980) provides a chronological account of French architecture, sculpture, and painting, and discusses the history behind each period and trend. *Cathedral School of Laon, from 850 to 930: Its Manuscripts and Masters* by John J. Contreni (Munich: Bei der Arbeo-Gesellschaft, 1978) is a very good account of Laon's religious history. *The Cabinet Cyclopaedia: The History of France* by Eyre Evans Crowe (London: Longman, 1830) is an old reference work that explores French history in three detailed volumes.

—Kim M. Magon

Laval (Mayenne, France)

Location: Situated on the Mayenne River in the Loire country, 175 miles southwest of Paris and 42 miles east of Rennes.

Description: A town rich in medieval history and Renaissance architecture, situated in the historic province of Maine, which changed hands between England and France throughout the Middle Ages and Hundred Years War; a battleground in the Vendean and Chouan rebellions of the 1790s.

Site Office: Office de Tourisme, Syndicat d'Initiative
Alle du Vieux St. Louis B.P. 614
53006 Laval, Mayenne
France
43 49 46 46

Laval, now a quiet provincial town, was an important crossroads in medieval times. It was founded as a feudal stronghold in the Middle Ages and is still dominated by its medieval castle. The town's old quarter, characterized by narrow twisting streets, is situated on a hill on the west bank of the Mayenne River. A wealth of the medieval architecture survives, including an eleventh-century castle, a thirteenth-century stone bridge, once fortified, and two eleventh-century churches in the northern Romanesque style. The modern town first developed on the east bank and now surrounds the old town on all sides.

The earliest records of the area go back to Roman times and show that the site was occupied by the Diablintes and the Cenomani, Gallo-Roman groups. In the mid–fifth century, these groups merged to create the single district of Le Mans, to which Laval also belonged. The Gallo-Romans were followed by Merovingian and Carolingian kings. After the death of Frankish emperor Charlemagne in 814, the district emerged as the province of Maine (or Le Maine), one of a number of semi-independent feudal states that made up "West Francia," later France. Le Mans, situated to the east of Laval, became the capital. Maine was close to both Brittany and Normandy and was subject to both Breton and Norman invasions in the ninth century. Local authority rested chiefly with the bishopric until the late ninth century, when a warlord, Roger, usurped the bishops' power, ruling as perhaps the first hereditary count of Maine.

About 1020, Count Guy II, one of the first seigneurs de Laval, began construction on what is now known as the Vieux Château on the bank of the Mayenne. The castle proved insufficient against the might of the Norman duke, William the Conqueror, however. Maine fell to the Normans before they invaded England in 1066. William's suzerainty over the province lasted only a short time: the Norman ex-pedition to England gave Maine its opportunity to regain independence.

The castle of Laval also played a frequent role in the wars between local counts of the eleventh and twelfth centuries. The powerful counts of Blois fought over the Maine territory with the rival counts of Anjou, until the Angevins, annexing the territory bit by bit, finally established themselves as the local rulers in the twelfth century. The great accomplishment in these centuries—despite the feuding—was the clearing of the land for agriculture. The region gained enormously in wealth, and Laval became an important market town.

When Henry Plantagenet of the house of Anjou was crowned as King Henry II of England in 1154, the province of Maine (and Laval with it) became part of the English domain. Philip II (Philip Augustus) of France conquered Maine, Anjou, and Normandy in 1204, after which Laval belonged to the French crown, at least for a while. In 1232, Louis IX of France gave Maine and Anjou to his brother, the future Charles I of Naples. His son, Charles II of Naples, in turn transfered Maine to Charles of Valois in 1290. When the Valois kings assumed the French throne in 1328, Maine was again united to France. The province again passed to the dominion of Naples when John II of France granted it as an appanage to his son Louis (later Louis I of Naples) in 1356.

Meanwhile, the successive lords and counts of Laval—among whom were the Montmorencys and Montforts, famous families of the French nobility—also played a prominent role in history. In 1218, the first house of Laval expired with the death of Guy VI. The lordship then passed into the house of Monmorency with the marriage of Emma, daughter of Guy VI, to Mathieu II de Montmorency, a war hero who was appointed constable of France in 1219. Their son, Guy VII, was the ancestor of the second house of Laval. The senior line of the Montmorency-Laval succession continued to Anne de Laval, who married Jean de Montfort. In the absence of male Laval descendants, Anne inherited the city in 1412 and the lordship passed into the house of Montfort.

Anne's most famous relative was her cousin, the notorious child-murderer Gilles de Rais, source of the Bluebeard legends. One of the wealthiest, most cultivated men of his time, Gilles fought during the Hundred Years War, when Maine was often held by the English. After English forces captured Maine in 1425 and then the town of Laval in 1428, Gilles became companion-in-arms to Joan of Arc in the struggle to reunite France under Charles VII. He fought by her side at the siege of Orléans and other battles in 1429 and was made a marshal of France in the same year. He was also with Joan of Arc when she was taken at the siege of Compiègne in 1430. In 1434 he refused to obey royal orders and descended into a life of profligate spending and debauchery, squander-

The Vieux Château and Pont Vieux in Laval
Photo courtesy of Mairie de Laval

ing the entire family fortune in a matter of years. Faced with poverty, he first became an alchemist in hopes of regaining his wealth and then became involved with black magic and satanism. He was hanged at Nantes in 1440 for the crimes of sorcery, sodomy, and murder, having ritually tortured and murdered more than 200 kidnapped children, mostly boys.

The English were expelled from the province of Maine in 1448, after the house of Laval allied itself through marriage to the ducal family of Brittany. Maine passed into the French domain once more in 1481, after Louis I's great-grandson Charles died childless. The family of la Tremoille held the countship of Laval from 1605 until the French Revolution.

During the Renaissance, Laval prospered as the market town of a rich farming area and as a center of textile production, particularly of its famed linen. Agriculture and linen weaving formed the backbone of the Laval economy into the twentieth century. The linen weaving industry dates back to 1289, when Beatrix de Gavre married into the house of Laval and brought weavers from her native Flanders. The city commemorates its linen heritage with a statue of Beatrix, on the old castle's terrace overlooking the Mayenne.

The Maine province, especially the Laval area, figures most prominently in French history late in the eighteenth century, when it was the stage of violent Royalist counter-revolutionary insurrections following the French Revolution. The peasant-led uprisings were known as the Wars of the Vendée (or the Vendean Rebellion) and La Chouannerie. Although the organized Vendean army and the guerilla Chouan bands formed separate revolts and for different reasons, they often joined in each other's battles against Republican forces. Once they even captured Laval's castle. The risings are a little-known aspect of the history of the French Revolution and deserve special telling here; after all, far more people died in the Vendean rebellions than during the Reign of Terror.

The town of La Vendée in western France became a center of resistance to the revolutionary Republic from 1793 to 1796, following the execution of Louis XVI. When the Republic moved to ban the Roman Catholic church and to seize its holdings, many peasants of western France rebelled throughout most of 1793. Peasant leaders were soon joined by Royalist nobles, forming the 50,000-man Vendean army. They ravaged the quiet Maine countryside. The Chouans had first revolted in Brittany and Normandy in 1792, but took part in a major Vendean uprising in the Laval area the next year. The Chouans owe their name—meaning screech-owl in Breton—to the fact that they adopted the hooting of the owl as a night signal. Led by four brothers of the Cottereau family, many of the original Chouans were smugglers and dealers in contraband salt. Their reason for revolting was less a devotion to monarchy and Catholicism than a resentment at the Republicans for abolishing their illicit trade and enforcing conscription. Always a motley group, the Chouans were barbarous fighters, operating mainly from the fields and forests to strike at farms and villages.

The Royalist cause had some initial success in 1793. An alliance of the Vendean army and a Chouan band called the Petit Vendée was under the command of the la Tremoille family's Antoine Philippe, prince de Talmond, lord of Laval and a general in the Vendean cavalry. The Petit Vendée defeated Republican forces and took the castle of Laval. Later that year, however, the Vendean army was forced to give up Laval and retreated south to the Loire. It was following this retreat that the Republicans executed thousands of insurgents, including the prince de Talmond, whom they guillotined in front of his own Laval castle gates. Another Vendean-Chouan insurrection gathered force in 1795 but was quickly put down by the Republican army. Peace did not come to the Maine region until 1800, when Napoléon sent forces to La Vendée and at the same time offered compensations, freedom from conscription, and liberty of worship. Sporadic violence, however, continued into the nineteenth century.

After the French Revolution the old Maine province was divided into the present-day *départements* of Sarthe, Mayenne, Orne, and Eure-et-Loir. Mayenne forms the western part of the historic province. Centrally located, Laval, Mayenne's largest town and the seat of the bishopric, was naturally made capital of the new *département*.

Laval was the birthplace and boyhood home of two of the most famous figures of France's belle époque—the era of elegance and gaiety that characterized fashionable Parisian life in the period preceding World War I: the painter Henri Rousseau (1844–1910) and the playwright Alfred Jarry (1873–1907). These two pioneers of the French avant-garde met in the early 1890s in Paris, where they happened to be neighbors. Rousseau, nicknamed "Le Douanier" (customs-inspector), had previously lived with his parents in the fifteenth-century Beucheresse gate at Laval. The ancient tower still bears a plaque in memory of the tinsmith's son who became a famous painter. The forerunner of the "naive" school of artists and a pioneer of a studied primitivism, Rousseau helped pave the way for surrealism a few decades later. Laval's Vieux Château now houses the Museum of Naive Painting, whose collection centers on some canvases by Rousseau. A generation younger, the visionary poet and playwright Alfred Jarry, son of a Laval wool factory manager, is perhaps best known as the creator of the of the drama *Ubu roi*. This political farce, which mercilessly satirized French government, caused riots when it was first performed in Paris in 1896. Today, the play is generally regarded as the first work of the Theater of the Absurd.

Laval remains an important regional and commercial center. Agricultural development again accelerated in the nineteenth century, when more woodland areas fell to the axe to make room for wheatfields, pasture land, and orchards. The once-thriving linen manufacture has been replaced by cotton and synthetic textile production. Periodic industrial growth, however, has left the old quarter of the town largely untouched. The oldest preserved parts of the Vieux Château, the crypt and keeps, date to the twelfth century. Even older is the Pritz church, the original parish church of Laval, which dates back to about 1000. The Basilica of Notre-Dame d'Avenières, about a century older, was more frequently altered but nevertheless preserves a unity of style that makes it a valuable example of the Romanesque. The Pont Vieux, a thirteenth-century stone bridge, is still in use. The most noteworthy Renaissance monuments include the Beucheresse Gate, which once made part of the defensive perimeter walls of Laval, and the Nouveau Château, which became the home of the counts of Laval in the sixteenth century and now houses the Palais de Justice.

Further Reading: Jean Dunbabin's *France in the Making: 843–1180* (Oxford and New York: Oxford University Press, 1985) covers the period from the collapse of the Carolingian Empire of West Francia to the rise of the French monarchy. Henry Myhill's *The Loire Valley: Plantagenet and Valois* (London and Boston: Faber, 1978) is a historical guidebook for tourists and armchair travelers alike. The *Companion to the French Revolution* by John Paxton (New York and Oxford: Facts on File, 1988) is a general reference guide to the people, places, and events of the Revolution.

—Jeff W. Huebner

Le Mans (Sarthe, France)

Location: In northwest France, on the River Sarthe (a tributary of the Loire) in the district of Sarthe.

Description: Famous for Le Grand Prix d'endurance de Vingt-Quatre Heures du Mans, an automobile race held every June, Le Mans is today an attractive industrial town. It was the site of several bloody battles during the Hundred Years War, the Wars of Religion, and the Franco-Prussian War. Formerly the capital of Maine, its economy is for the most part based on the manufacture of textiles, car parts, and tobacco-related goods.

Site Office: Office de Tourisme, Syndicat d'Initiative
Hotel des Ursulines rue de l'Etoile
7200 Le Mans, Sarthe
France
43 28 17 22

The city of Le Mans dates back to the third or fourth century B.C. During the Roman occupation the city's ramparts were built to defend its inhabitants against possible invasion. The Christian conversion of the population of Le Mans by St. Julian occurred during this period.

During the eleventh and twelfth centuries, Le Mans was a favorite residence of the Plantagenet family. Geoffrey V Plantagenet, the count of Anjou, added the town to his domains in 1128 when he married Matilda, the granddaughter of William the Conqueror (William I of England). He quickly became enamored with the town, and upon his death in 1151 was buried in Le Mans's Cathedral of St. Julian, built during the eleventh and twelfth centuries. Geoffrey's son Henry, the future King Henry II of England, was born in Le Mans and spent most of his life there. From 1173 until his death in 1189, Henry II and his sons, prince John and Richard Coeur de Lion (the Lion-Hearted), were engaged in almost continuous warfare against each other; warfare instigated by Henry II's wife, Eleanor of Aquitaine. In 1189, Henry II lost Le Mans and the chief castles of Maine, but the city became part of Eleanor of Aquitaine's acquisitions. Richard Coeur de Lion succeeded to the English throne, as Richard I, on his father's death. Richard's widow was awarded the city after her husband's death in 1199.

The history of Le Mans has been one of war and plunder. The Hundred Years War, fought sporadically by England and France between 1337 and 1453, led to an almost continual besiegement of the city. It finally fell into French hands in 1481.

Le Mans continued to be a battleground in succeeding centuries. In 1562, the first year of the Wars of Religion, a Huguenot (Protestant) army captured the city and occupied it for three months before being driven out by Catholics. Warfare would continue to rage around France until 1598, when the Edict of Nantes granted religious toleration to Protestants. Then in 1793, the Vendeans, a royalist group that opposed the French Revolution, attacked and occupied Le Mans briefly before being driven out by revolutionary forces in a bloody battle.

In 1871, Le Mans was the site of a decisive action in the Franco-Prussian War. French general Antoine-Eugène-Alfred Chanzy had been retreating through western France after a tough winter campaign in which he had been frequently defeated by the Prussians, but remained a determined opponent. He had brought his forces to Le Mans in December 1870; there the troops rested while the French government and its generals drew up new plans of battle. The Prussians, however, were out to deliver a final blow to Chanzy. In January, the Prussian Prince Frederick Charles, with four corps and four cavalry divisions, arrived at Le Mans and engaged Chanzy's forces in a stubbornly contested battle over three days, January 10 through 12. The eventual French defeat has been blamed on Breton troops who panicked and retreated in a disorderly fashion, in the process breaking the resolve of their fellow soldiers. Chanzy retreated toward Laval, where fresh troops joined him, and was preparing to attack the Prussians at Le Mans again when news of the armistice reached him.

A few years later, events took place in Le Mans that would bring the city fame for something other than warfare. Economically, Le Mans before the Industrial Revolution was largely dependent on candle- and textile-related industries for its income. However, the introduction of new technology following the scientific and technological revolutions of the late eighteenth and nineteenth centuries brought great change, and Le Mans became a leading industrial town, focused mainly on the automobile industry.

The rise of this new industry in Le Mans resulted from the achievements of the Bollée family, former bell founders. In 1873, Amédée built *L'Obéissante (Obediant)*, one of the first horseless carriages. Ten years later, Bollée and his son, also named Amédée, were commissioned by the marquis de Broc to build a huge steam coach; this was one of the first private commissions of a horseless carriage. The younger Bollée invented the first gasoline-powered car in 1896, and by 1898 he had built a sports car model with a streamlined exterior.

Almost simultaneously, Léon Bollée, the other son of the elder Bollée, invented the Léon Bollée Voiturette, a tricycle with a gasoline engine, in December 1895. The vehicle quickly became the favorite mode of transport for leading sportsmen.

The Bollée Company continued to produce innovative, mainly four-cylinder, automobiles until the outbreak of

The choir in the Le Mans cathedral
Photo courtesy of The Chicago Public Library

World War I in 1914, and Léon Bollée's name could be seen on many cars until the 1920s. His widow sold the company to William Morris in 1924. Morris, however, encountered many problems trying to produce Bollée's vehicles as cheaply as his competitors' similar models.

The first proposal for a motor racing circuit at Le Mans came from the Automobile Club de la Sarthe, founded in 1905. The French Grand Prix was held near Le Mans, over a 64-mile circuit, on June 26–27, 1906. It was won by Ferenc Szisz in a ninety-horsepower Renault car. Six years later the Grand Prix de France was held 33.5 miles south of Le Mans. The course was eventually incorporated into the circuit used for the twenty-four-hour race. In 1919 the Automobile Club de l'Ouest mapped out a new permanent circuit of 10.726 miles, south of Le Mans; this incorporated the 1911 circuit. It was won by a Bugatti car.

The most important motor race in France prior to the twenty-four-hour circuit was the Grand Prix de l'ACF in 1921. This was the first race to be held after World War I and was won by U.S. driver Jimmie Murphy in a Duesenberg car.

In 1922 the Automobile Club de l'Ouest devised a race course that would test the durability and efficiency of touring cars and involve night driving. The idea of a race that would continue throughout the night was not original—the first had taken place in Indianapolis in 1904. Annual twenty-four-hour races also took place in Brighton Beach and Coney Island, New York, and the first European twenty-four-hour race was held in Bol d'or St. Germain, near Paris, in 1922.

"Le Grand Prix d'endurance de Vingt-Quatre Heures du Mans," which is held today, owes much to Georges Durand, the secretary of the Automobile Club de l'Ouest, who discussed the idea of a twenty-four-hour race with Charles Faroux, editor of *La Vie Automobile*, and Emile Coquille. Faroux was originally keen to hold an eight-hour race but was eventually convinced of the merits of holding the longer, more challenging event. Coquille put up a purse of 100,000 francs and promised to present a trophy, the Rudge-Whitworth cup. Cars could only qualify for the 1923 race if there were at least thirty identical models being manufactured.

It was decided that the winner of the first race, held in 1923, would not receive the cup outright; rather, the circuit would act as a qualifying round for the Triennial Cup, which could be won in 1925. This cup was only awarded once; a Biennial Cup was given in 1924, and this replaced the Triennial.

The original circuit of 10.726 miles mapped out in 1919 was used only six times. The race course contained highly dangerous roads; thus, for the 1929 race the Automobile Club de l'Ouest built the Rue du Circuit to cut across the difficult Pontlieu, one of the most notorious parts of the track. The layout and length of the race led leading manufacturers, such as Porsche and Bugatti, to regard the track as a testing ground for their inventions. The first race was won by Lagache and Leonard in a Chenard et Walcker. They covered 1,372.94 miles at an average speed of 57.21 miles per hour.

Since 1923 the race has attracted some of the most prestigious drivers in the world. The course has undergone major renovations to make it more interesting and to make driving conditions less dangerous. Two pedestrian bridges shaped like Dunlop tires span the circuit at the Tertre Rouge (Red Hillock) section, and an amusement arcade has grown around the course to cater to the thousands of spectators who come to share the excitement of Le Mans. Today the course is 8.5 miles (13.528 km) in length.

If one examines the history of the races at Le Mans, there are three distinct eras: first, the pre-World War II races, dominated by Bentley, Alfa Romeo, and Bugatti and Simca; second, the golden age of the 1950s and 1960s, dominated by Jaguar, Aston-Martin, Mercedes-Benz, and Ferrari; and last, the modern era, which began in the mid-1960s, during which major technical innovations occurred and the regulations governing motor racing changed substantially.

The Bugatti circuit, which is between 2.5 and 2.7 miles in length and includes part of the Tertre Rouge and several difficult bends and turns, is available for public use; motorcycles can often be seen using the track. It was used for the 1967 French Grand Prix. Unsurprisingly, a car museum, which houses more than 150 impressive cars, is situated near the track.

Le Mans is the site of several important and beautiful historic sites. The former bishop's palace is today the Musée de Tesse. It contains several wonderful treasures, including an outstanding example of Philippe de Champaigne's still lifes and an impressive collection of northern European and French paintings.

Probably the most interesting and visited site is the Cathedral of St. Julian du Man, constructed in remembrance of the former bishop of Le Mans. The exterior is a mixture of architectural styles from the eleventh to fifteenth centuries. Although predominantly Gothic, the building possesses a Romanesque nave, decorated with beautifully illuminated stained glass windows. Inside the baptistry, the count of Maine's vault can be found. The Gothic chancel is one of the finest examples in France; it is extremely high—112 feet. A fine selection of tapestries decorates the choir stalls.

Le vieux Mans (the old town) is situated on a hill within the Roman-built fortifications. It is a picturesque town with tiny winding streets, wooden houses, and many charming shops and restaurants. In addition to the cathedral, the old town includes several other historical places, such as the oddly named Maison de la Reine Bérengèe, a house named after Richard Coeur de Lion's wife, Berengaria of Navarre, but built at the end of the fifteenth century, well after her death. Today the building houses a museum of history and ethnology. It contains interesting examples of the pottery and ceramics of the region, as well as paintings and other art.

Further Reading: *Blue Guide France* (London: Benn, and New York: Norton, 1984; third edition, London: Black, and New York:

Norton, 1994) is a good general introduction to the history and sites of the country as a whole, and the town more specifically. *Le Mans* by Anders Ditler Clausager (London: Arthur Barker, 1982) is a good brief history of Le Mans motor racing, with a detailed breakdown of individual races, winners, and cars, illustrated with photographs. *The Le Mans 24-Hour Race* by Michael Cotton (London: Stephens, 1989) is an illustrated history of the famous course.

—Aruna Vasudevan

Letchworth (Hertfordshire, England)

Location: Thirty-six miles north of London, just off the A1(M).

Description: The world's first Garden City, as defined by Ebenezer Howard in 1898. Begun in 1903; now covering 4,598 acres, of which 2,500 acres is agricultural land; the town was originally designed by architects Barry Parker and Raymond Unwin, based on plans by Howard.

Site Office: The First Garden City Heritage Museum
296 Norton Way South
Letchworth Garden City, Herts SG6 1SU
England
(462) 683149

Letchworth was Britain's—and the world's—first Garden City, as defined by Ebenezer Howard in his book, *Tomorrow: A Peaceful Path to Real Reform,* first published in 1898. Before Letchworth, Chicago had been known as "The Garden City," and Christchurch, founded in 1850 in New Zealand, was also known as that country's Garden City. The New York suburb on Long Island, started in 1869 by Alexander T. Stewart, is named Garden City, and there are now several more villages and towns in the United States bearing that name. However, none of them conforms to Howard's definition of a garden city: a town surrounded by an agricultural green belt, the whole of the land "being in public ownership or held in trust for the community."

Ebenezer Howard, born in 1850 to a London baker, started working at the age of fifteen in a city stockbroker's office. Within three years he had taught himself Pitman's shorthand, which he was able to perform so well that a famous preacher, Dr. Joseph Parker, employed Howard as his secretary. In 1871 Howard emigrated to America to seek his fortune; he passed the better part of five years working in Chicago as a stenographer to the law courts. Although he did not make a fortune, he did see the rebuilding of Chicago after the 1871 fire, including the building of its pioneer "garden suburbs," in particular those taking shape in Riverside.

After returning to England, Howard invented the Jordan-Howard printing press while working as a parliamentary stenographer. Then in the 1880s he read the American writer Henry George, whose book on land reform, *Progress and Poverty,* impressed him deeply. He also read Edward Bellamy's *Looking Backward,* which developed his Utopian ideals of a society run on cooperative principles. Howard became interested in the early industrial "villages," including George Cadbury's Bournville and W. H. Lever's Port Sunlight. All these influences appear in Howard's 1898 *Tomorrow,* the revised edition of which, entitled *Garden Cities of Tomorrow,* was published in 1902. Such was the impact of

Howard's ideas that by October 1903, the first Garden City was begun.

While there were precedents for Howard's ideas (among them the Old Testament Levitical Cities, Jerusalem, Plato, Aristotle, and Sir Thomas More's Utopia), Howard's garden city idea was intended as a practical solution to the serious problems facing his country since the explosion of city populations during the Industrial Revolution of the late eighteenth and nineteenth centuries. In his book, Howard contrasted city life with country life, outlining the benefits and drawbacks of each. His Garden City was to combine the best of both worlds: fresh air and water with plenty of jobs; low rents with high wages; fields and parks with plenty of social activity and culture; and gardens and pleasant housing with factories.

Howard proposed a town surrounded by a large agricultural area. The farmers from the country would find a market and social life in the town, and the townspeople would have the advantage of easy access to the countryside. The town, Howard felt, should be planned in its entirety, including the roads, buildings and gardens, landscaping, amenities and industry, and its size should be limited. Within each town he proposed there should be neighborhoods or wards, communities within the community. His diagrams dividing the town into zones—a very new idea at the time— show a central park, allotments, convalescent homes, and asylums. The underlying tenet of the plan was that the whole area, including the agricultural green belt, should belong to the people living there, and that any income and profit from the land should be used for the inhabitants' collective benefit. Howard also proposed to develop a cluster of towns around the main town.

One year after Howard's *Tomorrow* was first published, T. W. H. Idris founded the Garden City Association to publicize Howard's ideas and formulate plans for the building of a garden city. Within two years, the two men's enthusiasm had attracted many influential supporters, including George Cadbury and W. H. Lever. The Garden City Pioneer Co. Ltd. was formed in July 1902 to find a site for the first garden city and set Howard's plan in motion. By July 1903 the site had been chosen, and on September 1 of the same year First Garden City Ltd. was registered, with both Cadbury and Lever on the board. The stated aim was to develop the Letchworth estate on the principles outlined in Howard's book.

The site chosen, covering 3,822 acres (about two-thirds of the 6,000 acres Howard had envisioned), lay thirty-five miles north of London near a railway and the Great North Road (now the A1[M]). More land was soon purchased, bringing the area to 4,598 acres, and taking in the three small villages of Letchworth, Willian and Norton. The area now

Letchworth founder Ebenezer Howard, center, at 1911 ceremony

May Day parade at Letchworth, 1909
Photos courtesy of the First Garden City Heritage Museum, Letchworth

known as Letchworth previously had been settled during the Iron Age, around 600 B.C. The combined population of the three villages in 1903 was only 450. A few structures remain today, including the fifteenth-century Letchworth Hall, the twelfth-century St. Mary's Church, and some seventeenth-century cottages.

No precedent existed in England for the creation of a town plan, so after a careful survey of the area developers chose a plan submitted by two little-known architects, Barry Parker and Raymond Unwin, based largely on the "neighborhood principle." Unwin, a Fabian who supported William Morris' Socialist League, later wrote an innovative book, *Town Planning in Practice* (1909). Parker was involved with the Arts and Crafts Movement, and in 1927 he was to plan the Wythenshawe satellite town of Manchester. Parker and Unwin were enthusiastic supporters of Howard's ideas and pursued his ideals of the marriage of town and country and low-density housing, which allowed for a maximum of twelve houses per acre.

In April 1904 the Estate Office opened; on July 7 the first new house was occupied, and later in the same month the waterworks began operation. However, First Garden City Ltd. was severely hampered by a shortage of capital. The constitution of the company meant that shareholders had no equity—they would never actually own the land—and they were to receive, at some time in the future they hoped, a maximum of 5 percent per annum dividend; not a great inducement to any investor save those with the highest of ideals. Having bought the land, the company was already too heavily mortgaged to build. Determined not to compromise by selling freeholds, the company manager, Thomas Adams, persuaded the *Spectator*'s editor, J. St. Loc Strachey, to run a "Cheap Cottages Exhibition" at Letchworth in 1905. The editor was keen to promote the building of cottages for agricultural workers as a means of slowing down their migration from the country to the overcrowded cities.

As a result of this joint venture, approximately 120 cottages were built at a cost of about £150 per cottage, and the publicity generated attracted 60,000 visitors to the exhibition. The railway company had opened a station with regular service just three months before the exhibition; the gasworks were in operation, and Heatly Gresham Engineering Works had opened the first Letchworth factory. The exhibition drew some enterprising young architects to the new town, some of whom leased sites and built their own houses there. A cottage building society was started and a subsidiary building company was set up by First Garden City Ltd., which guaranteed 4 percent on capital. Letchworth Garden City was under way.

In 1907 an *Urban Cottage Exhibition* was held, producing another burst of building. Each subsequent year left a trail of significant firsts. In 1906 the first public building, the Mrs. Howard Memorial Hall, was built to commemorate Ebenezer Howard's first wife, who had died in 1904, and the local newspaper and first Dramatic Society were launched in the same year. In 1905 Howard himself had moved to Letchworth, where he stayed until 1920, the year he moved to his second garden city, Welwyn Garden City. When Howard died in 1928, he was buried in Letchworth. In 1907 The Skittles Inn a temperance pub, opened. At that time there was strong feeling against alcohol in England, similar to the sentiment that led to prohibition in the United States. The directors decided there would be no premises in Letchworth licensed to sell alcohol, and the center of the town remained dry until 1962, although three old village pubs on the outskirts of the estate continued to sell alcohol. Also in 1907, electricity was provided for the Letchworth residents, and in 1909 the Palace Cinema opened, one of the first purpose-built cinemas in England. In 1911 the census revealed a population of 5,324, and in 1913 First Garden City Ltd. paid its first dividend, of 2 percent.

World War I put the brakes on further development at Letchworth. However, the population was dramatically increased by about 3,000 Belgian refugees who descended on the town, most of them working in the new Kryn and Lahy Metal Works, helping the war effort. After the war, peace was celebrated by the building of the Peace Memorial Hospital, St. Paul's Church, and a new gate at All Saints' Church, in the village of Willian. By 1919 the population had reached 10,000, and an Urban District Council was formed. That same year Welwyn Garden City was begun, only thirteen miles from Letchworth.

Some residents felt that First Garden City Ltd. was not doing enough to promote the growth of Letchworth. The Urban District Council considered taking over the estate, but the legal and financial barriers proved too complicated. Despite this contention, Letchworth continued to grow in the interwar period (1918–39). New industry was attracted, a new parade of shops was built in 1924, and an open-air swimming pool was opened in 1935. A second public hall, The Icknied Halls, had been built in 1924–25, a Theosophical College in 1919–24 (extended 1934–39), and a Grammar School in 1931. In 1935 a Town Hall was provided, and in 1938 a library was built. By 1936 the town had acquired its second cinema, and in 1948 the Urban District Council started building a new housing estate. In 1939 the population had reached nearly 16,000, a number that was again increased during World War II by evacuees, troops, and workers who were training at the Government Training Centre in Letchworth.

1953 was Letchworth's Golden Jubilee Year, and the town held a Jubilee Fair of Industry and Trade. The 1950s and 1960s saw full employment in the town. By 1961 the population had topped 25,000, but Letchworth threatened to become a victim of its own success. The postwar growth had made its shares highly attractive to speculators: the first garden city was in danger of a takeover, first by Raglan Property Trust Ltd. and then by a company called Hotel York, which won control of First Garden City Ltd. at the end of 1960. The Urban District Council and the people of Letchworth were faced with the prospect of the breakup of the garden city estate and the end of Ebenezer Howard's vision. Battle commenced with a "Save Letchworth Garden City" campaign, in which the council and community fought to present a private bill to parliament that would set up a public body to take over the estate and preserve it, and Howard's principles, intact.

The campaign against Hotel York began in November 1960, and the bill received royal assent on August 1, 1962. The Letchworth Garden City Corporation Act transferred the estate from First Garden City Ltd. to Letchworth Garden City Corporation, which took control on January 1, 1963. The town began the second phase of its existence safe in the knowledge that the garden city concept had been preserved and the estate was protected intact.

Under the terms of the act, any profits arising from the operations of the corporation must be used for the benefit of the community. The corporation, as landlord of the estate, has ultimate control over the development and use of the land and its buildings. Hotel York had sold off three freeholds during its brief period of control, but all other land and buildings were still leasehold. When the Leasehold Reform Act was passed in 1967, Letchworth attempted unsuccessfully to get exemption status. The result was that freeholds could then be bought in the garden city. However, the corporation retains the power of giving "landlord consent" under a special "Scheme of Management" both for new buildings and for any alterations or development.

The Corporation and the Council have taken their responsibilities to preserve standards of appearance very seriously. When the town center was redeveloped from 1971 to 1975, architects used traditional brick and tiling details in an effort to help the new buildings blend in with the rest of the area. In 1973 the corporation went into surplus for the first time and, after consultation with the community, built several new facilities. These included the Standalone Farm, an educational model farm, a Day Hospital (the only privately funded U.K. hospital to take only nonpaying National Health Service patients), a community and leisure complex, and the excellent First Garden City Heritage Museum.

In 1974 a large Conservation Area was defined, which included the town center and the pre-1914 original garden city buildings, along with some older areas. The North Hertfordshire District Council, which took over from the Urban District Council in 1974, has responsibility for these areas. The Conservation Area designation means that all new buildings must blend with the "established character of the town" in terms of design and materials. In 1979 some 400 properties were listed as buildings of special architectural or historic interest, including many fine examples of the original cottages. Several of these cottages were designed by Parker and Unwin, along with others, designed by many different architects, which had been entered in the 1905 and 1907 Exhibitions. While their building methods varied (ranging from steel-framed to prefabricated and concrete block structures), all generally conform to the original garden city style defined as early as 1904 in the Building Regulations of First Garden City Ltd. Buildings were to be in the vernacular

tradition, using simple but beautiful materials. They had steep roof pitches and used handmade clay tiles. Because the local brick was considered to be aesthetically inferior, the walls were usually roughcast cement that was then finished. Considerable attention was paid to such details as the gables, dormer windows, doors, and porches. Many of the cottages had small-paned timber casement windows in cottage style, although some of the larger ones had leaded casements. Most featured tall shafted chimneys, in Yeoman Tudor style. The overall color scheme was white or cream with green and black. Although a great variety exists among the buildings up to 1925, most do observe these general principles, and guidance has been published for owners on how to conserve these qualities.

In addition to man-made structures, great importance is placed on the preservation of trees and hedges. It is said that when Letchworth was built only one tree was felled, and a great number and variety of trees, hedges, and grass verges were put in place. Originally each road was planted with a single species of tree, and in the early days the town would hold "Arbor Days" when the trees of a street came into bloom. Whereas the farms in the green belt area were once all cultivated independently, now the corporation farms the entire area (called the "Rural Estate"), making a good profit while at the same time employing nonintensive methods of farming and decreasing the use of pesticides.

At the time of this writing, Letchworth Garden City is in the process of another major change. When the Letchworth Garden City Corporation Act was passed in 1962, the private First Garden City Ltd. was replaced by a public body, the corporation, in keeping with Ebenezer Howard's thinking. The corporation has six board members, four of whom are appointed by the government, while the board itself remains independent of it. The new corporation took ten years to pay off the large compensation bill incurred by the change, but now the estate is valued at more than £50 million. The corporation's affairs were administered by the Department of the Environment's "New Towns Division," which has now been disbanded. Letchworth was, and is, different from the other new towns in not having been funded by the government.

Since Letchworth's official position needs clarification, the corporation, at the Secretary of State's instigation, has been reviewing the situation. In 1992 the first steps were taken to dissolve the corporation and replace it with the Letchworth Garden City Heritage Foundation. This is to be a society with charitable status, completely independent of the government but able to retain its tax-exempt status. Despite much debate, the change appears to be regarded as generally positive by the people of the town. The Letchworth Bill has already gone through the preliminary stages in the House of Commons. There are to be up to thirty Governors of the foundation, of whom two-thirds or more could be Letchworth residents. The chief aim of the change is "to preserve the continuity of Letchworth's garden city heritage, and the associated benefits to the community . . .".

As Letchworth approaches the beginning of the third

phase in its history as it nears its centenary, its importance remains. The garden cities represent the first real steps taken anywhere toward the concept of town planning. Although the only two garden cities, properly speaking, to be built in England, the examples of Letchworth and Welwyn Garden Cities have inspired others all over the world: Many of Howard's ideas have been applied to garden suburbs in England and elsewhere. Ironically, the extension of large cities is exactly what Howard was seeking to prevent.

Further Reading: *Tomorrow: A Peaceful Path to Real Reform* by Ebenezer Howard (originally published 1898; revised as *Garden Cities of Tomorrow,* 1902; third edition, edited by F.J. Osborn with an essay by Lewis Mumford, London: Faber, 1946) is the book that made Letchworth possible. Howard's plans and ideals for a garden city provide an invaluable backdrop to any history of Letchworth. *The Building of Satellite Towns* by C.B. Purdom (London: Dent, 1925; revised, 1949) is a well-documented account of the development of Letchworth and Welwyn Garden Cities by one who was involved in the creation of both. *The Letchworth Achievement,* also by C.B. Purdom (London: Dent, 1963), was written just after the passing of the Letchworth Corporation Act in 1962; it gives a full account of the battle leading up to the act and includes a copy of the act itself. *Green-Belt Cities* by F.J. Osborn, with a foreword by Lewis Mumford (London: Evelyn, Adams and Mackay, 1946), offers another account of the two garden cities by a writer also involved in the planning of both. *Letchworth: The First Garden City* by Mervyn Miller (Chichester, West Sussex: Phillmore, 1989) is the first comprehensive history of the town and surrounding area. A wide range of materials relating to Letchworth is also published by The Town and Country Planning Association, 17 Carlton House Terrace, London SW1Y 5BD.

—Beth F. Wood

Liège (Liège, Belgium)

Location: On the rivers Meuse and Ourthe in eastern Belgium, close to both the German and Netherlands borders.

Description: Third largest city in Belgium; major economic center whose importance dates back to the Middle Ages, when it was granted major privileges as a prince-bishopric; described as "Walloon Venice," it is the largest French-speaking town in Belgium.

Site Office: Liège Tourist Office
En Feronstree 92
4000 Liège, Liège
Belgium
(41) 32 24 56

Liège (Lüttich in German, Luik in Flemish) is an old and important city, dating to the eighth century A.D., when St. Hubert built a chapel in memory of St. Lambert, the bishop of Teongeren and Maastrict, who had been murdered there in 705. The chapel rapidly became a popular place for pilgrims to visit, and sixteen years later Liège became a bishopric.

One of the most influential of the bishops of Liège was Notger, who, at the close of the tenth century, turned the bishopric into a principality, beginning a tradition that lasted until 1794. Notger's kingdom was large, more than half of the size of modern Wallonia, and included land from the German Holy Roman Empire. Under Notger's guiding hand the city became a cultural and educational center. Its ties to Germany allowed it to benefit from the Ottonian Renaissance, the cultural and artistic revolution during the reign of Holy Roman Emperor Otto I in the tenth century. The Mosan school of art first developed during this period, and would produce some of its greatest works in ivory, gold, and silver between the tenth and thirteenth centuries. Ever since, the city of Liège has been linked to art.

Called the Athens of the north, medieval Liège became the site of many feuds and battles between the ruling classes and the populace. Quick to remonstrate against injustice, the Liègois have always fought for their rights. The Perron, a column topped with a pine cone and cross, stands in the central marketplace, and appears on the city's coat of arms; it symbolizes the freedom and liberty of the town's people. Originally erected in the thirteenth century, the Perron was destroyed during a storm and replaced in 1697. It was temporarily stolen by Charles the Bold, who later returned it to the city.

For most of the fifteenth century the city was dominated by the extravagant Burgundian family, who influenced the town's cultural, political, and intellectual life, but curbed most of the citizens' freedom. John the Fearless, the first Burgundian ruler, defeated Liège at Othee in 1408. Philip the Good also defeated the town at Montenaken in 1465, and in 1465, Charles the Bold, in response to a revolt by the Liègois, executed all of the town's able-bodied men and carried off the Perron to Bruges. He destroyed most of the city, except for the churches. Nine years later Mary of Burgundy restored the city's privileges and gave the Liègois permission to rebuild their city. For three centuries after this the town attempted to maintain its neutrality in the French wars with the Holy Roman Empire. The de la Marck family were important in restructuring the city, although William de la Marck, the black sheep of the family, nicknamed the Wild Boar of Ardennes, murdered the prince bishop Louis de Bourbon in 1482; he was sent to his death at Maastricht by the bishop. Evrard de la Marck brought Liège both peace and wealth; however, on his death in 1538 war broke out again.

Liège's policy of neutrality was born of its inability to defend itself. However, its strategic location meant that it was often forced by other powers into a defensive posture. The town raised only one army of any great significance during the War of the League of Augsburg, when it had 6,000 men in its army. However, this was completely inadequate for effective defense. The political alliances of the prince-bishops, and their relations to many of the important ruling houses of Europe made a policy of diplomacy almost impossible. During the seventeenth century the principality was ruled by a series of younger sons of the Bavarian Wittelsbachs who negotiated alliances for their own ends and not those of their people.

The ambivalence felt by the Liègois who wanted a democratic government came into conflict with the elitism of the ruling bishops. This is best illustrated by the conflict that arose between Ferdinand of Bavaria, piece of Liège, and the Liègois during the Thirty Years War (1618–48). The Liègois disagreed with the principle of absolute monarchy. A series of revolts made the citizens' displeasure obvious, and the prince invited the Spanish and imperial armies to help put down any insurrections. In contrast, France and the Netherlands were willing to help Liège for political gain.

While Ferdinand overtly supported the neutrality of the city, he accepted financial aid from the emperor. In the second half of the seventeenth century, his successors gave their support to the French instead, which led to a greater military presence in the city for the rest of the seventeenth century and most of the eighteenth century.

However, this was also a period of great cultural progress. Liège has always been a city of enlightenment and progress, breaking new ground both politically and culturally. Over the centuries it has attracted such artists as Lambert Lombard (1505–66) and Jean Lecour (1627–1707), and such composers as César Franck (1822–90) and André-

A view of Liège
Photo courtesy of Belgian National Tourist Office

Ernest-Modeste Grétry (1741–1813). During the eighteenth century several outstanding buildings were constructed and some of the city's best examples of cabinetmaking were produced.

During the same period the era of the prince-bishopric ended. Liège became involved in the revolutionary activity sweeping France in the late eighteenth century, and was annexed by that country in 1794; the last prince-bishop was expelled. Liège remained part of France until Napoléon's downfall in 1815, when the Congress of Vienna assigned it to the Netherlands.

The Liègois had always shown great courage and bravery during battle and had always fought for their freedom. In 1830 the working class and the bourgeoisie began to revolt against the Dutch government. A small group of journalists tried to direct the rebellion, and the clergy even became involved. However, the reasons behind the insurrection became confused—it was unclear whether the attack was aimed at the constitution, the country as a whole, the sitting government, or something else. The Dutch deemed it to be an attack on their country.

The Belgian economy was depressed at the time, and this exacerbated the situation. The winter had been severe and crops had failed. There was severe unemployment and rising discontent with the situation. Thus, the first signs of revolt were seen in Brussels on August 25, 1830, and rebels attacked the houses of leading dignitaries in the city, who were for the most part Belgians. During the next few days, the working class in the suburbs and in Liège followed suit, wrecking factories and equipment. The response was that the bourgeoisie decided to set up a civil guard to quell the rebellion. The revolt, however, resulted in the formation of the Kingdom of Belgium, independent of the Netherlands. The Liègois had shown their courage in supporting their right to freedom and to directing the future of their country.

Similarly in 1914 the townspeople of Liège were important in defending the city's citadel until the French and Belgian forces could rearm to fight the Battle of the Marne. When the city fell into German hands, fifty men were executed. Again, during World War II, the Germans attacked and occupied the city; much of it was destroyed during the blitz of 1944–45.

Historically, Liège's economy has been based on coal- and steel-related industries, ever since a blacksmith discovered coal seams, near Publement, in 1198. For centuries the Liègois have been nicknamed "coalheads," a fact of which they are proud. The population has historically been employed as blacksmiths, artisans, steelworkers, and armor makers—Liège was once famous for its armor production. Following the Industrial Revolution, Liège's location on the river made it a popular site for industry. The first Continental European locomotive was manufactured in the city, and the Bessemer steelmaking process was tested here. The construction of a canal in 1939 allowed the region to prosper—no mean feat considering the heavy bombardment that Liège suffered during the war. Today, metalworking, chemicals,

and plastic manufacture are among the city's most important economic activities.

Despite its heavy dependence on industry, Liège is an attractive city. Large areas that were once covered by the Meuse River are now streets and parks. Although many fine buildings, including some lovely art nouveau structures, were destroyed during the heavy bombing of the last two years of World War II, the architect Charles Vandenhove led efforts to reconstruct and renovate many eighteenth- and nineteenth-century houses and buildings in the city. In addition, Vandenhove brought in other important architects, famous artists, and interior designers to help him restore the city to its former glory.

Around the verdant area of Sart, the university became the catalyst for a series of innovative, and sometimes shocking, architectural pieces, including the Physical Education Institute, which has a roof made of undulating concrete, designed by Vandenhove.

Among the city's main historic sites is the Palais des Princes-Eveques (Prince-Bishops' Palace), a large Gothic building originally constructed for Notger in the late tenth century. The palace was restored completely in the sixteenth century, and today is used as an administrative building. The writer Victor Hugo once described the palace as "an architectural complex so strange, so sullen and so splendid." The palace has two courtyards—the larger one is celebrated for its sixty massive columns and galleries.

A seventeenth-century Mosan school friary has been converted into the popular Musée de la Walloone (Museum of Walloon Life). This houses an impressive collection of regional art and marionettes, and even has a puppet theater. A character frequently depicted in puppet theaters is Tchantchès (Francis or François), Liège's most famous folklore character. Local legends have the homely but lovable Tchantachès fighting alongside Charlemagne and his nephew Roland in the eighth century. Tchantchès has become a symbol of the city because he epitomizes the Liègois citizen who is "very obstinate . . . very independent, but at the same time [possesses] a heart of gold and [is] always ready to fight for a good cause."

One of the finest architectural examples of the Mosan School is a mansion built by the Curtius family. Jean Curtius's house was constructed for the princely sum of 1 million gold francs, which he had earned selling arms to Spain. The building has been converted into the Musée Curtius (Curtius Museum), and houses such masterpieces as the twelfth-century *Mystery of Apollo*.

Liège is the site of many museums, in keeping with the city's interest in the arts. The Musée d'Art religieux et d'Art Mosan (the Museum of Religious and Mosan Art) houses a brilliant collection of work from the school, including the *Virgin and Butterfly*. The pieces held in this museum trace the development of religious art in the region from the fourteenth century onward, and include work from other north European countries. The Walloon Art Museum contains pieces by leading artists from all over Belgium. The

Modern Art Museum, built in 1905, features art from the mid-nineteenth century to the present date, and focuses on modern French and Belgian art. The city also possesses a museum dedicated to its iron and coal industry, based in a nineteenth-century workshop.

Because of its episcopal past, Liège is full of beautiful churches; in fact, one can probably count at least 100 different spires on the city's skyline. Probably the most beautiful is the Romanesque Eglise St.-Barthélemy (St. Bartholomew's Church), which originated between 1010 and 1015 and was built in sandstone. The church was renovated and enlarged in the late eighteenth century, and today it contains one of Belgium's great art treasures—the bronze baptismal font by Renier du Huy. The Church of St.-Denis, like St.-Barthélemy a former collegiate church, possesses a sandstone tower previously used as a lookout in the city's defense. Liège also possesses some beautiful examples of Gothic religious architecture, including the Cathédrale St.-Paul, originally built in the thirteenth century, with its beautiful stained glass windows, and the impressive church of St.-Jacques, constructed during the sixteenth century.

Liège also contains an almost separate enclave—Outre-Meuse (Across the River Meuse)—situated on the right bank of the Meuse. Reputedly the birthplace of Tchantchès, Outre-Meuse is also called the Free Republic, as along with the borough of St.-Pholien-des-Près it is one of two independent areas. These areas originated from the existence of the weavers of St.-Nicholas and the Tanners of St.-Pholien. Each year on 15 August the "Republic" of Outre-Meuse celebrates its national day, and on the Festival of the Blessed Virgin the front of each house is made into an altar to celebrate the Black Virgin.

Further Reading: *Introducing Belgium* (London: Harrap, 1981) by Anthony Barnes is a good introduction to the country and its history. *Belgium and Luxembourg* (Harrow, Middlesex: Michelin, 1994) is part of the Michelin Series, which always provides comprehensive histories of the regions they cover. *Modern Belgium* (Brussels: Modern Belgium Association, 1990), edited by René Bryssinck et al., is composed of very good individual essays on the culture, economy, and society of the country.

—Aruna Vasudevan

Lincoln (Lincolnshire, England)

Location: Lincoln is situated in the county of Lincolnshire, an area bordered by Nottinghamshire to the west and the North Sea to the east. East Anglia stretches farther to the south and east, and the Midlands, farther west. Lincoln is approximately 150 miles due north of London. The nearest cities of Nottingham, Sheffield, Leeds, and York, all larger than Lincoln, cluster in an arc to the north and west.

Description: Lincoln is a mid-sized city, the county's seat, set among some of England's most productive farmland. Dominated by its early English Gothic cathedral, the city rests on one of the fens' few steep hills. It has retained its medieval aura, a place where it almost seems as if time stood still after the fifteenth century.

Site Office: The Tourist Office
9 Castle Hill
Lincoln, Lincolnshire LN1 3AA
England
(522) 529828

Before the Romans arrived on the shores of Kent, a local tribe named the Coritani established a town enclosed by earthworks, called Lind-coit, in the fen country. After a second invasion, during the reign of Augustus, the Romans pushed their way north toward Scotland, capping their campaign near the present-day border between England and Scotland by building Hadrian's Wall. Lindum Colonia, as the Romans named what would become Lincoln, was established about A.D. 100 as one of nine colonies made up of veteran legionaries, who held land in exchange for providing their military services when needed. Lindum clung to the side of a steep hill, rising dramatically above a wide pool and surrounded by the flat and swampy fens. "Lyn" was the British name for a clear, still pool; "dun" signified a hill fortress.

Situated along the Witham River, Lindum was a northern garrison town at the crossroads between two major military roads, rechristened Ermine Street and Fosse Way by the Saxons. It was lesser in importance than London to the south or York to the north, but still a strategic stronghold against the British tribes who strove to reclaim the land. The fortified camp was enclosed by walls and ringed by a ditch. As space within city walls ran out, growth extended south, down the hill. Newport Arch, which still spans a major road in Lincoln today, represented the northern gate of the city (though the city has risen over the centuries by almost ten feet, dwarfing the once-impressive height of the arch.) By the early fourth century, Lindum had achieved provincial capital

status. The Romans engineered a series of channels to drain parts of the fens, enabling large-scale farming in the area, and they introduced many new crops, such as rye, oats, flax, and cabbage. Lindum also became an important center for the wool trade and a seat of Christianity. When Constantine the Great presided over the "Golden Age" of Roman power in Britain, between 306 and 337 A.D., Christianity gained even more support as pagan practices, previously tolerated, were made illegal. It was during this age that a church was first built in Lincoln. And, when the Council of Arles was held in France in 314, three bishops from Britain attended: one from London, one from York, and one, Adelfius, from Lincoln.

But the death of Constantine in 337 mirrored the decreasing power of the Roman Empire as a whole. Security and prosperity were waning. By the late fourth century, after Roman troops were recalled to the continent to defend against Saxon attacks there, many of the towns stood empty. In Lincolnshire, much of the engineering progress was left to deteriorate, and the waters began to reclaim the farmland. Military rebels proclaimed power, sparking fierce battles as Saxon raids intensified along the coast and the Scots attacked from the north. The depleted Roman militia left in Britain struggled to cope with these fierce raids. By 410, Roman rule was in disarray, and the Saxons had succeeded in separating much of Britain from the empire. The last known appeal by Britons for military help from the Romans came in 446, but the plea went unheeded. The Saxons ultimately won control over most of Britain, and Lincolnshire became part of the Saxon kingdom of Mercia. In taking over Lincoln, the Saxons destroyed much of the city in their attempts to obliterate all traces of Christianity and the Romans. The Saxons, from the cold and windswept eastern shores of the North Sea, were not merely fierce fighters and eager colonizers; they were also expert farmers, and they succeeded where the Romans had failed in Lincolnshire. With their iron axes and deep ploughs, the Saxons cleared the heavy forests of the eastern midlands and turned marshy valleys into terraced fields, creating the now-familiar English countryside. Most Saxons were freemen, independent of taxes and tithes, instead owing their community leaders their warrior skills, when necessary. The Saxon kingdoms were held together by a sense of kinship, duty, and loyalty.

But that bond was not strong enough to withstand the powerful Danes, who swept into the northeast of England to challenge the Saxons toward the end of the eighth century. The Danes' attacks on Lincolnshire were successful due in part to the area's proximity to the North Sea and to inland waterways such as the Witham River. When the Danes threatened Wessex to the west, however, the Saxons were victorious, thanks largely to the military prowess of Alfred the Great in the late ninth century. Saxon Wessex soon became

Lincoln Cathedral, top, and Lincoln Castle, bottom
Photos by Kevin Newton, courtesy of City of Lincoln

the predominant economic and political force in Britain, due to Alfred's foresight and wisdom, which helped create a framework in which all of England's disparate peoples could live together and prosper. The Danes, who grew to respect Alfred as both ruler of Wessex and overlord of Mercia, quietly began to assimilate into Britain and slowly gave up fighting for farming and trade. Hardworking and shrewd, they took to commerce and made Lincoln, among a handful of other towns, a large trading town and prosperous port on the Witham River. In fact, the Danes have been credited with planting the seed of Britain's importing and exporting traditions, interested as they were in international exchange. Lincoln eventually became a major town under the Danelagh (Danish authority) and was home to many Norse settlers, as evidenced by the word "gate" in many of the street names, derived from the ancient Scandinavian word *gata,* meaning thoroughfare. The towns under Danish law replaced the kingdom of Mercia; each eventually was ruled by its own earl and twelve administrators who enforced the law, reinforced by a Common Court of Justice.

Toward the end of the tenth century, however, a new wave of hostile Vikings again threatened invasion. The line of strong rulers, Alfred and his descendants, had ended, and feuds between large landowners were fragmenting the country. In 1016, King Canute, a Dane, began his rule of England and succeeded in bringing a short spell of peace and prosperity as ruler of both England and Denmark. Known not only for his strength and military prowess but also for his piousness, King Canute raised and gave money to restore the Christian churches that had been destroyed over the years of pillaging. Lincoln's Christian tradition had survived, albeit weakly, in spite of the dominating pagans. When Canute died in 1035, a confusing dance of inheritance ensued, eventually leading to the invasion and conquest of England by William, Duke of Normandy, in 1066.

The Norman Conquest brought dramatic changes to a country hitherto ruled by fragmented powers. The Normans introduced fortified castles, one of which was built in Lincoln in 1068. Lincoln Castle, sited on the crest of the town's hill, most likely was designed to emphasize the Norman ascendancy over Lincoln's resistant inhabitants. More than 160 medieval homes were destroyed to make room for the castle, no doubt punctuating that point. The Normans also established the social and military system of feudalism, whereby a person's rank was determined by the amount of land he held. In 1072, Lincoln's imposing cathedral was begun. William the Conqueror, owing his friend Remigius the first vacant bishopric for his services during the invasion, granted Remigius his request for a cathedral at Lincoln. The cathedral started by Remigius underwent expansion until 1185, when a strong earthquake split the structure in two. At the time there was no bishop in Lincoln, thanks to Henry II's lack of interest, and the diocese had suffered from neglect.

In 1186, Hugh of Avallon was appointed the new bishop and handed the task of restoring the cathedral. Hugh was a Frenchman of a noble family from Avallon in Bur-

gundy. He became a Carthusian monk at the Grande Chartreuse and was sent to be a prior in Somerset. Devoted, learned, and energetic in his beliefs, he was greatly admired by Henry II, who selected him to take over the huge Lincoln diocese, which stretched from Grimsby to Eton and from Oxford to Huntingdon. The only Carthusian monk to occupy an English see, Hugh retained many of his old habits of life and shunned politics. His biographers record his piety, humility, love of solitude, and pragmatism, as well as his physical stamina. With these qualities, he immediately set about rebuilding the ruined church. In 1192, work began in a different architectural style than before, introducing Gothic elements and the intricately carved, Early English West Front. Hugh died of a sudden illness in London in 1200, but his work was continued. When he was canonized in 1220, the saint's shrine was constructed in a large new retrochoir that eventually became the Angel Choir, so named for the series of twenty-eight angels sculpted in stone under the top windows. St. Hugh's remains were moved there when it was finished in 1280, and it became a place of pilgrimage for many years.

The Normans had encouraged many Jews to settle in Lincoln throughout the late twelfth century, mainly for the purpose of financing trade in the town. Two stone buildings, among the oldest surviving examples of Norman architecture, testify to the importance of Jews in Lincoln as moneylenders. The House of Aaron the Jew stands below the castle, and the other, simply named Jew's House, sits farther down the hill. By the 1250s, the immediate area around Jew's House was self-contained, the lower boundary enclosed by a gateway that was secured at night; no Jew was permitted past this gate after sunset. This anti-Semitism was not new; St. Hugh had sheltered Jewish fugitives to protect them from armed and angry mobs half a century earlier. In 1255 this hatred was further aroused when the Jewish community of Lincoln was accused of murdering a young boy named Hugh as part of a mock crucifixion. As the Jews' banking contributions became less essential, their position became increasingly precarious. In 1290, Edward I expelled all Jews from the country; none would return until 1656.

Between the thirteenth and sixteenth centuries, Lincoln prospered as one of the principal market towns north of London (along with York). The wool trade, in particular, was strong. Then, in the mid-sixteenth century, the English Reformation of Henry VIII quite literally burst through Lincoln Cathedral's doors. After Henry had broken with the Roman Catholic Church and established the Church of England, he ordered the destruction of all monasteries and shrines in the land. Throughout the country, religious statuary was smashed and artwork despoiled. In Lincoln, the main shrine of St. Hugh was destroyed in 1540 and its treasures taken away, though not without a struggle, to the Tower of London to rest under the king's domain.

A century later, the city was torn apart again, this time by the English Civil War between Charles I and Parliament. Lincolnshire withstood many battles. Lincoln itself was taken and pillaged by Parliamentary troops, the castle stormed, and the cathedral desecrated and ransacked. Calmer times arrived with the Restoration. However, Lincoln's significance as a trading town had begun to wane as newer towns such as Leeds and Sheffield ascended. The city still revolves around its cathedral, which one can see from miles away rising impressively from flat fenland, now rich farmland. Lincoln Castle and numerous other historic buildings also mark the town's landscape. Among the displays at the castle is a copy of the Magna Carta, recently relocated from the cathedral.

Further Reading: Arthur Bryant's *The Medieval Foundation of England* (London: Collins, and New York: Doubleday, 1967) is an informative and interesting account of Britain, including Lincolnshire where pertinent, from the first invasion of the Celts to the fifteenth century. David Knowles sheds more light on St. Hugh of Avallon in his book, *Saints and Scholars* (Cambridge: Cambridge University Press, 1962; reprint, Westport, Connecticut: Greenwood, 1988). For a running commentary on Lincoln's historical roots, there is *Forgotten Lincoln* (Wakefield, West Yorkshire: E.P. Publishing, 1974), which reprints a series of essays written in the late nineteenth century by Herbert Green. Though at times difficult to piece together, it provides in-depth information about selected buildings and covers the Roman history especially fully. Christopher Daniell's *A Traveller's History of England* (New York: Interlink Books, 1991) provides an overview of England's turbulent political history, from Roman to modern times.

—Christine Walker Martin

Lindau (Bavaria, Germany)

Location: Southeast of Friedrichshafen, at the southeast corner of Lake Constance in Bavaria, southern Germany, approximately two miles from the border with Austria.

Description: Linked to the mainland by two bridges, Lindau is an island city with origins dating to Roman times. It was a commercial center during the Holy Roman Empire. Its medieval and baroque appearance, spectacular setting at the fringe of the Alps, and location in Germany's warmest region make Lindau a popular tourist destination.

Site Office: Verkehrsverein
Am Hauptbahnhof
Ludwigstrasse 68
8990 Lindau, Bavaria
Germany
(8382) 26 00 30

A charter from the year 822 is the earliest known record of the city's name, in which it appears as Lindoua. It is derived from the German for lime tree, *Linde,* which grows in the humid lakeshore area. The lime tree is also included in the city's coat of arms. Lake Constance, or Bodensee in German, is not in fact a lake, but a huge swelling of the Rhine River. Carved out by glacial activity, its water surface covers 163 square miles, making it the largest inland body of water in the German-speaking world. The lake forms the border to three nations: Germany, Switzerland, and Austria. The Rhine flows into the lake's southeast corner, where Austria and Switzerland meet, and proceeds out of its western end, where it continues just within Swiss territory before flowing north into Germany.

Archaeological finds dating back to 500 B.C. point to the presence of Celtic inhabitants in the area, though it is unlikely that they settled the island or the immediately surrounding lakeshore. The city's site was made up of two larger and two smaller islands until the beginning of the nineteenth century. Its present size is the combined effect of a gradual reduction in the water level and land reclamation. Tools from the period have been found here, though not in sufficient quantities to prove the past existence of an island settlement. It is likely that prehistoric peoples found the islands of the time inhospitable. Their location without the cover of natural features exposes them to the *Föhn,* the hot wind from the slopes of the Alps to the south, which would have blown away their lightweight shelters.

Roman troops arrived here in 15 B.C. under the leadership of Tiberius, who became Roman emperor in the following year. He had advanced from Gaul, present-day France, while his brother Drusus was marching up the valley of the River Adige in present-day Italy on his way to the Rhineland. A military fort was erected and named Tiberii— "of Tiberius." In due course two roads were built which provided quick access to Italy. One of these led to Como and Milan, the other led to Chiavenna in the Italian Alps, over Bregenz and Chur in present-day Switzerland. The freestanding part of the island in the harbor area settled by the Romans is today referred to as the Römerschanze, or "Roman entrenchment." Numerous archaeological discoveries, most importantly the foundations of the fort, confirm the Romans' presence.

Lindau and its surrounding areas, together with the Bavarian-Swabian highlands, the Tyrol in present-day Austria, and the Grisons in present-day Switzerland, constituted the Roman province of Rhaetia. It is known from an early fifth-century source that the Romans kept a flotilla at Lake Constance, for its admiral, the *praefectus barcariorum,* was based in nearby Bregenz, some four miles east along the shore from Lindau.

Incursions made into Roman territory by Germanic tribes first made themselves felt in Tiberii in A.D. 271. Aurelian pushed them back, returning a measure of stability to the area. Nonetheless, toward the end of the third century these tribes—the Alamanni—had secured an expanse of territory from south of the Swabian Mountains to the east bank of the Rhine. This development brought the frontier of the Roman Empire closer to Tiberii, which went from being an auxiliary fort to a border stronghold. Its function was changed to that of first-line defense to supplement the military post at present-day Bregenz, which stood in a more easily defensible location within a narrow pass. Valentinian I, Roman emperor from 364 to 375, ordered the improvement of additional border defenses, and it is thought by a number of historians that the *Heidenmauer,* or "heathen wall" dates from this time; others argue that it dates back to Carolingian times. Thirty feet high and up to twelve feet thick, it would have stood at the water's edge at the time of its construction.

Defensive measures were effective for a time but could not prevent the end of Roman rule. The empire collapsed definitively with Attila the Hun's invasion of Gaul in 451, which earned him the title of "Scourge of God." He died two years later when he overindulged himself at his wedding feast. The Alamanni invaded Italy in 458 and settled around Lake Constance at this time. They came under Frankish rule within half a century, when the Merovingian king Clovis I defeated them in 496 at the Battle of Zülpich (Tolbiacum), twenty miles east of present-day Bonn. The Merovingian dynasty ruled for two-and-a-half centuries until Carolingian Pepin III the Short acceded to the throne in 751. His son Charlemagne expanded Frankish dominion over the territories of the present-day Low Countries, Switzerland, Ger-

The thirteenth-century prison tower at Lindau
Photo courtesy of German National Tourist Office

many and France, as well as parts of the present-day Czech state and Italy. This empire was the most dominant European power at the time and stood as the protector of the Christian Church; Christian territories were at the time surrounded by Moslems to the south and east and heathens to the north. In recognition of this role, Pope Leo III revived the title of Roman Emperor, which had lapsed in the fifth-century, to grant it to Charlemagne in 800 adding ''Holy'' to the title. By that year, Lindau had grown into a fishing community. A chapel for Christian worship was in existence at this time.

This was the church of St. Peter, patron saint of fishermen. Today's St. Peter's Church in the Schrannenplatz—now a war memorial chapel—dates from the eleventh century and is recognized as the oldest church in the region of Lake Constance. Its interior is decorated by a series of wall paintings dating from the thirteenth and sixteenth centuries, of which the Passion scenes are by Hans Holbein the Elder.

A Benedictine nunnery was founded in Lindau in 810 through the patronage of Adalbert, Count of Rhaetia, who endowed it with many gifts. It was granted a number of rights by the crown in 839. First of these was the right to all revenues from Lindau's mint, customs, and navigation taxes. It was also determined that the abbess should be elected from among the nuns. In case there be no one suitable for the post, a nun from another convent could be chosen, as long as this met with royal approval. These freedoms, called immunities, were intended to remove convents from the jurisdiction of territorial lords, who often transgressed against these otherwise unprotected institutions. Because the nuns were barred from occupying themselves with worldly matters, the administration of the above revenues was placed in the hands of a *Vogt,* or guardian, who thus came to hold a powerful position within the city.

One of the gifts received by the convent during the Carolingian period was the Lindau Book of Gospels. A much-celebrated work of art, its cover was of finely crafted gold encrusted with pearls and precious stones. Following the convent's secularization in 1802, the book was sold to the Earl of Ashburnham, England, in 1901 for 200,000 marks. Today it belongs to the Morgan Library in New York City.

At the Treaty of Verdun in 843, and according to the Carolingian tradition, Charlemagne's grandchildren divided the empire into three parts. One of these was the eastern kingdom, containing Lindau, which was received by Louis II the German. This is the first time that Lindau can be said to be part of Germany, for it was Louis, a Frank by birth and a German by nurture, who cultivated German language and literature to create a distinct political and cultural entity out of his new kingdom.

Lindau's economy began to grow in the late ninth century, mainly as the result of two developments. In 887, Arnulf ascended the German throne and carried out repeated campaigns against Italy. During these campaigns, Lindau was used as a convenient transit point on the way south, greatly benefiting the city's economy. The other development was the creation of the Upper Burgundian Kingdom in 888 to the west of Lindau. This stimulated trade with Swabia, much of which also passed through Lindau, again enlarging the city's economy. This nascent prosperity was, however, threatened by frequent attacks on German territory by Magyar forces from the east. These attacks continued until 955, when Emperor Otto I decisively defeated the Magyars at the Battle of Lechfeld near Augsburg, eliminating the threat for good. Another major event brought about the emergence of Lindau as a trading center. This was the royally authorized transfer of the Aeschach market, on the mainland lakeshore,

to the island of Lindau in 1079. Historical sources indicate that the move was made to afford greater protection for out-of-town traders, which was more easily accomplished on the island. A further impetus for growth came during the reign of Frederick I, also known as Barbarossa. Duke of Swabia from 1147 before acceding to the German crown in 1152, he engaged the cities of northern Italy in a prolonged struggle for nearly three decades, which created even greater traffic through Lindau than Arnulf's ninth-century campaigns south. Lindau's connection with Italy endured for many centuries. Even in the eighteenth and nineteenth centuries, the city was the base of one of the most important stagecoach services between Germany and Italy. It was known as the Lindauer Bote, or "Lindau Messenger," on which Goethe traveled during his first visit to Italy in 1786 (described in his work *Italienreise*).

The city underwent great physical change during the twelfth century. Lindau was fortified during this time. The year 1180 saw the construction of the Stephanskirche in the Marktplatz. Little of its original Romanesque character is evident today. It underwent one major alteration in 1506 and was struck by lightning in 1608, which melted its 300-year-old bell. A second alteration took place from 1781 to 1783. Next to it and dating from the same century is the Stiftskirche, dedicated to the Virgin Mary. It was destroyed by fire on several occasions and finally rebuilt during the period from 1748 to 1752. Its present appearance is marked by baroque features.

One consequence of Lindau's wealth was greater political power. A major historical event was Lindau's elevation to the status of an imperial free city in 1275 by Rudolf I, two years after his coronation as Holy Roman Emperor. The citizenry came under the direct lordship of the emperor, without any intermediate authority. Reason for this change in status was Rudolf I's need to guarantee the city's loyalty to his own interests, which were in conflict with those of Count William de Montfort, who at the time held nearby Bregenz. The proclamation meant an enlargement of the citizen's rights and a greater degree of self-government by the city. Lindau was now set for the high noon of its history, the fourteenth and fifteenth centuries, when the city reached the pinnacle of its wealth.

Of particular importance was the establishment of a measure of law and order. Emperor Louis IV, the Bavarian, established the 1331 Swabian-Bavarian King's Peace in Ulm. This was a set of laws that prohibited theft, arson, and unjust taxation. It was acknowledged by Lindau, as well as a host of other cities. The resulting protection of trade and travel was a great benefit to these cities in general, and to Lindau in particular. The activities of robber barons were severely curtailed; Lindau and its neighbor cities Constance, St. Gall, and Zurich benefited especially from the destruction of one Count Albrecht of Werdenberg's castle.

The city's government was at first the exclusive domain of the civic patriciate. The tradespeople, who were organized into guilds, accounted for a significant portion of the city's economy, were not represented within local government. This situation changed when the resulting tensions culminated in a revolt by the guilds in 1345. The conflict was settled by an enlargement of the city council. Eight guild masters joined the fourteen council members. Council meetings appear to have been the only occasions when patricians and commoners came together.

The city played a small role in the Reformation, which was set in motion by Martin Luther in Wittenberg in 1517. Although the Lindau convent was left unaffected, the citizenry leaned toward the doctrines of another influential figure of the Reformation, the Zurich-based Huldrych Zwingli (1484–1521). The freedom to follow Zwingli's teachings had been secured at the 1526 Diet of Speyer but was revoked three years later under renewed Catholic pressure. This action in turn raised a protest from groups of Lutheran supporters, of which the city of Lindau was one. It was this event that gave its name to the new denomination of "Protestants."

With the weakening of the Holy Roman Empire around the beginning of the nineteenth century, Lindau lost its status as a free imperial city. It was ruled by Austria for a short time in 1804 and was incorporated into Bavaria in 1805. Bavaria became part of the German empire in 1871. In the midst of all these changes in political control, and subsequent events in Germany, Lindau has remained a popular resort town and has maintained many of its historic buildings.

The Altes Rathaus, or "Old Town Hall," was built during the period 1422–36 in what was then a vineyard. It was remodeled in 1578 and today has a stepped gable, adorned with graceful volutes, leading to a belfry. Emperor Maximilian held a diet, an assembly of the empire's constituent authorities, in Lindau in 1496, an event depicted in a wall painting on the south facade. An inscription bids the reader to "Take leave of evil, and learn to do good." This injunction was apparently for the benefit of those who were held in the city jail, which at one time was to be found on the building's first floor.

A famed landmark is Lindau's harbor entrance, which is guarded by a nineteenth-century Bavarian symbol, a seated stone lion, which looks out over the lake and towards the Alps from the top of a plinth at one end of the harbor walls. At the other end of the walls stands the "new" lighthouse, the Neuer Leuchtturm, built in 1896. Its old counterpart, the thirteenth-century Mangturm, also referred to as the Alter Leuchtturm, constructed in the 1200s, stands at the edge of the inner harbor on the remains of the thirteenth-century city walls. The Pulverturm, or "gunpowder tower," at the westernmost point of the harbor was formerly referred to as the Little Green Tower and served as the city's gunpowder store.

The Barfüssekirche, the Church of the Barefoot Pilgrims, was built between 1241 and 1270. This church was once part of a monastery, which in the fourteenth century was the home of a certain Brother Marquard, a German mystic. Very little is known of him apart from his writings. A few historical sources give an indication of his existence, but say little else about his life. He was part of a religious movement pioneered by the Franciscan monks David of Augsburg and Berchtold of Regensburg. Its proponents were called *Göttesfreunde,* "Friends of God," whose aim was to achieve spiritual peace through introspection. It was a spiritual orientation that dwelt on the mysteries of the world and of the heart, and as such acquired the name of mysticism. The church has served in different capacities in the course of the city's history, variously serving as jail, army barracks, armory, and riding school. It became a concert hall in 1868 and a theatre in 1887, a function it still serves today as Lindau's main theater, the Stadttheater, which is celebrated as being the most beautiful theater building of the Lake Constance area.

The Haus zum Cavazzen, facing the Marktplatz, is the city's museum. The baroque building was constructed in 1728 and 1729. It houses art and exhibits covering Lindau's history.

Further Reading: There is no complete history of Lindau in English. A work in German, *Geschichte der Stadt Lindau im Bondensee,* edited by Dr. K. Wolfart (Lindau: Kommissionsverlag Joh. Thomas Stettner, 1909) provides detailed coverage of the city's history in three volumes.

—Noel Sy-Quia

Loches (Indre-et-Loire, France)

Location: Loches is 25 miles southeast of Tours on the River Indre as it flows northwest to join the Loire. It is about 140 miles southwest of Paris.

Description: A small town dominated by its château, which was used as a fortress, royal residence, and prison and which was associated with many royal intrigues during the medieval and Renaissance periods.

Site Office: Office de Tourisme, Syndicat d'Initiative
7 Place de Wermelskirchen
37600 Loches, Indre-et-Loire
France
47 59 07 98

Unlike Saumur, where the château that had once been a palace became a fortress, Loches has a château that was built as a fortress but became a palace. The tall medieval defensive fort with its turrets, towers, sentry walk at the base of the roof, and crenellations, is now the northern wing of the building; the lower southern portion has square chimneys, gabled windows with pointed Gothic stone ornamentation above large oblong windows, and, in place of crenellations on the roof, a decorative surround of stone latticework. The whole building is the focal point of one of the most picturesque towns in central France, and was one of the principal residences of the Plantagenet family, counts of Anjou, one of the great feudal dynasties that took control of France in the ninth century, after the death of Charlemagne in 814.

According to Bishop Gregory of Tours, Loches is the site of a monastery founded about 500 by St. Ours. The Fulk branch of the counts of Anjou took control of Loches and its sixth-century fortress in 886; the greatest of these counts, Fulk III Nerra, whose robust military exploits and bouts of penance have become legendary, used Loches as a center from which to conquer Touraine. It was he who built the *donjon* tower to protect Loches against attack from the south. It was a massive three-story square stone building, the architecture of which patently derives from previous such edifices constructed of wood. The defenses were strengthened again when, with Henry II, the counts of Anjou came into possession of the English crown in 1154, initiating the Plantagenet dynasty of English kings.

Henry died in 1189. His son, Richard the Lionheart, who had lived at Loches, was imprisoned in Austria when returning from the Third Crusade (1189–92). The French King Philip II (Philip Augustus) had conspired with the Lionheart's brother, John, to sequester Loches and, on his release in 1194, Richard took the château in a surprise attack that lasted only three hours. The counts of Anjou lost Loches for good when Philip Augustus took it back again in 1205,

but only after a year-long siege in which most of the defenses were destroyed.

Philip repaired the château and its defenses, and the Loches château became the residence of the kings of France. The collegiate church of Notre-Dame, now the church of St. Ours, scarcely 100 yards south of the château, had already been started, and the fresco in the crypt apparently dates from the eleventh century. The church itself dates chiefly from the twelfth century. Between the two dissimilar square towers at each end of the nave, each with an octagonal pyramid spire, the two shorter octagonal hollow pyramids above the vaulted ceiling were constructed in the twelfth century. The richly ornate Romanesque door is now badly damaged.

In the thirteenth century the *porte royale* or ceremonial entry was built in the ramparts on the west side of the ensemble. Construction of the new, domestic wing of the château, the *nouveau logis* portion of the *logis royaux* (royal lodgings), dates from the very late fourteenth century, and its pointed Gothic gable windows and miniature decorative stone spires are now mimicked in the single gable window of the fifteenth-century porte des cordeliers at the northeast corner of the enclosed area, where it touches the Indre River. The porte des cordeliers, still showing the apparatus for retaining the drawbridge, has round turrets and a sentry walk below the top story. It was the principal medieval entry to the town, through which traffic passed on the route to Spain.

Loches also figures significantly in the reign of Charles VII. When his father, Charles VI, died in 1422, Charles VII controlled only part of central France—Touraine, Orléanais, Berry, Auvergne, and Dauphiné—the rest being under English domination. On the death in 1422 of Henry V, the ten-month old Henry VI, son of Henry V and Catherine (the daughter of Charles VI of France) was proclaimed king of England and France in accordance with the 1420 Treaty of Troyes, although the French dauphin was also proclaimed king of France at Mehun in Berry, and made Bourges his capital. It was at Loches, in the old part of the château, that Joan of Arc, having succeeded in lifting the siege of Orléans and freeing the city's inhabitants, returned to Charles VII on June 5, 1429, to urge him to have himself crowned king in the traditional way at Reims.

Loches is now permanently associated with Charles VII and his relationship with his mistress, Agnès Sorel, known as the Dame de Beauté, and so described on her tomb, on account not only of her well-attested radiant good looks, but also because of the estate of Beauté-sur-Marne, which Charles VII gave her. Agnès was born in the Loches area, bore Charles VII four daughters, and is buried at Loches. Her black marble sarcophagus is surmounted by a full-length reclining figure of her with hands joined in prayer. The two lambs at her feet are an allusion to her name—*agneau* is

The north wall of Loches' château
Photo courtesy of The Chicago Public Library

French for lamb—and whispering angels support the pillow on which her head rests. The tomb was originally placed in the church of St. Ours, but was mutilated during the 1789 revolution when her tomb was mistaken for that of a saint. Her remains were dispersed. The tomb, now restored, has been at the château since 1809. The tower at the southeast corner of the château was named after her.

The *nouveau logis* also contains the oratory of Anne of Brittany, from 1491 wife of Charles VIII, by whom she had no offspring, and from 1499 of Louis XII, by whom she became pregnant eight times, but produced no live male children. Anne brought Brittany into France, but in the absence of a male heir the succession passed to the Angoulême family, and the unity of France was preserved only by the marriage in 1514 of the heir presumptive, François de Valois (later Francis I) to Anne's eldest daughter, Claude de France. Anne's oratory at Loches contains a late Gothic stone prieudieu and is decorated with her emblems, ermine and a tasselled rope cord.

Loches, however, is best known for its role in French life during the reign of Louis XI, the rebellious son of Charles VII who achieved the unity of France by unsqueamishly crushing the feudal opposition. Under Louis, who reigned from 1461 to 1483, the château took increasingly important roles as a defensive outpost and as a prison. A four-story round tower and a structure called Le Martelet, with its layers of underground dungeons, were built to update the defenses and to provide cells for torture and imprisonment. The cells, according to the chronicler Philippe de Commynes, contained iron cages enclosed in wooden trelliswork. The cages were only two yards wide, and could be dangled from the ceiling.

Among the more distinguished prisoners was Cardinal Jean Balue, who betrayed Louis XI and was imprisoned from 1469 until Pope Sixtus IV achieved his release in 1480. Ludovico "the Moor" Sforza, duke of Milan, was imprisoned in Le Martelet in 1499 and is said to have died at the shock of seeing sunlight on his release. Other notable prisoners were the bishops of Autun and Le Puy, caught up in a conspiracy against François I in 1523. There was also St.-Vallier, father of Diane de Poitiers, embroiled in the same conspiracy, but too ill to undergo the torture to which he had been condemned in 1524. It is unlikely that Diane de Poitiers obtained her father's reprieve by becoming the king's mistress, a story that inspired Victor Hugo's play *Le Roi s'amuse* and Giuseppe Verdi's opera Rigoletto. She was, however, mistress to François's successor, Henry II.

The pleasant little town of Loches also has an attractive sixteenth-century town hall, and a belfry, known as the Tour St.-Antoine, of the same era. The poet Alfred de Vigny was born here in 1797. Recently the Institut collégial européen has held annual summer conferences at Loches. Less than a mile away, at Beaulieu-les-Loches, are the remains of an abbey built by Fulk III Nerra as penance while he was under threat of excommunication.

Further Reading: As with most other Loire châteaux towns, there is nothing in English devoted specifically to the town of Loches except the often adequate tourist guides, gazeteers, and brochures. Readers are forced to have recourse to often technical works on the history or France, or on the history of its architecture, or general works, some magnificently illustrated, on the dozens of châteaux in the Loire valley. Among general works in English is Jean Martin-Demézil's *The Loire Valley and its Treasures* (London: Allen and Unwin, 1969).

—Anthony Levi

London (England): City of London

Location: The City of London is the heart of Greater London, some 677 acres—slightly more than one square mile—on the left bank of the River Thames, in the southeastern part of England. The city includes the area from the Temple Bar at the Royal Courts of Justice to the Tower of London.

Description: The City of London is formed by the boundaries of the old Roman enclosing wall. The country's financial center is located here along Fleet Street, including the Bank of England, the Royal Exchange, and the Stock Exchange. In the western part of the borough is the legal center: the Temple and Law Courts. Other famous sites within the city are the Tower of London, St. Paul's Cathedral, and many fine churches designed by Sir Christopher Wren.

Site Office: London Tourist Board
26 Grosvenor Gardens
London SW1W 0DU
England
(71) 730 3488

Known as "The Square Mile" for the area within the ancient enclosing walls, the City of London has stood along the River Thames for almost 2,000 years. The capital of England, it has a rich and diverse history. From its early days as a small Roman trading village to the twentieth century, the city has survived wars, fires, and disease to dominate the country as a center of international commerce, finance, and banking. With its own city corporation and legal courts, London has long been unique as a self-governing entity, through the various periods of English history.

The English capital began as a small village in pre-Roman days. The name London probably derives from the Celtic "Lyn-dun," meaning "fortified village on a lake." The Romans arrived in A.D. 43, inhabiting the area for the next 400 years. They called it Londinium. The village developed along the River Thames at a gravelly bank, which was stable enough for the early bridges, yet with deep tidal water for Roman ships. At first Londinium was mostly a trading center, gradually growing into a busy port town. Londinium burned completely in A.D. 61 during a raid by Queen Boudicca and her Celtic tribe. In the 120s, another fire leveled about 100 acres in the heart of the city, from which it barely recovered. Not until the late second century did the city reach its former size. This was probably when the enclosing wall was added. The length of the wall was about two miles, consisting of substantial stone, in places almost twenty feet high and six to nine feet wide. (Most of this wall, with some alteration, survived to the late eighteenth century; in

1980 an intact section nine feet high was discovered in Crosswall house, and another section was found built into the Tower of London in 1977.) The large stones used in its construction had to be carried downriver on the Medway, along the seacoast, and up the Thames. The wall enclosed an area of 326 acres, from the current site of the Tower of London to the old junction of the Fleet and Thames Rivers.

From this point Londinium quickly grew as a center of commerce, and, by the third century, it was also a center of government and finance, with a population of 30,000. Inside the walls was a thriving city, with paved streets, public baths, small huts, and larger houses of wood, brick, and stone. The wealthy inhabitants enjoyed wine, spices, fabrics, and other luxury items imported from around the known world. Six main gates opened into the Roman roads leading to all civilized parts of the country, making Londinium its focal point. By the fourth century, the city's importance earned it the title Londinium Augusta. It was one of the biggest cities in western Europe, until the departure of the Romans in A.D. 410. Not much is known about its history from this point until the sixth century. Under Anglo-Saxon rule, tribal groups controlled the country, and there followed a long period of savage battles.

In the late sixth and early seventh centuries, London again grew as a port city with many trading enterprises. The first St. Paul's Cathedral was built in the reign of Aethelberht (ruled 560–616), when St. Augustine, first archbishop of Canterbury, brought Christianity to England. London was again destroyed by fire, in 851, during the Viking raids. After a series of battles with the Danes, King Alfred the Great of Wessex briefly regained control of the city, and the country, in 883. Then called Lunduntown, the city was rebuilt and prospered. Its citizens had extraordinary independence for the times. Living in an international trade center, they had the wealth and power to govern themselves apart from the strife of the rest of the country. In 1016, they elected Edmund Ironside as their king. With his death the same year, the country united under the rule of the Dane King Canute. But the citizens of London had established their autonomy, and were respected and feared by succeeding rulers. William, Duke of Normandy, who became king in 1066, fortified his London residence, the White Tower, to protect himself from the influential townspeople.

The White Tower, completed in 1097, was used by William and his successors both as a castle and prison; it remains today as the base of the Tower of London. Located on the Thames in the southeast part of the city, the rectangular "tower" (the word meant a castle at the time) was used as a residence by various reigning monarchs until the seventeenth century. The Royal Mint and Royal Menagerie also were kept here until the nineteenth century. Various sections

The Tower of London
Photo courtesy of H.M. Tower of London

held royal quarters, prisons, armories, and storerooms, including that of the Crown Jewels. The Chapel of St. John is a rare example of early Norman architecture.

From the early Middle Ages, London had gained its independence in matters of trade, religion, government, and law. Londoners functioned democratically through the city corporation, formed in the late twelfth century, by which time London was the country's capital. In 1215 King John signed a charter affirming the citizens' right to govern themselves and to vote for their own mayor. During the Middle Ages the craft guilds, or trade and merchant associations, held much power and wealth. Tradespeople lived and worked in specific areas of the city: for instance, the fishmongers' market was on Fish Street Hill; Lombard bankers, Italian immigrants, did their business on Lombard Street. The first Guildhall, from which the Lord Mayor and council governed (as they still do today), was built in the fifteenth century. The Royal Exchange was established as a place for the city's merchants to meet; well-established trading and export businesses flourished from the sixteenth century.

A number of London churches and monasteries were built in the Middle Ages, particularly after the arrival of various European orders of friars, such as the Black Friars in 1221 and the Franciscans (called Grey Friars) in 1224. Rich citizens also donated money for the construction of churches and chapels, some in the plain Saxon style, others more ornate, with stained-glass windows and marble tombs. St. Paul's Cathedral was rebuilt in the twelfth and thirteenth centuries, after a series of disastrous fires.

The legal quarter, called the Inns of Court, formed in the western part of the city by the end of the fourteenth century. Lawyers first conducted business within St. Paul's Cathedral (where there was even a law school), until the practice was banned by Henry III in 1234 to encourage use of the schools he had established in Oxford. The lawyers moved into buildings owned by the Bishops of Lincoln and the Lord Grey de Wilton, among others, located along Chancellor's Lane (now Chancery Lane); these were called inns, meaning houses of wealthy men. Another group of lawyers leased buildings originally used by the order of the Knights Templar. These houses developed into barristers' chambers, for conducting business and training new lawyers. The four Inns of the Court then established were Middle Temple, Inner Temple, Lincoln's Inn, and Gray's Inn; rebuilt in the eighteenth century, all are still in existence. Across the street are the Royal Courts of Justice, a nineteenth-century Victorian addition.

Through the Middle Ages, fires continued to be a great problem. Fire-damaged sections of London were continually rebuilt with wood from the northern countryside. Disease was another danger of city life. City streets and the ditch around the enclosing wall became polluted with human refuse and dead animals; fresh water had to be imported from cleaner regions outside the city. Plague outbreaks were common. One of the most severe, the Black Death in 1348–49, killed as many as 50,000 people in London.

The turmoil of the fifteenth and sixteenth centuries, first with the Wars of the Roses, then with the frequency of religious changes as the country became Protestant during the reign of Henry VIII, Catholic under Mary I, then Protestant again under Elizabeth I, had little effect on the City of London. A city only one square mile within its walls, it continued to govern itself. However, by the middle of the sixteenth century, the city had grown well beyond the old wall. The Reformation of 1535, which led to the dissolution of the monasteries, provided some release from the overcrowded conditions of the city; church land and buildings were granted to wealthy citizens and noblemen, who built large mansions. This was a time of significant demographic changes in society, as a deep chasm grew between the rich and the poor. By the end of the century London's population numbered about 200,000, but the wealthiest citizens began moving into suburbs outside the city walls, to escape overpopulation and disease. Royalty had already moved out of the city by the time of Henry VII, living lavishly in various palaces outside the city, first at Westminster, founded in the time of Edward the Confessor's reign, then at such favorite palaces as Eltham, Greenwich, and Hampton Court.

This flight of rich merchants and nobility to outlying suburbs increased as industries developed within the city walls. The level of pollution surged; coal, in particular, polluted everything in the area around its use. Neighborhoods around the river, especially, deteriorated into slums. The divisions between social classes grew larger, and, with a depressed economy, unemployment rates soared. This led to anti-Royalist sentiments and, ultimately, the Civil War.

Physically, London was changing. Coffeehouses, shops, and taverns replaced empty mansions. Inigo Jones, the king's surveyor and architect, and the first London city planner, influenced the radical change in architectural style from that of the Tudor and Jacobean Gothic period to a more classical design, based on Italian Palladianism. A fine example of his design is the Banqueting House in Whitehall. Much of the new construction of this period reveals his influence. Then, restoration and new building came to a standstill with the Civil War of 1642–48 and its aftermath. After the execution of Charles I in 1649 (at Jones's Banqueting House), Oliver Cromwell forbade further new construction, and growth of the city stalled until the mass destruction of the mid-seventeenth century.

The two major hazards of London life, fire and disease, together severely affected every aspect of the city, its population and economics, even its social structure. The Great Plague peaked in 1665, killing so many that burial pits were used to dispose of the 20,000 plague victims. After a hot summer, the Great Fire of 1666 started as a small fire in a bakery near London Bridge. During the four days of the fire, two-thirds of the city, including large areas of wood tenements and slums, as well as St. Paul's Cathedral and the Royal Exchange, burned. Reconstruction led to a slightly more ordered city plan, with straighter streets and more solid structures, building codes, and major development along the two rivers (the Fleet River canal was not enclosed until in the mid-eighteenth century). The famous architect Sir Christopher Wren was commissioned to help design the new city plans, rebuild St. Paul's Cathedral and other buildings, including fifty-one churches, all of different designs. Craftsmen who helped Wren included Grinling Gibbons and Jean Tijou, as well as Wren's pupil Nicholas Hawksmoor. Working together, they renovated the city, giving it a new foundation and facade.

The renovation of St. Paul's Cathedral, begun in 1672, was completed in 1710. Even before the fire, the cathedral, which had seen service as a stable during Cromwell's government, was in need of repairs. Wren spent much time trying to obtain approval for his final designs, which kept the thirteenth-century Gothic Choir intact and added the high dome. When Wren died in 1723, he was the first to be buried in the cathedral.

After the Great Fire, higher rents and taxes, and a preference for the better living spaces in the suburbs, kept most of the city's former citizens—those who could afford it—from returning to the city, with its inadequate water supply and poor drainage systems. Thus began a major shift in London's population that would continue through the twentieth century. Development centered in such areas outside the city walls as Piccadilly, Pall Mall, and St. James's, as well as in more distant regions. By the end of the century, London's population had grown to about 650,000, due in large part to increasing immigration.

The establishment of London banks began in the mid-seventeenth century. Wealthy Londoners had always lent money to reigning monarchs to fund wars, new construction and renovations, and navigation. Financial transactions eventually became more sophisticated and therefore more regulated. The first banks served pawnbrokers and goldsmiths, whose business involved trading bills of exchange, like an exchange of credit. This type of brokerage was conducted at the Royal Exchange, originally created mostly for stock trading. The first official Royal Exchange, designed by George Samson, was completed in 1773 on Threadneedle Street, becoming the financial center in the city. (In the nineteenth century the Royal Exchange burned and was rebuilt again, eventually to be replaced by a larger building.) The first place of business for the Bank of England, in 1694, was in the Grocer's Hall on Poultry Street; it quickly flourished because of speculation in investment on trade in the New World. Private banks also began to open, through the eighteenth century. The London Stock Exchange first opened in 1698 in a coffeehouse in Change Alley, after business grew too large to be accommodated in the Royal Exchange.

In the eighteenth century the outer regions of London continued to grow rapidly. New buildings erected inside its walls included the classical Bank of England (later replaced by a new building in 1940, six years before it was nationalized), and the Mansion House (designed by George Dance the Elder), which is still the residence for the Lord Mayor of the city. The Mansion House, completed in 1752, was modeled after a Palladian villa with a front portico and six Corinthian columns.

From 1839 to 1862, rail lines were built to connect various portions of the large city. The first underground opened in 1863. With the cheap fares, even members of the working class could afford to live outside the city, increasing the numbers leaving the city. Remaining citizens lived in unsanitary conditions, and there were serious cholera and typhus outbreaks by the mid-nineteenth century.

The City of London's population continued to decline into the twentieth century. Most people lived outside the old city and commuted to work within the city limits. In 1901 there were 27,000 people living within the city walls; many of these were shopowners living over their stores. By 1931 inner London's population had declined to 11,000.

During World War I air raids, several buildings were damaged or destroyed. This, however, was negligible compared to the destruction of World War II, which decimated entire neighborhoods. St. Paul's Cathedral, so many times destroyed through the centuries, survived the heavy bombing of the city. One bomb destroyed a high altar; another shattered all the stained glass. But the cathedral stood mostly intact. Even so, there was a vast amount of wartime damage around the city. The rebuilding after the war was a massive and slow undertaking. Most of the other, more heavily damaged, churches eventually were restored. The London Bridge, built in 1831 and itself a successor to earlier bridges, was replaced, with the new bridge opening in 1973. The Barbican project, completed in the 1970s, spans one of the bombed-out areas, a modern complex with residences, offices, shops, an art center, and theatres. It is also the home of the London Symphony Orchestra and the Royal Shakespeare Company. Newspaper offices on Fleet Street and other modern office blocks now fill the skyline of the city. There are a few surviving features of London's history: some churches, and notably St. Paul's Cathedral, the Guildhall, the Mansion House, the old Royal Exchange, the Bank of England, the Inns of the Court. In 1965, area outside the city was divided into the thirty-two boroughs of Greater London. The City of London retains its own Lord Mayor and City Corporation. Less than 10,000 people now live within the city borough; most of them are government employees who are residents of the Barbican.

Further Reading: Among the numerous books on London, there are several good ones that focus on particular periods in the city's history. These include: *London Life in the 14th Century* by Charles Pendrill (London: Allen and Unwin, and New York: Adelphi, 1925); *London and the Reformation* by Susan Brigden (Oxford: Clarendon Press, 1989; New York: Oxford University Press, 1991); *Eighteenth-Century London Life* by Rosamond Bayne-Powell (London: Murray, 1937; New York: Dutton, 1938); and *Plague and Fire: London 1665–66* by Leonard W. Cowie (London: Wayland Publishers, and New York: Putnam's, 1970). An important and inclusive series is *The Survey of London* by Sir Walter Besant (London: A. and C. Black, 1902–1925). The nine volumes, spanning each major period and region of the city, have hundreds of illustrations, maps, and city plans. *London: The City and Westminster* by Arthur Mee (London: Hodder and Stoughton, 1937; revised edition by Ann Saunders, 1975), is divided into chapters that cover distinct regions within the city and includes much architectural and historical detail. *The City of London: Its History, Institutions and Commercial Activities,* by Mary Cathcart Borer (New York: Pitman, and London: Museum Press, 1962), is a concise (less than 150 pages) synopsis of the city's history. Another good book by Borer is *The City of London: A History* (London: Constable, 1977; New York: McKay, 1978), which covers various aspects of the city's development from Roman times to the twentieth century. *London: The Biography of a City* by Christopher Hibbert (London: Longmans, 1969; New York:

Morrow, 1970) provides a good general introduction and includes many illustrations. *Studies in London History,* edited by A.E.J. Hollaender and William Kellaway (London: Hodder and Stoughton, and Mystic, Connecticut: Verry, 1969), is a "miscellaneous" symposium of essays on subjects ranging from London's pre-Norman bridge to the Holloway Prison. Two encyclopedias are worth mentioning here for their specificity of detail on a wide variety of subjects relating to the city: *An Encyclopaedia of London,* edited by William Kent (London: J.M. Dent, 1937; revised edition edited by Godfrey Thompson, 1970); and *The London Encyclopaedia,* edited by Ben Weinreb and Christopher Hibbert (London: Macmillan, 1983; Bethesda, Maryland: Adler and Adler, 1986).

—Noelle Watson

London (England): Greenwich

Location: On the southern bank of the River Thames; about eight miles east of London, toward the sea.

Description: Site of the prime meridian, also known for its architectural treasures, including the Old Royal Observatory, the Royal Naval College, the National Maritime Museum, and the Church of St. Alphege; two ships, the *Cutty Sark* and the *Gipsy Moth IV,* rest in dry dock here.

Site Office: Tourist Information
46 Greenwich Church Street
London SE10 9BL
England
(81) 858 6376

Greenwich is probably most famous as the site of the international zero of longitude. Rich in naval history, Greenwich was the place where Henry VIII built his dockyards, from which ships later sailed to the New World, and the Royal Observatory was established here in the late seventeenth century to further the science of navigation. The National Maritime Museum is located in what originally was the Queen's House, built by Inigo Jones. And the Royal Naval College, built over part of the old palace, once housed retired seamen in the eighteenth and nineteenth centuries.

Inhabited since the Bronze Age, the area around the River Thames that became Greenwich probably first developed as a fishing village. Because of its strategic location both on the river and along a Roman road from London to Dover and Canterbury, the village grew as an important center of trade and commerce. The remains of a Roman villa were discovered in the nineteenth century, indicating a site of some size. In the invasions of Saxon England during the early eleventh century, the Danes camped in Greenwich, using it as their base for attacks on London. When the marauders burned the Canterbury Cathedral in 1012, they took Archbishop Aelfheah (Alphege) back to Greenwich, where he was murdered. (A church, designed in 1718 by Nicholas Hawksmoor, was built on the supposed site.) Alphege was later canonized. Until the thirteenth century much of the land around Greenwich belonged to the Abbey at Ghent; a group of Ghent monks lived in a small manor, which later passed into the king's hands. The name Greenwich derives either from ''green village'' in Anglo-Saxon or ''green reach,'' a Scandinavian name.

The late thirteenth and early fourteenth centuries formed a turbulent period in English history. During the rebellions led by Wat Tyler in 1381, Richard II attempted to escape to Greenwich, but mobs of poor people along the river forced him to turn around his barge. Tyler himself may have marched through the area. Later the manor house was the site of intense royal negotiations. Henry IV greeted Emperor Manuel II Palaeologus of Constantinople here in 1400, giving him funds for the war against the Turks. And in 1416 the Holy Roman Emperor met there with the French and English to attempt to end the Hundred Years War.

In 1417 Humphrey, Duke of Gloucester, Henry V's brother, inherited the manor and converted it into the castle that would become a favorite royal residence for the next 200 years. Situated along the River Thames, the castle, which Humphrey called Bella Court, had a moat and battlements. He enclosed about 200 wooded acres to create Greenwich Park and built a watchtower on a hill by the river, to watch for invaders heading upriver toward London. With Humphrey's death in 1447, Bella Court was taken over by the Tudors. Henry VI and Margaret of Anjou renamed it Placentia (''pleasant place'') and made it a palace.

The heyday of Greenwich occurred during the reign of Henry VIII; he was born and grew up in the palace, and there he married Catherine of Aragon, Anne Boleyn, and Anne of Cleves. Henry added new halls and towers, a royal armory, and naval dockyards, where he watched his ships carrying supplies to and from London. He held extravagant hunting and sporting parties in the park, which was well stocked with deer. After Henry divorced Catherine, he kept her imprisoned here. Both of his daughters, Mary I and Elizabeth I, were born and lived at Greenwich. Like her father, Elizabeth spent much time entertaining, holding court, and governing here. According to legend, in 1581 Sir Walter Raleigh threw down his cloak for Elizabeth to walk over a puddle outside the palace. It is known with certainty that Raleigh visited frequently, and Elizabeth saw his ships off to the New World from the docks at Greenwich.

The history of the palace at Greenwich, from its use as a royal residence in the fifteenth century to its conversion in 1873 to the Royal Naval College, is a complex one. As was common, residing monarchs from the times of Henry VI altered and added to the palace, and they lived part of each year here. Thus, the buildings represented a combination of pre-Tudor and Tudor styles, with the old road to Deptford and Woolwich running down the middle of the complex. The Commonwealth government stripped its furnishings and used it as a prison and military barracks. After the Restoration in 1662, Charles II began renovation in the gardens and commissioned a King's House from John Webb. Much of the old palace complex was destroyed during this work. Because of a lack of funds, only one wing was completed (now the King Charles Block of the Royal Naval College). The palace was never finished.

During the reign of William and Mary, construction began again, but not to update or personalize a royal resi-

The National Maritime Museum
Photo courtesy of Greenwich Tourist Information Centre

dence. The new monarchs preferred the palace at Hampton Court, so Mary decided the buildings begun by Charles II should be converted into the Royal Naval Hospital for retired seamen. Christopher Wren undertook the monumental project, starting in 1694; before it was finished, more than fifty years later, Sir John Vanbrugh, Nicholas Hawksmoor, Sir James Thornhill, and others had all become involved.

The Royal Naval Hospital is considered one of Wren's masterpieces of baroque architecture. Wren designed the southwest wing in the King William Building, which includes the Painted Hall, its walls and ceiling done by Thornhill, and a refectory next to the chapel. Vanbrugh, assisted by Hawksmoor, later finished the western facade. (Vanbrugh also built a miniature, story-book style brick fortress called Vanbrugh's Castle in the park, where he lived as he worked on the hospital; it is now a school.) The stately buildings became the Royal Naval College in 1873, shortly after the new steamships replaced clippers.

Today the Royal Naval College stands between the river and the National Maritime Museum, originally built as the Queen's House. In 1615 Inigo Jones designed it as a residence for Anne of Denmark, wife of James I, based on the Italian Renaissance architectural style; it was the first such building in England. Its simple, classical style includes

carefully spaced windows, a simple exterior, inward-curving stairs, and, on the south side, Ionic pillars over a loggia. Jones added a bridge that passed over the road, connecting the old palace with the new Queen's House. The building was finally completed for Henrietta Maria, wife of Charles I, in 1635. For various reasons, Greenwich ceased to be a favored royal village, and its buildings became seldom used. In 1807 the Queen's House was made into the Royal Hospital School, then in 1937 the National Maritime Museum. The museum, restored to seventeenth-century conditions, now houses Elizabethan and Stuart exhibits and a fine collection of old royal barges, charts, maps, and other naval relics showing Britain's seafaring history.

Another famous building in Greenwich is the Old Royal Observatory. In 1675 Charles II commissioned Wren, a former professor of astronomy, to build an observatory on the site of the original watchtower. The first such observatory in the world, it was designed for the study of longitude, navigation, and astronomy. John Flamsteed became the first Astronomer Royal in 1675; others include Edmund Halley (appointed 1720) and Sir George Airy (appointed 1837). Map and chart makers only began to use the Greenwich meridian as a standard base in determining longitude in the late 1700s. In 1884, at the international Meridian Conference in Washington, D.C., Greenwich became the prime meridian for measurement. At the Royal Observatory a brass rail set in concrete marks the spot of the international zero of longitude. The observatory functioned until 1958, and it is now a museum.

Despite the grandeur of its buildings, Greenwich by the end of the nineteenth century reflected the worst of the industrial age, its rivers polluted, factories and tenements taking over the country landscape and parks. No longer used as a royal residence, the town developed as a suburb of London. With the first railway connecting the city to Greenwich in 1836, the population of Greenwich increased steadily, although many of its major attractions closed. The local fairs, closed by a vote of townspeople because of the undesirable city crowds, once had been popular enough to be described by Charles Dickens in *Sketches by Boz* (1836). Even Greenwich's important shipping industry collapsed with the invention of the ironclad battleship; the dockyards closed in 1869, including those in nearby Deptford and Woolwich, the British Navy's main naval dockyard. The Royal Hospital also closed in 1869. A cumbersome electric power station built alongside it in 1906 overshadows the older, majestic building.

During World War II, most of Greenwich was evacuated; those remaining worked in factory production for the war effort. Although it had suffered no major damage during the war, Greenwich began an era of reconstruction in the postwar years, as preservation societies formed to renovate many of the old buildings, including the Queen's House, St. Alphege's Church, and historic Georgian houses. As part of the general restoration, in 1957 the tea clipper *Cutty Sark* was put in dry dock along the river. The *Cutty,* one of the last

clippers built, in 1872, once had been the greatest and fastest of the trade ships from China to Britain, made obsolete by the steamboat. The *Gipsy Moth IV,* in which Sir Francis Chichester circumnavigated the world alone in 1966, is docked beside the *Cutty Sark.* (Chichester was knighted by Elizabeth II at Greenwich in 1967.)

Today the fashionable neighborhoods of Greenwich have rows of restored Georgian houses. One such house, the Grange, said to have been given by a daughter of Alfred the Great to Ghent Abbey in 818, still has timbers from the early twelfth century. The famous view from the hill at Greenwich Park offers a panorama of London behind the River Thames, with the National Maritime Museum and the Royal Naval College in the foreground.

Further Reading: Many general historical books on London include sections on Greenwich. Suggested books include *The City of London: A History* by Mary Cathcart Borer (London: Constable, 1977; New York: McKay, David, 1978); *London: The Biography of a City* by Christopher Hibbert (London: Longmans, 1969; New York: Morrow, 1970); and *Studies in London History,* edited by A. E. J. Hollaender and William Kellaway (London: Hodder and Stoughton, and Mystic, Connecticut: Verry, Lawrence, 1969). A general overview of Greenwich's history can be found in *A History of Greenwich* by Beryl Platts (Newton Abbot, Devon, and North Pomfret, Vermont: David and Charles, 1973). The best source is *Royal Greenwich* by Olive and Nigel Hamilton (London: Greenwich Bookshop, 1969), an illustrated guide to the history of the city, and the buildings in particular. The information on the buildings is condensed in a smaller booklet, *Nigel Hamilton's Guide to Greenwich* (London: Greenwich Bookshop, 1972).

—Noelle Watson

London (England): Hampstead and Highgate

Location: Hampstead and Highgate are in the western part of Camden Borough, about four miles northwest of the City of London. Hampstead is on the southern side of a high hill called Hampstead Heath, Highgate on a hill to the east of Hampstead.

Description: Hampstead and Highgate share a unique atmosphere and elegance, with distinctive seventeenth- and eighteenth-century buildings, including many fine Regency and Georgian houses. The Hampstead Garden Suburb is a good example of the first neighborhoods built for better integration of social classes. Highgate is known as well for its old Victorian cemetery.

Contact: London Tourist Board
26 Grosvenor Gardens
London SW1W 0DU
England
(71) 730 3488

Never of much political significance, Hampstead and Highgate are today exclusive residential neighborhoods, known as quiet places of retreat from the growing city of London. With similar geography, Hampstead and Highgate also share a history. From the Middle Ages, the small villages, atop twin hills, grew slowly as people moved out of London, escaping overcrowding and disasters like the plague and fires, to seek healthy country air. As late as the seventeenth century Highgate and Hampstead were still rural farmland, when surrounding areas had been quickly absorbed into the city. In large part, this was because the area, a sandy heath and heather-covered hills around dense forests, at first proved inhospitable to the development of civilization.

Thus the land, so unsuitable for farming, remained unchanged for centuries. Even during the Roman period, the Middlesex forest, particularly around the heath, was unpopulated. Although some Roman artifacts were discovered here in the late 1700s, there is little evidence that Romans spent much time in the immediate area other than to build a road to St. Albans that passed across the heath between the two hills. This road became a focal point of development in later years. The area that is now Hampstead grew into a small farming village with about twenty families, and a building for a few monks. Its name obviously derived from "hame" and "steede," Hampstead, probably was originally the homestead of a Saxon farmer. The village is first mentioned in official records in a tenth-century charter granting the land to a nobleman; it later passed to Westminster Abbey. The abbots sold or leased parcels to noblemen or let out to tenants and feudal peasants. In 1349 the abbot and several monks came here to escape the Black Death, but they died anyway, along with about one-third of the country's population. An abbot of St. Albans rebuilt the old Roman road over the Hampstead hill early in the fourteenth century and hired men to protect pilgrims on their journey from London. Much of the land remained for years in the possession of either Westminster or the bishops of London, and the woods were preserved as royal hunting grounds for the king and his bishops. During the Reformation, Henry VIII dispersed sections among wealthy landowners. Even then, the Highgate and Hampstead forests were uncleared for several hundred years.

The area around the small hamlet of Highgate, once part of the Hornsey estate, also belonged to the bishops of London. In the fourteenth century, after a dispute between the monks of Westminster and St. Albans, the Westminster monks decided to build a new road enclosed with a gate. One of the bishops allowed this private road, leading from London, to pass through his hunting park near the hill of Highgate. The road later was enlarged and toll gates, from which the name Highgate derived, were added, by permission of Edward III. A series of local hermits lived by the road to collect toll. The holes from which road gravel was taken formed small ponds (these ponds, long popular with London vacationers, became polluted by the mid-1800s and were covered over and made into Pond Square). A small village grew up around the old Highgate hermitage where the toll collectors had lived, abandoned after the Reformation.

The legend of Dick Whittington comes from the old Highgate road, dating from the fifteenth century. According to an old nursery rhyme, Richard Whittington rested at the foot of Highgate Hill and heard the Bow Bells telling him to return to London, where he would become lord mayor of the city. The Whittington Stone along the road now marks the site.

Inns and pubs were established along the roads to serve travelers on the way to and from London. One still surviving near Hampstead is Jack Straw's Castle, a tavern named after the contemporary term for farm laborers, possibly the meeting place for local farmers joining Wat Tyler's march to London in the Peasants' Revolt of 1381. Pub customers popularized a ceremony called the "Swearing on the Horns"—they grasped the horns of animals such as bulls, stags, or rams, and repeated the Highgate Oath. Tavern owners probably began the "swearing" as an amusement for their customers, many of whom were farmers bringing cattle to London. The custom, described by Lord Byron in *Childe Harold* (1812), lasted well into the nineteenth century, when Highgate then had about twenty inns.

Hampstead Heath was one of the ancient commons to the north of London, part of the hundreds of open acres that ringed the city, used by everyone for wood collection and

Kenwood House
Photo by Muriel Emanuel

pasture. Local inhabitants slowly took over, but only after years of constant contention between the locals and people of the city, who wanted to preserve the large green areas with their fresh spring water and country air. One of the main sources of business in the area during the reigns of Charles II and Henry VIII took place on this heath: the royal courts employed laundrywomen, who washed and dried clothes here. Most of the dense forest had been cleared by the end of the sixteenth century, the timber used in the fast-growing industry of shipbuilding and to rebuild London after the Great Fire of 1666. Almost 400 acres in the city had been destroyed by the fire. Just as Londoners fled to the area from the Great Plague of 1603, thousands escaped the fire by moving to the small outlying villages in the north.

In the sixteenth and seventeenth centuries, London aristocrats and rich merchants left the expanding city and began to build their manors in the northern countryside. Because much of the heath was still owned by manor lords, development centered in the east, near the villages in the foothills. Unforested land continued to be used as hunting grounds, and Hampstead and Highgate remained fairly iso-

lated villages. Even so, the ponds along the heath were favorite places of recreation, especially the large fairs held here at Easter and in spring and summer. Water parties were popular ways of coming down river to the villages, safer than taking the road because of dangerous highwaymen. In the early eighteenth century, Hampstead became a resort area famous for its spas. For centuries the people of London had used the cleaner water from northern regions: water flow from Hampstead to the city was regulated by government by the mid–sixteenth century, the reservoirs and ponds controlled by the Hampstead Water Company; its spring water was bottled and sold in the city because of its supposed therapeutic values and the polluted condition of London's water. By the eighteenth century, this belief in the medicinal values of waters and spas was widespread. The richest people traveled to such areas as the hot springs of Bath. Others stayed closer to London. One of the most celebrated chalybeate springs was at Hampstead, only four miles from London. In the busy season from March through November, a stagecoach ran several times a day. Although Hampstead was still a rural village, it offered a Great

Room and Long Room for dancing and music, card rooms, taverns, gardens, and all the necessities for tourists, including a race course, bowling green, and a dueling forest. In addition to shops, inns, and other businesses, rows of new houses were built during this growth spurt. However, scandals were frequent, and the reputation of the spas deteriorated. By the middle of the century the wells stood unkempt and unclean, out of fashion.

Nonetheless, Hampstead remained a popular holiday retreat and residence for many who worked in the city, overcrowded even in the eighteenth century. Merchants, doctors, bankers, publishers, and other wealthy professionals moved onto the top of the hill—the only available land—and built row houses, as permanent or summer homes, in the Georgian style of the day. Many still exist, such as Church Row, with its small cottages and three-story brick townhouses. The yellow-brick St. John's Church, completed in 1747, stands at the end of this row. By the end of the century, Hampstead had become one of the most desirable places to live in London. The heart of the old village was already fully developed. With little room for further growth, many houses were destroyed so that roads could be enlarged; the largest of the old Victorian mansions vanished as their owners moved to suburbs farther away. The railway brought cheap, fast travel to and from the city, and Hampstead continued to grow as a suburb in the early nineteenth century, especially after the London tube connection in 1907.

Mainstream development of Highgate, a much smaller village than Hampstead, also began in the eighteenth century. But the road from London to Highgate was worse than that from London to Hampstead. Eventually attempts were made to improve the main road up the hill; a tunnel was cut through the hill, which collapsed, but the way was now clear for construction of the Archway Road in 1812. The Highgate Archway, made of iron, replaced the stone and brick bridge in 1900. The first cable car in Europe went up the hill here in 1884. As in Hampstead, the wealthy moved here to build their mansions atop the hill; shops opened for business at the foot. Some Queen Anne houses remain today, mostly Georgian row houses, some with wrought-iron gates, along tree-lined streets; Regency, Edwardian, and Victorian houses also survive.

Famous buildings in Highgate include Ye Olde Gate House Tavern, dated from 1380, which originally stood by the tollgate. Frequent visitors included Byron, George Cruikshank, and Charles Dickens, who wrote frequently of Highgate—at one point, four houses claimed to be the Steerforth abode from *David Copperfield* (1849–50). Lauderdale House, built around the mid–seventeenth century, is said to be the house from which Nell Gwynn dangled her six-year-old son out the window in front of his father, Charles II. Samuel Pepys visited friends in Lauderdale House several times in the mid–seventeenth century. (In the late nineteenth century, it became part of St. Bartholomew's Hospital.)

Highgate Cemetery, on the western part of the hill, was created in 1839. With its grandly landscaped grounds, and even an Egyptian Avenue, it became one of the most popular cemeteries in Victorian England. Those buried here include the Rossettis, Karl Marx, George Eliot, William Friese-Green, William Foyle, and John Galsworthy. Members of Dickens's family, including his parents, his wife, and his daughter, also are buried here. St. Michael's Church, which overlooks the cemetery, is the burial place of Samuel Taylor Coleridge, who lived in the Grove from 1823 until he died in 1834.

Hampstead attracted its own famous inhabitants. Artists, musicians, writers, actors, scientists, and others have lived here since the seventeenth century, when Pepys frequently visited friends in the village. The British jurist William Murray, first earl of Mansfield, lived here in the eighteenth century in a house called Kenwood, designed by Robert Adam; the home is now a museum. Samuel Johnson and his wife had a house here. Kate Greenaway lived in a gabled house designed in 1885 by Norman Shaw. Galsworthy lived in the Grove Lodge, part of the Admiral's House, from 1918 to 1933, when he died. Rudyard Kipling visited his daughter here from 1934 to 1937. Sigmund Freud moved to 20 Maresfield Gardens with his daughter Anna in 1938; he died the following year and Anna stayed until her death in 1982. (The house is now a museum, complete with Freud's couch and exhibits on the history of psychoanalysis.) John Keats lived in a small Regency building called Wentworth Place, now known as Keats's House; Fanny Brawne and her mother lived in the connecting house. Keats wrote such poems as *Ode to a Nightingale* in the garden. Other well-known residents included Anna Pavlova, Sir Julian Huxley, Joshua Reynolds, H. G. Wells, Wilkie Collins, George Romney, Gerald du Maurier, Robert Louis Stevenson, and J. B. Priestly.

In the early twentieth century the flow of Londoners to Hampstead led to the establishment of an innovative residential neighborhood. Dame Henrietta Barnett, a local resident, was active in the early social consciousness movement. After the London underground extended into the village, she realized the potential for rehousing London slum dwellers—slum clearance already had begun in the Victorian era. Barnett conceived the idea of the Hampstead Garden Suburb, one of the first such projects. In 1905 she found two architects already working in garden city planning, Raymond Unwin and Barry Parker, and she established the Garden Suburb Trust to fund the development. The Trust purchased part of the Wyldes estate, still open country located between Golders Green and East Finchley, and 243 acres were set aside for social integration.

The architects designed tree-lined streets and a variety of house styles. Their guiding principles included housing residents from all social classes, with a certain number of rents proportioned to be more affordable to the poor; housing a mixture of old, young, disabled, and others in need; creating diverse house plans and gardens, separated by small hedges only; and placing no more than eight houses per acre.

In 1907 the first cottages were built. Unfortunately, Barnett's ideal never was fully realized; the suburbs today house mostly middle-class families.

Today Hampstead Heath still encompasses acres of open and unforested land and lakes, even though virtually every area around the heath has been built upon, with only a small portion in its natural, pre-Saxon state. Main attractions in Hampstead and Highgate include restored seventeenth- and eighteenth-century houses, several of which are museums. For example, one of Hampstead's finest and oldest houses is Fenton House, built in 1693. Of brown brick, it is a typical William and Mary house, with seven bays, the original pine staircase and doors, eighteenth-century furnishings, and a keyboard instrument collection dating from 1540 to 1805. Such areas as Keats Grove, Church Walk, and Flask Walk remain quiet little streets and alleys with period rowhouses, pretty courtyards, and gardens. And, of course, the old villages continue to attract the rich and famous.

Further Reading: The most significant work on the area is still *The Annals of Hampstead* by Thomas J. Barratt, three volumes (London: A. and C. Black, 1912). *The Book of Hampstead,* edited by Mavis and Ian Norrie (London: High Hill Press, 1960; revised, 1968), collects sixteen essays by writers living in Hampstead, about topics of general interest, including anecdotes about local writers and artists and history of buildings. *Hampstead: Building a Borough, 1650–1964* by F. M. L. Thompson (London and Boston: Routledge and Kegan Paul, 1974) traces the urban history of Hampstead, focusing on the development of local topography and populations. *Hampstead and Highgate: The Story of Two Hilltop Villages* by Mary Cathcart Borer (London: W. H. Allen, 1976) offers a good historical perspective of the social and political aspects of the villages' history from medieval times to the twentieth century. It has illustrations and a bibliography. An account of the last years of Samuel Taylor Coleridge's life, spent at Highgate, appears in *Coleridge at Highgate* (London and New York: Longmans, Green, 1925) by Lucy Watson, the granddaughter of Coleridge's doctor in Highgate.

—Noelle Watson

London (England): Hampton Court

Location: Hampton Court is on the north bank of the River Thames in the London Borough Richmond upon Thames, fifteen miles southwest of the City.

Description: The palace buildings of Hampton Court stand on open land in a riverbend; the grounds, with Bushy Park, are about 1,900 acres. The famous red-brick palace, with its Tudor turrets and chimneys, some 1,000 rooms, the State Apartments, and art collections, was the favored home of centuries of British royalty.

Site Office: Hampton Court Palace
East Molesey, Surrey KT8 9AU
England
(81) 781 9500

One of the most popular sites in England, Hampton Court is both artistically and historically of major significance. In many ways, the history of Hampton Court is the history of England itself during the fourteenth to seventeenth centuries; it was the place where members of the British royalty were born and died, imprisoned, or divorced. Diverse royal initials and arms on the palace gateways and ceilings represent the changes and additions of centuries of generations. There are even ghosts; two of Henry VIII's wives, Jane Seymour and Catherine Howard, are said to wander the palace.

Known for its architecture, the palace consists of a traditional series of three square courts, with red-brick turrets and distinctive chimneys. There were two main construction periods. In the early sixteenth century, Cardinal Wolsey first built the palace, later altered by Henry VIII; the Base Court and chapel remain from this period. The second major phase occurred during the reign of William and Mary, in the late seventeenth century. The Fountain Court, to the east, was added under the direction of Sir Christopher Wren and William Kent, who succeeded Wren. The third of the courts, now called the Clock Court, where Wolsey lived, represents a combination of work commissioned over the years. As a result, Hampton Court is an incomparable example of Tudor and Jacobean Gothic styles. Noteworthy architectural hallmarks of the palace include its tall brick chimneys (not the originals), archetypal gateways and courtyards, large terracotta medallions of Roman emperors by the Italian Giovanni da Maiano, the sixteenth-century Astronomical Clock by Nicholas Oursian, and the Great Hall with its ornate hammerbeam roof.

The site of Hampton Court was first settled by Saxons; its name probably derived from a Saxon word describing a farm by a riverbend: "ham," or "home;" "ton," meaning several houses close together; "court," because there was a common yard. The manor house belonged to the St. Valerie family until the thirteenth century, when the land was taken by Henry III. The Knights Hospitaller of St. John of Jerusalem, a powerful crusading order, held part of the land and an old manor, listed in the Domesday Book (1086). By this time, the area had developed into a village, much of it enclosed within walls, both the center of a secular government and the parish.

John Wolsey, a chaplain in the house of Henry Tudor, became archbishop of York in 1514, the same year he leased the manor and surrounding land from the Knights of St. John. Under Henry VIII, Wolsey became lord chancellor, cardinal, then papal legate; one of the most powerful and richest men of the time, he had much influence over the king. It was said that Wolsey's wealth surpassed even the king's. He had amassed a superior collection of art treasures, including tapestries, his great pride, and Flanders carpets, which were to be housed in his new home. Most likely the health-conscious Wolsey chose the site because of its clean waters and country air. Planning a residence to rival the royal palaces, he enclosed the land, then some 1,800 acres. When completed, his palace had about 1,000 richly furnished rooms, 280 designated as guest rooms for visitors arriving from throughout Europe. A staff of 500 served the palace. There were impressive innovations: water closets, lead pipes carrying spring water from Coombe Hill three miles away, and brick sewers draining into the Thames (the sewers survived until 1871). An eight-turreted Great Gatehouse is the most distinctive of the remains from this period.

Wolsey soon lost his palace. In 1529 Henry VIII plotted to divorce Catherine of Aragon and to marry Anne Boleyn, for which he needed Wolsey's help. Wolsey was a member of the court presiding over the case to validate the marriage to Boleyn. They quarreled over the lengthy court proceedings, and Wolsey presented Henry with Hampton Court as a gesture of goodwill. Although Henry was not appeased, he promptly moved into the palace. Wolsey, out of favor with both the king and Boleyn, died on his way to London to face a charge of treason.

Henry VIII destroyed or enlarged many sections of Wolsey's palace, adding new wings, the Great Hall, the chapel, moat and drawbridge, gardens, courts and other annexes. The covered Tennis Court was built in 1532, mostly for Henry to indulge not in playing tennis, but in gambling. (According to legend, however, he was engaged in a tennis match there when he was informed that Anne Boleyn's execution had been carried out.) The Astronomical Clock, designed by Nicholas Oursian in 1540, was added to the Clock Court. This famous clock showed tide times and moon phases as well as the hours and days.

The Great Hall of Hampton Court Palace
Photo courtesy of Historic Royal Palaces

Hampton Court was only one of several palaces used by both Tudor and Stuart royalty. The court moved from season to season, changing palaces to avoid the plague, and to live for awhile in improved sanitary conditions. Reigning monarchs and their courts traveled to the palace by barge, carrying large loads of household items and furniture. Guests also arrived from London by river transport. From the time of Henry VIII, who lived at the palace with five of his wives, Hampton Court became a favorite royal residence, a status that lasted until the eighteenth century. Here Jane Seymour gave birth to Edward VI and died soon after. Edward grew up in the palace, along with Henry's daughters Mary and Elizabeth. Mary Tudor spent several months here during her short reign. Even Elizabeth, who had been imprisoned briefly in the Water Gallery, favored Hampton Court. In the mornings she worked alone in the gardens among plants such as the New World potato and tobacco plants brought to her by Sir Walter Raleigh and Francis Drake. It was customary for the court to celebrate Christmas here, with hundreds of guests in attendance. Popular pastimes included masquerades, dancing, tennis, and plays, which were presented in the Great Hall, the oldest existing Elizabethan theater in England. Inigo Jones designed the settings, and the Companie of Comedians performed Shakespeare.

With monarchs so frequently in residence, many important events occurred at the palace. In 1604, for example, James I took part in the Hampton Court Conferences held to settle the differences between the Church of England and the Puritans. Petitioning for reforms in the church, the Puritans wanted several fundamental changes, including alterations in The Book of Common Prayer and a new translation of the Bible. James refused most of their demands, but he allowed for a new Bible translation. The Authorized (King James) Version of the Bible was completed in 1611.

Over the years, further minor alterations were made on the palace. Using the palace as a country estate, Charles I redesigned some buildings, adding a deer park and many ponds and fountains. Later he was held prisoner here in 1647, during the Civil War. After he was executed, the new Parliament tried to sell Hampton Court, but Oliver Cromwell took it for himself, living there till he died in 1658. The furnishings were preserved, but the palace began to deteriorate even after the Restoration. Charles II did not spend as much time there as his predecessors, but he upgraded the neglected gardens, added fountains, and restored many of the furnishings that had been sold. During the Great Fire of 1666 he removed his large fine art collection to Hampton Court for safekeeping.

Even with royalty in residence at least some of the time, the palace and grounds fell into a state of neglect that lasted for almost 150 years. Then the second phase of massive reconstruction of Hampton Court began with William and Mary, who both delighted in the palace, their main residence in England. In 1689 they commissioned Sir Christopher Wren (who worked for a time on the palace for Charles II) and a team of craftsmen, including Grinling Gibbons,

Antonio Verrio, and Jean Tijou, to work on the interior. About one-third of the old Tudor palace, mostly the Cloister Green Court where Henry VIII had lived, was torn down, replaced by the Fountain Court, in the classic style of seventeenth-century French Renaissance brick and stone. Wren redesigned and rebuilt the east and south fronts. He added the well-known State Apartments and private rooms for the king and queen, with traditional separate courts, and the grand staircases. The Cartoon Gallery to the west of the Fountain Court held seven Raphael cartoons that had belonged to Charles I (copies are there now). William and Mary also wanted the gardens enhanced in the Dutch style, and an orangery added. Mary stayed frequently in the Water Gallery, although reconstruction was not completed until 1699, five years after her death from smallpox.

Although this second phase of extensive renovation concluded during William's reign, small changes continued through the eighteenth century. Queen Anne, who liked to hunt, improved the grounds, and added the famous hedge maze, by the Lion Gate, which still exists. Wren was commissioned to redecorate the State Apartments and chapel, superbly finished by Gibbon carvings. (Wren himself retired in his mid-eighties and lived in Hampton until he died in 1723.) George I spent much of his time in England in quiet retreat at Hampton Court. George II held much livelier parties at the court, and the State Apartments were finally completed during his reign. He was the last royal to reside there. Since his death in 1760, no other reigning monarch has lived at Hampton Court. (Legend has it that George III as a child had his ears boxed by his grandfather at Hampton Court, and so he spent little time here as an adult.) In any event, Queen Victoria opened the State Apartments and gardens to the public in 1838.

The State Apartments, including some of Wren's additions in the south wing of Fountain Court, were closed after serious damage by a fire in 1986. A government project has restored this section to much of its original state from the period of the sixteenth century, and Hampton Court is again open to the public.

Further Reading: The first major work on the palace, and a principal source of information on it, is *The History of Hampton Court Palace* by Ernest Law (London: George Bell and Sons, three volumes, 1885, 1888, 1891). Both *Hampton Court Palace* by Olwen Hedley (London: Pitkin Pictorials, 1971; New York: British Book Center, 1975) and *The Gardens and Parks at Hampton Court and Bushy* by David Green (London: Her Majesty's Stationery Office, 1974) offer good general descriptions of the palace itself and its history, and include many photographs. Another good introduction for the general reader is *Hampton Court: The Palace and the People* by Roy Nash (London: Macdonald, 1983; Topsfield, Massachusetts: Merrimack, 1984). A more extensive account of the history surrounding the palace, from its beginnings, with a focus on the people who lived there, can be found in *Hampton Court Palace* by June Osborne (Kingswood, Surrey: Kaye and Ward, 1984).

—Noelle Watson

London (England): Southwark

Location: Inner borough of Greater London covering approximately eleven square miles and extending south from the Thames to Sydenham Hill, between the boroughs of Lambeth and Lewisham; composed of three metropolitan boroughs from the days of the London County Council: Bermondsey, Camberwell and Southwark.

Description: Site of a Roman bridge built across the Thames in A.D. 43; historically important crossroads and provider of necessary but unpleasant industrial services for the rest of the region, such as tanning, weaving, printing, and all manner of entertainment; home to Shakespeare's Globe Theatre; starting point for Geoffrey Chaucer's fictional Canterbury pilgrims.

Contact: London Tourist Board
26 Grosvenor Gardens
London SW1W 0DU
England
(71) 730 3488

Southwark is situated on the southeast side of what is generally known as Central London. Directly north of the borough lie many of London's most familiar landmarks, including St. Paul's Cathedral, the Old Bailey Central Criminal Court, and the city of London itself, now one of the most important financial districts in the world. Southwark is perhaps lesser known to the world at large, lacking the monumental period embellishments common in those areas of London favored by modern tourists; but it has, since its very beginnings, played a pivotal role in the development of the city that was for nearly two hundred years the most important in the world.

To the modern visitor Southwark may seem indistinguishable from the other south-of-the-Thames communities that surround it. Bomb damage from World War II and subsequent rebuilding schemes destroyed much of the area's original architecture and infrastructure, leaving the area with an eclectic combination of new—and generally bland—buildings and surviving Victorian structures. Furthermore, Southwark was historically a perpetually blighted area, filled with taverns, brothels, theaters, and unregulated industry, none of which lent itself to the construction of first-quality structures. Wood was the material of choice, and fires were common.

However, Southwark's role in the development of London had little to do with monument-building or the trappings of empire. Rather, Southwark was—and still is—a crossroads. For hundreds of years, Southwark claimed the only bridge over the Thames and was the only bastion of

civilization in the Greater London area south of the river. Until Westminster Bridge was built in 1750, every person who wished to enter the capital from the south had no choice but to pass through Southwark (as did residents of London desiring to head south). Some kept moving, some stayed in the borough temporarily, and some stayed permanently. It is this constant turnover of people and ideas that makes up Southwark's contribution to history.

Southwark's initial manifestation as an outpost of civilization can be marked by the construction of the first bridge across the Thames by the Romans in A.D. 43. Arriving at what is now Richborough, on the eastern coast of the county of Kent, the invading Romans were faced with a dilemma: a bridge across the Thames was an absolute necessity, but geography prohibited building such a structure anywhere near the sea, where a bridge would have been most convenient. The Thames estuary, a wide, brackish waterway leading into southeastern Britain from the North Sea, was both tide-prone and swampy, and provided few natural locations for a sturdy bridge with solid foundations on either side. The decision to build the bridge at Southwark was based almost entirely upon the fact that it was the easternmost location at which a structurally sound bridge could be built. All of what is now London was to grow from that single decision.

Nothing remains of that pivotal first bridge today. In fact, most of the particulars of its very existence have been extrapolated from studies of the growth of the city subsequent to the Roman invasion. It seems likely that it was wooden, and it probably also featured a drawbridge due to the fact that large Roman boats of more than sixty tons have been discovered upstream in the Thames. Its presumed location was further evidenced by the discovery in the nineteenth century of a large number of Roman coins and other common artifacts in the area during dredging undertaken to construct a later bridge. Additionally, remains of an enormous quay have been found on the north side of the Thames stretching approximately from London Bridge southeast to the Tower of London. Presumably, the quay was built on the north bank of the river to take advantage of its more stable geology, since the south bank, where Southwark now sits, was notoriously flood prone.

The combination of the bridge at Southwark and the commercial quay on the opposite side of the river provided London with its first engines of growth: a place to cross the river and a place to do business. In fact, Emperor Tacitus, in describing the Boudiccan Rebellion of A.D. 60, in which native British tribes, led by Druidic religious leaders, fought to oust the Roman invaders, stated that the population consisted primarily of merchants. The bridge and the quay made London, then called Londinium, the first port of call for merchants entering Britain and the only market for British

London Bridge
Photo courtesy of British Tourist Authority

merchants interested in exporting their wares to other outposts of the empire. It is not surprising, then, that London had become the capital of the British province of the Roman Empire by the middle of the first century A.D.

Whether Britain flourished or foundered under Roman rule is subject to debate. Certainly many indigenous British cultures were either irreparably altered or stamped out completely, but affiliation with the Roman Empire, however forcible, made Britain a charter member of what was then the most advanced society in Europe. Furthermore, the presence of a common enemy—the Romans—presented the tribes of Britain with their first opportunity to unify, however haphazardly or temporarily, into a single entity.

The great city of London, eventually the capital of an empire greater than even that of Rome, was born during the Roman occupation of Britain. London was the seat of Roman government in Britain, host to the representative of the emperor, his palace and staff, all provincial bureaucratic headquarters, and the only official mint in Britain. Much remains today of Roman London, primarily in the form of foundations, fortifications, and other masonry discovered over the centuries during the constant process of rebuilding the city.

Southwark's role as a crossroads was firmly entrenched even at this very early stage. The first Roman commercial district and seat of government was established on the north side of the bridge across the Thames due to its higher elevation and lack of floods. Known as the City, the one-square-mile district was surrounded by a protective wall that extended to just north of the quay and its adjoining bridge. Within the wall, merchants, government officials, and the well-to-do conducted their business. But just as all modern city centers have their satellite slums, red-light districts, and industrial corridors, so did early London. Those who wished to be near the conveniences and connections of the City, but could not or would not live there, set up shop in what is now Southwark.

It is not known when or how the original bridge across the Thames met its demise. However, the infrastructure that grew up on both sides of the river around the original bridge, in addition to the geographical and geological convenience of the site, led successive generations to continue to build bridges in almost exactly the same spot. By the tenth century a structure formally known as London Bridge had been built at the site, first mentioned in records of the period describing the drowning of a woman convicted of witchcraft.

The name "Southwark" also made its first formal appearance in the tenth century. According to the *Burghal Hidage,* a cartographical document from the tenth century, two towns are listed in what was then known as Surrey: "Eschingum" and "Suthringa geweorc." The latter translates as "the fort of the men of Surrey," and has evolved into the modern Southwark.

The bridge at Southwark gained immortality in 1014 when King Olaf of Norway, leading an invasion of Danish-controlled London in alliance with the English King Aethelred, vanquished the Danish defenders of the bridge by sailing up the Thames, roping his Viking vessels to the structure and rowing away, pulling the fragile bridge to pieces in the process. This heroic effort was celebrated in a Norse saga of the twelfth century, which eventually became the nursery rhyme "London Bridge is Falling Down."

Both Southwark and London Bridge played important roles in the Norman Invasion of 1066. Following the Battle of Hastings in what is now the county of East Sussex, located south-southeast of London, the victorious William the Conqueror marched to London, determined to finish what he had begun by killing King Harold Godwinson in battle. With approximately 15,000 inhabitants, London was by then the largest and wealthiest city in Britain, in addition to being the seat of Harold's government.

William may have been a great military leader, but he, too, had to pass through Southwark in order to enter the city from the south. Unfortunately for Southwark, William met with great resistance at London Bridge, forcing him to reconsider his plans. Deciding that isolating London would be a more effective means of capturing it than assaulting it directly, William opted for a scorched-earth approach, starting with Southwark. After leaving Southwark in ruins, William and his troops headed westward to Hampshire, where they turned north in order to cross the Thames at Wallingford. Within three months of the Battle of Hastings, the Normans had taken London and the conquest was over.

Although devastated by war, Southwark was given the opportunity to grow anew following the invasion. Twenty years after their victory, the Norman conquerors took an exhaustive inventory of their newest domain. Known as the Domesday Book, this pivotal census is the most complete and thorough chronicle of Britain in the eleventh century, and is, in many cases, the only quantitative record of the period. According to the Domesday Book, there were fifty-two houses in Southwark in 1086, and, apparently, a thriving fishing port. The area was also home to a church, located either on the site of the present Southwark Cathedral or in the place of St. Olave Church, an ancient church that once sat on the banks of the Thames where a warehouse now sits.

Between 1176 and 1209 a stone bridge was built across the Thames. Spanning 900 feet on a series of 19 arches, the bridge, designed by Peter of Colechurch, was destined to stand until 1831. In a move surprising to modern thinking, the bridge was considered to be real estate like any other and was soon covered with houses and shops precariously perched on either side and often reaching across the bridge itself, reducing the traffic lane in the middle to a close, dimly lit, and often dangerous passageway. By 1460 there were no less than 129 tenements on the bridge, all of which paid rent toward its upkeep.

By the fourteenth century what is now the Greater London area consisted primarily of three loosely joined cities: London, Westminster, and Southwark. The power lay in London, but the majority of the people lived in Westminster and Southwark.

Southwark in the fourteenth century was a jumble of people and ideas, some respectable, many not. It was still the only crossroads to the south of London and, as such, attracted people from all walks of life and industries. Both the Bishop of Winchester and the Abbot of St. Augustine's Canterbury maintained residences in Southwark because of the availability of land there and the borough's proximity to London. As at all crossroads, innkeeping flourished in the borough.

Southwark was also home to an important agricultural sector, providing an important market for cattle raised in Kent and, on its southern fringes, maintaining a large number of gardens that provided fruit and vegetables for the London market. Other respectable industries tended to be of the unpleasant kind, such as tanneries, kilns for smelting and lime burning, and printing. Necessary ventures, to be sure, but not enterprise of the variety welcomed in the city, where bureaucrats in fine clothing were held in higher regard than skilled tradesmen covered in the muck of their vocation.

Because it was an independent municipal entity, Southwark was not bound by the strict moral laws that mandated the behavior of city residents. It made sense for the local authorities in such an important crossroads for London-area commerce, to please as many people as they could. In any given day Southwark played host, for at least a few hours, to a wide variety of merchants, tradesmen, farmers, thieves, swindlers, prostitutes, politicians, and every other kind of person imaginable. Southwark became the place where all could find what they wanted.

Naturally, this presented problems. Compared to

London, Southwark was a lawless slum and, as such, attracted a large criminal element. The best pickings for such predators lay across the river, but Southwark provided an ideal refuge following a night of thievery in the city.

In 1327 King Edward III was granted a charter that gave the city financial control over Southwark, but left a gaping hole in the city's ability to maintain order, which remained partly in the hands of the bishop of Winchester and the sheriff of Surrey—the same office that had let things get out of hand in the first place. Southwark was not completely under the jurisdiction of the courts of the city, thereby giving legal refuge to criminals who were wanted for crimes committed there.

As the laws of London became ever more strict, that city's undesirables had little choice but to cross the bridge and set up shop in Southwark, where they were, for all intents and purposes, still beyond the law of the city. Prostitution became a booming industry in Southwark, benefiting from the high concentration of inns and travelers, and the proximity to city clients eager to spend an evening away from the prudish confines of the capital. In fact, the greatest concentration of brothels and associated enterprises lay along the riverbank within sight of London. Because the majority of this land was owned by the bishop of Winchester, who apparently turned a blind eye to the proceedings, the prostitutes of the area were often referred to as "Winchester geese." Near the river, the bishop also ran a prison known as the Clink, source of the slang term to describe a prison.

By 1376 Londoners had had enough of the two-tiered system of laws that allowed known felons from the city to live unmolested in Southwark, and petitioned the king to put Southwark entirely under the control of London. Because Edward was the only real authority in Southwark and would have lost a great deal of his control by handing the borough to the Londoners, he refused, and instead added to the facilities he had already placed there. In 1394, however, Londoners demanded and received an act that gave them complete jurisdiction over the nearby ward of Farringdon, setting the precedent for future battles over the sovereignty of Southwark.

Despite this flurry of activity, Southwark was in the fourteenth century a small hamlet of hardly more than 1,000 people. Its dense concentration of slums near the river gave way within a matter of a few hundred yards to fields in the southern portion of the borough. By 1377 the population of the entire Greater London area was only about 40,000, making the metropolis smaller than Paris or Milan and about the same size as Brussels or Cologne.

Southwark was forever immortalized in 1387 by Geoffrey Chaucer in *The Canterbury Tales*. The 17,000-line poem describes the meeting of twenty-nine pilgrims at the Tabard Inn in Southwark, where they begin their pilgrimage to Canterbury. Although the pilgrims, such as the Knight, the Cook, and the Miller are fictional, the Tabard Inn was not; it was a favorite haunt of Chaucer when he was employed in the king's court in London. Chaucer held a variety of positions in court prior to his death in 1400, from diplomatic posts to controller of customs for the port of London. In 1386 he became knight of the shire for Kent and probably lived there for the rest of his life, indicating that he passed through Southwark on a regular basis.

In 1550, during the short but turbulent reign of the boy-king Edward VI, Southwark was finally declared Bridge Ward Without, giving London authorities complete control to arrest and try in Southwark those who had broken city laws. This new status did not, however, give London authorities the power to create or enforce the laws of Southwark, which remained a freewheeling satellite of the prudish but powerful city across the river.

The puritanism of London made Southwark a mecca for entertainment and other "frivolous" pastimes, especially theater. In the mid–sixteenth century the repressive policies of the city's authorities were leveled against theater, which regularly drew larger crowds than the church. "Will not a filthy play, with the blast of the trumpet, sooner call thither a thousand, than an hour's tolling of the bell bring to the sermon a hundred?" went the reasoning of the day. Although they were not outlawed in London, theaters were not welcome there, either.

As with so many undesirable professions before it, theater had no choice but head for the unfettered borough of Southwark, which soon became London's undisputed theatrical district. The burgeoning business attracted many young actors and playwrights, including no less than William Shakespeare, who, at the time, was both an actor and writer. Shakespeare had no settled residence in London, and he moved about the area frequently, but he did live in Southwark at least once, and the erection in 1599 of the Globe Theatre, in which Shakespeare owned a one-tenth share and at which his plays were performed, suggests that he spent the majority of his time in London at Southwark.

Shakespeare's plays were also performed at several other theatres in the area, including the Rose (1587) and the Swan (1595), which seated 3,000 patrons. Depite official opposition to the theaters, by the late sixteenth century London was home to seventeen of them, with the greatest concentration in Southwark, while rival Paris claimed only two public playhouses, both of which were controlled by the government.

In 1592 the plague made an appearance in London, as it had done from time to time for centuries. Strictly, the plague never went away, but was only officially recognized as a menace when deaths from the disease reached critical levels. When plague deaths met a certain number, "plague orders" automatically went into effect. In June all "profane spectacles" were prohibited until September 29, 1592. Theatre was not necessarily a profane spectacle, but as of September 7 the plague had become such a problem that all theaters were closed. Banning the close assembly of large numbers of people made some bacteriological sense, but the reasoning behind the move was an entirely different matter. "The cause of plagues is sin, if you look to it well," said a London preacher during a previous manifestation of the dreaded disease. "And the cause of sin are plays; therefore the cause of plagues are plays." The theaters stayed closed until June 3, 1594.

The plague again visited London in 1665 and hit hardest in the areas of the filthiest slums, such as Southwark, leading the moralists of the day to once again associate the sin of the borough with its virulence. This would prove to be the last outbreak of the plague in London, for in 1666 London was almost completely destroyed by fire.

The fire proved to be something of a blessing, as the flames and intense heat destroyed most of the rats that spread the plague. It also gave the city an opportunity to reinvent itself, solving many of the problems of overcrowding and poor sanitation that had afflicted it for more than a thousand years. Southwark was essentially spared from the flames, although embers brought across the river by the wind did start small fires at a stable and St. Thomas' Hospital.

Because it escaped the fire of 1666, Southwark remained a blighted area. In 1748 an observer wrote, "a quarter of the town [is] ill-built, having but two streets in its breadth, and almost entirely occupied by tanners and weavers." Things had not improved much by the nineteenth century, when a critic described the industrial corridor that surrounded central London, including Southwark, as follows: "A Polynesian savage in his most primitive condition [did not endure conditions] half so savage, so unclean and so irreclaimable as the tenant of a tenement [there]." Still, it was areas like Southwark that provided the rest of the capital with many of its most important products and services.

Southwark received its own bridge, dubbed the Southwark Bridge, in 1819. Designed and built by John and George Rennie, the bridge was one of several important public bridges built by the duo, including the new London Bridge of 1831. The current London Bridge was erected in 1973 and, curiously enough, was financed by a surplus of tolls and rents collected since 1282.

Always under the control of the sovereign or, later, Parliament, London in 1889 finally received an elected government of its own. The London County Council exerted direct control over the entire London area, with the exception of the city itself, which maintained certain special privileges. The council's powers were relatively limited, but they did serve to tie together what for almost 2,000 years had been a haphazard collection of disparate elements, sometimes growing together and sometimes growing in different directions. Southwark, and the other constituent wards of the city, were finally united under a truly common government at this time. In 1899 this government was subdivided into twenty-eight separate metropolitan boroughs, of which Southwark was one. With powers shared between the boroughs and the council, inefficiency reigned.

Southwark was among the more grievous casualties of the Blitz during World War II, suffering between 20 and 30 German bombs per 100 acres during the first two months of nighttime raids that began on June 10, 1940. The Blitz destroyed much of Southwark's architecture and infrastructure,

leaving gaps that have since been filled with generally bland postwar buildings.

In 1965 the London County Council was abolished and replaced by the Greater London Council, which enlarged constituent boroughs and formalized the separation of powers, leaving housing and planning issues in the hands of separate borough councils, but maintaining a large central bureaucracy, the need for which was questioned from the start. The London County Council boroughs of Southwark, Bermondsey, and Camberwell were combined at this stage into a single borough named Southwark, which continues to exist. In 1986 the "Municipal Monster" was abolished completely and all powers were either devolved to individual borough councils or dispersed amongst citywide public entities.

In the 1980s Southwark was witness to one of the most controversial developments undertaken in the city's history. The docks of the eastern Thames, underutilized since before World War II, became the subject of the largest urban renewal project in Europe. Generally referred to simply as the Docklands Project, the development aimed to put the antiquated docklands area to new use in the modern, service-oriented economy. Approximately 8.5 square miles of commercial and residential property were slated for redevelopment along the banks of the Thames east of London Bridge. The primary concentrations of new development were located in Surrey Docks, the Isle of Dogs, and the Royal Docks, all located east of Southwark. Southwark, however, found its way into the plan, with many a waterfront warehouse being converted from empty space to expensive "loft" apartments.

The project was lambasted by some, including Prince Charles, who claimed that the American-style high-rise development is out of keeping with the area's heritage. But Southwark, the bridge, and the docks have always been a crossroads, and not prone to maintaining a single image for very long.

Further Reading: *A History of London* by Robert Gray (London: Hutchinson and Company, 1978; New York: Taplinger, 1979; revised, 1987) is the standard text on the history of the capital. The history of the borough of Southwark is given a more full historical context when treated as part of the history of London as a whole. *Shakespeare of London* by Marchette Chutte (London: Secker and Warburg, and New York: Putnam, 1951) tells the story of Shakespeare's life in London, where he completed most of his work before retiring to Stratford-upon-Avon. Although Shakespeare lived in Southwark for a comparatively short time, most of his early works were performed there. *Three Tours Through London in the Years 1748, 1776, 1797* by Wilmarth Sheldon Lewis (New Haven, Connecticut: Yale University Press, 1941; Oxford: Oxford University Press, 1942) is a well-researched account of what a visitor to London would have seen in three different years in the eighteenth century. *London in the Age of Chaucer* by A. R. Myers (Norman: University of Oklahoma Press, 1972) does for Chaucer's London what Chute's book does for Shakespeare's London.

—John A. Flink

London (England): Westminster Abbey/Parliament Square

Location: Along the north bank of the River Thames in the City of Westminster, a borough of Greater London.

Description: Westminster Abbey, a working church that has existed in one form or another for nearly 1,000 years, is the site of royal weddings and coronations and the burial place of many of England's most notable citizens. Across Parliament Square from the abbey is Westminster Palace, more popularly known as the Houses of Parliament, once a royal palace and now the meeting place of England's Parliament. Over the palace rises a clock tower commonly referred to as Big Ben (strictly speaking, the name applies only to the bell that tolls each hour); this is perhaps the most famous timepiece in the world.

Site Offices: Westminster Abbey
Broad Sanctuary
London SW1P WPA
England
(71) 222 5152

Westminster Palace
London SW1A 0AA
England
(71) 219 4272

The area around Parliament Square in the London borough of Westminster is of great spiritual and secular significance in England. Westminster Abbey is both a working church and a monument of national importance in its role as the site of royal weddings and coronations, as well as a burial site for many of England's most famous people. Westminster Palace is the meeting place of the country's Parliament, having evolved from its original role as a royal residence. Rising above the palace is the clock tower known as Big Ben, one of the most recognizable landmarks in the world and a symbol of stability to the English people.

Witness to countless royal unions and coronations, the Collegiate Church of St. Peter in Westminster, to give the abbey its official title, is, in many ways, the national church of England. Located in the heart of the city, this impressive structure has undergone many architectural and functional changes over its long life; however, its existence has always been closely tied to the history of England and particularly the monarchy. Now a much-visited tourist attraction, Westminster Abbey remains a functioning church, and noisy visitors are constantly reminded that such a structure demands more reverence than it is usually given. Within its walls are buried many of the country's notables, and those who are not actually buried there are often remembered with plaques and statues. So important is the church in the history of England that it has been called "the Valhalla of the British People."

Many legends exist concerning the origin of the abbey, but perhaps the most repeated concerns a church being built on an island in the Thames by Sebert, king of the Saxons. When the church was nearly completed, St. Peter appeared miraculously to a group of fishermen on the shores of the river. They ferried him across to the church and then witnessed the entire structure fill with light. Peter told them that he had personally consecrated the building. Furthermore, he told them that they would be blessed with a wonderful catch of salmon but that they must never again fish on the Sabbath. So powerful was this legend during the medieval period that in 1308 a tomb was set up that is said to contain Sebert's remains. The tomb still exists today in the south ambulatory of Westminster Abbey. This myth also helped form the basis for the abbey's claim to an exemption from Rome. Such an exemption was indeed granted, after which the monastery answered directly to the pope and not to local ecclesiastical leaders.

Other myths and legends concerning Westminster include one that says a temple to Apollo once stood on the site. There is no evidence to support this belief; however, fragments of Roman pottery as well as a stone sarcophagus from the Roman period indicate that at the very least the site was used in ancient times as a place of burial. The sarcophagus can still be seen in the entrance to the chapter house of the abbey.

Whatever the true story of the foundation of Westminster may be, it seems likely that the first structure on the site was probably a church and the location was once an island formed by the flow of the Thames. Much of the credit for the creation and the propagation of the many myths concerning Westminster can probably be ascribed to the monks themselves, who sought as early an origin for the church as possible. However, the first certain reference to the site appears in a land grant made in 693. Some 100 years later in another document, reference is made to the monastery "in that terrible place Westmunster."

The next references to Westminster come from the period around 980, when an Abbot Wulsin was appointed to lead the monastery. This figure went on to become the Bishop of Sherborne. Finally, the reliable history of the site begins with the accession of Edward the Confessor to the throne in 1042. It was during his reign that an entirely new church was built on the site. Upon establishing his rule, Edward desired to fulfill a vow he had made many years earlier to make a pilgrimage to Rome. It was deemed unwise for the king to leave his country, so the pope released Edward from his vow on the condition that he construct a church and

Westminster Palace
Photo courtesy of House of Commons

dedicate it to St. Peter. Legend has it that a monk informed the king that St. Peter had appeared in a vision instructing the monk to tell the king that he should construct the new church at the site of the earlier structure that Peter had already consecrated personally. Apparently Edward heeded the monk's message and began constructing a church alongside the monastery that already existed at the site.

Having spent much of his life in France in exile, Edward sought to build a church in the style of the new structures of Normandy; thus the church he constructed was the first in England in the Norman Romanesque style. Edward died shortly after the consecration of the church in 1065, and the only image that remains of that building appears in the Bayeux Tapestry.

Edward was buried at the church and has long been considered its most important patron. There exist several documents, now thought to be forgeries, that were purportedly signed by Edward granting the church special privileges. One of those documents names the church as the coronation site for all future British monarchs.

Just one year after the consecration of the church, the political tide in England took a dramatic turn, and William

the Conqueror was crowned at Westminster on Christmas Day, setting a tradition that has been followed by every monarch in England except Edward V and Edward VIII. The oft-told tale of the Christmas Day coronation of William is worth repeating. The tumult and shouts of joy within the church at the coronation so alarmed the Norman soldiers outside the church that they began burning the thatched-roofed buildings adjoining the structure. Thus the coronation took place amid the noise and flames coming from outside the walls of the church.

The tomb of Edward at Westminster was purported to have been the site of many miracles, and it was not long after his death that a campaign was begun to have him named a saint. In 1161, Edward was canonized and his remains were moved to a new and more elaborate shrine within Westminster. Now the church had a special saint, and many miracles, both before the canonization and afterward, have been attributed to St. Edward.

The current appearance of the church, or at least of the interior, dates back to the reconstruction undertaken by Henry III, an inept political leader but a devoted patron of the arts. Beginning in 1245, the old church was demolished bit by bit and was reconstructed. The new structure owed much to the Gothic style, which was beginning to overtake continental Europe. The model for the abbey church is believed to have been the cathedral of Rheims. Unfortunately, Henry died before the completion of the rebuilding, and the church remained a peculiar mix of old and new for many years.

One of the most curious episodes in the history of Westminster occurred in 1303, when King Edward I was informed that the royal treasury, which was stored in a crypt below the chapter house at the abbey, had been robbed. So displeased was the king that he sent all of the monks, including the abbot, to the Tower of London. After a thief was identified, the abbot and some of the monks were released, although others remained imprisoned for more than two years and two of the clerics were implicated in assisting the accused robber.

The rebuilding and completion of the Gothic church begun by Henry III was not completed until the start of the sixteenth century. At roughly the same time an important element was added to the church: King Henry VII's chapel. This structure is also known as the Tudor chapel, for it is decorated throughout with the Tudor badge and houses the remains of all the Tudor monarchs. This chapel is also home to one of Europe's largest collections of sculpted saints, who look down from above on all who enter the structure.

In 1540 as the result of intense pressure from the monarchy, the monks gave the church to King Henry VIII and dissolved the order. Following a tumultuous religious period including a brief restoration of the monastery, the accession of Queen Elizabeth I to the throne helped to finally define the future of Westminster as a collegiate church headed by a dean. The seventeenth century saw another period of religious upheaval in England, and for a time daily services at Westminster were discontinued and instead Presbyterian ministers preached from its pulpit. By 1660 the dean had been restored, and the church has continued as a collegiate church from that point until the present. As in earlier times when the abbey had been subject only to the pope, the abbey is now a "royal peculiar," subject only to the monarch.

While the interior of the church had remained in relatively good condition over the centuries, by the end of the eighteenth century the exterior had suffered serious deterioration, much of it due to the heavy use of coal in London. In order to correct the damage, Sir Christopher Wren was hired to oversee the reconstruction and restoration of the exterior. He was able to direct some of the work himself, but died before its completion. He left detailed plans for the remainder of the restoration and for the construction of the western towers, which are sometimes referred to as the Wren Towers. The towers were completed by 1740 with considerable variations by Wren's student Nicholas Hawksmoor. Additional renovation work was done in the nineteenth century by Sir Gilbert Scott, particularly to the exterior of Henry VII's chapel, which had not been rebuilt by Wren. The renovation work done from 1875 to 1890 is referred to as the great restoration, when much of the exterior was significantly altered.

The abbey suffered minor damage in World War I and more serious damage in World War II. The fires caused by German bombardments in the 1940s required restoration work that took several years, but the church was ready for the coronation of Elizabeth II in 1953. In the 1950s Winston Churchill launched an international fundraising campaign that provided monies for the complete cleaning of the interior of the abbey. In addition, a great deal of repainting and regilding has taken place during the past several decades. Although the church no longer resembles the original structure built on the site, it remains a truly national monument to the British people.

While the church has a long history in legitimating kings and queens through coronations, it has also served to elevate and canonize the dead through their burial within its walls. Also, many British notables who are not actually buried there are commemorated with statues and plaques. One of the most famous sections of the church has come to be known as Poets' Corner, and in it are buried some of Britain's literary luminaries. The first commoner to be buried in the abbey was Geoffrey Chaucer, the author of the *Canterbury Tales,* who was interred in 1400. Other notables have followed, including Edmund Spenser and Ben Jonson. Conspicuous by his absence is William Shakespeare, but he is commemorated with a plaque. It is perhaps testament to the great importance the British have placed on burying their favored sons in the abbey that the tomb of poet Alexander Pope, who did not desire to be buried in the church, stands in Twickenham with the epitaph "For one who would not be buried in Westminster Abbey."

Curiously, buried in the church are some individuals one would not imagine being highly desirable candidates to the church, most notable among them being Charles Darwin. Other notable nonroyals buried in Westminster include Isaac Newton, John Gay, Samuel Johnson, Charles Dickens, and the explorer David Livingstone. That the abbey has seen its share of funerals and burials is sure, and that they were among some of the most lavish events in British history is without question. Westminster, throughout its long history, has become synonymous with British nationalism and the glorification of national heroes. Legend has it that before entering into a decisive battle Lord Nelson declared "Victory or Westminster Abbey!" Its character as a national monument and symbol was only heightened when, in 1920, it was chosen as the site for the tomb of the unknown soldier.

The religious dimensions of the church have given way in large measure to its museum qualities. Today it is mainly visited by those desiring to browse and admire its monuments, but rarely by those who want to worship or pray. Unlike many other such buildings, Westminster Abbey is not associated with only one particular time period or person but indeed with the entire historical panorama of Britain.

Many patriots have been inspired by Westminster Abbey, and one of them, the Elizabethan playwright Francis Beaumont, penned these lines in tribute:

> Morality, behold and fear!
> What a change of flesh is here:
> Think how many royal bones
> Sleep within these heaps of stones:
> Here they lie, had realms and lands,
> Who now want strength to stir their hands.
> Here, from the pulpits seal's with dust,
> They preach, "In greatness is no trust!"
> Here's an acre sown indeed,
> With the richest royallest seed,
> That the earth did e'er drink in,
> Since the first man died for sin.

Westminster Palace's proximity to the abbey was no accident. Edward the Confessor decided to live in the area, known then as Thorney, during the abbey's construction. He wanted to be near the church and have easy access to river transportation. Nothing is left of Edward's living quarters, but his decision to locate there is responsible for the existence of Westminster Palace today. William the Conqueror lived at Westminster for a time, and the oldest extant part of the palace, Westminster Hall, dates from the reign of his son, William Rufus, in the late eleventh century. The hall took two years to build, beginning in 1097, and was huge for its time: more than 239 feet long and 67 feet wide, with walls rising almost 20 feet to a gallery that encompassed the four sides of the hall. These walls still stand today; the hall as a whole has been remodeled and embellished extensively over the centuries. Despite the hall's impressive size, William Rufus was disappointed; arriving there after a trip to Normandy in 1099, he called the hall "a mere bed-chamber."

William Rufus died the next year, and his brother and successor, Henry I, made Westminster his primary seat. He held a Great Council—a meeting of all the prominent clergy and laity of England—in 1102 at Westminster, more than likely in the hall. The hall definitely was the site in which Henry approved arrangements that established England's judicial system, an act considered his most important achievement. Successive kings expanded and redecorated the hall, built royal apartments adjacent to it, and held numerous important meetings there. In the twelfth century King Stephen built the Chapel of St. Stephen in honor of his namesake. A religious order, the Canons of St. Stephen's, held services for the royal family.

In Westminster Hall in 1275 Edward I convened what is considered the first truly representative Parliament, giving commoners a voice in national affairs. Their inclusion in government became even more firmly established with Edward's "Model Parliament" of 1295. The hall and the other buildings that made up Westminster Palace continued to be altered. Edward I, II, and III oversaw extensive remodeling of the Chapel of St. Stephen, adding stained glass and other works of art, and under Richard II in the late fourteenth century, Westminster Hall had its windows replaced, a hammer-beam roof installed, and its appearance generally changed from Norman to Gothic.

During the next few centuries Westminster Hall was the scene of many great trials, including that of Sir Thomas More in 1535. More had refused to recognize the validity of Henry VIII's divorce from Catherine of Aragon, or the king's right to head the Church of England. More was found guilty of treason and beheaded.

As a result of Henry's English Reformation, the order of the Canons of St. Stephen's was dissolved in 1547, and the chapel became the meeting place of the House of Commons, which since the fourteenth century had been meeting separately from the nobility, or House of Lords. (Another effect of Henry's reign was the abandonment of Westminster as a royal residence, after a fire in 1512.) The chapel's appearance remained unchanged until the late seventeenth century, when Sir Christopher Wren was given the task of making it more appropriate to use by the House of Commons. Wren took out the stained glass, covered over other religious art, removed the clerestory, and lowered the roof.

Westminster Hall continued to be the site of sensational trials. In 1605 English Catholics intended to blow up Parliament with stores of gunpowder hidden under the buildings at Westminster. After their so-called Gunpowder Plot was exposed, they were put on trial at Westminster Hall in January 1606; several conspirators were executed. Later in the century, during the reign of Charles I, a struggle for supremacy between the king and Parliament ensued, finally escalating into the English Civil War and resulting in Charles's trial for treason in 1649. He was found guilty and executed. Eighteenth-century trials included several related to the Jacobite uprisings in Scotland.

On October 16, 1834, a fire, started by an overheated stove, engulfed Westminster Palace. No one was killed, and only a few people suffered major injuries, but the palace incurred heavy damage and required extensive rebuilding. The new design by Charles Barry incorporated Westminster Hall, as well as a crypt and cloisters that represented the other surviving medieval construction. Land was reclaimed from the river and the palace site expanded to cover eight acres. The new Gothic-style palace, covering four levels, was built from 1840 to 1870. It encompasses the meeting chambers of the House of Commons and House of Lords, libraries, offices, and rooms with various other functions. The hall continues to be used for major public ceremonies.

One of the most notable features of the new design is the clock tower, which is 314 feet high. The highest standards of accuracy were required for the clock in the tower, and the manufacturer E. J. Dent won the contract to make this timepiece. The clock was installed in 1859 and has come to be known as Big Ben, although technically that is the name only of the bell that tolls on the hour. Big Ben has tolled to commemorate major events in English history, including the end of World War I, and it continued to be heard even during the bombing of London in World War II.

Bombing destroyed Westminster Palace's Commons

Chamber in 1941; a new chamber was constructed by 1950. The House of Commons staff has grown to such a point that it has taken over other, nearby buildings. Throughout the twentieth century the exterior of Westminster Palace has required frequent cleaning and restoration because of damage caused by air pollution.

Today the palace remains one of the most imposing landmarks of London, and it has been immortalized in paintings by such masters as the French impressionist Claude Monet. Complete tours are available by permits obtained though members of Parliament (for British residents) or through written request to the public information office (for overseas visitors). Debates in the House of Commons and House of Lords are open to the public.

Further Reading: A great deal has been written about Westminster Abbey over the centuries, but many of the sources concentrate on only one element of the architecture or one incident relating to its history. One of the nicest little texts on Westminster Abbey is an undated book from a series on important sites in Great Britain entitled *Classical Art Tours: Westminster Abbey* by Claudio Gorlier. Another book with a distinctly British tone is *The Glory of Westminster Abbey* (Norwich: Jarold, n.d.). Although no author is mentioned, the book is dedicated to the Dean and Chapter of Westminster. W. R. Lethaby's *Westminster Abbey and the King's Craftsmen* (New York: Blom, 1971) is a detailed account of the various artistic elements in the abbey, given in a narrative chronology. Finally, one of the most interesting accounts to be found is *Westminster Abbey* by Beatrice Home (London: Dent, 1925) from the series *Cathedrals, Abbeys and Famous Churches*. Certainly much of the reason this book is so enjoyable has to do with the fact that many of the incidents and stories that it relates are of dubious origin and questionable authenticity. An extensive treatment of Westminster Palace is available in *Westminster Palace and Parliament* by Patrick Cormack (London and New York: Frederick Warne, 1981).

—Michael D. Phillips

Lourdes (Hautes-Pyrénées, France)

Location: At the foot of the Pyrénées on the banks of the Gave de Pau in the southwestern part of France; in the Alps-Côte d'Azure region, southwest of Toulouse.

Description: Best known for its Marian shrine, Lourdes attracts millions of pilgrims each year. The Virgin Mary reportedly appeared to a young peasant girl, Bernadette Soubirous, in 1858 requesting her to build a chapel and begin a procession.

Site Location: Office de Tourisme, Syndicat d'Initiative
Place du Champ Commun
65100 Lourdes, Hautes-Pyrénées
France
62 42 77 40

Lourdes is internationally known for its shrine to the Virgin Mary, who, according to tradition, appeared to Bernadette Soubirous, a young peasant girl, in February 1858. Almost instantly, the remote village was transformed into an international pilgrimage center attracting millions of the devout, the sick, and the aged annually. The present-day landscape of Lourdes is marked by the Rosary Basilica and the fourteenth-century château, both symbols of Lourdes's diverse past.

Excavations at the grotto site at Massabielle, near the present site of the Marian shrine, suggest that the valley of the Gave de Pau has been continuously inhabited for more than 14,000 years. Since the nineteenth century, it has been a source of stunning archaeological finds. Hundreds of prehistoric tools made of stone and animal bones have been discovered, indicating that the valley of the Gave de Pau was abundantly supplied with water, game, and natural defenses. The most spectacular finds to date have been the sculptures, the most famous being a sculpted horse made of mammoth ivory. The archaeological record also suggests that a more settled population lived in the valley during the Neolithic Era. Some burial sites have been dated to 2500 B.C.

An ancient Roman shrine was built in the third century A.D., and the château occupies the site of an earlier Roman structure whose purpose is still not known. With the fall of the Roman Empire, the grotto at Massabielle became a haven for refugees from invaders, and the Roman shrine was converted to a Christian site, one of the earliest known in France.

Lourdes's written history does not begin until the Middle Ages. In the eleventh century, the monastery of St. Pé de Genestus was founded. But the medieval château, on the right bank of the Gave de Pau River, would determine Lourdes's fortunes throughout much of the pre-modern era.

Its fourteenth-century tower, la Tour de Garnabie, still dominates the landscape of Lourdes's old quarter. A strategic stronghold, the fortress and its outlying village were often victims of the dynastic disputes that gripped medieval Europe.

Built in the tenth and eleventh centuries, the château passed to the French crown in 1304 and was ceded to the English Plantagenets under the terms of the Treaty of Brétigny in 1360 during the Hundred Years War. A garrison of 250 to 300 men, made infamous as the "Company of Lourdes," was installed by Edward the Black Prince, son of Edward III of England. The natural defenses of the château were so impenetrable that the small force was able to withstand successive French attempts to win back the stronghold. Since the English garrison relied heavily on the produce of the local inhabitants for food, the unprotected village was frequently burned and destroyed by retreating French troops. A French force under the Duke of Berry finally ousted the English garrison in 1406–07 following a prolonged eighteen-month siege.

The château was used again as a stronghold during the religious wars of the sixteenth century when Protestant troops laid yet another siege against it. The military and strategic importance of the château at Lourdes waned throughout the seventeenth and eighteenth centuries. During the reign of Louis XIV, it was used primarily as a "Bastille of the Pyrénées," a prison for the king's political foes. Napoléon Bonaparte made similar use of the château during his imperial mastery over France. Despite the general decline in the fortunes of the château, the village of Lourdes prospered throughout these years of peace. Local agricultural production flourished, and by the end of the Napoleonic era, eight grist mills had been constructed to fill the needs of the local peasants.

It was against this mundane commercial backdrop in 1858 that Marie-Bernarde Soubirous saw the visions that would transform Lourdes into one of the major Catholic pilgrimage sites in the world. Bernadette was born on January 7, 1844. She was the eldest daughter of François and Louise Soubirous, a poor milling family. As a child, she was often sickly, afflicted by asthmatic attacks. While she was still quite young, Bernadette's father lost the use of his left eye due to a milling accident. Unable to keep the old family mill operating, he took work as a day-laborer, and the family was forced to move to poorer quarters. Her mother frequently took on work as a charwoman and Bernadette, as the eldest, looked after her younger brothers and sisters.

On February 11, 1858, Bernadette, her sister, and a friend went in search of some firewood. Falling behind her two companions, she paused at the grotto site at Massabielle where villagers often retrieved old bones for their kindling.

The Lourdes Grotto

As she paused to catch her breath, a beautiful lady appeared to her. It was the first of eighteen apparitions that the fourteen-year-old girl would witness from February 11 to 16 July, 1858.

Little was known about the lady in the visions, and though she only appeared to Bernadette, people began to assemble at the grotto site. The week following her first apparition, the lady spoke to the young girl for the first time. She asked Bernadette to return each day for fifteen consecutive days. Despite the initial scepticism of her family and the reticence of local church officials, Bernadette obeyed the lady's wishes.

A crowd of 350 people witnessed her bizarre behavior on February 25 as she followed instructions to drink and wash from a non-existent spring. As soon as she had begun to dig, the spring came forth. Many pilgrims to Lourdes believe that the spring possesses miraculous curative properties, and today, the water (which now flows at a rate of 32,000 gallons a day) is used for the bath at Lourdes.

Local authorities became alarmed by the increasing number of people gathering at the Massabielle grotto. Church officials were themselves concerned by the persistent rumors that the beautiful lady was in fact the Virgin Mary, even though Bernadette claimed she did not know the identity of the apparition. In an effort to curb public curiosity, town officials restricted access to the grotto, had all the religious statues and icons removed, and posted guards to prevent further trespassing on the site.

In spite of such opposition, Bernadette fulfilled her promise and visited the site for fifteen consecutive days. Often praying beyond the barricades barring entrance to the site, she would wait for the apparition and her instructions. On February 27, she was instructed to have a chapel built on the site and to have people come there in procession. The message was repeated again on March 2.

During the fifteenth day of her visits, a crowd of more than 8,000 had gathered in anticipation of a significant event, but were disappointed when the lady did not appear. Nor did she appear for the following three weeks. Bernadette, nonetheless, continued her daily visits. The absence of further apparitions had led to increasing doubt in the minds of the local officials and the local priest, Father Peyramale. As the officials looked for legal means of removing the child from the site, Father Peyramale, who had known Bernadette for most of her childhood, was troubled. Who was this lady? Why would she appear only to an illiterate girl?

On the feast day of the Assumption, March 25, the lady appeared again to Bernadette. "I am the Immaculate Conception," she told the girl. The dogma of the Immaculate Conception had only recently been defined by Pope Pius IX in 1854. It was unlikely that Bernadette would have known of such dogma. Adding to the uproar were the rumors of other spontaneous visions and reports of miraculous cures. The first such claim came from a woman named Catherine Latapie, who claimed to have been healed on March 1. In response to the hysteria, the local bishop, Monseigneur Laurence, announced the creation of a church commission to investigate Bernadette's story, as well as the many other claims. For their part, local officials closed all access to the spring or the grotto from June to October 1858 for "hygienic" reasons.

Increasingly, Bernadette and her family were besieged not only by the faithful, but also by those seeking to meet her. In the fall of 1858, the Soubirous family was forced to leave their home in Lourdes. Still fearing for Bernadette's safety two years later, Father Peyramale suggested she board with local nuns for protection. Meanwhile, the church commission continued its investigation. Bernadette held steadfast to her story throughout the ordeal.

In January 1862, the clerical commission officially recognized the authenticity of Bernadette Sourbirous's claim to have been visited by the Virgin Mary. It sanctioned the creation of a cult of Our Lady of the Grotto of Lourdes and announced that the church would undertake the building of a chapel on the site in fulfillment of the Virgin Mary's wishes. Already, Bishop Laurence had taken steps to purchase the grotto property for the church, and ministerial approval for construction was being sought.

Yet for Bernadette, such a vindication was short-lived. She was soon cut off from all the activities originating from the grotto. Still protected by the nuns, she became gravely ill in the spring of 1862, so ill that she was administered the last rites, but she recovered. She made her last appearance at Lourdes in May 1876 at the dedication mass for the church built on the grotto sites. Six months later, she said good-bye to her family and friends and joined the Sisters of Notre-Dame of Nevers, where she would spend the rest of her life in service to the order. She died on April 16, 1879, at the age of thirty-five. In 1933, she was canonized by Pope Pius XI in recognition of her religious commitment and service to others.

Within months of the report of the clerical commission in 1862, Lourdes was transformed into a major Marian shrine. Wooden baths were built on the spring uncovered by Bernadette. Construction was started on the church and Bishop Laurence commissioned a new marble statue of the Virgin Mary to be placed in the grotto. It was officially unveiled in 1864 before 10,000–20,000 people who had assembled for the first official procession of Lourdes. With the arrival of railway lines to the region in 1866, Lourdes's popularity began to reach well beyond the Côte d'Azure region.

Although the Franco-Prussian War of 1870 interrupted the flow of visitors to the shrine, Lourdes became one of the largest pilgrimage sites in the Christian world by the end of the nineteenth century. In 1871, the long-awaited church was finished and the first mass celebrated. By 1872, 50,000 people from all over France participated in the annual procession. The church was declared a minor basilica by the Vatican in 1876 and was consecrated in the presence of 100,000 worshippers, including Bernadette, who also saw the papal nuncio to France crown the statue of the Virgin Mary. In 1883 the first stone was laid for another, larger church, the Rosary Basilica, which with its fifteen chapels, was built to accommodate the thousands who flocked to Lourdes annually. It too, was later declared a minor basilica in 1926. In 1958, an immense underground church capable of seating 20,000 was consecrated as part of the centenary anniversary of Lourdes. The Vatican elevated Lourdes's place in the larger Catholic community as well. In 1892, the Vatican designated February 11 (the date of the first apparition) as the feast day of Our Lady of Lourdes for all the churches in France. This was later extended to all Roman Catholic churches worldwide in 1907. Lourdes eventually became so popular that Pope Leo XXI had a replica of the Lourdes grotto built in the Vatican gardens, and in 1983, Pope John Paul II came to the Pyrénées to pay homage to Our Lady of Lourdes.

Lourdes's three million pilgrims each year include all nationalities, classes, and ages, but the site has a particular significance for the sick. Since the first reported cure in 1858, thousands claim to have been healed there. Since 1882, a permanent medical bureau has investigated the validity of such claims. Reported cures are first studied by the medical bureau. Approved cases are then sent to a canonical commission in the person's own diocese. The bishop of that diocese then pronounces on the miraculous nature of the cure. Since 1858, sixty-five cures have been designated as miraculous. The most recent reported cure was on June 2, 1989.

Apart from the millions of faithful pilgrims, Lourdes has continued to attract considerable interest from writers, artists, and filmmakers. Émile Zola, a well-known nineteenth-century French author, visited the shrine in 1892 and subsequently published his work, *Lourdes,* as part of a trilogy satirizing religion. He was vilified in many Catholic circles for his caricature of Lourdes. Similarly, Joris-Karl Huysmans in his 1906 *The Crowds of Lourdes* called the procession "a paragon of the ecclesiastical turpitude of art." In 1941 Franz Werfel, a Jewish writer and refugee from Nazi Germany, published *Das Lied von Bernadette* as a tribute to those who had helped him flee the Nazis. Hollywood turned the novel into a motion picture in 1943, and although the accuracy of the story had since been challenged, it proved popular with wartime audiences. Even in today's popular culture, the story of Bernadette of Lourdes has held the public imagination; it inspired songs from poet-songwriter Leonard Cohen, and the eclectic Ken Russell as a young man filmed the images and crowds of Lourdes, to which he added a modern music score.

Further Reading: Recent works such as *St. Bernadette* by Leon Cristiani (New York: Alba, 1981) attest to the continued popularity of the story of Bernadette. Those interested in a more complete history of Lourdes should consult Stéphane Beaumont's *Histoire de Lourdes* (Paris: Privat, 1993).

—Manon Lamontagne

Louvain (Brabant, Belgium)

Location: In the Flemish-speaking province of Brabant in central Belgium; sixteen miles east of Brussels on the Dijle River.

Description: Founded in the ninth century by a German emperor seeking to defend himself against invasions by Norsemen; early capital of Brabant under the dukes of Brabant; a center of cloth weaving in the Middle Ages; since 1425 home of the Catholic University of Louvain, a center of learning and culture and a bastion of Flemish Catholicism.

Site Office: Tourist Office
Naamsestraat 1-a
B-3000 Louvain, Brabant
Belgium
(16) 23 49 41

On the north side of the town of Louvain (in Flemish, Leuven) sits a Benedictine abbey, modern but with some medieval traces, known as the Keizerberg. It occupies the site of an earlier fortified castle built in the ninth century by Arnulf of Carinthia for the purpose of defending himself against the Norsemen, who, after the death of Charlemagne, threatened the stability of the Franks. The territory west of the Schelde River was under the control of Charles the Bald; that on the right bank, known as Lotharingia, was a kingdom under Arnulf, who enjoyed a great deal of independence from both the French and German empires. The Norse armies were attracted by the fertility of the Brabant area, and one of the armies set up headquarters at Louvain along the Dijle River. Arnulf's army resisted their advances in 891, but the Norsemen still did not leave the area until a year later, when famine hit and they were obliged to move on to England and Normandy.

Soon afterward, the kingdom of Lotharingia was split into principalities governed by noblemen. Prince Lambert I set up a dynasty at Louvain. His castle and the surrounding town became the capital of the Hesbaye region in the tenth century. Maintaining the autonomy of the region, he resisted all attempts by the emperor and the bishops to subjugate his authority. It was Lambert who built a church where Sint Pieterskerk (St. Peter's Church) now stands.

Louvain's location on a navigable waterway close to the road leading from Bruges to Cologne made it one of several mercantile centers to develop by the end of the eleventh century in what is now Belgium. Artisans were organized by the local burghers into guilds whose function was to produce goods for export. The burghers developed into a wealthy haute bourgeoisie who profited from the labor of the artisans and enjoyed the protection of the aristocracy. As the town became an increasingly prosperous center for the cloth trade, the dukes of Brabant used Louvain's economic resources, along with those of Antwerp and Brussels, to build wealth throughout their territory and to ensure the security of the main trade routes.

It was not until the fourteenth century that the artisans began to rebel against this system. Economic hardships caused by the long conflict between France and England spurred a democratic movement among the tradesmen. Peter Coutereel, the bailiff of Louvain, took up their cause against the patrician oligarchy. In 1360 the artisans rose up against the aristocracy. Some of the aristocrats were imprisoned; others, refusing to cooperate in reforms, went into exile. Duke Wenceslas of Luxembourg, who through marriage had become ruler of Brabant, tried to end the conflict by ousting Coutereel, but this only made matters worse. During the course of the riots in 1379, the citizens threw seventeen noblemen out of the windows of the Stadhuis (the town hall). Wenceslas took revenge upon the citizenry, who were forced to submit to the nobles in 1383.

These events began the decline of the town of Louvain, which by then had reached a population of 50,000. Many of the weavers, unhappy with conditions here, left for England. The dukes moved their residence to Vilvoorde, and Brussels soon became the new capital of Brabant.

Although Louvain lost its stature as a prosperous trade center, it soon became a center of learning and of art. The fifteenth century saw the construction of a three-story town hall, the decorative ornamentation of which seems to overpower its architecture. A new church was built on the site of Lambert's original one and was decorated with stone tracery. Plans for the building originally called for a pair of Gothic towers to flank the facade, but they were never completed for lack of a proper foundation. The interior was made equally rich in ornamentation by the work of local sculptors and goldsmiths.

Art flourished in the fifteenth and sixteenth centuries, due to the spread of humanism and the influence of the Italian masters. Sculpted ironworks by Louvain-born artist Quentin Massys and paintings by Dirk Bouts can be seen today in the Sint Pieterskerk. Both artists were able to imbue religious subjects with human emotion.

Perhaps the most significant event in the history of Louvain was the establishment of the Catholic University of Louvain in 1425. Duke John IV of Brabant, with the support of the local magistrate, secured approval from Pope Martin V to open a *studium generale* to train doctors, men of law, and eventually, theologians. A joint effort of the church, the ducal authority, and the local officials, the university would help to unify the Belgian provinces. Scholars from the Low Countries no longer had to study at Paris or Cologne. In its

The Town Hall in Louvain
Photo courtesy of Belgian National Tourist Office

first year there were only fifteen professors and four faculties: arts (humanities), canon law, civil law, and medicine. The faculty of theology was added in 1432.

The town suffered several plagues during the fifteenth century, and the growth of the university was slowed by economic hard times. In addition, the warfare between the Habsburgs and the Burgundians brought a period of instability to the Low Countries after 1477.

The sixteenth century was a period of prosperity and glory for Louvain. While the ideas of Martin Luther were taking hold in Antwerp and other cities, Louvain became involved in the debate over humanism. Desiderius Erasmus of Rotterdam, who was on the faculty from 1517 to 1521, dominated the side of the humanists. He sought to reform society by enlightening the ruling classes so that they, in turn, could educate the masses. In the interest of promoting individuality of thought and tolerance of ideas, he opened the Trilingual College at Louvain. Here Hebrew, Greek, and Latin would be taught as part of a liberal Catholic education. Unfortunately, the conservative leaders of the university saw this school as a threat to orthodox teaching and as an open invitation to heresy. His adversaries felt that only the Latin Vulgate could be a source of dogma. They attacked his proposal for an original-language edition of the New Testament. Erasmus wanted theology to be more critical and to refer to the original texts. Accused of heretical ideas, Erasmus was forced to leave Louvain for Switzerland.

Erasmus was not the only thinker to meet resistance at Louvain. Gerardus Mercator, the renowned cartographer, a student and later a professor of geography, was driven out of the country in 1544. He and Andreas Vesalius, who modernized the study of human anatomy, taught side by side with classicists. Most of the teaching at Louvain was devoted to the study of antiquity and opposed to the idea of making knowledge accessible to the lower classes. Convinced that educating the masses would undermine the authority of the princes and the clergy, the university continued to perpetuate an intellectual elite. One of Louvain's most famous history professors, Justus Lipsius, dedicated himself to studying Seneca and Tacitus and to writing on many subjects. Unfortunately, his writings show more interest in form than in substance.

During the Reformation, the great questions of the age had to do with free will versus predestination, the meaning of original sin, and the infallibility of the Pope. The theologians at Louvain clung steadfastly to the ideas of Aristotle and Thomas Aquinas. Their contributions to this period included the creation of an index of censured books that became a model for later versions, and the publication of a new Latin translation of the Bible.

By the end of the sixteenth century, Belgium's population decreased as Calvinists emigrated rather than face the policies and practices of Philip II of Spain, who ruled the region and sought to wipe out opposition to Catholicism. Louvain's population went from 20,000 to 9,000 as commerce slowed. The university no longer enjoyed its former prestige, and intellectual decay set in. Part of the problem stemmed from the reorganization of the bishoprics, whereby Louvain became part of the diocese of Mechelen (Malines) instead of Liège. In addition, it now had to compete against universities at Douai (under the Spanish Netherlands) and Leiden (a Calvinist university established to oppose Spanish Catholicism). Economic and political troubles marked the end of the century. The university at Louvain encountered difficulties in meeting professors' stipends. Following an attack on the town by William the Silent, leading a revolt against the Spanish rulers, in 1572, a garrison was sent to keep order. When the soldiers were not paid, they turned to vandalism. The plague of 1578 claimed the lives of many on the faculty. Fewer and fewer students came to Louvain during these years.

A period of relative tranquility followed the appointment of the Habsburg archduke Albert of Austria as governor of the Low Countries. Shortly after his death, however, Louvain became embroiled in a dispute between the Spanish Netherlands on one side and Louis XIII of France and Frederick Henry of Orange on the other. In 1635 the townspeople and the students joined forces to take up arms against the Spanish governor.

The seventeenth and eighteenth centuries produced only mediocre scholarship. In 1652 the leaders of the university officially rejected the Cartesian philosophy popular in France in favor of more traditional Aristotelian philosophy.

During the eighteenth century, Austrian authorities sought to make reforms at Louvain in order to raise its standards. They withdrew the right of the university to appoint its own rector. After the French Revolution control of Louvain alternated between France and Austria until 1797, when the university was closed after 372 years of existence. Many priests who had been teaching there refused to take the oath to the new French republic, and they were exiled. Students continued their studies either in Brussels or in France.

In 1814 some of the former professors suggested reopening the university in its original form. The local authorities preferred a university based on the French model. Still others opposed the idea of reopening the school altogether, citing Louvain's opposition to the Enlightenment. But two years later, William I, king of the new union of Belgium and the Netherlands, established three universities in the southern Low Countries: at Ghent, at Liège, and at Louvain. The university was reopened in 1817.

The liberals and the Catholics in the community did not like William's authoritarian government, and many students joined the rebellion of 1830, part of a larger revolution ending the Belgium-Netherlands union and giving rise to the independent kingdom of Belgium. Disorder marked the next few years at Louvain. Some wanted a single state university for all of Belgium; others wished to retain only Ghent and Liège. In 1835, it was decided that only Ghent and Liège would be supported by the state. The bishops, who had established a Catholic university at Mechelen in 1834, took advantage of this opportunity to move their university to Louvain.

The new university was to continue the traditions of the old one, except that there were no ties to the state, as provided by Belgium's new constitution. Funds came from tuition and charitable contributions. Discipline was rigorous, and professional training was more important than scholarship. While French replaced Latin as the language of instruction, there were classes in Flemish language and literature that were not offered at other Belgian universities.

Some students felt that Louvain had become too conservative, and they transferred to Brussels. Some conservatives on the staff felt that the administration was too liberal, and they sought positions at other universities. Internal struggles continued through the next few decades, with the conservatives gaining many concessions, such as the power to appoint faculty and oversee lectures.

A new law of 1876 allowing the university to create its own exams encouraged experimentation in science courses. New schools opened in engineering and agronomy, with the intent of improving Belgium's agricultural economy and preserving Catholicism among the peasants. As the student body began to draw from the middle classes, student life became more relaxed. The university had to expand to meet the needs of its growing population, and this often resulted in financial strain.

Toward the end of the century, there was less emphasis on teaching and more on scholarly research. Study groups formed in theology and humanities classes, and student organizations began to develop. One group that called for promoting the Flemish language encountered resistance from the Walloon (French-speaking) students. Nevertheless, more courses were offered in Flemish and more documents were published in both languages as a result.

By 1900 more capital became available to resolve some of the financial difficulties and to build new facilities. New institutes were set up in bacteriology, pathology, and geology. Petitions for duplication of courses in both languages, however, were denied on financial grounds. This increased the tension between the two groups of students. Finally in 1911 more classes were added in Flemish, along with new courses in art history, dentistry, and banking.

When the Germans invaded Louvain in 1914, about one-third of the existing buildings were burned down. The university was forced to close for the duration of the war, since many of its professors had left to teach abroad, and the administration did not want to submit to German censorship. Professors who remained in Louvain busied themselves with doing research or with looking for books to replace those lost when the library was burned. When the university opened once again in 1919, Belgian bishops sent out an international appeal to raise money for the library. The Treaty of Versailles stipulated that the Germans had to compensate Louvain for any lost treasures.

As funds were secured from the bishops' appeals and from Belgian and American citizens, the university began to rebuild. More students matriculated than ever before, and women were accepted as students for the first time in 1920. Under pressure from Flemish groups, the duplication of classes in two languages was increased. Nevertheless, tensions between the two groups over the language question continued to mount. When the administration tried to suppress the militant groups, the Flemings rebelled by boycotting the 500th anniversary celebration in 1927.

The town of Louvain had to be evacuated in May 1940 when Belgium was invaded by the Germans a second time. Once again fire ravaged many of the buildings, including the rebuilt library. The rector reopened classes at the university in July, but the community suffered many problems during the occupation. There was a large increase in enrollment following the closing of the University of Brussels in 1941 and because many young people sought to register for classes rather than submit to forced labor under the Germans. The increase in the number of students from 4,600 at the start of the war to 7,700 by 1943 put a strain on the university's resources. Students had to work under uncomfortably crowded conditions and endure shortages of materials, food, and heat. When the rector refused to release the first-year students to work in the German factories, he was condemned and sent to prison. The vice-rector set up clandestine lectures and exams for students who were not allowed to register officially.

When the town was repeatedly hit by Allied bombs in 1944, many of the buildings were destroyed and some of the students and faculty killed. More damage was done as the Germans retreated, making it necessary for the town and the university to go about the task of rebuilding once again.

After the war, the university responded to the increased social needs of the students by assigning advisors and counselors to give them guidance. Academically, new institutes were set up in archaeology, art history, African studies, and other disciplines. In 1954 funds were allocated to establish a sister center of learning, Lovanium University in the Belgian Congo, in order to meet the needs of African students.

The language issue came to a head in the 1960s. Despite the duplication of courses and the hiring of Flemish faculty, the administration continued to hold its meetings in French, and French culture prevailed. When the number of Flemings outnumbered the Walloons, the Flemish students and professors began to press for more autonomy. While some were willing to accept bilingual status for Louvain on the model of Brussels, the majority felt that this was unacceptable, since Louvain was in Flemish territory. But the French-speaking professors wanted their children to learn French in school.

When the issue could not be resolved by dividing the university into two separate sections under separate deans, pressure mounted to transfer the French section to Wallonia. The medical school had already moved its facility to a suburb of Brussels in 1965. Proponents of the move argued that Louvain was much too small a town to accommodate the increasingly large student body. Opponents feared that Louvain would lose its status as a world-class university. The debate divided students, faculty, bishops, and eventually the Christian Social Party. In 1968 the plan was approved to separate the Katholieke Universiteit te Leuven, which was to remain in Louvain, from the Université Catholique de Louvain, which was to be relocated in French-speaking territory fifteen miles southwest of Louvain in Louvain-la-Neuve, a community to be built between Wavre and Ottignies.

The division had some advantages. The new university, while a Catholic institution, was led primarily by laypeople who abandoned the former authoritarian policies in favor of a new open-mindedness. State subsidies were increased, allowing for the expansion of the facilities at KU Leuven and the building of a well-planned university community at UCL. Each section was able to serve more than 16,000 students by 1975. On the negative side, the decision to divide the existing library in half by allocating books with even-numbered call letters to one facility and books with odd-numbered call letters to the other has devastated the historic collection.

Today's visitor to Louvain will find the University Library, built in the Flemish Renaissance style in 1928 and later restored, near the Herbert Hooverplein. Its high (280-foot) belfry contains a 63-bell, 5-octave carillon used today by students taking lessons from carillonneurs of the nearby Royal Carillon School in Mechelen.

The earlier library was housed in the Gothic Hallen (Cloth Hall) from 1629 until World War I. Originally a center for the cloth workers' guild in the fourteenth century, the Hallen has been rebuilt and now serves as a university administration building.

Also near the Grote Markt (main square) are the Stadhuis and Sint Pieterskerk. The Stadhuis was built in flamboyant Gothic style by Mathys de Layens in the fifteenth century. Its three stories of windows alternate with niches containing statues, added in the nineteenth century, of prominent citizens, artists, and noblemen. The bases of the windows are carved with Biblical scenes.

Sint Pieterskerk stands on the site of two earlier Romanesque churches, both destroyed by fire. The existing church was begun in 1425 in the late Gothic style. An attempt was made to replace the original Romanesque towers on the west facade with ambitious Gothic towers, but the foundation was too weak and the towers crumbled to roof level in the seventeenth century. The nave was destroyed during World War I; the choir was severely damaged by World War II. The interior is lighted by ninety windows and contains many significant works of art, including a black marble statue of Duke Henry I of Brabant and a reduced copy from 1440 of Roger van der Weyden's *Descent from the Cross*.

Alongside the Dijle River is the Groot Beginhof, founded in the thirteenth century as a residence for béguines (secular nuns). Acquired and restored by the university in 1961, it boasts many examples of local architecture from the fourteenth to the eighteenth centuries.

Much of Louvain's history has been retold in its buildings. The Hallen speaks of the evolution of the town from a mercantile center to a seat of learning. Sint Pieterskerk, with its collection of paintings of martyrs, is a symbol of the fervent Catholicism of the people of Louvain, who numerous times had to rebuild their church after it had been ravaged by fire or war. The statues adorning the Stadhuis show the importance to the town of citizens of all classes, not only the nobility and the clergy. The Flemish names of these and all the other buildings of Louvain are a testament to the people's will to uphold the Flemish culture in the face of divisions caused by war, religion, and language.

Further Reading: *The University of Louvain 1425–1975* by R. Aubert, et al. (Louvain: Leuven University Press, 1975) is a complete, balanced, and beautifully illustrated history of the university from its beginnings through the years of separation, written on the occasion of its 550th anniversary. It provides a detailed account of its most famous students and faculty and of the movements that shaped its history. *Belgium: The Making of a Nation* by H. Vander Linden, translated by Sybil Jane (Oxford: Clarendon, 1920) is a valuable, if generalized, history of Belgium. Linden examines the social, economic, religious, and political movements that helped to shape modern Belgium.

—Sherry Crane LaRue

Lübeck (Schleswig-Holstein, Germany)

Location: In northern Germany, on the Trave and Wakenitz Rivers, about nine miles from the Baltic Sea.

Description: City and major seaport of the state of Schleswig-Holstein; a leading city in the Hanseatic League in the Middle Ages.

Site Office: Verkehrsverein
Breite Strasse 75
23552 Lübeck, Schleswig-Holstein
Germany
(451) 7 23 00

Lübeck's history corresponds to the history of the Hanseatic League, a union of medieval trading towns that extended from Russia and Poland to England and Flanders. For nearly four centuries, Lübeck commanded this vast mercantile empire, and even though the last Hanseatic meeting occurred in 1669, the city still takes pride in the appellation *Hansestadt*. Lübeck's meteoric rise to Hanseatic leadership was based on the mercantile prowess it developed during the thirteenth century, a development fostered by the free imperial city's privileged status within a culture that was still largely feudal.

The original settlement was Liubice, site since the mid-eleventh century of a Slavic citadel near the mouth of the Trave River. Liubice was destroyed in wars between German and Slav lords that occurred as the Germans migrated east into Slavic territory, but was refounded in 1143 as the German town Lübeck by Adolf of Schauenburg, the count who ruled Holstein. In 1157 the town was destroyed once again, this time by fire. Aware of the advantages of the location, Henry the Lion, duke of Saxony, founded the town anew in 1159. The very next year, his efforts began to pay off as Lübeck replaced Oldenburg as the diocesan town for Holstein.

Even as Lübeck was being refounded by Henry, the engine of the town's eventual greatness was conceived in Visby, a trading post of Sweden on the island of Gotland in the Baltic Sea. From 1163 Visby became increasingly important as a base for German merchants trading with other Baltic ports; to promote their common interests, the Germans formed a *universitas mercatorum Romani imperii Gotlandiam*, a "company of German merchants of the Roman Empire visiting Gotland." The German influence soon also extended to Russia, where a German *Kontor* (trade delegation) was founded near Novgorod.

The new Lübeck grew rapidly as the main trading point between the countries of northern and eastern Europe that produced raw materials and the manufacturing centers in the west that were the destinations for those materials. Lübeck's strategic location made it a natural location for the

first German port on the Baltic Sea. Nestled securely at the southwestern corner of the Baltic region and only eleven miles from the mouth of the Trave River, Lübeck connected the Baltic with the land bridge to the North Sea, and to all points south on the Continent. This allowed traders to bypass the frequently dangerous straits between the Baltic and North Seas—such as the Øresund between Sweden and Denmark—which were dominated by the Danes. Though Lübeck was granted further rights in 1188 by the Holy Roman Emperor Frederick I after he had banished Henry the Lion following a quarrel, Denmark—which the Lübeckers have viewed throughout their history as both friend and foe— controlled the town from 1201 to 1226. But in the latter year, not even a century after its founding, Lübeck received its charter as a free city within the Holy Roman Empire. In 1227 Lübeck decisively defeated the Danes in battle at nearby Bornhöved.

To accommodate its new status, Lübeck devised a form of self-government with its own laws and constitution. The Laws of Lübeck notably addressed personal liberties, assuring every resident of Lübeck the right to be tried under the law of Lübeck, even in the ducal court. German merchants united with the Teutonic Knights in promoting settlements farther east, and as Lübeckers contributed substantially to the Germanization of these territories, the town's example consistently informed the new developments. In particular, the Laws of Lübeck were granted in charters to more than 100 cities in the Baltic area, among them Rostock (1218), Wismar (1229), Stralsund (1234), Demmin (1249), Greifswald (1250), and Anklam (1264). It was under these governing precepts that merchants and landowners rose to prominence as Lübeck's new aristocrats, rapidly supplanting the feudal systems of economy and politics. Lübeck's novel governance paid off; it was patrician control that orchestrated the city's ascendancy in, and ultimate domination of the monolithic Hanseatic trade network. Indeed, the patrician class embodied the very essence of mercantile ambition and efficiency.

In addition to its system of law, Lübeck strongly influenced the look of many Hanseatic towns through its *Backsteingotik*—brick Gothic architecture. Most notable was the style of church architecture found in the Marienkirche, which had been built by the patrician class from the early thirteenth century through the mid-fourteenth century to be larger than the cathedral. This building's influence can be found in Mecklenburg, Pomerania, Prussia, and Brandenburg, among other locations. The use of brick compelled the churches' builders to simplify their forms, allowing internal structural concerns such as vaulting to take precedence over decoration.

The Hanseatic League had begun as a merchant association of seventeen Flemish towns to promote and protect their trade in cloth; at the height of its development the league

A view of Lübeck
Photo courtesy of German Information Center

comprised 166 members, with Lübeck as the undisputed center. Cologne monopolized the Rhenish trade route, while Lübeck and Hamburg handled goods flowing in from the Baltic and North Seas, respectively. In about 1230, in the first of many alliances that would bind various Hanseatic cities, Lübeck signed a pact with Hamburg, a city on the Elbe River to the south. Both ports served as points where wares were transferred from ships to wagons, to be delivered on the land route that connected the Baltic region and the North Sea.

In 1248 the Norwegian king, Haakon IV Haakonsson, began to formalize trade relations between Lübeck and Norway. Lübeck's merchants, mindful of the economic advantages to Norway of importing foodstuffs from eastern Europe, lost no time in promoting exclusive arrangements for the Germans at the expense of their rivals on the North Sea. In 1276 the Germans trading in Bergen won the right to purchase or rent their own yards there, as Norway became permanently dependent on grain imports that only the Baltic Hanseatic towns could deliver. This so-called Deutsche Brücke (German Bridge) exemplified the closed type of trading settlement that contributed to the profitable German monopoly, with Bergen, on Norway's southwestern coast, serv-

ing the Germans as a base on the way to England's eastern ports.

Not content to settle for moderate success, Lübeck aggressively strengthened its ties both to the Baltic and to the Continent. Sweden had been a longtime trading partner, for example, and in 1251 extended its treaties with Lübeck. The Germans profited in particular from the trade in Swedish iron and copper, much desired throughout the rest of Europe. As well, Lübeck incorporated Flanders into the Hanseatic power base by according privileges to the merchants there in 1252–53. After all, by the mid-thirteenth century Lübecker merchants were the most frequent visitors to Bruges and other thriving Flemish towns. For its part, within half a century Bruges enjoyed preeminence as terminus of a thriving trade axis from Russia to Flanders.

Meanwhile, a league of Wendish towns developed between 1256 and 1264. Lübeck's relationships with the Wendish towns to the east, in particular with the busy ports of Rostock and Stralsund, were more competitive than its relationship with Hamburg. But even so Lübeck agreed with Rostock and Wismar in 1259 to unite against sea piracy. In a further agreement in 1264 the three towns arranged for mutual assistance in the event of war, and established legal parities for citizens. A dispute between Stralsund and Greifswald was resolved in the *Landfriedensbündnis* in 1283, which was a treaty of alliance to maintain the peace that was signed by the key towns: Lübeck, Wismar, Rostock, Stralsund, Greifswald, Stettin, Demmin, and Anklam.

While it was cooperation on a large scale that allowed the Hanseatic trading system to be so successful, the flow of goods and wealth across widely disparate regions was bound to lead to divergent interests on occasion. In 1280 Lübeck spearheaded the first effort within the Hanseatic sphere to resolve a trade conflict through sanctions. Local merchants in Bruges had objected to foreigners dealing directly with each other there, without using Flemish middlemen, but two years of trade sanctions against the Flemish town worked in the Germans' favor. The trading concessions they won from Bruges in 1282 exceeded the terms they had originally enjoyed in the town. This tactic also proved effective with the Norwegians, who had sought to circumvent the Hanseatic grain monopoly, and suffered a blockade in 1284–85. By 1294 Norway accorded to the Germans concessions that surpassed those of the Norwegians themselves in their own country.

Significantly, at about the same time a shift was occurring that transferred the nature of the league from an organization whose members were merchants to an organization comprising towns. Signaling this change was the abolition of Visby's special status in 1293. It was then that the Wendish towns voted to remove control of the German trade delegation at Novgorod from Visby and grant it to Lübeck.

So it was that Lübeck rose to the pinnacle of the Hanseatic system. As long as the land route through Lübeck remained preferable to the dangerous sea route around Denmark, the city's prosperity remained assured. Unfortunately for Lübeck, much of surrounding Schleswig-Holstein continued to be controlled by the Danes; the town's vital link to Hamburg was threatened when Danish King Erik VI won from Holy Roman Emperor Albert I control of all lands north of the Elbe River, including the territory between Lübeck and Hamburg. From 1307 to 1319 the proud city that had defeated the Danes so soundly in the previous century was forced into the role of a Danish protectorate.

More serious disaster befell the city when the Black Death struck in 1350; statistics suggest a large portion of Lübeck's population died during the plague, which seriously disrupted Hanseatic activities in general. But trade activity rebounded, and the first general Hanseatic diet took place soon afterward in Lübeck, in 1356. Just two years later, in 1358, the Hanseatic League made Lübeck its administrative headquarters. In all, Lübeck hosted fifty-four of the seventy-two Hanseatic diets that brought the trading towns together between 1356 and 1480.

A Hanseatic embargo against Flanders ensued from 1358 to 1360, with a repeat occurring in 1388. The latter embargo continued until 1392, and trade with England and Novgorod suffered during this time, too. These conflicts were costly, as could be Hanseatic administrative business in general, leading to inevitable hikes in taxes. Lübeck's populace began to express dissatisfaction with the patrician rulers, as indicated by the butchers' risings in 1380 and again in 1384.

The discontent festered until 1408, when a constitutional crisis forced the ruling town council to address popular demands. Some members of the council went into voluntary exile, new members were elected, marking the beginning of eight tumultuous years of split rule, the old councillors and new councillors repeatedly in conflict. When the new council came into conflict with the Holy Roman Emperor Rupert in 1410, he shocked the Hanseatic world by putting Lübeck under an imperial ban. Order did not return to the city until matters were resolved in favor of the old, patrician system in 1416, which nonetheless retained some democratic elements from the nearly ten years of instability.

Lübeck and the empire continued at cross-purposes, however, and the town's other external relations were contentious during the fifteenth century. Lübeck and the Hanseatic League engaged in repeated military actions with rivals of one sort or another. In 1419 a long-standing dispute marked mostly by piracy began between the league and Castile, and was not resolved until 1441. From 1426 to 1435 one of the ubiquitous wars between Lübeck and Denmark raged once more, followed almost immediately by a war with the Dutch from 1438 to 1441. Lübeck promoted yet another embargo against Flanders from 1451 to 1457, although this was the last such embargo against that region. Decades of disputes with England led to war with that country from 1470 to 1474. France was no exception to Hanseatic hostilities, necessitating the signing of a perpetual peace treaty between the league and France in 1483.

As well, by the end of the fifteenth century, the increasing power of the Scandinavian countries and an emergent Russia began to infringe on the hegemony of Lübeck and the Hanseatic League. Much to its detriment, Lübeck was largely shut out of the lucrative new trade with the Americas and the Indies, markets that were served mostly by naval powers such as England and Holland, the latter of which—under the astute leadership of Amsterdam—was successfully encroaching in traditional league markets, too.

As occurred in many trading cities, the Protestant Reformation induced further disruption. Lutheranism swept northern Germany in the 1520s. The politics across Europe became increasingly volatile, especially because of the Catholic faith of the Holy Roman Emperors and the eventual dominance of Catholic Spain over Protestant northern territories. Lübeck's council vehemently opposed Lutheranism, but the new doctrine gained popularity despite that; by 1530 the council had capitulated, to the point that Catholic ceremonies were outlawed, and gold and silver church ornaments were removed to help finance Lübeck's war chest. The revolutionary Jürgen Wullenwever became burgermeister, engaging the city in a rash and unjustified war against Denmark (1531–35) that led to his downfall. Allied with Denmark, for once, Lübeck suffered a decisive defeat in a war waged from 1563 to 1570 against Sweden for Baltic supremacy. At about the same time, the political and religious tumult in the Low Countries interfered with Hanseatic operations there, such as at the new mercantile house the league had just completed in Antwerp. Back east, even as Sweden was overrunning Prussia and other parts of northern Germany, Lübeck itself remained neutral during the Thirty Years War (1618–48). Nonetheless, the widely devastating conflict marked the end of the Hanseatic League as it had existed. With only three signatories, the alliance in 1630 between Lübeck, Bremen, and Hamburg signaled the de facto end of the league. Lübeck remained the most important harbor on the Baltic Sea, however, and in their individual ways each of the three cities continued to conduct business according to the entrenched Hanseatic principles. The last official Hanseatic diet occurred, appropriately enough, at Lübeck in 1669.

Until the ruinous disruptions in trade during the period of the French Revolution and Napoleonic Wars, from 1792 through 1815, Lübeck had remained prosperous enough to construct and maintain a series of new bastions and walls around its Gothic core. In a sorry turn of fate, however, Lübeck demolished its fortifications in 1803, only to be sacked by Napoléon's troops in 1806. Foreign domination was the order from 1811 to 1813, when the French ruled Lübeck. After 1815 Lübeck became a member state of the German confederation.

During the nineteenth century, Lübeck experienced the general advantages and disadvantages of the Industrial Revolution, including the enhancement of its land transport routes through railway connections (although the line to Hamburg opened only in 1864, after a Prussian defeat of Denmark secured the region). Until 1848 Lübeck's politics continued to reflect the city's entrenched patrician heritage: the richest families and merchants controlled its government. Even so, Lübeck was not exempt from the wave of liberalism that swept Europe during the 1840s. Lübeck acknowledged the inevitable in 1848, when the elite rulers of the essentially oligarchic city-state began to share political power with the average citizenry. From 1866 the city belonged to the North German Confederation, precursor to the eventual German Empire, of which Lübeck became a part after German unification in 1871.

Driving a much-needed boost for Lübeck's economy was the construction of the Elbe-Lübeck Canal in 1900, which enhanced the land routes with a water connection directly to the well-traveled Elbe River. Lübeck's proud status as a separate, self-governing city—in force since the imperial charter of 1226 despite the city's gradual decline—ended ignominiously in 1937, when the National Socialist government incorporated the city into the province of Schleswig-Holstein.

A large portion of the historic inner city was destroyed in World War II, but the distinctive medieval ensemble comprising the town hall and the adjacent Marienkirche have since been restored. Several other historic churches, some Gothic, some baroque, remain standing, as do numerous commercial buildings, now maintained as museums, from the Hanseatic era. There also are museums, the Buddenbrookhaus and the Drähaus, with exhibits on the life and work of two of Lübeck's most famous sons, the writers Thomas and Heinrich Mann.

Further Reading: The Hanseatic League is so integral to Lübeck's early history that accounts of that entity inevitably involve events in Lübeck. *The German Hansa* by Philippe Dollinger, translated from the French and edited by D. S. Ault and S. H. Steinberg (Stanford, California: Stanford University Press, and London: Macmillan, 1970; originally published as *La Hanse,* Paris: Aubier, 1964) is an authoritative and wide-ranging survey that includes maps, a list of Hanseatic League towns, a chronology of the league, and economic tables. *The Hansa: History and Culture* (Leipzig: Edition Leipzig, with Dorset: Dorset Press, 1988) by Johannes Schildhauer, translated from the German by Katherine Vanovitch, is an attractive, large-format book that includes detailed footnotes, exquisite woodcuts, illustrations and photographs, many full-page and in color. This book discusses the broad social and cultural issues of the mercantile milieu of the Hanseatic League cities. Karl Jordan's *Henry the Lion,* translated from the German by P. S. Falla (Oxford: Clarendon Press, and New York: Oxford University Press, 1986; originally published as *Heinrich der Löwe, eine Biographie,* Munich: C. H. Beck'sche Verlagsbuchhandlung, 1979) offers a fascinating narrative about the ambitious prince who set the wheels in motion for Lübeck's second founding in the late twelfth century. *A Portrait of Two German Cities: Lübeck and Hamburg* (Lubbock, Texas: Caprock Press, 1980) by David Rodnick is a subjective account with an emphasis on personal anecdote and sociology. *Buddenbrooks,* translated from the German by H. T. Lowe-Porter (New York: Knopf, 1976), an epic novel by Thomas Mann about the decline of a merchant family during the nineteenth century, captures in its narrative the essence of patrician Lübeck. The book has been published a number of times in German and in English since its debut in 1901, but enhancing this edition are Mann's own remarks, "Lübeck as a Way of Life and Thought," delivered in Lübeck in 1926, the 700th anniversary of its institution as a free imperial city.

—Randall J. Van Vynckt

Lucerne (Lucerne, Switzerland)

Location: Central Switzerland, in the Lucerne canton, at the point where the Reuss River issues from Lake Lucerne; 60 miles southeast of Basel and 179 miles north-northwest of Milan.

Description: Well-preserved and picturesque city that grew up around a Benedictine monastery; a stronghold of Swiss Catholicism and a major tourist center.

Site Office: Tourist Information Office
Frankenstrasse 1
CH-6002 Lucerne, Lucerne
Switzerland
(41) 51 71 71

Lucerne was named for the Benedictine monastery of St. Leodegar, or Luciaria, founded in the eighth century near the site of the present-day Cathedral of St. Leodegar, or Hofkirche. The monastery was later attached to the Murbach monastery in Alsace. The nearby fishing village grew into a town, chartered about 1178. Its inhabitants were originally the monastery's serfs.

Lucerne became a major center of trade between the upper Rhine and Lombardy when the St. Gotthard Pass opened about 1220. The citizens drew up their first municipal constitution in 1252. In 1291 the German king and Holy Roman Emperor Rudolf I, the founder of the Habsburg dynasty, purchased the town and monastery. This did not please many of Lucerne's citizens, who desired independence.

There was much political instability under Rudolf's successors, and in 1332 the canton of Lucerne formed an alliance with the Swiss cantons of Uri, Schwyz, and Unterwalden, which had joined forces in 1291. Lucerne thus became part of the Swiss Confederation; before joining, it actually had sent troops to assist the Habsburgs in putting down uprisings of the young confederation. In 1386 the confederation defeated the Habsburg army at Lake Sempach, ten miles from Lucerne, and gained independence.

By 1415 the town had annexed most of the territory of the present canton of Lucerne through treaties, armed occupation, or purchases. In 1480 Pope Sixtus IV granted Lucerne the right to depict the Apostles on the Mount of Olives on its city banner. This motif is still found in both chapels and private homes in Lucerne.

When the Protestant Reformation began, Lucerne became the leader of Switzerland's Catholic cantons and served as the seat of the papal nuncio from 1581 to 1848. To the present day, most of the residents of Lucerne are Catholic. In 1654 St. Francis Xavier was made the patron saint of Lucerne. A triumphal procession held in celebration featured a citizen playing the role of the saint drawn through the streets in a chariot modeled on those used by senators in ancient Rome.

In addition to being a center of Catholicism, Lucerne became one of the centers of opposition to the aristocrats who controlled Swiss towns. These patricians had gained power and wealth through mercenary service in the armies of foreign countries and through pensions granted by foreign royalty. Lucerne was a major participant in the unsuccessful peasant revolt against the aristocracy in 1653.

The event that finally stripped the patricians of their power, at least for a short period, was the invasion by Napoléon's armies in 1798. Lucerne's resistance was brief, and it became a protectorate of France as the Helvetic Republic, with Lucerne as its capital from 1798 to 1803. During this period Vinzenz Rüttimann, the son of a Lucerne burgher, gained favor with Napoléon, who made him the mayor of Lucerne and then the president of the Tagatzung, the governing body of Switzerland. When Napoléon was defeated, Rüttimann quickly repudiated his revolutionary ideals to gain favor with the Bourbon rulers of France, and became the leader of the aristocratic regime, which again gained control of Lucerne after 1814.

Between 1831 and 1841 the patrician regime was replaced by a more liberal government. In the early 1840s, however, the anti-Catholic Radical Party came to power in Lucerne and elsewhere in Switzerland. The Radicals proposed, among other things, putting seminaries under state control and making bishops' appointments subject to government approval. The rise of Catholic opposition gave the Radicals an excuse to close all eight convents in the canton of Aargau in 1841. This so outraged the Catholics in Lucerne that they overthrew their canton's Radical government. As a further act of defiance, Lucerne invited the Jesuit order, which had been suppressed by the Radicals, to take control of the canton's schools and colleges.

Lucerne's defiance did not go unnoticed; in 1844 and 1845 the canton was raided by pro-Radical forces from other cantons, and at least one of Lucerne's Catholic leaders was assassinated. Lucerne and other Catholic cantons therefore formed the Sonderbund, a defensive alliance, in 1845. In 1847 the federal diet, in which the Radicals held a majority, declared war on the Sonderbund, and the city of Lucerne was defeated and occupied by federal troops that year. The war ended after less than a month, with the federals victorious and with fewer than 100 casualties on either side. One result of the war was the adoption of a new federal constitution in 1848; this constitution provided for a stronger central government and limited the power of the cantons. While this may not have pleased the defeated cantons, they accepted the new arrangement, and peace was restored.

Lucerne and the rest of Switzerland likewise have

A view of Lucerne in the late nineteenth century
Illustration courtesy of Swiss National Tourist Office

remained peaceful in their relations with the rest of the world; Switzerland has maintained a policy of neutrality in all major international conflicts. The city of Lucerne has come into the twentieth century as a center of education—it has many institutions of higher learning—while its most important industry is tourism. It hosts numerous annual events, including the International Music Festival, featuring choral and symphony concerts, and a pre-Lenten carnival (Fasnacht) featuring masked processions, held since the fifteenth century.

The city is divided into two parts by the Reuss River, which is spanned by five bridges within the town's boundaries. On the right bank is the old town area, distinguished by charming alleys and squares and houses from the medieval, Renaissance, and baroque periods, with part of the old city walls on the northern border. This area has several churches, a cathedral, and museums.

The oldest of the bridges across the Reuss is the Chapel Bridge (Kapellbrucke), a covered wooden bridge completed in 1333 and decorated with seventeenth-century paintings of Lucerne's history. It spans the Reuss at an angle

and marks the point where the river and lake meet. It adjoins the thirteenth-century Water Tower (Wasserturn). The second-oldest bridge is the covered wooden Spreuerbrucke, completed in 1407 and decorated with seventeenth-century panels representing the Dance of Death, inspired by the fourteenth-century plague that affected all of Europe. The city's main bridge, the Seebrucke, connects the Station Square (Bahnhofplatz) on the left bank and the Swan Square (Schwanenplatz) on the right bank.

The remains of the fortified city walls, the crenelated Musegg walls, dating to the fourteenth century, have nine watchtowers that were built between 1350 and 1408. A clock installed in one of the towers in 1365 still keeps time.

Lucerne's historic churches include the Cathedral of St. Leodegar (Hofkirche), on the site of the eighth-century monastery. The present cathedral is a late-Renaissance style structure built in 1639 but incorporating Gothic towers from about a century earlier; the rest of the earlier structure was destroyed by fire in 1633. St. Peter's Chapel, built in 1178, is the oldest church in Lucerne. The Franciscan Church (Franziskanerkirche) is a Gothic building from the fourteenth century, and the Jesuit Church (Jesuitenkirche) was built in the baroque style from 1667 to 1678.

Lucerne has a wide variety of other historic attractions. The Glacier Garden (Gletschergarten), excavated from 1872 to 1875, consists of bedrock that has been dramatically marked by Ice Age glaciers. The Bourbaki-Panorama is a huge conical structure created between 1876 and 1878 as a tourist attraction and houses a lifelike wraparound painting, covering more than 118,000 square feet, depicting the French Eastern Army retreating into Switzerland, where the army was given asylum in 1871 during the Franco-Prussian War. The Lion Monument (Lowendenkmal), was carved from a sandstone cliff between 1819 and 1821 as a memorial to more than 700 Swiss Guards killed in the French Revolution while defending the Tuileries in Paris in 1792. It depicts a dying lion with his chin sagging on his shield and the broken stump of a spear in his side. Mark Twain described it as ''The most mournful and moving piece of stone in the world.'' The Weinmarkt, one of several fountain squares, was once the site of the city's wine market, and between the fifteenth and seventeenth centuries was the site of passion plays that drew visitors from throughout Europe.

The city's public buildings represent several architectural styles. The canton's government building, the Renaissance-style Ritter Palace (Rittersche Palast), was built from 1557 to 1564. The Old Town Hall (Altes Rathaus), a late-Renaissance building erected between 1599 and 1606, has been the meeting place of the Lucerne town council since 1606. The Lucerne library occupies a house that was built about 1674.

Lucerne has some notable museums. The Fine Arts Museum (Kunstmuseum), housed in the Palace of Arts and Congresses, displays works by Swiss artists and by such foreign artists as Maurice Utrillo and Maurice de Vlaminck. The Renaissance-style Am Rhyn House, built in 1617, contains an important collection of Picasso paintings. Utenberg Costume and Folk Museum (Schweizer Trachten- und Heimat-museum Utenberg) features life-size dolls in Swiss regional costumes. The Richard Wagner Museum, on the outskirts of Lucerne, was established in 1933 in the Villa Tribschen, a country house Wagner occupied from 1866 to 1872 while composing the *Meistersinger*. The Natural History Museum (Natur-Museum), which emphasizes the natural characteristics of central Switzerland, is known for its modern display techniques. The nearby Historical Museum (Historisches Museum) was built in 1567 as an armory and exhibits Swiss arms and flags, city sculptures, and reconstructed rooms. The Swiss Transport Museum (Verkehrshaus der Schweiz), built in 1959, treats all forms of transport, including space travel, and also includes a planetarium. It draws more visitors than any other museum in Switzerland.

Further Reading: In *The Cradle of Switzerland* (London: Hollis and Carter, 1952; New York: British Book Centre, 1953), Arnold Lunn provides insight into the cultural background of the Swiss and the role Lucerne played in the country's development. A good general overview of Swiss history can be found in *A History of Switzerland* by Charles Gilliard (Westport, Connecticut: Greenwood Press, 1955).

—Phyllis R. Miller

Ludlow (Shropshire, England)

Location: In Shropshire, in western England, on the Welsh-English border, on the A49, eleven miles north of Leominster.

Description: Medieval English border town, noted for its castle, which housed English royalty and Marcher lords, who administered law on the English frontier. The town's architecture is also reflective of the various historic eras in Ludlow's past.

Site Office: Ludlow Visitors' Centre
Castle Street
Ludlow, Shropshire SY8 1AS
England
(584) 875053

Located on the English-Welsh border in Shropshire, Ludlow now commands less cultural, economic, and political prestige than in its heyday in the Middle Ages, when the town provided the administrative seat for the Council of the Marches of Wales, a body that administered a sort of English frontier legal justice. Ludlow Castle, the town's focus and protector, both harbored and accommodated British royalty and future kings. The town also had a thriving market and small industrial base. In more recent times, Ludlow has, because of its well-preserved Norman, Gothic, Tudor, and Georgian architecture, become a resort town, one that allows visitors to experience the evolution of English history within the confines of one town.

Late in the twelfth century, Ludlow was established as a recognizable community by the unification of the manor Stanton Lacy and the surrounding territory. The town sometimes referred to as Lude, Lodelowe, Ludelawe, Luddelow, or Ludlowe evolved around an imposing castle, whose Norman design can be attributed to the builders' ancestral ties to Normandy. Ludlow was an early example for England of a town planned on a grid—also the result of Norman influence. Though smaller than the original plan, the town appeared much the same in 1800 as it did in 1270.

Ludlow was home to a wide range of economic classes, but wealthy merchants were numerous and truly poor individuals few, as suggested by the quality of housing throughout the town. While farmland was located outside the town walls, farmers usually resided in town, and stored their goods and property there. In addition to farming, Ludlow's principal industries involved sheep raising, cloth and textile manufacturing, and related industries such as glovemaking. According to early records, the fledgling Ludlow had fifty plows, making it one of the richest areas in Shropshire.

Wool processing and clothmaking (and related industries such as sheepherding) created Ludlow's wealth. The first mill, established in the mid-twelfth century, led to a boom in cloth production and mill establishments that thrived until the late sixteenth and early seventeenth centuries. The cloth trade and manufacture supported a whole range of craftsmen such as dyers, fullers, and weavers. Cloth products included derkegrenemedle (green cloth) and blue medle, and a multiplicity of other colorful cloths. However, Ludlow was most famous for the production of Ludlow whytes, a coarse broadcloth. Trade with Flanders and within England was brisk until the late sixteenth century, when the manufacture of a finer grade of cloth became possible. Unlike their Flemish counterparts, Ludlow's clothmakers continued milling coarse cloth, and trade declined.

Because of its agricultural industries, Ludlow also developed as a market town, providing a stop on the north-south trading route. The earliest recorded market was held in the thirteenth century, but it is certain that one existed prior to that time. Different markets were located throughout town. The long-enduring market for trade in cattle, sheep, and other livestock was located at present-day Bull Ring. The central marketplace, now the High Street, began with temporary stalls, which were succeeded by sturdier buildings. Markets served the local population, but fairs attracted peddlers and other merchants from a distance. Held less frequently than the weekly markets, fairs were granted for the benefit of an individual lord, local organization, the town itself, or for the crown, which received rent from the booths. In an early form of trade protection, Ludlow even charged out-of town vendors an entrance toll to sell goods in the market or fair.

The people of Ludlow were an interesting mix of immigrants and locals. Significant numbers of Welsh and Irish immigrants kept the town's population constant, despite the usual death rate from plague and other diseases. Until 1587 Ludlow had a surplus of births over deaths, due in part to this immigration.

Not all of Ludlow's inhabitants were well-to-do, however, and, like most medieval towns, Ludlow did offer universal, basic public services. Monasteries and religious houses not only took care of the spiritually poor but also provided for the educational needs of the community. The Houses of Austin Friars of Ludlow, built in about 1254, administered to the spiritual needs of the early parish. Established approximately a century later, the Carmelite Friars of Ludlow heard confessions and provided a means of education. Founded in the 1220s, the Hospital of the Holy Trinity, the Virgin Mary and St. John the Baptist provided relief and care for the sick and the poor, and perhaps a resting place for travelers. In the seventeenth and eighteenth centuries parts of the hospital became a residence and tenement housing for laborers.

Charity usually provided for the sick or for widows,

Ludlow Castle
Photo courtesy of Shropshire County Council

but not for economic and natural disasters such as crop failure and fire. Fortunately, because of the economic structure of Ludlow, masses of laborers were seldom out of work. It was not until the crash of the textile and cloth industry in the sixteenth industry that large numbers of the poor required assistance. In the mid-sixteenth century John Hosyer, a Ludlow draper, established through his will Hosyer's Almshouse. Admission was only to aged and honest poor residents of Ludlow who were of good character, although the criteria changed over time. Vagrants and the unemployed—considered the undeserving poor—were barred from admittance. In the eighteenth century, however, many potential residents preferred to live elsewhere because of the building's dangerous state of repair. Other forms of charity included loans from wealthy merchants and tradesmen to young men beginning in business or a trade.

The Palmers' Guild, a religious organization founded in 1284 by Ludlow's burgesses, was a form of economic and spiritual insurance company. Through dues from its members, the guild supported priests who would say prayers before and after the deaths of its members. Primarily local until about 1390, the society also provided public relief to lepers, the blind, injured, and ill. The guild also operated a legal defense fund. After 1390 the guild's membership began to spread beyond local boundaries. By the fifteenth century only half of the guild's membership lived in Shropshire. Membership consisted of about 50 percent merchants, tradesmen, and craftsmen, 25 percent clergy, and 13 percent nobility or gentry. Throughout the town's history the Palmers' Guild supported a grammar school, established sometime in the thirteenth century.

Medieval towns frequently were dominated and protected by a castle and its lord. Ludlow was no exception. Built in about 1086, probably by Roger de Lacy, Ludlow Castle and its buildings stand on a limestone cliff at the convergence of the Corve and Teme rivers. The castle physically dominated the landscape of the medieval countryside. The nearby cathedral-like St. Laurence's Church, made of pink sandstone, and completed in the 15th century, was the largest medieval church in Shropshire. Financed by a wealthy wool merchant, the church is 203 feet long and has a tower that stands 135 feet tall. The castle is now in ruins; the church, conscientiously restored, still stands.

It was the political position of the castle's lords that made Ludlow important. The English frontier required both protection from Welsh attack and a system for maintaining English law and order. Feudal lords maintained order and justice by protecting their towns' inhabitants through the castles' fortifications, and through usually benevolent dictatorship. Beginning in the twelfth and thirteenth centuries, the kings of England were forced to accept a franchise jurisdiction of justice, whereby they had to trust the lords to mete out English justice within their jurisdictions. This situation invested a great deal of power in the local lords, detracting from the king's control. The five westernmost manors, or hundreds, essentially created their own dominion. In the late

15th century the Council of the Marches, as their association was called (after the local name for the border country), ruled from Ludlow Castle, and for the next 200 years Ludlow in effect was the capital of Wales and the border counties.

Lacy and his descendants often found themselves in opposition to the crown, which summarily would confiscate their lands or appoint another Lacy male as lord. Often the castle became fighting ground for the crown. Two early tales about the castle reveal the chaos of manorial life, the frontier, and English politics. A fourteenth-century Norman poem and minstrel song, "The Romance of Fitz-Warin," relates a plot by Joce Dinan to take the castle from one Walter Lacy. The romance centers on a woman in Dinan's household named Marian and her lover. Ludlow Castle supposedly was lost because of Marian's liaison, which resulted in a murder-suicide. Perhaps a more historically accurate tale tells of King Stephen laying siege to the castle in 1138–39 and saving Prince Henry of Scotland.

The male line of the Lacy family declined, and caused the castle and its lands to be inherited by a nephew, Gilbert, who assumed the Lacy name. Gilbert's son Hugh became governor of Ireland, which allowed much of the Irish immigration and emigration to and from Ludlow to occur. Eventually passed on through the female line, the castle was inherited through marriage by Roger Mortimer. Mortimer not only was lord of nearby Wigmore, also in Shropshire, but also inherited the Lacy's Irish holdings in County Meath, becoming lord lieutenant.

Roger Mortimer was politically powerful in both England and Ireland. During the Scottish Border Wars he opposed some of Edward II's loyal lords, for which he was imprisoned in the Tower of London. Escaping to France, Mortimer established both a political and romantic liaison with Isabella, the king's wife. Isabella and Mortimer invaded England in 1326, forcing Edward II to abdicate in favor of his son, Edward III. Edward II's mysterious murder was avenged by his son through the arrest and subsequent hanging of Mortimer. The last of the male Mortimer line died of the plague in 1427, paving the way for the crown to seize the castle after the Wars of the Roses ended in 1485.

Through the convoluted bloodlines of his mother's family, Richard Plantagenet, Duke of York, also became lord of Ludlow Castle. Until Henry VI's wife bore the king a son, the Duke of York had claims to the English throne. In 1459 the duke's forces rallied at Ludlow but lost the subsequent battle, and the opposing Lancastrian forces sacked Ludlow. Eventually the Duke's son Edward defeated the Lancastrian army near Ludlow at Mortimer's Cross. With his accession to the throne in 1461 as Edward IV, Ludlow Castle became a royal home.

In 1472 Edward IV sent his two young sons to Ludlow Castle, where they remained until the death of the king ten years later. Much of the rest of the story of the "princes in the tower" is fraught with legend and mystery. Allegedly, the young boys were captured, placed in the Tower of London, and killed on orders of the Duke of Gloucester, who became

Richard III. The death of Richard III in 1485 at the hands of troops led by Henry VII in the Battle of Bosworth returned the monarchy to the house of Lancaster. Ludlow Castle again served as a royal home for Henry's son Arthur, who had recently married Catherine of Aragon. When Arthur died in 1502, his body lay in state in the castle for three weeks. Catherine subsequently married Arthur's brother, who became King Henry VIII.

In the sixteenth century Ludlow Castle ceased to be a royal residence, and was taken over by a series of presidents of the Council of the Marches. No longer running an outpost of independent justice, the Marcher presidents fairly enforced English law. Some lord presidents had ties to the Privy Council in London, and English kings and queens from Henry VIII onwards managed to administer a regular system of law enforcement. Sir Henry Sidney of Kent, as lord president in 1559, built housing for justices adjoining the castle, and created a courthouse and two offices for record keeping. In 1689 the Council of the Marches was abolished in favor of more centralized control from London.

The demise of the Marcher lords, the economic decline of the sixteenth century, the English civil war in the mid-seventeenth century, and the Industrial Revolution brought drastic change to Ludlow. The Council of the Marches, still in existence during the civil war, and the fact that Ludlow Castle was a royal residence made Ludlow a royalist town. The townspeople billeted the king's troops and defended themselves with the assistance of Welsh and Irish soldiers. In the spring of 1646 parliamentary troops laid siege to the town and castle, burning and destroying large portions of Ludlow. Thereafter Ludlow became an unimportant military backwater. In 1650 the castle's contents as well as the king's furnishings were sold. During the eighteenth century the Earl of Powis leased Ludlow Castle from the crown, and one of his successors finally purchased the property in 1811.

Economically Ludlow never recovered fully from the decline in wool and cloth production. While the Industrial Revolution had its origins in Shropshire, it was centered around Shrewsbury and made little impact on Ludlow. The eighteenth-century Ludlow economy was based on corn, cattle, provisions, and glovemaking. No longer broadly middle-class, its population was polarized between rich and poor. Nearly two-thirds of Ludlow's women were engaged in textile-related production, such as the manufacture of but-

tons, garments, and gloves. Compared with similar towns, Ludlow possessed an inordinate number of households headed by unmarried women.

In the eighteenth century Ludlow became a social and cultural center similar to Bath, but on a smaller scale. Because roads had improved greatly, and because of England's late-eighteenth-century war with France, travel to domestic destinations such as Ludlow preempted continental travel. Owners of many of the medieval wood-framed houses installed brick facades to create a more Georgian and "modern" look.

Ludlow also had artistic ties. In 1643 John Milton's masque *Comus* was performed in the castle. Milton created a precedent for writers to visit Ludlow that carried on through the nineteenth and into the twentieth century. The novelist Henry James visited in the late 19th century. The twentieth-century poet A. E. Housman wrote of Ludlow and his native Shropshire countryside. The annual Ludlow Festival of Arts, begun in 1959, features an open-air Shakespeare production in the inner bailey of the castle.

Today Ludlow retains pieces from each era of its past. The castle lies in ruins, but examples of Norman, Gothic, and Tudor architecture abound. Built by a secretary of the Marches council in 1558, the half-timbered Feathers Hotel displays carved heads and other ornamentation on its exterior. The Readers House, a three-story timber dwelling, may have been a prison for Mary Queen of Scots. Parts of Hosyer's Almshouse still stand. The eighteenth-century town hall, known as the Butter Cross, houses the town museum. In 1976 the Ludlow Historical Research Group was formed to further investigate Ludlow's eventful past.

Further Reading: *Ludlow 1085–1660: A Social, Economic and Political History* by Michael Faraday (Chichester, West Sussex: Phillimore, 1991) is a fascinating, well-written, and easy-to-read survey of medieval Ludlow. The focus is on the socio-economic development of the town. *Historic Ludlow Castle and Those Associated With It* by A. Lowndes Moir (Ludlow, Shropshire: Ludlow Adventiser, 1955) is a very detailed but enjoyable account of the intrigues of the castle. It includes good photographs. *Ludlow: A Historic Town in Words and Pictures* by David Lloyd and Peter Klein (Chichester, West Sussex: Phillimore, 1984) provides a brief historical overview, and then illustrates that history through diaries, personal accounts, woodcuts, and early recorded histories.

—Jenny L. Presnell

Lyons (Rhône, France)

Location: In east-central France, 291 miles southeast of Paris, 170 miles north of Marseilles, 58 miles northwest of Grenoble.

Description: Capital of Rhône Department and the third-largest city in France; one of the oldest cities in France, with a history that dates from 43 B.C.; site of repeated religious, social, and political unrest; the headquarters of the French Resistance during World War II.

Site Office: Pavillon du Tourisme
Place Bellecour
B.P. 2254
69214 Lyons, Rhône
France
78 42 25 75

Lyons, or Lyon, is the capital of Rhône Department and one of the largest cities in France, ranking only behind Paris and Marseilles. A major regional center because of its convenient access to trains and ships and its ability to attract various professionals and crafts workers, Lyons has played a key role in its country's history. Two rivers, the Rhône and the Saône, which flow into the city, aided its growth. The Saône River, which flows from the north, wraps along two hills, Croix-Rousse and Fourvière; the Rhône flows from the northeast. The rivers then run parallel, creating what was at one time an island that supported a wealthy residential district as well as a commercial district for warehouses and workshops. Today the island is a peninsula and supports the heart of the city. South of the peninsula, the two rivers join together and flow south. The center of the city runs north and south between the Croix-Rousse and Terreaux quarters. It is home to Lyons's famous silk district. In this area are also located the business district, the Bellecour quarter, which includes hotels, cafes, and shops, and the Perrache district. The latter in the site of the Perrache train station and was reclaimed from the rivers in the eighteenth century. Vieux Lyon, the city's old quarter, is located on the west bank of the Saône and at the foot of Fourvière; more recent developments and buildings are found on the east bank of the Rhône.

The recorded history of Lyons dates back more than 2,000 years, to October 10, 43 B.C., when Lucius Munatius Plancus, a former lieutenant of Julius Caesar, founded a colony on the Fourvière hill. He called the settlement Lugdunum: *Lug* denoted an early deity of the Celts, who inhabited the area 600 years before the Romans, and *dunum* meant hill. Lugdunum soon grew to be a major city and served as the capital of the three Roman provinces of Gaul. In 12 B.C., the altar of Rome and Augustus was dedicated here. Between 16 and 14

B.C., Augustus built a 4,500-seat theatre about halfway up the Fourvière hill, near what would become Vieux Lyon. Later, Hadrian enlarged it to seat 10,000. A smaller, 3,000-seat theatre was also built nearby for music and speeches. The ruins of these theatres can be found today near the ruins of the Temple of Cybèle and other archaeological sites.

Lugdunum was also home to a circus for chariot races and the Ampithéâtre des Trois-Gaules. The latter, discovered in 1958, was built in a part of Lyons now called Condate and was located across the Saône on the Croix-Rousse hill. When the Romans divided the three provinces into sixty Gallic tribes, delegates from the tribes would meet in the amphitheatre. It is believed these council meetings continued a tradition started in pre-Roman days, an assembly of Gauls held near Chartres by the Druids. In A.D. 48, Emperor Claudius gave a speech in the amphitheatre to the Roman Senate. The speech is now inscribed in bronze in Musée Gallo-Romain, a museum adjacent to the Roman theaters. The Ampithéâtre des Trois-Gaules was also used for public executions, and even animal and gladiator combats. During the second century, the amphitheatre was also a site of the martyrdom of the early Christians, who were there thrown to the lions.

The end of such persecution and the official adoption of Christianity by the Roman Empire brought increased religious importance to the city. Lyons became the first center of Christianity in Gaul. It is believed that the city's first cathedral, which served the city until the sixth century, was located on the site now occupied by Eglise St.-Nizier.

The political importance of Lyons declined with the collapse of the Roman Empire, but its religious influence increased. In 1245 Pope Innocent IV called the First Council of Lyons, at which he excommunicated and condemned the Holy Roman Emperor Frederick II. The Second Council of Lyons, held in 1274, dealt with matters of clerical morals, discipline, and the fight against heresy. It was particularly fitting that Lyons was chosen as the site of such a discussion; a century earlier, in 1174, the city had been the birthplace of the Poor Men of Lyons, a group of lay preachers who challenged the authority of the church. They taught in the style of Christ and the Apostles, preached the virtues of poverty, and maintained that the Bible was the only true source of knowledge and faith. Because they were organized and led by Peter Waldo, they eventually became known as Waldensians. The church declared them heretical in 1184, but they survived persecution and later joined the Protestant Reformation. The Waldensian controversy was among the earliest outbreaks of the defiance, unrest, and outright rebellion that would occur in Lyons throughout its subsequent history.

Until the early fourteenth century, Lyons and the surrounding area were ruled by archbishops. Then, in 1307, Philip IV incorporated the city into French crownland, mak-

Building facades in Lyons

Detail of a fountain sculpture by Frederic-Auguste Bartholdi
Photos courtesy of Ville de Lyon

ing Lyons part of the French kingdom. In 1320, Lyons's citizens gained self-rule.

With the coming of the Renaissance, Lyons became the center of banking, printing, and the silk trade for all of France. In fact, Lyons was granted a monopoly on the sale of silk. Charles VII secured the city's economic future in 1420, when he authorized the first trade fairs there. Initially, there were two such fairs each year, then four a year. By the late fifteenth century, Lyons attracted merchants and bankers from all areas of Europe. It was a wealthy Italian banking family, Gadagne, that purchased what became known as the Hôtel de Gadagne, a Renaissance mansion in Vieux Lyon. The mansion was built during the sixteenth century for two brothers; each had a wing of his own.

In the late fifteenth century, German printers set up shop in Lyons. Free of censorship, the city soon became a major printing center. As the print shops flourished and attracted many writers and scholars, Lyons became known as an intellectual center. Among the scholars to spend time in Lyons was Rabelais. He practiced medicine there and published two volumes of *Gargantua and Pantagruel* during the trade fairs held in 1532 and 1534. In the late sixteenth century, Lyons' printing industry was overtaken by that of Paris.

When Francis I allowed Lyons to make silk free of taxes, silk weavers flocked from Genoa to Lyons. Quickly, Lyons became Europe's leading silk producer. In 1506, France's first stock exchange opened in Lyons. The city was on its way to becoming a major financial center. What is now called Vieux Lyon became the new center of the city. Here, merchants and bankers built grand townhouses. Vieux Lyon now includes three quarters, St.-Paul, St.-Jean, and St.-Georges, in which there are approximately 200 buildings that date from the late Middle Ages through the seventeenth century. The Cathédrale St.-Jean was started in the twelfth century in the Romanesque style, but was completed in 1420 in Flamboyant Gothic. The cathedral features four towers and several stained glass windows, some of which date to the thirteenth century. It was the site of the two Councils of Lyons and the consecration of Pope John XXII during the fourteenth century. In 1600, Henry IV and Marie de Médicis were wed here. Richelieu received his cardinal's hat there twenty-two years later.

During the sixteenth century, a family named Morelli, who later shortened their name to Morel, arrived from Tuscany to start a silk weaving shop. Their success led them to invest in real estate, and soon they owned much of the suburb of Ecully. To help renovate the Chartreux neighborhood, Joseph Morel donated enough land to create a spacious square at the intersection of Tourette and Chartreux Streets. The area became known as the Place Morel. In 1945, the name was expanded to the Place Théodose Morel, after a resistance hero who was killed on the Plateau des Glières in March 1944.

History's first real labor disputes occurred in Lyons, when journeymen printers held strikes in 1540. The silk weavers staged a strike in 1786. Once again, the unrest in Lyons was to be a prelude to larger revolution.

In 1793, Revolutionary France entered the period known as the Reign of Terror. Nine government officials including Maximilien-François-Marie-Isidore de Robespierre, were elected to a Committee of Public Safety. When the committee members began acting as dictators, the people of Lyons rebelled. The counterrevolution was fueled primarily by socially conservative republicans, though a few liberal royalists took part as well. In addition to Lyons, the resistance movement affected such cities as Bordeaux, Marseilles, Nîmes, Montpellier, Toulon, and Caen. In response to the revolutionary cry of "Liberty, Equality, Fraternity," these leaders of the upper middle class struggled to defend liberty, order, and property, which they felt were jeopardized by proposals for redistribution of land and wealth. In Lyons and Marseilles, desperate federalists appointed royalists to command their armies, a move that resulted in brutal retaliation.

Commissioners responsible for suppressing counterrevolutionary actions in the provinces collaborated with the local Jacobin clubs and revolutionary committees to bring the Terror directly into their districts. When the revolutionary committees decreed that "Lyons be no more," the city's residents were nearly wiped out. In Place des Terreaux, 1,684 Lyonnais were executed by guillotine. The city's population was saved only by the overthrow and execution of Robespierre following the July 27, 1794, Revolution of the Ninth Thermidor.

The invention of the Jacquard loom in 1802 helped save the city economically, as it quickly made the silk business important again. Its success brought renewed unrest, however; as elsewhere in France, workers began demonstrating for higher wages, a shorter workday, and less employment-threatening mechanization. In November 1831, workers seized the city of Lyons, declaring that its people must "live working or die fighting." But their demands went unheard and their uprising was brutally repressed. Nevertheless, the efforts of the silk workers became an important milestone in the history of the labor movement. In the 1870s, the invention of the power loom did away with the need for thousands of hand looms.

Guignol, the famous hook-nosed puppet modeled after a nineteenth-century silk worker, was created in Lyons. Guignol, his wife Madelon, and their friend Gnafron were popular puppets, almost cultural heroes, who disliked work and authority but enjoyed Lyons' famed Beaujolais wine. Guignol expressed the resentment of the workers that caused them to strike and revolt: "they didn't want the herring's tail, and they didn't want the herring's head, they wanted the entire herring."

In 1840, rows of houses along the Fourvière hill fell victim to mudslides. Several Lyonnais marched up the hill and vowed to replace its chapel with a cathedral, in hopes that their religious zeal would prevent another mudslide. Thirty years later, during the Franco-Prussian War, the people of Lyons would again march up that hill. Led by the archbishop of Lyons, they prayed to the Virgin Mary to keep Prussian troops away from the city and the church. The archbishop vowed to replace the cathedral with a basilica so that Lyons would be spared from invasion. The invasion was postponed. In return, the cathedral was transformed into the Basilique Notre-Dame de Fourvière. The church also includes a chapel, built between the twelfth and eighteenth centuries, which houses a statue of the Virgin.

During World War I, Lyons attracted many industries from northern France. In 1916, an International Manufacturers' Sample Fair, which would become an annual event, was established. Industries in the area included automobiles, brass, chemicals, confectionery, dyes, embroideries, fertilizers, furniture, gelatin, glassware, gloves, glue, hats, iron and steel, paper, perfume, plastics, shoe polish, silks and artificial fabrics, watches, and wine. Following World War I, the water power of the Rhône River was finally harnassed. On July 14, 1918, Lyons dedicated the Wilson Bridge over the Rhône River, in honor of U.S. president Woodrow Wilson. When the French Parliament created the Loucheur project to offer state aid to build moderately priced houses, Lyons was one of the first cities to build the needed housing.

If World War I brought new industries and construction to Lyons, World War II brought terrible devastation. Germans invaded Lyons for the first time on June 19, 1940. Prefect Émile Bollaert wore a black tie as a sign of mourning. In Limonest, a northern suburb, Senegalese troops fired at the Germans in protest. The Germans saw this conduct as an excuse to retaliate for the conduct of the Sengalese in a previous war, when the Sengalese had cut off German ears and other appendages for trophies. This time around, the Germans reportedly massacred the Sengalese.

With the signing of the Armistice in June 1940, the Germans pulled out of southern France. As the major industrial center of the unoccupied zone, Lyons was a natural base for the resistance movement, especially given the city's long history of defiance and rebellion. In addition, Lyons was only eighty miles from the Swiss border. The resistance movement was supported by Lyonnais from all walks of life and social strata. The city's reoccupation by Germany two years later seemed only to stiffen their resolve.

Hitler ordered the reoccupation of southern France on November 10, 1942; the next day, German tanks rode quietly across the Place des Terreaux and a swastika hung above the entrance to the city hall. The Germans occupied Grange-Blanche Hospital, Montluc Prison, and Part-Dieu barracks. The KDS (Kommando der SIPO und SD), a German police unit, was formed in Lyons and twelve other unoccupied cities to deal with "disturbers of the peace." The KDS included six sections. Section Four, Repression of Political Crimes, was known as the Gestapo and was responsible for fighting the resistance and arresting Jews. Those who met with the Gestapo were usually subjected to brutal treatment and torture, then sent to concentration camps. Section Four was headed by Klaus Barbie, who had enlisted under Hitler in 1933, at the age of 29.

In response to the crackdown, the resistance grew bolder. On January 10, 1944, a group of young resistance fighters shot and killed two German policemen in the middle of the afternoon. Klaus Barbie's superior, Werner Knab, ordered the SD to take prisoner anyone caught near the area of the shooting. Although the resistance fighters had already fled the scene, the Germans randomly pulled twenty-two innocent people from the streets, put them in a single jail cell in Gestapo headquarters, and shot them to death with over 180 rounds of submachine-gun fire. Later, Klaus Barbie would claim that the victims, who included a postman still wearing a bag of mail, were actually prisoners trying to mutiny and escape.

Of course, for many of the city's 700,000 residents, life went on as before, and business remained a priority. Many Lyonnais continued to profit, despite the fact that much of their money came from business relations with the Germans. The silk industry sewed parachutes for the Luftwaffe; the vineyards of Rhône shipped wine to Berlin.

Marius Berliet was one Lyons-based businessman determined to prosper in these times. Throughout the war, Berliet, who owned the largest such company in France, made trucks for the Nazis, who supplied him with needed raw materials. He even deported "undesirable" workers to Germany. Because of the assistance he offered the Germans, the Allies twice bombed his factory. When he went to trial following the war, a judge questioned him about his actions: "Don't you think it would have been preferable to lay off your workers or put them on half pay, rather than continue to make goods that helped the enemy?" Berliet replied simply, "Those are subtleties I didn't consider." In fact, Berliet had learned a lesson during World War I, when he offered to help his country by converting his factory to the production of tanks. In 1918, with victory already assured, the government canceled its final order. Berliet was left with 800 tanks, for which the government never paid him.

On May 26, 1944, 800 American bombers flew from Italy to the Lyons area in a desperate attempt to rid the area of the Germans. While 200 bombers were sent to Chambéry, and 200 to Saint-Étienne, 400 headed for Lyons. For Lyons, this event became the worst disaster of the war. Several thousand incendiary bombs landed in the city. And, although sirens warned the people, many were too curious to seek protection in time. The bombs hit civilian and military targets alike. This action by the Allies killed and wounded more civilians than had the Germans during all of their occupation. When the bombing was over, 700 civilians were dead and 1,129 were wounded. Destroyed buildings included the SD headquarters in the École de Santé. The Resistance leadership viewed the bombing as an outrage. Alban Vistel cabled the Allied command in Algiers, exclaiming, "LYONS BOMBARDMENT, MORAL RESULT CALAMITOUS, POPULATION PAINFULLY INDIGNANT, REPEAT, ENORMOUS SACRIFICES FOR INSIGNIFICANT RESULTS."

On August 13, Lyons was bombed again, this time as a prelude to the Allied landing in Provence. Klaus Barbie and his men worked hard to destroy archives, knowing they had little time to conceal their actions before the invading Allies arrived. Many of the remaining prisoners were killed; a few fortunate ones survived and were released. Lyons was finally liberated on September 2, 1944, but it would take forty years for Barbie to be put on trial for his actions. When he was finally brought to justice during the mid-1980s, the proceedings were held in Lyons.

After recovering from the war, Lyons again established itself as a leading commercial city. The manufacture of silk still contributes to the local economy, as do the newer industries of synthetic textiles, metallurgy, and petrochemicals. The trade fairs begun in 1420 by Charles VII continue to be held. Lyons is perhaps best known, however, for its food. The city has a long tradition of producing world-renowned chefs, including Paul Bocuse, George Blanc, and Pierre Troisgros, and many connoisseurs consider the food in Lyons to be the best in all of France. Local specialties include gratinée dishes, quenelles, chicken with truffles, sausages, and braised trout.

Further Reading: *A Concise History of France* by Roger Price (Cambridge and New York: Cambridge University Press, 1993) describes French history from the early Middle Ages through François Mitterand's presidency, paying close attention to the relationship between state and society and the probable causes of the French Revolution. Ted Morgan's *An Uncertain Hour: The French, the Germans, the Jews, the Klaus Barbie Trial, and the City of Lyon, 1940–1945* (New York: Morrow, and London: Bodley Head, 1990) reads almost like a novel, and provides a detailed account of the people and events of France in the 1940s. *Birnbaum's France, 1993*, edited by Alexandra Mayes (New York: Harper, 1992) describes the places and the history of France, with much attention given to Lyons.

—Kim M. Magon

Mainz (Rhineland-Palatinate, Germany)

Location: On the left bank of the Rhine River where it turns west and meets the Main River, twenty miles southwest of Frankfurt am Main, Germany.

Description: One of the oldest cities in Germany, founded by the Romans; now capital of the state of Rhineland-Palatinate, formerly the capital of Electoral Mainz, a sovereign state ruled by its Archbishops; home of Johannes Gutenberg, the first printer in Europe.

Site Office: Verkehrsverein Mainz e. V.
Bahnhofstrasse 15, Postfach 4140
D-6500 Mainz, Rhineland-Palatinate
Germany
(6131) 23 37 41

The name "Mainz" derives from "Castrum Mogontiacum," the name of the Roman army camp established on the site over 2,000 years ago. The fact that this name in turn derives from that of Mogon, a Celtic god, indicates that there was human settlement here even earlier. Two sandstone statuettes of women and other artifacts from the Old Stone Age have been discovered in Mainz. During the New Stone Age, the fertile terraces above the Rhine were planted with crops; and there is evidence that the peoples of the Hallstatt and La Tène cultures, probably Celts, also settled on both banks of the Rhine.

By the time the Romans first crossed the Rhine, in 38 B.C., the Celtic people known to them as the Germanii were living on its banks. The army camp, housing two legions, was established 15 B.C., and became the base for campaigns against the Germanii. At one time, the camp was run by Drusus, a stepson of the Emperor Augustus, who is said to have come to Mainz to attend his funeral. Later, Mainz became the capital of the Roman province of Germania Superior (upper Germany), around the end of the first century A.D., and of Germania Prima (Germany No. 1) in the year 300. By then the city was protected by stone walls and had docks on the river and a stone bridge over it. Soon afterwards, the first Christian bishop of Mainz was appointed, beginning the city's long history as a center of religious worship and ecclesiastical administration.

The Romans abandoned Mainz as the legions were called home during the first half of the fifth century, but they left a great deal behind. Parts of their city walls and their aqueducts are still standing. The Roman ruins in the grounds of the Zitadelle, a fortress built between 1659 and 1661 include the Eichelstein, probably a monument to Drusus. Reconstructions of the Jupiter Column, from around 58 A.D., and of the third-century Arch of Dativius Victor stand in the city center. The originals are in the Landesmuseum (Middle Rhine Regional Museum), which also displays the remains of four boats and a sailing ship, uncovered on the riverbank in 1981 and 1982. The Römisch-Germanisches Zentralmuseum has exhibits covering the whole range of Roman activities in northwest Germany over 500 years. The Romans also left a tradition of vine-growing and wine-making, which has played a crucial part in the economy and culture of the region ever since.

Vandals, Alamans, Huns and other migrating peoples are known to have passed through the area, albeit during centuries for which little evidence remains. The recorded history of Mainz starts again in the middle of the sixth century, when Bishop Sidonius is said to have re-established the Christian church in Mainz. The church was most certainly in existence in 747, when the English missionary St. Boniface was appointed as archbishop. Boniface was under the protection of the Franks. Their conquest of most of what are now Germany, the Low Countries, France, and northern Italy culminated in the crowning of their leader, Charlemagne, as Emperor of the West in 800. From then on, the archbishops of Mainz were also heads of the Christian church in Germany, sending out missionaries throughout the ninth and tenth centuries and governing the city and its hinterland. By the middle of the ninth century, Mainz, located on the busiest commercial routes between northwest Europe and Italy and between France and Asia, was trading in spices, slaves, and other luxuries. It was probably the largest city in the German part of Charlemagne's empire, which was divided after his death.

The archbishops sometimes had to suppress rebellions by the city's merchant leaders; at other times they even led troops into battle. During the 880s, Archbishop Liutbert helped to keep the invading Vikings at bay; during the 890s, his successor, Sunderolt, died while fighting them. The archbishops also developed close relations with the kings of Germany. Mainz was a center of scholarship, best represented by Hrabanus Maurus, archbishop from 847 to 856 and a prolific commentator on the Bible and writer of religious treatises. This scholarship alone allowed the archbishops to serve as advisers in a society where even most landowners, let alone the peasant masses, were illiterate. From 965, the archbishops of Mainz were automatically arch-chancellors of what was to become the Holy Roman Empire, a largely honorary but usually influential position. In addition, they were among the select group of rulers who were given the title of elector (*Kurfürst*) and the duty of choosing the emperors, at meetings in Frankfurt am Main, which they chaired from 1257 onward.

Mainz thus became not only the religious capital of Germany but also the political capital of a state, known as Electoral Mainz, comprising the city itself and its hinterland,

A view of Mainz
Photo courtesy of Mainz Tourist Association

known as the *Unterstift* (lower electorate) and several other separate areas, mostly in northwest Germany. As a *Reichsfestung* (imperial fortress), the city received funds from the emperors to maintain its fortifications. Known as *Aurea Moguntia* or Golden Mainz, the city was the site of a number of meetings of the Reichstag (the imperial parliament) during the Middle Ages. From 1254 Mainz was also a leading member of the League of Rhenish Cities, created to defend merchant interests, especially in the river trade, against the noble landowners.

The magnificence of the city's major buildings reflected its status. Its oldest religious building, the Johanniskirche, was begun around 900 but was replaced as the main church of the city under Willigis, archbishop from 975 to

1011, who began the building of a Romanesque cathedral, dedicated to Sts. Stephen and Martin of Tours, in 1009. The Cathedral was renovated and expanded, in red sandstone, beginning in 1100 and further extended between 1190 and 1319, although its central tower, begun in 1239, did not reach its present height of 270 feet or acquire its ornate baroque crown until 1774. Three Gothic churches were also built in Mainz: St. Quintin's, begun in the thirteenth century; St. Stephan's, built in the fourteenth century, which now has stained glass windows by Russian artist Marc Chagall, installed in 1978; and the Carmelite Church, founded around 1360.

Golden Mainz also achieved prestige as the home of Johannes Gensfleisch Gutenberg, the first person in Europe to print books using movable type (although whether he was an independent inventor of the process or had adapted it from Chinese traditions is still open to debate). Between 1440 and 1450, Gutenberg printed a number of religious texts, including his world-famous Latin Bible and the Psalterium Moguntinum, an edition of the Psalms. The Gutenberg-Museum, inside a baroque building completed in 1664, has exhibits on the history of printing, a copy of his Bible, and a reconstruction of his workshop.

While Mainz was flourishing, it was involved in numerous disputes over trading rights and landholdings with neighbouring states. The Palatinate (Pfalz) region, ruled for six centuries by the Bavarian Wittelsbach family, was a frequent rival. In 1273, both states were allies of sorts when they joined in the campaign led by Emperor Rudolf I of Habsburg to suppress the minor nobles who used their Rhineland castles as bases for levying illegal customs duties on travelers. However, they ultimately faced off in an all-out war between 1459 and 1461. Mainz surrendered and then plunged into a civil war between supporters of two rival archbishops, Diether of Isenburg and Adolf of Nassau. Adolf had been appointed by the pope to replace Diether and was eventually victorious. A side effect of this conflict was that in 1462 Mainz, which had been declared bankrupt in 1429 and again in 1456, lost its status as a free imperial city. The city's right of immediate access to the emperor was revoked and Mainz and was "mediatised," that is, taken over, by Archbishop Adolf. Diether was permitted to succeed Adolf after the latter's death and is primarily remembered as the founder of the University of Mainz, established in 1477.

Trade on the Rhine was disrupted by these conflicts until 1468, when delegates from both the Palatinate and Mainz reduced their tolls on the river traffic. After mediatisation, the city benefited from the archbishops' insistence on their "right of staple," meaning that any goods transported through the city had to be offered for sale there before going farther. Since religious institutions were exempt from taxes, merchants trading on behalf of the archbishops or the many local monasteries had extra advantages. However, the rise of Frankfurt am Main as a rival commercial center and the general decline of the Rhine trade in the sixteenth and seventeenth centuries brought about a decline in the city's prosperity.

After Martin Luther initiated the Protestant Reformation in 1517, the Electoral authorities banned his writings and expelled his followers, but they could not hold off the peasant armies who claimed to be inspired by Luther (although he repudiated and condemned them) and who invaded the city during the Peasants' War of 1524–25. Mainz surrendered on May 7, 1525, and it briefly appeared that the city would become a center of social revolution, until the nobles and the cities united to crush the peasantry and restore the feudal order. After the rebellion, Mainz was embellished with a number of new features: a fountain erected by Archbishop Albrecht of Brandenburg in 1526 to mark Emperor Charles V's victory at the Battle of Pavia; statues of St. Willigis, St. Boniface, and the Virgin Mary (the last nicknamed the *Schöne Mainzerin* beautiful woman of Mainz), installed in the cathedral to indicate the continuity of the Archbishopric; and, in 1618, a new building, the Domus Universitatis, to house the university, which was placed under the control of the Jesuits in 1561. The lingering prosperity of Electoral Mainz is also suggested by surviving townhouses of wealthy citizens such as the Knebelscher Hof, built in 1598 or the Schönborner Hof, built between 1647 and 1673.

Mainz renewed its defenses during the religious upheavals of the period, building the Altes Zeughaus (old arsenal) in 1604, strengthening the city walls, fortifying the Kurfürstliches Schloss (Electoral Palace), begun in 1628, and creating a standing army. The army, however, never had more than 3,000 troops. Nevertheless, during the Thirty Years War from 1618 to 1648, the city was occupied first by the Swedes from 1631 to 1636 and then by the French from 1644 to 1648. Its population fell by one third to one-half of its previous level through violence, disease and emigration.

In 1658, Electoral Mainz joined the League of the Rhine, which brought many of the northwestern states, Catholic and Protestant alike, under the protection of France. The alliance did not last beyond 1668, when many of the members realized that France mainly wished to keep them neutral during its invasion of the Spanish Netherlands (now Belgium and part of northern France). During 1688 and 1689, the city was once again occupied by the French.

The eighteenth century saw one last revival of Electoral Mainz, concentrated as before in its capital city. The baroque buildings that still dominate much of the city center were constructed during this time, although almost all are reconstructions, following Allied bombing during World War II. The architect Anselm Franz von Grünstein added new wings to the Electoral Palace and, between 1730 and 1738, built the Deutschhaus, the former headquarters of the Order of Teutonic Knights which has been the meeting place of the *Landtag* (state legislature) of Rhineland-Palatinate since 1950. Other baroque buildings from these years include the Neues Zeughaus (new arsenal), completed in 1740 and now the headquarters of the state government; the church of St. Peter which has two "onion" domes; and the Erthaler Hof, the Osteiner Hof, and the Bassenheimer Hof. This era also saw the construction of other residences of the Imperial Knights (Reichsritter) whose country estates lay near the city.

By 1789, the city of Mainz had a population of at least 30,000 and was experiencing a minor boom based on the trade in wine and other produce from the surrounding countryside. Yet the city still had an essentially medieval economic structure, dominated by guilds and by the twenty-seven religious houses of monks, nuns, and priests. In 1781, Archbishop Friedrich Karl von Erthal abolished three of the monasteries and gave their wealth to the university, which was freed from Jesuit control in 1784 and encouraged to add history and politics to its traditional teaching of theology, philosophy, medicine, and law.

It was Archbishop von Erthal who unwittingly began the destruction of Electoral Mainz. In 1790, he had to call for aid from neighboring states to suppress an outbreak of violence between university students and city apprentices. In 1791 he expelled the radical university professor, Anton Dorsch, who went to Strasbourg to join in the French revolution. Thus, von Erthal made it obvious that his state was both opposed to the revolution and too weak to fight it alone. In 1792 the city was invaded by 13,000 French soldiers who easily defeated the Electoral troops. Professor Dorsch and other German radicals, organized as the Society of the Friends of Liberty and Equality, or the Mainz Club, organized elections in February, 1793, to a Rhenish-German National Convention. The convention was boycotted by most of the citizens. The convention, predictably, voted for union with France in March, but in June, Mainz was bombarded by the Austrian and Prussian armies, who drove the French out, abolished the short-lived Mainz Republic and restored the archbishop in July.

In 1794, the French besieged Mainz again, having conquered the rest of the German left bank of the Rhine, and Archbishop von Erthal fled to his residence at Aschaffenburg, forty-eight miles away. From 1797 to 1814, Mainz, under its French name Mayence, was the capital of the French dependency of Donnersberg or Mont Tonnerre (thunder mountain). Its university was abolished in 1798 and its legal and commercial codes were revised to eliminate the influence of the former Imperial Knights, the guilds, and the religious houses. Meanwhile the Archbishop, the eighty-fourth in succession, died in 1802; the Archbishopric and the Electorate of Mainz were both moved to Regensburg in 1803; and Emperor Francis II abolished the Holy Roman Empire in 1806.

After the second overthrow of Emperor Napoléon of France in 1815, the city of Mainz was incorporated into the Grand Duchy of Hesse-Darmstadt, which was itself absorbed into the Kingdom of Prussia during the process of unification that ended with the foundation of the German Empire in 1871. During the nineteenth century, industrial suburbs grew up around Mainz, producing glass, chemicals, cement, machinery, and books, all of which are still leading industries in the area today. Some of the problems created by industrialization were addressed by, among others, Bishop Emmanuel von Ketteler, a Christian socialist who died in 1877 and for whom the city's Ketteler Institute is named.

The expansion of Mainz required the removal of the medieval fortifications in 1904 and later involved the extension of the city boundaries to take in various suburbs in 1930, 1938, and 1969. The suburbs on the right bank of the Rhine were separated in 1945.

After the defeat of the German Empire in World War I, Mainz was occupied yet again by French troops, who used the former Domus Universitatis as the headquarters for their occupation of the Rhineland from 1918 to 1930. The resentment that this caused contributed to the (varying and still debated) levels of support that the Nazi regime received from Mainzers between its inception in 1933 and its total defeat in 1945. Mainz was designated as the capital of still another French occupation zone in 1946. A plan by the French architect Marcel Lods and the German architect Adolf Bayer to widen the streets, reduce the number of buildings in the center, and then surround it by skyscraper apartment blocks, was strongly opposed and ultimately abandoned after the occupation ended in 1949. Mainz was reconstructed to resemble its prewar layout and appearance as closely as possible. Even the university was refounded, as Johannes Gutenberg University, in 1946.

In 1950, the government and legislature of Rhineland-Palatinate, one of the new states created in 1945, were transferred from Koblenz to Mainz. In addition to the industries already mentioned, contemporary Mainz, a city of more than 175,000 people, is also a center of the computer software industry and houses the offices of three television companies: the regional branch of Südwestfunk; the national headquarters of Zweites Deutsches Fernsehen (ZDF), a station controlled by the sixteen states; and SAT-1, Germany's first private television station, established in 1985. After centuries of rule by archbishop-electors, as well as periods of foreign occupation, imperial control, and brutal dictatorship, Mainz, the city of a long-forgotten Celtic god, retains its Roman, medieval and Baroque heritage in a liberal democratic and technologically advanced setting.

Further Reading: There is usually some information on both the city of Mainz and the state of Electoral Mainz in general histories of Germany, but there are few up-to-date specialized studies in English. Henry J. Cohn's *The Government of the Rhine Palatinate in the Fifteenth Century* (Oxford and New York: Oxford University Press, 1965) belies its restrictive title by covering the relations between the Palatinate and Mainz through the Middle Ages. T.C.W. Blanning's *Reform and Revolution in Mainz 1743–1803* (Cambridge, and New York: Cambridge University Press, 1974) prefaces a sardonic account of eighteenth-century Mainz with a survey of the city's history from Drusus onwards.

—Patrick Heenan

Malbork (Elbląg, Poland)

Location: On the Nogat River, the eastern branch of the Vistula River, about twenty-five miles southeast of the port city of Gdańsk in northern Poland.

Description: Springing up around Malbork Castle, Malbork functioned as the capital of the Order of Teutonic Knights from the early fourteenth through the mid-fifteenth century.

Contact: Centrainy Osrodek Informacji Turystycznaj
Ulica Heweliusza 25133
Gdańsk, Gdańsk
Poland
31-43-55

Malbork played a pivotal role in conflicts involving Poland, Lithuania, and the Teutonic Order from the thirteenth to the fifteenth centuries. Founded by the Teutonic Knights, who had been invited into the region by Polish nobles requiring military assistance in their battles with pagan tribes, Malbork served as an important military outpost for more than 150 years. From 1309 until 1467, it was also the residence of the Grand Masters of the Teutonic Order. Malbork Castle, as William Urban writes in *The Prussian Crusade,* is "one of the largest and finest castles in the world."

In 1226 Duke Konrad of Mazovia, a province in northern Poland, invited the Teutonic Knights to join his war against neighboring pagan tribes who were raiding his territory. The Teutonic Order was a German military and religious order that had fought Crusades in the Holy Land on behalf of Christianity. The order was founded in the Holy Land in 1198 by crusaders from the German town of Bremen. With patronage from the pope and the German emperor, the Teutonic Order grew dramatically in power and prestige during the thirteenth century. Grand Master Herman von Salza had even been asked to arbitrate when Danish King Valdemar was kidnapped. In addition to their participation in Crusades in the Holy Land, the knights, in 1211, had been sent into Hungary to assist the Christians there in battles with pagans. When the knights established their own autonomous state within his country, the Hungarian king grew alarmed, and in 1224 he ordered the knights to leave. Konrad's request for assistance, then, gave the order a new chance to set up their own state.

In exchange for the order's military assistance, Konrad promised the knights a share of the conquered territories and the Chelmno region of Poland. Part of the agreement was that the knights would recognize Konrad as their sovereign. This agreement was soon forgotten, however, as the order quickly took power in Prussia and established an independent state of its own, stretching between the Vistula and Nieman Rivers. In alliance with the Order of the Sword—a German order of knights which had already conquered neighboring Livonia and Courland along the Baltic Coast—the Teutonic Knights controlled the only access to the sea from Poland and Lithuania. The two nations tried separately to oust the order, but the battle-hardened knights proved resistant to their efforts.

In 1275 the Teutonic Order began construction of a castle on the Nagot River, the eastern branch of the Vistula River. In 1276 the order received a charter authorizing the construction of a town at the site as well. Unlike most military structures in Poland, which were made of timber and dirt, Malbork Castle (originally called Marienburg and later renamed by the Poles) was built of brick. Building stone was scarce in coastal Prussia, while a brick industry was thriving. The castle as it stands today consists of three separate structures enclosed within double defensive walls of brick and a moat filled with water. These structures are known as the Lower Castle, the Middle Castle, and the High Castle. The Lower Castle, completed in the fourteenth century, includes the Chapel of St. Lawrence, the Main Gate, storage areas, and stables. In general terms, it was where the auxiliary services of the castle complex were housed. The Middle Castle, completed about 1400, contains the Great Refectory (a forty-eight-by-ninety-eight-foot meeting hall) and the Palace of the Grand Master. The palace later became the permanent living quarters of the order's leader. The High Castle, the oldest of all the castle's structures, completed in the mid-fourteenth century, houses the castle tower, a church, bedrooms, the chapter hall (used for large meetings) and a rectangular courtyard. The knights lived and ate in the High Castle.

In 1309 the Grand Master of the Teutonic Order, Siegfried of Feuchtwangen, moved to Malbork from Venice, making the castle his official residence, headquarters for the order, and the repository of the order's wealth. The move was hastened by several developments both within the order and within several southern European countries. Within the order there was dissent from the Prussian knights, who had long been concerned that their interests were not being heard at the highest levels of the order's hierarchy. In 1301, they had forced Grand Master Gottfried of Hohenlohe to resign at a meeting held in Elbing, resulting in a schism within the organization that did not fully heal until Gottfried's death in 1309. At the same time, political troubles in Italy endangered the order's headquarters in Venice. The King of France, in a dispute with the church, had kidnapped Pope Boniface VIII in 1303. In 1307, he arrested the membership of the Knights Templar, a rival military order that had extensive commercial dealings throughout Europe, on charges of heresy. The Templars' wealth was quickly seized by the French king, and the

Malbork Castle

knights were put on trial and condemned to death. Other European nations followed suit, arresting the Templars as heretics and seizing their extensive holdings. The incriminating evidence against the Templars consisted mostly of confessions obtained under torture, and their guilt has been a matter of historical debate ever since.

Watching these developments carefully, Grand Master Siegfried decided to leave Venice. He had no powerful military allies in the area and the Teutonic Order possessed considerable wealth of its own, which might well prove an irresistible temptation for an unscrupulous monarch. Venice had been the order's headquarters only because its location on the Mediterranean allowed sea access to the Holy Land during Crusades. With his move to Malbork, the Grand Master decisively committed the Teutonic Order to fight the enemies of Christendom in the frontier regions of eastern Europe, abandoning further Crusades to free the Holy Land from the Moslems. The order's official policy was that, only if a successful Crusade managed to regain a foothold in the Holy Land, the Teutonic Order would assist in efforts to consolidate the gains.

Although the Teutonic Knights did not keep their bargain with Duke Konrad of Mazovia, they did engage in battle with local pagan tribes that Konrad considered to be his en-

emies. These tribes, including the Pogesanians and Scalovians, had been attacking Christian areas of northern Poland, burning castles and settlements, and using captive men and women as slaves or as victims in human sacrifices. Throughout much of the thirteenth and fourteenth centuries, the Teutonic Knights fought these tribes. Many pagan prisoners were resettled in the Malbork area, where they could farm under the watchful eye of the grand master himself. As the knights won victories over local pagan tribes, they annexed new territories until Prussia was a formidable power in the Baltic region. This prominence in the area was augmented by Teutonic estates in Germany and Italy, which supplied the order with fresh men and income.

King Wenceslas of Poland died in 1305, and his son was murdered shortly thereafter, leaving no heir to the throne. An intense struggle for the crown broke out between Duke Henry of Silesia and Duke Władysław of Kujavia. In the power vacuum, the northern province of Pomerania fell into chaos as the dukes of neighboring Brandenburg claimed Pomerania as their own. Aided by rebel Pomeranians hoping to gain power for themselves, the Brandenburgers seized most of the province in 1306. Only the citadel at the city of Gdańsk held out against the rebels. Twice the Gdańsk military officials asked Władysław—who by then had won the battle for the crown—for reinforcements, but he could not afford to send any. Finally he suggested they call upon the Teutonic Knights for assistance. Teutonic Grand Master Eberhard sent a garrison to Gdańsk to drive the Brandenburgers out of the city. The knights did more than that however; in 1308 they seized control of Gdańsk and then recaptured the entire province, attacking every fortress that did not immediately surrender to them. When the bloody campaign was finished, Eberhard presented King Władysław with a bill of 10,000 marks, expecting to be paid for his mercenary services. But Władysław refused to pay, partly because he did not have the money and partly because he felt the Teutonic Order owed him the service as king. Realizing that Władysław was unable to muster an army because of his tenuous hold on the newly won throne, Eberhard decided to keep Pomerania for the order. The decision left a lasting rift between the Polish monarchy and the Teutonic Order which eventually led to the order's demise.

In 1386 a union between Poland and Lithuania was achieved with the marriage of Lithuanian Grand Duke Jagiełło and the Polish princess Jadwiga. Lithuania had been searching for an ally in its struggle against the Teutonic Order; a Lithuanian overture to the Russian kingdom of Muscovy led the Polish to propose a merger of their own. The Lithuanians hoped through this union to gain a military ally. More importantly, by joining a Catholic nation and publicly converting to the religion, the pagan Lithuanians hoped to take away the Teutonic Order's primary justification for a holy war as well. Lithuanians not only converted to Catholicism, but adopted many Polish customs and institutions also. Their aristocracy even adopted the Polish language. The new nation, called the Commonwealth, was ruled jointly by the royal couple, each one overseeing his or her homeland.

In 1409 a revolt among subjects of the order in the province of Samogitia, aided in part by neighboring Lithuania, led to a war between the Teutonic Order and the Commonwealth. The Grand Master Ulric von Jungingen, labeling the war as a fight against infidels and pagans, called on knights throughout western Europe to rally to his defense. In the last gathering of its kind, knights came to fight with the Teutonic Order from all over Europe.

The pivotal battle of the war was staged on July 15, 1410, in fields near the city of Grünwald (also called Tannenberg and now known in Polish as Stębark). Here, in a day-long struggle, the Polish and Lithuanian armies, aided by a few Tatar troops, bested the Teutonic Knights in what is often considered one of the largest battles of the Middle Ages. The Polish-Lithuanian forces, led by King Jagiełło of Poland and Grand Duke Witold (whose name is also given as Vytautas) of Lithuania, at first suffered a setback as the knights ran them from the field. But the second stage of the battle ended with the Polish nobles crushing the knights in an overwhelming victory. Grand Master von Jungingen and many of the order's top officers were killed on the field of battle.

Following the battle at Grünwald, towns in Pomerania and Prussia quickly surrendered to Polish troops. It seemed that the Teutonic Order's territories would be divided between Poland and Lithuania. But in the castle at Malbork, Henry von Plauen held out against a siege. The exhausted Polish and Lithuanian troops, plagued by disease, were forced to lift their siege of the castle in September of 1410. Von Plauen then rallied the remaining Teutonic Knights and reconquered much of the territory they had lost. In 1411 the peace treaty of Torun gave Samogitia to the Poles and Lithuanians, but only until the present rulers died; then the territory reverted to the Teutonic Order again.

Despite the unsatisfactory conclusion of the 1409–10 war, the conflict nonetheless marked the end of the Teutonic Order as a formidable military threat to Poland or Lithuania. In addition, the Teutonic Order lost its credibility with the Catholic Church as a crusader against pagan tribes on the frontiers of Europe. At the Council of Constance, held from 1414 to 1418, the Poles were charged with having joined with pagan Lithuania in a war against a Christian military order. However, the Poles successfully rebutted the charges by arguing a case that conversion by military force was immoral. When the council accepted the argument, the Teutonic Order no longer had the church's blessing for waging war against non-Christians.

Following the war of 1409–10, sporadic engagements between the powers and increasingly frequent uprisings of the Prussian population eventually led to a new war in 1422. This conflict resulted in the order giving up its claim to Samogitia altogether. Continued deterioration of Teutonic rule led in 1440 to the founding of the Prussian Union, a group of towns and gentry wishing to protect their interests in an increasingly chaotic situation. In 1454 Poland and Lithuania were asked by the union to intervene against the Teutonic Order on its behalf. At the same time, the union pledged its allegiance to the Polish monarchy and King Casimir IV declared Prussia a Polish state. This intervention led to the outbreak of the Thirteen Years War. Despite initial losses at Chojnice in 1454, which lost them the area around Königsberg, the Poles won major battles in 1462 under Piotr Dunin and a mercenary army and a sea battle in 1463. Malbork was besieged in 1457. That same year it was sold to the Polish crown by mercenary troops that the Teutonic Order had hired but failed to pay. The war ended with the Treaty of Torun in 1466. Under the terms of the treaty Poland received western Prussia, including Malbork, Gdańsk, and Chelmno, while eastern Prussia was left to the Teutonic Order as a fiefdom of the Polish monarchy. The grand master was made a vassal of the Polish crown. Although the Teutonic Order lingered on for a time, it was essentially finished. At the time of the Reformation, the order was transformed from a religious organization into a hereditary fief by the last grand master, Albrecht Hohenzollern of Ansbach. In 1605 it passed into the hands of the Hohenzollern family of Brandenburg.

From 1466 until 1772 Malbork was an official residence of the Polish monarchy. It also served as the capital of the province of Pomerania. In 1772 King Frederick the Great of Prussia, seeking to increase his territory, captured the city and turned the castle into warehouses and barracks. By the early nineteenth century, the castle's condition had deteriorated to the point where demolition was seriously considered. A movement to restore the site, however, gained enough momentum to save the historic complex. In 1887 Conrad Steinberg was appointed to oversee the restoration. His efforts on the project continued until the outbreak of the World War I. In 1945, Malbork Castle was seriously damaged by retreating German troops. A fire in 1958 caused yet further damage. Reconstruction of the castle and Malbork's Old Town by the Polish Ministry of Culture began in 1959 and was only finished in the 1980s. Today, Malbork Castle is a museum housing collections of medieval weapons and valuable amber. It is one of the best examples of medieval defensive architecture still in existence.

In Polish literature, Malbork enjoys a romantic association with the glorious battles of medieval knights. Nineteenth-century Polish writers such as Sienkiewicz and Gustaw Freytag wrote historical novels based on the battles between the Teutonic Knights and the Polish nation. Malbork figured prominently in these propagandistic tales. In Sienkiewicz's *Krzyzacy,* for example, the Teutonic Knights dress in black armor and are depicted as evil incarnate, while the Polish hero Zbyszko stands for the good and virtuous. In this novel the German Kaiser makes an appearance at the order's headquarters at Malbork, as, in fact, he was wont to do in real life, surrounded by his loyal courtiers and ladies of the court.

Further Reading: *A History of Poland* by O. Halecki (New York: David McKay, 1976; London: Routledge and Kegan Paul, 1978) gives a thorough account of the struggles between Poland and the Teutonic Order, while *The Prussian Crusade* (Washington, D.C.: University Press of America, 1980) and *The Livonian Crusade* (Washington, D.C.: University Press of America, 1981), both by William Urban, give an overview of the history of the Teutonic Order in the Baltic region. *The International Dictionary of Architecture,* edited by Randall J. Van Vynckt, (Detroit and London: St. James Press, 1993) provides a short history of Malbork Castle itself.

—Thomas Wiloch

Melk (Lower Austria, Austria)

Location: Located in the region known as Lower Austria, Melk is a small town approximately fifty miles west of Vienna. The Melk Abbey is located on a rocky bluff overlooking the Danube River.

Description: A small town of approximately 5,000 inhabitants, Melk is dominated by and primarily known for the imposing buildings of the Benedictine abbey perched on a rock ledge above the Danube. The architecture of the Abbey Church and particularly the beautifully preserved building interiors are considered to be among the finest examples of high Austrian baroque.

Site Office: Tourismusbüro
Babenbergerstrasse 1
A-3390 Melk, Lower Austria
Austria
(2752) 2307–32

In the beautiful Danube Valley of Lower Austria, high on a rocky bluff overlooking the river, stands a commanding and impressive structure, the Benedictine abbey of Melk. Considered by many to be the architectural high point of the high Austrian baroque, the abbey at Melk consists of a group of buildings surrounding the dominating central structure, the Abbey Church. Visited by thousands of people every year, the art and architecture of these structures have been carefully preserved and tend toward the lavish and overdone. High Austrian baroque is characterized by a style known as rococo. Mainly an interior style, rococo is considered a continuation and modification of the earlier baroque style and is filled with a sense of theatricality. In the case of the Abbey Church at Melk, the interior was actually designed by the Viennese theatre architect Jakob Prandtauer.

Melk marks the start of a twenty-mile stretch of the Danube known as the Wachau. The Wachau is famous as a wine-producing region and is considered by many to be the most beautiful stretch of the famous Blue Danube. Also on this stretch of the river one can see the ruins of castles perched precariously on the cliffs above. This whole experience is best viewed from the deck of a paddleboat steamer, and that is certainly the best way to approach the Monastery of Melk. It is from the river that the abbey is most impressive, and clearly the architects had that vista in mind.

The modern history of this site goes back more than 1,000 years. It is first mentioned as the site of the Roman fort known as Namare. The history of the transformation of Melk from a mountain stronghold to the seat of the first Christian rulers of Austria is somewhat unclear. It has long been believed that sometime in the tenth century Melk was established as the residence of the ruling Babenburg family.

Shortly after the arrival of the Babenburgs at Melk, a group of Benedictines was invited to establish a convent. The arrival of the Benedictines in 1089 marked the start of the influence of this order at the site, an influence that continues today. Thus, in its early years, Melk served the threefold purpose of cloister, court, and citadel; perhaps it is that unique foundation that has made the site such a rich source of art, architecture, and music.

During the first decade of the twelfth century, the Babenburgs moved their court from Melk to Klosterneuburg. When they left, the ruling family granted to the Benedictines not only the residence, but also vast tracts of surrounding lands, and thus Melk became Stift Melk, from the German word *Stiftung,* which means endowment. During this same period the abbey was granted an exemption from Rome. This meant that the abbey answered directly to the pope and was not subject to local ecclesiastical supervision. Although these exemptions became increasingly common in later periods, Melk was one of the first to receive one.

Melk was somewhat neglected during the early years after the departure of the ruling family. A small community grew around the abbey, and the town was granted market rights in 1227. The town was completely destroyed by fire in 1297 and was not rebuilt until 1307. As a center for the Benedictine order, Melk's influence was felt throughout Austria and, indeed, Europe. After a period of great expansion, the Benedictines undertook a significant tightening of discipline within their order in a move that has come to be known as the Melk Reform, which took place in 1418. For the next two decades, Melk continued its role as a cultural and religious center. In spite of repeated serious fires, the records indicate that unlike other religious enclaves Melk did not suffer from financial problems. It appears that the original endowment and the lands belonging to the abbey were sufficient to provide for both its needs and for considerable growth.

Increasing hostilities between the Ottoman and Holy Roman Empires reached their climax when, in 1683, the armies of the Turkish sultan marched into Austria. For two months the Turks laid siege to the countryside. The abbot of Melk made the decision to arm both his monks and the local citizenry to protect the abbey. In spite of the successful defense of the abbey and the eventual retreat of the Turks, the properties and farms surrounding the abbey stood in ruins.

During this tumultuous period the abbey also suffered extensive damage due to a fire. The abbot of Melk from 1679 to 1700 was Gregor Müller, and he is widely credited with successfully leading the abbey through this trying period and seeing to the repair and rebuilding of the abbey and its property. However, it was the abbot who succeeded Müller who is the most famous of Melk's Benedictine leaders. Berthold

Melk Abbey
Photo courtesy of Austrian National Tourist Board

Dietmayr served as abbot of Melk from 1700 to 1739, and was elected to the position when he was only thirty years old.

The apparent end of the Turkish threat and a sense of renewed confidence throughout all of Austria made this a period of rebuilding throughout the region. Immediately upon becoming head of the abbey, Dietmayr commissioned architect Jakob Prandtauer to refashion and replace the structures at Melk in order to create a unified whole. The construction of the new Abbey Church clearly symbolized the important and dominant role that Melk had achieved by the eighteenth century.

Dietmayr and Prandtauer worked together in the design of the abbey, and in 1702 the work was begun. Prandtauer died in 1726, but the work continued under the direction of his student Franz Munggenast. Parts of the new monastery were destroyed by fire in 1835, but Munggenast rebuilt the destroyed portions and in particular left his mark on the newly constructed towers. Most of the project was completed by 1740, and the current appearance of the monastery dates from that period.

The eighteenth century saw the abbey reach its high point and it also saw the most serious challenge to its continued existence. Abbot Dietmayr was a personal friend of two Austrian emperors and was in large part responsible for bringing Melk to its height of glory. However, the reign of Emperor Joseph II near the end of the century was marked by his intolerance for what he believed were the rather idle lives of monks. Many abbeys were dissolved and the number of monks at Melk itself was greatly reduced. During the last years of the century Melk saw its school transferred and theological seminary dissolved. In spite of the movement against monasteries throughout Austria, Melk was somehow able to survive. Its survival is surely related in no small way to its unique historical status and its close relations to the crown.

In succeeding generations the schools at Melk were restored, and the structures and site itself remained central in the history of Austria and Europe. In 1805 and 1809, Napoléon I stayed at Melk and established his headquarters there in his campaign against Austria. Perhaps Melk's most enduring legacy during the past two centuries has been the preservation of its art treasures and structures, and its continued role in the education of generations of Austrians.

Some have described the abbey at Melk, high above the beautiful Danube, as the first example of organic architecture because its buildings are so closely related to the actual landscape in which they were constructed. Seemingly a continuation of the rock cliff on which it sits, Melk seems to be an essential part of the surroundings. Certainly much of the credit for the remarkable appearance of Melk goes to Prandtauer, who clearly had a sense of the theatrical from his experience as a theatre architect and designer in Vienna. Melk is his masterpiece, and this glorious structure seems to command the entire Wachau.

The abbey buildings run from east to west, with the most ornate facade on the Danube side. However, visitors enter the abbey from the opposite side, where the outer gateway that leads to the inner court is flanked by statues of St. Leopold and St. Coloman, the patron saints of the abbey. Within the gates and beyond the vestibule lies the Court of Prelates. Lined with statues representing the prophets, the court gives way to a fountain, with the church visible beyond.

Among the most significant of the structures at the site is the emperor's Gallery. Originally used to house guests, many of the chambers have now been converted to museum galleries that contain historical items from the abbey. The Marble Hall is another important building known primarily for its ceiling painted by Paul Troger. Completed in 1732, this allegorical painting is entitled *Reason Guiding Humanity from the Darkness of Obscurity toward the Light of Civilization and Culture*. Opposite the Marble Hall stands the library, among the most important for scholars of early Christianity. It houses more than 80,000 books and 2,000 manuscripts, many dating from the early years of the abbey. The library also has a ceiling painted by Troger, and at its entrance stand four gilded statues representing the four faculties.

The pinnacle of Melk is without question the stunning Abbey Church. With symmetrical towers at the west end and a great octagonal 210-foot-high dome, the exterior is indeed impressive, but it is the interior for which the church is known. The color scheme of the interior is dominated by reddish-brown, grey, and orange combined with gold ornaments and marble. The interior of the church is extremely light both because of the color scheme and the many windows.

Throughout the nave and into the dome and altars, the church is filled with painted and sculpted figures. Within the dome itself one can see the Father and the Son as well as the Evangelists. In the middle of the high altar are gilded statues of St. Peter and St. Paul, to whom the church is consecrated. On the vaulting of the nave the figure of St. Benedict is depicted being received into heaven. The vaulting throughout the church is intended to be a vision of the heavens, and one's eyes are drawn upward to this vision by the red marble columns and pilasters that adorn the walls of the structure. Characteristic of the rococo style, the decoration of the church is replete with undulating lines and artistic planes that are broken by figures who appear to be moving from one area to another.

One of the most striking elements of the church is the organ gallery, which is filled with gold pipes bound together with garlands of silver laurel leaves. The gallery is also filled with delightful carved cherubs playing musical instruments. Melk was an important center for the development of religious music, and its library is filled with manuscripts and records attesting to the prodigious musical output of the abbey.

In addition to its beautiful architecture and interior decoration, Melk is known for three important relics. The first is the *Melker Kreuz*, a processional cross created to house a fragment of the True Cross. This artifact dates from 1362 and is the work of a Viennese goldsmith. The cross is gilded in silver and covered with precious stones, pearls, and a Roman cameo. The second treasure is the *Tragaltärchen Swanhilds,* which is a small, portable altar made of ivory. This altar apparently belonged to Swanhild, who was the wife of one of the early rulers residing at Melk. It is considered the oldest item definitely belonging to the Babenburgs. The third treasure, which dates from the thirteenth century, is a female head reliquary with crown.

Notable both for its setting and its design, Melk is an architectural jewel in the Danube Valley. With a long history tied closely to the monarchs of Austria and the Catholic Church, the Benedictine abbey is a fine example of the fusion of arts intended to create mystical-emotional excitement in the wake of the Counter Reformation. Still a functioning church and monastery, Melk is now regarded by much of the world as a museum that symbolizes Austria at its peak. In its current state, the abbey reflects the high Austrian baroque style and bears the imprint of Abbot Dietmayr and architect Prandtauer, who worked to rebuild and refashion the abbey in the early eighteenth century. It is like stepping back in time 200 years to wander through the abbey's grounds and buildings.

Further Reading: Although Melk is consistently mentioned in larger texts about Austria, baroque, or rococo, few books have been published on the site in English. Most of the available scholarship is in German, especially academic articles concerning the immense holdings of the Melk library. Of the books available in English, one of the most interesting is *The Practice of Music at Melk Abbey* by Robert N. Freeman (Vienna: Österreichischen Akademie Der Wissenschaften, 1989). Although primarily concerned with music, it does contain a concise history and discussion of some of the architecture. A nice introduction to the historical sites of Austria is James Reynolds's *Panorama of Austria* (New York: Putnam, and London: Hale, 1956). Among the many guidebooks that are available, one of the most interesting and complete in terms of art and architecture is *Austria, a Phaidon Cultural Guide,* edited by Franz Mehling (Englewood Cliffs, New Jersey: Prentice-Hall, and Oxford: Phaidon, 1985).

—Michael D. Phillips

Melrose (Borders, Scotland)

Location: On the banks of the Tweed River in the Borders region of southeast Scotland, 37 miles south of Edinburgh.

Description: An ancient town that developed at the entrance to Melrose Abbey, built in 1136. The abbey is the final resting place of several saints and monarchs including Alexander II. Sir Walter Scott lived on a farm here, and his remains are buried in nearby Dryburgh Abbey.

Site Office: Melrose Tourist Information Centre
Priorwood Gardens
Melrose, Borders
Scotland
(89) 682 2555

Melrose has survived even though Melrose Abbey, its original reason for existence, has not. The abbey ruins and town are just south of the Tweed River, on high enough ground to be safe from floods in a land of rolling hills, winding rivers, and green and gold fields in the central Borders region of Scotland, so called because it borders England. Tourism and farming are the main industries.

The area was first occupied some 2,000 years ago, when the ancient tribe of Selgovae had a settlement of 300 huts on top of one of the Eildon Hills to the south. In the first century A.D., the Romans built a signal station and established a frontier post called Trimontium on the lee side of the Eildon Hills, which they occupied for about 200 years. Twentieth-century excavations have uncovered many artifacts of the Roman fort and nearby settlements. Some of the artifacts are displayed at the Ormiston Institute in Melrose.

In 635, St. Aidan started the monastic community of Mailros, about two and one-half east of the current town of Melrose. One of the original monks was St. Cuthbert, whose body was believed to have miraculous powers. His remains were buried at Melrose for a short time, but were moved several times to prevent desecration by invading Danes. St. Boniface, the Apostle of Germany (also known as the Apostle of the Rhine), also spent time at the monastery. Kenneth MacAlpin, king of the Scots, destroyed the abbey in 839, but it was later rebuilt and was used as a retreat for several more centuries. Turgot, who was later confessor to St. Margaret, the patron saint of Scotland, is said to have visited there between 1073 and 1075.

In 1136, King David I of Scotland invited a group of Cistercian monks from Rievaulx in Yorkshire (the colony was originally from Clairvaux Abbey in Eastern France) to come north. They first stopped at Old Melrose, but decided the site was unsuitable, and came farther up the Tweed. The new abbey was founded on March 23, 1136, and the church was consecrated on July 28, 1146. It was the first Cistercian community in Scotland, and was dedicated to St. Mary, as were all Cistercian churches. David I granted much land, along with wood and fishing rights, to the monks.

During the twelfth and thirteenth centuries the various kings continued their patronage. Consequently, Melrose became one of the richest monasteries in the kingdom. Numerous land endowments in Carrick, Eskdale and Teviotdale; property in Edinburgh, Haddington, and Lanark; fishing rights in the Don and Tweed Rivers; salt-marshes at Turnberry and Prestonpans; pasture on the Lammermuirs and a peat-moss in Fairnington provided Melrose with its main sources of revenue. Philip, Count of Flanders, provided protection to the abbey: the community did not have to pay tolls and was protected in war time.

In the early fourteenth century, King Robert I (Robert the Bruce) of Scotland, although excommunicated because of a murder, gave very generously to the abbey and ordered that his heart be buried there. During that tumultuous century, Melrose was sacked more than once by England—in 1322 by Edward II and in 1385 by Richard II, and the abbey was largely destroyed. The ruins at the site today are of the abbey church that was built after that time, primarily from 1385 to 1505.

All that remains of the first church are a fragment of the west wall with the lower part of the entrance doorway, the foundations of the square nave-piers and their connecting screenwalls, and a few pieces of the north wall. The plan for the new church resembled the earlier one, albeit with more elaborate architecture. A series of chapels opening off the south nave were added, and the transepts and presbytery were made larger and extended further east. Rib vaults were constructed, except in the presbytery and the eastern chapels where the vaults were highly ornamental. Yellow, green, or brown tiles set in geometric patterns paved the floor. The existing front was retained. The construction continued throughout the fifteenth century and into the sixteenth. Although eleven successive abbots oversaw the rebuilding, the coats of arms of only two, Andrew Hunter (1444–1471) and William Turnbull (1503–1507), are evident on the church wall.

Local stone was used in building the monastery. Prior to 1385, agglomerate stone was used, obtained from Quarry Hill, southwest of Melrose. Construction after 1385 was of sandstone from the Eildon Hills.

Gradually the abbots became intensely involved in politics, and many of the leading churchmen became statesmen and advisers to the monarchy. This involvement may be the reason that the abbeys became an increasingly strategic element in defense of the realm against the English. For

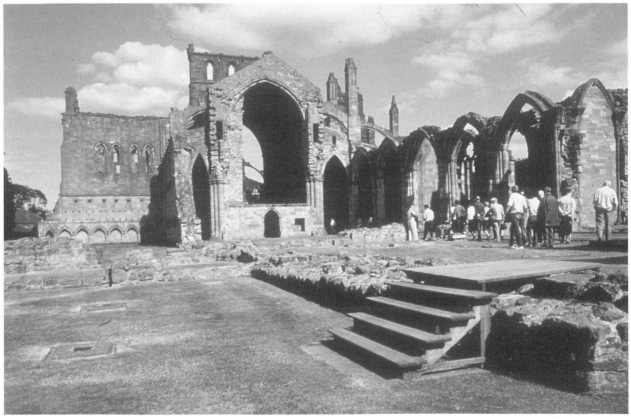

Two views of Melrose Abbey
Photos courtesy of Scottish Borders Tourist Board

instance, in 1496 James IV made the abbey his headquarters during a raid on Northumberland; in 1502, he received the English ambassador there. In a sense, the abbeys became castles.

The Border wars continued into the sixteenth century. A particularly destructive period followed the death of James V in 1542. In 1544 and 1545, Melrose town was set afire, and the church and tombs were pillaged. During the remainder of that century little was done to repair the buildings, and the number of monks steadily declined. When the last monk died in 1590, the story of Melrose Abbey effectively ended.

When an abbot was not in residence, the Pope often appointed a lay person to live at the abbey and reap the benefits of its luxurious services. Such appointees were called commendators. Melrose Abbey's last commendator was James Douglas, who became Earl of Morton, acted as regent for James VI, and was eventually executed for his involvement in the murder of Lord Darnley, husband of Mary, Queen of Scots. Even after the monastery declined, he continued to reside in the commendator's house, and retained part of the ruined cloister. After his death, the lordship of Melrose passed to the Earls of Haddington. Later, Duchess Ann of Buccleuch purchased it.

By the nineteenth century, neglect and frequent plundering had desolated the abbey. The writer Sir Walter Scott convinced the Duke of Buccleuch to undertake a major cleanup and repair of the remaining ruins. The Buccleuch descendants placed the property in the guardianship of the H.M. Office of Works in 1919, and it is now maintained by Historic Scotland.

In the abbey church, eight aisle-chapels survive; each had an entrance through a doorway in a wooden screen. Altars with shelves stood against the east walls, and in the fifth chapel high up on the wall the masonry is scorched where the shelf was burned. The fourth chapel may have been dedicated to St. Michael, as his image can be seen in the vaulting. This chapel is the burial place of the Pringles of Woodhouse and Whytebank, ancestors of South African poet Thomas Pringle. The fifth chapel has been appropriated as a burial place by the Scotts, lairds of Gala, and the Pringles of Galashiels.

The three bays of the nave that had constituted the monks' choir were adapted for use as the parish church in the 1620s. Extensive work was done to make it usable. When a new parish church was built to the west of the abbey in 1810, some of the new walls were taken down and the structure left as it appears today.

The collapse of the central tower damaged both transepts. The vaulting in the south transept is now limited to two bays and an eastern chapel. The south wall and window were not damaged. All the vaulting in the north transept was destroyed except for that in one of the eastern chapels, dedicated to Sts. Peter and Paul. The statutes of the saints still stand on the west wall.

A perpendicular window in the east wall and windows in the north and south walls fill the presbytery with light. An intricate pattern composed of ribs and bosses adorns the vaulted ceiling. A carving of the Holy Trinity and two angels is on the central boss over the spot where the high altar once stood. Carvings of St. Andrew, St. Bartholemew, St. Peter, St. Thomas, St. James the Less and the Greater, St. Paul, and St. Matthias also grace the presbytery walls.

Elaborate carvings of the Virgin, coats of arms, saints, gargoyles, angels, ordinary workers (a cook and a mason), and foliage cover the abbey's exterior. High on the roof is a carving of a pig playing a bagpipe.

The town of Melrose grew up at the gate of the abbey and eventually provided the villages of Darnick, Gattonside, Newstead, Newtown, Danieltown, Eildon, and Blainslie with shops, tradesmen, postal services, doctors, lawyers, bankers and insurance agents. As the center of the parish, it contained the parish church and school.

An attempt was made to develop weaving as a cottage industry in the seventeenth century, but failed because not enough flax was grown locally and using imported flax was not economic. Some woolen weaving was done in the early nineteenth century, but that trade eventually moved to Galashiels.

Tourism began to be an important part of the economy in the nineteenth century. Also, a growing number of wealthy families chose to build homes in and near Melrose because of its beauty and proximity to Edinburgh.

The region has always been famous for its good soil and therefore farming was and is a significant component of the economy. In the nineteenth century, Melrose was one of four market towns in the county, with a market every Saturday. Grain and corn were sold weekly, and periodic cattle fairs were held.

The red stone Town Hall on Abbey Street was built in 1822, and was given to the town in 1896 by the Duke of Buccleuch. The Market Cross in Market Square is dated 1645. Its shaft supports a unicorn bearing the royal arms. To the left of the Market Cross in the East Port section is a house containing a lintel and vaulting boss from the abbey, dated 1635. Another house in the East Port, distinguished by a rounded gable, was a municipal building and home to the sergeant of police until 1855.

The Masonic Lodge on High Street near the Market Square has been on its present site since 1791, and is one of the oldest in Scotland. The George and Abbotsford Hotel, across the road, is a former coaching inn and was mentioned by Sir Walter Scott in *The Monastery*. Another former coaching inn, located on the same side of the street, the Kings Arms Hotel, dates from 1830.

Abbotsford, Sir Walter Scott's magnificent estate, is on the banks of the Tweed three miles west of Melrose. The house was designed in the Scottish baronial style. Scott lived there from 1812 to 1832. Dryburgh Abbey, where Sir Walter Scott is buried, is on the Tweed a few miles to the southeast of Melrose.

The ruins of two other abbeys remain in the Borders area. Jedburgh, south of Melrose, has ruins of an abbey destroyed in 1544–45 by the English Earl of Hertford. Only ground patterns and foundations remain. The Kelso Abbey ruin, on the Tweed River about five miles northeast of Jedburgh, is the least intact of the abbeys. It was largely destroyed in 1545.

Further Reading: *Melrose Abbey* by J.S. Richardson and Marguerite Wood, revised by C.J. Tabraham (Edinburgh: Her Majesty's Stationery Office, 1981; revised, 1989) contains an excellent summary of the story of Melrose Abbey, including the abbey's history and a complete physical description. Two earlier accounts of the abbey are useful primarily because they provide nineteenth-century viewpoints. They are *History of St. Mary's Abbey, Melrose,* by James A. Wade (Edinburgh: Thomas C. Jack, 1871) and *Description of the Abbeys of Melrose and Old Melrose* by John Bower (Edinburgh: self-published, 1827). Insight into the town of Melrose during the nineteenth century is available in *Melrose 1826,* edited by D.M. Hood (Melrose: Melrose Historical Association, 1978). The book is based on maps of Melrose done by John Wood, a surveyor from Edinburgh, in 1826, and additional research by the historical association.

—Patricia Ann Shepard

Mont-St.-Michel (Manche, France)

Location: An islet off the coast of Normandy in the Bay of Mont-St.-Michel near the border with Brittany, about forty miles north of Rennes; in the *département* of Manche in the region of Basse-Normandie; joined to the mainland by a causeway.

Description: Once a peak rising above the forest of Scissy, now surrounded by the sea at high tide and moving sand banks at low tide; a sanctuary for Christian pilgrims dedicated to St. Michael the Archangel, who inspired the Bishop of Avranches, St. Aubert, to build an oratory near its summit in 708; crowned by a Benedictine abbey and church built over a period of more than 500 years; a citadel often attacked but never taken, damaged several times by fire, collapse, or vandalism; a site of frequent restorations, additions, and fortifications according to the needs of the times; a Catholic sanctuary during the Wars of Religion; a government prison after the French Revolution; a national historic monument and popular tourist attraction today.

Site Office: Office du Tourisme
Place Aristide Briand
50116 Mont-St.-Michel, Manche
France
33 60 14 30

Several hundred yards from the Normandy coast a massive granite rock rises out of the bay to a height of 285 feet, as if to defy the earth, the sea, and the heavens. As strong and as courageous as St. Michael himself, the rock and its abbey have protected their citizens for more than ten centuries from the perils of war and of nature. Like an ark for the new era, the abbey crowns the mount as a witness to the past and a symbol of hope for the future.

The islet was not always surrounded by water. Before the eighth century, the region was known as the forest of Scissy. Missionaries came to convert its pagan inhabitants to Christianity. When eventually the high tides of the sea washed away the forest and replaced it with sand banks, they left visible and isolated the three highest points: Tombe, Tombelaine, and Dol. Mount Tombe, now known as Mont-St.-Michel.

By this time people of the Normandy region had moved from the mercurial cults of Gallo-Roman times to a devotion to St. Michael the Archangel. He was God's messenger, a defender and protector of the human race. It is not difficult to see why he was chosen by the Normans to be their patron saint.

A manuscript believed to have been written in the ninth century tells of the origin of the church dedicated to St. Michael. Early in the eighth century, the archangel appeared to Aubert, the bishop of Avranches, designating Mount Tombe as the site for a temple to be built in his name, similar to the one that already existed at Mount Gargan in Italy. The location was remote, accessible only twice a day at low tide. At first hesitant, the bishop began his mission in 708 after receiving several signs that this was indeed what he was supposed to do. The location for the temple had been designated by the presence of a tethered bull, an unusual sight at the top of a mountain. The original oratory, built as a cavern hollowed out of the rock, was large enough to hold about 100 people. The first inhabitants are widely believed to have been canons, and the summit of the mount soon became a popular pilgrimage place. A monk named Bernard described in 870 the sea receding at low tide on the feast of St. Michael to allow the pilgrims to pass through.

In 966 Duke Richard I expelled the canons and founded a monastery as a way of consolidating his territory to the west toward Brittany. The twelve original Benedictines from St.-Wandrille led a contemplative existence, balancing prayer and study, manual labor and intellectual thought, silence and chant. Hospitality and almsgiving became important functions of everyday monastic life.

On the site of the original oratory, the monks built a church consisting of two parallel naves separated by the wall of two arcades. Known today as Notre-Dame-sous-Terre, this church was to become the crypt for the Romanesque church constructed later.

When a fire destroyed the wooden roof of the church in 992, construction began anew, and stone vaults replaced the wood. Duke Richard II's marriage to Princess Judith of Brittany was blessed here in 1017. He must have thought the conditions to be too crowded, because he decided to rebuild the abbey church. Built in early Romanesque style, it borrowed elements from the Romans and the Middle East. The choir was begun in 1020; the transepts and the nave, which crossed at the level of the summit, were completed between 1060 and 1080. Crypts were built as foundations under the choir and the transepts, and the original sanctuary was preserved as a crypt under part of the nave. The shape of the church was cruciform, with a porticoed entrance and terrace overlooking the sea to the west. Originally, the nave had seven bays; the circular choir, surrounded by a simple ambulatory, rose gradually toward the east to a level three yards higher than the level of the façade. The nave was covered by a flat roof with fishbone vaults over the side aisles. The capitals show a minimum of ornamentation. Although the four great pillars of the crossing begun in 1048 are still visible today, the eleventh-century choir collapsed in 1421 when the

Mont-St.-Michel
Photo courtesy of French Government Tourist Office

buttresses proved to be too weak to support the structure. The three westernmost bays of the nave were demolished after being destroyed by a fire in 1776. Only four remain today. The north wall of the nave collapsed in 1103, not long after its construction, destroying the monk's dormitory contiguous to it. It was reconstructed and reinforced under abbots Roger I and Roger II early in the twelfth century.

To accommodate the devout life of the monks, the monastery was built to the north side of the church, outside the mainstream of pilgrim activity, but allowing the abbot easy access to the church and the pilgrims. Visitors were welcomed in the Aquilon room, named for the north wind, on the lower level. The middle level had the Promenoir (promenade), and the dormitory was on the upper level, opposite the entrance to the church.

For the period, the style was ahead of its time. It used the broken arch and intersecting ribs, both imported innovations that were to influence some of France's later Gothic cathedrals. This early church was immortalized in the Queen Mathilda tapestry at Bayeux showing William the Conqueror with Harold before the expedition into England.

To protect against attack and to ensure peace, the abbey was fortified with defense corridors. The death of Wil-

liam the Conqueror had led to struggles for succession among his sons. Henry I took refuge at Mont-St.-Michel and eventually won out over his brothers. His grandson, Henry II, named Robert de Torigni as abbot in 1154. The choice was a judicious one, for Robert turned out to be an exceptional leader.

Robert's abbacy at Mont-St.-Michel was an active one: he wrote the chronicles of Henry II, acquired relics, and settled various disputes among the priories in his district. The scriptorium reached its peak during Robert's time, and the library was enriched to reflect the intellectual and material resources of Mont-St.-Michel during the twelfth century. The continuous flux of pilgrims and their concerns meant that the abbey had to be enlarged to meet their growing needs. As the number of monks increased, Robert doubled the existing structure on the west side, and he built another three-storied building facing south to receive the pilgrims. The abbot's quarters were positioned to allow access to the church, the monastery, the infirmary, and the guest quarters. Bell towers were added to the facade. Unfortunately, Robert's architecture was much too ambitious for the existing foundations, and most of what he built in the twelfth century has since crumbled.

A fire set in 1203 by the Duke of Brittany to the town below caused serious damage to the abbey. King Philip II Augustus compensated the abbey so that repairs could be made. With these funds Abbot Jordan began the construction of the group of buildings on the north face called La Merveille (The Wonder). From 1211 to 1228, while the cathedrals of Notre-Dame de Paris, Reims, and Chartres were still under construction, Frenchmen worked alongside Normans to build La Merveille.

In the tradition of the earlier structures, La Merveille was built on three levels. It was divided vertically by a wall into two independent structures. On the first level there was the reception area on one side, and the storage cellar on the other; on the second level, the inn with a capacity for 200 guests, mirrored by the Hall of Knights; above, the monks' refectory and the cloister.

The reception room and the storage cellar are simple but strong in design, belonging to the early transition style between Romanesque and Gothic. The inn is divided into two naves by a row of elegant columns ending in intersecting ribs. It allowed for the communal life of the pilgrims, which included eating and sleeping in the same large room. The Hall of Knights is also a commons room. It became the intellectual center of the abbey, where copyists and illuminists could work or think in quiet. Ninety feet long and sixty feet wide, it boasts two fireplaces and round windows in the arching of the vaults. Its style is masculine and military.

The third level belonged to the monks. The cloister is a calm place between the sky and the sea. Granite columns are arranged in triangular formation to frame a courtyard. The frieze decorating the gallery contains an image of the peace-loving Saint Francis of Assisi. The refectory is light and airy, due to the huge twenty-foot windows on the side walls. The Benedictines ate their meals here in silence, listening to a reader. The church could be reached by walking through the cloister.

Between the second and third levels off the Hall of Knights is a thirteen-foot-square library with a fireplace, accessible from the cloister by a spiral staircase and from the hall by a few steps.

A visitor viewing La Merveille from below would see fifteen long buttresses reinforcing the interior arches and the walls along the 235-foot length of the structure. The duality of its construction seems to echo the unity of priest and soldier, of secular and spiritual, of church and state that was so important to this sanctuary. All are welcomed within its walls.

The abbey continued to flourish throughout the thirteenth century and to attract pilgrims of all social classes. Even the kings of France came to visit on occasion: Louis IX, his son Philip the Bold, his grandson Philip the Fair, and their descendants all made pilgrimages to the mount. To allow for the increasing number of judicial cases under the abbot's jurisdiction, a new justice hall and court were added to the existing buildings.

As the civil duties of the abbot gradually took up more of his time than did his religious obligations, which he delegated to the prior, he was obliged to move his quarters out of the monastery and closer to the entrance to the administrative section. His lodging had to be simple, but attractive, reflecting his status as a spiritual leader.

Pilgrims continued to journey to the mount in great numbers during the fourteenth century, despite the ever-present threat of turbulence during the Hundred Years War. A system of fortification begun with the financial support of King Louis IX in 1256 was completed. The village below the abbey was protected by ramparts surrounding the islet. To increase the security of the mount, a garrison was provided to fend off any possible attacks by the English. The only entrance to the abbey was through the Châtelet, a model of military architecture, with its defense towers, parapets, and loophole openings. A reinforced wall was constructed below the Châtelet for further protection. Under the abbot Robert Jolivet early in the fifteenth century, the ramparts protecting the village were reinforced, and the abbey was supplied with enough provisions and water to continue its work in the event of an attack.

Although Jolivet surrendered to England, the monks refused to submit to the control of Henry V. They replaced Jolivet with a new leader appointed by the pope. With the help of the garrison and a number of Norman knights who preferred to fight, they successfully resisted the British attacks. Although the English took control of nearby Mount Tombelaine in 1423, they did not take Mont-St.-Michel. Two years later, Breton sailors came to their aid and broke the English blockade of the bay so that provisions could once again be secured. Inspired by the victories of Joan of Arc, those defending the mount turned the tiny islet into a symbol of French resistance.

In 1433 the village below the abbey caught fire. The English took advantage of this opportunity to break through the ramparts and launch a large-scale attack. Once again, they were pushed back by the determined Normans. The attacks and counter-attacks continued until the decisive battle of Formigny in 1450, which put an end to the Hundred Years War.

The successful resistance of Mont-St.-Michel focused even more attention on the powers of St. Michael and the dual function of the mount as a sanctuary and a citadel. The strategic importance of the islet did not escape the king of France, who appointed a new abbot, Guillaume d'Estouteville, to undertake the reconstruction of the buildings. He began with the collapsed choir, using the plan of the eleventh century choir, but adding a series of chapels that radiate from the ambulatory. The result is one of the most successful fusions of Gothic and Romanesque architecture. The simple and energetic style of the nave contrasts with the light and graceful lines of the Gothic choir built 500 years later, but somehow they complement each other. This time the choir was built on a solid foundation and reinforced by flying buttresses on the outside. The result of generations of evolution in the Gothic style, the choir represents the achievement of a nearly perfect form: high, elegant, and full of light that has been gently filtered by the delicate pinnacles of the buttresses.

The next two centuries were marked by a gradual decline in both the physical and spiritual allure of Mont-St.-Michel. Although the pilgrimages continued throughout the religious wars, the role of the mount as a sanctuary was eclipsed by its more temporal function as a fortress and prison of the state. The abbots were no longer elected by the monks, but instead were clergymen who had found favor with the king. Little was done to maintain or embellish the existing buildings. To their credit, the monks imposed the reforms of the Benedictines of St. Maur on themselves in an effort to revive their spirituality. During the French Revolution, however, they were forced to leave. The entire abbey became a state prison, ironically named Mont Libre (Mount Free). The church, neglected and vandalized, was partially burned in the nineteenth century. The scandalous situation lasted until 1863, when public pressure convinced Napoléon III to remove the prisoners and to declare Mont-St.-Michel an historic monument. Restoration of the site continued well into the twentieth century.

Although some of the abbey's manuscripts were lost after the Revolution, many of them were taken to the town library at Avranches where they can be seen today. Interest in Mont-St.-Michel was revived for the millennium celebration in 1966. Soon afterward, the order of St. Benoît, the Benedictines, were invited to return to the mount. They say daily mass in the restored church and conduct frequent retreats for the faithful.

Today's visitors to Mont-St.-Michel, much like their pilgrim predecessors, may find the ascent to the summit difficult, especially during the summer season. More than 2 million tourists visit the monument each year. The narrow streets of the village are crowded and busy with vendors selling souvenirs, and the road winds its way slowly to the top. But one must not be discouraged from making the journey. There is a kind of harmony of granite, sea, and sky that makes the climb well worth the effort. It is to this summit that one comes to find peace.

Further Reading: *Le Mont-Saint-Michel: mille ans au péril de l'histoire* by R. P. Michel Riquet (Paris: Hachette, 1965) is a scholarly, in-depth account in French of Mont-St.-Michel from Gallo-Roman times to the millennial anniversary of the abbey. It includes drawings of the elevation and floor plans at three levels, as well as selections from original manuscripts. More concise and readable is *Le Mont Saint-Michel* by French historian Yves-Marie Froidevaux (Paris: Hachette, 1965). It contains a collection of photographs showing details of the architecture of the church and abbey. *Mont-Saint-Michel and Chartres* by Henry Adams (Boston and New York: Houghton Mifflin, 1905; London: Constable, 1914) contains a delightful description of the mount that explains the difference between the Romanesque and Gothic as similar to the difference between masculinity and femininity. He relates the shift in style to the change in devotion from St. Michael to the Virgin. He also makes a thoughtful comparison of the eleventh century buildings to the *Chanson de Roland*. *Norman Illumination at Mont Saint Michel: 966–1100* by J. J. Alexander (Oxford: Clarendon, and New York: Oxford University Press, 1970) describes in words and photographs the illuminations done at the Benedictine monastery during its early years. The first chapter gives a brief history of the site.

—Sherry Crane LaRue

Moscow (Moscow, Russia)

Location: In northeastern Russia, between the Moskva and Neglinnaya Rivers, about 400 miles southeast of St. Petersburg; outer limit delineated by the Moscow Ring Road.

Description: City of about 9 million people; capital of Russia, former capital of the Soviet Union; fourth largest city in the world.

Site Office: Ministry of Culture
Office of the Minister
Kitayskiy Proyezd 7
Moscow 103693, Moscow Oblast
Russia
(95) 925-11-95

Moscow, the capital of Russia, is surprisingly young; it first appears in a historical record dated 1147, when Grand Prince Yury Dolgoruky wrote to a relative, inviting him to "Moscow." In 1156, the prince built an elevated fortress in Moscow, the Kremlin, to protect the strategically placed town, which lay between the Moskva and Neglinnaya Rivers. The Kremlin sat at the center of town, with the wooden homes of artisans and traders outside its walls and other homes, a church, and stables inside. When invading Mongol-Tartars burned the fortress down in 1238, the Russians rebuilt the structure atop Borovitsky (Pine Grove) Hill, and Moscow became the capital of a small principality. The Kremlin, from the Russian word *kreml*, meaning citadel or fortress, continued to expand concentrically.

Rich, prominent towns overshadowed poorer Moscow until the early fourteenth century. It was then that Ivan Danilovitch, nicknamed Kalita or "Moneybags," collaborated with the Tartars and so protected the city from much of the pillaging suffered periodically by most other towns. During Kalita's reign (1325–40), Moscow emerged as the political center of Russia and also became the official seat of the head of the Russian Orthodox Church in 1322. In 1339, Kalita added to the Kremlin and constructed new walls of oak. Owners of small businesses moved their shops within and just beyond the Kremlin's protecting walls. The boyars (noblemen) also lived in and around the Kremlin. In 1367, the wooden walls of the Kremlin were torn down and replaced by white stone. Fortified monasteries were constructed at strategic sites near Moscow to strengthen its defenses. The first stone church, Assumption Cathedral, was constructed in 1326 and 1327, in an area of the Kremlin now known as Cathedral Square.

The second half of the fifteenth century saw the official formation of the united Russian state, with Moscow as its center. The city flourished, particularly after 1480, when the Tartars suffered a crushing defeat that put a stop to their invasions. Grand Prince Dmitry Donskoi's army not only routed the Tartars, but also successfully withstood Lithuanian attacks. Moscow became the capital of the largest empire the Slavic peoples had ever assembled.

In 1475 the first Assumption Cathedral was dismantled to make room for a more grand ensemble. The partially built new structure collapsed in an earthquake, upon which the Italian architect Aristotle Fioravanti was commissioned to redesign and build the cathedral. Fioravanti completed the white five-domed Italianate-Byzantine Cathedral of the Assumption in 1479. Until the October Revolution of 1917, the cathedral served as a major religious and political center, in which czars were crowned and married, metropolitans and patriarchs were ordained and buried, and acts of state were proclaimed.

Many other great buildings were erected at this time, and the walls of the Kremlin were once again torn down in 1485. Over the next ten years, they were rebuilt and extended, this time in red brick. The walls, rising in some places to a height of sixty-five feet at a thickness ranging from ten to twenty feet, was laid out in a triangle. Eighteen towers—most of them designed by Pietro Solario, one of the Italian architects invited to Moscow by Ivan III, the "Great"—were built to strengthen the triangle, five on each side and one at each corner. These fortifications survive virtually unchanged today. The Savior Tower, which leads to Red Square, holds chimes in its belfry that are broadcast over radio and served as the nation's time signal in the twentieth century. Red Square itself had begun as an outdoor market area mostly built in wood. The area was destroyed by fire in 1493, following which Ivan III ordered the area left open as a firebreak to protect the Kremlin. The half-mile-long square was named Fire Square during its first five decades.

In 1547 Moscow was struck by disaster. A fire destroyed most of the city, including the food supplies, which led to famine among most of the city's common people. Rioting, incited by the boyars, against the czar and his family caused such upheavals that the Supreme Duma was formed, new legislation written, and the army expanded to control the unhappy populace. In spite of the unrest at home, the country continued to expand in the 1550s. In 1552, Russian soldiers captured Kazan and further expansion up to the Urals in the east was accomplished in 1556 and 1557.

Just outside the Kremlin walls stands the Church of the Holy Trinity, also known as the Protecting Veil of the Most Holy Mother of God, or, more commonly, St. Basil's. The cathedral complex was built between 1555 and 1561, to commemorate the capture of Kazan. The church's location outside the Kremlin walls but within the bustling market area of Red Square, was chosen to signify the national character of

St. Basil's Cathedral, seen in silhouette from Red Square
Photo courtesy of General Tours, Inc.

the victory. Russian craftsmen Barma and Postnik built the cathedral without blueprints or mathematical calculations. The complex contains nine separate churches, the central one, rising above the rest of the group, having ten domes, each of a different color and design.

In the 1560s Ivan IV, the Terrible, turned his attention once again to the continuing internal unrest, seeking to con-trol the boyars as much as Moscow's starving poor. In 1564, the czar unexpectedly left the city, taking the government with him. He threatened not to return unless the townspeople switched their allegiance from the boyars to the crown. As soon as he had gained the support of the people, Ivan re-turned to Moscow and divided the city and the country into two parts. One part was ruled by the crown, the other part by the boyar-controlled Duma. This divided government lasted from 1565 to 1572 and was punctuated by an invasion of the Crimean Tartars, who captured Moscow and burned everything but the Kremlin. Only 30,000 of the city's 200,000 residents survived, only to be plunged into the Reign of Terror of the 1570s that gave Ivan his nickname. An extensive secret police sought out leaders of popular unrest, and executions became the order of the day.

In spite of brutal government repression, the city continued to expand as a market town, also drawing craftsmen from far and wide. Blacksmiths, potters, tanners, and armorers lived and built their shops just outside the Kremlin walls, where they were vulnerable to attack. To protect them, a stone wall encircling the first was built toward the end of the sixteenth century. This second ring of defenses was later strengthened by earthen ramparts, with wooden walls and ramparts constructed on top. The fortress monasteries built earlier stood along these fortifications.

The second line of fortifications contains the second-oldest section of Moscow, Kitai Gorod. Although *kitai* means China in Russian, this area actually takes its name from the old word *kita,* which is the term for the earthen baskets used to reinforce the wooden ramparts. The area, east and northeast of the Kremlin, was Moscow's main commercial and crafts center from the fourteenth century.

In the wake of Ivan's death in 1582, a power struggle took place among the boyars that was eventually won by Boris Godunov. Czar Fyodor, who officially succeeded Ivan, was unable to assert himself and became a mere figurehead of government. The people of Moscow once again rebelled in 1586, a situation that allowed Godunov to consolidate his de facto authority. Godunov was crowned czar upon Fyodor's death in 1598, the official beginning of an ill-fated reign that led to what is known as Russia's Time of Troubles.

From 1601 through 1603 famine struck all of Russia. A total of 120,000 deaths occurred in Moscow alone, 50,000 of those coming in a seven-month period. Again starvation spurred the populace to rebellion, a crisis compounded by an invasion of Polish forces in 1604. Czar Boris Godunov died suddenly shortly after the Polish invasion. A royal impostor, known as the False Dmitry I, temporarily managed to take over the government with the backing of the Poles in 1605, but he was ousted in a popular uprising the following year. A second uprising accompanied by extensive streetfighting in Moscow took place later in 1606. The government was so unstable that another impostor, the False Dmitry II, was able to raise an army and lay siege to Moscow in 1608 and 1609.

In 1610 Czar Vasily Shuysky was ousted and the Poles took control of the city. Although they maintained a presence in Russia for some fifteen years, the Polish forces were unable to establish themselves firmly in the capital. In 1611 the people of Moscow rose up against the Poles unsuccessfully. The main result was a fire that laid waste to large parts of the city. The next year, however, no longer able to control their position, the Polish army retreated from Moscow. But the city's troubles were not over yet. In 1626, a fire devastated the wealthiest part of town, many of the boyars and wealthy merchants losing their mansions to the blaze. Rebuilding was begun quickly, this time in brick and stone at the czar's orders, to prevent fire damage in the future.

The second half of the seventeenth century was marked by a series of popular revolts. The salt riot of 1648 and the copper riot of 1662 were ultimately brutally repressed with the help of the Strelitz regiments that were permanently garrisoned in Moscow. The loyalty of the Strelitzers could not be counted on forever, though. In fact, 1682 was marked by the so-called Strelitz revolt, when soldiers ran amuck throughout the city.

The eighteenth century brought a major setback for Moscow, when in 1712 Peter I, the Great, finding Moscow too barbaric, moved the seat of government to St. Petersburg. Moscow's population initially declined, and its economy stagnated upon the departure of much of the nobility and many wealthy merchants for St. Petersburg. But Moscow remained a major commercial and industrial center and a kind of second capital, being the seat of the Senate and many ministries. After 1720, the city's textile mills expanded dramatically and produced silks and other fabrics that could compete with textiles from western Europe.

An epidemic of plague hit the city in 1771 and once again led to popular unrest. Those among the working people who fell ill were initially quarantined under ghastly conditions. Treatment of even the most rudimentary nature was in many cases not made available. An uprising against the church followed; this was brutally put down by the army with a massacre in Red Square. In 1773 a Yemelyan Pugachov, also known as the False Peter III, managed to raise an army of peasants and threatened Moscow for a while, but he was captured and executed in 1775. These uprisings were followed by a time of greater internal unity.

Moscow was again devastated in 1812, at the time of the French invasion. The Russian army had been unable to resist the advance of Napoléon's army until the Battle of Borodino, not far from Moscow, on August 26. In spite of the victory of Borodino, however, General Kutuzov decided to give up Moscow on August 29. The army's retreat from Moscow was followed by a mysterious fire, which some speculate may have been set by the Russians to thwart Napoléon. All provisions were burned. When Napoléon marched into the city, he found it virtually empty, both of people and of food. With winter fast approaching, the French army was in a difficult position. The troops were starving. Foraging parties venturing into the countryside surrounding Moscow invariably met with resistance and never returned. On October 7, after 30,000 officers and men had died in the city, Napoléon began an ignominious retreat with the Russian army following close on his heels.

The French invasion was followed by a period of increasingly organized resistance to the institution of serfdom. By 1861, the czar decided that he would rather abolish serfdom himself than see the country wrested from his control over the issue. Moscow fared well by the abolition of serfdom: capitalist expansion brought many new industrial enterprises to the city, not only in textiles but also in steel. By about 1900 Moscow had 1 million inhabitants, many of whom labored in the textile and steel mills. Overpopulation, high unemployment, and housing shortages caused horrific living conditions in Moscow's slums, however. The twentieth century opened with an economic depression that left many workers in peril. In the midst of the economic downturn, Russia went to war with Japan and lost, exacerbating conditions for the urban poor.

Strikes broke out in Moscow and elsewhere, which finally led to the unsuccessful Revolution of 1905. As the strikes spread throughout the country, the army was called in to force the people back to work. Streetfighting broke out in Moscow in December of 1904 and continued for several days. On January 22 of the following year, hundreds of laborers were killed at the czar's Winter Palace in St. Petersburg while protesting working conditions. The army was victorious in 1904 and 1905 and managed to suppress but not wipe out socialist organizing.

As Russia entered World War I, resentment of the government intensified as it sent 15 million Russians to fight in a war that most neither understood nor supported. Antiwar sentiments flourished, and the revolutionary Russian socialist Lenin (Vladimir Ilich Ulyanov), exiled in Switzerland at the time, urged Russian soldiers to "vote with their feet" and leave the fighting to return home. At the same time, "peace at any price" became a slogan throughout Russia. The poorly equipped Russian soldiers petitioned for peace. At home, unrest among the workers continued, particularly when the war led to food and fuel shortages. Once again workers struck in all the major industrial centers in Russia, but this time the army sided with the people. In March 1917, Czar Nicholas was forced to abdicate and was placed under house arrest.

Prince Georgy Lvov first assumed power, but on July 21 he was replaced by the provisional government of attorney Aleksandr Kerensky. Kerensky continued the war, in which Russia suffered more defeats. The socialists, under the leadership of Lenin, had gathered sufficient momentum that the Kerensky government could not stop them. On October 25 (under the old style or Julian calendar) or November 17 (the modern, Gregorian calendar date) of 1917, the Russian Social Democratic Labor Party, known as the Bolsheviks, overthrew Kerensky and took over the government.

Based in St. Petersburg, the Bolsheviks negotiated an armistice with Germany and abolished all private ownership of land. Within two days, however, Moscow became the center of a power struggle between Kerensky supporters, called the Whites, and soldiers in the Kremlin, the Reds, who supported the Bolsheviks. Martial law was declared in the city on October 27. For six days, a civil war raged with fighting both inside and outside the Kremlin. The Whites were initially victorious, killing the Red soldiers stationed in the Kremlin. The Bolshevik Revolutionary Committee of the Moscow Soviet of workers and soldiers responded with a general strike in the factories and joined the Red soldiers in

the streets of Moscow in building barricades. Finally, with outside support and shelling of the Kremlin, the Red Guard (demobilized soldiers) and the workers broke into the Kremlin through the Nicholas Gate. They overcame the White Guards and established the Bolshevik Council of Commissars. The council immediately abolished all privileges of title, ancestry, and religion. All Russian citizens were proclaimed equal.

The new government excavated what became the Brotherhood Grave between the Savior and Nicholas gates, where some 500 workers and soldiers who had died in the fighting were buried amid the first of many major public events held in Red Square. Later the Brotherhood Grave became the place where Russian Communist leaders, several Russian cosmonauts, and others of prominence were buried. The American journalist John Reed, who wrote *Ten Days that Shook the World,* an eyewitness account of the revolution, was also laid to rest there. Newlyweds now traditionally visit the grave after their marriage ceremony.

In March of 1918, the Soviet government was established in Moscow, which thus once again became the capital of the Russian empire, this time under the name of the Union of Soviet Socialist Republics. The red flag, the hammer-and-sickle banner, replaced the double-headed eagle of the czars. The Kremlin became synonymous with the Soviet government in the minds of many. In July of the same year, former Czar Nicholas II and his family were executed.

Civil war raged in the countryside for the next three years, between the White Guards, who received extensive aid from western European governments, and the Red Army. In 1919 Moscow briefly came under siege of the counterrevolutionary forces under Denikin. The Bolshevik leaders remained in the Kremlin during this time, living a spartan life. Lenin lived on the third floor of the Council of Ministers, previously the Senate Building. (His quarters are now a museum.) In March 1921, Lenin officially announced the end of the civil war at the Tenth Congress of the Communist Party in the Kremlin.

Following the civil war, Moscow undertook a massive effort to rebuild the city and the country as a whole. Lenin died in 1924, in the midst of the reconstruction efforts. His body was preserved and is still displayed in a mausoleum on the west side of Red Square as part of the Brotherhood Grave. Following Lenin's death, Stalin took over as head of the Communist Party and of the Soviet government. By 1925 economic rehabilitation was said to be complete, the economy more or less returned to the state it was in at the outbreak of World War I. Nevertheless, there were about 120,000 jobless in Moscow, and there were severe housing shortages. A five-year plan was launched, the first of a series of such plans that laid out the transformation of the Soviet economy. The plans set goals for industrial development, the expansion of cooperative farming, and housing and other construction.

In Moscow itself endless complexes of high-rise apartments were built, and construction of an underground railway system was begun. At the same time, the city became the cultural center of the entire U.S.S.R. The most notable universities, theatres, and museums were to be found in Moscow. Moscow, which had for long been a center of the book trade, became the premier center of publishing, always under the auspices of the Communist Party.

The Soviet Union did not enter the war until June 22, 1941, when unexpectedly the Germans launched an offensive. The country was immediately mobilized, and the people of Moscow turned the city into a veritable fortress, building defensive lines around the city and inside the city around the Kremlin. On October 20, the Germans had advanced to the outskirts of Moscow, and the city was declared under siege. By early December of 1941, the volunteer units and the Red Army defending Moscow had brought the Nazi offensive to a halt. On December 5, a counteroffensive was launched that managed to drive the German army out of the Moscow area by January of 1942. The defeat at Moscow was the first major setback suffered by the Germans in World War II. For the remainder of the war, Moscow became the center of the war industry, its factories retooled for the production of armaments. In November 1942, the grueling siege of Stalingrad came to an end, and the Soviet army began a relentless pursuit of the Germans, first sweeping the Nazi forces out of the Soviet Union and then pushing them back all the way to Berlin in 1945.

In the wake of World War II, the Soviet Union gained control over Eastern Europe, establishing communist regimes in what became satellite nations. East Germany, Poland, Hungary, Czechoslovakia, Romania, Albania, and, to a lesser extent, Yugoslavia all turned to Moscow for economic aid and cultural enrichment. Moscow's sphere of influence was enlarged with every addition to the Communist Bloc.

By the late 1980s, however, the Soviet Union was in disarray. Not only were many eastern European countries deserting the Communist fold, but there was also widespread dissension within the Soviet Republics. Civil war broke out between Azerbaijan and Armenia. The economy was unstable. The people pushed for democratization and for a market economy under the leadership of the political maverick Boris Yeltsin. In February of 1990 the Communist Party voted to end the one-party system that had made it synonymous with government for seventy-two years. In the following month Lithuania declared independence, followed later by Latvia and Estonia. Half-hearted attempts to stop the independence movements in the Baltics were unable to repress the desire for freedom in the Soviet Republics. Later the Ukraine, Georgia, and Moldavia also pursued secession.

Meanwhile, economic policy reforms caused power struggles within the government. Boris Yeltsin, elected president of the Russian Republic, pushed himself forward as a radical reformer, calling for more extensive economic reforms. Communist Party hard-liners opposed reform of any kind, however. On August 19, 1991, these hard-liners staged a short-lived coup, which, ironically, precipitated the dissolution of the Soviet Union. While President Gorbachev was away on vacation, the Kremlin was seized with the backing of the army. Under the leadership of Yeltsin, the people of Moscow turned out for massive demonstrations against the coup. They prevailed, in large part because the army refused to open fire on the demonstrators. The leaders of the coup yielded control of the Kremlin on August 21. Three days later, Gorbachev resigned as chief of the Communist Party, and the Supreme Soviet voted to remove all government functions from the party on August 29. A week later the Soviet Union was officially dissolved.

Moscow is once again capital of Russia. A market economy flourishes, but discontent is widespread. The people complain of unemployment, inflation, crime, and other ills. The return to capitalism has not been a miracle cure for the economic malaise of the 1980s, but Moscow with all its cultural and historical riches has been opened up to the rest of the world as it perhaps has never been before.

Further Reading: *History of Moscow: An Outline*, edited by S. S. Khromov and translated by Yuri Shirokov (Moscow: Progress Publishers, 1981) is particularly good on Moscow's early history, less successful as an account of its twentieth-century history. *The Kremlin: Citadel of History* by Mina C. Klein and H. Arthur Klein (New York: Macmillan, 1973) traces this symbol of Russian history over a period of 800 years, detailing how the fortress complex came to stand for all that is Russian. John Reed's *Ten Days that Shook the World* (1919), available in a variety of editions, gives a highly readable eye-witness account of the October Revolution. The *Moscow* edition of *The World's Cities* series, edited by Nicholas Wright (London: Cavendish, and Secaucus, New Jersey: Chartwell, 1978) delves into the cultural, political, social, and historical aspects of Moscow, giving a glimpse of the psyche of the people. The richly illustrated book presents a unique view of the city.

—Sharon Bakos and Marijke Rijsberman

Mtskheta (Georgia)

Location: In the Caucasus Mountains in the center of eastern Georgia, which lies in the center of the Caucasian Isthmus between the Caspian Sea and the Black Sea; approximately 12.4 miles from the capital, Tbilisi.

Description: State historic preserve. Former capital of the Kingdom of Iberia, a predecessor to Georgia, and former seat of the catholicos (patriarch) of the Georgian Orthodox Church. The Republic of Georgia was part of the Union of Soviet Socialist Republics until the Soviet Union ceased to exist in 1991.

Contact: Tourist Information
Metekhi Palace Hotel
Tbilisi 380003
Georgia
(8832) 744 556

Mtskheta, located at the confluence of the Kura (Mtkvari) and Aragvi rivers, was once the capital of the Kingdom of Iberia and an important center for first the pagan and Zoroastrian religions, and then the Georgian Orthodox Church. Its two rivers and the roads that paralleled them (east/west along the Kura and north/south along the Aragvi) gave this location significance as a trading center, at the crossroads where Europe met Asia.

The origin of the name "Mtskheta" is alternately thought to be from the city's legendary founder Mtskhetos, son of Kartlos, the demigod forefather of the Georgians, or, more probably, the name means "city of the Meskhians." (The Meskhians or Moskhoi were one of the ancient tribes that formed Iberia.) The Georgians call themselves *Kartvel-ebi* and their homeland *Sa-Kartvel-o,* both names linked to Kartlos. The name "Georgians," applied to them by Western Europe, and erroneously said to spring from the veneration of St. George, is derived from the name used by the Arabs and the Persians, *Guri (Kuri)*. The Russians call the Georgians *Gruzians.*

Twentieth-century archeological finds indicate that Georgia was home to some of the earth's first humans. Some details of the country's ancient past are cloudy, fact intermingled with myth. There were two main cultures in what is now the Republic of Georgia. In the west was the Koban-Colchian culture—the ancient Colchis of the Greek myths of Jason and the Golden Fleece, Prometheus, and the tragedy of the Colchian princess/enchantress Medea; the Ibero/Meskho/Albanian culture in the east developed into the ancient Kingdom of Iberia. (Although classical geographers confused this Iberia with the Iberians of Spain and invented theories to link the two, there is no evidence of any connection.) King

Midas of the Greek legend was actually King Mita of the Meskhians. Under the name Meshech, the Meskhians are numbered among the sons of Japhet in the Old Testament, and are associated with Tubal of Genesis.

This region, within the area of the Old Bronze Age, was subject to successive waves of nomadic looters, beginning with Phrygians, Mysians, and Bithynians, and ending with Scythians, Cimmerians, Medes, and Persians five centuries later. By the fifth century B.C. the tribes of Georgia were under the protection and control of the Persians. The Persian influence was probably the cause of Iberia's early development of a civilization out of the chaotic tribal society of the time. Later the Persian Empire weakened and was overthrown by the Macedonians, who sent a tyrannical relative of Alexander the Great to administer the area. (It does not appear that Alexander himself was ever in Iberia.) In the third century B.C., under the first Iberian king, the half-legendary Farnavazi (Parnabazus or Parnavaz), the tribes revolted and overthrew the Macedonians. Farnavazi was recognized by the kings of Syria and Armenia as the legitimate ruler. He and his descendants ruled until 93 B.C., when they were driven from the throne by the Armenian Arsakid dynasty, for supposed infractions against the religion practiced in Iberia and Armenia at the time. The Arsakids then ruled the country until about A.D. 226.

The precious objects found in the graves of local royalty indicate Iberia was a rich land, one that various invaders would find attractive. To protect the area, an ancient stronghold, the Armaz-tsikhe (Harmozika) fortress, the seat of the princes of Mtskheta, was built on Bagineti hill in the late fourth to early third century B.C. This was also the site of shrines to the moon god, Armaz (the primary god) and a secondary god, Zadeni, probably the god of vegetation. A new Persian-influenced religion, that of Ahura-Mazda, spread throughout Georgia during Farnavazi's time. This new cult was not accepted by everyone, however, as many, including the Armenian Arsakids, still believed in the old animist cults.

In 66–65 B.C. Pompeius invaded and brought both Iberia and Colchis under Roman dominance. Unlike the more passive Colchians, the Iberians resisted, but once conquered they rapidly adapted themselves to the rule of the Romans. By the end of the first century B.C. Iberia had detached itself enough from Rome to be more an ally than a subject land.

A turning point came in about A.D. 330 with Iberian conversion to Christianity. The effects on art, literature, culture, and the entire way of life among the Iberians cannot be overestimated. Although there is some confusion about the exact date of conversion, it coincided with the beginning of a new dynasty in eastern Georgia under King Mirian. St. Nino, a slave woman from Cappadocia who was said to have

A view of Mtskheta

miraculous healing powers, reportedly cured Queen Nana of a mysterious ailment. Shortly thereafter, King Mirian was enveloped in total darkness (probably an eclipse) while on a hunting trip. He called upon the pagan gods to no avail, but when he prayed to St. Nino's Christian God, the light was restored. He converted and was baptized along with Queen Nana and all of Mtskheta. Christianity was declared the state religion, and the king ordered pagan temples destroyed and

churches built. These new churches were the precursors to Sveti-tskhoveli and Samtavro. Historical circumstances favored the conversion, as neighboring Armenia had converted a generation earlier, and, by St. Nino's time, other missionaries had been actively proselytizing at the Greek colonies along the Black Sea.

In the first century A.D., Mtskheta had replaced Armazi and the Armaz-tsikhe fortress as the capital of Iberia.

Mtskheta remained capital until 458, when King Vakhtang Gorgaslan (Gorgasali) moved the capital to Tbilisi. Although no longer the political capital from that time forward, Mtskheta remained the religious metropolis and residence of the head of the Georgian Church, the catholicos (patriarch). At first the Georgian Church was subject to the patriarchs at Antioch, but it was given autonomy during the reign of Vakhtang Gorgaslan.

Georgia excelled in developing original designs and techniques for construction of churches. This was particularly true after the kingdom was united under Bagrat III (975–1014), leading to the advances in architecture, with the creation of cathedrals and monasteries. Because Mtskheta was the religious capital, the history of the city is inextricably intertwined with the history of its churches. Fragments remain from the earliest wooden churches, such as King Mirian's at Sveti-tskhoveli, and there are surviving examples from various periods of Iberia's history. Mtskheta has been made a state historical preserve to protect its treasures.

The Cathedral of Sveti-tskhoveli (Life Giving Column), built between 1010 and 1029 in the center of Mtskheta, is one of the most sacred places in Georgia. Legend says that a Georgian Jew, who had converted to Christianity, bought Jesus' robe from the Roman soldiers in Jerusalem. As he was bringing it home to Mtskheta, he showed it to his sister. She was so overwhelmed, she died clutching the robe so tightly that it had to be buried with her. King Mirian chose her gravesite as the place to build the first Christian church in Georgia. A column, made for the church from a cedar tree growing on her grave, was reputed to have magical properties and supposedly rose in the air of its own accord. It was later reported to have healing powers as well.

The current church is the third to be built on the site. The first was King Mirian's wooden building of the legend, the second a basilica built in the fifth century by Vakhtang Gorgaslan. When this basilica fell into disrepair, Melkhisedek, the catholicos of Georgia, commissioned a new cathedral. Elements of the fifth-century church were incorporated into the new church, and some traces of the foundation from the earlier church remain on the site. The cathedral is considered a masterpiece of the early Georgian Renaissance, demonstrating a high degree of technical and artistic achievement. Some elements, such as the two bulls' heads remaining from the fifth-century basilica, show the typical blending of pagan and Christian iconography of the period. The cathedral was damaged by an earthquake in 1283 and during the invasion by Timur in the beginning of the fifteenth century, but was subsequently restored. Despite the ravages, some original elements of the church remain intact.

The kings of Georgia were crowned at this cathedral, which also served as their burial place. Only three of the ten royal tombs have been located: those of Vakhtang Gorgaslan, Herekle II, who was responsible for placing Georgia under Russian protection in 1783; and Giorgi XII, Herekle's son, who died in 1800, the last king of Georgia. The church also contains a baptismal font thought to have been used to baptize King Mirian and Queen Nana. Inside the cathedral is a small stone church, a late-thirteenth- or early-fourteenth-century copy of the Church of Christ's Sepulchre in Jerusalem. The original is said to be the holiest place in the world; the copy marks Sveti-tskhoveli as the second most holy place. Since this cathedral was also the seat of the catholicos, there is a throne on one side of the church. Its use was discontinued when the Georgian Orthodox Church began to require the catholicos to be seated in the center of the church.

Nearby is Samtavro, an enormous ancient necropolis, where Scythian artifacts and other early objects have been found. The road leading to the site is bordered with tenth- and eleventh-century tombs. The name Samtavro ("Place of the ruler Mtavari") refers to King Mirian's garden, where he built a church in the fourth century to honor St. Nino, who prayed there. In addition to this tiny church, the site has three other edifices: the main church of Samtavro, the adjacent three-story bell tower (thirteenth century or later), and a building (eleventh century) that served as a nunnery and main theological seminary of the Georgian Orthodox Church.

The main church, dedicated to St. Nino, demonstrates the architectural advances of the eleventh century, particularly in decoration of the facade. In a reflection of their simple piety, King Mirian and Queen Nana chose to be buried in this humble church rather than in the Sveti-tskhoveli. The cupola of the church collapsed in the thirteenth-century earthquake and has been reconstructed.

On a high hilltop, overlooking the Mtkvari valley next to Mtskheta, is the Dzhvari Monastery. To commemorate the victory of Christianity over paganism, St. Nino erected a cross which made the site a place of pilgrimage. There are two churches on the site, both built in the latter part of the sixth century: Prince Guaram's, a small Dzhvari (Jvari, "the cross") church now in ruins, and a larger church built by his son, Duke Stephanos I, in which the foundation of St. Nino's cross can still be seen. This building is a classic, cruciform, domed church, and its remarkable exterior carvings include a lifelike portrayal of the duke, Jesus Christ, St. Stephen, and the Archangel Michael. The Arab invasions of the tenth century damaged the church, but it is generally well preserved. It was closed in 1811 by the Russians, and served as an historic monument until 1988, when it was reconsecrated.

Under Queen Tamara, whose reign began in 1184, Tbilisi was one of the most important cities in the Middle East, and the church flourished in Mtskheta. That was not to last, however. Over the next 600 years the country alternately suffered invasion and subjugation by Mongols, Turks, and Persians. Power passed back and forth as the Georgian rulers occasionally but temporarily won back territory and authority. The importance and power of Mtskheta declined with the fortunes of the country under its non-Christian conquerors, beginning with the domination by the Mongols in the thirteenth century. The country fragmented into principalities, grouped together along the old Colchian/Iberian lines. As the country was no longer united, a rival catholicos was established in western Georgia, but it was generally subject to the catholicos of Mtskheta.

In the late fourteenth century the Mongol invasions, led by Timur, became particularly brutal. During this period of aggression, they destroyed the Sveti-tskhoveli Cathedral at Mtskheta. The cathedral was restored, in 1656, by the Muslim King Rostom, who ruled on behalf of the Persians.

Only in the reign of Teimuraz II and his son Herekle II (Heraclius or Irakli) did the Georgians gain significant independence, first from Turkish and then from Persian control. The Georgians reunited most of the country in 1762, with Tbilisi once again the capital. The Georgian Orthodox Church played an important part in the intellectual life of the country.

At the end of the sixteenth century Russia had begun to extend its sphere of influence in Georgia on the pretext of protecting the Christian monarchy from Moslem domination. Caught between the Russians and the Turks, Herekle fought on the side of the Russians. Finally, under a 1783 treaty, the beleaguered king had the country placed under the protection of Russia, voluntarily giving up his sovereignty in return for the assistance of Georgia's powerful neighbor. Russia failed to save Georgia from a devastating blow in 1795, however.

That year, the Persian shah Agha Mohammed ravaged the entire area, sacked Mtskheta, and destroyed Tbilisi. He leveled churches, put priests to death, deported the whole population, and then withdrew. Herekle soon died, and his son Giorgi XII, the last king of Georgia, tried to preserve his line of succession by appealing to Russia for help. Another Persian attack was imminent, and again, rivals for the throne were conspiring with the Turks. But instead of merely providing assistance, Russia annexed Georgia shortly after Giorgi's death.

Although Russia had political authority over eastern Georgia from 1801 on and western Georgia from 1810, the church remained independent until 1811. In that year, fearing the church as the center of nationalism, Russia abolished the Georgian Orthodox Church. The catholicos was replaced by a representative of the Russian Orthodox synod, who along with his successors tried to subjugate the Georgian church the way the Russian Orthodox Church had been made subject to the czar. They were never completely successful, and the church remained a symbol of popular solidarity.

After the 1917 revolution, the church regained its autonomy for a time and appointed a new catholicos. After the Bolshevik victory, however, the Georgian Orthodox Church was again merged with the Russian Orthodox Church until 1943, when autonomy was restored due to the intervention of Joseph Stalin, himself a Georgian. The Georgian church was, of course, subject to all of the religious repressions of the Soviet government. In 1988, under the Gorbachev reforms, the catholicos was able to reconsecrate many churches that had been put to other uses under Soviet control, and with the 1991 dissolution of the Soviet Union, the church regained full freedom and autonomy, with its official seat at Sioni Cathedral in Tbilisi. Despite politics and the official change of the seat of the metropolitan, however, Mtskheta remained and still remains the spiritual center of the Georgian Orthodox Church, drawing both Georgian and foreign visitors.

Further Reading: *A History of the Georgian People: From the Beginning Down to the Russian Conquest in the Nineteenth Century* by W. E. D. Allen (New York: Barnes and Noble, 1932; reissued, 1971) is a scholarly work providing a wealth of information on the Georgian people and their religion, language, and monarchy. *The Georgians* by David Marshall Lang (New York: Frederick A. Praeger, and London: Thames and Hudson, 1966) traces the history of Georgia from prehistory to the thirteenth century. Although also a scholarly book, it is very readable. It includes sections specifically addressing art, architecture, literature, and learning.

—Julie A. Miller

Munich (Bavaria, Germany)

Location: In southern Germany, in the heart of the state of Bavaria, near the foothills of the Bavarian Alps and along the banks of the Isar River.

Description: Bavaria's capital and Germany's third-largest city features a diversity of architecture and culture found virtually nowhere else in the country. Its architectural palette ranges from medieval to modern and includes a variety of reconstructions built to replace structures lost to World War II bombing. Munich is today considered one of the most cosmopolitan cities in all of Europe.

Site Office: Verkehrsamt
Sendlinger Str. 1
80331 Munich, Bavaria
Germany
(89) 2 39 11

Like an innocent bystander caught in the crossfire, the German city of Munich has regularly found itself thrown into the midst of history-making events. Almost from the time of its founding in the mid-twelfth century until the present day, Munich has routinely been caught up in the maelstrom of change that has defined Bavarian, German, European, and even world history. Yet despite the significant role that Munich has had in helping to shape history, it has often been a reluctant participant, frequently getting drawn into the fray against its will. And with good reason, for Munich has often been hit by the bullets—and bombs—unleashed by these events.

Munich is located in the heart of the south German state of Bavaria, once an independent duchy. It sits on a glacial moraine at the foothills of the Bavarian Alps along the banks of the Isar River, a tributary of the Danube. Arguably, Munich's location is one reason why the city has often been drawn into history-making events. Situated at the crossroads of Europe, where north-south and east-west transit routes cross the continent, Munich literally has found itself in the middle of things. Thus as opposing forces in national, continental, or global conflicts have taken up arms through centuries, their campaigns have often taken them through Munich and its surroundings.

As the capital of Bavaria and a prominent German city, Munich's very nature has also frequently made it susceptible to the prevailing political and diplomatic climate. In general, as went the history of Bavaria (or Germany), so went the history of Munich. As a consequence, Bavaria and Munich have often come under the influence—or even control—of foreigners at various times through the centuries.

But despite its battle scars, Munich has nevertheless managed to survive and even prosper. It has always managed to retain (or at least restore) the architectural and cultural treasures that have made it one of the most diverse, colorful, and interesting cities in Germany, if not all of Europe.

Among the first to visit the site where Munich now stands were the Romans, who arrived about 15 B.C. But because of the rocky soil in this glacial moraine area, the Romans found the area unsuitable for agriculture ("barren," as one author describes it) and did little to establish anything of permanence in the area. Indeed, the Romans looked upon the whole of Bavaria with similar disinterest, using the area primarily as a transit corridor between Rome and its outer provinces, with only a few thousand occupants in the region at any one time.

With the fall of the Roman Empire, Bavaria was easily overrun by the barbarians (due to the limited Roman presence). But the barbarians, in turn, were quickly overrun by the Franks. Frankish control over the region gradually increased, despite attempts by would-be monarchs to assert local control, until Bavaria eventually became part of Charlemagne's Carolingian Empire, which maintained control until the early tenth century. With the decline of the Carolingian Empire, a number of local dukes attempted to assume leadership over Bavaria, but they often ended up fighting among themselves. They also faced frequent challenges from outside invaders, such as those from Hungary, Ostmark (Austria), and the German Empire.

In the midst of all these struggles for the control of Bavaria, a small monastic community was established along the banks of the Isar River. It is believed to have been founded in the tenth century by monks from the nearby monastery of Schäftlarn seeking to flee invading Hun forces. This settlement, which managed to survive unscathed (unlike the monastery of Schäftlarn), became known as zu den Mönchen, which loosely translates as "at the monks' place." The name was eventually shortened to München, the German name for the city of Munich. A monk even appears as part of the city's coat of arms.

Germans came to power in Bavaria in 948 and held it for roughly a century. When they ceded control in the late eleventh century, various foreign dukes vied for control of the duchy (or portions thereof) over the next 100 years. One of those dukes who held power for a time was Henry the Lion (Heinrich der Löwe) who ruled over both Bavaria and Saxony. During his reign from 1156 to 1180, Henry formally established the capital city of Munich near the site of the old monastic settlement. He also did much to help foster the growth of the city. Henry decided that Munich would be an ideal spot to build a bridge spanning the Isar River. To ensure maximum toll revenue, he destroyed the only other nearby bridge, at Vohring. Munich thus became an increasingly important center of commerce, and the city grew.

Munich's Rathaus, left, and Frauenkirche, right
Photo courtesy of German Information Center

Henry's rule came to an end in 1180, when the Wittelsbach family assumed control over Bavaria. The Wittelsbachs would maintain sovereignty for the next 738 years, one of the world's longest-reigning dynasties. Bavaria, and Munich in particular, prospered under their rule. The Wittelsbachs began holding court in Munich in 1253, which increased its importance as a political and commercial center.

Later, when the Reformation swept through other German states, the Wittelsbachs kept Bavaria as a Catholic stronghold. Dukes Albert V (1550–79) and William V (1579–97) zealously fought ''the new heresy,'' while their successor, Maximilian I (1597–1651), founded the Catholic League in 1609. These efforts all helped to solidify Catholicism's position in Bavarian life, where it remains the dominant religion to this day.

But the Wittelsbachs also encountered their share of problems during their reign: Opponents of Wittelsbach rule split off their territories from Bavaria, forming independent states that would never again belong to the duchy; internal squabbles led to the parceling of Bavarian territory among various Wittelsbach family members, and these lands would remain divided for centuries, becoming reunited only after much difficulty; the eventual dying-out of the Bavarian branch of the Wittelsbach family led to the ascendancy of relatives from other Germanic states, who often looked on Bavaria as a commodity (one such ruler, Charles Theodore, even tried to market Bavaria to the highest bidder); and political entanglements with powerful neighbors France and Austria (and later, Prussia and Germany) led to Bavarian involvement in several continental wars, the results of which included everything from military defeats to the capture of Munich to the surrender of some territory. But there were successes, too, such as the war against Napoléon, in which Bavaria sided with the victorious Germans, a move that made Bavaria a prominent European state for a time.

Later Wittelsbachs also looked to Bavaria as a means to finance personal projects. Perhaps the most notable of these rulers were King Louis I (reigning from 1825 to 1848) and his grandson, King Louis II (reigning from 1864 to 1886). Both appropriated vast sums from the Bavarian treasury to build grand palaces and other architectural splendors in and around Munich and throughout Bavaria. They also spent state funds extravagantly to indulge the whims of their romantic interests, the countess of Landsfeld (a Spanish dancer better known as the infamous Lola Montez), in the case of Louis I, and composer Richard Wagner, in the case of Louis II. Their excesses eventually led to their downfalls, however; Louis I was deposed in open revolt (a rare act among the usually peaceful Bavarians), and Louis II was incarcerated and shortly thereafter committed suicide.

Ironically, their extravagances helped establish Munich as a gleaming and impressive city in nineteenth-century Europe. Despite its commercial importance and its continued role as Bavaria's capital, Munich's prominence had declined somewhat in the previous century. But with the infusion of new building projects initiated by these free-spending monarchs, Munich again assumed center stage.

With the emergence of a unified Germany under Bismarck in the late nineteenth century, the role of individual Germanic states like Bavaria began to decline. The emergence of a unified Germany as a European power also contributed to the outbreak of World War I. Bavarians, by now well accustomed to the ravages of such international conflicts, were not willing supporters of this devastating and unpopular war. In fact, they made their sentiments known with the ouster of the Wittelsbach monarchy in 1918. Bavarian Socialists protesting Germany's continued involvement in the war (and the Wittelsbach family's support of it) overthrew the long-running dynasty and set up a new provisional government. The Socialists, in turn, fell out of favor and were quickly replaced by a series of fractional and coalition governments, including a short-lived Bolshevik regime.

From this point onward, Munich (like all of Germany) began playing an increasingly important role in the European and world politics. The first signs of this role came in the early 1920s, when a little-known radical named Adolf Hitler formed the National Socialist Party in Munich. In 1923, in a bold if somewhat uncoordinated move, Hitler and his band of Nazi followers initiated an attempt to seize the Bavarian state government. Hitler's short-lived plan was amateurish and ultimately led to his surrender and imprisonment, but it nearly embroiled Munich in the throes of a Bavarian civil war. What's more, it thrust Hitler to national prominence and allowed him to build a power base in the ensuing years.

The citizens of Bavaria and Munich were uncomfortable with their land and city having been the birthplace of Nazism. In part this was because Hitler was Austrian by birth, and Austrians were viewed as having been responsible for much of Bavaria's woes through the centuries. Consequently, Bavarians and Müncheners distrusted the führer after his rise to power.

In an attempt to defuse potential international hostilities, a 1938 conference involving Hitler and leaders from Britain and France was held in Munich. But the agreement produced during that conference—drafted under the principle of appeasing the German dictator—was ultimately not enough to contain the führer or prevent the outbreak of World War II. Many Bavarians and Müncheners openly defied the Reich during the war by refusing to engage in military service. Despite such resistance, Bavaria was not spared bombardment by Allied forces in their march across Germany. Munich was intensely bombed, and much of the city destroyed.

In the years after the war, Munich, like much of Germany, set about the process of rebuilding. Many of Munich's older buildings were restored as faithfully as possible to original specifications, but new, modern designs, symbolic of a new Munich and a new Germany, were added as well.

One of the crowning achievements in this rebuilding effort was Munich's being awarded the right to host the 1972 Summer Olympics. Germany in general and Munich in particular viewed this event as an opportunity to showcase a new, modern, civilized city and country. However, the event was marred by a massacre of Israeli athletes by Arab terrorists. Munich's desire to shine on the world stage had been severely tarnished.

Today, Munich is a center of commerce, culture, and entertainment that has become Germany's most popular tourist destination. Yet despite its modern, contemporary image, it is proud of its past and features many architectural and cultural elements that reflect the rich, long history of the city. Perhaps Munich's most significant role has been as Bavaria's capital and the one-time site of the Bavarian royal court. A number of structures used for such purposes can still be found in Munich today.

The first structure occupied by the Wittelsbach royal court was the Alter Hof, located near one of the oldest sections of the city. The Wittelsbachs resided here from 1253 to 1385, when they moved to a new palace to escape the encroaching tenements of the growing city. The inner courtyard of the Alter Hof is all that remains today.

Upon leaving the Alter Hof, the Wittelsbachs moved into the Residenz, which served as the family's royal palace for more than 600 years. The Residenz was begun as a small castle in 1385. Over the years, successive Wittelsbach monarchs added to it, creating an eclectic complex of facilities that at one time or another has included a church, a banquet and festival hall, and several theaters. In the process, these additions have also yielded a design that illustrates a variety of architectural styles, including Renaissance, baroque, rococo, and Neoclassical. While some portions of the Residenz have been damaged or destroyed over the years and others have been adapted to new purposes, the palace complex endures as a legacy of the Wittelsbach family. Treasures of this long-reigning dynasty can be found in the Schatzkammer (Treasury) and Residenzmuseum.

One monarch who did much to shape the nature of the Residenz (and indeed much of Munich) was King Louis I. He launched a series of ambitious architectural programs, many of which were executed under the direction of court architect Leo von Klenze. Many of these projects employed neoclassical styling, which Louis loved and which helped Munich earn the nickname Athens on the Isar. Perhaps the most notable of these projects was Königsplatz (Kings Place), a large city square featuring Grecian-style exhibition halls (now museums) and a massive neoclassical arch.

Louis I's son, King Maximilian II, also sponsored massive public architectural projects, including an ambitious city planning program. While he certainly had a flair for grandeur, he did not share his father's enthusiasm for the neoclassical style. To make his own mark, he sponsored the development of Maximilianstrasse, a broad boulevard lined by majestic buildings, many used for government purposes. Perhaps the most notable structure along this boulevard was the Maximilianeum, a lavish palace that has been the home of the Landtag, the Bavarian state government, since 1877.

Carrying on the building traditions started by his father and grandfather, King Louis II also launched a number of architectural projects. While he may be best known for his Bavarian countryside castles, such as Neuschwanstein, Louis II sponsored some ambitious projects in Munich as well. Perhaps the most prominent of these was the Neues Rathaus (New Town Hall), built on the perimeter of Marienplatz, a huge public square in the heart of Munich. The imposing neo-Gothic structure was built by Georg Hauberisser in stages between 1867 and 1908. Its ornately detailed facade and tall central tower with arched windows have been compared to the English Houses of Parliament. But perhaps the most distinctive feature of the Neues Rathaus is its charming, fanciful glockenspiel (added in 1904), which features jousting medieval knights and dancing coopers.

The Neues Rathaus is one of Munich's most popular tourist attractions, as is the square to which it is adjacent, Marienplatz. The square was named for a gilded statue of the Virgin Mary, which was erected in 1638 as an act of thanks for Munich having been spared the devastation of the Thirty Years War.

Given Munich's frequent involvement in military conflicts, the city has erected numerous war memorials over the years. Perhaps the best known is another of Louis I's projects, the Feldherrenhalle (Hall of Generals), an open structure on Odeonsplatz featuring statues of Bavarian military leaders, patterned after Florence's Loggia dei Lanzi. Other such memorials include Karolinenplatz, which features a monument (unveiled in 1812) in honor of Bavarians killed during the war against Napoléon, and the Siegestor (Victory Arch), patterned after Rome's Arch of Constantine, which honors the efforts of the Bavarian army during the Wars of Liberation (1813–15).

Authentic examples of medieval architecture are rare, but some of the best are the fortified gates that once served as entrances to the city. The Karlstor, adjacent to the city square Karlsplatz, is said to date to 1302. Another surviving fourteenth-century gate is the restored brick Sendlingertor, adjacent to the city square Sendlingerplatz. A third such gate, Isartor (adjacent to Isartorplatz), serves as an entrance to one of the oldest streets in Munich, the Tal (literally translated as "Valley"), which had been an early route into the city. The

buildings that line the Tal perhaps come closer than any others in Munich to reflecting the look and feel of the city's medieval past. Another example of medieval architecture is the Altes Rathuas (Old Town Hall). This charming structure was built by Jörg Ganghofer in 1474 and restored after extensive damage during World War II.

Munich is also rich in religious architecture. Peterskirche (St. Peter's Church) is Munich's oldest parish church, dating to the eleventh century. It has undergone various architectural changes over the years and today reflects those different styles, as evidenced by its late Gothic altar, baroque interior, and eighteenth-century statues of the apostles. Its 300-foot tower provides visitors with a panoramic view of the central city.

Perhaps Munich's most impressive church is its cathedral, the Frauenkirche (Church of Our Lady). This late Gothic brick structure features twin towers that exceed 300 feet, capped by distinctive onion-shaped domes. The church, designed by Jörg von Halspach, was built over a twenty-year period, with its consecration in 1494; the towers were added thirty years later. Extensive damage during World War II prompted an ambitious ten-year restoration program completed in 1957.

Both Peterskirche and Frauenkirche have interesting folklore associated with them. Each was allegedly subjected to clashes with the devil. In the case of Peterskirche, Satan was said to have wanted to rid Munich of its oldest and most devout parish by subjecting it to all sorts of spells, apparitions, and violent weather phenomena; however, the worst that he could do was bend the church's steeple.

Frauenkirche's encounter with Lucifer allegedly began with a wager between Satan and architect Jörg von Halspach. Satan was said to have challenged von Halspach to design a nave that could be illuminated without windows. When von Halspach succeeded in meeting the challenge, the devil was supposedly so outraged that he left his mark in the center of the nave—a footprint in the stone floor, the Teufelstritt (devil's footprint).

Another of Munich's prominent churches is the Renaissance-styled Michaelskirche (St. Michael's Church), built in the late sixteenth century by Duke William V. And like Peterskirche and Frauenkirche, it, too, has some colorful folklore associated with it. In 1590, shortly before its consecration, its tower collapsed. William interpreted this catastrophe as a sign from God that the church was not big enough, and so he embarked on a program to enlarge the structure, expending whatever sums the Bavarian treasury could afford to create a church more befitting of the Lord.

Despite William's belief that his parish was somehow not impressive enough, the design of Michaelskirche is remarkably conservative, at least compared to the decorative excesses of some of Munich's other churches. Its restrained facade and plain white stucco interior set it apart from many of Munich's other parishes. Michaelskirche is also the home parish of the Wittelsbach family; more than forty of its members (including William and Louis II) are entombed in its crypt.

One of Munich's smaller but most richly decorated parishes is the eighteenth-century Church of St. Johann Nepomuk. Built in honor of a fourteenth-century Bohemian monk by Cosmas and Egid Asam, the parish also sometimes goes by the name of Asamkirche (Asam Church). The parish, built between 1733 and 1746, was entirely financed by the Asam brothers. The exterior of their creation features a raw rock facade with giant pilasters. Its late-baroque interior features red stucco and rose-colored marble, adorned with frescoes, statues, and gilding.

Rich baroque styling can also be found in the Bürgersaal, built in 1710. Its interior features fanciful stucco foliage and paintings depicting parishioners in devout pilgrimages.

The interiors of both the Asamkirche and Bürgersaal reflect Bavarian sensibilities toward faith, which encompass a love of nature, a pious devotion to spirituality, and a harmonious relationship between the two. Many consider this outlook "earthier" (and more approachable) than the revered or austere views held by parishioners elsewhere throughout Germany and northern Europe.

Throughout the centuries, Munich (like Bavaria as a whole) has developed a reputation as a festive place, its inhabitants readily willing to partake in celebrations of all kinds. One of Munich's most raucous good times is the annual Oktoberfest, which has been held since 1810 in Theresienwiese, an enormous exhibition ground just outside the city center. The festival was first held to celebrate the engagement of Princess Therese von Sachsen-Hildburghausen to Bavarian crown prince Ludwig Wittelsbach (who would later become King Louis I). What began as a one-night party has since grown into a sixteen-day celebration, including carnivals, singing, dancing, parades, and, of course, beer drinking.

As noted earlier, Oktoberfest was by no means the only way Louis I liked to have a good time. He also derived great enjoyment from spending the Bavarian treasury's money, as evidenced by his many architectural endeavors. One of the last such projects he undertook is adjacent to Theresienwiese, the Bavaria statue. This bronze monument of the maiden Bavaria stands more than 100 feet tall and is regarded by many as Munich's Statute of Liberty.

Further Reading: A comprehensive but succinct book on Munich and its homeland is *Bavaria* by James Bunting (London: Batsford, 1972). In addition to describing sites of importance, the author provides a concise history of the city and state. *Legends and Tales of Old Munich* by Franz Trautmann, translated by Amelia Curtis Stahl (Munich: Lentner, 1910) provides a colorful history of the city's folklore. Written before two World Wars, the book presents an interesting historical perspective. An excellent recent accounting of one of the city's pivotal twentieth-century events can be found in *Munich 1923: The Story of Hitler's First Grab for Power* by John Dornberg (New York: Harper, 1982; as *Putsch That Failed: Munich 1923—Hitler's Rehearsal for Power,* London: Weidenfeld and Nicolson, 1982). Dornberg presents a detailed and insightful telling of Hitler's unsuccessful 1923 grab for power, which foreshadowed events to come.

—Brent Marchant

Nancy (Meurthe-et-Moselle, France)

Location: Northeastern France, 170 miles east of Paris in the Meurthe Valley.

Description: Capital of the Meurthe-et-Moselle department and the former Lorraine province, once the home of the dukes of Lorraine, known primarily for its collection of baroque and rococo architecture. The last duke, Stanislas, reigned in the eighteenth century and commissioned architect Emmanuel Héré du Corny to create a city center. Héré's design, Place Stanislas, is considered a classic example of urban planning. Nancy is also known for its contributions to the development of art nouveau through the work created by a group of artists and artisans known as the École de Nancy.

Site Office: Office de Tourisme
Syndicat d'Initiative
14 Place Stanislas
54011 Nancy, Meurthe-et-Moselle
France
83 35 22 41

Nancy's history begins as early as the twelfth century when it became the seat of the Lorraine duchy. The site was chosen by Gerard d'Alsace more for its proximity to the varied property holdings of the duchy than for its bucolic setting in the Meurthe River valley. In its infancy, Nancy was hardly more than the ducal castle and a monastery. A fire destroyed the original castle in 1228, but another one was built almost immediately.

By the fourteenth century, a tiny medieval town with twisting paths and alleyways had grown around the castle. Most of the businesses existed to serve the duke and his court. Heavy walls with mighty gateways surrounded the town to protect it from invading forces. The Porte de la Craffe, one of the city's oldest surviving gateways, was built in 1382 by Duke Jean I and until the French Revolution was used to house and torture prisoners. Architecturally, the structure is notable for its Renaissance-style exterior and its late medieval statues. In 1463, the Porte de la Craffe was further reinforced by the addition of two rounded towers.

In the fifteenth century, Duke René II founded the Church and Monastery of the Cordeliers, so named because of the white cords the monks wore around their robes. The two buildings are prime examples of the period's Gothic architecture. The Ducal Chapel, built over the tombs of the dukes of Lorraine, was inspired by the Medici Chapel in Florence.

Charles the Bold (Charles le Téméraire), the Duke of Burgundy, coveted the Lorraine region because it would link his territories in Burgundy to the south and Flanders to the north. In 1476, Charles successfully wrested control of Nancy from René II. The following year, René returned to rally the citizens to oust Charles. During a bloody battle that has grown through centuries of retelling to mythic proportions, Charles was slain. Legend holds that his body was later found, half-eaten by wolves, in a frozen pond.

The battle is considered one of Nancy's most historic moments, and René II is still revered as a hero. According to the legend, the cross that was used to inspire Nancy's soldiers was a relic from the crucifixion of Jesus Christ. The cross of Lorraine, with its two crossbars of unequal lengths, has since become a symbol of patriotism and was adopted by French naval forces in World War II as the combat symbol of France. In 1496, René II oversaw the initial construction of a new ducal palace. The building was completed by René's son, Antoine. A statue of Antoine astride a horse, built by Mansuy Gauvain in 1512, towers over the palace entryway, or Porterie.

Cardinal Richelieu, Louis XIII's chief minister, invaded Nancy in 1633 in his quest to extend France's royal holdings. The occupation was short lived, however, and the city soon returned to the Dukes of Lorraine.

Nancy was expanded during the reign of Duke Charles III in the late sixteenth century. Charles's creation was called New Town, an area of Renaissance-style buildings and straight streets crossing at right angles. Charles III also envisioned a city center that would connect the medieval and Renaissance sections of the city. However, it was not until the rule of Stanisław in the eighteenth century that the two sections were united by the magnificent architecture that made Nancy known worldwide.

Duke Leopold, who reigned from 1690 to 1729, commissioned the building of two churches. The Church of St. Sebastien, built between 1720 and 1731, was designed by an architect named Jennesson, who modeled it after the Church of St. Agnes in Rome. Germain Boffrand, a noted French architect of the time and author of a highly acclaimed tome called *Book of Architecture,* began work on the baroque-style cathedral during Leopold's reign. It was completed in 1742. Boffrand also designed the Hôtel de Fountenoy and Hôtel Ferrari, the homes of two of Nancy's wealthiest families.

The story of how Stanisław came to be Duke of Lorraine is typical of the peculiar manner in which the whims of the ruling classes often dictated daily life. In the eighteenth century, Duke François III of Lorraine married Austrian empress Maria Theresa and thus became an emperor. His new status apparently gave him more bargaining power, and he traded the Lorraine region to French king Louis XV in exchange for Tuscany. Louis was married to a Polish princess named Maria Leszczyńska, whose father, Stanisław, had

The fountain on Place Stanislas
Photo courtesy of Ville de Nancy

been dethroned in 1736. Louis, being a good son-in-law, gave Lorraine to Stanisław with the understanding that the region would become part of France when Stanisław died.

Stanisław was an educated and cultured man who enjoyed the fine things in life. In Lorraine, he created a stylish and brilliant court which attracted many talented and distinguished artists in the vanguard of the rococo and baroque movements. Stanisław had a particular passion for architecture and building projects. He loved to immerse himself in blueprints and to visit workshops. It is not surprising, then, that he devoted the remainder of his life to making Nancy an architectural wonder.

Arts and culture had always been influenced by reigning monarchs, so it was not unusual that when Louis XV ascended to the throne in the eighteenth century, architects and designers turned away from the majestic and imposing structures that had characterized Louis XIV's monarchy. The new style was called *rococo*, a combination of the French words *rocaille*, meaning rocks, and *coquille*, meaning shells. Rococo style was light, airy, and dainty. Whites and pastel colors were used in interiors. Ribbons, scrollwork, wreaths of flowers, and spiral ornamentations were popular. The popularity of rococo also signaled a shift in the center of architectural influence from Italy to France.

Stanisław commissioned two local artisans, architect Emmanuel Héré du Corny and wrought-iron master Jean Lamour, to create a city center to replace the moat that separated the Old Town and the New Town. Little is known about Héré and Lamour before they became part of Stanisław's court. It is known, however, that Héré studied with Germain Boffrand and that Héré used Boffrand's book as a guide in designing the town center in Nancy.

Héré's masterpiece was originally called Place Royale and is remarkable for its asymmetrical combination of diverse open spaces. At the southern end stands the Town Hall, onto which Lamour crafted a 120-foot wrought-iron bannister. To the right and left of the hall are four tall buildings. Facing the hall are two single-story structures embellished with arcades and terraces lined with balustrades decorated with stone vases, mythical figures, and trophies. On each of the four street corners, Lamour created gilded wrought-iron gateways. A set of gateways also provides the background for a fountain sculpted by Barthelemy Guibal honoring the mythical figures of Neptune and Aphrodite. The citizens of Nancy were so pleased with the results that they changed the name of Place Royale to Place Stanislas (Stanisław).

To link the Old Town and the New Town, Héré transformed a horse exercise yard into the Place de la Carrière, a linden tree–lined promenade that bisects Place Stanislas. At the north end is a rounded area bordered with curved colonnades and known popularly as the Hérmicycle. To the south is an Arc de Triomphe erected in honor of Louis XV. Two of Lamour's elaborate wrought-iron gateways stand on either end of the Place de la Carrière.

In 1750, Héré designed the smaller Place d'Alliance, east of Place Stanislas. It is surrounded by private homes and the ever-present linden trees. The square also contains a fountain designed by sculptor Paul-Louis Cyfflé to commemorate the 1756 treaty between the houses of Bourbon and Austria.

Héré's designs were a departure from traditional city squares that were surrounded and enclosed by tall buildings. Héré's squares were unique not only in their openness but also in their ability to be reached from all of the town's principal streets. Héré's design was also decidedly French rather than German, thus making a statement about the nationality of the town's rulers. This nationalism is evident in Lamour's incorporation of the symbols of French royalty—the sun, the lily, and the rooster—into the gates of the Arc de Triomphe.

Here also designed a church for Stanisław. Erected between 1738 and 1741, the Church of Notre-Dame de Bonsecours (Our Lady of Refuge), replaced the chapel built by René II in 1477 to commemorate his victory over Charles the Bold. The Church of Notre-Dame de Bonsecours holds the tombs of Stanisław and his wife, Catherine Opalinska. As agreed by Stanisław and Louis XV, Lorraine came under French rule upon Stanisław's death in 1766.

During the Franco-Prussian War of 1870–71, Nancy was captured by the Prussians, but ultimately freed. After the war, an influx of immigrants from Alsace and other parts of Lorraine bolstered its population and economy.

Nancy continued to attract artists and at the turn of the century played a large part in the development of the art nouveau style. An ornamental style of art noted for its long sinuous lines, the art nouveau movement began in England and quickly spread to the Continent. It was known by several names throughout Europe; however, the coining of the term art nouveau is attributed to a Parisian gallery.

Art nouveau's most famous names include Henri de Toulouse-Lautrec, Paul Gauguin, and Louis Tiffany. Within art circles, the members of the École de Nancy are also well known. Composed primarily of architects, cabinetmakers, and artisans working with wrought iron, ceramics, and glass, the École de Nancy also included designers specializing in tapestries, embroidery, jewelrymaking, and bookbinding. The group was headed by Émile Galle, a glass designer; Louis Majorelle, a furniture and ironwork designer; and Jacques Gruber, a stained-glass window designer. The art nouveau influence can be seen in Nancy's Chamber of Commerce building, banks, restaurants, shops, and private residences.

Nancy's participation in the world wars was limited. In August and September of 1914, the German army occupied several spots in the eastern heights, but, despite an attempt to enter Nancy on the north, was never able to fully take over the town. The citizens did suffer several bombing raids and a temporary loss of communication with Paris.

Although the city was occupied by German troops during World War II, Nancy was left virtually untouched by the bombing that devastated much of Europe. In August

1944, the German troops retreated from Nancy, and on September 11, the city was liberated by Allied forces and members of the French Resistance.

Today Nancy relies on tourism for much of its income. In addition to its historic architecture, Nancy is home to several museums. The Musée de Lorraine (Lorraine Historical Museum) is housed in the former Ducal Palace. Its exhibits include paintings, tapestries, sculptures, stain glass, and weapons from the Gothic and Renaissance periods. The Porte de la Craffe is now part of the Lorraine Historical Museum where visitors can see ancient instruments of torture and wall engravings scratched out by prisoners. The collection of paintings at the Musée des Beaux-Arts focuses on French and Italian masters from the fourteenth to the twentieth century.

Further Reading: *France at its Best* by Robert S. Kane (Lincolnwood, Illinois: Passport Books, 1986) is a lively travel guide with five pages on Nancy, including a short history, sightseeing, restaurants, and lodging. The *France* edition of the *Blue Guide* (London: Black, and New York: Norton, 1988) takes a considerably more serious look at the art and artifacts of Nancy. A diagram of Place Stanislas and a brief description of Nancy's significance in the history of architecture can be found in the *International Dictionary of Architecture*, edited by Randall J. Van Vynckt (Detroit and London: St. James Press, 1993).

—Mary F. McNulty

Nantes (Loire-Atlantique, France)

Location: At the confluence of the Loire and Erdre Rivers, about 35 miles inland from the Bay of Biscay and 240 miles southwest of Paris.

Description: Nantes is the largest city of France's historic Brittany region and capital of the Loire-Atlantique department. Its castle was home to the dukes of Brittany in the Middle Ages and in 1598 was the site of the signing of the Edict of Nantes, which granted religious freedom to Protestants in France. While much of Nantes has been altered in the twentieth century to improve river navigation and land transportation, some medieval landmarks remain, including the castle and the Cathedral of St. Peter.

Site Office: Office de Tourisme
Place du Commerce
44000 Nantes, Loire-Atlantique
France
40 47 04 51

Nantes is the first city of historic Brittany. The town's tenth century castle, which was rebuilt in the fifteenth century, was once the residence of the dukes of Brittany; the castle was also where the Edict of Nantes was signed by King Henry IV of France, in 1598. The famous edict granted religious and civil liberties to French Protestants, also called Huguenots, following the Wars of Religion; it was later revoked. During the French Revolution, Nantes was the scene of mass executions of royalists by drowning. A center of the resistance movement against the Germans during World War II, the city was badly damaged but has since been repaired. Nantes has undergone extensive change in the twentieth century.

Brittany's earliest recorded inhabitants were Celtic tribes from Britain, though prehistoric peoples had built stone monuments in the region. Nantes derives its name from the Namnetes, a major Gallic tribe of Brittany who built the town and made it their capital prior to the Roman conquest of Gaul. The original city center, Condovicnum, was situated atop low hills away from the Loire; the port, Portus Namnetum, was on the river. Roman soldiers established a settlement here in Caesar's time, and in the 50s B.C. the town was a Roman road terminus from the Rhine. It became an important commercial and administrative center under the Romans, who called the region Armorica. The town's two parts didn't join, however, until as late as the fourth century. Nantes withstood attacks by the Huns in 450, and in 560 the city fell into the hands of Frankish king Clotaire I, who placed it under the government of the bishop St. Felix. After the Romans withdrew from Armorica in the fifth and sixth centuries, there was a further influx of Celts into the area. Many Celtic missionaries came

at this time, following in the footsteps of St. Clair, who had introduced Christianity to the region in the third century.

The region was divided into several small lordships, without a strong central government. Charlemagne successfully imposed his authority, but Brittany revolted against his successors. Nominöe, who was proclaimed king of Brittany in 842, conquered Nantes and nearby Rennes, and razed Nantes's fortifications because the city had sided with Charles II, the Bald, king of the West Frankish kingdom. Viking raiders pillaged the city repeatedly in the ninth century, and they occupied it from 843 to 936. The successors of Nominöe rallied their people against the Norse invaders, until Count Alan (Alain Barbe-Torte) finally drove them away. Breton supremacy was established in 937, and Alan established his capital at Nantes. It was also about this time that feudalism took hold in Brittany, and the moated fortress castle of Nantes was built and became home to the dukes of Brittany.

In the Middle Ages, when Nantes began trading with northern Europe and became a local market center, the city saw a long struggle between the counts of Nantes and Rennes for control of Brittany, and the rulers of England sometimes became involved in the fights. When Philip Augustus, king of France, declared Pierre de Dreux (Pierre Mauclerc) the duke of Brittany, the duke made Nantes his capital; he fortified it, and defended it against King John of England in 1214. In the fourteenth century, France and England contested for Brittany, with the wars of succession fought between the partisans of two contenders for the rule of the region: Jean de Montfort, supported by King Edward III of England, of the Montfort faction; and Charles de Blois, duke of Brittany and nephew of of Philip VI of France, of the Penthièvre faction. Nantes first sided with the Montforts, but in 1342 the city was besieged and taken by the Blois-Penthièvres. The defeat and death of Charles at Auray, in 1364, secured the duchy for the Montfort house; Nantes didn't welcome the Montforts through its gates until their English allies had withdrawn from the region. The succeeding dukes of Montfort, aware of Brittany's strategic geographic position, tried to secure the duchy's neutrality between France and England during the Hundred Years War (1337–1453). In the fifteenth century, under Francis I and II, the great dukes of Montfort, Nantes was firmly established as the administrative center of Brittany. From the castle, the dukes governed Brittany as an independent country.

This era saw extensive building in Nantes. The first stone of the city's gothic Cathedral of St. Peter (Cathédrale de St.-Pierre) was laid by Duke John V in 1434; nothing remains of its twelfth-century Romanesque predecessor except a crypt that houses a museum of religions. The cathedral—higher than Notre-Dame in Paris—was not finished

Nantes Cathedral
Photo courtesy of French Government Tourist Office

until 1893, and the choir and chapels were not added until 1938 (the edifice suffered heavily from World War II bombing six years later.) The cathedral houses the magnificent Renaissance tomb of Francis II, the last Montfort duke of Brittany, who ruled from 1458 to 1488. The original University of Nantes (Université de Nantes) was established by papal bull in 1460, but was abolished during the Revolution. A new University of Nantes was founded in 1962, and its successor, an autonomous, state-supported school, was established in 1970, under a 1968 law reforming higher education. Francis II began reconstructing the ducal castle—called the Château des Ducs de Bretagne—in 1466. Though the exterior is noted for its high curtain walls, huge crenellated towers, and stout round bastions, the inner courtyard is in a typically flamboyant Gothic Renaissance style. Now Nantes's major historic building, the castle houses three museums.

The Nantes castle is notorious for another reason: Gilles de Rais (or Retz), a satanist who inspired the gruesome Bluebeard legends, was imprisoned in the original fortress before and during his 1440 trial for heresy and murder. Gilles had joined the Montfort clan, supporting Duke John V of Brittany against the Blois-Penthièvre house. He fought against the English in the late 1420s, and was made responsible for the safety of Joan of Arc in 1429–30, until she was captured. Gilles was made marshal of France in 1429. He became a patron of literature and music, and his court was more lavish than the king's. Though his parents had left him great wealth and possessions, Gilles rapidly squandered his inheritance. He became obsessed with alchemy, necromancy, and satanism, even while building a chapel near Nantes. In time, Gilles resorted to kidnapping children, mostly boys; he tortured and murdered about 200 children in grotesque sacrifical rituals. In 1440, he attacked the brother of a Brittany treasurer during Mass, which led Jean de Malestroit, bishop of Nantes, to investigate his activities. Gilles was arrested, condemned for heresy by the bishop, and sentenced to death for murder by a civil tribunal. Gilles confessed to the murders, and was hanged in the autumn of 1440.

Duke Francis II's daughter, Anne of Brittany or the Duchess Anne—twice queen of France—was born in the new chateau in 1476 or 1477. When Francis died in 1488, he left his young daughter as his only heir. King Charles VIII of France, accompanied by an army, came to the impregnable castle to woo Anne, but the king's men could enter only after Charles bribed the garrison commander. When Charles married Anne in 1491, Nantes, as well as the rest of the duchy, became French. Charles died in 1498. Anne's subsequent marriage to King Louis XII of France at the castle in 1499 solidified Brittany to the crown. Anne died in 1514. Her daughter Claude became queen of King Francis I of France, under whom the treaty of 1532 formally bound Brittany to France, yet granted the duchy certain local liberties. This provincial autonomy survived until the Revolution. King Francis II of France granted Nantes a communal constitution in 1560.

During the Wars of Religion (1562–98), when all of France was plunged into a civil war between the Roman Catholics and French Protestants (Calvinists, or Huguenots), Nantes backed the Catholic League against Henri de Bourbon, king of Navarre, heir to France and leader of the Protestant forces. At this time, Brittany—governed by Philippe Emmanuel de Lorraine, or the duke of Mercoeur—was the scene of many fights against the Huguenots. The duke tried to revive the independent duchy, beginning in 1588, but ten years later was defeated by King Henry IV of France, the former Henri de Bourbon. Religious peace was finally restored when Henry signed the Edict of Nantes at the castle on April 13, 1598. This charter granted limited religious and civil liberties to his Protestant Huguenot subjects, until the edict was revoked in 1685.

The Edict of Nantes was the first official recognition of religious toleration by a major European country. While the edict made Roman Catholicism the established religion of France (and reestablished the Catholic religion wherever it had been suppressed), it also permitted Protestants a large measure of freedom of conscience, a certain degree of freedom of public worship (except inside Paris), and a social and political equality with the Catholic majority. The decree provided special courts to adjudicate disputes between the faiths. The edict also allowed Protestants to hold government office and have parliamentary representation, to be admitted to colleges and academies, to trade freely and inherit property, and to hold synods and even political meetings. Although Huguenots represented a small percentage of France's total population, they had held more than 100 fortified towns; the edict formally gave them control of these and other specified towns, such as La Rochelle, which became the chief Protestant stronghold, for a period of time.

The edict was seriously enforced only until Henry IV's assassination in 1610, when the religious and political power of the Protestant Huguenots once again became precarious. The edict was resented by Pope Clement VIII, the Roman Catholic clergy in France, and by French governmental bodies. Cardinal Richelieu, especially, regarded the edict's political clauses as a danger to the state. Following a series of Huguenot revolts, Richelieu put down the political power of the Protestants by taking La Rochelle in 1628. With the Peace of Alais (Ales), in 1629, Richelieu modified the edict; while leaving intact the Huguenots' religious and civil rights, he took away their political rights and armed strongholds. King Louis XIV tried to force the Huguenots to con-

vert to Catholicism, and finally revoked the edict on October 18, 1685, depriving Protestants of all religious and civil liberties. The act led to more bloodshed and deprived France of its most prosperous commercial class: within a few years, several hundred thousand Huguenots emigrated to England, Prussia, Holland, and America. (The Edict of Toleration in 1787 restored Protestants' right to worship; the Declaration of the Rights of Man in 1789 once again guaranteed full liberty to Huguenots.)

Events surrounding the nine-decade duration of the Edict of Nantes did not have a substantial bearing on the city itself; except for an unsuccessful uprising against new taxation—''the revolt of the stamped paper''—in 1675, Nantes, and Brittany, remained mostly peaceful throughout the seventeenth century. An interesting episode occurred in 1661, however, when Louis XIV arrested his finance minister, Nicolas Fouquet, in Nantes.

Though a wealthy and powerful noble, Fouquet—who became France's last *surintendant des finances* in 1653—also enriched himself from the national treasury through irregular accounting practices. Between 1656 and 1660, he built a magnificent castle at Vaux-le-Vicomte, near Paris. A patron of writers such as Molière (who lived in Nantes for a time), Fouquet also acquired Belle-Isle, a fortress off the Breton coast, to which he could retreat in the event of prosecution. A plot was organized against Fouquet by his enemies, who informed the king of Fouquet's practices. In August 1661, Louis was honored with one of the most sumptuous fetes in French history at Fouquet's castle. The splendor of the setting and entertainment, which exceeded all that the king had ever spent on his own palaces, further aroused the king's suspicion; but he was dissuaded from arresting Fouquet on the spot. In September, Fouquet accompanied the court to Nantes. While the finance minister was being carried through town on his sedan chair, the lieutenant of the king's musketeers, d'Artagnan (of *Three Musketeers* fame), arrested Fouquet on charges of embezzlement. The sensational trial, which was a parody of justice, lasted nearly three years. Though initially hostile to Fouquet, public opinion turned in his favor when the maneuvers of his enemies were exposed; the charges did not turn out to be as serious as alleged. In any event, Fouquet was condemned to banishment in 1664, but the king commuted the sentence to life imprisonment. He died at a Piedmont fortress in 1680—the day before he was to be granted clemency.

Nantes expanded greatly as a trade center for sugar, rum, and cotton with the development of the first French colonial empire in North America. From the sixteenth through the eighteenth centuries, Nantes was the major European link in the slave trade, which also allowed the city to prosper; at one point, it was France's largest and busiest port. Nantes produced trinkets, ribbons, and fabrics that were traded for human cargo in Africa; the Africans were then brought to the New World. There, shippers would load sugar cane in the Indies and return it to Nantes for refining. Cotton shipped to Nantes would be woven into cloth sold in Africa.

Sea merchants made wealthy by the slave trade built lavish houses on the Ile Feydeau, a section of Nantes once surrounded by canals, which are now filled in and paved; most of the houses still stand. When the Revolution abolished slavery, sugar was made from beet roots; Nantes' economy declined, and became supported by canneries, cookie factories, and metal production.

In 1789, at the beginning of the French Revolution, Nantes declared its alliance with the future First Republic. In 1793, the city successfully resisted attacks by a royalist Catholic army in the 1790s Wars of the Vendée (although the royalists took the town and held it briefly in 1799). Named for a region south of Brittany, the Vendée was a peasant-organized counter-revolutionary insurrection. For four months in 1793, under the Reign of Terror, Nantes was the site of the infamous *noyades,* or drownings. Jean-Baptiste Carrier, an envoy of the revolutionary Committee of Public Safety, considered guillotine executions too slow, so he replaced them with wholesale drownings. Hundreds of men and women suspected of being for the *ancien regime*—or opposed to the Revolution—were bound together in pairs and forced into the Loire.

Famed science fiction writer Jules Verne—one of the most frequently translated of all French novelists—was born in Nantes in 1828 and lived here until he went to Paris to study law. It is said he dreamed of the romantic islands across the ocean, but never journeyed far on his sailboat. His house still survives in the Ile Feydeau area, although a separate museum houses Verne memorabilia.

In 1830, Nantes was one of the first provincial towns to revolt against Charles X, the last legitimate French king; the Bourbons were deposed, once for and all, in July, and Louis-Philippe was proclaimed "Citizen King." Two years later, in 1832, Nantes saw another episode of political intrigue. Charles X's daughter-in-law, Marie-Caroline-Ferdinande-Louise, duchess de Berry, had secretly gone to Marseilles in hopes of organizing a rebellion on behalf of her infant son, the legitimate Bourbon heir against Louis-Philippe. The attempt failed, and she stole into Nantes disguised as a peasant and hid in the house of faithful royalists. Armed with a printing press and a staff of two men, the duchess issued letters, dispatches, and proclamations to points throughout Europe from an attic room for six months. Her hiding place was revealed late in that year, however, and the authorities came to the house on rue Mathelin-Rodier looking for her, to no avail; the duchess and her staff had hidden in a secret compartment behind a fireplace. When two gendarmes posted to the attic lit a fire to keep warm, the duchess managed to snuff out the fire on her dress. The next morning, the guards lit another fire. Nearly asphyxiated, the duchess and her party finally opened the secret door and gave themselves up. She later served time at the Nantes castle, which had been turned into a state prison.

Railways, improved river navigation, and industrial diversification led to an economic revival in the Nantes area in the latter half of the nineteenth century. The nearby port of Saint-Nazaire—located thirty-five miles downstream on the Loire on the Bay of Biscay—was developed as a seaward terminal after 1856 to accommodate larger, deep-draft vessels; shipping and shipbuilding activities—Nantes's major industries—are still concentrated there. Between 1887 and 1891, a ship canal was built from Nantes to Saint-Nazaire. In 1914, during World War I, Saint-Nazaire was a major supply base for the British; Canadian forces disembarked there in 1915, and Americans landed there June 26, 1917, staying until 1919. Nantes was extensively modified by an urban renewal plan adopted in 1920, and harbor and port facilities were further improved between 1924 and 1935.

The Nantes area was occupied by the Germans during World War II, from June 22, 1940, to August 1944. Nantes was of strategic importance because a supposedly bomb-proof Nazi submarine base was located at Saint-Nazaire; the base grew to have fourteen bays that could accommodate more than fifty submarines. Nantes was also a center of the French Resistance movement; many of the town's citizens were taken hostage by the Germans and shot. Five days prior to German occupation, on June 17, 1940, Saint-Nazaire made naval history when it was the scene of the greatest loss of life on any single ship during a wartime engagement: the British ship *Lancastria* was dive-bombed by German forces, and 3,000 British troops were killed. The German submarine pens were damaged in March 1942 when the *Campbeltown,* an old British destroyer, was sent into the base loaded with time bombs planted by commandos. Allied bombing, which began in September 1943, caused extensive damage to the Nantes–Saint-Nazaire area.

After the war, the town center was replanned: river branches were filled in—wiping islands off the map—and made into roads; the railway was rerouted, made to run underground for most of its length; the Erdre River was diverted into a subterranean canal; and the port was substantially rebuilt again. Newer settlements have been built on former or existing river islands, and famed French architect Le Corbusier built flats in the Rézé-les-Nantes section in the 1950s.

Most recently, a national planning scheme has made Nantes into a major economic development center: transportation systems have been extended and vast industrial zones have been built, and much that was once unique or picturesque about the city has been sacrificed for a better flow of traffic. As Richard Wade writes in *The Companion Guide to the Loire:* "It has been neglected too long and allowed to become dowdy, and now suddenly with an aggressive, progressive vigour the authorities have awoken to a fury of futuristic rebuilding with all the pompous accompaniment of wide, neat boulevards in which the pedestrian feels lost and insignificant and only a mad motorist could be happy. The place is being torn apart in order to be remoulded to a style which itself will look shabby and out of date in little more than a generation."

Further Reading: *The Loire* by Vivian Rowe (New York and Washington, D.C.: Robert Luce, 1970) is a detailed, informative travel and history book about the Loire River and the surrounding region. *The Loire* by James Bentley (Topsfield, Massachusetts: Salem House, 1986) is profusely illustrated and evokes the unique flavor of the region, ranging from the city of Nantes to the Bourges area. *France in the Making: 843–1180* by Jean Dunbabin (Oxford and New York: Oxford University Press, 1985) covers the period from the collapse of the Carolingian Empire to the rise of the French monarchy, including Nantes's role in these developments. Another source worth consulting is Richard Wade's *The Companion Guide to the Loire* (Englewood Cliffs, New Jersey: Prentice-Hall, and London: Collins, 1983).

—Jeff W. Huebner

Nantwich (Cheshire, England)

Location: Nantwich lies in the rich dairy farming area of south Cheshire, in the northwest of England. It is sixteen miles from Chester, near the Welsh border, sixteen miles from the famous potteries of Stoke-on-Trent, and only eight miles from Junction 16 of the M6 motorway. Both the River Weaver and the Shropshire Union Canal flow through the town.

Description: Nantwich, a small market town, is one of Britain's listed towns of national historic importance, designated for conservation. It has a wealth of historic buildings, in particular timber-framed Elizabethan structures, and a good selection of Georgian buildings as well. It also has a thirteenth to fourteenth century church and a good deal of medieval woodcarving. The most important of the three Cheshire salt towns (the other two being Northwich and Middlewich), it has been an important trading center since Roman times. The city was the site of a siege during the English Civil War.

Site Office: Nantwich Tourist Information
Church House
Church Walk
Nantwich, Cheshire CW5 5RG
England
(270) 610983

Before the days of freezing and canning, salt was the only means of preserving food, as well as providing flavor. In fact, it was so important that Roman soldiers were paid part of their wages in salt, their *salarium,* from which we derive the word "salary." There are huge deposits of salt below the Cheshire plain, an area that was once a sea. When Britain had a tropical climate, millions of years ago, this sea evaporated and its salt eventually dissolved in underground water, producing brine. It was the discovery of these sources of brine during, or possibly before, Roman times that led to the establishment of Nantwich as a thriving community and trading center. Salt from Nantwich supplied the Roman garrisons of Stoke-on-Trent and Chester, and the original brine spring, called Old Biot, can still be seen today and now provides brine for the outdoor swimming pool. The Nantwich salt was particularly pure and therefore in great demand, not only for preservation but also as a seasoning and for use in cheese production. Cheshire cheese is one of England's best known cheeses.

In the past, wells and springs, as sources of life-giving water, were often treated as sacred places, and water cults were widespread all over the world. In Britain, holy wells had a special day when they were decorated with flowers and greenery, and rituals of thanks and blessing took place. The salt springs or brine pits in Cheshire were once decked out with green boughs, flowers, and ribbons on Ascension Day, but today well-dressing takes place mainly in the two adjoining counties of Derbyshire and Staffordshire, with some activity in Gloucestershire.

The various names by which Nantwich has been known underline its importance as a salt town. The Welsh, who first raided and then traded with the town for salt, called it *Hellath Wen,* meaning the white salt town; nearby Northwich was called *Heledd Dhu,* the black salt town. *Namet Wich* is a Saxon name referring to a center of industry, "the most famous wich."

Because of the demand for salt, roads known as the "salt routes" were built, connecting Nantwich with Wales and Ireland, and the north, south, and east of England, roads that can still be distinguished today. The River Weaver flows through the town, and later the Chester Canal was built, running initially from Chester to Nantwich and eventually extended to become part of the Shropshire Union Canal. Eight salt houses are recorded for the town in the Domesday Book, all of them being beside the river. However, roads have always been the most important form of communication for the town.

When the Normans came to Britain in 1066 Nantwich resisted; it was one of the last towns to succumb to the invaders. In the process it was nearly destroyed, with only one house left standing. It is from Norman times (1066–1189) that the earlier parts of the Church of St. Mary date, although there was a previous building dating from Saxon times (978–1066). In the twelfth century, Nantwich or Wich Malbank was part of the manor of Hugh Malbank, who in about 1130 was responsible for the founding of the nearby Combermere Abbey, one of the first Cistercian foundations in England. At this time he gave the Church of St. Mary to the abbey, along with part of the town. In the mid–thirteenth century a large church was built in the early English style of the time, over the Norman foundations, and in the fourteenth century it was rebuilt in the later decorated and perpendicular styles, incorporating the older parts of the building. In 1704 a library of theological books was established at the church; this library contains the Church Registers from 1539 onward and the only complete edition of Wynkyn de Worde's *Sarum Hymns and Sequences,* printed in 1506.

In the church there are some excellent examples of medieval woodcarving, including twenty misericords. Misericords are widely distributed in churches all over Britain, France, and most other European countries. The word originally referred to small indulgences or acts of mercy, such as allowing a sick monk or member of the clergy, bound to the

A view of Nantwich
Photo courtesy of Crewe and Nantwich Borough Council

strict rules of a monastic order, to have an extra blanket in winter or a little meat during Lent. The early church did not allow clerics to sit during services, but by the twelfth century churches began installing underseat ledges on which clergy could rest a little, half-sitting, half-standing. By the thirteenth century these ledge-seats, carved from a single block of wood, were being carved on the underside; the most surprising, even startling thing about the carvings is that they depict secular subject matter, including domestic and work scenes from everyday life. There are images of music and dancing; lovers; people dressing and undressing, in the bath; drinking beer, playing games, engaged in various professions and trades; and there are many depictions of animals. There are portraits and there is humor; there are even portrayals of sex.

No one is sure, but the subject matter of the carvings may be related to the fact that the clerics were to sit on them, and that to place one's bottom on a religious image would seem sacrilegious. In St. Mary's the misericords include images of St. George and the Dragon, a pair of wrestlers, and a woman beating her husband's head with a ladle while a pig eats the chicken she has prepared and a dog eats the stew. There is a mermaid, a virgin with a unicorn, and there are several dragons and other beasts.

There are also many other woodcarvings, including several of the Green Man or Jack-in-the-Green, a rather mysterious figure whose image appears in a great number of churches and cathedrals, often being the only decoration in early medieval churches, both in England and France. His image has also been found in the Rhineland of Germany and in Rome. Some think that he is a popular ancient pagan fertility god, his presence in churches being evidence of Christianity's flexibility in absorbing the pagan religions of the countries converted to it. The Green Man is often taken to be a symbol of spring and rebirth but, as he rarely looks happy, it has been suggested that he may represent lost souls. There is usually just a head with no body, almost always male, with foliage entwined around the head and growing out of the mouth and nose. He is often to be found on roof bosses, but in Nantwich he is to be seen on one of the bench ends, in the crossing, around the base of the vault, and on the exterior of the church, both in wood and stone. Throughout the town there are many examples of woodcarvings to be found on the inside and outside of buildings dating from medieval times onward.

During Edward I's Welsh Wars, which ended in 1284, Nantwich was particularly important to the crown, being a major route to North Wales. The town received many royal grants for the upkeep of a bridge and its roads because of this. In 1283, Edward I granted Nantwich the right to have an annual three-day fair at the Feast of St. Bartholemew, and with the conquest of Wales came a period of peace and prosperity for the area, in which farming and trade increased.

By the fourteenth century the street layout of Nantwich was established: a simple, radial plan still evident today. The names of the streets reflect the town's history. For example, Welsh Row refers to the Welsh farmers who drove their cattle along the street when bringing them from Wales to sell them for Nantwich salt. This street, the widest in the town, was also known as Le Frog Row after Frog Channel, which ran down the middle, once full of frogs. Hospital Street is named after The Hospital of St. Nicholas, founded by the first Baron of Nantwich in 1083. The several Wood Streets near the river are the streets where the wood needed to produce salt from brine was stored, while Monks Lane lies on land that belonged to the monastery.

There is also Castle Street, as a castle dominated the river crossing the twelfth and thirteenth centuries.

The oldest complete residence in Nantwich dates from about 1450, and is known as Sweetbriar Hall, on Hospital Street. It is one of the many in the town that are timber-framed. The Unitarian minister Dr. Joseph Priestley lived there for some time. He later became famous for his discovery of oxygen. John Gerard, the famous herbalist who published his *Herball or Generall Historie of Plants* in 1597, was born in Nantwich in 1545.

By 1548 the town's population was 1,800, and at the end of the sixteenth century there were 216 wich-houses, or salt-houses. This was the maximum number that had been agreed upon in order to keep the price of salt up. At this time other sources of salt were being discovered and developed, and slowly, from the seventeenth century onward, salt production in Nantwich declined. By 1792 there were only 3 salt-houses left. Fortunately, the town had developed other industries, such as cheese making and leather tanning.

In 1583 Nantwich suffered its Great Fire, which is said to have lasted twenty days. Fire fighting was hindered by the presence of four bears roaming the streets after being freed from their cages by the fire. In those days bear-baiting and cockfighting were still practised. The disaster left few buildings standing. Apart from Sweetbriar Hall, structures that survived included The Lamb Inn, built in 1552, and Churches Mansion, dating from 1577. The latter is perhaps the finest of the town's half-timbered buildings, and in 1930 there was a move to export it to the United States in its entirety. This was prevented and the restored building now serves as a restaurant.

After the Great Fire there was a surge of rebuilding. The town's importance to royalty as a supplier of salt, cattle, and services to the nearby army was confirmed by Queen Elizabeth I, who mounted a national relief collection, as she had done in Portsmouth two years earlier; she made a generous contribution to the fund herself. It is from the Elizabethan period that many of Nantwich's most beautiful buildings date, including many inns, almshouses, and schools, as well as residences. At this time the town was beginning to develop as a center of hospitality for travelers, and by the 1580s there were eight important inns and a postal service. By 1792, at the height of the coaching era, there were thirty-four inns and public houses.

The seventeenth century brought the English Civil War. Nantwich was alone in Cheshire in supporting the Parliamentarians, who had their headquarters in the Lamb Inn.

One mile out of Nantwich there is a Jacobean country house, Dorfold Hall (built 1616 between and 1621), which was the Royalist headquarters in the area and from which attacks on Nantwich were mounted. Many Civil War battles took place nearby, and on January 24, 1644, the Royalist siege of the town ended with the Battle of Nantwich, which resulted in a victory for the Parliamentarians. Today the battle is reenacted every year on the nearest Saturday to January 25, called Holly Holy Day because in 1644 the local people celebrated victory by wearing sprigs of holly in their hair and hats.

The Great Fire of the sixteenth century, and the growing scarcity and resulting price increase of wood led to an increase in the use of brick in construction as the seventeenth century progressed. Dorfold Hall was an early example, but by Georgian and Regency times (1714–1830) brick was the main building material in many areas, unless there was a good supply of local stone. There are many excellent Georgian buildings in Nantwich dating from the eighteenth century, a time when the town was becoming known as a spa town, the salts in the water being reputed to have healing properties. As a spa town and a stopping point for coaches, Nantwich experienced another period of expansion. New trades were introduced, including cotton spinning, wigmaking, tobacco, thread making, and straw plaiting and dyeing; the importance of agriculture to the market town has continued to the present century.

Today Nantwich is still a flourishing market town.

While its history is evident in the names of its streets, its layout, and its buildings, Nantwich is no museum piece. The timber-framed buildings are carefully preserved, but their present owners and occupants are computer firms, antiques and craft shops, restaurants and nightclubs. There is a marina on the canal and a museum is housed in the 1888 building that was originally the first ''free'' library. In Northwich, one of the other two salt towns about fifteen miles to the north, there is an excellent museum, the Salt Museum, displaying the history of the salt industry in the whole area. To the south of Nantwich are the Stapeley Water Gardens, the world's largest water garden center, set in sixty-five acres of countryside. There are two acres of water gardens, a one-and-one-third acre glass pavilion housing a tropical environment, complete with parrots and fish from all over the world, and a museum.

Further Reading: *A History and Guide to Nantwich* by J. J. Lake, fourth edition, (self-published, 1988) is the standard twenty-page outline history of the town. *The Great Fire of Nantwich,* also by J. J. Lake (Nantwich: Shiva Publishing, 1983), reconstructs the sixteenth-century town in detail, including accounts both of the fire and the rebuilding. *The Treasures of Cheshire* by Norman Bilsborough (Swinton, Manchester: North West Civic Trust, 1983) covers the history and noteworthy sites of the whole county. The NWCT was founded in 1961 with the aim of promoting environmental improvement in the northwest area of England.

—Beth F. Wood

Narbonne (Aude, France)

Location: Ninety-one miles east-southeast of Toulouse in southern France, twenty miles from the Mediterranean Sea; situated in the region of Languedoc and department of Aude.

Description: Former capital of Roman Gaul; also, with the decline of the Roman Empire, a capital of the Visigoth kingdom in southern Gaul; a principal center for the heretical movements of the Albigensians and the Franciscan Spirituals and, consequently, a center of Inquisitions in the thirteenth and fourteenth centuries; home for a significant Jewish population in the Middle Ages, from which they were expelled in the fourteenth century by the king of France. Retains much medieval architecture, including St. Just–St. Pasteur Cathedral and Archbishops' Palace.

Site Office: Office de Tourisme, Syndicat d'Initiative
Place Roger Salengro
11100 Narbonne, Aude
France
68 65 15 60

Narbonne, located in the southern French region of Languedoc, was at one time the capital of Roman Gaul. Known then as Narbo, it was the greatest city in Gaul in the first century A.D., the main port of exit to Spain and Italy for Gallic goods. Apollinaris Sidonius, famed Gallic bishop and poet of the fifth century, acclaimed Narbonne's flourishing city life, its promenades, markets, forum, and theater. In Sidonius's day, it could boast of one of the finest universities in Gaul.

Though conquered by Rome in 125 B.C., developed as a colony and center of trade by the tribune Caius in 124–23 B.C., and, in later centuries, regarded as nearly a colony by a more prosperous northern France, Narbonne has fought to maintain a tradition of independence from the powers that be. It enjoyed periods of material prosperity, from the days of Caius to the Middle Ages, when development in shipbuilding produced smaller vessels that could reach inland ports such as Narbonne, and therefore increased trade. However, Narbonne's material prosperity was inevitably threatened from time to time. In A.D. 406, Vandals, along with Alani and Suevi tribes from the east, devastated Gaul, plundering Narbonne. Later, beginning in 414, the Visigoths would control southern Gaul, designating Narbonne as its capital. Placidia, the wife of Autaulf, brother-in-law of the great Visigoth leader Alaric, ruled it steadfastly for twenty-five years. Perhaps a woman leader symbolizes in one way what sets Narbonne and its region, Languedoc, apart: not prosperity, but unorthodoxy.

In the early Middle Ages, knightly courts in Languedoc were noted for their divergence from traditions of the crown. Narbonne's was one of the most noteworthy in this regard. There, members of the clergy and nobility, men and women, gathered and experienced new delights in dress, cloth, and spice, challenged themselves with intellectual games like chess, and heard poetry recited. In these courts, love songs emerged as well, the theme of which was in sharp contrast to the overriding military preoccupation of northerners. These songs, as refined by troubadours, resulted in the formulation between 1100 and 1140 in Languedoc of a new ideal for relations between men and women: courtly love.

Courtly displays of unorthodoxy, however, pale next to the behavior of the Narbonnese during the period for which the city is best known, the time of the Albigensian Crusade in the thirteenth century and subsequent Inquisitions. Southern France in 1200 was composed of essentially independent principalities with only theoretical allegiance to the king. One of the by-products of this independence was religious tolerance. Narbonne, as a place of trade accustomed to contact with outsiders, had a significant Jewish population. In addition, it became one of the first centers of the Albigensian heresy, perhaps as a result of contact with Moslems and Jews, as well as merchants from heretical towns in Bosnia, Bulgaria, and Italy. From the French town of Albi, the Albigensian movement, also known as Cathari, from the Greek for "pure," rejected the Catholic Church while embracing the precepts of the Sermon on the Mount. An austere life was the Albigensian ideal, grounded in a belief that all matter was evil. Its clergy emulated the "perfect" life: abstinence from sex (Adam and Eve's sin was coitus), meat (with diet restricted to fish and vegetables), and the use of force (capital punishment was criminal). Albigensians believed that God would eventually triumph over this world of evil.

Most important, in view of later events, was the movement's harsh criticism of the Catholic Church. Its adherents did not regard the Catholic Church as the Church of Christ, nor did they believe that St. Paul founded the papacy; popes succeeded emperors, not the apostles in their view. The clergy, rich and worldly, were seen as latter-day Pharisees. Pope Innocent III, just after assuming the papacy in 1198, denounced the movement in a letter to the archbishop of the Languedocian town of Auch. He decreed that the heresy must be eliminated, by force if necessary. The archbishop of Narbonne, however, rejected these demands when legates of Innocent were sent there to enforce them. Eventually, Innocent called for a crusade against the Albigensians, summoning Christians from all lands to the holy war on Languedoc that began in 1209.

Narbonne's resistance to the crusading forces led by Simon de Montfort was initially nonviolent. With their town under threat of attack, Viscount Aymery and Archbishop Berenger, the two leading powers of Narbonne, negotiated its

Narbonne, with the Archbishops' Palace and the cathedral at center
Photo courtesy of Ville de Narbonne

surrender, including regulations against heresy. Minimal prosecution of heretics followed, however. Though one condition of the surrender was support for the crusade, this proved to be lukewarm at best. In one campaign led by Montfort against the knight Guirard de Pépieux in 1209, the Narbonnese men deserted Montfort and returned home under Aymery. Narbonne's resistance to the crusade became more aggressive in 1212, when the presence of Montfort's son and brother in the town sparked an anti-French riot resulting in the death of several crusaders. One year later, Montfort himself was prevented from entering Narbonne.

Nonetheless, Montfort possessed greater military resources than Aymery, who was essentially Narbonne's military chief, and in 1215, the viscount was compelled to pay homage to Montfort, who had proclaimed himself Duke of Narbonne. In addition, Montfort demanded the dismantling of Narbonne's walls. In 1218, Montfort agreed to their reconstruction on condition that no heretics or their supporters be admitted into the town. Aymery's acquiescence to the forces of the Albigensian Crusade, while not necessarily reflecting the wishes of the citizens, seems to have spared Narbonne the death and destruction suffered by the nearby towns of Béziers and Carcassone. While the Narbonnese cherished their independence, they were evidently a practical people.

With the death of Simon de Montfort later in 1218 and his son Amauri's subsequent failure to hold on to his father's conquests, Amauri left in 1224 and ceded all of his claims in Languedoc to the king of France. A year later, Pierre Amiel became Archbishop of Narbonne. Amiel, after securing an agreement with the king to confiscate from rebels what were now lands claimed by the crown, turned against his own people as no other viscount or archbishop of Narbonne ever had. His archiepiscopal court was empowered to make arrests and pronounce sentences following inquisitions led by the Dominican friar Ferrier. Ferrier would eventually become one of the leading inquisitors in the region of Languedoc.

Inquisitorial activity in Narbonne was actually mild at this time. However, an incident in 1234 gives evidence of the Narbonnese love of independence, even in confronting an archbishop aligned with the king of France. When one of their citizens, Raymond d'Argens, was arrested on suspicion of heresy and his home was targeted for confiscation, fellow Narbonnese forced his release and, the following day, defended him against another attempt at his arrest. This led to an interdict by Archbishop Amiel against that portion of Narbonne, the bourg, he deemed responsible. He excommunicated all members of the Armistance (a society sworn to defend the rights of the people), which he believed impeded the Inquisition. A letter written by consuls of the bourg to the consuls of Nîmes explaining their position reveals their attitude regarding the Inquisition: it abused the ordinary rights of the citizens; arrests and confiscations of property were frequently conducted without any justification; and properties were confiscated *before* trial. Eventually, three years of ne-

gotiations between consuls of the bourg and consuls of the cité, the section of Narbonne where the archbishop resided, resolved this conflict. The excommunications were lifted; the Narbonnese had not capitulated. After 1237, there were only ten recorded cases of heresy in Narbonne during the remainder of the thirteenth century. The conflict of 1234 to 1237 may have dampened the resolve of Amiel and subsequent archbishops of Narbonne.

However, in the fourteenth century, Narbonne experienced a resurgence of the Inquisition, this time at its worst. The town had become home to various mendicant (monastic) orders such as the Dominicans and the Franciscans. Within the Franciscans, there developed a group called the Spirituals who decried the order's increasing materialism, and Narbonne became an important center for the Spirituals. When Pope John XXII declared Spiritual doctrine heretical, his wrath was keenly felt in Narbonne, where the movement had enjoyed a large lay following. In 1318, two friars from the Narbonne convent were burned at the stake; by 1328, forty-nine citizens of Narbonne, men and women from all classes, were brought before the Inquisition. This time, the Narbonnese were in no position to resist, for by now the king of France had established his power in the region; with his immediate support, the archbishop could challenge any uprising.

This royal power was felt in another way as well. In 1306, King Philip IV called for the expulsion of Jews from Narbonne and other locations in southern France. While this may have been done to acquire their land and other possessions, it is likely that the king pursued this course mainly to distinguish himself as a defender of the Catholic faith. Though succeeding Capetian kings readmitted the Jewish population, Charles IV, the last of the Capetian monarchs, renewed the expulsion, and after 1322, there were virtually no Jews in France.

The Narbonnese most certainly did not embrace this policy, as they are noteworthy for a history of toleration of different faiths. Narbonne had been a home to a large Jewish community. Medieval scholars such as Joseph Kimchi and his sons Moses and David thrived there. David's *Compendium* served as an authoritative grammar of Hebrew for centuries and a constant aid in translating the King James Bible. As a gesture of its feelings toward the Jewish population, Narbonne formally regularized relations between Christians and Jews through an archiepiscopal charter of 1284. Narbonne emerged as a safe haven for Jews in the region. From there, Meir ben Simeon felt free to denounce the crown while praising life in his city, contrasting it with Frankish oppression in other domains. Until the expulsion of 1306, the citizens of Narbonne resisted pressures from the French monarchy to discriminate against Jews. But, as in the case of the Inquisition against Franciscan Spirituals, the Narbonnese were no match for a superior military supporting a definitive policy.

The crown's political influence in Narbonne and the region of Languedoc would not wane from the fourteenth century until the French Revolution. In the seventeenth century, it was evident in the institutions of the Estates and the person of Archbishop Bonzi. The Estates were a form of assembly convoked by the crown once a year to negotiate between king and province appropriate taxes. This assembly consisted of archbishops, bishops, barons, and representatives of the towns. Yet, while the Estates were ostensibly a means of negotiation, in fact the king had ultimate power to enforce his wishes.

The archbishop of Narbonne presided over the Estates. Pierre de Bonzi, one in a long line of bishops of neighboring Béziers who were essentially clients of the crown, quickly rose from that appointment in 1659 to Archbishop of Narbonne in 1673. Bonzi's sympathies were so aligned with the monarchy that he was called the "king of Languedoc." He even obtained a royal decree that he alone had the right to have an embroidered cushion under his arms and a carpet under his feet when Catholic mass was heard during sessions of the Estates. Thus the Estates, theoretically an assembly for negotiation, was headed by a man loyal to the king, who had benefited from a system of royal patronage.

Despite the power of the French monarchy in the seventeenth century, the Narbonnese spirit of independence and justice persisted. In 1635, citizens of the town angrily protested what they considered an unfair system of taxes, demanding its revision at a council meeting. A group of townspeople met secretly in 1640 to draw up a lawsuit against the tax system. This was followed by a similar lawsuit in 1641. Narbonnese disrupting an election in 1647 voiced the same grievance. And in 1651, many of its citizens even took up arms protesting corruption in the Estates, such as its extortion of taxes. While the people of Narbonne were relatively powerless to reform this system, they would at least be heard.

In the eighteenth century, the king appointed all bishops, and the higher clergy were nearly all from noble stock; 129 of the 130 bishops in France in 1789 came from titled families. When these men entered the church, they retained their habits of privilege. Narbonne's Archbishop Dillon told King Louis XVI why he hunted while prohibiting his clergy from doing so: "Sire, my clergy's vices are their own; mine come from my ancestry." However, revolution in the eighteenth century altered the balance of power in favor of the common person of Narbonne and other towns and cities throughout France.

In the nineteenth century, Narbonne continued its proud tradition of defending the individual's rights. In what is termed the red revolution of 1871, a Paris Commune, anarchist and opposed to state power, led the way to the formation of similar communes in other French cities. Narbonne was one of the few that had already organized revolutionary communes of their own. The 1871 communes considered themselves autonomous and free to devise their own reforms without interference from the central government. The communes' ultimate success was limited by the fear their anarchist approach engendered in some provincial cities (though not Narbonne), and by the communes' own realiza-

tion of the value of a centralized state in opposing one's opponents; still, the communes were responsible for an 1882 law that restored the right of municipal councils to elect mayors. Previously, the central government had appointed mayors in towns with populations exceeding 20,000, which included Narbonne.

In the twentieth century, there have been other instances of Narbonnese defiance of northern power. In 1907, wine producers in Languedoc revolted against falling prices they believed to be the result of fraudulent overproduction in the wine market. A massive demonstration of 100,000 people took place at Narbonne. Their leaders were Marcellin Albert, a wine producer from a town near Narbonne, and Ernest Ferroul, the radical-socialist deputy-mayor of Narbonne. Both evoked the ideals of Louis-Xavier de Ricard, a defender of Languedocians, whom he called "freethinkers." Ferroul equated industrialists of northern France, the sugar-beet producers who competed with the wine makers in alcohol production, with feudal barons of the thirteenth century who had ruined Languedoc and other southern regions. Albert called on the demonstrators to be as courageous as the people of Carcassone were in defending their city against the Albigensian Crusaders. Unfortunately, the wine producers suffered a fate similar to that of the thirteenth century Carcassès; Clemenceau's troops crushed their movement, firing on the crowd at Narbonne.

In the 1940s, all of France faced a threat far greater than the abuses of a central French power. When the Germans occupied southern France in late 1942, they confronted in Languedoc an opponent with a long history of freedom fighting. Some of the most effective Resistance fighters of the war came out of Languedoc, many from Narbonne.

Today, Narbonne is a quiet, unassuming town of some 40,000 and seems typically "country" French. There, the tourist might sit down to a bottle of Languedoc-Roussillon wine accompanied by a sampling of the region's famous goat cheese and a loaf of bread. Narbonne's Archbishops' Palace and its thirteenth- and fourteenth-century Gothic cathedral are much visited. The palace is actually two building complexes, the Old Palace and the New Palace, which are joined by an arcade. The oldest parts of the Old Palace date from the ninth century; the New Palace was begun in the fourteenth century. There have been extensive additions and renovations to both complexes since their original construction. The cathedral dates to the thirteenth century. Other historic sites include the remains of a Roman bridge that links the bourg district to the site of the original Roman settlement. But the pleasant atmosphere of this town belies its history of various struggles for independence. It is for those efforts that Narbonne should be remembered.

Further Reading: *Heresy and Inquisition in Narbonne* by Richard W. Emery (New York: Columbia University Press, 1941; reprint, New York: AMS, 1967) provides an in-depth study of Narbonne's involvement in the Albigensian and Franciscan Spiritual movements and the subsequent Albigensian Crusade and Inquisitions, arguing that archiepiscopal success in suppressing heresy there depended on the support of the French monarchy. *The French Monarchy and the Jews* by William C. Jordan (Philadelphia: University of Pennsylvania Press, 1989) devotes a lengthy section to the expulsion of Jews from Languedoc in the fourteenth century, an often overlooked event in French history. *The Age of Faith,* the fourth volume in Will Durant's eleven-volume *Story of Civilization* (New York: Simon and Schuster, 1950; Sydney and London: Angus and Robertson, 1956), includes a clear description of the philosophy and behavior of Albigensian believers, along with some information on the Jews of Narbonne. William Beik's *Absolutism and Society in Seventeenth-Century France* (Cambridge and New York: Cambridge University Press, 1985) is a thorough study of the workings of the French monarchy in Languedocian provincial government, including extensive discussion of its relationship with Archbishop Bonzi of Narbonne.

—Robert M. St. John

New Lanark (Strathclyde, Scotland)

Location: In the valley of the Clyde near the old market town of Lanark.

Description: One of the world's most important and influential industrial heritage sites; cotton manufacturing community established in the 1790s; turned into a model industrial community in 1799 by Robert Owen, who ran the facilities until 1828.

Site Office: New Lanark Conservation Trust
New Lanark Mills
Lanark, Strathclyde ML11 9DB
Scotland
(555) 665738

Cotton manufacturing was the first large-scale factory industry in Scotland. The largest and most successful "cotton town" was the model industrial village of New Lanark, which reached its peak population of 2,300 in 1821. A major tourist attraction in its day, New Lanark was admired throughout the world as much for its enlightened labor policies as its beautiful setting near the Falls of Clyde.

New Lanark earned its greatest fame under Robert Owen, an Anglo-Welsh entrepreneur whose brand of benevolent paternalism formed the foundation of his social and educational experiments. Community building and formation of character were among Owen's major concerns.

The cotton-spinning industry in Scotland was centered around Glasgow and its neighboring towns and villages. New Lanark was the perfect locale for a cotton mill. A steady supply of water was provided by the River Clyde, and the neighboring town of Lanark had a ready supply of well-skilled linen weavers, stonemasons, and clockmakers.

New Lanark was founded in the early 1780s by David Dale, a Scottish banker, to house the laborers who worked in the cotton mills. Construction on the first mill began in 1785; the mill commenced operations in 1786. By the end of 1793, four mills had been built and housing to accommodate as many as 200 families had been erected. For a short period of time, Richard Arkwright, the famous English inventor and entrepreneur, was a partner. Arkwright had expressed an interest in establishing a cotton mill in Scotland. In 1784, he visited Dale at New Lanark and came away immensely impressed with its desirable location adjacent to the Falls of Clyde, but the partnership ended on a sour note, apparently as the result of a dispute over the design and position of a wooden cupola.

Most of the workers were outsiders. In general, factory work in Britain was known for its strict discipline, harsh conditions, and long hours—hardly an incentive for most people. Indeed, factories were often compared to Dickensian workhouses. Despite its enlightened reputation, the owners of New Lanark were forced to look elsewhere for employees.

By far, the greatest number of employees at New Lanark were from the Scottish Highlands, mostly from Inverness and Argyllshire as well as from the Caithness area in the far north of Scotland. Displaced by economic and social turmoil in their native glens, Highlanders sought work in the Lowlands and the industrial centers. Highland ministers, exerting their influence, persuaded members of their flock to stay and work in Scotland. Dale, for his part, recruited potential employees by some rather unconventional means. In 1791, for example, bad weather forced an emigrant ship, the *Fortune,* bound for North Carolina with 400 passengers, to stop at the Scottish seaport town of Greenock for repairs. Dale convinced many to cast their lot with him rather than leave for the great unknown in North America.

Women and children formed a large proportion—as great as two-thirds—of New Lanark's workforce. Dale recruited children as young as five or six years old to work in the mills; the majority were orphaned children from the Charity Workhouse in Edinburgh.

By the turn of the century, New Lanark had apparently become too much of a financial burden for Dale to bear. In late 1799, Dale sold New Lanark to his son-in-law Robert Owen. While Owen had business partners who shared in the ownership, he was to be sole manager. Dale died seven years later, at the age of 67.

Robert Owen was born in Newtown, Montgomeryshire, England, in 1771, the son of a hardware merchant and saddler. He left home at the age of ten to move to London, where he worked in the drapery business. For fifteen years he worked in Manchester, learning factory management. In 1794, he was made a partner in Manchester's Chorlton Twist Company, which had business links in Glasgow. It was in Glasgow that Owen met Ann Caroline Dale, David Dale's eldest daughter. Around this time he first visited New Lanark and met Dale. Owen and Caroline Dale eventually married, after overcoming Dale's initial reservations, in September 1799.

Upon assuming control of operations at New Lanark, Owen introduced a series of changes that were unpopular with his employees. Discipline was strictly enforced, and every aspect of factory production supervised closely. Owen was also intent on improving the social conditions and social behavior of the workforce. He introduced evening patrols of the village streets, even going so far as issuing fines and threats of dismissal for public drunkenness. He improved housing and sanitation, had the streets cleaned regularly and, where necessary, paved, established a company store, and provided free medical service for his employees. He improved the working conditions through better organization

New Lanark, 1818, from an original by John Winning

A dancing class in Robert Owen's school, 1820
Prints courtesy of New Lanark Conservation Trust

and an appeal for cooperation. He also ended the practice of hiring children under the age of ten.

Word continued to spread about the famous community in the Scottish Lowlands. It became a particular favorite of middle- and upper-class visitors. New Lanark had certainly attracted its share of visitors in David Dale's day, but now people of the caliber of the English poets William and Dorothy Wordsworth and Samuel Coleridge included New Lanark on their traveling itinerary.

The growth of Britain's working-class population frightened many members of society's elite. Owen's solution to the labor problem was to promote a brand of social reform that would produce a healthy and content workforce—some argue that Owen really meant a docile workforce—who would willingly conform for the benefit of the community. He wanted to turn New Lanark into a model community, not by forcing his workers to do his bidding but by convincing them through education and behavior modification that his way was the right way. With some gentle coaxing and persistence, he was certain they would eventually come around to his way of thinking. His goal was the greatest happiness for the greatest number of people. In short, the perfect society would be based on a new moral order of mutual cooperation and harmony. Through proper education and improved working and social conditions, it could be attainable within a relatively short time span.

Owen was a businessman with a social conscience and he played the role of the visionary entrepreneur to the hilt, much to the chagrin of his increasingly irritated partners. The partnership was under severe strain; they had little patience for Owen's grandiose plans, nor did they share his essentially philanthropic view of conducting business. Owen had written a series of works declaring his philosophy of life. *A New View of Society,* published in 1812–13, called for improved social conditions and the establishment of a new world order. It received considerable attention as far away as London.

Meanwhile, the relationship with his partners continued to deteriorate. Conditions became so bad in fact that the partners—who, together, held majority control of New Lanark—insisted that the mills be sold by public auction. Owen was a formidable opponent, however. Although not a quarrelsome man, he did have many friends in high places and had become known as a social and educational reformer of the first order.

By this time, Owen had begun searching for more appropriate partners. He found six men who shared his ideas: Joseph Foster, a Quaker philanthropist; Joseph Fox, a dentist and religious dissenter; Michael Gibbs, who later became lord mayor of London; John Walker, a wealthy and cultured man; William Allen, a businessman, Quaker, and philanthropist; and the utilitarian philosopher Jeremy Bentham. But Owen's estranged partners refused to be intimidated. New Lanark was put up for sale in Glasgow in December 1813 with his rivals haggling over the sale price; they may have wanted to sell New Lanark but they did not wish for Owen to

be the one to purchase it. With the considerable economic assistance of his new partners, however, Owen regained control of New Lanark.

Owen's primary vehicle for reform was education. He felt it was vital for people to receive a proper education at a young age. The Institute for the Formation of Character, which opened its doors on January 1, 1816, was New Lanark's showpiece. It consisted of a school for the children who worked in the factory, public halls, and a nursery school. Owen took particular pride in his nursery school. Attendance was free, not just for children with parents employed in the mills, but for any child who lived in the community.

Owen was influenced by the Swiss educator Heinrich Pestalozzi. Pestalozzi promoted a gentle approach to teaching. For this reason, abuse of any kind—verbal or physical—toward the children was strictly forbidden. Voices could not even be raised. Pestalozzi also emphasized the importance of observation and experience, which suited Owen, who did not particularly care for the use of textbooks—he stressed hands-on experience when given a choice. Students were also encouraged to participate in field studies. In a departure from traditional teaching methods, especially with students so young, Owen encouraged instruction by lecture, discussions, and debates and offered such subjects, unusual for the time, as geography, science, and ancient and modern history.

The children also received lessons in dancing and singing. They were taught to sing in harmony in 200-member choirs. Military-style exercises helped instill a sense of discipline, as did the school uniform—a tartan cloth worn like a Roman toga. In short, Owen believed that education should be spontaneous and natural, it was something that children should enjoy, not dread—quite a radical philosophy in his day in Britain. Many visitors to New Lanark came away impressed by the children's grace and intelligence, hailing the Institute for the Formation of Character as a modern wonder.

Eventually, Owen spent more time lecturing and giving speeches than addressing the daily concerns of life at New Lanark. He was an opinionated man who did not attempt to hide his feelings. He also held some some fairly controversial views on established religion. So when he publicly attacked the usefulness of religion in modern society, his partners—some of whom were Quakers—balked. While he was on a trip to Ireland, they decided to take advantage of his frequent absences by changing policy. They insisted that religious instruction be taught in school; that singing, with the exception of psalmody, be discontinued; and that all boys six or older should wear trousers, not kilts, since the latter, they believed, helped to promote an atmosphere of impropriety. Owen reluctantly accepted these concessions; by this time, he had lost interest in New Lanark and had turned his sights to the New World.

Owen had known about the Rappite community of New Harmony, Indiana, for some time. An English silk manufacturer named George Courtauld had visited New Lanark and told Owen of Father George Rapp and the 1,000 German immigrants who had founded the Harmony Society in Indiana in 1814. Courtauld and others praised the community's high moral character and well-behaved inhabitants. New Harmony appealed to Owen, who, by this time, was looking for a ready-made community to continue the experiments of New Lanark. Father George Rapp, the community's founder, knew of Owen's interest. In 1824, Rapp and his followers returned to their original home in Pennsylvania and sold the town to Owen. Owen left New Lanark in 1824, although he did not actually relinquish management of the mills until the following year. Three years later, in 1828, he sold the mills and his remaining shares. Owen died in 1858.

After Owen left New Lanark, the Walker family took over. In 1881, they sold the works to Henry Birkmyre. Birkmyre introduced a number of changes at New Lanark, including the manufacturing of fishing nets, the installation of new windows in houses, the introduction of electric light, and a new drainage system.

By 1951, New Lanark's population had dwindled to 550. Mill buildings and machinery were modernized during the 1950s and 1960s, but the condition of the housing stock proved the most expensive undertaking. Finally, the village was faced with a painful dilemma—either update the housing or risk the very real possibility that the houses would be condemned for safety and health reasons. To relieve themselves of severe financial strain, the owners—the Gourock Ropework Company—offered the village to the Lanark Town Council for a nominal amount. The city fathers considered the proposal until they realized that upgrading the community was a more ambitious and expensive undertaking than they were willing to assume. By this time, though, the historic community's plight began to attract the attention of the government and the public.

On November 21, 1963, the New Lanark Association was formed. Among its members was Kenneth Dale Owen, a Texas businessman and a descendant of Robert Owen. The association eventually purchased New Lanark. Throughout this troubled period, New Lanark was still a working village. When the mills closed in 1968, however, the community continued to deteriorate even further, so much so that by the early 1970s the population of New Lanark had dropped to eighty or so people. With further renovation required, no single agency—certainly not the New Lanark Association—was willing or able to finance the enormous expenditures.

On the verge of demolition, New Lanark found new life in 1974 when the New Lanark Conservation and Civic Trust was formed and a full-time manager appointed. Other agencies also participated in the massive restoration project. Today, New Lanark is a revitalized community and one of the largest heritage conservation efforts in Scottish history, attracting up to 350,000 visitors a year. The village contains restored mills, examples of workers' housing, a cooperative store, and the Institute for the Formation of Character. In one of the mills a multimedia exhibit recreates nineteenth-century life in New Lanark. In 1990 New Lanark won the coveted Best in Britain trophy for the year's most outstanding tourist attraction.

Further Reading: The standard biography of Robert Owen is Frank Podmore's two-volume *Robert Owen: A Biography* (London, 1907). Also worth reading is Margaret Cole's *Robert Owen of New Lanark, 1771–1858* (London: Batchworth Press, and New York: Oxford University Press, 1953). Probably the best recent account of New Lanark is Ian Donnachie's and George Hewitt's *Historic New Lanark* (Edinburgh: Edinburgh University Press, 1993). For a history of New Harmony, Indiana, see D. F. Carmony and Josephine M. Elliott's "New Harmony, Indiana: Robert Owen's Seedbed for Utopia," in *Indiana Magazine of History* (Bloomington), volume 76, 1980, pages 161–261. In "New Harmony's First Utopians," from *Indiana Magazine of History* (Bloomington), volume 75, 1979, pages 24–300, D. E. Pitzer and Josephine M. Elliott examine the community's German Rappite origins.

—June Skinner Sawyers

Nijmegen (Gelderland, Netherlands)

Location: Gelderland province, eastern Netherlands, on the Waal River (southern arm of the Rhine)

Description: Largest city in Gelderland province; with Maastricht, one of the two oldest cities in the Netherlands.

Site Office: VVV Nijmegen
St. Jorisstraat 72
6511 TD Nijmegen, Gelderland
Netherlands
(80) 225440

Nijmegen enjoyed two heydays, first as a Roman outpost in the "barbarian" northern provinces, and again in the Middle Ages, when it served the Holy Roman Empire as an imperial seat. Nijmegen's strategic location on the Netherlands' Rhenish frontier with Germany has thrust the town into the history books since those earlier times too, mainly for several peace treaties between the Dutch and French signed here in 1678–79, and for battles fought here toward the end of World War II.

The Roman conquest of Gaul, which was completed by Caesar from 59 to 52 B.C., stopped short at the Rhine. The Low Countries formed part of the provinces of Belgica and Germania Inferior (later Belgica Secunda and Germania Secunda), which themselves were subdivided into *civitates;* in Germania Inferior, the *civitates* included that of the Batavi, a Germanic tribe that lived in the fertile Isle of Batavi between the Meuse and Waal Rivers in what is now the eastern Netherlands.

From the mid-first to the mid-third century A.D., Gallo-Roman culture from the more settled areas penetrated the outer provinces of the Roman Empire. The Romans extended their famous road network throughout the northern areas, and constructed garrisons along the Rhine, which comprised the empire's northern boundary. Some of these fortifications ranged along the southern branch of the Rhine known as the Waal, including at the site of present-day Nijmegen. Roman legions had been stationed at the site since the time of Augustus (27 B.C.–A.D. 14), who had determined that this convergence of major land and water routes was of strategic significance. Nijmegen was connected by fortlets to other garrison stations.

Despite the strong fortifications on the Roman frontier and the widespread assimilation by native tribes of Roman culture, some Germanic and Celtic tribes occasionally came into conflict with the Roman forces. In A.D. 69–70 the Batavi, led by Civilis, were incited to revolt against the legion. Because of obligations elsewhere in Europe, the Roman forces were at an ebb in the region, and the Batavi were initially successful. Superior organization eventually gave the Romans the upper hand, however, and the Batavi retreated. Following the revolt, the Romans replaced their wooden fortress with a more impregnable, stone structure for the Tenth Legion to occupy. This was built beside the Oppidum Batavorum, the former stronghold of the Batavi tribe on the site of Nijmegen. As well, the Roman naval patrol, the Classis Germanica, enhanced its control of the Waal and other nearby waterways from its depot at Nijmegen.

While the fortifications were being reinforced, the civil settlement—the *oppidum*—was moved to a site that was west of the present town center. This area became known as the Ulpia Noviomagus: the new market. To provide the Roman forces with a market, a quarter also grew near the harbor at the base of the Valkhof Ridge, a promontory that overlooks the Waal.

As incursions by the surrounding tribes abated, the Romans bestowed on Germania Inferior the status of a civil province, and most of the legionnaires departed the camp by A.D. 105. Noviomagus, which had grown to have a population of about 3,000 persons, declined notably following the military withdrawal. In an effort to stimulate the economy, the Romans granted the town special market privileges, followed by an urban upgrade to Municipium Batavorum in about 170.

By the mid-third century, however, Roman power in the Low Countries was seriously beginning to wane. Frankish tribes from the west took advantage of this weakness and invaded the city from 260 to 270. The Romans attempted to reinforce the border along the Rhine during the next century, especially through wars instigated between 355 and 360 by Julian, caesar of Gaul. Despite the short-term promise of these expeditions, the Germanic tribes decisively ended Roman rule in the Low Countries in 406–7. Without the Romans' administrative skills, the sophisticated urban systems that had been developing in the Low Countries fell victim to the violent Frankish incursions. The cities declined dramatically during the following three centuries as the infrastructure for procuring food, organizing trade, and providing protection disintegrated.

A reversal of this decline began with the ascendancy of the Pépin Dynasty following a long period of ineffectual Merovingian rule. The Pépin ruler Charles Martel finally managed by 719 to unite the various Frankish lands under one administration. This progress culminated in the reign of his grandson, Charlemagne, king of the Franks (768–814) and emperor of the West (800–814), who through additional conquests established the beginnings of the Holy Roman Empire. The expanded empire included all of the Low Countries, and they were administered as were the other Frankish lands: by

Bridge over Waal River at Nijmegen
Photo courtesy of Netherlands Tourist Board

the king as supreme authority, who traveled from place to place with his servants.

In 777 Charlemagne restored importance to Nijmegen—which, despite its decline, was still strategically located—by choosing the Valkhof Ridge overlooking the Waal as site for an imperial residence. Unlike the original Roman fortifications, and despite its commanding location, this structure was a simple wooden hall, and not a fortress at all. The Nijmegen residence, though simple, remained one of Charlemagne's favorites; he added to the compound thirty years later. His son and successor, Louis the Pious, eventually rebuilt the Valkhof compound to include a great audience chamber and chapel, in addition to a residential wing.

The presence of the court brought new life to Nijmegen, priming the town for further growth. In fact, few towns at the time were as large or important as Nijmegen. While the death of the great administrator Charlemagne in 814 spelled new problems for the vast empire in general,

Louis the Pious continued to favor Nijmegen. In addition to the usual military strategy and fealty sessions that the emperor conducted during the summer at the various imperial seats, including Nijmegen, he began to request additional assemblies for the purpose of preparing reforms or discussing general organizational tactics. Louis was an explicit promoter of unity throughout the empire, and the increased frequency of these convocations directly contributed to his goals. As one of the favored locations for these conventions, Nijmegen figured more prominently in this newfound sense of order and unity.

Louis eventually developed irreconcilable differences with his sons, who were his ostensible heirs, and especially Lothar, who sought to usurp his father's power by having him kept "in supervised freedom." One especially dramatic session in October 830, soon after Lothar had made his move, occurred when Louis called a general assembly at Nijmegen, where he could still count on loyal support despite Lothar's contentions. The aging emperor, by then considered largely powerless, stunned the assembly by defying his adversaries and demanding the exile of certain of them—and receiving it. Though Lothar's partisans continued to agitate for an in-house revolution, the son momentarily deferred to his father, who had a difficult time convincing the excited crowd that all was well.

The empire's disintegration accelerated as the Treaty of Verdun in 843 divided the government into three kingdoms, of the West Franks, the East Franks, and Lotharingia. The last of these was notably unworkable in both political and geographic terms, and soon splintered into small divisions.

Adding to the woes in the Low Countries were the Viking raiders from the north who plagued the area from several directions in the late ninth century. Plundering and ransacking their way south, the Vikings took pride in spending each successive winter at a point deeper in Frankish territory than the previous one had been. From one direction, for example, they worked their way from Ghent in the winter of 879 to Paris in the winter of 885. In the eastern Netherlands, the Vikings spent the winter of 880 at Nijmegen, where they destroyed the royal residence.

With the need for a genuine fortress obvious after the sacking, the complex was eventually rebuilt in stone and gained protective features such as a wall and a moat. The Valkhof was again destroyed in 1047, but Emperor Frederick I (Barbarossa) rebuilt it in 1155.

Although Nijmegen had already served for centuries as a de facto imperial locus, the town did not receive its imperial charter until 1230. The town had thrived commercially during the stable period following the Viking invasion, and actively competed with nearby Arnhem (a former market village) for mercantile activity, especially from the Rhineland. The chief products traveling through the area were cloth, grain, timber, fish, salt, and, especially, wine from the Rhineland. Nijmegen also dealt in the metal trades of the Maas Valley. Despite the towns' rivalry, the volume of trade flowing through the area allowed both cities to profit. By 1364 Nijmegen was conducting enough trade to become a member of the Hanseatic League, the huge trade network centered in Lübeck that monopolized trade from Russia to Flanders.

Despite the regional prosperity, Gelderland, like Holland and the other provinces of the Low Countries, became enmeshed in the political and religious strife that characterized the era of Spanish rule that began when Philip II of Spain inherited rule of the Netherlands from his father, Emperor Charles V, in 1556. The Spaniards' unabashedly pro-Catholic policies conflicted seriously with Calvinist sympathies in the northern provinces, especially in urban centers such as Nijmegen.

In 1579 Nijmegen subscribed to the Union of Utrecht against Spain, a contract that united the seven northern provinces under the leadership of Prince William of Orange, who became stadholder. The Spanish forces retaliated where they could, taking a number of towns in the eastern Netherlands

by force, among them Nijmegen in 1586. In 1591, however, Prince Maurice of Nassau, son of William (who had been murdered in 1584) reconquered several of these towns, including Zutphen, Deventer, Delfzijl, Hulst, and Nijmegen.

Nijmegen provided frontier protection for the prosperous Dutch Republic during the latter's most powerful period, the seventeenth century. During the Thirty Years War (1618–48) Nijmegen's fortifications housed Dutch forces; the disadvantages of a town's economy being tied to a military outpost surfaced again (as it had when the Roman garrison departed) when most troops were withdrawn from Nijmegen in 1648 upon the end of war. The sudden exodus had a serious impact on the town's population and economy.

The United Provinces' power on the seas and in trade inevitably led to conflicts with other countries such as England and France. After two wars with England in the mid-seventeenth century, the Dutch Republic went to war with both England and France, whose king, Louis XIV, had designs on Dutch territory. During the last campaign, Nijmegen was taken by the French in 1672. Appropriately, the treaties begun in 1676 between Louis XIV, the Netherlands, Spain, and the Holy Roman Empire that ended the hostilities were signed in Nijmegen in 1678–79.

Nijmegen served as capital of Gelderland province until the town's capture in 1794 by the French, who moved the capital to Arnhem. The French emancipated the long-suppressed Catholic majority of Nijmegen, and almost overnight the garrison town was transformed from a bastion of Protestant Orangist sympathies into a stronghold of republican Catholicism. The castle rebuilt by Frederick Barbarossa in the twelfth century was largely demolished in the revolutionary fervor of 1796–97; only a couple of chapels survived.

Many once-thriving towns fell upon hard times in the late eighteenth and early nineteenth centuries. Arnhem and Nijmegen were just two among a number of towns in the eastern Netherlands that suffered a decrease in population. The latter town nonetheless remained important as a fortified point on the frontier throughout much of the nineteenth century. Its fortifications were dismantled only in 1878. Soon after, in 1882, the Nijmegen-Venlo rail line provided a link between the Netherlands' two oldest towns, Nijmegen and Maastricht.

In the twentieth century, Nijmegen's frontier position put the city center stage once more. It was late in World War II, in 1944, and the Allied command in Europe was anxious to force the definitive retreat by the Germans that would bring the European war to a close. German strength on the western front depended on the so-called Siegfried Line of defense, and various Allied strategies sought to breach this line. In simple terms, the Allies needed to create various corridors that would allow ground and air forces to move eastward from already liberated areas across the Rhine and into the Ruhr Valley, Germany's industrial heartland.

On September 17, 1944, the Allies set in motion the hastily planned and optimistic Operation Market-Garden. The "market" component was an air operation requiring three and one-half airborne divisions to drop in the vicinity of the cities of Grave, Nijmegen, and Arnhem to capture key bridges that would allow land forces to cross key canals and rivers, including the Waal. The intent was to create a fifty-mile-long corridor. The "garden" component comprised British ground troops pushing from the Belgian-Dutch border north to the IJsselmeer (formerly the Zuider Zee) north of Amsterdam.

Operation Market-Garden was a bold plan, not least because of the inclement weather possible at that season. One especially bold maneuver required the ground forces to rush across the 400-yard-wide Waal River at Nijmegen—through a swift current under an unknown quantity of enemy fire—and capture the north ends of the rail and road bridges that crossed the river. Assisting in this risky maneuver, as usual, were Dutch resistance forces, who were eager to assist the Allies.

The failure of Operation Market-Garden has been blamed variously on adverse weather, delays of reinforcements, and faulty intelligence. The effort succeeded only insofar as a limited number of bridges, including those over the formidable Waal at Nijmegen, were indeed made available to the Allies; the Germans seem to have decided not to blow the bridges up as their command retreated, perhaps in the hope that the vital routes could still be defended. But even with the capture of these bridges, the lives lost in the whole operation were a high price to pay.

The Allied plan otherwise went awry in the Arnhem sector beyond Nijmegen, where the Germans had solidified their defenses and routed the Allied forces trying to take the Arnhem bridge. In practical terms, this meant that there would be no early incursion by the Allies into the German heartland.

Nijmegen suffered severe devastation in World War II, including the destruction of the Renaissance town hall, which has since been rebuilt. Further rebuilding has enabled the city to regain its industrial prosperity and become once again a vital rail junction and inland shipping center. Though the city has been largely reconfigured since World War II, some remains from the Carolingian buildings can still be seen in a park on the Valkhof, a reminder of the city's distinctive past.

Further Reading: For a useful account of Nijmegen in the context of broader Dutch developments, especially in its initial phases, see Audrey M. Lambert's *The Making of the Dutch Landscape: An Historical Geography of the Netherlands* (New York: Academic Press, and London: Seminar Press, 1971; second edition, London and Orlando, Florida: Academic Press, 1985). J. F. Drinkwater's *Roman Gaul: The Three Provinces, 58 B.C.–A.D. 260* (Ithaca, New York: Cornell University Press, 1983) and *Gallia Belgica* (London: Batsford, 1985) by Edith Mary Wightman both provide scholarly discussions of the Roman milieu that defined the territory at the northern boundary of which Nijmegen (Noviomagus) lay. *Charlemagne and the Carolingian Empire,* volume 3 of the series Europe in the Middle Ages, translated from the French by Giselle de Nie (Amsterdam, New York, and Oxford: North-Holland Publishing Company, 1977; originally published as *Charlemagne et l'empire carolingien,* Paris: Editions Albin Michel, 1947) by Louis Halphen is a clear, accessible account of the important developments that affected the medieval court at Nijmegen. Rosamond McKitterick's *The Frankish Kingdoms under the Carolingians, 751–987* (London and New York: Longman, 1983) deals with many of the same issues, with a thorough look at the realm that is based on an impressive roster of original sources. *The European Theater of Operations: The Siegfried Line Campaign,* part of the United States Army in World War II series (Washington, D.C.: Office of the Chief of Military History, Department of the Army, 1963) by Charles B. MacDonald presents a detailed account of the Allied airborne strategy to liberate Nijmegen and surrounding areas from the Germans toward the end of World War II. Considering the auspices under which this series was published, this history offers an admirably balanced account of events, drawing on German as well as Allied sources, and noting failures as well as successes. Narratives drawn from copious diaries, journals, and field reports provide the book with an impressive sense of immediacy.

—Randall J. Van Vynckt

Nîmes (Gard, France)

Location: To the northwest of the Rhone delta, between Avignon and Montpellier.

Description: Famous in history as an important center of Roman Gaul, today boasting some of the best preserved and most impressive Roman monuments anywhere. The city played a central role in the Wars of Religion in the sixteenth and seventeenth centuries.

Site Office: Office de Tourisme, Syndicat d'Initiative
6 rue Auguste
30000 Nîmes, Gard
France
66 67 29 11

Although Nîmes lies in the historic region of Languedoc-Rousillon, its Roman roots are so strong and its similarity to other nearby Roman towns such as Arles and Orange is so pronounced, that it can be considered an "annex" to the region of Provence.

The extraordinarily well-preserved Roman buildings that dominate the city's center make Nîmes one of the principal tourist attractions of the region, the highlight being the incomparable Pont du Gard—the aqueduct that spans the River Gardon about fifteen miles from Nîmes. Apart from its Roman legacy, Nîmes's second claim to historical significance is as a focus of religious conflict between Catholics and Protestants from the Reformation up to the nineteenth century, largely due to its proximity to the impenetrable Cévennes Mountains. The city was a Protestant stronghold throughout the Wars of Religion.

Prior to the arrival of the Romans, Nîmes was the capital of a local Gaulish tribe, the Volcae Aremocici, who traded with the fringes of the Greek colony of Masalia, modern Marseilles. From the fourth century B.C. the Gaulish hill forts of the Lower Rhone region began to assimilate the forms of Greek defenses, one of the main features being the construction of large lookout towers. One such tower stood on the highest of Nîmes's seven hills, later to be rebuilt by the Romans and still standing today as the Tour Mange.

With the Roman conquest of Southern Gaul in 121 B.C., most of the fortified sites in the Lower Rhone were abandoned in favor of undefended settlements in the valleys. Only Nîmes remained occupied to become a highly important Roman city. The significance of the site derived from its proximity to a strong natural spring. The Romans called the god of this spring Nemausius, and hence the name of the Roman settlement—Nemausus. The site of Nemausus was also an important military post on the Via Domitia, the first Roman road to be built in Gaul, running from the Rhone to the Pyrénées.

It is not clear exactly when the transition from garrison town to Latin "Colonia" took place. It is commonly held that the walls of the city were built by Caesar Augustus, Roman emperor from 27 B.C. to A.D. 14, but A.L.F. Rivet suggests that the discovery of pre-Augustan coins at the site bearing the inscription COL NEM (Colonia Nemausus) point to a date earlier than 27 B.C. Nemausus soon became one of the most distinguished cities in the flourishing province of Gallia-Narbonensis with a population of 25,000. The city and lands around Nemausus were settled by veterans from some of the eastern legions. Augustus distributed land to the legionnaires in recognition of their loyalty during his campaigns in Egypt. The presence of these colonists from the east is reflected in the coins issued at the time, which bear the emblem of a crocodile chained to a palm tree. The emblem commemorates the victory of Emperor Augustus over Antonius and Cleopatra in Egypt and symbolizes both Egypt's demise and Rome's dominance.

Nemausus was an important economic and social center. The city minted coins for the armies campaigning in Gaul and Germany during the Augustan period. From here, in A.D. 39, came the second Narbonensian to reach the consulship—the great orator Ca Domitius Afer—and it seems probable that the city was the home of Trajan's wife Platina.

The importance and prosperity of the city are illustrated by the monuments and defenses that still dominate the center of Nîmes. Most of the Roman buildings date either from the time of Augustus or his son-in-law Agrippa. The most imposing of Nîmes's antiquities is the amphitheater (or Arènes). Built by Agrippa in A.D. 50, it ranks twentieth in size among the seventy Roman amphitheaters known to have existed, with a capacity of about 21,000. It is 146 yards by 110 yards in area and up to 69 feet high. The amphitheater is notable because it is regarded as being the best preserved in the entire empire. It is still in use today as an arena for bullfights and theater performances. The upper part of the building is particularly well preserved, still bearing its ornate Doric columns, and the brackets for the wooden mast of the awning which would have shaded spectators from the sun can still be seen. On occasion, the arena was flooded to accommodate mock sea battles and marine animals. As well as the amphitheater, Nemausus also boasted a circus within the city walls, but while much of this site was still in evidence in the nineteenth century, little remains of it today.

The other monument of major significance is the Maison Carrée, considered one of the most outstanding monuments in Gaul. This temple, with a rectangular shape that belies its name of Square House, would have overlooked the forum in Roman times. Although small, it is perfectly proportioned, displaying a very strong Greek influence. Thirty pillars surround it—twenty of them engaged in the wall and

Roman amphitheatre in Nîmes, still in use today
Photo courtesy of French Government Tourist Office

ten forming the portico, each topped with Corinthian capitals. Thomas Jefferson was so impressed with these capitals that he had them copied for the Virginia State Capitol in Richmond. Apart from its original function as a temple, this building has been variously employed as a stable, residence, town hall, and church. Today it is a museum.

Nîmes's aqueduct tapped a source some thirty miles away at Uzes. Structures of ground level channel and tunnels still remain, but the most impressive feature is the famous Pont du Gard—the highest surviving bridge structure in the modern world. The harmonious tiers of arches are justifiably regarded as among the most outstanding of all Roman remains in Gaul. The actual aqueduct itself represented a remarkable feat of ancient engineering and surveying. An estimated flow of more than 50,000 gallons of water per day was carried at an average fall of just twenty-one inches per mile. Historian Anthony King estimates that the construction of the Nîmes aqueduct may have cost up to 100 million sesterces, equivalent to $100 million, a cost reflecting the difficulty of its construction and the high priority based on water provision.

Nimes remained a Roman city for more than 400 years, until the final disappearance of Roman authority from

Gaul in the 460s. The town was overrun in turn by Vandals, Visigoths (who used the amphitheater as a fortress), and Saracens. It was wrested from them in turn by the Franks and became part of the County of Toulouse.

Nîmes reemerged as a historically significant site during the Wars of Religion—a place where the ebb and flow of tensions between Catholics and Protestants dictated events, sometimes with tragic consequences. The city was the center of Protestantism for all of France. "Its hinterland, most notably the Cévennes mountains to the northwest, constituted the most impregnable citadel in the entire realm for Calvinism," as J.N. Hood notes in the *Historical Dictionary of the French Revolution*.

Nîmes had a mid-sixteenth-century population of about 10,000 and was an early center of the Reformed Church. Pastor Guillaume Mauget arrived from Geneva, the spiritual home of Calvinism, in 1559 and quickly established a church and consistory. The role of the consistory was to enforce the puritanical standards and conduct of the new faith and to eradicate the vestiges of medieval Catholicism. Moral control was exerted over the swelling congregations by meting out punishment on offenses ranging from blasphemy and relapse into popery, gluttony, fornication, and dancing.

By the outbreak of the Wars of Religion in 1562, Nîmes had a Protestant majority and the town council, responsible for local military functions, was under Protestant control. The town was a Huguenot stronghold and Protestant domination of the four member consulate, the leading organ of traditional city government, was to last until the end of the century. The strength of Protestant support in the town is illustrated by the fact that following the aggressive spread of Calvinism throughout much of France in the mid-sixteenth century, the consequent Catholic backlash against the political power of the reformed faith did not have much of an impact on Nîmes. While reprisals against Protestants culminated in Paris in 1572 in the bloody massacre of St. Bartholomew's Day, during which 5,000 to 6,000 Protestants were killed, Nîmes remained relatively calm.

Indeed, Protestants continued to consolidate their power in Nîmes and were not above violent acts themselves. September 29, 1567, witnessed the Michelade—a massacre of 100 Catholics, mostly priests and prominent laymen, at the hands of the Huguenot majority. The national synod of the Reformed Churches met in Nîmes in 1572, just prior to the St. Bartholomew massacres, and the provincial synod met there numerous times from the 1570s on. Political assemblies of Protestants also took place in Nîmes over the same time period, and the town eventually became a *place de sureté,* one of fifty-five fortified Protestant towns whose status was confirmed by the Edict of Nantes in 1598. Whereas in most of France the edict gave Protestants security of persons and property and a share in political life, in Nîmes it gave the outnumbered Catholics these same benefits.

Over the seventeenth century the prosperous Huguenot majority was increasingly challenged by a growing Catholic minority. Political factions formed around the incumbent regime, favoring the Protestant status quo, and around the forces advocating the revival of Catholic institutions. The year 1685 brought the Revocation Edict, bringing to an end the protection that Protestants had been afforded by the Edict of Nantes. Catholicism was back in the ascendant and Nîmes experienced massive forced conversions. As J. N. Hood writes, "Protestants were excluded from public offices and the free professions unless they fulfilled forms required for a certificate of Catholicism. Suffering under numerous other disabilities, their numbers declined to less than half the population of the city" ("Protestant-Catholic Relations").

A Protestant revival in 1689 was suppressed by the royal authorities, but in 1700, just before the outbreak of the War of the Spanish Succession, a group of young peasants and artisans from Nîmes and the Cévennes, claiming charismatic powers, organized a new revival opposing the tyranny of the Catholic Church. These militant Calvinists, known as Camisards on account of their characteristic short shirts, were treated as rebels by the crown. Robert Louis Stevenson's "Travels with a Donkey" provides a stirring account of the Camisard uprising.

The war with Spain diverted troops away from Nîmes and the surrounding hills, giving the new guerilla army enough scope not only to protect Protestant assemblies, but also to inflict heavy losses on the remaining Catholic forces. A cycle of violence spiraled on both sides, with the army razing and depopulating numerous mountain villages and the Camisards taking revenge by killing priests and burning Catholic villages. The Camisards held Nîmes in a virtual state of siege until 1705, when the city authorities negotiated a settlement and executed numerous merchants in Nîmes itself who had been accused of supplying the guerrillas and of planning to butcher the Catholics of the region. Even after this the Camisard resistance continued in the impenetrable Cévennes. The return of troops after the peace with Spain in 1713, however, closed the power vacuum that the Camisards had filled, and Protestants were subsequently much more cautious about when and where they held their assemblies.

Although never reaching the frenzied peak witnessed at the beginning of the century, persecution of Protestants in and around Nîmes continued throughout the eighteenth century. A fresh edict was issued in 1787 that again made it impossible for non-Catholics to worship in public and to hold high public office.

The tensions associated with the revolution in 1789 caused conflict to erupt once more along the lines of old sectarian hostilities. Civil war broke out in the Gard region in 1790 between Patriot (read Protestant) advocates of reform and Royalist (largely Catholic) opponents, these parties being identified with the leadership of the Calvinist bourgeoisie and the Catholic oligarchies respectively. Violence culminated in a period of rioting, commonly referred to as the Bagarre de Nîmes, in May and June of 1790. The counter-revolutionaries took a decisive beating from the Patriots, who then took the town's public offices, positions from which they had previously been largely excluded. The conflict was characteristically brutal, and great numbers of citizens lost their lives.

During the nineteenth century and especially after the arrival of the railway, Nîmes flourished as an industrial and agricultural center, developing its strength in the manufacture and trading of textiles. In 1860 an Austrian, Levi Strauss, began exporting heavy cloth from Nîmes to California to be used for tents in the gold diggers' camps. This tough wearing material was adapted for workers' overalls and was given the name of its town of origin—de Nîmes or denim. Today, aside from being a center for tourism, Nîmes is an important center for textile manufacturing and the wine trade.

Further Reading: For information on the Roman era in Nîmes, A. L. F. Rivet's *Gallia Narbonensis: Southern France in Roman Times* (London: Batsford, 1988) provides a detailed study of the town's archaeology and status. Also, Anthony King's *Roman Gaul and Germany* (London: British Museum Publications, and Berkeley: University of California Press, 1990) puts Nîmes and its neighboring towns clearly in their contemporary political and cultural context. For an examination of Nîmes during the Calvinist era, Howell Lloyd's *The State, France and the Sixteenth Century* (London and

Winchester, Massachusetts: Allen and Unwin, 1983) is useful. The most vivid account of Calvinism and the physical environment from which it developed is to be found in Robert Louis Stevenson's inspired work *Travels With a Donkey in the Cevennes,* available in various editions.

—Daniel D. Collison

Nizhny Novgorod (Nizhny Novgorod, Russia)

Location: Capital of oblast of the same name, spread across the Djatlovy hills and surrounding meadowlands, at the confluence of the Oka and Volga Rivers, 265 miles northeast of Moscow.

Description: Founded by Prince Vladimir in 1221; strong point along the eastern Russian frontier during the Middle Ages; medieval and early modern center of trade, particularly after the establishment of a trade fair in 1817; major industrial center since the start of the Communist era; renamed Gorky from 1932 to 1990, in honor of author Maksim Gorky.

Site Office: Nizhny Novgorod Oblast Administration
Office of the Mayor
Nizhny Novgorod, Nizhny Novgorod Oblast
Russia
(8312) 39–1506

Prince Vladimir of the Vladimir grand principality founded Nizhny Novgorod in 1221 as a fortified post for Suzdal, an important city in northeastern Russia during the late twelfth and early thirteenth centuries. The fort's location on the right bank of the Volga River was most likely chosen because it is higher than the left bank, and thus less susceptible to flooding. Along with Volga Gorodets, another of Suzdal's fortified posts first mentioned in a chronicle dated 1172, Nizhny Novgorod would eventually form the Nizhegorod grand principality.

In 1238, the Mongols invaded Russia, ushering in a prolonged occupation that would last until 1480. The years following the Mongolian conquest have been referred to as the "appanage period" of Russian history. It was an era of fratricidal warfare and little national unity.

Russian princely dynasties were allowed to continue under Tartar control so long as they obeyed the Mongolian Khan's orders. In the mid-thirteenth century, Prince Sviatoslav granted his nephew, Prince Andrei, the domain of Nizhny Novgorod and Volga Gorodets. Most scholars have speculated, however, that by 1256 Andrei traded this domain for the domain of Suzdal; documents after 1256 refer to him only as the grand prince of Suzdal. Other documents support this speculation by showing that Andrei's sons never inherited Nizhny Novgorod or Volga Gorodets from their father.

Between 1238 and 1328, two princely dynasties—the houses of Moscow and Tver—battled for hegemony over northeastern Russia. Evidence is unclear, but it is believed that one of the areas of contention was Nizhny Novgorod. Moscow and Tver ended their struggles in 1328, when Ivan Kalita of Moscow became grand princely ruler of Russia, and Alexander of Suzdal, son of Mikhail of Tver, was given the title of ruler of "Vladimir and the Volgaland." This year was also significant because it officially defined the Nizhegorod grand principality.

In 1343, Alexander moved the capital of his principality from Suzdal to Nizhny Novgorod. Four years later, the city was granted its own bishop. The 1340s and 1350s witnessed the rise of Nizhegorod as a power capable of rivaling Moscow for regional hegemony. Its rise was made possible by the decline of the two once-grand principalities: Tver and Novgorod (in northwest Russia). Tver's stature within Russia declined because of civil war; Novgorod slipped because it had lost territory.

Alexander attempted to take advantage of the situation by seizing control of the grand principality of Vladimir in fact as well as title. If he had been successful in annexing the territory, he would have controlled much of eastern Great Russia. But death cut short his plans. The territory of Nizhegorod passed to his brother Constantine, and Vladimir went to Grand Prince Ivan Kalita of Moscow.

When Constantine assumed control of the Nizhegorod principality in the mid-fourteenth century, its capital city of Nizhny Novgorod had become an important strategic and administrative center within Russia. Trade going up and down the Volga River made the city a vibrant commercial center. Nizhegorod was also on the border between Russia and its often hostile neighbors to the east, making it an important buffer zone. Moscow was forced to rely upon Constantine because it was not capable of protecting or absorbing Nizhegorod. Constantine used his great organizational abilities to carve out Nizhegorod as an independent center of power within Russia. The city was used as a staging point in expansionary attempts toward the east, as well as in battles against outside invaders. Constantine was an active colonizer, ordering his subjects to settle outside the original confines of the principality. Territory expanded further down the Volga, Oka, and Kudma Rivers.

Constantine also ordered, in 1350, the construction of the Church of Transfiguration in Nizhny Novgorod, to replace an earlier church that had been destroyed. Constantine supposedly brought icons to the church all the way from Greece.

Nizhegorod's growing importance in Russia led Constantine to challenge the grand principality of Moscow for princely rule of all Russia. The opportunity arose with the death of Grand Prince Semeon the Proud of Moscow. To be successful in his challenge, Constantine needed the explicit consent, also known as yarlik, from the Mongol Tartars—still the occupiers of Russia.

Prince Constantine received support in his challenge from both Lithuania and the principality of Great Novgorod. Together, the three petitioned the Mongol Horde, asking it to grant yarlik for all of Russia to Constantine. Instead, the

Nizhny Novgorod as it appeared in the nineteenth century

Tartars gave yarlik to Prince Ivan of Moscow, brother of Semeon. Determined to rule over Russia, Constantine, along with the Lithuanians and the Novgorodians, engaged in battle with Moscow for supremacy. Moscow won and Nizhegorod was forced to make peace with Prince Ivan.

When Constantine died, his son Andrei received yarlik from the Tartars to rule over his father's domain. Simmering hostilities between Nizhegorod and Moscow subsided under Andrei's rule because he was willing to recognize Prince Ivan II of Moscow as grand prince of Russia and as his superior.

For more than 100 years, the Golden Horde brought stability to Russia. But in 1359, Khan Berdi-Beg was murdered leading to chaos within the Tartar Empire and in Russia. The murder caused the formation of two competing centers of Tartar authority.

Andrei's brother, Dmitri, began ruling Nizhegorod in 1360. He headed the principality through the confusing years following the khan's murder. Exacerbating an already difficult situation was the return of the Black Death to Russia in 1364–66 (it previously struck in 1352–53) and the outbreak of hostilities between Dmitri and his brother, Boris of Volga Gorodets, for control of the principality. The situation in Nizhegorod was only made worse by the involvement of the warring Tartar khans. One supported Dmitri while the other supported Boris, who seized the throne at Nizhegorod.

In addition to the support received from one of the two competing Mongol khans, Dmitri sought and received aid from Grand Prince Dmitry Donskoi of Moscow and Metropolitan Alexei, the head of the Russian Orthodox Church located in Moscow. Metropolitan Alexei ordered Boris's control over the Suzdal Orthodox church taken away and commanded Boris to appear before him in Moscow. Boris refused. Dmitry Donskoi of Moscow and Dmitri of Nizhegorod marched together on Nizhny Novgorod and forced Boris to submit. Boris returned to Volga Gorodets and Dmitri to Nizhegorod as Boris's superior.

The battle between Dmitri and Boris reconfirmed Moscow's superiority in northeastern Russia and demonstrated Nizhegorod's subordinate position to the city. Dmitri could not have gained rule in the principality without Dmitry Donskoi's aid.

A marked increase in commercial activity occurred along the Volga River under Dmitri of Nizhegorod's reign. It was both a blessing and a curse: a blessing because it meant more money and increased political and economic importance for the principality, a curse because it made the Russian border principality even more attractive than before to raiders. In the 1360s, Nizhegorod was invaded and looted by various Tartar bands, the Bulgars, and the Mordvians.

Increased military activity along the Russian eastern front led to the construction of a stone fortress in Nizhny Novgorod and the establishment of the new dependent town of Kurmyshan along the Sura River in the 1370s. Nizhegorod became more closely bound to Moscow, working no longer as an independent power, but rather as a front-line force engaged in protecting Russia from invasions from the east.

Like the 1360s, the 1370s were a violent period for the principality. Twice it fell victim to large-scale sackings by the Tartars and the Mordvians between 1376 and 1378. The sackings came about because local princes began working closely with Dmitry Donskoi to strengthen Nizhegorod's defenses and rebuild its citadel. The Tartars viewed the cooperation between Nizhegorod and Moscow as a direct threat to their power over the region and as an attempt to halt the Tartars' right to collect tribute from Nizhegorod. Tartar forces arrived in 1374 to end the Moscow-Nizhegorod alliance, but were instead massacred and imprisoned by the Nizhegorod troops. The Tartars returned again in 1377 and were victorious, savagely sacking the principality.

Exhausted from almost constant battle and from Tartar and Mordvian invasions, Nizhegorod once again submitted to a Tartar khan, Tokhtamysh. Khan Tokhtamysh claimed the principality and took Dmitri's two sons, Vasili and Semeon, as well as his brother hostage. Moscow was unable to protect Nizhegorod from any of Tokhtamysh's actions since it, too, was bracing for an impending clash with the Mongol leader.

When Prince Dmitri died, his sons and his brother were still being held captive by the Tartars. Instead of granting yarlik to any of them, Tokhtamysh gave it to Dmitri's uncle, Boris. The principality fell apart under Boris's reign. It was once again at odds with Moscow. Boris also had to contend with Vasili and Semeon, now free from Tartar imprisonment and determined to regain a throne they believed to be rightfully theirs. Boris eventually yielded to the brothers, then returned to the throne again, just to be evicted once more—this time by the combined forces of Vasili, Semeon, the grand principality of Moscow, and the Tartars. But the principality's independence was gone. In 1392, it was annexed by Dmitry Donskoi's son, Vasily I, and placed under Moscow and Tartar subjugation.

The Nizhegorod grand principality's lessening of importance within Russia was a product of the conditions on the country's frontier. Nizhegorod was able to maintain power and political prestige only when it was engaged in the defense of Russian lands, involved in the capture of new lands for colonization, or active in protecting Russian trade routes. When these duties were taken over by Moscow, Nizhegorod's independence and importance declined.

Despite decreased power and prestige, Nizhny Novgorod remained an important Russian city throughout the Middle Ages, thanks to its position along three trade routes. East-west travel along the Oka River was the first trade route. The second went north-south on the Volga River. The final trade route went east-west across land and through both the Ural Mountains and the city of Nizhny Novgorod.

The city became one of the leading economic centers within the country by the sixteenth century. Trade peaked during the nineteenth century after an annual trade fair was established in the city in 1817. The fair continued more than 100 years, until it closed in 1930. From its inception, the annual fair was under the centralized control of the Russian state. The physical site for the fair, a low-lying swamp on the bank of the Oka River, was chosen in 1816. Centralized planners made sure that only the best were hired for the fair's construction. Andreyan Zakhanov, a St. Petersburg architect known for having designed that city's admiralty, designed the layout of the fairgrounds. Augustin Bethencourt, a well-respected engineer, was in charge of the actual construction. Initial costs for the fairgrounds were estimated at 6 million Russian rubles.

When construction was completed in 1822, the marketplace consisted of a massive rectangular structure surrounded by an amusement arcade, a giant wooden theater, along with plenty of coffeehouses and taverns. The fairgrounds were encircled on three sides by a canal that flowed into the Oka. To facilitate easier land travel to and from the fair, a highway was built in 1847 linking Nizhny Novgorod to Moscow.

During the fair's heyday, more than 60 percent of all wares sold within Russia came from Nizhny Novgorod. The fair transformed the city from a small town during the early 1800s into one of Russia's most populous cities by 1867, with an estimated 41,000 people.

For the remainder of the nineteenth century, Nizhny Novgorod grew at a steady rate. When the Communist revolution came, and Russia became part of the Soviet Union, Nizhny Novgorod experienced a massive population explosion unmatched by any other Russian city. Between 1920 and 1940, the city's population ballooned from 70,000 to 644,000. Massive growth was attributed to the city's increasing industrialization.

Nizhny Novgorod became a key center in European Russia for industry and transportation. Most of the city's new industrial complexes were located near the fairgrounds. The city's numerous industries included shipbuilding, sawmilling, oil refining, chemical and textile production, engine building, and woodworking. One of the Soviet Union's principal producers of chemicals was located west of the city, in the suburb of Dzerzhinsk. Southeast of the city was Kstovo, home to one of the country's largest oil refineries. Across the Volga River was the suburb of Bor, a major glassworking center. During the Cold War years, the city's important role in military manufacturing led the Communist government to close off the city to tourism. The city remained off limits to tourists until 1990.

In 1932, Joseph Stalin, then leader of the Soviet Union, ordered the city and oblast's name changed from Nizhny Novgorod to Gorky. Stalin wanted the name changed to honor one of Russia's most famous writers, Maksim Gorky.

Maksim Gorky was born into a merchant family in Nizhny Novgorod in 1868, and was named Aleksey Maksimovich Peshkov. He eventually dropped his first name and took the pseudonym "Gorkii," which means bitter, as his surname. After Gorky won fame with his story "Chelkash" in 1895, he moved gradually from travel writing to social criticism in such works as *Forma Gordeyev* and *Na dne*. His opinions led to his association with the Social Democratic (Marxist) movement and necessitated his flight from the country in 1906, following a brief imprisonment at the hands of the czarist government the previous year. While abroad, Gorky became friends with Vladimir Lenin, and upon Gorky's return to Russia in 1913, he became editor of socialist journals. Although he became increasingly skeptical of the Bolshevik cause, which he viewed as hostile to the intelligentsia, he continued to work with the regime once it assumed power. Gorky died in 1935 at the age of sixty-eight.

Some historians have speculated that he was poisoned on Stalin's orders; others believe his death to have been of natural causes.

Increased industrialization around the city led to the establishment of two prominent institutions of higher learning—the University of Gorky and Gorky Polytechnical Institute. The former was founded in 1918 and is also known as N. I. Lobachevskii University of Gorky. The latter was established to train individuals for careers in the industrial, transportation, and communications sectors. It was established in 1930 and received status as a polytechnical institute in 1950. Gorky Polytechnical Institute is also known as the A. A. Zhdanov Gorky Polytechnical Institute.

From 1980 to 1987, Gorky was the home-in-exile for Andrei Sakharov, a famous Soviet dissident. After 1990, the city was no longer off limits to foreigners. That same year the city's name was changed back to Nizhny Novgorod.

Today, Nizhny Novgorod contains two distinct districts. The first district, the old town area, is spread out across the hilly southern bank of the Oka River, and contains the city's kremlin. Many structures in the old section of town date back to the sixteenth century. The Archangel Cathedral (1631), housed within the kremlin, is one of the more famous buildings. It is now used as a museum. The newer section of town, called Kanavino, is located on the flatter northern bank of the Oka.

With the fall of Communism and the gradual introduction of market-driven reforms, Nizhny Novgorod has witnessed dramatic change. It is now home to the country's largest conversion bank. A joint-stock company has also formed for the purpose of reopening the city's long-closed trade festival.

Further Reading: While A. E. Presniakov's *The Formation of the Great Russian State* (Chicago: Quadrangle Books, 1970) does not deal wholly with Nizhny Novgorod, it does present a good deal of information on the city and surrounding area in northeastern Russia. J. L. I. Fennell's *The Emergence of Moscow 1304–1359* (Berkeley and Los Angeles: University of California Press, and London: Secker and Warburg, 1968) deals mostly with the rise of Moscow, but also touches upon the history of the city during this limited period. Robert O. Crummey's *The Formation of Muscovy 1304–1613* (London and New York: Longman, 1987) is written in a straightforward manner and is easy to understand.

—Peter C. Xantheas

Novgorod (Novgorod, Russia)

Location: Northwest Russia, approximately 120 miles south of St. Petersburg, on both sides of the Volkhov River.

Description: One of the oldest and most historically significant cities in Russia and the administrative center for the surrounding region; in the Middle Ages, was the site of one of the few democratic governments ever to exist in Russia, as well as an important trade center.

Site Office: Novgorod Oblast Administration
Dom Sovetov, Sotiyskaya Ploschad
Novgorod, Novgorod Oblast 173005
Russia
(8160) 9–2514

Novgorod, formerly called Lord Novgorod the Great, is today an unexciting provincial city, as well as the administrative center for the surrounding region. It has, however, a rich history dating back more than 1,000 years. It was at one time one of the most—if not the most—important city in all of Russia, and for more than 300 years, it boasted a distinctive political and social system.

While the exact date of Novgorod's founding is unclear, historians agree that it was in existence by the eighth century. It was mentioned in the early *Russian Chronicles* as the town where the first Russian ruling dynasty started. Rurik, a Varangian (Scandinavian) leader, settled there in 862. According to the *Russian Chronicles,* produced in the twelfth century, Rurik was invited into Novgorod, although some historians now argue that he took advantage of a power vacuum and gained control of the town through force.

Rurik's successor (and possibly a descendant), Oleg, extended the dynasty's power base by pushing south, expelling the Khazars, a Turkic-speaking people, from Kiev, and shifting his seat of government to that town. This unified all of the various tribes in the country and allowed for the formation of the first Russian state, which became known as Kievan Rus. A few historians now argue that the local tribes were already becoming unified before Rurik and Oleg, and that some type of Russian state would eventually have formed without their leadership; but even these historians agree that Rurik's and Oleg's actions helped speed the process.

Kievan Rus was at its most powerful between the tenth and twelfth centuries, and during this period, Novgorod continued to play an important role. For instance, it was the home to many princes before they ascended to power as the ruling grand princes of Kievan Rus, including Yaroslav the Wise, grand prince from 1019 to 1054. Yaroslav contributed greatly to Russian society as a whole and to Novgorod in particular. He saw to the development of the first formal legal code in Russia, called *Pravda russkaya.* This code originated as the local law for Novgorod when Yaroslav was the prince of the city, and he then applied it to the rest of the country once he became ruler of all Kievan Rus. It was then used by Russians in the following centuries as the model for the development of later laws.

During Yaroslav's reign, one of Novgorod's best-known buildings—and an outstanding example of the architecture of the Kievan Rus period—was constructed: the Cathedral of St. Sophia, which was named after a cathedral in Kiev that Yaroslav had built. Novgorod's Cathedral of St. Sophia boasts six domes and the famous Plotzk Door, which is notable for its forty-eight bronze plates featuring scenes from the Old and New Testaments as well as from mythology. This door was brought to Novgorod by local merchants and was probably crafted in Magdeburg. Novgorod's cathedral became a focal point of life in the city and the surrounding area.

Even with these great contributions, Yaroslav also laid the foundation for the decay of Kievan Rus. His two sons, upon his death, split the country in half. This, as well as attacks on Kiev by enemy forces, led to the gradual disintegration of the country. Consequently, Novgorod's bond with Kiev started slipping in the eleventh century.

By 1136, the tie between the two cities became so weak that the citizens of Novgorod—with a population of approximately 30,000—expelled its prince and formed the Novgorod Feudal Republic, which lasted until 1478. One should note that Novgorod always maintained a degree of independence during the dominance of Kiev on Russian life and that its citizens had been known to refuse to accept the authority of some of the princes assigned to them. They even had a saying for this: "If the prince is bad, into the mud with him." What is important, however, is the direction that Novgorod took after the expulsion, forming what was one of the few democratic republics in Russian history. It was during the period of the republic's existence that Novgorod boasted its most interesting developments.

Shortly before this break with Kiev, from 1119 to 1130, another of Novgorod's landmark buildings was constructed: the St. George Monastery, located just outside the city. This monastery, a striking example of the architecture of the Kievan Rus period, features three apses, three cupolas, and white stone walls, conveying impressive grace and simplicity.

After 1136, as the historian Nicholas V. Riasanovsky points out, Novgorod's prince became essentially an employee of the republic, and his authority was severely limited. The town council (called a veche) would select a prince to serve Novgorod and negotiate a contract with him. The

The Cathedral of St. Sophia

The walls of Novgorod's Kremlin
Photos by Laurence Minsky

prince's main responsibility was to serve as the local military leader. As a sign of his limited authority, his residence was located in the outlying parts of the town.

With the prince being so powerless, the veche also served as the highest governing body. In addition to recruiting and dismissing princes, this assembly elected the chief administrative officials, declared the law, and carried out other functions as needed.

Citizens found it relatively easy to convene the veche. Anyone from the prince or an official to the people could call a meeting. One simply had to ring a special bell that was located in the main marketplace to bring the veche together.

Novgorod's form of democracy operated with many different layers, starting with each street (called a *ulitsy*). Each *ulitsy* was to be a self-governing unit and had its own elected leader. A group of these units formed a *sotina;* and, in turn, a group of *sotinas* was combined into a *kontsy* (a borough). In addition to running its own veche and electing its own officials, each borough also oversaw part of the land that was just outside of the city limits; more distant lands that were allied with Novgorod were governed by the city as a whole.

This form of government did not always function smoothly. Because of factional quarrels and the need in most cases for a unanimous decision, the archbishop often had to appear at the veche and attempt to restore order.

In 1156, the veche also began selecting Novgorod's archbishop. This further established the city's independence. The veche identified three potential archbishops, and each candidate drew lots, the winner of which was then elevated to his rank by the head of the Russian church.

To participate in the veche, one simply had to be a free householder. Through their respective guilds, which were comparatively powerful, merchants and artisans also participated in governmental affairs. Women as well as men had a voice in the affairs of the city.

Another governmental body, the Council of Notables, saw to the city's day-to-day business, which the veche could not handle efficiently. This group was led by the archbishop and included persons who had previously held administrative offices. Because of its makeup, this council in effect protected the interests of the boyars, or nobility. As it grew more powerful, conflict within the city resulted, contributing to the republic's eventual downfall.

Novgorod's judicial system observed high legal and humanitarian standards. Although receiving the death penalty was a possibility, most punishments consisted of fines or, in extreme cases, banishment. As Riasanovsky notes, "in contrast to the general practices of the time, torture occupied little, if any, place in the Novgorodian judicial process."

One important factor that helped Novgorod thrive and further separate its development from the rest of Russia was its location along the Volkhov River. This location fostered commerce and communication, and provided a natural barrier against invasion. Novgorod was able to control its own trade routes, which led to the Baltic Sea and beyond, as well as to the Volga River and trade routes east. The town grew along both sides of the river, one of which became known as the Sophia Side, for the cathedral, and the other the Commercial Side.

With its access to all of these routes, Novgorod became a crafts, trading, and cultural center. Exports included fur, wax, flax, and hides. Because of these activities—combined with its distinctive form of government—Novgorod was more like the towns of the Hanseatic League than like the other principalities in Russia.

Even with Novgorod's success on the international stage, its existence was not peaceful. In 1237, the Mongols, led by Genghis Khan's grandson, Batu, invaded Russia. They were very successful in capturing and then plundering Russian cities. In 1238, Batu turned his attention to Novgorod; luck, however, was on Novgorod's side because the spring thaw had come, making the ground swampy and impossible.

Batu rested in 1239, but he captured additional cities in 1240, including Kiev. This made the Mongols overlords of Russia, a status that lasted nearly two and one-half centuries.

Meanwhile, the Teutonic Knights as well as the Swedes also invaded Russian lands, both attacking Novgorod in 1240. Novgorod's prince, Alexander, beat back these invasions, defeating the Swedes near the Neva River; because of that victory, he became known as Alexander Nevsky. Other attacks followed, however. Between 1142 and 1446 Novgorod fought twenty-six battles against the Swedes, eleven against the German knights, fourteen against the Lithuanians, and five against the Norwegians.

Alexander Nevsky took a different approach with the Mongols, because he believed that if he fought them, he would lose. Therefore, he gave in to the Mongols and offered them tribute. He became a favorite of the khan and grand prince of Russia from 1252 until he died in 1263. This did more than just help his career. Because Novgorod was not sacked by the Mongols, the city was able to preserve—and further develop—Russian culture and to continue thriving in its trading endeavors. The rest of Russia was not so fortunate; the Mongol invasion disrupted commerce and depressed the economy.

Although Novgorod was saved from ruin by the Mongols, not many buildings remain from this period because most were constructed of wood, a practice that continued until the seventeenth century. Because Novgorod was located near thick forests, the city constructed wooden streets and built an extensive wooden aqueduct system. Stone and brick were used only for defensive structures, churches, and the homes of the upper class. One can see examples of the brick construction techniques in the style of the walls of Novgorod's Kremlin. Encompassing St. Sophia and the marketplace where the veche met, these walls featured thirteen towers, of which nine remain today.

During the Middle Ages, Novgorod also became noted for its school of art. Novgorodian religious icons were known for their expressiveness, precision, and use of bright colors, with an emphasis on red and white. As for the city's architecture, starting in the thirteenth century, an occasional Romanesque element was incorporated into its buildings. In addition to its art and architecture, Novgorod was famous for its music; this city, along with Kiev, developed cycles of epic songs (called byliny), which remained popular through the sixteenth century.

Novgorod's eventual loss of importance was a direct result of the rise of Moscow, which at the start of the Mongol invasion was an insignificant trading outpost. Moscow grew under the leadership of princes who were ambitious and shrewd. They worked with the Mongol invaders and collected "tribute" for them; this enabled them to gain control of more and more land. Unable to compete with this new power center, Novgorod was forced to choose sides between either Moscow or Moscow's only major rival, Lithuania. The boyars, who took more and more control of governmental affairs, chose Lithuania, but the ordinary citizens felt closer in cultural traditions to Moscow and were unhappy with the increasing power of the boyars. With most of the population on Moscow's side, its prince, Ivan III, was able to take control of Novgorod in 1478. As a sign that Novgorod was no longer independent, Ivan III had the bell that was used to convene the veche carted away.

While this takeover was relatively bloodless—150 people were put to death—Ivan III's grandson, Ivan the Terrible, was uncomfortable with Novgorod's independent past, planted a forged letter that made Novgorodians appear traitorous, and had approximately 60,000 people executed.

Once Novgorod was incorporated into the Moscow-controlled Russian state, it never again achieved prominence. Other factors also contributed to its demise, including an invasion in 1611 and subsequent occupation by the Swedes until 1617, as well as an uprising in 1650.

The final blow to Novgorod's importance was the founding of nearby St. Petersburg in the eighteenth century.

Many of Novgorod's financial and commercial activities shifted there, leaving flax growing as its only important economic activity.

While no longer important, Novgorod has been recognized for its contributions to Russia's past. In 1862, a monument was erected in Novgorod's Kremlin commemorating the thousandth anniversary of Russia's existence.

Novgorod faced its worst devastation during World War II. German forces occupied the city from August 15, 1941, to January 19, 1944. When they left, only forty of Novgorod's 2,532 apartment houses remained standing. The city, however, has been rebuilt and today, as an industrial center, it is responsible for one-third of the region's economic activity. A few historic buildings, including the Kremlin and St. Sophia, remind visitors of Novgorod's glorious past.

Further Reading: *Bol'shaia Sovestskaia Entsiklopediia* (*Great Soviet Encyclopedia*), A. M. Prokhorov, editor in chief, volume 18 (New York: Macmillan, and London: Collier Macmillan, 1978; translation of third edition, originally published in Moscow: Sovestskaia Entsiklopediia Publishing House, 1974) is a good starting point for information about Novgorod, but suffers from Communist-era propaganda. *A History of Russia* by Nicholas V. Riasanovsky (New York and Oxford: Oxford University Press, 1969) is probably the most comprehensive book on the subject, but is also the slowest read. Faster and easier to read is *Soviet Union: A Country Study* by the Federal Research Division, Library of Congress, Raymond Zickel, editor (Washington, D.C.: Department of Army/Government Printing Office, 1989). Also helpful are *Hippocrene Companion Guide to the Soviet Union* by Lydle Brinkle (New York: Hippocrene, 1990); *Invitation to Russia* by Yuri Ovsianikov (London: Conran Octopus, and New York: Rizzoli International Publications, 1989); and *The Cambridge Encyclopedia of Russia and the Soviet Union* by Archie Brown, John Fennell, Michael Kaeser, and H. T. Willetts (Cambridge and New York: Cambridge University Press, 1982).

—Laurence Minsky

Nürnberg (Bavaria, Germany)

Location: On a craggy outcropping of the Franconian Basin, about 2,600 feet north of the Pegnitz; almost exactly halfway between Würzburg and Regensburg, or 68 miles from each; about 10 miles above the confluence of the Pegnitz and the Regnitz; connected to the Main, Rhine, and Danube Rivers.

Description: Site of a fortress to protect the Holy Roman Empire; center of trade with Italy, Northern Europe, and the Orient in the late Middle Ages; location of the Nazi Reichsparteig rallies, including the 1935 rally at which the Nazis formulated the Nürnberg Laws; site of the postwar international war crimes tribunal; renewed tourist attraction following the recreation and restoration of its medieval architecture, which had been severely damaged during World War II.

Site Office: Tourismus-Zentrale
Frauentorgraben 3
90443 Nürnberg, Bavaria
Germany
(911) 2 33 60

With a population of more than 500,000, Nürnberg is the second-largest town in Bavaria, the administrative center of Middle Franconia, a port on the Rhine-Danube canal, and the most important industrial town in north Bavaria, with electrical goods and electronics accounting for more than 40 percent of industrial production. Heavy industry and food processing remain important, and the service industries are growing fast. Each year the town provides for more than a million visitors, and Nürnberg has a year-round center for industrial fairs. The old town is roughly oblong in shape, and measures about one mile by a third of a mile. The Pegnitz flows across the center, separating north from south, and the old town is surrounded by a three-mile wall with towers, erected in 1400. Many older buildings use the characteristic local red-brown sandstone. There is an underground railway system.

Nürnberg was once one of Germany's most beautiful Gothic towns. Medieval blocks of half-timbered artisans' and merchants' houses lined cobbled streets with flamboyantly decorated facades and gables. The streets were narrow and crooked. The town suffered greatly in World War II, but still contains many imposing buildings, including its two famous churches, St. Lawrence and St. Sebaldus, and many charming streets and houses, especially near the river, as well as much impressive late medieval and Renaissance architecture. Modern Nürnberg, except in the old town and for tourist purposes, has attempted to distance itself from the image constituted by its late medieval buildings and atmosphere and from its association with Nazi Germany.

It seems more likely that Nürnberg was founded for strategic reasons and later became a crossing point for international trade routes than that the town was established on account of pre-existing routes. The first building, which dates from about A.D. 1000, was a fort on the rocky outcrop at the northeast edge of the present old town. The location was ideal to protect the Holy Roman Empire against attack from the northeast. It took 200 years for the town to extend 550 yards southward toward the Pegnitz. In 1050 the use of the name "Norenberc" is attested, and by 1062 its coinage was being accepted. The Burggrafenburg, almost completely destroyed by fire in the course of a dispute in 1420, was built for the German king and Holy Roman emperor Henry III (1017–56). It was unsuccessfully besieged by Lothair II, emperor from 1125 to 1137 and, after Conrad III, king of Germany from 1138, had lost control of it to the military governor, he built the adjacent Kaiserburg, which Frederick I Barbarossa again extended. Burggrafenburg and Kaiserburg, both restored since World War II, constitute an imposing miscellany of buildings dating in their present form from the fifteenth and sixteenth centuries.

Restored from 1854 to 1856, the complex covers the northwest corner of the old town. The present inner courtyard to the west is surrounded by the large palace building to the south, the court ladies' quarters to the west, and two high-roofed walls. To the east under an arch is a forecourt on whose south side is a fine two-story chapel with a tower to the east. The larger and better decorated upper story of the chapel, with its arched ceiling supported by gracefully decorative columns, was reserved for the court. The pillars are stockier in the lower story, part of which dates from the twelfth century.

To ensure viability for what was a mixture of palace and fortress, workmen during the twelfth century sunk a well in the courtyard to an extraordinary 55 yards. It provides for an average water level of ten feet, and is housed in a well-house built in 1563. The northeast corner of the courtyard is formed by superb half-timbered Gothic buildings that, although designed as treasury and secretariat, are still of nearly domestic proportions and charm. To the east of the courtyard is the 100-foot round Sinwell Tower, and in the castle's main building there is also a Knights' Hall with a wooden ceiling supported by thirty crossbeams with trusses held in place by oak supports. Above it, echoing the arrangements for the chapel, is the Emperors' Hall, more graceful and with a lighter ceiling.

The Kaiserburg is now connected by a covered passage to the Burggrafenburg to the east. Its courtyard was known as the "Freiung" because it bestowed right of sanc-

A view of Nürnberg
Photo courtesy of German Information Center

tuary, and of the original buildings only the Walburgis chapel, first restored in 1892, the guardian's cottage, the pentagonal tower, and the 1377 watchtower (the Luginsland), have remained. The pentagonal tower is the city's oldest building and the incidental result of an unsuccessful attempt to reinforce one of the original six-and-one-half-foot-thick walls. It was erected in 1200, although its upper part was not added until 1560. The massive building between the towers was constructed in 1494 and 1495 as stables, with the six upper stories used as a grain store.

It was from Nürnberg that in 1147 Conrad III set out on the second crusade. In 1192 the Burggrafenburg passed by marriage into the Hohenzollern family, and relations between the emperors and the Hohenzollern burgraves became generally hostile. Early in the thirteenth century Nürnberg was granted the privileges that gradually made it into an imperial free city. That meant that, although it belonged theoretically to the empire, it was in practice autonomously governed. The constitutions of the free cities differed, but the cities were generally invited to attend and sometimes to vote at imperial diets. Formal oaths of allegiance and feudal dues were not generally onerous.

In 1320 Nürnberg was granted independent supreme

jurisdiction. In 1332 Louis the Bavarian, who stayed in the town on seventy different occasions, confirmed the tax exemption. In 1356, the Golden Bull was drawn up on the instructions of Emperor Charles IV, king of Bohemia, to regulate the election of "Roman kings" (i.e., emperors); the bull stipulated that all new emperors should hold their first diet in Nürnberg, where Emperor Sigismund, son of Charles IV, decreed in 1424 that the imperial regalia were to be kept. They remained there until 1796. All previous emperors had stayed in the town, and Charles IV himself made some forty visits. In 1349 he also allowed the destruction of the Jewish quarter to make way for a marketplace, today's Hauptmarkt. Doors were sealed off and the houses set on fire. Those of the 600 residents who escaped were beaten and burned, and the marshy ground was then filled in and the ground level raised nearly twenty feet.

Between 1350 and 1358, on the site of the destroyed synagogue, Charles IV oversaw the erection of the Frauenkirche by master mason and architect Peter Parler, a Swabian partly responsible for Prague cathedral. The Frauenkirche—its gable with its pinnacles and niches was the work of Adam Kraft—was until 1361 the court chapel. In 1323, by which time the right-bank town had reached nearly as far as the river and the left-bank town stretched about 430 yards to its south, the present old town was physically constituted by the erection of the town walls, with roofed sentry walk. As early as the 1350s a Jewish community reassembled in the town, although it continued to suffer under severe discriminatory legislation and, forced into banking by prohibitions preventing them from becoming craftsmen or artisans, could scarcely hope to lessen the hostility toward them. In 1385 the Jews were rounded up all over Swabia and forced to bargain their way out of imprisonment. The town raised a ransom of 80,986 guilder in the process. The Jews were finally exiled again in 1500.

As in many German towns, in Nürnberg there was civic unrest in the mid–fourteenth century when the skilled craftsmen demanded more power in the city's affairs. They governed briefly and encouraged the formation of craftsmen's guilds, but they were dissolved when Charles IV restored a patrician governing council in 1349. The town, which maintained an armed militia in the Kaiserburg, also erected a wall against the Grafenburg in the fourteenth century and finally affronted the burgrave by erecting the Luginsland watchtower in 1377. The Grafenburg was finally occupied by the town in 1388–89, then later returned to the Hohenzollerns. In 1427 the town was financially strong enough to buy the ruin, and its independence from the Hohenzollerns, for 127,000 guilder. The Hohenzollerns attempted to regain control of the Grafenburg from 1449 to 1453 with an army of 7,000, but the town was able to resist them.

Metalwork and other trades were now flourishing. Nürnberg, like Augsburg, profited from its role as intermediary between Italy and both northern Europe and the Orient. Despite such appalling setbacks as the loss of 13,000 victims—possibly a third of the population—to the plague in 1437, and indeed partly on account of the plague's ultimate economic effects on the balance of urban and rural populations and the value of property, Nürnberg entered the high point of its prosperity and cultural achievement shortly after 1470. It was not surrounded by good agricultural land, and may not have had its economic development retarded as much as some other cities by rural populations that were forced by plummeting prices to move into towns. The plague was again to strike severely in 1533, 1562 (9,000 deaths), 1585 (5,000 deaths), and 1634 (2,000 deaths).

The earlier of Nürnberg's two most famous parish churches is that of St. Sebaldus, who, before his death in 1070, had built a chapel where his church now stands. Building on the church began in 1230, and the church was constructed in various stages. By the time it was completed in 1345, the town needed a larger monument to its patron, and the architectural shell of the present structure was standing by 1372. The thirteenth-century construction incorporated the subsequently removed original church, and there is now an early Gothic thirteenth-century west front with twin towers, made higher in 1345, and two Romanesque doorways on each side of a protruding chancel at the end of an elaborately

decorative fourteenth-century late Gothic church lavishly adorned with pinnacles and pointed arches. Substantial restoration work was undertaken from 1888 to 1906, and again from 1946 to 1957.

The tympanum above the south door to the west front is by Adam Kraft, also responsible for the *Carrying of the Cross* on the south side of the nave and the *Passion and Resurrection* under the first northern window of the east chancel. Outside the residual transept near the west chancel are two doors, each with an impressive tympanum. The church contains a central nave, high in appearance because separated from side aisles of half the height by thick square pillars, and enclosed above by solid walls with only minuscule windows. The interior of the church contains superb work by Veit Stoss and the family of Peter Vischer the elder, whose younger members Hermann, Peter the younger, and Hans derived their style from Italian work with which they had become familiar. The whole family worked on the tomb of St. Sebaldus, impressive and fascinating in spite of the bewildering variety of styles it incorporates.

The building of the parish church of the settlement on the Pegnitz's left bank, the Lorenzkirche, or church of St. Lawrence, as a three-aisled Gothic basilica modeled on the Sebalduskirche, first began in about 1250 on the site of a series of earlier Romanesque chapels. The impressive west front—with its tall twin-steeple–topped towers reaching twice as high as the nave, its decorative sculpted portal in the Bohemian style of Peter Parler (Prague's chief architect) under a carved pointed arch richly adorned with sculpted stone figures, its large rose window, and its filigree gable with a silver spire, a triangular embellishment of stone tracery in which it culminated—must have been added about 1353 because it contains the intertwined arms of Charles IV and his third wife Anna von Schweidnitz und Jauer, whom he married in that year. The elaborately ornamental architectural style of the Bohemian court had replaced the austerity of early Gothic. The basilica was finished in 1360, but later in the fourteenth century the nave walls were extended to the outer edge of the flying buttresses, with the consequence that much larger windows could be installed. The church's interior became broader and much lighter.

The chancel was started in 1439, erected around the altar of the abbot Deocarus, Charlemagne's confessor, whose relics, awarded to the town by Louis the Bavarian in 1316, were put in 1437 into a silver shrine reminiscent of St. Sebaldus's grave in the older parish church. The chancel was completed in 1477. No further changes were made until the town became part of Bavaria in 1806, when many of the more valuable works of art were sold off, particularly if they were in precious metal and could be melted down, like a medieval bronze font and the Deocarus shrine that used to stand on the altar. Restoration work started quite soon, in 1825, but was set back when lightning burned out the upper part of the north tower in 1865. Extensive restoration after World War II made the church usable again by 1952.

The interior contains superb work by Veit Stoss in his

later, post-Kraków manner, most examples of which are in limewood, often monochrome, like the high altar crucifix here, and generally consist of no more than one or two figures. The church contains his celebrated *Annunciation* of 1517–18, a bronze chandelier cast by Peter Vischer in 1489, and the 1493–96 stone tabernacle by Adam Kraft, who sculpted an image of himself in working clothes beneath it. The stained glass of the chancel windows, with individual windows given by donors in the interest of furthering the attainment of salvation for themselves and members of their families, has been almost entirely preserved. Four of the five apse windows come from the workshop of Michael Wolgemut, intimate friend of the scholar Wilibald Pirckheimer, where Dürer learned to paint and design woodcuts.

Elsewhere in Nürnberg the late medieval house just west of the Kaiserburg, purchased by Albrecht Dürer in 1509, is now a museum, one of the more imposing relics in a town of winding, narrow streets and busy marketplaces of the prosperous late medieval town. Property taxation made street frontage expensive, and the often half-timbered houses with sharply pitched roofs and dormers had to be built to great heights, with craftsmen's houses seldom having fewer than four stories, on top of which further living quarters were situated just beneath the roof.

In October 1518 Martin Luther stayed briefly in Nürnberg, where a following soon built up for his views. The governing council was embarrassed but did nothing. Of all the south German towns, the debate over Reformation remained most purely religious at Nürnberg, where there was little social unrest. On the one hand, the council could scarcely resist popular feeling, which was inclined strongly to Luther's doctrines, if not immediately to his liturgical reforms, and the council was not at all anxious to continue to allow the town's religious life to be regulated externally, any more than it would have relished the abandonment of political autonomy. On the other, the free city had too much to lose by breaking with the emperor.

It temporized, secularizing the monastic foundations, removing clerical exemptions from taxation and subjecting the clergy to ordinary civil jurisdiction, without, however, giving the laity any greater role in the administration of ecclesiastical affairs or joining the Schmalkaldic League of Protestant princes, although it made a modest contribution to the league's treasury. By the 1530s war between the Protestant princes and the emperor appeared likely. The Nürnberg council did not want to find itself obliged to go to war against the empire on which its status as a free city and a prosperous commercial center entirely depended. No mere city could have retained independence in any Germany governed by the Protestant princes.

In fact, Nürnberg's decline had become inevitable with the discovery of the sea route to India in 1497. Then in 1552 the margave Albert Alcibiades of Brandenburg switched his allegiance from the emperor to the Protestant princes. He used the religious divide as an excuse to terrorize Nürnberg into ransoming itself for 200,000 guilder and the

restoration of the feudal privileges his family had lost in 1427. Nürnberg yielded, but a year later the margrave, having switched back to the emperor's side, mounted another attack, this time purely for booty. The destruction was enormous, but eventually the imperial princes and cities rallied, and Albert was defeated at Schweinfurt in June 1554. Still, Nürnberg had suffered much damage and was 3.5 million guilder in debt.

During the Thirty Years War (1618–48), Nürnberg remained officially neutral, although Albrecht Wallenstein of Austria defeated Gustavus Adolphus of Sweden near the city. Nürnberg had barricaded itself, but during the blockade some tens of thousands died of the plague, dysentery, and typhoid. Debt increased, and the town's decline accelerated. In 1792 and 1796 the town had to yield parts of its territory to Bavaria and Prussia, and offered itself to Prussia in return for the payment of its debts. Prussia turned down the offer. In 1803 Nürnberg was allowed to retain its nominal position as a free city, but in 1806 it lost its free status and was incorporated into Bavaria. In the nineteenth century it began to develop again as an industrial city, and the first German railway was opened in 1836 between Nürnberg and Fürth.

It is not impossible that Hitler's choice of Nürnberg for his Reichsparteitag, or annual Nazi rallies, was influenced as much as his attested reaction to Wagner's *Die Meistersinger,* which is set in Nürnberg and ends in a celebration of German art, as by Nürnberg's historical freedom from non-Germanic "cultural contamination." The first Reichsparteitag to be held there was in 1927, but Hitler and his supporters had first demonstrated there in 1923. By 1927 he could stage in the city a vast military pageant, with resounding marches, banners, and 20,000 party members, of whom 8,500 were in uniform. Vaguely, but ominously, Hitler's speech on this occasion first publicly linked the concepts of *Lebensraum* (living space) with Hitler's already fierce anti-Semitism.

The first great showpiece rally was on Zeppelin field in 1934. Alfred Speer was the architect and stage manager who erected a stone structure 400 meters (1,312 feet) long and 24.5 meters (80 feet) high, inspired by the Pergamum altar of Zeus in Berlin, to underline the pagan and religious nature of the rally. The stadium was crowned by an eagle with a 30-meter (100 foot) wingspan, hung with thousands of swastikas, and it was choreographed with the aid of 130 anti-aircraft searchlights. Leni Riefenstahl used a crew of 120 to film it, with cameras on planes, cranes, roller skates, and a flagpole, and the phrase "the thousand-year empire" was publicly used for the first time. The rallies, wrote the British ambassador, were superior in grandiose beauty to anything the Russian ballet had to offer. The spectacular massed bands, the perfectly synchronized marching, the flag-waving, and the meticulous coordination of everything during the six days were successfully designed to produce a hypnotic sense of power, force, and unity.

The rallies grew in popularity. The last rally before World War II was attended by 1.6 million people.

The Nürnberg laws against the Jews were the product of the rally of 1935, by which point Hitler's anti-Semitism had seemed to have become milder, perhaps under economic pressure from abroad. But on September 13 he ordered that a decree be prepared within forty-eight hours to be entitled "Law for the Protection of German Blood and Honor." After a draft prohibiting marriage and extramarital sex between Jews and Aryans had been prepared, further orders were issued to produce a Reich citizenship law. The authors ran out of paper and had to use old menu cards. It was 2:30 A.M. on September 15 before it was agreed that only those "of German or related blood" could be citizens. Hitler announced the new laws at 9:00 P.M.

After the war, Nürenberg was chosen as site of the war crimes trial. The principle of a trial for war crimes was agreed upon as early as 1942 by the United States, Great Britain, and the Soviet Union at the instigation of nine of the occupied countries. In 1943 a United Nations War Crimes Commission was established. The gestation of the form that retribution should take for the major Nazi survivors was slow, even when fifty-two, of whom fifteen would eventually be chosen for trial, were collected at Bad Mondorf between May and August 1945. Agreement on Nürnberg as the site of the trial was largely determined by America's ability to provide better rations and comfort there than the Russians could have produced for a Berlin trial in their sector, which they would have preferred. Everything about the trial—judges, charges, prisoners, and procedures—was the hard-fought result of compromise. The trial started on November 20, 1945. Twenty-one of the twenty-two accused Nazis actually appeared in court. Verdicts and sentences were pronounced on October 1, 1946, and executions took place on 16 October. Twelve were sentenced to be hanged, seven to imprisonment, and three were acquitted.

Until 1945 Nürnberg had produced roughly half the airplane, submarine, and tank engines in Germany and was in consequence heavily bombed. On January 2, 1945, some 90 percent of the old town was destroyed, with the estimated loss of nearly 2,000 lives. The city has been lovingly and meticulously restored and exudes again the atmosphere of the prosperous late medieval town. The real industrial city, where the majority of the population lives, is outside the walled old town, where assorted relics of a colorful past, so often painful for those who lived through it, are visible in the streets or carefully labeled in the many museums. The old town overflows with priceless and sometimes intensely moving artistic treasures, with magnificently restored medieval architectural masterpieces, and with picturesque reconstructions of patrician dwellings. But, even more than most historic towns, Nürnberg has sadly had to rely more on replication than on restoration. The present center, with its completely new replication of the old Rathaus and its Tradesmen's Court in which stall holders in medieval dress purport to be making the traditional wares they sell, has not totally avoided the atmosphere of a theme park. Life in medieval Nürnberg was a great deal more unpleasant for everyone than the flawlessly maintained modern old town would like the tourists it attracts to think. Even the famed instruments of torture, carefully laid out, are somehow sanitized by their display.

Further Reading: Apart from the usual guides, brochures, encyclopedia articles, and the mass of recent scholarly material in German, there is a popular but informed book in English by Gerald Strauss, *Nuremberg in the Sixteenth Century* (Bloomington: Indiana University Press, 1966; Chichester: Wiley, 1967). For those with a serious interest in south German art history, it would be difficult to recommend a better book in any language than Michael Baxendall's *The Limewood Sculptors of Renaissance Germany* (New Haven and London: Yale University Press, 1980). For the history of Nürnberg, the second edition of the *Handbuch der Bayerischen Geschichte*, edited by Max Spindler, four vols. in six (Munich: C. H. Beck'sche Verlagsbuchhandlung, 1968) contains indispensable information, but there are more specific treatments of the town and various aspects of its history, including E. Mulzer's *Die Nürnberger Altstadt: Das architektonische Gesicht eines historischen Großtadtkerns* (Freiburg: Rombach, 1984) and the series of volumes edited by G. Pfeiffer, *Nürnberg: Geschichte einer europäischen Stadt* (Munich: C. H. Beck'sche Verlagsbuchhandlung, 1982).

—Anthony Levi

Odense (Fyn, Denmark)

Location: In north-central Fyn Island, east of the Jutland Peninsula.

Description: Third-largest city in Denmark after the capital Copenhagen and Arhus; a garden city, famous for its cathedral, as the birthplace of the writer Hans Christian Andersen, and as a former place of pilgrimage.

Site Office: Odense Tourist Association
Råhuset
DK-5000 Odense C, Fyn
Denmark
66 12 75 20

Odense is one of the oldest cities in northern Europe. Its name, which means "Odin's Shrine," is derived from two Norse words: "Odin/Woden/Wotan" and "*vi*"; the former is the name of the head Norse god, and the latter means "sanctuary." Odense is thought to have been a place of pilgrimage before Christianity arrived in Europe.

The first recorded mention of the city is in a religious document sent by the German emperor Otto III to the bishop of Bremen in 988. The emperor shared the bishop's desire to make Odense the site of a bishopric. In recognition of this significant date Odense celebrated its millennium in 1988.

Large amounts of coin were struck in the city during the eleventh century, particularly during the reign of Canute the Great (1018–35), who founded a convent in the town. However, it was Canute the Holy, who reigned from 1040 to 1086, who brought Odense to public recognition. Although Canute had been rebellious in his youth, he proved to be a strong-willed king who took his duties seriously. He was particularly interested in taxes, and introduced a 10 percent tithe on harvested corn. He also worked to free indentured laborers.

A fierce enemy of William I (William the Conqueror), Canute the Holy planned in 1085 an expedition against the former Danish colony of England, envisaging that it would once again form part of the Danish empire. To this end, he enlisted the support of the Norwegian king, Olaf III, and the count of Flanders. The war fleet on the eve of the planned invasion numbered more than 1,500 ships (about 1,200 Danish and 300 Norwegian).

When rumors about the planned attack began to circulate throughout Denmark, however, the tribes south of the border of Schleswig began to revolt. While Canute was calming the rebellion, William I sent envoys to Denmark to bribe Canute's fleet. At the same time, the peasants who had been mobilized for the planned attack on England, grew restless; they had exhausted their supplies, and their land suffered from neglect. The peasants demanded Canute's immediate return. Canute returned to Aggersbog to find that the Danish forces had deserted him, and only the Norwegian ships were left. Unfortunately, Canute imposed heavy fines on the peasants for interfering with the war, and this resulted in a revolt north of Jutland, from which Canute had to flee. He ended up in Odense, in the wooden church of St. Alban, where he was killed by a group of marauding peasants.

In the following years the country was stricken with famine and poor harvests. A series of miracles were reported to have taken place at the site of Canute the Holy's murder, and the Danes began to regard him as a saint. In the eleventh century Canute was canonized by the pope, and today he is known as St. Canute. His remains were moved to the church he had built in 1084. Fifty-five years later this church was consecrated as Skt. Knuds Domkirke. Unfortunately, little remains of the original cathedral; a series of fires resulted in its destruction by the thirteenth century. As a result of Canute the Holy's canonization, Odense again became a popular place of pilgrimage during the Middle Ages. Today the town is home to six convents and monasteries, and three churches.

In the early fourteenth century, St. Canute's original stone building was replaced by one of the finest examples of the High Gothic style that exists today. It took almost 200 years to complete. Today the whitewashed brick structure houses shrines to St. Canute and St. Alban. It is a bright, spacious building. The altar, a stunning triptych by the German artist Claus Berg, was installed in the 1880s; it is probably one of the most memorable in Denmark, with its magnificent depictions of the Crucifixion, the church fathers, and members of the Danish royal family, and its heavy use of gold. In the early nineteenth century the remains of King John and his family were also moved to the cathedral.

During the sixteenth century, Odense became a popular place of residence for the Danish and Pomeranian aristocracy. Destroyed by fire in 1529, three years after the religious Reformation in Denmark, the city suffered further when it became caught between the warring factions involved in the Count's War. By the 1550s, however, Odense was benefiting once more from its position on the river Little Odense, and foreign trade revitalized the economy.

Unfortunately, the outbreak in 1654 of war with Sweden, which was then seeking independence from the Danish empire, and the subsequent Swedish occupation of the city from 1654 to 1657 affected the economy adversely. With the imposition of absolute monarchy by the Danes in 1661, after the Swedes had been driven out, the great assemblies of the aristocracy no longer took place in the city, and the economy declined further.

King Frederick IV and his wife popularized Odense with the aristocracy again in the mid-eighteenth century by

Part of the collection at Hans Christian Andersen's home
Photo courtesy of Danish Tourist Board

renovating the Sanki Hans Monastery into a castle. The king loved the city, and died in the castle there in 1730. The royal patronage brought new industry to the area and soap and sugar were among the commodities produced during this period. This period of prosperity was temporarily interrupted by the Napoleonic Wars, but in 1804 a canal was constructed, linking the river to the Odense Fjord. A harbor and the construction of railways across the district helped to establish Odense's position as a central town in Danish commerce. The city began to expand south of the river as trade and industry prospered.

Odense was also the birthplace of the writer Hans Christian Andersen. The son of a cobbler, Andersen was born into relative poverty in 1805 and resided in the city until he was fourteen. Today the area where Andersen lived has been restored. In 1930 his home (H. C. Andersen Barndom-shjem) was turned into a museum, which forms part of the much larger Hans Christian Andersen Museum along with

several other houses. The museum contains Andersen manuscripts and other memorabilia.

The people of Odense have always banded together in times of strife. Never has this been more evident than during the Nazi occupation of Denmark. During World War II Denmark developed an impressive Resistance network. Originally a neutral country at the outbreak of war, Denmark quickly and effectively mobilized people willing to fight against the Nazis once Germany invaded Norway, and Denmark capitulated to the Germans in 1940. Support for the Nazis was quickly in evidence as local fascists came out onto the streets wearing uniforms. But Nazi support increased in Odense and the Resistance movement grew. Great quantities of arms and explosives came into Denmark in the early 1940s and were used to sabotage Nazi mobilization plans—the main targets included bridges, armament factories, and boat manufacturers. Young and old alike were used to fight the fascists, and a highly skilled team of paratroopers were used to hold back the Nazis' plans for the eventual conquest of Europe.

Each night during the summer of 1943, there were open clashes between the Nazis and local people at the shipyards in Odense. Although the manufacture of German U-Boats and warships had been delayed by the slow work of Danish employees, the minesweeper *Linz* was finally ready for use. To prevent it from being launched, shipyard workers smuggled in a limpet mine and blew the ship up. The German commander in chief, General von Hanneken, sent in armed troops to quell the situation, and as a result the whole city of Odense went on strike. This situation was repeated all over Denmark.

After the war, Odense became a university town. The University of Odense, founded in the 1960s, is an internationally respected educational institution, featuring some of Denmark's best modernist architecture, made of concrete and rusted iron.

Since the beginning of the twentieth century, the character of Odense has changed significantly. The city was important in Denmark's industrialization, and its population has more than tripled since 1900. Today Odense is a remarkably pretty city, despite its dependence on heavy industry, and only the Cathedral spire and a relatively new university hospital break the relatively low skyline. Every year thousands of tourists go to Odense to admire its attractions.

Further Reading: *Baedeker Denmark* (London: Automobile Association, and Englewood Cliffs, New Jersey: Prentice Hall, 1987) provides brief but basic information on the country's major sites and the history of Denmark; *Blue Guide Denmark* (London: Black, and New York: Norton, 1992) is a sound guide to the country with a comprehensive overview of its history; *A History of the Kingdom of Denmark* by Palle Lauring (Copenhagen: Host and Sris Forag, 1960) is a detailed history of the country from its beginnings.

—Aruna Vasudevan

Omaha Beach (Calvados, France)

Location: The landing beaches for the Allied invasion of Normandy stretched eastward nearly fifty miles along the coast from Quineville on the east shore of the Cotentin Peninsula to the mouth of the Orne River. Utah, the westernmost invasion beach, covered about five miles along the Cotentin Peninsula in front of the town of La Madeleine. About twelve miles east of Utah was Omaha Beach, stretching about four miles from Vierville to just east of Colleville-sur-Mer. These beaches were assigned to the American forces. Continuing eastward to the Orne River were the British beaches: Gold, Juno, and Sword. Inland about eight miles from Sword Beach is the city of Caen, one of the primary objectives of the D-Day attack. The modern French roads D514 and N13 run westward along the invasion coast from Caen to St.-Laurent-sur-Mer, then northwest along the east side of the Cotentin Peninsula.

Description: Utah Beach is a stretch of sloping sand between the sea and marshes (flooded deliberately by the Germans at the time of the invasion); several causeways led up to the town of St.-Mère Église, a key D-Day target. Conditions were more difficult at Omaha Beach, where wide stretches of sand faced stone shingle shelves and steep bluffs more than 150-feet high in places, covered by dense scrub and grass. Gold, Juno, and Sword Beaches, in contrast to Omaha, sloped gently and had no cliffs or bluffs baring the way inland. Today, a number of museums, monuments, and cemeteries dot the coast where the invasion took place.

Site Office: American Cemetery at Omaha Beach
14710 Colleville-sur-Mer, Calvados
France
31 22 44 60

Operation Overlord, the Allied invasion of the European continent, began in the early hours of June 6, 1944, with British and American airborne assaults intended to secure the east and west flanks of the invasion area. These assaults occurred between 1:30 and 2:30 A.M., and at 6:30 A.M. the main seaborne attack by American forces began on the westernmost beaches: Utah and Omaha. These initial attacks were followed by British seaborne landings on the three eastern beaches: Gold, Juno, and Sword. Utah Beach, lightly defended, was secured by late morning, the British beaches by afternoon. The Omaha Beach assault, facing steep bluffs and cliffs, developed into a hard-fought, bloody battle that came close to failure. The U.S. First Division finally overcame the beach defenses, paying a price of more than 4,000 casualties.

The massive Allied invasion across the English Channel marked the beginning of the end for Nazi Germany. The invasion itself was preceded by a series of dramatic military and political events that stretched back for nearly two years of planning, preparation, and conflict. In August 1942 British and Canadian forces with a small contingent of U.S. Rangers staged a seaborne hit-and-run raid on the small French coastal resort of Dieppe. The Germans were alerted by intelligence sources and inflicted a terrible defeat on the raiders. Some 3,500 of the 5,000 who made it ashore were lost. The Dieppe failure was a warning to Allied planners that an immense concentration of forces would be required to gain a foothold on the heavily defended coast.

In May 1943 the Anglo-American Conference in Washington, D.C., considered opening a second front in Europe in the context of overall Allied plans for operations against Germany, and in relation to the war in Russia and against Japan. The Combined Chiefs of Staff had already established the parameters for a cross-channel invasion: concentration on a single site close to a major port, near a good road network leading inland, and within range of fighter planes based in Britain. The Pas de Calais area was closest to Britain but for that reason was heavily fortified, as was the area around Dieppe. In March 1943 the planners decided on the coast of Normandy between Caen and Cherbourg, along the base of the Cotentin Peninsula. In May 1943 a target date for the invasion, May 1, 1944, was set by Winston Churchill, Franklin D. Roosevelt, and their military advisers. The Russians agreed with the timing and with the invasion plans during conferences at Quebec and Tehran later in the year.

Dwight D. Eisenhower had headed General George Marshall's War Plans Division since the Japanese attack at Pearl Harbor, and in 1942 he was named to command Operation Torch, the Allied invasion of North Africa. Roosevelt decided at Teheran that Eisenhower would command the cross-channel invasion, Operation Overlord. Eisenhower first met with the high Allied planning staff in Britain on January 21, 1944, and took formal command of Supreme Headquarters, Allied Expeditionary Forces (SHAEF) on February 13, 1944. Planning for D-Day involved preparation for the largest amphibious military operation in history. Plans included military coordination among the forces of several nations and enormously complex logistical arrangements for operations on D-Day itself and for an extended time thereafter. In addition, counterintelligence activities were carried out to fool the enemy about the location and timing of the assault. Finding sufficient ships to transport troops was in itself a major problem. Huge numbers of personnel and massive amounts of materiel were assembled. By June there were more than 1.5 million U.S. troops, 2.5 million tons of supplies, 1,200 assembly camps, and 100 marshalling centers in the United

The American Monument at Omaha Beach
Photo courtesy of French Government Tourist Office

Kingdom. More than 4,000 ships of all kinds stood by, including substantial portions of both the British and American fleets. The Allied air forces also were ready.

Allied policy was to convince the Germans that the invasion would come at the Pas de Calais, the coastal area nearest Britain. Phony intelligence reports from captured German spies were radioed to Europe. An entirely imaginary army, "Army Group Patton," was created on the British coast near Dover. Dummy tanks, trucks, and landing craft were constructed, false radio traffic was broadcast, and aerial attacks were concentrated on the Pas de Calais area. Until well past D-Day, the Germans were fooled and kept a large army pinned down near Calais waiting for an attack that never came.

On the German side plans were also being made to repel the invasion. The commander-in-chief in the west was Field Marshal Gerd von Rundstedt, one of the most respected officers in the German army, the victor in 1940 over French and British armies. Hitler, with a chronic distrust of army officers, divided the command responsibilities in the west, a policy that had proven disastrous in Russia. Late in 1943 Hitler put Field Marshal Erwin Rommel, famous for his North African victories, in charge of planning

the defeat of the invasion. He was made responsible to Hitler's headquarters rather than to Rundstedt, and was placed in command of two armies forming Army Group B along the Channel. Rundstedt, however, was left in command of all armored units in France. Rommel believed the invasion needed to be defeated on the beaches, while Rundstedt wanted to hold back a mobile reserve that could be rushed forward to counterattack. Allied air superiority made this a dangerous policy that would prove disastrous for the Germans on D-Day.

Though Hitler had bragged for years that the "Atlantic Wall" was heavily fortified and impregnable, Rommel found the actual coastal defenses skimpy and in places non-existent. Some fortresses had been built at major ports in the aftermath of the Dieppe raid, but much remained to be done. With characteristic energy Rommel moved to complete the coastal defenses. On every possible landing site soldiers and labor battalions erected lines of anti-invasion devices. There were jagged triangles of steel, gatelike structures of iron, slanted wooden stakes, and cone-shaped concrete barriers, all covered or tipped by mines, on which Rommel had learned to rely in North Africa. Mines were laid on obstacles, in the shallow beach water, in the sands, on the bluffs, and in the gullies and pathways leading off the beaches. More than 5 million mines were laid down and only the coming of D-Day prevented Rommel from planting 6 million more. Behind the mines German troops waited in pillboxes and concrete bunkers linked by communication trenches and layers of barbed wire. Artillery pieces were sighted to give overlapping fields of fire on possible invasion beaches. Batteries of rocket launchers, mortars, tanks, and flame throwers rounded out the German arsenal. But the defenses were still unfinished on D-Day, and the German divisions manning the defenses were far from Germany's best. The enormous losses on the Russian front had caused Germany to siphon off the best personnel from the western forces. The German garrisons in the west were made up to a great extent of young, partially trained boys, old men, units shattered on the Russian front, "volunteers" from occupied countries, including Russians, and soldiers still recuperating from wounds. The most powerful German forces in the west, the Panzer and SS divisions, were held far back from the coast in Hitler's and von Rundstedt's mobile reserve and would play little or no part in the D-Day fighting.

Rommel, who had prepared the defenses that would almost defeat the invaders on Omaha Beach, himself played no part in the D-Day action. Stormy conditions on the Channel on June 4 (which convinced Eisenhower to postpone D-Day from June 5 to June 6) encouraged Rommel to take several days' leave. He decided to spend his wife's fiftieth birthday with her at Herrlingen, then go on to Hitler's Bavarian headquarters to confront the dictator. It was understood at the time that he wanted to demand of Hitler more weapons and personnel for the Channel defenses; after the war it was revealed that Rommel was heavily involved in the German army officers' plot to remove Hitler and make peace with the western Allies. The plotters wanted to assassinate Hitler, but Rommel favored arresting him, trying him before a military court, and arranging a limited surrender with Britain and the United States while continuing to fight the Russians. According to Rommel's chief of staff, Lieutenant General Hans Speidel, Rommel's June trip was planned as a last attempt to reason with Hitler before taking action against him. In the event, Rommel turned back to the coast on receiving news of the invasion. His role in the anti-Hitler conspiracy was revealed by arrested plotters after the failure of the July 20 bomb attempt on Hitler's life. On October 14 Rommel was forced by Hitler's emissaries to commit suicide by taking poison in order to save his family from a concentration camp.

The long-awaited invasion began a little after midnight on June 6, with paratroopers of the British 6th Airborne Division and the U.S. 82d and 101st Airborne Divisions dropping behind the invasion beaches. Pathfinders dropped first, followed by the main airborne units. Stormy weather caused many of the planes to drop their men miles from their landing zones. Some fell into German-controlled areas and were killed or captured; some fell into swamps or the sea and were drowned. The survivors rallied and moved on to their objectives: to cut off roads, bridges, and rail lines leading to the landing beaches.

Omaha Beach was a 7,000-yard stretch of sand with up to 200 feet of ground exposed at high tide and 400 feet at low tide. Behind the sand an 8-foot bank of gravel covered the western end of the beach. A line of scrub-covered bluffs 100 to 170 feet high were behind the middle part of the beach, merging at either end with rocky cliffs that enclosed Omaha. Two parallel bands of mined obstacles 50 to 75 yards wide and about the same distance apart had been set up by the Germans, backed by a dozen artillery strongpoints. Machine gun nests and trenches were located between the bunkers. On the Pointe du Hoc, a cliff about 5,000 yards west of Omaha, there was a battery of six howitzers said to be sighted on the landing beaches. A U.S. Ranger Battalion was assigned to scale the cliff and silence the guns. The Rangers assaulted the Pointe shortly after the landings began, taking heavy losses. They found the howitzers near the cliff and fought desperately for the rest of the day to keep them from being recaptured.

By dawn on June 6, with the 5,000-ship invasion fleet approaching, nearly 18,000 American, British, and Canadian airborne soldiers were already on the flanks of the Normandy battlefield. The Germans, still believing that the main attack would come at the Pas de Calais, were slow to respond to the airborne landings. Their confusing command structure, compounded by the absence of Rommel, continued to lull the Germans into inactivity, although local units were already fighting skirmishes with groups of Allied paratroopers.

At about 6:30 A.M. the first wave of soldiers of the U.S. First Infantry Division, in various kinds of landing craft and floating tanks, assaulted Omaha Beach near Colleville and in the Vierville area west of Colleville. Though British decrypters had cracked the German's Enigma code and were reading German radio transmissions, they missed the arrival at Omaha Beach of the 352d Infantry Division, one of the best German units in Normandy. High seas racked the coast, and numerous craft were swamped or destroyed by shellfire or mines before reaching shore. As the surviving boats closed on the beach a hail of machine gun fire and mortar shells rained down on them. The Allied air attacks meant to soften up the beach happened too far behind the shore; the Allies had been too cautious, for fear of hitting friendly troops. Infantrymen leaving their landing craft were cut down by gunfire and shelling, or blown up by mines. A few managed to reach the shelter of the sea wall, but most of the first wave became casualties. In many units every officer was killed or wounded within minutes. More soldiers of the second wave made it across the beach, but by 8:00 A.M. there were more than 3,000 casualties on the sand.

By 11:00 A.M. the commander of the German 352d Division thought he had won. But the Americans were gradually beginning to move forward off the beach. Numerous individual acts of courage and heroism took place. Individuals, what remained of units, moved up the bluffs. By noon small parties had captured a road that ran along the bluffs a few hundred yards inland. Vierville was captured by the 5th Ranger Battalion by noon. After fierce fighting, by the end of the day American forces held a beachhead at Omaha up to a mile and a half deep, and almost all the coastal villages were taken. Men and supplies continued to land and a regiment of the 29th Division came ashore to help extend the beachhead. Total Allied casualties on D-Day came to more than 9,000, but more than 100,000 Allied troops were ashore. Omaha Beach came to be called "Bloody Omaha" by the invaders, but the overall assault was successful.

In the days that followed, Rommel's predictions came true: strong German Panzer and SS divisions held back from the shore were shattered by Allied air power while trying to reach the coastal battles. Nevertheless, there was hard fighting around Caen and in the Cherbourg Peninsula before German forces finally collapsed and the Allies were able to burst out of Normandy. Cherbourg surrendered on June 27, and the Allied breakout began July 25 near Saint-Lô. By August 12 the Germans were retreating, and suffered a disastrous defeat near Falaise. On August 25 the Allies liberated Paris, and Operation Overlord was declared over. Germany surrendered unconditionally on May 7, 1945, and the war in Europe officially ended at 11:01 P.M., May 8, 1945.

Today, the battles fought on the D-Day beaches are memorialized by a host of museums, monuments, and cemeteries. Among the better known are the American Cemetery at Omaha Beach, located in Colleville-sur-Mer; the Musée du Débarquement in Arromanches, which focuses on the British contributions to the invasion; the Monument du Débarquement in St.-Laurent-sur-Mer, near the dunes that overlook Omaha Beach; and Le Mémorial, a museum in Caen that focuses not only on the D-Day invasion, but also on the war's place in the larger context of the quest for peace. Less formal, but equally poignant, reminders of the fierce fighting that took on D-Day are the scattered remains of bunkers, trenches, and barbed wire still visible from the dunes and bluffs overlooking the beaches.

Further Reading: *Omaha Beachhead: 6 June–13 June 1944* (Washington, D.C.: U.S. War Department, Historical Division, 1945) is a thorough, well-illustrated report based on unit reports, interviews, and enemy records of the landings. *Builders and Fighters: U.S. Army Engineers in World War II,* edited by Barry W. Fowle (Fort Belvoir, Virginia: U.S. Army Corps of Engineers, 1992) is a long, well-illustrated account of the Army Engineers in the war, including their activities on Omaha Beach. David W. Hogan Jr.'s *U.S. Army Special Operations in World War II* (Washington, D.C.: Department of the Army, Center of Military History, 1992) includes a dramatic account of the U.S. Rangers' operations on D-Day. Cornelius Ryan's *The Longest Day: June 6, 1944* (New York: Simon and Schuster, 1959; London: Gollancz, 1960) is probably the most widely read history of D-Day and was made into a popular movie. Charles B. MacDonald's *The Mighty Endeavor* (New York: Oxford University Press, 1969) is an exhaustive study of American forces in Europe, including much material on the Normandy landings. The German side is discussed in Samuel W. Mitcham Jr.'s *Rommel's Last Battle* (New York: Stein and Day, 1983), which probes German actions on D-Day and includes a dramatic account of Rommel's subsequent death at Hitler's orders. A more personal account is provided by Rommel's chief of staff, Hans Speidel, in *Invasion 1944* (Chicago: Regnery, 1950). Current information about the invasion beaches can be found in *Fodor's France* (New York: Random House, 1991).

—Bernard A. Block

Orange (Vaucluse, France)

Location: Southeastern France, on the left bank of the Rhône River; twelve miles north of Avignon, on the route from Lyons to Marseilles; situated in the French department of Vaucluse in the Provence-Alpes-Cote d'Azur region.

Description: Market town that manufactures brooms and glass, and processes food; town originally grew with the rise of its Roman theater and the triumphal arch that still stand; was an independent principality from the eleventh century until King Louis XIV conquered it in 1672; was formally ceded to France in 1713 by the Treaty of Utrecht.

Site Office: Office de Tourisme, Syndicat d'Initiative
5 cours Aristide Briand
84100 Orange, Vaucluse
France
90 34 70 88

Twelve miles north of Avignon in southeastern France, the triumphal arch and semicircular theater of Orange hark back to the town's ancient Roman roots. Throughout the centuries, Orange has developed into a market town.

Many centuries B.C., a new civilization of Celtic tribes migrated into the valleys of the Rhine and Rhône rivers from the Danube area. The Iron Age bolstered the Celtic civilization's development from about 900 B.C., and encouraged the tribes to move beyond their boundaries. Starting in 500 B.C., the long-haired Celts began roaming Europe and launching quick, furious battles to expand their territories. They were considered the strongest power north of the Mediterranean, and were a constant threat to Greece and Rome centuries before Caesar added Gaul to the Roman Empire.

In the fifth century B.C. Rome and the Greek colony of Massilia, now known as Marseilles, became allies, signaling the beginning of a power struggle between the Romans and the Celts—whom the Romans considered barbarians—over northern Europe. In 390 B.C. a group of Celtic wanderers crossed the Alps, triggering the first open conflict between Celts and Romans. The Celts defeated a Roman army at Allia, only twelve miles outside of Rome. Slowly the Celts penetrated the lands now known as France and Spain, but the Romans eventually conquered the Celts, forcing them into the area between the Alps and the Po River that became known as Cisalpine Gaul.

The submission to the Romans came in two stages. In 121 B.C. the Celts led an unsuccessful campaign against the Romans, who won under the leadership of Domitius. The Romans took over southern Gaul, having responded to pleas from Marseilles for protection against pirates along the coast. That victory marked the beginning of the Roman province of Gaul. By the end of the second century B.C. the Romans had overtaken the Mediterranean region extending from the Alps to the Pyrénées. The whole Mediterranean shore of Gaul created a bridge between Italy and Spain, which the Romans had already conquered.

Eventually Roman troops were sent to the Gaul area to guard travelers against attacks along the trade routes. The Romans engineered a system of roads and settlements in southern Gaul that were designed as military arteries. The goal was to create a passageway for the legions to move quickly from one point to another, assuring peace and order. The focal point of the network of roads was Lyons, which was the capital of Roman Gaul at that time. The oldest Roman road, the Via Domitia, stretched between the Alps and the Pyrénées.

The vast countryside, thick forests, and great river systems of Gaul—the Rhône, the Loire, the Seine, and the Rhine—all made the region a rich source of trade and agriculture. The southern region was a particularly good source of wine and olive production.

At first, the country seemed too big and the people too rustic for the Romans to form an organized network of cities. But the land's material wealth, Roman organization, and eventual peace with the Celts spurred on the growth of the Celtic marketplaces into towns. Before the Romans' arrival, Gaul was marked by a number of native *oppida,* or towns, that were actually fortresses. Many were at the sites of defensive hilltops or were protected by surrounding rivers or marshlands. Eventually, the *oppida* took on new importance as centers of administration, manufacturing, religion, and trade.

The Romans quickly urbanized the area, paving the way with bricks, mortar, stone, and tiles. The results were in stark contrast to the primitive wooden bridges of the Celts. The "cities," which were essentially rural territories, drew their water supplies from giant aqueducts built by the Romans. Orange was built in Roman style in a valley known as Saint-Eutrope, rather than on top of a hill.

In 58 B.C. Julius Caesar was made proconsul of the Roman Gallic provinces of Cisalpine and Transalpine Gaul. Caesar arrived in Gaul, taking advantage of the divisions among tribes, and among the cities' aristocracies and the plebeians. At this time Celtic Gaul included what is now known as France, Belgium, the Rhineland, and much of the Netherlands and Switzerland.

For about eight years, Caesar and his lieutenants waged battles throughout the area. Eventually, a rebellion led by the Arvernian chief, Vercingetorix, tried to cut the lines of communication of the northern Roman legions, but Caesar led 20,000 men on a forty-eight-mile march to the threatened point. Vercingetorix tried to starve the Romans by destroying their crops and villages, and held them in check for several

Arch of Orange
Photo courtesy of Office de Tourisme, Orange

months. But Caesar fought ruthlessly, stealing the Gauls' gold and silver, destroying entire populations, and selling any survivors into slavery. A million prisoners were deported and sold; many rebels' right hands were cut off.

Vercingetorix chose the site of the final battle, the hilltop field of Alesia (known today as the town of Alyse-Sainte-Reine) in Burgundy. While Vercingetorix had thought he could fight below the walls of the fortress, Caesar had constructed siege works surrounding the base of the hill. The Celtic troops soon ran out of garrison supplies. In the end, Caesar's united Roman armies soundly defeated the tribes of the Gauls, and set the path toward the Roman domination that lasted for five centuries. Vercingetorix was obliged to surrender, and his soldiers delivered him to Caesar. The Roman general kept Vercingetorix in prison for six years before exhibiting him in a parade through Rome. At the foot of the Capitol, Vercingetorix was publicly executed for high treason against Rome. His death signaled the end of unity among the Celts.

At the time of the Roman conquest, the people of Gaul were living in primitive conditions in the wilderness. They were an illiterate people who had not developed arts and sciences. The Romans eventually taught the people of Gaul the material arts, the crafts of building, furnishing, weaving, metalwork, tanning, and glass-making, among others. The Romans were also a source for knowledge of the alphabet, the calendar, law, and architecture. They replaced marble with stone and created a strong cement out of baked brick bound by mortar. The Romans knew how to build vaulted roofs that would not collapse, a technology that was used to erect arches and domes of enormous strength.

The sixty-one-foot-high triumphal arch in Orange is one of the largest constructed by the Romans, and an unusually early example of the triple arch. Built in honor of the first Roman emperor, Augustus, it is among the pagan monuments of the Midi, or south of France, that provided models for later architects embellishing Christian churches with classical elements. Sculptures carved into the arch depict Caesar's victories in the first century B.C.

In about 27 B.C. Augustus established a city at Orange for his veterans of the Second Legion, naming the site Arausio. At the time, the thriving town housed as many as 80,000 to 100,000 people. Gone are the Roman baths and forum that served the residents of Orange, but the town's semicircular theater is still the site of shows. Estimated to have been built during Augustus's reign (27 B.C.-A.D. 14), it later served as part of a fortress. Rising up from the slopes of a hill are tiered benches that originally seated an audience of about 10,000. The back wall of the theater stands 124 feet high and 334 feet long. Ensconced in the center of the theater is a 12-foot-tall statue of Augustus (placed there in 1951). Today the theater can still seat audiences of more than 7,000.

Many of the monuments and other displays that the Gauls learned to build from the Romans were symbols of convenience: centers of health treatment, and aqueducts that brought water from afar. Even the houses were built with comfort as the focus. Many of the homes of the rich had central heating systems, walls and floors decorated with mosaics, and furnishings that included couches.

Gaul was rapidly assimilated into the Roman world. The swiftness of the Romanization stemmed from the Gauls' appreciation for Greco-Roman culture and from the way in which it was presented to them: Roman citizenship was not forced on the Gauls, but rather was treated as a privilege. Cities were given the freedom to govern themselves.

It was in the Gallo-Roman cities that the religions of the two races were notably fused. The Romans and the Gauls worshipped both the Eternal City, the goddess Rome, and her divine ruler on earth, the emperor. Yet the town of Orange derives its name from Arausio, a Gaulish god.

While the Gauls could integrate well their religion into Roman life, the Celtic language gradually gave way to Latin. It was not until 1539 that French became the language of royal justice and administration. Latin, however, remained the language of science, higher education, and diplomacy up until the seventeenth century. In fact, the earliest Gallic authors wrote in Latin. In the south of France are still to be found many ancient towns with Greek or Latin names. There is Orange, for instance, in the department of Vaucluse, along with the towns of Avignon, Carpentras, Vaison, and Cavalion.

The Roman colonies for military veterans were one of three categories of Gallo-Roman towns that were developed. Arles was first, later joined by Beziers, Narbonne, and Orange. The colonies exemplified Roman planning, with their checkerboard-grid layouts, amphitheaters, public baths, and Roman theaters. Grid patterns of centuriation divided parts of Roman Gaul into frontier districts. Each century was separated into 100 lots for veterans and their families who settled there. These squares of centuriation can still be seen in aerial photographs of the landscape of southern France.

With populations of up to 6,000 veterans with families, the Roman colonies were much smaller than what were known as Latin colonies, which were planned to accommodate both indigenous residents and emigrants from rural Italy. The Latin colonies of Nîmes and Vienna had large populations, possibly with as many as 60,000 people each. The third type of colonies were purely indigenous areas that were reestablished as Roman capitals by Augustus.

While Gaul was enjoying newfound peace, prosperity, and culture in a well-organized society, Roman authority soon began to dwindle throughout Europe. Civil strife in Rome, the restless masses of barbarians, abuse of power by the military, and the greed of Romans all contributed to the demise of the Roman Empire.

During the third century, all of the problems came to a head. The Romans were driven by a policy of exploitation, stealing the land and resources of entire classes of people. The downfall of Rome did not stem from military defeat but from internal woes of impoverishment, anarchy, and demoralization. Had the government been more cohesive, the Romans could have held the Germans at bay. But in 257 the Franks and the Alamanni attacked Gaul, destroying monuments and burning down an important sanctuary. After the fall of the Roman Empire, during the fifth century Orange was pillaged by the Visigoths, who came from Spain to conquer Aquitaine.

Although the origins of the House of Orange (royal family of the Netherlands) are obscure, tradition holds that the lineage began in 793, when Charlemagne gave William, whose surname was Cornet, sovereignty over the small principality of Orange. It is not clear who succeeded William, but eventually the principality was passed to Gerald Adhémar, who ruled there in 1086.

In 1174 Orange was willed to Bertrand des Baux, a prince whose descendants ruled until 1393. The principality changed hands through a line of princes of the House of Orange-Châlons until 1530, when Philibert died without heirs. It was during Philibert's reign that the Netherlands became associated with the House of Orange: the Holy Roman Emperor Charles V rewarded the prince for loyal service by giving him title to large parts of the Netherlands. Philibert's nephew, René of Nassau-Châlons, inherited Orange and added it to the German and Dutch possessions of his father's family. René passed his title of prince to his cousin, William of Nassau-Dillenburg, later known as the *stadholder* William the Silent.

William the Silent protested the persecutions of Protestants and established the Dutch Republic, which he linked with the history of the House of Orange-Nassau. He was murdered in 1584. His brother led a long line of descendants who succeeded as prince of Orange through the mid-1600s. Among them was Maurice of Nassau, who incorporated the

Roman arch into a fort, and the theater into the ramparts. Louis XIV of France captured Orange in 1672 during one of his wars against the Netherlands, and removed those fortifications. It was not until 1713, however, that Orange was ceded formally to France through the Treaty of Utrecht.

Early in the 1800s, Orange's Roman theater was refurbished so that plays could be presented. In 1888 *Oedipus Rex* was performed for a capacity audience, which at the time was 8,000. The theater was again restored in 1894 and became the national open-air theater of France.

As the French department of Vaucluse evolved into a prosperous center of marketing for agriculture products during the nineteenth century, Orange flourished, handling fruit, vegetables, wine, and cereals. In the past few decades Orange's economy has also been based on the fiberglass industry and tourism.

Further Reading: *The Heart of Provence* by Amy Oakley (New York: D. Appleton-Century Co., 1936) offers a colorful description of this French region, including vivid details of Orange. The book includes a diary of the author's trip through the area, focusing more on the people and plays of the time and less on the history of the town. *The Birth of France: Warriors, Bishops and Long-Haired Kings* by Katharine Scherman (New York: Random House, 1987) provides a more updated, readable history of the entire nation. *France: A Modern History* by Albert Guerard (Ann Arbor: University of Michigan Press, 1959; revised edition by Paul A. Gagnon, 1969) offers a concise, though at times textbook account, of France from its early history through the 1950s.

—Laura Duncan

Oxford (Oxfordshire, England)

Location: Oxford lies in England's West Midlands between the Chiltern and the Cotswold hills, 56 miles northwest of London and 111 miles southeast of Birmingham. Branches of the Thames River, known as the Isis within Oxford, flow through the city, which is situated just north of the confluence of the Thames and the Cherwell.

Description: The town's main attraction is Oxford University, England's oldest university. No longer merely "that sweet city with her dreaming spires," as Matthew Arnold called it, Oxford successfully combines a working university of 14,500 students with a shopping and administrative center, while allowing public access to college buildings that, although privately owned and run, are also national monuments.

Site Office: Oxford Information Centre
St. Aldates
Oxford OX1 1DY
England
(865) 726871

Modern Oxford developed from a cluster of twelfth-century scholars and teachers, who came together at a convenient crossing of routes on the site of an old ford and trading settlement and satisfied needs no longer being met by the abbey and cathedral schools. The city's central crossroads is still known as Carfax, from the Latin *quadrifurcus* (French *carrefour*) meaning four-forked crossing. Nearby were the priory of Saint Frideswide's on the site of the present cathedral, founded in 1122 to house the remains of the saint, who died in 735; the priory of Osney; and the church of St. George in the Castle. A Saxon settlement on the ford, first mentioned in the 912 Anglo-Saxon Chronicle, Oxford became a burg on the frontier with Mercia to defend northern Wessex. The site was repeatedly raided by the Danes in the tenth and eleventh centuries, assumed considerable political importance as a location for negotiations between Danes and Saxons, and was captured by William the Conqueror's Norman forces in 1068.

Coins of King Alfred found, although not at Oxford, carrying the imprint of "Oksnaforda" suggest that there may have been an ninth-century mint there. A few fragments of Saxon remains can still be seen, and the tower of St. Michael-at-the-Northgate at the top of Cornmarket, which is of Saxon origin and part of which dates from the early eleventh century, formed a portion of the city's northern fortifications. Signs of the Norman occupation dating from 1071 can be found in the castle tower and mound on the site of what until very recently was Oxford prison.

Under the first Norman governor, Robert d'Oilly, in the early twelfth century, prosperity returned to the town. He built the first three bridges, Magdalen, Folly, and Hythe, and erected a stone wall around the settlement, enclosing some ninety-five acres. A few sections of the wall still remain, principally within the grounds of New College and along Brewer Street.

It was d'Oilly's nephew, also named Robert, who founded the priory in 1129. Later it became the abbey of Osney. In its time it was a magnificent building, but only fragments now remain. About 1130, Henry I, who already had a residence a few miles away at Woodstock, built himself a palace in what is now Beaumont Street, and granted the town a charter. It was in 1142 that Henry I's daughter, the Empress Matilda, escaped from the castle over the frozen river and through the snow to the neighbouring town of Abingdon when the city was besieged by Stephen in 1142.

The university's first charter of 1214 was little more than a settlement imposed to resolve a dispute between the clerics and the townspeople following a series of incidents in 1208 and 1209, in which a townswoman was killed and two or three clerics were hanged in revenge. The townspeople were eventually forced to apologize to the papal legate and to accept his decision that substantial privileges be bestowed on the scholars, some of whom were also suspended from teaching for three years. In addition, a chancellor was appointed by the bishop of Lincoln, the diocese which at its outer edge contained Oxford. The judgment implied recognition of the body of scholars as forming a corporation separate from the town. Part of the reason for the divergence between the universities of Paris and Oxford is that the Paris masters, originally subject to a bishop, put themselves under immediate papal jurisdiction, whereas the Oxford teachers or "regents," although in the diocese of Lincoln, in practice depended directly on the king for patronage and protection.

The history of late medieval Oxford chronicles the increasingly dominant part played by the university in the life of the town. This role was not achieved without antagonism and often resulted in riots, from which the scholarly community, protected by the king, generally emerged with enhanced privileges. The academic corporation had been strengthened by the founding of new mendicant orders in the early thirteenth century: the Dominicans in 1221 and the Franciscans in 1224. They were soon joined by nonmendicants such as Carmelites, Benedictines, and Cistercians. By the end of the thirteenth century the earliest of the present colleges were under construction, replacing the semi-private halls where handfuls of students had hitherto lodged with regent masters. In 1444 there were sixty-nine halls, generally housing between twelve and twenty students.

The colleges were originally religious foundations.

Oxford University
Photo courtesy of Oxford University

They eventually became autonomous academic institutions, governed by teaching masters, with their own chapels, libraries, dining halls, and more recently, student accommodations outside college precincts. Some of the older buildings belonging to the university have interiors as impressive as those of the best college buildings, and Oxford also contains some fine examples of late-twentieth-century architecture.

While it is scarcely possible to walk through the center of Oxford without being overwhelmed by the antiquity and grandeur of some of the buildings, the town's real treasures are not immediately visible to the casual stroller. It is the interiors of the quadrangles, the older libraries, the chapels, and the dining halls that are most impressive.

There are twenty-eight colleges that admit undergraduates for a first degree, as well as other colleges and halls that admit only graduate students. Most of the colleges, with their lawns and quadrangles, are a few minutes' walk from the center of the town and from either the Isis or the Cherwell. Their street fronts are interspersed with shops, institutes, banks, restaurants, and domestic buildings and by tradition, the names of the colleges are not displayed on the street front. Oxford's planning authorities have been criticized for allowing the town to sprawl, with large areas of housing as well as parking lots, chain stores, and a large covered market. However, for over a century there has been little tension between town and gown.

The medieval and early modern university's concentration of monastic foundations, its dependence on the teaching of theology and consequent vulnerability to accusations of heresy, and its position in the vanguard of intellectual speculation and educational development are all reflected in the history of the town. Corpus Christi was founded specifically to promote theological studies involving a knowledge of Greek, and was, like Christ Church, a product of the suppression of the monasteries and of a new Renaissance approach to theological study. It is probable that developments at Oxford were arrested in the fifteenth century in favor of those at Cambridge because of Oxford's residual association with the heterodox teaching on the eucharist of John Wycliffe, an early leader of the Protestant Reformation. A schism occurred in the sixteenth century, when the university acquiesced in Henry VIII's repudiation of papal primatial jurisdiction. This decision eventually led to the burning in Oxford of Nicholas Ridley, bishop of London; Hugh Latimer, bishop of Worcester; and Thomas Cranmer, archbishop of Canterbury, when Mary Tudor, a Catholic, ascended to the throne in 1553.

During the English Civil War Charles I made Oxford his headquarters from 1642 to 1646, when on June 24 the city surrendered to the Parliamentary forces under Fairfax. Although the city appears on the whole to have sympathized with the anti-royalist forces, there were no disturbances. Oliver Cromwell made himself university chancellor from 1651 to 1657. Charles I had held his Oxford parliament in 1644, and Charles II held the last Oxford parliament in 1681. Later on the city, having at first opposed James II, became strongly Jacobite. Later still, in opposition to the university, which remained Jacobite, the city became Hanoverian. By 1785, however, when George III visited Oxford, both university and city welcomed him. During the eighteenth century, a canal connecting Oxford to the new network of English waterways was built. Not until 1929 were the city's boundaries extended to include the new industrial area that had developed at Cowley, some three miles from the city center.

A thirteenth-century legacy from William of Durham makes University College, on the south side of the present High Street, the university's oldest collegiate foundation. However, none of its buildings were constructed before 1634. It was this college that expelled Shelley for atheism. Balliol College on Broad Street was founded as an act of penance imposed by the bishop of Durham on John Balliol in the thirteenth century and is probably the oldest college to

function as an independent institution. Its principal buildings date from the nineteenth century, which was not the most distinguished period for Oxford architecture. There are nonetheless remains of the medieval buildings in Balliol's old hall and library.

The oldest college to establish the pattern of its successors was Merton, which looks south over the great water meadow toward the Isis. The foundation grew out of an institution to administer estates at Malden on behalf of the family of Walter de Merton, twice chancellor of England, and bishop of Rochester from 1274. The chapel choir, with its splendid windows and delicate stone tracery, dates from the period 1289 to 1311, as do the library and muniment room, which form part of Oxford's oldest quadrangle. The massive tower dates from the fifteenth century. There were four fourteenth-century collegiate foundations: Exeter (1314), where Edward Burne-Jones first made the acquaintance of William Morris in the nineteenth century; Oriel (1326), which gave birth to the "Oxford Movement;" The Queen's College (1340–41), with buildings by Christopher Wren and Nicholas Hawksmoor and wood carving by Grinling Gibbons in the library, and New College (1379), the twin foundation with Winchester College of William of Wykeham (1323–1404). In New College, the chapel, a choir with transepts as an ante-chapel but no nave, stands in line with the dining hall. The magnificent reredos was plastered over in the sixteenth century.

The original High Street front and the first quadrangle of All Souls (1437) remains much as it was when the college was founded, and the eighteenth-century extensions by Hawksmoor include two magnificently proportioned towers. The structure of the governing body permits All Souls to provide an Oxford base for a number of distinguished academics who have taken up political or legal careers, and for a period in the twentieth century provided an academic forum for political discussions, now often criticized for the policies to which they led. Both of the other fifteenth-century foundations, Lincoln (1427) and Magdalen (1458), were early to confirm the probability that by the mid-century Oxford had surrendered academic and educational primacy to Cambridge, where new foundations and exciting developments continued to happen. Lincoln College, one of a compact group in Turl Street between High Street and Broad Street, has a hall dating from 1436, although the wainscoting was added in 1701. Magdalen, with a long frontage at the far end of High Street and a splendidly graceful bell-tower built from 1492 to 1505 and overlooking the Cherwell, may be Oxford's most beautifully proportioned college. Its charms are not diminished by a deer park, sixteenth-century panelling in the hall, and the large meadow next to the Cherwell enclosed by "Addison's Walk," named after Joseph Addison, founder of *The Spectator,* who used to walk there.

The largest college of all is Christ Church, originally founded by Cardinal Wolsey and known as Cardinal College, then refounded by Henry VIII after Wolsey's disgrace and death. The great front of the college on St. Aldate's is surmounted by a bell-tower from a design by Wren. The gate opens to Oxford's largest quadrangle, "Tom Quad," named after the bell, dedicated to Saint Thomas of Canterbury. It originally hung in Osney Abbey, but was recast in 1680. The fan tracery roof of the staircase leading to the hall dates from 1640. The hall itself has an ornate timber roof from 1529 and the best portrait collection of any college hall in Oxford.

The most architecturally and decoratively interesting of the university buildings are in the central complex comprising the Sheldonian Theatre, the Bodleian library, and the Radcliffe Camera. The Sheldonian Theatre, built between 1664 and 1669 by Wren, is notable for its flat wooden ceiling of some seventy feet by eighty feet. The spanning problem was solved by a system of wooden beams and trusses in a roof space where the University Press once stored the books it printed in the basement. The original cupola was replaced in 1838. The principal university ceremonies, including graduations, are held here.

The Bodleian library, which receives a free copy of all books published in the United Kingdom, contains some 4,500,000 volumes. It boasts one magnificent late Gothic room from 1450, much restored and now furnished principally with ancient folio-sized volumes and the basic furniture required to read them. It surmounts the divinity schools with a groined stone roof, widely splayed arches, and carved pendants, also from 1480. The Radcliffe Camera is an eighteenth-century circular reading room rising from a strong stone base, with cupola and lantern.

Yet it almost betrays the city merely to catalogue some of its architectural and artistic treasures if that means becoming distinctly selective after about 1530 or omitting to mention the beautifully designed gardens, the immaculate lawns, the sports grounds, rivers, and large areas of spacious green meadow, or, more importantly, forgetting the intellectual and religious history of the university. Starting in the late Middle Ages, Oxford was second only to Paris in its contribution to the development of the western church. In 1277 Robert Kilwardby condemned scholastic propositions discussed at Oxford that cast doubts on the survival of the individual soul after death. In 1323 John Lutterell, the former chancellor of the university, denounced the famous Franciscan theologian, William of Ockham.

The changes that the nineteenth-century Oxford Movement brought about in English ecclesiastical life are perhaps the best known of any emanating from or otherwise centrally involving Oxford. The Oxford Movement insisted on such Catholic doctrines as apostolic succession and on Catholic practices, fasts, and feasts, and in so doing it strained the theological foundation that had come to underpin the Anglican separation from Rome. Nevertheless, the movement was essentially spiritual rather than theological in origin and was expressed more clearly in the temper of religious feelings experienced by individuals than in the spiritual development of its principal figures, such as John Henry Newman, who became a Catholic in 1845 and was made a cardinal in 1879; the mystically inclined Edward Bouverie Pusey; and John Keble, who was elected the Oxford professor of poetry in 1831. The movement is often dated from Keble's assize sermon of July 14, 1833, and culminated in the publication in 1841 of Newman's *Tract 90*, the final installment in the series *Tracts for the Times*. Intellectually, the Oxford Movement had something in common with the general European cultural reaction against eighteenth-century rationalism. At Oxford a college was chartered in 1870 in memory of Keble, and a non-university theological college in memory of Pusey was chartered in 1884.

Immense skill, much money, and great good will has gone into making modern Oxford capable of combining the responsibilities of keeping one of the world's great universities functioning in a heterogenous mixture of buildings spread over several square miles within a busy town and without any central campus; retaining the city's viability as a market town and the administrative center of a tier of local government; preserving and protecting its ceremonies, markets, traditions, and rituals; and maintaining accessibility to its architectural and artistic treasures for the hundreds of thousands of tourists who visit it. The task is not made easier by Oxford's convenient location on what is colloquially known as "the milk round," taking tourists in busloads from London to Windsor and Eton, Oxford, and Stratford.

But the traffic does continue to flow, and the city's life is virtually dominated by a common consciousness that its character is determined by the balance it preserves between these different functions. It is as if, after all those centuries, town and gown have finally come to realize that they depend on one another. Oxford is one of the two or three genuinely university towns left in Europe.

Further Reading: There are innumerable guides, maps, encyclopedia articles and books about Oxford life, Oxford buildings, and the history of each of the colleges. There is a rather old-fashioned and not unprejudiced but very readable *History of Oxford University* (London: Batsford, 1974) with a good reading list by a former fellow of Lincoln College in history, V. H. H. Green. Equally pleasant reading is the nostalgic *Oxford Now and Then* by Dacre Balsdon (London: Duckworth, and New York: St. Martin's, 1970). For more serious reading, all previous work has been surpassed by the new series *The History of the University of Oxford* in eight volumes (Oxford: Clarendon, and New York: Oxford University Press, 1984–94).

—Claudia Levi

Paris (France): Île de la Cité/Quartier Latin

Location: Île de la Cité is one of two small boat-shaped islands in the River Seine; it lies partly in Paris' first and fourth *arrondissements*. The Latin Quarter (Quartier Latin) district is on the Left (south) Bank of the Seine, directly opposite the Île de la Cité; it lies mostly in the fifth *arrondissement*, with Montagne Ste-Geneviève at its geographic center.

Description: The ancient and medieval heart of Paris, the nucleus around which the city—as well as the French nation—developed. The Île de la Cité had become by the Middle Ages the city's administrative and religious center, while the Latin Quarter became the seat of the powerful University of Paris, as well as the student quarter. Much—but not all—of the medieval character of the Cité and the Latin Quarter was lost when town planner Georges Haussmann demolished the area in the latter half of the nineteenth century.

Site Office: Office du Tourisme et des Congrès de Paris
127 avenue des Champs-Elysées
75008 Paris
France
(1) 47 23 61 72

Île de la Cité, a River Seine islet dominated by Notre-Dame Cathedral, along with the Latin Quarter (Quartier Latin), site of the Sorbonne and other educational institutions, form the ancient and medieval heart of Paris. The Île de la Cité, Paris's original Gallic-Celtic settlement, and the Roman-established Latin Quarter, on the Left (south) Bank of the Seine just opposite the Cité, have been inhabited for more than two millenia. These two city districts were the nucleus around which the modern capital—and nation—developed.

While the Cité had become Paris's center of religious and administrative life by the Middle Ages, the Latin Quarter developed into a center of academic and intellectual activity. Its name derives from the fact that students and their teachers spoke Latin as the everyday *lingua franca,* a tradition which disappeared during the French Revolution (1789–99), when the schools were temporarily closed. Although the Cité and the Latin Quarter are the oldest parts of Paris, their rich medieval character was largely lost when Napoléon III's town planner Baron Georges Haussmann, the creator of modern Paris, remodeled the antiquated, overcrowded city between 1853 and 1870. Some small, crowded streets still remain in the Latin Quarter, which still has the largest concentration of colleges in France.

The Seine played a vital role in the birth and growth of the city of Paris, providing food, water, and protection for the early peoples who established settlements along its banks. While traders occupied what is now the Île de la Cité even as early as Neolithic times, the first inhabitants to leave their names were boatmen, the Gallic *Parisii,* a fierce and warlike tribe who settled there around 250 B.C. and made a living as fishermen and farmers on the small, boat-shaped island. Julius Caesar's Romans conquered the island in 52 B.C., naming it "Lutetia," from a Celtic word meaning "dwelling surrounded by water." The Romans built a garrison on the island, defeated the *Parisii,* and the prosperous river trade colony spread to a "suburb" on the Left Bank by the first century A.D. While the Romans' public buildings and bath houses were located on the Left Bank, their palace of governors and temple to Jupiter were later built on the island.

Barbarians began raiding Lutetia in the late second century; the Roman Left Bank was burned to the ground by the end of the third century, and would not be developed again for centuries. The Gallo-Roman inhabitants took refuge on the island, building a defensive wall from the fire-blackened stones transported across the river to the Cité. This wall, as well as the earliest bridge to the Left Bank, the Petit Pont, was rebuilt numerous times over the next several centuries.

The town, renamed Paris by the early fourth century, miraculously escaped a sacking by Attila's Huns in 451, before the fall of the Roman Empire. The Romans fled in terror to Orleans as the Asian Huns approached Paris. But the *Parisii* decided to stay when Geneviève, a devout Christian girl, predicted that God would spare the city. At the last minute, the Huns changed course and decided to sack Orleans—and the Romans—instead. Geneviève later became patron saint of Paris.

The Franks conquered the Gauls toward the end of the fifth century, and King Clovis made Paris—or just the "Cité"—the capital of his Frankish kingdom in about 508. The old Roman palace of governors became the seat of his Christian monarchy. The Merovingians (as Clovis's dynasty was called) ruled from 511 to 741. Their successors, the Carolingians, notably Emperor Charlemagne (742–814), centered the crown more to the east, leaving Paris (now a collection of small distinct villages clustered around abbeys and chapels, with the old city on the island at its center) in the hands of counts and bishops. It fell into slow, chaotic decline, and was raided several times throughout the ninth century by the Vikings, but was successfully fortified and defended by Eudes, Count of Paris, in 885. One of Eudes's heirs, Hugh Capet, elected to the throne in 987, moved into the royal palace, which grew more magnificent as the kings grew wealthier. Hugh Capet's house would reign throughout most of the next 800 years. The Capetian kings brought power, political stability, and public order to the capital. Paris became the most important city in Europe during the

Notre-Dame Cathedral, Paris
Photo courtesy of The Chicago Public Library

Middle Ages: it was the major commercial metropolis of northern Europe, the cosmopolitan center of learning and intellectual Christendom, as well as the birthplace of Gothic architecture, the building style that defined medieval culture.

The Notre-Dame Cathedral, located on the eastern end of the Île de la Cité, is perhaps the most enduring symbol of Paris, the city's geographic and historic heart. Although an earlier Notre-Dame church had been erected on the Cité site of the Roman palace to Jupiter in the sixth century (rebuilt in the ninth century after a Viking raid), the great cathedral of Notre-Dame was begun in 1163 under the reign of King Louis VII (1137–80). Considered the consummate expression of French Gothic, Notre Dame Cathedral had a profound religious and architectural influence on Europe. When Maurice de Sully became bishop of Paris in 1159, he decided to replace the decaying sixth-century church with a cathedral in the new Gothic style, characterized by pointed arches, flying buttresses, vaulted roofs, and stained-glass windows. Construction on the new cathedral began in 1163 and continued until 1345.

While the cathedral's original designers remain unknown, records show that master architects Jean de Chelles and Pierre de Montreuil worked on it during the thirteenth century. For nearly two centuries, the construction site was a permanent workshop where thousands of the continent's most skilled stonemasons, sculptors, carpenters, and glassmakers plied their trades. Plans constantly changed as new problems and ideas arose. By the end of the thirteenth century, two dozen chapels and churches had been constructed in the vicinity of Notre-Dame. Parisians damaged the cathedral during the 1789 Revolution, destroying many of its statues, and Notre-Dame fell into disrepair. It was about to be demolished when Napoléon ordered it redecorated for his coronation as emperor in 1804. After Victor Hugo spurred a revival of interest in the cathedral's medieval roots with his novel *Notre-Dame de Paris,* serious restoration of the edifice was carried out under architect Eugène Viollet-le-Duc from 1845 to 1864. The cathedral regained its past splendor, and was capable of holding 9000 worshippers.

While the Île de la Cité has changed considerably since the Middle Ages (Baron Haussmann gutted more than three-fourths of it in the late nineteenth century), it still contains many other buildings of historical value—notably, the old royal palace, which eventually became France's seat of justice embracing the medieval-era La Conciergerie and

Sainte-Chapelle, as well as the nineteenth-century Palais de Justice.

The kings of France abandoned the Cité's vulnerable royal palace after 1358, when Charles V's ministers were assassinated in a bourgeois revolt led by Etienne Marcel. The monarchs first preferred the better-fortified Louvre (originally a castle built about 1200) on the Right Bank, and then other royal residences outside Paris during the height of the Hundred Years War (c. 1390–1420), when the desolated city was besieged by English forces. The royal palace compound, situated on the western end of Ile de la Cité, evolved into France's seat of the supreme court of justice, as well as its prison.

The ancient Roman palace of governors (now the Palais de Justice) was first rebuilt by King Louis IX in the thirteenth century, and later enlarged by King Philip IV. It was damaged or destroyed by a series of fires in 1618, 1776, and 1871. The palace's Great Hall, once the king's parliamentary meeting-place (now a court waiting room), contains a chamber where the Revolutionary Tribunal sentenced hundreds of people to the guillotine beginning in 1793, a year after the First Republic had been proclaimed. The modern, classical Palais de Justice was built by Baron Haussmann beginning about 1860 and finished after the 1871 fire.

The oldest parts of the former royal palace complex—the Conciergerie and Sainte-Chapelle—were spared in Haussmann's near-wholesale "sanitation" of the Île de la Cité's lesser medieval buildings to make way for wider streets and squares, and newer government offices. The grim, gray-turreted Gothic Conciergerie, which became a public monument in 1914, was added to the palace during the reign of Philip IV, and soon became a notorious prison. The name of the building is derived from the royal *concierge* (caretaker), a powerful lord who was appointed by the king to look after the palace and who enjoyed the privilege of levying taxes and collecting the rents of ground-floor boutiques and workshops, later cells and beds. During the Revolution, the building held such famous prisoners as Robespierre, Marie-Antoinette, and Danton; they were among more than 2,500 taken from here to the Place de la Concorde to be publicly beheaded during the Reign of Terror (1793–94).

Sainte-Chapelle ("Holy Chapel"), a luminous Gothic masterpiece, was built by architect Pierre de Montreuil from 1245–48, under the direction of later-canonized King Louis IX. The king constructed it to house what he thought was the Crown of Thorns from Christ's crucifixion, as well as other holy relics such as fragments from the True Cross. He had purchased the relics in 1239 from the impoverished Emperor Baldwin of Constantinople at phenomenal expense, and immediately decided to build a special shrine to house them in the courtyard of the royal palace. After the Revolution, Sainte-Chapelle was no longer used as a church, and Louis's precious relics were placed in the Treasury of Notre-Dame. Sainte-Chapelle's tall, glittering thirteenth-century stained-glass windows—which depict more than a thousand scenes from the Old and New Testaments—are the oldest in Paris.

The Latin Quarter began in the early twelfth century when some students and teachers broke away from the strictly orthodox church school of Notre-Dame and established their own university on the relatively unsettled Left Bank, across the Seine from the Cité. It quickly drew scholars from throughout Europe; by the end of the thirteenth century, several schools and colleges had been established in the Latin Quarter, including the Sorbonne (1257). When King Philip II (Philip-Auguste) and the pope finally granted the scholars the right to form and govern their own academic community—the University of Paris—in 1200–15, the act was also a recognition of Paris's natural division into three parts: the Right Bank formed the busy, river-trading mercantile quarter; the Left Bank contained the university and the academic quarter; and on the island was the Cité—home of the palace, government buildings, and the cathedral.

One of the most legendary—and tragic—love affairs of all time eventually led to the establishment of the University Quarter (as it was called before Rabelais gave the quarter its present name in the sixteenth century) and the University of Paris. Renegade theologian Peter Abelard, one of the greatest thinkers of the Middle Ages, left his native Brittany in 1099 to attend the school of Notre-Dame. He mastered medieval-era dialectics, the art of reasoning through debate, and eventually formulated his own "conceptualist" doctrine. Abelard's brilliant lectures on logic and theology attracted students and scholars from all over Europe. About 1118, the thirty-eight-year-old teacher-philosopher became the tutor of eighteen-year-old Heloise, the niece of Canon Fulbert, for which office Abelard received free lodging. Abelard and Heloise fell in love; she became pregnant, and the couple fled to Brittany where she gave birth to a son. When they returned to Paris, Fulbert punished Abelard by having him castrated in 1118. Abelard became a monk at Saint-Denis, and Heloise obeyed his command to become a nun. Their love survived and they continued writing passionate letters to each other.

Meanwhile, 3,000 of Abelard's disciples converged on the Notre-Dame school, demanding the return of the famous scholar. The bishop expelled them in 1127, and they took refuge on the Left Bank's Montagne de Ste-Geneviève, which was to become the Latin Quarter's center of studies. Abelard rejoined his followers in the 1130s. He lectured in the open air and daringly applied Aristotlean principles of dialectic and logic to sacred Church dogma, sparking an intellectual revolution. Condemned as a heretic in 1140, Abelard appealed to the pope; he died in 1142 en route to Rome to defend himself. Abelard and Heloise were later buried together in a cemetery outside Paris.

One of the oldest and most influential learning institutions in Europe, the University of Paris was informally founded about 1170. Abelard's high standards of inquiring Christian scholarship had made the Left Bank's Mont-Ste-Geneviève area a haven for students and prominent professors from across Europe. They first used what buildings they could, and then private citizens and religious orders established college-hostels to house the scholars by nationality.

King Philip II—who beginning in 1180 built walls around the University Quarter that remained standing until the eighteenth century—formally recognized the growing master-student community by incorporating it as *Universitas Magistrorum et Scolarium Parisiensium* in 1200. Pope Innocent III confirmed the charter in 1215, freeing the university from the Paris bishop's authority and making it accountable only to Rome's ecclesiastical courts. This action also widened the scope of the curriculum. The university was originally divided into four faculties: theology, canon law, medicine, and the arts and sciences. It was also at this time that Latin began to be the only language permitted in the university precincts, even outside classes.

In the thirteenth century, Paris's explosive economic growth surpassed that of every other northern European city (its population rose to about 100,000), permitting the kings to solidify their rule across what is now France. The thirteenth century was also the city's golden age of building and teaching. The University of Paris's growth reflected that of the Latin Quarter. The first retail shops in the Western world began here with the selling of books and art objects. Illuminators and parchment makers also moved to locations near the university—the beginning of the Quarter's leading role in French publishing. By the end of the thirteenth century, the university was the most celebrated teaching center of all Christendom, particularly in theology; the works of Aristotle, too, unread since the Dark Ages, were made mandatory in 1255. The university attracted the greatest thinkers of the day; its famous professors included Albertus Magnus from Germany, St. Thomas Aquinas from Naples, and Roger Bacon from England.

The Sorbonne was the first and most famous college built to accomodate the growing student population. Founded between 1253 and 1257 by Canon Robert de Sorbon, chaplain-confessor to Louis IX, the Sorbonne became the center of the theological faculty. While a number of other colleges soon grew up around it, the Sorbonne served for centuries as the administrative seat of the powerful university, becoming almost synonymous with it. But soon the University of Paris's educational program—a proxy for Roman Catholic orthodoxy and Scholastic dialectics—became stiflingly fixed and formulaic, and many serious scholars were driven away. The reactionary institution reluctantly accepted advanced theories, and was resistant to the rising tide of French nationalism as advanced by Joan of Arc; it was a Sorbonne judge who sent the Maid to the stake in 1431. (Though the Sorbonne installed France's first printing presses, in 1470, this happened before the free press had come to be seen as dangerous.) Sorbonne theologians also approved the 1572 St. Bartholomew's Day Massacre, during the Wars of Religion (1562–98), when 10,000 Protestant Huguenots were murdered by Roman Catholic mobs. Consequently, the university contributed little to the humanistic studies of the Renaissance, and it declined in stature during the Reformation and Counter-Reformation.

In the early sixteenth century, humanists petitioned

King Francis I for a modern curriculum "republic of scholars" and, in 1530, the more liberal College de France was founded across the street from the Sorbonne to teach "pagan" Greek and Hebrew texts, as well as suppressed classical Latin authors. Professors received salaries from the royal purse—unlike the case at the university, where students paid their teachers. The College de France, which has been considerably extended since the eighteenth century, still retains its scholastic independence, and continues to give free public lectures on a wide range of topics without granting degrees.

King Louis XIII's devoted minister Cardinal Richelieu, appointed university chancellor in 1622, attempted to revive the Sorbonne's moribund status by enlarging the college through the 1620s and '30s. But this move did not stop the Latin Quarter from becoming an infamous slum, full of gambling houses, bordellos, and drunken students (humanists, heretics, Huguenots, and scoundrels of all sorts were put to death in the nearby Place Maubert, once an outdoor classroom). When Cardinal Richelieu died in 1642—the same year as Louis XIII—he was entombed in his beloved Sorbonne chapel, which was built from 1635 to 1642. The college itself was extensively rebuilt into its present undistinguished classical style in 1885–1901.

The Revolution swept away all institutions of King Louis XVI (who reigned from 1774–92), as well as the monarchy itself. Deemed an instrument of the king and the church, the University of Paris's colleges and four faculties were dissolved in 1793. Revolutionists deconsecrated its Gothic abbeys and churches, and even considered demolishing the Latin Quarter altogether. The use of Latin in the student quarter, another tradition of the *ancien regime,* was also banished. Napoléon, however, at the beginning of the nineteenth century revived and reorganized the university, which became an academy of the newly created University of France. The new University of Paris had a secular program independent of political and religious doctrine; theology, however, was not abandoned until 1886.

The great student uprising of May 1968 led to major educational reform. Police entered the Sorbonne for the first time to arrest demonstrators, which provoked considerable unrest and repression. The student riots developed into a serious national outbreak of labor strikes and protests, precipitating the resignation of Charles de Gaulle. The upheaval also led to the university being split into thirteen autonomous, numbered (I–XIII) universities, in 1968–71; seven of the faculties, schools, and institutes occupy old Latin Quarter buildings (two in the Sorbonne), while the remainder were decentralized to more spacious suburban sites in an effort to ease student overcrowding in the Quarter. Other specialized colleges and prestigious *lycées* (secondary schools) still inhabit the area.

The Latin Quarter contains two other sites of deeply historic note: the Panthéon, a seldom-used church that became a burial place for France's "Great Men"; and the Musée de Cluny, an exceptional museum—second in importance to the Louvre—that contains one of the world's greatest collections of medieval art.

A physically overwhelming Neoclassical basilica, the Panthéon (and its 10,000-ton dome) dominates the summit of Montagne Ste-Geneviève. It was commissioned by King Louis XV in 1744, in gratitude for his recovery from a grave illness, as a beautiful replacement for the ancient church of the Ste-Geneviève abbey, and to hold the relics of Paris's patroness. Work began in 1755, but the structure was not finished until after architect Jacques Soufflot's death, in 1792. Though intended as Paris's principal church, the building was renamed the Panthéon by Revolutionists, who planned to make it the final resting place of Revolutionary heroes. Napoléon, however, turned it into a church again until, after the death of Victor Hugo in 1885, it was restored to a national mausoleum. Today it houses the tombs of Hugo, Voltaire, Rousseau, Zola, Louis Braille, and Resistance leader Jean Moulin, among many others.

The Musée de Cluny is situated near the intersection of the Latin Quarter's two main thoroughfares: the Boulevard St-Michel (the district's principal promenade, the students' original "Boul' Mich") and Boulevard St-Germain; originally Roman roads to Spain and Italy, these boulevards were straightened and widened by Haussmann from 1855–59. The museum is housed in one of only two Gothic medieval mansion-residences left in Paris. The Hotel de Cluny was built in the late fifteenth century for the abbots of the famous Cluny Abbey on the same site of the ancient Roman baths that survived a fourth century Barbarian sacking; the remnants of one of three baths can still be seen. The Musée is entirely devoted to decorative arts of the Middle Ages, including the renowed fifteenth-century tapestry *Lady and the Unicorn (La Dame à la Licorne.)* Off the Boulevard St-Michel, near the Seine, are a number of narrow, medieval-era side streets and alleyways that escaped Haussmann's bulldozing, such as those around the St-Severin Quarter.

The Left Bank—most notably, the St-Germain-des-Prés and Montparnasse districts—became a celebrated gathering place for artists, writers, and philosophers in the late nineteenth and early twentieth centuries. But the Latin Quarter's academic-intellectual ambiance and lively cafes also nurtured Paris's cultural arts revival; the city's main publishing houses and bookshops are still located here. From about 1945 to the mid-1950s, the Latin Quarter was the hub of the existential movement, and it became a center for artistic bohemianism; however, this characteristic of the Latin Quarter has declined in recent decades due to gentrification and the fact that many students and artists can no longer afford to live there.

Further Reading: *Cadogan City Guides: Paris* (London: Mercury House, and Old Saybrook, Connecticut: Globe Pequot, 1993) by Dana Facaros and Michael Pauls is one of the most extensive and literate travel guidebooks to Paris, written by two professional travel writers who currently live in France. The book is particularly strong on in-depth historical information, often imparted in a witty, anecdotal manner. *Fodor's 94 Pocket Paris* (New York: Random House, 1993) edited by Suzanne De Galan and Katherine Kane, and *Passport's Illustrated Travel Guide to Paris* (Lincolnwood, Illinois: NTC Publishing, 1993) by Elisabeth Morris, present essential district-by-district tourist information in a practical, selective, easy-to-use, and up-to-the-minute fashion. Each guide has sections on the Île de la Cité and Latin Quarter, with concise, generalized historical information. *Paris* by Sean Jennet (New York: Hastings House, 1973), a travel book rather than a traveler's guide, is a thorough introduction to the city's historic sites, arranged by *arrondissement*. The chapter devoted to the fifth *arrondissement* treats the Latin Quarter and neighboring areas.

—Jeff W. Huebner

Paris (France): The Louvre

Location: In the first *arrondissement* of Paris; at the Quai des Tuileries and Quai du Louvre, on the north bank of the Seine River. The Tuileries gardens to the west reach to the Place de la Concorde.

Description: Originally a stronghold built in the 12th century as part of the defenses of Paris against invaders, it became both the symbol and residence of the French monarchy, and is now the foremost museum of art in France.

Site Office: Musée du Louvre
75058 Paris
France
(1) 40 20 50 50

Over the last eight centuries the Louvre has seen its role change repeatedly. It started as a garrison fortress of the walls defending the city of Paris; later, it become the state treasury and royal palace of kings and emperors; it has housed artists and artisans and academies of science and culture; the French Revolution was debated inside its halls; and now it is the greatest French museum of art and antiquities. Excavations in the mid-1980s revealed the original foundations to parts of the building, and these are now on display. The architecture of the palace itself reflects almost every period of its construction and includes some stunning late-twentieth-century additions.

The origins of the name Louvre are not clear, although it is widely held to come from the Latin *luperia,* meaning the kennels where wolfhounds were kept. Paris in the twelfth century had expanded from its origins as a settlement on the Île de la Cité in the Seine River; its inhabitants now populated the left and right banks. In 1190 King Philip II (Philip Augustus) gave orders for defensive walls and a fortress to be built preparatory to embarking on the third Crusade. The fortress was finished in 1202. Surrounded by an almost square outer wall, defended by ten towers, and with two gates and a moat, the huge circular keep was intended to reinforce the defenses of the city, and was additionally protected by an inner dry moat, crossable by a draw bridge. It housed soldiers and artillery, and was also used to hold important prisoners and hostages.

The French court, state archives, and administrative bodies in the thirteenth and fourteenth centuries remained on the Ile de la Cité, although occasional council meetings were held in the keep. By the mid–fourteenth century, however, the Louvre was beginning to lose its strategic importance as the size of Paris increased, and new walls were required, farther afield. These were built in the reign of Charles V, and at the same period additions and improvements were made to the Louvre so that the king could use it as a residence. Windows were pierced in the outer walls, pointed lead roofs topped with tall weathercocks were added to the towers, existing buildings were heightened by a story, and two additional royal apartments were built by the architect Raymond du Temple and decorated with statues, frescoes, and tapestries. The king's famous library of rare manuscripts was installed in one tower, and its name was changed to the Library Tower, and gardens were planted, a larger one to the north, and a smaller one to the south on the Seine.

During the Hundred Years War in the fifteenth century, the English occupied Paris from 1420 to 1436, and the Louvre became neglected; succeeding French kings adopted a peripatetic existence, preferring to use their castles in the valley of the Loire in the Touraine and Berry districts in western France. The Louvre continued to be used as an arsenal and a prison. It was not until Francis I declared in 1527 that he would make Paris his permanent base with the Louvre as the center of his court that it was once again architecturally transformed.

First, the ancient keep was demolished and its moat filled in 1528. The resulting square court was paved, and the interior facades could receive light and air. Preparation for an official visit in 1540 by Holy Roman Emperor Charles V saw much renovation and decoration, including the regilding of the fleur-de-lis weathercocks and the installation of a large sculpture of Vulcan in the court. Work began in 1546 on the construction of a new west wing; the project, under the direction of the architect Pierre Lescot, replaced the old buildings but retained their foundations. A southern wing was begun in the reigns of Charles IX and Henry III; Jean Goujon created allegorical sculptured decorations on the facades honoring the fame and reputation of the monarchy, justice, peace, and piety. These two wings housed a large ballroom, apartments for the king, queen, and Queen Mother Catherine de Médicis, and in 1566 a small Italianate gallery was also begun, stretching down to the river. These new wings were the setting for such festivities as the celebrations for the wedding between the future Francis II and Mary, Queen of Scots, in 1558, and in 1572 for that between the future French king Henry IV and Marguerite de Valois, Charles IX's sister. Many Protestants were gathered in Paris for the latter event; their presence gave Charles IX and Catherine de Médicis the opportunity to order their slaughter. This act became known as the St. Bartholomew's Day massacre and led to the resumption of the Wars of Religion.

Queen Mother Catherine de Médicis had grown tired, meanwhile, of living in a palace of which one-half was a medieval fortress and the other a building site. She wanted to have a palace of her own, and chose the area just west of the Louvre called the Tuileries. The architect Philibert Delorme

The Louvre Pyramid
Photo courtesy of French Government Tourist Office

drew up plans for a vast construction around a square court, but between 1564 and 1572 he and his successor, Jean Bullant, had only managed to erect a small part of it. The main attraction of the intended palace, however, was its gardens, and the plans for these, including a labyrinth and a grotto, were drawn up by a Florentine. Polish ambassadors who came in 1573 to offer the crown of that country to the duke of Anjou, Catherine's favorite son, were given a magnificent banquet there.

When Henry IV came to the throne in 1594, converting to Catholicism in order to do so, the Louvre had been abandoned for six years. He determined to complete and enlarge the palace with ambitious plans to pull down the remaining medieval wings, quadruple the size of the court, and link the palace to the Tuileries with two long galleries running parallel to the Seine. First to be completed were the southern wing and the small gallery where portraits of kings and queens of France were hung, in order to emphasise the continuity of the monarchy through the Valois and Bourbon dynasties. It was decorated with Henry's cipher and royal emblems, and columns with fleur-de-lis capitals. Next built was the long gallery running along the river side and joining the two palaces, with a promenading passageway on the up-

per floor, and workshops and lodgings beneath. A gallery for the royal collection of antiquities was included on the ground floor, and beginning in 1608 painters, sculptors, goldsmiths, clockmakers, and other artists were housed there, together with their families and their workshops. The tapestry workers were to remain until 1671, when they were relocated in the Gobelins factory. Upon Henry's assassination in 1610, work stopped once more, but the mint and the royal printing press were installed when Richelieu came to power as first minister.

The quadrupling of the court was finally begun, with architect Jacques Le Mercier employed to redesign and build the north wing facing the Lescot wing that had been constructed some fifty years earlier. Three quarters of the old wing was removed to accommodate the plan, but the rest of the scheme would have to wait until the 1660s. Meanwhile, the painter Nicolas Poussin, after much persuasion, was brought from Rome to decorate the long gallery; given lodgings in one of the pavilions of the Tuileries, he was at first delighted by his flattering reception in Paris. He returned to Rome, however, at the end of 1642, less than two years after his arrival, driven away by the difficulties of the kind of large-scale paintings he was commissioned to produce (in particular the difficulties he faced in preparing a series on the life of Hercules for the long gallery, a room some 1,400 feet long by only 28 feet wide), together with the jealousy demonstrated by the Parisian artistic community. Although he promised to continue to direct the project from Rome, he never returned. Only drawings and documents remain of this project.

During the Regency the court left the Louvre, but Louis XIV and his mother later returned. From this time until the court finally removed to the much more grandiose setting of the newly and vastly enlarged palace of Versailles outside Paris, the Louvre experienced another twenty-five years of refurbishment and rebuilding. Initially, in the mid-1650s, the plasterwork and decoration of the queen mother's and king's suites of rooms were improved, using artists such as Le Sueur, Le Brun, and Guérin. Then the *grand dessein,* or great design, was reintroduced. The remaining medieval buildings were at long last to be brought down, the court was to be surrounded according to its new size, and in the late 1660s and early 1670s a colonnaded front was added to the east wing. Louis Le Vau was the architect responsible for the plan, and private citizens were forbidden to build in Paris in order for the necessary colossal work force to be available for its fulfillment. The north wing and the Pavillon de Beauvais at its junction with the west wing were completed in 1660, the south wing and central pavilion in 1663. A fire that gutted the small gallery in 1661 offered the opportunity to modernize its facade, and at the same time a gallery named after the interior decorations by Le Brun of the life of Apollo was added. The facade of the Tuileries was also completed, with a gallery and a pavilion symmetrical to the existing south wing.

While this work was going on, the Louvre was the stage for spectacular entertainments. Ballets, masquerades, plays (Molière's company used the hall known as the Salle des Cariatides from 1658 to 1661, staging the *Docteur amoureux* there for the king), and in 1662, in the Tuileries court, a famous two-day Carrousel was held, a huge equestrian event in which Louis XIV was acclaimed in his costume of a young Roman emperor. An international architectural competition was held for the completion of the eastern colonnade, and after producing a second design, the Italian sculptor and architect Giovanni Lorenzo Bernini was invited to Paris. Although the king laid the foundation stone in 1665 of his essentially un-French design, the project was shelved two years later and the way was open for a French team to build a colonnaded facade, the work largely of Le Vau and Claude Perrault, brother of the famous writer, Charles. Before it was completed, however, Louis had decided to move the court to Versailles, originally a small mansion built by his father and used for hunting expeditions. Louis was to enlarge the mansion on a vast scale, turning it into one of Europe's most monumental palaces.

In the last two decades of the seventeenth century, and in the eighteenth century before the French Revolution, the Louvre housed in various parts the different Academies of Paris. These included the Académie française, and the academic bodies devoted to painting and sculpture, the sciences, architecture, and politics. Upon Louis XIV's death in 1715 the new king and the regent came to the Tuileries, but little renovation was done there apart from work in the gardens and reorganization of the collections of sculptures. Political councils were held in the Louvre, however, a reminder of the palace's initial role.

Equally important was the Salon du Louvre, an exhibition of contemporary paintings and sculpture that was reestablished there after an initial Salon in 1725. From 1737 it became a regular event, at first annually then biennially, in the last week of August. Such artists as Boucher, Nattier, and Carle van Loo were frequent exhibitors, and the Salons attracted extensive public criticism, both partisan and hostile. The writer Diderot was one of many who wrote articles reviewing the exhibitions, in his case those of 1759 to 1781. In the middle of the century the Louvre became more and more crowded with artists, tradespeople and entrepreneurs, royal employees, and their families, all seeking studios, shops, and lodgings. The feeling began to emerge that the Louvre should be decolonized and made into a setting for the public enjoyment of the royal collections. These collections were open to the public from 1750 to 1777 in the Luxembourg Palace, but plans were put in place for their eventual installation in the Long Gallery, with better lighting and a new staircase. The Revolution of 1789 interrupted this work, but ultimately led to its realization as a museum.

In 1789 Louis XVI and his family were brought back to the Tuileries Palace, and eventually he and his queen, Marie Antoinette, were executed. The Tuileries, renamed the *Palais national,* housed the revolutionary councils and assemblies, until Napoléon Bonaparte became first consul, and, in 1804, emperor. In 1793 the museum, long anticipated, was opened in the Long Gallery and the Square Salon in the south wing. A museum of antiquities was opened in 1800 on the ground floor, capable of sustaining the weight of the marble statues. The policy of acquiring additional works went on briskly, with, for instance, the purchase of the Borghese collection belonging to the emperor's brother-in-law. Italian primitives, German masterpieces, and other European paintings were added. In order to accommodate this influx, restoration and completion work on other parts of the palace was undertaken. The rue de Rivoli on the north side of the Louvre was laid out, and the western part of the north wing built.

The Arc de Triomphe du Carrousel was begun in 1806 to commemorate the victories of Napoléon of the previous year. It was based on the design of the Arch of Septimus Severus in Rome, about two-thirds its size, and was the main entrance to the courtyard of the Tuileries from the Cour du Carrousel, where Louis XIV had held his celebrations in 1662. The years of the Second Empire, the 1850s and 1860s, saw further building and rebuilding, mostly in the wings leading from the Cour Carrée. Various state ministries occupied some of these new buildings, including the Finance Ministry in the Richelieu wing. The Tuileries Palace, burned down by the Paris Communards in 1871, was never rebuilt, although various projects were put forward, and it was finally demolished in 1882.

In the twentieth century the end of the Third and Fourth Republics, the establishment of the Fifth in 1958, two world wars, and the invasion of Paris by the Germans in June 1940 have all influenced the development of the Louvre as one of the world's major art galleries and museums. The Pavillon de Flore at the western end of the south wing on the Quai des Tuileries housed successive ministries and was returned to the museum in 1961; the Richelieu wing on the rue de Rivoli only in 1989. It has therefore now become possible to rearrange the Louvre's vast holdings on a more permanent basis. The items in this collection are in some cases of royal or imperial origin, some have been bequeathed or donated, and many are the result of careful acquisition on behalf of the nation. The Louvre is also the center of educational, cultural, and commercial enterprises. The École du Louvre, opened in 1882, trains specialists in different artistic disciplines; a research laboratory was founded in 1931, and engraving and metal casting studios had long been there, too, though this last transferred elsewhere in 1930. The directorate of national museums and its commercial department have been in operation since 1895, with its printing and reproductions an important source of funding for further acquisitions. With the opening of other, specialized museums in Paris, such as the former railway station d'Orsay in 1986, the Louvre has been able to concentrate on its major fields of importance in the history of art.

When the decision was taken in 1981 to integrate the Richelieu wing into the museum, work began on the whole future of the Louvre, its accessibility and its layout. Archaeological excavations of the medieval origins of the palace were made from 1983 to 1985 in the Cour Carrée. A superficial exploration of the site had been made in 1866, but this time something more thorough was undertaken to show the public not only the early structure of the Louvre but also the fabric of life as it was then. The work was done fairly quickly to meet the pressing need of completing the reorganization of the museum by the end of the century. A glass pyramid designed by the architect I. M. Pei was built in 1989 in the Cour Napoléon to the west of the Cour Carrée, and was greeted in some quarters with the same negative reaction that had followed the erection of the Eiffel Tower in 1889; it will, no doubt, be treated eventually with the same reverence that Parisians feel for the tower today. It dominates the underground concourse that links the Louvre museum with the underground station, parking, restaurants, shops, auditorium, and exhibition space. This impressive, marble-lined hall has another, inverted glass pyramid, also the work of Pei, which imparts an airy, light quality to what otherwise might have become a gloomy and spartan entrance. The new wing of the museum, which has had a glass cover added to its three courtyards, allows the museum's collection of eastern and Islamic art and sculptures to be shown properly for the first time, and has meant that a satisfying and logical distribution of the diverse collections of the Louvre can at last be achieved.

Further Reading: The Louvre publishes its own catalogues in English editions and in French; recent works include *Mémoires du Louvre* by G. Bresc (Découvertes Gallimard: Réunion des Musées Nationaux, 1989) which is well illustrated. On the early Louvre and the recent excavations, M. Fleury and V. Kruta have produced *Le Château du Louvre* (Paris: Editions Atlas, 1990) which is also illustrated, and P. Quoniam and L. Guinamard's *Le Palais du Louvre* (Paris: Nathan, 1988) is lengthy but recommended. The Pelican History of Art series includes volumes specifically devoted to France: Anthony Blunt's *Art and Architecture in France: 1500–1700* (London: Penguin, 1953; Baltimore, Maryland: Penguin, 1954; second edition, 1970) and Wend G. Kalnein and Michael Levey's *Art and Architecture of the Eighteenth Century in France* (London: Penguin, 1972; New York: Penguin, 1973).

—Honor Levi

Paris (France): Place de la Bastille

Location: Place de la Bastille is on the eastern boundary of Paris's 4th Arrondissement (district), part of a section of the city called the Marais. The square is at the intersection of a number of streets, including the boulevard Beaumarchais, boulevard Henri IV, and the rue St. Antoine.

Description: Site of a 14th-century fortress that was later used as a prison for political prisoners. The storming of that fortress by Parisians on July 14, 1789 marked the beginning of the French Revolution. The building itself was destroyed at the end of 1789, and until 1988 the square remained little more than a traffic circle containing a monument known as the July Column. Today Place de la Bastille is the site of a new opera house, and the surrounding neighborhood has undergone a renaissance of redevelopment and resurgence of nightlife.

Contact: Caisse Nationale des Monuments Historiques
Hôtel de Sully
62, rue Antoine
75004 Paris
France
(1) 44 66 21 50

The Bastille—properly the Bastille St. Antoine—was built by King Charles V (1337–80) in the late fourteenth century as a fortress to guard the eastern entrance to the city of Paris. The Bastille had eight towers and a water-filled moat seventy-five feet wide. Its walls were ninety feet high. It was built in Paris's Marais district on the north side of the River Seine, an area that had once been a swamp ("marais" in French means marsh or swamp). Charles V moved his court to this area from the Ile de la Cité. When, in the early seventeenth century, Henry IV laid out the Place Royale, now the Place des Vosges, the Marais soon became an exclusive enclave filled with the grand residences of the French aristocracy.

The Bastille as a fortress increasingly became an anachronism, and by the reign of Louis XIII (1610–43), the Bastille was used almost exclusively to house political prisoners, those with whom the crown had some quarrel. So it would remain until its destruction.

Among these prisoners was the writer/philosopher Voltaire (1694–1778). As did many of his fellow prisoners, Voltaire expounded philosophies that challenged the existing order. In writings such as *The Spirit of Laws* and *An Essay on the Customs and Spirit of Nations,* he focused on the progress of civilization and the development of society, and he argued for a social order favoring the bourgeoisie, to the detriment of a privileged class. His deistic religious views challenged the power of the institutional church.

It was such thinking that helped to create the intellectual environment in which the French Revolution would topple established institutions. That prisoners in the Bastille were held there by order of the king made it a symbol of royal oppression to Parisians increasingly discontented with the monarchy.

In 1789, Louis XVI, who reigned 1774–93, occupied what had become over the years an increasingly unpopular throne. Attempts at tax reform by his grandfather, Louis XV (reigned 1715–74), had failed, and there were vast discrepancies between the economic status of the common people and that of the upper classes. Louis XVI's queen, Marie Antoinette, had a growing reputation as an insensitive snob who spent huge sums on a lavish lifestyle at the royal palace at Versailles while the poor in Paris lacked bread. The famous, although probably untrue, statement attributed to her, "let them eat cake," typifies her reputation.

France at the time was beset by severe financial problems precipitated in part by its involvement in the American Revolution—an effort that was financed from loans made by bankers, most notably the Swiss-born financier Jacques Necker (1732–1804). The extent of the hardship faced by the French people is now almost inconceivable. It has been estimated that in 1790 10 million out of France's total population of 23 million, almost half the population, were in need of some kind of economic relief; 3 million of these people were no more than beggars. During the years 1775 to 1788 riots over food shortages and tax increases were common occurrences.

Yet the aristocracy continued to insist that any insurrection was the result of outsiders attempting to overthrow the existing order. Blinkered and determined to preserve the status quo, they refused to acknowledge the genuine grievances of the common people.

On July 12, 1789, the news broke that Necker had been dismissed as controller-general of finance, a position he had held for two years. Necker, who was attempting to make the tax system more equitable, was something of a folk hero, and his dismissal seemed yet more evidence that there was an aristocratic conspiracy afoot to stop at any cost any relief for the common people. Most damning, it was rumored throughout Paris that the king had assembled 60,000 foreign troops who would march on Paris and squelch once and for all any hope of liberty. Terrified that the Court was plotting their destruction, the merchant and artisan classes took the leadership without which revolution would not have been possible.

At the Hôtel de Ville in Paris, a committee had been organized to plan a defense against expected attack; another group, journalists and pamphleteers, congregated at the Palais Royal, and from there attempted to persuade members of the French National Guard to desert to the people's militia.

The July Column
Photo courtesy of French Government Tourist Office

As rioting over Necker's dismissal took place on July 12 and 13, Parisians, now a mob, began a non-systematic and often hysterical search for arms and supplies. Shops were looted, grain stores sacked, and even churches and monasteries were searched for hidden stores. On the morning of July 14, a crowd of some 7,000–8,000 people took some 30,000 guns from the Hôtel des Invalides. Spurred on by their stunning success, the crowd next looked to the Bastille, known to be equipped with gunpowder and arms recently sent there from the arsenal. To secure those weapons and gunpowder was the real goal of the siege of the Bastille.

A delegation was dispatched to the prison to demand that its governor, the Marquis de Launay, release these arms and withdraw the cannon that had been moved into the prison's enclosure and aimed into the streets of Paris. The governor did meet with the delegation, while an increasingly restless crowd began to gather in the Bastille's outer courtyard awaiting word of the meeting's disposition. A second delegation led by Thuriot de la Rozière was sent inside when no word seemed forthcoming.

Thuriot returned to say that the governor refused to surrender, although he would agree not to fire unless attacked. Half an hour later, some of the crowd forced down a small undefended drawbridge that led into the inner, or Governor's, court. The crowd, which now numbered many hundreds of people, immediately swarmed through this entrance. It was at this point that de Launay panicked and ordered his troops to fire, even though they were protected by the prison's thick walls. At least ninety-eight of the crowd were killed and seventy-three wounded. Even two delegates carrying white flags were fired on. Word of the carnage reached troops stationed at the Hôtel de Ville under the command of former National Guard noncommissioned officer Hulin and a lieutenant named Élie who quickly came to the defense, bringing with them two detachments of the National Guard along with five cannon removed from the Invalides that morning. Joined at the Bastille by several hundred armed civilians, this new force fought its way to the inner courtyard of the fortress and aimed the cannon at the main gate. De Launay offered terms at this point.

He offered surrender—if the mob would spare the fortress. The mob, in no mood for concessions, continued the siege. De Launay in panic lowered the second drawbridge leading into the prison, and the garrison fell. The mob stormed the fortress, and although he escaped, the governor was later killed as he tried to take refuge at the Hôtel de Ville. His head and the head of a man named de Flesselles, who had refused to give arms to the crowd, were paraded on pikes through the streets of Paris.

The mob released the seven prisoners remaining in the Bastille; ironically, none of them was in any way a political prisoner. One was an insane young man kept in the Bastille by his family, two others were also mentally disturbed, and four were disreputable characters being prosecuted for forgery.

According to a census in 1790 of those who partici-

pated in the storming of the Bastille, people from all classes of society were involved, although most were artisans from the surrounding neighborhood. The most reliable records show that *the vaingueurs de la Bastille,* as they were called, numbered between 800 and 900 people, and names, addresses, and occupations were established for 662 of the survivors.

The storming of the Bastille, although a fairly minor incident in and of itself, particularly militarily, was a potent political symbol: it came to be regarded as the beginning of the French Revolution. And, despite many setbacks to come, in a real sense it was. The king had lost control of Paris and did not have the resources to take the city back by force. The people had been energized; for them, there was no turning back. The monarchy was eventually overthrown and a republic established. July 14 became a French national holiday—Bastille Day—that is celebrated today in much the same manner that Americans celebrate the Fourth of July, their Independence Day.

The Bastille itself did not survive the Revolution. It was razed later that same year, 1789; some of the stones from the building were carved into replicas and sent throughout France as mementos of the event. General Lafayette presented George Washington with the key to the Bastille; it remains at Washington's home at Mount Vernon, Virginia. The outline of the prison is today marked by stones set into the Place de la Bastille.

Turmoil in France did not end with the French Revolution. In 1815 the Emperor Napoléon was overthrown and the French monarchy restored under the Bourbon king Charles X, who reigned 1824–30. In July 1830, Charles passed the Ordinances of St. Cloud restricting the right to vote to a relatively small group of wealthy landowners. That action provoked another revolt, and Charles was overthown during what became known as the Three Glorious Days. To commemorate this event, the *Colonne de Juillet* (July column) was erected at the Place de la Bastille. The names of 500 people who were killed in the revolt were inscribed on the column by Charles's successor, the constitutionally-elected monarch Louis-Philippe (reigned 1830–1848).

Louis-Philippe (the July Monarch) was himself ousted by a revolt in 1848. Again, the French economy had fallen on hard times, and antiregime propaganda flourished. In February, the government cracked down on this type of activity, and insurrection was the result. The king abdicated on February 24 and fled to England with his queen. These events were also commemorated on the Colonne de Juillet: the names of those killed in the insurrection were added to those killed in the 1830 revolution.

The area around the Place de la Bastille in subsequent decades went into decline, and the site itself became little more than a traffic circle until 1988. It was then, in anticipation of the bicentennial of the French Revolution, that the French government began construction on a new opera house on the south side of the square. The Opéra de la Bastille, designed by Argentinian architect Carlos Ott, opened July

14, 1989. It seats more than 3,000 people, has five moving stages, and is notable for its state-of-the-art equipment.

The Marais area also has seen a renaissance, beginning with the opening of the Pompidou Center in 1977, the subsequent restoration of the Hôtel de Ville, and the development of the rue de Lappe, where there is now a burgeoning nightlife with cafes, nightclubs, and fine restaurants.

Further Reading: *The French Revolution, Volume I* by George Lefebvre, translated by Elizabeth Moss Evanson (New York: Columbia University Press, and London: Routledge, 1962), is a translation of the work of one of the greatest authorities on the French Revolution. Although highly readable, this book goes into great detail and depth about every aspect of the causes, history, and legacies of the French Revolution. Another similar, although less comprehensive, book is *The Coming of the French Revolution* by the same author, translated by R. R. Palmer (Princeton, New Jersey: Princeton University Press, 1979). This work deals with only the events leading up to the revolution and not the revolution itself, although it does give a full account of the storming of the Bastille. For a concise yet informative synopsis of the revolution, *A Traveller's History of France* by Robert Cole (New York: Interlink Books, and Moreton-in-Marsh, Gloucestershire: Windrush, 1988) is excellent. It has the advantage of putting the revolution clearly within the context of France's overall history. *Fodor's 94 Paris* (New York: Random House, 1993) contains an excellent description and history of the Bastille itself. *The Crowd in the French Revolution* by George Rudé (London and New York: Oxford University Press, 1959) gives an interesting perspective on the events of July 1789 by focusing on the make-up and psychology of the crowds responsible for those events and ultimately for the French Revolution itself.

—Linda J. King

Paris (France): Place de la Concorde

Location: Public square situated at a central crossroads in Paris, eighth arrondissement; looking east, across the Jardin des Tuileries, to the Louvre; west, via the Avenue des Champs Elysées, to the Arc de Triomphe; north, via the Rue Royale, to the Place de la Madeleine; and south, across the Seine via the Pont de la Concorde, to the Palais Bourbon.

Description: Designed by Ange-Jacques Gabriel from 1756 to 1773 as a setting for the royal equestrian statue commemorating Louis XV after the Peace of Aix-la-Chapelle in 1748; has since been a significant public square and site of executions after the French Revolution; originally called Place Louis XV (until 1792), then Place de la Révolution (1792–1795), and now Place de la Concorde.

Site Office: Office de tourisme; Syndicat d'initiative
127 avenue des Champs Elysées
75017 Paris
France
(1) 49 52 53 54

Arguably the most impressive single square in Paris, the Place de la Concorde is at the same time the site of some of the modern world's grimmest scenes. The guillotine was erected there in 1793 during the Reign of Terror that followed the French Revolution; more than a thousand people were decapitated in the square (at that time called the Place de la Révolution), including, most famously, Danton, Robespierre, King Louis XVI, and his wife Marie-Antoinette, all executed in 1793 and 1794. Only a few years earlier, the royal couple might have been given cause to fear the newly created square: in 1770, during a fireworks display to celebrate the dauphin Louis's marriage to Marie-Antoinette, panic broke out in the Place Louis XV (the original name of the Place de la Concorde), killing some 130 people.

The Place de la Concorde's three successive names reflect the site's importance, both physical and symbolic, in French history, particularly during the period surrounding the French Revolution. In 1757 the area, at the time a vacant space to the west of the Jardin des Tuileries, was chosen to be the new site for Edmé Bouchardon's bronze equestrian statue of Louis XV. Numerous similar statues had been erected in cities across France to commemorate the Peace of Aix-la-Chapelle, which in 1748 ended the War of the Austrian Succession. Bouchardon's statue for Paris was unveiled in 1763, and the square, completed ten years later by Louis XV's *premier architecte* Ange-Jacques Gabriel, was named after the king. It remained Place Louis XV until 1792, when revolutionary forces toppled the royal statue and replaced it with an allegorical statue of Liberty; the square itself was renamed Place de la Révolution. In 1795, under the rule of the Directoire, the name was again changed to its present name, Place de la Concorde.

Gabriel's original plan did not actually call for a square but rather an octagonal area surrounded by moats. The south end facing the Seine was left open to maintain the vista across the river to the Palais Bourbon. Likewise, the east-west axis remained unobstructed to provide expansive vistas across the entire city; the Tuileries gardens lie to the east and the Elysées gardens to the west of the Place. Only on the northern section of the Place did Gabriel provide anything like a clear border. Even here, however, the emphasis is on openness: Gabriel abandoned his original plan to close off the northern edge with a huge palace, in favor of a plan that maintained a perspective from the north, down the rue Royale from where now sits the Eglise de la Madeleine, to the royal statue. He flanked the rue Royale with two mansions (designed by Gabriel from 1763 to 1772), one serving today as the Ministry of Naval Affairs and the other as the Hôtel Crillon and the Automobile Club de France. Together these splendid structures remain an important component in the overall effect accomplished by the Place de la Concorde.

If Gabriel's two mansions represent the architectural triumph of his design for the Place, the square's most distinctive feature to modern visitors is perhaps the Obelisk of Luxor, a gift from the Egyptian Viceroy Muhammad Ali to Louis-Philippe in 1831. A monolith weighing 220 tons and rising 72 feet from the ground, the Obelisk dates from the reign of Ramses II (thirteenth century B.C.). The installation of the Obelisk in the center of the square was the first in a series of additions to the Place de la Concorde initiated by Louis-Philippe and carried out from 1836 to 1854 by the architect Jacob Ignaz Hittorff, who had been asked by the sovereign to alter the square in such a way that its central position in the city be accented.

Hittorff, respectful of Gabriel's original design, made relatively few and conservative changes to the square: aside from the Obelisk, his most important additions are the two fountains reflecting the importance to France of sea and river, and the eight statues that were placed on Gabriel's sentry boxes at each angle of the original octagon. These statues are allegorical female figures representing the major provincial cities of France: Marseilles, Bordeaux, Nantes, Brest, Rouen, Lille, Strasbourg, Lyons. Hittorff's changes transformed the Place de la Concorde into a symbol of national unity. In fact, the hold exerted by the statues on the French populace was so great that the statue representing Strasbourg was draped in black crêpe and wreaths from 1871, when France lost Alsace during the Franco-Prussian War, until

Place de la Concorde
Photo courtesy of French Government Tourist Office

1918, when France recovered the territory at the end of World War I.

The Place de la Concorde is a magnificent example of urban architecture without any clear definition. Gabriel's original plan for an enclosed garden has not been preserved (Napoléon III ordered the place's surrounding moats filled to facilitate traffic circulation), yet the square's highly effective combination of urban and natural elements is still evident to the viewer today. Gabriel gestured back to the classical seventeenth-century style of Claude Perrault's Louvre colonnade in his design for the twin palaces on the rue Royale, thus maintaining a link to the architectural style of his surround-

ings as well as to what was considered France's heroic past. But the square as a whole points to the future, with its open spaces subject to infinite change as the city around it is transformed.

Further Reading: *Le deuxième centenaire de la Place de la Concorde* by Pierre Lavedan (Paris: n.p., 1956) provides an overview of the place's history. Any of several city guides to Paris contains a detailed section on the Place de la Concorde; Ian Robertson's *Blue Guide Paris and Versailles* (eighth edition, London: Black, and New York: Norton, 1995; first published as *Blue Guide to Paris and Environs,* London: Muirhead, 1921) is perhaps the finest of these.

—Paul E. Schellinger

Paris (France): Place de l'Étoile/Arc de Triomphe

Location: At the point of convergence of twelve major avenues on the right bank of Paris, at the west end of the Avenue des Champs-Elysées; sixteenth arrondissement.

Description: The "Étoile" is named for the star shape created by the avenues circling the area upon which stands the great Arc de Triomphe (commissioned by Napoléon Bonaparte, designed by Jean-François-Therèse Chalgrin, and built from 1806 to 1836); site of important national occasions, such as Victor Hugo's state funeral in 1885; in 1970 it was officially renamed Place Charles-de-Gaulle.

Site Office: Caisse Natìonale des Monuments Historiques
Hôtel de Sully
62, rue Antoine
75004 Paris
France
(1) 44 66 21 50

The Place Charles-de-Gaulle (formerly Place de l'Étoile, the name by which it is still commonly known), like so much of Paris, owes much of its present character to the urban reform initiative of Baron George-Eugène Haussmann, prefect of Paris under Napoléon III. Haussmann created the star-shaped effect that has given the Place de l'Étoile its name. He added eight avenues to the four existing ones to form the radiating flow of traffic around the Arc de Triomphe, the monument left by Napoléon III's more famous uncle, Napoléon Bonaparte. Louis-Napoléon (or "Napoléon le petit," as Victor Hugo was to call Napoléon III) was understandably eager to cash in on the great emperor's glory in order to justify the coup d'état he himself had accomplished in 1852. While the Place de l'Étoile may not carry much significance today as the site of one ruler's capitalizing on the reputation of another, the Arc de Triomphe certainly stands as a powerful symbol of French identity and one of the signature monuments for foreign visitors to Paris.

In 1920, after World War I, the Place de l'Étoile became the site for the Tomb of the Unknown Soldier, thus solidifying its status as a national symbol. The tomb lies beneath the Arc de Triomphe and contains a flame which, thanks to the daily ministrations by a small group from the Old Soldiers' Association, has burned constantly since November 11, 1923. On each November 11, the anniversary of the 1918 Armistice, services are held at the Tomb of the Unknown Soldier to commemorate the dead in both world wars.

The Place itself is today hardly distinguishable from the great Arc de Triomphe, the largest triumphal arch in the world. Rising 164 feet from the ground and standing 148 feet wide, the Arc de Triomphe dominates the Avenue des Champs-Elysées and has a splendid eastern perspective on its axis with the Place de la Concorde and the Louvre. However, when Napoléon Bonaparte commissioned the Arc in 1806 to commemorate his victory over the Austrians at the Battle of Austerlitz on December 2, 1805, its current setting on the Chaillot Hill was not his first choice. Napoléon wanted his great triumphal arch to stand on the site of the Bastille, no doubt in large part because of its revolutionary significance, but also because of the area's central location. His minister of the interior, Jean-Baptiste Champagny, persuaded Napoléon that the relatively removed area at the top of the Champs-Elysées, while not so prominent as the Bastille location, would offer a more impressive setting for his monument. Napoléon in the end agreed, and few today would argue with Champagny's preference for Chaillot Hill as the ideal spot for the Arc. The site's elevation serves to emphasize the monolithic nature of the structure. What was once a large pasture with a commanding view of the city is now an important part of the urban fabric.

Jean-François-Therèse Chalgrin, along with Jean-Arnaud Raymond, began designing the Arc de Triomphe in 1806. Chalgrin's design to some extent drew on that for the Arc du Carrousel at the entryway to the Palace of the Tuileries, near the Louvre. The Roman influence in the Arc du Carrousel, also commissioned by Napoléon in 1806, to celebrate his victory at Marengo, and built by Percier and Fontaine on the model of the Arch of Septimius Severus in Rome, reflects the emperor's wish to draw upon imperial Roman imagery in order to lend historical weight to his own exploits and set himself among the great rulers of the past. In fact, Napoléon at one time envisioned a string of triumphal arches across Paris that would give the city a Roman aspect. For the Arc de Triomphe, Chalgrin originally planned to maintain the strong Roman flavor by including freestanding columns. However, he abandoned the idea in favor of a more austere, simpler facade, in line with the emperor's own wishes.

From the date construction began in August of 1806 to the time of Chalgrin's death in 1811, very little progress had been made. When Napoléon married Marie-Louise in 1810, only a small portion of the arch, rising five yards high, had been completed. Napoléon had the arch finished to scale, in wood, so that he could incorporate his great monument into his imperial wedding procession. Already in that unfinished state, the Arc de Triomphe's sheer dimensions, as well as its impressive setting, overwhelmed all onlookers.

Upon Chalgrin's death, one of his students, Goust, took over the construction of the arch. Work halted four years later after Napoléon's defeat at Waterloo and did not resume

The Arc de Triomphe by night
Photo courtesy of French Government Tourist Office

until 1823, under the Restoration of Louis XVIII. Once construction resumed, with Goust now assisted by Jean-Nicholas Huyot, the plans for the arch began to take on a new character. Louis XVIII had appointed Huyot to check the imperialist tendencies in Goust's design. Huyot proceeded to alter the starkness of Chalgrin's design (a design that recalled the official aesthetic, linking beauty to simplicity, that characterized Napoléon's reign) by adding Corinthian columns and other decorative features. The changes led Goust to resign in 1830. However, under Louis-Philippe and the architect he appointed, Abel Blouet, plans for the arch reverted to Chalgrin's design. The columns were once again rejected, and the arch was built very much according to its original creator's final plans.

Since its inauguration on July 29, 1836, thirty years after its date of commission, the Arc de Triomphe has become one of the most powerful symbols of French unity. Its sculptural decorations promote a fairly limited range of national concerns—battles, funerals of generals, and the like. Perhaps most important are the four colossal high-relief groupings on the main facades. These depict the Triumph of Napoléon in 1810 (by Cortot), the Resistance of the French in 1814 and the Peace of 1815 (both by Étex), and, most famously, the Departure of the Army in 1792, also known as "La Marseillaise" (by François Rude). Also included on the vaults are the names of more than 650 generals who took part in battles under the republic or the empire.

If, during the 1830s and 1840s, the "Citizen King" Louis-Philippe's political agenda caused him to commemorate primarily those occasions that glorified the republic and the empire (enompassing the years from 1792 to 1814), it is not as a cogent reminder of either period that the Arc de Triomphe stands today. In fact, its meaning has been dispersed across the many efforts to involve it in shaping national identity, starting with Napoléon's own attempt to align himself, his city, and his country with the glory of Rome. The arch finally served the purpose for which it had been designed—welcoming victorious armies—when Allied troops marched through it at the end of World War I. With the renaming of the Place de l'Étoile as Place Charles-de-Gaulle in 1970, the site continues to be identified with that aspect of French identity having to do with war and the involvement of great men in battle.

Further Reading: A detailed account of the construction of the Arc de Triomphe, and the monument's significance in French life, may be found in the *International Dictionary of Architecture*, edited by Randall J. Van Vynckt (Detroit and London: St. James Press, 1993).

—Paul E. Schellinger

Passau (Bavaria, Germany)

Location: Just east of the confluence of the Ilz and the
Danube Rivers, on a tongue of land narrowing
toward the east, at the tip of which the Inn River
flows northeast from the Swiss Alps into the
Danube; roughly 118 miles from Munich, 155
miles from Vienna, 56 miles from Salzburg, and
118 miles from Prague.

Description: Celtic settlement and site of Roman fortifications;
early regional center of Christianity and center of
trade in the Middle Ages, both of which centers
led to the creation of numerous medieval
structures, many of which were rebuilt in an
Italian baroque style following fires in the
seventeenth century.

Site Office: Fremdenverkehrsverein
Rathausplatz 3 (C1)
94032 Passau, Bavaria
Germany
(851) 3 34 21

Known as Dreiflussestadt (Town of the Three Rivers), Passau owes its prosperity largely to the various commercial
advantages it has derived from the confluence of the three
rivers, which nearly doubles the flow of water through the
Danube and for a short distance triples its width; this dramatic
increase in size makes the river more easily navigable but less
easy to cross. Passau's picturesque situation offers spectac
ular views from the wooded elevations on the left bank of the
Danube over the rocky promontory between the Danube and
Inn, on which the finest private houses and public buildings
were constructed. The rivers made the town easy to defend;
Passau was on an important intersection of north-south and
east-west trade routes; it controlled extensive facilities for
cheap river transport in four directions, and it could also levy
tolls for use of these facilities and charge rights of storage;
and it has magnificent tourist potential, not least on account
of its proximity to the Bavarian Forest National Park. After
fires in 1662 and 1680 destroyed the medieval city, Italians
rebuilt and largely redecorated its principal buildings, giving
Passau the character of an eighteenth-century Italianate town
with extensive baroque features.

Passau is now chiefly a resort, largely pedestrianized,
with the principal buildings erected hierarchically on and just
beneath the high point of the tongue of land on which the
cathedral stands, and which slopes steeply toward the rivers
on each side as it narrows. Passau is still also a river port, rail
junction, and center for light industry. It contains some fine
public and private patrician houses, and has overflowed into
small, village-type suburbs on the left bank of the Ilz (which
used to be the Jewish quarter), the right bank of the Inn (the
site of the Roman settlement), the spit of land between the

Danube and Ilz, and the left bank of the Danube. The town's
new university opened in 1978. The population is just over
50,000, and the diocese is now dependent on Munich.

The Danube was a natural barrier for imperial Rome
to use as a frontier against invading barbarians, but before the
Romans came, there was a Celtic settlement known as
Boiodorum (Beiderwies) on the site of the present Innstadt,
on the right bank of the Inn, at the end of the first century
A.D. From A.D. 139 to 141, the Romans established a colony
of Batavian veterans from the ninth cohort, the *castra batava,*
from which the name Passau (Batavis) derives. When the
Roman administration collapsed in the fifth century, the peripatetic monk Severinus strengthened the population's newly
established Christianity; established a monastic institution in
Passau; negotiated with Gibuld, leader of the Alamanni; and
organized the withdrawal of the population no longer able to
defend itself against marauding Germanic tribes.

Exact information from this period is sparse, but by
the early eighth century, on the arrival of Boniface, "the
apostle of Germany," from England on a commission
awarded in 719 by Pope Gregory II to counteract the heterodox influence of Irish monks, there were already four politically organized geographical areas in the region—Regensburg, Freising, Passau, and Salzburg. Boniface elevated
these four areas into bishoprics when he went to Rome in
738. It is Boniface's personal career, as archbishop in 732,
papal legate in 741, bishop of Mainz, and metropolitan over
the Rhine bishoprics and Utrecht, that accounts for the preeminence of Mainz in the histories of such towns as Erfurt,
Regensburg, and Würzburg, as well as of Passau. In the
eighth century, Duke Hucbert transferred St. Valentine's relics to St. Stephen's church in Passau, and it seems likely that
Charlemagne established a monastic institution in Passau
dedicated to St. Florian, perhaps in 791. By 815 the town was
walled, and a century later monks were clearly distinguished
in Passau from nonmonastic canons, and there was some
acknowledged distinction between ecclesiastical and civil jurisdictions, which the bishops later for a period successfully
amalgamated.

In 977 the town, taken by the Bohemians for its strategic importance at the edge of the Holy Roman Empire, was
retaken and razed by the emperor Otto II. Ecclesiastical authority quickly increased under Bishop Pilgrim, who governed the diocese from 971 to 991 and, with the emperor's
aid, replaced the church with a new cathedral, consecrated
probably in 985. By means that appear to have included
forgery, the bishop unsuccessfully attempted to extend his
status to that of metropolitan, and his jurisdiction to Salzburg
and Hungary. Passau had by that time become the mother
church of the eastern Danube basin. Its eighth-century church
had given the name Stephen to the first king of Hungary, and

A view of Passau
Photo courtesy of German Information Center

to the cathedral belonging to Passau's archdiocese which was consecrated in Vienna in 1147, ecclesiastically independent of Passau only from 1469, and an archbishopric only from 1723. Until 1783, when Linz and Sankt Pölten became independent, Passau's cathedral was the episcopal church of the largest diocese on German soil.

As early as the ninth century Passau was known as the front-line fortress of Rome against Byzantium. Exemption from customs dues had been granted to its merchants as early as 886. In 976 the emperor Otto II granted additional financial exemptions and privileges, and in 999 Otto III further enhanced the status of Pilgrim's successor, Christian, by entrusting him with full juridical and administrative authority, except over the exempt abbey of Niedernburg, which was freed from taxation in 1010. A bridge over the Inn was built in 1143, and over the Danube in 1278. In 1161, under Bishop Konrad, the jurisdictions of abbey and diocese were united, giving Passau a single administrative and financial authority. It was ecclesiastical: the officials who kept order, exercised the jurisdiction, and administered the town were appointed by and responsible to the bishop.

Not surprisingly, friction arose between the merchant community and the ecclesiastical authorities, who from the

time of Wolfger von Ellenbrechtskirchen, bishop from 1194 to 1204, held all territorial overlordship. It is probable that it was in his circle that the celebrated tragic epic poem, the *Nibelungenlied,* was composed in more than 2,300, four-line rhyming stanzas. There are several narrative layers, but in its final form the poem was certainly written on the Danube, most probably between 1200 and 1204. The protagonists pass through Passau and other towns along the river. The poem's ethic is courtly, and it is pessimistic and emotionally sensitive.

In 1217 the bishops were created princes of the empire, and two years later they began to build the present Veste Oberhaus fortress overlooking the left bank of the Danube to protect themselves from the citizenry. The Veste Oberhaus is linked by a fortified road to the Niederhaus fortress beneath it, at the river's edge on the point at which the Ilz flows into the Danube. Meanwhile, the merchant community unsuccessfully sought to achieve for Passau the status of a free imperial city, but the first serious confrontation did not take place until 1298. Although the bishops saw no need for a merely civic building where the merchants could meet, feast, and conduct business, in that year the merchant community acquired from one of its number, Christian Haller, a house with a tower in the Fischmarkt (now the Rathausplatz) to the northeast of the cathedral. The prince-bishop, Wernhard von Prambach, who governed from 1284 to 1314, regarded the ringing of its bell to summon a council as a harbinger of revolution, and retired to his Oberhaus fortress. The merchants were quickly obliged to hand over the house, the bell, and the town seal, although Prambach issued a conciliatory decree the following year, granting the merchant community some autonomy and the right to create its own council.

These events in Passau, mirrored elsewhere in the area, were important. The conflict between Pope Gregory IX and Emperor Frederick II resulted not only in the defeat of the emperor (who was excommunicated in 1227), but also in demonstrating the manifest inefficacy of spiritual sanctions. In 1239 the pope excommunicated not only the emperor again, but also a string of bishops and their chapters, including Rüdiger of Passau, bishop from 1233 to 1250, who was conducting an ordination when he received the brief. His reply was to strike the messenger. Rüdiger was removed in 1250, but power was shifting as stronger and stronger sanctions proved less and less effective. Real power would in the future be exercised not by emperor or pope, but would be fought for at a lower level. The struggle would be between, on the one hand, the feudal overlords, dukes, and princes, and, on the other, the bishops; the struggle would also increasingly be between overlords and their subjects, and between bishops and their chapters. Passau profited during the dispute from the religious and administrative energy of the prince-bishop Otto von Lonsdorf, who governed the diocese from 1254 to 1265.

After Prambach's death in 1314, the Passau see remained vacant for seven years, during which the merchants acquired new rights. Subsequent bishops initially acquiesced, and in 1322 the prince-bishop Albrecht I, prince of Saxony, even bought back the house on the Fischmarkt for the merchants. In 1345 King Ludwig of Bavaria incorporated the citizens' rights in a Golden Bull, partially abrogated by Charles IV, king of Bohemia, in 1348. But prince-bishop Albrecht von Winkel again attempted to acquire a monopoly of authority, provoking a revolution in 1367 that led to a battle in which 200 members of the citizens' army were killed. The rest were excommunicated and had their property expropriated. In 1368, however, a constitution for the town was drafted. There was further unrest in 1387, but two years later the citizenry was again allowed its house, tower, and bell. There was another confrontation four years later, during which the town hall, as the house became, was enlarged, although the bell had to be taken down; the merchants entered into full and uncontroverted possession only in 1396.

They never achieved full medieval autonomy, since the bishop retained the right to appoint or renew the presiding magistrate, and two lawyers appointed by him attended the sittings. All members of the council had to pledge allegiance to new bishops, who could at any time dismiss them. The tower contained the town jail; the eastern house, the floor on which taxes were levied on wine, beer, meat, and other commodities; and the south house, the court and council room. The council consisted always of an odd number of elected members and a secretary. It met on Tuesdays, Thursdays, and Saturdays, summoned by two rings of the bell, and sealed important documents with the archbishop's seal and lesser ones with Passau's wolf sign.

The magistrate was responsible for policing and was also the town's financial officer. It was he who appointed, with episcopal approval, the directors of institutions, guardians of the gates, and all other civic employees. The prince-archbishop owned the town's fortifications and gates. The town hall was much enlarged in 1408, but these administrative arrangements survived almost unaltered until the secularization of the bishopric in 1803 and the incorporation of its lands into Bavaria. The town hall—a grand, richly ornamented building 200 or 300 yards down the tongue of land east of the cathedral—was decorated with tapestries, had a red and white marble staircase, finished in 1446, and contained a large council room, finished in 1425; the council room doubled as a theater for passing entertainers. Construction work continued sporadically, and several stories and a sundial were added to the tower. On the river side to the north was the fish market square, with a wall dropping down to the river. All Danube traffic had to stop there to pay dues.

The cathedral had been damaged in a fire in 1181, repaired, and had apparently again fallen into decrepitude. A tax was imposed by Otto von Lonsdorf in 1264 for its restoration. Only Vienna was exempted, since Vienna was having to rebuild its own cathedral after a fire in 1258. Prambach, profiting from Lonsdorf's administrative skills, carried the restoration through, creating in 1291 ornate tombs for the relics of the cathedral's two patrons, saints Valentine and Maximilian. The church, extended to its western towers, was completed by 1380. Prince-bishop Georg von Hohenlohe, who governed from 1390 to 1423, then had the choir rebuilt in the new, more elaborate Gothic style. It was also enlarged, and had to be bent 44 inches away from its true axis at the tip of the chancel in order to leave a passage between cathedral and episcopal residence. The enlargements and improvements, including an unusual octagonal central cupola, were not completed until 1570. Prints show that the central nave was rib-vaulted, with pointed arches between the pillars giving access to the side aisles; arches repeated over the windows—both above the openings to the side aisles and in the aisles' exterior walls—unified the architectural ensemble.

On April 27, 1662, medieval Passau was destroyed by fire. Both of its principal buildings were reduced to shells. The cathedral lost its roof, vaulting, both towers, the nave, north aisle, and cupola. The structural damage to the town hall must have been less but the need functionally to repair it more urgent: the master builder, with seventy laborers, was ready to put the roof on by July. The town secretary could use his offices again by September, and the tower had undergone necessary repair by December. It is likely that the architect was the Italian Carlo Lurago, who had helped to introduce Italian baroque into Prague, where he still had important commitments. He was soon engaged to rebuild Passau's cathedral, and did so in an exuberant Italian baroque style that also respected the Gothic ground plan and retained the eastern aisle, which was still standing. The stuccowork on the town hall was probably by Giovanni Battista Carlone, who was also responsible for the richly ornamental stucco decoration in the cathedral. The town hall appears to have avoided serious damage in the 1680 fire, although Lurago was working on its tower in 1683.

At the time of the 1662 fire the newly elected prince-bishop, Archduke Carl Joseph of Austria, was thirteen years old. The cathedral chapter appears to have intended to reinstate the Gothic cathedral, but the archduke died, and on March 27, 1664, the Bohemian cathedral provost of Salzburg, Wenceslaus, a count of Thun, became prince-archbishop. He decided immediately to rebuild the cathedral "in the Salzburg style," but to retain the east aisle and the cupola. The first of seven contracts with Lurago is dated March 12, 1668, and Wenceslaus himself made a sketch for the facade. During Lurago's absences to finish work in Prague, two pillars and three side chapels collapsed, killing five workmen. Lurago appointed a deputy, Francesco Torre, a fellow Italian who had also worked in Prague, and by 1675 the two three-story towers of the west facade were finished. Lurago's technical innovations relate the cathedral to his work in Prague. With a length of 330 feet, it is one of the biggest churches in south Germany, and the first German baroque church to replace cross beams with elongated paintings on the ceiling.

The fire of July 29, 1680, did little structural damage to the cathedral. The building as it now stands retains the strict, formal Gothic lines of the building's sides, which contrast with the stone latticework and tracery of the choir in a much later Gothic style. The late-nineteenth-century octagonal stories added to the towers of the west facade are now generally deplored. Wenceslaus had incarcerated the town council until the 2,000-gulder tax was collected for the rebuilding, but he died before the interior decoration could be undertaken. The new prince-archbishop engaged Carlone, whose team of some dozen plasterers, stucco experts, and sculptors all came from the same Italian district as Lurago, had already worked on Passau's Jesuit church, and were being employed in Austrian and Bavarian monasteries and castles. There is a reredos by Carlone in the seventeenth-century Jesuit church overlooking the north bank of the Inn; the church was begun in 1655 and is notable for its early-eighteenth-century gilding and stuccowork. Some of the work in the cathedral was already virtually rococo. Carlone accepted thirteen contracts for work in the cathedral from 1677 to 1693, finishing in 1695.

Most of the painting was undertaken by Italian artists, of whom the most important were Carpoforo Tencalla from the region around Lake Lugano and, after Tencalla's death in 1685, his pupil Antonio Giovanni Galliardi and son-in-law Carlo Antonio Bussi. Many of the paintings are thematically linked in some way, and the cathedral contains fine twentieth-century work by Josef Henselmann, including the

high altar. There are four side chapels with admired paintings by the Austrian Johann Michael Rottmayr, and a magnificently decorated Viennese pulpit dating from 1722 to 1726. The eighteenth-century principal organ together with the four organs connected to it constitute the world's biggest church organ, with 231 sound registers and 17,720 pipes. The chancel of the cathedral is an impressive and slender ornate late-Gothic construction, jutting eastwards into the Residenzplatz next to the episcopal residence.

In the details of its luxuriant ornamentation, its Corinthian columns (on which stand figures connected by a cornice that support the rounded ceiling vault), in its occasional gilding, its patches of color, and the contrast between unbelievably extravagant decoration in white and its allegorical ceiling paintings, the church achieves daring feats of visual imagination. Although one guidebook austerely dismisses the church as "overloaded with frescoes and stuccowork," the overall impression, strengthened by the comparatively economical use of color, is rather one which communicates overwhelming power, making Passau's Italianate cathedral one of the great south German churches.

There are still late-medieval features to be found in other Passau churches, but the buildings were all largely decorated or renewed in the dominant baroque style. The most important is the twin-towered church of Niedernburg abbey, whose late-Gothic silhouette is still preserved. The early-baroque pilgrimage church of Mariahilf on the west bank of the Inn, built between 1624 and 1627, combines traditional Inn ridge roofs behind facades, a fundamentally late-Gothic outline, and baroque decorative features and ornamentation.

The Passauer Vertrag or Treaty of Passau (1552) was an agreement to allow Protestants the free exercise of their religion pending a diet to be held in Augsburg in 1555. Passau nonetheless remained Catholic in the sixteenth century, a fact no doubt reflected in the ease with which it threw off the sharp angles of Gothic architecture for the elaborately expressed and confident flamboyance of the cathedral's decoration—and its no doubt inappositely joyous exploitation of classical figures and motifs.

In 1886 Ferdinand Wagner undertook the series of paintings and frescoes in the Rathaus, and more lighteartedly painted scenes in the rathskeller from Passau's history. The new glockenspiel was installed in 1991.

Today's town, at least on the central tongue of land, contains few buildings that predate the eighteenth century, although glimpses of this history are still available, and the attractive small squares and quiet streets, sometimes with gradients steep enough to break into flights of steps, give it an atmosphere to be found nowhere else. The rivers and surrounding woods do not allow the visitor to feel as near the Mediterranean as the topography and generally relaxed layout and atmosphere may suggest. The architecture, in spite of the Italianate features, announces a German-speaking region, but in the south. This is clearly Danube country, as beautiful north of Passau as, more famously, south of it.

Further Reading: Apart from the material contained in guidebooks, encyclopedias, works of art and architectural history, and works on the history of the German Middle Ages, there is very little material in English specifically devoted to Passau. There is a good German brochure by Gisa Schäffer-Huber, *Rathaus Passau* (Passau: Stephanus Verlag, 1993), and a better one by Gottfried Schäffer and Gregor Peda, *Der Stephansdom zu Passau* in the Pannonia series (Freilassing: Pannonia Verlag, 1980). The fundamental work to which recourse must be had, and through which further reading on all aspects of the town and its history can be discovered, is the *Handbuch der Bayerischen Geschichte,* edited by Max Spindler (four volumes in six, second edition, Munich: C.H. Beck'sche Verlagsbuchhandlung, 1968). The word "*Handbuch*" in the title should not mislead, as the six volumes contain nearly 5,000 octavo pages.

—Anthony Levi

Périgueux (Dordogne, France)

Location: Seventy miles northeast of Bordeaux and sixty-three miles southwest of Limoges, on a bend in the Isle River; capital of the *département* of Dordogne on the western edge of the Massif Central in southwestern France.

Description: A city that developed around two separate centers: on the western plain, originally a settlement of Gauls known as the Petrucorii, later a prosperous Roman oppidum called Vesunna in the first and second centuries; on a hill to the east, a flourishing center of trade by bourgeois artisans in the Middle Ages and a stopping point for pilgrims en route to Santiago de Campostela during the Crusades.

Site Office: Office du Tourisme
26, Place Francheville
Périgueux, Dordogne 24000
France
53 53 10 63

While the Périgord region is known to have been inhabited for over 15,000 years, little is known about the first settlers of the Périgueux. The Petrucorii, named for the four clans of Celts who first made their home on the banks of the Isle River, left few traces. Several objects retrieved from excavations at the nearby hill of Ecorneboeuf point to the existence of settlements dating to the beginning of the Bronze Age (1800 B.C.), but most of the objects date from the late Bronze Age (1200 to 650 B.C.). It is believed that they lived by fishing, hunting, and raising animals.

When the Petrucorii were unsuccessful in helping Vercingetorix hold back Julius Caesar's army in 52 B.C., the settlement became a Gallo-Roman agglomeration called Vesunna, in honor of its tutelary deity. The Romans built villas with floors made of pink stone and decorated with mosaics around the area where the Tower of Vesunna would be built several generations later. They traded with Limoges, Bordeaux, and other cities in southwestern France. An amphitheatre, baths, and aqueducts for the fountains were all built in the first century A.D.

In the late second century, the original structures were replaced by dwellings forming a religious quarter around the Tower of Vesunna, which stood eighty-nine feet high and fifty-nine in diameter.

Invasion by the Franks and Alamanni in the third century forced the Gallo-Romans to build a wall to fortify their city. They used stones from demolished buildings to construct a wall 2,592 feet long with towers sixty-five to ninety-eight feet apart. The wall, however, did more damage to the town than did the enemy, choking the growing popu-

lation into a few acres of space. The wealthier aristocracy moved outside the walls and built houses in the surrounding countryside. Vesunna as such ceased to exist and became instead the Civitas Petrucoriorum, or the Cité, an administrative center for the region. Once the community became Christianized in the fourth century, there was more organized opposition to the Visigoths. Nevertheless, the barbarians took control of the region for most of the fifth century, burned many of the buildings, and massacred or enslaved some of the population.

Led by their bishops, the Christians persisted in building churches. A church dedicated to St. Front, a legendary disciple, was built in 511 on the Puy-St.-Front overlooking the Isle, only to be destroyed later by the Normans. St. Front is believed to have been the disciple who evangelized Périgord and built the original church of St. Stephen on the site of the former Mars temple. Since he had asked to be buried in a crypt on top of a hill surrounded by his martyred friends, Bishop Chronopius chose the *puy* (hill) to the east of the Cité overlooking the river, now known as the Puy-St.-Front.

Despite the efforts of Bishop Chronopius to attract Christians to Périgueux, the city's population continued to decrease. Weak leadership under Clovis's descendants in the face of persistent attacks by the Goths, chronic famine, and pestilence marked the sixth century. The city then experienced a period of relative calm under the Frankish king Charibert II until invasions by Islamic groups and a territorial struggle between the duke of Aquitaine and Charles Martel disturbed the peace in the eighth century. This battle continued between Charles Martel's successor, Pépin the short, and Waïfre, the duke of Périgueux, until the latter was assassinated in 768. To compensate citizens who were loyal to him, Pépin invested them with fiefdoms and titles, a tradition that lasted until the eighteenth century.

Pépin's son, Charlemagne, divided Aquitaine into nine provinces, one of which was Périgord. Each province was assigned a count to serve as a governor. Périgueux developed as two parallel cities. The Cité on the west side remained an enclave of canons, knights, and aristocrats such as the count. A new settlement of artisans, merchants, and laborers formed to the east on the Puy-St.-Front. The monastery here served the same function as the feudal château. The burg of artisans began to prosper from trade by the end of the tenth century. Frotaire, the bishop named by Hugh Capet, began construction of the abbey church of St.-Front, which was consecrated in 1047.

The two cities vied with each other until well into the thirteenth century. There was animosity between the successive counts of Périgord and the citizens of Puy-St.-Front, who pledged allegiance to the king of France.

Norman invasions continued to be a threat to the re-

The Tower of Vesunna at Périgueux
Photo courtesy of Office du Tourisme, Périgueux

gion, but Périgord remained under French rule until Eleanor of Aquitaine divorced Louis VII and married Henry Plantagenet, the duke of Normandy. When he became Henry II of England, Périgueux fell under English rule for the latter part of the twelfth century. Henry's son, Richard I, the Lion-Hearted, inherited the kingdom of Aquitaine and Poitou.

With the ascension to the French throne of Louis VIII and Louis IX shortly thereafter, the citizens of Périgueux proclaimed their support of the king of France, partly because they needed his protection, but also because they wished to protest against the count of Périgord. The mayor of Puy-St.-Front made accords with neighboring noblemen so that citizens could pass more freely through the surrounding territory. This, too, strained relations with the count.

An accord ending the bloodshed between the two cities was finally reached in 1240. The treaty provided for the cessation of all squabbles and the forgiveness of all wrongdoing on both sides. The Cité and the Puy-St.-Front were to become one unified city, Périgueux, governed by a mayor and twelve elected councilmen. A wall was to be built to encircle the two sections. Each section was to have a similar set of bells that would ring simultaneously to announce the closing of the city gates. Not everyone was satisfied with the treaty, however. To the inhabitants of the Cité, it seemed that it was they who had to make the larger sacrifices.

Several generations of counts continued to do battle with the citizens of Puy-St.-Front over the next hundred years, until the imminent threat of war with the English forced the two sides to reach a second accord in 1353. Nevertheless, the English took control of Périgueux and occupied it in 1356. During this time, Count Roger-Bernard fell victim to the plague, and his son, Archambaud V, swore allegiance to England. The city remained under English domination until 1370, when Charles V sent Bernard du Guesclin to take back the cities of Périgord for France. The citizens saw this as an opportunity to take revenge on Archambaud V. They treated him as a criminal, deprived him of his possessions, and banished him from his land.

The city continued to prosper from its commerce by manufacturing cutlery, raising animals, and selling game and animal hides throughout the Middle Ages. Pilgrimages and festivals brought throngs of visitors who supported the tourist business. Although the wars had ceased, the population continued to decrease as many of the citizens fell victim to plagues and leprosy. Many of Périgueux's dark, narrow streets were rebuilt during the Renaissance. Elegant new lodgings were built and the streets were widened.

The city suffered again during the Wars of Religion in the sixteenth century. Huguenots dressed as peasants going to market took the city's guards by surprise in 1575. They burned the Château Barrière, destroyed St. Stephen's cathedral, desecrated the tomb of St. Front, and occupied Périgueux for six years (1575 to 1581) in spite of several attempts by Catholics to overthrow them. Finally, in July 1581, a group of Catholics turned the tables on the Protestant guards,

whose commander escaped over the walls. The Catholics were then able to reclaim their city.

Périgueux achieved notoriety in 1650 when, tired of shouldering the expense of playing host to the many regiments who passed through their city, the citizens destroyed the quarters that had housed the regiment from Picardy for over ten weeks. This civil unrest was the first of a series of disturbances that came to be known as La Fronde, after the name of a popular children's game. The marquis of Chanlost retaliated by tyrannizing the citizens of Périgueux, causing even more rebellion. Louis XIV granted the city amnesty in 1654 and appointed as governor Joseph Bodin, who had been sympathetic to the cause of the citizens.

Peace reigned in Périgueux for the next century, and restorations were begun to the city's buildings. The Church of St. Front, which had become the Cathedral of Périgueux in 1669 when St. Stephen's was no longer serviceable, hosted the last meeting of the Estates General in 1789 just before the French Revolution. For several years afterward, the nearby Place de la Clautre was the scene of dozens of executions by guillotine.

During the nineteenth and twentieth centuries, many improvements were made to allow for a growing population. An elaborate system of locks was designed to make the Isle navigable from Périgueux to the Gironde estuary. Rail lines were built connecting Périgueux to Bordeaux.

Today Périgueux attracts visitors who come to sample the region's internationally famous pâté de foie gras, truffles, and wine. It also serves as a center for those who come to visit the Dordogne's many prehistoric caves.

The city's main attraction is the Cathedral of St. Front, an architectural curiosity since its restoration in the late nineteenth century under the architect Abadie, who later used it as an inspiration to design Sacré-Coeur in Paris. Only the Romanesque bell tower was kept in the original style. Rectangular at the base, square in the middle, and circular at the top, it stands 210 feet high. The church has a Byzantine appearance reminiscent of Constantinople or San Marco in Venice. It is built in the form of a Greek cross, and each arm of the cross is crowned by a cupola. A fifth cupola dominates the center. Built in the twelfth century, it stands on the site of the earlier abbey church, which was destroyed by fire in 1120. The five cupolas are supported on the inside by twelve massive pillars. The overall impression is one of austere strength and volume.

The tomb of St. Front, destroyed by the Huguenots, was placed in the crypt of the earlier church, which can be visited along with the half-Romanesque, half-Gothic cloister, dating back to the eighth or ninth century.

Four blocks north of the cathedral near the gardens called the Allées de Tourny is the Musée du Périgord, which contains one of France's finest collections of prehistoric and Gallo-Roman artifacts. Besides statues, mosaics, and jewelry from ancient Vesunna, it also contains one of the oldest human skeletons ever found.

The Tour Mataguerre on the Place Francheville is all

that remains of the city's medieval walls. Named for the captain who was detained there by the English for years, it affords a spectacular panoramic view from its top. Restored several times over, it has been standing in its present state since 1477. Polygonal at the base and round from the top of its doors to the parapet, the tower has the French royal symbol, the fleur-de-lis, carved into its crown.

In the old Cité section of Périgueux, on the other side of the Place Francheville, stand the Tower of Vesunna, the ruins of an ancient gladiator amphitheatre built to hold 20,000 spectators, and the remains of the first-century Roman villa, Pompeiius, originally built in a quadrangular form around a courtyard containing a fountain. Frescoes (now restored) are painted in geometric designs and plant and animal motifs. The house was complete with kitchens, baths, and even a fireplace. Many examples of pottery from the Vesunna period have been excavated near this site.

The first Church of St. Etienne (St. Stephen) was built on the site of the ancient temple of Mars and dedicated by the first bishop of Périgueux, St. Front. The present-day church is the oldest example of Romanesque architecture in Périgueux, dating back to the twelfth century. It originally had four cupolas; since the Wars of Religion, it has had only two, balanced upon the two remaining bays. The westernmost bay contains a baptismal font framed by a very finely sculpted arch outlining the tomb of the soldier-bishop, Jean d'Asside. This bay is impressive for the solidity of its square columns, the simplicity of its style, and the harmony of its lines. The eastern bay, built years later, shows a movement toward the Gothic style. Its arches are slightly more broken, its cupola is smaller, its pillars are decorated with columns, and its stained-glass windows allow for a diffused light.

Less than twenty miles from Périgueux in the valley of the Vézère are the cave paintings at Lascaux, painted over 17,000 years ago by masterful artists who mixed pigments of black, yellow, red, and white with various minerals in order to achieve variations of color. They used stone tools as paintbrushes and sticks for scaffolding so that they could reach the high places on the walls. Their drawings were sure and without erasures, indicating a certain amount of experience. The Hall of the Bulls, fifty-five feet long, has paintings of unicorns, bulls, black horses, and red and black deer. The art-

ists, called Magdalenians after one of the prehistoric sites, La Madeleine, were adept at incorporating the shape of the rock on the limestone walls to fit the animal. For example, a bulge in the rock might designate the shoulder of a bull. They had an astounding sense of perspective, not widely used in European art until the latter part of the Middle Ages. While the purpose of the caves is uncertain, most archaeologists think they were related to some hunting-magic ritual, given the use of animals as subjects.

The Grotte de Lascaux was discovered in 1940 by a group of schoolboys who were exploring a hole in the ground. As tourists began to crowd the site, it became necessary to enlarge the entrance and to install electricity. The constant flux of visitors caused the carbon dioxide levels to increase, which resulted in condensation on the paintings. A humidity control system was installed in 1958, but when algae spots started forming on the paintings, experts realized that the complex ecosystem was not going to be easily controlled. The French cultural minister ordered the cave closed to visitors in 1963. Today only scholars and scientists are allowed to visit the original site, but an exact replica of the cave, Lascaux II, was built several hundred feet away in Montignac to serve the thousands of visitors who come annually to see the majesty of Lascaux.

Further Reading: *Histoire de Périgueux des Origines à Nos Jours* by Guy Penaud (Périgueux: Pierre Fanlac, 1983) is a complete account, in French, of the history of the city and its leaders from the time of the Petrucorii to the present day. It includes maps and engravings of the city at various stages in its history, a chronological list of its mayors and bishops, and an extensive bibliography. Each period or event is titled to facilitate access. A more concise overview of the history is provided by *Connaissance de Périgueux* by J-L Galet (Périgueux: Pierre Fanlac, 1972), written in French as a guide for visitors to the city. It begins with a brief account of the history, offers a description of each of the city's historical monuments, and suggests excursions to nearby points of interest. Its photos and engravings highlight the enduring charm of Périgueux. "Art Treasures from the Ice Age: Lascaux Cave" by Jean-Philippe Rigaud, in *National Geographic* (Washington, D.C.) October 1988, is a fascinating description of the prehistoric cave drawings and the problems inherent in preserving them. Its color photos bring the story to life.

—Sherry Crane LaRue

Peterhof (Leningradskaiya, Russia)

Location: On the Gulf of Finland, about eighteen miles west of St. Petersburg (formerly Leningrad), Russia.

Description: Complex of palaces, gardens, and fountains begun by Czar Peter the Great. Palaces served as summer residences for the imperial family until the revolution in 1917.

Contact: Leningradskaiya Oblast Administration
Shvorovskiy Prospekt 67
St. Petersburg 193311, Leningradskaiya
Russia
(812) 274-3563

Peterhof, a major park and palace complex near St. Petersburg dating from 1707, was begun by Czar Peter the Great. Peter also built St. Petersburg, the former capital of the Russian Empire, and was responsible for the westernization of Russia. Initially called Peterhof (or strictly transliterated, Petergof), the name was changed to the Russian Petrodvorets (Peter's Palace) in 1944, when many German names were eliminated. It resumed the name of Peterhof in the 1990s.

When Czar Fyodor III died in 1682, his mentally and physically feeble fifteen-year-old brother, Ivan, was his legitimate successor. The patriarch of the Russian Orthodox Church convened a council of nobles that proclaimed Fyodor's younger half-brother the nine-year-old Peter, rather than Ivan, czar. The *streltsy* (soldier-traders) of Moscow revolted and made Ivan and Peter co-czars, giving Ivan senior status and making Fyodor's sister Sophia regent. In 1689 Peter's supporters executed a coup d'état, and put Sophia in a convent. Although now clearly the ruler, the seventeen-year-old Peter chose to leave affairs of state to his mother while he learned and played at the art of war. When his mother died in 1694 (and Ivan two years later), Peter finally took on his full duties as czar.

In 1697 Peter embarked on an eighteen-month great embassy to western Europe, supposedly incognito, but fooling no one. In addition to observing and learning European ways, he recruited foreign specialists. Upon his return to Russia, he set out on a westernization/Europeanization campaign, refusing to let the *boyars* (nobles) defer to him in the Oriental way, and insisting they cut off their beards and abandon their wide, flowing, Oriental-style sleeves. He also adopted the Julian calendar, which was in use by the rest of Europe. (Since the old Russian calendar began counting from the "beginning" of the world, Russian dates were 6,508 years ahead of the western world.) In addition to westernization, Peter sought to reform society but used a repressive bureaucracy and political police, with the additional threat of army intervention, to implement his ideas. He firmly established his authority by killing the rebellious *streltsy*. Peter's reforms and westernization further separated the gentry, who spoke Western European languages and wore western clothes, from the bulk of the Russian population.

In 1703 Peter founded a new city, St. Petersburg, on land on the Baltic coast claimed by both Russia and Sweden. He finally won the Baltic area from the Swedes in 1709, in a war that dragged on until 1721. In 1712 he moved the capital of Russia from Moscow to St. Petersburg, as he had always hated Moscow, especially since the revolt of the *streltsy* there early in his reign. St. Petersburg remained the capital until 1918. (The name of St. Petersburg was changed to Petrograd in 1914 and then to Leningrad in 1924. The name St. Petersburg was resumed in September 1991 after the dissolution of the Soviet Union.)

In 1707 Peter was building the Kronstadt fortress on Kotlin Island, a few miles off the coast of St. Petersburg. He went to visit Kronstadt often and found it easier to get to the island by boat from a place about thirteen miles west of St. Petersburg rather than from St. Petersburg itself. He initially ordered two wooden houses and a building for workers constructed there for the sake of convenience. He loved the sea and the view across to Kronstadt from the deserted coast, and decided to build a permanent home on the site; the home came to be known as Peterhof. Although it ultimately became a grand complex of palaces, Peter initially wanted a very simple house.

In 1713 he hired J. F. Braunstein, an architect and landscape designer. Using Peter's own sketches, Braunstein began to plan the layout of the park and the building of the first house, the Dutch House. Peter later called it Monplaisir. (Peter's original sketches are now in the Hermitage Museum in St. Petersburg.) The French architect Alexandre-Jean-Baptiste Leblond's work had impressed Peter on his visit to Paris. He engaged Leblond not only to be chief architect of St. Petersburg, but also to teach his craft to Russian architects. Three years after Braunstein started work on the Dutch House, Leblond arrived and laid the foundation for the Peterhof that exists today. Monplaisir is considered the joint work of Braunstein and Leblond.

The house consisted of a square central structure with a heightened roof, flanked by straight galleries on the east and west sides, each ending in a small pavilion. These galleries, with round-topped French windows, gave Peter a place to take exercise while being protected from the weather. The central block contained a Great Hall, which was the first picture gallery in Russia, housing Dutch and Flemish paintings acquired during Peter's Grand Embassy abroad. The remainder of the main portion of the house contained six small rooms: Peter's bedroom, the Sea Cabinet (Peter's study), and his secretary's office on one side and the kitchen,

Fountains at Peterhof
Photo courtesy of General Tours, Inc.

pantry, and the Lacquer Cabinet on the other. The Lacquer Cabinet was Peter's one purely decorative room, paneled with Chinese scenes and adorned with precious Chinese porcelain.

Peter lived several months a year at Monplaisir, conducting business and receiving ambassadors. He particularly loved the house because he could see and hear the sea from its open windows. He found tranquility and lived there even when his wife Catherine was in residence in the Grand Palace at the top of the hill.

In addition to Monplaisir, Leblond created two other summer pavilions. The first was the Hermitage, which sat at the edge of the sea. Its very name meant privacy and peace to the royals, who were always surrounded by courtiers and servants. It was a tiny, elegant, two-story house, surrounded by a moat with a drawbridge leading to a single entrance. The ground floor contained a kitchen and an office; the upper floor was an airy dining room with large windows opening onto balconies. To ensure the privacy of the diners, the table had an innovative mechanical center that could be lowered to the kitchen for each change of courses. There were also Delft soup plates in special holders that could be lowered to the kitchen. The obsession with privacy was so great that the house initially had no stairway to the second floor. Visitors had to be winched up in a special chair.

The other pavilion was Marly, named for a house belonging to Louis XIV of France. Another simple Dutch house, Marly bore no resemblance to its namesake. Originally called the Small Seaside House, it was begun in 1719 and completed in 1724, just a year before Peter's death. It housed some of his paintings and had an oblong lake at its front which was the site of water spectacles and fireworks.

Leblond exploited the physical properties of the Peterhof site, a flat coast broken by a ridge sixty feet high, about one-quarter mile back from the sea. Peter had soon realized he needed a larger Grand Palace to receive foreign dignitaries properly, and the top of the ridge was the perfect location. Since it was to stand exactly above a Grand Cascade, Peter himself, with his ministers, went out to the ridge of hills behind Peterhof armed with hoes and found the necessary water supply. While construction of the Grand Cascade was under way, Leblond started work on the palace, to be exactly the same width as the cascade. Many projects could progress at the same time, as resources were essentially unlimited for "public," i.e., "imperial" work. Even the army could be ordered to help in peacetime.

In 1719 Leblond died of smallpox at the age of thirty-six, three years after his arrival at Peterhof, and Braunstein again took charge of the building. Nevertheless, Leblond is considered the principal author of the complex. Since he was a more brilliant landscape designer than architect, Petrodvorets appeared to be gardens set off by palaces, rather than the other way around. The real glory of the complex was the use of water, which soared, plumed, sprayed, and splashed. There were mirror-like pools, basins, and canals. Water flowed around statues of men, gods, mythical creatures, and animals. The Grand Cascade consisted of cascades, sixty-four fountains, and thirty-seven statues. Water flowed down two marble staircases, flanked by gilded statues. The center-piece was the gilded Samson Fountain, which commemo-rated the Russian victory over the Swedes on St. Samson's day, which won the Baltic coast for Russia. The Samson Fountain was destroyed in World War II and has been re-placed with a modern replica.

Peter soon realized that the original Grand Palace, a simple, two-story building with a central pavilion and two wings, was inadequate. In addition, parts had been damaged by fire in 1721. Niccolo Michetti, an architect with special gifts as a decorator, enlarged the rooms and enriched the decoration. Wood panels for the czar's study were one of the great artistic creations of the period, designed by sculptor Nicolas Pineau and carved by Louis Rolland. In 1723 M. G. Zemtsov (an important Russian architect) was commissioned to draw up plans to enlarge the palace. He was assisted by P. M. Yeropkin, the first Russian architect to study in Italy. Work started in 1725, but Peter died that very year, and work stopped for fifteen years.

Ironically, Peter died without naming a successor, after he himself had enacted a law requiring the current ruler to name his own successor. Peter's wife, Catherine, was named empress by a caucus of nobles. Her reign, which was supported by the threat of intervention by the guards, set precedents in two ways. She was the first woman to rule in her own right in Russia, and the threat of the guard would play a part in future dynastic conflicts. Her heir, empress Anna, was detested because of her lover and her German followers. This incited a coup d'état, which put Elizabeth, daughter of Peter the Great and Catherine, on the throne.

Elizabeth had initially deserted Peterhof for another palace at Tsarskoye Selo, but on her accession to the throne in 1741, she started partial renovations at Peterhof. In 1746 she decided a complete reconstruction was needed and chose Bartolomeo Francesco Rastrelli as the architect. Rastrelli, an Italian who spent most of his life in St. Petersburg, is con-sidered the outstanding St. Petersburg architect of the ba-roque period. Elizabeth and Rastrelli agreed that Leblond's central building should remain intact, and Rastrelli designed additions that provided a unified appearance. He raised the building to three stories, and added new three-story wings, increasing the width of the Grand Palace to twice that of the Grand Cascade. He also added one-story galleries on the east and west sides of the palace, each ending in a pavilion. At one end was the Church Pavilion with one dome, which soon had to be altered, since Orthodox Church tradition required five domes. (When renovations were done after World War II, the original Rastrelli single-dome design was restored.) At the other end was the Heraldic Pavilion. Rastrelli's exterior was restrained out of respect for Leblond's original design. It was in the interiors of the palace that his baroque style was truly showcased. Much of his work was lost in the damage during World War II.

In the 1770s Jegor Velten, an architect of the Gothic Revival style, redecorated a number of the rooms and, in the style of the times, toned down Rastrelli's exuberant decora-tion. He also made radical changes in the floor plan, parti-tioning some rooms.

Peterhof continued to play a role under subsequent czars. Elizabeth selected as her successor the inept Peter III, her nephew. Peterhof was the site of his abdication, barely a year after his accession. His wife, who would later become the Empress Catherine the Great, and her lover, Grigori Or-lov, supported by a regiment of guards, had Peter brought to Peterhof from the nearby Palace of Oranienbaum and forced him to sign an Act of Abdication.

Under Catherine's successor, Czar Paul, the Neptune Fountain was added to the Upper Park. At the end of the Thirty Years' War, two Nürnberg sculptors had designed a fountain for the Nürnberg Marktplatz. The parts were cast, but the fountain was never assembled because there was not enough water available for it to operate. In 1799 the Nürn-berg Council sold the fountain to Paul.

Under the czars Nicholas I and Alexander I, Peterhof was the site of a yearly *fête champêtre* to which all the in-habitants of St. Petersburg were invited. The celebrations were on a gigantic scale, the gardens and fountains ablaze with lights. People of every class participated, bejeweled nobility mingling with local rustics.

Just to the east of the main complex lies Alexandria Park, the former location of the Lower Palace, the site of some important historic events. Built in 1885 and expanded as the Summer Palace of the last czar, Nicholas II, it stood on the shore, an easily defensible, isolated retreat for the worried ruler. It was there that he signed the "October Manifesto" in 1905, granting a *duma,* or parliament. On disastrous advice of his ministers, troops had fired on a peaceful procession of workers moving toward the Winter Palace in St. Petersburg. A revolution followed, although it was not a concerted at-tempt to take power, but rather a breakdown of authority, with scattered violence and strikes. This revolution was put down by military force, but as a concession, Nicholas per-mitted formation of an elective parliament, the state *duma.* In 1907 the Lower Palace was also the site of the signing of a subsequent decree by Nicholas, limiting the *duma's* powers. In 1914 the Russian Empire was drawn into World War I on the side of the Allies. It was in the Lower Palace that Nicho-las signed the order to mobilize for war. On March 2, 1917,

Czar Nicholas II gave up the throne, and the Communists took power. The Grand Palace was later converted into a Museum of Historical Art.

In September 1941 Hitler began his march north to take Leningrad. The first line of defense for Leningrad ran in a semicircle twenty-five miles from the city, starting at Peterhof and ending at the Neva River. The second line of defense was between Peterhof and Leningrad, about fifteen miles from the center of the city. Soldiers and civilians, including children, built these defensive barriers, an extensive system of fortifications, by working day and night. After intense fighting between the German and Russian armies, German bombers came to support the ground troops, and the first line of defense fell on September 11, 1941. The Germans occupied Peterhof from September 23, 1941, to January 19, 1944, damaging and looting the complex. Some of the buildings were left as burned-out shells. Only a pile of rubble remained of the Lower Palace.

Since World War II much reconstruction work has been done at Peterhof. Monplaisir, Marly, the Hermitage, and many rooms in the Grand Palace have been fully restored. The State Rooms in the Grand Palace that have been completed include Peter's study (with some of its original wood panels); the Blue Dining Room, with its eighteenth-century dinner service; the Kavalerskaya (Room for Gentlemen-in-Waiting), designed by Rastrelli with raspberry silk wall coverings; the study used by the imperial family, decorated in a white silk floral design; the East and West Chinese Studios, with modern reproductions of the original 1760s Chinese lacquer panels; the Partridge Room, designed by Rastrelli, modified by Velten, and named for the pattern on its silk wall coverings; and the White Dining Room, also by Rastrelli and Velten, decorated with white stucco reliefs.

Of particular note is the Throne Room, which is more Velten's design than Rastrelli's. Between the upper windows are portraits of the czars and their families. At one end of the room is a full-length portrait of Catherine the Great on horseback, making a triumphal entry into St. Petersburg after seizing the throne.

The original State Bedroom, designed by Rastrelli, was divided in two by Velten, forming the Crown Room, which housed the crown when the imperial court was in residence, and the Divan Room with a Turkish divan. The original Chinese silk wall panels were removed for safekeeping in 1941 and have been reinstalled. (Dutch marquetry commodes and Chinese vases were also removed and saved from war damage.)

Another interesting room is the Cabinet of Modes and Graces, also called the Rotari Room. It is a superb example of Rastrelli's decor, with exuberant white and gold woodwork. The walls are papered with 368 portraits painted by Count Pietro Rotari from 1757 to 1762. He illustrated various types of dress, much like a twentieth-century fashion photographer.

Further Reading: *The Palaces of Leningrad* by Victor and Audrey Kennett (London: Thames and Hudson, 1971; New York: Putnam, 1973) is a beautifully illustrated book (including some color photographs) which provides detailed history and description of many important palaces in and around Leningrad (St. Petersburg), including a section on Peterhof (pp. 202–217). For a room-by-room tour of the reconstructed rooms of the Grand Palace, see *Blue Guide Moscow and Leningrad* by Evan and Margaret Mawdsley (New York: W. W. Norton, 1980; reprinted with corrections, London: Black, 1989). This comprehensive tourist guidebook to Russia's two main cities, part of a series of books covering many destinations, provides more detail on the interior than does the Kennett book.

—Julie A. Miller

Poitiers (Vienne, France)

Location: In the department of Vienne, in western France, fifty-six miles south of Tours.

Description: Religious center of Gaul in the fourth century; sixth century battle near the town between armies led by Frankish king Clovis and Alaric II, leader of the Visigoths; Merovingian queen Radegund, later saint, founded convent in 559; battle near the town in 732 between forces led by King Charles Martel of the Franks and Arab leader Abdel-Rhaman; twelfth-century court of Eleanor of Aquitaine; Battle of Poitiers in 1356 during Hundred Years War between England and France; also site of questioning of Joan of Arc in 1429 during the same war; Edict of Poitiers signed in 1577 during Wars of Religion, granting religious freedom to Protestants.

Site Office: Office de Tourisme, Syndicat d'Initiative
8 rue des Grandes Écoles
86000 Poitiers, Vienne
France
49 41 21 24

Several centuries before Christ, an immigrant Germanic tribe, the Pictones, founded a settlement on a hill overlooking the Clain River in what is now western France. Poitiers, its name derived from the name of the tribe, would become an important site in the history of the country, beginning in the fourth century with the town's emergence as a religious center of Gaul.

By the third century, Poitiers was already distinguished as one of the earliest Christian towns in Gaul. Its first church, the Baptistry of St. John, is the oldest Christian building in France. This octagonal Romanesque structure of the fourth century featured a deep baptismal font, now closed off, originally used in a total immersion ritual. Candidates, almost always adults, disrobed, walked down the stone steps into the cold water, and received the sacrament from the priest standing above. Upon emerging from the water, the new member of the church would be clad in a white tunic.

This religious community found its greatest leader in the fourth century. Hilary of Poitiers, born in 315, was a man instrumental in the spread of Christianity throughout Gaul. One of the earliest Latin theologians and the first Gallic Christian both born and educated in Gaul, Hilary was unanimously elected Poitiers' first bishop in 353. His diocese was enormous (his closest peer in southern Gaul was the bishop of Arles), but his influence was to extend even beyond its boundaries.

A dispute between Hilary and Arles's bishop, Saturn-

imus, served as the impetus for Hilary's greatest work. Saturnimus was one of the favorites of Emperor Constantius, and they both embraced the teachings espoused by Arius, a priest in Alexandria during the reign of Constantine the Great. Arius believed that God was indivisible, and thus Christ could not be considered God in the strict sense. Instead, he had been created by God from nothing before the rest of the world came into being. In essence, Christ was a lesser god, a demigod, an idea that was attractive to pagans. Hilary regarded this Arian theory as nothing short of heretical, setting himself at odds not only with Saturnimus, but also his emperor. At the Council of Milan in 355, Hilary staunchly defended the consubstantial indivisibility of the Trinity, and went so far as to write a letter to Constantius that was years ahead of its time in political thought: it protested state intrusion in religious affairs.

The emperor's response was predictably undemocratic: Hilary was exiled to the eastern province of Phrygia for civil interference. Ironically, this exile proved to be his blessing. In Phrygia, with time for study and reflection, he composed the treatise, *On the Trinity,* continued to administer his diocese through correspondence, and wrote another challenging letter to the emperor, this time declaring the West the seat of the true creed. He thus became a hero to the Poitiers see. Constantius eventually relented, sending Hilary back to Poitiers in hopes that a bishop occupied with the affairs of his people would be less bothersome to him.

That hope was not to be. Despite his busy schedule in Poitiers, the indefatigable Hilary wrote *Invective against Constantius.* After Constantius, coincidentally or not, died in the same year of its publication, 361, his cousin Julian, upon assuming the throne, showed prudent respect for Hilary's influence among Gauls by removing Bishop Saturnimus from the church. Hilary had become a champion for Christian orthodoxy; his efforts would inspire the later ascendancy of Catholic Frankish Gaul as well as subsequent battles against heresy in the West.

Hilary died in 367 and was canonized. The citizens of Poitiers honored Hilary with the church of Saint-Hilaire-le-Grand, a Romanesque structure built in the eleventh century over the chapel that houses his tomb. St. Martin, a patron saint of France, was a disciple of Hilary, and founded a monastery near Poitiers in 360.

Early in the sixth century, King Clovis of the Franks fought his own battle against Arian heresy near Poitiers. In 507, he marched against Alaric II, the king of the Arian Visigoths, at the time a dominant force in southern Gaul. Just outside Poitiers, in the village of Vouillé, the Franks dealt a crushing blow to the Visigoths, and Clovis himself slew Alaric. Although clashes followed at other sites, the defeat of the Visigoths near Poitiers marked the end of their rule in the

The Cathedral of Notre-Dame-la-Grande
Photo courtesy of French Government Tourist Office

vast expanse of southern Gaul. A memorial in Vouillé recalls the pivotal battle.

Clotaire I, in 558, initiated the unimpressive, and at times brutal and shameful, Merovingian dynasty of twenty-eight kings. None of them could match Clovis in greatness. While Clotaire is not significant in Poitiers's history, his wife, Queen Radegund, is. Preferring the life of a nun to that of a king's wife, she founded the St. Croix convent near St. John's Baptistry in 559. Initially, her husband was reluctant to let her go. In the end Radegund went with his blessing, along with a generous portion of his wealth.

The St. Croix convent resembled a luxurious villa, spacious and elegant, with gardens and baths, whose members, mostly from wealthy, noble Gallo-Roman families, occupied themselves with pleasantries such as sewing, embroidery, reading, and the copying of manuscripts. Curiously, in the midst of what was essentially a retreat, Radegund denied herself these comforts. The convent's founder even refused the title of abbess, conferring it on another, and remained only a plain sister. Unlike St. Croix's other nuns, she chose a rigorous, austere existence. Radegund subsisted on vegetables, bread, and water, wore a hair shirt, and performed menial chores such as cleaning the nuns' shoes and scrubbing the church floor. She even branded her skin with a heated metal cross.

She was not, however, merely self-absorbed in her penitence. Her charity towards the less fortunate was extraordinary. One day a week, the poor and diseased of Poitiers were admitted inside the walls of the convent. Radegund bathed them herself, then provided them with fresh clothing and a meal of the exquisite convent cuisine she herself refused.

Her generosity inspired the poet and composer Venantius Fortunatus, who became her chaplain at the convent, and later bishop of Poitiers. Fortunatus's hymns would in turn inspire Thomas Aquinas; two of them, "Vexilla Regis" and "Pange Lingua," became enduring pieces of the Catholic liturgy.

Radegund died in 587 and was canonized soon after. Her body rests in St. Radegund's Church, Poitiers' beautiful part-Romanesque, part-Angevin Gothic structure erected over the original sixth-century chapel of the nuns' burial ground. Beneath its Romanesque choir, distinguished by thick foliage-ornamented columns highlighted with Bible stories, lies the crypt, at whose entrance stands a seventeenth-century stone statue of the saint. It was the gift of Anne of Austria, whose prayers to St. Radegund for her ailing son, the future Louis XIV, were answered. While the church in her name remains, Radegund's convent is today St. Croix Museum, housing Celtic, Roman, and Merovingian artifacts, as well as Radegund's oak reading desk.

In the eighth century, Poitiers was the site for yet another battle between the forces of orthodox Christianity and heresy, not far from the very place where Clovis met Alaric II. In 732, the Frankish king Charles Martel challenged the invading Arab army of Abdel-Rhaman. The battle would determine the fate of France, perhaps all of Europe: would it remain Christian or become Islamic? The armies faced one another for a week without incident, then on an October Sunday, Abdel-Rhaman and his men attacked. The stout armor of the Franks proved to be the deciding factor, and the Arabs were forced to retreat. Abdel-Rahman himself was slain. Not only was the battle at Poitiers critical in preserving Christianity in Europe, but it propelled Charles Martel into a series of successful campaigns to increase the Frankish dominion. His growing kingdom would come to be known as France.

Through war and political maneuvering, Poitiers eventually fell under English rule for many years, becoming part of Eleanor of Aquitaine's dowry. In the twelfth century, Eleanor established one of the most brilliant courts in medieval Europe in Poitiers. Her husband, Henry II, king of England, had become involved in an extramarital affair and was eager to have Eleanor out of England. In 1168, he gladly allowed her to move to Poitiers, where she spent five blissful years at court. During her stay, an enormous decorative hall was added to the ducal palace; it still stands, now as the Palais de Justice of Poitiers. The hall was part of an impressive display of Romanesque architecture in the town, which included the twelfth-century church Notre Dame le Grande, with its richly decorated facade.

At her court, Eleanor entertained such noted troubadours as Rigaut de Barbezieux, Bertran de Born, and Chrétien de Troyes, along with such dignitaries as Marie of France. Tournaments were conducted and plays performed; there were romantic song contests, later described as courts of love, presided over by Eleanor herself. The queen's eldest daughter, Marie of Champagne, sometimes assumed her mother's place at the contests. A literary patroness, Marie encouraged Chrétien de Troyes to write *Lancelot*.

The song contests of the troubadours, their *gai saber* (joyous art), were the principal form of entertainment at the court of Poitiers. The troubadour's extraordinary devotion to his lady, analogous to a vassal's loyalty to his overlord, was described by a witness to Poitiers's contests, Andrew the Chaplain, in his *Treatise on Love and Its Cure*. This was apparently an attempt to publish Marie of Champagne's theories on love, one of which is that it teaches good manners. But she added that love cannot exist for a married couple, as the institution compels them to submit to one another's wishes; in love, all is given gratuitously. Andrew does write, however, that the knight appears to become the lady's slave.

In this atmosphere of liberation for and devotion to women, the appearance of the courtiers of Eleanor's world at Poitiers reflected a similar freshness: slashed cloaks, flowing sleeves, long hair, and pointed shoes. Gifts presented to the ladies included fine handkerchiefs, brooches, mirrors, purses, combs, and gloves, almost anything for the toilet or dressing table. As Eleanor herself was the lady of many troubadours, it is no wonder that her days at Poitiers were so pleasant.

Two years before Eleanor began her stay at Poitiers, the town initiated construction of a cathedral in the breathtaking new style of Gothic architecture. Although work on St. Pierre commenced in 1166, it would not be completed until 1271. The completed cathedral featured three naves of equal height. The highest parts of the church, soaring above, appear to crush the lowest tier of the west facade, on the ground level. While its sculptures are simple, St. Pierre's interior is impressive in its spaciousness, with light streaming in through stained-glass windows of the twelfth and thirteenth centuries. St. Pierre Cathedral stands with the other churches of Poitiers—the Baptistry of St. John, Saint-Hilaire-le-Grand, Notre-Dame-la-Grande, and St. Radegund—to form a veritable museum of church art and architecture over the centuries.

In the century following the completion of St. Pierre, Poitiers, along with the rest of France and England, entered a darker chapter of history, the Hundred Years War. Once again, Poitiers would serve as the site for a crucial battle. In 1355 and 1356, Edward, prince of Wales, later known as the Black Prince, led his army on a campaign that ravaged towns and the countryside in south and southwest France. The king of France, John II, or John the Good, gathered an army of some 40,000 to 50,000 men, 26 dukes and counts, and nearly all the baronage of France to move against the prince. Hearing of the advance of the French forces, the prince and his relatively small army retreated to a plateau just south of Poitiers. Two legates from the pope arrived to attempt a negotiation of peace as the armies at last faced one another. The prince of Wales, cut off from Bordeaux by superior forces, agreed to return all that he had conquered to that point and to refrain from bearing arms against the king of France for seven years. When John would accept nothing less than the prince's surrender along with 100 of his knights, all hope for peace was lost; the Black Prince would not endure such humiliation.

On the morning of September 19, 1356, the Battle of Poitiers erupted. Surprisingly, the French army was soundly defeated by the smaller numbers of the English. Though John fought gallantly, his troops were undisciplined, moving without order and cooperation, and were poorly commanded. John himself was taken prisoner, and his defeat triggered a popular uprising against the nobility.

John the Good may have been brave and chivalrous, but he was self-destructive. His refusal to accept the prince of Wales's offer at Poitiers proved fatally arrogant. Yet the wrath of the people turned not against him, but against the French nobility, which was denounced in ballads as responsible for the defeat; they had dishonored their country. The people's fury exploded in riots throughout the land. Poitiers suffered in another way; the disastrous battle near its walls resulted in the Treaty of Brétigny four years later, which gave Poitiers, along with other parts of western France, to the English.

With a new king, Charles V, and his reform-minded military advisor, Bertrand Du Guesclin, the French were able to take Poitiers back in 1372. Du Guesclin understood that the feudal system of warfare, which led to defeat at Poitiers in 1356, was outmoded. Frontal attacks by undisciplined forces were replaced by methodical and scientific approaches, including guerilla tactics using paid soldiers. Upon regaining Poitiers, Du Guesclin proudly presented it to the king's representative, Duke Jean de Berry, brother of Charles V.

In 1422, Charles VII, repudiated son of Charles VI, proclaimed himself king of France, having already established his court and parliament in Poitiers in 1418. Poitiers could hardly be proud of its new king. There were doubts about the legitimacy of his birth. In addition, although Charles was extremely religious, attending Mass three times a day, he enjoyed a succession of mistresses. He was no leader during this time of war with England, timidly leaving that struggle to his ministers and generals, who frequently quarreled among themselves. When the English moved south to lay siege to Orléans in 1428, no concerted effort was made to stop them.

Orléans and France, and the reputation of Charles VII as a king, would be saved by a village girl. Yet, before Joan of Arc emerged as the great leader of France in the Hundred Years War, she would have to face her doubters in Poitiers in 1429. Despite Joan's assurance of allegiance for the insecure Charles in Chinon, he sent her to Poitiers to face the questions of theologians regarding her faith and of matrons on the issue of her virginity. There, in the Gothic Great Hall of the town's recently rebuilt law courts, Joan impressed all with her uprightness, while at the same time revealing her annoyance with the delay caused by this inquiry. Frère Seguin, a professor of theology, asked Joan what language her voices spoke. To this learned man whose French was in the dialect of Limousin, she replied, "A better language than yours." Undaunted, Frère Seguin asked her if she believed in God. "Yes, and better than you," was her bold response. Then he informed her that they could not supply her with men-at-arms merely on her word; they needed some proof that she was deserving of their trust. At this point, Joan could not contain her anger. She said, "I have not come to Poitiers to perform signs. Lead me to Orléans, and I will show you the signs for which I am sent." In the end, she convinced her examiners of the appropriateness of her cause, although Frère Seguin's record neglects to explain precisely how she managed to persuade them.

Unfortunately, the record of the exchange between Joan and Frère Seguin is one of the few remaining documents from the inquiry at Poitiers. The official record of her examination, the so-called Book of Poitiers, no longer exists; it was either lost, destroyed, or suppressed soon thereafter. The absence of this document from her trial at Rouen in 1431 may well have been Joan's undoing. During the trial, she repeatedly asked her judges to refer to it, but they were unable, or unwilling, to produce it. Joan must have set great store by her responses to the questions at Poitiers, and in their effect on her audience. We can only guess at the contents of the Book of Poitiers, and at what its presence at Rouen might have meant for the outcome of the trial.

While Joan of Arc sacrificed her life for her cause, her bold leadership saved France and inspired Charles VII to right himself in the eyes of his people. He thereafter ruled wisely, choosing competent ministers and disbanding the private companies of soldiers in France, reorganizing them in 1439 into the first standing army in Europe. This army was responsible for driving the English from France. Charles was important to Poitiers in yet another way. In 1431, he founded its university.

In the sixteenth century, Poitiers again became the site of a historic battle. By 1530, as France's third-largest city, it had become a residence for many Protestants. By 1559, Poitiers was predominantly Huguenot. It would inevitably become embroiled in the approaching conflicts between Catholics and Protestants—the Wars of Religion.

In 1569, during the Third Religious War, Gaspard de Coligny assumed the leadership of the Huguenots in France with the death of Louis Condé during the Battle of Jarnac. Coligny's troops joined a force composed of German, French, and Flemish soldiers under the leadership of William of Orange and Louis of Nassau and, on June 12, marched against Poitiers, which was defended by Henry, Duke of Guise. A siege of the town ensued and lasted seven weeks, to be broken only upon the arrival of the forces of the Duke of Anjou (later Henry III).

Although Coligny and his allies were driven from Poitiers, he was eventually successful in gaining more rights for the Huguenots than they had ever enjoyed. The Peace of St. Germain, signed in 1570, promised them freedom of worship except in Paris or near the court, eligibility for public office, and, as a guarantee for these terms, the right to hold four cities—La Rochelle, Montauban, Cognac, and La Charité—under their rule for two years.

However, peace was not lasting. Animosities were too deep seated, and King Henry III proved incapable of understanding the political complexities of the day. A series of wars punctuated by temporary cessations of fighting en-

sued. Four treaties of peace, from 1576 to 1580, were unable to ensure a lasting resolution to the conflict. One of these treaties, the Edict of Poitiers in 1577, granted Protestants freedom of worship in specified castles of Calvinistic lords and in certain towns. In addition, this edict recognized the holdings and heritages of Protestants, union by marriage with Catholics, and admission to public office.

Not until the reign of Henry IV would these rights take hold. The Edict of Nantes in 1598 authorized the exercise of the Protestant faith and freedom of the Protestant press in all but a few overwhelmingly Catholic French towns. While religious liberty was restricted to Catholics and Protestants, Henry IV was responsible for the most advanced religious tolerance in Europe.

With the passing of the sixteenth century, Poitiers assumed a less prominent role in its country's history. In the eighteenth century, under the rule of governors, known as intendants and appointed by the central government, Poitiers became a quiet provincial capital. Although the area near the train station was damaged in air raids during World War II, Poitiers, as a site of important religious and military conflicts in the development of France, remains basically intact. Its monuments, from the Romanesque St. John's Baptistry to the Gothic Great Hall, remind us of those struggles.

Further Reading: *The Birth of France* by Katharine Scherman (New York: Random House, 1987) offers detailed profiles of two prominent religious figures in Poitiers' early history, Hilary and Radegund. François Guizot's *History of France* (New York: Bolles, 1869) examines in depth several battles, from the clash between Clovis and Alaric II in 507 to the Battle of Poitiers in the Hundred Years War, which shaped Poitiers's and France's development. Desmond Seward's *Eleanor of Aquitaine* (New York: Dorset, and London: David, 1978) devotes a chapter to the twelfth-century Poitiers court of this fascinating individual. *Saint Joan of Arc* by Vita Sackville-West (London: Quartet, 1973; Boston: G. K. Hall, 1984) tells in refreshing prose the story of this famous saint, including her examination at Poitiers in 1429.

—Robert M. St. John

Prague (Czech Republic)

Location: Central Europe; near the center of the Bohemian plateau, situated along both banks of the Vltava River, a tributary of the Elbe River; 160 miles northwest of Vienna and 175 miles southeast of Berlin.

Description: Capital of the Czech Republic; dominant city of Bohemia from the ninth century until the end of World War I; central European intellectual and commercial center that has withstood Habsburg Austrian, Nazi German, and Communist Russian subjugation; emerging from political repression and disillusionment after the Prague Spring of 1968, the country underwent a democratic revolution in 1989 and elected the celebrated playwright Václav Havel as president.

Site Office: Czech Tourist Authority
Staromětské nám 6
110 15 Prague 1
Czech Republic
(2) 24 89 72 78

Since its beginnings in the ninth century, Prague has been a leader in political and intellectual change in central Europe. The development of Prague moved in concert with the formulation of the Bohemian state. The city started as a network of ancient Slavic hill forts that protected a grouping of small communities on the banks of the Vltava River. During the ninth and tenth centuries, the hill fort village settlements diminished with the construction of the Prague Castle. Rising over the countryside as the protective fortress, it became the home of Bohemian kings and continues to this day as the seat of Czech leadership.

Bohemia, which had previously been under the rule of Moravia, became a duchy in the tenth century, and royal and administrative power began to centralize in Prague. Bohemia became Christianized at this time, due in part to the influence of the Bohemian prince-duke Wenceslas (in Czech, Václav). Wenceslas encouraged neighboring German missionaries and made peace with the aggressive German rulers, but he made many enemies among Bohemian nobles, including his brother, Boleslav I, who assassinated Wenceslas in 929. His remains were entombed in a church in Prague. Wenceslas was subsequently canonized and is the patron saint of Bohemia. He is also the "Good King Wenceslas" of the Christmas carol.

After Boleslav became prince of Bohemia, the German rulers were still seeking to expand their empire, and he was forced to accept German suzerainty in 950. Bohemia subsequently became part of the German-controlled Holy Ro-

man Empire, founded in 962. Bohemia and Prague flourished under imperial control, however, and in the late twelfth century Bohemia was made an electorate and hereditary kingdom within the empire. By this time Prague was already a well-developed and attractive city, with much fine Romanesque architecture. Additions to Prague Castle were begun in this period. In the thirteenth century Prague grew into one of Europe's largest cities. Much construction was carried out in the Gothic style then coming into fashion.

Prague enjoyed a golden era in the fourteenth century, under the Luxembourg dynasty of Bohemian kings. King John, known as John of Luxembourg, initiated the construction of St. Vitus' Cathedral, part of the Prague Castle complex, in 1344. The cathedral was built on the site of the church where St. Wenceslas was entombed, and his remains were reinterred in the cathedral. Additions to the cathedral, which still stands, continued into the twentieth century. It houses the Czech crown jewels and the remains of many rulers. John's son, Charles of Luxembourg, became King Charles I of Bohemia in 1347 and Holy Roman Emperor Charles IV in 1355. He made Prague the empire's capital during his reign, and the city achieved unparalleled prominence. A skilled urban planner, he brought architects and artisans to Prague from throughout the empire. The city expanded, with the New Town (Nové Město) growing up around the former city center, which became known as the Old Town (Staré Město). A Gothic-style stone bridge, the Charles Bridge, was built across the Vltava River, beginning in 1357. Charles also established a university.

Upon Charles's death in 1378, the spirit of exhilaration and creation that emanated from Prague gave way to an extended period of religious and political agitation. During the reign of Charles's son Wenceslas IV, a church reform movement commencenced. In 1401, Jan Hus became dean of philosophy at the university and initiated attacks on religious doctrine and church organization. He spoke out against the sale of indulgences, questioned the concept of papal infallibility, and generally sought to bring the church closer to the people. His teachings met heavy resistance by the foreign faculty that were in the majority at the university.

In 1409, King Wenceslas IV sided with Hus, allowing the radical reformer to head the university. The makeup of the university began to shift as the faculty soon reflected a Czech majority. In his new position, Hus continued his criticism of the church and drew the attention of the papal authority in Rome. Hus traveled to a church conference in Konstanz in 1414 to defend his position on reform. When Hus arrived he discovered he had been deceived: he was confronted by a tribunal. Acting to discourage heresy and repress the widening movement, church leaders condemned Hus and burned him at the stake.

Worshipers in St. Vitus' Cathedral
Photo courtesy of Czech and Slovak Service Center

The death of Hus became a rallying point for the Hussite movement in Bohemia. Their emergence as a religious and nationalistic party began to challenge imperial control over the Bohemian kingdom. The Hussite wars consumed the Bohemian region for decades during the fifteenth century. Catholic and imperial forces eventually quashed the Hussites. During the fifteenth and sixteenth centuries the Habsburg dynasty came to power within the Holy Roman Empire. From 1526 to 1918, Bohemia was absorbed into a grand German-speaking supranational empire that the Habsburgs ruled. The empire came together, not from conquest on the battlefield, but through Habsburg marriages into monarchies across Europe. The Catholic Habsburg combined their hereditary possessions from Austria, the Netherlands, Spain and Italy.

As the Reformation upheaval swept across Europe, the Catholic-Protestant conflict came to a head in the Habsburg crownlands. In both Hungary and Bohemia many citizens turned to Protestantism to resist Habsburg rule. In 1618 an incident in Prague, in which two advisers to Holy Roman Emperor Matthias, a Catholic Habsburg, were thrown from the windows of Prague Castle, touched off the Thirty Years War. Although the two landed in the moat and suffered only minor injuries, the "Defenestration of Prague" was sufficient to inflame the emotions of both Protestants and Catholics. In 1619 Protestant Bohemian nobles overthrew Ferdinand II, the Catholic king of Bohemia, who also had become Holy Roman Emperor that year. The Protestants chose Frederick V, elector of the Palatinate, as the new Bohemian king. Frederick became known as "the winter king" because his reign lasted only through the winter of 1619–20. He and his army were decisively defeated by forces loyal to the Catholic Habsburgs at the Battle of White Mountain, near Prague, in 1620. This battle effectively ended Bohemia's part in the war, which finally ended in a draw in 1648.

After White Mountain, Ferdinand II resumed control of Bohemia and worked at imposing Catholicism upon Bohemia and eliminating any vestige of Protestant religion. Protestant nobles were killed or exiled and their lands portioned off to aristocrats from Catholic territories. The new nobility was loyal to their imperial sponsors rather than the native population. Jesuits were brought in to reconvert the Czech population to Catholicism, and German became the official language of the kingdom.

Czech culture and language were nearly snuffed out. Bohemian society went underground as some 35,000 of Bohemia's leading families went into exile. Bohemia, while nominally a separate kingdom, was completely under the Habsburgs' Vienna-centered authority. In the Bohemian cities, Prague and Bratislava, the middle classes were largely German in thought and origin. The peasants of the Bohemian countryside were the main source of hope for the future in kindling Czech customs and language.

In Prague, the resurgent Catholic leaders worked at placing their stamp on the city's architecture. This is most evident in Prague's many baroque Catholic churches, de-

signed by some of the era's most prominent architects. Two of the churches, built in the first part of the eighteenth century, bear the name of St. Nicholas; one was designed by Christoph Dientzenhofer and his son Kilian Ignatz, and the other by the son alone. St. Clement's Church, designed by F. M. Kaňka and built from 1711 to 1717, is part of the Clementium, a complex of buildings begun in the seventeenth century to house a Jesuit college. The Clementium is now the headquarters of the Czech National Library. The Church and Monastery of St. James, originally built in the Gothic style in the thirteenth century, were remodeled in the baroque style between 1682 and 1702.

In the middle of the eighteenth century, Prague became an international battleground once again, in the Seven Years War. The war was a struggle among several European powers; the primary issues were the contest between France and England for colonial supremacy in North America and India, and the battle for European power between the Habsburg Austrian empress Maria Theresa and the Prussian king Frederick II, known as Frederick the Great. Prague was the site of a major victory for Frederick in 1757. The war ended in 1763 after extensive peace negotiations; there was little immediate effect on Bohemia, which remained a Habsburg possession, but the war did help to establish Prussia as a significant power. Similarly, Bohemia remained under Austrian control through the Napoleonic Wars of the early nineteenth century, even though the Holy Roman Empire was dissolved. In 1813 Prague hosted the Congress of Prague, at which European powers attempted but failed to reach a negotiated peace with Napoléon.

By the late eighteenth and early nineteenth centuries, several factors began to create a resurgence in Czech identity in Prague. First, the Industrial Revolution sent masses of peasants into the city, sparking a change in its character from German to Czech. Second, the Catholic aristocracy, over the course of generations, became interested in their own autonomy and less devout in their allegiance to Vienna. Third, Germans began to step back from leadership and allow the increasingly impassioned Czechs to play greater roles in Bohemian politics. Finally, a revival of Czech culture was permitted in the intellectual and academic communities. The Royal Bohemian Academy of Science was introduced to elevate Czech culture and the first professorship in Czech language was established at the university in Prague. In 1836, the *History of the Czech People* was published by the nationalist historian František Palacký.

Bohemia was not immune to the revolutionary fervor that spread across Europe in 1848. An assortment of Czech nationalists, under the direction of the historian Palacký and the journalist Karel Havlíček, called for the independent status of Bohemia and other historic crownlands within the Habsburg Empire. After spurning the opportunity to go to an all-German conference in Frankfurt, they convened a Pan-Slavic Congress in Prague that summer. The assembly took an anti-German tenor and was fueled by the chronic local resentment between Czechs and Germans. While the conference was underway, pro-Slavic demonstrators took to the streets across the city. Soon a Habsburg detachment of troops stormed Prague and suppressed the uprising.

The Czech nationalists' calls for autonomous rule were swiftly silenced by counter revolutionary forces that went to work in Bohemia. The new Habsburg Emperor Francis Joseph I masterfully allowed one group to play off another and thereby maintain the status quo. Whether he used Bohemian conservatives and moderates to offset extremist nationalists or Czechs to counterbalance Germans, Francis Joseph worked the political situation in Bohemia to his favor. At first, it appeared that the Bohemians would receive some reasonable measure of autonomy from the Austrian Habsburgs. Then, Francis Joseph tightly closed off that possibility by repressing the Bohemians for decades to follow.

The only concession Francis Joseph made within his empire was to the Hungarians. After the revolutions of 1848–49 in Hungary nearly toppled Habsburg hegemony in the region, he agreed to formulate the Austro-Hungarian Monarchy. The historic compromise of 1867 created a divide between the Czechs on the Austrian-controlled side and the Slovaks on the Hungarian-ruled portion. The ruling systems of this dual empire developed in different directions. The Austrians slowly allowed larger numbers of Czechs to be a part of the political process, while the Hungarians granted very few Slovaks any voice in governmental affairs.

By the end of the nineteenth century, a dynamic political and cultural transformation was settling upon Bohemia. Prague's Czech population between 1856 and 1886 exploded from 60,000 to 150,000 while German inhabitants of the city dwindled in the same time period from 73,000 to 30,000. Czechs came to be well represented in both local government and in the imperial ruling council. New political parties began to spring up as industrialization created greater class consciousness among workers.

The intermingling of the Czech and Slovak nationalism leading to the Czechoslovakian state began in Prague in the 1890s. Tomáš Masaryk, who was part Slovak, started to bring the two communities together through his university professorship in Prague. He generated such a nationalist and activist spirit within his students that they formed the Czechoslakian Union in 1896 and in 1898 launched the radical publication *Hlas* (The Voice). When World War I broke out in 1914, Czech and Slovak interests became closer. As the Habsburgs' moribund hold over the Austro-Hungarian Empire began to slip, the Bohemians lost confidence in Habsburg rule. The goal of Czech and Slovak nationalism shifted from autonomy within the empire to independence from it.

The coupling of the Czechs and Slovaks was due primarily to the efforts of Masaryk. Exiled during the war, Masaryk and his associates traveled widely to court western leaders' support for Czechoslovak independence. When they met with U.S. president Woodrow Wilson in Pittsburgh in 1918, the resulting Pittsburgh Declaration opened the door to a postwar independent Czechoslovakia. Germans in the Sudetenland, a border region of Bohemia, were strongly op-

posed to the plan for a new Czechoslovakia. They acted out their opposition with protest marches and intermittent armed violence. Later that year, however, the Allies gave their full approval for a provisional Czechoslovakian government that occupied the Prague Castle.

The First Czechoslovak Republic was modeled after western democratic governments. Its Czech founders devised a liberal constitution and a freely elected parliament with authority to select the nation's president. A dynamic spirit fostered the simultaneous existence of cooperation and competition as the country developed numerous political parties. For two decades, from 1918 to 1938, the First Republic relied on external support through a series of treaties with its Slavic neighbors and a defense pact with France. Masaryk became the first president of the republic and Eduard Beneš was foreign minister. When Masaryk retired in 1935, Beneš succeeded him. In the mid-1930s, however, Czechoslovakia began to be threatened by Nazi Germany.

As the Austrian-born Adolf Hitler seized power in Germany, he made known his deep-seated resentment toward Czechs. The very existence of Czechoslovakia agitated Hitler because its creation was a result of the Allied peace treaty of Versailles, which many Germans found deeply humiliating. Hitler's effort to annex Czechoslovakia began when he secretly attempted to make a nonaggression pact between the two countries. When his request was rebuffed by the Prague government, he waited until his reoccupation of the Rhineland in 1936 to endorse self-determination for Germans in the Sudetenland, and to criticize the Czech policy of granting asylum to anti-Nazi refugees from Germany. Berlin funneled money to Bohemian Germans to construct political parties that were similar to the German Nazi party. His strategy began to succeed as the Sudeten German Party won more votes than any other party in the 1935 Czechoslovakian elections.

In 1937 Hitler began secretly making plans for a military takeover of Czechoslovakia. In the 1938 Munich Pact, Czechoslovakia's allies, Great Britain and France, granted Hitler the Sudetenland. The pact was widely regarded as a surrender to Hitler and an abandonment of Czechoslovakia. President Beneš had not been invited to the negotiations, and he resigned shortly afterward. In early 1939 the Nazis launched their military invasion of Prague, giving as excuses internal disorder in Czechoslovakia and Czech oppression of Germans and Slovaks. The German troops met no military opposition in the Czech capital. After the German army occupied Prague Castle, Hitler entered it to celebrate with a snack of Prague ham and Pilsner beer. The Czech cabinet, meeting on the other side of the castle, did not even know of his presence.

The invasion marked the beginning of a prolonged German occupation of Prague. Czechs, especially university students and intellectuals, tried to mount some opposition to the Nazis, but on November 17, 1939, Hitler ordered the "Special Action Prague," in which 9 student leaders were executed and 1,200 students were sent to the Sachsenhausen concentration camp. All Czech universities were closed and the faculty thrown out of work. After this, no effective Czech resistance developed until the final weeks of the war, when their was a major demonstration in Prague.

Beneš came back to Prague in 1945 to assume the Czechoslovakian presidency. He tried to work with the Soviets and the Czech Communist party but resigned in 1948 when Communist rule was imposed on the Czech Republic. As the Czech economy went through a severe depression in the 1960's, a restless mood swept the country. Prague became a center for the arts, counterculture, and student activism. Playwright Václav Havel and film director Milos Forman drew international renown, and Prague even hosted an appearance by the American beat poet Allen Ginsberg in 1965; Ginsberg was deported within a few days by the Czech authorities, however. From 1960 to 1967, the government vacillated between censorship and tolerance of intellectual and artistic expression. The year 1968 brought a major confrontation between outspoken students and the Czech government. That spring, 1,500 students from Prague Technical College marched toward the castle to protest the lack of heat and light in their living quarters. They were met by police, who entered the campus and attacked students in their path, including nondemonstrating students in their dorm rooms.

The event so angered the country that the protests took a political turn and students at other colleges and universities demonstrated their outrage toward the government and contradictions of Czech society. The Soviets invaded Prague in August 1968, sending hundreds of thousands of Warsaw Pact troops across twenty border points into the country. Unarmed Czechs confronted the armies in the streets and were no match for the well-equipped Soviet troops. In all, 70 were killed and 1,000 wounded.

The Soviets imposed a period of "normalization" on Czechoslovakia in the years after the Prague Spring of 1968. They forced Czech president Alexander Dubček to accept the Moscow Protocols, which allowed the Soviets to station troops in Czechoslovakia indefinitely and reversed various negotiated reforms that were agreed upon earlier in 1968. In April 1969, Dubček himself became a casualty of the normalization; he was removed from his top post in the Czech Communist Party. The post-Dubček government had no legitimacy in the eyes of the people and relied on the presence of Soviet troops to enforce normalization, which consisted of strict central control of the economy, the media, and other Czech institutions.

One hero who emerged from the failed revolution of 1968 was the playwright Václav Havel. For the next twenty years, he became a leader in the growing dissident movement. He maintained a high profile in spite of being frequently arrested and harassed for his activities to push for democratic reform.

In the fall of 1989, with the fall of the Berlin Wall and the rise of democratic movements in various Soviet bloc countries, Czechoslovakians began to push for dramatic change. On November 17, the fiftieth anniversary of Special Action Prague, a demonstration in the city not only commemorated that event but called for a liberalization of Czech society. Police attacked the demonstrators, but opposition to the government continued to grow. Again, students and the intelligentsia were the forces behind the movement for democracy, but they eventually drew support from industrial workers. The umbrella group for the movement was Citizens Forum, which Václav Havel helped found. The government had few means of countering the opposition; for example, it did not have the option of calling in Soviet troops. The Soviet government, under Mikhail Gorbachev, considered the unrest in Czechoslovakia as part of the democratization process and not a matter for outside interference. The Czech regime also recognized that the use of force would be unlikely to help its cause, and it finally collapsed in what became known as the "Velvet Revolution" because of its general lack of violence. Free elections were held in 1990, and Havel was elected president.

By the mid-1990s the Czech Republic—Slovakia, with significant ethnic and cultural differences, having broken off in 1993—was one of the most stable and prosperous of the former Soviet bloc countries. Prague continued to attract millions of tourists each year to view its castle, cathedral, historic churches, other architecturally significant build ings, and many monuments to victims of various forces of oppression throughout Czech history.

Further Reading: *The Fall of the House of Habsburg* by Edward Crankshaw (London: Longmans, Green, 1945; New York: Viking, 1963; reprint, London: Macmillan, 1981; New York: Penguin, 1983) offers insight into the developments in history that sustained and eventually ended the Habsburg rulers' dominance. Harry Schwartz's *Prague's 200 Days: The Struggle for Democracy in Czechoslovakia* (New York, Washington, and London: Praeger, 1969) takes an interesting look into the roots of the Prague Spring of 1968. *The Velvet Revolution: Czechoslovakia, 1988–1991* by Bernard Wheaton and Zdenek Kavan (Boulder, Colorado, San Francisco, and Oxford: Westview, 1992) offers a compelling view of the 1989 revolution that toppled the Czech Communist government.

—Andrew M. Kloak and Trudy Ring

Pskov (Pskov, Russia)

Location: In northwest Russia, 100 miles southwest of Novgorod, 160 miles southwest of St. Petersburg, on the confluence of the Velikaya and Pskov Rivers. The city of Pskov is located in Pskov district (oblast), which borders Latvia on the west, Belarus on the south, and the St. Petersburg district on the northeast.

Description: One of Russia's oldest cities, probably first settled in the seventh century. Famous as a merchant republic in medieval times. Pskov and Pskov district are a historical treasure trove of medieval Russian art and architecture. Pskov's historic downtown is richly medieval, graced by the ancient Kremlin (fortress), Trinity Cathedral, and imposing homes of the city's medieval merchants. Nearby towns contain ancient fortresses and monasteries (Pskov cave monastery; Mirozhsky monastery); estate museum of Russia's most famous poet, Alexander Pushkin; and the birthplace museum of composer Modest Mussorgsky.

Site Office: Pskov Oblast Administration
Ulltsa Nekrosova 23
Pskov 180001, Pskov Oblast
Russia
(811-22) 2-2444

The ancient Russian city of Pskov and surrounding territory formed a satellite city of Novgorod the Great (as Novgorod was called in the Middle Ages) for several centuries. Together, the two cities are perhaps the oldest in Russia, far older than Moscow. They lie in the heartland of Russia, with Pskov's having the distinction of being the westernmost outpost of Russian civilization during medieval times. Pskov's glory as an independent merchant city-state lasted for a mere century and a half, from the time of its independence from Novgorod in 1348 until it was annexed to the Muscovite state (or Muscovy) in 1510. Despite its loss of independence, Pskov remained an important commercial and trade center until St. Petersburg was established on the Gulf of Finland in the early eighteenth century. Thereafter, trade and commercial routes bypassed Pskov, which sank into provincial oblivion. Its subsequent obscurity has proven fortunate to modern historians or tourists, for little was torn down or altered. Pskov remained a medieval gem, shabby and decidedly Russian, impervious to the winds of westernization blowing out of St. Petersburg. Even today, when so many Russian cities have returned pre-Revolutionary names to their streets and squares, Pskov stubbornly clings to Communist-era names such as October, Red Partisan, Soviet, and Lenin. It is in-

teresting to note that Pskov was home to one of Russia's most notorious nineteenth century terrorists, Sofia Perofskaya, hanged in 1881 for masterminding the bloody assassination of Czar Alexander II, who had tried to westernize his government.

Pskov mirrors ancient Russian contempt and distrust of foreigners as do few other cities in Russia. Geographically, Pskov was situated close to territories peopled by its non-Russian, non-Slavic enemies, such as the Germans. Nearby Poland was populated with Slavs, but they were Roman Catholics, which therefore made them foes and rivals. One of the biggest, most destructive invasions of Russian territory, in which Pskov was a target, occurred in the late sixteenth century. King Stefan Báthory of Poland led an army of thousands of Poles and allied Lithuanians against the city. Eyewitness accounts cited the role that Pskovian women played in repelling the invaders, who managed to scale the outermost wall of the city. Leaving the safety of their homes, the women threw boiling water and heavy stones at the advancing Poles, who gave up and fled.

Throughout its long history, Pskov was under constant attack by these and other neighbors. More than two hundred sieges were recorded by the late sixteenth century. Ironically, Pskovians traded with their foes and even adopted many foreign words into their Russian vocabulary, while also learning to hate and fear the invaders. Even today, the heart of Pskov is dominated by a massive fortress, or Kremlin, situated behind thick walls. The oldest part of the city was contained within the Kremlin and adjacent to it, hugging the banks of the Pskov and Velikaya Rivers.

The Pskov River gave the city its name. According to Russia's most ancient chronicle, *A Tale of Bygone Years,* the ancient Slavic Krivichi tribe settled along this volatile river in the early 600s. Because of the river's constant splashing and lapping against the shore, they called it *pleskova,* an old Slavic word connoting splashing, and their settlement, *Pleskov,* evolved into the medieval Pskov. In that heavily wooded region, they probably built fortifications of wood.

This Slavic tribe and others eventually bowed to the rule of Prince Rurik and his brothers in the ninth century. One of the brothers, Truvor, became Pskov's first prince. Rurik, probably a Scandinavian, was considered to be the first political ruler of the nascent Russian state, the capital of which was distant Kiev. In 988, his descendant, Vladimir I, Grand Prince of the Kievan state or principality, which included Pskov and Novgorod, adopted Christianity. It was of the eastern, or Byzantine, suasion, which would decisively distinguish the emerging Russian identity from western Europeans. In due course, Pskovians converted to Christianity, probably in the way that was common then. The Prince

Pskov in winter, with Trinity Cathedral at left
Photo courtesy of Pskov Oblast Administration Office

would be baptized publicly, followed by mass baptism in the nearest river, whether the populace was ready or not. By the mid-1000s, two spiritual havens had grown up in the vicinity, the monasteries of St. John's and Mirozhsky, both still in existence. These, too, were heavily fortified and in fact, were part of Pskov's outer defense.

By then, Pskovians had also carved out the foundations of their most famous landmarks—Trinity Cathedral and the Kremlin. Both were initially made of wood. The cathedral, securely situated within the Kremlin walls, would be-

come the focal point of both civic and religious life. Its windows were mere slits, to foil enemy attacks. Its basement was spacious; it held the town's inhabitants during times of siege. The belfry, a separate structure unique to Pskov's churches, called the faithful to worship, and warned of impending attacks.

The other conspicuous symbol of Pskovian life, the Kremlin, was eventually converted from wood to stone, becoming a tremendous structure surrounded by a moat on one side and two rivers on the others, making it a virtual island.

A ring of five sixteen-foot-thick fortified stone walls, only two of which have survived, enclosed the adjacent medieval city as well. All of Pskov's pathways led to the Kremlin and to the safety of the cathedral within. In times of danger, the Kremlin and the town walls would be manned by citizen soldiers. The town itself was subdivided into *kontsy* or districts, with residents of each district responsible for defending their section of the walls.

Despite this grim existence, Pskov was a remarkably vital and vivid city. Until St. Petersburg was built in the early eighteenth century, Pskov was the closest Russian city to the Baltic Sea, though it was decidedly inland. Because it was also at the juncture of various important trade routes, commerce was the lifeblood of Pskov. Business was more important than farming, perhaps accounting for the relatively high level of literacy among Pskovians (along with Christianity had come the alphabet). In times of peace, foreign merchants were welcome in the city. Commercial wealth was flaunted in the dozens of beautiful churches, whose architects and artists attained fame throughout the Russian lands. These masters of iconography and church architecture established Pskov's reputation as a great locus of Russian medieval culture.

After the Kievan state's fragmentation in the twelfth century, Great Novgorod ruled Pskov. By the fourteenth century, Great Novgorod had overextended itself, and signed a treaty in 1348 with the citizens of Pskov, granting them their independence. That independence marked the heyday of Pskovian commercial life and cultural achievement. Pskovians evolved their own unique styles of architecture and icon painting, and grew even wealthier, mainly from the trade with the Hanseatic cities on the Baltic coast.

Rich details of Pskov's inner life are revealed in the medieval Pskov chronicles, written from the fourteenth to the seventeenth centuries, and carefully stored away in the vaults of Trinity Cathedral. From them it is known that Pskov was ruled not by its prince, but by its citizens, who voted on town matters en masse in the "veche," or town assembly. There was no voting requirement other than to be male. The veche was mandated to elect the city's prince who in turn had to pledge allegiance to the veche. This vigorous democratic institution, on what was called the veche square, was located within the formidable Kremlin walls. In addition, each district of the city boasted its local veche. Rich merchants from the 1200s onwards, until the advent of Muscovite rule in the 1500s, were often outvoted by the Pskovian "rabble," craftsmen and the like. This nascent democracy, also evident in medieval cities in western Europe, was possible because of the absence of central power and authority in Russia. Actually, it was not called "Russia" then, although inhabitants were conscious of being "Rus" (pronounced "roose"), that is, Russian.

There was much social tension throughout Pskov's stormy history, leading at one point to betrayal at the hands of the wealthy elite, known as boyars, and Pskov's brief occupation by German knights in 1240. After the invaders were expelled with the help of Novgorod, tension between merchants and the poor was constant.

Social tensions, invasions from outside enemies, and even heresies, came and went. The Mongols, conquering the Russian heartland in the twelfth century and ruling Russian towns until the early 1400s, demanded tribute from Pskov, but interfered little with town life. Of course, the Mongols were not popular, and Pskovians were among the Russian forces of Muscovite prince Dmitry Donskoi when he inflicted the first crushing defeat on superior Mongol forces in 1380. Thereafter, there was no stopping Muscovite expansion and the political centralization that accompanied it, all in the name of defending "Holy Russia" from foreign, non-Orthodox Christians and Moslems.

Pskovians were unwitting allies in Muscovy's expansion, mainly because of the city's dependence on outside help for its own defense. By the fifteenth century, the Pskov veche no longer was electing its princes; that right now belonged to Moscow's Grand Prince. When Muscovite armies, which included Pskovian volunteers, crushed Novgorod's independence and annexed the city to the centralized Muscovite state, Pskov's days as a merchant republic were clearly numbered. The day came, in 1510, when Grand Prince Vasily III took control of Pskov, disbanded the veche, and went even farther than the pro-Muscovite wealthy boyars thought necessary by forcing them to pay taxes to the new state. Belatedly, the citizenry realized what had happened to them, and great unrest ensued. Muscovy would have none of it. Three hundred wealthy families were summarily arrested, deported, and their estates were confiscated. As a symbolic gesture of who was in charge, the new government dismantled Pskov's great belfry and packed it off to Moscow.

It took a hundred years for Pskov to recover commercially from these blows to its independence and from the extinction of its leading merchant gentry. When the Muscovite dynasty died out in 1598 with the demise of Czar Ivan the Terrible, chaos reigned. Pskov's citizens took advantage of the anarchy and declared the city's independence, resurrecting its veche in the process. Independence was short-lived, and Pskovians found themselves once again crushed under the heels of Muscovite absolutism when a new dynasty, the Romanovs, and a new czar, took charge in 1613.

The Muscovite state, however, was no Soviet Union, and trade and commerce continued to rest in private hands. Pskov was still an important trade crossroads, and the commercial spirit was very much alive. Pskov's central market was transferred from within the Kremlin walls to the adjacent city. A continuous trade fair existed with hundreds, if not thousands, of wares on display, most of which were produced by Pskov's leading tradesmen. Such famous seventeenth-century merchant families as Pogankin, Menshikov, Pechenko and Guryev left imposing stone edifices that survived the ravages of time and two world wars. The houses resemble mini-fortresses, and served as good places of refuge during times of siege. Some of them even had indoor water supplies. The home of the millionaire Pogankin family on Nekrasov

Street is not only evidence of great private fortune, but also proof that even in that day and age in Russia, there was some economic opportunity. The founder of this family, Sergei Pogankin, had begun life as a common laborer and managed to accumulate a fortune.

Pskov's commercial wealth and promise declined precipitously when Peter the Great began the construction of his new capital. St. Petersburg was located directly on the Baltic shore, on territory newly wrested from the Swedes. Henceforth, Pskov's identity as the gateway to the Baltic Sea was lost forever. Two hundred years or more of slow stagnation and decay followed. The pristine medieval splendor of Pskov lay undiscovered in the nineteenth century by all but the hardiest and wealthiest travelers from abroad. The town's economy shifted away from commerce to agriculture, especially the production of flax, as well as the felling of trees and the processing of timber. Pskov's huge repositories of medieval manuscripts and icons were of interest only to learned academicians.

Somewhat miraculously, Pskov kept its name and its medieval identity virtually intact during the seventy-year Soviet period. Although no new life was injected into Pskov, many of the city's churches and museums were preserved. Under the Soviets, Moscow once more became the capital of a Russian empire. Pskov struggled on as a backwater.

While the city managed to escape the ravages of World War I, even though the front lines were only a hundred miles away and the last czar had abdicated in Pskov, it was not so fortunate during World War II. Pskov lay directly in the path of invading Nazi forces in June of 1941. The city and the surrounding villages were wholly occupied by German forces. The Nazis enslaved the Pskovians and herded them off to Germany. Many of the city's glorious medieval monuments that had survived even the brutalities of the Stalinist regime were utterly destroyed. Hundreds of nearby villages suffered similarly, including Mikhailovskoye, the estate museum of the revered poet, Alexander Pushkin, who lay buried there.

The postwar years witnessed an astonishingly quick restoration and rebuilding of the city, including every seventeenth-century merchant's grandiose stone residence. Entire medieval libraries were restored and lovingly catalogued by the brilliant Pskovian scholar and museum curator, Leonid Tvorogov. Even in the drab postwar Soviet era, it was possible for Pskov to begin, albeit tentatively, a life as a museum city.

The chaos resulting from the demise of Soviet rule has threatened the financial well-being of Pskov's monuments, including the ancient Kremlin. Nonetheless, the city's greatest asset still lies in its identity as one of the major cultural and spiritual repositories of Russia's medieval past, and therein lies its promise. Pskov's surroundings are as rich as its past with the fortress of Izborsk a few miles distant, the ancient monastery of Mirozhsky a stone's throw from the city limits, and the house museums of Pushkin and of the renowned musician-composer Modest Mussorgsky within an easy drive. For the first time in 250 years, Pskov's location as Russia's westernmost city will pay dividends, as western tourists begin to discover the legendary beauty of Russia's architectural and artistic heritage.

Further Reading: *Pskov, A Guide* by Yelena Morozkina, translated into English (Moscow: Raduga Publishers, 1988) is still the most detailed, authoritative, and richly illustrated source on Pskov's history, heritage, and surroundings. It is a delightful book, marred occasionally by obligatory pro-Communist jargon. *Fodor's Russia and the Baltic Countries* (New York: Fodor's, 1993) is updated annually, and has a very adequate and useful section devoted to Pskov with the usual practical tips and information. Arthur Voyce's *The Art and Architecture of Medieval Russia* (Norman: University of Oklahoma Press, 1967) is an excellent introduction to this complex subject.

—Sina Dubovoy

Regensburg (Bavaria, Germany)

Location: In southeastern Germany; at the northernmost point of the Danube, and at the Danube's confluence first with the Naab and then the Regen Rivers; 76 miles from Munich, 62 miles from Nürnberg, and 71 miles from Passau.

Description: Site of a Celtic settlement and Roman fortification; principal seat of Louis the Pious; flourishing commercial center during the thirteenth and fourteenth centuries, when much of its medieval architecture was constructed.

Site Office: Tourist-Information
Alten Rathaus (B2)
93047 Regensburg, Bavaria
Germany
(941) 5 07 44 10

Regensburg reached its high point as a prosperous commercial town in the thirteenth and fourteenth centuries and—despite modern efforts to resurrect ancient glories, like the foundation in 1962 of a university which opened in 1967—its 1,400 historical monuments, relatively undamaged during World War II, destine it to remain a focus for cultural tourism. The old town is compact, itself a living museum of medieval urban development. It extends for about one and one-quarter miles along the right bank of the Danube in the form of a half-ellipse whose axis is the river and whose furthest point is barely one-half mile away, still separated from the railway station by parks and tree-bordered avenues.

Although the old town has a modern shopping center, tourism is itself a growing industry, registering more than 400,000 "visitor-nights" in 1989, and depends on the careful preservation of the relics of the past. These relics have been carefully restored, and the town, now dominated by the cathedral's twin steeples, erected between 1859 and 1869, still contains an extraordinary concentration of ecclesiastical buildings. Some of the dazzling, gilded rococo interiors of the old town's churches communicate a powerful euphoria, but the exteriors still exhibit deposits from all periods of the town's development. There are remains from the Roman fort that replaced the Celtic settlement, and reminders of Regensburg's roles as the first Bavarian capital, then for nearly 600 years the most easterly free city in the empire, and then for 250 years an evangelical citadel with a mostly Catholic population. Regensburg's position as permanent seat of the imperial diet from 1663 to 1806 has left magnificent and often exquisite monuments of architecture and decoration, and even its capitulation to Napoléon and subsequent incorporation into Bavaria in 1810, while inaugurating a period of relative decline, have left its series of characteristic build-

ings. The old town offers a section through the cultural fortunes of 1,800 years of the history of one of the more important trading towns and administrative capitals of central Europe. Regensburg has a population of nearly 130,000.

While it is certain that there were Celtic encampments several centuries B.C. on the site of Regensburg, there is no evidence of any lasting settlement. The "Radaspona" mentioned in the *Life of Saint Emmeramus* by the bishop Arbeo von Freising in about A.D. 772 can indicate only a region that housed small communities from time to time. The recorded history of Regensburg starts with the Romans, who conquered the territory south of the Danube a decade before the Christian era began. Archaeological evidence suggests that until the middle of the first century, Roman military concerns were limited to the Danube west of the Lech, which enters it from the south near Donauwörth, about fifty miles directly south of Nürnberg, and fifty-six miles upriver from Regensburg. The Danube was a natural frontier for imperial Rome at its easternmost extent, and the site of Regensburg was ideally suited for a fort to defend Roman territory against the Germanic tribes. In the second half of the first century, a row of forts down the Danube as far as Passau was built, and there was a Roman camp known as Castra Regina with its dependent village on the present site of Regensburg. There was a road to Augsburg (Augusta Vindelicum), established earlier and about sixty-eight miles southwest of Regensburg.

The emperor Domitian's nearby fort at Kumpfmühl, established in about A.D. 90, fell to plundering Germanic tribes in about A.D. 170, and Marcus Aurelius decided to station his third *legio italica* opposite the mouth of the Regen. That meant building quarters for some 6,000 people in the camp itself, and an inscription, still preserved in the Stadtmuseum with many other relics from Regensburg's history, tells us that by A.D. 179 the walls with gates and towers had been completed. The Porta Praetoria, one block north of the cathedral and one block south of the river, is still standing, and much archaeologically valuable material has been discovered only quite recently, notably the gold and silver unearthed in 1989.

The Porta Praetoria was near the center of the north, shorter wall of the oblong camp, which measured nearly 1,800 feet by 1,500 feet, and whose southwest corner was between the Obermünster and the present-day Church of St. Emmeram. Other parts of the twenty-three-foot fortification walls can be seen next to the Niedermünster, 650 feet south of the river. There were originally at least twenty-two towers twenty-six feet high, and probably eight more twice as high, still standing in the late eighth century.

In A.D. 180 a peace was made with the tribes, and trade flourished until the Alamanni twice overran the fort in the third century. Evidence from the cemetery shows that the

The old town of Regensburg, with cathedral at its center
Photo courtesy of German Information Center

richer Romans were withdrawing from the luxury villas that they had constructed, that living standards were falling, and that Rome was withdrawing troops in order to defend the alpine passes. The fort was again destroyed in 357, but was certainly still in Roman use as late as 408. The conquering tribe, the Bajuwarii, who were ruled from 555 to 788 by dukes from the Agilolfing family, had settled on the site by about 500, retaining the Roman walls and making it their principal settlement. Saints Rupert, Erhard, and Emmeram, who was martyred, attempted to convert them, and a church was built to house the coffin of St. Erhard in about 700. Regensburg must have been an episcopal see before Boniface had it formally erected into a diocese in 739. It became a principal focus from which Christianity spread to southern Germany.

When, in 771, his brother died, Charlemagne seized the inheritance and sought as king of the Franks to impose his personal sovereignty on the other tribes of what is now Germany. Tassilo III, the last Agilolfinger and Charlemagne's brother-in-law, was forced to become his vassal and, in spite of an appeal to the pope in 787, to abdicate and enter a monastery in 788. Charlemagne visited Regensburg, and his son Louis the Pious, king of the Franks and emperor of the West from 814, made Regensburg his principal seat; Louis rebuilt his citadel on the present Alter Kornmarkt, where the Alte Kapelle, its high altar the wildly exuberant, gilded triumphal gate for the entry of Mary into heaven, now stands, almost mischievously extending late-rococo play with space and light to the outer edge of its possibilities, if not even straining to move beyond them. Louis used stones from the Roman remains, some with still incongruously visible inscriptions; across the market square his Roman Tower, with some of the original masonry, still stands. Louis's choice of Regensburg as capital was determined by its superb geographical position as a trade center.

Louis had held three imperial diets in Regensburg, which his chancery referred to as the royal city (*urbs/civitas regia*), a term otherwise used only for Pavia. His great-grandson Arnulf, king from 887 and emperor in 896, held four such diets. Arnulf's rule was repeatedly threatened by Conrad I, king of Germany from 911 to 918; with Arnulf's defeat of Conrad I, Regensburg was firmly fixed as the imperial capital, offering the key to the control of Bavaria. It was regarded as the mighty and majestic seat of power, and attracted in consequence numerous monastic foundations from the tenth to the thirteenth century. It also gestated such ambitious projects as the famous romanesque Jakobskirche, with its twelfth-century "Schottenportal" (the door of the church attached to the monastery of Irish Benedictines, the "Scoti"); the All Saints Chapel, built for episcopal tombs probably by Italians in about 1160; and the celebrated stone bridge, all of which are still there.

With the construction of the Danube bridge in the early Middle Ages, Regensburg became a point of intersection for routes between the Italian peninsula and the north, and from France and the Low Countries to Bohemia, Russia, the Middle East, and India. The bridge, which now rises some sixteen feet in the middle and bends to lessen the strain imposed by the river's flow, was started in 1135 and finished in 1146. It is the oldest German bridge, and for 800 years Regensburg's only bridge, replacing a wooden structure erected 330 feet downstream during Charlemagne's reign. Advantage was taken of a particularly hot summer, when water levels were unusually low. The emperor Frederick I Barbarossa decreed in 1182 that no tolls should be levied, but his decision was revoked by later emperors in 1310 and 1331. The bridge was originally 1,100 feet long, with fifteen supports—four of which were originally of wood—for sixteen round arches, of which buildings on each bank have left only fifteen now visible, with towers at each end and one over the twelfth support, erected in 1200 and destroyed in 1784. Each support is mounted on an ice-breaking structure that protects the supports from erosion. From the fourteenth century until the flooding of 1784, various types of mills on these structures utilized the water flow to drive machinery. The four wooden supports were replaced by stone between 1583 and 1586. Technically the engineering was advanced, using the water pressure against the supports to help strengthen the bridge.

The aspect of the bridge has changed much in the course of history, particularly through the disappearance not only of the midstream tower, the small chapels on the carriageway, and the buildings at the base of the supports, but also of the north tower and its surrounding buildings when Napoléon's troops stormed the town in April 1809; changing the bridge as well were the construction of the salt lofts (1551 and 1610–12), and the "sausage kitchen" at the south end, replacing a 1616 eating-house. A new north tower that could accommodate trams was built in 1901–03. The fourth of the original supports was blown up in 1633 to hinder the advancing Swedish army in the Thirty Years War, and it was not properly rebuilt until 1790–91. The second and eleventh supports were blown up to hinder the U.S. advance on April 23, 1945. They were rebuilt in 1967.

In 920 Arnulf, duke of Bavaria, had walled the city to include the basilica of St. Emmeram, which houses the tombs of many great figures associated with the city, including nineteen of the first twenty-one bishops. The successor of Conrad I as king of Germany was Henry I, whose attacks on Regensburg were repulsed. His son, Otto I (the Great), succeeded in achieving a political reorganization that separated kings from dukes in Bavaria. The kings kept largely to the St. Emmeram quarter, leaving the heart of the city, the Alter Kornmarkt, to the dukes, who began the construction of the Niedermünsterkirche and of the original Alte Kapelle in the tenth and eleventh centuries. Otto made his brother Henry duke of Bavaria, but both Otto and Duke Henry were ousted by Arnulf's son and the Bavarian nobles in 953; in the ensuing power struggle Regensburg began to be left in the shadow of Bamberg, when the emperor Henry II began systematically to build that town up at Regensburg's expense. The situation remained unstable until the end of the twelfth century, by which time Regensburg seemed again to be the natural capital of Bavaria, and the obvious place to convoke the imperial diets. It was also a convenient point to assemble crusading armies.

In the late tenth century the kings, who had striven for the monopoly of indistinguishably spiritual and temporal power, began to find the support of their bishops necessary, although no formal arrangement was made before the agreement of Duke Louis I of Bavaria and Bishop Conrad IV of Regensburg in 1205. Wolfram, bishop of Regensburg from 972 to 994, had accepted the separation of Bohemia from Bavaria, allowing the institution of a bishopric in Prague; he reformed Regensburg's monastic institutions, strengthening their libraries and educational activities, and encouraging more daring forms of manuscript illumination.

From the thirteenth century date the first extant city seal (1211), the first mention of a mayor (1243), and the first artisans' guild (the shoemakers' in 1244). In 1245 the emperor Frederick II allowed Regensburg to elect its own mayor and council, the principal prerogative of a free imperial town, paying taxes to neither king nor bishop. It was difficult to deny such privileges to the rich trading centers, but it meant that until 1810 Regensburg was not part of Bavaria. The town was known for its silk from the Levant, cloth from France and Italy, pelts from Russia, and metals from Hungary. There was also trade in commodities such as wine, oil, spices, and salt. A five-year dispute between bishops and council over primacy of jurisdiction was resolved in 1253 and must be regarded as part of the overall attempt to regulate the relationship between spiritual and temporal sovereignties in Europe from the eleventh to the thirteenth centuries.

In 1284 construction began on the new town walls, which now enclose some 235 acres, more than Nürnberg or Frankfurt, although much less than Cologne. It had become fashionable for the big merchants to advertise their importance by erecting tall towers, often built by Italian workmen, to heights up to twelve stories. Regensburg still has some twenty towers, of which the highest is more than 130 feet tall. The vast romanesque porch of St. Emmeram dates from the twelfth century, and the eleventh-century sculptures next to it are among the oldest in Germany. Building of the Gothic cathedral began in 1250, and a Rathaus was constructed in 1350.

The cathedral drew on French building designs and techniques, and has three vaulted naves. There had been an episcopal church on the site of the present Niedermünsterkirche since about 700, a Carolingian cathedral with a facade facing the Via Praetoria, and on the present site an eleventh-century building with two towers on the western facade, of which only the "donkey tower" (Eselsturm) on the northern part of the transept remained. In 1273 one of the town's many fires (fires had previously occurred in 954, 1020, 1132, 1153, 1166, and 1176) delayed work, but the towers were ready in the fourteenth century, with the northern tower markedly more elaborate than the southern, which was finished later in the century and is later in style. The

simplicity of its top third was due to the need to economize. With the exception of the high altar, the baroque decoration added in the seventeenth and eighteenth centuries was removed during the reign of Louis I of Bavaria (1825–48), and the towers were then given their Gothic steeples. The cathedral is considered to be the finest in southern Germany, at least of those derived from the French models. Its magnificent west front, work of the Roritzer family, has an unusual triangular porch, with fine statues.

Inside, the tall cathedral is 280 feet long by 115 feet wide, with the central nave 105 feet high, virtually twice as high as the 55-foot-tall aisles. The redecoration was finished in 1989. Almost all the medieval stained glass has been preserved intact, although nineteenth century neo-Gothic glass has been added to it. The stained-glass decoration was completed in 1988 by Josef Oberberger, and in front of the west transept pillars there are remarkable statues dating from about 1280 and depicting Gabriel and Mary. The other statues were created later, and the style of the more recent pieces is less formally religious. Statues of the devil and his grandmother correspond to a late-medieval religious sensibility and warn against the evils of the world outside.

By the end of the fifteenth century strenuous remedies were needed to reform the city's monastic institutions. Pressure for reform came largely from the city council, which, like the rest of Regensburg's population, resented the clerical exemptions from taxes. Regensburg had, with Vienna, been the only large German town to protect the Jews in 1348 during the plague, and the council protected the Jewish community again in 1384 before turning against the Jews in the late fifteenth century. The emperor, who needed the tax income the Jews brought him, involved the town in a number of legal actions to retain the Jews, but they were driven out in 1519.

The Neupfarrkirche, Regensburg's first Lutheran church, was erected on the site of the old synagogue. The Jewish gravestones were broken and spread around the town. Luther's repudiation of Roman primacy, fueled by anticlericalism, was supported above all by the patricians and the artisans seeking an interior, evangelical religion to replace a late-medieval spirituality based on the observance of apparently arbitrary practices not intrinsically connected to religious perfection. It was the opposition to Luther by Austria and Bavaria that kept Regensburg's citizens Catholic, and although the town did not sign the famous "Protest" at Speyer of April 1529, its official religion was Lutheran from 1542.

Trade had dropped significantly with Turkish victories, and from the end of the first third of the fifteenth century Regensburg could maintain financial solvency only by borrowing. There was a danger that the Habsburg emperors would regard Regensburg as a buffer against Bavaria, and the citizens were generally relieved when the town gave itself into the hands of Albert IV, duke of Upper Bavaria, in 1486. Unhappily, the emperor would not consent to the arrangement, saying he would prefer to lose all of Austria. The town was handed back to the emperor, who reorganized its administration.

The city's artistic life remained vigorous. The library at St. Emmeram contained 700 works in 1500, making it one of the biggest in Germany. The first Regensburg book, a missal, was printed in 1485. The painter Albrecht Altdorfer, famous for the relationship he established between his figures and their background landscapes and deriving his style partly from Mantegna, was town architect and a member of the council. Johannes Kepler, the astronomer who completed the system of Copernicus, spent much of his life in the town.

From 1663 until 1806 Regensburg was the perpetual seat of the imperial diet, which met in Augsburg only in 1713, when 8,000 Regensburg citizens died of the plague. Napoléon reorganized the Holy Roman Empire at a diet from 1801 to 1803. Despite Regensburg's political importance, the town's trade continued to diminish. Regensburg was insolvent by 1800. Of 22,000 inhabitants, only 1,300 paid taxes, and comparatively little secular building work was undertaken in the eighteenth century, although the Alte Kapelle was splendidly redecorated between 1747 and 1761, and the same team created a rococo interior from 1749 to 1760 at St. Kassian, a romanesque church first mentioned in 885. The rococo conversion of the Benedictine Monastery of St. Emmeram between 1731 and 1733 by the Rome-trained Asam brothers, Cosmas Damian and Egid Quirin, is not generally considered to have been so successful, despite their delight in light and color. The difficulty was in balancing the extreme range of styles from the eighth century to the eighteenth century in different parts of the complex. The cloister became the home of the Thurn and Taxis family, who maintained monopoly rights over postal services in Germany until 1866.

Regensburg has many other buildings of importance. The Rathaus is a complex of buildings from the fourteenth to the eighteenth centuries, and on the Haidplatz, some one thousand feet west of the cathedral and half that distance from the river, is the decorated five-story former inn "Das Haus zum goldenen Kreuz," with tower; it was there that Emperor Charles V met Barbara Blomberg, who was to bear him Don Juan of Austria, the victor of Lepanto in 1571. Near its east gate the city built a castle between 1854 and 1856 for Maximilian II, with a facade facing the Danube and with an informal English garden on the town side. It is now an administrative building.

In April 1809 the Austrian army, defeated by Napoléon, retreated to Regensburg, where they were followed by the French, who looted and destroyed much of the southeast of the town and the buildings on the north side of the stone bridge. In 1810 Regensburg was incorporated into Bavaria as capital of the Upper Palatinate. The prince-archbishop, Karl

Theodor von Dalberg, the last elector of Mainz, had attempted from 1802 to maintain Regensburg's independence. He had reformed the school system, created public buildings, and established a forty-year debt-repayment program. After handing over Regensburg in 1810, he was to return modestly to the town in March 1814.

Regensburg's more recent history has not been happy. The Versailles Treaty ruined the town, and Regensburg underwent the experiences common to German towns during the rise of national socialism. In the later stages of World War II air raids killed more than 1,000 people, and more than 1,000 buildings were destroyed. The Obermünsterkirche was obliterated, and all the bridges were blown up. An attempt to surrender to advancing U.S. troops in order to avoid useless bloodshed led to the shooting of the commander, Michael Lottner. Happily, after the twenty-five years it took after the war to reestablish equilibrium, the subsequent quarter century has seen many encouraging signs of a revitalization of Regensburg's economy.

The town is immensely rich in historical monuments. Those surviving monuments not standing on the streets are housed in the town's half-dozen museums. Like so many other towns with historic centers, Regensburg's central area is being kept functional. It is accessible to motorized vehicles, but the tendency has been to pedestrianize as much of the small old town as possible; as well, new light industry has been relegated to the fairly flat land to the south and west beyond the girdle of parks and avenues, and the town encourages its tourist industry. For tourists, Regensburg has charm and is adequately equipped with the usual facilities, but the town does not yield itself easily to the casual visitor. Almost anywhere in the old town one may encounter individual buildings and architectural features extending over 1,800 years, and an acquaintance with the political, social, architectural, and general cultural development of the town is useful indeed.

Further Reading: Recent and reliable work on Regensburg is available only in German, notably the well-illustrated *Regensburg: Geschichte in Bildokumenten,* edited by Andreas Kraus and Wolfgang Pfeiffer (Munich: C.H. Beck'sche Verlagsbuchhandlung, 1979). Most of the photographs are in black and white. The 1,000-page *Regensburg* by Karl Bauer (Regensburg: Mittlebayerische Druckerei und Verlagsgesellschaft, 1988) is popular, prize-winning, well illustrated, and reliable, and the *Regensburger Taschenbücher* books by the same publisher are informative, illustrated, and accurate. The fundamental modern work is the *Handbuch der Bayerischen Geschichte,* edited by Max Spindler (four volumes in six, second edition, Munich: C.H. Beck'sche Verlagsbuchhandlung, 1968).

—Anthony Levi

Reims (Marne, France)

Location: In the Marne department of Champagne, northeast France; seventy miles east-northeast of Paris, and fifty miles from the Belgian border; situated on the Vesle River, a tributary of the Aisne, and on the Marne-Aisne canal.

Description: Originally a settlement of the Gallic Remi tribe, Reims was conquered by the Romans, under whom the town flourished. It became an important Christian center, and from very early times most French kings were crowned in the cathedral, known as Reims Cathedral or the Cathedral of Notre Dame. The town was very badly damaged in both World Wars, particularly in the first, but the main buildings have been restored. The town is surrounded by vine-growing country, in which champagne wine is produced. The major champagne wine houses have their cellars and headquarters there.

Site Office: Office de Tourisme et Syndicat d'Initiative
1 Rue Jadart
51100 Reims, Marne
France
26 88 37 89

Although based on the site of a prehistoric settlement, Reims was given its name by the Remi tribe of Gaul. Taken by the Romans in Caesar's conquest, the site was then called Durocortorum; by the third century it was the capital of the Belgian Roman province, housed its governor, and was the starting point for roads leading to Boulogne, Cologne, Metz, and elsewhere. It became a center of the wool trade, and textiles have remained since those times an important industry. Of the Roman city two monuments remain, the Porte de Mars and the Cryptoportique.

The Porte de Mars, to the north of the original city, is one of the best preserved Roman triumphal arches in France. Uncharacteristically, its three bays are almost the same in width and height, and the four rectangular towers or pylons are of identical shape. The vault of the central bay has a unique frieze symbolizing the months of the year. It was probably built towards the end of the second, or beginning of the third century, and was one of four in the city. It was not, strictly speaking, a gate at all as the city boundaries lay further away. The Cryptoportique's date of construction is also uncertain, though of the same period, or just later. It is an underground and vaulted hallway, 318 feet long by twenty-nine feet wide, with shorter branches at right angles on either end. Each hall is divided down the center by a series of pillars that create, in effect, twin corridors. The walls, pillars, and vaulting retain a few traces of their original red

coating; niches in the walls were colored green. The purpose for which the Cryptoportique was intended is also unclear, but probably changed over the centuries. It may have been a series of shops where fragile stocks such as grain and cereals could be stored safely away from light and damp. However, given its central position in the city and the fact that it was almost certainly related to another building and forum that lay on top of it and have long since disappeared, it is more likely that it had some religious or public use.

Christianity was first introduced to the city in the second half of the third century, and developed with more vigor in the fourth. A cathedral and associated episcopal buildings were erected in the center of the city by Bishop Niçaise at the beginning of the fifth century. He was murdered in its forecourt in 406, in an invasion by the tribe known as the Vandals. During the rest of the century, the Franks pushed forward in this part of France, and it was in Reims Cathedral on Christmas Day, probably in 498, that Clovis, king of the Franks, and by now married to a Catholic princess, was baptized by Bishop Remi. In the following centuries, the city was to become one of the leading Catholic cities in the kingdom, and by the middle of the eighth century more than twenty churches and chapels had been built inside and outside its walls. In 816, Louis the Pious was crowned in the cathedral by the pope, creating a precedent. The coronation of Henri I in 1027 cemented the tradition, stoutly defended by the archbishops of Reims against the claims of their Parisian counterparts. In all, thirty-seven French kings took their coronation oaths there, ending with Charles X in 1825. One of the most famous of these ceremonies, of course, was that of Charles VII in 1429, with Joan of Arc in attendance.

By the mid-thirteenth century the population was over 20,000, making it one of the largest cities in France. The textile industry was important, but it was also a market center for the sale and distribution of agricultural produce. It was as both an ecclesiastical and scholastic center, however, that Reims was supremely influential during this period. One master of the school attached to the cathedral in the eleventh century went on to found the monastery of the Grande Chartreuse in the Alps, and the Carthusian order. Two of the principal churches of the period which still stand, despite frequent war damage, are the Abbey church of St. Remi, and the Cathedral of Notre Dame.

The church of St. Remi stands on the site of an earlier one, but was renamed in honor of Archbishop Remi, whose tomb lies behind the high altar. It is the oldest church in the city, having been largely rebuilt in the eleventh century in the Romanesque style, with the vaulting of the nave and the Gothic choir completed at the end of the twelfth. Pope Leo IX dedicated the church on the feast of St. Remi, October 1st, in

Reims Cathedral
Photo courtesy of French Government Tourist Office

1049. It was particularly renowned for housing the ampulla containing the holy oil used in the annointing of French kings at their coronation. The nave of St. Remi stretches to 400 feet, and is almost as long as that of the cathedral itself. Two statues, of St. Peter and St. Remi, decorate the main facade. Valuable monuments and the ampulla were damaged or removed during the French Revolution. Some twelfth-century stained glass remains in the choir in spite of the fact that the church suffered enormously from bombardments during World War I, leaving it almost entirely roofless. Excavations in 1931 allowed archeologists and historians to determine much of the original, pre-Romanesque structure and the church has been well restored. The abbey buildings, attached to the church, are now used as a museum of medieval art and archaeology.

The Cathedral of Notre-Dame, built on the site of the original fifth-century structure, burned down in 1210. The following year, work began on a new cathedral. Louis VIII was crowned there in 1223, as was Louis IX in 1226. This magnificent Gothic building, completed within the span of 100 years (apart from the two west towers which were not finished until 1428), was a major architectural influence far beyond the boundaries of France. It was designed to be a majestic setting for the coronation of kings, and is both longer and taller than the cathedral of Notre-Dame in Paris.

The west facade is richly and gloriously decorated. Three doorways with pointed arches dominate the ground level, the central one with sculptures of the life of the Virgin Mary, the left with Christ's passion, and the right with a scene from the Apocalypse. All three doorways are deeply indented, with sculptured figures of saints at the bottom. The huge rose window in the center of the middle level is flanked by tall pairs of clerestory windows; statues on the plinths and jutting pillars that separate these three architectural sections include one of Christ in medieval clothing. The third tier consists of a series of statues within gothic arches, and they continue around the sides of the two lateral towers, which rise to a height of 269 feet from the ground. The central figure in this series, of which there are seven above the rose window alone, represents Clovis at his baptism. He is surrounded on either side by sculptures of successive kings of France. This profusion of sculpted figures continues on the other facades of the cathedral, and the carvings of angels in particular have wonderfully mischievous smiles. The choir is supported by double tiers of flying buttresses, and there is a profusion of carved pinnacles. The impression is exuberantly gothic and of an almost overwhelming, architectural vivacity.

Inside, the west front has a series of seven rows of niches on either side of the main doorway, each filled with figures or scenes from the testaments, and filled between with decorative carved foliage. Within the cathedral, carving is apparent in the capitals of the pillars of the nave, which have extravagantly varied foliage and small groups of figures. Otherwise, the interior is austere and majestic. Not a great deal of the original stained glass remains; the lower windows were replaced in the eighteenth century, and the rose window of the south transept was blown out in a hurricane in 1580.

The upper windows in the choir and apse commemorate the life of the Archbishop Henri de Braine, who died in 1240, and who was responsible for much of the rebuilding of the cathedral. Modern stained glass, designed by Marc Chagall, has been put into the chapel behind the altar. Despite fires, the destruction inflicted during the French Revolution of 1789, and various wars, the cathedral has been well restored, and remains much as it was initially intended to be.

On the south side of the cathedral is the Palais de Tau, built at various stages from the twelfth century onwards. It was the archbishops' palace, and traditionally used by the kings who were crowned in the cathedral as their residence during the coronation celebration. The inhabitants of Reims had to pay a special tax in order to accomodate the king and his retinue on each occasion. The palace is now a museum, of necessity much restored after war damage, and holds collections of chalices and ecclesiastical ornaments, tapestries, and the robes worn by Charles X at his coronation.

Reims's modern reputation, apart from its historical buildings, lies first in its connection with the wine trade, and with sparkling champagne wine in particular. It shares its position as headquarters of this industry with Epernay, a town about fifteen miles south. Vines have long been cultivated in the district, which encompasses the northernmost vineyards in France, even before the Roman invasion of the area. The discovery of how to make the sparkling wine is said to have been made by Dom Pérignon, a monk who was cellar master of the abbey of Hautvillers north of Epernay in the late seventeenth and early eighteenth centuries. He is alleged to have exclaimed, "I am tasting stars!" after trying the fermented wine.

The process of turning grape juice into champagne is a complicated one, involving careful blending, fermentation, and the removal of sediment from the bottle while it is upside down. The ideal temperature (eleven degrees Celsius, or fifty-two degrees Farenheit) necessary for the various stages of its production is to be found in the *caves,* or cellars, made in the chalk deposits under the city. These are mostly in the Champs de Mars quarter north of the city center, and in the St. Niçaise quarter to the southeast. Many of these cellars were originally quarried by the Romans to extract the chalk for building purposes, and are shaped like a pyramid with a hole at the top only wide enough to allow one block at a time to be taken out. This arrangement was designed to prevent too much moisture from entering the chalk face. The best example of this type of cellar is that of the House of Ruinart. Now the major champagne houses use miles of these underground cellars to hold their stocks of maturing wines, and those belonging to Mumm and Pommery are themselves both over eleven miles long. The other major champagne houses in Reims are Piper-Heidseick, Taittinger, Veuve-Cliquot-Ponsardin, Roederer, and Lanson.

The second reason for the importance of Reims in modern times is military; it has been invaded or attacked repeatedly over the centuries. In the Franco-Prussian War of 1870, Prussian troops took the city, and in World War I the German army invaded, then bombarded, the city. In 1914, the cathedral roof was set on fire, and three-quarters of the rest of Reims was also destroyed, including the Town Hall, in 1917. The cathedral was restored, with a celebratory mass attended by the president of the French Republic and other members of the government and army in July 1938, to mark its completion. Two years later German troops again entered the city and stayed until August 29, 1944. The Second Regiment of the Fifth American Infantry Division arrived the following day. On May 7, 1945, the German surrender was signed in General Eisenhower's headquarters, which had been a technical college, north of the railway station. The *Salle de la Reddition,* or War Room, has been preserved, and is open to the public. Although the destruction incurred during World War II was relatively light, enormous efforts were required to restore Reims's prosperity and economy in the postwar period.

Despite an increase of about 80,000 inhabitants in the first four decades after the war, Reims has seen the near termination of its woolen weaving industry. Clothes manufacturing and food, pharmaceutical, and light engineering companies have take its place. At the same time, however, the champagne trade has increased, with rising markets both abroad (some firms such as Veuve-Cliquot and Mumm exporting more than two-thirds of their output), and at home. The area in the vicinity devoted to grape growing has more than doubled. Reims remains, therefore, what it has always been; an ecclesiastical, cultural, intellectual, industrial, and above all, invigorating city.

Further Reading: *Saint-Remi de Reims: l'oeuvre de Pierre de Celle et sa place dans l'architecture gothique* by Anne Prache (Paris: Flammarion, 1978) and *Gothic cathedrals: Paris, Chartres, Amiens, Reims* by P. Clemen (Oxford: Oxford University Press, 1938) are two sources of further information about these churches. The history of the city is given in a series of detailed essays in *Histoire de Reims,* edited by P. Desportes (Toulouse: Privat, 1983).

—Honor Levi

Rostov (Yaroslavl, Russia)

Location: On the northwestern shore of Lake Nero, 35 miles southwest of Yaroslavl and 120 miles northeast of Moscow.

Description: The city of Rostov is one of the most ancient cities in Russia. Settled originally by Finns and named in honor of Prince Rosta, the city was first mentioned in Russian annals in 862. Later, it became one of the first Slavic towns to appear in Russia. The city became the capital of an appanage principality and a key center for the development of the modern Russian state, later superseded by the cities of Vladimir and Moscow. Throughout the Middle Ages, Rostov retained an important position within Russia as an international trading center. Today, the city is noted for its traditional crafts industry.

Contact: Yaroslavl Oblast Administration
Sovetskaya Ploschad 5
Yaroslavl/50000, Yaroslavl
Russia
(85–2) 22–2328

The first known settlers of Rostov were the Meria, a Finno-Ugrian people. (Most lakes and rivers in and around the city still have their original Finnish names.) In 862, Rostov was mentioned in the Russian annal *Primary Chronicle*. During that same century, trade increased between Europe and Asia, turning Rostov and other small settlements in Russia into commercial centers with an active merchant class.

Over the next two centuries, many different ethnic groups settled around Rostov. Some archaeologists have speculated that Arabian coins found in Rostov were brought to the region by Norman (or, as they were known in Russia, Varangian) settlers. Three Slavic tribes, the Slovenes, the Krivichi, and the Viatichi, are known to have colonized Rostov during these two centuries. The Slavic colonists often lived communally in small fortresses with their immediate family, extended family, and strangers. They actively practiced polygamy.

The Slavs took advantage of the area's rich soil by cultivating rye, flax, fruits, vegetables, and hops. They also raised cattle, tended sheep, trapped beaver, collected honey, fished Lake Nero, and were active traders with both Scandinavia and Arabia. Commodities traded included amber, silver coins, honey, fur, and grain.

The Rostov-Suzdal principality formed in the tenth century on territory located between the Volga and Kliaz'ma Rivers. Rostov was the principality's capital, an honor it retained until the second quarter of the twelfth century.

In 989, residents of the city were Christianized, at least nominally, by Byzantine Orthodox priests. Locals were divided into groups of ten to fifteen individuals and baptized in the waters of Lake Nero. Despite these attempts, paganism continued in Rostov for sometime thereafter.

Kiev was the preeminent Russian city from the ninth to the eleventh centuries. The district of Rostov-Suzdal came under the authority of the grand prince of Kiev during either the late tenth or early eleventh century. During part of the eleventh century, while under Kievan control, Rostov and the surrounding area remained ungoverned. The grand prince of Kiev, Vladimir Monomakh, took interest in the district and in 1108 established the city of Vladimir on the shores of the Kliaz'ma River. He placed his youngest son, Yury Dolgoruky, in charge of Vladimir.

After Grand Prince Vladimir Monomakh died in 1125, the Kievan state was rocked by numerous Turkish raids. A power vacuum was created in Russia due to Kievan troubles and three power centers formed within Russia to fill the void. The first center of power was in the north, around Novgorod. A second power center formed to the west and southwest. It was eventually taken over by Lithuanians and Poles. The final region originated to the north and northeast, between the Oka and Volga Rivers. It later became the political, economic, and cultural center of Russia and gave rise to the Moscow Grand Principality and the modern Russian state. Rostov was found in this last region.

Upon his death, Grand Prince Vladimir Monomakh deeded to Yury Dolgoruky the newly formed principality of Rostov-Suzdal-Vladimir. Dolgoruky severed Rostov's dependence on Kiev, named Suzdal as the princedom's capital, and expanded and defined the principality's boundaries.

As the first independent ruler of Rostov, Dolgoruky ran the principality well. He worked to strengthen his frontiers against Bulgar attacks and fought hard against princes to the south and the Russian aristocracy, or boyars, in an attempt to become the grand prince of Kiev in 1155. His desire to capture Kiev and rule his domain from there gained him the nickname "Long Arm." Dolgoruky ordered the construction of many new cities and fortresses to protect his holdings, and also ordered the building of many new churches. He actively encouraged settlement into his princedom with promises of land and exemption from taxes. Many of those who moved into the region were military personnel, merchants, and artisans. During his rule, the principality of Rostov-Suzdal-Vladimir became an important center for commerce and industry. Local merchants traveled from the princedom down the Volga River, to the Caspian Sea, and to the cities of Novgorod, Kiev, and Constantinople.

Upon Dolgoruky's death in 1157, his son, Andrei Bogoliubsky, became prince. In his first year as prince he transferred the capital from Suzdal to Vladimir. The move

Rostov's Church of the Resurrection and Cathedral of the Dormition

changed the name of the principality from Rostov-Suzdal-Vladimir to the Vladimir-Suzdal. The principality was the largest princely domain in all of northeastern Russia. Andrei pursued the policies set forth by his father and attempted to further increase his own power over the domain. He further enlarged and fortified the principality, led a successful campaign against the city of Kiev, and established hegemony over the land of the Rus. Industrialization also increased under his reign, especially in Vladimir. Artistic achievements reached new heights.

In 1207, the city of Rostov became independent from Vladimir-Suzdal. It became the capital of a domain that consisted of the city of Rostov plus the towns of Yaroslavl, Uglich, Mologa, Beloozero, and Ustiug. Rostov's first prince was Konstantin Vsevolodovich, who ascended to the throne in 1212 by challenging his elder brother, Yury, for control of the domain upon his father's death. Fighting between the brothers lessened the importance of Rostov and weakened the region. Four years later Konstantin defeated Yury and became the grand prince of Vladimir. After Konstantin's death in 1219, Yury briefly restored unity.

Unity was once again shattered when the domain was divided between two brothers, Vasil'ko and Vsevolod. Vasil'ko received as his domain the city of Rostov and possibly the town of Beloozero. Beloozero later became an independent principality in 1238, along with Yaroslavl and Uglich.

For most of the remainder of the thirteenth century, Rostov remained in the hands of Vasil'ko's eldest direct male descendants. This was accomplished for three primary reasons. First, his heirs chose not to follow a long-held Russian tradition of horizontal succession, in which family holdings passed from brother to brother until one generation of brothers died off. Then, in theory, but not always, the throne would pass to the sons of the deceased eldest brother. This practice often caused problems with succession. Instead, Vasil'ko's descendants followed vertical succession, passing holdings down family lines from father to son, keeping the principality within the immediate family's possession. Second, his heirs tended to have small families, lessening the likelihood for large-scale fraternal discord. Lastly, a high infant mortality rate further thinned competition between brothers.

In 1238 the Rostov principality was captured by the invading Mongolian Tartars, but occupation by the Tartars did not bring about an end to the principality's existence. Continued internal fragmentation led Rostov to fall increasingly within the Grand Principality of Moscow's sphere of influence. Moscow's grand prince Ivan Kalita purchased Uglich from the Rostovian princes during this period. Later, Grand Prince Dmitry Donskoi of Moscow turned the princes of Rostov into his ''service princes.'' Rostov joined other Russian towns in 1262 in an unsuccessful uprising against the Mongol Tartar yoke, and briefly managed to unify itself with Beloozero in 1277–78, and with Uglich in the 1280s.

The fourteenth century saw the envelopment of Rostov by Moscow. In 1328, Rostov was divided into two halves, Sretenskaia and Borisoglebskaia. Sretenskaia was sold to Moscow in the middle of the fifteenth century. Borisoglebskaia remained quasi-independent until it too was sold to Grand Prince Ivan III of Moscow in 1474 and became part of the Muscovite state.

Rostov prospered during the sixteenth century as an important stop on the Moscow-to-White Sea trade route. Rostov also became the residence of the Russian Orthodox metropolitan between the years of 1589 to 1788.

In the seventeenth century the city was sacked and ruined twice by Polish and Lithuanian invaders. An earthwall was built between 1631 and 1633 to protect the Rostov Metropolia, or kremlin as it is called today, from further attacks. Inside the earthwall the kremlin was further protected by an additional wall and ten round towers. Inside the walls were a complex of houses, administrative structures, and religious buildings used mostly by the Russian Orthodox Church. These buildings were linked to one another by a series of arched passageways.

The shape of the kremlin was rectangular, occupying five acres in the center of town, with the rest of Rostov spread around it in a radial-semicircular manner. The kremlin's most unique feature was that it did not possess a dominant central cathedral, unlike most other Russian cities' kremlins.

Rostov was named district capital for the Yaroslavl vicegenerency in 1777. In 1796, it became the district capital for the Yaroslavl province. The city established an annual trade fair in the seventeenth century. It eventually grew into Russia's third largest trade fair, behind Nizhny Novgorod and Irbit, by the first half of the nineteenth century.

Much of the industry and manufacturing found within Rostov today can be traced to the late nineteenth century. Enterprises such as a flax spinning mill, a coffee and chicory factory, and a molasses plant have been in Rostov for more than 100 years. These industries were all later updated during the twentieth century after Rostov came under Soviet control. Other key industries found in the city include a clothing factory and an automotive parts plant. Rostov is also home to an agricultural technicum, a pedagogical school, and is a well-known center for traditional crafts—specifically, linen weaving and the painting of enamel onto copper objects.

In 1870, the city was linked to Yaroslavl and Moscow by rail, turning Rostov into an important railway junction. Today, a number of major highways also converge upon the city, making it an important highway transportation center as well.

In 1970, Rostov was declared a national preserve. Today, one finds many surviving architecturally significant structures both inside and outside the city's historic kremlin walls. Structures found within the walls of the kremlin are excellent examples of sixteenth- and seventeenth-century Russian architecture. Historic churches in the city include the Church of Ascension built in 1566. Also known as the Church of Isodore the Blessed, it is a five-domed building in the Moscow style of architecture. Uspenskii Cathedral, also called the Church of the Resurrection, was founded in 1214 and consecrated sixteen years later. Its icons date to the eleventh century. The stone exterior was updated in 1589 and frescoes were added in 1659 and 1670. There is also a bell tower from the 1680s. One of two gate churches found inside the kremlin's walls is the five-domed Voskresenie Church constructed in 1670. Frescoes for the church were completed five years later. The Church of St. John the Evangelist is the second kremlin gate church. Both its construction and its frescoes were completed in 1683. Spas na Seniakh Church, Church of St. Gregory, Church of Smolensk Mother of God, and Odigitriia Church were constructed in 1675, 1680, 1693, and 1698, respectively. Churches constructed between 1667 to 1691 fell under the direct supervision of the Russian Orthodox metropolitan Ion Sisoyevich.

Famous secular structures located inside the kremlin include Byelata Palata, or White Palace, constructed in 1670 for the Russian czars. It possesses a 3,230-square-foot hall and is now used as a museum. The Red Palace, also known as the Metropolitan's Palace, was built between 1672 and 1680. Ierarshaia Palace was completed that same century. Continuing farther out from the kremlin walls are additional churches dating from the sixteenth through nineteenth centuries, along with many stone homes from the late eighteenth and early nineteenth centuries.

Rostov was one of just a handful of key towns that helped form the modern Russian state. It was an important principality during a period in Russian history when the nation was affected by conflicting centripetal and centrifugal forces. Centripetally, the country was brought together for two centuries by one ruling entity, the Mongolian Tartars, and by a professed allegiance to the Orthodox Christian faith. Centrifugally, Russia was split apart by several sovereign and self-contained principalities, often engaged in political and military intrigues. Amid the confusion, Rostov became a hegemonic power in northeastern Russia and established a fort in Moscow. Moscow later took over Rostov and the rest of Russia and brought unity to a previously disunified state, leading to the formation of modern Russia.

Further Reading: J. L. I. Fennell's *The Emergence of Moscow, 1304–1359* (Berkeley and Los Angeles: University of California Press, and London: Secker and Warburg, 1968) deals with the rise of the grand principality of Moscow, but also does a good job of explaining how the succession process worked for the Rostovian princes. It also deals with some of the accomplishments of the individual princes. Paul Miliukov's *History of Russia: From the Beginnings to the Empire of Peter the Great* (New York: Funk and Wagnalls, 1968) is an excellent source for ancient Russian history.

—Peter C. Xantheas

Rouen (Seine-Maritime, France)

Location: Capital of the Seine-Maritime department, Upper Normandy, situated on the Seine River about thirty-one miles from its mouth at the English Channel, and about eighty-seven miles northwest of Paris.

Description: Developed from a Gallic settlement by the Romans into a major European city, and until the seventeenth century, a more important one than Paris. Now the capital of the Seine-Maritime department, an important port and industrial center. The old part of the town was extensively damaged in bombing raids during World War II but has been largely rebuilt.

Site Offices: In person:
Office de Tourisme
Syndicat d'Initiative
25 Place de la Cathédrale
76000 Rouen
France
35 71 41 77

By mail:
Office de Tourisme
Syndicat d'Initiative
B.P. 666
76008 Rouen Cedex
France

The original Celtic name for Rouen was Ratuma or Ratumacos; it was changed by the Romans into Rotomagus, and then by later medieval writers of Latin into Rodomum, from which the present name derives. Under the Roman emperor Diocletian (A.D. 245–313) it became the capital of a Gallo-Roman province and an episcopal see; much of its importance during that time was owed to its early bishops, from St. Mello (c. 260) to St. Remigius (d. 772). After the Romans withdrew from what is now known as France, the city was under frequent attack from the Normans (originally "Northmen"), but in 911 Rollo, the Norman chief, ratified a treaty with King Charles III of France at St. Clair sur Epte, thirty-one miles to the southeast of Rouen. He subsequently converted to Christianity, married the king's daughter, was created the first duke of Normandy, and made the town his capital. Its prosperity had always depended on river traffic, and trade now increased. After the conquest of England in 1066 by Rollo's grandson, William the Conqueror, the links with England were firmly established. William himself died in Rouen in 1087 after being wounded at Mantes.

Throughout the twelfth century, English kings were invested as dukes of Normandy in the Cathedral, after coronation at Westminster in London, but although Rouen was sometimes neglected by these kings in favor of nearby Caen and Poitiers, it maintained a flourishing trade with England, extending later to Flanders, Champagne, and the Hanseatic towns of Germany. With the marriage of Henry II of England to Eleanor of Aquitaine in 1152, Rouen became the administrative centre of a huge Anglo-Norman territory.

During the following century the rebuilding of the Cathedral of Notre Dame began. Its construction had been started in the twelfth century but it had been extensively damaged by a fire in 1200. It is now a fine example of French Gothic architecture, although it was not until the fifteenth and sixteenth centuries that the building took on its final shape, under the direction of the master builders Guillaume Pontifs and, later, Roulland le Roux, responsible for the central doorway of the West Facade. This facade of the Cathedral is distinctive because of the two quite different towers that flank it and its doorway. The northwest tower, the St. Romanus Tower, is the older, and its base rests on some of the few remains from the pre–twelfth century church; it rises from the Romanesque, through early Gothic, to the Flamboyant architectural styles (the last of which is a late Gothic term meaning "flamelike" and refers to the wavy window divisions, or tracery). The other tower, the Tour du Beurre (Butter Tower), was given this nickname in the seventeenth century because, it was believed, its construction had been funded by the sale of dispensations to people who did not want to forgo butter during the Lenten fast. It is topped by an octagonal crown, and houses a carillon of fifty-six bells. The Cathedral's most striking architectural feature, however, is its spire built atop the central lantern tower over the transept crossing. Rising to 495 feet, it is the tallest church spire in France, and dominates the city skyline. It was started in the thirteenth century, and in 1544 a wooden structure covered in gilded lead was raised; when this was, much later, struck by lightning, it was replaced in 1876 by the modern-day, cast-iron landmark.

Inside the Cathedral, similarly mixed architectural styles harmonize with one another. The early Gothic nave leads through the transept to the thirteenth-century chancel, which is light and simple. In the ambulatory beyond are the tombs of Rollo, his son, and Richard the Lionheart. The Lady Chapel contains the early Renaissance tombs of the Cardinals of Amboise, carved from drawings by Roulland le Roux. There are stained-glass windows from the thirteenth century to the Renaissance in the transept and the ambulatory, and the chancel chapel later dedicated to Joan of Arc has modern stained glass windows by Max Ingrand. It is this exuberant building with its towers, spire, and intricate openwork carving that the artist Claude Monet painted in a series of thirty canvases in the last decade of the nineteenth century, showing the Cathedral in various lighting at different times of day. These paintings are now dispersed in European galleries and

Rouen street scene, with the Gros Horloge at center
Photo courtesy of French National Tourist Office

American collections including Boston, Philadelphia, and the Smith College Museum of Art. There are other fine medieval churches in Rouen. The choir of the beautifully proportioned Eglise St. Ouen was built between 1318 and 1339, when work was stopped by the Hundred Years War (1337–1453) caused largely by disputes and complications arising from the possession by English kings of French territories. The nave was built during the latter part of the fifteenth century, and although the church was ransacked by Protestants in the Religious Wars in 1562, and by revolutionaries two centuries later, there still remains fourteenth-century stained glass in the great windows, together with seventeenth-century stalls, eighteenth-century choir grilles, and a nineteenth-century organ, one of France's largest. St. Ouen was originally a monastery, and its dormitory was made into the Town Hall following the revolution of 1789.

The Eglise St. Maclou was built between 1437 and 1517 in the Gothic-Flamboyant style; because of the speed of construction, it is more homogeneous than many other buildings of this time. Some of the panels of the west porch entrance are attributed to the sculptor Jean Goujon, as are the marble columns inside that support the organ case, decorated with remarkable Renaissance wood carving. Near the church are the St. Maclou cloisters that housed a medieval plague cemetery, and the upper floor of the half-timbered buildings was used as a charnel house until the eighteenth century.

During the Hundred Years War Rouen suffered much damage. In 1418 the town was besieged by Henry V and capitulated from starvation after six months. Plots against the successful English forces, known as the Goddons (from their habit of swearing "God damned") were thereafter frequent, and it was in Rouen that one of the major acts in the whole drama took place. Joan of Arc had inspired Charles VII, then uncrowned, to renew resistance to the English invaders. Although victorious in several battles, and despite Charles's coronation in Reims Cathedral, she was captured by the Burgundians at Compiègne and handed over to the English, after threats of economic reprisals, in return for a reward of 10,000 gold ducats. Starting Christmas Day 1430, she was held in the thirteenth-century fortress in Rouen, of which one tower now remains: the plain, circular Tour Jeanne d'Arc, the former keep of the castle built by King Philippe-Auguste. It was in this tower that she was tried by an ecclesiastical court over three months starting on February 21, 1431. On May 9 she was tortured and imprisoned in another tower, the Tour des Champs (Tower of the Fields), of which only the foundations are visible today. On May 24 she was tied to a scaffold in the cemetery of the Abbey of St. Ouen and forced to recant; she was reprieved, but condemned to life imprisonment. The English were, naturally, furious; the guards removed her women's clothes, and, being therefore obliged to wear male clothing, she was deemed to have broken the conditions of her confinement and was burned at the stake on May 30 in the Place du Vieux Marché. Her unburned heart was thrown into the Seine. Joan of Arc was rehabilitated in 1456, however, and in 1920 she was finally canonized and named the Patron Saint of France. The last weekend in May is now celebrated in Rouen as the festival of Joan of Arc.

Rouen was an early trading and manufacturing city and was the port upstream from Paris at the nearest point from the sea at which a bridge could be constructed over the Seine. It was therefore inevitable that the city was to become not only the major administrative and ecclesiastical capital of Normandy, but also a seafaring and industrial center. Rollo had had the river dredged and narrowed, reclaiming marshland and reinforcing the river banks with quays. This work ensured an extremely efficient port, and for many centuries, indeed, it was France's main one. Together with Le Havre and Dieppe from the fifteenth to seventeenth centuries, Rouen was known as the embarkation point for voyages of exploration and discovery; it was the Rouennais la Salle who, in 1682, sailed down the Mississippi and claimed possession of Louisiana in the name of the king, Louis XIV. Trade was always worldwide, not simply with Britain; as Rouen was a manufacturing town of linen and later woven silk, textiles formed a large part of its exports. Already in 1550 a Colonial Exhibition could be put on in the town. Tapestry was woven there, and in the early eighteenth century a cotton cloth, dyed indigo blue and called Rouennerie, became very popular; a decade or two later other textiles such as velveteen and twill were also produced. Associated industries such as the dyeing, finishing, bleaching, and printing of cloth, particularly with advancing technology, ensured Rouen's continued prosperity.

While trade flourished in the late fifteenth and sixteenth centuries following the defeat and expulsion of the English in 1449, so too did domestic and civic building. Rouen is famous for its half-timbered houses, and more than 700 of them have been restored after the damage caused by the bombs of World War II. The oak-framed houses, completed with lath and plaster, usually have projecting upper stories roofed with slates or small tiles. In the sixteenth century, inspired by the Italian Renaissance style that was introduced to Rouen by Cardinal Georges d'Amboise, minister to Louis XII and archbishop of Rouen, the facades became more highly decorated. Later, in the eighteenth century, they were to become plainer, and slightly later still they tended to be covered with plaster to resemble stonework. The streets around the Cathedral in the old part of the town have, of course, many of the most interesting of these buildings, and the rues St. Romain, Martainville, and Damiette are lined with well-restored examples of merchant and bourgeois houses. The rue St. Romain, bordering the Cathedral and leading to the Eglise St. Maclou, is one of the most fascinating. The Hôtel de Bourgtheroulde in the Place de la Pucelle is an early-sixteenth-century house built by Guillaume le Roux, a counsellor to the exchequer; now, appropriately, it houses a bank. Its architecture is part Gothic and part Renaissance, and in the inner court friezes surround the southern gallery; the lower frieze shows scenes from the Field of the Cloth of Gold, when Henry VIII of England and François I of France met in 1520 in great splendor. Guillaume le Roux's son, the Abbot Aumale, was present at this meeting.

The Gros Horloge, or Great Clock, was moved from the nearby Gothic belfry in 1527 to a specially constructed arch. Made in 1389, the clock has scarcely needed any modifications; it has a single hand telling the hours, a central section indicating the phases of the moon, and an inset showing the weeks of the year. In the belfry from which the clock was removed, a curfew bell still rings every night at nine o'clock.

During the sixteenth and seventeenth centuries, Rouen was a focus for political opposition to courts and governments based on the Loire or in Paris, and the city played an important part in both the sixteenth-century Religious Wars and the seventeenth-century unrest that led to the civil wars known as the Fronde.

The town's literary associations are numerous. Among the most famous of French authors to be born there were seventeenth-century playwright Pierre Corneille, at 4, rue de la Pie, where he lived for fifty-six years. The house, known as the "Petite Maison," is now a museum commemorating the dramatist, often considered the father of French classical tragedy. Although little of the original house remains (the cellar and stairs, the two side walls and part of the back wall), there is a reconstruction of an early-seventeenth-century study on the first floor, and on the second a library devoted to the dramatist. His play *Mélite* was first performed in Rouen in 1629. The Collège Bourbon that he attended from 1615 to 1622 was renamed the Lycée Corneille in 1873. Pierre's brother Thomas, also a prolific playwright, was born in the adjoining house, which has been demolished. A nephew of the two brothers, Fontenelle, was also born in Rouen and lived there for the first thirty years of his life. A writer of great learning, he was a member of the Académie Française, the Académie des Sciences, and the Royal Society of London. One of the best known of nineteenth-century novelists, Gustave Flaubert was born in the hospital Hôtel-Dieu, where his father was chief surgeon. Educated in the same school as Corneille, as also were the writer Guy de Maupassant and the painter Jean-Baptiste Camille Corot, Flaubert studied law unsuccessfully in Paris and spent most of the rest of his life at Croisset, six miles outside Rouen in the Normandy countryside. He was buried in the city. In *Madame Bovary,* his first novel, published in 1856, an important assignation between protagonists Emma and Léon is set in Rouen Cathedral. The Flaubert Museum (and Museum of the History of Medicine) is housed in the pavilion of the Hôtel-Dieu in which he was born.

Increasing industrial productivity in the nineteenth and early twentieth centuries required the modernization and enlargement of the port. Urban development ensued, and the town's population—90,000 in 1831 (the same as Bordeaux and surpassed only by Paris with 700,000, and Marseille and Lyons, each with 115,000)—now stands at around 400,000, including the conurbation that spread to the other bank of the river and the tributary valleys and hillsides. The port is now France's fifth busiest, and the third most important river port. It is the prime port for exports, especially for agricultural produce, and trades in wood pulp, chemical products, and the by-products of refined oil. The industries this trade supports include mechanical engineering, metallurgy, and chemical and wood processing. The industrial zone on the south bank has been rebuilt after the damage of World War II, and a semicircular Préfecture (regional administrative offices for the Seine-Maritime department) with a 262-foot skyscraper tower to hold archives was opened in 1966. The Seine is crossed at that point by the modern Pont Boieldieu (named for the opera composer François-Adrien Boieldieu, who was born in Rouen in 1775) below the Pont Corneille. Close by this bridge on the other, older side the Halle aux Toiles, or Linen Hall, has been partly rebuilt in modern style as a conference and exhibition center. The university, previously part of Caen University, has become independent, with strong scientific and medical faculties and a large student population. The campus has been built on the Mont-aux-Malades and the Mont-Saint-Aignan hillsides, part of the suburbs that overlook the town, port, and river.

Today, the historical center of Rouen can easily be explored on foot. In addition to the museums already mentioned there is the Musée des Beaux Arts near the Eglise St. Godard. The Musée de la Céramique houses a very fine collection of Rouen faience, a type of ceramic made of compound clay covered with a tin-based enamel. In the seventeenth century the decoration was mainly in blue and white, but later Dutch influence added red, and in the eighteenth century polychrome coloring appeared on the tableware. This museum is housed in the seventeenth-century Hôtel d'Hocqueville, close to the Musée des Beaux Arts. The Palais de Justice (Law Courts) again are not far from this central area; originally built to house the Exchequer of Normandy, the building became the Parliament under François I. It is said to be the work of Roulland le Roux and was rebuilt in the nineteenth century and again after World War II. In this part of the city, one final, and modern, memorial to Joan of Arc was completed in 1979: the Eglise Jeanne d'Arc in the Place du Vieux Marché, close to the spot where she was burned at the stake. The roof of this church evokes the flames that consumed her.

Further Reading: *Michelin Guide to Normandy* (Clermont-Ferrand, London, and Milltown, New Jersey, Michelin, 1993) is one of the best general guidebooks to the region. *Portrait of Normandy* by Derek Pitt and Michael Shaw (London: Robert Hale, 1974) is a more personal account of traveling in Normandy. *Guide Littéraire de la France* (Paris: Hachette, 1964) provides information for self-guided tours throughout France, arranged regionally. *Rouen: images et écrits* (Rouen: Imprimerie Rouennaise, n.d.) is a splendid collection of black-and-white photographs accompanied by extracts from the various writers associated with the city.

—Honor Levi

Runnymede (Surrey, England)

Location: In the northwest corner of the county of Surrey, on the banks of the River Thames just west of London. Runnymede is near the village of Egham and the city of Windsor.

Description: A largely rural portion of Surrey, one of the Home Counties (so named because they are adjacent to the County of London), Runnymede is known primarily as the location of the signing of the Magna Carta (Great Charter) between King John and the disenchanted barons of England. Generally considered to be the first constitutional document to effectively limit the power of a sovereign ruler, the signing of the Magna Carta in 1215 is among the most prominent events in English history. Fittingly, Runnymede has been bestowed with a succession of modern memorials to the principles of freedom and the rule of law, including the American Bar Association Memorial, the Commonwealth Air Forces Memorial, and a one-acre memorial to John F. Kennedy, bequeathed to the United States by Queen Elizabeth II.

Site Office: Warden of Runnymede
Langham Farmhouse
Langham Place
Egham, Surrey
England
(784) 470–194

A bucolic sweep of meadowland along the banks of the Thames, just far enough from London to escape the ravages of modern urbanity but just close enough to benefit from its cosmopolitan sophistication, Runnymede is to the evolution of democracy in Britain what the Big Bang is to the evolution of everything else: the beginning.

Composed primarily of Windsor Great Park and a long stretch of meadow along the Thames, the thirty-square-mile borough of Runnymede was historically one of the playgrounds of royalty, which kept the area well stocked with deer and other recreational game. As aesthetically pleasing as it may be, the topography of Runnymede is of little interest, consisting primarily of Coopers Hill, which overlooks the Thames and hosts two of the area's modern memorials. Unlike much of the British Isles, Runnymede is not known for its archaeological significance, either, bearing neither significant earthworks nor particularly revealing artifacts.

But unlike notable topography, which means little to anyone who has not been there, or fascinating archaeology, which interests only a small number of people, Runnymede symbolizes something very important to everyone who lives in a democratic society.

Today, Runnymede serves as a public park, allowing visitors to roam the very meadows in which King John and the angry barons hammered out their Great Charter in 1215. Picnicking is the favored pastime in the hallowed meadow, and modern monuments to freedom draw visitors from around the world.

King John, who ruled from 1199 to 1216, like most English kings up to his time, rarely, if ever, bothered to consult the barons and associated aristocracy from whom he collected the taxes he required to maintain his kingdom and lavish lifestyle. This tendency did little to create loyalty among the most influential of John's subjects.

Still, John was to reign over the opening years of a pivotal period in the history of England, a period that would include everything from John's warring, feudalistic domain at the dawn of the thirteenth century all the way to the first "united kingdom" under Henry VII by the beginning of the sixteenth. This period saw the transition of England from a feudal state with a nascent parliament, at best, to a near-modern nation-state with a parliament that, if not extremely powerful, was certainly there to stay.

Although this process of democratization began at Runnymede in 1215, events leading up to the signing of the Magna Carta began much earlier. John was the third of the Plantagenets to ascend to the throne, succeeding his brother Richard I in 1199 under less than ideal circumstances. Although backed by the barons of England and Normandy, John's claim to the throne was seriously disputed by the leaders of the lands of Anjou, Maine, and Touraine, who favored the accession of Arthur I of Brittany, son of Geoffrey (the late brother of Richard and John), and then a small child. John's behavior prior to accession only fueled the concerns of his critics.

Born on December 24, 1167, John was the youngest son of Henry II and Eleanor of Aquitaine. Born without lands of his own, John relied upon the favoritism of his father to gain a degree of power after coming of age. For this purpose he was chosen to succeed to the earldom of Gloucester and granted the lordship of Ireland.

In 1185 Henry sent his favorite son to Ireland, and it was during this brief tenure that John first gained a reputation for selfishness and immaturity. "There met him at Waterford a great many of the Irish of the better class in those parts," wrote the contemporary historian Giraldus Cambrensis, at the scene for the entirety of John's visit. "Our newcomers and Normans not only treated them with contempt and derision, but even rudely pulled them by their beards, which the Irishmen wore full long, according to the custom of their country."

In a prelude of things to come, John's insolent attitude caused several Irish chieftains to avoid him, and caused the

American Bar Association monument to the Magna Carta

Memorial to John F. Kennedy at Runnymede
Photos courtesy of Royal Borough of Windsor and Maidenhead

kings of Connacht, Thomond, and Desmond to actually form a defensive league against him. His seemingly capricious award of lands did not make him any friends, either, and Henry called him back to England after only eight months.

John's prominence increased dramatically following his brother Richard's accession to the throne in 1189. Although a popular, heroic king, Richard I, known as the Lion-Hearted, spent all but a few months of his reign abroad, either fighting in the Crusades in the Holy Land or battling for his territories in France, where he was killed in 1199.

During his brother's absentee reign, John was confirmed as lord of Ireland and granted extensive lands in England. He also married Isabella, the heiress to the earldom of Gloucester that he had been promised (their marriage was dissolved in 1199 because John and Isabella were blood relations; both being great-grandchildren of Henry I). John was also forbidden to enter England during his brother's absence, for fear that he would try to usurp power for himself.

In October of 1190, Richard formally recognized the three-year-old Arthur as his heir, infuriating John and forcing him to return to England, where he immediately took up opposition to Richard's chancellor and de facto king, William Longchamp. When Richard was imprisoned in Austria in 1193, John went so far as to ally himself with Philip II (Philip Augustus) of France for the purpose of seizing control of all England for himself. His original plan failed, but John still made arrangements with Philip to split up Richard's possessions and foment rebellion in England.

Richard returned to England in 1194 and immediately banished his brother and took away his ill-gotten lands. By 1195 John had managed to recover some of his lands, including Ireland, but still was not happy until Richard recognized him as his heir after Arthur was surrendered to Philip II by the Bretons in 1196.

John was crowned king in May 1199, one month after his brother's death, and began a reign that many historians have branded inept. Some have even called it evil. However, the facts indicate that John was an involved administrator with a solid understanding of law and management, even if his egocentricity was legendary.

In all fairness, John was doomed to a difficult reign. The extraordinary expense of Richard's campaigns in the Holy Land, along with the ransom extracted by Austria, were taking their toll on the landed aristocracy upon whom John depended for financial and political support. The barons grumbled about what they perceived as the growing inefficiency of the royal government, which seemed always to want more and more money, and had begun to threaten their independence. The culprit, of course, was inflation—a phenomenon commonly understood today, but not yet considered in the thirteenth century. Furthermore John was duty-bound to try his best to maintain British possessions in France, despite the fact that Philip Augustus obviously wanted them back.

Beset from the beginning with significant problems that would only grow more complex with time, John attacked his situation with characteristic bluster and immediately made things worse. In 1200 he married Isabella of Angoulême in an effort to placate the French territories, but instead fomented a war that resulted in the loss of all of England's continental possessions by 1206.

Determined to put his continental disaster permanently behind him, John turned his attention to domestic matters, namely, replenishing his nearly empty treasury. To restore a positive balance, John resorted to every fundraising device at his command, including new taxes, customs duties, forest-law fines, taxation of Jewish moneylenders, and all manner of feudal dues and obligations. From the barons' perspective, John, who had lost the French possessions, was asking a great deal.

John further alienated the barons through his complete indifference toward the Catholic Church. In 1205 Archbishop Hubert Walter died, prompting the monks and bishops of Canterbury to send their own delegations to Rome to seek confirmation of their own archbishops, assiduously bypassing John. John protested to Pope Innocent III, who declared that election of the new archbishop was in the hands of the monks of Canterbury, but that they were duty-bound to choose a successor pleasing to their king. The monks chose their prior, Reginald, but John chose John Grey, bishop of Norwich.

When the pope bypassed the matter by appointing English scholar Stephen Langton, who was then living in Italy, and consecrating him in 1207, John simply refused to recognize Langton. This act led to six years of tension with Rome, during which time England was placed under interdict and John was excommunicated. John also took the opportunity to seize church lands and revenues. To make matters worse, Philip Augustus was preparing to invade England, with the backing of the pope.

In 1213 John capitulated, going so far as to surrender his kingdoms of England and Ireland to the pope in order to receive them in return for his pledge to be a papal vassal. It was the price John felt he had to pay in order to secure the support of the pope in his long-running battle with Philip.

John may have been faced with serious problems from the very beginning, but to the barons he was a greedy, inept, and self-centered ruler who cared little for their welfare. The barons also were offended by John's overt disregard for time-honored rules and the role of the church in English life. Something had to give.

By 1212 the state of relations between John and the barons had deteriorated to such a degree that John had begun to take possession of castles and demand proof of goodwill from barons, usually in the form of hostages. In 1213 a group of northern barons refused to provide their share of manpower for a planned French campaign, arguing that they had already fulfilled their obligation for the year and that John's demands were, therefore, excessive and unlawful. Only intervention by Stephen Langton prevented civil war. In February 1214 John advanced upon France anyway, but returned to England by October, defeated again.

It was at this point that the barons decided to take collective actions against John, who they felt was running roughshod over their most cherished liberties. The first blow came when John sent his justiciar, Peter des Roches, to collect scutage—a tax levied in lieu of military services—from those barons who had not assisted with the French campaign. The attempt was met with open opposition. Furthermore, the basis of the barons' argument, according to them, was the coronation charter of Henry I, which in 1100 expressly spelled out the rights of barons regarding obligatory "work-service" in defense of the realm.

The confrontation grew progressively more hostile. In April 1215 a group of the most outraged barons took their displeasure directly to John, who recommended that they submit their grievances to a special council consisting of four members from each side of the issue, with the pope as president. The barons rejected the offer, leading John to muster mercenaries in Ireland in preparation to crush the rebellion by force, if necessary. Only intervention by the more moderate barons, led by Stephen Langton, prevented calamity.

Langton argued that the matter should be discussed rationally, with the two sides meeting at a predetermined place. While John and his advisers worked from Windsor, the barons set up shop first at London, and later at Staines, in what is now suburban London, near Windsor and Runnymede.

At Langton's behest, the parties met at Runnymede on June 15, 1215, where John agreed to accept the forty-nine clauses of what was then called the Articles of the Barons. Lawyers assigned to the negotiations rewrote the hastily prepared documents, adding fourteen clauses and making the sometimes vague language more precise. This document they called the Charter of Liberties, which John signed. It is not known exactly when John signed the charter, but it is known that the conference was concluded on June 23.

Although the original text of the document was written in continuous form, modern scholars have generally divided the document into a preamble and sixty-three separate clauses roughly divided into nine groups. The first group affirmed the right of the English church to be free from royal intervention. The second group defined the rights of those holding lands directly from the crown, while the third section passed similar rights on to subtenants. The fourth, fifth, and sixth groups dealt with trade, law, and administration of justice, and with the behavior of royal officials. The seventh group, which dealt with the use of forests, was eventually turned into a separate document, and the eighth dealt with purely immediate issues. The ninth and final section established a council of twenty-five barons charged with keeping an eye on the king, and making sure that he adhered to the covenants of the charter.

Of the sixty-three clauses enshrined in the original charter, it is clause 39 that has taken center stage in history. As translated into modern English, the clause reads:

No freeman shall be arrested, imprisoned, dispossessed, outlawed, exiled or in any way deprived of his standing, nor shall we proceed against him by force or send others against him, except by the lawful judgment of his equals and according to the law of the land.

Clause 39 was the first explicit guarantee of the rights of the individual. To a modern reader, this may seem revolutionary within the context of 1215, but at the time, the clause meant relatively little. Few people were free and the subclause about "judgment of his equals" meant exactly what it said, but within the highly stratified, class-conscious society of its day. The Magna Carta was not written by barons honor-bound to defend the freedoms of the masses. It simply put into writing the personal freedoms and responsibilities commonly accepted in English society at the time, so that they could not be arbitrarily dismissed by the sovereign.

At first, John changed his ways and followed the charter, issuing several writs to address pressing complaints. However, after the angry barons who had first forced his hand disbanded their army, John began to disregard the Charter. He was supported in his behavior by Innocent III, who called the charter "shameful, debasing, illegal and unjust." Innocent even forbade John to follow it or his barons to enforce it. When the pope summoned Langton to Rome to discuss the matter, the rebellious barons prepared to wage war against their king, even seeking help from the French and offering the English throne to Philip's son, Louis. John reacted quickly to put down the rebellion, but his death in October of 1216 was what really saved the day.

John's heir, Henry III, was not of age at John's passing, but his regent, William Marshal, reissued the charter shortly after John's death, and again in 1217, with minor revisions. It was in 1217 that the charter received the name by which it has been known ever since. Lawyers removed clauses 44, 47, and 48 to use as the basis for a new Charter of the Forest, and, to distinguish the two charters, named the first one Magna, from the Latin for "great."

In 1225 Henry III published the Magna Carta under his own seal, omitting several clauses that were no longer relevant or, as in the case of clauses 25 and 61, severely limited the power of the crown. It was this version that was adopted as statute by Edward I in 1297 and became something of a constitution: not so much a collection of laws as a sourcebook of legal principles.

Copies of the Magna Carta were sent out to all major population centers in 1215. Only four of these survive: two in the British Museum and one each in the cathedrals of Lincoln and Salisbury.

The Lincoln copy was actually in the care of the Library of Congress in the United States during World War II. Having first been sent to New York in April 1939 to be put on exhibit at the World's Fair, the threat of German bombing six months later convinced British leaders of the need to keep the document in a safe place until after the war. Upon handing the document over to U.S. officials, the British ambassador Lord Lothian said:

The principles which underlay Magna Carta are the ultimate foundation of your liberties no less than ours. Samuel Adams appealed to the rights of Magna Carta. . . . It was in their name that your ancestors threw the tea into Boston Harbor and rejected the claim of King George III to tax the colonies for defense.

The meadow of Runnymede has not changed significantly since 1215, but is home to several memorials to the principles enshrined in the Magna Carta. In 1953 the British government erected the Commonwealth Air Forces Memorial in honor of the commonwealth airmen who fought for Britain in World War II.

In 1957 the American Bar Association erected a monument to the Magna Carta at Runnymede in recognition of its status as the first document to officially uphold the sanctity of law. In May 1965, 750 years after the signing of the Magna Carta, Queen Elizabeth II bequeathed an acre of land at Runnymede to the people of the United States in honor of President John F. Kennedy.

Further Reading: A wealth of information exists on the Magna Carta and its ramifications, much of which is purely legalistic in nature. However, *A Documentary History of England,* volume one (Harmondsworth: Penguin, and Magnolia, Massachusetts: Peter Smith, 1966) by J. J. Bagley and P. B. Rowley collects fourteen of the major defining documents of English history between the years 1066 and 1540, including the Magna Carta. Explanations are short and cursory, but valuable to the lay reader. Summary explanations of the Magna Carta can also be found in most encyclopedias. *The High Middle Ages* (London: Routledge and Keagan Paul, 1986) by Trevor Rowley covers the years 1200 to 1550 and is part of the acclaimed five-part *The Making of Britain* series, which, together, can put any aspect of British history into an identifiable context.

—John A. Flink

St. Albans (Hertfordshire, England)

Location: An old cathedral town twenty miles north of London, southeast England.

Description: St. Albans is the third important town built on its site on the River Ver, and includes landmarks dating back more than 2,000 years. Dominating the landscape is the Cathedral and Abbey Church of St. Albans, built on the spot where England's first Christian martyr was executed.

Site Office: Tourist Information Centre
St. Albans Town Hall
Market Place
St. Albans, Hertfordshire AL3 5DJ
England
(727) 864511

The early history of England is a history of incursion. Hardly an isolated island kingdom, the country endured one invasion after another. From 8000 B.C., when Stone-Age people were crossing a land bridge, which disappeared into the North Sea about 5000 B.C. through the invasions by Neolithic tribes from the south, Celts, Picts, Belgics, Romans, Angles, Saxons, Jutes, Vikings, and others fought for domination of the island. Most were assimilated, adding their works, their cultures and their languages to the character of the land. The hills along the River Ver, north of modern London, have attracted these peoples, and the current city of St. Albans is the third important town to be built on the site.

Little is known of the first of these settlements, though the Belgic tribes that spread through the area in the first century B.C. left earthworks in nearby Prae Wood. They settled in the south and east of England, using the more efficient farming implements developed on the Continent to extend agriculture from easier-to-cultivate areas with lighter, more easily drained soils to heavier lands characterized by forests or marshes.

Julius Caesar led the first Roman raids on Britain in 55 and 43 B.C. as part of his effort to secure Rome's hold on Gaul. For the next four and one-half centuries, Rome held varying degrees of control over the country.

Even before Claudius conquered southern Britain in A.D. 43 and initiated Rome's attempts, never completely successful, to conquer the whole of Britain, the second settlement, Verulamium, had begun to grow by the Ver. By modern standards, it and other Roman settlements were not large, but Roman techniques of planning and construction made them the first settlements we would recognize as towns. Verulamium grew to become the third largest city in Roman Britain, covering 200 acres. London, the largest Roman regional center, would cover 330 acres.

In about A.D. 7, coins were minted at the site of Verulamium by Cunobelinus, who was immortalized by Shakespeare as Cymbeline. With the military attention of Rome elsewhere, Cunobelinus established a kingdom centered in Colchester.

Verulamium developed into a self-governing settlement composed to a large extent of members of the Roman legions whose time in military service had expired and who had retired to civilian life. It became the first, and the only certain, example of a British town given the privileged status of Municipium. Municipia were native cities occupied by Rome and raised to chartered status, largely governed by Roman citizens and with inhabitants having rights as citizens of Rome.

Small shops were among the first buildings put up in Verulamium and the oldest included a wine shop, a bronzesmith and a carpenter. Extensive ranges of small shops, rented from larger landlords, soon flourished.

In A.D. 61, Verulamium was sacked by the warrior queen of the Iceni, Boudicca, whose forces also destroyed the commercial and administrative centers of Colchester and London, effectively paralyzing for a short time Rome's government and cutting its communications with the Continent. Boudicca's revolution began after the death of her husband, Prasutagus, a client king of the Romans. When he died, he left his kingdom jointly to his two daughters and the emperor. Though this split was meant to protect his family, local officials treated the territory as if it had been left completely to the empire and proceeded to plunder the kingdom, evicting its nobles from their ancestral homes and treating members of the royal house as slaves. When Prasutagus's widow protested, she was flogged and her daughters were raped. She roused her own tribe and the neighboring Trinovantes into rebellion.

The Britons, led by Boudicca in her chariot, were initially confronted by a portion of Rome's Ninth Legion under the command of Q. Petillius Cerialis. Boudicca's forces cut the opposing infantry to pieces and Cerialis escaped with only his calvary intact. The governor, C. Suetonius Paulinus, decided to retreat to London, mistakenly believing that troops waited there to help turn back the attackers. A Roman writer of the period, P. Cornelius Tacitus, wrote that the governor "decided to sacrifice the one town (Colchester) to save the general situation. Undeflected by the prayers and tears of those who begged for his help, he gave the signal to move, taking into his column any who could join it. Those who were unfit for war because of their sex or too aged to go or too fond of the place to leave, were butchered by the enemy."

The Roman towns at the time were largely defenseless, without walls and with few regular troops. Boudicca's

The Abbey and Cathedral Church of St. Albans
Photo courtesy of St. Albans City and District

force, estimated as high as 230,000, quickly overran Colchester and proceeded toward London, a city without administrative status but a flourishing trade center. Tacitus's account continues: "A similar massacre took place at the city of Verulamium, for the barbarian British, happiest when looting and unenthusiastic about real effort, bypassed the forts and the garrisons and headed for the spots where they knew the most undefended booty lay. Something like 70,000 Roman citizens and other friends of Rome died in the places I have mentioned. The Britons took no prisoners, sold no captives as slaves and went in for none of the usual trading of war. They wasted no time in getting down to the bloody business of hanging, burning and crucifying. It was as if they feared that retribution might catch up with them while their vengeance was only half complete."

Boudicca's forces were finally met in the Midlands by Roman legions and were defeated by the more disciplined troops. That battle also turned into a massacre in which, Tacitus reported, 80,000 Britons—men and women—were killed, as were 400 Romans. Boudicca escaped, but died a short time later, from illness, according to some accounts, or by poison, according to others.

To the north, unconquered tribes remained a threat, as they would until the empire crumbled and the tribes moved toward the south. Hadrian's Wall, the first and most ambitious of the bulwarks intended to hold back the Picts, was built between 122 and 126, even as towns in the well-settled south flourished.

At Verulamium reconstruction was, at first, sluggish. It was fifteen years before rented shops reappeared, but after they resumed operations, they continued to operate with little change for the next seventy-five years. Then, in approximately 150, the shopkeepers began to buy up their leases and, when they needed new quarters, to rebuild in separate buildings, though with a common covered arcade.

In 155, the city was swept by a disastrous fire. An estimated fifty-two acres of buildings were consumed as the flames spread through the mostly timber-framed wattle-and-daub private houses and shops. Rebuilding, drawing on the resources of what had become a prosperous city, began immediately. Construction, as was normal in the empire, drew heavily on private funds both for public works and individual enterprises. The destruction actually may have assisted Verulamium's development by clearing away old buildings and leaving the way clear for new infrastructure, including new streets laid in some areas.

Much of the town was rebuilt as large town houses, suggesting that the retail trade moved away from its center, even though demand for goods and services increased as the gentry moved in. Smiths and jewelers flourished, and a municipal market building, or macellum, was built. Retailers who had acquired their leases sold their now prime sites at a substantial profit and set up business elsewhere.

At the same time, construction methods changed. Builders relied more on fire-resistant masonry and solid clay walls. The wood-framed houses had been far from spare, featuring boldly painted interior plaster walls, but in the new buildings, wooden floorboards were replaced with pictorial or geometric mosaics, which became very popular and continued to be laid into the fourth century. More important from a technological standpoint, the new construction allowed the old brazier system of heating to be replaced by a system called a hypocaust, which used flues to distribute heat generated underground.

The center of the city boasted a theatre and directly behind it was an elaborate Romano-Celtic temple, making up what in other parts of the Roman Empire was a common grouping: a temple or sanctuary, a theatre, and a piazza, making up a Forum. Verulamium's temple was deserted in about 380.

While no longer sluggish overall, development of some amenities remained relaxed and unhurried. With its citizens perceiving that conditions in the surrounding area were largely peaceful, they built grand gates toward the end of the second century but delayed construction of the stone walls that would have joined them until the third century. In the interim, earthworks topped with timber ramparts were developed as temporary defenses.

Verulamium became grander through the second century. Expansion continued despite periods of civil war, economic recession, and political upheaval that took place well into the third century and eventually led to a crumbling of the city's infrastructure and support systems.

Toward the end of the third century, the threat to lands controlled by Rome expanded. To the hazards of the northern frontier was added the danger of Frankish and Saxon pirates on the east and south coasts and attacks by the Scots tribes of Ireland on the western shores of Wales and Dumnonia.

At the beginning of the fourth century, Verulamium was extensively rebuilt with government assistance as part of an effort by Constantine to streamline provincial government and revitalize the towns. By A.D. 350, however, it had reverted to the disrepair and squalor that had been evident a century earlier. The city, except for the area around its civic center, was deserted, street gutters were blocked, and the theatre was used as a garbage dump.

Still, Verulamium outlived many other towns. Standards of Roman culture and taste were for a time maintained in isolated rural villas which exploited an increasingly barbarous and uncooperative peasantry. After 360, marauding Picts, Scots, and Saxons held sway, and town populations were swelled by refugees fleeing the villas as they became less secure. Despite changes in appearance and standards, organization and functions; despite the recall of Roman forces by the Emperor Honorius in 410, effectively ending Rome's presence in Britain; and despite the demise of sophisticated commerce by 450, by which year minting of new money had long since ceased and even worn coinage was no longer circulating, Verulamium remained populous and active. Excavations have uncovered a fifth-century corn-drying plant built atop one of the town houses. It was used long enough to need repairs before being replaced by a large stone barn or public hall. This last building came down and a new water main was laid across the site in about 450. Major municipal improvements under an active city council were evident as late as 460.

In addition to its secular role, Verulamium figured prominently in the development of Christianity in Britain. In 209, on the hill outside the city now crowned by St. Albans Cathedral, a Roman soldier was executed, becoming the first Christian martyr in the country. Alban, a Roman solider and a pagan, gave shelter to a fugitive Christian priest during a persecution of Christians and was baptized by him. When soldiers came to Alban's house, he exchanged cloaks with the priest and was taken by the soldiers to the local magistrate. Questioned about his new beliefs, Alban replied, "I am called Alban and I worship and adore the true and living God who created all things." He was taken out of Verulamium, across the river and up the hill, and beheaded. The tale took on mythic proportions as the centuries passed. In *Ecclesiastical History* written about 730, the historian Bede reported that a church had been built on the site of the execution and that numerous works of healing had occurred. By the thirteenth century, when Matthew Paris, a monk at the Norman Abbey later built on the site, wrote his account, the priest was identified as Amphibalus (cloak) the heavens were filled with wondrous signs, and the headsman who killed Alban was smitten down and then restored to life by touching Alban's body.

In 312, the Emperor Constantine made Christianity the official religion of the Roman Empire, but belief ebbed as the empire declined and heresies attempted to push English belief away from the Roman church. By 500, pagan Saxons, Angles, and Jutes had migrated from northern Europe and spread over eastern and central England, and by 600 they occupied much of the island. Thus, while Christianity flourished in Ireland, Wales, and southwestern England during the fifth and sixth centuries, it became almost extinct in the rest of England, except for a few strongholds, such as at the site of the Saxon village that succeeded the decayed Verulamium.

In 429, a mission from Gaul was sent to England by Pope Celestine I to combat British heretics. Led by Germanus, bishop of Auxerre and later sainted, the mission debated proponents of the heresy at a public meeting in Verulamium and, after swaying the crowd to orthodoxy, became part of the flock of pilgrims who, over the years, visited the shrine of St. Alban. Germanus's visit effectively reinforced an activity—the veneration of martyrs—which the heretics opposed. In 601, Pope Gregory I strengthened Christianity's position in the southeast, installing Augustine as the first archbishop of Canterbury, home of the king of Kent and his Christian wife.

King Offa II of Mercia founded and endowed with land a Benedictine abbey on the site of Alban's martyrdom in 793. He had Alban's bones moved into a more splendid sepulchre.

Christianity suffered again under the next wave of invaders, the Vikings, who between 876 and 937 developed a distinct kingdom of their own. However, they, too, adopted Christianity and by the tenth century were developing new monasteries.

In about 990, St. Albans Abbot Wulsin laid out a new town north of the abbey, centered on a long, tapering marketplace. The abbot also built three churches, including St. Michael's on the site of Verulamium's basilica. This tenth-century church retains much of its Saxon character and contains the tomb of Sir Francis Bacon, whose family lived nearby.

In 1077, the Normans started construction of an abbey on the site of Offa's church, using, in the material-poor region, Roman bricks gathered from ruins in Verulamium. The first Norman abbot, Paul of Caen, rebuilt the entire monastery except the bakehouse and buttery. It was finished in 1088 and dedicated in 1116. His abbey church makes up most of the central part of today's cathedral. The interior was plastered and painted, though paintings done in the thirteenth and fourteenth centuries were later covered with whitewash by the Puritans.

St. Albans Abbey rose to wealth and power in the Middle Ages, with the monastery numbering 100 monks during the thirteenth century. Starting in 1154 and lasting until 1396, its leaders were the foremost abbots in England, decreed so by a papal bull issued by Adrian IV, born Nicholas Breakspeare in St. Albans and the only Englishman ever to become pope. During that time, the church building was expanded. The abbey had become rich enough to import stone, which was used between 1200 and 1235 to build the east end and west bays of the nave.

Late in the period, when Thomas de la Mere was abbot, the abbey's influence was substantial. The abbot was a friend of Prince Edward, called the Black Prince because of the color of his armor. The prince was the son of King Edward III, whose aspirations to the French throne led to the Hundred Years War between England and France. The abbot was also a privy councillor and friend of the French king John, who was captured by the Black Prince at the battle of Poitiers in 1356 and sent to St. Albans where he was held for ransom.

With the dissolution of the monasteries in 1539, the abbey's buildings, including St. Alban's shrine, were surrendered to King Henry VIII and destroyed. The church was sold for £400 to its parishioners who, for the next 300 years, struggled to keep in repair a structure far too big for their needs. They gradually withdrew from much of the building until they were using only the presbytery and part of the north transept.

Parts of the shrine were used to patch up the church and, during the major restoration begun in 1856, these were recovered. In 1872, more than 2,000 fragments were pieced together to reconstruct the elaborately carved marble base of the shrine that had been a major destination of medieval pilgrims. During this period, a new west front and transept windows were added and whitewash was removed from the paintings.

In 1877, the enlarged building became the cathedral of a new diocese, which includes more than 300 churches in the counties of Bedfordshire and Hertfordshire.

Starting in 1955, the Norman paintings were restored and additional paintings were found and, in 1982, a new chapter house, built in brick like the Norman church, was opened by Queen Elizabeth II.

Further Reading: A variety of books are available on Roman Britain and the development of England's towns, including *The Oxford Illustrated History of Roman Britain,* by Peter Salway (Oxford and New York: Oxford University Press, 1993), and *The Making of English Towns, 2000 Years of Evolution,* by David W. Lloyd (London: Gollancz, in association with Peter Crawley, 1984).

—Richard Greb

St. Andrews (Fife, Scotland)

Location: St. Andrews lies on a bay on the east coast of Scotland, flanking the North Sea to the east and the Eden river to the north. The town is twelve miles southeast of Dundee and just under fifty miles from Edinburgh.

Description: A town with medieval origins, individual buildings of historic interest, spacious seascapes, and ample green park areas that are now carefully preserved. Until well into the twentieth century St. Andrews was a medium-sized fishing port. Scotland's oldest university, St. Andrew's University is located here. The town claims to be the original home of the game of golf. It has four publicly owned eighteen-hole courses and is the seat of the sport's rule-making body outside North America and Mexico. The twentieth century has seen the university expand and the town become a resort.

Site Office: St. Andrews and North-East Fife Tourist Board
70 Market Street
St. Andrews, Fife KY16 9NU
Scotland
0334-472021

Between St. Andrews and the Eden estuary to the north is the forest of Tentsmuir, now a bird sanctuary, where excavation has shown inhabitation by nomads in about 8000 B.C. By 4500 B.C. there were farming, pottery, and stone axes, and by 2500 B.C. weaving and metal-working was performed in the area, followed by carved designs on stone around 2000 B.C. Traces of cremations and burials suggest continuous habitation of the town site by Celtic-speakers for most of the last millenium B.C. From about A.D. 80, the local Venicones tribesmen appear to have lived amicably with the Roman invaders, who constructed a row of forts across Fife. The Picts appear to have gained control of the St. Andrews site until about A.D. 850.

The Picts were converted by St. Columba's Irish monks, who had settled at Iona around 565. They then professed allegiance to the missionaries sent by Gregory I in 596 under Augustine of Canterbury, who had been granted authority over all the Celtic churches, but had had difficulty in imposing the Roman Easter. The Celtic monks were expelled by the Picts in 717, and a link was formed between St. Andrews and the Northumbrian church, which followed Roman usage.

A religious community nonetheless grew up out of an assemblage of Celtic anchorites in the early ninth century, and by 921 they had become a canonical body of thirteen. A church was built for them at St. Andrews before 877. It was later replaced by another structure on the rocky headland known as Kirkhill, where ruins of a twelfth-century nave and thirteenth-century choir can still be seen. By 1144 there appear to have been two bodies of clergy, the Celtic Culdees and the canons regular, following Roman usage and a relaxed Benedictine rule. It was only in 1273 that the Culdees, less and less Celtish, were excluded from episcopal elections, although individuals among them, like William Wishart, bishop in 1272, continued to hold high positions in the thirteenth century. Their last church was pulled down in 1559.

The first known name of the site is Cennrigmonaid or Headland of the King's Mount, which was Latinized into Kilrimont, changing "Kin" (headland) into "Kil" (church). Various versions of the legend in which a Saint Regulus or Rule miraculously brought to St. Andrews the relics of the apostle Andrew, brother of Simon Peter, suggest a relatively commonplace attempt to provide apostolic origins for a local church. This was done to guarantee the orthodoxy of the doctrine transmitted to it and, where relics were involved, to establish a justification for pilgrimages, which could be lucrative and certainly conferred prestige. St. Andrews excelled both Canterbury and Iona in eminence. The 108-foot-high, eleventh-century tower in the cathedral grounds at St. Andrews was not referred to as St. Rule's or St. Regulus's until around 1500. At one time, there was a small church to house the relics, with the enormous square tower with fourteen-foot sides to proclaim their presence.

By 975, the diocese of St. Andrews included land stretching as far as the Tweed, and the bishop was the seniormost in Scotland. Only scraps of information about tenth- and eleventh-century bishops were recorded before Malcolm III Canmore surrendered to William the Conqueror in 1072. Thereafter, the Scottish church was under the ecclesiastical authority of the Norman archbishop of York. By 1107 St. Andrews had a Norman bishop; and by 1238 the see was the largest and wealthiest in Scotland. Robert, the prior of Scone, was nominated bishop in 1124 and consecrated in 1127. By 1144 he had endowed the priory, whose canons regular were made into the electoral chapter for the see in 1147. The present ruins date from the middle of the thirteenth century.

The cathedral itself was founded in 1160 and consecrated in the presence of King Robert the Bruce in 1318. With a nave shortened to 358 feet after the first west gable blew down, it was the biggest church in Scotland. However, it was abandoned in 1559 as a result of the schism. The whole ecclesiastical complex extended across the Pends over the land now occupied by St. Leonard's School buildings and grounds. The remaining walling is from the sixteenth century, part of a continuous wall reconstituted by Prior John Hepburn and enclosing some thirty acres. It rose at places to twenty feet, was three feet thick, and was fortified with thirteen towers. Pope Sixtus IV elevated the see into an arch-

The Royal and Ancient Clubhouse

St. Mary's College
Photos courtesy of St. Andrews and North-East Fife Tourist Board

bishopric in 1472. The religious sentiment of John Knox was used by the nobility to excite the citizenry and to provide an excuse to loot the cathedral, the priory, and especially the graves, of all that had not already been removed by their custodians.

The castle, part fortress and part episcopal palace, was probably begun in 1200, although the extant ruin dates mostly from the sixteenth century. The castle was taken several times by English forces and burned down. It was rebuilt in 1385 with its famous bottle dungeon, which may have been a grain store. James I was educated in the castle presided over at its moment of greatest glory by Archbishop James Beaton and his nephew and successor, David Beaton.

Catholic forces aided by the French recaptured the castle after a siege in which the present mine and counter-mine were created. By 1559, the castle was again in Protestant hands, and in 1587 it passed to the crown. It was a prison for a while, but was then dismantled in the seventeenth century, and the materials used to repair the harbor and pier.

James VI made St. Andrews into a royal borough in 1620. Of the two main roads, both running from east to west, South Street led directly to the cathedral. The other, North Street, led to the castle area. The merchants' houses of South Street had long strips of land leading to the burgh's edges, and, although the town was never walled, all the riggs and wynds were closed with ports or gateways. The only surviving gate, the West Port, was rebuilt, partly to a new design, between 1843 and 1845.

The university arose from the need for immense numbers of educated clerics, hitherto trained in France, required to administer the huge diocese. The need did not become acute until the beginning of the fifteenth century, when the senior clergy, under bishop Henry Wardlaw with the support of James I, informally established a body of teachers. They were chartered in 1412 by Wardlaw to keep their earnings free of tax and to protect them from civil jurisdiction. The original bulls enabling the corporation to grant teaching licences in the usual undergraduate discipline of philosophy and the four graduate disciplines of civil law, canon law, theology, and medicine were issued by the antipope Benedict XIII in 1413. The corporation originally met in the priory, but gradually acquired property of its own, with the result that St. Andrews is today one of the few genuine university towns in Europe. There is no campus. With some exceptions, the casual visitor may not easily discern which buildings in the town form part of the university, and which do not.

The university was refounded in the mid–fifteenth century, when Bishop James Kennedy established the college of St. Salvator, which still occupies its North Street site. The foundation was confirmed by a bull of Pope Nicholas V in 1451 and by a new charter of Pius II in 1458. Most of the present buildings are of nineteenth-century construction, although the street frontage is medieval.

The university's most attractive premises are in South Street, where St. Mary's College incorporates a fine range of buildings originally erected in the sixteenth century. Half a dozen large student residences are scattered throughout the center of the town and on the outskirts. Much of the science teaching takes place in the large buildings on the North Haugh, on the outskirts of the town and opposite the golf courses, although there are distinguished science departments in the town center. Not surprisingly, given its position, the university has established a reputation for marine biology. Unfortunately, it has not established a reputation for any architectural distinction in its modern buildings.

The town has become increasingly dependent economically on its golf courses, which are the source of some controversy. Golf as a sport emerged from a number of games played with sticks and balls, and its origins can be traced to 1413. It was almost certainly first played elsewhere, as the game was not mentioned at St. Andrews until 1552. James II had forbidden it in 1457. It did, however, become popular at St. Andrews, where it was played on common ground, in coarse grass growing on sand dunes on which sheep roamed in search of nourishment.

The Old Course fascinates players not only because of its history, but also because it is exceedingly difficult to play. Even on a fine, windless day it strongly favors the play of long, low, and accurate shots. However, the course is entirely unsheltered from the sea, and the weather's mood changes are sudden and extreme. Quite savage winds and storms give only minutes' warning, so that a dedicated championship player who is more or less independent of mere luck can often score very differently by starting as little as an hour earlier or later.

In 1754 a number of local patricians formed a club to play an annual challenge, and in 1834 the patronage of William IV allowed the club to be known as The Royal and Ancient. The members met at local inns, until a clubhouse was built in 1854. Although it is an entirely private club limited to 1,800 members, all of whom must be male, and who join by invitation, the regulations of the rules committee set up in 1897 have been accepted throughout substantial parts of the golfing world. The Royal and Ancient now legislates for the sport and negotiates rules where its writ does not run, runs the Open and Amateur of the British Isles, and built the New Course in 1895. The town built the Jubilee course in 1897, the Eden course in 1912, and the new Strathtyrum, opened in 1993. The nine-hole Balgove course was built in 1971. The famous Old Course was reduced from twenty-two to eighteen holes in 1836, and was free until 1913, when fees for visitors were instituted. Only in 1946 were charges first made for local rate-payers. In 1993, visitors played 80,000 rounds, while locals played 100,000. The Links Trust revenue for that year was £2.25 million.

The current controversy stems from a feeling on the part of the citizens of St. Andrews that they have been deprived of their common ground, as much their own public amenity as the beaches. The common ground is controlled by a Links Trust, although ownership is vested in the district council. The Links Trust has been accused locally of eroding the privileges of residents, but also of allowing the five eighteen-hole courses and the nine-hole course to be overplayed, damaging the turf and jeopardizing the course's suitability for hosting the Open tournament every five years.

The effect on the local economy may be exaggerated but cannot be insignificant, and since many local residents are now dependent on the influx of golfing visitors for their livelihood, the local debate is complex. The danger is not so much the erosion of local rights as the over-dependence of so small a community on the sport, even with an internationally famous university and the town's manifest attractions as a center for historical tourism.

The university students hold an annual procession through the town dressed as characters from the past, and have other arcane ceremonial occasions on which their presence is allowed to disturb the business of the town. A noisy annual Lammas Day Fair takes place on the second Monday of August.

Further Reading: Apart from brochures, guides, gazeteers, and encyclopedia articles, St. Andrews figures largely in histories of Scotland and of golf, and in memoirs of prominent persons connected with both. The history of the university has been treated by formal historians such as R. G. Cant in *The College of Saint Salvator* (Edinburgh: Oliver and Boyd, 1950) and *The University of Saint Andrews,* (Edinburgh: Scottish Academic Press, 1971) has plundered the records. Recommendable and undemanding general works include *Saint Andrews: City of Change* (1984), *A Visitor's Guide to Saint Andrews and the East Neuk* (1985), and *The Life and Times of Saint Andrews* (1989), all written by Raymond Lamont-Brown and published by John Donald of Edinburgh.

—Claudia Levi

St. David's (Dyfed, Wales)

Location: On the southwest peninsula of Wales, the village of St. David's lies on the Alun river about one mile from the Irish sea

Description: Site of one of the earliest Christian monasteries in Wales, St. David's was founded by Wales's patron saint in the sixth century. The smallest cathedral city in the United Kingdom, it was and remains the seat of the bishopric in Wales.

Site Office: The St. David's Peninsula Tourist Association
The City Hall
St. David's, Dyfed SA62 6SD
Wales
(437) 720392

Symbol of the evangelizing work of the Celtic saints, St. David's in south Wales was founded by the monk and bishop Dewi (David is the nearest English approximation of his name). Since the twelfth century, he has been regarded as the patron saint of Wales. Throughout the Middle Ages, his shrine was one of the most sacred sites and was frequently visited by pilgrims; two pilgrimages to St. David's were considered equivalent to one pilgrimage to Rome. David is the only Welsh saint to be canonized and culted in the western church. According to legend, David was the son of a Welsh chieftain and St. Lon. He established his monastery of Mynyw (Menevia) sometime in the sixth century. The monastery was said to be one of the strictest and David was known as the "Waterman," perhaps for his abstinence from wine or other alcoholic beverages. Irish monks of the eighth and ninth centuries were the first to record the name of Dewi and the monastery and church at Mynyw in their martyrologies.

In the ninth and tenth centuries, St. David's became the ecclesiastical center of Wales. Its monks were renowned scholars; when King Alfred the Great sought to revive learning in Wessex in the 880s, he turned to Asser, a monk of St. David's, and the cult of St. David became a national one. Despite its reputation as a sacred site, it was not spared the ravages of the Viking raids. It was abandoned for seven years after its buildings were plundered and destroyed. On the eve of the Norman invasion in the eleventh century, St. David's was again inhabited. It was the cultural center of Wales and many traditions, such as permitting Welsh churchmen to marry, separated it from its Anglo-Saxon counterparts.

William the Conqueror came to worship at St. David's shrine following the end of hostilities between the kingdom of South Wales and the Norman invader. It seems unlikely that William came only to worship at the shrine. His visit enabled him to establish peaceful relations with Rhys ap Tewdr, ruler of southwest Wales, as well as to oversee the Norman crossing to Ireland.

Norman encroachment upon Welsh lands began almost immediately. It was only a matter of time before the bishopric was placed in Norman hands. When the bishopric became vacant in 1115, it was given to a Norman, Bernard, the chancellor of the queen.

Bernard, the episcopate from 1115 to 1148, displayed a great deal of respect for Welsh traditions. He was primarily responsible for the reorganization of the Welsh diocese into archdeaconries and deaneries. Careful not to offend local sensitivities, he organized them along the boundaries of the ancient Welsh kingdoms. He also sought to protect St. David's status and ancient privileges. In 1119 he secured from the papacy the canonization of St. David and the formal recognition of his cult. He argued for St. David's independence from England as an archbishopric and suggested it be the center of an independent province.

Bernard's successors did not continue his attempt to establish St. David's independence from Canterbury. Many of them derived much of their power from connections with the English marcher lords along the English–Welsh borderlands. It is not surprising, then, that the bishopric itself was recognized as a marcher-lordship and its fate became closely associated with the English baronage.

The dream of Welsh independence, however, was not entirely abandoned. It was renewed by Gerald of Wales, nephew of Bishop David fitz Gerald, Bernard's successor. Gerald was a man of letters, an archdeacon of the church, and historian. A descendant of Welsh princes and Anglo-Norman marcher baronage, he was educated in England and later in France. He was appointed archdeacon by his uncle. Having failed in his bid to succeed his uncle he served Bishop Peter de Leia as commissary before being appointed as court chaplain to King Henry II in 1184. With the death of Peter de Leia in 1198, he was elected bishop by the chapter of St. David's. His selection, however, was opposed by the new king, John, and by Hubert Walter, archbishop of Canterbury, unleashing a bitter struggle.

Gerald opted to take his claim for Welsh ancient rights directly to Pope Innocent III. He attempted to persuade the pope to make St. David's independent of Canterbury. Gerald believed that his uncle and Peter de Leia had been prevented from upholding their ancient rights when they were coerced into recognizing the authority of Canterbury. After four years of struggling, Gerald of Wales eventually capitulated to the English and recognized Geoffrey of Henlaw, Walter's nominee, as the new bishop of St. David's. He retired to Lincoln where, for the remaining twenty years of his life, he devoted himself to his writing.

In spite of Gerald's attempt to gain greater autonomy for St. David's, the Welsh see greatly benefitted from its close association with the kings of England. In the thirteenth

The interior of St. David's Cathedral
Photo courtesy of Preseli Pembrokeshire District Council

and fourteenth centuries, a succession of royal officials were appointed as episcopates. The rebuilding of the cathedral at St. David's was the main preoccupation during this period. Parts of the present cathedral date back to Peter de Leia's first undertaking in 1180. The original design called for a cruciform building with an aisled nave and presbytery, a flat east end and a lower tower rising above pitched roofs. Subsequent reconstruction has left only the nave intact. In 1220, the tower collapsed and was replaced with a new one in a more pointed style. Successive additions include the chapel of St. Thomas, the ambulatory, and Lady Chapel by Bishop Martyn, all of which extended the cathedral plan to its present-day size.

In 1328, Henry de Gower was elected to the episcopate. He belonged to a noble family and distinguished himself with his knowledge of several languages. He is better known, however, for his building activity. In addition to rebuilding the aisle walls of the nave, he built the close wall, added the lantern stage of the tower, and constructed the pulpiturum in which he is buried. He is best remembered for the Bishop's Palace constructed under his auspices.

The Bishop's Palace consists of three ranges of buildings, enclosing a rectangular courtyard. The open arcade and parapet that run atop the whole building on the outer side are characteristic of Bishop Gower's influence. The entrance to the palace was through a vaulted gateway marked by buttresses. To the east of the gateway lay the Bishop's Chapel with its familiar arcaded parapet. Farther east along the wall lie the Bishop's Hall and the Solar, both of which were originally designed for an open roof. To the south of the Bishop's Hall lies the kitchen and the Great Hall, which was generally used for public purposes.

Bishop Gower died in 1347, and a number of his successors held royal appointments along with the bishopric. John Thoresby was keeper of the privy seal and, later, chancellor. Reginald Brian had been a royal clerk; his successor Thomas Fastoff had served as ambassador, and Adam Houghton, also renowned for his building activity, though not at St. David's, was chancellor. John Gilbert was not only royal confessor but also served as treasurer. Guy Mone was keeper of the privy seal and, later, treasurer. The close association between the English crown and the bishops of St. David's ended with Henry Chicele, who served from 1408 to 1414.

Much of the decline of St. David's after the fifteenth century may be attributed to the defeat of the Welsh independence movement. In 1406 Owen Glendower led a Welsh revolt against English rule. A descendant of the Welsh princes of Gwynedd, he sought to revive the hopes for Welsh independence. In addition to the restoration of the Welsh kingdom, Owen sought to restore the Welsh church to its former glory. This meant that the Welsh church should be independent of the English church. In 1406 the Great Schism within the Catholic Church, with competing lines of popes, enabled him to pursue such a claim. In return for his recognition of the Avignon pope Benedict XIII, St. David's was to be recognized as a metropolitan center, and as such could administer its own affairs free of the interference from Canterbury. The defeat of Owen in 1409 by Henry, then Prince of Wales, and later King Henry V, not only heralded the end of the Welsh dream of independence, it also heralded a general decline for the Welsh church. In the fifteenth century, the appointment of Welshmen to Welsh sees, or to royal offices, became very rare. Having utterly defeated any Welsh resistance, the English crown tended to appoint local men from the English marches whose connections tended to be local rather than national. St. David's was to suffer a period of decay and decline, and not until the rise of the Tudor monarchs was it again to have any favor with the English crown.

Some construction continued at St. David's. In the sixteenth century, the unique wooden oak ceiling of the nave was completed. At the same time Bishop Vaughn, who served from 1509 to 1522, was building his Trinity Chapel, which was to serve as his burial place. Much of what had been constructed in the medieval period was allowed to decay following the Reformation. The lead from the roof of the Great Hall was stripped away by Bishop Barlow, who served from 1536 to 1548 and wanted to abandon St. David's in favor of Carmarthen. Slowly, St. David's was entirely abandoned. The Bishop's Palace was last used in 1633.

During the English Civil Wars (1642–48) the buildings were further damaged when Commonwealth troops vandalized the windows, organ, bells, and library. By the end of the seventeenth century, the palace was roofless and derelict. Steps to fully restore the palace and cathedral to their former state of grandeur did not come until the nineteenth century, when Sir Gilbert Scott undertook the task. The main arcading and aisle walls of the cathedral were dangerously out of line and the tower about to collapse at any time, so Scott regarded the restoration of St. David's as the most hazardous commission he had ever undertaken. Today the preservation of the restored cathedral and the palace ruins is under the guardianship of the Ministry of Public Buildings and Works, and the village of St. David's has become a tourist center. The cathedral is the only one in the United Kingdom in which the reigning monarch has a permanently reserved space. Each year in the last week of May the cathedral hosts an acclaimed classical music festival. The remains of St. David are interred in the cathedral. Nearby attractions include the saints birthplace and numerous holy wells and preaching stations associated with him and his followers.

Further Reading: David Walker's *Medieval Wales* (Cambridge and New York: Cambridge University Press, 1990) provides a good discussion of the role of the Welsh church in the Middle Ages. Readers also may wish to consult David W. James's *St. David's and Dewisland: A Social History* (Cardiff: University of Wales Press, and Mystic, Connecticut: Verry, Lawrence, 1981). Nona Rees's *St. David of Dewisland: Patron Saint of Wales* (Llandysul, Dyfed: Gomer, 1992) is a study of the ancient Welsh cult. J. Wyn Evans's *St. David's Cathedral* (London: Pitkin, 1991) is an overview of the building of the cathedral.

—Manon Lamontagne

Saint-Denis (Seine-Saint-Denis, France)

Location: In the northern suburbs of Paris, about seven miles from the center of the capital, on the Seine and the Canal de Saint-Denis.

Description: A Gallo-Roman settlement from the first to the third centuries A.D., now an industrial suburb; the site is particularly known for its ties to the kings of France and for its twelfth- and thirteenth-century basilica, formerly the abbey church.

Site Office: Office de Tourisme
1 rue de la République
93200 Saint-Denis
France
42 43 33 55

The religious buildings and town of Saint-Denis took their name from St. Denis, patron saint of France, and their long history has been intimately linked from its beginnings with the history of the nation. Christianity gave the place its identity when Gaul was still ruled from Rome; later a close, durable spiritual link was forged with the French royal dynasties. As the settlement grew and prospered, Saint-Denis added wide economic and political power to the preeminence it drew from its religious and royal connections. In modern times, by an ironic contrast, the town of the royal abbey became the anticlerical "red Saint-Denis," a citadel of French socialism and Communism. Its role in national history is now augmented by a pleased awareness of its multicultural present.

The legendary St. Denis we are concerned with here—several other martyrs of that name appear in early church history—is the result of a conflation of three different historical men, living at three different periods, all referred to in early texts by the Latin form of their name, Dionysius. The first of these, known as Dionysius the Areopagite, is mentioned in the Acts of the Apostles as having been converted to Christianity in Athens by St. Paul. Secondly, Gregory of Tours in his sixth-century *History of the Franks* refers to an Italian Dionysius, Christian missionary to Gaul and first bishop of Paris, who was tortured and executed in the middle of the third century. In the late fifth or early sixth century, another Dionysius was the author of the *Celestial Hierarchy,* a mystical treatise on the heavenly host and divine light. In the ninth century these three personages were fused by Hilduin, an abbot of Saint-Denis, in his *Historia Sancti Dionysii,* and since then the legendary French St. Denis, the bringer of Christianity to Gaul, combines in himself the traits and deeds of these three men.

This compound St. Denis is supposed in older versions of the legend to have been sent by Pope Clement I, after the death of St. Paul, on a mission to convert the pagans in first-century Gaul, and to have become the first bishop of Paris, although modern scholarship places the first Christian settlement in Paris and the first bishop in the third century. When the Roman emperor—either Diocletian (emperor from 284 to 305) or Valerian (emperor from 253 to 260)—ordered the persecution of Christians, Dionysius, with his legendary companions Rusticus and Eleutherius, is supposed to have been arrested in Paris, imprisoned and tortured and, still refusing to deny his faith, executed on the slopes of what is now Montmartre. According to the legend, Dionysius picked up his severed head and walked with it, while chanting psalms, to a spot two miles away where he fell to the ground; he was buried in a field, under a simple monument, by Catulla, a Christian noblewoman.

Excavations have shown that a Gallo-Roman village, Catulliacum, once stood where the abbey of Saint-Denis was built, with a large pre-Christian and Merovingian (roughly mid-fifth to mid-eighth century) cemetery on and adjacent to the site of the future abbey church (now the cathedral). Archeological remains and early texts indicate the third-century burial in this cemetery of a Dionysius probably martyred not on Montmartre but at Catulliacum. A series of chapels and churches dedicated to St. Denis were built through the centuries on the spot where oral tradition had located the original burial site and relics of the saint.

The first authenticated structure was the mortuary chapel built at the end of the fifth century or the beginning of the sixth (the year is conventionally given as 475) on a site probably abandoned, by order of St. Geneviève, patron saint of Paris, who died in 512. This was a simple rectangular basilica-style edifice with masonry walls, using Gallo-Roman stone blocks in its foundations and setting the east-west orientation followed in all subsequent churches at Saint-Denis. Traces of St. Geneviève's chapel have been uncovered in modern excavations of the present building. It measured roughly 65 feet by 33 feet. (The present cathedral is 354 feet long, 121 feet wide, and 95 feet high.) Within the next century, a religious community grew up around the chapel. The shrine of the saint became the most venerated one in the north of the realm; various miracles were said to have taken place there. From the sixth century on, noble, then royal, burials were carried out in the church, until by the end of the tenth century it had become unusual for a French king to be interred elsewhere.

In the seventh century the Saint-Denis community's association with the kings of the Franks became very close. St. Denis was adopted as the royal patron saint, and royal gifts and benefactions brought wealth, power, and privileges to his abbey. This rise in importance was particularly marked during the second quarter of the seventh century, with the

The basilica at Saint-Denis
Photo courtesy of Office de Tourisme, Saint-Denis

reigns of the Merovingian king Dagobert I (of nursery rhyme fame) and his son Clovis II. Dagobert enlarged St. Geneviève's church at Saint-Denis, where he eventually died (c. 638) and was buried. Clovis and his wife, St. Bathilde, founded the abbey of Saint-Denis in about 650 by making the existing community into a regular monastic order following the rule of St. Benedict and St. Columba. The custom was established of depositing copies of royal documents at the abbey, thus endowing its library with the annals of early French history and perhaps helping to form the basis of the school where, among others, the young kings Pépin the Short and, much later, Louis IV studied. The royal regalia (scattered or melted down during the Revolution) were held at Saint-Denis, along with the *oriflamme,* the red banner of France, mentioned in the *Song of Roland* and carried into battle by the marshal of the armies of France. In 1152 St. Denis would be named by proclamation the patron saint, the leader and protector, of all France.

Of the abbots of Saint-Denis who acted as advisers and ambassadors to the Carolingian kings (751–987), the most interesting is Fulrad (died 784), who constructed, keeping to the original rectangular basilica form, a larger and completely new church at Saint-Denis (begun in 750 and consecrated in 775 by Charlemagne) to replace the old one, found to be too small for the then recently adopted complete Roman liturgy. Fulrad's church, measuring roughly 206 feet by 75 feet, appears from documents to have had round arches resting on marble columns, but nothing now remains of it above the present level of the paving (which is a couple of feet above the level of his floor, so that some of his column bases have been discovered during excavations). The abbey's prosperity suffered a setback with the Norman invasions in the second half of the ninth century, when several times the monks had to take the treasures and relics and seek refuge elsewhere. Important domains and privileges were lost and not fully regained until the twelfth century.

The most notable architectural development at Saint-Denis was the move away from Romanesque style with the construction by Suger (abbot from 1122 to 1151) of what is generally considered to be the first Gothic church. Suger's spacious building embodied in stone, with soaring pointed arches and vaults and large stained-glass windows, the theme of divine light suggested by the *Celestial Hierarchy* attributed to the composite St. Denis. The Gothic style introduced in Suger's new *abbatiale* would dominate Western church architecture until the Renaissance.

Starting at the west end, retaining the nave of the old Carolingian basilica, Suger gave the abbey church a crenellated façade with a rose window and three splendid carved portals that are still standing, though much degraded by eighteenth-century alterations that included the removal of a figure of St. Denis, and by the nineteenth century repairs and restorations of Debret, Viollet-le-Duc, and others. Suger's three pairs of bronze doors were melted down in 1794. The western end as planned by Suger had two towers, of which only the southern one survives: when, for his restoration of

the north tower, Debret used stone much heavier than the original, it became unstable and in 1847 had to be dismantled. At the eastern end, parts of Suger's semi-circular nine-bay choir, with some of their stained glass, remain, and, supporting them below, the round-arched crypt, now heavily restored, that he built over and around the ancient shrine. In the thirteenth century a new nave and transepts with large rosc windows were built in the *rayonnant* or refined Gothic style by Pierre de Montreuil; the chapels on the north of the nave are fourteenth-century work. The fabric of the Gothic abbey church was by that time substantially complete, and further changes would result only from modifications, accidental or deliberate damage, repairs, and restorations.

Surviving all the church's architectural avatars has been the national interest in its royal tombs and effigies. Since the end of the tenth century all but four of the kings of France have been buried here (the absentees are Philip I, Louis XI, Louis-Philippe, and Charles X). During the Revolution, among other mutilations of the building, the royal tombs were opened and their human remains thrown into a pit. These bones, including those of the wives of the Stuart kings Charles I and James II, are now in an ossuary in the crypt. The monuments were removed to Paris and after their return on the orders of Louis XVIII they suffered restoration.

Among the most beautiful and interesting of the many tombs and memorials to be seen at Saint-Denis are those of Dagobert I (died c. 638) in the ambulatory; Philip "the Bold" (died 1285), by Pierre de Chelles and Jean d'Arras (in the west end of the choir); Louis XII (died 1515) and Anne of Brittany, by Giovanni di Giusto (north aisle); Francis I (died 1547) and his first wife Claude de France, by Philibert Delorme and others (in the form of a triumphal arch; south transept); Henry II (died 1559) and Catherine de Medicis (died 1589), by Primaticcio, Germain Pilon, and others (north transept). The resting places of the hearts of Francis II (died 1560) and Henry III (died 1589) are marked by tall free-standing columns. After the restoration of the Bourbons (1815), Louis XVIII, making up burial arrears caused by the Revolution, had the remains of Louis XVI and Marie-Antoinette ceremonially interred in the crypt, where he too lies.

Many other historical personages and events are associated with the abbey, its church, and the town. In the twelfth century, Peter Abelard took refuge at the abbey. In 1429 Joan of Arc dedicated her armor in the church, and in 1593 Henry IV abjured the Protestant faith in whose name during the Wars of Religion the first Prince de Condé had taken the town in 1567, only to be defeated some months later by connétable Anne de Montmorency on the Plaine Saint-Denis in the south. The town of Saint-Denis was already accustomed to war, having been over the centuries besieged, sacked, or occupied by Normans, Burgundians, and Englishmen. In 1871 it was bombarded, besieged, and occupied by the Prussians (the Musée d'Art et d'Histoire has a unique collection of documents on the Commune), and in 1944 it was bombed by the Allies in the course of the Liberation struggle in which its own citizens participated. It was the birthplace in 1895 of Paul Éluard, the surrealist, Resistance, and Communist poet.

While its church was preserved in the Gothic form, the abbey's other buildings, and the monastic community, went gradually into decline. By the beginning of the eighteenth century the abbey dependencies were old and dilapidated. Demolition and reconstruction of the chapterhouse, refectory, dormitories, and cloisters were undertaken in slow stages, and completed in 1786 with a large semicircular front courtyard, only for the whole soon to be converted by Napoléon I to its present use as a school for the daughters of members of the order of the Légion d'Honneur. The privilege of wearing its blue uniform is still keenly sought after.

As the church and abbey gradually assumed their modern forms, the little Gallo-Roman religious community had grown and spread to include a secular population. Saint-Denis's position always favored such expansion: it is situated near the capital, on the main routes, whether Roman road, river, canal, railway, or motorway, to the north; and to the south, above the flood-level of the Seine, lies a plain. The fertile soil here, as in the rest of the Île-de-France, was suitable for agriculture, to which in the Middle Ages was added cloth manufacture and dyeing with natural colorants—trades helped by the regular fairs and the availability of clear rivers and streams running through the town. At the end of the eighteenth century there were still about fifteen water-mills strung out along the River Croult flowing westward to the Seine. The Industrial Revolution came slowly to France but, by the middle of the nineteenth century, to the textile trades and the production of associated chemicals were added (facilitated by the existence nearby of the abattoirs of La Villette) tanneries and factories for making soap and stearine. The Canal Saint-Denis was opened in 1821.

Then between 1850 and 1880 came the construction of the railway network linking the region to the coal mines of Belgium and northern France. Textiles rapidly gave way to heavy industry, with large factories producing iron and steel, military and naval equipment, boilers, gas, chemicals, leather, glassware, pianos (Pleyel), silverware and jewelry (Christofle), etc. The mushrooming of large factories and small workshops and the influx of low-paid labor without the provision of suitable infrastructures produced industrial squalor for much of the population, which had risen from less than 4,000 in 1801 to more than 60,000 by 1901. Urban sprawl encroached upon the rural setting and gradually filled the space that had separated Saint-Denis from Paris and from neighboring communities that made up the girdle of suburbs outside the ring of the old fortifications.

The politicization of the work force that started seriously between about 1889 and 1892 resulted at first from a two-way resentment between Paris and its suburbs more than from a conflict between capital and labor. Class solidarity developed later, after years of intermittent struggle and protest, with Communism taking hold in the late 1920s. Jacques Doriot, Communist mayor of Saint-Denis in 1931, was expelled from the French Communist Party in 1934; drifting more and more toward fascism, he founded the French Popular Party (PPF) in 1936, and was a collaborator during the German occupation of France in World War II. Industrial Saint-Denis, its economic and social problems now addressed systematically and with long-term planning, continues to be the symbol of the workers' movement and is still sometimes called "la Ville Rouge."

Further Reading: *Saint-Denis la ville rouge 1890–1939* by Jean-Paul Brunet (Paris: Hachette, 1920); *The Royal Abbey of Saint-Denis from its Beginnings to the Death of Suger 475–1151* by Sumner McKnight Crosby, edited and completed by Pamela Z. Blum (New Haven, Connecticut, and London: Yale University Press, 1987); and *Roissy-Express* by François Maspero (translated by Paul Jones, London: Verso, 1994), all provide valuable treatments of various aspects of the town's history.

—Olive Classe

St. Peter Port (Guernsey, Bailiwick of Guernsey)

Location: On the eastern side of the island of Guernsey, forty-two miles from the Cherbourg and seventy-five miles from the English coastline.

Description: The chief town and administrative center of the Island of Guernsey, a self-governing commonwealth owing allegiance to the British Crown.

Site Office: Guernsey Tourist Board Information Centre
P.O. Box 23
St. Peter Port, Guernsey GY1 3AN
(481) 723552

St. Peter Port is the main town of the Island of Guernsey and the administrative center of a group of the Channel Islands— Guernsey, Alderney, Sark and Herm—which collectively form the political unit called the Bailiwick of Guernsey, an autonomous commonwealth with its own unique legal and legislative system dating from the Middle Ages. Until recently the island also possessed distinctive cultural features such as its own old Norman dialect, which remains in the names of streets and houses. The bailiwick is geographically part of a larger archipelago known as the Channel Islands. St. Peter Port is also one of ten Guernsey parishes, each of which has its own administration.

Since Roman times St. Peter Port has existed on its present site, the best natural harbor in the Channel Islands. Christianity arrived in the sixth century from Brittany. Later, in the eleventh and twelfth centuries, the Channel Islands formed part of the Duchy of Normandy, which had passed to the English Crown. Guernsey's unique status began with the loss of Normandy to the French in 1204. The island, with its larger neighbor Jersey, continued to offer allegiance to the English king John. In return, the king gave the inhabitants the privilege of electing their own magistrates, while the monarch continued to exercise sovereignty as de jure duke of Normandy. The special status of Guernsey was enhanced in 1483 when Pope Sixtus IV granted the Privilege of Neutrality in time of war, which enabled the island to continue to trade with both France and England while they conducted a war against each other. Guernsey remains a self-governing commonwealth offering formal allegiance only to the British Crown and not the British government.

St. Peter Port is the commercial and administrative center of Guernsey, deriving its importance from the sheltered position of its harbor. It was first referred to a "ville" in 1275; Edward III ordered that a town wall be erected in 1350, but because of the cost this command was not obeyed. At the king's insistence, however, some kind of defensive wall was almost certainly built. The only visible sign of the medieval walls that remain are Les Barrières de la Ville. These are six stones erected in 1700 on the orders of the Royal Court to mark the spots of the original gates. Their erection was probably prompted by the laws of inheritance, which required firm boundaries to establish property held within the town. These stones then established a firm legal boundary for the town. In 1840 the boundaries were changed to accommodate the extensive growth of the town and a change in the inheritance laws.

Since its inception the town has been an important shelter for shipping along the English Channel. The sea brought trade and opportunities that were reinforced by the Papal Bull of 1483, which granted the Privilege of Neutrality. Guernseymen could freely operate as smugglers and privateers, exploiting their location to prey on foreign shipping. Real prosperity only came later during the American Revolution and Napoleonic Wars.

The harbor is the raison d'être of the town, giving it an ongoing strategic importance. Edward I ordered the construction of a new pier there in 1275; although he stipulated how the money for its construction should be collected, the collectors held onto the funds themselves. Elizabeth I sought to circumvent this problem in 1563 by ordering the Royal Court of Guernsey to be responsible for raising the money. These funds were used to build the South Pier of the Old Harbor in 1590. When Guernsey became a center for storing wine, the North Pier of the old Harbor was also built. Quays were constructed in 1775 and an arch built over Cow Lane, from which to load cattle onto ships. But the harbor gradually became incapable of sheltering all visiting ships, and in the 1840s it was decided to implement a plan proposed by the engineer James Rendel to expand the facilities of the port. Following intense pressure from merchants, building began on the present harbor in 1853 and lasted until 1909; among the new facilities were a bathing point constructed at Les Terres in 1863. The entire cost of the harbor expansion was £360,000, which included the creation of the North and South Esplanades to replace shingle beaches.

The one outstanding example of the earlier St. Peter Port is Castle Cornet, which was built as the medieval garrison to guard the harbor and protect the sea route from England to France. Undoubtedly Castle Cornet's impregnability was a considerable factor in the rise of St. Peter Port's commercial success. It was built on an islet to one side of the harbor and is cut off from the town at high tide. First constructed in 1206, the castle was enlarged by the French when they captured the stronghold in 1338 and occupied it for seven years. It was later renovated around 1540 to accommodate cannon. The longest period of siege and conflict took place during the English Civil War, when the garrison remained loyal to the king, but the island supported Parliament.

The Hauteville House, where Victor Hugo lived

German soldiers who occupied Guernsey in World War II
Photos courtesy of States of Guernsey Department of Recreation and Tourism

Parliamentary forces put the castle under siege from 1643 to 1651; it was the last Royalist stronghold to surrender.

The greatest destruction to the castle occurred in 1672, when the magazine was struck by lightning and exploded. The original keep was not repaired. The most interesting feature of the castle is its complex plan, designed to confuse any intruder, and as the Civil War siege proved, the castle was remarkably easy to defend. It continues to offer excellent views of St. Peter Port, and its seventeenth-century gardens are still maintained. They were planted and designed by General John Lambert, Cromwell's deputy, who was imprisoned here after the restoration of Charles II.

The other important building that remains from the town's earliest days is the Town Church, which stands overlooking the harbor in the center of town. The existing building dates from the fourteenth and fifteenth centuries. The spire was built in the eighteenth century. The original church was given by Duke William in 1048 to the Benedictine Abbey of Marmoutier, to which it belonged until the Reformation. The nave is the earliest part of the existing building, to which were added the north and south aisles and the chancel and central tower. It is probable that the church was frequently damaged during the Civil War. Later the building housed the militia's artillery and the island fire engines brought from London in 1768. The monuments in the church are an important guide to the history of Guernsey families over the past 200 years, and some of the inscriptions are written in French. During World War II a bomb dropped by the Royal Air Force blew out almost all the windows of the town, including those of the church.

St. Peter Port evolved because of its excellent harbour, but the nearby terrain is impractical for building because of its steep bank. Houses climb in tiers up the hillside overlooking the port. The town church marks the earliest place of settlement and the oldest quarter. In the eighteenth century many of these early buildings were demolished to make way for a more planned town. Inevitably, because it was built on the same terrain, the new town retains its cluttered character. But on the plateau above the town the newly affluent could build their comfortable villas.

The town hospital to house the poor was built in 1743 around a quadrangle. Its completion established a new precedent for the town. St. Peter Port is unusual in that there is no record of any guild there during the Middle Ages, and so no almshouses exist on the island. But even with the new town hospital, social problems continued. In the 1830s there was a cholera epidemic, made worse by the tight lanes and streets in which the residents lived. This calamity prompted important changes to the structure of the town and a more active approach to town planning. Houses were demolished to widen streets, and drains were built. As the town expanded, the old town became more of a commercial district and the area surrounding the town church was cleared to make a square.

The great era of Guernsey development and opportunity stemmed from the prosperity of the privateering and free trade era between 1770 and 1820, which coincided with the French Revolution and continental wars. With growing affluence, the residents could also afford to fund civic improvements. The best example is the Assembly Rooms, erected in 1780 with finance drawn from twenty of Guernsey's leading families. Here John Wesley preached on his visit to Guernsey in 1787. More commonly, balls and parties were held in these rooms; membership was restricted to about sixty families, the island aristocracy who came to be known as the Sixties. Membership requests had to win two-thirds approval, and officers of the garrison were the only non-natives admitted. Subsequently the nouveaux riches wanted to join this social elite but were excluded. These people came to be known as the Forties, and a running social feud persisted through the nineteenth century between these antagonistic groups. The Assembly Rooms were designed as a traditional market house with nine granite bays, and beneath was an arcade intended for use as a market. They were purchased by the States (Guernsey's parliament) in 1870 and then sold to Mr. Guilles and Mr. Alles, who turned the space into a library to house the books that the two men had collected. The arcade below continues to function as a marketplace.

In 1799 the Plaiderie, meeting place of the States, was replaced by the existing building, the Royal Court House, to house both the parliament and the Royal Court. The building is plain and discreet. The States meet in the Royal Court Chamber, which has a small public gallery. To commemorate Queen Victoria's Diamond Jubilee, the town added symmetrical wings to accommodate the Police Court and the Greffe, Guernsey's record office. At the close of the eighteenth century, citizens first requested reform of the representative assembly to allow greater representation for the parish of St. Peter Port, which then provided two-thirds of all tax revenues. In 1846 St. Peter Port was awarded more proportional representation in the States, giving it six representatives to its former one.

An important event in the economic circumstances of the town in the late eighteenth and early nineteenth centuries was the erection of the town markets. Formerly, the fish and meat markets had been held in the vicinity of the town church. In 1780 the meat market moved beneath the Assembly Rooms to the arcade later known as Les Halles. In the early nineteenth century a distinct commercial district was created and new market buildings were constructed. In 1820 the States decided to build a covered market. Designed by John Wilson, the Doric market hall was completed in 1822, and in 1830 a southeast extension was added. The creation of this market area reflected and in turn generated a growth of economic activity; a visitor in 1847 commented that "marketing seems here the grand object of people's lives." Later in the 1870s, two fine High Victorian buildings, the fish and meat markets, were added to the complex by John Newton.

It was also during the period from the 1780s to the 1820s that the old medieval lanes were replaced with new roads and terraces, and large houses were built of stucco and stone with gardens in the front or rear. It is this period of expansion that continues to give the town its prevailing character, and little remains of the town before this wave of planned urban design.

The town's suburbs are perhaps its most distinctive feature, however, with some of the most outstanding examples of Regency and Early Victorian design outside of Cheltenham and Sidmouth on the English mainland. Funded by the gains of smuggling and privateering, this new residential district developed on the plateau above the town, allowing the construction of terraces and houses standing in their own gardens. These buildings might be described as a kind of vernacular stucco. Following the French Revolution and the turmoil of the 1840s, French refugees brought their own influences, which can be discerned in local buildings. An eclectic range of styles such as Gothic, Tudor, and Jacobean were frequently borrowed during the period of peak building activity between 1770 and 1890. This expansion also allowed the town to create new municipal facilities such as a cemetery, a fire station, and a public park, Candie Gardens.

John Wilson is the most distinguished architect to have practiced on the island. His work illustrates the local interest in different styles; his markets are neoclassical, whereas his design for the island's principal school, St. Elizabeth's College, is Gothic. He also constructed a range of market buildings of the 1820s and St. James the Less Church, built in 1817 as an English-language church to serve the new district beyond the old town.

With the new construction and better communications with England and France, St. Peter Port began to attract new visitors. The first royal visit occurred in 1846, when Queen Victoria and Prince Albert landed on the island from the royal yacht. To commemorate the event, the government in 1848 erected a 108-foot tower west of the town. St. Peter Port's most distinguished resident, Victor Hugo, lived in town as an exile from the government of Emperor Napoléon III from 1855 to 1870. It was during this period that he wrote some of his most famous works, including *Les Misérables*. Hugo's home, Hauteville House, remains as he left it. The house stands on a steep slope with views of the harbor and the neighboring islands. Up the road lived Hugo's mistress Juliette Drouet, a young actress who held her own salon. Even after returning to France in 1870, Hugo kept the house; he had lost all his possessions in France, and wrote to a friend that "I have no longer a country, but I want a home." It was

bequeathed to the people of Paris by his heirs. The house is designed in an eclectic and idiosyncratic style.

Much of the fabric of St. Peter Port has remained the same since the late nineteenth century. The major event in St. Peter Port during the twentieth century was the German occupation of the Channel Islands during World War II, from 1940 to 1945; it was the only area of the entire British Isles to suffer invasion. The invasion had no lasting effect on the fabric of the town, but there was an inevitable impact on the civilian population who had to endure the harsh nature of enemy military occupation. People not born on the island were deported to concentration camps in Germany, where many died.

During the twentieth century St. Peter Port became a major center of offshore banking owing to Guernsey's favorable tax laws. Banking is now one of the principal sources of income for the residents. The major constitutional issue has been negotiation of special terms for the island when the United Kingdom entered the European Economic Community (EEC). Guernsey was exempted from European legislation and fiscal rules but allowed to trade freely with the United Kingdom and other Common Market nations, thereby enabling St. Peter Port to maintain its traditional role of being an open port for economic activity.

Further Reading: *The Bailiwick of Guernsey* by C. P. Le Huray (London: Hodder and Stoughton, 1969; New York: International Publications Service, 1971) remains the standard history of St. Peter Port and Guernsey. It contains a sixty-nine-page chapter on the town. *Buildings in the Town of St. Peter Port* by C. E. B. Brett (Guernsey: National Trust of Guernsey, 1975) is the most detailed and scholarly study of individual streets and buildings in St. Peter Port. It describes the town's development and planning, and also provides information on the most important architects who practiced in St. Peter Port. *Guernsey* by G. W. S. Robinson (Newton Abbot, Devon, and North Pomfret, Vermont: David and Charles, 1977), part of a series on world islands, has a short but useful chapter on St. Peter Port. *A Short History of Guernsey* by Peter Johnston (Guernsey: Guernsey Press, 1978) is no more than a booklet, but is especially informative about how the expansion of trade during the eighteenth and nineteenth centuries affected the development of St. Peter Port. *An Historical Look at Guernsey and the Bailiwick* by G. H. Mahy (Guernsey: Guernsey Press, 1977) has only a small chapter on St. Peter Port but does have separate chapters on the harbor and Castle Cornet, and provides general information on such subjects as smuggling and shipwrecks.

—Joshua White

St. Petersburg (Leningradskaiya, Russia)

Location: Northwest Russia on the delta of the Neva River at the east end of the Gulf of Finland, an arm of the Baltic Sea; 400 miles north of Moscow, the center of the city is the south bank of the Neva, containing St. Petersburg's business district and most of the famous historical buildings.

Description: Capital of Russian Empire from 1712 to 1918; founded by and named after Peter the Great (Peter I); official residence of the ruling Romanov family (the Winter Palace was modeled in design from Versailles); site of the overthrow of czarist and provisional government power by revolutionaries in 1917. Known as Petrograd from 1914 to 1924, then as Leningrad from 1924 until its name change back to St. Petersburg in 1991, the city withstood a devastating 900-day siege by the German army from 1941 to 1944 during World War II.

Site Office: Leningradskaiya Oblast Administration
Suvorovskiy Prospect 67
St. Petersburg, Leningradskaiya
Russia
(7812) 274-3563

St. Petersburg is a city that has held Russia's dreams to create new beginnings against great odds. It began as a fortress that Peter I established as a strategic stronghold at the mouth of the Neva River to protect the Russian interior from the open waters of the Baltic Sea. Although the city is relatively young by European standards, St. Petersburg's name has changed several times over the course of just under 300 years of existence.

In response to continual Swedish attacks during the Great Northern War (1700–21), Peter I (the Great) saw the need to counteract Sweden's drive to control the Baltic region. After a bitter defeat at the hands of Charles XII of Sweden in 1700 at Narva Region, Peter battled back to capture the fort of Nienschantz at the very mouth of the Neva from the Swedes in May 1703. Peter wanted to control Russia's newly won access leading out to the Swedish-dominated Gulf of Finland and Baltic Sea. From the beginning, the czar was involved in transforming strategically located Hare Island from a seasonal Finnish fishing village into a fortress-city.

His first impulsive action was to cut two strips of turf with a bayonet on Hare Island and lay them in the form of a cross to dedicate the foundation of St. Petersburg on May 16, 1703. The island was situated on low, swampy ground that made it unhabitable for year-round use due to flooding. The fortress of Sts. Peter and Paul, covering the entire Hare Island, began to come together under Peter's coercive mandate.

Russian laborers worked to raise the level of the is-land above the Neva River's threatening reach. The oblong ramparts of Sts. Peter and Paul were first constructed of earth and timber, but then Peter had the walls made thicker and higher with stone. Just outside the fort, army carpenters built a simple, three-room log cabin that Peter lived in during the summer months as the work on his rising citadel progressed. The cabin, now surrounded by an outer stone covering to protect it from the elements, is the oldest building in St. Petersburg to this day.

St. Petersburg began to develop beyond just a military center into an eighteenth-century European city. Peter started to petition provincial governors from regions across Russia for 40,000 men per year to work in the island city. Both skilled tradesmen and unskilled peasants toiled at fashioning whatever they were told to in an environment that had deplorable living and working conditions. Peter began to establish shipyards near Lake Ladoga, which flowed into the Neva delta. Under his direction, significant strides were taken to modernize Russian army and naval equipment, firepower, and operations.

Peter had grown to dislike much of what the Russian capital of Moscow represented. He felt that Moscow, as the old residence of the czars, was backward, with its archaic traditions, court intrigues, and landlocked outlook. As time passed he started to see that he could create a city and palace that would offer something completely new. St. Petersburg began to represent Peter's vision for a new Russia that turned toward the West. He wanted to create a city with a window to the sea, such as the Dutch had in Amsterdam or the Italians in Venice.

Even before the year 1712, when St. Petersburg was made the capital of Russia, Peter issued decrees that made resettlement to the city compulsory. Merchants, artisans, peasants, and nobility were forced to settle in the artificial urban center from other Russian towns. Moscow began to be emptied of its government ministers and administrative offices as the population of the new capital began quickly to swell.

Peter, in describing his new city, called St. Petersburg ''Paradise.'' At first, it fell far short of that description because there was no specific plan besides Peter's changing whims. Construction of the city's main buildings was halted on several occasions because of flooding. Peter's impractical attempt to build on the city's lowest site at Vasilevsky Island could not defy nature, as the project was forced to move to higher ground. It was almost a certainty when the autumn storms violently hit St. Petersburg that buildings would be swamped with icy water on their ground-floor levels. The flood of November 1721 produced the most widespread destruction, as nearly every house in the badly damaged city was hit.

The rise of St. Petersburg as a commercial port came

St. Isaac's Cathedral and the Neva River
Photo courtesy of General Tours, Inc.

about through Peter's conscious efforts to overcome two key realities that stood in his way: the Swedes' control of the waters along the Baltic trade route, and the resistance of Russian merchants to give up the established Russian port of Arkhangel'sk. Peter did everything possible to encourage the use of the port of St. Petersburg, not only by his own countrymen, but also by foreign merchants. In a 1713 decree, he prohibited the shipping of the main goods of the Russian export trade between interior ports and Arkhangel'sk. Instead, the goods were to be sent to St. Petersburg. For foreign traders, he offered tolls at his island port that were less than half what the Swedes charged at their Baltic Sea ports, and he guaranteed the British that he would offer Russian goods at low prices if they picked them up at St. Petersburg, rather than at Arkhangel'sk.

While St. Petersburg was twice as close to western Europe as was Archangel'sk, Peter's decree of 1713 actually crippled Russian trade for more than a decade, because the lingering Swedish bottleneck of the Baltic made it virtually impossible to traverse the sea safely. St. Petersburg was to become Russia's major commercial anchorage, but only until well after the Great Northern War's final peace treaty was

struck at Nystadt in 1721. The Admiralty was established, at first, as a simple shipyard, then was expanded to become a well-protected fortress that served as the headquarters for the Russian naval fleet. When the Admiralty was rebuilt in the early nineteenth century, its impressive spire facing the steeple of the fortress-like St. Isaac's Cathedral dominated the skyline, much as the landmark does to this day.

Peter saw the need to have his new northern capital follow an architectural plan similar to those of the great cities he admired, at the time, in western Europe. He brought in a group of the best architects available from the West. A Swiss-Italian architect, Domenico Tressini, came in to design large-scale, expensive undertakings. Tressini engineered the development of the Cathedral of Sts. Peter and Paul within the original fortress on Hare Island. In this classical and baroque-style church, Peter the Great and his successors are buried, with the exceptions of Alexander I, whose body was mysteriously removed from his tomb, and Nicholas II, who was killed in Sverdlovsk (Exaterinburg), Siberia, with his family after the 1917 overthrow of czarist imperialism.

St. Petersburg's world-famous street, known as Nevsky Prospekt, was part of the early master plan for the city's central district. The three-mile-long boulevard that stretches east from the Neva was lined with private mansions of Russian princes and dignitaries. At the end of the Nevsky, Peter founded the Alexander Nevsky Monastery in 1710 on the spot where the legendary Novgorod prince and saint Alexander Nevsky reportedly won a glorious military victory over the Swedes and Teutonic Knights in 1241. Just one year later, in 1711, Peter had the Winter Palace constructed on a street that runs parallel to the Neva. This original Winter Palace was austerely crafted to conform to the other homes in its vicinity during the early days of the emerging capital. The imperial residence had a great reception hall and wings of narrow rooms, which served Peter's sense of practicality in architectural function and style.

As the city of St. Petersburg grew, Peter and top Russian nobles began to consider escaping to country palaces outside the capital. Luxurious summer estates began to be constructed in the suburban environs of Petrodvorets, (Peterhof), Lomonosov, Tsarskoye Selo (Pushkin), and Pavlovsk. In the year 1716, Peter brought in the French architect Alexandre-Jean-Baptiste Leblond as his architect general at a handsome guaranteed salary with privileges. Leblond was well respected in France and a student of André Lenôtre, the designer of the palace of Versailles outside Paris. At Petrodvorets, eighteen miles from the center of St. Petersburg, Leblond designed the thirty-seven-acre garden park at Peter's summer home. He created for Peter a true French formal garden with ornate designs of trees, flowers, waterfalls, statues, and trails. The Great Cascade of three waterfalls and sixty-four fountains has a water system that has remained largely the same since 1721.

Leblond worked with Peter to establish the Chancellery of Building, which required any new building plan to be submitted for approval before construction could begin. Leblond also began to draft an overall plan to direct any major development for St. Petersburg in the future. While Leblond instituted these guidelines as something of a visionary, Peter utilized them as a harsh form of control to assert his own power. He decreed that all buildings were to be built strictly to specific architectural plans that regulated the kind of material used, color, and size, or face the consequence of having them torn down. Likewise, if Petersburgers did not build on or choose to move to Vasilevsky Island, the czar threatened in a decree of 1724, they would face the same result of having their homes ripped down.

Six miles beyond Petrodvorets at Lomonosov, the powerful noble Prince Alexander Menshikov built his own summer mansion at Oranienbaum between 1710 and 1725. The name "Oranienbaum" (German for "orange tree") traces to the heated orangeries that only those with exalted status during Peter's rule enjoyed. Menshikov, who was Peter's favorite, also owned a great palace on Vasilevsky Island. His summer residence faced the sea and had its own formal garden constructed by his own Prussian architect, Gottfried Schädel. When Leblond was brought in by Peter, Menshikov, as the governor general of the city, did his best

secretly to sabotage the architect's work. Leblond's series of grand canals for the new city were made narrower and shallower than he wanted them to be, due in large part to Menshikov. The jealous prince, who was responsible for all construction while Peter was away for several months, made Leblond's St. Petersburg canal plan so unsuitable that it eventually had to be scrapped.

While Peter's corps of architects and craftsmen did much to immortalize and symbolize his reign, St. Petersburg's true architectural radiance came with the reigns of Elizabeth Petrovna, who ruled from 1741 to 1762, and Catherine the Great, who ruled from 1762 to 1796. The Italian architect Bartolomeo Francesco Rastrelli greatly upgraded the architectural designs begun under Peter to create the elegant Russian baroque and classical styles that have remained to this day. His work on the Winter Palace between 1754 and 1762 created an impressive main imperial palace with more than 1,000 halls and rooms. It was used primarily for ceremonial events, receptions, and grand balls and is, today, part of the overall museum complex called the State Hermitage. Rastrelli went on to reconstruct or build anew other unparalleled structures: the Smolny Convent, the fairy tale-style summer palaces at Petrodvorets and Pushkin, and the massive bazaar of Gostinny Dvor on the Nevsky Prospekt.

These splendid architectural embellishments provided an appropriate setting for the emergence of culture and industry in St. Petersburg, but the nineteenth century saw stirrings of a different sort, too, as the capital city witnessed the beginnings of Russian revolutionary fervor. In a revolt in December 1825, the so-called "Decembrist" group of military officers attempted a coup d'etat against the new czar, Nicholas I. That effort failed, but as the paternalistic, repressive rule of Russian czarist tradition began to prove ever more ineffectual at the beginning of the twentieth century, St. Petersburg again took the lead toward revolutionary change. Following months of unrest during a revolution in 1905, Czar Nicholas II finally granted a *duma* (parliament), although he limited its authority at the first opportunity.

Before World War I, St. Petersburg changed its name to the Russified Petrograd. It would stay Petrograd until 1924, when it was renamed Leningrad after the Bolshevik revolutionary leader Vladimir Ilich Lenin. By the beginning of World War I in 1914, the Bolshevik movement, directed by the exiled Lenin, began to gain ground in Petrograd. The Bolsheviks' revolutionary cry was to stop the war, overthrow the czar, and form a new government. As the war against the Germans dragged on, there were an increased number of workers' strikes, protests, and demonstrations that broke out in the streets of Petrograd.

The clamor for revolution started to build up to a fever pitch when nearly 90,000 men and women went on strike in the city's heavily industrial Vyborg district on February 23, 1917. A large group of them pushed their way past the police to the Petrograd municipal *duma,* demanding bread. The next day a massive strike of half the city's industrial workers culminated with emboldened crowds pouring into the streets across the city. The best that the imperial authorities could throw at the demonstrators was not enough, as police, cossacks, gendarmes, and even the most battle-hardened military units, direct from the front, were completely overrun. More than any other oppressed groups in Petrograd, peasants and workers were most responsible for carrying out the people's revolution against the czarist government. On February 27, Czar Nicholas II and his family were expelled from the capital, and the czar formally abdicated March 2. The leaderless "February Revolution" was a spontaneous outburst that was over amazingly fast. The final casualty count was 1,443 people dead or wounded in the cold Petrograd streets.

A new, coalition-based provisional government developed from the aftermath of the February Revolution. The real source of power during this period came from the Petrograd Soviet of Workers' and Soldiers' Deputies, which was remotely supported by a Russia-wide organization of regional soviets. The soviets comprised three parties: the Mensheviks, the Social Revolutionaries, and the Bolsheviks. From the Social Revolutionaries came Aleksandr Kerensky, who became prime minister of the provisional government in July 1917. With the support of the majority of the executive committee, the new Russian government chose to continue the war against the Germans. The government tried to bolster an already-teetering Russian military machine by restoring a semblance of discipline in the army and by taking out a new war loan.

The entire nation supported the redoubled military effort, except for the small party of Bolsheviks. Then in April of that revolutionary year, the German kaiser's government purposefully allowed Lenin to pass from Switzerland through Germany and into Russia on a sealed train. He arrived at Petrograd's Finland Station, and from then on began to coalesce the Bolshevik movement. In July Lenin tried to overthrow the provisional government, but was unsuccessful, and briefly went into exile in Finland. Lenin secretly came back to Petrograd in September to prepare for an armed insurrection against the government. Through the formation of military revolutionary committees, he set the stage to overpower the weakened provisional government.

The Bolshevik center of power started to gather in the Petrograd district of Smolny. Practically overnight, Smolny became an insurrection command center and well-stocked arsenal for the Bolshevik movement. By the last few weeks of October, large amounts of weapons, vehicles, and supplies were amassed there by revolutionary organizers. On the evening of October 24, 1917, the Red Guard began to take up positions around Smolny to defend it from possible attack from government forces. Like clockwork, the Bolsheviks launched active operations to capture the capital. During the night the Red Guard forces captured the Baltic railroad station, cut off all telephone and telegraph communication, and sealed off major thoroughfares and bridges around the Winter Palace. The cruiser *Aurora* on the Neva served the special revolutionary committees as a way to back their demands

with force and as a radio station to broadcast messages to the companies of the Red Guard.

Russian Prime Minister Kerensky called in his ministers for an emergency meeting at the Winter Palace. The junkers, the main group left to defend the Winter Palace, received a demand from the Bolsheviks to leave the palace. The next day the leader of the Military Revolutionary Committee, Leon Trotsky, explained what he knew would become reality when he announced that the provisional government ceased to exist. Feeling the raging storm closing in on him, Kerensky was forced to abandon the Winter Palace in a getaway car followed by another car, from the U.S. embassy, that flew a small American flag. In almost no time, the palace was surrounded by troops of the Red Guard. The Military Revolutionary Committee delivered an ultimatum to government authorities inside to surrender the Winter Palace or face bombardment from cannon at the Peter and Paul Fortress and aboard the *Aurora*.

The gun batteries from the fort and the ship opened fire on the Winter Palace, but their purpose was to threaten more than to do any real damage. Out of a total of thirty-five shots that were reportedly fired in a span of two hours, just two hit their mark, and the only damage done was to the plaster walls. The six-inch guns of the *Aurora* were silenced when it was learned that the Winter Palace had fallen to the Bolsheviks. The well-planned October Revolution ended with the Red Guard storming into the room where the ministers of the provisional government were gathered and arresting them in the early morning hours of October 26. Lenin came out of hiding at the last possible moment to become the new head of the Russian Bolshevik government. In order to protect Petrograd from a German threat to the city in March 1918, Lenin decided to move the capital of Russia to Moscow. This "temporary" transfer to Moscow became permanent when the Baltic states of Finland, Estonia, Lithuania, and Latvia broke away from Russia that same year, leaving the Russian-Finnish border just twenty miles from Petrograd. Also in 1918, Lenin signed the peace treaty of Brest-Litovsk with Germany, following his vow to get his country out of the war. When Lenin died in 1924, Petrograd was named Leningrad in his honor.

When Russia signed the Nazi-Soviet pact in August 1939 to guarantee that each country would not aggress against the other, the city of Leningrad was lured into a false sense of security. With the strategic city so close to the Finnish border, the Russian leader Joseph Stalin forcibly captured buffer territories from neighboring nations to protect Russia against the growing threat. The Russians took the offensive with the winter war of 1939–40 that served to capture bordering Finnish lands, and then annexed Poland in 1939 and the Baltic states in 1940. Leningrad started to move toward a state of military alert in mid-1941 with the knowledge that the Germans continually had been violating eastern air space borders and had landed troops in Finland. The mobilization of Leningrad was slowed because of the tremendous damage Stalin himself caused. During the terror and repressions of 1937–38 in Leningrad, hundreds of party members and high-ranking military officials had been purged, leaving a severe leadership vacuum that could not be filled quickly enough.

Party leaders began to work around the clock to create a network of defense works around the city to slow the German advance and convert nearly every industry into mass-production centers for military supplies and weapons. Many Leningrad families, fearing that the Nazi attack would come from Luftwaffe bombers, such as had occurred in 1940 in England, sent their children out of the city. Between late June and early July 1941, more than 200,000 children were sent to country areas that were southwest and west of the city. This evacuation proved to be a fatal mistake for thousands of them, as they were sent right into the path of advancing German units. The Soviets braced for air raids on the city by fireproofing factories and residential buildings. Camouflage nets and brown, green, and gray paint began to cover the city to deceive and confuse German planes. The Military Council ordered those not drafted into the army (largely women) to work extra hours each day to dig fortifications outside the city or to construct shelters. More than 30,000 of these women were working to dig trenches, minefields, and tank traps along the Luga line. They faced daily attacks by German planes that bombed and strafed them while they were working or traveling along the roads to their destinations.

The Luga line saved the city from total collapse as it slowed the German advance for nearly a month. It threw the Germans' timetable off and bought the Russian forces valuable time to prepare to defend the city. When Hitler's forces captured the town of Mga from the battered Soviet army, all of Leningrad's rail connections to the rest of the country were cut off. Nazi panzer tanks forced the Soviet Forty-Eighth Army to retreat toward Leningrad, and it was at that point that the army's chief of staff, General D. N. Nikishev, severely began to doubt Moscow's real desire or ability to provide the Leningrad forces with adequate resources. Nikishev soon after was mysteriously relieved of his command. Then Germans drove ahead to surround the city completely. By the second week of September 1941, German Field Marshal von Leeb was confident that Leningrad would soon fall when the German Supreme Command finally came through with armor and air reinforcements. The Soviet armies held out, but a choking blockade of the city had begun.

As the cold Russian winter began to settle in on both entrenched armies in October 1941, Leningrad survived on the brink of despair and defeat. With the reserves of the city food supply giving out, hundreds of thousands died of starvation, and those who survived were too weak to bury many of the dead until springtime. Only in March of 1942 could city and party officials begin to bury the mountains of frozen, decayed bodies that lay in the streets of Leningrad. The only way to get food into the surrounded Leningrad was by the less-than-reliable Lake Ladoga shipping route. Ships loaded with food supplies took the perilous sixteen-hour trip and were continually bombarded by German planes. When the lake froze in winter, trucks drove across the last available link to the outside world, but the Germans' aerial accuracy made even the trucks' arrival into Leningrad an occasion for celebration. The Lake Ladoga ice road became of such strategic importance that members of the Young Communists staffed traffic control posts every 200 to 300 yards in extreme, below-zero temperatures. Every mile of the road was guarded by antiaircraft guns, and the last remaining planes of the Seventh Russian Air Corps made it a priority to provide the essential air cover.

As the siege of Leningrad continued, the city started to see small signs of hope on the horizon. In January 1943 the Russians broke through a small corridor that allowed trains to pass back and forth under murderous German gunfire. It was not until later in the year that passage along the corridor began to pay off for Leningrad. The Soviets introduced an elite military train detachment, Special Engine Column No. 48, which effectively began to bring freight into the city with reliability. American food supplies began to flow into the city. The 900-day siege would continue until 1944, when the Russian and Allied victories elsewhere loosened the German grip on Leningrad.

Leningrad devoted considerable energy in the postwar years to reconstruction, with concentrated efforts on restoring its historic architectural properties. With new freedoms emerging and old barriers falling in Russia, the city was allowed to take one more step in reclaiming its heritage. In September 1991 voters in the proud Russian city decided to change the name of their city from Leningrad back to the original St. Petersburg.

Further Reading: *The Reforms of Peter the Great: Progress Through Coercion in Russia* by Evgenii Anisimov; translated by John T. Alexander (New York: M. E. Sharpe, 1993) is an excellent look into Peter the Great and his ironfisted intention to fashion St. Petersburg into a military and imperial capital to rival and imitate the best of other European nations. Leon Trotsky's *The Russian Revolution: The Overthrow of Tzarism and the Triumph of the Soviets* (New York: F. W. Dupee, 1932; reprint, New York: Doubleday, 1959) offers a surprisingly balanced analysis of how the revolutionary events of 1917 developed in the streets of Petrograd, but spends more time looking at the October Revolution. Harrison E. Salisbury's classic *The 900 Days: The Siege of Leningrad* (New York: Harper, 1969; London: Pan Books, 1971) is a fascinating historical account of the Battle of Leningrad and how the city's people heroically struggled to survive in the face of death and starvation.

—Andrew M. Kloak

Salisbury (Wiltshire, England)

Location: In southern England, on the Avon River, twenty-two miles northwest of Southampton.

Description: City that developed around its distinctive thirteenth-century cathedral; the cathedral has many notable architecture features, including the tallest spire in England, and is one of the best examples of the English Gothic, or "Early English," style. The cathedral's close—the area that houses its ecclesiastical community—is the largest in England.

Site Office: Salisbury Tourist Information Centre
Fish Row
Salisbury, Wiltshire SP1 1EJ
England
(722) 334956

The town of Salisbury, besides being the site of Buckingham's execution in Shakespeare's *Richard III*, is notable for its magnificent Gothic Cathedral of St. Mary, better known as Salisbury Cathedral. Salisbury Cathedral is unusual for several reasons. It is the one medieval English cathedral completed according to its original designs, built throughout in a uniform style. Also, it was the only English cathedral built on a virgin site in the Gothic period.

The town is also known as New Sarum. Its predecessor, Old Sarum, was an ancient fortress town situated on a hilltop. In 1086 William the Conqueror held a council at Old Sarum. One of the outcomes of this meeting was the Oath of Salisbury, which established the principle that a vassal's obligations to his immediate lord could not supersede his allegiance to the king. Old Sarum was a mile north of the present site of Salisbury. Medieval records indicate the old town suffered from shortages of water and exposure to the winds. The cathedral of Old Sarum, built by St. Osmund, was severely damaged by lightning just five days after its dedication in 1092. Old Sarum dwindled in size after the cathedral was moved, but remained an official entity until 1831. The foundations of the old cathedral, and the ruins of the castle that surrounded it, can still be seen.

The move of the cathedral came during a great period of church building in England. This activity began with the reconstruction of the choir of Canterbury Cathedral 1174 to 1184. Following Thomas Becket's death in 1170, English clergy saw him as a martyr in the cause of protecting the church against royal tyranny. This led the religious community to take a new pride in the church, and this pride was expressed through the building of new shrines. There was also a surge in English nationalism, resulting in the development of the English Gothic, or "Early English," architec-

tural style, of which Salisbury Cathedral is a great example. This atmosphere was strengthened by the upbeat, reformist mood of English bishops following the Fourth Lateran Council of 1215, which is regarded as the high point of the Catholic Church in medieval times.

Bishop Richard Poore was the driving force behind Salisbury Cathedral's construction. It was he who got permission to move his see from Old Sarum. Bishop Poore wanted the cathedral and its community moved because of constant friction with the soldiers based at Old Sarum and because of the cramped living conditions. He laid out a town in the meadows by the river Avon, not Shakespeare's Avon but another one flowing southward to the English Channel.

The move brought about new problems, as the land was low and swampy and the cathedral was near enough to the Avon that in times of flood the river waters lapped into the church, leading some pessimists to label the cathedral close "the sink of the city." Nevertheless, in normal weather the location and the building combine to offer the picture of purity and peace. Mrs. Schuyler Van Rensselaer writes about the view of Salisbury in her book *English Cathedrals* and offers this adoring description: "The men who built and planted here were sick of the temples of Baalim, tired of being cribbed and cabined, weary of quarrelsome winds and voices. They wanted space and sun and stillness, comfort and rest and beauty, and the quiet ownership of their own; and no men ever more perfectly expressed, for future times to read, the ideal that they had in mind. The cathedral stands upon a great unbroken, absolutely level lawn which sweeps around it to west and north and east, while close beyond it to the south rise the trees of the episcopal garden."

Bishop Poore did not live to see the cathedral finished, but he did lay the foundation stones for a cathedral that, incredibly, was complete enough to be consecrated after less than half a century. Bishop Poore's labor of love began in 1220, and Bishop Giles de Bridport consecrated the cathedral in 1258. The west front of the cathedral was finished in 1265, and the roof completed a year later. Nicholas of Ely was the cathedral's master mason, and Elias of Dereham was supervisor of building. The design to some degree recalls the old Romanesque cathedral at Old Sarum. For the most part, however, it was the epitome of English Gothic.

As Van Rensselaer writes, "If Salisbury had been built with the express desire to show what, in its simplest form, the Early English style implied, its witness could not be fuller or more precise." The prime characteristics of the style are a relatively long and narrow nave and choir, shallow buttressing, comparatively low walls, elaborate moldings, little ornamental sculpture, a low-pitched roof, and lancet windows.

Some features of the new cathedral were influenced

Salisbury Cathedral
Photo courtesy of Salisbury Cathedral

by the designs for Lincoln and Wells Cathedrals, then under construction. These included the use of double transepts, as at Lincoln, and the adherence to simple external contours without flying buttresses, as at Wells. Balancing verticals and horizontals within a series of rectangular blocks that direct attention toward the center, the cathedral's appearance changes with every variation in light, shadow, or viewpoint.

The interior of Salisbury displays the most unity of any English cathedral. There is a variation from eight-shafted piers in the choir to four-shafted piers in the nave, but this seems primarily a subtle expression of the lesser importance of the publicly accessible areas in relation to the choir area. Extensive use of marble, which was available from the nearby Isle of Purbeck, intensifies the sense of decorum and restraint begun by the plain capitals and vaulting. This austerer style is similar to that of Canterbury's Trinity Chapel, the resting place of Becket's relics. The interior design creates a sense of great height, accentuated by the equal-length aisles, lancet windows, and enormous marble pillars.

There were regional variations among English cathedrals of this period. Northern churches were characterized by a straight east wall while southern ones featured projecting chapels and squared ambulatories. Salisbury is a prime example of the latter style, with its Lady Chapel on the east end of the building. In fact, some consider the Lady Chapel to be the best thing about the Cathedral's rather disappointing interior.

The decision not to build a towered facade on the west end of the cathedral probably arose from considerations of speed and economy, for a rather plain facade made it possible to complete the project quickly. A grandiose facade also could have been too much of a distraction from the tower that evidently was already planned to rise above the central crossing. The west facade has five levels of niches filled with figures of saints. The cloisters and chapterhouse were completed by the end of the century, as was a separate, freestanding bell tower, which survived until James Wyatt made alterations to the cathedral in the eighteenth century. Richard Farleigh added the the crowning touch, the central tower and spire, in the fourteenth century.

The cost of Salisbury Cathedral was 27,000 pounds. This sum included the construction of the chapter house and cloister, which were built at a later date. For perspective on this price, the average annual income of the crown was 35,000 pounds.

Although the basic construction and decoration of Salisbury Cathedral was accomplished in a very short time, like other cathedrals and much like public documents, it was amended over time. The cloisters were built between 1240 and 1270, but there was never a monastery on the grounds. The chapterhouse was built between 1263 and 1284 in a style similar to Westminster Abbey's. The east side of the cloisters was built over in 1445. A library has been there since 1756, containing an Anglo-Saxon liturgy and one of the four copies of the Magna Carta. Gothic arches in the crossing were built under Bishop Beauchamp in the fifteenth century. They were

necessary because the weight of the great spire on the original arches threatened their collapse. The wall around the close was built in the fourteenth century, when Edward III gave permission to "embattle" the cathedral and to use stones from the church at Old Sarum. Some of the stones have Norman carvings.

A window in the south aisle dates from the fourteenth century. A clock in the north aisle dates from 1386, and is one of the oldest in Europe. A memorial slab from 1375 honors Bishop Wyvill, and the Shrine of St. Osmund commemorates the Bishop of Old Sarum, who died in 1099.

The tower and spire, however, form the really important centralizing unit of the building. Various accounts of the tower and spire's construction differ concerning dates. Most likely, however, the tower was completed in the early fourteenth century, and the spire later in the century. The work is variously attributed to Richard Farleigh or an unknown master mason. At any rate, the tower-spire unit has always been universally acclaimed. It is integrated with the earlier building. The additional decorative pinnacles that were inserted in the angles between the corners of the tower and the spire prepare the observer's eye for the great leap of the spire upward to a height of 404 feet, the highest in all of England.

In part, the lowness of the roof and the height of the spire combine for an impressive effect that is enhanced by Salisbury's garden-like location on a river meadow in the Salisbury Plain. Also, the appearance of wide-spreading limbs—the transepts, porch, chapel, sacristy, cloister, and chapter house—help to sustain the vertical lines of the spire. Compare Salisbury to the Amiens Cathedral in France: Amiens's tower rises twenty-two feet higher than that of Salisbury, yet Amiens does not appear as impressive because its roof ridge is more than 100 feet higher than Salisbury's.

The overall effect of Salisbury's architectural elements is one of grace rather than power, of serenity rather than severity and force. The splendid result, however, did not come without some difficulties. Chief among the aesthetic dilemmas is the west front, which is seen by critics as incongruous to the other parts of the building and to the interior, chiefly because it appears too mechanical a collection of features that are not visually unified.

Another problem was the enormous load of the spire on the walls. Initially, there were piers only six feet square to support more than 6,400 tons of masonry. Flying buttresses were added so that the weight could be distributed to the outer walls. Abutments were added to the upper stages of the church before the tower was completed. In the fifteenth century bracers were added to the supporting piers inside the church. Reversed arches in the opening of the choir transepts were added, creating a dramatic perspective when looking from one side of the transept to the other. At the main crossing another vault was added underneath the steeple. One can observe a bulge near the top of the northwest pier at the crossing as evidence of the need for reinforcement. Even now the point of the spire is twenty-three inches out of perpendicular, though this is scarcely perceptible. No additional settling or leaning has occurred in recent centuries, and modern engineering has helped guard against further problems.

Christopher Wren and, later, George Gilbert Scott helped strengthen supports of the spire. In 1950 it was necessary to rebuild the top thirty feet of the spire because the elements had rusted and weathered it.

More changes to the cathedral came in the eighteenth century at the hands of architect James Wyatt, known as the "licensed vandal." In 1788 and 1789 Wyatt rearranged monuments and ordered replacement of the pleasingly colored ancient glass with plain clear glass. Wyatt also deemed the bell tower on the north side of the church of no importance and had it destroyed. Sir Gilbert Scott endeavored to improve the building in the late nineteenth century with vault paintings, choir stalls, and the bishop's throne.

Salisbury Cathedral's idyllic, sylvan setting among ancient elms is complemented by houses representing periods from medieval times to the seventeenth and eighteenth centuries. Beyond the west wall is a row of homes in which the canons of the church lived. Most of these houses are Elizabethan, some Queen Anne or Georgian. There are examples throughout the town of medieval and Georgian architecture. Mompesson House, built in 1701, still has much of its original plasterwork and paneling. Malmesbury House, another notable structure, dates from the fourteenth century. The Salisbury and South Wiltshire Museum in the center of town has a collection of prehistoric artifacts as well as some fine models of Stonehenge, which is just seven miles away.

Nearby are Old Wardour Castle and Wardour Castle. The old castle, built in the 1390s by the fifth Lord Lovel, was modeled after a French chateau, with a hexagonal courtyard and a great hall that occupied the upper stories on the east facade flanked by two high towers. This castle was badly damaged during the Civil War, when it was under siege twice, once by each of the opposing sides. The "new" Wardour Castle is a Palladian mansion designed by James Paine and is presently the property of the Carnbourne Chase School.

Wilton House, a few miles west of Salisbury, is one of the finest of England's stately homes by virtue of its design and interior decoration. Its present form is the result of three stages of construction: the original house constructed by William Herbert; a later design by Inigo Jones, whose major influence was Palladio; and the latest reconstruction, done by James Wyatt in the early 1800s. Wilton House was a Benedictine abbey that Henry VIII gave to his brother-in-law General William Herbert. Herbert's Tudor house burned in 1647 but was rebuilt in 1653 as a simple classical mansion by Inigo Jones and John Webb. There is fine baroque styling in the seven staterooms. The showpiece of the house is the Double Cube Room, which has Van Dyck paintings on white walls embellished by gilded representations of garlands of fruit and flowers. Other paintings of the old masters are in the house, along with fifty-five gouaches of the Spanish Riding School by Baron D'Eisenberg.

Salisbury hosts a two-week festival in September of each year. One of the leading cultural affairs in southern England, the festival includes performances of orchestral works in the cathedral, concerts by candlelight in neighboring churches, and various artistic and outdoor events.

Further Reading: Three books that offer insightful and detailed descriptions about Salisbury Cathedral are *English Cathedrals,* by Mrs. Schuyler (Mariana) Van Rensselaer (New York: Century, 1914), *Cathedrals of Europe,* by Ann Mitchell (London: Hamlyn, 1968), and *The Gothic Cathedral: The Architecture of the Great Church 1130–1530* by Christopher Wilson (London and New York: Thames and Hudson, 1989). Van Rensselaer's book is at various times gushing in its praise, florid in its phrasing, and supercilious in tone; but it is still quite useful in its architectural analysis and contains some lovely line drawings of the cathedral and its environs. Mitchell's book is quite lucid and descriptive, also with some fine illustrations. Wilson's book is much like Van Rensselaer's in that it is an analysis from the standpoint of architectural criticism. An excellent visitors' guide to the cathedral is *Salisbury Cathedral* by Canon A.F. Smethurst, revised by Canon I.G.D. Dunlop and Roy Spring (Andover and London: Pitkin Pictorials, 1976; reprint, 1993). Although a slim volume, it contains much historical background on the cathedral and numerous beautiful color photographs. For an entertaining read about the construction of cathedrals, David Macaulay offers *Cathedral: The Story of Its Construction* (Boston: Houghton Mifflin, 1973; London: Collins, 1974). He describes an imaginary town and the cathedral that rises within its walls over a period of two centuries. The methods described correspond closely to those used in the actual construction of a Gothic cathedral.

—Bob Lange

Salzburg (Salzburg, Austria)

Location: In central Austria, near the border of Southern Germany; Salzburg stands on both sides of the Salzach River, on the northern edge of the Alps.

Description: Capital of the province of the same name, Salzburg is Austria's fourth largest city, renowned for its architectural and scenic beauty and as the birthplace of Wolfgang Amadeus Mozart. The city is today a major travel destination and, since the end of World War I, the site of the famed Salzburg Festival.

Site Offices: The Salzburg City Tourist Office
Auerspergstr. 7
A-5024 Salzburg
Austria
(662) 8072-0

For the Salzburg Festival:
Salzburger Festspiele
Postfach 140
A-5010 Salzburg
Austria
(662) 8045-0

When Salzburg was liberated at the end of World War II by the Rainbow Division of the Allied Forces, its commander, U.S. Army General Harry J. Collins, was so captivated by the extraordinary sight of the city and its Alpine surroundings that he asked to be buried there upon his death. His wish came true twenty-one years later in 1966, when he was given a grave of honor in St. Peter's Cemetery.

St. Peter's was the first church to be built in the town that later became Salzburg. Prior to that, the region had been known as Iuvavum, an administrative center during Roman times. It lay in the Roman imperial province of Noricum, which included the present-day territories of Upper Austria, Lower Austria below the Danube, Eastern Tyrol, Styria and Carinthia. Its site provided several advantages. Bordered by a river to the north and east and cliffs to its south and west, the area was protected by natural defenses; deposits of salt and mineral ores in the surrounding area added an economic dimension to its attractions. "Salzburg," literally translated, means "Salt Castle." The Roman occupation in Iuvavum came to an end with the disintegration of the Roman Empire, brought on by barbarian invasions. Two tiny chapels dating from the Roman period can be seen in the catacombs today. Carved out of the rock, they date from the third century, when they were used by Christian Romans to celebrate mass. In A.D. 471 St. Maximus tried to hide some 400 of his parishioners in his chapel, deep within the cliffside catacombs, from the advancing hordes. They were discovered, mur-

dered, and their bodies tossed into the cemetery. Those who managed to escape the massacre fled into the mountains and the forest. The town lay in ruins, uninhabited, for two centuries until the arrival of Bishop Rupert of Worms at the beginning of the eighth century.

Rupert had come to the area with the missionary aim of re-establishing Christianity among the old Romano-Christian communities that had survived in the frontier provinces of Noricum. Rupert had come from Regensburg to the north, where, in 696, he had baptized the Bavarian Duke Theodo and his family. He had left the town and ridden south until reaching a lake, the Wallersee, which seemed to him a suitable location for establishing a new diocese. He remained at this location, today the village of Seekirchen, for several years. After sending his servants to explore the surrounding area, he soon learned of ruined Iuvavum, some two hours' ride to the southwest. Though desolate, the town still offered a number of advantages over Seekirchen as the site of a new diocese, including the natural defenses of the town as well as its location at the intersection of major lines of communication. This last factor was particularly important as the town's central placement would make the process of reaching the scattered Christian populations easier. There was also a legal advantage to setting up in old Iuvavum. Canon law requires that a bishopric be based in a town; in legal terms, Iuvavum still was a town, which would simplify the proposal for a see here as opposed to Seekirchen. Duke Theodo approved Rupert's proposal, and granted him full possession of the town, including the castle on the hill, the temple below it, and the surrounding land. A church was soon built on the site and dedicated to St. Peter. A Benedictine monastery was constructed on the Mönchsberg (the "Monks' Hill"), and a nunnery was founded on the Nonnberg (the "Nuns' Hill"), in which the bishop's cousin Erentrudis was installed as abbess, and which is today considered to be the world's oldest convent. These structures were soon followed by a Capuchin monastery on the Kapuzinerberg.

Priests traveled from here to locate the old Christian communities, and Rupert himself was continuously on the road. Many heathen tribes were converted, an activity he found more satisfying than dealing with the Christians, whose faith he considered to have been irredeemably neglected. The need for such religious restoration must have been a frustrating experience for Rupert, who was eventually expelled from Salzburg by a populace infuriated by his attempted reforms. His body was returned to the town as Salzburg's patron saint when the Englishman St. Boniface (675–754) succeeded in establishing the church throughout Bavaria. He is entombed in St. Peter's, the church he had built.

Salzburg was to become an important city; its sphere

Salzburg by night
Photo courtesy of Austrian National Tourist Office

of power and influence would cover a large part of present-day Austria, from the River Inn to the River Drava. The first indication of the city's ascendance came with its elevation to an archdiocese in 798, when Rupert's successor Bishop Arno was appointed as Salzburg's first archbishop and papal nuncio by Pope Leo III. A friend of Charlemagne's counselor Alcuin, Arno was well connected to the Frankish court, and this tie became significant in the extension of Salzburg's jurisdictions. With the defeat of the Avar Empire by Charlemagne in 796, Frankish rule had expanded beyond the borders of western Hungary; Salzburg's domain grew accordingly as a result of Arno's connections.

The Magyars invaded Austria in the tenth century, and Salzburg's extended territory was lost. This was not the only loss to Salzburg: Dittmar, its archbishop, died in battle as the Frankish armies were defeated in 907. The Magyar occupation lasted half a century until 955, when the troops of Emperor Otto I crushed the Magyar forces at the Battle of Lechfeld near Augsburg, a decisive event that would come to mark the beginnings of Austrian history. Following the victory at Lechfeld, Emperor Otto established the Bavarian Ostmark, an "Eastern March" or border province, under the command of the house of Babenberg in what is today Austrian territory. It was to serve as a bulwark against future Magyar attacks, which never occurred. The name Ostarichi, from which the modern Österreich (German for Austria) is derived is first used in records of 966 to refer to the Ostmark.

Salzburg's power and wealth continued to grow under the leadership of its archbishops, who displayed their abilities more notably as statesmen and soldiers than as priests. Eberhard II (1170–1246) in particular was one of many archbishops to extend the reach of Salzburg's power and influence through skillful use of political circumstance and the practice of legal opportunism. One of the reasons for a prolonged rivalry between Salzburg and Bavaria was the political alignment of Salzburg's archbishops. These consistently sided with the emperor, who preferred a centralized form of government. Opposing this alignment was the feudal nobility, which included the Duke of Bavaria, who preferred an arrangement of decentralized power. Appreciative of Salzburg's support, the emperor granted the archbishops numerous gifts including market, customs, and even coin-minting privileges, as well as tracts of land and hunting rights. The archbishops also enlarged the archdiocese's holdings by exploiting a legal situation. Most of the local nobility held its lands in fief from the Duke of Bavaria and the emperor. These properties were hereditary, but only according to the German Lehnrecht, a law of inheritance that recognized transfer of ownership only from father to son. Because many of the male descendants of local noble families died in battle during the period of the Crusades (1095–1270), the archbishops took the opportunity to declare many inheritances invalid and claim them for the archdiocese. The pinnacle of Salzburg's power was reached in 1278, when Archbishop Friedrich II, after helping Rudolf of Habsburg gain the im-

perial throne, was given the title of Imperial Prince. This gave the Prince Archbishop precedence over other German princes in the election of the emperor, making Salzburg one of the Empire's leading cities.

The archbishops' relations with other powers, like those with the townspeople of Salzburg themselves, were often strained. As the wealth of Salzburg's citizens grew during the Middle Ages, so did their demands for a greater say in the running of the city, which meant a reduction in the number of the archbishop's rights and privileges. The Archbishopric of Leonhard von Keutschach (1495–1519) is remembered to this day not so much for the rebuilding of the Hohensalzburg fortress (many parts of which bear his emblem, a turnip) but for his utter disregard for the townspeople's wishes in matters of local government. Emperor Frederick III granted to the burghers of Salzburg the right to elect their own mayor without having to consult the archbishop, as well as giving them the rights of Reichsunmittelbarkeit. The latter greatly reduced the archdiocese's say in city matters, as it entitled the city burghers to self-government under the direct lordship of the emperor, not of the archbishop. Seeming to accept this new arrangement with unaccustomed grace, Leonhard invited the town's dignitaries to a banquet, where a rude surprise awaited them. Almost as soon as they arrived, the guests were bound and dumped into sledges that took them to Radstadt, fifty miles away. They were made to feel the cold of the winter night without blankets or coats, although their discomfort must have been greater still as they noticed the archbishop's executioner riding among the guards. No doubt these must have been powerful incentives for agreeing to the archbishop's demand to relinquish their newly acquired rights and privileges. This they did under oath and were thereby allowed to return to their families. Cardinal Matthäus Lang von Wellenburg, who succeeded Archbishop Leonhard, was not as adept at controlling his subjects. Matthäus had to face the widespread peasant rebellion of 1524–25, in which the townspeople opened the city gates to let in the mob, forcing the archbishop into the fortress, where he rode out the artillery storm that came from below.

Salzburg's present-day appearance is due in large part to the building zeal of Archbishop Wolf Dietrich von Raitenau (1587–1611). Born to a family of the old Swabian nobility, he received his education in Rome, where his immersion in and consequent love of the arts of the Italian Renaissance were to result in the city that came to impress General Collins some three and a half centuries later. As with many large-scale projects, Wolf Dietrich's planned changes caused some upheaval. Before his architectural interventions could begin, he had to raze the area then occupied by the cathedral squares and the chapter—a medieval maze of narrow lanes in which buildings were clustered around the large cathedral. Fifty-five houses and a cemetery were destroyed to make way for the work of the Italian architect Vincenzo Scamozzi, who firmly believed that a perfect city should have five squares. Wolf Dietrich also wanted Scamozzi to build a new cathedral, in the grand, Italian style, to replace the old one. Expecting to meet with stern opposition, Wolf Dietrich was helped when a fire broke out in the oratory. The archbishop apparently looked so pleased when he learned of the fire that rumors implicating him immediately spread. While the fire caused little damage to the actual structure, it provided sufficient justification to press ahead with the archbishop's ambitions. However, there was a price to pay for the fulfillment of such grand architectural dreams. The people who were displaced by his clearing of the town's central area had to be re-housed, a task the archdiocese's treasury lacked the funds to carry out. Wolf Dietrich attempted to absorb into Salzburg's jurisdiction the Augustinian monastery at Berchtesgaden in neighboring Bavaria, with the ultimate aim of securing its wealth, derived from the town's salt mine. This attempted appropriation angered the Duke of Bavaria and brought about the end of Wolf Dietrich's archbishopric. Fearing the duke's retribution, Wolf Dietrich lost his nerve and fled into the mountains, only to be captured and imprisoned, first in Werfen Castle (some thirty miles south of Salzburg) and then in the Salzburg fortress, where he died after five years.

Salzburg's cathedral, the Dom, is thus the second cathedral building to stand on this site. The cathedral is dedicated to St. Virgil (who as bishop had ordered the construction of the first cathedral in 770), and to St. Rupert. Bishop Conrad III had the structure enlarged considerably. With its five aisles and 395-foot long nave, the Dom was one of the largest churches north of the Alps. Archbishop Wolf Dietrich asked Scamozzi (who was also working on the new layout of what is today referred to as the old town, the area lying between the cliffs of the Mönchsberg and the river) to work on the design and construction of the new cathedral. Construction had barely begun when Wolf Dietrich's days as an ecclesiastical official came to an end. His successor, Archbishop Markus Sittikus, employed another Italian architect, Santino Solari, and ordered him to draw up simpler and less ambitious plans. Built between 1614 and 1628, the new cathedral was styled on the Church of Il Gesù in Rome and was the first to be built in the Italian style north of the Alps. The first side chapel on the left contains a bronze font from the original cathedral dating from 1320. The architect Solari was laid to rest nearby, in St. Peter's Cemetery; Crypt XXXI contains his grave. The richly decorated façade of white Untersberg marble provides the impressive backdrop for the performance of Jedermann ("Everyman"), held annually in front of the cathedral during the Salzburg Festival.

The Residenz is another major landmark of Wolf Dietrich's tenure as archbishop. As with the town's cathedral, this replaced an older building, a much smaller episcopal residence dating from the twelfth century. Enlarged substantially in 1708, it remained the archbishops' residence until 1803, and became imperial property in 1816. In 1867, the Emperor Franz Joseph received Napoleon III here. In 1871, the German Emperor William I was also a guest here.

Another noteworthy detail of Wolf Dietrich's arch-bishopric is Mirabell Palace, built in 1606 and originally named Altenau, after Salome Alt, the archbishop's mistress and seventh child of Salzburg merchant and town councilor Wilhelm Alt. Salome was said to be the most beautiful woman of her time, and she soon aroused the archbishop's attention. According to local lore, the archbishop seduced her and installed her in a wing of his palace. Later he built Altenau for her and their children. When the crisis of Wolf Dietrich's downfall began, she was sent, with the youngest children, to Carinthia. He meant to follow, but was never allowed to join her. She lived out her days in peace in Wels, Upper Austria.

Though not as grand in his architectural ambitions, Wolf Dietrich's successor Markus Sittikus also left his architectural mark in Hellbrunn, a summer palace some four miles south of Salzburg. Built between 1613 and 1619, at about the same time work on the new Cathedral was underway, Hellbrunn was also under the supervision of the architect Santino Solari. Among its gardens and artificial ponds, its main attraction is the Steintheater, "Stone Theater," a natural cave in which the first performance north of the Alps of Monteverdi's opera Orfeo was staged in 1617.

The eighteenth and nineteenth centuries were uncertain times for Salzburg, particularly during the Napoleonic Wars of 1789–1815. During this period, Salzburg featured in two treaties. The first of these, the Treaty of Pressburg (present-day Bratislava), was signed between Austria and France in 1805 following Napoléon's victories at Ulm and Austerlitz. Austria was forced to give up territories which, among other things, resulted in a sharp reduction of its influence in Germany. In return, Austria was allowed to annex Salzburg. Four years later, the Treaty of Schönbrunn was signed after Austria had failed in its military initiative at liberation from Napoléon. Further territory was lost; Salzburg was handed to Bavaria.

The period of 1848–49 was a time of revolutionary unrest in Austria, during which the province of Salzburg achieved independence. In the following years, the city became the site of many events of national importance. These included the betrothal of Franz Joseph and Elizabeth of Bavaria, Franz Joseph's meeting with Napoléon III and a visit by the German Emperor William I of Bismarck. In March 1938, German troops under Hitler's command marched into the city. The Germans remained until the arrival of the Allied forces in 1945.

Salzburg is widely known as the birthplace of Wolfgang Amadeus Mozart, born on January 27, 1756 on the third floor of Getreidegasse 9, where his family lived until 1773. Mozart was employed as Archbishop Colloredo's concertmaster from 1771 until 1781 when, prompted by the poor treatment he received from his patron, he moved to Vienna, where he settled at the Emperor's court. In 1787, he was appointed Court Composer of Emperor Joseph II. He remained there until his death in 1791. The Salzburg Festival, which takes place annually during July and August, was founded after World War I by Max Reinhard. Salzburg achieved further popular renown with the filming of *The Sound of Music*. Many of the city's landmarks were featured in the film, which can be viewed as part of special Sound of Music tours.

Further Reading: *Fodor's 94 Austria* (New York: Random House, 1994) is a well-researched guide for the visitor to Salzburg. *Nagel's Encyclopedia-Guide, Austria* (Geneva, Switzerland: Nagel Publishers, 1970) contains a detailed summary of Salzburg's history in general, and of its buildings in particular. *Austria and the Aus-* *trians* by Stella Musulin (London: Faber, 1971, and New York: Praeger, 1972) includes an engagingly written essay on the city's history, with interesting anecdotal detail. *Salzburg under Siege, U.S. Occupation, 1945–1955* by Donald R. Whitnah and Florentine E. Whitnah (Westport, Connecticut, and London: Greenwood, 1991) offers an account of Salzburg's first decade of postwar history written from first-hand experience. The authors worked in the U.S. War Department in Salzburg during 1946 and 1947.

—Noel Sy-Quia

Saumur (Maine-et-Loire, France)

Location: On the Loire after it has been joined by the Cher, the Indre, and the Vienne, between Tours (40 miles to the east) and Angers (30 miles to the west). Saumur is 100 miles, by river, southwest of Orléans.

Description: A small town best known for its château, a residence of royals and nobles in the Middle Ages and Renaissance; a Protestant stronghold in the seventeenth century; also noted for its wine-making, its cavalry school, its equestrian museum.

Site Office: Office de Tourisme, Syndicat d'Initiative
Place de la Bilange B.P. 241
49418 Saumur, Maine-et-Loire
France
41 51 03 06

The Saumur caves along the Loire and on both sides of the valley of its tributary, the Thouet, must have been occupied in prehistorical times, and the present town was certainly the site of a Roman settlement, but the town's recorded history starts in the tenth century, when its Tour du Tronc, on the site of the present château, became a place of refuge for the region's inhabitants during invasions by hostile feudal sovereigns. The second syllable of the town's name almost certainly derives from the Latin *murus*, a wall, from which it may be inferred that Saumur probably came into existence as a defensive enclosure. A monastery was built on the site of the Tour, and in the eleventh century a fortress replaced it.

In 1026 the settlement, hitherto in the territory of the counts of Blois, fell into the hands of Fulk III Nerra, the warrior count of Anjou, whose robust military exploits and subsequent penititential feats were to enter the realm of folklore. Saumur, however, continued to be raided frequently by the Normans and was much disputed for two centuries between the feudal overlords of Blois and Anjou. As part of Anjou, Saumur came under the control of the English crown in the twelfth century, with the rise of the Plantagenet dynasty of English kings, whose line originated with the counts of Anjou. The French king Philip II (Philip Augustus) won Anjou and other territories back from the English in 1205.

The oldest surviving structure in Saumur is the eleventh-century nave of the church of Notre-Dame-de-Nantilly, south of the château and on the southeast edge of the town. The church was frequently visited in the fifteenth century by King Louis XI, who built the southern aisle, and whose oratory is now the baptistry chapel. The doorway and choir are of twelfth-century construction, and the enameled copper crozier of Gilles, archbishop of Tyr and guardian of the seals to King Louis IX (St. Louis), is fixed to one of the pillars.

The church also contains a thirteenth-century wooden statue of Notre-Dame de Nantilly and a series of fifteenth- and sixteenth-century tapestries, completed by a series of eight seventeenth-century Aubusson tapestries on the life of Christ.

The chronicler Jean de Sire Joinville, adviser to Louis IX, gives an account of the sumptuous festivities and lavish entertainment provided by the king at Saumur in 1241 on the occasion of the marriage of his brother Alphonse to the daughter of the comte de Toulouse. In 1246 Saumur passed to Charles, brother of Louis IX, and became royal property again only in 1328. In 1356, after the French King John II (the Good) was defeated and imprisoned in England by Edward, the Black Prince, John handed the property to his second son, Louis I, duke of Anjou, who began the construction of the present château. The work was continued by his son, Louis II, king of Sicily and duke of Anjou, and then by his grandson, René I (the Good), titular king of Naples, duke of Anjou, Bar, and Lorraine, and count of Provence. Both Louis II and René lived in the château.

In the early-fifteenth-century *Très riches Heures,* (Book of Hours) commissioned by Jean, duc de Berry, there is an illustration of the Saumur château in its then completed state. Much smaller than today's building, the château was highly decorative in outline, bedecked with miniature towers, bell turrets, crenellations, and all manner of spires, ornate chimneys, pointed stone dormer windows, circular protuberant rooms mounted on stepped stone supports, and a large number of fleurs-de-lis. René much admired the effect of the light on the gilded weather-cocks and called the building "the castle of love."

On René's death it was seized by Louis XI, who turned it from a simple palace into a fortress. Anjou remained a royal domain, passing through the hands of Francis I, his mother, Louise de Savoie, and then Henry II and his sons Henry III and François d'Alençon. The château of Saumur was further fortified by Philippe de Mornay, seigneur du Plessis-Marly, the moderate Huguenot leader, when in 1589 he was appointed governor of Saumur by Henry of Navarre, the future King Henry IV. Duplessis-Mornay was also in 1599 a founder-patron of the city's celebrated Protestant Academy, the principal theological citadel of moderate Calvinism in the seventeenth century. From the academy emanated a strong challenge to the less liberal views on the theology of predestination held at the seminary of Sedan. The château, in addition to serving as residence of the town's governors, was used as a prison and then a barracks in the seventeenth and eighteenth centuries. Among those imprisoned there were Nicolas Fouquet, Louis XVI's finance superintendent, who was accused of embezzlement, and the Marquis de Sade.

On March 3, 1810, the château formally became a

A view of Saumur, with the château at top left
Photo courtesy of Office de Tourisme, Saumur

state prison, and from 1830 to 1890 was used as a military arsenal and barracks. In 1908 it was acquired by the city of Saumur and restored with state subsidy. It now contains three museums, the Musée des arts décoratifs, with chiefly seventeenth- and eighteenth-century enamels, ceramics, sculptures, liturgical vestments, silver, paintings, and furniture; the Musée du cheval, containing items connected with the history of horseback riding; and the Musée de la figurine et du jouet, containing a collection of miniature and toy figures of all sorts.

While Saumur was a Protestant center there were up to twelve publishing houses in operation, and the town's whole life came to revolve round religious activities. The town's fortifications were dismantled in 1623. However, some two-thirds of the population left when the Edict of Nantes, which had granted widespread and in some areas complete toleration to the Huguenots, was repealed in 1685, and the population of Saumur today—about 35,000—is smaller by several thousand than it was in the seventeenth century. It contains further churches of some historical interest and much characteristic domestic architecture of the late medieval and early modern eras. The church of St.-Pierre, twelfth-century in origin, has a Renaissance nave and a seventeenth-century facade, and the architecture of the partly

sixteenth-century town hall is explained by its function as part of the town's fortifications when the building was still touching the river. The part of the building away from the river is modern.

Saumur was once the home of the national cavalry school, founded in 1768. The school's officers and cadets heroically defended the town with obsolete weapons against advancing German troops in June 1940. Saumur, however, was heavily damaged in the fighting. The school is now devoted to training in the use of armored vehicles, but the equestrian role of the cavalry school has merged with that of the national riding school, founded in 1789 and famous for its Cadre Noir riding team.

Saumur is also rightly celebrated for its characteristic red wines in a region where the predominant product is white, although Saumur does produce white and rosé wines as well. The mushroom crop, cultivated in the caves, constitutes another important local industry and amounts to about 70 percent of France's total. The town also has been associated since the seventeenth century with the manufacture of religious medals, rosaries, and other small religious items.

Further Reading: The only works in English devoted to Saumur, apart from guidebooks and encyclopedia articles, are histories of architecture or political, religious, or military histories of France. Saumur also figures in books on French wine-making and in general works on the Loire châteaux. A good introduction to the Loire châteaux can be found in Jean Martin-Demézil's *The Loire Valley and its Treasures* (London: Allen and Unwin, 1969) and in François Lebrun's *Histoire des pays de la Loire* in the series on French towns issued by Privat of Toulouse (no date).

—Anthony Levi

Schleswig (Schleswig-Holstein, Germany): Danevirke and Hedeby

Location: *Danevirke:* Southern Jutland peninsula, between the Schlei fjord in the east and the Treene River in the west, in the Schleswig-Holstein region of Germany. *Hedeby:* Western edge of Haddeby Noor, the southern branch of the Schlei fjord; approximately twenty-five miles inland from the Baltic Sea on the Jutland peninsula, south of modern Schleswig, Germany.

Description: *Danevirke:* Monumental earthwork fortification nineteen miles long; built in phases between A.D. 737 and the early thirteenth century by Danish rulers. Now located in northern Germany, it originally protected Denmark's southern border. *Hedeby:* Viking-age trading town in ancient Denmark, encompassing 68 acres within a semicircular rampart. Established in the early ninth century by Danish King Godfred, Hedeby prospered for 250 years until its destruction in the mid–eleventh century.

Site Office: Haithabu Viking Museum
Am Haddebyer Noor
24837 Schleswig, Schleswig-Holstein
Germany
(4621) 813 300

The Danevirke is a complex system of earthworks built to defend Denmark's southern border. This series of fortifications, one of the largest ancient defenses in northern Europe, was constructed in multiple phases over a period of 500 years. Its earliest section was built during the Viking age, and repairs and extensions continued to expand the Danevirke until the early thirteenth century. The Danevirke also played a strategic role in military conflicts of the nineteenth and twentieth centuries.

At the peak of its defensive capability, the Danevirke encompassed approximately nineteen miles of ramparts and ditches stretching across the southern Jutland peninsula, from the inland tip of the Schlei fjord in the east to the Rheide and Treene Rivers in the west. The fortifications were probably not extended all the way to the North Sea because the two rivers and surrounding swampland provided an effective natural barrier to any military threat.

The Danevirke evolved as a series of individual earthworks, some connecting to each other and some standing alone. Although the complicated story of its construction has not yet been fully resolved, historians have concluded that it was not built according to any uniform plan; rather, it was rebuilt or extended according to the defensive needs of Danish rulers over the 500-year period.

Recent excavations have established that the Danevirke's earliest phase of construction began in A.D. 737. Such precise dating is possible through a scientific process called dendrochronology, which determines the exact year a piece of timber was felled by comparing the width of its annual growth rings to those in a piece of wood of known age. Construction in 737 produced the north wall, located between the Schlei fjord and Thyreborg, and part of the main wall, which extended the north wall southwest past Thyreborg. Together, these two ramparts formed a defense four miles long, thirty-three feet wide and seven feet high. The east wall, a freestanding defense southeast of the Schlei fjord, may date from the same time.

From the beginning, the Danevirke was designed as a formidable, and permanent, barrier. The ramparts were constructed of earth faced with a wooden palisade and lined on the south side by a ditch. In marshy areas, a specially designed foundation provided stability for the wall.

Although no written sources remain that allude to the Danevirke's origin, historians associate it with conflicts between the eight-century Danes and their neighbors, the Saxons (to the south), the Franks (to the southwest), and the Slavs (to the southeast). The power and prestige needed to build such a fortification suggest that the Danevirke was initiated by a central ruler, perhaps King Ongendus, who rejected the attempts of a Frankish missionary to convert him to Christianity.

Historians attribute the next major phase of the Danevirke's construction to King Godfred, who ruled Denmark at the beginning of the ninth century. At this time, conflicts between the Danes and the Franks, who had conquered Saxony under Charlemagne, raged along the border. Historical sources describe King Godfred's order in 808 to build a defensive wall along the Danish border from the Baltic to the North Sea.

Despite such specific historical information, scientists and scholars have so far been unable to determine precisely which part of the Danevirke resulted from King Godfred's decree, as no section has been dated to 808. One theory proposes that Godfred's wall was simply a repair of the earlier main wall. Godfred had ordered that his wall contain only one gate, and the main wall does include a single gate through which passed Denmark's main north-south route, known as the Army Road or Ox Road.

Another theory suggests that the Kovirke may be Godfred's wall. The Kovirke, a straight wall positioned south of the main wall, stretched four miles from the southern tip of Haddeby Noor (an inlet of the Schlei fjord) westward to the swampland near the River Rheide. Like the main wall, it was faced with a timber barrier and was bordered by an outer ditch. A gate also allowed passage along the Army Road, which the Kovirke intersected southwest of Hedeby. The date of the Kovirke's construction remains unknown.

The Hedeby Viking settlement

Viking artifacts
Photos courtesy of German National Tourist Office

The Danevirke underwent major expansion in the tenth century, during the reign of King Harold Bluetooth. A new earthwork called the connecting wall, constructed in 968, joined the main wall to the semicircular wall and fore wall that surrounded Hedeby. This linkage created a solid defense from the Schlei fjord to the western edge of the main wall, which was then significantly extended by the crooked wall. This curving rampart was forty to forty-three feet wide and stood ten feet high, its sloping sides topped by a wooden barricade.

Harold's conflicts with the German Empire at this time probably necessitated a strengthening of the Danevirke border defense. Even so, when war broke out in 974, German forces captured the Danevirke and occupied the border area until the Danes won it back in 983.

The Danevirke was modified again in the 1160s by King Valdemar the Great. Lauded as "the mighty conqueror of the Slavs" on his gravestone, King Valdemar is credited with extending the Danevirke and facing the section with stone and later with bricks. This formidable brick wall, embellished with buttresses and battlements, was considered so noteworthy that it is also listed among Valdemar's chief accomplishments on his gravestone.

The Danevirke again played a defensive role in military conflicts of the modern age. In the mid–nineteenth century, it became a landmark in the nationalistic disputes involving Denmark, the German Confederation, and the Duchies of Schleswig and Holstein. Both Denmark and its southern neighbor, Prussia, sought to control the Duchies, which fought vigorously for independence. The Danevirke, which stands in Schleswig territory, was the site of a battle in 1850, during which the small Schleswig-Holstein army lost decisively to Danish forces. Denmark regained the Danevirke, and the Duchies fell largely under Danish control.

Tensions continued, however, and in 1864 the Danes extended the Danevirke with trenches in the expectation that Prussia would invade the Duchies. Despite this preparation, Prussian and Austrian forces easily overcame Danish opposition when they invaded Holstein and then Schleswig. The Danes abandoned the Danevirke and retreated far up the Jutland peninsula. As a result, Denmark gave up control of Schleswig and Holstein, which were eventually annexed by Prussia.

The Danevirke's strategic importance was realized most recently during the German occupation of Denmark in World War II. In early 1945, Nazi forces dug deep antitank trenches in the Danevirke's ancient walls to strengthen their defense against an anticipated British invasion.

Today, portions of the Danevirke are disturbed by newer construction or lost amid Schleswig's low-lying hills. Some segments, however, remain intact and highly visible. Although it stands in what is now northern Germany, the Danevirke is a formidable reminder of Denmark's ancient southern border and the conflicts that raged there.

Hedeby, an ancient Danish settlement on the southern Jutland peninsula, was the preeminent trading town of Viking-age Scandinavia. Located on the western shore of Haddeby Noor, the southern arm of the Schlei fjord, the town was lost to history until the twentieth century, when site excavation began. Archaeological work at Hedeby has contributed to a new understanding of Viking culture and commerce.

Transcontinental trade routes during the Viking age fostered the exchange of goods and currency between Asia, Russia, and Western Europe. Previous theories viewed the Vikings as middlemen who profited by transporting luxury goods from Asia to Western Europe and back. The discovery of Hedeby and other ancient trading towns in Scandinavia, however, has broadened this view significantly. Rich and well-preserved archaeological finds reveal a town the wealth of which was not solely dependent on the exchange of foreign goods. Industry flourished at Hedeby, where local artisans refined raw materals into products traded at colonies and commercial centers throughout the North Atlantic, British Isles, Western Europe, and the Baltic-Russian region.

The origin of urban Hedeby dates to the early ninth century. In 808, Danish King Godfred destroyed the Slavic trading station called Reric—he had taxed its income very profitably—and moved its merchants to "Sliesthorp" (later known as Hedeby).

Godfred's decision to establish a trading community on the Schlei fjord was farsighted. The earliest portion of the Danevirke had been constructed just west of Hedeby, and the new wall would provide further protection against invasions from the south. Hedeby's accessibility to major land and water trade routes also contributed to its growth and prosperity. The major north-south transportation route through Jutland (the Army or Ox Road) passed through the Danevirke just west of Hedeby, and a road beside the Danevirke provided east-west access across the peninsula to the North Sea. Additionally, Hedeby had direct access to the Baltic via the Schlei fjord, and rivers that flowed into the Eider River and North Sea were also nearby. Hedeby's location made it an instrumental link in the network of trade routes across the North Atlantic and Scandinavia.

Despite the proximity of the Danevirke, however, Hedeby must still have been considered vulnerable. During Hedeby's early history, the Hochburg, a natural hill just north of the town, was fortified with a rampart of stone and earth as a refuge in case of attack. Hedeby itself remained unprotected until the mid–tenth century, when a defensive rampart was built to encircle the town. A short time later, in 968, Hedeby's rampart was joined to the Danevirke's main wall, providing a formidable defense that was further strengthened by the fore wall and a series of ditches to the town's southwest.

Hedeby's rampart was constructed in phases, beginning with an earthen rampart ten feet high. Subsequent additions to the rampart were built with earth and perhaps stones and may have been faced with sod or timber. Wooden palisades and towers may also have been constructed on the rampart's summit. Three wood-lined gates, two parallel to the water's edge and one opening inland, allowed access to the town. In its final stage, Hedeby's semicircular rampart stood 33 to 36 feet high and 4,250 feet long.

Hedeby's open shoreline, which left the town unguarded to attacks from the Baltic, was also fortified. A semicircular barrier of wooden pilings erected in the harbor afforded protection for the shore and the ships moored there. Remains of jetties suggest that the Vikings may even have attempted to control shore erosion.

Hedeby grew within the rampart's boundary. A stream running east-west through the settlement was channeled and bridged. Fresh drinking water came from wells, which were wood-lined shafts often created by sinking containers, such as wine barrels, into the ground. Hedeby's streets, paved with wood, separated small fenced plots on which stood rectangular dwellings and sometimes outhouses.

Excavations have revealed detailed information about Hedeby's houses. An average house, including a living space with hearth and a storage and working area, measured 13 to 20 feet by 16 to 33 feet. The exterior walls, supplemented by sloping outer posts, bore the weight of the roof and made interior support posts unnecessary. The walls were generally constructed of timber frames filled with either wooden staves or wattle and daub.

Most Viking-age houses have long since deteriorated, leaving only foundations of sunken wood beams or stone. At Hedeby, however, the water level has risen four feet since the ninth century, creating damp soil conditions that preserve organic materials exceptionally well. Excavations have uncovered the intact walls and gable of one house, dendrochronologically dated to 870. This house, sixteen by thirty-nine feet in size, had a timber foundation, wattle-and-daub walls, and exterior buttress posts supporting the roof. Study of this house has revealed that the structural elements of Viking homes (such as upright staves in the wall) were often replaced after fifteen to twenty years because of rot.

Using grave counts from burial grounds, scholars estimate that Hedeby's population averaged about 1,500 people throughout its 250 years of occupation. The contents of these graves reveal class distinctions and varying levels of wealth among its inhabitants. A merchant class most likely resided in Hedeby alongside a permanent population of craftsmen.

Hedeby emerged as Denmark's most prominent trading town at a time when trade was increasing throughout Europe. By the early ninth century, a network of trading centers was established: Quentovic near Boulogne; Dorestad on the Rhine River; Hamwih (present-day Southampton), London, and York in England; Ribe, Hedeby, Kaupang, and Birka in Scandinavia; Ralswiek, Wolin, Truso, and Grobin on the Baltic coast; and Staraja Ladoga in northern Russia.

Hedeby was a port for the exchange of goods from all over Europe and the Orient. Prized imports included metal, sword blades and armor, salt, wine, silk, spices, and other luxury items. Scandinavian products traded in return included furs, walrus tusks, iron, and even slaves seized on

Viking raids. Trading within Scandinavia itself flourished as well, and archaeological remains at Hedeby offer evidence of the town's role as a producer of typical Scandinavian goods.

A wide variety of industries coexisted at Hedeby. Tools such as spindle-whorls, loom weights, shears, and needles found there are evidence of textile production. Excavation of the harbor has even uncovered intact bales of cloth and rags that may have been produced on site. Leather footwear was also manufactured at Hedeby; remains include tools and wooden vats for tanning. Metalworking shops where molds and scrap metal were excavated probably produced iron tools and jewelry from precious metals.

Horn, antler, and bone were plentiful raw materials in Scandinavia. Craftsmen fashioned horn from cattle, sheep, and goats, as well as antler from red deer, reindeer, and elk, into practical and decorative objects such as combs, pins, game pieces, dice, and flutes. Discarded horn cores and burrs were found in large quantities at Hedeby. An excavated furnace for melting glass proves that glass beads and finger rings of green and yellow glass were manufactured there. Craftsmen at Hedeby refined local amber into jewelry, and potters created a dark gray, unglazed cookware that was traded locally.

With so much production and trade occurring in Hedeby, it is not surprising that the first mint in Scandinavia was established there in the ninth century. Early silver coins were based on those struck in Dorestad, Frisia, for Frankish Emperor Charlemagne. The Vikings often treated these coins as bullion rather than currency, a fact underscored by the scales and weights found in most northern European trading towns.

By the time of King Harold Bluetooth's reign in the late tenth century, Danish coins bore a cross symbolizing the kingdom's Christianity. In Hedeby's early period, though, pagan religion still predominated. Christians were not persecuted, but they were forbidden to build churches or worship publicly. Hedeby played a role in Denmark's eventual conversion to Christianity, for a Frankish missionary named Ansgar persuaded Danish king Horik to build Denmark's first church there around 850. Horik may have recognized that building a Christian church in Hedeby would promote trade with Christian nations.

The wealth and prestige that Hedeby commanded made it a desirable property. Despite the various fortifications erected by the Danes, Hedeby fell into enemy hands. Swedish forces controlled the city sometime during the late ninth and early tenth centuries, until German emperor Henry the Fowler and the Saxons invaded in 934. The Danish Gorm dynasty regained power thereafter, but faced another German threat in 974, when the Ottonian forces managed to occupy part of southern Jutland. In 983 Harold Bluetooth's son, Sven Forkbeard, captured the area and regained control of Hedeby.

The numerous foreign invasions and foreign visitors to Hedeby account for the variety of names by which the town has been known: Sliaswig, Sliaswich, Sliasthorp, Heidiba, æt Hædum, and Haithabu among them. These place names link ancient Hedeby with modern Schleswig, Germany, located on the north shore of the Schlei fjord.

Schleswig's growth coincided with Hedeby's decline. Hedeby suffered several devastating attacks in the mid–eleventh century, and was at last destroyed by King Harold Hardrada of Norway. Trade and population then shifted to Schleswig. One theory to explain this shift proposes that Schleswig's deeper harbor better accommodated the larger ships that developed after the Viking age.

Ancient Hedeby's remains were long thought to lie under modern Schleswig. In 1896, Danish archaeologist Sophus Müller suggested looking for Hedeby within the semicircular rampart on Haddeby Noor. Excavations commenced in 1900 and immediately confirmed Hedeby's location. Although only about 5 percent of the area within the rampart has been excavated, the archaeological finds have supplied a great deal of information about the Viking-age community. Underwater exploration of Hedeby's harbor in 1953 discovered not only intact bales of cloth but also the wreck of a Viking ship almost sixty-six feet long. This prime example of Viking craftsmanship was raised in the 1970s and is now on display in the Boat Hall of the Haithabu Viking Museum. Hedeby's rampart, thirty-three feet high in places, survives to this day, evidence of the city's 250-year dominance in regional and international trade.

Further Reading: *Towns in the Viking Age* by Helen Clarke and Björn Ambrosiani (New York: St. Martin's, and London: Leicester University Press, 1991) offers an up-to-date, extremely detailed discussion of Viking-age settlements in the Baltic, North Atlantic, western Europe, and British Isles, incorporating recent archaeological scholarship. Equally informative, Else Roesdahl's *The Vikings*, translated by Susan M. Margeson and Kirsten Williams (London: Allen Lane/Penguin, and New York: Viking Penguin, 1991; first published as *Vikingernes Verden* by Copenhagen: Gylendal, 1987) presents an incisive portrait of Viking history and culture that also draws on current scholarship. Two serviceable, though less recent, histories of Denmark are W. Glyn Jones's *Denmark* (New York: Praeger, and London: Benn, 1970) and Stewart Oakley's *A Short History of Denmark* (New York: Praeger, 1972; as *Story of Denmark,* London: Faber, 1972).

—Elizabeth E. Broadrup

Sergiev (Moscow, Russia)

Location: A Moscow suburb, Sergiev is situated forty miles northeast of the capital, connected to it by electric rail, highway, and bus.

Description: Sergiev is a small city that developed around the huge, sprawling monastery complex of Holy Trinity-St. Sergius, founded in the mid-1300s. The city was called Sergiev, after the founder of the monastery, St. Sergius, until 1930. It was then renamed Zagorsk after a Bolshevik hero. It resumed its original name after the breakup of the Soviet Union.

Contact: Ministry of Culture
Office of the Minister
Kitayskiy Proyezd 7
Moscow 103693, Moscow Oblast
Russia
(95) 925-11-95

Holy Trinity-St. Sergius monastery has been a spiritual nucleus of immense prestige and influence in Russia for six centuries. Under the dynamic leadership of its founder, the monk Sergius, the monastery radiated its influence throughout northern Russia, sending out scores of missionary monks to found monasteries of their own in the indomitable wilderness. Some 100 monasteries were founded in this time, extending Christianity and bringing the alphabet to regions and peoples that had not known either. At the same time, Holy Trinity-St. Sergius was a witness to and a participant in the rise of a centralized Russian state in the fourteenth and fifteenth centuries. For better or worse, the monastery supported that state, which protected and nurtured it. The atheist Communist government that supplanted the czars in 1919 could not suppress the monastery's influence. Deprived of the state's traditional support and nurture, it nevertheless survived Communist rule and continues to flourish.

Sergius took on this name after he became a monk in his early twenties. He was born Bartholomew around 1321, at a time when Mongols controlled the heartland of Russia. Their power was waning, and the dynamic city-state of Moscow was challenging that power, at the expense of the independence of neighboring cities. Rostov, the town in which the future monk Sergius was born, suffered economic ruin after being annexed forcibly to Moscow, and his family's fortunes declined along with it.

He and his older brother, Stephen, grew up in Radonezh, nine miles from the future site of the monastery. Both were intensely religious, and in young adulthood they decided to distance themselves from the troubled world. They left family and friends to seek spiritual consolation in the dense forests around Radoslav. Russians habitually sought refuge in the forests, which Mongols were known to avoid. While Sergius took easily to a life of isolated prayer and self-denial, his widowed brother soon had enough of the wilderness. Abandoning his younger brother to lonely piety, Stephen returned to the world—to Moscow—where he attained influence in high political circles. According to his biographer, Epifany the Wise, Sergius's sole friend and companion at this time was a tame bear.

Word of Sergius's astounding piety reached others. By the end of his two-year forest sojourn, he had a small circle of disciples. He became a tonsured monk, and, together with his spiritual brothers, he built a chapel of wood and called it Holy Trinity. This was the foundation of their monastery. He also adopted a set of rules by which all the monks were to live and which incorporated both work and prayer. The work was hard, and Sergius labored alongside his monks clearing woods, building a residence, and planting vegetables (the most important component of their diet). The monastery included a library, known to have been filled with many works of Greek mysticism that stressed an individual approach to God. This was the approach of Sergius, which he implanted in his monks.

So great was Sergius's humility that he refused to become the abbot or *igumen* of the monastery he founded. If anything, this heightened his reputation. He became known in Moscow (thanks in part to his brother), and pilgrims began heading in the direction of Holy Trinity to seek Sergius's blessing. After the death of the first abbot, Sergius bowed to the pressure to become his successor, however, and remained head of his monastery until his death, more than twenty years later.

It was a dramatic time in Russian history. The Grand Prince of Moscow sought Sergius's counsel, and it was recorded that in 1380 Grand Prince Dmitry Donskoi and his troops made their way to Holy Trinity to be blessed by Sergius on their way to do battle with the Mongols. Soviet historians maintain that Sergius even appointed two of his monks to lead the troops in battle. Others, however, insist that the two monks were sent along to provide spiritual consolation in the frightening ordeal that lay ahead. After the tremendous battle, the first in which the Mongols suffered serious defeat at the hands of the Russians, Dmitry Donskoi returned with his surviving troops to Holy Trinity to give thanks and to pray for the dead. Sergius became the spiritual mentor of the Grand Prince's sons.

Perhaps inadvertently, the monastery was drawn to the side of Muscovite expansion. Even in Sergius's lifetime, pious *boyars,* or noblemen, donated tracts of land and money to the monastery in exchange for prayers and masses in their honor. Peasants began tilling the monastery's tracts, while others derived a livelihood from the growing pilgrimage

Icon to St. Sergius, at his monastery

trade. The town that developed around the monastery was called Sergiev, after the founding monk. By the early fifteenth century, vast tracts of land had been cleared and tilled, a stone church replaced the wooden structure built by the hands of the founder, and tiny hamlets surrounded the monastery.

Sergius died a *starets,* or holy elder, around 1392, when he was over seventy years old. He was canonized after his death, having refused to become head of the Russian Orthodox Church. It seems he preferred to be the spiritual conscience of his countrymen and is remembered as such. His influence was visible in the scores of monasteries that his own monks established throughout the wilderness of northern Russia. Some twenty-five years after his death, his friend and fellow monk Epifany the Wise wrote a *Life of St. Sergius.* It was an auspicious time to record the life of this remarkable man, since by that time there were very few remaining monks who known the founder. Furthermore, the monastery had been burned to the ground by the Mongols and was slowly being rebuilt.

In 1408, a marauding Mongol detachment under Khan Edigei had attacked the monastery and burned it, including the church painstakingly built in timber by Sergius himself. The monks must have been forewarned, since they had all fled to safety, taking the monastery's most valuable possessions with them. The rebuilt monastery was decidedly more grandiose and impressive than what their founder had left when he died. Holy Trinity became Holy Trinity-St. Sergius monastery and was elevated to a *lavra,* or monastery of the first rank, the only monastery after the ancient Monastery of the Caves in Kiev to be so designated. (Two others were raised to that status in more modern times.) The monastery was becoming a beautiful complex of structures, centered on the Church of Holy Trinity. The whitewashed stone walls of the new stone building were decorated with the finest frescos and icons depicting biblical subjects.

The paintings on the walls were intended to help the illiterate worshiper to grasp the basic Christian doctrines. The icon of the Trinity, of course, took pride of place among the other decorations. The famed iconographer Andrey Rublyov, who had known Sergius, was commissioned to do this icon. A monk for most of his life and presumably literate, Rublyov masterfully revealed the essence of the Trinity by using, surprisingly, Old Testament material. The icon, four and one-half feet high and nearly four feet wide, at first glance appears to depict the story in Genesis of the divine visit to Abraham and Sarah's tent. The three visitors are angels who will deliver a message from God; and while Abraham hurriedly gives orders to prepare a meal for the guests, they sit at a simple table with a chalice of wine set before them. All three are dressed differently, but their faces are identical. While the lamb for the guests is being slaughtered—a symbol of Christ's coming crucifixion—the three drink the wine which suggests the blood of Christ. The table becomes an altar. Completely unlike other icons of the time, Rublyov's

Trinity is fluid and the background uncluttered. It is bathed in a graceful, luminous light. His graceful simplicity and masterful use of light had many imitators, but no rivals. His artistry had been touched by the mystical individualism of St. Sergius. Today, all of Rublyov's paintings have been replaced by copies, the originals, including frescoes, secured in the hands of the Tretyakov Gallery in Moscow.

When the cathedral was completed in 1422, set on the same spot as the original wooden church, the monastery was becoming one of the richest in Russia. It was a monastic fortress, ringed with walls and a dozen towers. What a horde of Mongols could do in 1408 was impossible for 20,000 Polish troops in 1608, when they laid siege to the monastery for nearly sixteen months. It was at the height of the "Time of Troubles" in Russian history, when one dynastic line had died out and a half dozen men contended for absolute rule. A moving opera by the Russian composer Modest Mussorgsky, set to the words of Russia's greatest poet, Alexander Pushkin, centers on one of these contenders, Boris Godunov. Acclaimed czar in 1598, he ruled for a short time from behind the Kremlin walls until his sudden death, possibly by poisoning, in 1605. His wife and son were murdered to get them out of the way, and all three lie buried in Holy Trinity-St. Sergius. Godunov is the only czar in Russian history to be buried in a monastery.

The monastery withstood the siege of 1608 as well as a minor siege in 1618. It was then secure until the time of the Russian Revolution. Building after building was erected in the monastery compound, among the most notable being the czar's palace, built in the late seventeenth century. The seventeenth-century hospital and the Church of Sts. Zosima and Savaaty and the sixteenth-century Assumption Cathedral, the largest church in the compound, are also particularly worthy of attention. In 1740, when Peter the Great's daughter Elizabeth ascended the throne, the monastery's famous four-storied Bell Tower was built, quickly becoming a landmark for the monastery and the surrounding region. The Empress Elizabeth's private generosity also laid the foundation for the monastery's theological seminary. Dozens of settlements developed near the monastery. In 1782, these were designated a *posad,* or suburb of Moscow.

By the mid-eighteenth century, the monastery had become the largest private owner of serfs in Russia, owning approximately 100,000 men (together with their wives and children, considerably more than that). Since serfs could be bought and sold, serfdom differed little from outright slavery. Manumission was practiced by some landowners in Russia, as in the American South. The monastery was not one of these. Not all the monks and abbots of the approximately 1,000 monasteries in Russia were unanimous in defending serfdom, nor did most of them own serfs. Donations of land and the "souls" that accompanied the land continued unabated, until Empress Catherine the Great put a stop to the acquisition of further property, including serfs, by monasteries. In 1763 she deprived the monasteries of their serfs and made them the property of the state. In 1858 no Russian

Orthodox monastery possessed a single serf. Serfdom and great wealth in no way diminished the prestige or credibility of Holy Trinity-St. Sergius for believers, rich and poor alike. Pilgrimages were constant, and on orthodox holy days the religious filled the town of Sergiev to capacity.

Down to the present day, the monastery has remained the number-one industry of the town, along with handicrafts. The population of Sergiev remained fairly small, reaching 100,000 only by the 1970s, despite the presence of a railroad in the town since 1862. It is noteworthy that the town square was a center of trade and commerce, a veritable market place of hundreds, if not thousands, of goods.

With the Communist takeover, Lenin decreed in April 1920 that the town was to become a museum city, perhaps to emphasize the "obsolescence" of religion in the new era. The monastery was nationalized, although religious services were permitted to continue. A museum dedicated to medieval art was to be constructed in the town (the present-day State Museum–Preserve of History and the Arts). In official Soviet history, Sergius was honored not as a religious figure but as an early proponent of Russian nationalism. Lenin's successor Stalin intensified religious persecution in the Soviet Union overall, and the situation perhaps became threatening enough to warrant the town to change its name in 1930 to Zagorsk, derived from the name of the Bolshevik revolutionary Vladimir Zagorsky, who was assassinated by rival socialists in 1919.

The ancient monastery of Holy Trinity-St. Sergius, in which its founder lies buried, has survived the vicissitudes of Communism and has outlived an atheist government. Now returned to its original name, Sergiev continues to attract thousands of tourists and pilgrims annually to its incomparable monastery, the spiritual heart of Russia during medieval times. In 1992 a year-long series of events was held at the monastery and throughout Russia to commemorate Sergius on the 600th anniversary of his death.

Further Reading: A good description of Sergiev, of the founder and its subsequent history and buildings, is contained in the relevant volumes of the multi-volume *Modern Encyclopedia of Russian and Soviet History,* edited by Joseph L. Wieczynski, 58 volumes (Gulf Breeze, Florida: Academic International, 1976–1994). There is also the Soviet guidebook of the Trinity-St. Sergius monastery, translated into English, and richly illustrated: *Zagorsk. Trinity-St. Sergius Monastery* (Moscow: Sovietskii Khudozhnik, 1967). While its perspective is decidedly ideological and antireligious, it is a useful guidebook to the art and architecture of the monastery. Arthur Voyce's *The Art and Architecture of Medieval Russia* (Norman: University of Oklahoma Press, 1967) is an excellent introduction into this complex subject.

—Sina Dubovoy

Sibiu (Sibiu, Romania)

Location: Sibiu is set on a hill 1,400 feet above sea level, above the Olt Valley along the northern expanse of the Făgăraş Mountains (part of the southern Carpathian Alps), roughly halfway between Cluj and Brasov. Sibiu sits in the center of Romania, along the lower rim of the Transylvanian region, approximately 150 miles northwest of the capital of Bucharest.

Description: Sibiu, capital of the county of the same name, is divided into upper and lower halves connected by stairway and by an intricate network of tunnels (now mostly sealed off) intended for evacuation during wartime. The upper level of the city retains much of its medieval charm, marked by roads too narrow for motor vehicles and ancient buildings that articulate the craftsmanship of Saxon settlers. Five of Sibiu's seven ancient defense towers remain standing, as do many of the city's historic cathedrals and cultural monuments. Sibiu's lower half sprawls into a suburban/industrial district representative of twentieth-century Romania.

Site Office: Prima Ardeleana S.A.
Piaţa Unirii 1
Sibiu, Sibiu
Romania
(924) 11-788

For a relatively small city, Sibiu has a rich history. Located along the Carpathian mountain range in the often tumultuous Transylvania region, Sibiu has maintained its largely German heritage through centuries of eastern European infighting. To the Germans, the city was known as Hermannstadt; to the Hungarians, it was Nagyszeben. For a time, Sibiu was the military capital of Transylvania, with its *Cetatii,* or fortified brick walls, providing testimony to the confrontational nature of surrounding peoples. For much of the eighteenth century and again during the mid-nineteenth century, Sibiu served as capital of the Voivodate. Sibiu is the home of the one of Europe's oldest museums, the Brukenthal, which today is the site of an extensive folk art collection and a comprehensive library documenting the history of the region. Romanians often refer to Sibiu as the city of "mystical sadness and impossible beauty."

The first settlers of the Carpathian-Danube area, Bronze Age peoples from north of the Black Sea, arrived in the early second millenium B.C. The extraordinary fertility of the region encouraged farming, and subsequent generations left a legacy of skilled craftsmanship. Local blacksmiths became societal members of particular standing, as their vocation was considered indispensable to the achievements of the community.

Of the first Europeans to settle this part of eastern Europe, the Illyrians settled along the Sava River toward the Adriatic Coast, while the Thracians moved east toward the Carpathians. Between the ninth and fifth centuries B.C., the people called the Dacians by Latin scholars (Getians by the Greeks) began to develop their singular tribal characteristics; they are the ancestors of modern-day Romanians.

At this time, the so-called "Hallstatt" Iron Age culture was developing in and around the Carpathians. The Dacians, who spoke in an Indo-European dialect from which a few words survive in contemporary Romanian, absorbed much of their lifestyle from Greek societies living along the Black Sea. Additionally, the Dacians learned to ride horses from the nomadic Scythian tribe, who arrived in the Carpathians by way of the Ukraine around the sixth century B.C. The Dacians improved their ironwork and pottery through the teachings of the "La Tène" Celts, who moved into the Transylvanian Alps during the fourth century B.C.

During the fourth and third centuries B.C., the Dacian people opposed both Phillip II of Macedonia and his son, Alexander the Great. Later, during the first century A.D., the Dacians endured periodic intrusions by Roman forces. In coalition with other groups, Dacians killed the governor of the Roman province in A.D. 85 or 86. However, the Romans held the Province of Dacia between the years 106 and 271. A long-standing calm among the various ethnic groups accompanied this period of Roman rule in the Carpathians until groups of Germanic peoples banded together to mount a fierce attack on the Romans. Rather than battle, the Roman Emperor Aurelian chose to retreat south of the Danube, thus evacuating Dacia.

Activities in the Dacian region lack historic record from the year of the Roman departure through the year 602, a void of information that has contributed to the centuries-old argument between Romanians and Hungarians over the true lineage of the territory. In 896, the Carpathian Basin was occupied by the Magyars, newcomers of Finnish-Hungarian descent. The Magyars were a nomadic tribe who named the region Transilvania, "the land beyond the forest". Led by Prince Árpád, the Magyars used their settlement as a home base for banditry in other parts of Europe, a practice they continued until their bloody defeat, near Augsburg in 955, at the hands of the Holy Roman Emperor Otto the Great. By the year 1000, the Magyars had converted to Christianity, and had renounced their malevolent tendencies.

Hungarian historians hold that the area now known as Transylvania had already been inhabited by Slavic tribes united under rulers called *voivodi,* tribes who fell under Magyar influence upon the arrival of the latter group. A couple of centuries after the arrival of the Magyars, German settlers began entering the region, an immigration documented as

Street in Sibiu, near the city walls
Photo courtesy of Romanian National Tourist Office

early as 1143. Hungarian leaders invited German families to share the Transylvanian farmland in order to bolster their numbers against invading Mongols, and they made the coalition official with King Andrew II's charter "Andreanum," drafted in 1206. Saxon Germans from Flanders, Luxembourg, and the Rhineland, as well as Lower Saxony, created the *sedes* (seats) of the *Sieben Burgen,* the Seven Fortresses of self-governing city-states including Hermannstadt (Sibiu), Kronstadt (Brașov), and Bistritz (Bistrita). In 1195, the newly arrived Germans won a decisive court case before the Roman Curia in which the bishop of Transylvania had disputed the legitimacy of their Hermannstadt clerical chapter.

Hermannstadt's rise as an important commercial hub can be attributed to its position along the roadways linking northern Europe with the Black Sea and beyond. With access to the Red Tower pass (*Turnu Roșu*) in the Carpathians (so named because it is said to be permanently stained with the blood of invading Turks), the city provided a midway point along the trade route, making it a source of some prosperity for its German merchant class.

Another ethnic group with significant ties to the Carpathians was the Szeklers, a Turkish people who spoke Hungarian. Hungarians see the Szeklers as Turkish-Bulgarian collaborators with the nomadic Magyars; Romanians prefer to describe them as Turks driven back into Transylvania by Hungarians after having occupied the Roman province named Pannonia. In either view, the Szeklers add another piece to the compositional puzzle that is modern-day northwest Romania.

By the twelfth century, recorded history shows that Transylvania was divided into counties, leading Hungarians to believe that this marks the assimilation of the region into the Hungarian nation, which was also organized by county. Romanians, on the other hand, assert that their voivods retained rule over much of Transylvania. Romanian voivods, however, were gradually absorbed into the Hungarian nobility, who ostensibly created the first statehood in Transylvania, a feudal society that extended privileges to neither Romanian nor Hungarian tenant farmers. Unlike these disenfranchised serfs, the Saxons and the Szeklers in Transylvania enjoyed statutory rights as the second and third "nations" of the area, electing their own counts, selecting their own adjudicators, and realizing tax exemptions in exchange for military service.

Romanians seeking status as a fourth Transylvanian nation were repeatedly stifled. In September 1437, and again in January of the following year, a union of the Hungarians, Saxons, and Szeklers reiterated their commitment to the regional social tier system that relegated Romanians to a politically powerless peasantry. Although Romanians, with approximately 500,000 inhabitants in fifteenth-century Transylvania, constituted a majority of that region's population, they were perceived as a "people without a history". They were treated as second-class citizens, and the few remaining members of the Romanian upper class generally assimilated themselves into Hungarian traditions and into Catholicism.

In 1438, Sultan Murad II of Turkey joined with the voivod Vlad Dracul of Romanian Wallachia to conquer Transylvanian villages as far as Sebeş and Orăştie, before being turned back at Sibiu. This march precipitated a century of antagonism between Transylvania's noblemen and Turkey's Ottoman Empire. During this struggle, the Transylvanian voivod John Hunyadi emerged as a heroic figure, earning a hard-won victory over the Turks in July of 1456. Two decades later, the voivod Stephen Báthory once again staved off a Turkish advance, but by 1521, Süleyman the Magnificent had taken southeastern Europe as far as Belgrade and Budapest.

During the 1450s, an unusual alliance was struck between the Transylvanian governor Hunyadi and the second son of Vlad Dracul, Vlad Dracula, also known as Vlad the Impaler, the "prince of darkness" whose life story inspired Bram Stoker to write his widely influential novel, *Dracula.* In an era marked by crippling concessions on the part of even the most combative leaders, Hunyadi offered an army appointment to Vlad Dracula, a man who was already a veteran of five tours of duty in the enemy Turkish military. For his cooperation, Dracula was ensconced in a command post at Sibiu, where the German locals were obliged to accommodate him because he and his Romanian mercenary army protected the Transylvanian passes from the advances of the Turks.

As his political power increased with successive victories, Dracula's once-strong relations with the Germans of Transylvania began to sour. His imposition of strict trade customs on the Saxons led to political betrayal and in response, Dracula laid waste to several Saxon towns in a furious attack, mercilessly murdering their citizens. Dracula's rampage was curtailed when his ally, Mihály Szilágy, failed in his attempt to take Sibiu on October 9, 1458; by this time, the city was one of the best-fortified in the land.

The relentless Dracula stormed Transylvania once again in 1460, again destroying entire colonies using the methods of gruesome warfare that would pepper his legend. This time his mission was to avenge Saxon support in Sibiu for Vlad the Monk (his own half-brother, whom Dracula considered a threat to his second reign as Prince of Wallachia). A peace treaty, guaranteed by the Hungarian king Matthias, required that Dracula restore German commercial privileges in exchange for "maintenance money" he would receive from the cities of Sibiu, Brașov, and others to keep a standing army for their protection against the Turks. With Transylvania back in his fold, Dracula once more turned his attention to the more threatening warriors of the Ottoman Empire.

During a brief third rule as prince of Wallachia, Dracula, his Hungarian wife Ilona (Matthias's cousin), and her son lived in a Sibiu mansion before moving to Buda as guests of the Hungarian court. In December 1476, the lifelong fighter was killed during a battle with a Turkish battalion twice as large as his own. Thirty-four years later, his first-born and illegitimate son, Mihnea the Bad, a ruthless ruler cast in his father's mold who had also achieved the Wallachian throne, was murdered by a hired assassin as he attended Mass in the Roman Catholic church of Sibiu. Mihnea's body is entombed in the church (now of Evangelical denomination), which stands in the center of the Piata Grivitei. Most of his father's official correspondence can be found today in the archives of Sibiu and Brasov.

Following the Turkish conquest of Transylvania during the 1520s, the territory was buffeted throughout the century by repeated transfers of power among the Austrian and Polish dynasties and the Ottoman Empire. Michael the Brave, another Wallachian prince whose name is commonly heard in the histories of Romanian development, was ousted from Transylvania in September 1600 by Hungarian and Saxon leaders. His short-lived union of the three Romanian provinces—Wallachia, Transylvania, and Moldavia—was an early attempt to declare some form of statehood for descendant Romanians.

By 1686, Transylvania had aligned itself with Vienna's Habsburg dynasty. At the same time, however, Hungary's royalty considered Transylvania an independent Hungarian principality. Furthermore, the sultans of Constantinople, treating it as a vassal, continued to tax the region for an annual monetary tribute of 10,000 ducats and retained the right to confirm Transylvania's princes, upon election by the Diet of Cluj. The toll in human life was heavy as Transylvania provided the battleground for disputing factions on many occasions.

Sibiu, the most affluent of the Saxons' "Seven Fortresses," came to be known as the City of Seven Towers. Five of the Towers survive and stand as a kind of living monument to the city's graceful but embattled past. Houses and public structures that date to the long era of instability in Sibiu are equipped with tiny windows on their uppermost levels that were used as watchposts. Their visual effect has inspired an old local reference to "the eyes of Hermannstadt".

The Reformation of the sixteenth century did little to improve the religious standing of Romanians, whose adherence to Eastern rite of Catholicism did not afford them the privileges enjoyed by the clergies of Saxon Lutherans, Szekler Unitarians, Calvinists, and Roman Catholics. The Reformation did, however, lead to greater freedoms in education. The first book published in Romania, a catechism, was printed by a Romanian in Sibiu in 1544. Saxon Lutherans had established a college in Sibiu by 1600, joining schools of higher learning in Cluj, Braşov, and Bistrita.

On December 4, 1691, the signing of the Leopold Diploma reiterated the Union of the Three Nations in Transylvania as recognized under Habsburg rule; a subsequent seat of the diet convened in Sibiu. Despite this confirmation of the status quo, residents of Romanian heritage were gradually shedding the shackles of serfdom as they volunteered for armed service in exchange for exemption from labor obligations.

Baron Samuel Brukenthal, governor of Transylvania between 1777 and 1787, courtier of the notorious Empress Maria Theresa of Habsburg, and inventor of implements of torture such as the rack, was also a distinguished patron of the arts. His residence in Sibiu became a gallery for some of the best works by master painters of the day. In accordance with his will, his palace at Piaţa Revoluţiei 3 was converted to a museum space in 1817. The Brukenthal boasts an extensive collection of masterworks by artists such as Rubens, Van Dyck, and Titian; folk art and ancient remains; sculpture; religious artifacts; nineteenth-century Romanian paintings inspired by French artists; a comprehensive library of texts on Sibiu and Transylvania; and, lastly, the baron's own designs for his instruments of torture, all of which make the museum one of the finest in eastern Europe.

Throughout the eighteenth century, Habsburg rulers expanded the role of production on behalf of the state, employing miners of iron, copper, salt, gold, and silver. Sibiu, besides being a leading manufacturer of gunpowder, employed much of its citizenry in consumer industries. By 1792, the city claimed six glass factories, six for the treatment of wax, one for cloth, and two paper mills. The miners' rebellion of 1784 resulted in the abolition of Transylvanian serfdom by the reformist ruler, Joseph II. The official Hungarian language was replaced with German, and the autonomous Saxon and Szekler districts were eradicated. In turn, Magyar resentment of these reforms led to the transfer of the Gubernium, or government seat, from Hermannstadt to Kolozsvár. Romanians, meanwhile, were increasingly free to marry, settle, and learn a trade as they chose.

Joseph II's death in 1790 brought on the reign of Leopold II, an ardent supporter of the nobility. By this time, the Romanian people had fostered their own intellectual subculture and were determined to end their inferior social standing. Consequently, Leopold's reactionary measures unleashed vehement Romanian opposition. After decades of increasing protest, the Romanians saw the Paris revolution of 1848 and the fall of Metternich in Vienna as their cues to revolt in the name of national consciousness. The intellectual leader George Baritiu led the way by publishing the first Romanian newspaper, the *Transylvanian Gazette,* in response to resurgent Hungarian nationalism. On May 16, 1848, Romanian planners presented a sixteen-point petition demanding political representation; free trade; and liberties of religion, education, language, and the press. A permanent committee was created at Sibiu to ensure application of the petition.

The diet at Klausenburg voted down the proposal. The Romanians convened a committee of six at Sibiu, who in turn enlisted commander Avram Iancu to organize a corps of soldiers for the expulsion of the Hungarian dictate in Transylvania. Magyar estates were pillaged and national guardsmen were attacked. The Germans of the Seven Fortresses remained neutral. On December 16, the Sibiu Romanians delivered another petition to Francis Joseph I, the new emperor of Austria and Hungary, but in six months they were driven back to Wallachia by the Hungarian revolutionary army. An impending truce between the two sides was rendered meaningless when the Czar Nicholas of Russia sent troops to quash the Hungarian rebels at Francis Joseph's request. The emperor then proceeded to instigate reform legislation, but Romanian accession to statehood had been thwarted.

The Transylvanian Diet returned from Kolosvár to Sibiu in 1863, where protesting Magyar representatives refused to sit, creating a "rump Parliament" in which the Saxons and Romanians shared an open floor. During the absence of the Magyars, the Romanians passed legislation demanding equal rights for their language and church, but these amendments were declared illegal upon the return of the Magyars. As an indirect result of this effort, however, the Romanians' Bishop Şaguna secured an archbishopric at Sibiu for the Romanian Orthodox Church, independent of the patriarchate in Serbia.

Though gradually attaining justice, Transylvania's Romanian population would have its voice substantially oppressed until the culmination of World War I. During the war, their two newspapers in Sibiu, the *Romanian* and the *Luceafărul,* were silenced. By 1918, however, the decline of Habsburg power had paved the way for Romanian national unity. With the approval of the National Assembly of Saxons, a Unified Greater Romania became a reality on January 11, 1919.

In 1940, the new country watched as its northern Transylvania border was rolled south by the Axis powers, a revision that awarded Hungary a 26,000-square mile parcel of land inhabited by 2.6 million people, half of whom were of Romanian descent. In an effort to appease the increasingly powerful Adolf Hitler, the Romanian government began using the existing National Center for Romanization to strip Romanian Jews of their privileges. World War II severely damaged the young nation of Romania, and after 1945 the country came within the sphere of Soviet influence. However, its distinct culture and the independent will of its Ceauşescu-family leadership eventually set it apart from other Communist nations.

In recent decades, the German percentage of Sibiu's population has dwindled (only 7000 of 170,000 residents in 1990) and yet the city is indelibly stamped with a characteristically Saxon high standard of efficiency and a culturally pervasive Saxon atmosphere. Activity in Old Sibiu, the upper portion of the town, generally centered around Piata Revolutiei until recently, when that square began to share its traffic with the newly renovated Piata Unirii.

Off the Piaţa Republicii stands the renowned Hotel Imparatul Romanilor, built in 1800. Its dazzling Baroque appointments made the hotel an attractive destination for such eminent guests as Franz Liszt and Johann Strauss. Several historic structures stand on Piaţa Mica (Little Square). One is

Podul de Fier, commonly called the Liars' Bridge. Local legend has it that if one tells a lie while standing on the bridge, it will collapse. The Turnul Sfatului, or City Council's Tower, a bastion-like structure dating from the fourteenth century, houses seven floors of medieval relics and artifacts. The Muzeul de Ethnografic Universala Franz Binder loans Romanian treasures in exchange for exhibits of art and artifacts from far-off lands such as Zaire, Cuba, and Tahiti.

The Orthodox Cathedral, southwest of Piaţa Revoluţiei, was built in 1905 as a replica of Istanbul's Hagia Sofia. The Catholic Cathedral, on Piaţa Revoluţiei 2, was built by Jesuits in 1726. The interior shimmers with gold-laced walls and pink marble colonnades that belie its drab exterior, the intricate frescoes of which were covered by concrete at the order of Ceauşescu. The Town Hall, known as Primaria Veche, lies one block south of Piaţa Griviţei and houses a history museum on the site of a one-time Nazi administrative office. Rising alongside Strada Cetatii are the well-preserved remains of Sibiu's archaic fortifications, their archers' portals reflecting on the city's long history of conflict.

During the 1989 revolution, which saw the execution of Nicolae and Elena Ceauşescu, skirmishes lasted longer in Sibiu, the base of operations for Nicu, the Ceauşescus' son. In an account of that uprising, Sibiu-born National Public Radio correspondent Andrei Codrescu writes that the ancient city, threatened by the Ceauşescus' intent to raze it to make way for agro-industrial complexes, was spared by Nicu's protests. "Sibiu is a place for the ambitious sorrow of youth," writes Codrescu, and he quotes the Romanian-French writer Cioran: "There are only two cities in the world—Paris and Sibiu.''

Further Reading: *A History of the Romanians* by Georges Castellan (Boulder, Colorado: East European Monographs, distributed by Columbia University Press, New York: 1989) is an exhaustive study of the fitful, under-documented history of the people and their land, with multiple references to Sibiu's role in their development. *A History of the Roumanians, from Roman Times to the Completion of Unity* by R.W. Seton-Watson (Hamden, Connecticut: Archon, 1963) contains less factual information concerning Sibiu but serves well as a secondary source. A little surprisingly, one of the best reference materials on Sibiu is *Dracula, Prince of Many Faces: His Life and His Times,* a meticulous account of the real Vlad Dracula's life as a royal leader and warrior, written by Radu R. Florescu and Raymond T. McNally (Boston: Little, Brown, 1989; and London: Little, Brown, 1990); Vlad Dracula spent considerable time in Sibiu, and the authors spent considerable time researching the era. *The Hole in the Flag: A Romanian Exile's Story of Return and Revolution* by Andrei Codrescu (New York: William Morrow, 1991), as it pertains to Sibiu, is useful primarily as background information, given its descriptive reminiscences written by a former resident.

—James Sullivan

Skara Brae (Orkney, Scotland)

Location: Orkney consists of about seventy islands and is situated six miles north of the Scottish mainland; about twenty of the islands are inhabited. The population of Orkney is approximately 21,000. Kirkwall is the capital and the largest town. Skara Brae is located nineteen miles northwest of Kirkwall on the west coast of Orkney's main island, overlooking the Bay of Skaill.

Description: Ruins of a neolithic fishing and farming village that was inhabited between 3100 and 2500 B.C. Skara Brae is one of the best preserved Stone Age villages in Europe.

Contact: Orkney Tourist Board
6 Broad Street
Kirkwall, Orkney KW15 1NX
Scotland
(856) 872856

Many peoples have populated the Orkney Islands during the past 6,000 years, from neolithic settlers to the Picts to the Vikings to the Lowland Scots. Not surprisingly, the islands, located off the northern coast of the Scottish mainland, contain many prehistoric monuments; indeed, few other places in Britain have as many archaeologically significant sites. The neolithic village of Skara Brae on the west coast of Orkney's main island is considered the best-preserved prehistoric village in northern Europe. Continuously inhabited for 600 years from about 3100 to 2500 B.C., Skara Brae existed before the pyramids of Egypt were built and before construction began at Stonehenge.

Orkney separated from the Scottish mainland around 10000 B.C. At that time, Orkney consisted of a treeless landscape battered by the incessant winds. The islands are mostly composed of flagstone and sandstones—the building blocks of Orkney's Neolithic villages. Around 1900 B.C. the climate turned colder and wetter. It worsened again around 1200 B.C. By this time, peat bogs had developed. Only the sturdiest crops were able to withstand Orkney's fierce climate.

The first permanent settlers arrived in Orkney around 3500 B.C., crossing the Pentland Firth from the Scottish mainland. The oldest recorded settlement appears to be at Knap of Howar on the island of Papa Westray. Archaeological findings reveal that life within Orkney's Neolithic communities was fairly sophisticated—the settlers employed mixed farming techniques and made their own pottery and tools. The lack of trees on the islands forced the inhabitants to consider other means of construction besides wood. (Wood from the forests of North America, in the form of driftwood, did come ashore, but not in sufficient quantities to construct wooden homes.) The islands may have lacked timber but they had an ample supply of stone. Consequently, Orcadians became master stone builders.

Skara Brae was a self-sufficient village. By all accounts, the inhabitants enjoyed a stable community life for more than 500 years. The village's design indicates that Skara Brae was essentially an egalitarian society. The surrounding land was available for pasture while the sea yielded a rich harvest of marine life. The inhabitants were farmers who raised sheep and cattle and grew wheat, barley, and grain. The most important form of sea food was fish and marine mollusks, but the settlers also hunted whales, seals, seabirds, and otters, most likely for their skins, oils, feathers, and pelts rather than as a source of food. Unlike today, sand dunes separated the village from the sea; thus, in Neolithic Orkney, the village was located a considerable distance from the water's edge.

The settlers apparently lived in harmony with their environment. In addition to mastering the necessary survival techniques, they also were able to enjoy the fruits of an artistic life. Pottery, pins, beads, and pendants as well as carved stone artifacts have been discovered at Skara Brae, in addition to what appears to be a workshop area, where craftspeople created jewelry and other arts. Skara Brae was a small community, according to some historians, probably with no more than ten or fifteen families, certainly no more than twenty families at any one time.

The descendants of the original settlers improved the dwellings, building a new village on the same site in a process that may have taken a generation to complete. The plan of the second village, however, was somewhat different from that of the first village. The earlier and smaller houses were built around a central hearth, whereas the later homes were larger, with beds situated in the main living area. The present remains are mostly of the village from its second phase of habitation. At one point, Skara Brae consisted of at least six houses that were connected by a narrow passageway covered with turf. Beams supported the roof.

The inhabitants of Skara Brae lived simply. Individual houses consisted of a single room, mostly similar in size, made of thick drystone walls. The entrance was by a single, low, narrow doorway. The furniture, too, was simple—dressers and beds made of stone. Some of the stones were decorated with geometric designs.

A large central hearth was located between the door and the dresser with box beds on both sides of the wall. There were also cupboards and a number of cells of various sizes and shapes. The cells were used for various purposes, possibly as storage spaces, as safes, or as lavatories. Boxbeds were made of stone, mattresses of bracken. Sheepskins served as blankets. The settlers used various natural resources for fuel, such as mixtures of animal dung, dried seawood,

The interiors of dwellings at Skara Brae
Photos courtesy of Orkney Tourist Board

heather, and whalebone. Ventilation was primitive, at best, consisting as it did of a hole in the roof that allowed smoke to escape. The villagers probably collected wild plants and herbs as well as fruits and nuts. At least one historian has suggested that the diet of the villagers was a diverse one, including cow's milk, fresh water from nearby streams, and even such items as venison and oysters.

Unlike most settlements in Britain at the time, which consisted of a single farmstead with surrounding buildings, Skara Brae formed a closely knit community. It was also one of the first known farming villages in Britain. The villagers shared a large central space with further divisions within the settlement based on a family unit. Despite this pattern of communal living, the villagers did apparently make arrangements for a degree of privacy. Houses were designed as manageable units, best suited to accommodate small groups of people rather than extended families.

The semi-subterranean houses were surrounded by a mound of midden material (a mixture of the waste of everyday material such as bones, animal dung, rotting vegetables, and shells). Since the mound essentially functioned as a foundation for later construction, the creation of the midden heap was the first step in the construction process. Villagers collected their daily rubbish and accumulated it on a nearby site. When sufficient amounts of midden formed, the villagers deposited it wherever they wished to build. They then dug holes and put into place the stones for the new structures.

The descendants of these early Neolithic peoples were the Picts, an enigmatic people who left no written records behind, although they did leave carved symbol stones, containing the cryptic *ogam* inscriptions. Orkney was a part of this Pictish kingdom. The Pictish era began in A.D. 297 and ended with the union of the Picts and Scots under the Scottish king Kenneth MacAlpin in A.D. 843.

Some historians believe that the emergence of regional communities eventually replaced the self-contained villages and, consequently, may have led to the ultimate collapse of Skara Brae and similar sites. Others maintain that the depopulation resulted over a long period of time, as younger people began to move to single farmsteads or leave the community entirely in search of greater fortune. Gradually, the population of the village decreased until there came a day when no one remained. Sand from the nearby beach buried the community after it was abandoned. The once thriving village of Skara Brae lay dormant, covered under a layer of sand, for many years.

Finally, in 1850, a violent storm partially exposed a huge refuse heap and the ruins of ancient structures. William Watt, the landowner at the time who lived nearby in the seventeenth-century mansion called Skaill House, examined the remarkably well preserved ruins that the great storm uncovered. By 1868, four houses had been excavated and various objects and artifacts removed to Skaill House. The site remained in essentially the same condition until a violent storm in December 1925 damaged part of the previously cleared structures.

The ruins were placed under government protection. To prevent further damage, a sea wall was built. Around this time, in 1927, additional structures were discovered beneath the refuse heap. Dr. V. Gordon Childe was hired to supervise the archaeological excavations. Between 1928 and 1930, the rest of the buildings were excavated. Further excavation was done in 1972 and 1973.

Initially, Childe concluded that the remains at Skara Brae were of Pictish origin, that is, circa early in the first millennium A.D. After further deliberation, he realized his mistake and determined that they were, in fact, Neolithic—some 2,000 years older. Radiocarbon dating has since indicated that Skara Brae is actually 3,000 years older than Childe originally thought.

Skara Brae is only one of four well-preserved Neolithic settlements to be excavated in Orkney. The others are the Links of Noltland and Knap of Howar on Westray Island and Rinyo on Rousay. The early Neolithic peoples who built Skara Brae were also responsible for other prehistoric monuments in Orkney, including the magnificent chamber tomb at Maes Howe, the Standing Stones of Stenness, and the henge monument of the Ring of Brodgar. The latter two structures are the most northerly example in Britain of henge monuments. These and other structures suggest that the early inhabitants of Orkney were a deeply spiritual people. Important social and religious functions were probably held at these sites, each with their own peculiar rituals and customs.

Further Reading: *The Prehistory of Orkney BC 4000–1000 AD* (Edinburgh: Edinburgh University Press, and New York: Columbia University Press, 1990), edited by Colin Renfrew, is a comprehensive examination of 5,000 years of Orkney life from the Neolithic era to the age of the Vikings. The book, academic in both tone and approach, also contains information on the latest findings gleaned from archeological digs. *Skara Brae: Northern Europe's Best Preserved Prehistoric Village* (Edinburgh: Historic Scotland, 1989) by David Clarke and Patrick Maguire is the official guide to the village. The history of Orkney does not end with Skara Brae, however. For a later view of Orcadian history—including the Viking era—the best source is *Orkneyinga Saga: The History of the Earls of Orkney,* translated by Hermann Palsson and Paul Edwards (London: Penguin, 1978). Written by an anonymous Icelander around 1200, the saga chronicles the lives of the earls of Orkney over 300 years from the ninth to the thirteenth centuries. One of the best ways to get a feel for a place is to read fictional works set there. The finest writer to emerge from Orkney in this century is arguably George Mackay Brown. Brown's novels, short stories, and poetry are imbued with the feeling of the rich Orcadian past. Among his recommended fictional works are a novel, *Magnus* (Glasgow: Richard Drew, 1987), about the murder of St. Magnus by his cousin Hakon in the twelfth century, and *Hawkfall* (London: Triad/Panther, 1983), a collection of short stories. Mackay Brown is also a thoughtful prose writer. His *Portrait of Orkney* (London: Hogarth, 1981) is a loving tribute to his beloved islands.

—June Skinner Sawyers

Slagelse (Vestsjælland, Denmark): Trelleborg Viking Fortress

Location: Confluence of the Tudeå and Vårbyå Rivers on the western coast of Sjælland (Zealand), Denmark, approximately three miles west of Slagelse and one and three-quarters miles inland from the Great Belt.

Description: Tenth-century Viking fortress encompassing approximately seventeen and one-half acres. A circular rampart enclosed roads, courtyards, and houses built on a precise geometric plan; an outer fortification protected the rampart and additional buildings. Constructed in A.D. 980, probably by order of King Harald Bluetooth, and abandoned or destroyed less than fifty years later. The excavated site is now open to the public.

Site Office: Vikingeborgen Trelleborg
Trelleborg Alle, Hejninge
4200 Slagelse, Vestsjælland
Denmark
53 54 95 06

Trelleborg, an ancient Viking fortress built in A.D. 980, was abandoned after only fifty years in use and remained unknown until its discovery in the early twentieth century. Extensive on-site investigations have not only clarified Trelleborg's original purpose but have also enriched and refined modern views of Viking society and culture.

Encompassing approximately seventeen and one-half acres, Trelleborg consists chiefly of a main ward, enclosed within a circular earthen rampart, and an outer ward of structures shielded by a narrower embankment and ditch. The fortress's most striking characteristic is its precise geometrical design. The rampart surrounding the main ward was perfectly circular, broken only by gateways placed at the four points of the compass. The two wood-paved roads connecting the gateways crossed at the ward's center, dividing it into quadrants. In each section stood four identical buildings clustered around a central courtyard.

In addition to the sixteen buildings located within the circular rampart, fifteen similar structures stood beyond it to Trelleborg's eastern, landward side. This outer ward also displayed a geometrical, mathematical precision. The fifteen buildings were aligned radially with the circular rampart, their gables pointing toward its center and forming a precise arc. The radius measured from the inner gables of these structures to the main ward's center was twice the radius to the rampart's inner boundary. A narrower embankment and ditch formed Trelleborg's outermost defense on its eastern side, which was not protected by the rivers. This fortification, too, followed a concentric arc until veering east to enclose a burial ground.

Excavations at Trelleborg have given scholars a clearer idea of how the fortress looked during its active period at the end of the tenth century. The circular earthen rampart, already formidable at six and one-half yards high, was further fortified on both sides by palisades, rows of pointed stakes set in the ground to form a wall. On its eastern side, the rampart was bordered by a moat nineteen yards wide and four yards deep.

The four gateways providing entrance to the main ward cut through the rampart at its base, which was eighteen yards thick. The interior sides of each gateway passage were lined with semicircular wooden posts. These posts helped support the roof of the gateway, made of wooden planks covered with clay. A massing of boulders stretched six and one-half yards into the rampart behind the wooden walls on both sides of the passage, helping to stabilize them against the pressure of the huge earthwork. Gate-rings and keys discovered near some of the gateways revealed that double-doored gates enclosed the passageways.

Although none of Trelleborg's wooden buildings survived, clues to their original appearance remained in the earth itself. These buildings were constructed from timber planks, or "staves," sunk upright into the ground to form walls and support the roof. Additional posts, sunk outside and parallel to the wall, leaned against it to buttress the entire structure. The holes in which these posts were sunk left discernable traces in the soil, allowing historians to gauge the dimensions and groundplans of the buildings. This information, combined with known Scandinavian building styles of the period, produced a fairly detailed portrait of the buildings.

The buildings at Trelleborg were standard in shape and layout. Their general shape resembled that of a ship whose ends have been blunted, with convexly curved sides tapering to truncated gables. In the main ward, the building length from gable to gable measured 32.3 yards, or 100 Roman feet. At its midpoint, each building was 8.7 yards wide. Broad oak planks formed the walls as well as the roof, which was lined with shingles. Given modern knowledge of other Viking structures from the same period, scholars have suggested that some of Trelleborg's buildings may have been highly decorated with detailed carvings and painted with vivid colors.

The interior of each building was divided into three areas, a central hall 19.5 yards long and two small gable rooms. Many buildings in the main ward featured sunken hearths in the central hall. Four doors opened from the central hall: two to the outside and one to each gable room. Doors also opened from the gables to the outside. The fifteen buildings in the outer ward shared the same shape and construction, although they were slightly smaller and contained no hearths.

An aerial view of Trelleborg
Photo courtesy of Vikingeborgen Trelleborg

The exact use of these buildings cannot be determined with certainty. Many of the buildings in the main ward, especially those with hearths, probably served as dwellings for the people who lived at Trelleborg, including soldiers. The smaller buildings in the outer ward may have been used as storehouses. Other potential uses include barns, stables, and workshops where gold, silver, and iron products were crafted.

The precise geometry of Trelleborg's plan, the strength of its fortifications, and the selection of its defensible position all attest to the extreme efforts of planning and construction that produced the fortress. Although its location at the headland of two rivers makes excellent use of its natural defenses, the site required extensive leveling and filling-in to create a large, stable area on which to construct the fortress. The river banks were fortified with a stone foundation. Excavations have also revealed that a village formerly on the site was demolished to make way for Trelleborg.

The preparations and planning for Trelleborg, most likely carried out by a powerful central ruler, were part of a larger campaign that produced three other similar fortresses in Denmark during the same period. The other fortresses, all discovered in the mid–twentieth century, are Fyrkat, in the northern part of central Jutland; Aggersborg, the largest of the four, on the Limfjord in northern Jutland; and Nonnebakken (Nun's Hill) in the center of the island of Funen (or Fyn). Each of these locations had access to land and watershed transportation routes.

The four fortresses, which were built concurrently, share the same striking geometry and precision of design. Each includes a perfectly circular rampart partially bordered by a concentric ditch; four gateway passages at exactly north-south and east-west points on the rampart; streets that, connecting the gateways, form perpendicular axes; four quadrants in which equal numbers of houses were grouped around central courtyards; and an outer ditch fortification on its landward side. Although the fortresses differ in size, they all use the Roman foot (11.5 inches) as the standard unit of measurement.

Historians seeking architectural models for the fortresses' design have examined other ancient monuments throughout Europe and the East. Several fortresses of circular design have been discovered on the coasts of the Netherlands and Belgium. Although these structures have circular walls with four equidistant gates and axis-streets, they do not share many other characteristics of the Danish fortresses; the general geometric pattern common to these sites and the Danish forts may have been derived from Carolingian or Ottonian monumental architecture in Europe.

Another potential model, however, is found in the East. The circular pattern that characterized Assyrian military camps as early as the first millenium B.C. was still in use in the tenth century A.D., when Firuzabad was built in central Persia. This royal residence is almost completely circular and is bordered by a concentric rampart and moat with four gates at the points of the compass. This model could have influenced Danish fortress design, since Viking trade routes are known to have reached the Middle East. Large numbers of silver coins minted there have been found in Scandinavian Viking sites, evidence of the commercial ties between the two cultures.

The most central question concerning Trelleborg and the other Danish fortresses, however, is their original purpose. Their unity of design and fortification implies a common builder and mission. The vast quantities of materials and human labor needed to erect four such impressive strongholds across the country suggest a powerful builder, most likely a Danish king. Theories as to which ruler may have ordered the construction, and to what purpose, have been examined and revised in recent decades.

Poul Nørlund, the first excavator of Trelleborg, proposed that the fortresses were built as military training camps for the Viking troops who launched a series of raids against England under Sven Forkbeard between 994 and 1013. The fortresses were large enough to have housed a substantial army, it was reasoned, and money collected during the attacks on England may have helped to finance their construction. This theory was generally accepted from the time that excavations began at Trelleborg, in the 1930s, until it was disproved by recent scientific advances.

One of the most crucial factors in determing Trelleborg's purpose is its date of origin. Until recently, estimates of the Danish fortresses' years of origin ranged from 900 to 1050. A new dating technique called dendrochronology, however, has provided information that clarifies Trelleborg's beginnings. Dendrochronology identifies the precise year that timber was felled, by comparing the width of the annual growth rings with those in another piece of wood whose age is known (given that the outer growth rings remain intact). Dendrochronological methods used to date wood recovered from different parts of the Trelleborg fortress revealed that all the timber had been felled between August 980 and April 981. Timber examined at Fyrkat was dated to precisely the same time.

Dating the fortresses to 980 effectively disproved the theory that they served as training camps to support raids on England, as these attacks did not begin until the mid-990s. Construction in 980 instead dates the forts to the reign of King Harold Bluetooth, who is traditionally credited with uniting the Danish Kingdom. A new theory incorporating this information explains the four fortresses as part of the process by which Harald gained and sought to maintain a centralized power in Denmark.

Harold Bluetooth faced a number of political and economic problems around 980. His financial problems included a shortage of Arabic silver, which played a vital role in Viking trade, and the lack of tribute money won from Norway, over which he had lost power. His political problems included tensions with the German emperor, to whom Danish forces had lost an important battle in 974. He also faced a potential rebellion from his Danish subjects, whom he had taxed substantially to fund a large number of public building projects during his reign.

Harold may have intended the construction of four large, impressive fortresses on Jutland and the islands of Funen and Sjaelland as a way to revive his prestige, consolidate his power, and regain some lost territory. These symbols of military might and royal prestige, built at disparate points around the kingdom, would help maintain his control over the population and squelch the local powers of petty kings and chieftans. Soldiers stationed at the fortresses could organize quickly to face foreign threats or to plunder foreign territories to collect tribute for the king. Finally, the fortresses may have served as administrative centers for the king, playing a role in trade, coinage, and craft production.

Many characteristics of Trelleborg's design support this theory. The strong fortifications—rampart, moat, ditch—would have been important in skirmishes with local chieftains. They would have provided protection for the king's administrative interests and, on a basic level, would be solid regional reminders of the central ruler's power. Such fortifications would have been unnecessary if Trelleborg's main purpose was to quarter and train troops for raids across the sea.

The fortresses' inland locations also support the central-power theory. Coastal locations would have been more favorable to the launching of attacks on England as well as defending against seaborne invaders from abroad. Instead, the fortresses were built inland, close to land and water trade routes by which communication and contact could be more easily maintained across the kingdom. The fortresses were also constructed near regions that were not controlled by the south Jutland dynasty that preceded Harold Bluetooth. They may indeed have played an important part in Harold's historic claim to uniting the Danish kingdom.

Archaeological finds at Trelleborg further validate the central-power theory. The earlier theory portrayed the fortresses as military barracks for as many as 10,000 men. This estimate, however, has been drastically lowered by the examination of buildings in the fortress. While many of them were living quarters, many were not, serving instead as storehouses or workshops for craftsmen. A stable community of craftsmen, such as potters, goldsmiths, coppersmiths, and weapon-makers, existed alongside the military population and participated actively in trade with Baltic and other Scandinavian countries.

Despite a tremendous effort of organization, funding, and labor, Trelleborg and the three other fortresses had a very short life span. They survived less than fifty years, perhaps less than ten. Although wood buildings needed frequent repair and replacement to survive, there is no evidence that Trelleborg's timbered houses and rampart palisades were ever repaired. Shortly after construction, Trelleborg and the other royal fortresses were abandoned and fell into ruin.

These events may well coincide with Harald's own fall from power. Although in 983 he was successful in recapturing the border area lost to German forces nine years earlier, he again faced rebellion from within in 986. His son Sven Forkbeard, future conqueror of England, overthrew Harold and claimed the crown as his own. Harold Bluetooth was exiled and died in November 987. Whether Sven ordered the destruction of his father's fortresses or simply abandoned them is unclear.

These fortresses remained unknown until the twentieth century. Trelleborg, the first of the fortresses to be discovered, was the proposed site of a motorcycle racecourse in the 1930s. A preliminary investigation of the area led to the discovery of the Viking fort. Aggersborg, in northern Jutland, came to light in the 1940s, followed by Fyrkat in the 1950s. Nonnebakken, the fourth fortress, has been excavated least because subsequent building on the site severely disturbed its Viking Age appearance.

Visitors to Trelleborg today find its fortifications largely intact, thanks to restorations that cleared debris and repaired damaged areas. The massed boulders that once helped support gateways through the circular rampart have been replaced at the eastern gate. Outlines demonstrating the arrangement and shape of the buildings have been marked permanently in the ground. At the entrance to the outer ward stands a full-scale reconstruction of a Trelleborg house, built in 1942. The house replicates Viking building techniques: its walls are stave-built from oak, and its roof is lined with oak shingles. Although the house contains some features that are now known to be inaccurate (an exterior roofed gallery instead of support buttresses), its general depiction of a tenth-century Viking dwelling remains valid.

The Vikings have traditionally been viewed as a crude, ruthless warrior race who preyed physically and financially on their poorly defended European neighbors. The idea of a Viking civilization, in this view, was almost a contradiction in terms. The discovery of Trelleborg and the three related fortresses forced scholars to reexamine long-held notions and to accept a broader view of what was, in fact, a Viking civilization.

The fortresses themselves, with their precise geometrical design and skillfully constructed fortifications and buildings, evince a culture with formidable organizational capability and technical knowledge. Excavations at Trelleborg have revealed that it was a center not only of military force but also of craftsmanship and peaceful, prosperous international trade. According to the current theory, the four Danish fortresses may also have fostered the emergence of a central national government, unifying the power once divided among local petty kingdoms. The findings brought to light through archaeological excavations at the fortresses continue to challenge and refine modern understanding of the Vikings.

Further Reading: *The Vikings* by Else Roesdahl, translated by Susan M. Margeson and Kirsten Williams (London: Allen Lane/ Penguin, and New York: Viking, 1991; first published as *Vikingernes Verden*, Copenhagen: Gyldendal, 1987) offers an up-to-date, incisive portrait of Viking history and culture that incorporates findings from recent archaeological excavations. Although dated in some respects, P. V. Glob's *Denmark: An Archaeological History from the Stone Age to the Vikings,* translated by Joan Bulman (Ithaca, New York: Cornell University Press, and London: Faber, 1971; first published as *Danske Oltidsminder,* Copenhagen: Gyldendal, 1967), offers a thorough, lively description of prehistoric Danish history written by a renowned researcher. A useful introduction to Danish history presented in brief, explanatory essays is the *Dictionary of Scandinavian History,* edited by Byron J. Nordstrom (London and Westport, Connecticut: Greenwood, 1986).

—Elizabeth E. Broadrup

Soissons (Aisne, France)

Location: In northern France, on the bank of the Aisne River, eighteen miles southwest of Laon, sixty miles northeast of Paris.

Description: Ancient city that played a role in the formation of the medieval Frankish kingdom and was much contested throughout centuries of warfare, most notably in the twentieth.

Site Office: Office de Tourisme, Syndicat d'Initiative
Cour St. Jean-des-Vignes
02200 Soissons, Aisne
France
23 53 08 27

The history of Soissons dates back to Roman times, when northern France was settled primarily by Celtic tribes. Through the campaigns of Julius Caesar, Gaul fell under Roman control in the mid–first century. After the Roman conquest, Soissons became a garrison town of strategic importance to the continued presence of the conquerors. The Romans divided Gaul into four administrative provinces: Narbonensis, Lugdunensis, Aquitania, and Belgica, the last of which included the town of Soissons. The Celts were receptive to early Roman rule, and northwest Gaul quickly assimilated into the Greco-Roman world. Further administrative division and organization eventually laid the foundation for the political formation of medieval France.

The Celtic tribes inserted themselves into the Roman system at the level of local government. The city of Soissons derives its name from the Suessiones, a tribe that chose the city as its capital in the third century. At the time, the town, situated on the bank of the Aisne River, was surrounded by dense forest, few clearings, and vast woodland swamps populated by many large wild animals. Located at an important crossroads, however, the city saw many travelers passing through its gates.

Soissons was important not only in the Romans' worldly conquest, but also in the establishment of Catholicism. When Christian missionaries gained a foothold in the region in the third century, the city was made a bishopric. Throughout the centuries, Soissons retained its importance as a religious center.

Politically, Soissons had its share of troubles. High Roman Gaul suffered from several foreign invasions during the years of the late Roman Empire of the third and fourth centuries. In 260 and again in 276, Gaul was plundered by two Germanic confederations, the Alamanni and the Franks, and their incursions led to a string of revolts against Roman rule. Rome managed to reform the empire and restore Gaul in the century following, but Germanic invasions continued, progressively weakening Roman control.

Toward the end of the fourth century the western empire collapsed, and Gaul was unified under Frankish rule by the Merovingian and later the Carolingian dynasty. The first leader of the Merovingians was Childeric, who only ruled a short time before passing the throne to his son Clovis in 482. To secure his kingdom, Clovis had to defeat several other German factions. His most formidable opponent was Sygarius, who also controlled a substantial part of Gaul and had established his capital at Soissons. In the four years after his accession to the throne, Clovis conquered several small regions of Gaul, and, in 486, he finally defeated Sygarius in a battle fought at Soissons. Sygarius's defeat secured the area as the new Frankish kingdom.

Clovis established Paris as the capital of this new kingdom in 507. When he died in 511, he divided France among his four sons, giving each a capital region. Soissons was assigned to Chlotar I, who outlived his brothers and went on to reunite the kingdom under his own rule. At Chlotar's death, France was again divided, the Soissons region falling to Chilperic I. During the time of Chilperic's rule, a new political configuration arose out of internal struggles within the kingdom. A section of the kingdom of Soissons was joined with a section of the Paris kingdom to form Neustria, a broad coastal area with the lower Seine valley as its center. Soissons was made Neustria's first capital.

With political configurations constantly in flux, Merovingian dominion was weakening. Another Frankish family, the Carolingians, were consolidating power in Gaul by exploiting their positions as mayors of the palace. Their leader, Charles Martel, led a small but loyal following to victory over the Neustrians at Soissons in 719, which enabled the Carolingians to take control over northern France. Martel then moved on to establish his power in the northeast, while also strengthening his influence in Germany. Upon the death of the Merovingian King Theodoric IV, who had occupied the Frankish throne at the time, Martel seized control over the entire Frankish kingdom. When Martel died in 751, his son, Pepin III, the Short, assumed the throne. He was crowned king at Soissons's St.-Médard Abbey in 752. Later, Pepin handed the crown to his son, Charlemagne, who was to introduce feudalism to France.

For most of the following four centuries, France was wracked by territorial conflicts between the country's nobles. The monarchy's efforts to contain the local warfare were often futile, until Louis VII was able to use the influence of the Catholic Church to create a truce. To establish peace in the realm, Louis assembled the most important nobles and most influential church dignitaries of northern France at Soissons in 1155 and negotiated a ten-year peace with them. The king took it upon himself to make sure that all parties lived up to the agreement. In fact, the truce secured a cessation of

The refectory of the Abbey of St. Jean-des-Vignes
Photo courtesy of Ville de Soissons

hostilities well beyond its own expiration and was not seriously violated until the French crown itself undertook a new war. King Philip II (Philip Augustus) invaded Flanders and provoked the famous Battle of Bouvines in Flanders in 1214. Raoul, Count of Soissons, fought in the French army, and the city also sent several bodies of trained men to participate in the invasion. By then, the peace agreement of 1155 was all but forgotten.

For more than a century after the Battle of Bouvines, the city of Soissons remained quiet and undisturbed. Then, in the fourteenth century, with the approach of the Hundred Years War, Soissons woke from its rest, to suffer hundreds of years of devastation. The city is estimated to have been subject to more than thirty sieges and major battles through the centuries.

The Hundred Years War began with the refusal of Edward III of England to pay feudal homage to the French crown for the French territories that were among his possessions. On the basis of complicated ancestral connections, Edward also laid claim to the French throne and, in 1336, mounted an expedition to France to realize his claim. In the first phase of the war that ensued, England won many battles, successfully took the French king Philip hostage, and forced France to give up considerable territory. However, the French persisted and, in time, won back their territories and ultimately expelled the English from French soil.

Soissons played a major part in the Hundred Years War in 1414, when the city was taken by force by Charles VI of France, along with the neighboring towns of Compiègne and Bapaume. The towns had allied with the duke of Burgundy against the king. A horrendous massacre followed, causing such a scattering of inhabitants that in the after-years not one family was to be found in the city whose ancestors had lived in Soissons before the siege. Soissons survived the war, but barely.

Soissons spent the next few centuries unsuccessfully trying to rebuild itself. Despite the ravages, the city remained an important religious center. And although the town made efforts to improve its image and standard of living, Soissons suffered again in the Wars of Religion in the sixteenth and seventeenth centuries and from many other small battles and sieges throughout the years.

Soissons makes a rather unappealing entrance in the historical record of the eighteenth century with a project in social engineering. In the 1780s, the city built an institution to rehabilitate beggars with discipline and hard work—the prototype of the workhouse. Despite high escape and death rates, the project was widely acclaimed and imitated, particularly when industrialization created a so-called surplus population in the nineteenth century.

The one major event of the nineteenth century in Soissons took place during the Franco-Prussian War. In 1870 the town was taken by the Prussians and was almost continuously under bombardment during several years of hostilities. The destruction of these years, however, pales in comparison with the ravages caused by World War I. France entered the war

with a great sense of national unity, if not with a great deal of military competence. In the first four months of the war, France suffered 850,000 casualties, while the Germans advanced apace on Paris. It was in September 1914 that the massively destructive Battle of Soissons—perhaps the town's most famous historical event—was fought. The battle, which lasted four days, is often seen as a decisive moment in the allied efforts to break the German offensive.

The battle began on September 15. The German forces had then established a front north of Paris that ran from Compiègne in the east to Reims in the west. Soissons lay right in the middle of the front and offered, just north of the city, a natural stronghold that would have solidified the German line tremendously. Both the French and the British were deployed to defend Soissons, and the British, stationed just north of the river, engaged the German forces in particularly fierce combat, suffering severely. Despite several small attacks and defenses by the Allies, the German front at Soissons proved strong, delivering harsh bombardment and taking many prisoners. Nevertheless, and somewhat to the surprise of the Allied troops, the French War Office announced after four days of fighting that the Germans had failed to hold their lines on the river's north banks. Unable to secure the Soissons stronghold, the Germans were forced to pull back along the entire length of the front.

Weeks later, Soissons suffered a double bombardment at the hands of the Germans, causing almost the whole population to flee the city. Later that year, Soissons was caught in the Battle of the Aisne. After several years during which the front moved away from the city, Soissons was captured again toward the end of the war, on May 29, 1918, as the Germans made their way through the Aisne-Marne region. However, American troops were advancing from the west with the French Sixth and Tenth Armies and launched a counteroffensive in the Soissons area. Twenty-four divisions were thrown into the assault along with more than 2,000 guns, 500 tanks, and 1,000 aircraft. The Germans were forced to withdraw. In early August, another battle was fought just south of Soissons, again forcing the Germans into the defense. In early September, the German troops had to pull back further toward the Aisne River as a result of the Oise-Aisne offensive. On this battlefield, about fourteen miles from Soissons, now lies the Oise-Aisne Cemetery, the resting place of the battle's numerous casualties.

As the war drew to a close in the following months, Soissons was left in almost complete ruin. In some sections of the city, not a single structure was spared. By the end of the war, the population of the city had dropped from 18,000 to 500. The interwar years were devoted to rebuilding the city and restoring its ancient architectural treasures.

During World War II, Soissons again found itself in the path of an advancing German army. This time the city was freed soon after the Germans captured it, however. After the war, Soissons continued to develop as an industrial and technological center. Today, Soissons is a producer of iron and copper goods, boilers and agricultural implements, rubber products, and glass.

Ravaged by war again and again, the city has lost much of its ancient architecture, but a few monuments have been restored. The Abbey of St. Jean-des-Vignes, founded in 1076, was destroyed in large part under Napoléon, but the facade was spared. The abbey is known for its towers, imposing and asymmetrical, that support a large cloister and refectory. In the twelfth century, Archbishop Thomas Becket lived at the abbey for nine years. In the north part of the city stands the cathedral, which dates to the twelfth century and is famous for its beautiful transepts, its arcades and galleries, and its lofty clerestory. Near the main square of the city is the former Abbey of St. Lèger, which now houses a museum of paintings and sculpture. To the east stands the Crypt of St.-Médard, which contains the tombs of several Merovingian kings.

Further Reading: *A Traveller's History of France* by Robert Cole (New York: Interlink Books, and Moreton-in-Marsh, Gloucester: Windrush, 1989) gives a brief but thorough description of the history of France from its birth up to contemporary times. The book also includes a chronology of events, a glossary, and a reference list of governmental figures and religious sites. *The Martyred Towns of France* by Clara E. Laughlin (New York: The Knickerbocker Press, 1919) dedicates each of its chapters to the story of a different French town, including Soissons. *Merovingian Military Organization 481–751* by Bernard S. Bachrach (Minneapolis: University of Minnesota Press, 1972) tells the story of France under the rule of the Merovingian dynasty. *History of the Franks* by Ernest Brehaut (New York: Octagon Books, 1973) also recounts the history of the Merovingians, in a style similar to the Bible's. *Birth Of France: Warriors, Bishops and Long-Haired Kings* by Katherine Scherman (New York: Random House, 1974) again traces the roots of the Merovingians but goes back further to include Celtic occupation of Begian Gaul. *The Legend of the Bouvines* by Georges Duby, translated by C. Tihanyi (Los Angeles and Berkeley: University of California Press, and Oxford: Polity Press, 1990) recounts the Battle at Bouvines in Flanders. *The Campaign of 1914* by G. H. Perris (New York: Henry Holt and Company, 1915) is the journal of a French soldier's observations and accounts during early World War I. The journal provides much solid historical fact while capturing an inside perspective of the war. *American Armies and Battlefields in Europe*, prepared by the American Battle Monuments Commission, (Washington, D.C.: U.S. Government Printing Office, 1938) describes European battles from a geographic standpoint. *To Win A War: 1918, The Year Of Victory* by John Terraine (London: Sidgwick and Jackson, 1978; Garden City, New York: Doubleday, 1981) tells a detailed story of the last year of World War I and how the conflict ended.

—Cynthia L. Langston

Speyer (Rhineland-Palatinate, Germany)

Location: South of Ludwigshafen and Mannheim, on the left bank of the Rhine River, at the mouth of the Speyer River.

Description: One of Germany's oldest cities, with origins dating back to the Roman period. An episcopal city until 1294, it was closely linked with the Holy Roman Empire, especially during the Salian dynasty. The medieval core of the city is today surrounded by modern residential and industrial areas.

Site Office: Verkehrsamt
Maximilianstrasse 11
67346 Speyer, Rhineland-Palatinate
Germany
(6232) 14395

In Speyer's historical museum, the Historisches Museum der Pfalz, there is an exhibit dating from Speyer's prehistory. It is a large bone, which until the eighteenth century was believed to be the bone of a giant. According to local lore, the giant had fallen into a ditch while storming the city and died there when Speyer's defenders poured boiling tar over him. The bone is, in fact, the vertebra of a mammoth. Another valued exhibit in the museum dates from the Bronze Age (c. 2000–1200 B.C.) and is known as the "Golden Hat." Crafted in gold, this "hat" is an ornate conical object with a flat rim at its base. Though little is known about it, it is thought to have been used in fertility rites, when it was mounted on a carriage and taken through fields in procession. It is known that the area around Speyer was inhabited by Celts, but little else is known of Speyer's prehistoric times.

The city celebrated its 2,000th year of history in 1990. The beginning of Speyer's documented history in 10 B.C. was marked by the erection of a fort by Drusus, stepson of the Roman emperor Augustus and commander of the forces that occupied this area of the Rhine during the period from 12 to 9 B.C. The fort was part of a series of border strongholds, fifty in total, that were established along the Rhine and the Maas rivers. Civil engineering excavations in the present-day Kleine Pfaffengasse at Speyer in 1927 led to the discovery of this fort's location. This building was enlarged during Emperor Claudius's reign, early in the first century A.D., further developing the site as a permanent settlement. The next major impetus for growth came in A.D. 74 with the Roman conquest of the east bank of the Rhine. The Roman troops were moved on to secure the new frontier, which extended as far as present-day Frankfurt. This territory was included in the new province of Germania Superior, which also included territories west of the Rhine. The aban-

doned fort was taken over by the local inhabitants, the Nemetae, who gave rise to the civil settlement of Noviomagus, or "Newtown," which from the third century came to be known as Nemetum. The settlement's location on the road running along the west bank of the Rhine drove the city's economic development and facilitated the establishment of Roman culture. The remains of pottery, dating from A.D. 100, were discovered under the street surface near the present-day museum, and a chance find in the late nineteenth century yielded a fourth-century glass bottle containing wine that had remained preserved since Roman times. An element of social life was provided by an amphitheater, the presence of which is indicated by an exhibit in the historical museum, a stone parapet inscribed with a seat reservation for an unknown party. The practice of various pre-Christian forms of Roman religious worship is attested to by various artifacts, which include altars to Jupiter, Diana, Hercules, Minerva, and Mercury.

The town's Roman way of life was seriously threatened in 260 with the incursion of the Alamanni into Roman territory. The invaders quickly seized the entire east bank of the Rhine, which had been in Roman hands for 186 years. Nemetum was once again part of frontier territory. It was an uncertain time for the city, especially as the Alamanni made frequent skirmishes across the Rhine. Nemetum's commercial life suffered as a result, and it is thought that during this period some of the wealthier citizens moved into more securely held Roman territory in central Gaul, present-day France. With the enemy so near, the city shifted toward the southeast. Parts of the city were abandoned and leveled, and the resulting rubble was used for the construction of a city wall and new buildings.

There was a brief lull in the barbarian threat from the east in the 370s, during the reign of Emperor Valentinian I, when border defenses along the Rhine were strengthened and pre-emptive skirmishes were made into Alamanni territory. Despite such efforts, the Rhine border proved to be increasingly indefensible in the face of Germanic advances. After numerous incursions into various parts of the Roman Empire, Rome recalled its troops in 406 in response to the growing barbarian threats, which resulted in the sacking of the imperial capital in 410. Nemetum came under Alamanni control and remained so for nearly a century. In 496, the Alamanni were defeated by Clovis I (the Frankish king of the Merovingian dynasty) at the Battle of Zülpich (Tolbiacum), some twenty miles west of present-day Bonn.

Clovis I was the founder of the Frankish kingdom that came to be a dominant force in western Europe during the early Middle Ages. He was also the first barbarian king to convert to Christianity. Historical sources from the mid-fourth century already refer to the city as a diocese, reestab-

Interior of the imperial cathedral at Speyer

lished in the sixth century, after the defeat of the Alamanni. It is also during Clovis's reign that the name Spira first appears, though it was not to come into common usage until the eleventh century. It was during the Merovingian period that a church, dedicated to the Virgin Mary and St. Stephen, was built by King Dagobert I (628–683) on the site now occupied by the Cathedral.

The year 751 saw the beginning of the Carolingian period, when Pépin III, the Short, acceded to the Frankish crown. It was Pépin III's son, Charlemagne, who laid the foundations for the Holy Roman Empire in which the city of Speyer, in the course of time, was to play a major role.

Charlemagne's Frankish empire originally included the territories of present-day eastern France, Germany, Switzerland, and the Low Countries, as well as parts of present-day Italy, Austria, and the Czech state. It was the dominant European power of the time and, as a Christian one, stood against the well-established Moslem forces in Spain and the heathen peoples to the north. Charlemagne's de facto role as defender of the Christian faith was acknowledged by Pope Leo III. The title of Roman emperor had lapsed with the disintegration of the eponymous empire in the fifth century, but Pope Leo III reestablished the title to bestow it upon Charlemagne in 800.

In 843, Charlemagne's grandchildren partitioned the empire at the Treaty of Verdun. Lothair I kept the title of emperor and received the Italian lands, as well as a strip of territory that ran from Provence to Friesland. Charles II the Bald claimed the territory west of the Rhine, much of what is currently France. Louis II the German was given the lands between the Rhine and the Elbe. This created the Eastern Kingdom, an event that effectively marked the beginning of German history. Louis II cultivated German language and literature in a way that would lead to the emergence of a distinct cultural and political entity. He became the first of a long line of kings that held the German crown until 1918. In the partition, Louis II also succeeded in claiming for his domain the western archdiocese of Mainz as well as the bishoprics of Worms and Speyer. These cities, with their Roman origins, formed an important triad as the ancient nucleus of a new kingdom.

The name of Dagobert I's church in Speyer was changed during Louis II's reign. With a charter by Louis II, dating from 859, St. Stephen's name was removed from the church, which was to be known simply as the Church of the Blessed Virgin Mary until the late Middle Ages. Little is known about the cause of this change, but it is likely that the small Merovingian church was replaced by a larger Carolingian cathedral at this time.

Louis II's line died out in 911, and the German crown passed to Conrad, duke of the Franks, who reigned until his death in 918. He was succeeded by the kings of the House of Saxony. The second of these, Otto I the Great, was crowned German king in 936. The Roman imperial title, which had been handed on to Charlemagne's son Lothair I at the 843 partition, had lapsed again with the demise of the Carolingian

line. In 962, the imperial crown was conferred on Otto I and was to be borne hence by successive German kings until the abolition of the Holy Roman Empire. The House of Saxony gave way to the Salian dynasty in 1024. It was this dynastic change that signaled the most illustrious period of Speyer's history.

The first Salian king was Conrad II the Elder, who was particularly fond of Speyer, where he held a large estate. This attachment to the city also earned him the nickname of Conrad the Speyerer. In 1027, he was crowned emperor in Rome by Pope John XII. In his role as protector of the faith he decided to build the greatest cathedral of the contemporary Christian world in Speyer. The foundation stone for this ambitious undertaking, which was to occupy three Salian generations, was laid in 1030, following removal of the preceding Carolingian cathedral. The new cathedral's proportions, 438 feet long and 180 feet wide, were massive for the time. The new cathedral was also intended to serve as the final resting place of the Salian kings, but construction was only half completed when Conrad II died in 1039 in Utrecht. His remains were placed in front of the partially erected choir screen in what at the time amounted to a huge construction site. The cover to the grave was attached with iron bands to prevent its accidental removal by workmen. Conrad's son, Henry III, continued with the project. The crypt was consecrated in 1041. The world-renowned *codex aureus,* a beautifully decorated book of the Gospels, was given to the church in 1046 to mark the consecration of the high altar. Henry III also endowed the cathedral with a number of relics from his voyage to Rome, among them the head of the martyr pope St. Stephen. Henry III died in 1056, and his wife, Agnes, and his six-year old son, Henry IV, were left with the task of completing the cathedral, accomplished in 1060. The three decades of construction had wrought other major changes to the city. The city's population had grown considerably to include hundreds of craftsmen and all manner of tradespeople, whose accommodation mandated the expansion of the city. The consequent increase in the city's demand for goods and services also stimulated vigorous growth in Speyer's economy. Henry IV ordered various improvements to the cathedral in 1081. One of these was the vaulting of the originally flat ceiling, a ground-breaking and audacious feat in the annals of German architecture. Changes were also made to the apse. Side chapels as well as a roof gallery were built, resulting in Christendom's most splendid structure of the time.

Henry IV's successor, his son Henry V, granted two significant privileges in 1111, beginning Speyer's change in status to that of imperial free city. The first of these, extraordinary at the time, was the populace's exemption from death duties. Payable to landlords, these duties could frequently amount to half the inheritance, which had resulted in great poverty and misery among common people. The second exemption was from all manner of customs, such as those levied on operators of market stalls and other traders. Another expansion of citizens' rights, to the further reduction of the bishop's privileges, occurred in 1198. Duke Philip of Swabia, to be crowned German king in the same year, granted Speyer's citizens the right to elect from their ranks a council of twelve representatives. The councilmen would be accountable to the citizenry and were charged with the government of the city, "to the best of their ability." It was a freedom granted to few contemporary German cities, and eventually led to the replacement of episcopal representatives in urban administrative posts by townspeople. The city's own seal, introduced in 1208, symbolized this newly established authority. Evidence of the power shift within Speyer came in 1226, when the city council entered into an accord with the cities of Binge, Frankfurt, Friedberg, Gelnhausen, Mainz, and Worms to form the First Rhenish League of Cities. Such transfer of authority was, more often than not, the result of constant confrontation with the bishop. This uneasy state of affairs found a parallel with the bishop's own relations with Rudolf I, Holy Roman Emperor from 1273 to 1291. Historical sources place the origin of Rudolf's antipathy in a visit his young wife made to Speyer. Bishop Frederick, eager to welcome her to the city, personally lifted her out of her carriage and kissed her on the lips. The fifteen year-old queen did not appreciate this greeting and told her husband as much. Rudolf's discontent was such that, capitalizing on the townspeople's own displeasure, he forced the bishop's departure from Speyer. Bishop Frederick was allowed to return in 1292, after Rudolf's death. By 1294, episcopal authority had been curtailed to such an extent that Speyer had become an independent city. Its loyalty was to the emperor and not to any intermediate bishop, prince, or king.

Speyer is also famed for being the venue for imperial assemblies. Referred to as diets, these were meetings of the legislature of the Holy Roman Empire from the twelfth century onward. They would take place as convened by the emperor, in either episcopal or imperial cities. Fifty diets were held in Speyer; after 1524 they were held in the Rathof, or city council court, which was built that year. Two diets stand out as having had a bearing on the history of the Reformation. Martin Luther had issued his attack against the church in Wittenberg in 1517. In 1521, at the Diet of Worms, Catholic Emperor Charles V's first diet, Luther was summoned to abjure his heretical doctrine, which he refused to do. This resulted in the Edict of Worms, which placed Luther under the ban of the empire, which meant that he was convicted as a heretic. However, Charles V had to leave Germany to see to other imperial matters and was unable to enforce the edict. His inaction actually weakened Charles's authority and strengthened Luther's position, and Luther's supporters were able to organize themselves more effectively. The Edict of Worms was suspended at the 1526 Diet of Speyer, when the states and cities of the empire agreed that the conduct of religious affairs, guided by conscience, should be left to each constituent authority. With renewed Catholic pressure, this arrangement was repealed at the 1529 Diet of Speyer, prohibiting interference with Catholic religious practices. The removal of the 1526 concessions prompted a protest from the Lutheran princes and cities, which gave rise to the term Protestant. This event is memorialized in the neo-Gothic Gedächtniskirche, "Memorial Church," in Speyer's Landauer Strasse, which was constructed between 1893 and 1904.

Another role served by Speyer was as venue for the Reichskammergericht, the Imperial Court, which was based here from 1527 to 1689. The year 1689 was an unfortunate one for the city. It was destroyed by French troops in the War of the Grand Alliance, also known as the War of Augsburg (1689–97), in which French King Louis XIV staked his territorial claims in the Palatinate area of Germany. The cathedral by then contained the graves of eight German kings and emperors: those of the Salians Conrad II, Henry III, Henry IV and Henry V; that of the Habsburg Rudolf I; and those of Philip of Swabia and Albrecht of Austria. These were plundered, along with the cathedral and the rest of Speyer. The population scattered into villages in the surrounding countryside, and some people crossed the Rhine to make their way to Heidelberg and Frankfurt. The city council met in Heidelberg a few weeks after its enforced departure from Speyer. It then transferred to Frankfurt, where it remained the city government in exile for the next nine years. On January 6, 1698, the city council was back in session in Speyer, in the inn Zum Riesen, "The Giant"—today's Kaufstätte—which had remained relatively unaffected in 1689. The council called for the return of the population, and in January of 1699 decided to deny citizenship rights to those who had not returned by St. Martin's Day (November 11). Many stayed where they had resettled, and many had died in the intervening period. Others, such as craftsmen who hoped to find long-term reconstruction work, returned to the city. The cathedral's full reconstruction did not take place until the period 1772–78.

It was only a brief time after the cathedral's reconstruction that the city was again subjected to the ravages of war. War was declared by France on the Holy Roman Empire in April, 1792. French troops took the city in September of the same year. Damage was suffered on several occasions, but most notably in 1794, when the cathedral was desecrated and its flammable contents burned. Speyer was reclaimed by the Holy Roman Empire on several occasions but in 1797 was absorbed into the French Republic, where it stayed, known by the name of Spire, until 1815.

This period saw the formal dissolution of the Holy Roman Empire, which had effectively ceased to exist following the Peace of Westphalia in 1648. This treaty acknowledged the territorial sovereignty of the empire's member states, which were henceforth empowered to enter into accords with each other and with foreign states as long as these were not to the empire's detriment. Thus the institutions of the Holy Roman Emperor and the diet continued only in greatly diminished form. The empire's nominal end came about with Napoléon I's self-coronation as French emperor, with which he aimed to eliminate the Holy Roman Emperor's age-old superiority among European sovereigns. Francis II, resigned title as the last Holy Roman Emperor on August 6,

1806. In 1812 Napoléon began his ill-fated campaign against Russia; in 1813 he was defeated by an alliance of European nations at Leipzig. France was invaded the following year, and Speyer became part of Bavarian territory in 1815, where it remained as the capital of the Rhenish Palatinate section until 1945.

The present-day city center has retained much of its medieval layout. The cathedral was reconsecrated in 1961 following completion of its most recent restoration program, begun in 1957, which returned the cathedral almost exactly to its original condition. In the park behind the cathedral stands the Heidentürmchen of 1280, which was once part of the old city wall. Near it stands the Mikwe, Speyer's Jewish baths, built in 1104. It is one of the few complexes of its kind remaining in Germany. Also part of the town's defenses was the Altpörtel, a thirteenth-century gate tower that is considered the most beautiful of its kind in Germany. The baroque Dreifaltigkeitskirche, or "Church of the Holy Trinity," in the Grosse Himmelsgasse, was constructed between 1701 and 1717 and counts as one of the most important Protestant churches in the Rhenish Palatinate.

Further Reading: There are no detailed English-language histories of Speyer. In German, *Speyer, Kleine Stadtgeschichte* by Fritz Klotz (Speyer, Bezirksgruppe Speyer des Historischen Vereins der Pfalz, 1988) is a comprehensive work on the history of Speyer, describing events up to World War II with interesting detail. It contains a short chronology covering the period 1948–87. For a more exhaustive history of the city, consult the three-volume *Speyer: Geschichte der Stadt Speyer*, edited by Wolfgang Eger and published by the City of Speyer (Stuttgart: Verlag W. Kohlhammer, 1982) which covers all aspects of Speyer's history with encyclopedic thoroughness. *Der Kaiserdom zu Speyer, Geschichte und Führer* by Franz Klimm (Speyer: Verlag Jaeger Druck, 1978) is a brief summary, with illustrations, of the cathedral's history.

—Noel Sy-Quia

Stirling (Central Region, Scotland)

Location: South central Scotland, along the Forth River, thirty-six miles northwest of Edinburgh.

Description: Much contested during the Wars of Independence, Stirling was the site of the Battle of Stirling Bridge in 1297 and the Battle of the Bannockburn in 1314. Stirling Castle was often used as a fortress and was for many years the residence of the Stuart monarchs.

Site Office: Loch Lomond, Stirling and Trossachs Tourist Board
Tourist Information Centre
41 Dumbarton Road
Stirling FK8 2QQ
Scotland
(786) 475019

Stirling Castle and the surrounding town have been flashpoints in Scottish history because of their strategic location as the gateway between the Highlands and Lowlands. Situated on a high rock at the southern opening to the Highlands, the site became a point of contention between the Scots and the English in the twelfth century. It has been called "the Key to Scotland," in part because the Old Bridge of Stirling, built about 1400 and still used as a footbridge, was the only route leading north from Stirling until Charles I's reign in the seventeenth century.

According to legend, King Arthur captured the castle from the Saxons. Actual records indicate that there was a castle at Stirling for more than a century prior to 1124, when Alexander I died there. Alexander's son refers to his "burgh of Stirling," suggesting the town had established some degree of importance. For the next 600 years Stirling was in the path of English and Scottish troops seeking the upper hand in the Wars of Independence. The castle was also the site of various schemes and intrigues as Scottish clans vied for power.

The stage for English-Scottish struggles was set late in the eleventh century, when the Scottish king Malcolm III of Canmore married Margaret, a granddaughter of England's Edward the Confessor. This match resulted in considerable English influence in Scotland. In 1174, when William the Lion of Scotland sought to aid English rebels in hopes of regaining some Scottish lands, he was defeated in battle, captured, and forced to swear allegiance to England's King Henry II. This was the first time Scotland lost its independence and thus began the quest to recover it.

In the late thirteenth century, a crisis of royal succession arose between Robert the Bruce and John de Baliol. When England's King Edward I also claimed the throne, he crossed the border, occupied southern Scotland, and in 1292 made Baliol king—Baliol, however, was just a figurehead. Baliol became unpopular with the Scottish people; frustrated with his position, he revolted against Edward, who came storming back into Scotland, defeating the insurgents at Dunbar, and left Scotland under military occupation.

At this time a nobleman named William Wallace led a revolt, and in 1297 he defeated troops under the Earl of Surrey at the Battle of Stirling Bridge. This bridge predates the current one, and its exact location is unknown. Wallace became a national hero and ruled for a year in the name of John de Baliol. Edward returned in 1298, this time defeating Wallace at Falkirk and slaughtering 15,000 Scots. Wallace kept up the fight with guerilla warfare that required Edward to invade southern Scotland three more times. Wallace captured Stirling in 1303, and in 1304 Edward began his great siege of Stirling Castle, then the last bastion of the Scottish insurgents. The Scottish forces held out for three months until they ran out of food. Wallace spent years in hiding but was finally captured by Edward in 1305. In London, Wallace was tried and condemned for treason. In 1306 he was executed by being dragged by a horse, then hanged, disemboweled, beheaded, and quartered. His severed head was displayed on London Bridge.

That same year, Robert the Bruce, namesake grandson of the Robert who had vied with Baliol for the throne, was crowned Robert I, king of Scotland, shortly after stabbing to death his rival John Comyn at a meeting that had been called to plan strategy against England. Over the next several years Robert consolidated his position, capturing most of the castles that had been held by the English, and eventually defeated Edward II in the historic Battle of the Bannockburn near Stirling in 1314. At the time of the battle Robert's brother Edward Bruce had been besieging the English-occupied Stirling Castle, and Edward II sent troops to assist the garrison. The English army consisted of some 2,000 armored knights and 17,000 foot soldiers. The Scots had about 5,000 foot soldiers and 500 light horse troops. In preparation for the battle, Robert instructed his men to dig man traps all over the grounds and disguise them with branches, and to scatter iron spikes, or calthrops, over the fields. Although Robert was outnumbered, he selected a strategically strong position that required the English to make an arduous crossing of the Bannockburn, a narrow but deep tributary of the Forth River. The crossing, on makeshift bridges, took most of the night of June 23. The next morning the Scots slowly, methodically forced the English back against the Bannockburn. When the English saw Edward II's banner leaving the field and the Scots camp followers streaming down the hill toward them, they broke ranks and the rout began. Edward and his small entourage escaped, but the victory was decisive. After the battle and according to his usual procedures, Robert destroyed the for-

The Wallace Monument
Photo courtesy of Loch Lomond, Stirling and Trossachs Tourist Board

tifications at Stirling Castle, so that it could not be used as a stronghold, but repairs were made several years later. The victory brought Scotland its independence, but the country suffered from internal strife and intermittent conflict with the English for years to come.

When the Stuarts came to the throne, Stirling Castle became their royal residence. James I became King of Scot-

land in 1424. Dismayed at the power his nobles were wielding, and James attempted to strengthen his own authority by executing Lord Murdoch, two of Murdoch's sons, and Murdoch's aged father-in-law. His actions outraged many nobles and led to James's assassination in 1437 by Sir Robert Graham, who was consequently tortured to death over a period of three days at Stirling.

James II was no less involved in royal intrigues. In 1452, believing that William, eighth earl of Douglas, was involved in a plot against the crown, James II invited the earl under a letter of safe conduct to come to Stirling Castle to discuss the matter. After the earl refused to dissolve his group of conspirators, James stabbed the earl with a dagger. In due course his courtiers threw the corpse out of a window. The earl's brother, the ninth earl of Douglas, retaliated by attacking the castle and the city of Stirling, causing extensive damage, but James II continued to rule.

Despite these intrigues, there was some measure of stability generally in Scotland and particularly in Stirling during the reigns of the Stuarts. The town grew in importance because of the castle's status as residence of the kings. James III founded the Chapel Royal of Stirling and James IV made many improvements to the castle. James V finished many additions to the palace complex and improved its fortifications. Mary, Queen of Scots, lived in the castle for four years as a child, was crowned there, and visited at other times during her stormy reign. And James VI rebuilt the Chapel Royal for the christening of his son Prince Frederick Henry, who was the last prince of Scotland brought up in Stirling. When James VI became King James I of Great Britain in 1603, Stirling Castle became far less important, and one of its main uses for many years thereafter was as a prison for persons of dissident political and religious views.

Stirling Castle was the site of a battle between Scottish forces and one of Oliver Cromwell's armies in 1651. English soldiers garrisoned it following their victory. Stirling, rather ironically, served as a military stronghold for the English against attempts to restore the Stuart monarchy in 1715 and 1745–46. The strategic importance of the city and castle ended after the latter uprising.

Stirling Castle is now the regimental headquarters of the Argyll and Sutherland Highlanders. The oldest extant portions of the castle date from James III's time in the fifteenth century. He commissioned the Great Hall to hold his parliaments and state ceremonies, and is credited with the gatehouse and flanking towers. James IV also put his own touches on the castle, building a new palace behind the previous building. After his death his son James V continued work on the Royal Palace, a remarkable building with beautiful early Renaissance sculptures.

The palace has changed little since completion. It has recessed ornamental panels in the Renaissance style situated between the windows of the royal apartments, which are protected with wrought-iron grilles, and the entire building is filled with sculptures. The rooms surround an open court dubbed the Lion's Den because both James III and James IV kept lions.

The Chapel Royal, built by James VI in the early classic Renaissance style, is unchanged except that the coats of arms and royal badges were removed during Cromwell's time. It was in the Chapel Royal that Mary was crowned Queen of Scots when she was nine months old.

The oldest part of the town is located along Spittal Street and Broad Street, which lead up the hill to the castle. Atop Broad Street is Mar's Wark. Little more than a ruined facade now, it was to be a Renaissance palace that John Erskine, the first Earl of Mar and Regent of Scotland when James VI was a child, began in 1570. The palace was never finished, and most of the building was destroyed in 1746 during the Jacobite uprising. Other points of interest in this area include the Cross, where Archibishop Hamilton was tried and hanged the same day for his part in the murder of Henry, Lord Darnley, in 1571. Darnley was the unpopular husband of Mary, Queen of Scots. On Corn Exchange Road are the Municipal Buildings, which show the town's history in stained glass.

Argyll's Lodging is located on Castle Wynd. Originally an elegant quadrangular townhouse, it was constructed on preexisting foundations in the 1630s by Sir William Alexander, first Earl of Stirling. Archibald Campbell, ninth Earl of Argyll, lived there in the 1670s and added a screen wall with an entrance doorway. The building's ornamentation is a good example of the Scottish Renaissance style. In recent years Argyll's Lodging has served as a youth hostel.

Also on Castle Wynd is the Church of the Holy Rude, which dates from the early fifteenth century. It was here that James VI was crowned in his infancy, with John Knox participating in the ceremony.

Another significant landmark is the Wallace Monument, situated one and one-half miles northeast of Stirling. The monument, erected in the 1860s by a Scottish nationalist group, is a huge Victorian tower standing 220 feet high. An imposing bronze statue of Wallace is on the wall above the door. Inside is the Hall of Heroes, containing busts of eminent Scotsmen, and the Hall of Arms, containing a collection of weapons. There is a two-handed broadsword five feet four inches long, said to be Wallace's, but swords of this type, called claymores, were not noted in use in Scotland until the late fifteenth or early sixteenth century. One can climb 246 steps to a parapet, with the reward of an excellent view of the castle and Stirling town. Also at the monument, there is an audiovisual presentation on the Battle of Stirling Bridge.

Another tourist attraction is the Bannockburn Heritage Centre. The location is not on the battle site, which has been overtaken by the mining industry, but southwest of it, where Robert the Bruce reportedly set up his standard on the night before the battle. Nearby is an equestrian status of Robert.

Another nearby battleground is Sauchieburn, where rebel lords defeated James III in 1488. As the king fled from the battlefield he fell from his horse. Knocked unconscious and then carried to shelter, James III was assassinated by one of the rebels disguised as a priest administering the last sacrament.

James III was buried at Cambuskeneth Abbey just east of Stirling. Founded in the twelfth century by King David I, not much of the structure survives today because the Earl of Mar built Mar's Wark from the Abbey's stones. However, there is a complete campanile, or tower, different from anything else from medieval Scotland. The campanile stands sixty-seven feet high with high narrow windows reflecting the First Pointed style of the thirteenth Century.

The modern part of Stirling is late Victorian in appearance; it dates from the coming of the railroads. It is a commercial center for the agriculture of the surrounding countryside. The latest significant addition to Stirling has been its university, founded in 1967, the first completely new university in Scotland since the university at Edinburgh was established in 1583. The University of Stirling stands on 300 acres of land from the Airthrey estate, including 63 acres of woodlands and a 23-acre loch. Adding to its cultural attractiveness is the University's MacRobert Centre, which offers both the public and the students exhibitions of theatre, music, art, and dance.

Further Reading: Susan Ross's *The Castles of Scotland* (London: George Philip and Son, 1973; Newfoundland, New Jersey: Haessner, 1974) has a good chronology of Stirling's history, another chronology of important historical dates in Scottish history, and an introduction that examines how the construction of castles developed. For more information on Stirling and Scottish history see *The History of Scotland* by Peter and Fiona Somerset Fry (London and Boston: ARK Paperbacks, 1985); *Scotland: Forever Home* by Geddes MacGregor (New York: Dodd, Mead, 1980; as *Scotland: An Intimate Portrait,* Boston: Houghton Mifflin, 1990); and *The Making of the Highlands* by Michael Brander (London: Constable, 1980). These last three books provide the religious and political contexts within which Scotland developed. Travel guides that provide useful information and anecdotes about Stirling Castle and the town of Stirling include *Cadogan Guides: Scotland* by Richard Miers (Chester, Connecticut: Globe Pequot, 1987); *The New Shell Guide to Scotland,* edited by Donald Lamond Macnie (London: Ebury, 1977); *Around Scotland: A Touring Guide* by Ken and Julie Slavin (London: Cadogan, 1983); and *The Intelligent Traveller's Guide to Historic Scotland* by Philip A. Crowl (London: Sidgwick and Jackson, 1986; New York: Congdon and Weeds, 1986).

—Bob Lange

Stockholm (Sweden)

Location: Where Lake Mälar meets the Baltic Sea.

Description: Capital city of the Kingdom of Sweden.

Site Office: Stockholm Information Service
Box 2542
S-10393 Stockholm
Sweden
(8) 789-24 00

By around the middle of the ninth century the region around Lake Mälar in central Sweden had come under the control of the Svear people, whose kings resided at Gamla Uppsala. During the period when Norwegians and Danes, and some Svear too, were becoming known to western Europe as Vikings, Svear traders and warriors journeyed east as far as Baghdad and the Caspian Sea and governed Novgorod (now in Western Russia) and Kiev (now in Ukraine), where the name "Rus," formerly applied to the Svear, came to be used for the Slavs. Around 945 other Svear formed the Varangian Guard of the Emperor at Constantinople (now İstanbul).

The trade routes out of Lake Mälar were controlled first from Birka, twenty miles west of modern Stockholm, then from Sigtuna, twenty miles to the north. As the land rose and the shape of the lake gradually changed, the harbor at Sigtuna became clogged with silt while the lake's outlet to the Baltic became more navigable. It is not clear what settlement there may have been at the outlet before around 1250, when a castle was built on the island of Stade by Birger Jarl, then ruling Sweden, including Finland, on behalf of his infant son King Valdemar. This castle was to become known as the Tre Kronor, in reference to the three crowns on the royal coat of arms (representing the Svear, the Goths of Västergötland and Östergötland to the South, and the Wends who lived on the eastern coasts of the Baltic). In 1252 Birger signed a treaty with the German city of Lübeck, giving its merchants exemption from taxes and half the seats on the council of his new town, Stockholm. Within a hundred years 35 to 40 percent of Stockholm's residents were Germans, providing commercial links between Sweden and other countries on the Baltic Sea and making the town the main conduit for imports of cloth and salt and exports of copper and iron. Settlement spread to Helgeandsholm ("Holy Spirit Island") and Riddarholm ("Nobles' Island"), which, together with Stade, are now known collectively as Gamla Stan ("old town").

Two brick churches survive from this period. The older of the two, the Storkyrka ("great church"), is where most coronations took place between 1336 and 1628 and again between 1721 and the abandonment of the ritual in 1907. Alterations in the late fifteenth century and in the 1730s concealed its brick pillars and ceiling ribs and its white

walls, but these features were uncovered earlier in this century. The other, the Riddarholmskyrka, is the burial place of seventeen of Sweden's kings and queens, starting with Magnus Ladulås, who was interred in 1290, eleven years after his decision to exempt his leading advisers and warriors from taxation had created the nobility for whom the island is named.

In 1388 members of this nobility invited Queen Margaret of Denmark-Norway to take the Swedish crown. Within a year, the entire country, except Stockholm, had submitted to her armies, and in 1398 Stockholm also surrendered, but the Scandinavian union did not last long after her death in 1412. In 1434 the rebel leader Engelbrekt Engelbrektsson met her successor Erik of Pomerania at Stockholm to extract promises that taxes would not be raised and that foreigners would not be appointed to Swedish offices, but Erik soon broke both promises, and the rebels seized Stockholm in 1436. The murder of Engelbrekt and the expulsion of Erik ushered in a period of political chaos, as power over Sweden was passed around among rebel leaders, factions of the nobility, Kings of Denmark-Norway and the nobleman Karl Knutsson Bonde, who was King of Sweden for three separate periods up to his death in 1470. His nephew Sten Sture led the Swedish army that defeated an invading force sent by Christian I of Denmark-Norway at Brunkeberg (now in Stockholm) in 1471, an event commemorated by the wooden figures, in the Storkyrka, of Sture as St. George slaying the Danish Dragon, while a princess, symbolizing Sweden, watches from a separate pedestal.

Sture governed Sweden as regent until 1496, when the council of the realm removed him from office. A Danish army attacked Stockholm, Sture surrendered, and Hans of Denmark-Norway was crowned in the Storkyrka in 1497. But Sture carried on his fight, re-taking the castle in 1503, the year of his death. Svante Nilsson, his successor as regent, eventually submitted to Hans, but his son, who was (confusingly) also called Sten Sture, became regent in 1512 and started yet another rebellion. Christian II of Denmark-Norway launched another invasion in 1520. Sture died of battle wounds on his way to join the defense of Stockholm; his supporters surrendered on the promise of an amnesty. There followed the notorious "Stockholm bloodbath," when 82 of them were beheaded on the market square in front of the castle. Sture's corpse was exhumed to be displayed and then burned along with their remains.

Gustav Vasa, a relative of the Stures, took up the struggle and had himself made King of Sweden in 1523. He then signalled his decision to break with Rome in 1524 by putting the Lutheran Olaus Petri in charge of the city's only printing press. A Swedish translation of the New Testament was published in 1526, and in the following year the king

The entrance to the old section of Stockholm

replaced the pope as head of the Church in Sweden. Olaus Petri was tried for high treason in 1639 for failing to reveal what he had known of a conspiracy by German merchants to assassinate the king three years before. One piece of evidence brought against him at the trial, an extraordinary painting that he had commissioned of a parhelion ("mock sun") appearing over Stockholm at the time of the plot, can still be seen in the Storkyrka, of which he was appointed pastor in 1543, his death sentence having been commuted to a heavy fine.

Stockholm became the permanent capital of Sweden in 1634. Its population, which had been only around 7,000 in 1500, had already risen to 10,000 by 1620 and reached around 50,000 half a century later, an expansion eased by the demolition of Birger Jarl's city walls in 1640 and stimulated by a series of military victories and territorial gains. In wars with Denmark, Poland and Russia, which started shortly before Gustav's death in 1560 and went on until 1595, Sweden had already conquered most of modern Estonia and Latvia. From the 1620s onward King Gustavus II Adolphus and his chancellor Axel Oxenstierna expanded the bureaucracy and sponsored Stockholm's shipbuilding industry, though the largest warship, the *Vasa*, sank in the harbor just after being

launched in 1628. (Raised from the seabed in 1961 under the direction of Anders Franzen, this magnificent vessel and the thousands of items discovered inside it are now displayed in the Vasa Museum in the Djurgård park, to the east of the city center.) Intervention in the Thirty Years War, during the two years prior to Gustavus Adolphus's death in the battle of Lützen in 1632, led eventually to the acquisition of two Norwegian provinces and of Denmark's islands in the Baltic in 1645, and of Pomerania and other districts of North Germany in 1648.

In 1658 the Swedish empire reached its maximum extent when Skåne and other southern provinces were taken from Denmark. The grandeur of its capital was enhanced by the completion of the opulent Riddarhus (House of Nobles) on Stade in 1674, but the "Age of Greatness" ended soon afterward in a series of disasters. In 1697 the castle and many other buildings in Stockholm were destroyed by fire. Charles XII's campaigns against Russia ended in defeat at the battle of Poltava in 1709, with the loss of Estonia, Latvia and the Finnish province of Karelia. Then the plague, which reached Stockholm in 1710, caused as many as 18,000 deaths.

By the time that the present royal palace, designed by

Nicodemus Tessin the Younger, was completed on the castle site in 1754, Stockholm's relative superiority within Sweden was in decline. Ships built in the north and in Finland proved to be cheaper, the once unique glassworks on Kungsholm faced a growing number of rivals (and was closed down in 1814), and other ports took more and more trade away from the capital. After the death of Charles XII in 1718, the Riksdag (parliament) took effective control of the country until 1772, when King Gustav III mounted a peaceful *coup*, financed by the French, and ended the "Age of Freedom." After twenty years of almost absolute rule, during which he undermined the power of the nobility by increasing the influence of the non-noble sections of the Riksdag, Gustav was assassinated by a nobleman, Jacob Johan Anckarström, at a masked ball in the Opera House, in Norrmalm to the north of Gamla Stan, which he had founded in 1782. It was rebuilt in the 1890s and has since seen many performances of the opera which the incident inspired, Verdi's *Un Ballo in Maschera*.

In 1808 Russia launched a surprise attack on Finland. The ineffectual response of King Gustav IV Adolph and some of his officers provoked a rebellion by others. In 1809 Gustav was arrested inside his palace, imprisoned for nine months and then sent, on Christmas Eve, into an exile which ended only at his death, twenty-eight years later. His uncle was made king, as Charles XIII, only after he had accepted a new constitution drastically limiting his powers. The problem of the succession to the aging and childless Charles was solved when the Riksdag elected the French Marshal Bernadette as Prince Charles John (later King Charles XIV John). The new prince joined the alliance against his former master Napoléon, receiving Norway as his reward in 1813, though he had to give up his north German lands and Finland was lost to Russia.

Stockholm was at the forefront of Sweden's modernization in the nineteenth century. Its first paved streets were constructed in the 1840s, as also were its first stone apartment blocks, which then became standard after many of the city's wooden houses burned down in 1857. After outbreaks of cholera in 1834 and 1853, Stockholm's first water pipes were brought into use in 1858. As early as 1860 more than half of its population of around 100,000 had been born outside the city but had come there in search of work, mainly in the expanding manufacturing sector, in which mechanical engineering and printing were becoming prominent. Contacts with western Sweden and Norway, and through them with western Europe, increased after the completion of the railroad to Göteborg in 1864 and of the line to the Norwegian capital Kristiania (now Oslo) in 1867. By 1900 there were 300,000 people in Stockholm, which had spread far beyond the surviving medieval streets and merchant houses of Gamla Stan onto the surrounding islands and the mainland.

Political modernization accompanied these economic and demographic changes. Stockholm's modern city council was set up in 1863 as part of a nationwide reform of local government. Three years later the traditional four "estates" of the Riksdag (nobility, clergy, burghers, farmers) were replaced by a single franchise. Both reforms helped to undermine the domination of the city's affairs by the twenty or so wealthy families known as the *Skeppsbroadel* ("harbor nobility"). Stockholm's cultural life was also transformed, with the transfer of the royal art collections to the National Museum in 1866 and the realization in Djurgård of Artur Hazelius's projects for the Nordic Museum, in 1880, and for Skansen, the world's first open air museum, in 1891. From 1901 onward the Swedish Academy, founded by Gustav III in 1786, achieved international fame as the body responsible for awarding the Nobel Prizes (other than the Peace Prize). These awards attracted controversy as early as 1911, when an "anti-Nobel" Prize was awarded to the writer and painter August Strindberg, whose plays, including *Miss Julie*, are better known outside Sweden than the satirical stories about Stockholm that had led the Academy to refuse him a real Prize. (The house in Norrmalm where he died in 1912 is now a museum of his life and work).

In 1905 the Riksdag, meeting in its baroque building on Helgeandsholm, accepted the outcome of Norway's referendum on independence and ended the union between the two countries that had existed since 1813. Conservative politicians, defeated over Norway and alarmed by the general strike of 1909 in which the Swedish labor unions first showed their strength, rallied against the Liberal Prime Minister Karl Staaff's plans to reduce defense spending. Political conflict came to a head in February 1914, when King Gustav V greeted 31,000 protesters who had arrived in front of his palace after the so-called "Farmers' March" by declaring that he supported their demands for rearmament. Staaff resigned, and in the ensuing elections the Social Democrats became the biggest party in the Riksdag for the first time. But then all parties agreed on higher defense spending and neutrality for the duration of World War I; that was followed by the introduction of universal suffrage.

In the boom years of the 1920s Stockholm was at the center of the financial empire created by Ivar Kreuger, whose fortune derived from match factories but extended into banking, telephones, lumber, and mining. The opening of the Stadshus (city hall) on the island of Kungsholm in 1923 seemed to symbolize the optimism of the decade (though it had been designed before World War I). Its architect, Ragnar Östberg, successfully blended Swedish and Venetian traditions to create a monumental landmark next to a walled garden on the shore of Lake Mälar, as the venue not only for meetings of the city council but also for the annual Nobel Laureates' dinner.

Kreuger's bankruptcy, his suicide in Paris in 1932, and subsequent revelations of political corruption marked the onset of the Great Depression and brought down the government. From 1933 the Social Demorats and the Agrarians agreed to introduce a range of welfare, subsidy, and public works programs. Among these were the building of the cloverleaf intersection at Slusson ("sluice"), the meeting point of Lake Mälar and the Baltic, opened in 1935 to connect Norrmalm, Gamla Stan, and Södermalm, the district to their south. Together with the national agreement between employers and labor unions, signed in 1938, these acts formed the basis of the ''Swedish model'' of economic and social policy that lasted into the 1980s. They also underlay the consensus that kept Sweden neutral during World War II, when Stockholm became a base for numerous intelligence agents of both the Axis and the Allied countries.

After 44 years in office, alone or in coalitions, the Social Democrats were defeated in 1976, but they returned to power six years later, again led by Olof Palme, who had been Prime Minister from 1969 to 1976. His assassination in Stockholm in February 1986 was entirely unexpected and remains unexplained, for the only suspect in the case was eventually acquitted.

Such freak events apart, Stockholm has long been the peaceful and sophisticated capital of one of the richest countries in the world. Gamla Stan, the site of so many sieges and risings, intrigues and protests, now attracts tourists rather than invaders. Its role as the heart of the city has passed to Norrmalm, where most businesses have their offices and where there are numerous shops, restaurants, and other amenities for the 660,000 people of Stockholm and the 780,000 inhabitants of its suburbs. With Sweden now governed once again by a non-socialist coalition, which has applied to join the European Union, Norrmalm may be expected to become even more like the centers of other north European capitals. Yet the historic churches, public buildings, and houses in and near Gamla Stan—as well as the open spaces of Djurgård and the archipelago beyond it—will still ensure that Stockholm is not quite like any other city.

Further Reading: *The Early Vasas: A History of Sweden 1523–1611* (Cambridge and New York: Cambridge University Press 1968) and other books by Michael Roberts are the most rewarding guides to the history of Sweden and of Stockholm for the periods they cover. There are two informative but unexciting one-volume histories: Stewart Oakley's *The Story of Sweden* (London: Faber, and New York: Praeger, 1966) and Franklin D. Scott's *Sweden: The Nation's History* (Minneapolis: University of Minnesota Press, 1977). Eric Elstob's *Sweden: A Traveler's History* (Woodbridge, Suffolk: Boydell Press, and Totowa, New Jersey: Rowman and Littlefield, 1979) is probably harder to obtain than either of the other two, but it is much more readable, usefully relating historical events to places and to artistic achievements.

—Patrick Heenan

Strasbourg (Bas-Rhin, France)

Location: In northeastern France, in the province of Alsace, on the French border with Germany, 300 miles east of Paris and 220 miles west of Bonn. The city lies on the Ill River, two miles west of its confluence with the Rhine. It is the capital of the governmental department of the Bas-Rhin.

Description: Strasbourg is the sixth-largest city in France and France's largest river port. It is the terminus of the Marne-Rhine Canal, which links Strasbourg and the Rhine with the region around Paris, and Rhine-Rhône Canal, which connects Strasbourg with the city of Lyons and the Rhône River. The oldest section of the city, built on an island lying between two arms of the Ill River, contains the magnificent thirteenth-century Cathedral of Strasbourg and the Château des Rohan, the residence of the bishops of Strasbourg, built between 1731 and 1742.

Site Office: Office du Tourisme de Strasbourg
17, place de la Cathedrale
67082 Strasbourg-Cedex, Bas-Rhin
France
88 52 28 28

If one took a map of Europe and eliminated the colors and the thin lines that the cartographer used to define the different countries and to delineate their boundaries, what would remain would be only the physical features of the land, rivers and mountains and valleys, and the cities. With a map such as this, the truth of Strasbourg could be readily seen, that which has been both its good fortune and the source of its turmoil. The city, just east of the Vosges Mountains, lies in the great valley of the Rhine River, which divides the continent almost equally into eastern and western halves. Strasbourg is equidistant from the Atlantic to the north and the Mediterranean to the south. It is the heart of western Europe.

There are signs of permanent habitation in the area of Strasbourg dating back to the Stone Age. The Celts, who had settled there prior to the Roman occupation, built dams and walls to counter the floods of the Rhine, which at that time ran two miles east of its present course. In 12 B.C., the Romans conquered the area and turned the settlement into the city Argentoratum, one of fifty cities built by the Roman general Claudius Drusus, to guard the empire's easternmost frontier in Europe.

Argentoratum was destroyed by the invading armies of Attila the Hun in 455 and was eventually rebuilt by the Franks, a West Germanic people who were establishing themselves throughout the Roman province of Gaul and along the Rhine River. The restored city was renamed Strataburgum, or "fortress of the roads."

In 842, the empire of Charlemagne was divided among his three grandsons in an agreement called the Oath of Strasbourg. Louis received Germany. Charles reigned over the regions to the west of the Meuse and the Rhine rivers. Thus were the foundations of France and Germany laid. To Lothair went Italy and the lands along the Rhine. This area, much reduced, would eventually become the province of Lorraine. The Oath of Strasbourg itself remains important as the oldest surviving document of the language spoken in the ninth century in that area of Europe.

Strasbourg was linked to Germany in 923 through homage paid to Henry I of Germany by the Duke of Lorraine. At the beginning of the eleventh century, Strasbourg became an episcopal seat. Bishop Wernher laid the foundations for a Romanesque church in 1015. The church was destroyed several times by fire and rebuilt, and the great cathedral which stands today was begun on the site in 1240 and completed in 1439. During the eleventh and twelfth centuries, Strasbourg was in the total control of its bishops, who appointed the city's officials and who could claim the profit of five days of work per month from each of its citizens.

In the thirteenth century, however, Strasbourg was granted the privileges of a free imperial city by Philip of Swabia and began its growth as one of the centers of Europe. A bridge spanning the Rhine was built in 1388, greatly facilitating commerce and travel. No other bridge crossed the river between Strasbourg and the sea. From 1434 to 1444, Johannes Gutenberg resided in Strasbourg, and there, with the help of local artisans, he began his work on the design of movable type that led directly to the introduction of the printing press in Germany.

In the sixteenth century, Strasbourg played a major role in the Protestant Reformation. Joining in the movement begun by Martin Luther, the city embraced the Reformation and became a refuge for French Protestants from Catholic France. The initial cause of Luther's revolt was the abuse of indulgences by the papacy. An indulgence was a remission of part or all of the temporal and especially the purgatorial punishment for sins. Through long-standing practice, indulgences could be purchased. In 1476, Pope Sixtus IV extended these indulgences to souls already in purgatory, thus, in the words of historian A. G. Pickens, "exploiting for cash the natural anxieties of simple people for their departed relatives." Additionally, the secular finances of the papacy drew criticism. Pope Leo X realized huge personal profit from the sale of some 2,000 church offices, which were purchased as a form of annuity.

The most dynamic and vigorous leader of the Reformation in Strasbourg was Martin Bucer. Bucer was a former Dominican monk who left the order after he was first forbidden by his superiors to continue his studies, then forbidden even to read. Bucer was a moderate reformer, who attempted

The Château des Rohan, top, and Strasbourg Cathedral, bottom
Photos courtesy of Office de Tourisme, Strasbourg

to be a mediator between Luther and the more radical Zwinglian movement taking root in Switzerland. The focus of the Reformation in Strasbourg was on preaching the pure word of God, taken directly from the ancient texts. The newly invented printing press played an important part in the Reformation. The writings of Luther and other Protestants were printed and widely distributed. When Bucer urged Protestant theologians to go out into the world and preach, Luther replied "we do that with our books."

In 1528, the Catholic Mass was abolished in the city. In 1538, Johann Sturm founded the Scola Argentinensis. John Calvin was a member of the theological faculty. The school was the forerunner of the University of Strasbourg, which was founded in 1621 under an imperial charter. The Reformation also resulted in the establishment of city schools in Strasbourg and the start of a centralized system of welfare relief controlled by the municipal authorities.

The first known German newspaper, the *Relation,* began publication in 1609 in Strasbourg. Begun by Johann Carolus, it reported news from as far afield as Cologne, Vienna, Prague, Venice, and Rome.

Although badly wounded by the loss of its parishes, monasteries, and convents, the Catholic Church in Strasbourg was not totally erased by the Reformation. The chapters, the ecclesiastical courts, the episcopal administration, and the bishop—its basic structures and those most important from a financial point of view—remained. At the end of the Thirty Years War in 1648, Strasbourg found itself still an imperial free city, solidly Lutheran, German speaking, a small republic within the Holy Roman Empire.

The expansion of France under Louis XIV, however, soon led Strasbourg down a new path. Michel Le Tellier, Marquis de Louvois and minister of war to Louis XIV, was obsessed with the importance of the Rhine frontier, and with Strasbourg in particular, to the strength and the defense of France. After months of logistic planning, for which Louvois was famed, the armies of France began their strike against Strasbourg just after midnight on Sunday, September 28, 1681. The outlying forts were seized and their garrisons scattered. Observers sent out from the city saw the crushing mass of French troops, artillery, and supplies being amassed for an assault.

On Monday, September 29, representatives of the city rode out to meet with Louvois, who gave them the simple choice of submitting to the sovereignty of France or having their city leveled. After a day of fevered deliberation, the assemblymen of Strasbourg voted almost to a man to capitulate. The lone holdout in favor of resistance to the end was a seventy-year-old tailor. On Tuesday, September 30, with terms, including a general amnesty, deemed favorable to the citizens of the city, the assemblymen signed the capitulation, and the troops of France marched through the Butcher's Gate and occupied the city.

What Louvois had taken was now put into the hands of Sébastien Le Prestre de Vauban, an architect and the chief military engineer of Louis XIV. Vauban, arguably the most famous military engineer in history, is credited with the construction or repair of 150 "places de guerre" and is said to have directed 53 sieges. His genius was in the resourcefulness he demonstrated in adapting traditional means of defense to new situations, rather than inventing new ones. Some of his fortifications were in effective military use until the time of World War I.

Vauban, at the summons of Louvois, was en route to Strasbourg even before the city was taken. After careful study of the inadequacies of the current defenses, Vauban launched his refortification at a pace that amazed the townspeople. Vauban's fortifications, some of which remain to this day, resulted in a star-shaped complex of defensive towers and facades. An enemy could not assault one position without coming under withering protective fire from another. Vauban also expanded the fortifications east of the city, including the Kehl fort across the Rhine, and defensive positions on midstream islands in the Rhine, making any potential crossing much more difficult. Vauban transformed Strasbourg into a fortress for France, although not without provoking some criticism. One critic of the time declared that Vauban had converted Strasbourg into a "gigantic porcupine." In 1697, seemingly as an afterthought, Strasbourg was confirmed as a French city by the Treaty of Reswick.

The next hundred years saw Strasbourg's culture adapt, sometimes forcibly so, to its new identity as a French city. The Cathedral of Strasbourg, which had become a Protestant church during the Reformation, was returned to the Catholic Church in 1681. Louis XIV personally attended the rededication. Ordinances were passed requiring women to dress in "la mode françoise." Intermarriage between French and Germans aided in the transition. In 1725, King Louis XV married Maria Leszczyńska, daughter of King Stanisław of Poland, in the cathedral. An immense increase in travel occurred in the 18th century in Europe, and the citizens of Strasbourg not only traveled east to Germany, but westward, across the Vosges Mountains, to Paris.

Despite edicts from Paris, the populace clung stubbornly to the German language. Up until 1789, the press of Strasbourg was still mainly in German, although individual items in French were printed, and German sentences often contained French phrases. The weekly *Wochenblatt,* a trade newspaper, eventually adopted a double issue to resolve the situation, publishing in French on Wednesdays and German on Saturdays. Nowhere was the dichotomy of the city, the pulls of culture, and Strasbourg's dual identity as both French and German better exemplified than in its press, its literature, and its arts. In 1786, Jean-Chrétien Treitlinger's newspaper reading room at the Swedish Coffee House, on what is now Place Kleber, offered the reader, for a nominal fee, a choice of twenty-nine current out-of-town newspapers in French and fifty-four in German, along with the *London Evening Post,* the *Notizie del Mondo* from Venice and *L'Observatore Triestino.*

The composition of the student body at the University of Strasbourg also reflected the dual identity of the city and its growing position in Europe. In the years just prior to the

French Revolution, there was parity between students from France and from the Holy Roman Empire, with approximately 25 percent of the students from each. The remaining 50 percent were English, Swiss, Hungarian, Czech, and Russian.

In the latter half of the eighteenth century, after more than fifty years of existence, the Comédie Française de Strasbourg had become a successful enterprise, regularly featuring works by the playwrights Molière, Racine, Marivaux, Voltaire, and Beaumarchais. At the same time, the German literary movement called Sturm und Drang took root and flourished in the city. Characterized by rousing action and high emotionalism, and often dealing with the individual's revolt against society, the movement's most famous proponents include Johann Wolfgang von Goethe and Jakob Michael Lenz, who, as young men barely in their twenties, came to Strasbourg as students during the 1770s.

One continuous thread that ran through the social and civil life of the city was the Schwörbrief, or civic oath. Conceived in 1482 after more than 200 years of internal coups, revolts, and feuds following Strasbourg's emergence as an imperial free city, the Schwörbrief was a concise list of certain recognized municipal and civil procedures. For more than 300 years, until the French Revolution, all municipal officers swore obedience yearly to the Schwörbrief. For Strasbourg, the oath was a functioning symbol of the city's continuity, despite whatever pressures were exerted by forces beyond the city walls.

With the French Revolution, there was more upheaval in Strasbourg. At the height of the Terror, the cathedral was turned into a Temple of Reason, and its statues were ordered destroyed. In 1792, the victorious Jacobins considered destroying the cathedral's great spire. The statues were saved by a public works administrator, who had them removed and hidden. City legend has it that a quick-witted citizen proposed crowning the spire with a huge red cap, a symbol of the revolution, thereby saving it from destruction. Rouget de Lisle, a captain in the French Army of the Rhine, wrote the ''Marseilles'' in 1792 while garrisoned at Strasbourg. The war song was adopted as the national anthem of France in 1795.

Strasbourg had been under French control for two centuries when, in 1870, it was heavily bombarded by Germany during the Franco-Prussian War and sustained serious damage. By August 12, 1870, Germany had gained almost complete control of the French provinces of Alsace and Lorraine. Only Strasbourg and Metz remained in French hands. On August 13, the first shells fell on Strasbourg, and two days later the bombardment began in earnest, directed not against fortified military positions but toward public and private buildings. The German commander, von Werder, whom the people of Strasbourg grew to call Mörder (assassin), carried on a modern-style campaign of terror against the civilian population. In addition to directing bombs at schools, private homes, the art museum, and two libraries, the Germans also targeted and damaged the cathedral. On August 26, the roof caught fire, melting the copper tiles. Stone statues were pulverized and stained glass windows shattered. The bombing continued, destroying the railroad station, the Palais de Justice, theatres, the prefecture of police, and approximately 600 houses. Strasbourg finally surrendered on September 27. Estimates put the number of projectiles and bombs hurled into the city at 200,000. The city passed to the rule of Germany under the Treaty of Frankfurt in 1871 and remained under German control through World War I, until it was returned to France in 1919. Strasbourg and the Alsace-Lorraine region had remained culturally and spiritually tied to France during the period of German control; the citizens of Strasbourg still harbored bitterness against the Germans for the destruction of 1870. Alsace-Lorraine and its resistance to attempts to impose German culture became a rallying point for the Allied powers in World War I.

During World War II, from June of 1940 to November of 1944, the city was again under German control and suffered considerable damage, this time as a result of Allied bombing. After its liberation, the city once again was returned to France.

Since World War II, Strasbourg has taken its place as one of the centers of the greater European community. In 1949, the Council of Europe was established at Strasbourg, and in 1979 Strasbourg became the meeting place of the European Parliament. Strasbourg is also home to the European Commission on Human Rights and the European Science Foundation.

Strasbourg today remains one of Europe's most beautiful cities. Historic structures include the cathedral; the eighteenth-century Château des Rohan, once the home of bishops and today the home of an art museum; the Musée Historique, a sixteenth-century building displaying exhibits relating to Strasbourg's history; the Musée Alsacien, a museum of regional art and culture, housed in a historic inn; and the eleventh-century Church of St. Pierre le Jeune, which now belongs to a Protestant congregation. There are also many restored half-timbered houses from the sixteenth century and earlier.

Further Reading: *Strasbourg in Transition, 1648–1789* by Franklin L. Ford (Cambridge, Massachusetts: Harvard University Press, 1958) focuses on the social, political, and economic changes that Strasbourg underwent in its transition from an imperial free city of the Holy Roman Empire to a stronghold of France. *Reformation and Society in Sixteenth Century Europe* by A.G. Dickens (London and New York: Harcourt Brace, 1966) is a history of the Protestant Reformation, its root causes and spread throughout Europe, and the rise of the Reformation's more radical elements. *Strasbourg and the Reform* by Miriam Usher Chrisman (New Haven and London: Yale University Press, 1967) is a very detailed account of the events of the Reformation in Strasbourg and the role the city played in the events of a wider European stage. *Alsace-Lorraine under German Rule* by Charles Downer Hazen (Freeport, New York: Books for Libraries Press, 1917; reprint, 1971) was written in the turmoil of the days just preceding America's entry into World War I. The book explores in depth the ''question of Alsace-Lorraine'' as one of the causes of the Great War. Although well written and persuasive, the work must be viewed in the light of the times in which it was conceived.

—Rion Klawinski

Stratford-upon-Avon (Warwickshire, England)

Location: Southwestern Warwickshire County, ninety miles northwest of London, on the River Avon.

Description: A small town on the banks of the River Avon acclaimed as the birthplace of playwright William Shakespeare. It is now promoted as a major tourist attraction with medieval homes, inns, and churches. Shakespeare's remains are buried in the Holy Trinity Church.

Site Office: Tourist Information Centre
Bridgefoot
Stratford-upon-Avon, Warwickshire CV37 6GW
England
(789) 293127

For several centuries, the English countryside was a sparsely populated area dotted with large farms and isolated manors. In the 1500s, however, small towns began to be established. One such town was Stratford-upon-Avon, which prospered as a country market. Located literally on the banks of the Avon, the town was originally entered by means of a wooden bridge that was often treacherous to cross when the river crested during the summer months.

Stratford-upon-Avon was granted a borough charter in 1553 that provided for fourteen aldermen and fourteen burgesses to govern the town. Each September during Michaelmas, the combined group elected a bailiff or mayor and a sub-bailiff (also known as the head alderman) from among its members. Other town officials included two chamberlains to manage the finances, two tasters to oversee the quality of the bread, meat, and beer sold in the borough, four constables to maintain order, and leather sealers to approve and stamp the leather goods that were sold there. Two "serjeants-at-the-mace" were assigned to the bailiff and sub-bailiff for the purposes of issuing warrants.

In 1568, the bailiff was John Shakespeare, a successful glove maker and husband of Mary Arden, the daughter of a wealthy land owner. Four years earlier, in 1564, the first of John and Mary's five children, William, was born in the house they had purchased on Henley Street. The Shakespeare home was actually a row of connected dwellings with a separate space for John's glove business. The house was divided into three parts: the East House, the West House, and the Middle House.

Much of what is known about Stratford comes from church records and other official documents. The dearth of reliable information about William Shakespeare's life has produced myriad accounts that are often based on supposition and legends. Tradition holds that William Shakespeare was born in the Middle House of the block on Henley Street.

However, some scholars believe that the birth took place in the East House. The debate centers around the fact that, after John Shakespeare's death in 1601, William gave the Middle and West Houses to his sister Joan Hart and leased the East House to non–family members. Apparently, sentimental biographers could not imagine that the great Bard would lease his birthplace to strangers.

William was baptized in Holy Trinity Church. With its tall spire, the church is one of Stratford's most recognizable landmarks. The original section of the church dates to 1210 and was built on the site of a Saxon monastery. The main section was completed in 1480, and the spire added in 1763. The church's fifteenth-century stone altar, removed during the Reformation, was rescued and returned to the church and is now used as the High Altar.

In the sixteenth century, traveling acting troupes provided a vital communication link between London and the small towns of England. It is not surprising, then, that the actors' arrival was greeted with much anticipation and excitement, and this is no doubt when young William got his first taste of the theatre. The troupes would first call on the bailiff to receive permission to perform in the town. If the bailiff approved of the troupe's performance, the actors would set up a stage in the inn courtyard. Most of the townspeople stood on the ground below to watch the performances while the more prosperous citizens sat on the inn's balconies.

Stratford's most famous inns, the Swan and the Bear, figure prominently in Shakespeare's works. During his lifetime, they stood on opposite corners at the foot of the town's main thoroughfare and featured glazed windows with leaden panes. Inside, the walls were either wainscoted or hung with painted cloths. The open-beamed rooms bore such names as the Lion, Talbot, Dixon's, and the Cock, and were filled with the hand-carved furniture of the time: oak cupboards, tables and chairs, and four-poster beds.

Clopton Bridge, an arched stone structure that spans the river, was built at the end of the fifteenth century by a wealthy land owner named Sir Hugh Clopton to replace the wooden bridge that was finally swept away by floodwaters during the summer of 1588. (The same storms put a temporary halt to the advance of the Spanish Armada when the Spanish fleet was forced to hold its course off the coast of Cornwall while the English ships waited out the rains at Plymouth.) At the time that the Clopton Bridge was built, both banks of the river were thickly forested with willows, elms, and ashes. Upon crossing into Stratford, a visitor would first encounter the bankcroft, a grazing common for cattle and sheep. The bankcroft gave way to the butt close, an area reserved for archers to practice their skills. Further on, the town spread out with barns, gardens, and buildings. At the time of William Shakespeare's birth, Bridge Street was

The River Avon and Holy Trinity Church

The gardens at New Place
Photos courtesy of Stratford-on-Avon District Council

always filled with people—children ran through the town and shopkeepers displayed their wares in the town square.

The houses were typically half-timbered cottages with thatched roofs that hung over gables and dormer windows. The home of Shakespeare's wife's family, the Hathaway cottage, is a prime example of this architecture and is still standing today. Outside the town limits, Queen Elizabeth I's courtiers built lavish country homes hung with tapestries. One of these was Charlecote, the home of Sir Thomas Lucy. It was from this estate that a young William Shakespeare was reputed to have poached deer.

During Shakespeare's time, Stratford had a free grammar school for children ages seven through twelve. School was in session five days a week, year round. The school day typically ran from 7:00 A.M. to 5:00 P.M., with a two-hour break for the traditional midday meal.

William Shakespeare married Anne Hathaway in 1582. In the early years of their marriage, the Shakespeares lived with William's father in the house on Henley Street. The back of the Middle House was constructed as a separate residence, and it is here that William and Anne most likely began their family. Daughter Susanna was born in May 1583; twins Judith and Hamnet followed in January 1585.

In 1590, Stratford was forced to petition for tax relief—the rapid population growth in Europe after the Middle Ages put a strain on small town finances. Simultaneously, many young men were leaving the countryside to seek their fortunes in London. Stratford might very well have vanished from the face of the earth had William Shakespeare not been among those young men.

Shakespeare is thought to have left for London in 1587. Anne remained behind in Stratford with the children and continued to live in the house on Henley Street. A devastating fire swept through Stratford in 1594. Miraculously, though much of Henley Street was destroyed, the block on which the Shakespeares lived was spared. Further tragedy struck when Hamnet died in the summer of 1596 at the age of 11.

Shakespeare did indeed find fame and fortune in London. In 1597, he was successful enough to purchase the second largest home in Stratford, known as New Place. Its original owner was none other than Sir Hugh Clopton, the bridge builder. Anne and the two remaining Shakespeare children moved from Henley Street to New Place.

William Shakespeare died in 1616 and was buried in the chancel of Holy Trinity Church. A bust of Shakespeare, sculpted by Geraert Janssen, overlooks the tomb from a recessed space in the wall. Janssen's artistic rendering was never wholeheartedly received, and the bust has been repainted or whitewashed several times.

The rising influence of Puritanism during the final years of Elizabeth I's reign continued into the seventeenth century, and the popularity of traveling acting troupes waned. Shakespeare's former acting company paid a visit to Stratford six years after his death, but the visit was chiefly a sentimental one. By the mid–seventeenth century, Charles II was on the throne and the Restoration movement brought a renewed

interest in the theatre. People began to visit Stratford in increasing numbers. A benefit performance of *Othello,* perhaps the first of Shakespeare's plays to be performed in Stratford, was held in 1759 to pay for repairs to the bust at Holy Trinity. At the time, David Garrick, a renowned actor and producer, called Stratford "the most dirty, unseemly, illpav'd, wretched-looking town in all of Britain."

Ten years later, Garrick devised a plan to bring Stratford the recognition he believed it deserved as the birthplace of England's greatest poet. He organized an extravaganza in conjunction with the grand opening of the new Stratford Town Hall. Billed as the Jubilee of 1769, the festivities included singers, plays, lavish dinners, and firework displays. Although torrential rains caused the cancellation of most of the events on the second day, the Jubilee brought Stratford the publicity it needed to become an important tourist attraction.

David Garrick contributed to the birth-room controversy by declaring that Shakespeare was born in the front bedroom on the second floor of the West House. His promotion of the room led tourists to scrawl their names on the walls and pillage the room for souvenirs.

After Anne Hathaway Shakespeare died, Susanna continued to live at New Place with her husband, Dr. John Hall, and their family. Queen Henrietta Maria, the French-born wife of King Charles I, is said to have spent three days with the Halls in 1643 during the English Civil War. By the end of the seventeenth century, however, the home was back in the hands of the Clopton family and in desperate need of repair, prompting the Cloptons to direct a major renovation in 1702.

In 1753, the house was purchased by a wealthy vicar named Francis Gastrell. Evidently, Reverend Gastrell was unaware of his home's famous former owner and became vexed when tourists began arriving in vast numbers. Many of them came to view the mulberry tree that Shakespeare was rumored to have planted in the back yard. Gastrell decided to put a stop to the steady stream of unwanted visitors by chopping the tree down, thus causing a minor riot in Stratford. An enterprising man named Thomas Sharp purchased the wood and used it to manufacture relics and trinkets to sell to tourists. The astonishing array of trinkets included toothpicks, snuff boxes, eyeglass cases, goblets, and pieces of furniture. Sharp's inventory of "official" Shakespearean trinkets soon exceeded the wood that one mulberry tree could possibly produce.

Reverend Gastrell aroused further ire by protesting the monthly tax assessment imposed by the Stratford Corporation. When the Corporation refused to accept his argument that the assessment was unfair because he lived in New Place only during the summer months, an outraged Gastrell destroyed the home in 1759.

The house on Henley Street passed through several owners. The Hart family, descendants of Shakespeare's sister Joan, continued to reside there well into the seventeenth century, when the eastern portion was converted into an inn, the

Swan and Maidenhead, and its half-timbered facade was covered with bricks. The western portion was converted to a butcher shop.

Mary Hornby rented the butcher shop from the Hart family and remained a tenant even after the entire building was sold to the Court family in 1806. The Courts assumed management of the inn. Mrs. Hornby and Mrs. Court became bitter rivals in the fiercely competitive Shakespearean relics trade. Mrs. Hornby went so far as to claim a family relationship to the playwright. Mrs. Court prevailed, however, by evicting Mary Hornby in 1820.

After Mrs. Court's death in 1847, the house was placed on the market. Rumors were circulating that a group of Americans was planning to purchase the house, dismantle it, and reconstruct it in the United States. This led to the creation of the Stratford Birthplace Committee, which raised enough money through benefit theatre performances to successfully bid on the property. The committee's bid of £3,000 reportedly included £250 from Prince Albert.

The committee found that its new possession was in dire need of restoration. Ironically, what remained of the home's original timber was sold as relics in order to raise the funds necessary to return the house to a replica of what it was in Shakespeare's day. In 1866, the Committee became the Birthplace Trust and took over management of the New Place estate, the Hathaway cottage, Mary Arden's house, and Hall's Croft, the one-time home of Susanna and John Hall.

In 1864, Charles Edward Flower led fund raising efforts to create the Royal Shakespeare Company and to build a theater on the banks of the Avon. The first performance, *Much Ado About Nothing,* was held in 1879. The original theatre was destroyed in a fire in 1926. The current structure was built in 1932. The large but plain red brick building houses an extended stage that can be viewed on three sides. Adjacent to the theatre is a memorial statue of the playwright, surrounded by sculptures depicting various characters from his plays.

Today, Stratford exists primarily as a tourist attraction and has been restored to offer visitors a glimpse into history. The Elizabethan-era cottages and manor homes have been converted into museums. Shakespeare's birthday is celebrated annually. The passage of time has done little to lessen the fascination with Shakespeare and the town where he was born. The first official report of annual visits to Stratford was made in 1854 and show a figure of nearly 3,000. By 1900, the number had reached 30,000 and in the 1980s, Stratford reported over 1 million visitors a year.

Further Reading: *Shakespeare's Stratford* by Edgar I. Fripp (Freeport, New York: Books for Libraries Press, 1928; reprinted 1970) is written by one of the life trustees of Shakespeare's Birthplace Trust. It includes maps of the area as it appeared during Shakespeare's time and descriptions of the most prominent buildings. *Shakespeare's England* by the editors of *Horizon Magazine* in consultation with Louis B. Wright, director of the Folger Shakespeare Library (New York: American Heritage, 1964) is an easy-to-read narrative about life during the Elizabethan period and the life of William Shakespeare. It contains a one-page bibliography. *The Riverside Shakespeare* (Boston: Houghton Mifflin, 1974) is a collection of the complete works of William Shakespeare with introductions and explanatory matter by a panel of Shakespearean scholars. This edition also includes early criticisms of the plays and poems as well as an extensive bibliography. A highly recommended one-volume British edition of Shakespeare's plays is *The Complete Works* (Oxford Shakespeare), edited by Stanley Wells and Gary Taylor (Oxford: Clarendon Press, and New York: Oxford University Press, 1986). *The Englishman's England: Taste, Travel, and the Rise of Tourism* by Ian Ousby (Cambridge and New York: Cambridge University Press, 1990) is a slightly ironic look at the most popular tourist stops in England.

—Mary F. McNulty

Tallinn (Harjumaa, Estonia)

Location: On the southern shore of the Gulf of Finland.

Description: Capital city of the Republic of Estonia, with a mixed architectural heritage from centuries of Danish, German, Swedish, and Russian control.

Site Office: Estonian National Tourist Board
Pikk Tanav 71
EE0001 Tallinn, Harjumaa
Estonia
2 602 716

Today Tallinn is a city of about half a million Estonians and Russians living on and around a hill called the Toompea and the natural harbor below, but for most of its recorded history it was dominated by Germans, for whom the hill was the Domberg. From about 2,000 years ago there had been an Estonian settlement here, known as Lindanissa, the place where the hero Kalev was supposedly buried by his mother Linda. In 1219 the Christian king Valdemar II of Denmark defeated the pagan Estonians and established a castle on the Toompea. From that time forward, Lindanissa was known as Reval to the Danes and other non-Estonians, but was named Tallinn, which probably means "Danish castle," by the natives.

Reval's development as a trading center began in 1230, when 200 German merchants joined the Danish and Swedish settlers who were already there and brought the town into close relations with the German settlement of Riga, founded in 1201 by Albert, bishop of Livonia. In 1202 Albert founded a crusading order, the Brethren of the Sword, which took over the castle at Reval in 1227 and in 1232 began building the Domkirk, the cathedral next to the castle, which is the city's oldest church (though the present building, a Lutheran establishment, is largely the result of reconstruction in the 1680s). After the Lithuanians defeated the Brethren in 1236, those surviving joined the Teutonic Order, and Reval was returned to Danish control. (The castle, with its watchtower nicknamed Long Herman, was also to be extensively rebuilt, in 1780.)

In 1284 Reval joined the Hanseatic League, the association of trading cities that dominated the northern European seas from the late thirteenth century until the second half of the fifteenth century, when the Dutch gradually replaced it. The Hanse also included the Teutonic Order, whose High Master was its Beschermer (protector). Together with Riga and Dorpat (now Tartu), Reval came to serve as the main conduit for furs, timber, tar, honey, and rye from Russia. Most merchants from western Europe took the overland route from these towns to Novgorod, at that time the farthest west of the Russian states; the alternative route along rivers and lakes was obstructed by waterfalls. The German elite of Reval officially excluded non-Germans from this trade, while unofficially subcontracting to Estonians and others; nor were they averse to the profits that came their way from illicit trade with Finland, then a Swedish possession, where many farmers and even the dean of the cathedral at Åbo (Turku) preferred dealing with Reval, a boat-ride away, to obeying the distant Swedish authorities. The Lower Town, the area of Reval below the Toompea, became a busy trading settlement centered on the market square constructed in 1288.

Around this time the Germanized landowning nobles of Harrien and Wierland, the districts around Reval, formed a corporation that, in various forms, was to maintain a powerful political and economic position in city and region alike through to 1926, when independent Estonia redistributed the German minority's land to their tenants. Their mansions can still be seen on the Toompea, surrounding the castle, the Domkirk, as can their corporation's headquarters, the Nobility House (rebuilt in the 1680s and extended in 1840).

In 1347 the Danish king Valdemar IV sold his Estonian lands to the Teutonic Order, which now controlled two areas, Livonia and Prussia. These possessions were separated from each other by the powerful state of Lithuania, then still pagan, and depended on the Hanseatic League for supplies and communications. The Order's provincial masters, residing on the Toompea, collected rents from its land, as well as a share of the League's customs dues, and arbitrated between the merchants and the nobles and among the increasingly divided merchants themselves. The wealthier among them were responsible for the present-day appearance of the Lower Town, with its cobbled streets, its typically Hanseatic houses with their red-tiled roofs, and such buildings as the Town Hall, built on the market square between 1371 and 1374, and the Gothic headquarters of the Great Guild, built between 1405 and 1410.

The Teutonic Order declined after 1410, when it was decisively defeated by Poland and Lithuania at the battle of Tannenberg, and again after 1435, with the assertion of the rights of the Diet (*Landtag* in German), which represented the bishops, the Order, the nobles, and the towns of Livonia. The Order was divided in 1513, when the Prussian branch sold all its rights over the Livonian Knights to the Grand Master based at Riga. Reval also ceased to collaborate with other Hanse towns. Since 1422, when Dorpat took control of the Hanse base in Novgorod, the Livonian towns had monopolized trade with Russia, continuing to do so even after 1478, when the Novgorod Republic was destroyed and the city was absorbed into Muscovy.

By the fifteenth century only half of the population of Reval were Estonians, mostly day-laborers. About one-third were Germans, some monopolizing positions of power and

A view of old Tallinn
Photo courtesy of Estonian Consulate, New York

wealth and others engaged in skilled crafts and trade. The remainder were Swedes and Finns. In the 1520s, in spite of these social and cultural divisions, Reval joined with other Baltic towns in embracing the Lutheran Reformation, and Lutheranism gradually became the common faith not only of the Baltic Germans, the Estonians, and the Latvians, but also of the neighboring peoples of northeast Germany and Scandinavia.

In 1558, during an invasion of Livonia by Ivan IV (the Terrible), czar of Muscovy, Reval was briefly governed by a faction favoring a return to Danish rule. Three years later, however, further incursions by Ivan's troops, as well as a peasant revolt against taxes and conscription, frightened the merchants and nobles of Reval, Harrien, and Wierland into submitting to Erik XIV of Sweden instead. The rest of Livonia, under the now-secularized Teutonic Knights, acceded to Poland, while Lübeck and other Hanse towns abandoned their trade with Reval in favour of Narva, the only Baltic port controlled by the czar. Reval was attacked in 1565 by Estonians rebelling against Sweden, in 1569 by Lübeck gunships supporting the claims of its ally Denmark, in 1570–71 by Magnus, a Danish duke allied to the czar, and in 1576–77 by Ivan's own soldiers. But it remained true to its new Swedish

allegiance, which was reinforced in 1581 when the Swedes captured the rival port of Narva and ended its trading role. Thus, for almost 150 years, the Swedish province of Estland (the northern half of modern Estonia) was governed from the Nobility House by the chancellery of the nobility (*Ritterschaftskanzlei* in German), chosen by a diet from which church and city leaders were excluded. The nobles cooperated with royal officials from Sweden in overhauling local government, establishing secondary schools and enforcing the religious monopoly of the Lutheran Church.

The turn of the eighteenth century was marked by disasters. There was widespread famine in 1696–97; the plague, which arrived in 1709, killed perhaps two-thirds of the peasantry and more than one-quarter of the townspeople; and in 1710, following his decisive victory over Sweden at Poltava, the Russian emperor Peter I (the Great) seized Estonia, Latvia, and Lithuania, a transfer confirmed by the Treaty of Nystad in 1721. Peter was shrewd enough to leave the privileges of the local nobles unaltered, and, since Russia could not provide the close supervision that Sweden had, almost all administration passed into the hands of Baltic Germans working under Russian governors on the Toompea, which the Russians called Vyshgorod.

The population of Reval did not return to its seventeenth-century level until after 1800, though immigration did continue, with about half of the new citizens coming from Germany and half from Estland itself, where the ability of rural people to move to the towns was enhanced by the abolition of serfdom in 1819. In the late nineteenth century the czarist authorities sought to undermine the Baltic German elite, enforcing the use of the Russian language beginning in 1881 and building an Orthodox cathedral on the Toompea between 1894 and 1900, but the mainly German character of Reval and other towns was already threatened by the spread of industry and railways, which helped to increase the Estonian element in the population of Reval from 51.8 percent in 1871 to 88.7 percent in 1897. Thus there was an ever more receptive audience for the nationalist ideas of Konstantin Päts, editor of the newspaper *Teataja* from 1901. Estonians formed the majority on the town council beginning in 1904, while the industrial areas, mainly populated by Russian immigrants, became centers of left-wing agitation. In June 1917, three months after the first Russian revolution of that year, the Provisional Government made Reval the capital of a re-formed Estonia and the meeting place of the Maapäev, the new legislature, under a Russian commissar.

In November, following the Bolshevik Revolution, the city was taken over by a Military Revolutionary Committee chaired by a Russian, but the committee's decision to use force to disperse the Maapäev lost it whatever popular support it had possessed. On January 14, 1918, the Maapäev's Council of Elders declared Estonia independent. On February 24 the Bolshevik officials left the city and a Provisional Government was formed under Päts's leadership, just hours before the arrival of the German army, which had been invited into Latvia and Estonia by the Baltic Germans. Nine months later Päts and his colleagues seized the chance presented by the German surrender to the Allies to take control of the city they called Tallinn. By the end of January 1919 all the German and Soviet troops in Estonia had been removed, with help from a British naval squadron based at Tallinn and volunteers from among the Finns, who share a common cultural and linguistic heritage with the Estonians.

Tallinn was the seat of the Estonian Constituent Assembly from 1919 to 1920 and of the Riigikogu (parliament) from 1920 to 1940. The executive ministries were housed on the Toompea while the blue, black, and white national flag was raised on Long Herman. Tallinn was also, briefly, a center of Communist activity. The party was officially banned, but in 1923 it won 36 of the 100 places on the city council. In 1924, after 149 party members were tried for treason and the Communist legislator Jaan Tomp was executed, about 500 Communists attacked the railroad station, the government buildings, and other sites, but the uprising was defeated within hours.

Tallinn continued to develop as an industrial center, with paper and cellulose production adding to the existing industries in metal goods, cement, and railroad cars, but during the worldwide depression of the early 1930s unemployment rose, the currency (the *kroon*) was devalued, and the political culture deteriorated accordingly. In 1934 the Estonian Association of Freedom Fighters, a veterans' group transformed into a fascist party, won the elections for the councils in Tallinn, Tartu, and Narva and stepped up its violent activities. Konstantin Päts, now prime minister, banned the Association, postponed the presidential election, and dissolved Parliament. He then replaced all political parties with a National Association (Isamaaliit), obtained a referendum result in his favor in 1936, used the National Association to supervise elections to a new Parliament, and became president in 1938.

The Päts regime succeeded in neutralizing both communism and fascism within Estonia, but, like the governments of Poland, Latvia, and Lithuania, it was condemned by its geographical position to destruction by communist and fascist powers abroad. The wholly unexpected nonaggression pact between Germany and the Soviet Union, signed on August 23, 1939, threatened the existence of Poland and included a secret agreement to divide and conquer the three Baltic states. On September 19, sixteen days after the German invasion of Poland had triggered World War II, the Soviet government responded to the news that a Polish submarine had left the harbor at Tallinn by sending its own warships and airplanes to "defend" Estonia. By November all three Baltic governments had been forced to accept Soviet bases on their territories and the Nazis had arranged for the evacuation of all the Baltic Germans from Latvia and Estonia.

In June 1940 all three states were invaded by the Red Army, in July rigged elections produced People's Assemblies, and in August their applications to join the Soviet Union were accepted. There followed a year of terror and repression, in which about 60,000 Estonians and about 110,000 citizens of the other states were deported, conscripted, murdered, or made to disappear. Then came the German armies, which swept into Soviet-held territories on June 22, 1941, and reached Tallin in August. Throughout the three states about 250,000 Jews were murdered by the Nazis and their local collaborators; about 125,000 others were also murdered during their three years of occupation. In Estonia resistance movements based mainly in Tallinn and Tartu were brought together early in 1944 and even managed to operate a provisional government in Tallinn during the four days between the withdrawal of the Germans, on September 18, 1944, and the return of the Red Army and absorption into the Soviet Union.

Tallinn was closed to foreigners until 1959, but proximity to Finland made some contact with the outside world possible, both officially, as in 1964 when the Finnish President, Urho Kekkonen, visited Tallinn, and unofficially, through Finnish television, which could be received in the city. Meanwhile, Soviet industrialization programs increased the proportion of Russians in the populations of the three Baltic states and of their capitals, so that by 1979 only 51.3 percent of the inhabitants of Tallinn were Estonians. By the mid-1970s a new generation was prepared to protest against the Soviet empire. In 1975 the regime acknowledged the existence of the Estonian Democratic Movement by putting five of its members on trial in Tallinn; in September and October 1980 thousands of high school and university students demonstrated in Tallinn and in Tartu; and during 1981 there were several brief strikes in Tallinn and elsewhere in the Baltic states supporting the demand for a nuclear-free zone. All three peoples then took advantage of Gorbachev's limited reforms to begin demanding their freedom, which came faster than anyone could have expected. The anniversary of the 1918 independence was openly celebrated in 1987 and 1988. In May 1989 the three Popular Fronts met at Tallinn, and on the fiftieth anniversary of the Nazi-Soviet Pact, between 1 million and 2 million people joined hands to link Tallinn with Riga and Vilnius in an unforgettable demonstration. In March 1990 groups favoring independence won three-quarters of the seats in the Supreme Soviet of the Estonian SSR. At this stage Estonian leaders spoke of being in transition to independence, and, unlike in Latvia or Lithuania, nobody was killed by Soviet forces during 1991. The unsuccessful coup that took place in Moscow in August 1991 inadvertently gave all three states the opportunity to declare independence at last. On September 17 Tallinn became the capital city of a member-state of the United Nations.

The castle for which Tallinn is named still stands today, along with most of the Danish city walls, their towers and gateways, many of the buildings of the German elite, and the Orthodox cathedral and other remnants of two separate phases of Russian rule. It is surely to the credit of independent Estonia that, amid all the difficulties of reconstruction and development, most of this mixed endowment from Tallinn's fascinating and sometimes tragic history has been preserved.

Further Reading: Leaving aside the unreliable books on Tallinn and Estonia published during the period of Soviet occupation, the most interesting sources are such general histories of the Baltic states as David Kirby's *Northern Europe in the Early Modern Period: The Baltic World, 1492–1772* (London and New York: Longman, 1990), Georg von Rauch's *The Baltic States: The Years of Independence: Estonia, Latvia, Lithuania, 1917–1940* (London: Hurst, and Berkeley: University of California Press, 1974), and Romuald Misiunas's and Rein Taagepera's *The Baltic States: Years of Dependence, 1940–1990* (London: Hurst, and Berkeley: University of California Press, 1993).

—Patrick Heenan

Tbilisi (Georgia)

Location: In the Caucasus Mountains in the center of the eastern part of Georgia, which lies in the center of the Caucasian Isthmus between the Caspian Sea and the Black Sea.

Description: Since the late fifth century A.D., capital of the Kingdom of Iberia, predecessor to the Republic of Georgia. Important political and cultural center; has been damaged many times by invaders, but has retained some historic buildings.

Site Office: Tourist Information
Metekhi Palace Hotel
Tbilisi 380003
Georgia
(8832) 744 556

The historic city of Tbilisi, capital of the Republic of Georgia, was founded in 458. It sits on the banks of the Kura (Mtkvari) River, in a valley enclosed on three sides by the Caucasus Mountains. This modern, dynamic capital of a nation with an ancient culture stands at the crossroads between Europe and Asia, and exhibits a mix of the two continents' characteristics.

Archaeological finds show that Georgia was home to the earliest human cultures. The Tbilisi area has been inhabited at least since the neolithic period. Ruins and artifacts dating to 5000 B.C. have been discovered.

The first written records of Tbilisi date from A.D. 400, when the area was controlled by Persians, who had built a fortress there. Various legends surround the founding of Tbilisi, and its replacement of nearby Mtskheta as capital of the Kingdom of Iberia, in the fifth century. King Vakhtang Gorgaslan (Gorgasali) was hunting and, depending on the legend, wounded either a doe or a pheasant. When it plunged into one of the sulphur springs and was instantly healed, the king decided to found a town at the spot. The more plausible reason was that he recognized the benefits of the springs, and was impressed by the strategic advantages of the site. The town was named Tbilisi from the Georgian word *tbili* (warm), in recognition of the sulphur springs. Tiflis is a Persian name that was used widely until 1917, when the Georgian name Tbilisi was adopted. The name Tiflis held too many associations with past conquerors.

This site, where the road from central Asia to Asia Minor intersected with the road from Russia to Persia, was strategically located but also susceptible to attack. Outside forces sacked the city more than twenty-nine times from the Byzantine invasion in 626 through incursions by Arabs, Mongols, Seljuk and Ottoman Turks, tribes from the North Caucasus and, most devastating of all, the destruction of the city by the Persian Agha Mohammed Khan in 1795.

Although they had some degree of autonomy in the fifth Century, the Georgians were caught up in the growing fight between Byzantium and Persia. The struggle was concentrated in the Caucasus region. First, Persia abolished the Iberian monarchy after King Vakhtang Gorgaslan's death in 510; then, while the Persians were occupied on other fronts, the Byzantine Empire put Guaram, its chosen descendant of Vakhtang, on the Iberian throne as the reigning prince. Guaram's son, Duke Stephen I (Stephanos), regained political autonomy for Georgia, although Persia still held Tbilisi. The town fell under Byzantine control in 627.

As the Arabs under Prophet Mohammed began seeking to conquer other lands, Byzantium, weakened by internal religious strife, was unable to resist. After the Arabs succeeded in nearby countries, Georgia submitted to Arab rule, more or less voluntarily, in 654, and an emir, or viceroy, was installed in Tbilisi's Narikala Fortress to rule eastern Georgia. The Arabs governed until 1121, the longest period of foreign rule over Georgia. During this time Tbilisi prospered as a commercial and cultural center.

Georgia was finally able to develop its own feudal monarchy in the beginning of the tenth Century. Moslem power in the Caucasus was on the decline and the Byzantine Empire's imperialistic campaigns were temporarily stopped. The Bagration family, claiming to be descended from King David and King Solomon of Israel, was waiting in the wings to unify Georgia under one ruler. (In reality the family originated in the Caucasus region.) They won control of more and more of the principalities in Georgia so that by the year 1001 they controlled all of Georgia with the exception of a small amount of territory, including Tbilisi, which remained under the control of the emirs. By 1038 the Bagrations were strong enough to lay siege to Tbilisi, which fell to them two years later.

Alarmed by the rise of the Bagration family, the Byzantine Emperor Basil II invaded Georgia in 1021. After a temporary peace, fighting renewed in the 1040s. The Byzantines conquered Armenia and would have conquered Georgia, except for the fierce resistance of King Bagrat IV and the people. The Turks also were active in the area, ravaging the country until Bagrat's successor, King Giorgi II, capitulated in 1080. He later abdicated in favor of his sixteen-year-old son, King David the builder, who turned out to be an excellent ruler and military leader. During his reign, 1089 to 1125, he was able to discontinue payment of tribute to the Turkish sultan. Following a series of victories over the Turks, Tbilisi was finally back in Georgian hands by 1122 and the territory of Georgia was extended beyond its traditional boundaries. After David's death, however, the family sank into internecine warfare until the throne passed to David the Builder's great-granddaughter Tamara, who ruled from 1184 until the

Monument to King Vakhtang Gorgaslan

early thirteenth century. Under her rule the country flourished and Tbilisi became one of the most important cities in the region. The end of her reign, however, signaled the start of the decline of Georgia.

Waves of Mongol invasions began during the reign of Tamara's son, Giorgi IV, and continued until 1236 when his sister and successor, Rusudani, was forced to flee. She ordered the burning of Tbilisi to prevent its falling into Mongol hands. She finally capitulated in 1243. Georgia, forced to submit to Mongol rule and pay huge tribute to the khan in exchange for peace, continued to decline. The country was fragmented. David, "the clever," Rusudani's son, and David, "the big," the illegitimate son of Giorgi, each ruled a different part of Georgia. Ulu was ruler of Tbilisi, but was really no more than an administrator who reported to the Mongols. There was some strengthening of the power of the Georgian king under the reign of Ulu's son, Dmitri, until he got caught up in the internal struggles of the Mongol leaders. In 1289 he was tortured and beheaded, leaving the kingdom leaderless. Wakhtang II, a son of Narin, ruled

Tbilisi briefly, followed by David VI, the son of Dmitri who, facing increasingly repressive Mongol policies, fled to the mountains. The Mongols placed figurehead kings on the Tbilisi throne until 1318, when the Mongol strength was waning, and one of these figureheads, Giorgi V, began to take on real authority. Under Giorgi V and his successors, the kingdom was fairly stable until the late 1380s and early 1390s, when the Mongol power reasserted itself under Timur (Tamerlane). He pillaged Georgia for several years, attacking Tbilisi in 1393. The warfare was motivated to some degree by religious differences; Timur and his men were Moslems and the Georgians were Christians. In 1403 Timur accepted the allegiance of the king and left to conquer other lands.

In the fifteenth century, while many other countries in Europe were consolidating and merging their petty fiefdoms, Georgia took the opposite course. The country was split into principalities, and territorial princes wasted the land in continuing battles. This was a period of constant fighting back and forth among the Georgian factions, the Persians, and the

Turks. Even into the seventeenth century there was still disunity. The country was split into eastern and western segments that generally followed the boundaries of the ancient kingdoms of Iberia and Colchis-Laziea. The Iberian kingdoms were ruled from Tbilisi, under control of the Persians. The Colchian kingdoms were under Turkish control.

At the end of the sixteenth Century, Russia had made incursions into the Caucasus on the pretext of protecting Christians from Moslem domination. In 1722, after the collapse of Persia, Russia launched a full-scale invasion of the Caucusus region, including Georgia. Turkey also took advantage of the Persian weakness and conquered Tbilisi in 1723. In 1724, rather than provoke a war with Turkey, Russia signed an agreement recognizing Turkish control over Georgia. For the next several decades, Russia tried to avoid conflict with Turkey, but in 1768 Turkey declared war, urged on by France, which had its own differences with Russia. A treaty was negotiated in 1774, but it provided only a brief interruption of the continuing conflict between Russia and Turkey that went on from 1768 to 1829.

Caught between the Russians and the Turks, the Georgian King Herekle II participated in the fighting on the side of the Russians and often bore the brunt of the battles. Finally, in 1783, the beleaguered king gave up his sovereignty and had the country placed under the protection of Russia. This obligation was not taken very seriously by the Russians. In the larger context of the war with Turkey, the Georgian problem was not of prime significance to Russia.

In 1795 Georgia suffered a severe blow. The Persian Shah Agha Mohammed destroyed Tbilisi, deported its inhabitants, and then withdrew. Herekle died in 1798 and his son Giorgi XII, the last king of Georgia, tried to assure his own succession and that of his son by calling on Russia for help. The Persians were about to attack again and rivals for the throne were conspiring with the Turks. In 1801, shortly after Giorgi's death, Russia annexed eastern Georgia. In 1810, Russia annexed the remainder of the country, and there ensued a period of peace and some prosperity, at least for the Georgian elite, who fit well into Russian society; the peasants suffered poor conditions and revolted frequently. During this period Tbilisi was rebuilt, under the encouragement of Russian author-statesman Alexander Griboedov, and throughout the nineteenth century the city attracted many Russian artists and intellectuals. With its bazaars and baths, Tbilisi appeared exotic and inspiring to Russian sensibilities. Gribodoev, who sought to develop Georgia's economy, was the only Russian political official for whom Georgians bore much affection. Other Russian leaders regarded Georgia as merely a colony with resources to be exploited by Russia.

In 1918, the year after the Russian Revolution, Georgia was able to establish itself as an independent country. Independence lasted until 1921, when the Russians forcibly returned Georgia to Moscow's control. Georgia became the Georgian Soviet Socialist Republic, part of the Union of Soviet Socialist Republics, and remained such until 1991.

Outward manifestations of Georgian feeling against Russia began to have an effect as early as the 1950s, as demonstrated by the departure of Russian nationals from Georgia in increasing numbers. By 1987 the hostility escalated to protests, demonstrations and talk of secession. In April 1989 Soviet troops attacked a group of peaceful demonstrators in Tbilisi. The soldiers killed nineteen people, and hundreds more were injured. The demonstration that resulted in all the bloodshed sprang not only from resentment of the Soviets, but from Georgia's internal ethnic conflicts. The Republic of Georgia is composed of three states in addition to Georgia: Abkhazia, Adzharia, and South Ossetia. Abkhazia and South Ossetia were seeking annexation by Russia to protect them from the Georgian majority and as a first step toward gaining their own autonomy. Moslem Adzharia was seeking its own links with neighboring Turkey. The violence of April 1989 aroused the population, but there was no unity among the various dissident groups. In the 1990 elections there were thirty-four registered political parties, all espousing independence.

At first, there was a movement to restore the Georgian monarchy, but true to historical precedent, this was thwarted by squabbling among rivals for the throne. As protests continued, the statue of Lenin that dominated Tbilisi's Lenin Square was taken down, and the square was renamed Freedom Square. Later a statue of Stalin, who was formerly venerated in Georgia, was also taken down.

There finally emerged a leader, Zviad Gamsakhurdia, a longtime nationalist dissident and professor. In 1990 he was elected president with 87 percent of the vote, but was not the savior Georgians expected. He thought democracy less important than independence, and his nationalism increasingly turned into chauvinism directed against Abkhazia, Adzharia, and South Ossetia. He became increasingly authoritarian, detaining opponents and censoring the press. In April 1991 he declared independence from the U.S.S.R. with the rubber-stamp approval of the parliament. The U.S.S.R. dissolved in December 1991 and, at virtually the same time, Gamsakhurdia was violently overthrown. After two weeks of fighting left Tbilisi's central boulevard in ruins, he fled into exile.

An unlikely deliverer, Soviet Foreign Minister Eduard Shevardnadze, a Georgian, returned to "save" Georgia. This former KGB boss and Communist Party chief during Leonid Brezhnev's 1970s crackdown on Georgian nationalism was elected chairman of the legislature. His presence gave the country legitimacy and recognition in the international community, which had previously ostracized it. Initially repeating the pattern of his predecessor, he censored the media and arrested Gamsakhurdia's supporters. Later he announced an amnesty and appealed for reconciliation. The parliamentary elections in October 1992 again showed the historical disunity and lack of a common direction in the country. Although Shevardnadze won 96 percent of the vote, no single party won a majority. Ethnic strife and civil war in Abkhazia and South Ossetia continued, and Georgia's economy remained unstable.

More than 1,500 years of foreign invasions have left

a few buildings but no unified historical complex in Tbilisi. The buildings in the Old Town section, a typically Georgian amalgam of the Asian and European, were built in the nineteenth century. The area's street pattern, a maze of narrow alleys, dead ends, squares, and tiny courtyards, dates from the seventeenth and eighteenth centuries. Only the sulphur springs and Anchiskhati Church survived the ravages of the Agha Mohammed Khan in the Old Town in 1795. The unmistakable stone domes of the underground sulphur spring baths can still be seen today, as can the church, the first version of which was built shortly after the founding of the town of Tbilisi. It has been restored several times. A section of about 200 acres in the Old Town has been declared a historic zone.

Additional historic sites remain in other parts of the city. Metekhi Church of the Virgin, built in the thirteenth century, stands on Metekhi Hill. This was the site of both a fifth century church built by Vakhtang Gorgaslan, and David the Builder's royal residence. Both were destroyed in 1235 by the Mongols and subsequently rebuilt. After further damage to both in subsequent wars, the palace was finally destroyed in 1795. In the nineteenth century the Russians built a jail on the site; many dissidents, including Maksim Gorky, were imprisoned there. The jail was demolished in 1937. An equestrian statue of Vakhtang Gorgaslan stands next to the church.

Sioni (Zion) Cathedral, on the site of a church built by Prince Guaram in the late sixth century, also has been destroyed and reconstructed several times. It contains the cross of St. Nino, who converted Georgia to Christianity in the fourth century. Sioni is now the seat of the Catholicos (Patriarch) of the Georgian Orthodox Church.

Prominent above the Old Town, on Solalaki Ridge, are the ruins of Narikala Fortress, on the site of the first fortress built in the fourth century by the Persians. It was enlarged by Vakhtang Gorgaslan when he moved the capital to Tbilisi. The foundations visible today are from the eighth century when the palaces of the Arab emirs were built within the fort. Portions of the fort blew up in 1827 when lightning struck and ignited a store of gunpowder.

On a terrace halfway up Mtatsminda (the Sacred Mountain or David's Mountain) is the site of a chapel and monk's cell said to have been built by St. David, one of the thirteen Syraic saints. The current church on the site dates to 1855. Halfway up the mountain is a pantheon with tombs, statues and monuments commemorating writers and other famous Georgians. The site includes the tomb of Alexander Griboedov. At the top of the mountain is an immense park overlooking the city and the countryside.

These historic landmarks have survived the conflicts in the 1990s, which have caused further devastation in some parts of the city. Many cultural institutions were located near the parliament building, which was the focal point of the fighting. Museums, galleries, and a major research library have been destroyed, and some feel the city's cultural life has been ruined.

Further Reading: *A History of the Georgian People: From the Beginning Down to the Russian Conquest in the Nineteenth Century* by W.E.D. Allen (New York: Barnes & Noble, 1932; reissued, 1971) is a scholarly work providing very detailed information on the history of the Georgians and their religion, language, and monarchy. *The Last Years of the Georgian Monarchy: 1658–1832* by David Marshall Lang (New York: Columbia University Press, and London: Oxford University Press, 1957) covers the history of Georgia from the seventeenth century up through the nineteenth century and the annexation by Russia. This volume analyzes political history, social and economic conditions, and culture during that time period. For information on events in Tbilisi in the 1990s, see "Georgia Since Independence: Plus Ça Change" by Elizabeth Fuller, *Current History* (Philadelphia, Pennsylvania), October 1993. The article provides an analysis of the turmoil surrounding independence.

—Julie A. Miller

Tønsberg (Vestfold, Norway)

Location: Southeastern Norway, at the head of the Tønsbergfjorden, in the county of Vestfold.

Description: One of Norway's oldest towns, founded A.D. 871; prominent Viking and medieval trading center; one of the country's largest shipping and whaling towns in the eighteenth and nineteenth centuries.

Site Office: Tourist Information
Storgt. 55
N-3100 Tønsberg, Vestfold
Norway
(033) 10 220

Tønsberg first rose to prominence as a market town during the Viking Age, which began shortly before A.D. 800 and came to an end around the year 1000. The region's earliest market had been located at Bjerkøy, a small island off Nøtterøy, but it was moved to Tønsberg around the ninth century. Frisian and other foreign merchants sold their wares there, and eventually trade was opened with Novgorod.

Just outside Tønsberg, in approximately 850, one of the greatest known troves of Viking treasure was buried: the Oseberg funeral barge. Archaeologists excavated the burial ship in 1904. Two female skeletons, presumably a Viking queen and a servant sacrificed to care for her in the afterlife, were discovered inside the vessel, along with the remains of thirteen horses, six dogs, and a bull's head, presumably provided to allow the queen to hunt as she crossed from one world to the next. More important archaeologically, however, was the wealth of luxurious artifacts surrounding them. The artistic style, dubbed *Osebergstilen,* features exquisitely detailed ornamentation such as decorative animal heads, and suggests that the area was an important center for the development of Viking art. It is believed that much of the queen's jewelry and the fabric of her clothing were purchased at the market at Tønsberg.

Norway was divided among petty kings and chieftains until the reign of Harold I, better known in Norse folklore as Harold Fairhair (Hårfager). Harold belonged to the Ynglinger family who had come to Norway from eastern Sweden around 700. (It is believed that the queen buried aboard the Oseberg ship belonged to this family; she was most likely Aasa, wife of Yngling king Gudrød Storlatnes.) Around 900, in an attempt to conquer the northern regions of Norway and add them to his kingdom, Harold I defeated the chieftains and petty kings in battle at Hafrsfjord and declared himself king of Norway. For the seat of his administration he chose Tønsberg, named for the "berg" of Slottsfjellet, or Castle Hill. His parliament was held on the Haugar estate in the heart of town, and he soon gave control of the town's market to his son, Bjørn.

Following Harold's abdication in 935, Norway was again divided into smaller kingdoms ruled by his descendants and local chieftains. According to legend, Erik Bloodaxe (Blódøx), Harold's chosen heir, surprised and killed rival half-brothers Olav Digerbein and Halvdan as they plotted against him in Tønsberg. Eventually, Harold's youngest son, Haakon I, the Good, drove off Erik in 935 and temporarily reunited the kingdom. Haakon began the process of Christianizing Norway, but the bulk of this task was carried out, often ruthlessly, by his successors, Olaf I and Olaf II.

Following two centuries of political turmoil and civil war, Norway was again united by King Haakon IV, Haakonsson, in 1247. Under Haakon IV Tønsberg became a center of Norwegian affairs, both politically and militarily. For decades, Tønsberg was a favored royal residence, even though the official capital was at Bergen and, later, Oslo. The town also hosted numerous national assemblies and ecclesiastical conferences. It was at Tønsberg Harbor that Haakon IV assembled a fleet of 360 vessels to attack Denmark in 1253. When, the following year, the Castillian government sent ambassadors to request the hand of the Norwegian princess, they stayed through the winter at Tønsberg with Crown Prince Haakon. That spring, Princess Kristina set sail with a grand fleet from Tønsberg for Castile, where she would marry Don Philip, the king's youngest son.

Fortifications had been erected on the Slottsfjellet at Tønsberg in the 1160s, and a defensive wall was added around 1230. Haakon IV erected additional structures, including gate houses and a royal hall, Bredestuen, near St. Michael's Church. He visited the town for the last time in 1262, a year before he was killed defending the Hebrides and Isle of Man against invading Scots.

Norwegian prosperity continued during the reign of Haakon IV's son, Magnus VI Lagabøter (Lawmender), so called because of the municipal codes he instituted during his reign. (Those codes would govern Norway for 400 years.) Overseas trade at Tønsberg increased steadily, mostly in exports of timbers, skins, and butter to the British Isles. In 1260 trade was established with Rostock, and by the 1290s trade with German towns along the Baltic coast, especially Lübeck, began to outstrip trade with all other regions. In approximately 1275, Magnus VI added a 190-foot tower, called Teglkastellet, to the fortifications at Slottsfjellet, making the garrison the strongest in all of Norway until the construction of Akerhus in Oslo. Magnus also declared St. Michael's Church a chapel royal.

While Norway's urban centers thrived and Norwegian culture flourished under Haakon and Magnus, the rural communities were in trouble. The two trends were, in fact, related: the achievements of the age were made possible by a concentration of resources in the hands of the church and the

A view of Tønsberg, Norway's oldest town
Photo courtesy of Tourist Board of Norway

centralized government. Many of the rents and taxes were collected in the form of foodstuffs and raw materials that could be sold overseas, thereby spurring the growth of foreign trade that brought luxury goods into the country; the resulting revenues also funded large-scale architectural and artistic projects. As church and state each grew more powerful, they began contesting each other's positions. The church ultimately emerged victorious. In a concordat signed in Tønsberg in 1277, Magnus conceded to church demands by renouncing much of his legislative, judicial, and fiscal authority over ecclesiastical holdings.

With the close of the thirteenth century, Tønsberg entered a period of decline. In 1349, the Black Plague swept through Norway, killing an estimated two-thirds of the population. The fortress fell into disrepair and was easily captured by the invading armies of Knut Alvsson and Henrik Krummedike. The fortifications were ultimately destroyed

during an invasion by Sweden and an internal revolt in 1503. In 1536 a fire swept through the town. Most of Norway's buildings, from stave churches to manor houses and estates, were of timber construction; as a result, most were destroyed, including St. Olaf's Premonstratensian Abbey, which dated from at least 1191. Today, only the foundation of the structure remains.

Following the fire, the town was deserted for several years. It recovered as a center of trade in the seventeenth century, particularly during the conflicts between England and Holland that cut off much competition. The outbreak of the Great Northern War in 1709 brought renewed economic decline, however, and another fire in 1738 destroyed yet another of the town's architectural landmarks: the St. Laurence Cathedral, built in approximately 1120. Tønsberg would not rise again until the advent of the sealing and whaling industry in the nineteenth century.

Norwegians had been hunting seal in a limited capacity—mostly for food—since the early eighteenth century, but no large-scale attempts had been made until one of Tønsberg's most prominent citizens, Svend Foyn, embarked in 1847 on an arctic expedition that produced a catch of 6,000 seals. Soon Tønsberg Harbor was home to a fleet of sealing vessels. The growth of sealing and other commercial trade out of the city was spurred in 1849 by the repeal of England's Navigation Acts, which had impeded Norway's ability to trade with Britain.

The sealers were so successful, however, that they hunted the animals to near-extinction in the Arctic Ocean, and the hunters turned to the bottle-nosed whale in the 1870s. Svend Foyn developed many of the techniques used in the early days of modern whaling, and when Germany threatened to steal away the men he had trained, the Norwegian government granted him sole patent rights to his inventions. As a result, Norway maintained a near monopoly on European whaling until as late as 1883. Once again, however, overhunting led to a scarcity of game, and the whalers were forced to venture to far-off Antarctica for prey. By the early twentieth century, Tønsberg lost its preeminence in the industry to the town of Sandejford, south of Tønsberg.

Today, tourists visit Tønsberg to view the Oseberg ship and the many fine Viking relics on display at the Vestfold Museum, walk among the ruins on the Slottsfjellet, and tour the remains of stone and timber churches like St. Michael's Church and other medieval structures.

Further Reading: *Norway* by Ronald G. Popperwell (New York: Praeger, and Tornbridge, Kent: Benn, 1972) provides a detailed and well-researched history of the nation. *South Norway* by Frank Noel Stagg (London: Allen and Unwin, and New York: Macmillan, 1958) concentrates on the history of Vestfold County and Tønsberg, but Stagg's style is somewhat dense and chatty.

—Elizabeth Taggart

Toulouse (Haute-Garonne, France)

Location: On the banks of the Garonne River, centrally located between the Mediterranean Sea and the Atlantic Ocean, 150 miles north of the Pyrenees, southern France.

Description: Once capital of the Visigoth kingdom, and capital of Languedoc; frequently referred to as the *ville rose* because of the pinkish-colored brick with which much of the town is built. Its most famed monuments are the great Romanesque church of St. Sernin, and the Dominican mother church of Les Jacobins. It also has a cathedral, St. Étienne, reflecting the various styles of the long period of its construction. The many merchants who made their fortunes from pastel built fine town houses, a good number with towers, an indication of honor gained as a *capitoul,* or powerful town councillor. The seat of the *capitouls,* the Capitole, has a splendid facade from the middle of the eighteenth century. On the outskirts of Toulouse are the aircraft and space companies for which the town is renowned.

Site Office: Office du tourisme
donjon du Capitole
rue Lafayette
31000 Toulouse, Haute-Garonne
France
61 23 32 00

A site known as Vieille-Toulouse some six miles south of the present Toulouse grew early into a key town for the Celtic tribe of the Volques-Tectosages from the third century B.C. The wealth of the tribe became legendary, and in 106 B.C. the Romans pillaged its treasure. The story goes that the Roman general in charge, Servilius Caepio, left with seventy tons of gold ingots dedicated to Apollo, supplemented possibly by some treasures that the Volques themselves supposedly had looted from Delphi. Beaten by the Cimbre and Teuton tribes on his way home, he claimed on arrival back in Rome that his army had been robbed of the fabulous treasures; he was officially disgraced.

Known as Tolosa, the town flourished in Roman times, staying loyal to the empire to the end. From the first century B.C., trading of its wheat and Italian wine brought it prosperity. Tolosa was also adopted as the name for a whole colony administered from the town beginning in the second century A.D. Estimates have put the colony's population at approximately 20,000 people. The town grew on the right bank of the Garonne, covering an impressive 222 acres, protected by a high semicircular brick wall to the north, and bounded by the Garonne River to the south.

Christianity was spreading in the third century. The first man to be appointed bishop to Toulouse, Saturnin, also became its first Christian martyr in 250. Saturnin would pass in front of the Capitole temple devoted to Jupiter, Juno, and Minerva on the way to his church. The worshipers of the Roman gods felt that their prayers were not being answered and blamed Saturnin for the silence of the oracles. One day when they had gathered to sacrifice a bull to appease the gods, Saturnin passed by but refused to partake in the sacrifice. Enraged, the crowd tied him to the bull, which then dragged him along the street, killing him. Christians buried his remains on the spot where his body was found. Saturnin would later be canonized.

More than a century later, Bishop Hilaire built the first structure over Saturnin's grave. Pilgrims came to the site in increasing numbers. Toward the end of the fourth century, St. Silve built a basilica 330 yards to the north of the sepulchre. The basilica was completed by St. Exupère, and with imperial authorization, Saturnin's remains were transferred there in 402. The saint would come to be known as Sernin in French, and it was to him that the famed Romanesque church of Toulouse was built.

The fourth century Gallo-Roman poet-counsul Ausonius described the town in his writings, referring to it fondly as his nursemaid, and placing it as the third largest settlement in Gaul after Narbonne and Nîmes. It was a center of learning, with a famed university noted for its school of rhetoric. Its amphitheatre, of which only a few vestiges remain, was almost on the scale of that at Nîmes.

The Vandals invaded in 406. They were followed by the Visigoths around 420, who conquered Languedoc and Spain, and would rule a kingdom stretching from the Loire to Gibraltar. Toulouse knew glory as the capital of this kingdom from around 435 to 507. The Visigoths established themselves outside the Roman ramparts, building their palace, the Château Narbonnais, which would later be transformed into the seat of the counts of Toulouse. The learned counts spoke in Gothic, but taught and wrote in Latin. One of Toulouse's Visigoth queens was Ranachilde, wife of Theodore II. She became known as the *reine Pedauque (pé d'aouco)* meaning goose foot in Occitan, not only for the appearance of her feet but also for her predilection for water, which led her to build aqueducts in which to paddle around. The Daurade church was built in the fifth century, its name referring to the magnificent golden mosaic it contained.

After the Frankish victory of Vouillé in 507, the Visigoths were forced to make Toledo their new capital. The Franks entered Toulouse without having to fight for it. In 629, the Frankish king Dagobert gathered together the territories from the Loire to the Pyrénées and offered them to his brother Charibert as the kingdom of Aquitaine. Toulouse

Place Wilson at Toulouse
Photo courtesy of French Government Tourist Office

briefly became its capital. Charibert soon died, and a governing duke was appointed to Toulouse, giving the town a status independent of Aquitaine. The dukes were the precursors of the counts of Toulouse.

Invaders were unsuccessful in their attempts to capture Toulouse. The Moorish chief El Samah led an assault on Toulouse, but he was killed at its ramparts in 721. This was a significant defeat in the Moorish conquest plans, and the date of his death on June 9 was for a long time marked as one of grief in the Moorish calendar. The Toulousain country and Languedoc became a base for fighting the Moors later in the century. In the ninth century it was the Vikings who attacked Toulouse, laying unsuccessful siege to it in 844 and 874. The Hungarians were also repelled. The walls proved fine defenses.

The first counts of Toulouse were appointed by royal decree in the ninth century. Their authority extended far beyond the town. It would be contested by two powerful groups within the town, the church and the noble merchants. By the eleventh century, the counts of Toulouse ruled over a great region of land with boundaries on the Rhone, the Garonne, and the Pyrénées. The line was familiarly known as the Raymondine dynasty because of its predilection for the name Raymond. The first count to come to great prominence was Raymond Saint-Gilles, Count Raymond IV. While his brother, Guilhem, ruled Toulouse, he was renowned for the courage with which he fought the Moors in Spain. As a reward, he was married to the daughter of the King of Castile, and placed under the Cid. He was called on to become count of Toulouse on his brother's death in 1090.

In 1095, Pope Urban II visited Toulouse with two purposes. He came to consecrate the church of St. Sernin, and to place the saint's shrine there. However, Urban II's primary reason for visiting was to ask Raymond IV to head the First Crusade to free Christ's tomb from the Saracens. Although Raymond IV declined to lead the holy army, he did join the crusade and vowed not to return alive unless the tomb was liberated. He chose the twelve-pointed red cross for his arms, and these remain those of Toulouse. The well-liked count took leave of the people of Toulouse on the square of the church of St. Pierre-des-Cuisines. His son Bertrand assumed the role of count. In Palestine, Raymond IV turned down the throne of Jerusalem, leaving the title to Godfrey of Bouillon, although he did accept a county in Lebanon. The construction of the cathedral of St. Étienne was underway in Toulouse by this time.

Before he died in 1105, Raymond IV witnessed the birth of his fourth son. This son was baptized in the River Jordan, and hence given the name Alphonse-Jourdain. The son returned to Toulouse and became its ruler when his older brother departed on a Crusade. Alphonse-Jourdain's most important historical act was to create a body of councillors, the *domini de capitulo* (lords of the chapter), later shortened to the title of *capitouls,* to oversee justice in Toulouse.

Alphonse-Jourdain's son, Raymond V, continued his father's policies of granting further powers of self-government to selected nobles of the city. Poetry flourished in Toulouse in these times of the troubadour, the count himself playing an active creative part composing verse. The Benedictine monks of the Daurade followed more commercial pursuits by exploiting their rights to the use of the Garonne by creating the first-ever shareholding enterprise in the world: the watermills of the Moulins du Bazacle, which were granted a charter in 1192.

The neighboring powers around Toulouse posed serious threat to the count, who came to be known by some historians as "the weathervane" because of his shifting policies. Richard the Lion-hearted stole land holdings around Cahors from him, using the excuse that some pilgrims had been attacked on Toulousain land. Raymond V reacted by asking his allies on the Mediterranean not to offer Richard's fleet a harbor on its way to and from the crusade, an alliance that led to Richard's imprisonment by Leopold of Austria.

With the growth of Catharism, a Christian sect promoting a simple spiritual life, and denouncing the evil of the material world, in the twelfth and thirteenth centuries, Toulouse found itself involved in another Crusade, this time against it. The counts of Toulouse were fighting over territory with the counts of Barcelona in the second half of the twelfth century. One of the latter's supporters was the viscount of Carcassonne, a protector of Cathars. Rome was relatively slow to react, although in 1180 the then-count of Toulouse had proclaimed the heresy of the viscount of Carcassonne to the pope. In 1208 the papacy unleashed its responses. Pope Innocent II excommunicated Raymond VI for his lack of interest in rooting out the Cathars. Following the assassination of a papal legate by one of Raymond's officers, the pope insisted on Raymond's flogging. Still angered by the count's attitude, in 1209, he declared what became known as the Albigensian crusade against the Cathars, bringing bloody fighting to the entire region.

Armies from northern France descended to do battle, the Crusade becoming in great part a power struggle between regional lords. Those fighting the Cathars benefited from the promise of papal indulgences similar to those given to Crusaders who went to the Holy Land. Raymond elected to defend the people under his rule against the Crusade and called on his brother-in-law, Pedro of Aragon, count of Barcelona, to join him. The notorious Simon de Montfort was one of the protagonists leading the Crusade, seizing lands from Raymond. Pedro was keen to fight near Toulouse because it brought him closer to his love, Azalaïs de Boissezon. De

Montfort discovered a letter arranging a meeting between the two, and is said to have exclaimed: "How can I respect a king who marches against his God for the sake of a woman!"

Against numbers of possibly two to one, de Montfort won a remarkable victory at Muret, near Toulouse, in 1213. Pedro died in the battle. De Montfort entered Toulouse in triumph with the French king in 1215. The pope declared him the new count. He was joined in the rule of Toulouse by Folquet of Marseille, a troubadour turned church fundamentalist who as bishop of Toulouse led a violent hunt against so-called heretics. In 1217, Raymond, son of Raymond VI, led a successful revolt to oust de Montfort from Toulouse. Raymond VI used the opportunity to reenter the town to popular support. De Montfort returned to lay siege to the rebellious town, but he met his death at the city walls, slain by a woman's catapult shot in 1218.

De Montfort's successor, his son Amaury, continued to fight for the city against Raymond VII for another seventeen years. Amaury eventually gave up his claim in 1224, passing it to the king of France, Louis VIII, who continued to ravage the Toulouse area in another so-called Crusade, during which he took Languedoc. Raymond VII finally surrendered to King Louis IX in 1229. The Treaty of Paris, signed that year, effectively put an end to the possibility of an independent kingdom of Occitanie or Languedoc. Much of Languedoc was ceded to the king of France, the Comtat Venaissin to the pope.

While the fighting raged around Toulouse, the counts Raymond VI and VII encouraged the building of the cathedral. Raymond VI financed the great nave. Its execution has been viewed as the birth of the Gothic style of architecture in southern France. It is ironic that while they were erecting this important Catholic monument, they were also embroiled in the fight against the papal Crusade and its aftermath.

Two major medieval Catholic institutions emerged from the Albigensian crusade: the Dominican order and the Inquisition. Domingo de Guzman, a Spanish priest, founded the Dominicans in his efforts to convert Cathars. By 1206, he had enough female converts to form a convent. He gained the support of Toulouse's new bishop, Folquet, and the order of preaching friars was founded in the town in 1215. The order was sanctioned by the pope in 1216. The third Dominican church to be built in the town was Les Jacobins. A glorious building, begun in 1230, it has an austere style, bar its famed flamboyant vaulting and palm-tree column in the apse. The pope elected to place the shrine of the greatest Dominican, St. Thomas Aquinas, who died in 1274, in Les Jacobins.

Before that, Pope Gregory IX used the Dominicans to establish the Inquisition in 1233 to root out remaining Cathars. The first Grand Inquisitors were so unpopular that the *capitouls* arranged for them to be chased out of town. The Toulousains were forced to accept their return; but Cathars from Montségur came one night to exterminate them, leading to terrible reprisals in that town.

After the Treaty of Paris, Raymond VII was allowed to keep his position as count of Toulouse, but only to oversee the wider territory of Haut-Languedoc, and by usufruct. A more positive result of the new powers was that Raymond was ordered to found a medieval university, the second oldest in France. However, election of the masters of theology and canon law was tightly controlled by king and pope to uphold orthodox teaching and authority. Meanwhile, unrepentant southerners and Moors captured in Spain were sold into slavery in northern France. The *capitouls* offered asylum in Toulouse to any slave who escaped, regardless of origin.

Raymond VII had only one daughter, Jeanne, who married Alphonse de France, brother of the French king Louis IX. Alphonse and Louis set off on a Crusade, never to return, and therefore, in 1271, the line of counts of Toulouse came to an end, their territories falling under royal jurisdiction. The Toulousains feared the loss of many of the liberties they had acquired, but managed to have these reconfirmed in writing. The *capitouls* now governed the town under the king's *viguier*.

Toulouse University thrived in the fourteenth century, attracting students from across Europe and reaching some 5,000 in number. Three popes were educated there: John XXIII, Innocent VI, and Urban V. Troubadours also flourished in Toulouse. Their art was greatly encouraged by the founding in 1323 of the Consistoire du Gai Savoir by seven nobles of the town. Occitan poets were invited to compete for the golden violet in what became known as the "Jeux Floraux." The Consistoire has been called the oldest literary society in Europe. It would gradually accept French poetry, and was granted the title of an Académie de Jeux Floraux by Louis XIV in 1694.

In the mid-fourteenth century, Toulouse entered further rough times, hit by plague and the Hundred Years War. The town lost half its population to the plague. At the start of the Hundred Years War, the town ramparts were rebuilt with towers. But without, marauding groups halted supplies of food. The citizens were among the few to remain loyal to the French king, welcoming Charles VII in 1420. The city benefited from its fidelity, becoming the seat of the Languedoc region, when the *parlement* of Toulouse was granted a charter in 1443. The *parlement* was already in operation prior to this date, but Charles VII's letters patent made it the first recognized *parlement* of the French provinces. Of the twelve councillors, six were to speak the *langue d'oc* dialect, six the *langue d'oïl* dialect. Its seat was in the Château Narbonnais, its power extending over a vast area of southern France.

Toulouse's golden age came from the production of pastel, or woad, and its prized dye. The boom started in 1463, the year of a great fire that destroyed 7,000 dwellings. The pastel trade brought immense wealth to the area and the town. Splendid Renaissance architecture featured the area's pinkish-red brick, sometimes with additional detail in stone. Nicolas Bachelier was the main architect and sculptor of the town, and designed its Pont Neuf bridge. The Italian influence is evident. Many of the important private houses were further distinguished by towers, the building of which was an honor for the family and signaled a member who had been a

capitoul. An indication of the noblemen's wealth is evident in the tale of Jean de Bernuy. When the French king Francis I was held for ransom in Madrid, the royal treasury could not meet the payment demand. Jean de Bernuy was able to guarantee it out of his personal fortune.

The doctor and writer François Rabelais was a student in Toulouse during this period but left, apparently finding it dangerously conservative. Michel de Montaigne was another famous French writer to study at the university.

The arrival of indigo from the New World put an abrupt end to the pastel market. Toulouse's magnificent century of wealth came to an end around 1560, only eight years after the founding of the Toulouse stock exchange that was established to gather together the pastel business activities.

The deterioration in trade was exacerbated by new epidemics and by the Wars of Religion. In fact the Toulousain protestants were expelled early from the town, in 1562, the Toulouse authorities remaining staunchly Catholic. These were bloody times for the region. In 1629, the king removed the *parlement*'s right to distribute taxes. An uprising ensued in Languedoc, led by the region's governor, the duke of Montmorency, who mistakenly counted on the promised support of Gaston d'Orléans, Louis XIII's brother. Captured and beaten by Richelieu, Montmorency was sentenced to death in Toulouse in 1632, and decapitated despite pleas for clemency from numerous European dignitaries. His execution had to be carried out in the courtyard of the Capitole, such was the force of public outcry against his death. Around this time, the brilliant mathematician Pierre de Fermat, perhaps the university's most renowned son, invented integral calculus, advanced the laws of probability, and developed his last theorem.

A new period of prosperity came with the exploitation of the region's wheat crop, its export facilitated by new canals. The Canal du Midi, stretching from near Toulouse to Sète on the Mediterranean, was built with extraordinary speed from 1666 to 1681 under the direction of Pierre-Paul Riquet. The impressive building of the new Capitole from 1750 to 1760, its neoclassical facade by Guillaume Cammas, indicated how Toulouse had found renewed wealth and confidence. Strangely, soon after this, the *capitouls* ordered that all town buildings be whitewashed, in part to help citizens see around the town at night. It was not until 1946 that Toulouse set about restoring its pink facades.

The Toulousain, de Mondran, treasurer of France for a time, played an important role in encouraging commercial initiatives in the eighteenth century. Quays were built on the river bank to improve waterway trade. The Brienne canal linking Toulouse to the Canal du Midi was built from 1768 to 1776. When royal provincial power was placed more firmly in the hands of the *intendants* in the eighteenth century, the administrative center and *parlement* of Languedoc was removed from Toulouse to Montpellier. However, Toulouse retained its own *parlement*.

The town authorities' reputation was tarnished by the Jean Calas affair of 1762. He was found guilty of assassinating a son, supposedly to stop him from converting to Catholicism, in a rigged trial. Calas was cruelly tortured and then killed on the wheel. Voltaire championed his case, and campaigned against the intolerance and prejudice of the Toulouse councillors, leading to widespread condemnation across Europe, and Calas's posthumous pardon three years later.

During the French Revolution, Toulouse supported the radical Jacobins while most of the rest of the south stood against the full force of Parisian revolutionary zeal. Approximately fifty Toulousains lost their heads to the Terror's guillotine, including members of the *parlement*. This ruling body was disbanded in the Revolution. Louis XIII's equestrian status was brought down, and many works of art were destroyed. Alexandre Dumège managed to save a great deal, statuary in particular, and put the works on display in the Musée des Augustins. Many of the church buildings, which had occupied such a substantial proportion of Toulouse, were emptied and put to quite different uses, for example as arsenals or stables for the military. The towering 280-foot spire of Notre-Dame-la-Dalbade was shortened to reduce its religious symbolism. Toulouse came to be described in the nineteenth century as the "capital of vandalism."

With the creation of the system of *départements* in 1790, Toulouse lost much of its regional power, becoming simply the administrative center of the *département* of Haute-Garonne, reduced in size in 1808 by Napoléon's creation of the *département* of Tarn-et-Garonne.

Napoléon had ambitions for Toulouse, encouraging building; the prestigious Place Wilson is one example. But the empire soon came to a close. The Battle of Toulouse in 1814 took place three days after Napoléon's abdication. His General Nicholas-Jean de Dieu Soult was unaware of the empire's demise and, on Easter Sunday, April 10, he fought Wellington's coalition troops, which had followed his army from Spain. Defeated, Soult and his men retreated toward Carcassonne in the night.

After Napoléon's fall, wheat prices were greatly reduced, and Toulouse trade lost its dynamism. There was virtually no industrialization except for a tobacco factory, although the population continued to grow rapidly. New bridges were built in 1844 and 1850, but would be swept away by a great flood at the end of the century. The arrival of the railroad to Toulouse in 1856 improved its modern links with the rest of France, and a couple of large avenues in the Hausmannian style were opened up. But it would take twentieth-century wars to restore the town's fortunes.

Toulouse has become synonymous with the aircraft industry in France. As early as 1873, a local engineer named Clément Ader had made a machine with wings like a bat's and flew it over the Languedoc plain. In World War I, Toulouse's location far from the war zone was a great advantage, and much war work was carried out there. A powder factory employed some 20,000 by the end of the war. Most importantly for the future, in 1917, the Pierre Latécoère started building planes for the military.

Immediately after the war, Toulouse suffered from unemployment with the loss of defense contracts, but soon the national nitrogen industry was based there. Latécoère's business became the Compagnie Générale Aéropostale, pioneering civil flights. These journeys, to southern Europe and North Africa, then to elsewhere in Africa and South America, would be recorded by one of the daring pilots, Antoine de Saint-Exupéry, in his novels *Courrier-Sud* and *Vol de nuit*. Other aircraft builders, Dewoitine and Bréguet, came to Toulouse before World War II.

The town took in numerous refugees from the Spanish Civil War. The short-lived *Front populaire* government nationalized a number of the aircraft companies to create Sud-Aviation at this time. Early in World War II, Toulouse lay outside the part of France occupied by the Nazis, and, well situated on the transport network to Spain, became an important point for organizing escapes south. After the Allies invaded North Africa, Toulouse, like the rest of France, was occupied by the Germans, in November 1942. It suffered from Allied bombing in 1944. In August of that year the German soldiers fled by any means they could find, which apparently included carts and bicycles.

After World War II, the aircraft industry continued to expand. Thousands of dispossessed people from Algeria brought a second wave of new citizens to the town at the beginning of the 1960s. At the same time, a national program of decentralization and encouragement of regional industries placed Toulouse as one of the eight metropolises of France and capital of the Midi-Pyrénées region, gathering together eight *départements,* with the particular remit of developing the aircraft and space industries. Dassault Aircraft came to Toulouse in 1967. The former Sud-Aviation merged with further concerns to create Aérospatiale.

Attracting a large work force for its growing industries, Toulouse was a socialist-governed town for most of the twentieth century until 1971. In the 1960s, the city planner Georges Chandalis created a famous five-pointed star project at Le Mirail. It was intended as an ideal socialist residential cooperative for white-collar workers and a total of 100,000 people. The 1971 change of political direction put a halt to the project, and Le Mirail became something of a ghetto for immigrants.

The Caravelles and Concorde aircraft were born in Toulouse, followed by the Airbus. Great numbers of high-technology companies settled in the town. The European Ariane rocket and Hermès shuttle projects are based there. Toulouse now has its own radio and television stations, and since 1993, an underground rail service. It is reputed as a major European "technopolis" at the forefront of European technology and is considered France's center for space studies.

Further Reading: Philippe Wolff has overseen three serious historical publications on Toulouse in French, one concentrating on the town's general history, entitled *Histoire de Toulouse* (Toulouse: Privat, 1958: second edition, 1974), one focusing on the city's church history, entitled *Le Diocèse de Toulouse* (Paris: Beauchesne, 1983), and one on notable figures at particularly important moments in Toulouse's history, entitled *Les Toulousains dans l'histoire* (Toulouse: Privat, 1984). Fernand Cousteaux and Michel Valdigué are the authors of a shorter account in French of Toulouse's history, *Toulouse* (Toulouse: Privat, 1989), which provides a readable account, although passing over the Cathar period, and missing certain dates. At times it reads like a panegyric. The Blue Guides have published a section on Toulouse and its surroundings originally included in its title on the Midi-Pyrénées as a separate volume, *Toulouse et ses environs* (Paris: Hachette, 1991).

—Philippe Barbour

Tournai (Hainaut, Belgium)

Location: Southwestern Belgium in the province of Hainaut, on the Scheldt River; forty-five miles southwest of Brussels.

Description: Small city, one of the oldest in Belgium; a Roman settlement (Tornacum) until the fifth century; a Merovingian capital and birthplace of Clovis, founder of the Merovingian monarchy; seat of the bishopric and an important trade center during the Middle Ages; controlled at various times by the French, the English, the Spanish, and the Austrians; center of Calvinism in the mid-sixteenth century; known in earlier times for its sculpture and stonework, tapestries, copperware, carpets and porcelain; today renowned for the quarrying of stone and the manufacture of steel, leather, and hosiery; home to a Romanesque/Gothic cathedral (eleventh-twelfth century), a belfry (c.1188), the Tower of Henry VIII (1513–16), Renaissance guild halls, and two of the oldest houses in Europe (1175–1200).

Site Office: Tournai Tourist Office
14, Vieux Marché-aux-Poteries
7500 Tournai, Hainaut
Belgium
(69) 222045

Tournai (Doornik in Flemish) is a city with a colorful and tempestuous history that spans 2,000 years. Its location on the River Scheldt (Escaut in French) has made it an important center for trade since the early days of the Roman Empire.

The Nervii tribe of the Belgae were the inhabitants of what is now Tournai at the time of the Roman Conquest in 51 B.C. During the early years of the empire, commerce developed quickly; Tournai (Tornacum to the Romans) soon became a center of cloth-making, exporting wool and linen to foreign markets. The area surrounding Tournai was a rich source of stone used by the Romans to construct highways, aqueducts, and buildings. These stones were sent to other settlements along the Scheldt and its tributaries. Tiles excavated beneath the present rue de Pont have revealed the bed of a road that in Roman times led from Arras to Frasnes, thirty feet wide, dividing the city in two. A second road from Bavay to Courtrai intersected the main road on the left bank of the river. The rectangular-shaped settlement of stone houses with tile roofs extended along the river to the present rue Madame on the east side, and from the Vieux-Marché-aux-Poissons to the Grand Place on the west side. An aqueduct, discovered in 1960, supplemented by cisterns and a pipeline system, supplied the settlement with water. The houses, dating from the middle of the first century, were sturdy, built of thick stones on top of foundations deep enough to counteract the instability of the damp soil near the river.

Before Christianity took hold in the third century, the inhabitants of Tornacum were pagans. Archaeologists have discovered fragments of vases showing symbols associated with the Roman god Mercury (a purse, a cock, a snake with a ram's head) and some statuettes of Venus. They have also discovered, to the southwest of the city, an extensive necropolis dating from the first to the third centuries. In 1887–88 another Roman cemetery was uncovered on the right bank of the city, where seventy tombs were excavated. The tombs were of the cremation type, where the bones were burned in a clay pot and then placed in a trench, a common funerary practice in most of northern Gaul. To date the cemetery excavations have not revealed any tombs that are more elaborate than others. The most richly furnished tombs—containing vases, goblets, and urns—have been found separated from the cemeteries. Only one funerary inscription has been preserved, carved in French stone. Uncovered in 1943, it names Caius Domitius, the son of Caius.

Stone was a principal source of wealth for the early inhabitants of the region—not just the local stone, which was abundant, but other types as well. Stone from France, thought to be more suitable for sculpture, was made into columns and cornices. Clay pottery was imported from London and Cologne.

By the end of the second century, no more burials took place on the right bank of the river. Frequent Germanic raids discouraged the inhabitants from crossing the Scheldt. Two new cemeteries were established on the left bank, one near the rue Perdue, the other at the citadel. Most of these tombs used inhumation rather than cremation as a funerary practice. Excavations of these sites have uncovered jewelry, pottery and glass fragments, and coins; no weapons or Christian symbols have been found.

It was not until Frankish attacks on the town in the third and fourth centuries forced the settlers of Tornacum to defend themselves that weapons came to be buried with the dead. Tombs excavated at the cemetery near the Town Hall Park contained weapons and buckles decorated with dragons and other Germanic motifs, suggesting that there must have been some military battles during this period of invasion. At the same time, late in the third century, Christianity began to take hold in Tournai, having been brought to the area by missionaries sent by Pope Clement. St. Piat is said to have been Tournai's first evangelist.

During the fourth and fifth centuries, the Franks gradually took over the Roman territory. In 431 the Salian Franks established themselves in Tournai. The first Merovingian rulers, Childeric and later his son Clovis, tried to unify the

Tournai by night
Photo courtesy of Belgian National Tourist Office

territories. Childeric, whose greed and ambition had driven him to conduct many expeditions, died in Tournai in 482. His tomb, discovered in 1653 by a worker in the garden of the Saint-Brice church, Adrien Quinquin, was identified by the name and date inscribed on the seal containing his bust. It was an extravagant commission: he was buried in richly decorated cloth, and his treasure included elaborate weapons, gold coins, jewels, and a crystal ball.

The birth of Clovis at Tournai in 466 was the beginning of a new order for the Franks. By the time Clovis succeeded his father in 482, the Roman Empire had disappeared from Western Europe. Clovis set about the task of rebuilding Gaul. His victory over Syagrius at Soissons in 486 solidified his reputation as a powerful leader.

Upon leaving Tournai for Paris, Clovis left the government in the hands of Eleutherus, thereby making him the political as well as spiritual leader of the Tournaisiens. Before he died in 511, Clovis, who had converted to Christianity, named Tournai the seat of the bishopric, an area including all of Flanders and some of France. Eleutherus, the first bishop of Tournai, built a church on the site where the cathedral now stands.

Little is known about the sixth and seventh centuries under the Franks. Although Clovis had succeeded in reuniting the Frankish kingdoms, his successors could not keep the bishoprics together, and the area returned to barbarism. Not until the Carolingian epoch, beginning with the accession to the throne of Pepin the Short in 751, was order restored by unifying the heads of church and state. Tournai was one of many cities that began to prosper under the Carolingian dynasty. Monasteries and abbeys played an active role in the lives of the inhabitants of the Frankish region, with many monks and abbots keeping farms and participating in politics. The noble landowners, such as the family of Pepin, enjoyed administrative powers in their respective districts. The son of Pépin, Charlemagne, effectively organized all of the districts under one administration. His dream of a Christian empire extending over all of Western Europe ended, however, with his death in 814. The Norsemen attacked Tournai in 881, burning the cathedral that had been built there only thirty years before. The bishop of Tournai, who at that time controlled not only the bishopric but the administration of the city as well, raised the money necessary to restore the cathedral.

By the tenth century, a more austere form of monastic life evolved, less inclined toward temporal interests and more devoted to the spiritual realm. With the Middle Ages came a religious enthusiasm and talk of crusades and pilgrimages to the Holy Land. The cathedral of Tournai is a monument to the spirituality of the times. Destroyed by fire in 1060, it was restored by the year of the plague in 1090. On September 14 of that year, the bishop organized a procession at the cathedral to pray for an end to the deadly disease. For seven days people came from all over Flanders hoping for a cure. To this day the procession of Our Lady of the Sick takes place annually in September to commemorate the city's survival of the pestilence.

Tournai continued to flourish in the Middle Ages as a result of its trade in textiles and stone. The blue-gray stone from the region is not only durable but easily sculpted and polished to a shiny black. Some examples of the period sculpture have survived at the cathedral, but many works were destroyed during the religious insurrections of the 1560s. One of the finest works of art is the golden reliquary of Notre-Dame by goldsmith Nicolas de Verdun, on display in the cathedral treasury.

In 1187 the French took control of Tournai in an effort to strengthen their position in Flanders. French styles began to influence the Tournaisiens, and in 1243 the bishop began converting the cathedral to a Gothic style, beginning with the choir. Since the conversion was never finished due to a lack of funds, the nave is still Romanesque. The French were popular in Tournai, and the Tournaisiens loyally supported them through the Hundred Years War (1337–1453).

Art in the region continued to flourish. Two important painters of the Flemish school began their careers in Tournai in the late fourteenth and early fifteenth centuries: Robert Campin, otherwise known as the Master of Flemalle, and Rogier de la Pasture, who later called himself Rogier van der Weyden when he left to become the official painter at Brussels. The Tournai sculptors continued to carve church monuments out of the blue Tournai stone and export them to foreign countries. The fourteenth-century Tournai Mass was written (for three parts) to be sung at the cathedral services. The city was also an important center for tapestries during the fifteenth century.

Hostilities continued between France and England even after the Hundred Years War. In 1513 Tournai fell victim to an attack by King Henry VIII of England, who remained in control of Tournai for more than five years. By choosing Tournai over Calais (a strategic and fortified port on the Channel), Henry took a risk that turned out to be costly for the English. He may have been persuaded to keep Tournai by Cardinal Wolsey, who was named bishop of Tournai. Communication between Tournai and London proved difficult, and the people of Tournai were not at all cooperative. They showed their independence by appealing even minor decisions of the governor to London. In this way they managed to keep the upper hand. The expenses that Henry incurred by keeping a garrison at Tournai and by building the citadel, as well as the tower that still bears his name, were great. When he finally sold Tournai back to the French in 1518, he had lost more than £100,000 on his investment.

During this time, tension developed too between the citizens of Tournai and the occupying soldiers, as well as between the city authorities and the English governor. The English soldiers, underpaid and unhappy away from home, suffered the hardships of the cold winter of 1513, the plague of 1514, and, worst of all, the indignity of being refused credit by local merchants. When the garrison finally left in February of 1519, the people of Tournai celebrated their departure. Their joy was to be short-lived, however: only three years later Charles V took over Tournai for his Holy

Roman Empire, which was to be far less benevolent than Henry's occupation.

Despite Charles V's severe penalties for heresy, which he imposed in order to check the spread of the Calvinist movement, by 1543 Calvinism was already taking hold. Its leaders wanted to reform the state and make it subordinate to the church. Calvinist pastors won converts in Tournai. In 1545 the pastor of Tournai, Peter Brully, was executed. When Charles V abdicated in favor of his son, Philip II, matters became worse. Preferring to rule from Spain, the new king named his sister, Margaret de Parma, as regent, giving her little power. Instead, commissioners were sent from Brussels to carry out the king's policies. Many of the aristocrats—William of Orange and Lord Montigny among them—refused to give their support to the Brussels government. They wanted the States General, made up of nobles, clergy, and bourgeois, to have more importance.

Tournai had always had a tradition of strong local government, and its citizens resented the interference by the king's councils. Philip reduced the power and revenues of the Tournai government by restructuring the diocesan borders and giving Tournai a much smaller jurisdiction. The king, irritated by any opposition, was especially eager to suppress the Calvinist movement, since it posed a threat to his political order. He introduced *placards* or antiheresy edicts, and inquisition-like practices in an effort to gain control. Even those who had been loyal to Philip found the *placards* offensive because they encouraged secret accusations and allowed hearings based on shaky evidence. The definition of heresy was very vague, and the penalties were severe. Opposition to the Spanish regime grew as Philip became even more determined to eradicate heresy.

In 1566 the Calvinists revolted, destroying churches, monasteries, and other symbols of the Church. The citadel, which remained untouched, was used to house royal troops the following year to help put down the rebellion. The governor, Lord Montigny, was imprisoned and executed for his opposition to royal interference. The magistrates continued to lose power in Tournai since they neither espoused the Calvinist philosophy nor carried out the king's orders. The Calvinists temporarily took control of the city, but their wholesale destruction of art objects bred hatred among the city's Catholics. To settle the dispute, Margaret de Parma was obliged to launch a military attack on the Calvinist army in 1567. Those who did not leave had to face heresy charges. Catholics readily provided information to the commissioners and inquisitors. Many "heretics" were executed, imprisoned, or fined. The Council of Troubles extended the investigations and executions to those government officials who failed to suppress heresy and to enforce the *placards*. From this time on, Tournai no longer enjoyed the independence of government it once had.

In 1581 the States General deposed Philip II and affirmed the people's sovereignty. The Tournaisiens began to restore the damage done to their cathedral, commissioning a Renaissance-style arch between the Romanesque nave and the Gothic choir. The cathedral was to see more destruction following the French Revolution, when anticlerical French soldiers destroyed much of its art and auctioned off some of the treasures. Once again the bishop launched a campaign to restore the cathedral and return some of its art objects to their rightful place.

Another art industry sprang up in Tournai in the mid-eighteenth century: Tournai porcelain. In 1752 Joseph Peterinck, a porcelain manufacturer, was named to imperial status by the Austrian Empress Maria Theresa. His company became well known in Germany, France, Holland, Spain, and Russia for producing a fine product on a par with that of Sèvres in France. The company remained successful until the French occupation in 1794, when French protectionism proved fatal to the business. Peterinck's grandson was able to make the business profitable again from 1817 to 1830 under the Dutch regime. After that time, the production became commercialized.

There are three main periods of the Tournai porcelain designs. The first (1750–62), characterized by birds and flowers, was done by talented French artists. The second period (1763–74) shows an interest in cameo landscapes similar to Italian designs, although birds continued to be used in the patterns. Occasionally fruit designs adorned the plates. The third period (1775–1825) moved in a classical direction, white with gold in Louis XVI style, or wildflowers and herbs. In 1815 King William of Holland commissioned a service with a blue border decorated with a gold oak leaf garland. Examples of Tournai porcelain can be seen today in the city's Musée des Arts Décoratifs.

Today the city of Tournai retains a decidedly French flavor, with occasional references to other civilizations important to its history. Roman relics can be seen at the Museum of History and Archaeology on the rue des Carmes. An elaborately carved sarcophagus from a fourth-century tomb found in a nearby Roman cemetery is a well-preserved exhibit. Unfortunately, nothing remains in Tournai from the tomb of Childeric, whose treasures were pirated by other countries subsequent to the discovery of the tomb in 1653.

The Middle Ages are well represented in Tournai through its many examples of Romanesque and Gothic architecture. The Pont des Trous, built at the end of the thirteenth century, is a bridge made of Tournai stone with three Gothic arches. It was originally built to guard the river entrance to the city. The Belfry, 236 feet high, affords an exceptional view of the cathedral for those willing to climb its 256 steps. Built about 1200 and left untouched by the ravages of war, it continues its bell-ringing tradition with midday carillon concerts each Saturday in the summer. Two of Western Europe's oldest houses, dating from the late twelfth century, can be seen on the rue Barre Saint-Brice, along with other homes from the fourteenth and fifteenth centuries. An example of later architecture is the Henry VIII Tower, built in 1515 during the brief English occupation of Tournai. The massive cylinder is all that remains of the costly citadel that once housed 5,000 troops, the rest having been destroyed by Louis XIV. The tower now houses a weapons museum.

The flourishing art of the Renaissance can be seen in the Musée des Beaux-Arts, which has paintings by the Master of Flemalle, Van der Weyden, the Breughels, and Rubens. It also displays art from more recent periods, including works by nineteenth-century historical painter Louis Gallait, a Tournai native, and by the Belgian surrealist James Ensor. The French impressionists are represented by Manet, Monet, Seurat, and Van Gogh.

Tournai's Tapestry Museum, located next to the art museum, has a fine collection of fifteenth-century tapestries. Another important tapestry, showing the history of Tournai, is on display in the cathedral treasury. Made in Arras in 1402, it joins the golden reliquaries, the mantle of Charles V, and the chasuble of Thomas Becket to be among the cathedral's most valued treasures. The Musée du Folklore near the Grand' Place is a museum of Tournai history and traditions. It includes a reduced model of the city in the eighteenth century, and a later model of the Grand' Place after the destruction of 1940.

The city's most prominent landmark is its cathedral. Restored countless times, it is a monument to the fortitude of the citizens of Tournai. Its massive form dominates the city from every perspective. Four hundred and forty feet long, the Cathédrale de Notre-Dame de Tournai is the product of a series of constructions: the twelfth-century nave, facing west, is Romanesque; the central transept is transitional; and the thirteenth-century choir, facing east, is Gothic. The western facade displays a combination of styles. Figures carved of Tournai stone adorn the Romanesque doors. A statue of Our Lady of the Sick from the seventeenth century adorns the top and Old Testament scenes dating from the fourteenth century are visible at the bottom. A rose window was added much later to add light to the nave.

The three-sectioned nave has nine bays on each side covering a total length of 157 feet. The overall impression is one of ample space and durability. The two lower arcades are almost equal in height and follow the same pattern of repetition. The heavy square pillars supporting the arches are actually groups of eight columns. The arches at the tribune, also with grouped supports, are lighter in mass. A vaulted ceiling was built in 1640 to replace the original flat one. The triforium has twice as many arches as the two levels below it, and the arches rest flat against the wall. The windows above it throw ample light into the four-story nave, the first of its kind to succeed; it inspired later Gothic versions. The pillars in the nave are crowned by capitals carved from unpolished Tournai stone. Many have geometric or naturalistic motifs.

The exterior of the nave boasts important portals on either side. The Port Mantilus door (north side) recalls the miracle performed by St. Eleutherus on a blind man named Mantilus. Although badly eroded, the portal also depicts the battle between vice and virtue. The Port Capitol door (south side) is even more deteriorated. It appears to foretell the end of the world: a resurrection scene, the Last Judgment, and the four cardinal virtues are carved in stone. Near the south por-

tal, the Gothic Saint Louis Chapel, added in 1299, contains Rubens's *Deliverance of Souls from Purgatory*.

The transept is marked by a later architectural style. The capitals on the pillars show leaf clusters typical of the thirteenth century. Its vertical emphasis foreshadows the Gothic style. The arches are almost twice as high on the first level as in the tribune. Compared to the nave, the transept is huge; it is 220 feet wide, with an oblong bay at each end. Five towers dominate the central portion, each with distinctive features. The four corner towers are 272 feet high, while the central bell tower is slightly shorter. Completed in 1198 under Bishop Etienne, the Chapel of Saint Vincent is purely Gothic. The east walls of the transept, decorated in thirteenth-century frescoes, depict the legend of St. Margaret, who was attacked by a dragon: St. Michael the Archangel protecting heaven; and Christ and Mary accompanied by the apostles Peter and Paul.

Only a half century separates the building of the transept and that of the choir, but stylistically the difference is remarkable. The choir covers an area almost equal to that of the rest of the cathedral. Begun in 1243, the choir consists of six bays leading to an apse containing several chapels. It has been compared to the Sainte Chapelle in Paris because of its huge stained-glass windows, which nearly cover the walls at the base and at the triforium. The structural elements are minimal, giving an impression of lightness and elegance. The columns start at the base and continue unbroken to the ribbed vaults 118 feet above. When in 1359 it became apparent that the pillars were too thin to support the superstructure, a restoration was begun. Additional supports were added to the interior pillars and the exterior flying buttresses to stabilize the choir. The gables crowning the windows on the outside help to accentuate the vertical thrust which was to typify Gothic architecture.

As an example of Scaldian (Scheldt region) architecture, the cathedral at Tournai became a model for other Belgian cathedrals, such as Ghent and Bruges. But more importantly, it has remained the symbolic heart of the city. When in 1940 bombs burned some of the cathedral's art treasures and destroyed most of the Grand'Place, the citizens once again began to restore their city. The people of Tournai, like their prize landmark, are strong, durable, energetic, and resilient.

Further Reading: *Fielding's Belgium 1994* by H. Constance Hill (Redondo Beach, California: Fielding, 1993) is a comprehensive guide to the major cities of Belgium, Holland, and Luxembourg, intended for travelers. *Belgium: The Making of a Nation* by H. Vander Linden, translated by Sybil Jane (Oxford: Clarendon, 1920) is an old but complete history of Belgium through the nineteenth century, with limited references to Tournai. On the years of the English occupation, see *The English Occupation of Tournai, 1513–1519* by C.G. Cruickshank (Oxford: Clarendon, and New York: Oxford University Press, 1971), a detailed account of the social, military, political, and ecclesiastical events of that period. The Reformation years are well documented in *The Time of Troubles in the Low Countries: The Chronicles and Memoirs of Pasquier de le*

Barre of Tournai, 1559–1567 by Charlie R. Steen (New York: Peter Lang, 1989). Steen provides an historical overview of the period, as well as the original chronicles and memoirs of one of the Tournai magistrates. For an account of the Roman and Frankish periods based on archaeological excavations, see *Tournai de César à Clovis* by Marcel Amand (Gembloux: Editions Duculot, 1972). Published as the fifteenth volume of a series entitled *Wallonie, art et histoire,* it includes maps of the period and photographs of some of the artefacts discovered. The first volume of the same series, *La Cathédrale Notre-Dame de Tournai* by Luc-François Genicot (Gembloux: Editions Duculot, 1970), is a detailed description of the architecture, both past and present, of the Tournai cathedral. It includes a glossary of architectural terms and photographs of interior and exterior views of the cathedral. *Porcelaines de Tournai* by A. M. Mariën-Dugardin (Brussels: Musées d'art et d'histoire, 1959) describes the history of the manufacture of porcelain in Tournai. Its many photographs and drawings of porcelain patterns may be of interest to collectors for identification purposes.

—Sherry Crane LaRue

Trier (Rhineland-Palatinate, Germany)

Location: On the Moselle River near the Luxembourg border.

Description: Established by Romans just before the time of Christ; a market city with Roman ruins and medieval architectural influences, and the birthplace of Karl Marx. Famous structures include the Porta Nigra Roman city gate, several Roman baths, the Basilica, the Dom or Cathedral, and Marx's birth home.

Site Office: Tourist-Information
an der Porta Nigra
Simeonstift B 2
54290 Trier, Rhineland-Palatinate
Germany
(651) 9780 80

Trier, strategically situated on the banks of the Moselle River, was one of the major outposts of Roman civilization. Economically prosperous, it was home to several Roman emperors and a regional center for commerce and government. With the fall of the Roman Empire, the city experienced numerous economic fluctuations, never completely regaining its regional dominance. In the nineteenth century, Trier was the birthplace of Karl Marx, whose ideology would shape many of the revolutions of the twentieth century.

Trier has evidence of human habitation since 3000 B.C.. However, the city derives its name from a Gallo-Germanic people called the Treveri, who occupied the territory just before the Romans. The area was an important religious center for this tribe. While the Treveri sometimes supported the Roman emperor Augustus, the Romans required a more stable community at this strategic point on the Moselle River. Augustus founded Trier as a Roman fort, called Augusta Treverorum, in 15 B.C. Emperor Claudius made Trier a colony. Designation as a colony signified Trier's commercial importance and prosperity. As their empire increased, the Romans found Trier not only a beautiful location but also commercially and strategically valuable. Here, three major highways from Cologne, Koblenz, and Mainz intersected. The surrounding cliffs and mountains provided natural protection for the town, and wealthy landowners on its outskirts planted the vineyards that still remain an economic staple of the region. Pottery-making also became an important industry, and Trier flourished. Many of the ornate residences, the Santa Barbara baths, the amphitheatre, and various temples were constructed during the Roman period.

In the late second century Trier was enclosed by a wall, the northern gate of which was the largest ever built by the Romans. This gate, which still stands, is called the Porta Nigra (Black Gate) because the sandstone from which it was built turned black from weathering over the centuries. During the same period, the Romans built a bridge over the Moselle.

Trier suffered from invasions by the Franks and the Alemanni in 275, but it recovered. Its importance increased in 293, when Diocletian reorganized the Roman Empire's provinces into four regions and made Trier the capital of the western region, Belgica Prima, under the rule of Constantius Chlorus. Trier became one of the greatest cities of the western world, rivaling Alexandria and Antioch. A mint was established, the amphitheatre was expanded, and more elaborate baths built.

Constantius's son, the Roman emperor Constantine I (the Great), lived at Trier from 306 to 315 and added to the city's prestige. Constantine became a convert to Christianity, which became the empire's official religion, and Trier became a bishopric during his reign. Construction of the city's cathedral, St. Peter's, began in 326 on the foundations of a palace that is said to have belonged to Constantine's mother, Helena. The cathedral has been rebuilt extensively over the centuries, but still contains some of the original Roman construction. The cathedral houses many art treasures and holy relics, including the robe said to have belonged to Jesus Christ and for which Roman soldiers gambled at the time of his crucifixion. The robe is displayed only once every thirty years.

Another building dating from Constantine's reign, the so-called Basilica, was begun about 310. Although today it houses the Protestant Church of Our Savior, it actually began as a government building. Long, rectangular, and constructed of bricks, Constantine apparently used it as an imperial reception area, or perhaps as a judgment hall. The interior was decorated with marble and mosaics, some of which may have been removed to Charlemagne's cathedral at Aachen. The building has seen a variety of uses: as a seat of Frankish and Carolingian princes and of archbishops; as a wing of an elector's palace; and as a church. It began to be called the Basilica in the nineteenth century. Damaged extensively by bombing in World War II, the building has been meticulously restored.

Trier continued as a large and important city under Constantine's successors, but its splendor began to fade about 400, when troops were recalled to Rome to defend against barbarian invasions, and the provincial government moved to Arles. With the withdrawal of the Romans, Trier's importance as a commercial center faded, and many residents turned to subsistence farming, although evidence indicates that the pottery and wine industries continued to exist. The Franks took over Trier in the fifth century; this transition may not have been a wholly traumatic one, as people of Germanic

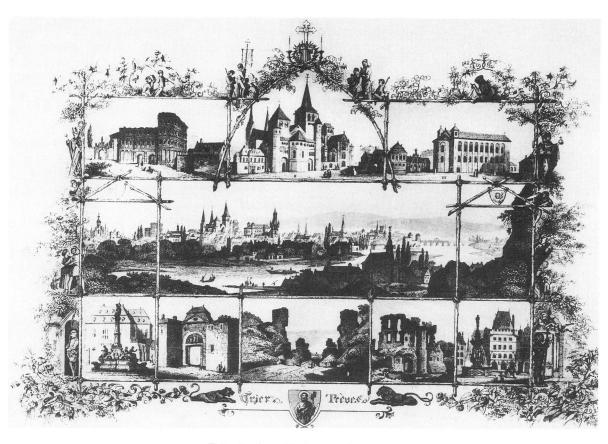

Trier in the mid-nineteenth century

Trier in the twentieth century
Photos courtesy of German Information Center

descent already living in the area assimilated easily with the Franks.

The Frankish king Charlemagne made Trier an archbishopric, and gradually the city regained prestige and prosperity. The archbishops governed Trier in secular as well as spiritual matters. In 958 Archbishop Heinrich granted Trier the right to hold a market, an event commemorated by putting up the Market Cross in the middle of the Hauptmarkt, or Great Market Place. The market, which continues to this day, boosted Trier's economy, and most of the city's new construction in the Middle Ages took place around the Hauptmarkt. Also enhancing Trier's status was the fact that its archbishops, from the twelfth century onward, also were prince-electors of the Holy Roman Empire, allowing them to participate in the election of emperors.

Trier's significant position in the empire and its progress in commerce helped the city prosper through the Middle Ages. Citizens began to chafe at the archbishops' rule, however, and began to seek the status of a free city within the empire. Through diplomatic maneuverings, the archbishops kept control. Such internal strife took its toll on Trier, and toward the end of the Middle Ages other cities in the Holy Roman Empire began to overshadow it in commerce and prestige. After centuries of stagnation some revival came in the eighteenth century, when the prince-electors tended to be patrons of the arts and encouraged the construction of many fine baroque buildings.

The prince-electors' rule finally came to an end with the French Revolutionary Wars toward the close of the century. In 1794 France captured and annexed Trier, and it became the capital of the French department of the Sarre. Church properties were secularized and large estates divided up; a substantial middle class began to develop, although many residents suffered extreme poverty. After the downfall of Napoléon, Trier came under Prussian rule in 1815.

Trier was a quiet, pre-industrial community when its most famous resident, Karl Marx, was born in 1818. Marx's father, Heinrich, was a prosperous lawyer in Trier; born Jewish, he had converted to Christianity because of an 1816 court ruling that prohibited Jews from practicing law in Prussia. Heinrich Marx was more of a follower of such Enlightenment freethinkers as Rousseau and Voltaire than of any religion, however, and this factor undoubtedly had an effect on his son. Another factor that likely influenced the development of Marx's socialist philosophy was the poverty he observed among the lower classes of Trier. His education, in a city that had retained much sympathy for the philosophies of the French Revolution, was a further influence.

Young Karl received his secondary education at the Friedrich Wilhelm Gymnasium in Trier. A Jesuit school when it was founded in 1563, by the time Karl Marx attended—1830 to 1835—it was a public institution and a center of liberal, anti-Prussian ideas. Its director, Johann Hugo Wyttenbach, encouraged intellectual freedom. Prominent among the teachers was Thomas Simon, who lectured extensively on the conditions of the poor and urged students to

pursue not "the possession of cold, filthy, printed money, but character, principles, reason and sympathy for the weal and woe of one's fellow men." Simon also may have encouraged Marx to read the works of Ludwig Gall, a Trier resident and Germany's first utopian socialist, who fled to France in 1832 to escape surveillance by the Prussian authorities.

Karl Marx left Trier in 1830 to study law at Bonn University, but left after a disastrous year spent mainly in drinking and carousing. He applied himself more seriously at Berlin University, then went on to develop his philosophy of Marxism, viewing all of society as a class struggle in which the working class would eventually seize control of the means of production from the exploitative capitalist class, with a cooperative, classless society the end result. In collaboration with Friedrich Engels, he wrote such works as *The Communist Manifesto* and *Das Kapital*, which influenced socialist and communist thought—and revolutions—throughout the twentieth century.

Marx's birth home at 10 Bruckenstrasse in Trier is a major tourist attraction today. He spent only the first two years of his life there; his family had rented the home but moved when they were able to purchase another one, near the Porta Nigra. Built about 1727, the baroque structure is a typical middle-class residence of its time. In 1928 the Social Democratic Party of Germany acquired the house and planned to make it a Marx-Engels memorial, but the National Socialist (Nazi) regime seized it in 1933. The Nazis used it as the editorial and printing office for their party newspaper. At the end of World War II in 1945, the house was returned to the Social Democratic Party. In 1968 Willy Brandt, party chairman and future chancellor of West Germany, gave the house to the Friedrich-Ebert-Foundation, which operates it as a museum.

Since the time of Marx, Trier has developed steadily but unspectacularly, with the vineyards and other light industries, plus tourism, forming the core of its economy. Trier was damaged heavily in World War II, but has been largely restored. Visitors are drawn to a diverse collection of sites—the Roman remains, including the amphitheatre, baths, and Porta Nigra; St. Peter's Cathedral and other historic churches; and the birthplace of Karl Marx.

Further Reading: *Roman Trier and the Trevri* by Edith Mary Wightman (New York: Prager, and London: Hart-Davis, 1970) provides the most comprehensive view of Trier's Roman history. It examines not only the artifacts and antiquities, but also the cultural and social life of the city's inhabitants. Another work, *Romans on the Rhine: Archaeology in Germany* by Paul MacKendrick (New York: Funk and Wagnalls, 1970), provides a larger picture of the Romans' influence in the region as well as the importance of the river valley. Many writers have examined the life of Karl Marx. *Karl Marx: An Intimate Biography* by Saul K. Padover (New York and London: McGraw-Hill, 1978) is a readable and detailed account of Marx's life and ideological development. Marx's own writings are available in a variety of editions.

—Jenny L. Presnell and Trudy Ring

Troyes (Aube, France)

Location: In eastern France, on the alluvial plain of the Seine; capital of the governmental department of the Aube; in the Champagne region, ninety-seven miles southeast of Paris.

Description.   Site of the marriage of Henry V of England and Catherine of France; birthplace of Pope Urban IV; site of several historic buildings, including the Gothic Basilica of St. Urban, several Renaissance churches, and the cathedral of Sts. Peter and Paul, whose tower bears an inscription commemorating the city's capitulation to Joan of Arc in 1429; scene of much devastation during World War II, especially in 1940 and in 1944, after which several of the city's districts were totally rebuilt.

Site Office: Office de Tourisme et Accueil de France
16 Boulevard Carnot, B.P. 48022
10014 Troyes, Aube
France
25 73 00 36

Prior to its conquest by Julius Caesar, the site of Troyes was occupied by the Gauls. Modern-day Troyes grew from the Roman city of Augustobona, located on a piece of high ground where the road between Lyons and Boulogne crossed the plain of the Seine River. During the reign of Emperor Augustus, the city became the capital of a Gallic tribe called the Tricasse, from whom the city derived its modern name.

The town turned to Christianity in the third century and became a bishopric in the fourth century. In the fifth century the city faced invasion by the Huns, but was saved through the defense mounted by its bishop, St. Loup. During raids by the Vikings, the walls of Troyes were rebuilt and refortified, but in 889, the city was sacked by the Normans, conquerors from Scandinavia who subsequently took control of Normandy.

During the tenth century, the church and the secular government fought for the control of Troyes. The bishop of Troyes, Anségise, attempted to gain control of not only the city but also the surrounding countryside. He was defeated by Count Robert of Champagne, the son of Herbert of Vermandois. As a result, the bishops retained their ecclesiastical authority, but secular power remained in the hands of the count of Champagne, who, until 1150, retained the privilege of selecting the bishop.

The major religious institutions were located within the fortified area of Troyes, called the *cité*, which was bisected by two main roads. The cathedral was located in the southeastern section of the *cité*. In addition to the cathedral, the *cité* was home to the bishop's palace, the monastery of St.

Loup, and various canons' houses and priories. Outside the walls of the *cité*, however, the suburbs that developed in the eleventh and twelfth centuries were totally under control of the counts of Champagne. It was here that the medieval fairs for which Troyes was famed were held. The two largest, held yearly, were the *foire chaude de Saint Jean* during June and July, and the *foire froide de Saint Remi* during October and November. Under the protection of the counts, the fairs grew in size and importance, becoming international events, and money-changing booths became central to the fairs. As a result, the city gave its name to a system of weights, the troy system, which was developed to standardize the measure of gold, silver, and other precious metals throughout Europe.

In 1188, a fire in the *cité* badly damaged the old pre-Gothic cathedral, thought to have been built under the auspices of Bishop Milo around the year 1000. Construction of the new cathedral began in approximately 1208 and continued for more than 300 years.

The Gothic cathedral of Sts. Peter and Paul is in the style of a basilica. The building is an oblong, with five parallel aisles ending in a curving semicircular apse at its east end. A transept crosses the aisles. The result is a cathedral in the shape of a Latin cross with a rounded top. The vaults of the central aisle rise to a height of ninety-two feet, twice the measurement of its width.

The use of defective materials and structural problems plagued the cathedral during its construction and led to numerous collapses. Chalky stone cracked when exposed to frost. The drainage system to remove rain water from the cathedral's roof was insufficient, frequently clogging and causing flooding. In addition, the initial design of the flying buttresses proved too weak to support the weight of the walls.

Construction of the cathedral was also impeded by a number of political, social, and financial circumstances. Work on Troyes cathedral was begun in a period of prosperity. However, the early thirteenth century found the counts of Champagne involved in the Crusades in the Holy Land, funneling resources away from the cathedral's construction. In 1274, the Champagne count, Henry IV, died. His daughter Jeanne wed the future King Philip IV. Upon Philip's ascension to the throne in 1285, the county of Champagne, including Troyes, lost its last vestiges of independence and was totally absorbed into the kingdom of France. Now belonging to the French king, Troyes became a battleground for the three separate powers competing for French dominance: the English; the Burgundians, who held territory in the north of France; and the house of Valois, whose influence extended over the regions to the south. Various armies overran the countryside of Champagne during the Hundred Years War. In 1358, the English attempted to capture Troyes but failed. In

The pedestrian section of Troyes
Photo courtesy of French Government Tourist Office

1380, the English tried again but were defeated by the forces of the duke of Burgundy, whose garrisons defended the city.

In 1415, Louis the Dauphin, eldest son and heir apparent of the French Valois king, Charles VI, died. Louis's brother, Jean, succeeded him, but in 1417 Jean also died. This left Charles VI's sole surviving son, fourteen-year-old Charles, as dauphin and new heir apparent to the throne.

John the Fearless, son of Philip the Bold of Burgundy and a cousin of Charles VI, seized this opportunity and attempted to gain power. He established an alternative Council of Regents at Troyes in 1417 and also established a mint in the city. Eventually, the Burgundians also seized Paris and took the king prisoner. The Dauphin Charles was unable to rescue his father or retake the city.

While these events ensued, Henry V of England was making his bid for the French throne. He won impressive military victories at Harfleur and Agincourt and finally laid a successful siege to the city of Rouen. In order to survive the English threat, France had to cease its internal conflicts. A meeting was arranged between John the Fearless and the Dauphin Charles. At the meeting, John the Fearless was struck dead by a blow to the head with a mace. The aristocracy in France condemned Charles, then sixteen years old, for his part in John's death. Leading the condemnation was Philip the Good of Burgundy, the son of John the Fearless. When the Council of Regents began peace talks with Henry V, it sent Philip the Good, not the Dauphin Charles, to represent Charles VI, now hopelessly insane.

These peace talks led to the Treaty of Troyes, signed in 1420. The treaty put the succession to the throne of France firmly into the hands of Henry V of England and his heirs, and totally excluded the exiled dauphin from any succession to the throne. To further seal the treaty beyond all question, Henry V married Catherine of Valois, the daughter of Charles VI. Henry and Catherine wed on June 2, 1420, in the church of St. Jean-au-Marché in Troyes. Philip the Good wed Michelle, Catherine's older sister.

Subsequent to the Treaty of Troyes, the Anglo-Burgundian alliance controlled the city for the next nine years, until 1429. Heavy taxes were levied in Troyes to finance the military campaign against the forces supporting the Dauphin Charles. This taxation contributed to the divided loyalty felt by the citizens of Troyes. The common people supported the Burgundians, but Bishop Jean Léguisé, along with certain leading families, supported the Dauphin Charles.

Upon his father's death, Charles proclaimed himself the king of France. His claim to the throne was tenuous; many considered him a bastard and not a true heir. Charles himself was plagued by doubt on this matter. He had established an exile government and was still a strong force south of the Loire and in the Languedoc, but in order to gain the throne, he would require great—some say divine—assistance. This assistance came in the person of Joan of Arc.

Almost all of the information regarding Joan's life is from the transcripts of her trial at the hands of the English at Rouen. Joan was born in the French village of Domrémy.

She was, as most peasants of her time, illiterate and learned to sign her name only in adulthood. The very early years of her life appear to have passed normally. At the age of 13, Joan began to hear voices sent from the Lord. In 1427, Joan was taken to see Robert de Baudricot at Vaucouleurs, the nearest stronghold of the dauphin Charles. At the meeting, Joan told de Baudricot that "her Lord intended that the dauphin should be the king, and would restore his kingdom to his control, saying that in spite of the dauphin's enemies he would be king and that she [Joan] herself would take him to be anointed" (translation by James Holland Smith). De Baudricot eventually became convinced of Joan's importance, and in February 1429, he supplied her with an escort to Chinon, where Charles was in residence. Once Joan had gained admittance to Charles, she won his confidence so quickly that the English eventually accused her of bewitching him.

Charles appointed Joan *Chef de Guerre,* or Director of the War. In April 1429, she marched from Blois to the relief of Orléans where, by May 8, she successfully raised the English siege of that city. On June 29, 1429, the march began to Reims, where Charles, again according to Joan's voices, was to be crowned king. On June 30, Joan captured the town of Auxerre. On July 4, she stood before the walls of Troyes.

Joan composed a letter to the citizens of Troyes, urging them to yield and recognize Charles as the true king of France. However, the gates of the town remained closed, the citizens fearing retribution by Charles on account of the treaty signed there, nine years before, which had excluded him from the throne. As the city did not respond to her letter, Joan urged an immediate attack. Prior to the actual assault, however, the citizens of Troyes sent out one of their preachers, a Brother Thomas, to investigate the rumors that Joan was a witch or, worse, a devil who had taken female form.

After meeting with Joan and being convinced that she was neither witch nor devil, Brother Thomas returned to Troyes. The citizens requested a parley. After negotiations, the Anglo-Burgundian garrison was allowed to leave the castle and the city was turned over to Charles's officers. On July 10, 1429, Joan and the dauphin entered Troyes peacefully. One week later, on July 17, Charles, accompanied by Joan, was crowned Charles VII in the cathedral at Reims. Joan would soon be captured and executed by the English, but the people of Troyes would not forget her. They added an inscription to the tower of the cathedral, commemorating Joan's entrance into the city in 1429.

Although consecrated on July 9, 1430, by Bishop Léguisé, the cathedral of Troyes remained unfinished. It was not until the middle of the sixteenth century, with the prosperity that came after the end of the Hundred Years War, that the cathedral at Troyes could be completed. Throughout the sixteenth century, the town prospered, despite a horrible fire in 1524 that destroyed a thousand houses in the city.

During the Protestant Reformation, Troyes strongly embraced the teachings of John Calvin. As a result, Troyes suffered greatly during the religious wars that soon consumed the nation. In 1562, Troyes's entire Huguenot (Protestant) population was expelled and forced to retreat to the relative safety of Bar-sur-Seine; in 1572, a number of these Protestants were put to death.

On April 3, 1598, King Henry IV of France signed the Edict of Nantes, designed to settle these religious wars and to improve the situation of the Huguenots. Although the Huguenots were still required to tithe to the Catholic Church, the edict allowed them to practice their religion freely, to reside in any part of the kingdom, to hold public office, and to control the religious education of their children. Special courts were created that included Protestant judges. The Edict of Nantes was deemed "perpetual and irrevocable" by Henry IV.

Henry IV was murdered in 1610, however, and the Wars of Religion were renewed during the 1620s. On October 18, 1685, King Louis XIV revoked the Edict of Nantes, ending any hope for religious toleration of the Huguenots in France. With the revocation of the edict, the Huguenots of Troyes were dispersed. As the Huguenots were among the leading citizens and merchants of Troyes, the revocation dealt a severe blow to the local economy; the economy declined still further as a result of almost continuous warfare and increased tax burdens.

But by 1789, the time of the French Revolution, Troyes had regained its preeminence as the economic center of Champagne. The city was favorably located at the intersection of trade routes between Switzerland and Germany to the east, and Paris, Flanders, and Great Britain to the west. Troyes was also a center of cotton production, with nearly two-thirds of its inhabitants employed in this industry. Raw cotton was imported from the Caribbean islands of Santo Domingo, Martinique, and Guadeloupe. The finished cloth was exported to southern France, Italy, and Spain. Merchants and royal officials dominated the social and cultural life of the city.

The revolution was particularly violent in Troyes. The soaring price of grain resulted in three days of food riots. Less than a week after the fall of the Bastille in Paris, a revolutionary municipal committee took over the functions of the town council. Less than two weeks after that, the mayor of Troyes was killed and publicly mutilated.

In 1790, Troyes was designated as the capital of the newly formed Department of the Aube. The city continued its prosperity during the development of the Industrial Revolution in the nineteenth century. The two world wars brought much devastation, and several of the city's districts needed to be completely rebuilt in the postwar years.

Today, Troyes is the chief center of the hosiery industry in France, along with the allied industries of dyeing, spinning, and textile machine and needle manufacturing. The manufacture of motor vehicles, tires, cigarette paper, and locks also takes place, along with flour milling.

Further Reading: *Building Troyes Cathedral* by Stephen Murray (Bloomington and Indianapolis: Indiana University Press, 1987) is a detailed description of the construction of the Cathedral of Saints Peter and Paul between the thirteenth and sixteenth centuries. Photographs and architectural drawings are included in the work. Due in part to the length of time it took to complete the cathedral, its construction is one of the best documented in France. *Joan of Arc* by John Holland Smith (New York: Scribner's, and London: Sidgwick and Jackson, 1973) is an account of the life of the peasant girl who aided a king of France to gain his crown. It is an objective, scholarly work, based largely on the transcripts of Joan's trial at the hands of the English at Rouen. *The Huguenots: A Biography of a Minority* by George A. Rothrock (Chicago: Nelson-Hall, 1979) chronicles the struggle and persecution of the French Protestants from the time of the Protestant Reformation to their dispersal subsequent to the revocation of the Edict of Nantes.

—Rion Klawinski

Tsarskoye Selo (Leningradskaiya, Russia)

Location: Approximately fifteen miles south of the city of St. Petersburg, at the edge of the St. Petersburg plain, in northwestern Russia.

Description: Site of two palaces used by generations of czars and czarinas, near the town they named Tsarskoye Selo (Czar's Village); site of Imperial Lyceum, opened in 1811, and attended by poet Alexander Pushkin; transformed into children's hospital and renamed Detskoye Selo (Children's Village) in 1918, following Russian Revolution; renamed Pushkin on centenary of author's death in 1937; home to the St. Petersburg Agricultural Institute and a population of 95,300. Resumed name of Tsarskoye Selo in 1990s.

Contact: Leningradskaiya Oblast Administration
Suvorovskiy Prospect 67
St. Petersburg, Leningradskaiya
Russia
(7812) 274-3563

Two palaces, one huge and one small, are surrounded by immaculately landscaped parks. The small palace has more than 100 rooms. There are statues, a theater, a pavilion, a Chinese pagoda, and a concert hall in the parks. There is an artificial lake, called the Great Pond, with a man-made island in its center, and near the island a heroic column rises out of the lake, high above the placid surface. A pyramid, a grotto, and a Turkish bath are set at various points along the water's edge. The lake can be drained or filled with the flip of a switch. It has been said that this estate was the first in Europe to be wired for electricity. All of this used to belong to the czar. It was called Tsarskoye Selo (Czar's Village) until 1918.

Tsarskoye Selo was built following Peter the Great's creation of the new Russian capital in nearby St. Petersburg. The czar felt that the old capital, Moscow, reflected the backwardness of Russia in the early eighteenth century. Despite the Kremlin and a few mansions, Moscow was little more than a collection of huts. Its residents, in comparison to other Europeans, were uneducated, uncultured, and uncouth. The young czar had been abroad and was all too aware of Russia's reputation. He resolved to reverse Russia's direction and push the country into the eighteenth century, by force if necessary.

To establish his new, modern capital, Peter fought Sweden for the land north of Moscow at the Neva delta, which had been lost by Russia in 1617. By 1704 he was calling his fort there his capital. By 1712, he moved his court from Moscow to the newly named St. Petersburg. St. Peters-burg would be a beacon of modern imperialism, and Peter I rounded up laborers from all around his empire to build it. There were 40,000 people there by 1725.

South of the growing capital were the remains of a Swedish village named Saaris Moiso (Island Farmstead). The land was given to Catherine, Peter's wife, in 1710, and the name was Russianized to Sarzkaya Myza. When Catherine officially became czarina in 1716, the land was given the name Tsarskoye Selo, and construction of a small stone palace was begun there the following year.

Though the first stone palace was built by Catherine I, the royal estate at Tsarskoye Selo didn't really take shape until after her death, when her daughter Elizabeth consented to take the throne after a coup in 1740. In the first decade of her rule, Elizabeth had the original palace enlarged, but then in the 1750s, she hired Italian architect Bartolomeo Francisco Rastrelli to rebuild the palace entirely. Rastrelli had the greatest reputation of any architect working in Russia at the time, but most of his work on the Catherine Palace (named by Elizabeth after her mother) would not survive the century.

The palace and the surrounding estate would be almost completely rebuilt by Elizabeth's daughter-in-law, Catherine the Great, a woman who strove to refashion her own identity and the identity of her country with the same fervor. Born Sophia-Augusta, the Prussian princess was a mere fifteen years old, and nearly penniless, when she was first invited to Russia to meet Elizabeth's son, Grand Duke Peter. By the time she assumed the throne in 1762, she had changed her religion to Russian Orthodox, her name to Catherine II, and her status to that of one of the most powerful rulers in Russian history.

Catherine seized power a scant six months after Elizabeth's death. During those six months she shared the throne with her husband, Czar Peter III. But Peter had never been a match for her. According to legend, Peter played constantly with toy soldiers, prepared mock executions of animals, neglected his wife, and generally played the fool, both publicly and privately. There is some evidence that their marriage was never consummated, and that Catherine's son, Paul, was fathered by Serge Soltikoff, one of Catherine's many lovers. (If true, Paul had not a drop of Romanov blood, and the dynasty was therefore broken long before the Russian Revolution.)

Catherine engineered a coup to remove and execute her husband, then set about remaking Russia, and, with it, Tsarskoye Selo. She hired a Scottish architect, Charles Cameron, to rebuild parts of the Catherine Palace in the classical style, which she preferred. She had the estate landscaped and filled with commemorative statuary. The artificial lake called the Great Pond was created, and the Chesma Column, honoring one of Russia's few naval victories, was placed next to an island in its center. The column is also called the Orlov

The Catherine Palace
Photo courtesy of General Tours, Inc.

Column, after the commander in the victory, who also was the brother of one of Catherine's alleged lovers. Pavilions were added around the pond, and a new palace was built for Catherine's favorite grandson, Alexander. She also had another park constructed, and just to be sure everyone, including her son Paul, knew for whom these places were intended, she named them Alexander Palace, and Alexander Park.

Catherine ruled Russia for thirty-four years, longer than any other Russian ruler in the eighteenth century. She had begun her rule as a liberal, loosening restrictions on free speech, encouraging the arts and sciences, but ended as a reactionary, having those who exercised their rights to criticize her (even mildly) exiled or put to death, and solidifying the rule of the nobility with the Charter of the Gentry.

Her son, Paul I, lasted only five years as czar before being removed by a coup that placed his son, Alexander I, on the throne. It was Alexander who established the Imperial Lyceum at Tsarskoye Selo. The Lyceum was Russia's first liberal arts school, based on English and French models, but without the strict discipline that marked them; the students were not beaten, and their daily regime was somewhat flex-

ible. The curriculum was meant to provide special training for future government officials, but it covered so wide a range of subjects that the students later complained that they had hardly learned anything. The school was Alexander's pet project, which was understandable, for the boys were practically moving in with him. They were quartered in the Catherine Palace itself, in a wing that had housed the grand duchesses during the reign of Catherine the Great. The Grand Duchess Anna Pavlovna still lived there.

When the first class of thirty students arrived in 1811, Tsarskoye Selo was a miniature Russia as portrayed in a fairy tale for the ruling family. All around the Catherine Palace, by now several times its original size and grandeur, were graceful gardens, heroic statuary, and triumphal arches; bands played in the gazebo on the artificial island, while guests of the czar glided past in rowboats. Clusters of lilacs lined the paths that wound through the gentle woods of Alexander Park. The boys might have been overwhelmed by the wondrous village when they first arrived, but they would soon find it all too familiar; Alexander had decreed that the students were not to leave for the entire six-year course of their studies, not even for a visit home or a ride into town. Tsarskoye Selo was like a precious, jeweled egg, with the boys stuck inside.

Alexander Pushkin, the famous poet whose name the town bore for a time, arrived, age twelve, as part of this first class. From the start, he was one of the more rebellious students at the Lyceum, spending most of his time on poetry and pranks until he became old enough to drink and pursue women. He wasn't alone in rebelliousness, though; the student body and even some of the faculty were an undisciplined bunch, particularly after the early death of the first director, V. F. Malinovski. Until a new director was chosen in 1816, the school was run by the assorted teachers, each with his own ideas on education. The first class of students was to be the only class, and for nearly four years, it was the only class in the world making its rules as it went along.

Pushkin's first run-in with Alexander I was the result of a misunderstanding on Pushkin's part. As a seventeen-year-old ladies' man, one of his first conquests was a chambermaid in the palace. Natasha worked for one of the empress' maids of honor, an old woman named Princess Volkonsky. One night in a dark corridor of the palace, Pushkin saw a dim figure in skirts and embraced her, roughly kissing the opened mouth of a woman who turned out to be the princess herself. The princess's brother complained to the emperor, who in turn, summoned the new headmaster, Egor Englehardt. At first, Alexander was all for having Pushkin whipped, but Englehardt pleaded for mercy; in the end, Pushkin had to write a letter of apology. The emperor added, "tell him this sort of thing must not happen again."

Pushkin wrote an apology, and a nasty poem about the princess as well. He was by this time already a renowned poet, both in the school and among the literati of the day. When he had read his *Recollections of Tsarskoye Selo* at the graduation from the lower school to the upper the year before the incident with the princess, many of the most important men in Russia were in the audience. The words of the poem had reduced Derzhavin, Russia's foremost poet and the poet of the court of Catherine the Great, to public weeping. "I am not dead," the old man whispered, seeking to hug Pushkin, who had run off the stage as soon as he had finished. Derzhavin did not find Pushkin, but fame did. The poem was published in one of the country's top literary magazines, *The Russian Museum*, and great things were expected from him in the future.

In 1816 the boys were finally given permission to leave the royal estate and go into the town of Tsarskoye Selo. Pushkin spent most of his time there, hanging around with the czar's hussars in the bars and barracks. There, some of the radical thought circulating in the school was tested in the world beyond the walls. Some of the teachers were freethinkers and one of Pushkin's teachers, David de Boudry, was the younger brother of Jean-Paul Marat, the French revolutionary. Boudry had changed his name to avoid notoriety, but venerated his brother's memory. And though the school was notably unsuccessful in producing good government servants, it became apparent after the death of Alexander in

1825 that Pushkin's class had produced a disproportionate number of revolutionaries. Several of the "Decembrists," the fomenters of the first, unsuccessful Russian Revolution in December 1825, were his classmates. Pushkin would have been among them, but he had angered Alexander with inflammatory verse and had subsequently been exiled to southern Russia in 1820. This exile saved him from execution or exile to Siberia.

In 1831, Pushkin moved back to Tsarskoye Selo and stayed for three months, virtually penniless. Finally, while out walking one day, Pushkin met Alexander's brother and successor, Nicholas I, who offered Pushkin a job with the National Archives in St. Petersburg. Pushkin never lived in Tsarskoye Selo again. In 1837, he was killed in a duel defending the honor of his wife. The same year, the first railroad line in Russia was constructed; it ran from St. Petersburg to Tsarskoye Selo.

Of all the major figures in Russian history who lived at Tsarskoye Selo, Nicholas II and his German-born bride, Alexandra, probably loved it most. They spent more time there than any other royal couple. Immediately after their marriage on November 26, 1894, Alexandra ordered a wing of the Alexander Palace redecorated, and they moved in soon after the coronation in 1896. That the royal family spent so much time in this gilded bubble had a bearing on the later events that finished the monarchy for good.

Nicholas II never wanted to be czar in the first place, but his father's sudden death at the age of forty-nine put him on the throne at the age of twenty-six. Bullied about by his uncles for the first ten years of his rule, he was less under their control after the Russo-Japanese War in 1904. Following a failed revolution in 1905, he established a representative ruling body, the Duma. The real influence on the czar, however, lay with a new visitor to Tsarskoye Selo: hypnotist and mystic Grigory Yefimovich, better known as Rasputin ("Debauchee"), a nickname acquired in his youth and one he would live up to during his stay at the palace. Rasputin gained control over the royal family, especially the czarina, by his seemingly miraculous curative powers. Nicholas's son Alexis was a hemophiliac who came close to death many times. But when Rasputin visited Alexis's room in the palace and prayed until morning, the child always recovered. When not at the palace advising the czar or caring for Alexis, Rasputin spent his time in St. Petersburg, drinking heavily, carousing, and womanizing.

Rasputin was murdered by nobles at the end of 1916, and after his body was discovered on January 1, 1917, Nicholas retreated deeply into the imperial cocoon at Tsarskoye Selo. His timing could not have been worse. Russia was strained terribly by its involvement in World War I. Though Nicholas had been warned that revolution was a strong possibility, he chose to ignore the threat. The uprising came that spring, while Nicholas was briefly away from Tsarskoye Selo on a rare excursion.

Nicholas was forced off the throne, to be replaced as czar by his brother Michael for a few days. Then he, too, was forced to abdicate. Nicholas returned under guard, by train to Tsarskoye Selo, where he was imprisoned along with his family and a few faithful employees. The Provisional Government had already sent troops to the estate, where they had battled briefly with the royal guards on March 13 near the Catherine Palace. The royal guards deserted two days later, and the royal family was imprisoned in the palace for the next five months.

They were limited to one wing of the Alexander Palace and a small patch of the park. The family went outside every day, picnicking and taking the sun inside a fence erected to keep out curiosity seekers. Townspeople and soldiers sometimes jeered at the hapless family, who were caged like zoo animals, but displayed grace nonetheless, chatting in friendly fashion with the soldiers and onlookers whenever they could. On August 13, 1917, Nicholas, Alexandra, and their five children were removed from Tsarskoye Selo by the Provisional Government; they were murdered in Ekaterinburg the following summer, on July 16, 1918.

That same year, the name of the estate was changed from Tsarskoye Selo to Detskoye Selo (Children's Village), and the mansions and palaces of the town converted into rest homes and schools for sick children from all over Russia. As the tumult of the revolution continued into 1919, White Guards captured the town briefly, but lost it again after a few days. The town underwent another name change in 1937, on the hundredth anniversary of Alexander Pushkin's death, and four years later, saw more fighting, as German soldiers cap-

tured the former czar's estates. The Germans used the Catherine Palace as a barracks while they attempted to starve St. Petersburg (then Leningrad) into submission. When the Germans were finally driven out, they blew it up. Restoration and conversion of the palace were begun in 1957.

The restoration is not entirely finished in the 1990s, but the entire estate is open to the public, and a museum to honor Alexander Pushkin was opened in the Catherine Palace in 1967. Also in the 1990s, the town reverted to its former name of Tsarskoye Selo.

Further Reading: *Pushkin* by Henri Troyat (New York: Doubleday, 1970) is the standard work on the life of Alexander Pushkin. Well written and researched, this is a big, entertaining, and informative book. A shorter biography, *Pushkin* by Ernest Simmons (New York: Vintage, 1938) is also well written by an enthusiastic admirer of the writer. Pushkin was hardly known outside of Russia when Simmons wrote the work. There are any number of biographies of Catherine the Great available—her original writings are well worth consulting. Several are available in *Catherine the Great,* edited by L. Jay Oliva (Englewood Cliffs, New Jersey: Prentice-Hall, 1971; Hemel, Hempstead: Prentice-Hall, 1972). *Nicholas and Alexandra* by Robert K. Massie (New York: Atheneum, 1967; London: Gollancz, 1968) was a huge bestseller, made into a popular television movie. Massie is obviously sympathetic to the czar and czarina, to a fault at times, but besides being documented history, this is a marvelous, tragic story, well told.

—Jeffrey Felshman

Ulm (Baden-Württemberg, Germany)

Location: In south-central Germany, along the left bank of the Danube River; its younger sibling, Neu-Ulm, or New Ulm, was created in the nineteenth century on the right bank of the river.

Description: Dating back more than 1,100 years, Ulm grew from a tiny hamlet to an economically prosperous town by the Middle Ages. This prosperity allowed the funding and subsequent construction of Ulm's most famous landmark, the fantastic Gothic cathedral of Münster. The town was largely destroyed during the Allied bombing raids of World War II, but miraculously the cathedral survived.

Site Office: Verkehrsbüro
Tourist-Information
Münsterplatz 51 (B 2)
89073 Ulm (Donau), Baden-Württemberg
Germany
(731) 6 41 61

The area of Ulm has been populated for centuries; it was an Alamannic settlement as far back as the seventh century. Its strategic location at the intersection of the Danube, Iller, and Blau Rivers made Ulm an important center of trading and commerce as early as the ninth century. The town is mentioned for the first time in 854 as a royal domain and called Hulma. Hulma was given a municipal charter from Holy Roman Emperor Frederick I (Barbarossa) and officially became a town in 1165.

The economic growth in thirteenth-century Europe led to the rapid development of industry and workshops throughout the Holy Roman Empire. By the fourteenth century, a number of towns with flourishing crafts and trade were granted special rights and liberties by the emperor and declared free cities. Ulm acquired this status in 1274 as a result of the growth and success of its workshops specializing in the weaving of fustian, a blend of linen and cotton. Ulm now was free from the rule of local princely authority (most importantly, from the prince's taxes); it was an independent jurisdiction, bound only to the emperor. Ulm also gained the power to produce its own currency and coins, and to grant its own democratic constitutions. Ulm's first such constitution came about in 1397 and was granted to the city's various trade guilds.

From its beneficial position on the Danube, Ulm enjoyed prosperous trade in these early years as a free city. The Danube, central Europe's longest river, snakes its way 1,725 miles until it empties into the Black Sea, and its length and navigability gave the river an important role in early trade. The bridge spanning the Danube at Ulm, combined with the river itself, provided a clear trade route between central Germany, Italy, the Mediterranean, the North Sea, and the Baltic.

It was at Ulm that the Danube River first became navigable for smaller boats. These boats were called *Ulmer Schachteln*, or Ulm boxes. The *Ulmer Schachteln* docked in Ulm as they were being loaded with cargo and passengers. Once underway along the then-treacherous Danube, these small craft were controlled by experienced and strong men at the oar until their docking as far downriver as Belgrade.

While Ulm enjoyed economic prosperity during the Middle Ages, many areas of Germany did not. The formation of free cities and the continuous division of duchies and other properties due to war or death created numerous tiny holdings. Many of these holdings were far too small to be economically viable. A number of the free imperial knights owning the small subdivisions eked out an existence farming the land; however, others situated on strategic locations turned to another livelihood: toll collection. These tolls became a severe financial hardship for the growing trade in the free cities and other areas.

In order to protect their citizens and commerce, a number of cities and towns in western and southern Germany banded together into a union known as the Swabian League of Cities. Ulm headed the union throughout the fourteenth and fifteenth centuries. This newly formed league found itself at odds with the church, as the Golden Bull of 1356 clearly prohibited the formation of such groups. The leaders of the Holy Roman Empire quickly expressed their displeasure with the league and immediately sent imperial forces to forcibly disband it. But the imperial forces were defeated by the cities, and peace was grudgingly declared in 1377.

In response, the imperial knights themselves banded together, forming the leagues of St. George, St. William, and the League of the Lion. The knights' leagues regarded the Swabian League of Cities with open hostility, and in 1382 war broke out between these two groups. Once again, the League of Cities proved victorious, and peace was declared. This truce was short-lived, and battle resumed five years later when a duke of Bavaria—in favor with the knights—was kidnapped by the archbishop of Salzburg—sympathetic to the cities. This time, in what was later called the South German City War, the cities were defeated and lost their political power.

Ulm began construction of its famous cathedral, Münster, in 1377. The period from the beginning of the fifteenth century to the start of the Reformation was a time of frantic building of churches in Germany. This tide swept through not only the larger cities such as Ulm, but through tiny villages as well, producing structures from enormous cathedrals to tiny but elaborate chapels. Towns spent hundreds of years

Interior of the cathedral at Ulm

and countless sums of money constructing buildings to honor the Lord—and to bring honor and recognition to the town's power, influence, and economic prosperity. Ulm's own Münster is a prime example of this type of undertaking. A chronicle of Ulm's cathedral tells of a small hut erected to serve as a place where pious citizens could bring their donations for its construction. "No apron, bodice, or necktie would be disdained," according to the manuscript. The offerings were then sold in the market and the money used to benefit the church. Other citizens dedicated their horses or themselves to labor on the construction for periods ranging from one month to an entire year. Popular legend tells that the cathedral's cornerstone, laid in 1377 and now covered with mortar, was encrusted with gold coins and precious stones by the town's mayor and citizens.

With this financial and personal commitment of Ulm's faithful citizens, Ulm's Münster was constructed, roofed, and furnished on the inside with fifty-two altars at the cost of nine tons of gold by 1488. This enormous endeavor was completely funded by the people of Ulm alone, with no outside help.

The Münster is one of the largest cathedrals in the world, its ceiling rising 137 feet above the central nave. The church remains one of Europe's finest examples of Gothic architecture and art. An undertaking of this scale attracted the finest craftsmen of medieval Germany, all dedicated to producing their greatest works for the glory of God. Münster's rich interior includes large, beautifully colored fifteenth-century choir windows and elaborately carved choir stalls, crafted by Jörg Syrlin the Elder from 1469 to 1471. Syrlin also created the immense eighty-five-foot-high limestone and sandstone tabernacle housed in the cathedral. On the cathedral's interior wall is a fresco of the Last Judgment, done in 1471 and considered one of the finest remaining representations of Gothic wall painting. In 1452 the sacristy was constructed with the funds donated by one single person, Claus Lieb, whose anvil and hammer are rumored to be buried in the sacristy's foundation. His name remains inscribed over the sacristy's door.

A popular legend can also be found in Münster, located on the roof of the church's central nave. According to folklore, medieval workers on the cathedral were struggling to carry a long log through Ulm's gates. On the verge of giving up, the laborers spotted a small sparrow carrying a blade of grass for its nest. The workers watched as the tiny bird carefully negotiated the blade through a small hole, turning the grass lengthwise to easily enter the hole and continue the nest. Inspired, the workers followed the sparrow's lead and brought the log successfully into town. The Ulmer Spatz (Sparrow of Ulm) became the town's mascot and can be found carved in stone atop the great cathedral.

While most work on Münster was completed during the fifteenth and early sixteenth centuries, construction and additions continued over the next few centuries. The two eastern towers and a 528-foot steeple—the world's tallest steeple and Ulm's most famous landmark—were not completed until 1890. By some miracle, the impressive finished product was undamaged during the fierce Allied bombing raids of 1944 and 1945.

By the fifteenth century, Ulm had become a very prosperous city, as shown by the extravagant cathedral. Cities during this period of the Middle Ages were particularly eager to acquire new lands on which to support their growing population. By the late fifteenth century, the city of Ulm had grown from the tiny hamlet of Hulma to a landed estate of more than fifteen square miles, quite large for a free city.

City life during this period differed greatly from the cosmopolitanism of later eras. While commerce played a large role in the cities, agriculture and animal husbandry were equally important. The city's farmland was generally tended by free farmers, while city tenants also kept their own livestock well within the city proper. Considering the potential number of animals, city governments were quickly forced to limit the amount each family could raise in town. Ulm laws dictated that no citizen was allowed more than twenty-four pigs, still a great number considering the swine grazed freely during the day.

Despite the number of humans and animals living within the city gates, life in these medieval towns was prosperous: food from the surrounding farms was plentiful and inexpensive, fresh meat and dairy products were in abundant supply, and the crafts and industry of each town produced products necessary for everyday life at a low cost. Ulm's own flax and hemp production was considerable, turning out 60,000 pieces of bleached linen and cotton per year.

The fruitful estates of Ulm, the large numbers of well-tended stock, and the successful industries of the town made Ulm one of the wealthiest cities in Germany. So great was Ulm's prosperity that, in the imperial levy assessed in 1520, Ulm was taxed at the same rate as the lands of the wealthy electoral princes.

To protect its citizens, industry and agriculture, Ulm and its surrounding territories were encircled by fortified walls. The first such wall dates back to the twelfth century and runs along the waters of the Danube. As the town grew, another line of defenses was placed around Ulm and its property, creating a protected area so large that the city did not outgrow its enclosure for 500 years.

These formidable walls protected Ulm from the unrest of the Middle Ages and allowed the city to survive the terrible battles of the Thirty Years War, battles that devastated a number of similar towns in Germany. Ulm's greatest involvement in the war that shook Europe came in 1620 with the Treaty of Ulm.

French ministers in Paris were eyeing the growing unrest in Europe between Catholics and Protestants with concern. Caught between two rising powers, Germany and Spain, the French wanted to maneuver safely between the two regions. With just cause, the French were concerned with the ties to the British throne of Frederick, king of Bohemia. Should the frail prince of Wales die, it was not impossible that Frederick's realm could be extended to include

Britain. On the other hand, should Spain use the growing unrest in Bohemia as an excuse to invade the Rhenish Palatinate, the French were too near that border to go unaffected.

Concerned, the French ambassadors left for Germany to seek peace between the Protestants and Catholics. When they arrived in Ulm in the summer of 1620, the ministers found the two armies—the Protestant Union and the Catholic League—gathering ominously on either side of the Danube. No one wanted to enter the war, but princes on both sides feared for their own lands. The French ambassadors put forward a plan, suggesting that if the Catholic League agreed to respect the neutrality of the Protestant states in regard to the affairs of Bohemia and not attack, the Union would guarantee that no aggressive Protestant forces would threaten the lands of the League. Relieved to have avoided needless confrontation, both sides agreed to the treaty on July 3, 1620.

While the treaty spared Ulm serious devastation, it could not completely end the hostilities, which raged across Europe for another twenty-eight years. Protected by its fortified walls, Ulm survived the Thirty Years War. Following the Peace of Westphalia, order reigned in the region for nearly 200 years until the approach of Napoléon and his Grand Army. In 1802, Napoléon declared that the Danube River would serve as the official dividing line between the provinces of Bavaria and Württemberg, effectively cutting Ulm in half. The older portion of Ulm lay in Württemberg, while the newer Bavarian town adopted the name Neu-Ulm, or New Ulm.

In 1805, the city of Ulm was the site of a strategically important victory for Napoléon and a crushing defeat for Austrian forces. Austria had joined with the English and the Russians fighting Napoléon in the summer of 1805. The Austrian Army had planned to fight most of its battles in Italy, and therefore kept the majority of its forces there.

In September of that year, commander Baron Karl Mack von Leiberich led some 72,000 troops into Bavaria, a territory then allied with France, to wait for Napoléon's entrance from France through the Black Forest. Mack's forces were centered along the Danube between Ulm and Gunzburg, a distance of approximately eighty miles. The Austrian army was to be reinforced by the Russians in their battle against the French, estimated to be approaching with no more than 70,000 soldiers.

Napoléon, however, had planned on Germany being the battleground and was advancing with 210,000 men, ready to crush Mack and his Austrian army before the Russians could arrive. The rapidly advancing Grand Army quickly cut off all of Mack's escape routes to the east. Grossly outnumbered and outmaneuvered, Mack's army suffered crushing defeats in several battles in the beginning of October. As the French continued their attack, the territory controlled by the Austrian army grew smaller and smaller, until October 15, when the bulk of the army was cornered in Ulm.

Napoléon's troops fired on the city on October 16, and with Russian help still 100 miles away, Mack saw little hope for his depleted army. Mack surrendered to Napoléon at Ulm on October 20, 1805; the French captured some 50,000 to 60,000 Austrian prisoners of war, while the French army suffered negligible losses.

Ulm next rose to historical prominence in the fall of 1930, when the city became the site of an investigation and subsequent trial of three German army lieutenants from the city's garrison. According to the prosecution, Lieutenants Scheringer and Ludin had been attempting to recruit officers of other army stations to join Adolf Hitler's National Socialist movement. Once the officers supported the Nazis, the army would be unable to prevent a National Socialist uprising. The case reached the German Supreme Court on September 23, 1930, and Hitler himself addressed the court in a long, impassioned attempt to persuade the court and the nation that the Nazis were not attempting to overthrow the Reichswehr. Yet, he also asserted that with National Socialist victory, "heads will most certainly roll in the sand." The defendants were found guilty of conspiracy to commit an act of high treason and were sentenced to eighteen months' imprisonment.

The city of Ulm was badly damaged in the fury of World War II. The Allied bombing raids of 1944 and 1945 destroyed more than 65 percent of the city. Fortunately, Ulm's most famous landmark, the cathedral of Münster and its 528-foot spire, miraculously escaped destruction.

Today's Ulm and Neu-Ulm boast a population of nearly 100,000 and combine historical landmarks such as Münster with modern industry and a university founded in 1967. As in its early days of trade on the Danube River, Ulm continues to be one of Germany's major transportation and communications centers.

Further Reading: Johannes Janssen's *History of the German People at the Close of the Middle Ages* (London: Kegan Paul, Trench, Trübner, 1905) provides a detailed look at a number of facets of daily life in medieval times. *Germany: A Short History* (Carbondale: Southern Illinois University Press, 1976) by Donald S. Detwiler offers a concise account of Germany from antiquity through the modern era. Those readers interested in the numerous intrigues of Europe during the chaotic Thirty Years War will enjoy C. V. Wedgwood's detailed *The Thirty Years War* (London: Cape, 1938; reprint, New York: Anchor, 1961). Accounts of both the Habsburg dynasty and modern German history are found in Ralph Flenley's *Modern German History* (London: Dent, and New York: Dutton, 1959) and Erich Eyck's two-volume *A History of the Weimar Republic* (Cambridge, Massachusetts: Harvard University Press, and Oxford: Oxford University Press, 1962 and 1964).

—Monica Cable

Uppsala (Uppland, Sweden)

Location: 40 miles northwest of Stockholm on highway E4.

Description: Religious capital of Sweden, site of the country's oldest university and home of the eighteenth-century scientist Linnaeus.

Site Office: Uppsala Turistkontor
Fyris torg 8
S-75310 Uppsala, Uppland
Sweden
(18) 27 4820

Uppsala is the capital city of the ancient province and modern district of Uppland, the original heartland of the Svear people for whom Sweden is now named (for *Sverige* means "land of the Svear"). Gamla Uppsala (Old Uppsala), two miles northeast of the modern city, appears to have been their political and religious center from around the year 300 or earlier. The three enormous artificial mounds at this site, which date from the sixth or seventh century, may well be the tombs of pagan kings. It has been suggested that they or their descendants may have invaded eastern England, where the buried royal ship excavated at Sutton Hoo in 1939 was found to contain weapons probably made in Uppland and where the poem *Beowulf*, which may well refer to people and events in Gamla Uppsala, was composed around 700.

By around the middle of the ninth century at the latest, perhaps even from before the building of the mounds, the Svear had extended their power south into other parts of Sweden and east into the Baltic and sent trading ships to Baghdad and the Caspian Sea. Some of them later settled in Novgorod (now in Western Russia), Kiev (now in Ukraine), and Constantinople (now Istanbul). Their kings were chosen at meetings of their chiefs in a meadow at Mora, which, with Gamla Uppsala and the surrounding area, formed the kings' personal estate. Their duties included dispensing justice on the *domarhög* ("mound of justice"), a fourth, flat-topped mound near the three tombs, also known as the *Ting* ("assembly") mound, and leading worship in the nearby aspen grove or, perhaps, in a wooden temple. It is hard to be sure which gods were worshipped or how, for the main evidence about paganism at Gamla Uppsala, apart from fragmentary archeological discoveries, comes from the hostile writings of the Christian missionary Adam of Bremen. The rituals may have included the hanging of men and animals from sacred trees in honor of Odin, the god of wisdom, who was said to have hanged himself.

Christian missionaries were active in Scandinavia from the tenth century onwards, but Uppland remained a pagan stronghold until the 1080s, when King Inge the Elder, having converted to Christianity and been briefly replaced by his half-brother Blot-Sven, reasserted his authority and began the process of Christianizing Uppland. One of Inge's successors, Erik, ruled for only five years, much of which he spent fighting pagans in Finland, up to his violent death in 1160. He was buried in the Romanesque church at Gamla Uppsala, built earlier in the twelfth century—either on the site of the pagan temple, if there ever was one, or on the ruins of a previous wooden church—and was soon being venerated as the patron saint of Sweden. Only four years after his death the church became a cathedral, the seat of Stefan, the first of the archbishops of Uppsala. It was damaged by fire in 1245 and reverted to being a church around 1276, when the archbishop took the saint's remains to the market-town and port that was to become modern Uppsala. Today only the tower, choir and crypt of the Gamla Uppsala church remain standing, in a field overshadowed by the four mounds of the pagan kings.

The Domkyrka, the cathedral built at Uppsala to house the saint's remains, is distinctive for having been constructed with locally-made bricks, though in other respects it follows the tradition of Gothic churches and cathedrals built in stone, for its chief architect after 1287 was a French immigrant, Etienne de Bonneuil. In the fourteenth century one of its chapels was decorated with wall paintings depicting the life of St. Erik. The cathedral was dedicated in 1435, but its external appearance was greatly altered in the 1890s when Hugo Zettervall carried out a restoration and built its double spire.

By the beginning of the sixteenth century the cathedrals, monasteries and other institutions together owned about 20 percent of Swedish land, and the archbishops of Uppsala, the leading priests in Sweden, were frequently drawn into political conflicts. In 1434, when Engelbrekt Engelbrektsson came to Uppsala to initiate the nationwide rebellion that would take the Swedish crown away from Erik of Pomerania, King of Denmark, Norway and Sweden, the nobles of Uppland joined him and replaced the archbishop appointed by the king with one chosen by themselves. In 1457, another archbishop, Jöns Bengtsson, led the uprising that expelled the regent Karl Knutsson and reunited Scandinavia under the Danish King Christian; it was Bengtsson again who led the Swedes against Christian's attempts to impose new taxes, and, in 1465, after a brief return by Karl, became administrator of the kingdom for two years, until Karl was brought back yet again.

Karl's nephew Sten Sture governed Sweden from 1471 to 1496 and supported the establishment of the first university in Scandinavia at Uppsala in 1477. It acquired one of the first four printing presses in the country, but it appears to have been closed around 1515. One of its graduates, Gustav Vasa, returned in 1521 to occupy the city and call for rebellion against the Danish King Christian II at a meeting on

Uppsala Castle

the Ting mound. Two years later he became king of an independent Sweden.

One of Gustav's first acts as king was to order the expulsion from Sweden of Gustav Trolle, who had been Archbishop of Uppsala since 1514 and had collaborated with Christian II in arranging the "Stockholm bloodbath" of 1520, in which eighty-two leaders of the resistance to Christian's invasion had been killed. The pope's refusal to recognize the new archbishop elected in Uppsala, and the startling assertion by Laurentius Andreae, the Dean of the Domkyrka, that the church's property belonged to the people, both helped to precipitate the separation of Sweden from the Catholic Church.

In 1524 Gustav re-established the Uppsala printing press, which was swiftly removed to Stockholm after it had been used to print anti-Lutheran books; there, it was put under the control of the Lutheran Olaus Petri (also a graduate of Uppsala University). In 1527 the king replaced the pope as head of the Church in Sweden, and began the transfer of most of the church's wealth into the hands of his family and his supporters; in 1528 he was crowned king in a Lutheran service in the Uppsala Domkyrka, where all his successors up to 1721 would also be crowned.

In 1531, needing an archbishop of Uppsala to officiate at his wedding, Gustav saw to it that an assembly of priests elected Olaus Petri's brother Laurentius to the post. In 1536 Laurentius presided over a council of the church in Uppsala, which ordered that all services should be in Swedish, not Latin, and that priests could marry if they wished. But another council at Uppsala in 1539 could not agree on the organization of the church, and Gustav intervened to make it a department of his national government.

Gustav Vasa ordered the construction of several castles, including Uppsala's, which still dominates the city from its hilltop site to the south of the cathedral and university buildings. He died in 1560 and was buried in the Domkyrka of Uppsala, between his first two queens, under effigies carved by a Dutch artist, Willem Boy. Seven years later his son Erik XIV, having begun the revival of the university in Uppsala, went mad in Uppsala Castle, murdering the nobleman Nils Sture, who was imprisoned there, ordering the killing of other prisoners and then running away into the nearby forest. Sture was buried in his family's chapel in the Domkyrka, and Erik was replaced in 1568 by his half-brother John III.

In 1571 yet another council at Uppsala approved a

church ordinance that reserved the right to elect bishops to local officials rather than to the monarch, reaffirmed the bishops' supremacy in religious affairs, and permitted the continued use of Catholic vestments and rituals. This ordinance was the last major contribution of Laurentius Petri, who died in 1573 after forty-two years as archbishop; it was effectively superseded only two years later by King John's *Nova Ordinantia,* which, among other measures, reintroduced the monarch's right to reject candidates for bishoprics. In 1577 John imposed a new liturgy, known as the Red Book, which shocked many Lutherans by its apparent incorporation of Roman practices, and also gave the Domkyrka the silver reliquary in which St. Erik's skull and crown are still kept.

John died in 1592. While Sweden waited for his son Sigismund, a Roman Catholic, to return from Poland, of which he had been elected king, John's brother Duke Karl organized an assembly of both clergy and laity which met in 1593 to produce the Uppsala Resolution, restoring the Church Ordinance, denouncing the Red Book and making the German Luthcrans' Confession of Augsburg the basis of official Christian doctrine in Sweden. Thus placed in an impossible position, Sigismund, crowned in Uppsala in 1594, found himself invading Sweden in 1598, only to be defeated, deposed, and replaced by his uncle Karl.

Meanwhile, John had been buried in the Domkyrka, though his elaborate tomb was not placed over his remains until 1782, having been kept in Danzig (now Gdańsk) until it was paid for. Beside it, oddly enough, is the tomb of his first wife, the Polish Catholic Katarina Jagiellonica, but not that of his second wife, the Swedish Lutheran Gunilla Bielke. She, however, has her own memorial elsewhere in Uppsala, the Gunillaklocka (Gunilla Bell), which she gave to the Castle in 1588 and which is still rung at six o'clock every morning and nine o'clock every evening.

A decisive revival of the university had also taken place in 1593, though it continued to lack funds and to be disrupted by theological disputes among its teachers until King Gustavus II Adolphus intervened. In 1620 and again in 1622 he increased the size of its faculty, adding the teaching of law, medicine, mathematics, history and politics to its original theological concerns, and in 1624 he endowed it with the revenues of 317 estates. He also donated to its library the books and manuscripts seized by his troops during his military campaigns in Germany. His benefactions are commemorated by the Gustavianum, a building to the west of the Domkyrka that was completed in 1625 and is now a museum of ancient Egyptian, Greek, Roman and Nordic artifacts. The university was well placed to provide training for many of Sweden's government officials and to attract young nobles who would previously have gone to Germany to study. It would thus remain the leading university in the kingdom even after the foundation of rival universities at Dorpat (now Tartu, in Estonia) in 1632, at Åbo (or Turku, in Finland) in 1640, and at Lund in 1668.

Gustavus Adolphus's daughter, Queen Christina, came to Uppsala Castle in 1654. She had always been a reluctant ruler and had decided to abdicate, not, as the Greta Garbo film *Queen Christina* would have it, because she had fallen in love with John Gilbert, but because she wished to become a Roman Catholic. The abdication ceremony took place in the vast Rikssal (Throne Room), which is inside one of the two wings between two towers that are the only parts of the Castle remaining after a fire in 1702 destroyed or damaged most of the city.

During and after the rebuilding of Uppsala, its intellectual prestige was enhanced by the international reputations of five scientists who lived in the city. The first of these was Olof Rudbeck (1630–1702), the anatomist who discovered how the lymphatic system functions. He added a dome to the Gustavianum and built the anatomy theatre that can still be seen inside it; somewhat less scientifically, at least by modern standards, he also argued that Sweden was the lost realm of Atlantis. The second was Anders Celsius (1701–44), who followed his father and both his grandfathers as professor of astronomy at the university and established the centigrade measure of temperature that was later to be named for him. The third and greatest was Carl Linnaeus (1707–78), who developed the theory that plants reproduce sexually and created the system of classification, still used today, that assigns every animal or plant to a genus and a species, both indicated by Latin names. Although his specimens and manuscripts are no longer in Uppsala (they are preserved in London), the university's Linnaeus Museum, in the house he occupied until his death, and its botanical gardens, which he reorganized in 1741 and which are now also named for him, form a fitting memorial. Like Linnaeus, Torbern Bergman (1735–84) was an innovative classifier, initiating modern chemical analysis with his attempts to define elements and compounds (which helped Jöns Berzelius to devise the periodic table in 1818). He was also the sponsor and research partner of Carl Wilhelm Scheele (1742–86), an apothecary who neither taught nor studied at any university but who improved the refining of iron and discovered manganese, glycerin, and chlorine.

In their own day, such scientists were decidedly in a minority within the university. In 1700, 57 percent of all university students in Sweden (then including Finland and Estonia) became Lutheran priests in the only legally permitted church. By the middle of the nineteenth century, however, only 20 percent were doing so. The university was given its own representative in the *Riksdag* (parliament) in 1823, but the house of the clergy in which he sat as a member of the "second estate" was abolished along with the other three estates in 1866, in a reform initiated by an Uppsala graduate, Louis Gerhard de Geer. Most legal restrictions on freedom of religion were removed seven years later.

The changing and growing university needed new buildings. The Carolina Rediviva, which stands to the south of the Gustavianum at the foot of the road up to the Castle, was designed by C. F. Sundvall and completed in 1842. Its Latin name refers both to King Charles XIV John and to its intended role in renewing the university by rehousing its main offices. Instead it houses its library, in which the most precious object is the sixth-century Codex Argenteus, a Gothic version of the New Testament, written with silver instead of ink. The main offices were eventually moved into the New University building, still farther south and higher up the Castle hill, which was opened in 1886 where the archbishops' palace had once been; it contains the only portrait of Gustav Vasa painted in his lifetime.

To a certain extent the university, traditionally the servant of church and state, now became a center of liberal thought. This trend began with such figures as Eric Gustaf Geijer, the professor of history, whose abandonment of conservatism in 1838 startled his colleagues both in Uppsala itself and in Stockholm, where he was the university's Riksdag representative. It was maintained well into this century by Verdandi, a student society founded in 1882 by Karl Staaff, later a Liberal Prime Minister. Its early members also included Hjalmar Branting, who was to be a founder of the Social Democratic Party, its first Riksdag member and its first Prime Minister.

The population of Uppsala rose during the nineteenth century from 5,100 at its beginning to 23,000 at its end; it has since risen to more than 150,000, making it Sweden's fourth-largest city (after Stockholm, Göteborg, and Malmö). Most of this growth has taken place on the east bank of the Fyris River, where factories, offices, and railroad stations make for a very different ambience than that of the ecclesiastical and academic buildings on the west bank. The film director Ingmar Bergman, who was born in Uppsala in 1918, has used the historic quarter both in his own film *Fanny and Alexander* and in *The Best Intentions,* directed by Bille August from a screenplay by Bergman based on his parents' lives there. Both films are set in the early years of this century, but surprisingly little has changed in the quarter since then. Now as then, Uppsala is a lively city in the midst of which the Cathedral, the University and the Castle commemorate Sweden's religious leaders, many of its most influential scholars and the kings and queens who sponsored them, just a few miles from the mysterious remnants of earlier rulers and an older faith.

Further Reading: The development of Gamla Uppsala and of modern Uppsala within the larger history of Sweden can be traced in Franklin D. Scott's comprehensive *Sweden: The Nation's History* (Minneapolis: University of Minnesota Press, 1977) and in Eric Elstob's entertaining and well-illustrated *Sweden: A Traveler's History* (Woodbridge, Suffolk: Boydell Press, and Totowa, New Jersey: Rowman and Littlefield, 1979), both of which, interestingly, treat Adam of Bremen's account of Gamla Uppsala as reliable. Michael Roberts, the doyen of Swedish studies in English, describes and explains the Swedish Reformation in *The Early Vasas: A History of Sweden, 1523–1611* (Cambridge and New York: Cambridge University Press, 1968). Ingmar Bergman's screenplay for *The Best Intentions* is available in an English translation by Joan Tate (New York: Harper, 1993).

—Patrick Heenan

Utrecht (Utrecht, Netherlands)

Location: Twenty-one miles south of Amsterdam.

Description: Capital of the Province of Utrecht; headquarters of powerful medieval bishops; site of a Gothic cathedral, a university, and museums of medieval art, rail transport, and mechanical musical instruments.

Site Office: VVV Utrecht
Vredenburg 90
3511 BD Utrecht, Utrecht
Netherlands
06-34034085

Two thousand years ago the coastal region to the north of the Rhine was inhabited by a tribe known to historians as the Frisians. For a few years they were drawn into the Roman Empire, but in A.D. 16 the Roman soldiers were withdrawn and twelve years later the Frisians drove out the remaining officials. In A.D. 48, during an unsuccessful campaign to regain the region, Roman soldiers set up a camp near the spot where the Domkerk (cathedral) of Utrecht would be built. They left behind them the place name Trajectum ad Rhenum (Ford over the Rhine), from which, via Oude Trecht (Old Ford), the city's name derives.

Utrecht does not reappear in the historical record for nearly 500 years. In the meantime the Frisians had been overrun by the Angles, who were on the way from Germany to their invasion of England, and a merged Anglo-Frisian people had created a kingdom along the North Sea coast. Utrecht, lying on its contested frontier with the Christian kingdom of the Franks, was the site of a Frankish church, built around 500 but demolished around 630 by the Frisians. It was then the site of a Frankish fortress that became, in 695, the base for a very different type of Roman soldiers, the missionaries of the Catholic Church. While the Franks set about the military conquest of Frisia, between 720 and 776, the two successive archbishops of Frisia, Saints Willibrord and Boniface (both English), began a religious conquest, starting with the foundation of a monastery and a school in Utrecht. The Archdiocese of Frisia was redesignated the Diocese of Utrecht after Boniface was murdered by pagans in 754; its Bishops were still engaged in rooting out pagan beliefs 200 years later.

Their task was made even more difficult by the fragility of the political structures in the region, for the Frankish empire created by Charlemagne was divided after his death in 814. From 843 to 870 Charlemagne's grandson Lothair, and then his son Lothair II, were kings of Middle Francia, which included modern Belgium, the Netherlands, Luxembourg, Alsace, and Lorraine (which is named for them); then, from 870 to 925, these lands were disputed between the West and East Franks. In practice they were parceled out among local lords and also subjected to repeated incursions by the Vikings, who destroyed Utrecht and expelled the bishop around 900. For about 125 years afterward, the bishops lived at Deventer, not at Utrecht itself.

In 925 these "middle" lands became part of East Francia, and in 953 King Otto I granted them to the archbishop of Cologne, his brother Bruno, who created two duchies out of them. Once Otto had been made Holy Roman Emperor in 962, it fell to the bishops of Utrecht to try to bring the Duchy of Lower Lorraine, including the modern Netherlands, fully under imperial authority. The emperors confirmed the bishops' control over the province of Utrecht and, between 1024 and 1040, granted them the provinces of Overijssel, Gelderland, and Drenthe and the town of Groningen, making them the most powerful lords in the Netherlands. However, their support of the emperors against the victorious popes in a dispute over control of church appointments cost them their remaining influence in Zeeland and Holland, where the local lords submitted to the counts of Flanders instead and Gelderland, which had the crucial advantage of control over the Rhine delta, was seized by their rivals the Counts of Gelders. Throughout the large areas left to them after these losses in the early twelfth century, the bishops sponsored the diking and damming of the rivers, leased out land reclaimed from the sea, and tried to arbitrate in the disputes among their rural vassals and among the growing merchant towns. In Utrecht, Bishops Adelbold, Bernold, and Konrad saw to the construction of the Domkerk (cathedral) and four parish churches between about 1025 and about 1080. These formed a cross, with the Domkerk, dedicated to St. Martin, in the middle, the Janskerk to the north, the St. Pieterskerk to the east, the Mariakerk to the west (it was pulled down in the nineteenth century), and the Pauluskerk to the south. These were in addition to the city's oldest church, the Buurkerk, first built in the tenth century and often rebuilt. (Since 1984 it has housed the remarkable and delightful Rijksmuseum van Speelklok tot Pierement, the national museum of music boxes, street organs, and other mechanical instruments. From 1080 to 1250 there was also an imperial palace in Utrecht, from which Emperor Henry V gave the city its charter in 1122.

Thus endowed with their own local administration, though still subordinate to the bishops, the city's merchant leaders founded the Vismarkt (fish market) in 1196 and about four years later completed the Oude Gracht, the first of the medieval canals that still run through Utrecht, connecting it to Zeeland on the west coast and to the mouth of the Elbe on the Baltic Sea. Utrecht became the most important trading city in the Netherlands, exchanging fish and salt for goods from Germany, England, and Sweden. But it declined during

Utrecht's cathedral
Photo courtesy of Netherlands Board of Tourism

the thirteenth century as the towns in Overijssel found prosperity through membership of the Hanse, the association of merchant towns that came to dominate trade around the Baltic.

Like the rulers of other Dutch provinces, the bishops found themselves less and less able to suppress the customary rights of their subjects. They were frequently forced to flee from peasant or urban revolts, and one of the bishops was even murdered by local farmers when visiting his northern lands in Drenthe. In the late fourteenth century another bishop, Jan van Arkel, who had fallen deeply in debt to the counts of Holland and Gelders, was expelled from the city by its council, which was elected each year by the merchants' gilds. It proceeded to take charge of the rebuilding of the Domkerk, begun in 1254 and not completed until 1517; the Domtoren, at 312 feet still the highest church tower in the Netherlands, was completed between 1321 and 1383.

By 1456 marriage, inheritance, and purchase had made Duke Philip II of Burgundy and Flanders the master of all the provinces that are now Belgium, the Netherlands, and Luxembourg—including Utrecht and its territories by ensuring that his illegitimate son was made bishop. By the time of the death in battle of his legitimate son Duke Charles (the Bold) in 1477, the provinces were loosely united under a States General, which successfully asserted its rights against Charles's daughter Mary, her husband Maximilian of Austria, and their son Philip III.

In 1506 Philip's son Charles became duke in his turn. In addition Charles inherited Spain, Naples, and America and was elected Holy Roman Emperor, as Charles V. The bishops of Utrecht were now confronted not only by the claims of their subjects but by those of their overlord, the most powerful individual in Europe. From 1522 to 1523 his former tutor Adriaen Florisz, born in Utrecht, was pope, as Adrian VI, but in 1527 Charles's troops invaded Rome, imprisoned Adrian's successor Clement VII, and secured his blessing on Charles's purchase of all the territories governed by Henry of Bavaria, the last of the prince-bishops. Charles then had a fortress, the Vredenburg, built in Utrecht and founded the convent that, since 1979, has housed the medieval art and other treasures of the Rijksmuseum het Catherijneconvent.

Emperor, bishop, and pope alike were already being challenged by the Protestant Reformation. It was in Utrecht in 1525 that Willem Dirks and Jan de Bakker became the first of at least 223 Dutch Reformers to be executed during Charles's reign. The persecution of Protestants continued under his son King Philip II of Spain, who decided in 1559 that the bishops of Utrecht and two other cities would become archbishops, that he himself would nominate and pay them, and that they would automatically sit in the provincial states assemblies. These proposals to integrate religious and political authority served to promote what became an equally integrated religious and political rebellion over the course of the Eighty Years War between Spain and the Netherlands, from 1568 to 1648.

In 1577 the transfer of Spanish troops from Utrecht to other rebel centers gave the townspeople their opportunity to pull down the Vredenburg. In 1579 representatives from most of what is now the Netherlands, including William, prince of Orange, the leader of the revolt, gathered to sign the Union of Utrecht in the Domkerk's chapter-house (now part of the university) and proclaim it at the Stadhuis, the city hall. This document was to serve as the constitution of the Dutch confederation for more than 200 years; it was also to be the model for the Articles of Confederation which preceded the U. S. Constitution. The Reformation in Utrecht culminated in 1584 with the secularization of all the buildings and lands of the Catholic Church.

Between 1586 and 1588 Utrecht was taken over by a group of devout Calvinists who wanted to push the process further. They supported the English earl of Leicester, who had been sent to aid the rebels and had been declared governor-general of the Netherlands, but he found little endorsement elsewhere and returned to England, leaving the Utrecht Calvinists isolated and easy to topple. The national revolt came under the leadership of Johan van Oldenbarneveld, advocate of Holland, and Maurice of Nassau, who had followed his father William the Silent as stadholder (military commander) of five of the seven United Provinces; they secured recognition from France and England in 1596 and the expulsion of the last Spanish troops in 1597.

It was against this background that the Calvinists in 1610 made another attempt to take over Utrecht. They and their cobelievers in other cities turned to Maurice, who had already taken up the orthodox Calvinist cause against the Arminians, or Remonstrants, who argued that God's determination of human destinies could be conditional on human acts. The religious controversy was quickly transformed into a political crisis over Oldenbarneveld's policy of toleration in Holland. In 1618, after the failure of negotiations in Utrecht, Maurice arranged to have Oldenbarneveld tried for treason: there were Utrecht Calvinists among the judges who condemned him to death.

For the Netherlands as a whole the seventeenth century was an era of colonial expeditions and wars with England, Portugal, and France, while struggles over the distribution of powers continued among the States-General, the stadholders, and the provinces. For Utrecht it was an era of economic development; the city became famous for its velvet cloth, then known as *velours d'Utrecht,* and for its university, established in 1636. This Golden Age ended with the European wars initiated by Louis XIV of France. From 1672 to 1673 Utrecht was occupied by French troops. Order was restored by appointees of the stadholder, who set aside the guilds' right to choose councillors. The storm of 1674, which destroyed the nave of the Domkerk, leaving a gap between the cathedral and its tower, and also damaged the Pieterskerk and the Janskerk, may have seemed just one more disaster. From 1690 to 1697 the Dutch and their allies fought Louis again; the War of the Spanish Succession, from 1702 to 1713, brought victory for the alliance against Louis but bankruptcy for the Netherlands, and was followed by peace talks and the signing of the Treaties of Utrecht in the Stadhuis in 1713 and 1714.

There followed seventy years of relative decline, during which the Dutch lost their leadership in trade and colonization to the British, who had taken Gibraltar, Newfoundland, and Nova Scotia from the French under the treaties. Many citizens of Utrecht, aggrieved at the provincial States Assembly's lack of interest in their problems, were increasingly attracted to more radical solutions and helped to revive the traditional militias, which quickly became a national network of Free Corps or Patriots, meeting in Utrecht from 1784 and issuing demands for greater democracy. In 1786 they forced the city council to resign and replaced it with citizens' delegates. Similar revolts in Amsterdam and elsewhere gave the Patriots control of the States Assembly of Holland. The stadholder William V ultimately had to rely on the army of his brother-in-law, the king of Prussia, to restore order in 1787 and 1788. Six years later the Patriots welcomed the French revolutionary invasion, which was followed in 1795 by the creation of a puppet Batavian Republic, in 1806 with the installation of Napoleon's brother Louis as king, and in 1810 by absorption into the French Empire. In 1815 the European powers recognized William V's son as King William I.

The nineteenth century saw the decisive transformation of Utrecht, beginning in 1818 with the abolition of the guilds and the freeing of market forces. Railroads soon connected the city to its suppliers and customers all over Western Europe. The Centraal Station still functions, but the Maliebaan Station, built in 1874, is now the Nederlands Spoorwegsmuseum, exhibiting both trains and streetcars. Liberalization also extended to religion: since 1851 there have been Catholic archbishops in Utrecht once again.

The Netherlands was neutral from 1815 to 1940, when the Nazis invaded and Queen Wilhelmina and her government went to London to organize resistance. Utrecht was the home of Anton Mussert, the leader of the Dutch Nazi party, which had little real power and disintegrated in September 1944, when the Allied advance panicked most of its members into fleeing to Germany. Mass starvation came to the Netherlands during the "Hunger Winter" of 1944–45 but was followed at last by liberation, celebrated in Utrecht by the ringing of the Domtoren bells on the day the Germans surrendered.

The face of the city—now, with about 300,000 inhabitants, the fourth largest in the Netherlands—was radically altered in the 1970s by the construction of the Hoog Catharijne, then the biggest shopping center in Europe, which, like some alien spaceship, confronts visitors trying to find their way from the railroad station, the bus station, or the parking lots in the west to the historic buildings in the east. The restoration and conversion of some of these into superb new museums has further enhanced the attractiveness of a city that for centuries was a focal point, not of shopping, but of European religion, trade, statecraft, and, occasionally, revolution.

Further Reading: Utrecht figures prominently in English-language histories of the Netherlands, among which G. J. Renier's *The Dutch Nation: An Historical Study* (London: Allen and Unwin, 1944), Bernard H. M. Vlekke's *Evolution of the Dutch Nation* (New York: Roy, 1945; London: Dobson, 1951), and the writings of Jan Huizinga and Pieter Geyl still stand out as uncondescending and thought-provoking. Sherrin Marshall's *The Dutch Gentry, 1500–1650: Family, Faith and Fortune* (Westport, Connecticut, and London: Greenwood, 1987) combines unconvincing statistical exercizes with interesting anecdotes. Simon Schama's *Patriots and Liberators: Revolution in the Netherlands, 1780–1813* (London: Collins, and New York: Knopf, 1977; second edition, London: Fontana, and New York: Random House, 1992) is, like his other books, informative but also verbose.

—Patrick Heenan

Versailles (Yvelines, France)

Location: Approximately fifteen miles west of Paris.

Description: The enormous palace of Louis XIV, the Sun King; renowned for its lavish architecture and decoration as well as its immense gardens and beautiful fountains.

Site Office: Musée et Domaine National du château de Versailles et de Trianon
78000 Versailles, Yvelines
France
30 84 74 00

Versailles today is a town of more than 95,000. The area originally developed as a dependency of the famous palace from which it has taken its name. However, when most people hear the name Versailles they think not of a small suburban area outside Paris, but of an expansive estate with palaces, gardens, and fountains; perhaps they even envision Louis the Sun King or Marie-Antoinette.

What was once the modest hunting lodge of Louis XIII became the most lavish and ostentatious of all European royal residences during the reign of his son, Louis XIV. It continued as the royal residence until the French Revolution, after which it fell into disuse and disrepair. Only during the past century have serious efforts been made to restore the palace to some of its eighteenth-century splendor. Now visited by more than 4 million people a year, Versailles is among the most important tourist attractions in all of France.

Mark Twain once visited the site and was so overwhelmed that he wrote, "You gaze, and stare, and try to understand that it is real, that it is on earth, that it is not the Garden of Eden. . . ." But Versailles did not come by its stunning appearance at all naturally; in fact, it was created through the efforts of thousands of workers and the expenditure of untold fortunes.

Louis XIII died in 1643, leaving a five-year-old heir to the throne. The queen regent, Anne of Austria, and her prime minister, Cardinal Mazarin, ruled France and saw it through a terribly tumultuous period that was to leave the young monarch, Louis XIV, with a strong distrust for life in Paris after the royal residence was invaded and his family forced to flee. The monarchy managed to survive this period, and when Louis reached his majority, he decided to settle outside the city, first at Tuileries and then at Saint-Germain-en-Laye.

Louis became involved with a woman named Louise de la Valleire, and when she became his mistress he housed her at the château of Versailles, which was not far from Saint-Germain-en-Laye. Through the influence of one of the royal architects, Louis approved a plan for the expansion of Versailles in 1668 and moved his residence there in 1682, even before the construction was completed. As many as 36,000 laborers and soldiers were involved in the construction project under the direction of Jules Hardouin-Mansart. Mansart's name has become synonymous with Versailles and particularly a certain type of roof, known as the Mansart roof, which is used throughout the compound. The modest château of Louis XIII thus became the palace of Louis XIV, with the west facade extending more than 2,000 feet. The work on Versailles gave new vitality to French architecture, and the style of Versailles was repeated throughout Europe as others tried to imitate Louis and his courtly lifestyle.

Versailles is a symmetrically balanced structure. The classical influence is felt not only in the architecture but also in the formal French gardens that were patterned after the formal gardens of Italy. Great care was taken to see that all the elements of design in the palace and the grounds worked together in harmony. Later this same attention was given to the creation of furnishings for the royal residence.

With the growth of Versailles, a more refined lifestyle came to the French court. Greater importance was placed on customs and manners, and each person knew his precise place in relation to the king. In order to tighten his control over the nobles, Louis enforced a strict social code and raised the prestige of the court to new levels. This was the high point of the French monarchy, and the purported divine right of Louis was embodied in Versailles itself. The king chose the sun as his emblem and Versailles became the palace of the sun, with corresponding mythological references to Apollo and Ovid's *Metamorphoses*.

With as many as 10,000 people living at Versailles, Louis demanded that his courtiers live in unheard-of luxury. In order to promote French trade, he also outlawed the importation of certain luxury goods such as mirrors from Venice and tapestries from Flanders, thus assisting the growth of French manufacture.

All the pomp and circumstance at Versailles awed the other royal courts of Europe, but still it was impossible to hide the overcrowding, disorganization, and filth that plagued the palace. It is no wonder that the king would often take leave of the estate to visit one of his smaller royal retreats.

Versailles remained central to France and indeed to all of Europe throughout most of the eighteenth century. In 1774 Marie-Antoinette, the most famous resident of Versailles after Louis the Sun King, became queen of France at the age of 19. With the monarchy already weakened, the lavish lifestyle of Marie-Antoinette and her never-ending projects at Versailles pushed the situation to the breaking point. She had hills and lakes constructed as well as a miniature Austrian village complete with farm. Her projects and expenditures

The Château de Versailles
Photo courtesy of French Government Tourist Office

not only alienated the masses, but also made her unpopular at Versailles itself. Marie did not enjoy the public life of the main palace and took up residence in a separate building known as the Petit Trianon. She also spent a great deal of time living in luxury accommodations at her miniature village. Her husband, Louis XVI, made little effort to control his wife, and she became farther and farther removed from life in Versailles and spent nearly all of her time living outside the main palace with a group of close friends at the Petit Trianon.

Marie was disliked by others in the court and was detested by the commoners in France, who blamed her for many of the country's ills and found her and her life at Versailles to be symbolic of all they wanted to change. Not long after the storming of the Bastille in 1789, an angry mob marched on Versailles and took the royal family to Paris, where the king and queen were both executed by guillotine in 1793.

After the storming of Versailles and the start of the French Revolution, the imperial palace was essentially abandoned. Feeble efforts were made by both Napoléon I and Louis XVIII to restore the palace as a royal residence, but the immense costs involved precluded any future use of Versailles as a residence. When Louis-Philippe became the last king of France, he proposed that Versailles be turned into a museum dedicated ''to all the glories of France.'' It was probably this action that saved Versailles from complete ruin, but in executing the plans to make the palace a museum, much of the interior was destroyed. Walls were torn down, doorways and fireplaces moved, and murals painted over. Luckily, Louis-Philippe's architect made careful drawings of each of the rooms before they were changed. The museum was dedicated in 1837 but the actual structures at Versailles continued to be neglected.

In spite of the alterations and neglect that Versailles had suffered, it continued to be an inspiration for wealthy individuals throughout Europe, who attempted to copy its style in their own palaces and châteaux. Even in its dilapidated state, Versailles was used from time to time to entertain visiting dignitaries. It also suffered during times of war: in 1870 the palace was used as the headquarters for the Prussians during the Franco-Prussian War, and it was there that King William of Prussia was named emperor of a united Germany. The following winter the palace was used as a hospital. Later that century the French government was forced to flee Paris and ultimately set up operations at Versailles.

Plans for renovating and restoring Versailles were put on hold by the two world wars, but it was at Versailles that the treaty ending World War I was signed. Early attempts to start renovations moved very slowly because of lack of funding.

Much of the preliminary research for the renovation was already complete by the end of World War I. It had been done by a man named Pierre de Nolhac, who came to work in the museum at Versailles and was its curator from 1892 to 1920. Through his meticulous research, Nolhac was able to recreate the original plans of the palace as well as the uses of many of the rooms that had been converted into museum space. He even discovered original decorative plans and set about the restoration of the palace. Nolhac's writing and lecturing about Versailles did much to create public interest in the site and helped raise funds for its restoration. Even the American John D. Rockefeller contributed $700,000 toward the cause.

It was not until after World War II that restoration at Versailles really began on a large scale. The Conservation Law of 1953 was instrumental in getting things started, and with additional funding and revenues, significant restoration and rebuilding has occurred, much of it since 1970. A visitor to Versailles today will see portions of the palace as they have not been seen in more than 100 years. The gardens and outbuildings have also undergone renovation, and the fantasy world of Marie-Antoinette is truly a sight to behold.

Versailles is perhaps the ultimate symbol of the power of the French monarchy and the conception of the Divine Right of Kings. Millions of visitors are attracted to this site every year to experience the lavish setting of the French court during the eighteenth century. They come to see the ultimate in power and prestige in this museum that is devoted ''to all the glories of France,'' but perhaps what they really see is the corruption of wealth and privilege and the somewhat hollow life of the eighteenth-century French elites. While Versailles may indeed represent a glittering page in the history of mankind, it is harder to determine exactly where such decadence fits in the history of humanity. That the palace is still used by French heads of state to receive world leaders and that important summits are held at the site is an interesting commentary on its role as a uniquely French symbol. That it has somehow become a national landmark cannot be attributed simply to the royal history of the place, but perhaps instead to the opulence and draftsmanship of this, the greatest imperial palace in European history.

Further Reading: Versailles has been widely studied, photographed, and sketched. A wonderful introduction to the site is *Versailles* by Christopher Hibbert (New York: Newsweek, 1972; London: Reader's Digest, 1975). This text is well illustrated and quite complete in its purview. A beautiful oversized text filled with color photos is Jean-Marie Perouse de Monclos's *Versailles* (New York: Abbeville Press, 1991). Gerald Van der Kemp's *Versailles* (New York: Park Lane, 1981) concentrates on some of the restoration that took place during the 1970s.

—Michael D. Phillips

Vézelay (Yonne, France)

Location: In north central France, about 135 miles to the southeast of Paris, Vézelay stands on the left bank of the north-flowing River Cure and at the southern edge of the department of Yonne. Yonne constitutes the northwest quarter of the administrative region—once the kingdom, then the dukedom, then the province—of Burgundy. Vézelay is reached by the D36, N151, N457, and N458 roads.

Description: A village of approximately 575 inhabitants, known as Vézéliens, Vézelay is famous for the beauty of Sainte-Marie-Madeleine, its largely Romanesque, partly Gothic basilica, the earliest remaining parts of which date from the twelfth century. Vézelay's history, however, goes back much farther. The original local Christian settlement grew up in the ninth century a mile to the southeast of Vézelay, on the site of the village of Saint-Père-sous-Vézelay. Two and one-half miles farther southeast, at Fontaines Salées, remains can be seen of Iron Age and Gallo-Roman constructions centering on natural saline springs.

Site Office: Office de Tourisme, Syndicat d'Initiative
Salle Gothique
rue Saint-Pierre
89450 Vézelay, Yonne
France
86 33 23 69

The focus of interest in Vézelay has gradually moved upward over the centuries, from the salt springs in the valley to the Celtic and Gallo-Roman baths built around them, then to the first abbey still down by the River Cure, then up to the abbey on the steep hill, with its church, and the town that grew about it. There is no abbey now, but the abbey church—today called the basilica—of La Madeleine shines from the top of its rocky spur over a broad calm sea of oaks and beechwoods, fields and meadows. It is a dazzling landmark, once for pilgrims on their way to Vézelay itself and often on to Compostella, now mainly for tourists. With the mighty *abbatiale* of Cluny gone, destroyed at the beginning of the nineteenth century, the Vézelay basilica is the most magnificent of the remaining Romanesque churches of Burgundy. It represents a past in which Vézelay was one of the centers in France of active, militant, and even militarist piety.

The immediate district had long attracted settlements. There was a Celtic stronghold, called Vercellas or Vercelai, on the hilltop. Evidence of a Celtic site in the neighborhood of present-day Vézelay exists at the spot to the southeast now suitably called Fontaines Salées. Excavations begun there in 1934 showed that the water from still-existing saline springs,

collected in wells made from hollowed-out trunks of oak trees and led along oakwood channels, had begun to be used for medicinal baths from about 600 B.C., or perhaps a century or two earlier, in the first (Hallstatt) Iron Age. By the first century B.C. the complex formed the heart of a large circular sanctuary dedicated to the supposed god of the springs. Fragments of these installations and other Iron Age remains, in the shape of salt-workings and a burial-ground, can be seen at the site, while others are kept in the museum, the former presbytery, at Saint-Père-sous-Vézelay, next to the beautiful late Gothic church, which was built in the thirteenth, fourteenth, and fifteenth centuries.

These same springs were utilized in the second century A.D., when the Romans built elaborate thermal baths over the old Celtic sanctuary. Remains of these baths can be seen near the springs. Gallo-Roman jewelry and other personal objects discovered on the site, known as Vizeliacum, are preserved in the Saint-Père-sous-Vézelay museum.

With his wife Berthe, who is commemorated with him in carvings in the church at Saint-Père, Girart de Roussillon, count of Lyons, regent of Provence, and hero of a famous twelfth-century *chanson de geste,* founded a convent for nuns at Saint-Père between 855 and 859. The place was subject to attacks by the Normans; this factor led to decisions to found a male Benedictine community and, in 887, to move the community up the hill to a new abbey. The monks, whose charter made them independent of the local spiritual and temporal powers, were in constant dispute with the bishops of the diocese, Autun, over church affairs; with the counts of Nevers, descendants of Girard de Roussillon, over their property; and with the townspeople, whose taxes and labor they roughly commandeered.

The church at Vézelay was originally dedicated to the Virgin Mary and called Notre-Dame, but in the early eleventh century, by the routine distribution of holy relics to churches, Vézelay acquired bones reputed to be those of St. Mary Magdalene, a traditional figure combining the identities of three different Biblical Marys: Mary of Bethany, the sister of Lazarus and Martha; the unnamed sinful woman in St. Luke who washed Jesus' feet with her tears; and Mary of Magdala, one of the first witnesses of the Resurrection. As one dear to Jesus, as a penitent sinner privileged to be among the first few to see the risen master, she was a potent symbol of erring humanity and its hope of salvation. On the strength of the rumor about the relics, the church was rededicated in 1050 and became Sainte-Marie-Madeleine or La Madeleine. For more than two centuries it was a great center of pilgrimage on account of its patroness and the favors and miracles produced at her shrine. Her feast day is July 22.

Vézelay was also an assembly point for pilgrims on the long road to the shrine at Compostella of St. James the

Tympanum at Sainte-Marie-Madeleine, Vézelay
Photo courtesy of The Chicago Public Library

Great. His mother, according to legend, had been one of the group of Christ's followers who escaped with Mary Magdalene from Palestine and took refuge in Provence. Lodging houses, shops, and hostelries were needed to serve these travelers. At the height of its fortunes in the Middle Ages, Vézelay had 10,000 inhabitants. The relics at La Madeleine had been authenticated by Rome, and the belief grew that the church at Vézelay had the entire remains of St. Mary Magdalene, but the church of St. Maximin in Provence had always maintained a similar claim. When St. Maximin's claim was given papal recognition, in 1295, Vézelay's fame declined.

The present church was constructed mainly in the twelfth century. The influence of Cluny was strong on what is, and typifies, an edifice designed in the first place for a monastic community rather than a town parish. At the end of the eleventh century the Carolingian abbey church on the site had seemed to the abbot, Artaud, insufficiently splendid, and from about 1096 he had begun to replace it; but he died in 1105 at the hands of the townspeople, who were oppressed by his taxes. The work of his successor, Abbot Renaud, was interrupted in 1120 by a fire that destroyed the old Carolingian nave. By about 1140, however, the new Romanesque nave was finished. Over the next twenty or so years the great enclosed porch, or narthex, was added to the west end of the building. With the construction, begun about 1185, of the new choir in the Gothic style, the rebuilding begun by Abbot Artaud was complete.

At Easter 1146, in the presence of the French king Louis VII and his wife Eleanor of Aquitaine, later the queen of Henry II of England, St. Bernard of Clairvaux, the "mellifluous Doctor," reformer of the Cistercians, preached a sermon at Vézelay in favor of the Second Crusade (1147–49), against the Turks. He had to speak out of doors because the church was not large enough to hold the huge and enthusiastic throng; a large wooden cross now marks the spot, less than a mile north of the town ramparts. Nearby is the twelfth-century chapel of Sainte-Croix, probably built by the masons who added the narthex to the *abbatiale*. It is known as La Cordelle because there was a community of Franciscan monks, called *cordeliers* for their cord belts, who lived there from 1217; they were massacred during the sixteenth-century religious strife, but successors came back in 1946. Kings Philip II (Philip Augustus) of France and Richard I, the Lion-Hearted, of England met in Vézelay before starting for the Third Crusade (1189–92), and King Louis IX of France—St. Louis—prayed there before leaving on the Eighth (his second) Crusade (1270), on which he died without reaching the Holy Land.

In the thirteenth century, not long, that is, after the new building had been finished, the fortunes of the abbey community, which was still quarreling with the people, had begun to waver and fall away, as did those of the church, with its relics demoted, and those of the dependent town. In 1537 the Benedictine monks were replaced by clergy not bound to monastic rules, and the church was then run by a chapter of canons. During the religious civil wars that wracked France between 1562 and 1598, Vézelay, birthplace of the Calvinist writer and reformer Théodore de Bèze, whose house may still be seen there, contained many Protestant inhabitants, some of whom attacked and damaged the church in the 1560s.

During the French Revolution, the canons were dismissed, practically all the remaining abbey buildings were demolished, and the abbey church became the property of the nation, to be eventually sold off. Much further damage was done to the church and its ornamentation. Years of neglect and a fire in 1819 reduced La Madeleine to a near ruin.

In 1840 the head of the national Commission for Historic Monuments, Prosper Mérimee (author of *Carmen*), appointed Eugène-Emmanuel Viollet-le-Duc, the architect and archaeologist, to restore the building. In twenty years he completed his mission. Some say he went beyond the line of duty, replacing, altering, and inventing instead of doing only the minimum necessary to keep the edifice standing in a form true to the work of the original artists. However, his many meticulous drawings and watercolors for the project, conserved in the former monks' dormitory, now the Musée Lapidaire, bear witness to his skills and seriousness; and the untouched great tympanum of the narthex, with other elements surviving in their original state in and around the basilica and in the Musée Lapidaire, provide a base for reassuring stylistic comparisons. La Madeleine is once again among the brightest jewels of French architecture.

The main direct influences on the design and construction of La Madeleine were, for the Romanesque parts, the builders of the church at Anzy-le-Duc, in the Brionnais part of Burgundy where Abbot Renaud grew up, and, for the Île-de-France Gothic of the choir, Suger's abbey church at Saint-Denis. The exterior of the Vézelay church is sometimes disparaged as uninteresting. It is true that the flying buttresses may seem heavy, and the assortment of roof levels capricious. The west facade, in most great churches a showpiece, may inspire reservations: the large statues, one representing Mary Magdalene, standing above the five lancets of the central gable, are replacements; the carved Last Judgement of its tympanum is also a nineteenth-century copy of dubious quality; and its towers are asymmetrical. When Vézelay is seen on its hilltop by an approaching traveler, however, the basilica's delicate variations of color, shape, and texture are enhanced by its dramatic and picturesque situation; and symmetry after all is not everything. The general effect is awe-inspiring.

The twelfth-century narthex was constructed a little later than the nave, to accommodate, even perhaps as a lodging, large numbers of worshipers and pilgrims. It is the same width as the body of the church into which it leads, and has three bays, three aisles, and a triforium. There is a narthex at Saint-Père, too, dating from the fourteenth century.

The famous main feature of the Madeleine narthex is the three-pillared double doorway opening on to the nave. This entrance is surmounted by a magnificent carved tympanum elaborately representing a theme in keeping with the story of St. Mary Magdalene, the church's patroness, and with the Christian mission—the spreading through the world of the gospel of redemption. From the central figure of Christ in majesty the Holy Spirit radiates toward the apostles, beneath his hands, and toward the actual and potential human recipients of their good news, depicted all around. The effect of the rushing mighty wind of Pentecost is indicated in the harmoniously swerving lines of Christ's draperies, and in those of the apostles and saints and even of the tinier personages in the semicircular tympanum and the supporting columns. To convey the impression of movement, in fact, seems to have been been a major preoccupation of the team of sculptors, possibly including Gislebertus of Autun, engaged to decorate Romanesque Vézelay.

Along the lintel over the double doorway of the main narthex tympanum, a frieze of simplified but amazingly mobile-looking figures portrays, on the left, an exotic soldiery, and on the right, various grotesque legendary races—pygmies and giants and men with large ears. The line of feet at the bottom edge is particularly eloquent and charming. The innermost of three concentric arcs around the main subject is divided into twelve roughly trapezoid sections, in the first of which apostles or evangelists are shown writing; in the others appear further representations of real and legendary peoples to whom the church's mission may be directed. The second half-circle consists of medallions depicting the signs of the zodiac interspersed with human figures performing the labors of the months. The three roundels above the head of Christ, however, show a mermaid, a dog, and an acrobat, who pose one of the many problems of interpretation presented by Vézelay's carvings. The vividly lifelike effect produced by all these little sculpted personages is enhanced when they escape the convention of their circular frame by leaning or standing on it or poking out in front. The outermost of the three tympanum arcs consists of a repeated foliage motif.

Leading out of the narthex into the north and south aisles of the nave are two smaller doors, one on each side of the main tympanum and each with a sculptured tympanum of its own, that on the left (north) showing the Resurrection, the other the story of the Nativity.

One of the most striking visual experiences offered by the basilica is the chiaroscuro contrast that meets the eye looking east from beneath the central narthex tympanum into the body of the church. The nave, which has a small-windowed clerestory but no triforium, is built of a local limestone that shines pale pink, brown, or buff, even with gleams from the other end of the spectrum, at each change in the dim illumination. Overhead the tunnel of ten rounded transverse arches, banded brown and white, opens out and upward into the brilliant white limestone and airy white light of the plain-glassed Gothic choir.

The bays of the nave have quadripartite groin vaulting, and the capitals of the columns (including those in the narthex there are more than 100) are mostly decorated with figures in the lively style of the central tympanum. These sculptures—although it must be remembered that some are not the originals or are not in their original state—are extremely beautiful and of the greatest interest. The range of their subjects is wide, and includes Biblical (Old and New Testament), classical pagan, legendary, and imaginary figures and incidents, with a number of amiable-looking animals and fearsome but to modern eyes comical demons, but not the life of Christ and only once, on the south side of a column on the north side of the narthex, St. Mary Magdalene. Two examples, among many, of artistic virtuosity and symbolism are: in the narthex, the angel with the oliphant on the capital of the first pillar in the south corner of the south portal; and, in the nave, at the lower level of the fourth column on the south side, the two figures processing the Word of God through the so-called "mystic mill." Some capitals are adorned with conventional carvings of flowers and foliage.

The dark vaulted crypt beneath the choir, originally ninth-century Carolingian, still contains the once controversial relics; in its present form, which retains much of its early aspect, it dates from the eleventh and twelfth centuries.

The south transept of the choir leads into a nineteenth-century chapter house. A wood and stucco jubé, or screen, erected by the canons between the nave and the choir, has been removed. The cloister gallery was added by Viollet-le-Duc. Almost all the basilica's windows are now of clear glass. The terrace beyond the apse, which looks out over the countryside, is on the site of what was once the abbot's house.

The little town of Vézelay stretches along its ridge within quite well preserved medieval ramparts with remnants of towers and gates, notably the Porte Neuve to the northwest. Its oldest houses date from the fifteenth century and include the former abbey guest-house.

Further Reading: The essay "Some Great Churches in France: Vézelay" by Walter Pater (*The Nineteenth Century,* June 1894, reprinted in *Miscellaneous Studies,* London and New York: Macmillan, 1899) is a detailed, perceptive, but biased account, marked by an amusingly partisan slant against Romanesque architecture and in favor of Gothic; Pater, although grudgingly admitting the harmony of the whole, much prefers the choir of the basilica to the glories of its nave and narthex, which strike him as decidedly sinister. *Vézelay* (in French), photographs and selected texts by E. M. Janet Lecaisne, introduction by Victor-Henry Debidour (Paris: Plon, 1962) is excellent in all respects. In *Vézelay ou L'Amour fou* by Jules Roy (Paris: Albin Michel, 1990), there is much obliquely conveyed fact; the tone is pious; the style may strike some readers as fanciful and sentimental. "Vézelay" in *Blue Guide: Burgundy* by Ian Ousby (London: A. and C. Black, and New York: Norton, 1992) is a very compact and helpful account. *Burgundy: Landscape with Figures* by Peter Gunn (London: Gollancz, and Mystic, Connecticut: Verny, 1976); *Burgundy* by Anthony Turner and Christopher Brown (London: Batsford, 1977); and *The Companion Guide to Burgundy* by Robert Speaight, revised and expanded by Francis Pagan (London: Collins, 1975; revised, 1990) all have very useful sections on Vézelay and indexed references that place it in a historical and architectural context.

—Olive Classe

Vienna (Lower Austria, Austria)

Location: Northeastern Austria, on the banks of the Danube River.

Description: The capital and largest city of Austria, covering about 160 square miles; population is approximately 1.5 million people. Important center of music and culture.

Site Office: Vienna Tourist Board
Obere Augartenstrasse 40
A-1025 Vienna, Lower Austria
Austria
(1) 211 14-0

Although no one can know when people first settled the area that is now Vienna, it is certain that humans have lived there for millennia. A stone-age relic found near present-day Vienna, a limestone sculpture of a plump woman known as The Venus of Willendorf, was carved between 30,000 and 25,000 B.C.

Celts moved into what is now southern and eastern Austria between 500 and 400 B.C., building a settlement they called Vindomina. Salt—used to preserve as well as flavor food—was precious, and the region's salt mines were a source of wealth for the Celts. (Trade in locally mined ore and salt began before the Celts arrived, and stretched back to at least 800 B.C.) Vindomina's situation made it an ideal trading center. A gap in the Carpathian Mountains just to the east allowed traders to travel between Europe and Asia. The settlement stood beside another profitable trade route: the Danube, Europe's second longest river. Nor did the Celts trade only salt along these routes. Valuable amber from the Baltic and the North Sea traveled overland through their territory to Italy or via the Danube to Greece and Asia Minor.

Celtic rule over the area ended when Roman forces moved north into Austria. By 15 B.C., Rome controlled land south of the Danube. The Romans used Vindomina, which they renamed *Vindobona,* to control the Danube. The region produced salt, gold, iron, and wine. Trade continued to thrive: grain and amber traveled along the Danube, silk and spices arrived from Byzantium, and pottery and olive oil came from the west. In the second century A.D., the defense of Vindobona, whose residents were now considered Roman citizens, was supervised by Roman Emperor Marcus Aurelius, who during his stay there wrote his *Meditations.* The emperor died in or near Vindobona in A.D. 180. Later, Roman control over the region weakened as a result of invasions from the north; Germanic tribes finally conquered it; Rome itself fell to invaders in 476.

The region was conquered again toward the end of the eighth century, this time by Charlemagne. As king of the Franks, Charlemagne already controlled what is now France, Belgium, the Netherlands, Luxembourg, and parts of western Germany. He used Vienna as a border fortress to protect his empire's Ostmark (Eastern March, later Österreich) from attack. Charlemagne's empire broke up after his death in 814.

By the tenth century, the settlement was known by its present name of Wien, or Vienna, and controlled by the Magyars, a people who came from Hungary. Germany's Otto I defeated the Magyars in 955, and in 962 he was crowned emperor of what would become the Holy Roman Empire. Austria was to become the most important state—and its capital a leading city—of the Holy Roman Empire, which continued to be ruled by German emperors until its dissolution in 1806.

In 976, Emperor Otto II gave control of northeastern Austria, including Vienna, to Leopold I of Babenberg. During 270 years of rule, the Babenbergs fortified the region with towers and castles, established monasteries, and encouraged the production of gold, silver, and salt. Vienna was rebuilt. By now the city's language was German. Its history of repeated invasion gave it an ethnically mixed population, including the largest Jewish population in medieval Europe.

By the twelfth century, Vienna was once again the main trading post between Europe and Asia. In 1137 Vienna earned the status of *civitus,* or self-governing city. The *Ostmark* gained prestige when Emperor Frederick I named it a *duchy* in 1156. Duke Henry II moved his capital from Leopoldsberg, a castle in the Vienna woods, to the city itself. The city expanded beyond its Roman walls, which at the end of the twelfth century were extended to encircle most of the Inner Stadt (inner city). In 1186, the Duchy of Styria to the south came under Austrian control. As a capital and trade center, Vienna became the Holy Roman Empire's chief city after Cologne. Under Duke Leopold VI, a public hospital opened, and in 1221 Vienna's citizens received a charter of legal codes that protected privacy and gave women the right to inherit property.

The last Babenberg duke, Frederick II, died childless in battle against Hungary in 1246. King Otakar (or Ottokar) of Bohemia took control of the duchies of Austria and Styria. In 1273, however, German princes elected Rudolf von Habsburg (or Hapsburg) of Switzerland the Holy Roman Emperor. Five years later, Rudolf defeated Otakar in the Battle of the Marchfeld. The Habsburg family was to rule Austria, with Vienna as capital, until 1918.

During their first two centuries of rule, the Habsburgs expanded their territory toward the south and west. They lost control of the Holy Roman Empire during the fourteenth century, but in 1438 the Habsburg Albert II was elected

The spire of St. Stephen's
Photo courtesy of Austrian National Tourist Office

emperor. In 1453, the duchy of Austria became an archduchy.

In Vienna, a university had been founded in 1365. The south tower of St. Stephen's Cathedral was completed in 1439. With its 553 steps, it was regarded as a technological marvel. The Hungarian King Matthias Corvinus captured Vienna in 1482, but was driven out three years later by Maximilian, who would become emperor in 1493. Maximilian obtained the Spanish empire for the Habsburgs by marrying his son Philip to the Spanish princess Juana. Their son became Charles I of Spain and later Holy Roman Emperor Charles V.

In 1529 Turkey, with an empire that stretched from Morocco to Persia and included Hungary, laid siege to Vienna but failed to capture it. Even so, the Turkish empire encircled the Mediterranean and Black seas, disrupting trade routes and hurting the Viennese economy.

Meanwhile, Charles V was finding his huge territory too unwieldy to govern. He eventually divided it, giving the Spanish portion to his son and the Austrian to his brother Ferdinand. Ferdinand, who became Holy Roman Emperor in 1558, obtained Hungary and Bohemia through marriage, although Turkey still occupied much of Hungary. Ferdinand I patronized the arts, beginning an impressive collection of paintings that would be expanded by his descendants. However, he was to clash with his Viennese subjects, many of whom had joined Protestant religions during the Reformation. Ferdinand, determined to uphold the Catholic faith, began a period of religious repression. Protestants were burned at the stake or forced to flee. Habsburg rulers also failed to respect Vienna's legal charter and eventually became absolute monarchs.

In 1618, Protestant Bohemians rebelled against Habsburg rule. They were defeated in 1620, but their uprising began the Thirty Years War, in which most of the European nations participated. The war was ended by the Peace of Westphalia in 1648, which stated that a nation's religion could be decided by its monarch.

In 1679 Vienna was devastated by plague, which cost 70,000 lives. Four years later, the city was again attacked by Turkish soldiers. The reigning emperor, Leopold, fled with his wife and 60,000 others. The Turks' attempts to breach the city walls by digging tunnels beneath them failed. The King of Poland, John Sobieski, came to the aid of Vienna and helped the Austrian Duke of Lorraine defeat Turkey. Austria's victory was seen by European rulers as a victory for Christianity over Islam.

The defeat of the Turks, who were driven out of Hungary toward the end of the seventeenth century, launched a building boom in Vienna. Some of the city's most beautiful buildings, such as the Belvedere Palace, the Church of St. Charles Borromeo, and the Schönbrunn, were completed during this time. The boom in turn fostered such crafts as furniture making, carpentry, and masonry. In 1684 a hero of the siege, Franz Koltschitzky, used coffee beans left behind by fleeing Turks and opened Vienna's first *Kaffeehaus* (coffee

house). By 1740, with fewer than 200,000 residents, Vienna was considered one of the most beautiful cities in Europe.

When the Habsburg King of Spain died in 1700, both Austria and France claimed the throne. This led to the War of the Spanish Succession, which lasted until 1714. Austria gained control of Belgium and parts of Italy as a result. The Archduke of Austria became Holy Roman Emperor Charles VI in 1711. Charles had no son, and salic law held that the Austrian crown could pass only to a male. In 1724, Charles proposed the Pragmatic Sanction, under which European leaders promised to recognize the succession of his daughter, Maria Theresa. After Charles died in 1740, several states broke this promise. As a result of the eight-year War of the Austrian Succession, Maria Theresa lost the duchy of Silesia to Prussia. While she failed to regain Silesia during the Seven Years War of 1756 to 1763, she solidified her rule over Austria, Bohemia, and Hungary.

During Maria Theresa's reign, Vienna began to acquire its reputation as a home for musical genius. The Habsburgs were music lovers; Maria Theresa herself sang soprano arias and brought up her five sons and eleven daughters to be musicians. Her heir Joseph played viola, cello, and harpsichord.

In 1754, German composer Christoph Willibald Gluck was appointed *Kapellmeister* of the Vienna Opera. Gluck's important opera reforms during his years in Vienna included balancing opera's musical and dramatic aspects. The future composer Franz Joseph Haydn was admitted to the choir of St. Stephen's Cathedral as a soprano at the age of eight. After working for the noble Hungarian Esterhazy family, Haydn in 1790 made Vienna his permanent home. Six-year-old Wolfgang Amadeus Mozart played for Maria Theresa in 1762; nineteen years later he moved to Vienna where, with Haydn, he became the leading composer of the classical style, producing masterpieces in every musical form. None of his compositions made much money, however, and he died in poverty in 1791.

German composer Ludwig van Beethoven had played for Mozart when he visited Vienna in 1787. After Mozart's death, Haydn praised Beethoven's work and encouraged him to come to Vienna. Beethoven did, and rarely left the city afterward. The music he wrote in Vienna—symphonies, concertos, overtures, chamber music, an opera, songs, and Masses—remains among the most influential ever written.

Beethoven's *Eroica Symphony*, first performed in 1804, was originally dedicated to Napoléon Bonaparte. Enraged when Napoléon declared himself emperor of the French, Beethoven withdrew the dedication. During the Napoleonic Wars, Napoléon seized large parts of the Holy Roman Empire. After winning the Battle of Austerlitz against Austria and Russia in 1805, Napoléon forced Francis II to dissolve the Holy Roman Empire. Francis was known as Emperor Francis I of Austria after that date. In 1809, Napoléon surrounded Vienna and bombarded it with the largest concentration of cannon fire ever used up to that time. He was able to capture and occupy the city without breaching its medieval walls. (The shock of the bombardment proved too much for Haydn, who took to his bed and died shortly afterward.) Emperor Francis settled the crisis by agreeing that his eighteen-year-old daughter Marie Louise should marry Napoléon. The couple's son, Napoléon II, was born in 1811. Two years later, Napoléon was defeated at the Battle of Leipzig by Austria, Prussia, and Russia. He abdicated and was exiled to Elba.

After Napoléon's defeat, victorious powers gathered in Vienna to carve up Europe. The Congress of Vienna opened in October 1814, and during its five months drew 100,000 visitors (including 247 members of royalty) to the city. Most of the decisions were made by Foreign Minister Prince Klemens von Metternich of Austria, Foreign Secretary Lord Castlereagh of Great Britain, King Fredrick William II of Prussia, Czar Alexander of Russia, and Prince Tallyrand of France. The Congress restored to power many European royal rulers ousted by Napoléon. Austria regained land lost to Napoléon in addition to parts of Poland and Italy, but lost Belgium. The Congress set up the German Confederation, a loose union of independent states that Austria and Prussia both sought to lead.

In addition to being a great diplomatic event, the Congress was a vast social gathering. Balls, banquets, plays, and dances were held throughout the city, night after night, to celebrate the new era of peace. The Viennese waltz became the most popular ballroom dance of Europe. Beethoven, whose opera *Fidelio* (with its theme of liberation) was revived, composed a cantata especially for the occasion, called *Der glorreiche Augenblick* (The Glorious Moment). In March 1815, Napoléon escaped from Elba and reentered France. But the Congress continued its work. Its last act was signed on June 8; on June 15 Britain's Duke of Wellington defeated Napoléon at Waterloo.

The end of the Congress began the Age of Metternich, during which the conservative foreign minister tried to suppress any nationalistic or democratic tendencies in the Habsburg empire. Metternich used a network of spies and strict censorship to keep revolutionary ideas at bay.

At this time Vienna was home to Franz Schubert, perhaps the leading composer of German *lieder* (songs). While his enormous output (including more than 600 *lieder*) won him some recognition, Schubert never achieved any real success during his lifetime. In 1826 he tried unsuccessfully to get a position at the emperor's court. He died, at the age of 32, in 1828, a year after Beethoven.

The deaths of Schubert and Beethoven left Vienna void of major composers, and the music climate darkened. German composer Robert Schumann wrote that the Viennese "fear everything new, every departure from the lazy old rut." Polish composer Frédéric Chopin visited Vienna in 1831, but chose to settle in Paris instead. In 1825, Johann Strauss, Sr. had organized an orchestra to play his waltzes. Strauss eventually wrote more than 150 waltzes, which became popular at home and abroad. His son Johann Strauss, Jr. followed suit, forming an orchestra in the 1840s that grew

as popular as his father's. As court music director from 1863 to 1871, the younger Strauss composed such famous waltzes as "Tales from the Vienna Woods," "On the Beautiful Blue Danube," and "Wine, Women, and Song."

In 1848, revolution began in France and spread to Vienna. Students and workers rioted, demanding Metternich's resignation and a constitutional government. Metternich fled to England, the Habsburgs to Innsbruck. The army shelled Vienna and regained control in the autumn of 1848. Ferdinand I abdicated in favor of his nephew, eighteen-year-old Francis Joseph. In 1850, Vienna became a self-governing municipality for the first time since Habsburg rule began.

In 1857, Francis Joseph issued a decree allowing Vienna's medieval fortifications to be destroyed, thus making way for the Ringstrasse, a stately avenue to be lined by imposing buildings. Buildings erected along the Ringstrasse included the Hofoper (State Opera House), university, parliament building, Rathaus (town hall), and Natural and Kunsthistorisches museums.

Unification movements in Italy and Germany began to threaten the Austrian Empire. Austria declared war on Sardinia in 1859 but was defeated by Italy and France. As a result, Austria lost control of Lombardy. In 1860, Vienna's flirtation with democracy ended as the Habsburgs once again became absolute monarchs. Austria again suffered defeat in the Seven Weeks War of 1866, this time at the hands of Prussia and Italy. The German Confederation dissolved, and Prussia set up a new confederation without Austria. In 1867, Hungary forced Austria to give it equal status in the Dual Monarchy of Austria-Hungary. Both countries remained loyal to the Habsburgs, but each had its own constitutional government. That same year saw the emancipation of Austria's Jewish citizens, who made up about ten percent of the population.

During the second half of the nineteenth century, Vienna's musical reputation revived. The State Opera rivaled the world's greatest opera houses. The Vienna Philharmonic Orchestra, founded in 1842, reached heights of excellence. Leading composers such as Johannes Brahms and Anton Bruckner made Vienna their home. Bruckner taught at the Vienna Conservatory and University of Vienna, and his students included composers Gustav Mahler and Hugo Wolf. Mahler would take charge of the State Opera in 1897.

Other disciplines also flourished. The artist Gustav Klimt led the Succession school of painting. In 1898, the Succession's first exhibition featured works by Auguste Rodin and John Singer Sargent. At the turn of the century, Viennese resident Sigmund Freud published *The Interpretation of Dreams,* which would revolutionize our understanding of the human mind.

Nationalistic fervor spread throughout the empire. In 1914, Serbian nationalist Gavrilo Princip assassinated Archduke Francis Ferdinand, the emperor's heir, and his wife Sophie, in Sarajevo. Austria-Hungary declared war on Serbia, an act that precipitated World War I. With Germany and other countries, Austria-Hungary fought Britain, France,

Russia, and eventually the United States. Emperor Francis Joseph died in 1916 and was succeeded by Charles I, who went into exile in November 1918 after a defeated Austria-Hungary signed an armistice.

The Treaty of St. Germain broke up Austria-Hungary, reducing Austria from 115,000 square miles (297,800 square kilometers) to 32,369 square miles (83,835 square kilometers). Much of its territory went to the new nations of Poland, Czechoslovakia, Hungary, and what later became Yugoslavia. Austria lost many industries, and its population fell from 30 million to 6 million people. Many Austrians wished to unite with Germany, but this possibility was forbidden under the treaty. In 1920, Austria became a republic, with Vienna as its capital, and adopted a democratic constitution.

The war's end brought economic chaos to Austria. The krone became worthless and inflation soared, wiping out the savings of many residents. Viennese stripped wood from the Vienna Woods to keep warm. Political chaos also reigned: the Christian Social Party and the Social Democratic Party each kept a private army, which clashed with each other and with the Austrian Nazi party.

Freud, who lost his life savings in the economic turmoil, continued to work in Vienna. Klimt died in 1917; his gifted disciple Egon Schiele died in 1918. Composer Arnold Schoenberg worked on a new twelve tone method of composing, which would become the twentieth century's most influential musical mode. Schoenberg left Vienna in 1926. He was not alone: increasing anti-Semitism made Vienna an uncomfortable place for such intellectuals as Schoenberg or Ludwig Wittgenstein, an influential philosopher who left two years later. Yet Vienna soon was attracting refugees from Nazi Germany, such as writer Arthur Koestler.

In March 1933, Chancellor Engelbert Dollfuss, a Christian Socialist, dismissed parliament. A four-day war between his party and the Social Democratic Party ended in victory for Dollfuss, who then ruled as dictator. Opposed to unity with Germany, Dollfuss was assassinated by an Austrian Nazi in 1934. Although his successor, Kurt Schuschnigg, also fought unity, German troops arrived on March 12, 1938. The streets of Vienna were lined with cheering crowds. Adolf Hitler arrived to a hero's welcome two days later. In an April election, 99.73 percent of voters favored unity. The country of Austria was abolished; once again it was known as the Ostmark.

Vienna was home to about 200,000 Jews at this time. Less than 10,000 remained in 1941. About half escaped the country before World War II began in 1939; the rest were sent to death camps. Freud was allowed to leave in 1938 after a ransom of 250,000 Austrian schillings was paid for his release. Many Christian intellectuals fled the country as well.

German troops remained in the city until Soviet soldiers drove them out on April 13, 1945. One-fifth of the city—including the Opera House and St. Stephen's Cathedral—lay in ruins. The city had fewer than 1 million people, half its prewar population. Its people were impoverished; at one point, the daily food ration was 800 calories per person. Vienna was divided into four zones, occupied by Britain, the Soviet Union, France, and the United States. A national coalition government was elected in November 1945.

The Austrian economy revived over the next decade and received about $1 billion under the Marshall Plan. Artistic vigor also returned. The Wiener Schule, founded by Surrealist artists, became influential. Ruined buildings were restored; the Opera House reopened in 1955. That year, Austria became a neutral, independent nation and joined the United Nations. Allied troops withdrew from Vienna.

Because of its historical position between East and West, Vienna became an important diplomatic meeting ground during the Cold War. It hosted some of the Strategic Arms Limitation Talks (SALT) that began in 1969 between the United States and the Soviet Union. A United Nations Center, known as the Vienna International Center, opened in 1979. Today, Vienna once again is regarded as one of Europe's political, economic, and cultural centers.

Many of Vienna's historic structures are remarkably intact. The city's medieval period is showcased in St. Stephen's Cathedral and other historic churches, including St. Rupert's Church and the church of Mary on the Banks; the oldest buildings of Vienna University, including one now housing the Austrian Academy of Sciences; and private homes, including some on the Judenplatz (Jews' Square), once the center of the city's Jewish ghetto.

The many structures from the baroque period of the late seventeenth and early eighteenth centuries include the Austrian National Library, designed by Johann Bernhard Fischer von Erlach and his son, Josef Emanuel; the Schönbrunn, the Habsburgs' summer residence, also designed by the elder Fischer von Erlach; the imperial stables, designed by his son; and the imperial parish church of St. Augustine. The great buildings of the Ringstrasse represent a wide variety of nineteenth-century architectural styles. And one building that encompasses the history of the Habsburg Empire is the Hofburg, the rulers' winter residence until 1918. Begun as a Gothic castle in the thirteenth century, it was expanded as the empire's power grew; additions in many different architectural styles were made through the nineteenth century.

Further Reading: *Vienna* by Frederic V. Grunfeld (New York: Newsweek, 1981) is a readable account of Vienna's history, culture, and people. *Vienna* by Ilsa Barea (New York: Knopf, and London: Secker and Warburg, 1966) provides historical information sprinkled with lively personal observations and reminiscences.

—Mary Feely

Vladimir (Vladimir, Russia)

Location: In the center of the western portion of European Russia; about 110 miles east of Moscow.

Description: City founded in the twelfth century; onetime capital of Kievan Rus, the cradle of Russia.

Site Office: Vladimir Oblast Administration
Oktyabrskiy Prospect 21
Vladimir 600000, Vladimir Oblast
Russia
(9222) 2-5252

Vladimir-on-the-Klyaz'ma was founded in 1108 in Suzdalia, the "land beyond the forests." Suzdalia, considered the cradle of Russia, is a convenient name for the federation of principalities in the northeastern portion of what is now European Russia. It lies in the basin of four major rivers: the Sheksna, Unzha, and Klyaz'ma, which all flow into the Volga, and the Volga itself, as it makes its way to the Caspian Sea and the markets of the East.

In the ninth century, Slavs from the east abandoned their nomadic lives and settled in the area that is now southern Russia. The Old Russian state was established in 862 when the Slavic people of Novgorod asked the Vikings to rule over them. The Viking chief, Rurik, became the first of the Rurikid Dynasty that ruled Russia until the seventeenth century. In 882 Oleg, regent for Rurik's son Igor, captured Kiev and established his capital there. Ruling a territory stretching from Novgorod to Kiev, Oleg was the first independent ruler of a unified Kievan Rus.

The term "Russians" was at first used to describe the Vikings, but the name "Rus" eventually came to be applied to Slavs and Vikings alike. These ancestors of the Russians, Ukrainians, and Belarussians (White Russians) called themselves and their lands Rus. The Kievan period, from the first Slavic migration into the area until the Mongol invasions in the thirteenth century, was a golden age. Kiev was the seat of the grand-prince from the ninth century and the seat of the metropolitan of the Russian Orthodox Church from 988, the year Prince (later, St.) Vladimir converted the country to Christianity.

The town of Vladimir sits on the central reaches of the Klyaz'ma in the southern part of Suzdalia. Vladimir derives its name from Vladimir Monomakh, great-grandson of St. Vladimir (who had married the daughter of Emperor Constantine IX Monomachus of Byzantium). Vladimir Monomakh, the first to appreciate the strategic position of the site, built a wooden frontier fortress there in 1108. At the end of the eleventh century Vladimir became an independent principality, and Monomakh bestowed it on his son, Yury Dolgoruky ("Long-in-the Arm"), although Yury waited to turn his attention to Vladimir until after Monomakh's death. Elected grand-prince of Kiev at the age of sixty, Monomakh ruled over most of the Kievan Rus for twelve years, maintaining a reasonable degree of law and order.

The nomadic Polovtsy exploited the turmoil that arose among the princes after Monomakh's death in 1125. Their invasion resulted in the flight of some of Kiev's population to the forest lands in Suzdalia. This fact played a part in the decline of Kiev, but the overriding cause was that no one dynasty ruled the principality. The last unified rule had been under Monomakh and his son Mstislav the Great. After that, rule passed back and forth between the various branches of Monomakh's family and Oleg's descendants. With the decline of Kiev, two strong principalities arose: Novgorod and Vladimir-Suzdal.

The development of Vladimir provides a sharp contrast with the fate of Kiev. Yury Dolgoruky and his sons ruled single-mindedly: they had no tolerance for rivalry and did not splinter their territory by letting parts of it slip into the hands of other relatives. Yury's son, Andrei Bogolubsky, realized that Vladimir was in a good location for the princes to consolidate and centralize their power. Against the wishes of his father, Andrei left secretly for Vladimir, taking with him an icon from Constantinople said to have been painted by St. Luke himself. Later named Our Lady of Vladimir, the icon became the most revered object in northern Russia. On Yury's death in 1157, Andrei became prince and chose Vladimir as his capital. He built Uspensky Cathedral, strengthened Monomakh's original kremlin, and built gates in the fortifications, including the Silver Gate on the east, and on the west the Golden Gate, copied from one in Kiev.

In 1169 Andrei's son captured Kiev, but Andrei did not move his capital there, preferring Vladimir. In 1174 Andrei was killed by rebellious *boyars* (nobles) in league with his wife (the daughter of a boyar living at the future site of Moscow who had been captured by Yury and brought back as a wife for Andrei). The assassination took place at Andrei's palace at Bogolubovo, just outside Vladimir. The assassins were caught, executed, and their bodies deposited in the nearby swamps in specially tarred coffins. It is said that their wailing and moaning can still be heard. Today the assassination staircase survives, as does part of a passageway that linked the palace to the adjacent church. The rest of the complex is gone.

Andrei was succeeded by his younger brother Vsevolod III. During Vsevolod's rule, which began in 1176, Suzdalia, with Vladimir as capital, reached the peak of its power. Vsevolod died in 1212 and was succeeded by his son, Yury II. A period of fierce warring among branches of the family followed. The throne of Vladimir passed briefly to Yury's brother Konstantin, but returned to Yury upon Konstantin's

Uspensky (Assumption) Cathedral in Vladimir
Photo courtesy of Vladimir Region Administration

death in 1218. A period of relative peace among the princes followed.

The Mongols' arrival in the thirteenth century took the Russians by surprise, as they seemed to come out of nowhere. The Russians had no idea what they were facing when they naively went to the aid of the neighboring Polovtsy and a terrible slaughter ensued. The Mongols, under Batu Khan, a grandson of Genghis Khan, arrived at Vladimir in 1238 and within days conquered the city; most of Russia soon fell to them. In addition to the greater size of the Mongol army, their superior tactics, reconnaissance, and efficient methods gave them enormous advantages. The Russians, with no unity, no central command, and an almost total lack of intelligence gathering, were no match for the Golden Horde. Yury II was beheaded; although accounts of the time are sketchy, there is reason to believe he may have been killed by his own troops as he tried to prevent them from fleeing. His three sons were also killed or taken prisoner in that same year. By summer of 1238, the Mongols had left Russia to rest and regroup for other conquests.

Most of Russia came under the control of the Mongol Empire. While the princes were made subject to their heavy tribute, the Mongols did not otherwise interfere in day-to-day life. Old chronicles are vague on the exact nature of the conquest, in a conscious or unconscious effort to downplay the extent of Russia's total submission. To maintain their dominance, the Mongols used periodic depredations of the area to keep Mongol violence fresh in Russian minds. Some damage was done purposely and some was incidental destruction as Mongol warriors returning from joint military missions looted the villages of their Russian allies.

After Yury II's death, his brother, Yaroslav, who already controlled Novgorod, succeeded as grand prince. Yaroslav had named his son Alexander ruler of Novgorod in 1236. Novgorod was not captured by the Mongols, but Alexander chose to pay tribute in return for protection. This left him strong enough that, when the Swedes invaded in 1240, he was able to defeat them at the Neva River, near the present site of St. Petersburg. For that reason he was named Alexander Nevsky. He also won battles against the German knights in league with the pope and the Holy Roman Empire. Alexander Nevsky is credited with keeping the country from the spread of Roman Catholicism, and was later canonized by the Russian Orthodox Church.

After Yaroslav died in 1246, there followed a period of bitter conflict over the throne of Vladimir. Yaroslav's brother, Svyatoslav Vsevolodovich, succeeded him, but within a year he had been ousted by Yaroslav's second oldest son Andrei, probably by means of a military coup. Andrei, Alexander (the older surviving son and prince of Novgorod), and Svyatoslav all separately made their way to the Mongols, as a "patent" was necessary for any one of them to rule. The Khan gave Andrei the throne of Vladimir and gave "Kiev and all the land of Rus" (i.e., southern Russia) to Alexander. Svyatoslav received nothing and sank into obscurity.

A crisis arose in 1252, the exact nature of which is unknown, resulting in Alexander again traveling to the Mongol headquarters. With an army supplied by the Khan, Alexander returned to Vladimir, defeated Andrei, and took over as grand prince of Vladimir. He was welcomed by the metropolitan of the Russian Orthodox Church and by the citizens at the Golden Gate. Fearing his brothers would back foreign invasions against him, Alexander immediately made peace with them.

A new era in Russia's subjugation began with Alexander's rise to power. Any organized resistance to the Golden Horde ceased for some time to come. Nevsky's collaboration with the Mongols has been seen as an embarrassment by some, but his willingness to cooperate may have saved Russia from further violence at the hands of the Mongols. He died on his way home from pleading the case of some of his subjects before the Khan.

After Nevsky's death, the Khan chose to make Nevsky's brother Yaroslav the grand-prince of Vladimir, rather than the older brother, Andrei, from whom they feared rebellion. In 1272, Yaroslav was succeeded by his younger brother Vasily Yaroslavich, and the civil wars temporarily stopped. There was, however, a sinister and significant dependence on the Mongols during this period, with a considerable Mongol army stationed in Russia. Due to Nevsky's relationship with the Mongols, his successors Yaroslav and Vasily were unable to take any positive steps toward independence. Vasily died in 1277 and was succeeded by Nevsky's eldest surviving son, Dmitry of Pereyaslavl'. About five years into Dmitry's rule, civil war again broke out. On three separate occasions, Andrei, the next eldest son of Nevsky, invaded Russia with armies supplied by the Mongols and seized the throne, only to be later defeated by Dmitry, first with Swedish troops and later with a rival Mongol force. Andrei's fourth and final invasion finally secured the throne; he survived as grand-prince for ten years after.

After the Mongol invasions, Vladimir had retained nominal seniority among cities. Grand-princes were crowned in Uspensky Cathedral, where the princes held their councils. The power of Vladimir was waning, however, and Tver and Moscow were emerging as the real power bases. After Andrei's death in 1304, a struggle arose between Tver and Moscow that was unlike the previous internecine battles. This time the contenders were dynasties with powerful hereditary bases.

The Mongols exploited the hostilities among the princes, using their control of the throne of Vladimir as leverage to neutralize princely power, in turn giving the throne to the Moscow faction. Yury Dolgoruky had realized the importance of the future site of Moscow, near the headwaters of the Oka, Volga, Don, and Dnieper Rivers, when he seized the area two centuries earlier. Moscow grew from a small principality to a grand-principality, eventually swallowing up Vladimir. It became the seat of the metropolitan of the Russian Orthodox Church, and soon the titles of metropolitan and grand-prince of Moscow both included the phrase "of all Rus."

In the first half of the fourteenth century, Moscow benefited from its alliance with the Mongols, since the continued raids weakened the power of the other princes. Throughout the period of Mongol rule, all of Suzdalia had benefited most from lucrative oriental trade. The Mongols, solicitous of trade, found it in their best interest to help rather than hinder Russia's economic recovery.

Eventually Moscow did begin to oppose the Mongols; in 1380 it achieved the first major Russian victory over the Mongols during their 140 year rule. This victory did little to change the status of Russia, however, and it did not break the Mongol "yoke." What brought down the Mongol Empire was the deterioration of relations among the Mongols themselves. Eventually Tamerlane attacked the Mongols, with Moscow fighting on the side of the Khan. In 1395 the Muscovites were convinced (wrongly) that Tamerlane intended to ravage the city. They brought the Virgin of Vladimir to Moscow and credited her intercession with the salvation of the city. (The icon is now in the Tretyakov Gallery in Moscow.) Mongol domination finally ended in 1480 when the weakened Golden Horde gave up its claims to Russia.

Vladimir and Suzdalis had enjoyed 125 years of prosperity from the accession of Vladimir Monomakh in 1113 to the death of Yuri II in 1238—a period during which Russian art and architecture flourished. Some buildings in Vladimir still survive from that time. The Cathedral of the Dormition (Uspensky Sobor), a one-domed church built by Andrei Bogolubsky in 1160, demonstrated a level of skill in carving virtually unknown at the time; its interior is embellished with frescoes, gilt, and painting, and it served as the principal cathedral of the Russians for centuries. The princes were crowned there until 1440, and from the beginning of the fourteenth century it was the seat of the metropolitan of the Russian Orthodox Church. After a fire, Vsevolod III restored the cathedral in 1185–89, greatly enlarging it. The single gold dome was changed to the traditional five domes of the Russian Orthodox Church. The building that exists today is largely the result of these renovations. Considered one of the great masterpieces of Russian architecture, the cathedral was studied as a model for the Church of the Dormition in Moscow's Kremlin. When the Mongols conquered Vladimir in 1238, they locked the royal family, the metropolitan clergy, and leading citizens in the cathedral and burned it. The people died in agony and the interior of the church was destroyed, but the exterior remained intact. The frescoes were clumsily restored in 1408, and the Chapel of St. George, added in 1862, destroyed the harmony of the interior. Further repairs in the twentieth century uncovered twelfth-century frescos.

Two other Assumption churches exist in Vladimir. The wife of Vsevolod III founded the Princess Convent and built Uspensky Church of the Convent, which served as the burial place for the grand-princesses. Re-built in the sixteenth and seventeenth centuries, it is now the only building remaining of the convent complex. Another Uspensky Church, built from 1644 to 1649 by the prosperous merchants of Vladimir, has five cupolas and a pyramid-shaped belfry.

Of historic note is the Nativity Monastery (Rozhdestvensky), built from 1191 to 1196. The monastery's massive walls correspond to the original kremlin (fortress) built by Monomakh. Further evidence that this was the site of the fortification is provided by the name of a nearby church, St. Nicholas-in-the-Kremlin. There had been a fine late twelfth-century cathedral on the grounds, but it was demolished. All that remains today is a jumble of buildings of varying age and architectural merit, most dating from the eighteenth century. Nativity was the largest monastery in Russia until the mid-sixteenth century. Alexander Nevsky was buried there, but when Peter the Great built the Monastery of St. Alexander in the Cathedral of the Trinity in his new city of St. Petersburg, Alexander's remains were moved to the new cathedral at the site of his historic victory on the Neva.

The Golden Gate, built 1158 to 1164 as part of the city fortifications, survives today. A masterpiece of defense architecture, it served as both a powerful guard tower and a ceremonial entrance to the city. According to legend, it collapsed on its dedication day, trapping many citizens. When Andrei prayed, the gate lifted itself and spared the people. The Golden Gate is the only remaining monument of twelfth-century Russian military art. On principal religious holidays, processions moved along a route lined with costly embroideries, from the Archbishop's Palace through the Golden Gate to Uspensky Cathedral. The gate, a gigantic cube pierced with an arch, was badly damaged in the Mongol invasion; it was reconstructed in the fifteenth century. The gates that filled the arch were of gilded copper, hence the name Golden Gate. On top of the gate, surmounted by a golden dome, is the small Church of the Rizopolozhenye (Deposition of the Robe), which was rebuilt in the eighteenth century. There are no traces of the four other gates Bogolubsky built.

Further Reading: *Russia: A Concise History* by Ronald Hingley (London: Thames and Hudson, 1972; revised edition, London and New York: Thames and Hudson, 1991) is a good general history of Russia from its ninth-century roots up through the changes under Gorbachev in the late 1980s and early 1990s. For a comprehensive study of thirteenth-century Russia, the effects of the Mongol conquest, and the civil wars, see *The Crisis of Medieval Russia: 1200–1304* by John Fennel (London and White Plains, New York: Longman, 1983). *Russia and the Golden Horde: The Mongol Impact on Medieval Russian History* by Charles J. Halperin (Bloomington: University of Indiana Press, 1985) gives an account of the history of the Mongol domination of Russia and provides an in-depth analysis of its effects on Russian government, politics, economy, and society. *Nagel's Encyclopedia-Guide: U.S.S.R.* (fourth edition, Geneva: Nagel, 1978), a very comprehensive guidebook to the former U.S.S.R., discusses the history, geography, economy, art, and literature of Russia and the Soviet Union, and provides detailed historical and architectural information on significant sites.

—Julie A. Miller

Warsaw (Warszawa, Poland)

Location: In eastern Poland, on the banks of the Vistula River.

Description: Capital of Poland and of Warszawa province; first settled in the tenth century; achieved prominence during fourteenth century under Janusz the Elder; city virtually destroyed during World War II, but most of historic buildings painstakingly reconstructed in 1950s.

Site Office: Tourist Office
Plac Zamkowy 1/13
Warsaw, Warszawa
Poland
635-1881

During the tenth century a group of Slav nomadic tribes joined together and formed roughly what is today Poland. Their Christian convert leader, Mieszko I, was the first of the Piast rulers, who governed the region until their demise in the fourteenth century. The Jagiellonians (1386–1572) succeeded them.

The Jagiellonians were descended from the union between the Polish queen Jadwiga and the Lithuanian prince Władysław Jagiełło. This marriage led to the *Rzezpospolita,* the political alliance between Poland and Lithuania.

Although there is evidence that Warsaw was a settlement during the tenth century A.D., the city only became an important center during the fourteenth century, when Janusz the Elder made it the capital of the Duchy of Mazovia, and developed it as a center of trade. The demise of the Jagiellonian dynasty in 1572 resulted in the election of the first Polish king by members of the aristocracy. During this period Warsaw grew as an important trade center, because the Vistula River was the main corn trading route. An inventory found from 1546 illustrates Warsaw's relative wealth during this period, detailing 700 brick buildings, 9 churches, and 3 hospitals.

King Sigismund III Vasa moved his capital from Kraków to Warsaw in 1596, following Poland's union with Lithuania. For the next hundred years Warsaw's economy prospered, although various wars with the Muscovites, Cossacks, and Swedes, among others, and the factional warfare among the Polish and Lithuanian nobility brought an end to the Polish-Lithuanian Commonwealth.

Destroyed by the Swedish king Charles X Gustavus in 1655, Warsaw was rebuilt in the late seventeenth century. Most of the reconstruction of the city coincided with a particularly vigorous burst of economic, cultural, and intellectual activity, promoted by Stanisław II Augustus Poniatowski, the last of the Polish kings, who reigned from 1764

to 1795. The painter Bernardo Bellotto, who called himself Canaletto after his famous uncle, Giovanni Antonio Canal, was among the artists who came to work in Warsaw during this period. Many fine buildings were erected during Poniatowski's reign.

Although Poland flourished culturally during the eighteenth century, it was weak politically, a factor that resulted in its partition between Austria, Prussia, and Russia in 1772, the first of three such divisions of the territory—the others occurring in 1793 and 1795. Following the abdication of Poniatowski and the final partition of the country in 1795, Prussia absorbed Poland into its lands, and Poland vanished from most maps of the world until 1918. These divisions adversely affected Warsaw, which in 1790 had 115,000 inhabitants; less than ten years later its population numbered 63,000.

Napoléon's France epitomized liberation to the stateless Poles at the beginning of the nineteenth century. The emperor's ties with the country were extensive. His mistress of almost thirty years, Maria Walewska, was a Pole; and many native Poles served in his military campaigns and could be counted among his intimate circle. Although in reality the restoration of Poland's independence was secondary to Napoléon's other territorial plans, he created the Duchy of Warsaw in July 1807, with the city of Warsaw as its capital. This move was the result of the Russian czar's refusal to oversee the Prussian administration of Poland. The duchy was created out of a small part of former Poland—from the Prussian land gained in the 1795 partition of the country. It originally included Mazovia and Wiełkopolska, but Kraków, Lublin, and Zamosč were added two years later. At its largest it was approximately 95,000 miles square, and had a population of almost 4.5 million people. One of the radical changes introduced into the region at this time was the abolition of serfdom, which placed most peasants in the ambiguous and unwanted position of choosing liberty or starvation, as most of the nobility were vehemently against this democratic action.

For the most part, the duchy served as a source of men and money for Napoléon's various military campaigns. In 1808, Napoléon introduced a compulsory six-year conscription for able-bodied men between the ages of twenty and twenty-eight. In addition to losing most of their young men, the inhabitants of this region were required to pay 25 million francs in taxes over a period of four years to reclaim property that had originally been their own.

After Napoléon's downfall, the duchy ceased to exist, although Warsaw continued to benefit from being the capital of the new Kingdom of Poland, under the patronage of Russia, created by the 1815 Congress of Vienna. Although this was a period of growth for Warsaw, it was also a time of great suffering and hardship for its citizens who were tyrannized by

A view of Warsaw
Photo courtesy of Polish National Tourist Office

the Russians. The subsequent unsuccessful rebellion of 1830 brought even greater suffering. Poland lost any semblance of independence from Russia, Warsaw was reduced to the status of a provincial city, and the city's intellectual institutions were closed. A second uprising in 1863 resulted in Czar Nicholas I razing part of the city and erecting a citadel in its stead. After these two revolts, many intellectuals and artists left the city. Even with this political oppression, there was some growth in Warsaw. The first railway between Warsaw and Vienna was built in 1845. The city's population increased to 690,000 people by 1900, more than five times its 1864 level.

The disintegration of the Austro-Hungarian empire, the Russian Revolution, World War I, and the subsequent Treaty of Versailles in 1919 led to Poland's reemergence as an independent state. Although the country was politically and socially unstable between 1919 and 1939, Warsaw's population continued to rise to 1.3 million by the outbreak of World War II.

In September 1939, Warsaw was bombed and occupied by Germany. Poland found itself again partitioned, this time between Russia and Nazi Germany as a result of the

Nazi-Soviet Pact. During the war the country developed one of the most effective and impressive Resistance networks in Europe. It also organized a secret education system after the Nazis closed down intellectual institutions in the country. Poland also had one of the largest Jewish populations in Europe, and this population suffered greatly under the German occupation. At the beginning of the war, it is estimated that there were between 380,000 and 400,000 Jews living in Warsaw, and 3.5 million living in Poland as a whole. The Jews had arrived in Poland during the eleventh century, working as artisans, merchants, financiers, and craftsmen.

Within months of invading Poland, the Nazis were discussing the possible partition of Warsaw, and the creation of a ghetto for the Jewish population. Early in 1940 a resettlement program was introduced by the Nazi authorities, which involved the mass transportation of Jews to preordained camps around the country. On October 2, the Nazis established an area in which the Jewish population would be forced to live; on November 15, 400,000 Jews were herded into a specific zone and effectively sealed off from the rest of the city. It is estimated that at its height, between 470,000 and 590,000 people lived in this cramped district—an average of 15 people per apartment.

By the summer of 1942, massive numbers of people were being deported to concentration camps. During this period the Resistance movement continued to grow, inside and outside the ghetto. In 1943, the inhabitants of the Jewish district staged a four-week uprising that ended in 56,000 deaths and the deportation of approximately 400,000 Jews to death camps.

The city of Warsaw staged a separate uprising in 1944. The sixty-three-day revolt began on August 1, and centered on the left bank of the city. This was ruthlessly put down by the Nazis, and resulted in the near-destruction of the city. The Germans killed more than 150,000 civilians, and approximately 26,000 of their own forces were killed or wounded. More than 85 percent of Warsaw was destroyed, including 782 out of the city's 957 historic sites, and most of the civilian population was deported from the city. This led Hitler to declare that Warsaw was no more than a name on a map. By the end of the war more than 850,000 of Warsaw's population were missing or dead; and out of the approximately 400,000 Jews living in the city in 1939, 200 were left. Poland lost one-sixth of its population in the six years of war.

Poland was eventually liberated by the Soviet Army in 1945. Once again the land mass of the country changed. It lost approximately half of its prewar lands to Russia, but also received German territory to the north and west of its borders. Postwar Poland became a Communist country.

The people of Warsaw participated in various uprisings against Communist rule, including the 1980s strikes and demonstrations in support of the Solidarity labor movement. Cardinal Stefan Wyszyński, the Warsaw-based primate of Poland from 1951 to 1981, was an outspoken advocate for human rights. A monument to Wyszyński was unveiled in Warsaw in 1987. Another activist clergyman, and the one most closely identified with Solidarity, was the young Warsaw parish priest Jerzy Popiełuszko, who in the early 1980s preached politically charged sermons in St. Stanisław Kostka church in the Żoliborz neighborhood. Popiełuszko was abducted and killed by four police officers in 1984; the event galvanized support for Solidarity. Poland's Communist government collapsed at the end of the decade, and Solidarity leader Lech Wałęsa was elected the country's president in 1990.

Today, as Poland seeks to establish its role in the world under a democratic government, its capital city remains a place of great historic and cultural interest. The decades following World War II saw the nearly complete reconstruction of Warsaw. The architects Roman Piotrowski and Jozef Sigalin are largely responsible for this achievement, which involved rebuilding some of the finest historical sites.

The completion of a concourse joining the Praga district to western Warsaw was one of the first major successes of reconstruction. From there new areas of the city, painstakingly rebuilt regions, and industrial zones have developed. Suburbs have grown around the city center.

Today Warsaw is the center of administration and government. It is also the artistic and cultural center of Poland, with many museums, galleries, and palaces. It hosts one of the most important international events for pianists—the Chopin competition.

Among the many historic sites rebuilt in the city center was the beautiful Zamek Królewski (Royal Castle). The castle was originally built in the fifteenth century for the dukes of Mazovia. King Sigismund II Augustus began expanding it in 1568 and King Sigismund III Vasa had it remodeled extensively in the early baroque style from 1598 to 1619. It underwent further alterations in the eighteenth century. It was destroyed by the Nazis at the end of the war, but many fragments of the building were preserved, and were used in the rebuilding of the new structure during the 1970s. Today the Zamek houses some of the original art collection held before the Nazis burned and looted the building.

Krakowskie Przedmieście is one of the most lovely boulevards in Warsaw, and like most of the city center was virtually destroyed during the war. The rebuilt street is lined with impressive palaces, houses, and churches, and is the site of the famous Radziwill Palace (now the Building of the Council of Ministers), where the 1955 Warsaw Pact was signed. The pact established a mutual defense alliance among Communist-ruled nations in eastern Europe. The palace also was where Frédéric Chopin gave his first public concert. Chopin grew up on Krakowskie Przedmieście; one of the homes he occupied now belongs to the Fine Arts Academy and memorializes the composer in a room called the Chopin Salon. Also on this street is a statue of Nicolaus Copernicus. The Nazis removed the statue in 1944, but it was returned after the war.

Łazienki, now a museum and park complex, is the former summer palace of Stanisław II Augustus Poniatowski, built in the 1770s, and designed by the architect Domenico Merlini. Set in 180 beautifully landscaped acres, the palace houses impressive paintings by European artists, such as Breughal, and Ferdinand Bols. Damaged by the Germans in 1944, most of the palace has been restored. The orangerie museum in the gardens houses one of the few remaining eighteenth-century court theaters in Europe. The park was also the site from which the rebels began the uprising of 1830. Near Łazienki is the eighteenth-century neoclassical Belvedere Palace, which today is the president's official residence.

One of the most famous churches in town is the eighteenth-century Church of the Holy Cross, which is the resting place of two famous Poles. Sealed in two pillars are the hearts of Chopin and of the Nobel Prize-winning novelist Władysław Reymont. The church was also extensively damaged during the war.

The Gothic Cathedral of St. John can be found within the ramparts of the old town. Used by the Resistance during the Warsaw Uprising, the old part of Warsaw, formerly the poorest area in the city, was completely gutted by the Germans in retaliation for the revolt. The cathedral was originally built during the Mazovian period in the fourteenth century, and was reduced to rubble in 1944. After the war, it was painstakingly restored to its former glory. The cathedral is the site of the tomb of the former president of Poland, Gabriel Narutowicz. He was assassinated just a few days after taking office in 1922.

The Historical Museum of Warsaw, originally housed in the only building still standing in 1947 in the market square, today lies in eight reconstructed and renovated houses that have been joined. The museum contains pieces by Canaletto, original manuscripts, armor, and other artifacts from the tenth century to more contemporary times.

Given that so many of Poland's Jews suffered and died during World War II, it is hardly surprising that modern Warsaw has a street dedicated to this community. Trakt Pamieci Meczenstwa Walki Zydow is named in memory of the people who lived and died in the Nazi-created ghetto. The Jewish Historical Institute has been built on the site of the former Judaic Library, destroyed in the 1940s by the Nazis. It contains clothing, manuscripts, and other artifacts from the ghetto and recounts the events that culminated in the mass murder of almost 400,000 people.

Further Reading: *Poland: Eastern Europe and the Commonwealth of Independent States* (London: Europa Publications, 1994) is a comprehensive book that details the history, economy, and other characteristics of individual countries in eastern Europe. *God's Playground: A History of Poland,* in two volumes, by Norman Davies (Oxford: Clarendon Press, 1981; New York: Columbia University Press, 1992) is a very detailed history of Poland. *A Cup of Tears: A Diary of the Warsaw Ghetto* by Abraham Lewin (Oxford and New York: Basil Blackwell, 1988) is a haunting memoir of a man who lived in the ghetto during the war. *Essential Handbook to Europe's Tribes* by Felipe Férnandez-Armesto (London: Times Books, 1994) looks at ethnic background and history of different indigenous peoples of Europe. *Political Culture in Vienna and Warsaw* by Hans-George Heinrich and Slawamir Wiatr (Boulder, Colorado: Westview, 1991) is a series of essays on the two cities.

—Aruna Vasudevan

Waterloo (Brabant, Belgium)

Location: Fifteen miles south of Brussels, on the Charleroi-Brussels highway.

Description: Small rural town; site of battle fought June 18, 1815, in which Allied armies under the Duke of Wellington defeated Napoléon's French forces.

Site Office: Braine-l'Alleud
Rout du Lion 252–254
Waterloo, Brabant B-1420
Belgium
021/385.19.12

The bucolic calm that pervades Waterloo today belies its history. On the surrounding fields and hills on June 18, 1815, two of Europe's greatest armies, led by two of Europe's most illustrious generals, clashed in a battle that would finally put an end to Napoléon's imperial designs and redraw the map of political power in the continent and beyond.

Waterloo stands with the battles of Trafalgar and Agincourt, not only as a great British military victory, but as an event that changed the course of European history. For Napoléon, a master tactician who had stamped his military genius across Europe from Austerlitz to Jena, Waterloo was a battle too far. The emperor was driven by his obsessional desire to humiliate the English and blinded by his confidence in his own powers and good fortune, but in Arthur Wellesley, the duke of Wellington, he finally met his match. The Iron Duke's strategic brilliance, combined with timely reinforcements from his Prussian allies, some costly blunders from Napoléon's frontline commanders, and the heavy rain that slowed the French attack, resulted in a defeat that destroyed the emperor's reputation and led shortly to renewed abdication and exile on St. Helena.

Waterloo was the first time that the two generals had met face to face in battle, in a war that had already lasted for twenty-two years. Although both were outstanding soldiers with reputations for being autocratic and dictatorial, their approaches to battle differed in important ways. Napoléon was the flamboyant Corsican, supremely confident in his own abilities as a commander and given to ambitious, aggressive campaigns that had often brought him glory, sometimes ignominy (e.g., his humiliating retreat from Moscow in 1811). In Wellington he met a stolid Dubliner who had been molded by the traditions of England's public schools—hence Wellington's assertion that Waterloo was won "on the playing fields of Eton." Wellington generally led smaller armies than Napoléon, in which command could be centralized and in which steady obedience of subordinates was the key ingredient. Wellington combined foresight with common sense, and his imagination seldom ran away with him. A prudent commander and master of defensive warfare, he was also capable, on occasion, of audacious offensives.

In 1815 Napoléon had determined to smash the Seventh Coalition and regain his dominant position in Europe. He raised an army of more than 70,000 men and revamped his command structure. Most commentators agree that it was Napoléon's poor choice of commanders that led to his defeat at Waterloo. He gave command of his right wing to Emmanuel de Grouchy, a cavalry general who had never led a corps, and assigned Michel Ney to his left wing—two inexperienced individuals who would ultimately let the emperor down very badly. Other omens were not good either. One of Wellington's spies described the disposition of the imperial army as "impressionable, critical, without discipline and without confidence in its leaders, haunted by the dread of treason and on that account, perhaps, liable to sudden fits of panic."

Meanwhile, the Seventh Coalition, based in Vienna, was raising its own armies, including an Anglo-Dutch army of 93,000 men under Wellington and a Prussian army of 117,000 men under Gebhard Leberecht Blücher, both positioned in Belgium. Other armies were built up along the Rhine and in Northern Italy. The Seventh Coalition's plan was simple—to march directly on Paris and crush Napoléon by force of numbers.

In April 1815 Wellington set out from Vienna for Brussels, rendezvousing with Blücher at Tirlemont in early May. Napoléon decided to seize the initiative, enter Belgium and beat the Prussians and English before they could unite their forces. Indeed, he found the allied forces unprepared for his arrival, and on June 15 established his headquarters at Charleroi, about fifteen miles south of Waterloo. Wellington was occupied in Brussels, attending a cricket match and a ball given by the duchess of Richmond, while Blücher was collecting his forces at Sombreffe.

The battle of Waterloo was preceded by two preemptive strikes by Napoléon against first Blücher at Ligny and then the English forces at Quatre Bras. On June 15 Grouchy engaged the Prussians at Ligny, about two miles southwest of Sombreffe, and eventually put Blücher's forces to flight, forcing them to withdraw north toward Wavre. This was a great victory for Napoléon, and it presented him with the opportunity to attack Wellington the following day without Prussian interference.

Breakdowns in the chain of command on both sides meant that when Ney attacked the English at Quatre Bras on June 16, each had only managed to assemble half his forces; 31,000 under Wellington and 22,000 under Ney. The battle ended in a draw at nine o'clock in the evening, both armies retaking the positions that they had held in the morning, and with equal casualties of between 4,000 and 5,000 on each side.

The Lion of Waterloo
Photo courtesy of Belgian National Tourist Office

Early on the morning of June 17 Wellington learned from a cavalry troop led by Colonel Alexander Gordon that the Prussians had been beaten and were retiring to Wavre. The duke sent word to Blücher that he was falling back to Mont-St.-Jean, only about a mile south of Waterloo, and that he would offer battle to Napoléon there if Blücher would support him with one army corps. That Wellington was able to retreat from Quatre Bras unmolested was entirely the result of Ney's inactivity. Although he had been ordered to attack the English, Ney did nothing. While the enemy were in full

retreat and thus at their most vulnerable, Ney's troops were preparing their lunch. The emperor was furious when he found his forces thus engaged and quickly sent them off in pursuit of the retreating English, but valuable hours had been lost and the French charge was hampered first by Lord Uxbridge's cavalry, who were protecting the English rear, and then by a torrential downpour. The storm probably saved Wellington because it made the ground so wet that the French were unable to advance across country, and were therefore tied to the Brussels road. The retreating forces were able to regroup with the entire Anglo-Dutch army just north of the heights of La Belle Alliance.

Napoléon fell back to his headquarters at the farm of Le Caillou, one and one-half miles south of La Belle Alliance, still confident that he would defeat the Allies the next day, and sure that Blücher would not dare to outflank Grouchy's men in order to join Wellington at Waterloo. However, just as Ney had been slow to pursue the English out of Quatre Bras, Grouchy had let Napoléon down very badly. He completely failed to press home the advantage of the victory over Blücher's Prussians at Ligny, following the Prussian rear guard with incredible slowness. Grouchy also failed to understand Napoléon's order to prevent the Prussians from moving west to join Wellington, a task that Grouchy could have quite easily achieved, as the Prussians had retreated in disarray and thinly spread out. Grouchy's inaction allowed the Prussian columns to regroup toward the end of the day on June 17, and Blücher replied to Wellington's request for reinforcements with the news that, at daybreak on June 18, Friedrich Wilhelm von Bülow's corps would march to his aid, immediately followed by George Dubislaw Ludwig von Pirch's.

Napoléon spent the night riding through torrential rain to inspect his outposts, and between 4 and 5 A.M. issued orders for the French troops to be in place to attack at 9 A.M. However, by daybreak it was clear that the heavy rain had made the ground too wet for rapid artillery movements. Napoléon was still convinced that the Prussians had been too weakened by their defeat at Ligny to intervene at Waterloo, and believed that "Wellington's polyglot army could be smashed at a single blow." Added to this, he was still under the impression that Grouchy would be able to head off any Prussian attempt to link with the Anglo-Dutch army. The emperor decided to delay his attack until 1 P.M. in order to give the ground a chance to dry, but this was perhaps his most fateful blunder of the whole campaign. Had he signaled the attack earlier in the morning, the English would have been routed before the arrival of the Prussians.

Napoléon aimed to employ his tried and tested tactic: break the enemy's center and then exploit this penetration. He began an intensive artillery bombardment at noon and instructed Ney to advance on Wellington's forces at 1 P.M. The battlefield was comparatively small, about two miles by four miles, being bisected by the Charleroi-Brussels road and flanked to the north and south by low ridges, the southern one being that of La Belle Alliance. Wellington's main line lay along the northern ridge, the two armies separated by a shallow valley containing two or three very small hamlets. As they lined up at 1 P.M., Wellington had 67,661 men under his command, and Napoléon, 71,947.

Just as Napoléon's artillery were hammering the English, Bülow's advance guard emerged from the woods at Chapelle St. Lambert, four or five miles to the east. Napoléon quickly deployed light cavalry and corps to meet the Prussians at the village of Plancenoit and prevent them from joining the fray. The French right flank thus secured, Ney sent forward infantry columns under Jean-Baptiste Drouet, Comte d'Erlon, to pierce the allied center. Through confused orders three of his columns adopted a huge, outdated formation that invited devastating artillery fire from the English at the top of the ridge. John Paget, earl of Uxbridge, launched his cavalry into the maelstrom, which not only repulsed d'Erlon's command, but then charged clear across the valley to attack the French grand battery, where it was decimated in turn by fresh lancers. Thus Wellington lost a full third of his cavalry—but the French grand attack had failed.

At about 3 P.M. there was a lull in the fighting, but Wellington, weakened by the loss of his cavalry, was now in desperate need of the Prussian reinforcements still pinned down in Plancenoit. Blücher's arrival behind the advance guard had been painfully slow. Napoléon ordered Ney to occupy the farmhouse of La Haye Sainte, which lay in the middle of the battlefield, but before the hamlet had been secured Ney launched a furious cavalry charge against Wellington's right center in the belief that he was pulling back. The steadiness of the English infantry squares and gunners resulted in another failed French attack, and Ney's cavalry was ultimately routed.

By about 5:30 P.M., the fighting around Plancenoit to the east escalated as Bülow's corps were joined first by Pirch and then by the seventy-three-year-old Prussian field marshal Blücher himself, and Napoléon had to commit increasing numbers of his reserves to this skirmish. However, at about 6 P.M. Ney finally managed to lead a properly coordinated attack against La Haye Sainte and began to punch holes in Wellington's center with a battery of guns. He needed Napoléon to send in the Imperial Guard to finish the job, but these crack troops were tied up with the Prussians at Plancenoit. Meanwhile Hans Ernest Karl Graf von Ziethen's First Corps arrived on Wellington's left wing, allowing him to divert Allied infantry units to shore up the center.

The battle entered its final stage after 6 P.M., with the arrival on the front line of almost a dozen battalions of the feared Imperial Guard. Ney led the two columns directly up against Wellington's right center, where his steadiest infantry lay waiting in the corn. The scarlet-coated British Guardsmen unleashed unremitting volleys of grapeshot into the French Guard, and soon after, Wellington signaled the bayonet charge that would finally break the French ranks, just as the Prussians came thundering out of Plancenoit to press the victory home. With the sun setting and the field swathed in plumes of black gun smoke, the duke sent about 40,000 men down the slope to finish the day. Napoléon made good his escape to Paris, leaving 25,000 of his troops dead on the battlefield along with 22,000 of the allied forces.

The victory was a narrow one, despite the ineptitude of Ney's command; had Napoléon triumphed, it is almost certain that the Seventh Coalition would have collapsed. However, Wellington won the day, giving England a victory of profound economic and political significance. While Trafalgar had assured British command of the sea, Waterloo opened to Britain the markets of the world.

Today the town of Waterloo commemorates Wellington's victory with a variety of monuments and historical exhibits, including the Lion Mound, a 150-foot monument on the battle site; the Wellington Museum, a collection of period uniforms and documents on the site of Wellington's headquarters; and the Ferme du Caillou Provincial Museum, a museum dedicated to Napoléon and placed on the site where the general and his staff made camp the night of June 17, 1815.

Further Reading: For a stirring military history of the Battle of Waterloo, J. F. C. Fuller's *Military History of the Western World* volume two (London: Eyre and Spottiswode, and Funk and Wagnalls, 1955) provides an excellent account, with vivid descriptions and close attention to tactics and strategy. *The Dictionary of the Napoleonic War* by David Chandler (London: Arms and Armour Press, and New York: Macmillan, 1979) provides useful background of all the people and events surrounding the battle, while R. J. White's *From Waterloo to Peterloo* (Middlesex: Heinemann, and New York: Macmillan, 1957) neatly analyzes the implications that the Battle of Waterloo had for policies in Europe, and more particularly in Britain, in the early nineteenth century.

—Daniel D. Collison

Weimar (Thuringia, Germany)

Location: In the middle highlands of Germany at the edge of the Thuringian forest; fifteen miles east of Erfurt, the economic and communication center of Thuringia; fifteen miles east of Jena, an important university town; fifty-four miles to the southwest of Leipzig, the closest major city.

Description: Intellectual and cultural center of Germany during the late eighteenth and nineteenth centuries; capital of the Weimar Republic from 1919 to 1933; center of avant-garde art scene known as the Weimar culture in the years following World War I.

Site Office: Weimarinformation
Marktstrasse 4
99421 Weimar, Thuringia
Germany
(3643) 20 21 73

Although this relatively small municipality (its 1989 population was estimated at 63,412) is an important rail center and is home to industries including agricultural machinery, chemicals, building materials, and furniture, it is not for these products that Weimar is known worldwide. Rather, the city is famous for the wonderfully rich artistic community that thrived there from the late eighteenth century up to the 1920s.

Weimar first appeared in a historical document dated A.D. 975. Three centuries later, in 1254, it was officially declared a town by the counts of Weimar-Orlamunde, under whose dominion it was located. In 1348, Weimar's status was upgraded to that of a municipality. The municipality later became the capital of the Saxe-Weimar Duchy in 1547, and the capital of the Saxe-Weimar-Eisenach Grand Duchy in 1815.

Weimar became an intellectual center within the various German kingdoms during the late eighteenth century. It was under the Dowager Duchess Anna Amelia and her son, Charles August II, that the town reached its peak as a cultural center. During their reign, great writers such as Johann Wolfgang von Goethe and Friedrich von Schiller lived and wrote in the city. The famous composer, Franz Liszt, was the city's musical director; Liszt was the first musical director to conduct Richard Wagner's *Lohengrin*. Friedrich Wilhelm Nietzsche, the influential philosopher, lived and died in Weimar. Johann Sebastian Bach, one of the world's greatest composers, was for a short while the court organist.

In 1775, King Charles August II, the young Duke of Weimar, invited Goethe to serve in his duchy as his companion and advisor. Goethe accepted the offer intending to remain in the city for only a short period of time. He ended up staying in the city for the rest of his life. While in Weimar, Goethe wrote prolifically and became a leading figure in the town's cultural and civic affairs. Although he was already an accomplished writer when he arrived in the city, it was in Weimar that Goethe blossomed into one of western civilization's greatest literary figures, on a par with Virgil, Dante, and Shakespeare. Some of the works Goethe completed while in Weimar include: *Iphigenie auf Tauris* (1779–88), *Torquato Tasso* (1780–89), *Romische Elegien* (1788), *Faust: Ein Fragment* (1790), *West-östlicher Divan* (1814–18), "Marienbader Elegie" (1823), *Novelle* (1826), and the completed version of *Faust: Part I and II* (1808–32).

Schiller lived in Weimar during much of this same period of time; he arrived in 1787 and remained until his death in 1805. While in the city, Schiller discovered a deep love of history, which motivated him to devote much of his energies, during his first years in Weimar, to writing historical works. Despite chronic illness, he wrote at least one major work during each year of his stay there. In addition to poetry and such plays as *Maria Stuart* (1800), he produced a series of important works dealing with the concept of aesthetic beauty, which included *Über Anmut und Würde* (1793), *Über die aesthetische Erziehung des Menschen* (1793), and *Über naive und sentimentalische Dichtung* (1795–96).

Die Horen, a literary journal, was started by Schiller in Weimar in 1794. Schiller's invitation to Goethe to write for the journal sparked a close association between the two men that lasted until Schiller's death.

When the journal proved to be a failure, these two literary giants collaborated on the work *Xenien* (1796). Over the course of the next nine years, the two worked closely preparing adaptations and translations of the works of Shakespeare, Gozzi, and Racine, thus helping to stage some of Weimar's best plays. Each man was also instrumental in pushing the other to pursue the completion of some of his greatest literary works. At Schiller's urging, Goethe again resumed working on *Faust*. Schiller, after being prodded by Goethe, completed *Balladenalmanach* (1797), which contains what many consider his finest ballads. Schiller's most famous play, *Wilhelm Tell* (1804), was completed one year before his death.

Throughout the rest of the nineteenth century, Weimar continued to be an important center for intellectual and artistic creativity, but it was not until the early twentieth century that Weimar fully bloomed once again. This period of the city's history introduced the world to the artistry and craftsmanship of Bauhaus; expressionist art, music, literature and film; the *Neue Sachlichkeit* artistic movement; and a newfound spirit of liberalism. This liberal, artistic, and intellectual renaissance achieved its greatest heights immediately following the end of World War I and lasted until the rise of Hitler's fascist regime in 1933. These were the years of the Weimar Republic.

Monument to Goethe and Schiller in Weimar
Photo courtesy of German Information Center

The Weimar Republic came into being when the German National Assembly met in February 1919, shortly after their country's defeat at the hands of the Allies in World War I, in an attempt to establish a republican and democratic Germany. By August 14 of that same year, the assembly's newly drafted constitution went into effect. The new constitution named Weimar as the seat of the National Assembly, rather than more traditional cities such as Berlin or Potsdam.

Weimar was chosen for two main reasons. First, Weimar could afford the National Assembly the security it required during the republic's turbulent first years. Many German cities were hotbeds of discontent, political demonstrations, and attempted coups des'état. Weimar was a smaller city, less caught up in political intrigues, and thus more able to insulate the assembly from the turbulence of the day.

The selection of Weimar was also supposed to represent a break from the domination in governmental affairs of Potsdam and Prussia, and a shift from the spirit of militarism and autocracy that many believed to be responsible for leading Germany down the destructive path of World War I. The city of Goethe and Schiller was elevated in the hope that this symbolic move would change Germany's character and destiny. But Germany's legacy of "blood and iron" would not be so easily repressed, and a dualist struggle emerged during the years of the Weimar Republic between those who wanted a more free and democratic Germany and those who favored a more conservative approach.

During the years of the Weimar Republic, representatives of what was to be termed "Weimar culture" created an art scene that was unabashedly modern and urban in its outlook. Those who helped to create the Weimar Republic's artistic scene felt that, with the end of the war, they were entering a new age, and they hoped to bring down the traditions and institutions of the past by creating a totally new system. Yet, these modernists were not the sole representatives of the art created within the confines of the city of Weimar, or in the whole of the Weimar Republic. Instead, these modernists represented only one of many different strands of thought, artistry, and expression found during the time of the republic.

In fact, the German cultural scene was highly fragmented and deeply divided. Avant-garde and modernity clashed head-on with tradition and antimodernism. With the government seated in Weimar, the city became closely associated with the more cutting-edge art, music, literature, and thought that was sweeping both the nation and world during the 1920's. It was this association that would cause so many conservatives to loathe the city throughout the 1920s, and into the 1930s, when Hitler and the Nazis rose to power. Those who deplored the cutting-edge art scene saw the government in Weimar as fostering artists who created nothing but garbage.

As more exhibits displayed the modern art of the Weimar Republic, as increasing numbers of museums added modern works to their collections, and as more expressionist and avant-garde artists and writers gained professorships at German universities, it might have seemed logical to assume that modern art would gain acceptance by the German masses. That was not so. A gulf between artists (and their supporters) and the masses remained throughout all of the Weimar years, separating those who appreciated the more modern works and those who loathed them. In fact, it seems that the more recognition the art received within the country, the greater the general public's hatred for it grew.

It seems odd, but in many ways the stylistic trends that swept through Germany, causing such an uproar from the traditionalists, strangely mirrored the hopelessness and fears of many within the Weimar Republic during the republic's formative years from 1918 to roughly 1923. The avant-garde and expressionist works of this period were disturbing, with scenes of seething unrest, apocalyptic visions, and almost universal hatred and fear of technology. This feeling was shared by many people throughout the country, whether they liked this kind of art or not. There was a general confusion about where the nation was headed. Germany had not only lost a war, but had been economically ravaged by it; its economy continued to suffer from runaway inflation and enormous debt; and now the citizens were forced to experiment with a federal and democratic system that appeared to many of them as strange and foreign.

By 1923, Germany had begun to move beyond these traumas and anxieties. The loss of the war was fading into memory. The economy was showing signs of recovery and stabilization. Massive demonstrations were becoming much less commonplace. Democracy and federalism were being given a chance. With this growing sense of stability came an end to the reckless and fatalistic visions associated with expressionism and some elements of the avant-garde. A change in the public's consciousness ushered in a new stylistic phase in the Weimar years: *Neue Sachlichkeit*.

Taken literally, *Neue Sachlichkeit* meant a "matter-of-factness." The movement treated technology as a fact of life rather than as a threat. Art was now less jarring than previous works. Phenomena, both in art and the real world, seemed to be more subdued. *Neue Sachlichkeit* had about it a respect for reality, while still retaining elements of the expressionist years, primarily abstractionism and certain elements of stylization.

Of all the artistic accomplishments of the Weimar Republic, it is the Bauhaus school that has made the most lasting impression both inside and outside of Germany. It helped introduce to the world modern architecture and modern furniture design.

In May 1919, six months after Germany's defeat in World War I, the architect Walter Gropius formed the Bauhaus in Weimar. It was more than just an art school, unlike any other art school in Germany. The school called for its students to cast aside everything they had previously learned and focus instead on their spontaneous impulses to develop a sense of touch, color, and space. Training both artists and craftsmen together helped to create a "cult of craftsmen-ship.'' Objects were designed to be both functional and simple in form. Objects created were quite often made from modern materials such as steel, glass, or concrete. Gone from designs were the superficial and artistic touches, associated with more traditional works, which served a solely ornamental purpose. In many ways the ideals of Bauhaus were linked to those of *Neue Sachlichkeit*.

Bauhaus had as its instructors some of the most talented individuals ever assembled at one school to teach architecture, design, and its related fields. The staff included painting instructors Lyonel Feininger, Paul Klee, Wassily Kandinsky, and Josef Albers; photography instructor László Moholy-Nagy; graphics instructor Herbert Bayer; textile instructor Anni Albers; stage designer Oskar Schlemmer; sculptor Gerhard Marcks; and architecture and furniture design instructors Marcel Breuer and Ludwig Mies van der Rohe.

Bauhaus would last in Germany until just before the rise of Adolf Hitler. His intense hatred of modern architecture hastened the school's demise. During its short existence, Bauhaus spent six of its years in the city of Weimar; in 1925 it was driven from the city following difficulties with the Thuringian government. From 1925 until its closing, Bauhaus would call the city of Dessau home.

It would not be an extreme statement to claim that the art associated with the cities of Weimar and Berlin during the years of the Weimar Republic was the most significant created thus far in the twentieth century. Virtually all forms of modern architecture, art, film, literature, and music can trace their roots to the Weimar culture. This artistic renaissance was able to sustain itself in an often hostile environment because it was given freedom and encouragement to do so by a democratic and republican government. It is ironic that many of the artists benefiting from these freedoms were hostile to the political system that made them possible. The artistic community's hostility toward the republic would help to bring about its downfall.

After 1929, the polarization between traditionalists and the modernists intensified markedly in Germany. The extreme rightist elements in society were gaining momentum. Weimar culture would fall victim not only to attacks from the right, but to changes within Germany's economy, the worldwide economy, and the advance of technology. A worldwide depression meant a curtailment of public grants funding the arts. Technological advances dealt the literary and theatrical worlds a severe blow as talking films and radio began to rob them of their audience. Affected were not only avant-garde works, but also more traditional forms of art such as the opera and symphony.

It has been reported that upon his ascension to power, Hitler was quoted as saying about the Weimar years, ''Fourteen years, a junkyard!'' Hitler never hid his hatred for the artistic community associated with Weimar, and he did everything he could to crush it. He was very effective. Those artists who continued to create influential works could only do so by leaving the country. Many artists did so as early as the late 1920s.

World War II also brought the city of Weimar some unwanted notoriety; it became famous for its proximity to Buchenwald, site of a notorious Nazi concentration camp. At Buchenwald, approximately 56,000 prisoners were starved and tortured to death in a facility whose body disposal plant could handle up to 400 dead bodies every ten hours. Victims at the camp were political dissidents, individuals from the occupied lands, and Jews.

During the war, Weimar was not spared from Allied bombings. Despite this bombardment, in no city in the east of Germany were so many historic structures spared from destruction.

After the World War II, Weimar came under the control of the Communist government of East Germany until 1989, when the two Germanies were united. Being under Communist rule for almost half of a century did little to change the character of the city. Walking through Weimar, one finds the city in much the same condition as it was during the time of Goethe and Schiller. Narrow cobblestone streets are surrounded by architecture dating back hundreds of years.

Important historical sites in Weimar include the Goethe National Museum, the home where Goethe lived from 1782 to 1832. The main attraction of Weimar, the house contains fourteen rooms that reveal the varied interests of the man. Many of the rooms are in nearly the same condition as they were upon Goethe's death.

The Goethe garden house, located in a small park on the Ilm River, is yet another structure devoted to the city's most beloved literary figure. Originally his first residence, it was later used by Goethe as a summer home.

The Schiller house is also preserved in Weimar. Schiller and his family lived in the home from 1802 until 1805. Currently a museum, it houses mementos of Schiller's life and his works.

Wittums Palace, the castle where the Dowager Duchess Anna Amelia and her son, King Charles August II, lived, is yet another landmark preserved for the public. The castle is currently devoted to displaying objects from the German classical Enlightenment.

Château Belvedere and Tiefurt Castle, both from the time of Anna Amelia and Charles August II, have been preserved. On Château Belvedere's baroque walls Goethe and Schiller once staged plays. Tiefurt Castle was a summer home for the imperial family.

Other areas of interest in the modern-day city of Weimar include the home of Franz Liszt, which houses many of the musician's personal effects. Nietzsche's Weimar residence, where he lived for three years, now houses the philosopher's archives. The National Research and Memorial Center of the Classical Writers of German Literature is also located in the city. Established in 1953, its purpose is to celebrate the literary works of German classical writers. It is housed in Weimar Castle. A national observatory, an agricultural school, the Franz Liszt College of Music, and the national headquarters for the German Shakespeare

Society round out the list of famous historical sites within the city. Outside of Weimar is the Buchenwald concentration camp.

Further Reading: Eberhardt Kolb's *The Weimar Republic* (London and Boston: Unwin Hyman, 1988) details the problems of the Weimar Republic. Very thorough in detail, it briefly touches upon the city of Weimar itself. Gordon A. Craig's *Germany* (Oxford: Clarendon, and New York: Oxford University Press, 1978) provides the reader with an unbiased and informative account of German history from 1850 to 1945. Allan C. Greenberg's *Artists and Revolution: Dada and the Bauhaus 1917–1925* (Ann Arbor: UMI Research Press, 1979) provides the reader with information about societal and political divisions within the Weimar Republic's artistic community.

—Peter C. Xantheas

Winchester (Hampshire, England)

Location: Winchester lies close to the south coast of England, north of the port of Southampton and to the southwest of London.

Description: Surrounded by undulating downs, the center of the town lies on flat lands. At its heart is the massive cathedral, together with most of the remnants of medieval Winchester; the fine domestic architecture of Stuart and Georgian periods is also to be found in this area. The college and its meadows are situated close to the cathedral. On the west hill, the Great Hall is all that remains of the castle; next to it is one of the former gates to the city.

Site Office: Winchester Tourist Information Office
The Guildhall
Broadway
Winchester, Hampshire SO23 9LJ
England
(962) 840500

Best known as the Saxon capital of England, Winchester was home to the West Saxon or Wessex kings. In particular, it is linked with the names of the powerful King Alfred and the legendary King Arthur. Most of its vast, imposing cathedral dates from Norman and Plantagenet times.

Before the Anglo-Saxon and Norman periods, the site was inhabited first by Iron Age people, and then by Romans. Evidence of an Iron Age fort has been unearthed on St. Catherine's Hill, to the south of the city. Precise dates for the Iron Age and Roman settlements are not known, but coins found depicting the Roman emperors Claudius and Nero indicate a Roman presence there from quite soon after the Roman invasion of Britain. The Roman name for the place was Venta Belgarum, or the market town of the Belgii, and it was apparently an important trading center. Excavations have revealed that much of the medieval city wall was built on top of a Roman one, while ornamental mosaic pavements attest to the settlement's prosperity.

Fiction has triumphed over fact in the tales of King Arthur resisting the Anglo-Saxon hordes who invaded the British Isles once the Romans had left. Winchester has been identified by some as Camelot, Arthur portrayed as the builder of the early Winchester castle in the sixth century, and its Great Hall depicted as the setting of his legendary Round Table. The stories hold no historical truth, although the tales may be a glorifying portrayal of a Roman-British chief resisting the invading tribes.

It was Cenwealh, second son of Cynegils, king of Wessex from 611 to 641, who, once he had converted to Christianity like his father before him, made the important decision to have a minster church built at Winchester. Bishop Haeddi of Wessex supposedly transferred his religious seat from Dorchester to Winchester in 676. From that time Winchester's history was tied to that of the West Saxon kings and their civilization. The minster, which came to be known as the Old Minster, was where most of the them were buried. It lay just north of the cathedral that replaced it. Thus, Winchester became a royal and religious center. Many have described Egbert, king of Wessex from 802 to 839, as the first king of a united England, his center at Winchester as the English capital. The bishops were close confidants of their kings, extremely powerful members of the establishment, and influential educators. Trade and the arts flourished.

England was being attacked by the Vikings in the ninth century, and one of the bishops of Winchester, St. Swithun, erected a defensive wall around the town. Alfred, known as Alfred the Great, became king of Wessex in this period. He is commonly portrayed as perhaps the most heroic king of English history, defeating the Danes, and reestablishing order and learning. He strongly encouraged the power of the church, and two further important religious establishments in Winchester stem from his reign. Nunnaminster was a convent he founded with his wife Alswitha; after his death she lived there until she died in 902. Donation of land by Bishop Denewulf around the start of the tenth century led to the building of the New Minster next to the Old Minster, fully established by Alfred's son Edward the Elder, with the aid of the French scholar St. Grimbold. Alfred and Alswitha were buried in the New Minster, which became the home of an extraordinary collection of holy relics.

Under King Edgar's rule in the late tenth century, Bishop Aethelwold placed monks in charge of the Old Minster. Important additions were made to it at the end of the century to give greater prominence to the resting place of St. Swithun's bones, his cult having flourished, and to vie with the New Minster. Legend has it that when the saint's remains were transferred, it rained for forty days and nights, hence the superstitious predictions should it rain on St. Swithun's day, July 15. The town's three important monastic institutions were highly influential in setting the tone of religious revival in England, and also produced fine illuminations, commonly termed of the Winchester school, one of the most elaborate examples being the Benedictional of St. Aethelwold. The prosperous town had a substantial defensive wall and was manned against the threat of invasion. It had its own mint and, with London, kept a set of standard measures for weights and lengths. But it was Winchester that was the effective capital of England at this time. Here Edward the Confessor was crowned in 1043, in one of the last great ceremonies of Anglo-Saxon England.

With the Norman conquest, Winchester was handed

*Winchester Cathedral, top, and the exterior of the Great Hall of the Castle, bottom
Photos courtesy of Winchester City Council*

over quite peacefully to William the Conqueror by Queen Edith, Harold Godwinson's sister, and widow of Edward the Confessor. She continued to reside in the town undisturbed. Her chaplain and adviser Bishop Stigand, however, spent the rest of his life in custody, while Earl Waltheof was executed outside Winchester for plotting against the new king. The Normans started building an imposing castle in Winchester, and the town remained a royal center, with a growing bilingual civil service. Although London's power was increasing, the records for the Domesday Book were sent to Winchester. The royal treasury was also established there, and the town prospered from the flurry of Norman activity.

The conquest affected the town's religious institutions. The New Minster's abbot was among those killed at the Battle of Hastings in 1066, and some of its buildings burned down the same year. Also, the new rulers decided to tear down the Old Minster. To replace the Old Minster, construction of a vast Norman cathedral began in 1079 under the first Norman bishop, Walkelin. Its west front went out much farther than it does now, but at 556 feet in length, it remains the longest of all medieval cathedrals. The first tower collapsed, and was rebuilt under the second Norman bishop, Giffard. Some blamed the disaster on the fact that King William Rufus, considered by many an immoral king, was supposedly buried under it. (Rufus had, however, granted Walkelin the right to hold a fair of St. Giles, after which a Winchester hill was named, and which would develop into one of the most important trading fairs in Europe in the Middle Ages.)

In 1093, with the consecration of the cathedral, the shrine of St. Swithun was ceremonially transferred to the new building, and on the same day destruction of the Old Minster began. The New Minster was transferred in 1110 by Giffard to the north of the city, and was renamed Hyde Abbey. Through the medieval period Winchester was a city where the bishops wielded great power, and ecclesiastical institutions were the town's main landowners.

King Stephen's reign, from 1135 to 1154, was troubled by civil war. Stephen's brother Henry of Blois became bishop of Winchester. A man of action, Henry helped fight Matilda, their cousin, who was trying to claim the throne. He fortified his bishop's palace at Wolvesey, but burned down part of Winchester, including the Norman castle, during the fighting. He also was a great patron of the arts and charitable institutions, and founded the Hospital of St. Cross in the village of Sparkford, then outside Winchester. In 1153 the Treaty of Winchester brought a conclusion to the war, establishing the succession of Henry Plantagenet to the English throne. Construction of a new royal castle, which included the kingdom's treasury, began at mid-century. Under King John's reign, from 1199 to 1216, the mayor replaced the alderman of the guild as the principal official of the town. Winchester also was granted the right to oversee its own finances without going through the county sheriff. This was an important element in securing the town's independence.

Henry III was born in Winchester castle in 1207, and would devote much attention to the city. A Great Hall was constructed beginning in 1222, and still stands as Henry III's lasting mark on Winchester. A dungeon was added, beginning in 1259. A few of the first English parliaments, instituted by Henry's brother-in-law Simon de Montfort during the civil war in this reign, were held in the hall, while in 1270 Henry's son and heir Edward gathered his troops at the castle to set out on a crusade. During these times, Winchester was characterized by disputes, both internally, among the town's business and governmental leaders, and externally, with its neighbor and trading partner, Southampton. King Edward I even had to intervene in 1274. The town was declining in importance, though. The mint was closed in 1279. The Winchester Jewish community, which dated from the Norman conquest, was persecuted during Edward's reign. Jews had made an important contribution to the city's trade, Winchester had not been affected by the pogroms that occurred elsewhere in England in 1189 and 1190, and one Jew, Benedict the Guildsman, had achieved the distinction of being made a member of Winchester's merchant guild in 1268. Still, the Jews' position in Winchester was precarious, and anti-Semitic sentiments led to Edward's expulsion of Jews from England in 1290.

Medieval Winchester was rich in church communities. There were the churches of the three great Benedictine foundations: the priory of St. Swithun surrounding the cathedral on all sides, the Abbey of St. Mary (the former Nunnaminster), and the Hyde Abbey church. Many parish churches were built in this period. The town contained five hospitals for the poor. William Wayneflete, the master of St. Mary Magdalene hospital at one time, would become bishop of Winchester, and found Magdalen College in Oxford. The city governing body met at St. John's—the oldest of the hospitals, perhaps dating back to the end of the tenth century—to conduct its business. From the 1230s onward the town had four friaries.

Winchester's economic activity in the Middle Ages centered on the wool trade. The main event of the trading year was the great fair of St. Giles, extended from its original three to sixteen days. The traders from nations such as France, Spain, and the Netherlands bore wine, spices, silk, and other luxury goods to exchange. The hill had a fixed grid of streets and stalls, and a special court, named the Pie Powder court, was instituted for the fair time. Both fair and its court were under the bishops' control.

In 1302 much of the castle was destroyed in a fire, from which Edward I and his wife narrowly escaped. The east end of the cathedral was altered in the late twelfth and early thirteenth centuries. The west front with two towers was added at a later date.

The middle of the fourteenth century was marked by the great plague, which killed perhaps three-quarters of the population. Poverty and decline ensued. In 1370, Bishop Wykeham had to threaten excommunication to persons caught stealing building materials from the abandoned parish churches. The wool trade had already been diminishing greatly in importance from the start of the fourteenth century.

In Edward III's reign (1327–77) numerous foreign weavers were brought there under royal protection to help the wool trade survive.

There were important new developments, however, in the town in the second half of the fourteenth century. Two forceful bishops, Edington and Wykeham, ordered important changes to the cathedral in the time when English Gothic architecture was at its high point. The great reconstruction of the nave was still not completed by the end of the century. Wykeham founded Winchester College, which through its existence has produced an extraordinary number of prominent figures of the establishment in Britain. The school's pupils are known to this day as Wykehamists. Teaching apparently started in 1373; in 1382 the school was granted a royal license by King Richard II. The original school was for seventy scholars, and the college buildings date from 1387 to 1400. Wykeham also founded New College, Oxford University, as a place where the scholars could continue their studies.

After Wykeham's death in 1404, the city suffered further steady decline through much of the fifteenth century. In 1449 a new fair was instituted beginning on the vigil of St. Swithun and lasting ten days; its emergence demonstrated the decline of the once-great St. Giles fair. The advent of the Tudor dynasty with Henry VII's accession in 1485, however, brought royal attention back to Winchester. Henry VII played on the ancient royal connections of the city as part of his policy of establishing the authority of his family to rule the country. His first son, named Arthur, was born in Winchester, and christened in the cathedral. A number of Wykehamists succeeded as eminent doctors in this time, some serving as royal physicians.

Henry VIII's reign brought the dissolution of the monasteries. In Winchester, many of the major institutions were destroyed, including Hyde Abbey and the Abbey of St. Mary. The Priory of St. Swithun's, next to the cathedral, survived, though in modified form, the prior and convent becoming dean and chapter. The cathedral was barely touched, except in 1538, when three royal commissioners destroyed St. Swithun's shrine and took down the high altar. One of the three, Thomas Wriothesely, would later take away many of the cathedral's treasures. The Holy Hole, through which relics had been exhibited, was filled in. The chapter continued as a wealthy establishment, but in 1545 its duty to support the Wykehamist scholars at Oxford and Cambridge was handed over to the crown.

The town's four friaries were all dissolved. Henry VIII handed over their sites to Winchester College as an exchange in 1543. The college's connection with Oxford University, as well as lavish attention to the king's retinue on a royal visit, protected it from dissolution.

When King Edward VI visited the town in 1547, he ordered that the Bible be read in English and banned prayers and anthems to the Virgin Mary. Catholicism returned, however, under Queen Mary I, whose marriage to Philip II of Spain in 1554 was marked by a mass in the cathedral. Winchester was a Catholic stronghold even after Elizabeth I restored the Church of England, and the execution of two Catholic priests only inflamed Catholic feelings. In 1564 Mayor Robert Hodson and other city officials were imprisoned in London for a time for their lack of zeal against Papists.

Under Elizabeth I, in 1588 the city received a new charter, which detailed roles of responsibility to a degree no previous royal charter had. Sir Francis Walsingham, the queen's secretary of state and a privy councillor, helped draw up this charter, and became high sheriff of the city for life. The city in the seventeenth century became less staunchly Catholic; the first bishops of the century were firm followers of Archbishop William Laud's Anglican revival. The cathedral underwent physical changes at this time. A screen designed by Inigo Jones was erected to separate choir from nave, and was adorned with bronze statues by Le Sueur of James I and Charles I.

Winchester was of course rent by the Civil War. Twice the royal garrison in the town surrendered to supporters of the Parliamentarian opposition. On Charles I's last visit to Winchester in 1648, the Royalist mayor gave him the keys of the city before being attacked by the guards. In 1649, the Royalists were expelled from the city government. Of the local institutions, only the college and a few hospitals emerged unscathed from the war. The college was fortunate to have a supporter in Nicholas Love, one of the most influential members of Parliament and a Wykehamist, and many of the cathedral manuscripts were transported to the college by his doing. Somehow Wykeham's chantry chapel in the college survived the purges of the Puritans; another Parliamentarian Wykehamist, Nathaniel Fiennes, apparently saw to its protection. The cathedral's dean and chapters were exiled and the bishop's palace became a ruin. The cathedral itself was spared important damage. The castle, however, was almost entirely dismantled by 1660, except for the Great Hall.

The restoration of the monarchy in the 1660s brought increased royal intervention in the town's affairs. This intervention became open when Edward Harfell was made a member of the city governing body, or corporation, in 1682 at the King Charles II's specific request. The following year, Charles instructed Christopher Wren to draw up designs for a royal palace in Winchester. The foundation stone was laid that year, and Charles donated a famous portrait of himself by Peter Lely to the town. By now, court control was firmly established. In 1684 the corporation surrendered the charter that had been issued by Elizabeth I; a coach delivered it to Charles while he was attending the fashionable Winchester horse races. The new charter of 1688 basically aimed to put an end to local independence, with central government choosing all the members of the corporation, but it was short-lived, as King James II's three-year reign ended that year, and under King William III the corporation's old privileges were restored to Winchester. One result of the royal interest in the city was that the court in Winchester attracted wealth and fine architecture. The bishop of the period, George Morley, rebuilt Wolvesey, made donations to various charitable homes, and collected a fine library, now in the cathedral. Doctors proliferated in the city and there was a movement to set up the first county hospital outside London.

Work on the Wren-designed royal palace, however, progressed slowly. Queen Anne intended to finish it, but after her death in 1714 her successor, George I, reversed royal interest in Winchester and began to have the building dismantled. The remains of the palace housed French prisoners of war in 1756, and were used as a prison in later wars as well, including the American Revolution. Much of the rest of the city decayed in the eighteenth century. The medieval city gates and walls were crumbling, and in the 1780s much of the new Wolvesey Palace was demolished. There was important new building, though, reflecting the changing social tide: the county hospital opened in 1737, a theater in 1785, and a prison in 1788. In 1792 a new Catholic chapel (now Milner Hall) was built; it was an important example of Gothic revival, planned by John Carter for Bishop John Milner and decorated by William Cave. Milner was an important figure in the town, not least because he wrote a standard history of Winchester.

The college had been in decline in the early part of the eighteenth century, seen by many as encouraging Jacobite ideas. The number of paying students, so-called Commoners, decreased drastically by mid-century. Its fortunes rose again under a new headmaster after this period, although the 1770s in particular saw a number of school riots, often caused by conflict between town and gown (so-called because the college scholars wore gowns). The riot of 1770 was sparked off when the landlord of the White Hart Inn requested students of the college to leave the premises. The school's reputation, apart from its academic one, was of a rough institution producing reactionary politicians.

Winchester was not fundamentally changed by the Industrial Revolution because of its lack of industrial resources. The town did contain a high number of men of the professions. The reforming acts of the 1830s and the vastly broadened franchise they brought focused elections on local issues. Winchester had suffered from lack of sewage facilities, and the life expectancy of its inhabitants was much shorter than in towns with better sanitation. The committee overseeing the matter expressed its "earnest hope that the time is not far distant when Winchester shall cease to remain a city of cesspits." The Garnier Street pumping station, named after a cathedral dean, came into operation in 1875. The mid-nineteenth century saw other advances in civic in-stitutions, with the opening of the Hampshire County Museum and the Public Library. Social reform advanced greatly thanks to the deans of the time. A keen sense of religious duty also led the Victorians to restore the churches of the town, and add such landmarks as the St. Thomas spire. Certain changes to the cathedral earlier in the century were less successful, much of the perpendicular work in the transepts being destroyed and replaced by neo-Norman.

With the opening of the railway line and the advent of the automobile, tourism became important. The town expanded rapidly. Saxon relics were discovered during construction of buildings on Winchester's eastern hill. On the western hill, the county council buildings were erected following the passing of the Local Government Bill in 1888. The remnants of Charles II's palace were destroyed by fire in 1894. Wolvesey Palace was rented out, and became a museum for stuffed birds for a time, while the bishops preferred to live in Farnham Castle. The city held a celebration honoring Alfred the Great in 1901, and ordered the creation of the imposing statue of Alfred by Hamo Thornycroft, which rises in the middle of the Broadway.

At the start of the twentieth century there were fears that the cathedral's foundation was no longer stable. A huge pageant was held in 1908 to help raise attention and funds. Workers completed the dangerous task of laying cement under the cathedral in 1912. The town suffered some bomb damage in World War II. In recent decades, a major development was the campaign was mounted by protesters against the route of the planned Winchester bypass, menacing some of Winchester's most beautiful undeveloped areas. The bypass has been built, avoiding the meadows that are said to have inspired Keats's *Ode to Autumn,* but cutting through St. Catherine's Hill. Today Winchester College continues to educate future leaders, while the cathedral, the Great Hall of the former castle, and other historic buildings provide reminders of Winchester's past.

Further Reading: Paul Cave Publications of Southampton have published solid, reliable histories of the town, the cathedral and the college. *Winchester* (first published 1980; revised, 1992, as *A History of Winchester*) by Barbara Carpenter Turner covers the town in general; *Winchester Cathedral, 1079–1979* (1979) by Frederick Bussby deals with the cathedral; and *Winchester College* (1981) by James Sabben-Clare concerns the college. *Winchester Cathedral: 900 Years,* edited by John Crook (Chichester, West Sussex: Phillimore, 1993), is another serious study of the history of the cathedral.

—Philippe Barbour

Windsor (Berkshire, England)

Location: Twenty miles southwest of London, on the Thames River.

Description: Site of Windsor Castle, the largest inhabited castle in the world, which still serves as the monarch's official residence. The Norman edifice is more than 900 years old, but the location may have been used by the invading Roman army.

Site Office: Tourist Information
Central Station
Thames Street
Windsor, Berkshire SL4 1PJ
England
(753) 852010

The story of Windsor begins many centuries before the construction of the castle commenced in 1066. Roman coins minted as early as A.D. 69 have been found in local gardens. Some historians suggest that the Second Legion "Augusta" moved up the Thames Valley to make camp at the hill-top location when the Romans made themselves the rulers of the countryside. The site, high above the river, lent the invaders an excellent view of the land.

After the Roman Empire declined, England was again conquered, this time by Anglian invaders. As early as the seventh century, Saxon sovereigns held court and had festivals at a royal manor in the settlement now called Old Windsor. The site on the chalk cliffs where Windsor Castle stands was not used by the Saxons. In January 1066, Saxon monarch Edward the Confessor died, signaling the next and final successful invasion of England. Windsor Castle was built by these Norman conquerors and has belonged to their descendants for more than 900 years.

After he returned from his victory at the Battle of Hastings in October 1066, William the Conqueror ordered the construction of a ring of defenses around London. The stronghold at Windsor, on the high cliffs overlooking the approach to London along the Thames, was one of these. It was situated outside of Windsor proper in the manor of Clivore, later Clewer, and took its name only for its proximity to the Saxon royal manor at Windsor. The first fortress possessed little of the grandeur and none of the elegance of the later castle. Simple in form and constructed of locally available materials by Anglo-Saxon labor, the structure was quickly built and was virtually impregnable at the time.

Conforming to Norman principles of military engineering, the nearly rectangular thirteen-acre site was enclosed with a wooden palisade and encircled with a ditch except on the northern side, where the cliffs were so steep that other defenses were superfluous. Within the palisade the site was divided into three areas called wards or baileys, and this division is still in evidence today. The Lower Ward took up the western half of the area and contained stables, storehouses, and other service buildings. The eastern Upper Ward, divided from the Lower Ward by a ditch and drawbridge, itself contained the Middle Ward, which was the center of the stronghold. The Middle Ward consisted of a huge earthen mound encircled by another ditch. The keep or *donjon* was built inside a stockade on top of the mound.

The fortress could be accessed only by the drawbridge and gate at the southwest end that gave onto the Lower Ward. The various ditches around and within Norman fortifications were usually filled with water, but the porous chalk cliffs of Windsor would not permit this, even if the Normans had known how to raise water to such a height. However, a well was sunk through the cliffs down to the level of the Thames to supply the stronghold with water. Rainwater may also have been collected at this time.

During the first decades of its existence, the fortress was not used as a royal residence. The Norman kings continued to use the Saxon manor at Old Windsor, more suitable for royal festivals and more convenient for hunting expeditions to the surrounding forests. The fortress was designed as a refuge of last resort for the king during times of siege and as a place to hold prisoners. The most famous prisoner during this period was Robert, Earl of Northumberland, who had rebelled and was captured by William II. Robert languished in the Windsor keep for thirty years.

Henry I was the first English king to assemble his court in New Windsor in 1110, and the castle was accordingly provided with royal apartments. There is no record of this first palace in Windsor, but it is believed to have been placed in the northeast corner of the Upper Ward, the more defensible half of the fortress. Henry would have had a magnificent view of the Thames Valley from his quarters. An assembly hall, a chapel, and further residential quarters were added in the Lower Ward, to accommodate the court.

Windsor Castle began its transformation from wood to stone under King Henry II, in the latter half of the twelfth century. Henry's four contentious sons—among whom were the later King Richard I, Coeur de Lion, and King John—had mounted a rebellion, and Henry sought to improve his fortifications. The wooden keep made way for a stone shell keep, that is, a tower surrounded by a stone collar from behind which bowmen could shoot at an enemy who had broken through the other defenses. The new keep, called the Round Tower (though it is in fact closer to a rounded square), was built out of heath-stone, quarried in the vicinity, and still dominates the Windsor skyline virtually unchanged. The palisades on the northern, eastern, and southern sides were replaced by walls of the same stone. Though restored and

The Upper Ward of Windsor Castle

The roof of St. George's Chapel

changed, these walls still stand. Three defensive towers were added to the walls, and these served as prisons throughout most of their history. By the time work on the western wall was to start, Henry had put down the rebellion, and the project was discontinued.

The king also replaced the original wooden royal apartments with a stone structure built around a cloister, or herb garden. The medicinal qualities of the herbs in the garden were thought essential for sanitary reasons. In the Lower Ward Henry added the first Great Hall, built out of stone. This was the first of a series of assembly halls for the king's knights, and, of all the buildings in the complex, the greatest political significance was attached to this structure. Since its magnificence symbolized the power and centrality of the monarchy, great care was lavished on this and later versions of the Great Hall.

The castle's defenses were first tested in 1193, during the reign of Richard I. While the crusading king was making his way back from the Holy Land, his younger brother John usurped the throne, garrisoned Windsor Castle, and arranged for Richard's imprisonment in Austria. An army of knights, still loyal to Richard, attacked the castle the following March. After a month's siege, John surrendered the castle to Eleanor of Aquitaine, the Queen Mother. The siege of 1194 was not John's only unsuccessful confrontation with the barons. After his accession to the throne upon Richard's death, John made himself unpopular with the nobility and common people alike. He was not only a glutton with a cruel and uncontrollable temper, but his tyrannical rule moved the barons to revolt in 1215.

Without baronial support and thus without an army, John was powerless, and he entered into negotiations with the barons in the meadow at Runnymede, some three miles downstream from Old Windsor. After several days of negotiations, John signed the earliest bill of rights in European history, the Magna Carta, which subjected the king to the rule of law and detailed the liberties of his subjects. Not content that his barons had been able to force his hand, John managed to persuade the pope to annul the charter about a month later.

The English knights again rebelled, this time calling upon France for aid. The French army invaded the country in 1216 and quickly took all strongholds, except Dover and Windsor. In May of that year, John left Windsor to meet his enemies, never to return. The castle was besieged by French and baronial forces but did not fall. The conflict came to an end when John died at Newark, on October 19, and a final settlement was made at Windsor. John was succeeded by Henry III, a more able and deliberate monarch. Realizing that he was so dependent on his feudatories that he could not govern without their consent, the new king reissued the Magna Carta in 1225 in its current form. It is the earliest enactment of the Statute Rolls of England and is considered the starting point of constitutional rule in the country.

Although John had liked Windsor and spent much time there, he did not improve the castle, adding nothing to it but records of scandalous gluttony and heartless brutality. Reportedly he imprisoned Lady de Braose and her son at Windsor, after her rebellious husband had fled the country. Locked up without water and provided only with some wheat and raw bacon, the two prisoners perished before their dungeon was reopened eleven days later. According to legend, Lady de Braose had vainly tried to keep herself alive by gnawing away her dead son's cheek.

Henry III, by contrast, made extensive improvements at Windsor and left a record of charity, regularly opening the castle to the poor, who were fed and clothed at his expense. In 1227, Henry walled the west end, providing it with the rounded towers that still form a marked contrast to the earlier square towers built into the other stone walls. The Lower Ward was extended across the old ditch, a new ditch was dug, and a barbican was added on the other side. A portcullis further fortified the main gate. To accommodate the queen, Henry also added to and beautified the royal apartments in the Upper Ward. After a violent thunderstorm did extensive damage to the queen's apartments, a separate queen's lodging around a second cloister was built in 1255. Henry's efforts also included a decorative program that, among other things, furnished the royal apartments with splendid colored glass. These improvements made Windsor the foremost royal residence in England and one of the most splendid in Europe. By this time, a sizable town had developed on the western side of the stronghold, but on all other sides the castle was surrounded by its parks. These were primarily hunting grounds, though over time houses for royal relatives and mistresses were built in both the Little Park and the Great Park.

Relative quiet reigned at Windsor for the next century, punctuated only by an elaborate tournament put on by Edward I in 1278. It was not until the warlike Edward III acceded to the throne that major changes once again took place at the castle. In 1344, at the close of a tournament that had drawn guests from all over western Europe, Edward announced his intention to revive the Arthurian Round Table. Immediately afterward, the construction of a great hall in the Upper Ward was begun to accommodate the revived order. By 1347, however, Edward gave a new direction to his chivalric ideals and instead created an order dedicated to St. George, the Order of the Garter. The order was formally instituted in 1348 with twenty-four founding knights in addition to the king and the Prince of Wales. The great new hall was dubbed St. George's Hall and became the site for the annual festivals of the order on St. George's Day. Edward also added a number of buildings for the clergy in the Lower Ward, which were granted to them as freehold within the castle domain, and he started reconstructions in the Upper Ward in 1359.

Little is to be reported of the last decades of the fourteenth century at Windsor, but Geoffrey Chaucer was clerk of the works for a time. He seems to have done little to maintain its physical splendor, but he did add to its cultural riches. Parts of the *Canterbury Tales* were apparently written for recitation at Windsor on Garter Day. By the mid-fifteenth century, the castle had been neglected for so long that it was visibly dilapidated. Edward IV undertook some reconstruction and began rebuilding St. George's Chapel in the Lower Ward in the Gothic Perpendicular style. It is particularly notable for its complex fan vaulting over the nave and choir. One of the chief attractions of present-day Windsor Castle, this gorgeous chapel was finished some fifty years later by Henry VIII. Henry, favoring convenience over considerations of defense, also built a wooden platform on the north side, outside the stronghold.

The reign of Henry VIII brought more than physical changes, however. When the king broke with Rome, he removed all associations with St. George from the Order of the Garter and decreed that the annual festivals could thenceforth be held wherever the king happened to be. Accordingly, Windsor began to take a less central place both in the actual workings and the mystique of English government.

Windsor's history once again accelerates during the Civil War in the mid-seventeenth century. Windsor Castle was taken over by the Parliamentary army under Colonel John Venn shortly after the outbreak of hostilities in 1642. Royalists attacked and pounded Windsor with heavy artillery but were unable to breach the walls. Parliamentary army headquarters were safe there for the duration of the war. Venn became governor of the castle, allowing the destruction and dispersal of its riches. Although some items were simply destroyed, many of the valuables were sold to subsidize the war effort. The army used Windsor both as a prison, a meeting place for the long prayer meetings presided over by Oliver Cromwell, and later as a training ground for new recruits. The most famous prisoner at this time was the defeated Charles I, who was brought in December 23, 1648, only to leave again for his trial and beheading at Whitehall in January 1649. Parliament had the body returned to Windsor Castle, where it was eventually entombed in a vault in St. George's Chapel together with the remains of Henry VIII and Jane Seymour. In 1652, Parliament considered selling Windsor Castle but refrained from doing so. Both parks were sold, but were recovered later.

In 1660 Windsor reverted to the crown, when Charles II was proclaimed king. Restoration and reconstruction began immediately. The most notable work was done on the royal lodgings, which were entirely redone and richly decorated with elaborate stone carvings, ceiling paintings, and gilding. St. George's Hall was also rebuilt in the baroque style, and the royal chapel was redecorated.

Windsor was fully restored to its previous state by the time William III was invited by Parliament to take over the government from the all-too-openly Catholic James II in 1689. William spent his time at Windsor until James finally left the country and the Glorious Revolution was accomplished without bloodshed. Although William and Mary frequented Windsor Castle, their preferred residence was the newer palace at Hampton Court. Before her accession to the throne, Queen Anne lived in a house in the Great Park and

relished her move to Windsor Castle in 1702. Her chief addition to the complex was the design of the gardens in the Great Park, just north of the castle. After her death, however, the castle was seriously neglected by her successors. Over the course of the next seven decades, the royal apartments became uninhabitable.

Charlotte, wife of George III, took a liking to a house in the Little Park and regularly stayed there with her family in the 1770s. Queenly favor for the Little Park gradually brought back life to the castle itself also. The ditch was filled in, and some repairs were undertaken, enough to allow a royal open house in 1780. In subsequent years, the royal family came to stay at Windsor more and more often and gradually transformed the State Apartments along the northern and eastern fronts of the Upper Ward. The outer northern facade was also changed, with new towers to relieve its monotony. Windows in the Gothic Revival style gave a new look to the previously somber and windowless defensive walls.

When George IV became king, however, a new program of major reconstruction was embarked upon. A new gate with towers provided an entrance in the middle of the south front, which still serves as the main entryway to the complex today. To suit romantic tastes, battlements and machicolations—which the castle had never had or needed— were added to many of the outer walls. The most prominent of these battlements are those added to the old Norman Round Tower, heightening it by some thirty feet. The old royal chapel and St. George's Hall were destroyed to make room for a larger and more splendid St. George's Hall in the Gothic Revival style.

Queen Victoria was essentially satisfied with the improvements undertaken by George IV. Only the Grand Staircase to the State Apartments was built, in 1866, to replace the less than splendid staircase constructed under the auspices of her predecessor. Victoria increasingly used Windsor to conduct state business when improvements in public transportation put the castle within easy reach of the capital. The queen went into relative seclusion and avoided Windsor after the death of her husband, Prince Albert, however. Edward VII used Windsor often as have his successors. Today, the royal family still lives at the castle on weekends, while large parts of the complex are open to the public in their absence. The architecture and furnishings are not the only points of interest: the castle's art collection, containing works by da Vinci, Michelangelo, and Raphael, is also quite splendid.

On November 20, 1992, a fire broke out in the queen's private chapel in the Upper Ward, reportedly because a light had been placed too close to a curtain. The most serious damage involved St. George's Hall, which was almost completely destroyed. The cost of reconstruction was estimated at $100 million. Since the castle had not been insured, the restoration would have to be financed with public funds. A public outcry resulted, and as a consequence the queen agreed to contribute to the project. However, Windsor Castle is so closely associated with the British monarchy and even the nation that the restoration of St. George's Hall was never seriously under debate, despite the prohibitive cost.

The centrality of Windsor Castle in the history of the British monarchy has been such that, when their German ancestry and titles became an embarrassment to the royal family during World War II, these Hanoverian monarchs chose to rename themselves the House of Windsor.

Further Reading: *The History and Treasures of Windsor Castle* by Robin Mackworth-Young (London: British Tourist Authority, and New York, Paris, and Lausanne: Vendome, 1990) offers an excellent and objective description of Windsor Castle and the people who occupied it throughout its long history. *Windsor Castle: In the History of the Nation* by A. L. Rowse (London: Weidenfeld and Nicolson, and New York: Putnam's, 1974) gives a good portrait of Windsor Castle's role in English and British history. Olwen Hedley's *Windsor Castle* (London: Hale, 1967) gives the most comprehensive and detailed description of events and improvements at Windsor.

—Thomas Cermak McPheron and Marijke Rijsberman

Wittenberg (Saxony-Anhalt, Germany)

Location: Forty-five miles southwest of Berlin, situated on the Elbe River.

Description: First settled in the twelfth century by immigrants from Flanders and Holland, the city became the official residence of the Elector of Saxony in 1486, which it remained until 1547. It was at the University of Wittenberg that the German monk Martin Luther became theology professor and set in motion the Reformation in 1517.

Site Office: Wittenberg-Information
Collegienstrasse 29
06886 Wittenberg, Saxony-Anhalt
Germany
(3491) 22 39

Wittenberg is one of Germany's oldest towns, but for the first 350 years of its existence, it excited no notice. No one would have guessed that it would become the spiritual heart of Germany—some would maintain, of Protestant Europe—in the sixteenth century. Germany (which was a nation only in a cultural, but not in a political, sense before the country was united in 1871) had no city on a par with Rome, Paris, or London. Consequently, Germany seemed the least likely place to inaugurate a revolution, and Wittenberg one of the least likely cities to host it. By the time Martin Luther died in 1546, his University of Wittenberg (established only in 1502) had the highest enrollment of students in the German states—many of which had far older universities—and one of the highest of any university in Europe. In fact, William Shakespeare's Hamlet had been a Wittenberg student. The city continued to flourish culturally and economically in the afterglow of the Reformation, declining precipitously only with the onset of incessant warfare from the eighteenth century onwards. Not until the late nineteenth century were serious efforts made to preserve the town's important cultural monuments and to commemorate its most prominent citizen.

Wittenberg is more than 700 years old. Its origins are so obscure that not even the monastic chronicles of the day recorded the founding of the settlement. By the time there was a record of its existence in 1180, the hamlet was no longer young. It was situated (as it still is) on the Elbe River, then the easternmost outpost of German civilization. The earliest inhabitants were probably immigrants from the Low Countries. Even in the 1000s and 1100s, the Low Countries of Flanders and Holland were becoming densely populated. In what are now the eastern parts of Germany, the Elbe River still had plenty of fish, especially valuable sturgeon and salmon, and offered a livelihood to those who could plumb its riches. At first the settlers called their village Wittenborch,

which was the Low German dialect version of "white mountain," referring to the white sand that was everywhere noticeable. In time, the name evolved to Wittenberg.

The town lay in the German state of Saxony, a frontier area exposed to constant invasions from the east, especially from the Slavs. Almost from the outset, Wittenberg had to be fortified. By 1293, there was enough commerce and population in the town, which even had two city gates, to warrant its receiving a city charter from the Saxon ruler, Albrecht II. Only eleven years later, under his successor, Rudolf I, Jews were expelled from the town and prohibited from even entering the city. While Jews were persecuted at this time all over Europe, there is little doubt that anti-Semitism has extremely long and deep roots in German history.

Aside from this disturbing milestone, Wittenberg's history was fairly uneventful until 1429, when a major invasion of Saxony occurred in which Wittenberg was directly threatened. Bohemians (i.e., Czechs) in the east had embraced the reformist doctrines of their native son, Jan Hus, who was condemned by the pope and burned at the stake in 1415 for the crime of heresy. The resulting Hussite wars, in which the German princes succeeded in crushing the heresy and occupying Bohemia, were a foreshadowing of the religious wars that would erupt in Germany after Luther's death. Luther himself acknowledged his indebtedness to the great Czech religious and political leader.

When the danger subsided for Wittenberg (the invaders never did take the town, although the suburbs were not so fortunate), little that was noteworthy occurred until 1486, when Wittenberg became the political seat of the German state of Saxony. In 1356 the princes of Saxony had had the good fortune to be appointed elector, or Kurfürst, a hereditary office. This was an extremely prestigious position, since there were only seven electors and only these seven "elected" the Holy Roman Emperor. Since there was no centralized government in the German states, neither the emperor nor the elector had much power. Nevertheless, the prestigious elector's move to Wittenberg gave the city a tremendous boost. The elector, Frederick III, nicknamed the Wise, was the future protector and mentor of Martin Luther.

Frederick found Wittenberg to be a rude town of no particular architectural distinction. The one famous architectural landmark that the town possessed, besides his ancestral fortress on a hill overlooking the town, was the castle chapel, built in 1338 by the pious wife of Rudolf I (the anti-Semite). The town was also known far and wide for its possession of one of Christ's original thorns. If a Christian traveled to Wittenberg and worshipped this relic, he could gain an indulgence for the remission of future sins. Wittenberg in this way attracted a steady stream of pilgrims, which boosted town revenues.

Wittenberg's castle church
Photo courtesy of German Information Center

Frederick III set about to raze the old fortress on the hill and to erect a new, "modern" residence in the Italian Renaissance style of the day; this residence still stands. Next, a church went up next to it, on the street side. Years later, Luther would nail his ninety-five theses on the door of this church. In it Frederick carefully installed his own awe-inspiring collection of relics, purportedly the second largest in the German states. Such was the credulity of the time that he, a very literate man, and others not so literate, actually seems to have believed that these were authentic relics: the Virgin Mary's breast milk, the body of one of the children murdered on King Herod's order, straw from baby Jesus' crib. These were among the most notable of the 5,000 items. The church became famous and drew far more pilgrims on All Saints Day (when the relics were put on public display) than the smaller chapel had. The new castle church also featured more masses said in the course of the day—more than twenty—than St. Peter's in Rome. Not to be overlooked was the fact that anyone worshipping the relics could gain remittance from either hell, purgatory, or other drastic divine punishment. It was just these kinds of entrenched attitudes and practices that Martin Luther sought to eradicate.

The church, which employed a staff of more than eighty, formed the impetus for Frederick III to establish a university in the town of Wittenberg. He wanted to improve the education of the male canons. He appointed his personal physician, Martin Pollich, as head or rector of the future university, the first building of which was completed in 1502. The castle church now had an added function as the university auditorium. The church's wooden entrance door became an announcement board. In October of that year, more than 400 students enrolled, mainly to study theology. All came from Wittenberg and the surrounding communities. The city itself had no more than 2,500 inhabitants.

Wittenberg was rising to some prominence as a political center and now as a university town. A building boom ensued, during which two new monasteries for the Augustinian and Franciscan orders were built. The monasteries also served as residence halls for the monks who were enrolled in the university. The chairs of theology and classical languages were filled by men from these two orders, one of whom was Martin Luther, an Augustinian monk.

By the time he arrived on campus to teach, in 1511, the university had expanded to include two other new buildings. But Luther, who had just returned from a sojourn in Rome, was extremely unimpressed with Wittenberg's provincialism and mania for relics. He took up his post as theology professor, at the same time completing his doctorate in theology, which he received from the new university in the following year.

The year 1512 also is a milestone in the town's history: the court painter and renowned artist of the Reformation, Lucas Cranach, completed building his imposing private home in the center of the town. To this day, it remains one of Wittenberg's most important landmarks. Eventually this versatile genius would acquire a printing press and publish Luther's new translation of the New Testament into German as well as many other Luther tracts and treatises. Wittenberg would be the publishing capital of Germany until the city of Frankfurt usurped this position in the seventeenth century.

What follows is the story of the birth pangs of the Reformation, which sent shock waves throughout Europe as far east as Poland. Luther, born in 1483, was thirty-four years old, a young Ph.D., when he did what was common practice at the university—nailing an announcement on the college "bulletin board," the wooden door of the castle church. His announcement was a challenge to a debate on the sale of indulgences. Luther eventually broadened his call for reform, and in 1521 he was excommunicated from the Roman Catholic church. (The spot where he publicly burned the pope's bull of excommunication has been turned into a memorial park.) Other changes followed swiftly. The monastery of his order was dissolved upon Frederick III's order. Luther remained its sole inhabitant, but not for long. In 1525 he married the former nun Katharina von Bora. They had six children in the former monastery, which became their home. Today, the Augusteum is the most important historic site in Wittenberg and contains the Lutherhalle, Luther's original rooms, including his sitting room, which was newly constructed in 1536. A few items that had belonged to Luther are on display. The Stadtkirche, or city church of St. Mary's, became Luther's church, where he preached regularly. The artist Lucas Cranach was so moved by his preaching that he captured several poses of Luther in the pulpit, beautifully rendered in an altarpiece that still stands in the church, among the city's prized possessions. The church also boasts an elegant, pre-Reformation baptismal font designed and constructed by the famous German artist Hermann Vischer, which also has survived intact to this day.

Luther's fame brought not only droves of new students to the university, but other famous people to the town as well. In 1518, a brilliant young former monk, Philip Melanchthon, arrived in Wittenberg to teach Greek at the university. He became one of Luther's staunchest allies and a major figure of the Reformation in his own right. The town council considered him such an asset that they ordered an imposing house built for him, in the center of town, which was not turned into a museum until 1967. In the nineteenth century, along with a statue of Luther, a statue in honor of Melanchthon was raised in the town square, in front of the old city hall (which was built in Luther's day). The courtyard of the castle church, on whose wooden door Luther unknowingly launched a revolution, contains the graves of both Luther and Melanchthon, lying side by side.

When Luther died in 1546 at age sixty-three, the German states were embroiled in a religious war between Protestant princes in the League of Schmalkald and the Holy Roman Emperor and his Catholic forces, which resulted in the ignominious defeat of the princes. Wittenberg was badly damaged in 1547: the new town hall and the Stadtkirche of St. Mary's were in ruins. As punishment for its part in the rebellion, Wittenberg was deprived of its status as the seat of the elector. Barely two decades later, an epidemic of plague descended on the city. Although this outbreak was not as savage as the one in 1636, it claimed the lives of half the town's residents.

War and pestilence were endemic in Europe at that time, and recovery was usually swift. In fact, in the dark days following Luther's death, the University of Wittenberg enrolled more students than before, and its endowment increased so that new buildings and a new library were added. The town's prosperity was evident in the expansion of the running water supply system that piped water into many buildings, including the university, and was in part due to the fact that Wittenberg had become a leading center of the European book trade. From a crude town that for hundreds of years had relied on meager livelihoods from fishing, agriculture, and the pilgrim trade, Wittenberg evolved into a small intellectual and cultural haven, which it remained until warfare engulfed it completely.

Wittenberg was spared serious harm during the Thirty Years War (1618–1648), and, though it was occupied by Swedes in the later Northern War (fought between Sweden and Russia), it remained unscathed. When Peter the Great of Russia stopped off in Wittenberg in 1711 to attend a wedding, he was shown the Luther landmarks, including the original wooden door of the castle church. Wittenberg was not so fortunate in the Seven Years War, waged between France and Great Britain and their respective allies, from 1756 to 1763. In 1760, Wittenberg suffered great devastation, and most of the landmarks of Luther's day, including the castle church with its wooden doors and Frederick III's palace, were burned to the ground. As in previous catastrophes, the city was soon rebuilt; the wooden doors, however, could not be salvaged, and were replaced by fireproof iron ones.

During the protracted conflict fifty years later with Napoléon Bonaparte, whose forces occupied Wittenberg and turned the Stadtkirche of St. Mary's into a military hospital, the town found itself once again amid warfare and ruin. When Napoléon was finally defeated in 1815, Wittenberg had been steadily sliding downhill economically. It was uneconomic to keep a university open when the nearby city of Halle boasted one as well. In 1816 the two universities were merged, with the Wittenberg campus closed for all practical purposes. From then on, Wittenberg housed a permanent garrison of soldiers, and fortifications around the city were modernized and strengthened. The castle hill, with its university buildings, became a military barracks for most of the nineteenth century. Wittenberg had lost its luster as the city of Luther and Saxony's intellectual gem and became something of a ghost of its former self.

It was in the nineteenth century, however, that serious attention began to be paid, for the first time, to preserving the historic landmarks that had made the city unique. Luther's house was restored and refurbished and, in 1883, opened as a museum. Statues were erected to the Reformation leaders in

the town square. Memorials and commemorations of important Reformation milestones were celebrated for the first time. After Germany was unified into one political entity in 1871, the fortifications surrounding the city became obsolete and were demolished, to be replaced by a lovely ring of parkland. On the eve of World War I, the Stadtkirche and castle hill were being renovated; "educational tourism" was returning to Wittenberg some of its former glory.

When World War I ended, the city continued the preservation of its heritage. In 1922, the town council took the unusual step of adopting a second name, Lutherstadt, or City of Luther. Historic preservation was the theme of the 1920s and 1930s, even in the backdrop of the Depression and the coming to power of Hitler. In a throwback to medieval times, Wittenberg's Jewish residents were subjected to outrageous attack and mistreatment on the infamous Kristallnacht in 1938.

Energy and attention were soon diverted to the armaments factories that were being built in the area. In Wittenberg, a new airplane factory went up in the mid-1930s, and women incarcerated in the nearby Ravensbrück concentration camp were forced to work there. All renovation and restoration work in the town ground to a halt in 1939, at the start of World War II.

The Soviet Red Army occupied Wittenberg in late April 1945, and several years later, Wittenberg found itself part of the new German Democratic Republic, officially an atheist state. Because Martin Luther had championed the use of the vernacular in church services and had translated the Bible into German, the new Communist state emphasized his German "patriotism" and entered him into the pantheon of national heroes. Heavy industry came to Wittenberg at this time, in the form of massive, inefficient chemical works that polluted the nearby countryside. Nonetheless, Wittenberg's reputation as the city of Luther was not disputed. An important milestone was achieved when Philip Melanchthon's residence was renovated, restored, and opened to the public as a museum for the first time in 1967.

In the heady days of Gorbachev's "perestroika" and the collapse of communism, Wittenbergians were in the forefront. Seven theses were affixed on the door of the castle church in 1989, calling for civil and human rights. Entrance to the church was blocked by secret police agents. Other churches were used instead as places of prayer and demonstration.

By 1990, with the end of Communist rule, the city could reassert its spiritual heritage. It lost no time in doing so, resurrecting religious conventions, symposia, and festivals. Since then, tourism has increased to an all-time high of 120,000 annually—more than twice the population of the city itself.

Further Reading: The best illustrated history of Wittenberg from earliest days to the present has been written by theologian and director of the Lutherhalle Museum, Martin Treu: *Wittenberg: The Town of Martin Luther* (Berlin: Nicolaische Verlagsbuchhandlung,

1993). It is a short but comprehensive book, filled with photographs. A good, readable biography of Martin Luther in the Wittenberg years, in fewer than 200 pages, is Mike Faeron's *Martin Luther* (Minneapolis: Bethany House Publishing Co., 1986). For those who want to delve into all the rich detail of the Reformation, there is the two-volume work by Joseph Lortz, translated by Ronald Walls, *The Reformation in Germany* (New York: Herder and Herder, and London: Darton, Longman, and Todd, 1968; as *Die Reformation in Deutschland*, Freiburg: Herder, 1939–1940; second edition, 1949).

—Sina Dubovoy

Würzburg (Bavaria, Germany)

Location: On the Main River in Bavaria, sixty-eight miles west of Nürnberg and seventy-three miles east of Frankfurt.

Description: A fifteenth-century town with archaeological artifacts dating to 500 B.C., Würzburg's architectural contributions are dominated by two principal buildings, the Marienberg fortress on a rise overlooking the Main's left bank, and the Residenz with its fine gardens. The construction of the Residenz between 1719 and 1744 synthesized French château design with Viennese imperial baroque in what became known as Würzburg rococo.

Site Office: Fremdenverkehrsamt
Am Congrecss Centrum
8700 Würzburg, Bavaria
Germany
(931) 3 73 35

Present-day Würzburg was in one of the four clusters of fourth-century Germanic settlements north of the Danube which, like those on the Rhine and the Danube themselves, the guardians of Rome's imperial boundaries had been unable to prevent from arising. The other three clusters were the north Bavarian, in the neighborhoods of Bamberg and Pegnitz; that at the confluence of the Rivers Regen, Naab, and Danube; and that near Ingolstadt. The northwest Bavarian group was situated in the triangle made by the Main when it turns south at Schweinfurt and north again to flow through Würzburg to Gemünden. Archaeological evidence suggests that the Germanic settlement, which may have gone back to the first century, had been preceded by a Celtic principality about 500 B.C. and then by small fishing colonies sheltered by the rise on which the Marienberg stands.

The name *castellum Virterburch* is attested in 704 A.D. as the capital of a Merovingian duchy. The recorded history of Würzburg, which was to become Germany's most important ecclesiastical center, begins with the erection in 706 of a round church whose center is probably the oldest preserved German ecclesiastical building east of the Rhine. The bishopric of Würzburg, covering all east Franconia, was established between 741 and 742 by Boniface, "Apostle of Germany" and archbishop of Mainz from 722 to 754, when he appointed Burkhard, who founded the monastery of St. Andreas, near the Romanesque-style Burkardskirche, consecrated in 1042.

In 788, a cathedral was constructed above the grave discovered in 752 of three Irish missionary monks, Kilian, Colman, and Totnan, martyred in about 689. Charlemagne attended the consecration. After a fire in 855 it was rebuilt,

nearby. The eleventh-century Neumünster, begun in 1042 and consecrated in 1189, was built on the original cathedral site. This site was the center from which the settlement spread, joined to the left bank at first by a ford, and then by a bridge erected before 1133. The third cathedral, St. Kilian's, erected under Bishop Bruno who held the see from 1034 to 1045, is Germany's fourth largest Romanesque church. The eastern towers date from 1237, the rococo facade and rich interior stucco decoration from 1701 to 1704, and the dome from 1731. The exterior was restored in 1882–83, and the whole edifice had to be rebuilt in 1966–67 after destruction incurred at the end of World War II. The Neumünster's choir and decorated tower were added in the thirteenth century, the cupola and rococo facade from 1710 to 1716. Walther von der Vogelweide, considered the greatest of the Middle High German lyric poets, died in Würzburg in 1230, and is said to be buried in the small garden, the Lusamgärtlein, north of the Neumünster chancel.

Würzburg acquired the rights to mint coinage, hold markets, and levy dues from Emperor Conrad III in 1030. It was in Würzburg that the Holy Roman Emperor Frederick Barbarossa married Beatrix, heiress of Upper Burgundy, on June 9, 1156, and that the emperors held several diets, notably that of 1180 when Barbarossa outlawed Henry the Lion, duke of Saxony and Bavaria. In 1168, Barbarossa had made the bishops of Würzburg, already princes of the empire, dukes of eastern Franconia, thereby resolving the medieval clash of jusrisdictions.

During the thirteenth century, in Würzburg as elsewhere, the merchants and craftsmen began their long struggle against the prince-bishops for civic independence. They supported the Emperor Henry IV (c. 1050–1106) against Gregory VII in the "investiture controversy," essentially a struggle to affirm the primacy of secular power that Henry IV humiliatingly lost, while the prince-bishops supported the pope. In spite of imperial protection afforded to the town by King Wenceslas in 1397, an uprising against Prince-Bishop Gerhard von Schwarzenberg was suppressed at the battle of Bergtheim in 1400.

By this date, the predecessors of the present churches of St. Stephan and St. Peter, in their eleventh- and twelfth-century forms, had been joined by the early thirteenth-century Franciscan and Dominican, later Augustinian, foundations. All were rebuilt and redecorated in the seventeenth or eighteenth centuries. In 1316, the town acquired the thirteenth-century Romanesque dwelling of a distinguished administrator, Graf Eckard, and made it into the Rathaus. The home was subsequently enlarged on several occasions.

The present market place was established on the ruins of the Jewish ghetto destroyed in 1349. The Marienkapelle (St. Mary's Chapel) on Würzburg's north side is a late Gothic

The Residenz at Würzburg
Photo courtesy of German Information Center

municipal foundation begun in 1377 and finished in 1481 with the construction of the tower. The chapel's original portal sculptures of Tilman Riemenschneider, mayor of Würzburg from 1520 to 1521, are now in the museum and have been replaced by copies. There is a tombstone by Riemenschneider inside the west front. To the east of the church is a splendid merchant dwelling, the Haus zum Falken, with elegant stucco work from 1752.

Würzburg played no important part in the religious disputes of the sixteenth century, but in 1525 the Würzburg citizens sided with the peasants in the Peasants' War, and under Götz von Berlichingen attempted in vain to storm the Marienberg, the episcopal palace from 1253 until 1719.

Between 1473 and 1543, the Alte Mainbrücke replaced the previously destroyed Romanesque structure. Statues were added in 1730. Two additional bridges were erected in the late nineteenth century: the Friedensbrücke downstream in 1887 and the Ludwigsbrücke upstream in 1894.

It was Prince-Bishop Julius Echter von Mespelbrunn who, profiting from the renewed stability after the religious disturbances, built the Juliusspital from 1576 to 1585, affording food and lodging to 600 persons a day. He also built the university from 1582 to 1594 to replace that which had been founded in 1492 and had existed for a few short years. Its accompanying Neubaukirche, a German Renaissance church, features a 1696 tower by A. Petrini. Petrini is best known for Würzburg's baroque Stift Haug with double towers and a cupola built from 1670 to 1691. The rich decoration was destroyed in 1945, but there is now a Tintoretto *Crucifixion* from 1583 above the high altar. Julius also extended the Marienberg palace-fortress.

In 1631, during the Thirty Years War, the city and the Marienberg were captured by Gustavus Adolphus, who installed a new government that lasted until 1634. From 1642 to 1673 Prince-Bishop Johann Philipp Franz von Schönborn renewed the town's fortifications and again extended the Marienberg, digging the 340-foot well. The fort's armory, built from 1702 to 1712, now contains the Main-Fränkische museum, with its large collection of historical relics, and also the Riemenschneider pieces and paintings by Giovanni Batista Tiepolo (1696–1770) and Lucas Cranach the Elder (1472–1553).

The elaborate complex that is the Würzburg Residenz was created through a near-miraculous linking of circumstances. It is possible to see the building as a collective work, and in detail to point to the contrasting plans and painful quarrels between the principal architects. The integration of the ensemble was primarily due to the efforts of Prince-Bishop Johann Philipp Franz von Schönborn, who acceded to the see in 1719, won 600,000 florins in a lawsuit, and immediately made plans to build. Von Schönborn's principal architect, Balthasar Neumann, was not given absolute control of the project, which consequently led to numerous confrontations with the other designers: Maximilian von Welsch, court architect to the prince-bishop's uncle, Lothar Franz von Schönborn, elector of Mainz; Johann Dientzenhofer, creator

of the elector's summer residence at Pommersfelden; Johann Lukas von Hildebrandt, primarily responsible for the great pavilion on the garden front at Würzburg and principal architect to the prince-bishop's uncle, Friedrich Karl, Arch-Chancellor of the empire in Vienna; and two French architects, Robert de Cotte and Germain Boffrand, who were concerned primarily with the facades.

The solution was to create two great hollow oblong wings protruding westwards toward the town, each wing formed by its four sides with a transverse block joining the longer sides in the middle to create two interior courtyards, and united by a single great transverse block at the rear, running from north to south and creating an impressive *cour d'honneur* in front of a great square. Each of the protruding lateral wings contains, on the insistence of Welsch, an oval pavilion in the middle of its outer facade, and the south wing of the south block contains the immensely graceful and richly ornate Hofkirche.

Johann Philipp Franz von Schönborn laid the foundation stone on May 22, 1720, and moved from the Marienburg to the immediate neighborhood of the site. His chief personal contribution to the architectural design was the spectacular staircase, two single flights that peel to the right and left, giving two parallel double flights to the upper landing, all in white marble, which Cotte wanted to do away with. Neumann successfully defended his designs for projections that flow inwards from the long sides of the oblongs enclosing the *cour d'honneur*.

Schönborn was much hated and little mourned on his death in 1724. His opponents were responsible for the election of his succesor, Christoph Franz von Hutten, who suspended building with only the north block completed. The elector of Mainz continued to protect Neumann and, when Hutten died in 1729, he was succeeded as prince bishop by Arch-Chancellor Friedrich Karl. Now, Neumann's chief architectural adversary was Hildebrandt, whose taste for imperial Viennese baroque was disregarded. Hildebrandt's main contribution was a stone and metal screen in front of the *cour d'honneur*, removed in 1821. The building was roofed by December 1744.

Friedrich Karl died in 1746, and his successor, Anselm Franz von Ingelheim, dismissed Neumann. Ingelheim died in 1749, and his succesor as prince bishop, Carl Philipp von Greiffenklau, resumed the tradition of princely patronage until his own death in 1754. Neumann was reinstated; Tiepolo was commissioned to execute the magnificent ceilings over the staircase and in the Kaisersaal; and Antonio Bossi took charge of the stucco work and of the whole team of gifted French, German, Italian, and Dutch craftsmen chosen from all over Europe. This was the group which created Würzburg rococo, the dominating architectural style of the Hofkirche.

It is a tribute to Neumann that his immense and daring single unsupported stone vault over the staircase, vestibule, and Kaisersaal, from which Hildebrandt had offered ''to have himself hanged at his own expense'' if it were to hold, not only did hold, but withstood both a fire in 1896 and the bombs of 1945, which left only the vault, the staircase, the Kaisersaal, and the Weisser Saal undamaged. The staircase, which has been called "one of the most magnificent spatial creations ever achieved in secular building" is decorated in a riotous medley of styles and appears to flout the laws of gravity.

In their way, the gardens constitute a complementary masterpiece, aided rather than impeded by the lack of space and the constraints of the available terrain. The individual rooms of the Residenz itself constitute a series of miniature masterpieces.

Between the luminous color of the staircase and the illusionism of the gilded draperies, figures, and fantasies in the Kaisersaal that move audaciously from sculpture to painting in false relief, is the Weisser Saal, formerly a *salle des gardes*. It is particularly daring in its total absence of any colors other than white on pale shades of grey. Even when Tiepolo did use allegorical subjects for his frescoes, or great scenes from Würzburg's history depicting imperial grandeur, the overall magnificence of conception makes the decoding of his narrative subjects seem almost an irrelevance to the point about decoration that his frescoes make. Not the least of Tiepolo's qualities is charm. He teasingly refuses to allow the beholder to know where painting merges into sculpture, and he obstinately makes fun of his grander subjects, which he will not treat with imperial earnestness. He gives a mediterranean background to his painting of Barbarossa's bestowal of the duchy of Franconia at the 1168 diet although it was actually held at Würzburg.

In 1753 Neumann died and Tiepolo finished his frescoes. In 1802, as a result of Napoléon's campaigns, Würzburg lost its independence, and became conditionally incorporated into Bavaria for the three years from 1802 to 1805. The bishopric became part of the province of Bamberg and Würzburg was governed by Ferdinand von Toskana from 1805 until 1814, when the town was definitively incorporated into Bavaria according to the terms laid down at the Congress of Vienna. The town was bombarded and taken by the Prussians in 1866, and its fortifications were removed in the last decades of the nineteenth century, and replaced by parks and alleys. Röntgen discovered X-rays here in 1895.

On March 16, 1945, some 85 percent of the old town was destroyed in twenty minutes in an air raid barely seven weeks before Germany's surrender. At least 4,000 people died, and as many buildings were demolished. Happily the furniture had been removed from the Residenz, the three principal rooms were intact, and much material recovered from the wreckage could be used again. The town has been totally and successfully restored, with a greater proportion of restoration to replication than might have been thought possible. The sole structural alteration was the closing of Schönbornstrasse to automobile traffic. Not even the fact that everything looks simultaneously new disturbs as much as it does elsewhere, since in Würzburg the important decorative gables and facades, the interior decoration of the churches, and the flamboyant exteriors of the merchants' houses, were once new almost together, in the first half of the eighteenth century, when what remained of Romanesque and Gothic was redecorated in Würzburg's no doubt historically conditioned but nonetheless unique brand of rococo.

Further Reading: While there is very little in English specifically devoted to the town of Würzburg, there is an official guide by Erich Bachmann and Burkard von Roda *The Würzburg Residence and Court Gardens,* ninth English edition, (Munich: Gärten und Seen, 1992). It is informed, but too dry and pedantic for its subject, on which its aesthetic judgements tend to insensitivity. For the art and architectural history a general reader is likely to find most help in the *Larousse Encyclopedia of Renaissance and Baroque Art,* edited by René Huyghe (Feltham: Paul Hamlyn, 1964; French edition, *L'Art et l'homme,* Paris: Larousse, 1958). On Tilman Riemenschneider, see Michael Baxendall's *The Limewood Sculptors of Renaissance Germany* (New Haven and London: Yale University Press, 1980). On Neumann, see Max H. von Freeden's *Balthasar Neumann, Leben und Werk* (Munich: Deutsche Kunstverlag, 1981).

—Anthony Levi

York (North Yorkshire, England)

Location: In the Vale of York at the junction of the Ouse and Foss Rivers, 194 miles northwest of London and 202 miles southeast of Edinburgh.

Description: County town of Yorkshire, seat of the archbishop of York, and historically a site of strategic importance; includes archaeological sites of Roman and Viking occupation, the great medieval cathedral York Minster, a number of historically significant buildings, well-preserved medieval streets, and many museums.

Site Office: York City Council
De Grey Rooms
Exhibition Square
York, North Yorkshire YO1 2HB
England
(904) 621756

"The history of York," observed King George VI, "is the history of England." It is difficult to fault his statement. Occupied for nearly 2,000 years, the city of York has witnessed the ebb and flow of England's past, from the invasion and influence of the Romans to the subsequent settlement of the Danes, from the devastation of the Norman Conquest to the prosperity of the Middle Ages, from the soaring triumphs of the church to the comfortable success of commerce and industry.

York is a city of layers, as archaeologists have learned. The first layer was put down by the Romans in A.D. 71 when they built a fortress for the Ninth Legion on the point of land at the confluence of the Ouse and Foss Rivers. Far from the established Roman towns of the south, the fortress, known as Eboracum, and its accompanying settlement or *colonia,* gave the Romans an outpost from which to monitor the unpredictable northern tribes. The site allowed for easy communication with the south by road or by water and was surrounded by land fertile enough to support the 5,000 soldiers stationed there. Although there is evidence of pre-Roman habitations nearby, the Romans chose an unoccupied and naturally defensible site for their fortress, although enemy attack was unlikely.

The experienced Legion engineers used the design proven again and again throughout the empire in laying out Eboracum. The primary purpose of a Roman fortress was to house a legion's men and equipment, and it was traditionally laid out in a large rectangle divided by crossroads giving access to the legion's headquarters, or *principia,* placed at the center. Barracks took up much of the remaining space, along with the baths, granaries, and workshops. The whole was surrounded by a palisade of timbers set upon a rampart, which was further heightened by an outlying ditch. The site

upon which the fortress of Eboracum was built is now occupied by York Minster.

Outside the fortress a settlement developed that eventually spread across the River Ouse and became a substantial civilian town with its own wall. By the early third century it had been named the capital of Britannia Inferior, one of the two Roman British provinces. Around this same time it was given the status of *colonia,* one of only four British towns so honored. Nor was it remote from affairs of state. Eboracum was visited by the Emperors Hadrian, Severus, and Constantius I. The latter two died in York, and Constantius's son, who would become known as Constantine the Great, was proclaimed emperor there, the only Roman emperor named in Britain.

The Ninth Legion apparently left Britain sometime around A.D. 110, and the fortress of Eboracum was reoccupied by the Sixth Legion, which arrived in A.D. 120 upon the visit of Hadrian to Britain. Reconstruction of the fortress defenses began at this time and was to continue for the next 150 years. From 209 to 211, York was visited by Emperor Septimius Severus, who, like Hadrian, was concerned with quelling the unrest of the northern tribes. Severus established Eboracum as his base of operations. In 211 he died at York, and the resulting political crisis in Rome diminished imperial concerns with the troublesome British tribes.

During the mid-third century, internal political wrangling and military problems on the continent occupied the empire. Britain seems to have been relatively calm, but probably suffered from interrupted trade. Much of the garrison at Eboracum may have been moved to serve the emperor elsewhere, for there is evidence that the fortress and defenses were not well maintained. But by the end of the third century and the beginning of the fourth, attention was being paid to the refurbishment of Eboracum's defenses. The transformation of the timber fortress into a stone one was completed at this time. The most ambitious project was a great polygonal tower at the southwest side of the fortress and six projecting interval towers. The great tower is still visible today and is known as the Multangular Tower. The changes made in the fortress reflect a changing attitude toward defensive strategy. Originally conceived as a base of operations, the fortress now took on a more medieval role as a defensive post to hold off attackers. There is no evidence that the fortress ever faced such a siege, and the towers were most likely intended to impress residents and state visitors as much as potential enemies.

The civilian area southeast of the fortress in the triangle of land between the Rivers Ouse and Foss was composed of shops and dwellings for those people who made their living supporting the soldiers. The production of building materials for the fortress, warehouses for food, and the

The York City Wall
Photo courtesy of Yorkshire and Humberside Tourist Board

shops of potters, smith, and other necessary craftsmen would have been convenient to the fortress here. The areas north and west of the fortress seem to have been less settled, used more for fields and cemeteries, and possibly kept clear of structures to allow the fortress a clear line of sight.

Across the River Ouse was where the *colonia* flourished. There was apparently a Roman wall about the settlement, later replaced by Anglo-Scandinavian defenses and the extant medieval wall, all situated upon the original rampart. The date of the origins of the *colonia* is now known, but a road laid down through the area in the mid-second century may be a strong clue. Its initial growth seems to have been rapid, involving a series of booms with subsequent expansions and reorganizations of existing neighborhoods. Archaeological evidence has been found for the existence of a variety of crafts in the *colonia*, including metalworking and leatherworking. Trade was carried out across the empire, with goods being imported from Gaul and the Rhineland. One important export was jet, traded in both raw form and finished pieces. The Ouse provided access to the English Channel and established Eboracum as an important trading center, although little is known about the details of shipping for the settlement. By the early third century, Eboracum's prosperity was reflected in the building of stone structures to replace timber ones. In addition to craftsmen and merchants, the occupants would have included retired legionnaires and Roman bureaucrats, all with their families. Eboracum was a cosmopolitan town, where Romans and Britons mixed with inhabitants from all over the empire. Although prosperous, it never attained the size of Roman towns in southern England. The *colonia* occupied an area of 67 acres and had no more than 3,000 civilian residents.

The decline of the Roman Empire led to the withdrawal of troops from Britain around the year 400. Little is known of York's history during this time. A drop in the population and in its prosperity likely came with the abandonment of the fortress. There is some evidence for British occupation in the late fifth and sixth centuries, when it may have been a part of a kingdom called Elmet. In 627 King Edwin of Northumbria conquered Elmet. Edwin converted to Christianity, was baptized in York, and founded the first minster or cathedral there the same year. It quickly became the ecclesiastical capital of the north, and under the leadership of the scholar Alcuin, the setting for a great flowering of Anglo-Saxon culture. Born in York around 732, Alcuin lived there for fifty years, serving as headmaster of the cathedral school where he had once been a student and influencing the shape of learning for generations to come. In 781, while traveling in Italy, he accepted the invitation of Emperor Charlemagne to assume leadership of the scholars being collected at the court of Aachen. Alcuin taught the emperor and his family, making literacy fashionable and establishing the study of the liberal arts as the foundation of scholarship. Eventually he served as abbot of the Abbey of St. Martin at Tours and promulgated the use of the Carolingian script for writing, which later served as a model for modern typefaces.

The fate of York took a dramatic turn again in 866 when the city was captured by the Danes. The Vikings, as we commonly call them, had begun raiding the rich and vulnerable monasteries along the English coast decades before. The raids gradually transformed into an invasion that ultimately defeated three out of the four major English kingdoms then in existence. York, the invaders' first target, was captured easily. The Northumbrians rallied with a counterattack the following year but were defeated with terrible losses. The city that had once been Eboracum became Jorvik, the capital of a Danish kingdom comparable to modern Yorkshire. The invaders settled in this newly won land and became its new farmers, craftsmen, and merchants. The city saw another boom unparalleled since Roman times. Recent archaeological work in the Coppergate section of the city has produced a wealth of Viking artifacts that present a picture of a busy and prosperous community.

By the end of the tenth century the Anglo and Danish elements of the York population had become comfortably integrated, and York had become a major center for trade and the second largest city in England after London. The city became a part of the new kingdom of England in 954, but retained the flavor of its Scandinavian heritage, even to its name, a derivation of the Danish Jorvik (pronounced Yorvik). An attempt was made to recapture York in 1066 by Tostig and the king of Norway. King Harold of England defeated them and reclaimed York in a battle at Stamford Bridge outside the city. Then, hearing of an invasion by William of Normandy, Harold and his troops hurried south to meet their fate at Hastings. The Norman conquerors dealt harshly with the Saxon stronghold of York when they arrived two years later. After his first motte-and-bailey castle was attacked in 1069, William returned and built another, stronger castle. His decisive action in the face of opposition quelled further rebellion. William's castle was severely damaged in 1190 when the Jewish citizens of York sought refuge there in the face of anti-Semitic riots. When the mob was joined by the sheriff and the castle constable the refugees set fire to the timber structure and committed mass suicide. The survivors were massacred as they emerged in surrender. The timber fortress was repaired and then replaced with a stone castle by Henry III. Only the castle keep, known as Clifford's Tower, survives.

The Middle Ages brought prosperity and relative peace to York, an opportunity for its citizens to exploit their natural trade advantages. York became a staple supplier of wool to Europe, trading from the Baltic states down to France. Craft guilds, churches, religious schools, and monasteries flourished, making it the commercial, political, and religious center of northern England. The Merchant Adventurers' Hall is a large timber-framed structure dating from the mid-fourteenth century. Built by the city's oldest guild, it is still in use today. Two streets dating from this period, the Shambles and Stonegate, are superb examples of medieval urban architecture. Though easily accessible, they are closed to traffic to assist in their preservation. The medieval walls, encircling the 260 acres of the old city, provide an excellent walking circuit of about three miles. Their height provides a good vantage point from which to view the York of the Middle Ages.

One consequence of the city's religious prominence was the construction, over the course of two centuries, of the great cathedral of York Minster, one of the finest medieval cathedrals in Europe. Built on the site of the original Roman fortress and two previous churches, the minster was begun in the thirteenth century by Archbishop Walter de Grey. When completed it stood 524 feet long and 249 feet across at the transepts. The transepts, constructed first, are in the Early English style. Services continue to be held in the nave, one of the widest Gothic naves in Europe. The choir, the last of the major sections of the minster to be built, is in the Perpendicular style.

In addition to the spectacular architecture of the minster, there is the stunning array of stained glass. More than 100 windows adorn the walls, dating from the twelfth to the twentieth centuries and including over half the surviving medieval stained glass in England. Eighty windows were removed for safekeeping during World War II, permitting caretakers the opportunity for cleaning and restoration. When serious structural problems were discovered at the base of the center tower of the minster in 1967, the resulting restoration and repair work uncovered a wealth of archaeological riches from the underlying Roman foundations and the minster's eleventh-century predecessor. Called the Foundations, the area is now open to viewing.

During the English Civil War in the seventeenth century, York served as Charles I's northern headquarters and a royalist stronghold. The city was besieged by Cromwell's army in 1644. After losing the battle of Marston Moor, six miles west of York, the royalists were forced to surrender the city. But the Parliamentarian leader Lord Fairfax, a Yorkshireman himself, ensured that the minster and the city's many other churches were spared desecration by the victorious Puritans. By the beginning of the eighteenth century, the city's wool trade had moved elsewhere, but York's location on the major coach roads led to its becoming a popular resort town. Much elegant Georgian architecture remains in the older parts of the city.

York was reborn as a commercial center in the nineteenth century, thanks to the coming of the railroad and the pointed efforts of George Hudson, entrepreneur, lord mayor of York, and manager of the York and North Midland Railway Company. Hudson rode the rails to riches and took York along for the ride, making it an important railway center and helping to generate the resulting industrial boom.

As befits a city rich in history and antiquities, York is also rich in museums. The York Heritage Center, located in the old Church of St. Mary, offers exhibits and an audiovisual display, the *York Story*, to introduce the rich variety of York's past. The Yorkshire Museum is dedicated to the history and archaeology of the shire. Its impressive collections include the Middleham Jewel, a fifteenth-century pendant with a large sapphire discovered in 1985 near the former home of Richard III. Remains of the Roman wall and the Multangular Tower can be seen from the museum gardens. The Jorvik Viking Centre is located at the site of the recent excavations in Coppergate. Artifacts from the excavation are also on exhibit. The history of York Castle, the Jewish massacre, and the Civil War are explained at Clifford's Tower, which also offers excellent views of the city. Across from the Tower is the York Castle Museum, which provides re-creations of life in several different periods of York's history. The National Railway Museum is the largest railway museum in the world and offers a variety of exhibits relating to the impact of the railway, from paintings to royal rail cars to a collection of locomotives.

In addition to its traditional industries, commerce, and tourism, York is again a center for scholarship, as is appropriate for the city of Alcuin. The University of York opened in 1963 and is housed in a modern campus on a 190-acre site.

Further Reading: For more detail of the archaeological investigations of the Roman and Danish inhabitants of York, the following accounts are both informative and well illustrated with photographs and plans: Patrick Ottaway's *English Heritage Book of Roman York* (London: Batsford/English Heritage, and North Pomfret, Vermont: Trafalgar Square, 1993) and Richard A. Hall's *The Viking Dig: The Excavations at York* (London: Bodley Head, 1984). An examination of the events leading up to the Jewish Massacre of 1190 is provided by R. B. Dobson in *The Jews of Medieval York and the Massacre of March 1190* (York: St. Anthony's Press, 1974). For a broader look at the history of the region see David Hey's *Yorkshire From A.D. 1000* (London and White Plains, New York: Longman, 1986).

—Elizabeth Brice

Ypres (West Flanders, Belgium)

Location: Approximately 10 miles north of the French border, 21 miles inland from the North Sea, and about 60 miles west of Brussels.

Description: One of the three great cities of Flanders in the Middle Ages, Ypres (also spelled Iepec) became an important trading center for artisans and merchants engaged in the manufacture and sale of cloth. The riches and commercial advantages of Ypres and indeed of all of Flanders attracted the envy of Europe and thus became the object of numerous conquests by foreign powers, primarily the French. During World War I, Ypres represented a critical strategic point on the front lines, and most of the town was destroyed in the fighting. Following the war, many of the old buildings were meticulously restored, including St. Martin's Cathedral and the great Cloth Hall, a monument to Ypres's textile industry built in 1214.

Site Office: Town Hall
8900 Ypres, West Flanders
Belgium
(57) 20 07 24

Before its destruction during World War I, Ypres, on the upper Yser River, was a long-celebrated city dating to the tenth century. It grew up around the ruins of an eighth-century fortified castle that was rebuilt around 958 by Baudoin, Count of Flanders. Located in the center of the maritime plain, the town prospered as a locus of artisans and merchants specializing in the manufacture and sale of cloth. From the twelfth through the fourteenth centuries, the counts of Flanders bestowed important privileges on Ypres, making it a major industrial center of western Europe and a rendezvous of merchants throughout the medieval world. The lure of riches in Ypres and other Flemish cities led French kings to launch numerous expeditions into Flanders. Ypres was seized in 1128 by Louis VI, in 1213 by Philip Augustus, and in 1297 by Philip the Fair, but suffered little damage during these wars. The destruction of the town in 1383 by the people of Ghent caused many of the weavers to flee, leaving as Ypres' only industry the manufacture of Valenciennes lace.

In the fourteenth century, Flanders came under French dominion, and under the moderate rule of a succession of dukes, the cities of Flanders, including Ypres, prospered. In 1481, Austria took possession of Flanders when Marie, heiress of Burgundy, married the Archduke Maximilian. In 1558, it came under the rule of Spain, and in 1559 Flanders replaced Thérouanne as the center of the diocese. By this time, however, much of Ypres's original splendor had vanished. In the sixteenth and seventeenth centuries, Ypres was ravaged by disease. It was then captured by the French, who slaughtered most of the inhabitants. In the seventeenth century, the French seized Ypres on four separate occasions: 1648, 1649, 1658, and 1678 when France took possession of the city under the Treaty of Nimègue. The Netherlands regained Ypres in 1715, but again lost it to France in 1792. Under the Napoleonic Empire, it became the capital of the Department of Lys. The treaties of 1815, which ended the Napoleonic wars, returned Ypres to the Netherlands, and after 1830 it became part of the Kingdom of Belgium.

By the start of the World War I in 1914, Ypres's population stood at about 18,000. Its main industries were textiles, the manufacture of woollen goods, printed cottons, linens, ribbons, and Valenciennes lace. Before its destruction, Ypres was graced with many rich and splendid buildings—the Cloth Hall, St. Martin's Cathedral, churches, mansions, houses, and other structures of the medieval and later periods. During World War I, however, the area surrounding Ypres was bitterly contested by both Allied and German forces attempting to gain control of the channel ports and the northern defensive line. For the French, the numerous battles at Ypres and across the region constituted one vast contest, the "Battle of Flanders." The Germans also viewed the battles as a whole, calling them the "Battle for Calais," or for the channel ports they sought to capture. For the British, the fighting at Ypres represented one of the most critical contests of the war, since it meant the defense of England itself.

The town of Ypres lies in a basin formed by a maritime plain intersected by canals and surrounded by low wooded hills. The canals and streams that run through the basin generally follow a southeast-northwest direction. The Yser Canal, which assumed considerable significance during World War I, is the most important waterway. The sides of the basin consist of low-lying and partly wooded hills, which form a line of crests running north to south through Houthulst Forest, Poelcappelle, Passchendaele, Broodseinde, Becelaere, Gheluvelt, St-Éloi, and other areas that became critically important during the war. Because of its marshy terrain, almost at sea level, and rainy climate, the war in this region assumed an entirely different character from elsewhere on the western front. Both sides found it virtually impossible to dig trenches or underground shelters where water was found immediately below the surface. The huge craters formed by shell fire became watery graves for many onrushing troops. Thus, the only possible defense-works were parapets. These conditions led both sides to establish defensive lines around wooded areas, farms, and villages, and made every hill or incline a critical tactical point. The crests surrounding the Ypres basin consequently became important as observation

Menin Gate, top, and Cloth Hall, bottom
Photo courtesy of Belgian National Tourist Office

posts and strategic points from which to establish defensive positions and launch attacks. As a result, the fighting along the crests and farms became the most fierce and bloody in the region. The numerous battles which occurred in the vicinity of the town between October 1914 and November 1917 are generally described as the three battles of Ypres, albeit each involved numerous separate contests for terrain.

With the outbreak of World War I, both the Allies and the Central European powers anticipated a quick and decisive victory. In the beginning Germany appeared close to defeating Allied forces in northern France and Belgium, and Germans themselves looked close to defeat on the eastern front. On every front, however, advantage passed to the defenders, and the war soon bogged down into years of bloody and indecisive trench warfare. Germany based its attack on the Schlieffen plan, named after the chief of Germany's General Staff from 1891 to 1905, Alfred von Graf Schlieffen, who anticipated fighting and winning a two-front war. Schlieffen's strategy first called for rapidly mobilizing forces against France and then turning on Russia, which he assumed would be the slower of the two in marshaling its forces. Since time was vitally important in order to overwhelm France, Germany would have to perform a rapid flanking movement through Belgium. The speed of the assault and numerical superiority would defeat the French and allow Germany to then marshal forces on the eastern front. Initially, Schlieffen's bold and daring plan proved successful. In August the German attack overwhelmed Belgium and Luxembourg, and forced French forces into rapid retreat, causing grave alarm in Paris. France tried to hold the line from Amiens to Verdun, but continued to cede ground on their left, while the British engaged German forces at Le Cateau. But in the last days of August, the German forces met fierce resistance, and in a series of battles collectively called the Battle of the Marne in early September 1914, the British and the French won a major success. The Germans were also repulsed at Verdun and were forced to retreat from Nancy to the Vesle.

In the next few months, the war on the western front shifted to the area between the Oise River and the coast, often called the "Race to the Sea." In fierce but inconclusive fighting in Autumn 1914, forces on either side tried to turn the other's flank and control the channel ports. Although the Germans made gains in the series of battles that came to called the First Battle of Ypres, they failed to achieve any breakthrough. At this point, the western front became stalemated for three years.

The First Battle of Ypres began on October 20 and involved a powerful German attack carried out by its Sixth and Fourth Armies along the whole front extending from La Bassée canal to the sea. The Germans attempted a breakthrough on the Yser front between Ypres and Nieuport in order to envelop and destroy the Allies' northern flank. So confident of success was the kaiser that he arrived at Courtrai for a triumphant entry into Ypres. But the British maintained their defensive line at La Bassée, Armentières, and Messines,

and the line at Ypres was secured by the timely arrival of forces led by Sir Douglas Haig.

Thus, Germany's first great attempt failed, and it soon began a second smaller attack between Messines and the northern end of Gheluvelt. The Germans introduced seven fresh divisions and, with the support of heavy artillery, attempted to break through the Allied line at Ypres; the offensive came within an inch of success. Haig met the attack with three divisions supported by General Edmund Allenby's cavalry.

On November 11, German forces broke through at Gheluvelt and had nothing before them but a British gun line. Failing to seize the critical moment, the German troops wavered and were driven back by counterattacks manned, in part, by cooks, and orderlies, and clerks. In the fighting, the Allies averted military disaster but lost Messines Ridge as well as some ground to its north. The British lost 50,000 dead, wounded, and missing at the First Battle of Ypres; figures for French and German losses are unavailable, but are assumed to be just as high or higher. Both sides now entrenched themselves more deeply, and open warfare on the western front would not reappear until 1918.

Following the near breakthrough of November 11, the battles around Ypres died down as the Germans turned their attention to the east. In response to Russian advances in East Prussia and Silesia, the Germans transported nine divisions from the west to the eastern front. They were thus left to maintain the defensive on the western front, and as an experiment decided to use chlorine gas at Ypres where the winds were said to be favorable. Since the use of gas was to be purely experimental, the Germans failed to hold any troops in reserve to exploit success, but around the salient were eleven divisions of the German Fourth Army. The Allies had received several reports of German intentions to use poison gas, but they ignored the intelligence at their own considerable cost.

On April 22, 1915, at 5:00 P.M., the Germans began a furious bombardment, releasing gas from steel cylinders buried in parapets. A light breeze from the northeast, wafting the bluish white mist toward the French, caused a panic that quickly spread throughout the lines. Behind the cloud of gas the Germans attacked under protection of artillery and intense machine gun fire as French colonial troops fell back toward Ypres. The French retreat exposed the left flank of the Canadians, who also fell back. The Germans advanced unopposed for about two miles until they were forced to halt after running into their own gas. On April 23, the Germans resumed the attack. The British tried to fill the breach with battalions from three divisions, but by April 25 the Germans gained the higher ground of Ypres Ridge and about half the breadth of the salient. Under heavy shell fire from three sides, further Allied defense of the salient was untenable. But when the Second Army commander, Sir Horace Smith-Dorien, proposed a two-and-one-half mile retreat to Ypres, he was summarily dismissed and replaced by Sir Herbert Plumer, who subsequently received permission to carry out the withdrawal as the only course of action. Like his predecessor, Plumer reasoned that nothing but a carefully planned offensive could dislodge the Germans from the heights around Ypres. Furthermore, to mount an offensive in the Ypres area without reinforcements not only would invite failure, but might also critically disrupt the planned joint British-French campaign at Artois. After several delays, the British carried out the withdrawal during the first three nights of May. Then, on May 8, the German Fourth Army attacked to overrun the Ypres Salient, and in six days of fierce fighting known as the battle of Fresenberg Ridge, the British lost valuable ground. On May 24, the Germans again bombarded the British with poison gas, followed by an infantry attack. The British maintained their positions, however, and the Germans failed to make further gains.

For its size, the Second Battle of Ypres was one of the bloodiest battles of the war, with total casualties exceeding 100,000. The Germans had used poison gas in defiance of the laws of war, but in doing so they had achieved their biggest success of the year in the west. Not only had they gained territory, but they had diverted seven Allied divisions away from the Artois offensive. Physicists and chemists were originally recruited to produce poison gas on the pretext that it would save lives by bringing the war to a quick end. Tear gas was first employed early in 1915, but it was the use of chlorine gas by the Germans at Ypres that caused widespread outrage. According to author Cyril Falls, through its use the Germans "reduced the Ypres salient to a flat curve just two miles east of the city and secured all the commanding ground. Yet, because their use of the chlorine gas had been experimental and they had so slender a reserve, they missed a far greater victory before the effect of surprise wore off." Their action also laid them open for retaliation, especially since the prevailing winds were westerly.

By 1917, the year of the Third Battle of Ypres, the disappointments of the previous year had produced a decidedly uncertain mood among the Allies. The U.S. president, Woodrow Wilson, called for an end to the hostilities, and throughout Europe, politicians oscillated between Wilson's call for peace and planning a major military breakthrough. Moreover, new European leaders had assumed power in political and military circles. Lloyd George became the new prime minister of England. In France, General Robert-George Nivelle, organizer of the successful counter attacks at Verdun, replaced General Joseph-Jacques-Césaire Joffre as the army's new commander in chief. And in Germany, General Erich Friedrich Wilhelm Ludendorff's star was rising and, together with Paul von Hindenburg, he urged unrestricted submarine warfare as a means of winning the war.

After conferring in Rome in January 1917, the Allies agreed on coordinating a major spring offensive. As events proved, however, misfortune made the year 1917 one of near disaster for the Allies. First came the crushing defeat of the great French spring offensive led by Nivelle. Nivelle planned an all-out offensive against the main German forces at the Noyon Salient along the Aisne River, which was to be preceded by a major diversionary attack by the British around the Arras. As it turned out, he was tricked by the Germans, who secretly withdrew their front line troops and then destroyed the advancing French armor and troops with artillery fire. The terrible slaughter ignited mutinies throughout the French army and resulted in the appointment of a new army commander, General Henri Phillipe Pétain.

The French defeat was followed that June by the start of a major British offensive in Flanders, known as the Third Battle of Ypres. A successful strike in this region not only would eliminate the destructive U-Boat bases on the Belgium North Sea coast, but would also turn the northern flank of the German army. The British government had long considered it imperative to expel the Germans from Belgium, but Nivelle's failure as well as other factors cast the situation in a different light. Following Nivelle's disastrous defeat, the British agreed to continue the offensive with the full backing of the French army, but General Pétain, on succeeding Nivelle on June 2, informed Haig that any support would be nearly impossible given the mutinies and unrest among his forces. To divert attention away from the French front as well as prevent the Germans from moving to the eastern front where the Allied situation was deteriorating, the British decided on an immediate offensive at Messines, a prelude to the major offensive.

On the front near Ypres, the battle of Messines opened on June 7 when the British simultaneously detonated nineteen huge mines containing almost a million pounds of explosives. Lloyd George, at his study in London, felt the tremor and heard the distant rumble. Then the British Second Army unleashed the full strength of its artillery, and nine divisions, with three in reserve, successfully made the assault, taking Messines Ridge. The British commander, General Haig, planned an interval between Messines and the major assault on the Ypres front, but continuing political hesitations in London cost the campaign six valuable weeks of fair weather. In the interim, the Germans anticipated the offensive and made preparations to strengthen their defensive lines across from the British.

On July 11, the Allied offensive began with an air assault, and in the heavy fighting that ensued little ground was gained. The bombardment destroyed the delicate network of drains and dikes, turning the area into a veritable swamp, a situation exacerbated by heavier than normal rainfall in August. Under these conditions, the Allies launched two additional attacks with only minor success. In the battle of Langemark (August 16–18), the Allies took a mere 500 yards, but at Lens (August 15–25), the Canadians routed five Germans divisions to capture Hill 70. In light of the limited gains, Haig assigned the principal role of the Ypres offensive to General Plumer, who led the successful attack on Messines Ridge. After three weeks of preparations, Plumer waged a three-step campaign in the battles of Menin Road (September 20–25), Polygon Wood (September 26), and Broodseinde (October 4) to take the crest of Ypres Ridge. Haig then pressed the campaign forward to seize the remaining part of the ridge at Passchendaele before the onset of winter. At Poelcappelle (October 9) and the first battle Passchendaele (October 12), where the weather completely broke, the Allies made little progress.

The Allies decided to suspend further operations on October 13 due to poor weather. Success depended on the laying of communications across swollen streams, mud, and water-filled craters. When the weather showed slight signs of improvement, Haig renewed the offensive on October 26. It was this last phase of the offensive in cold rain and mud that earned the terrible reputation for the Third Battle of Ypres. In the French sector, troops overwhelmed the villages of Draeibank, Hockske, Aschhoop, Merckem, and Kippe, as well as many fortified farms. With the aid of the Belgians, the villages of Luyghem and Vyfhuyzen were also taken. For their part, the British and Canadians advanced on the remains of the small village of Passchendaele, the main objective. Hindenburg issued orders that the town must be held at all costs and retaken if lost. The Canadian Corps led the attack on October 26 and, after renewing the assault from October 30 to November 6, the village was finally taken. On November 10, the Allies broadened their hold on Ypres Ridge before Haig stopped the offensive in order to proceed with the Battle of Cambrai. British losses from July 31 to November 10 totaled 244,897, the French, who played a minor role, lost only 8,525. Total German losses are estimated at 260,000.

Aside from the terrible death toll, the British offensive failed to achieve a major breakthrough. At the conclusion of the campaign, total British casualties exceeded 300,000 while German losses numbered about 275,000 along the entire western front. Although the Allies had gained ground, the Germans remained entrenched in the region. The campaign was subsequently criticized for its horrible attrition, but Haig defended its importance in giving the French enough time to restore their army. Moreover, had Haig been able to begin the campaign in April in better weather, it might have achieved greater success. Although both sides were temporarily exhausted, the Ypres experience convinced the Germans that a successful offensive could be made in Flanders. Consequently, in 1918 German forces made five attempts to achieve a major breakthrough before the Allies launched another offensive of their own in the fall. By the evening of the first day of the campaign on September 28, the Allies cleared most of the remaining portion of the notorious Ypres Ridge. Although the campaign soon bogged down, the Allies were on the offensive and marching toward victory when Germany agreed to sign the Armistice on November 11, 1918.

As a result of these long years of war, the once-picturesque town of Ypres, with its many historical medieval structures, was laid to waste. The Cloth Hall, begun by Count Baldwin IX of Flanders in the twelfth century, was badly damaged, and the Cathedral of Saint Martin, the Places des Halles, and the Hôtel de Ville, built in the sixteenth century, were completely destroyed. But, following World War I, the town was meticulously restored to its prewar state, including the seventeenth-century facades along many streets as well as the city's two great structures, St. Martin's Cathedral and the Cloth Hall.

Although Ypres never fell, the surrounding battles took an estimated 1 million lives. A memorial was built at Menin Gate in honor of the soldiers of the British Commonwealth who died defending Ypres. Most of the Ypres battlefield was plowed under by local farmers following the war, but approximately two acres of ground have been preserved since 1918; a museum was established at Sanctuary Wood in Zillebeke.

During World War II, Ypres, along with all of Belgium, fell into German hands. While under attack from the German Blitzkrieg, Allied army leaders met at Ypres on May 20, 1940, to plan a double counterattack. These plans subsequently failed, however, and on May 27th, the King of Belgium had surrendered. Ypres would not be liberated until the Allies crossed the English Channel and moved into western Belgium in Autumn 1944.

Further Reading: For a general history of Ypres see the illustrated Michelin Guide, *Ypres and the Battles for Ypres* (Clermont-Ferrand, France: Michelin, 1920). Ypres's medieval period is mentioned in several books on the history of Flanders, including *Medieval Flanders* by David Nichols (Harlow, Essex: Longman, 1992) and *Flemish Cities: Their History and Art* by William Gaunt (New York: Putnam, and London: Elek, 1969). Numerous books cover the battles of Ypres during World War I. The most comprehensive is Beatrix Brice's *The Battle Book of Ypres* (Stevenage, Hertfordshire: SPA, 1987; New York: St. Martin's Press, 1988). Other informative accounts are found in the following: *A Short History of World War I* by James E. Edmonds (Oxford: Oxford University Press, 1951; New York: Greenwood, 1968); *The First Word World War* by Keith Robbins (Oxford and New York: Oxford University Press, 1984); and *The Great War* by Cyril Falls (New York: Putnam, 1959; as *The First World War*, London: Longman, 1960).

—Bruce P. Montgomery

INDEX

Listings are arranged in alphabetical order. Entries in bold type have historical essays on the page numbers appearing in bold type. Page numbers in italic indicate illustrations.

NOTES ON CONTRIBUTORS

BAKOS, Sharon. Freelance writer and editor.

BARBOUR, Philippe. Freelance writer. Commissioning editor, Gale Research International, 1992–94; editor, St. James Press, 1991; editor, Wine Buyers Guides, 1988–90. Co-author of *Wine Buyers Guide: Saint Emilion, 1991.*

BLOCK, Bernard A. Freelance writer. Reference and documents librarian, Ohio State University, Columbus, 1969–92.

BRENNAN, Shawn. Editor, Gale Research Inc. Editor of *Newsletters in Print,* 1994; *The Resourceful Woman,* 1993; *Women's Information Directory,* 1993.

BRICE, Elizabeth. Special collections librarian, Miami University, Oxford, Ohio.

BROADRUP, Elizabeth E. Freelance writer and picture researcher.

CABLE, Monica. Product manager, Anixter. Recipient of Watson Foundation fellowship for independent research in southeast Asia, 1991–92.

CHEPESIUK, Ron. Associate professor and head of special collections, Winthrop University, Rock Hill, South Carolina.

CLASSE, Olive. Freelance writer and translator. Translator of *Mission to Marseilles* by Leo Malek, 1991; lecturer, then senior lecturer in French, University of Glasgow, 1965–90.

COLLISON, Daniel D. Project assistant, Horn of Africa desk, Christian Aid.

DEVINE, Elizabeth. Professor, Department of English, Salem State College, Salem, Massachusetts. Author of *European Customs and Manners,* 1984, 1992; *The Travelers' Guide to Middle Eastern Customs and Manners,* 1989; *The Travelers' Guide to Asian Customs and Manners,* 1986. Editor of *The Annual Obituary, 1983,* 1984.

DUBOVOY, Sina. Independent scholar and freelance writer specializing in history and biography.

DUNCAN, Laura. Reporter, *Chicago Daily Law Bulletin.*

ELLINGSON, Stephen. Project manager, Quality of Student Life Study, University of Chicago. Associate book review editor, 1989–91, and associate editor, 1991–92, *American Journal of Sociology.*

FEELY, Mary. Freelance writer and editor. Senior editor, World Book Publishing, 1992–94. Editor of *Study Power,* two volumes, 1993; winner of Chicago Women in Publishing award, 1994.

FELSHMAN, Jeffrey. Freelance writer. Recipient of 1993 Peter Lisagor Award for exemplary journalism, from the Society of Professional Journalists, Chicago Headline Club, for "The Fall and Rise of Anita Brick," Chicago *Reader,* December 3, 1993.

FLINK, John A. Freelance writer.

GOODMAN, Lawrence F. English instructor, Illinois Institute of Technology, Chicago.

GREB, Richard. Freelance writer. Reporter and editor, Reuters Information Services, 1976–88.

GURNEY, Judith. Research fellow, Oxford Institute for Energy Studies, Oxford University.

HEENAN, Patrick. Research student, University of London. Editor of *1992,* 1989.

HUEBNER, Jeff W. Freelance writer.

JAROS, Tony. Copy editor, *Vegetarian Times.*

KEELEY, Patrick. Freelance researcher.

KING, Linda J. Master of divinity candidate, General Theological Seminary, New York.

KLAWINSKI, Rion. Freelance writer.

KLOAK, Andrew M. Freelance writer.

LAMONTAGNE, Manon. Assistant director, National Film Board of Canada.

LAMONTAGNE, Monique. Research student, University of London. Co-editor of *The Voice of the People: Reminiscences of Early Settlers, 1866–1895,* 1984.

LANGE, Bob. Freelance writer.

LANGSTON, Cynthia L. Strategic planner, TBWA Advertising, New York.

LaRUE, Sherry Crane. Spanish instructor, Madison Area Technical College, Madison, Wisconsin.

LEDGER, Gregory J. Freelance writer; contributing writer, *Windy City Times,* Chicago.

LEVI, Anthony. Freelance historian. Buchanan Professor of French Language and Literature, St. Andrews University, 1971–88. Author of *Guide to French Literature, 1789 to the Present,* 1992; and *Guide to French Literature, Beginnings to 1789,* 1994. Editor of *The Collected Works of Erasmus,* *volumes 27–28,* 1986; and *Erasmus: Satirical Works,* forthcoming.

LEVI, Claudia. Freelance writer.

LEVI, Honor. Teacher and translator. Editor of *Inventaires de Cardinal Richelieu,* 1986; translator of *Pascal: The Pensées and Other Writings,* to be published 1995.

McNULTY, Mary F. Freelance writer and editor. Editor, American Association of Law Libraries newsletter, 1988–1993.

McPHERON, Thomas Cermak. News editor, *North Loop News,* Chicago.

MAGON, Kim M. Freelance writer and editor.

MARCHANT, Brent. Freelance writer and editor. Editor, *Interior Landscape/Interior Landscape Industry Magazine,* 1983–92.

MARTIN, Christine Walker. Freelance writer and editor. Author of *Breaking with Tradition,* 1994.

MILLER, Julie A. Freelance writer.

MILLER, Phyllis R. Freelance writer and editor. Author of *Keeping Kids Drug-Free,* 1993.

MINSKY, Laurence. Freelance copywriter. Co-author, with Emily Calvo, of *How to Succeed in Advertising When All You Have Is Talent,* 1994.

MONTGOMERY, Bruce P. Curator/head of archives, University of Colorado at Boulder.

PHILLIPS, Michael D. Instructor in humanities, Brigham Young University, Provo, Utah.

PRESNELL, Jenny L. Humanities/Social Sciences librarian, Miami University, Oxford, Ohio.

RIJSBERMAN, Marijke. Freelance writer and editor; English instructor, Chicago State University.

RING, Trudy. Commissioning editor, Fitzroy Dearborn Publishers.

St. JOHN, Robert M. Project director, Riverside Publishing.

SALKIN, Robert M. Commissioning editor, Fitzroy Dearborn Publishers.

SAWYERS, June Skinner. Associate editor, Loyola University Press. Co-author of *The Chicago Arts Guide,* 1994; author of *Chicago Portraits: Biographies of 250 Famous Chicagoans,* 1991.

SCHELLINGER, Paul E. Doctoral candidate in French, Northwestern University, Evanston. Editor of *St. James Guide to Biography,* 1991; co-editor of *Twentieth-Century Science-Fiction Writers,* third edition, 1991.

SHEPARD, Patricia Ann. Freelance writer.

SULLIVAN, James. Freelance writer.

SY-QUIA, Hilary Collier. Doctoral candidate in German literature, University of California at Berkeley. Translator of and author of introduction to Schiller's *Don Carlos and Maria Stuart,* to be published 1995.

SY-QUIA, Noel. Management consulting associate, Andersen Consulting, San Francisco, California.

TAGGART, Elizabeth. Freelance writer and editor.

VAN VYNCKT, Randall J. Master's candidate in architecture, University of Illinois at Chicago. Editor of *International Dictionary of Architects and Architecture,* 1993.

VASUDEVAN, Aruna. Freelance writer. Editor, St. James Press, London, 1991–94. Editor of *Twentieth-Century Romance and Historical Writers,* 1994 and *International Dictionary of Films and Filmmakers, Volume 1: Films,* to be published 1995.

WATSON, Noelle. Instructor in English, Clemson University, Clemson, South Carolina. Editorial coordinator, St. James Press, 1989–91. Editor of *Reference Guide to Short Fiction,* 1993; co-editor of *Twentieth-Century Science-Fiction Writers,* third edition, 1991.

WHITE, Joshua. Freelance writer.

WILKINS, Richard G. Executive speechwriter and editor.

WILOCH, Thomas. Freelance writer.

WOOD, Beth F. Freelance writer, researcher, and editor. Lecturer, University of Central London, 1984–90.

XANTHEAS, Peter C. Membership coordinator and researcher, Preservation Wayne, Detroit, Michigan.